# GRAPHS

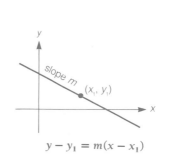

$$y - y_1 = m(x - x_1)$$

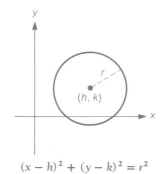

$$(x - h)^2 + (y - k)^2 = r^2$$

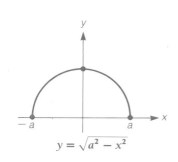

$$y = \sqrt{a^2 - x^2}$$

$$y = |x|$$

$$y = x^2$$

$$y = x^3$$

$$y = \sqrt{x}$$

$$y = \frac{1}{x}$$

$$y = [x]$$

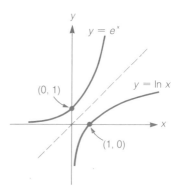

$$y = e^x \quad \text{and} \quad y = \ln x$$

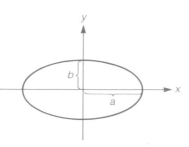

$$\frac{x^2}{a^2} + \frac{y^2}{b^2} = 1$$

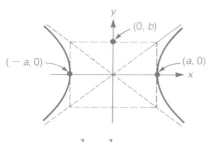

$$\frac{x^2}{a^2} - \frac{y^2}{b^2} = 1$$

# ALGEBRA AND TRIGONOMETRY

# Algebra and Trigonometry

**SECOND EDITION**

DAVID COHEN
Department of Mathematics
University of California
Los Angeles

**WEST PUBLISHING COMPANY**
St. Paul    New York    Los Angeles    San Francisco

**FOR MY DAUGHTER EMILY**

Text design:  Janet Bollow
Technical illustrations:  Rolin Graphics
Copyediting:  Susan Gerstein
Indexing:  Susan Cohen
Typesetting:  Syntax International
Cover image:  Bill Imhoff

**LIBRARY OF CONGRESS CATALOGING-IN-PUBLICATION DATA**

Cohen, David, 1942 Sept. 28–
    Algebra and trigonometry/David Cohen.—2nd ed.
       p.     cm.
    Includes index.
    ISBN 0-314-48117-6
    1. Algebra.   2. Trigonometry.   I. Title.
QD923.P44      1988
512′. 13—dc 19

88-28739
CIP

# CONTENTS

## ROOTS OF POLYNOMIAL EQUATIONS                                              568

## 11

## THE CONIC SECTIONS                                                         604

## 12

## ADDITIONAL TOPICS IN ALGEBRA                                              633

## 13

## APPENDIX                                                                   A-1

## TABLES                                                                     A-9

# PREFACE

This text develops the elements of college algebra and trigonometry in a straightforward manner. As in the first edition, my goal has been to create a book that is *accessible* to the student. The presentation is student-oriented in three specific ways. First, I've tried to talk to, rather than lecture at, the student. Second, examples are consistently used to introduce, to explain, and to motivate concepts. And third, all of the initial exercises for each section are carefully correlated with the worked examples in that section.

**AUDIENCE**  In writing *Algebra and Trigonometry*, I have assumed that the students have been exposed to intermediate algebra, but that they have not necessarily mastered that subject. Also, for many students, there may be a gap of several years between their last mathematics course and the present one. For these reasons, the review material (in Chapter 1 and in parts of Chapter 2) is unusually thorough.

**FEATURES**  1. *Word problems and applications.* Beginning with the review material in Chapter 1, there is a constant emphasis on learning to use the *language* of algebra as a tool in problem-solving. Word problems and strategies for solving them are explained and developed throughout the book. For examples of this, see any of the following sections.

Sec. 2.3 (word problems leading to linear and quadratic equations)
Sec. 5.2 (setting up equations that describe or model a given situation)
Sec. 5.3 (maximum-minimum problems relating to quadratic functions)
Sec. 6.5 (applications of the exponential function)

2. *Emphasis on graphing.* Graphs and techniques for graphing are developed throughout the text, and graphs are used to explain and reinforce algebraic concepts. (See, for example Sections 4.3 and 6.3.)

3. *Calculator exercises.* Many (perhaps even most) students entering a college mathematics course own a scientific calculator, and they are eager to use it. Calculator exercises, indicated in this text by the symbol Ⓒ, are designed to take advantage of this situation. (It should be pointed out, however, that this is not a calculator-based text.) There are four broad types of calculator exercises here:

(i) Exercises that reinforce and supplement the core material. (See, for example, Exercises 41 and 42 in Section 1.3.)
(ii) Exercises with surprising results that motivate subsequent theoretical questions. (See, for example, Exercise 71 in Section 1.6.)
(iii) Exercises that introduce supplementary material. (See, for example, Exercise 61 in Section 1.8.)
(iv) Exercises demonstrating that the use of a calculator cannot replace thinking or the need for mathematical proofs. (See, for example, Exercise 74 in Section 1.6 or Exercise 95 in Section 6.4.)

In addition to these specially marked calculator exercises, there are many routine problems, integrated throughout the text, for which either a calculator or tables will suffice. Such problems are not preceded by the symbol Ⓒ.

4. *End-of-chapter material.* Each chapter concludes with a chapter summary and an extensive review exercise set covering the entire chapter. In each review set, fifteen or so exercises are singled out with colored problem numbers; these constitute a sample chapter test for the student. In general, the review exercise sets serve several key functions. First, they provide a source of routine drill-type problems in a context that is not explicitly linked to a specific section of the chapter. Thus, just as on an exam, the student must decide for herself or himself which tools to bring to bear on a problem. Second, the review exercise sets provide a rich source of problems at various levels, ranging from the elementary to the challenging. Some of these approach the core material from perspectives that differ slightly from that in the exposition. This is important for students who are planning to apply their algebra skills in subsequent courses.

**CHANGES IN THIS EDITION**

Comments and suggestions from students, instructors, and reviewers have helped me to revise the text in a number of ways that I believe will prove helpful to both students and instructors. The major changes are in the following areas.

1. *Graded Exercise Sets.* As a convenience for the instructor, most exercise sets (except for the review sets) are now divided into three catagories (A, B, and C). The group A exercises are based directly on the examples and definitions in that section of the text. Moreover, these problems treat topics in roughly the same order in which they appear in the text. The group B exercises serve several key functions. Some of them require students to use several different techniques or topics in a single solution. Some group B problems, while not conceptually more difficult than their group A counterparts, require lengthier calculations or algebraic manipulations. Finally, some group B exercises are included simply because they are interesting or because they are intended to challenge the student. The group C exercises contain the more challenging problems.

2. *Trigonometry.* A third chapter has been added to the material on trigonometry. The topics in this additional chapter are vectors, polar coordinates, trigonometric form for complex numbers (including DeMoivre's theorem), and the laws of sines and cosines. (The material on the laws of sines and cosines and on trigonometric form did appear in the previous edition, but in different chapters.) In addition to this, the initial introduction to right-angle trigonometry in Chapter 7 has been expanded from one to two sections, and the number of worked examples and routine drill-type exercises has been increased. Finally, an additional section has been added (in Chapter 8) providing a detailed discussion of the graphs of the tangent function and the three reciprocal functions.

3. *Symmetry and graphing.* The use of symmetry in graphing is given greater emphasis and used more frequently in this edition.

4. *Maximum and minimum problems.* Word problems involving the maximum or minimum of quadratic functions are now developed in a separate section (Section 5.3), and more worked examples are included.

5. *The exponential function $y = e^x$*. The material introducing the number $e$ and the exponential function $y = e^x$ is now contained in a separate section (Section 6.2), and here too, the number of worked examples has been increased.

6. *Matrix inverses*. A new section on matrix inverses has been added to the chapter *Systems of Equations*. This new section develops the idea of an inverse matrix and shows how inverses are used in solving certain systems of linear equations.

7. *The conic sections*. The development of the conics has been expanded to include the focus definitions of the curves. This material now appears later in the book, in a separate chapter (Chapter 12).

**SUPPLEMENTARY MATERIALS**

1. *Students' Solutions Manual*, by Ross Rueger. This contains complete solutions for the odd-numbered exercises.

2. *Instructor's Solutions Manual*, by Ross Rueger. This manual contains answers or solutions for every exercise in *Algebra and Trigonometry*.

3. The *West MathTest*. This is a computer-generated testing program that is available to schools adopting *Algebra and Trigonometry*.

4. *Test Bank*. This contains two types of chapter tests for each Chapter in *Algebra and Trigonometry*. The first type (version A) consists of core problems that would be classified as group A in the graded exercise sets. The second type (version AB) contains problems from both the group A and group B levels.

5. *GraphToolz*, by Tom Saxton, is a software program for graphing and evaluating functions. This ingenious, easy-to-use program, along with drill and enrichment exercises by David Cohen, is provided free to qualified adopters. Available for the Apple Macintosh family of computers.

6. *Transparency Masters*. Transparency masters for many of the key figures and tables appearing in the text are available to schools adopting *Algebra and Trigonometry*.

**ACKNOWLEDGMENTS**

Many students and colleagues have made useful suggestions about the text and exercises, and I thank them for that. Also, particular thanks go to Professor Michael Dagg and Professor Charles Heuer for their conscientious work in reading and checking this new edition for accuracy. I am also grateful for the help and valuable suggestions given to me by the following reviewers for this second edition.

Karen Barker,
Indiana University, South Bend

Kathleen Bavelas,
Manchester Community College,
Connecticut

Clevis Carter,
Union College, Kentucky

Barbara Duch,
University of Delaware, Newark

Louis Gioia,
Nassau Community College, New York

Donald Gnirk,
South Dakota State University

Mark Hall,
Simpson College, Iowa

Linda Halligan,
Mohawk Valley Community College,
New York

George Hilton,
Loma Linda University,
La Sierra, California

Patricia Hirschey,
Delaware Technical Community
College

David Johnson,
Diablo Valley College, California

Charlotte Lewis,
University of New Orleans

Stanley Lukawecki,
Clemson University

Giles Maloof,
Boise State University

Richard Nadel,
Florida International University

Randolph Plotts,
St. Petersburg Junior College

Sherry Tornwall,
University of Florida, Gainesville

I would also like to thank two people in particular for their help and excellent advice: Ross Rueger of College of the Sequoias who wrote the supplementary manuals and created the answer section for the text, and Susan Gerstein, who served both as copyeditor and problem checker. Finally, I would like to express my appreciation to Mark Jacobsen of West Publishing Company for his work in bringing this second edition into print.

David Cohen
Lunada Bay, California, 1989

# ALGEBRA AND TRIGONOMETRY

# 1 FUNDAMENTAL CONCEPTS

**INTRODUCTION**   This first chapter sets the foundations for our work in algebra. The central topics here are the real numbers, polynomials, and operations with polynomials. In the last section of the chapter we describe the complex number system. The notation and the results in this chapter will be used repeatedly throughout the rest of the book.

## NOTATION AND LANGUAGE

**1.1**

*Perhaps Pythagoras was a kind of magician to his followers because he taught them that nature is commanded by numbers. There is a harmony in nature, he said, a unity in her variety, and it has a language: numbers are the language of nature.*

JACOB BRONOWSKI

*Ever since the time of Descartes, the terms "real" and "imaginary" numbers have been used in mathematics, despite their misleading associations.*

ALFRED HOOPER

Here, as in your previous mathematics courses, most of the numbers we deal with are *real numbers*. These are the numbers used in everyday life, in the sciences, in industry, and in business. Perhaps the simplest way to define a real number is this: A **real number** is any number that can be expressed in decimal form. Some examples of real numbers are

$$7 \ (= 7.000 \cdots) \qquad -\frac{2}{3} \ (= -0.\overline{6}) \qquad \sqrt{2} \ (= 1.4142 \cdots)$$

In algebra we often use letters to stand for numbers. This simple idea is what gives algebra much of its power and applicability in problem solving. By using letters, we can often see patterns and reach conclusions that might not be so apparent otherwise.

Arithmetic, and therefore algebra too, begins with the *operations* of addition, subtraction, multiplication, and division. Table 1 will serve to remind you that, for the most part, the notation for these operations in both arithmetic and

**TABLE 1**

| OPERATION | ARITHMETIC EXAMPLE | ALGEBRAIC EXAMPLE |
|---|---|---|
| Addition ($+$) | $5 + 7.01$ | $p + q$ |
| Subtraction ($-$) | $13 - \sqrt{5}$ | $r - s$ |
| Multiplication ($\times$) | $3 \times 4$ | $a \times b$ or $a \cdot b$ or $ab$ |
| Division ($\div$) | $5 \div 6$ | $x \div y$ or $\dfrac{x}{y}$ or $x/y$ |

algebra is the same. One exception: As the table indicates, the product $a \times b$ is often abbreviated to simply $ab$.

There are several other notational conventions that will seem familiar to you from previous courses. A sum such as $a + a + a$ is denoted by $3a$, while a product such as $a \cdot a \cdot a$ is written $a^3$. In general, if $n$ is a **positive integer** (i.e., one of the numbers 1, 2, 3, 4, ...), then

$$na \quad \text{means} \quad \underbrace{a + a + a + \cdots + a}_{n \text{ terms}}$$

and

$$a^n \quad \text{means} \quad \underbrace{a \cdot a \cdot a \cdots a}_{n \text{ factors}}$$

In the expression $na$, the number $n$ is called the **coefficient** of $a$. For example, in the expression $3a$, the coefficient is 3. When we write $a^n$, we refer to $a$ as the **base** and $n$ as the **exponent** or **power** to which the base is raised. The process of raising a base to a power is called **exponentiation**. The notation $a^2$ is read "$a$ squared," because the area of a square with sides each of length $a$ is $a \times a$ or $a^2$. Similarly, $a^3$ is read "$a$ cubed," because the volume of a cube with sides each of length $a$ is $a \times a \times a$ or $a^3$.

There are several important conventions regarding the order in which the operations of arithmetic and algebra are to be performed. Again, these will seem familiar to you from previous courses, and we summarize them in the box that follows.

---

| PROPERTY SUMMARY | The Order of Operations |
|---|---|

| CONVENTION | EXAMPLE |
|---|---|
| 1. If no grouping symbols (such as parentheses) are present: | |
|   **(a)** First do the exponentiations, working from right to left; | $3 + 4^2 = 3 + 16 = 19; \quad 5 \times 2^3 = 5 \times 8 = 40;$ <br> $2^{3^2} = 2^9 = 512$ |
|   **(b)** Next do the multiplications and divisions, working from left to right; | $9 - 2 \times 3 = 9 - 6 = 3; \quad 12 + 8 \div 2 = 12 + 4 = 16;$ <br> $4 + 2 \times 3^2 = 4 + 2 \times 9 = 4 + 18 = 22$ |
|   **(c)** Then do the additions and subtractions, working from left to right. | $1 + 2 \times 3 - 4 = 1 + 6 - 4 = 7 - 4 = 3$ <br> $15 + 6^2 \div 4 - 2 \times 3^2 = 15 + 36 \div 4 - 2 \times 9$ <br> $\qquad\qquad = 15 + 9 - 18$ <br> $\qquad\qquad = 24 - 18 = 6$ |
| 2. If grouping symbols are present: | |
|   **(a)** First do the operations within the grouping symbols; | $(1 + 2) \times 5 = 3 \times 5 = 15; \quad (3 + 4)^2 = 7^2 = 49$ |
|   **(b)** If grouping symbols appear within grouping symbols, the operations within the innermost set are to be done first. | $3[8 - 2(6 - 4)] = 3(8 - 2 \times 2)$ <br> $\qquad\qquad = 3(8 - 4)$ <br> $\qquad\qquad = 3 \times 4 = 12$ |

The four examples that follow serve to summarize the ideas discussed in this section. The fourth example (and the exercises like it) will also help you begin to review the geometric formulas given on the inside front cover of this book.

**EXAMPLE 1**    Rewrite each expression using exponential notation.

(a) $x \cdot x \cdot x \cdot y \cdot y \cdot y \cdot y$
(b) $(a + 3)(a + 3)(b - 1)(b - 1)(b - 1)$

Solution    (a) $x \cdot x \cdot x \cdot y \cdot y \cdot y \cdot y = x^3 y^4$
(b) $(a + 3)(a + 3)(b - 1)(b - 1)(b - 1) = (a + 3)^2 (b - 1)^3$

**EXAMPLE 2**    (a) Evaluate $x - 4y^2$ when $x = 19$ and $y = 2$.
(b) Evaluate $(x - 4)y^2$ using the values for $x$ and $y$ given in part (a).

Solution    (a) $x - 4y^2 = 19 - 4 \times 2^2$
$$= 19 - 4 \times 4 = 19 - 16 = 3$$

(b) $(x - 4)y^2 = (19 - 4) \times 2^2 = 15 \times 4 = 60$

**EXAMPLE 3**    Translate each of the following into algebraic notation.

(a) five more than twice the number $x$
(b) three times the sum of $x$ and $y^3$
(c) the sum of three times $x$ and $y^3$

Solution    (a)        $2x + 5$
twice $x$         five more

(b)        $3 \times (x + y^3) = 3(x + y^3)$
three times
the sum of $x$ and $y^3$

(c) $3x + y^3$
the sum of $3x$ and $y^3$

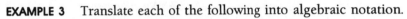

$V = \pi r^2 h$

**FIGURE 1**

**EXAMPLE 4**    The formula for the volume of a right circular cylinder is $V = \pi r^2 h$. (See Figure 1.)

(a) Suppose that in a right circular cylinder, the height is twice the radius. Write an equation relating $r$ and $h$.
(b) Express the volume of the cylinder described in part (a) in terms of the radius $r$.

Solution    (a) We translate the English into algebra as follows:

The height    is    twice the radius
$h$          =         $2r$

That is, we have

$h = 2r$

**(b)** The general formula for the volume is $V = \pi r^2 h$. Replacing $h$ with $2r$, we obtain

$$V = \pi r^2 (2r)$$

or

$$V = 2\pi r^3$$

as required.

## EXERCISE SET 1.1

### A

*In Exercises 1–10, rewrite the given expressions using coefficients and exponents.*

1. $x \cdot x \cdot x \cdot x \cdot x \cdot y \cdot y$    2. $x \cdot x \cdot y \cdot y \cdot z$

3. $(x + 1)(x + 1)(x + 1)$

4. $(a^2 + b^2)(a^2 + b^2)(a^2 + b^2)$

5. (a) $x + x + x + x$
   (b) $(x^2 + 1) + (x^2 + 1) + (x^2 + 1)$

6. (a) $a^5 + a^5$
   (b) $(a^4 + 1) + (a^4 + 1)$

7. (a) $a + a + a + b + b$
   (b) $a^2 + a^2 + a^2 + b^2 + b^2$

8. (a) $t^4 + t^4 + t^4 + z^2 + z^2$
   (b) $(t^4 + 1) + (t^4 + 1) + (t^4 + 1)$

9. (a) $(2a + 1)(2a + 1)(2a + 1)(2b + 1)(2b + 1)$
   (b) $(1 + 2a)(2a + 1)(1 + 2a)(2b + 1)(1 + 2b)$

10. (a) $(x + y^2)(x + y^2)(x^2 + y)(x^2 + y)$
    (b) $(x + y^2)(y^2 + x)(x^2 + y)(y + x^2)$

*For Exercises 11–26, evaluate the expressions using the given values of the variables.*

11. $x + 3y$; $x = 4, y = 6$    12. $a - 4b^2$; $a = 20, b = 2$

13. $a^2 + b^2$; $a = 3, b = 4$    14. $(a + b)^2$; $a = 3, b = 4$

15. $x^2 - 4y^2$; $x = 10, y = 4$

16. $x^2 - xy + y^2$; $x = 1, y = 1$

17. $x^2 - x + 1$; $x = 5$

18. $(x + 1)(x^2 - x + 1)$; $x = 5$

19. $2x^3 - 3y^2$; $x = 3, y = 2$

20. $2(x^3 - 3y^2)$; $x = 3, y = 2$

21. $1 \div a^2 + b^2$; $a = 1, b = 1$

22. $1 \div (a^2 + b^2)$; $a = 1, b = 1$

23. $\dfrac{1}{a^2} + \dfrac{1}{b^2} - \dfrac{1}{a^2 + b^2}$; $a = \dfrac{1}{2}, b = \dfrac{1}{2}$

24. $\dfrac{1}{1 - \dfrac{1}{x}}$; $x = 2$

25. $x^{y^z}$
    (a) $x = y = 2, z = 3$
    (b) $x = z = 2, y = 3$
    (c) $y = z = 2, x = 3$

26. $A^{B^C}$
    (a) $A = 3, B = 4, C = 2$
    (b) $A = 4, B = 2, C = 3$
    (c) $A = 4, B = 3, C = 2$

27. Which is larger: $(2^3)^4$ or $2^{3^4}$?

28. Compute $2^{3^2} - (2^3)^2$.

*In Exercises 29–42, translate each phrase into algebraic notation.*

29. Two more than $x$    30. Two less than $x$

31. Four times the sum of $a$ and $b^2$

32. The sum of four times $a$ and $b^2$

33. The sum of $x^2$ and $y^2$

34. The square of the sum of $x$ and $y$

35. The sum of $x$ and twice the square of $x$

36. Twice the sum of $x$ and the square of $x$

37. The average of $x$, $y$, and $z$

38. The square of the average of $x$, $y$, and $z$

39. The average of the squares of $x$, $y$, and $z$

40. The square of the average of the squares of $x$, $y$, and $z$

41. One less than twice the product of $x$ and $y$

42. The cube of three more than the product of $x^2$ and $y^3$

*For Exercises 43–58, make use of the geometric formulas that are given on the inside front cover of this book. In Exercises 49–52, which are calculator exercises, refer to the conventions for significant digits that are explained in Section A.1 of the appendix.*

43. Write expressions for the area and the circumference of a circle whose radius is $3y$.

44. Find an expression for the average of the areas of two circles whose radii are $r$ and $R$, respectively.

45. The radius of a certain sphere is $3x$. Find a formula expressing the volume V of the sphere in terms of $x$.

46. Write an expression for the surface area of a sphere whose radius is $x/2$.

47. Suppose that in a certain right circular cylinder, the height is equal to the radius.
    (a) Write a formula that expresses the volume of this cylinder in terms of the radius $r$.
    (b) Use the formula you found in part (a) to compute the volume of the cylinder when $r = 4$ cm.

48. Suppose that in a certain right circular cone, the height is six times the radius.
    (a) Write a formula expressing the volume in terms of the radius $r$.
    (b) Write a formula expressing the lateral surface area in terms of the radius $r$.

49. Compute the volume of a box with dimensions $l = 1.46$ cm, $w = 3.50$ cm, and $h = 2.12$ cm.

50. Which has the greater area, a circle with radius 5.15 cm or a square with side 9.12 cm?

51. Compute the volume of a right circular cone with radius 5.05 cm and height 3.25 cm.

52. The dimensions of a right circular cylinder and a right circular cone are as follows:

    cylinder:  $r = 3.76$ m; $h = 5.20$ m

    cone:  $r = 4.80$ m; $h = 6.50$ m

    Which has the larger lateral surface area, the cylinder or the cone?

## B

53. The accompanying figure shows a sphere of radius $r$ inscribed within a cylinder.
    (a) Find the ratio of the volume of the cylinder to the volume of the sphere.
    (b) Find the ratio of the total surface area of the cylinder to the surface area of the sphere.

54. The following figure shows a circle inscribed within a square of side $s$. Express the circumference of the circle in terms of $s$.

55. A certain rectangular box has length twice its width and height six times its width.
    (a) Express the volume of the box in terms of the width $w$.
    (b) Express the surface area of the box in terms of the width $w$.

56. Suppose that the length of each side of an equilateral triangle is $2s$.
    (a) Express the height of the triangle in terms of $s$.
    (b) Express the area of the triangle in terms of $s$.

57. (a) The dimensions of a right circular cylinder are $r = h = 4.48$ cm. Compute the ratio of the volume to the total surface area. Round off your answer to two decimal places.
    (b) By using more algebra and less arithmetic, you can compute the ratio in part (a) without a calculator. Do this and use the result to check your answer in part (a).

58. Assume that the Earth is a sphere of radius 4000 miles and that a string is tied around the equator.
    (a) Compute the length of the string to the nearest foot. (There are 5280 feet in one mile.)
    (b) If the radius of the earth were 1 ft longer, how much more string would be required (to the nearest foot)?

*For Exercises 59 and 60, evaluate the expressions using the given values of the variables.*

59. (a) $\dfrac{1000}{19 - \dfrac{12}{1 - 1/x}}$; $x = 3$

    (b) $\dfrac{1}{1 + \dfrac{1}{1 + \dfrac{1}{1 + 1/x}}}$; $x = 2$

60. (a) $[p(p^2 - 2^p)]^{p+1}$; $p = 3$

    (b) $\dfrac{\dfrac{a+b}{a-b} + \dfrac{a-b}{a+b}}{\dfrac{a-b}{a+b} - \dfrac{a+b}{a-b}} \cdot \dfrac{ab^3 - a^3b}{a^2 + b^2}$; $a = \dfrac{5}{4}$, $b = \dfrac{1}{4}$

# PROPERTIES OF THE REAL NUMBERS

**1.2**

*Today's familiar plus and minus signs were first used in 15th-century Germany as warehouse marks. They indicated when a container held something that weighed over or under a certain standard weight.*

MARTIN GARDNER, "Mathematical Games," *Scientific American*, June 1977

*I do not like × as a symbol for multiplication, as it is easily confounded with x; . . . often I simply relate two quantities by an interposed dot . . .*

G. W. LEIBNIZ in a letter dated July 29, 1698

In this section we first will list the basic properties for the real number system. After that we'll summarize some of the more down-to-earth implications of those properties, such as the procedures for working with signed numbers and fractions. (Section A.2 of the appendix, at the back of this book, presents a more theoretical treatment of some of these ideas.)

The set of real numbers is **closed** with respect to the operations of addition and multiplication. This just means that when we add or multiply two real numbers, the result (that is, the **sum** or the **product**) is again a real number. Some of the other most basic properties and definitions for the real number system are listed in the following box. In the box, the lowercase letters $a$, $b$, and $c$ denote arbitrary real numbers.

| PROPERTY SUMMARY | Some Fundamental Properties of the Real Numbers |
|---|---|
| Commutative properties | $a + b = b + a \qquad ab = ba$ |
| Associative properties | $a + (b + c) = (a + b) + c \qquad a(bc) = (ab)c$ |
| Identity properties | 1. There is a unique real number 0 (called **zero** or the **additive identity**) such that $$a + 0 = a \quad \text{and} \quad 0 + a = a$$ 2. There is a unique real number 1 (called **one** or the **multiplicative identity**) such that $$a \cdot 1 = a \quad \text{and} \quad 1 \cdot a = a$$ |
| Inverse properties | 1. For each real number $a$ there is a real number $-a$ (called the **additive inverse** of $a$ or the **negative** of $a$) such that $$a + (-a) = 0 \quad \text{and} \quad (-a) + a = 0$$ 2. For each real number $a \neq 0$, there is a real number denoted by $\frac{1}{a}$ (or $1/a$ or $a^{-1}$), and called the **multiplicative inverse** or **reciprocal** of $a$, such that $$a \cdot \frac{1}{a} = 1 \quad \text{and} \quad \frac{1}{a} \cdot a = 1$$ |
| Distributive properties | $a(b + c) = ab + ac \qquad (b + c)a = ba + ca$ |

On reading this list of properties for the first time, many students ask the natural question, "Why do we even bother to list such obvious properties?" One reason is that all the other laws of arithmetic and algebra (including the "not-so-obvious" ones) can be derived from our rather short list. For example, the rule $0 \cdot a = 0$ can be proved using the distributive property, as can the rule that the product of two negative numbers is a positive number. (See Section A.2 of the appendix for these proofs.) The distributive properties in a sense indicate the way in which multiplication and addition interact with each other. More specifically, each of the following examples relies on the distributive properties.

| EXAMPLE | REMARK |
|---|---|
| $12 \times 103 = 12(100 + 3)$ <br> $= 1200 + 36 = 1236$ | The distributive property can be used as an aid in mental calculations. |
| $3a^2(a + 2b) = 3a^2 \cdot a + 3a^2 \cdot 2b$ <br> $= 3a^3 + 6a^2b$ | An application of the distributive property, followed by a simplification. |
| $A(c + d) = Ac + Ad$ | A direct application of the distributive property. In the next example we are going to replace $A$ with the quantity $(a + b)$. |
| $(a + b)(c + d) = (a + b)c + (a + b)d$ <br> $= ac + bc + ad + bd$ | The distributive properties are used three times here. |
| $(x + 3)(x + 6) = (x + 3)x + (x + 3)6$ <br> $= x^2 + 3x + 6x + 18$ <br> $= x^2 + 9x + 18$ | This is similar to the previous example. Examples of this type will be discussed in greater detail in Section 1.8. |

The distributive and associative properties can be generalized. The **generalized distributive property** asserts that

$$a(b_1 + b_2 + \cdots + b_n) = ab_1 + ab_2 + \cdots + ab_n$$

The generalized associative property tells us that we can group the terms as we please in computing a sum or a product. For instance, we have

$$17 + 47 + 53 + 80 = 17 + [(47 + 53) + 80]$$
$$= 17 + [100 + 80]$$
$$= 17 + 180 = 197$$

**EXAMPLE 1**   Identify the property of the real numbers that justifies each statement.

**(a)**  $(17 \times 50) \times 2 = 17 \times (50 \times 2) = 17 \times 100 = 1700$

**(b)**  Since $\pi$ and $\sqrt{2}$ are real numbers, so is $\pi + \sqrt{2}$.

**(c)** $3 + y^2 = y^2 + 3$

**(d)** $[y + (z + 1)]x = yx + (z + 1)x$

**(e)** $(z + 1) + [-(z + 1)] = 0$

**(f)** $(a^2 + 1) \cdot \left(\dfrac{1}{a^2 + 1}\right) = 1$

**Solution**  **(a)** Associative property of multiplication

**(b)** Closure property with respect to addition

**(c)** Commutative property of addition

**(d)** Distributive property

**(e)** Additive inverse property

**(f)** Multiplicative inverse property

If you look back at the box on page 6, you'll see that the operations of subtraction and division are not mentioned. This is because addition and multiplication are actually the more fundamental operations. In fact, we can define subtraction and division in terms of addition and multiplication as follows.

---

**Definitions of Subtraction and Division**

$$a - b = a + (-b)$$

$$a \div b = a \cdot \left(\frac{1}{b}\right) = ab^{-1} \qquad \text{provided that } b \neq 0$$

---

Division by zero is not defined. To see why this must be the case, suppose just for the moment that division by zero were defined. Then we would have $\frac{1}{0} = k$ for some real number $k$. But then, multiplying both sides of the equation by 0 would yield $1 = k \cdot 0$; that is, $1 = 0$, a contradiction. In summary, if we wish the rules of arithmetic and algebra to be consistent, then division by zero must remain undefined.

In basic algebra you were introduced to a number of procedures that were useful in working with algebraic expressions. In the three boxes that follow, we summarize the properties of the real numbers that are behind those procedures you learned. In the statements of these properties, the lowercase letters $a$, $b$, $c$, and $d$ are used to denote arbitrary real numbers. In the properties of fractions, of course, we are assuming that none of the denominators is zero.

---

**PROPERTY SUMMARY**  Properties of Zero

| PROPERTY | EXAMPLE |
|---|---|
| 1. $a \cdot 0 = 0$ | $(x^3 - \pi) \cdot 0 = 0$ |
| 2. If $\dfrac{a}{b} = 0$, then $a = 0$. | If $\dfrac{x^2 + x - 1}{2x} = 0$, then $x^2 + x - 1 = 0$. |
| 3. If $ab = 0$, then $a = 0$ or $b = 0$. | If $5x = 0$, then $x = 0$. |

| PROPERTY SUMMARY | Properties of Negatives |
|---|---|

| PROPERTY | EXAMPLE |
|---|---|
| 1. $-(-a) = a$ | $-[-(x+1)] = x + 1$ |
| 2. $(-a)b = -(ab) = a(-b)$ | $(-3)4 = -12 = 3(-4)$ |
| 3. $(-a)(-b) = ab$ | $[-(x+y)][-(x+y)] = (x+y)^2$ |
| 4. $-a = (-1)a$ | $-(x + y - z) = -1(x + y - z) = -x - y + z$ |

| PROPERTY SUMMARY | Properties of Fractions |
|---|---|

(The denominators in each case are assumed to be nonzero.)

| PROPERTY | EXAMPLE |
|---|---|
| 1. $\dfrac{a}{b} = \dfrac{c}{d}$ if and only if $ad = bc$ | $\dfrac{8}{12} = \dfrac{2}{3}$ since $8 \cdot 3 = 12 \cdot 2$ |
| 2. $\dfrac{a}{b} = \dfrac{ac}{bc}$ | $\dfrac{4}{7} = \dfrac{4 \cdot 5}{7 \cdot 5} = \dfrac{(4 \cdot 5)\pi}{(7 \cdot 5)\pi}$ |
| 3. $\dfrac{a}{b} \pm \dfrac{c}{b} = \dfrac{a \pm c}{b}$ | $\dfrac{10}{3} + \dfrac{3}{3} = \dfrac{13}{3}; \quad \dfrac{10}{3} - \dfrac{3}{3} = \dfrac{7}{3}$ |
| 4. $\dfrac{a}{b} \pm \dfrac{c}{d} = \dfrac{ad \pm bc}{bd}$ | $\dfrac{2}{3} + \dfrac{3}{4} = \dfrac{2 \cdot 4 + 3 \cdot 3}{3 \cdot 4} = \dfrac{17}{12}; \quad \dfrac{2}{3} - \dfrac{3}{4} = \dfrac{2 \cdot 4 - 3 \cdot 3}{3 \cdot 4} = \dfrac{-1}{12}$ |
| 5. $\dfrac{a}{b} \cdot \dfrac{c}{d} = \dfrac{ac}{bd}$ | $\dfrac{5}{4} \cdot \dfrac{7}{9} = \dfrac{35}{36}$ |
| 6. $\dfrac{a}{b} \div \dfrac{c}{d} = \dfrac{a}{b} \cdot \dfrac{d}{c} = \dfrac{ad}{bc}$ | $\dfrac{3}{5} \div \dfrac{2}{7} = \dfrac{3}{5} \cdot \dfrac{7}{2} = \dfrac{21}{10}; \quad 1 \div \dfrac{1}{10} = \dfrac{1}{1} \cdot \dfrac{10}{1} = 10$ |
| 7. $-\dfrac{a}{b} = \dfrac{-a}{b} = \dfrac{a}{-b}$ | $-\dfrac{3}{4} = \dfrac{-3}{4} = \dfrac{3}{-4}; \quad -\dfrac{x-1}{x+1} = \dfrac{-(x-1)}{x+1} = \dfrac{x-1}{-(x+1)}$ |
| 8. $\dfrac{a-b}{b-a} = -1$ | $\dfrac{7-x}{x-7} = -1; \quad \dfrac{x^2 - y^2}{y^2 - x^2} = -1$ |

For the next example, you'll need to recall the definition of least common multiple. The **least common multiple** (l.c.m.) is the smallest positive integer that is a multiple of each of the given integers. *Examples:* The l.c.m. of 2 and 3 is 6; the l.c.m. of 3, 5, and 6 is 30.

**EXAMPLE 2**  Simplify the following expression (that is, write it as a single fraction).

$$\frac{\frac{5}{4} + \frac{7}{3}}{\frac{1}{6} + \frac{3}{2}}$$

**Solution**  The individual denominators are 4, 3, 6, and 2. Since the least common multiple of these numbers is 12, we multiply the entire original fraction by $\frac{12}{12}$ (which is equal to 1) to obtain

$$\frac{12}{12} \cdot \frac{\frac{5}{4} + \frac{7}{3}}{\frac{1}{6} + \frac{3}{2}} = \frac{12(\frac{5}{4} + \frac{7}{3})}{12(\frac{1}{6} + \frac{3}{2})}$$

$$= \frac{15 + 28}{2 + 18}$$

$$= \frac{43}{20}$$

## EXERCISE SET 1.2

### A

*For Exercises 1–10, use the distributive properties to carry out the indicated multiplication.*

1. $8 \times 104 = 8(100 + 4) = ?$

2. $17 \times 102 = 17(100 + 2) = ?$

3. $45 \times 98 = 45[100 + (-2)] = ?$

4. (a) $2a[3a + (-1)]$
   (b) $2a(3a - 4)$
   (c) $2a(3a + b - 4)$

5. (a) $(A + B)(A + B)$
   (b) $(A + B)(A - B)$
   (c) $(A - B)(A - B)$

6. (a) $(A + B)(C + D)$
   (b) $(A + B)(2A + 3B)$
   (c) $(A + B)(2A - 3B)$

7. (a) $(x + 2y)(x + 2y)$
   (b) $(x + 2y)(x - 2y)$
   (c) $(x - 2y)(x - 2y)$

8. (a) $(x + 3)(x + 3)$
   (b) $(x + 3)(x - 3)$
   (c) $(x^2 + 3)(x^2 - 3)$

9. (a) $A(x^2 - x + 1)$
   (b) $(x + 1)(x^2 - x + 1)$

10. (a) $(2x + 3)(x - 1)$
    (b) $(3x + 2)(x - 1)$

*In Exercises 11–24, identify the property or properties of the real numbers that justify each statement.*

11. $x + 3y = 3y + x$

12. $x(3y) = (3y)x$

13. $\frac{a^2}{b^2} + \left(-\frac{a^2}{b^2}\right) = 0$

14. $6(x + 8) = 6x + 48$

15. $(x + 1) + y = x + (1 + y)$

16. $(x + y) \cdot \left(\frac{1}{x + y}\right) = 1$

17. $(x + y)[(x + y)(x - y)] = (x + y)^2(x - y)$

18. $(\sqrt{2})(1) = \sqrt{2}$          19. $(5 + \pi) + 0 = 5 + \pi$

20. $[(x + y) + (x^2 + y^2)]x = (x + y)x + (x^2 + y^2)x$

21. Since $\sqrt{7}$ and $\sqrt{\pi}$ are real numbers, so is $\sqrt{7} + \sqrt{\pi}$.

22. $y(y + y^2 + y^3) = y^2 + y^3 + y^4$

23. $(x + 2)(x^2 + 4) = (x + 2)x^2 + (x + 2)4$

24. Since $\dfrac{1}{\sqrt{7}}$ and $\sqrt{5}$ are real numbers, so is $\left(\dfrac{1}{\sqrt{7}}\right)(\sqrt{5})$.

*For Exercises 25–50, simplify the given expressions.*

25. $1 + 2(-3)(4)$

26. $(97)(-98)(99)(0)$

27. $4 - (-3 + 5)$

28. $(-1)^4$

29. $-[(-1)^4]$

30. $-1^4$

31. $1 - \{1 - [-(-1 - 1)]\}$

32. $4^2 - (-4)^3$

33. $\frac{7}{5} + \frac{2}{3}$

34. $\frac{3}{4} - \frac{4}{3}$

35. $\dfrac{7}{x} - \dfrac{12}{x}$

36. $\dfrac{a+1}{a^2+1} + \dfrac{a-1}{a^2+1}$

37. $\dfrac{x^2 + y^2}{x + y} - \dfrac{x^2 - y^2}{x + y}$

38. $\frac{7}{12} \cdot \frac{5}{4}$

39. $\frac{7}{12} \cdot \left(\frac{-7}{3}\right)$

40. $1 \div \frac{1}{3}$

41. $\frac{2}{5} \div \frac{7}{3}$

42. $1 \div \frac{2}{3} + 1$

43. $1 \div (\frac{2}{3} + 1)$

44. $\dfrac{\frac{4}{5} + \frac{1}{2}}{\frac{5}{2}}$

45. (a) $\dfrac{\frac{4}{3} + \frac{1}{2}}{\frac{11}{6} + \frac{5}{4}}$   (b) $\dfrac{\frac{4}{3} - \frac{1}{2}}{\frac{1}{2} - \frac{4}{3}}$

46. (a) $\dfrac{b - a}{a - b}$   (b) $\dfrac{x - 7}{7 - x}$   (c) $\dfrac{3 - x^4}{x^4 - 3}$

47. $\dfrac{1 - (x + y)}{x + y - 1}$

48. $\dfrac{\frac{1}{5} - \frac{1}{4}}{\frac{1}{5} + \frac{1}{4}}$

49. $\dfrac{\frac{1}{2} - \frac{1}{3} + \frac{1}{4}}{\frac{1}{2} + \frac{1}{3} - \frac{1}{4}}$

50. $1 + \dfrac{1}{1 - \dfrac{1}{1 - \frac{1}{5}}}$

*In Exercises 51–55, evaluate each expression using the given value for x.*

51. $x^2 - 8x - 4$; $x = -2$

52. $2x^3 + 4x^2 - 1$; $x = -3$

53. $8x^3 - 4x^2 - 4x$; $x = -\frac{1}{2}$

54. $\dfrac{x^2 - 2x + 1}{x^3 - 3x^2 + 3x - 1}$; $x = -4$

55. $1 - x - x^2 - x^3 - x^4$; $x = -1$

56. If $m = \dfrac{y_2 - y_1}{x_2 - x_1}$, find $m$ when $y_2 = 3$, $y_1 = -1$, $x_2 = 1$, and $x_1 = 4$.

57. If $d^2 = (x_2 - x_1)^2 + (y_2 - y_1)^2$, compute the value of $d^2$ when $x_2 = -1$, $x_1 = -2$, $y_2 = 6$, and $y_1 = -6$.

**B**

58. In each case, give an example to show that the following are, in general, false.

(a) $\dfrac{1}{a + b} = \dfrac{1}{a} + \dfrac{1}{b}$   (b) $\dfrac{a}{b} + \dfrac{c}{d} = \dfrac{a + c}{b + d}$

(c) $\dfrac{a + b}{b} = a + 1$   (d) $a - (b - c) = a - b - c$

(e) $a^2 + a^4 = a^6$   (f) $(a + b)^2 = a^2 + b^2$

*Exercises 59–62 ask you to show that certain fractions are equal. To do this, make use of Property 1 in the summary box for fractions on page 9. (In each of the exercises assume that the denominators are nonzero.)*

59. If $\dfrac{a}{b} = \dfrac{c}{d}$, show that $\dfrac{a}{c} = \dfrac{b}{d}$.

60. If $\dfrac{a}{b} = \dfrac{c}{d}$, show that $\dfrac{a + b}{b} = \dfrac{c + d}{d}$.

61. If $\dfrac{a}{b} = \dfrac{c}{d}$, show that $\dfrac{a - b}{b} = \dfrac{c - d}{d}$.

62. If $\dfrac{a}{b} = \dfrac{c}{d}$, show that $\dfrac{a + b}{a - b} = \dfrac{c + d}{c - d}$.

☐ 63. If $y = x^3 - 3x^2 - 6x - 4$, find $y$ when $x = -0.75$.

☐ 64. If $y = \dfrac{x^2 - 4x + 1}{x^3 - x^2 - 3x - 1}$, compute $y$ when $x = -1.01$.

☐ 65. Let $y = \dfrac{x^2 - 1}{x^3}$.

(a) Complete the following table.

| $x$ | 1 | 1.250 | 1.500 | 1.750 | 2.000 | 2.250 |
|-----|---|-------|-------|-------|-------|-------|
| $y$ |   |       |       |       |       |       |

(b) Which $x$-value in the table corresponds to the largest $y$-value?

(c) It can be shown using calculus that the (positive) $x$-value that produces the largest possible $y$-value for the equation $y = \dfrac{x^2 - 1}{x^3}$ is $x = \sqrt{3}$. Use a calculator to compute this largest $y$-value and compare your answer with the $y$-values obtained in part (a).

☐ 66. Let $y = x^4 - 2x^3 - 12x^2 + 13$.

(a) Complete the following table.

| $x$ | 3.000 | 3.100 | 3.200 | 3.300 | 3.400 | 3.500 |
|-----|-------|-------|-------|-------|-------|-------|
| $y$ |       |       |       |       |       |       |

(b) Which $x$-value in the table yields the smallest $y$-value?

(c) It can be shown using calculus that the $x$-value that produces the smallest possible $y$-value for the equation $y = x^4 - 2x^3 - 12x^2 + 13$ is $x = \dfrac{3 + \sqrt{105}}{4}$. Which $x$-value in the table in part (a) is closest to this? Compute the $y$-value corresponding to $x = \dfrac{3 + \sqrt{105}}{4}$.

## SETS OF REAL NUMBERS

**1.3**

Certain collections of real numbers are referred to often enough to be given special names. These are summarized in the box that follows.

| PROPERTY SUMMARY | Sets of Real Numbers | |
|---|---|---|
| NAME | DEFINITION AND COMMENTS | EXAMPLES |
| Natural numbers | These are the ordinary counting numbers, 1, 2, 3, and so on. | 1, 4, 29, 1066 |
| Integers | These are the natural numbers along with their negatives and zero. | $-26, 0, 1, 1989$ |
| Rational numbers | As the name suggests, these are the real numbers that are *ratios* of two integers (with nonzero denominators, of course). In terms of decimal expansions, it can be proved that a real number is rational if and only if its decimal expansion terminates (e.g., 3.15) or is periodic (e.g., $2.\overline{4}$). | $4\ (=\frac{4}{1}), -\frac{2}{3}$ $8.7, 4.\overline{3}, 4.1\overline{73}$ |
| Irrational numbers | These are the real numbers that are not rational. It can be shown that any number of the form $\sqrt{n}$, where $n$ is a natural number that is not a perfect square, is irrational.* Also, any number that is the sum or the (nonzero) product of a rational number and an irrational number is irrational. (More generally, it can be shown that if the decimal expansion of a real number neither terminates nor is periodic, then the number is irrational.) | $\sqrt{2}, \sqrt{14}, \pi$ $1 + \sqrt{2}, 4\pi$ |

According to the comments in the box, the repeating decimal number $2.\overline{4}$ is rational. In other words, there is some fraction $p/q$ such that $2.\overline{4} = p/q$. How can we find this fraction? Example 1 shows one way to do this.

**EXAMPLE 1**   Express the repeating decimal $2.\overline{4}$ in the form $p/q$, where $p$ and $q$ are integers and $q \neq 0$.

Solution   Let $x = 2.444\dots$. Then $10x = 24.44\dots$, and we have

$$10x - x = 24.\overline{4} - 2.\overline{4} = 22$$

or

$$9x = 22$$

$$x = \frac{22}{9}$$

We now have $2.\overline{4} = \frac{22}{9}$, as required.

Actually, a rigorous justification of each step used in Example 1 would require certain techniques from calculus. Nevertheless, you can easily verify for yourself that the answer in Example 1 is correct; just divide 22 by 9 and see what you find.

---

* The fact that $\sqrt{2}$ is irrational (that is, not a ratio of integers) was known to Pythagoras more than two thousand years ago. Section A.3 of the appendix, at the back of this book, contains a proof of the fact that the number $\sqrt{2}$ is irrational. The proof that $\pi$ is irrational is more difficult. The first person to prove that $\pi$ is irrational was the Swiss mathematician J. H. Lambert (1728–1777).

FIGURE 1

In previous courses you learned that the real numbers could be viewed as points on a *number line*. Because this is such an important idea, let's take a moment to review the essential features. Beginning with a straight line, we choose two points, $O$ and $P$, as shown in Figure 1(a). We call $O$ the **origin**. With the points $O$ and $P$ we identify the numbers 0 and 1, as shown in Figure 1(b).

FIGURE 2

Now, using the distance from $O$ to $P$ as our unit of length, we mark off points at one-unit intervals on both sides of the origin. As indicated in Figure 2, the points that we've marked to the right of the origin are identified with the numbers 1, 2, 3, and so on; the points to the left of the origin are identified with $-1$, $-2$, $-3$, and so on. A fundamental fact is that there is a **one-to-one correspondence** between the set of real numbers and the set of points on the line. This means that each real number is identified with exactly one point on the line; conversely, with each point on the line we identify exactly one real number. The real number associated with a given point is called the **coordinate** of the point. As a practical matter, we're usually more interested in relative locations than precise locations on a number line. For instance, since $\pi$ is approximately 3.1, we show $\pi$ slightly to the right of 3 in Figure 3. Similarly, since $\sqrt{2}$ is approximately 1.4, we show $\sqrt{2}$ slightly less than halfway from 1 to 2 in Figure 3.

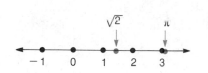

FIGURE 3

There are times when it is convenient to use number lines that show reference points other than the integers used in Figure 2. For instance, Figure 4(a) displays a number line with reference points that are multiples of $\pi$. In this case, of course, it is the integers that we then locate approximately. For example, in Figure 4(b) we show the approximate location of the number 1 on such a line.

FIGURE 4

Two of the most basic relations for real numbers are **less than** and **greater than**, symbolized by $<$ and $>$, respectively. For ease of reference, we define these and two related symbols in the box that follows. [Incidentally, the symbols $<$ and $>$ were invented by the English mathematician and astronomer Thomas Harriot (1560–1621).]

| PROPERTY SUMMARY | | Notation for Less than and Greater than |
| --- | --- | --- |
| NOTATION | DEFINITION | EXAMPLES |
| $a < b$ | $a$ is less than $b$. On a number line, oriented as in Figure 2, 3, or 4, the point $a$ lies to the left of $b$. | $2 < 3$; $\;-4 < 1$ |
| $a \le b$ | $a$ is less than or equal to $b$. | $2 \le 3$; $\;3 \le 3$ |
| $b > a$ | $b$ is greater than $a$. On a number line oriented as in Figure 2, 3, or 4, the point $b$ lies to the right of $a$. [$b > a$ is equivalent to $a < b$.] | $3 > 2$; $\;0 > -1$ |
| $b \ge a$ | $b$ is greater than or equal to $a$. | $3 \ge 2$; $\;3 \ge 3$ |

In general, relationships involving real numbers and any of the four symbols $<$, $\le$, $>$, and $\ge$ are called **inequalities**. One of the simplest uses of inequalities is in defining certain sets of real numbers called *intervals*. Roughly speaking, any uninterrupted portion of the number line is referred to as an **interval**. In the definitions that follow, you'll see notation such as $a < x < b$. This means that *both* of the inequalities $a < x$ and $x < b$ hold; in other words, the number $x$ is between $a$ and $b$.

**(a)** The open interval $(a, b)$ contains all real numbers from $a$ to $b$, excluding $a$ and $b$.

| Definition:   Open Intervals and Closed Intervals |
| --- |
| The **open interval** $(a, b)$ consists of all real numbers $x$ such that $a < x < b$. See Figure 5(a). |
| The **closed interval** $[a, b]$ consists of all real numbers $x$ such that $a \le x \le b$. See Figure 5(b). |

**(b)** The closed interval $[a, b]$ contains all real numbers from $a$ to $b$, including $a$ and $b$.

**FIGURE 5**

Notice that the brackets in Figure 5(b) are used to indicate that the numbers $a$ and $b$ are included in the interval $[a, b]$, whereas the parentheses in Figure 5(a) indicate that $a$ and $b$ are excluded from the interval $(a, b)$. At times you'll see notation such as $[a, b)$. This stands for the set of all real numbers $x$ such that $a \le x < b$. Similarly, $(a, b]$ denotes the set of all real numbers $x$ such that $a < x \le b$.

**EXAMPLE 2**    Show the following intervals on a number line, and specify inequalities describing the numbers $x$ in each interval.

$$[-1, 2] \qquad (-1, 2) \qquad (-1, 2] \qquad [-1, 2)$$

**Solution**    See Figure 6.

FIGURE 6

Although each interval depicted in Figure 6 is of finite length, this need not always be the case. For instance, in Figures 7 and 8 we show intervals that extend indefinitely in one direction or the other. These are called **unbounded intervals**. Also, the entire real number line itself is considered to be an unbounded interval. (We'll have more to say about unbounded intervals when we solve inequalities in Section 2.5.)

**FIGURE 7**  The set of real numbers $x$ such that $x \geq 2$.

**FIGURE 8**  The set of real numbers $x$ such that $x < 1$.

## EXERCISE SET 1.3

### A

*For Exercises 1–12, specify the category (or categories) to which the numbers belong. For convenience in writing your answers, use the following abbreviations: nat.—natural number; int.— integer; rat.—rational number; irr.—irrational number.*

**Examples**

(a) $-27$: int., rat.  (b) $\sqrt{2}$: irr.

1.  (a) 7  (b) $-7$
2.  (a) $-203$  (b) $\frac{204}{2}$
3.  (a) $\frac{27}{4}$  (b) $\sqrt{\frac{27}{4}}$
4.  (a) $\sqrt{7}$  (b) $4\sqrt{7}$
5.  (a) $10^6$  (b) $10^6/10^7$
6.  (a) 8.7  (b) $8.\overline{7}$
7.  (a) 8.74  (b) $8.\overline{74}$
8.  (a) $\sqrt{99}$  (b) $\sqrt{99} + 1$
9.  $3\sqrt{101} + 1$
10.  $\dfrac{\sqrt{5} + 1}{4}$
11.  $(3 - \sqrt{2}) + (3 + \sqrt{2})$
12.  $\dfrac{0.1234}{0.5677}$

*In Exercises 13–18, each of the given numbers is rational because the decimal expansion either terminates or is periodic. In each case, express the number in the form p/q, where p and q are integers and $q \neq 0$.*

13.  (a) 5.4  (b) $5.\overline{4}$
14.  (a) 2.8  (b) $2.\overline{8}$
15.  (a) 0.99  (b) $0.\overline{9}$
16.  (a) 0.66  (b) $0.\overline{6}$
17.  (a) 1.7  (b) $1.\overline{7}$
18.  (a) 9.001  (b) $9.\overline{1}$

19.  Draw a number line similar to the one shown in Figure 2 in the text. Then show the approximate locations of $\sqrt{5} + 3$ and $\sqrt{5} - 2$ on the line.  *Note:*  $\sqrt{5} = 2.24$, to three significant digits.

20.  Draw a number line similar to the one shown in Figure 2 in the text. Then show the approximate locations of $\pi/2$ and $\pi/4$ on the line.

21.  Draw a number line similar to the one shown in Figure 4(a) in the text. Then show the approximate locations of 6 and $-3$ on the line.

ⓒ 22.  Draw a number line similar to the one shown in Figure 2 in the text. Then show the approximate locations of $\sqrt{2}$, $\sqrt{3}$, and $\sqrt{5}$ on the line.

*For Exercises 23–28, say whether the statement is true or false.*

23.  $-5 < -100$
24.  $0 \leq -1$
25.  $-1 \leq -1$
26.  $\sqrt{2} + \sqrt{5} \geq 0$
27.  $\frac{15}{16} > \frac{17}{18}$
28.  $0.\overline{8} > 0.8$

*In Exercises 29–36, show the intervals on a number line. For Exercises 29–32, specify inequalities describing the numbers $x$ in each interval.*

29.  (a) $(-1, 3)$  (b) $[-1, 3]$
30.  (a) $[4, 5]$  (b) $(4, 5)$
31.  (a) $[-2, \frac{3}{2})$  (b) $(-2, \frac{3}{2}]$
32.  (a) $(-3, -2]$  (b) $[-3, -2)$
33.  All real numbers $x$ such that $x > -2$
34.  All real numbers $x$ such that $x \leq 5$
35.  All real numbers $x$ such that $-1 < x < 1$
36.  All real numbers $x$ such that $x \geq \sqrt{2}$

**B**

*In Exercises 37–40, each of the given numbers is rational because the decimal expansion is periodic. In each case, express the number in the form p/q, where p and q are integers and q ≠ 0.*

37. $0.\overline{19}$   *Hint:*   Let $x = 0.\overline{19}$ and consider $100x - x$.

38. $61.\overline{26}$   (See the hint given for Exercise 37.)

39. $0.3121212\ldots$   *Hint:*   Let $x = 0.3\overline{12}$ and consider $1000x - 10x$.

40. $0.\overline{142857}$   (Adapt the hint given in Exercise 37.)

$\boxed{\text{c}}$ 41. From grade school on, we all acquire a good deal of experience in working with rational numbers. Certainly we can tell when two rational numbers are equal, although it may require some calculation, as, for example, in the case of $\frac{129}{31} = \frac{2193}{527}$. The point of this exercise is to show you that, in working with irrational numbers, sometimes it's hard to tell (or even to show) that two numbers are equal. In each of the following cases, use a calculator to verify that the quantities on both sides of the equation agree, to six decimal places. (In each case it can be shown that the two numbers are indeed equal.)

(a) $\sqrt{6} + \sqrt{2} = 2\sqrt{2 + \sqrt{3}}$

(b) $\sqrt{3 + \sqrt{5}} + \sqrt{3 - \sqrt{5}} = \sqrt{10}$

(c) $\sqrt{\sqrt{6 + 4\sqrt{2}}} = \sqrt{2 + \sqrt{2}}$

(d) $\sqrt{\frac{3}{2} + \frac{2}{\sqrt{2}}} + \sqrt{\frac{3}{2} - \frac{2}{\sqrt{2}}} = 2$

$\boxed{\text{c}}$ 42. The value of the irrational number $\pi$, correct to ten decimal places (without rounding off), is 3.1415926535. By using a calculator, determine to how many decimal places each of the following quantities agrees with $\pi$.

(a) $\frac{22}{7}$   (b) $\frac{355}{113}$   (c) $\frac{63}{25}\left(\frac{17 + 15\sqrt{5}}{7 + 15\sqrt{5}}\right)$

*Remark:*   A simple approximation that agrees with $\pi$ through the first 14 decimal places is $\frac{355}{113}\left(1 - \frac{0.0003}{3533}\right)$. This approximation was discovered by the Indian mathematician Srinivasa Ramanujan (1887–1920). For a fascinating account of the history of $\pi$, see the book by Petr Beckmann, *A History of $\pi$*, 3rd ed. (New York: St. Martin's Press, 1974).

## ABSOLUTE VALUE

**1.4**

*There has been a real need in analysis for a convenient symbolism for "absolute value" . . . and the two vertical bars introduced in 1841 by Weierstrass, as in $|z|$, have met with wide adoption; . . .*

FLORIAN CAJORI, *A History of Mathematical Notations, Vol. 1* (La Salle, Illinois: The Open Court Publishing Company, 1928)

As an aid in measuring distances on the number line, we introduce the concept of *absolute value*. We will begin with a definition of absolute value that is geometric in nature. Then, after you have developed some familiarity with the concept, we will explain a more algebraic approach that is often useful in analytical work.

> **Definition:   Absolute Value**
>
> The **absolute value** of a real number $x$, denoted by $|x|$, is the distance from $x$ to the origin.

For instance, because the numbers 5 and $-5$ are both five units from the origin, we have $|5| = 5$ and $|-5| = 5$. Here are three more examples:

$$|17| = 17 \qquad \left|-\frac{2}{3}\right| = \frac{2}{3} \qquad |0| = 0$$

In dealing with an expression such as $|-5 + 3|$, the convention is to compute the quantity $-5 + 3$ first and then to take the absolute value. We therefore have, in this case,

$$|-5 + 3| = |-2| = 2$$

**EXAMPLE 1**  Evaluate each expression.

(a) $5 - |6 - 7|$      (b) $||-2| - |-3||$

Solution    (a) $5 - |6 - 7| = 5 - |-1|$      (b) $||-2| - |-3|| = |2 - 3|$
$$= 5 - 1 = 4 \qquad\qquad\qquad = |-1| = 1$$

As we said at the beginning of this section, there is an equivalent, more algebraic way to define absolute value. According to this equivalent definition, the value of $|x|$ is $x$ itself when $x \geq 0$, and the value of $|x|$ is $-x$ when $x < 0$. We can write this symbolically as follows:

$$|x| = \begin{cases} x & \text{when } x \geq 0 \\ -x & \text{when } x < 0 \end{cases}$$

By looking at examples with specific numbers, you should be able to convince yourself that both definitions yield the same results. We use the algebraic definition of absolute value in Examples 2 and 3.

**EXAMPLE 2**  Rewrite each expression in a form that does not contain absolute values.

(a) $|\pi - 4| + 1$
(b) $|x - 5|$      given that $x \geq 5$
(c) $|t - 5|$      given that $t < 5$

Solution    (a) The quantity $\pi - 4$ is negative (since $\pi \approx 3.14$), and therefore its absolute value is equal to $-(\pi - 4)$. In view of this, we have

$$|\pi - 4| + 1 = -(\pi - 4) + 1 = -\pi + 5$$

(b) Since $x \geq 5$, the quantity $x - 5$ is nonnegative, and therefore its absolute value is equal to $x - 5$ itself. Thus, we have

$$|x - 5| = x - 5 \qquad \text{when } x \geq 5$$

(c) Since $t < 5$, the quantity $t - 5$ is negative, and therefore its absolute value is equal to $-(t - 5)$, which in turn is equal to $5 - t$. In view of this, we have

$$|t - 5| = 5 - t \qquad \text{when } t < 5$$

**EXAMPLE 3**  Simplify the expression $|x - 1| + |x - 2|$, given that $x$ is in the open interval $(1, 2)$.

Solution    Since $x$ is greater than 1, the quantity $x - 1$ is positive, and consequently,

$$|x - 1| = x - 1$$

On the other hand, we are also given that $x$ is less than 2. Therefore, the quantity $x - 2$ is negative, and we have

$$|x - 2| = -(x - 2) = -x + 2$$

Putting things together now, we can write

$$|x - 1| + |x - 2| = (x - 1) + (-x + 2)$$
$$= -1 + 2 = 1$$

In the box that follows, we list several basic properties of absolute value. Each of these properties can be derived from the definitions. (With the exception of the *triangle inequality*, we shall omit the derivations. For a proof of the triangle inequality, see Exercise 71.)

| PROPERTY SUMMARY | Properties of Absolute Value |
|---|---|

1. For all real numbers $x$, we have
   (a) $|x| \geq 0$
   (b) $x \leq |x|$   and   $-x \leq |x|$

2. For all real numbers $a$ and $b$, we have
   (a) $|ab| = |a||b|$   and   $|a/b| = |a|/|b|$   $(b \neq 0)$
   (b) $|a + b| \leq |a| + |b|$      (the triangle inequality)

**EXAMPLE 4**   Write the expression $|-2 - x^2|$ in an equivalent form that does not contain absolute values.

**Solution**   We have

$$|-2 - x^2| = |-1(2 + x^2)|$$
$$= |-1||2 + x^2| \quad \text{using Property 2(a)}$$
$$= 2 + x^2$$

The last equality follows from the fact that $x^2$ is nonnegative for any real number $x$ and, consequently, the quantity $2 + x^2$ is positive.

Distance $= |5 - 7| = |7 - 5| = 2$

**FIGURE 1**

If we think of the real numbers as points on a number line, the distance between two numbers $a$ and $b$ is given by the absolute value of their difference. For instance, as indicated in Figure 1, the distance between 5 and 7, namely 2, is given by either $|5 - 7|$ or $|7 - 5|$. For reference, we summarize this simple but important fact as follows.

| PROPERTY SUMMARY | Distance on a Number Line |
|---|---|

The **distance** between $a$ and $b$ is $|a - b| = |b - a|$.

**EXAMPLE 5**   Rewrite each of the following statements using absolute values.

(a) The distance between 12 and $-5$ is 17.
(b) The distance between $x$ and 2 is 4.

(c) The distance between $x$ and 2 is less than 4.

(d) The point $t$ is more than five units from the origin.

**Solution**

(a) $|12 - (-5)| = 17$ or $|-5 - 12| = 17$

(b) $|x - 2| = 4$ or $|2 - x| = 4$

(c) $|x - 2| < 4$ or $|2 - x| < 4$

(d) $|t| > 5$

**EXAMPLE 6** In each case, the set of real numbers satisfying the given inequality is one or more intervals on the number line. Show the intervals on a number line.

(a) $|x| < 2$    (b) $|x| > 2$    (c) $|x - 3| < 1$    (d) $|x - 3| \geq 1$

**Solution**

(a) The given inequality tells us that the distance from $x$ to the origin is less than two units. So, as indicated in Figure 2(a), the point $x$ must lie in the open interval $(-2, 2)$.

(b) The condition $|x| > 2$ means that $x$ is more than two units from the origin. Thus, as indicated in Figure 2(b), the point $x$ lies either to the right of 2 or to the left of $-2$.

(c) The given inequality tells us that $x$ must be less than one unit away from 3 on the number line. Looking one unit to either side of 3, then, we see that $x$ must lie between 2 and 4 and $x$ cannot equal 2 or 4. See Figure 2(c).

(d) The inequality $|x - 3| \geq 1$ says that $x$ is at least one unit away from 3 on the number line. This means that either $x \geq 4$ or $x \leq 2$, as shown in Figure 2(d). [Here's an alternate way of thinking about this. The numbers satisfying the given inequality are precisely those numbers that *do not* satisfy the inequality in part (c). So for part (d), we need to shade that portion of the number line that was not shaded in part (c).]

(a) $|x| < 2$    (b) $|x| > 2$    (c) $|x - 3| < 1$    (d) $|x - 3| \geq 1$

**FIGURE 2**

# EXERCISE SET 1.4

## A

*In Exercises 1–16, evaluate each expression.*

1. $|3|$

2. $3 + |-3|$

3. $|-6|$

4. $-6 - |-6|$

5. $|-1 + 3|$

6. $|-6 + 3|$

7. $\left|-\frac{4}{5}\right| - \frac{4}{5}$

8. $\left|\frac{4}{5}\right| - \frac{4}{5}$

9. $|-6 + 2| - |4|$

10. $|-3 - 4| - |-4|$

11. $||-8| + |-9||$

12. $||-8| - |-9||$

13. $\left|\dfrac{27 - 5}{5 - 27}\right|$

14. $\dfrac{|27 - 5|}{|5 - 27|}$

15. $|7(-8)| - |7||-8|$

16. $|(-7)^2| + |-7|^2 - (-|-3|)^3$

*In Exercises 17–24, evaluate each expression, given that $a = -2$, $b = 3$, and $c = -4$.*

17. $|a - b|^2$

18. $a^2 - |bc|$

19. $|c| - |b| - |a|$

20. $|b + c| - |b| - |c|$

21. $|a + b|^2 - |b + c|^2$

22. $\dfrac{|a| + |b| + |c|}{|a + b + c|}$

23. $\dfrac{a + b + |a - b|}{2}$

24. $\dfrac{a + b - |a - b|}{2}$

*In Exercises 25–38, rewrite each expression in a form that does not contain absolute values.*

25. $\left|\sqrt{2} - 1\right| - 1$

26. $\left|1 - \sqrt{2}\right| + 1$

27. $|x - 3|$ given that $x \geq 3$

28. $|x - 3|$ given that $x < 3$

29. $|t^2 + 1|$

30. $|x^4 + 1|$

31. $\left|-\sqrt{3} - 4\right|$

32. $\left|-\sqrt{3} - \sqrt{5}\right|$

33. $|x - 3| + |x - 4|$ given that $x < 3$

34. $|x - 3| + |x - 4|$ given that $x > 4$

35. $|x - 3| + |x - 4|$ given that $3 < x < 4$

36. $|x - 3| + |x - 4|$ given that $x = 4$

37. $|x + 1| + 4|x + 3|$ given that $-\frac{5}{2} < x < -\frac{3}{2}$

38. $|x + 1| + 4|x + 3|$ given that $x < -3$

*In Exercises 39–50, rewrite each statement using absolute values, as in Example 5.*

39. The distance between $x$ and 4 is 8.

40. The distance between $x$ and 4 is at least 8.

41. The distance between $x$ and 1 is $\frac{1}{2}$.

42. The distance between $x$ and 1 is less than $\frac{1}{2}$.

43. The distance between $x$ and 1 is at least $\frac{1}{2}$.

44. The distance between $x$ and 1 exceeds $\frac{1}{2}$.

45. The distance between $y$ and $-4$ is less than 1.

46. The distance between $x^3$ and $-1$ is at most 0.001.

47. The number $y$ is less than three units from the origin.

48. The number $y$ is less than one unit from the number $t$.

49. The distance between $x^2$ and $a^2$ is less than $M$.

50. The sum of the distances of $a$ and $b$ from the origin is greater than or equal to the distance of $a + b$ from the origin.

*In Exercises 51–62, the set of real numbers satisfying the given inequality is one or more intervals on the number line. Show the intervals on a number line.*

51. $|x| < 4$

52. $|x| < 2$

53. $|x| > 1$

54. $|x| > 0$

55. $|x - 5| < 3$

56. $|x - 4| < 4$

57. $|x - 3| \leq 4$

58. $|x - 1| \leq \frac{1}{2}$

59. $\left|x + \frac{1}{3}\right| < \frac{3}{2}$

60. $\left|x + \dfrac{\pi}{2}\right| > 1$

61. $|x - 5| \geq 2$

62. $|x + 5| \geq 2$

**B**

63. In parts (a) and (b), sketch the interval or intervals corresponding to the given inequality.
    (a) $|x - 2| < 1$
    (b) $0 < |x - 2| < 1$
    (c) In what way do your answers in (a) and (b) differ? (The distinction is important in the study of *limits* in calculus.)

64. Given two real numbers $a$ and $b$, the notation **max$(a, b)$** denotes the larger of the two numbers. For instance, if $a = 1$ and $b = 2$, then $\max(1, 2) = 2$. In cases where $a = b$, then $\max(a, b)$ denotes the common value of $a$ and $b$. It can be shown (see Exercise 69) that $\max(a, b)$ can be expressed in terms of absolute value as follows:

$$\max(a, b) = \frac{a + b + |a - b|}{2}$$

Verify this equation in each of the following cases.
    (a) $a = 1$ and $b = 2$
    (b) $a = -1$ and $b = -5$
    (c) $a = b = 10$

65. Given two real numbers $a$ and $b$, the notation **min$(a, b)$** denotes the smaller of the two numbers. In cases where $a = b$, then $\min(a, b)$ denotes the common value of $a$ and $b$. It can be shown (see Exercise 70) that $\min(a, b)$ can be expressed in terms of absolute value as follows:

$$\min(a, b) = \frac{a + b - |a - b|}{2}$$

Verify this equation in each of the following cases.
    (a) $a = 6$ and $b = 1$
    (b) $a = 1$ and $b = -6$
    (c) $a = b = -6$

66. Show that for all real numbers $a$ and $b$, we have

$$|a| - |b| \leq |a - b|$$

*Hint:* Beginning with the identity $a = (a - b) + b$, take the absolute value of each side and then use the triangle inequality.

67. Show that

$$|a + b + c| \leq |a| + |b| + |c|$$

for all real numbers $a$, $b$, and $c$. *Hint:* The left-hand side can be written $|a + (b + c)|$. Now use the triangle inequality.

68. Explain why there are no real numbers that satisfy the equation $|x^2 + 4x| = -12$.

C

69. (As background for this exercise, you'll need to have worked Exercise 64.) Prove that

$$\max(a,b) = \frac{a + b + |a - b|}{2}$$

*Hint:* Consider three separate cases: $a = b$; $a > b$; $b > a$.

70. (As background for this exercise, you'll need to have worked Exercise 65.) Prove that

$$\min(a,b) = \frac{a + b - |a - b|}{2}$$

71. Complete the following steps to prove the triangle inequality.
    (a) Let $a$ and $b$ be real numbers. Which property in the box on page 18 tells us that $a \leq |a|$ and $b \leq |b|$?
    (b) Add the two inequalities in part (a) to obtain $a + b \leq |a| + |b|$.
    (c) In a similar fashion, add the two inequalities $-a \leq |a|$ and $-b \leq |b|$ and deduce that $-(a + b) \leq |a| + |b|$.
    (d) Why do the results in parts (b) and (c) imply that $|a + b| \leq |a| + |b|$?

## INTEGER EXPONENTS. SCIENTIFIC NOTATION

**1.5**

*I write $a^{-1}$, $a^{-2}$, $a^{-3}$, etc., for $\dfrac{1}{a}$, $\dfrac{1}{aa}$, $\dfrac{1}{aaa}$, etc.*

ISAAC NEWTON (June 13, 1676)

In Section 1.1 we defined the exponential notation $a^n$, where $a$ is a real number and $n$ is a natural number. Since that definition is basic to all that follows in this and the next two sections, we repeat it here.

---
**Definition 1: Base and Exponent**

Given a real number $a$ and a natural number $n$, we define $a^n$ by

$$a^n = \underbrace{a \cdot a \cdot a \cdots a}_{n \text{ factors}}$$

In the expression $a^n$, the number $a$ is the **base** and $n$ is the **exponent** or **power** to which the base is raised.

---

**EXAMPLE 1**  Rewrite each expression using algebraic notation.

(a) $x$ to the fourth power, plus five
(b) the fourth power of the quantity $x$ plus five
(c) $x$ plus $y$, to the fourth power
(d) $x$ plus $y$ to the fourth power

Solution  (a) $x^4 + 5$    (b) $(x + 5)^4$    (c) $(x + y)^4$    (d) $x + y^4$

CAUTION  Note that (c) and (d) differ only in the use of the comma. So, for spoken purposes, the idea in (c) would be more clearly conveyed by saying "the fourth power of the quantity $x$ plus $y$." In the case of (d), if you read it aloud to another student or your instructor, chances are you'll be asked, "Do you mean $x + y^4$ or do you mean $(x + y)^4$?"

MORAL  Algebra is a precise language; use it carefully. Learn to ask yourself if what you've written will be interpreted in the manner you intended.

In basic algebra the following four properties are developed for working with exponents that are natural numbers.

| PROPERTY SUMMARY | Properties of Exponents |
|---|---|

| PROPERTY | EXAMPLES |
|---|---|
| 1. $a^m a^n = a^{m+n}$ | $a^5 a^6 = a^{11}$; $(x+1)(x+1)^2 = (x+1)^3$ |
| 2. $(a^m)^n = a^{mn}$ | $(2^3)^4 = 2^{12}$; $[(x+1)^2]^3 = (x+1)^6$ |
| 3. $\dfrac{a^m}{a^n} = \begin{cases} a^{m-n} & \text{if } m > n \\ \dfrac{1}{a^{n-m}} & \text{if } m < n \\ 1 & \text{if } m = n \end{cases}$ | $\dfrac{a^6}{a^2} = a^4$; $\dfrac{a^2}{a^6} = \dfrac{1}{a^4}$; $\dfrac{a^5}{a^5} = 1$ |
| 4. $(ab)^m = a^m b^m$; $\left(\dfrac{a}{b}\right)^m = \dfrac{a^m}{b^m}$ | $(2x^2)^3 = 2^3 \cdot (x^2)^3 = 8x^6$; $\left(\dfrac{x^2}{y^3}\right)^4 = \dfrac{x^8}{y^{12}}$ |

Each of these properties is a direct consequence of the definition of $a^n$. For instance, according to the first property we have $a^2 a^3 = a^5$. To verify that this is indeed correct, we note that

$$a^2 a^3 = (aa)(aaa) = a^5.$$

Now we want to extend our definition of $a^n$ to allow for exponents that are integers but not necessarily natural numbers. We begin by defining $a^0$.

| Definition 2: Zero Exponent | |
|---|---|
| | EXAMPLES |
| For any nonzero real number $a$, | (a) $2^0 = 1$ |
| $a^0 = 1$ | (b) $(-\pi)^0 = 1$ |
| ($0^0$ is not defined) | (c) $\left(\dfrac{3}{1 + a^2 + b^2}\right)^0 = 1$ |

It's easy to see the motivation for defining $a^0$ to be 1. Assuming that the exponent zero is to have the same properties as do exponents that are natural numbers, we can write

$$a^0 a^n = a^{0+n}$$

That is,

$$a^0 a^n = a^n$$

Now divide both sides of this last equation by $a^n$ to obtain $a^0 = 1$, which agrees with our definition.

Our next definition assigns a meaning to the expression $a^{-n}$ when $n$ is a natural number.

---

**Definition 3:  Negative Exponent**

EXAMPLES

$$a^{-n} = \frac{1}{a^n},$$

where $a \neq 0$ and $n$ is a natural number

(a)  $2^{-1} = \frac{1}{2^1} = \frac{1}{2}$

(b)  $\left(\frac{1}{10}\right)^{-1} = \frac{1}{(\frac{1}{10})^1} = 10$

(c)  $x^{-2} = \frac{1}{x^2}$

(d)  $(a^2 b)^{-3} = \frac{1}{(a^2 b)^3} = \frac{1}{a^6 b^3}$

(e)  $\frac{1}{2^{-3}} = \frac{1}{\frac{1}{2^3}} = 2^3 = 8$

---

Again, it's easy to see the motivation for this definition. We have

$$a^n a^{-n} = a^{n + (-n)} = a^0 = 1$$

That is,

$$a^n a^{-n} = 1$$

Now divide both sides of this last equation by $a^n$ to obtain $a^{-n} = 1/a^n$, in agreement with Definition 3.

It can be shown that the four properties of exponents that we listed earlier continue to hold now for all integer exponents. We make use of this fact in the next three examples.

**EXAMPLE 2**  Simplify the following expression. Write the answer in such a way that only positive exponents appear.

$$(a^2 b^3)^2 (a^5 b)^{-1}$$

**Solution**   FIRST METHOD

$$(a^2 b^3)^2 (a^5 b)^{-1} = (a^2 b^3)^2 \cdot \frac{1}{a^5 b}$$

$$= \frac{a^4 b^6}{a^5 b} = \frac{b^{6-1}}{a^{5-4}}$$

$$= \frac{b^5}{a}$$

ALTERNATE METHOD   $(a^2b^3)^2(a^5b)^{-1} = (a^4b^6)(a^{-5}b^{-1})$
$$= a^{4-5}b^{6-1} = a^{-1}b^5$$
$$= \frac{b^5}{a} \quad \text{as obtained previously}$$

**EXAMPLE 3**   Simplify the following expression, writing the answer so that negative exponents are not used.

$$\left(\frac{a^{-5}b^2c^0}{a^3b^{-1}}\right)^3$$

Solution   We show two solutions. The first makes immediate use of the property $(a^m)^n = a^{mn}$. In the second solution we begin by working within the parentheses.

FIRST SOLUTION

$$\left(\frac{a^{-5}b^2c^0}{a^3b^{-1}}\right)^3 = \frac{a^{-15}b^6}{a^9b^{-3}}$$
$$= \frac{b^{6-(-3)}}{a^{9-(-15)}} = \frac{b^9}{a^{24}}$$

ALTERNATE SOLUTION

$$\left(\frac{a^{-5}b^2c^0}{a^3b^{-1}}\right)^3 = \left(\frac{b^3}{a^8}\right)^3$$
$$= \frac{b^9}{a^{24}}$$

**EXAMPLE 4**   Use the properties of exponents to compute the quantity $\dfrac{2^{10} \cdot 3^{13}}{27 \cdot 6^{12}}$.

Solution   
$$\frac{2^{10} \cdot 3^{13}}{27 \cdot 6^{12}} = \frac{2^{10} \cdot 3^{13}}{(3^3)(3 \cdot 2)^{12}}$$
$$= \frac{2^{10} \cdot 3^{13}}{3^3 \cdot 3^{12} \cdot 2^{12}} = \frac{2^{10} \cdot 3^{13}}{3^{15} \cdot 2^{12}}$$
$$= \frac{1}{(3^{15-13})(2^{12-10})} = \frac{1}{3^2 \cdot 2^2}$$
$$= \frac{1}{36}$$

As an application of some of the ideas in this section, we briefly discuss *scientific notation*. Scientific notation is a convenient form for writing very large or very small numbers. Such numbers occur often in the sciences. For instance, the speed of light in a vacuum is

29,979,000,000 cm/sec

As written, this number would be awkward to work with in calculating. In fact, the number as written cannot even be displayed on a hand-held calculator, since there are too many digits. To write the number 29,979,000,000 (or any positive number) in scientific notation, we express it as a number between 1 and 10, multiplied by an appropriate power of 10. That is, we write it in the form

$a \times 10^n$   where   $1 \le a < 10$   and   $n$ is an integer

According to this convention, the number $4.03 \times 10^6$ is in scientific notation, but the same quantity written as $40.3 \times 10^5$ is not in scientific notation. In order to convert a given number into scientific notation, we'll rely on the following two-step procedure.

---

**To Express a Number Using Scientific Notation:**

1. First move the decimal point until it is to the immediate right of the first nonzero digit.
2. Then multiply by $10^n$ or $10^{-n}$, depending on whether the decimal point was moved $n$ places to the left or to the right, respectively.

---

For example, to express the number 29,979,000,000 in scientific notation, first move the decimal point 10 places to the left so that it's located between the 2 and the 9. Then multiply by $10^{10}$. The result is

$$29{,}979{,}000{,}000 = 2.9979000000 \times 10^{10}$$

or, more simply,

$$29{,}979{,}000{,}000 = 2.9979 \times 10^{10}$$

Incidentally, in this last form, the number can be displayed on a scientific calculator by means of keys such as $\boxed{\text{exp}}$ or $\boxed{\text{EE}}$. For instance, using a Texas Instruments calculator, we would enter this number by pressing

2.9979 $\boxed{\text{EE}}$ 10

The display would then read

$\boxed{2.9979 \quad 10}$

As additional examples we list the following numbers expressed in both ordinary and scientific notation. For practice you should verify each conversion for yourself using our two-step procedure.

$$55708 = 5.5708 \times 10^4$$
$$0.000099 = 9.9 \times 10^{-5}$$
$$0.0000002 = 2 \times 10^{-7}$$

## EXERCISE SET 1.5

### A

*In Exercises 1–6, evaluate each expression using the given value of x.*

1. $2x^3 - x + 4; x = -2$
2. $1 - x + 2x^2 - 3x^3; x = -1$

3. $\dfrac{1 - 2x^2}{1 + 2x^3}; x = \dfrac{-1}{2}$

4. $\dfrac{1 - (x - 1)^2}{1 + (x - 1)^2}; x = -1$

5. $\dfrac{x^2 + x^3 - x^x}{2^x + 3^x - (x+1)^2}; x = 2$

6. $\dfrac{2^{x-1} - \dfrac{1}{2^{x-1}}}{2^{x-1} + \dfrac{1}{2^{x-1}}}; x = 3$

*For Exercises 7–14, rewrite each expression using algebraic notation.*

7. The square of the quantity $x$ plus $y$

8. The square of the product of $x$ and $y$

9. Three more than the square of the quantity $x$ plus $y$

10. The square of $x$, minus twice the cube of $y$

11. The square of half of the quantity $x^2 - 2y^3$

12. The cube of the average of $x^2$ and $-2y^2$

13. The cube of the absolute value of the quantity $x - 1$

14. The absolute value of the fifth power of the average of $-x^2$ and $y^2$

*In Exercises 15–26, use the properties of exponents to simplify each expression.*

15. (a) $a^3 a^{12}$    (b) $(a+1)^3(a+1)^{12}$
    (c) $(a+1)^{12}(a+1)^3$

16. (a) $(3^2)^3 - (2^3)^2$    (b) $(x^2)^3 - (x^3)^2$
    (c) $(x^2)^a - (x^a)^2$

17. (a) $yy^2y^8$    (b) $(y+1)(y+1)^2(y+1)^8$
    (c) $[(y+1)(y+1)^8]^2$

18. (a) $\dfrac{x^{12}}{x^{10}}$    (b) $\dfrac{x^{10}}{x^{12}}$    (c) $\dfrac{2^{10}}{2^{12}}$

19. (a) $\dfrac{(x^2+3)^{10}}{(x^2+3)^9}$    (b) $\dfrac{(x^2+3)^9}{(x^2+3)^{10}}$
    (c) $\dfrac{12^{10}}{12^9}$

20. (a) $(y^2y^3)^2$    (b) $y^2(y^3)^2$    (c) $2^m(2^n)^2$

21. (a) $\dfrac{t^{15}}{t^9}$    (b) $\dfrac{t^9}{t^{15}}$    (c) $\dfrac{(t^2+3)^{15}}{(t^2+3)^9}$

22. (a) $\dfrac{x^6}{x}$    (b) $\dfrac{x}{x^6}$
    (c) $\dfrac{(x^2+3x-2)^6}{x^2+3x-2}$

23. (a) $\dfrac{x^6y^{15}}{x^2y^{20}}$    (b) $\dfrac{x^2y^{20}}{x^6y^{15}}$    (c) $\left(\dfrac{x^2y^{20}}{x^6y^{15}}\right)^2$

24. (a) $(x^2y^3z)^4$    (b) $2(x^2y^3z)^4$
    (c) $(2x^2y^3z)^4$

25. (a) $4(x^3)^2$    (b) $(4x^3)^2$    (c) $\dfrac{(4x^2)^3}{(4x^3)^2}$

26. (a) $2(x-1)^7 - (x-1)^7$
    (b) $[2(x-1)]^7 - (x-1)^7$
    (c) $2(x-1)^7 - [2(x-1)]^7$

*For Exercises 27–48, simplify each expression. Write the answers in such a way that negative exponents do not appear.*

27. $64^0$    28. $(2^0 + 3^0)^2$

29. $(\frac{1}{25})^0$    30. $(\frac{2}{3})^{-1}$

31. $10^{-1} + 10^{-2}$    32. $(2^{-1} + 2^{-2})^{-1}$

33. $(5^{-1})^{-1}$    34. $(10)^{-1} + (10^{-1})^{-1}$

35. $(10^{-30})^0$    36. $(xy)^{-2}$

37. $(a^2bc^0)^{-3}$    38. $(a^3b)^3(a^2b^4)^{-1}$

39. $(a^{-2}b^{-1}c^3)^{-2}$    40. $(2^{-2} + 2^{-1} + 2^0)^{-2}$

41. $\left(\dfrac{x^3y^{-2}z}{xy^2z^{-3}}\right)^{-3}$    42. $\left(\dfrac{x^4y^{-8}z^2}{xy^2z^{-6}}\right)^2$

43. $\left(\dfrac{x^4y^{-8}z^2}{xy^2z^{-6}}\right)^{-2}$    44. $\left(\dfrac{a^3b^{-9}c^2}{a^5b^2c^{-4}}\right)^0$

45. $\left(\dfrac{a^{-2}b^{-3}c^{-4}}{a^2b^3c^4}\right)^2$    46. $(-2x)^{-3}(-2x^{-2})^{-1}$

47. $\dfrac{x^2}{y^{-3}} \div \dfrac{x^2}{y^3}$    48. $(2x^2)^{-3} - 2(x^2)^{-3}$

*In Exercises 49–52, use the properties of exponents in computing each quantity. (The point here is to do as little arithmetic as possible.)*

49. $\dfrac{2^8 \cdot 3^{15}}{9 \cdot 3^{10} \cdot 12}$    50. $\dfrac{2^{12} \cdot 5^{13}}{10^{12}}$

51. $\dfrac{24^5}{32 \cdot 12^4}$    52. $\left(\dfrac{144 \cdot 125}{2^3 \cdot 3^2}\right)^{-1}$

*For Exercises 53–62, express each number in scientific notation.*

53. The average distance (in miles) from Earth to the Sun:

    92,900,000

54. The average distance (in miles) from the planet Pluto to the Sun:

    3,666,000,000

55. The average orbital speed (in miles per hour) of Earth:

    66,800

56. The average orbital speed (in miles per hour) of the planet Mercury:

    107,300

57. The average distance (in miles) from the Sun to the nearest star:

    25,000,000,000,000,000,000

58. The equatorial diameter (in miles) of
    (a) Mercury: 3031
    (b) Earth: 7927
    (c) Jupiter: 88,733
    (d) the Sun: 865,000

59. The length of the "year," that is, the time to orbit the Sun once (in terms of Earth days), for the planet
    (a) Mercury: 86.688
    (b) Saturn: 10604.772
    (c) Pluto: 89424

60. The mass (in grams) of
    (a) Earth:

       60000000000000000000000000000

    (b) Jupiter:

       1900000000000000000000000000000

61. The time (in seconds) for light to travel
    (a) one foot:

       0.000000001

    (b) across an atom:

       0.000000000000000001

    (c) across the nucleus of an atom:

       0.00000000000000000000001

62. The mass (in grams) of
    (a) a proton:

       0.00000000000000000000000167

    (b) an electron:

       0.000000000000000000000000000911

## B

*For Exercises 63–66, use the properties of exponents to simplify each expression.*

63. $\dfrac{a^{3x+y}}{a^{2x} \cdot a^{x+y}}$

64. $\dfrac{a^{4p+2q}}{a^{3p}a^p(a^q)^2}$

65. $\dfrac{(x^{5n+1})^n}{(x^n)^{5n}} \cdot \dfrac{1}{x^{n-2}}$

66. $\dfrac{x^{3a+2b-c}}{(x^{2a})(x^b)} \cdot x^{3c-a-b}$

67. Which number is larger, $10^9$ or $9^{10}$?

68. (a) Which is larger, $11^{12}$ or $12^{11}$?
    (b) Which is larger, $18^{19}$ or $19^{18}$?

69. Use the following data to compute the time required, in minutes, for light to travel from the Sun to Earth. Round off your answer to the nearest minute.

    speed of light: $2.9979 \times 10^{10}$ cm/sec

    distance from the Sun to Earth: $92.9 \times 10^6$ miles

    1 mile = 160930 cm

70. Each planet in our solar system travels around the Sun in an elliptical orbit. Let $T$ denote the time for a planet to complete one orbit around the Sun. For example, for Earth, $T = 365$ days. (Actually, more precisely, $T = 365.26$ days.) Let $a$ denote the length of the so-called *semimajor axis* of the ellipse. (See the figure.)

In 1618 the German astronomer Johannes Kepler announced his Third Law of Planetary Motion, which stated that the ratio $a^3/T^2$ is the same for all planets in the solar system. Verify Kepler's Third Law empirically by completing the following table. Express the required values of $a^3/T^2$ in scientific notation, rounding off to three significant digits. (See Appendix A1 for the definition of significant digits.)

| PLANET | $a$ (KM) | $T$ (EARTH DAYS) | $\dfrac{a^3}{T^2}$ |
|---|---|---|---|
| Mercury | $5.791 \times 10^7$ | 87.95 | |
| Venus | $1.082 \times 10^8$ | 224.71 | |
| Earth | $1.496 \times 10^8$ | 365.26 | |
| Mars | $2.279 \times 10^8$ | 687.02 | |
| Jupiter | $7.783 \times 10^8$ | 4332.79 | |
| Saturn | $1.427 \times 10^9$ | 10759.72 | |
| Uranus | $2.869 \times 10^9$ | 30686.59 | |
| Neptune | $4.498 \times 10^9$ | 60191.20 | |
| Pluto | $5.900 \times 10^9$ | 90730.6 | |

## C

71. Suppose that $p = b^x$, $q = b^y$, and $b^2 = (p^y q^x)^z$, where all the letters denote natural numbers and $b \neq 1$. Show that $xyz = 1$.

72. Which pair of numbers is closer together: $10^{27}$ and $10^{98}$ or $10^{105}$ and $10^{106}$?

## *n*TH ROOTS

**1.6**

*Hindu mathematicians first recognized negative roots, and the two square roots of a positive number, . . . though they were suspicious also.*

DAVID WELLS, *The Penguin Dictionary of Curious and Interesting Numbers* (Harmondsworth, Middlesex, England: Penguin Books Ltd., 1986)

In this section we generalize the notion of square root that you studied in elementary algebra. The new idea is that of an *n*th root; the definition and some basic examples are given in the box that follows.

---

**Definition 1:   *n*th Roots**

EXAMPLES

Let $n$ be a natural number. If $a$ and $b$ are real numbers and

$$a^n = b$$

then we say that $a$ is an **nth root** of $b$. When $n = 2$ and when $n = 3$, we refer to the roots as **square roots** and **cube roots**, respectively.

Both 3 and $-3$ are square roots of 9 because $3^2 = 9$ and $(-3)^2 = 9$.

Both 2 and $-2$ are fourth roots of 16 because $2^4 = 16$ and $(-2)^4 = 16$.

2 is a cube root of 8 because $2^3 = 8$.

$-3$ is a fifth root of $-243$ because $(-3)^5 = -243$.

---

As the examples in the box suggest, square roots, fourth roots, and all *even* roots of positive numbers occur in pairs, one positive and one negative. In these cases, we use the notation $\sqrt[n]{b}$ to denote the positive, or **principal**, *n*th root of $b$. As examples of this notation, we can write

$$\sqrt[4]{81} = 3 \qquad \text{(The principal fourth root of 81 is 3.)}$$

$$\sqrt[4]{81} \neq -3 \qquad \text{(The principal fourth root of 81 is not } -3.\text{)}$$

$$-\sqrt[4]{81} = -3 \qquad \text{(The negative of the principal fourth root of 81 is } -3.\text{)}$$

The symbol $\sqrt{\phantom{x}}$ is called a **radical sign**, and the number within the radical sign is the **radicand**.* The natural number $n$ used in the notation $\sqrt[n]{\phantom{x}}$ is called the **index** of the radical. For square roots, as you know from basic algebra, we suppress the index and simply write $\sqrt{\phantom{x}}$ rather than $\sqrt[2]{\phantom{x}}$. So, for example, $\sqrt{25} = 5$.

On the other hand, as we saw in the examples, cube roots, fifth roots, and, in fact, all *odd* roots occur singly, not in pairs. In these cases, we again use the

---

* The radical sign ($\sqrt{\phantom{x}}$) was introduced in 1525 by the German mathematician Christoff Rudolff. Historian Howard Eves suggests (as did Leonhard Euler) that Rudolff may have obtained this symbol by modifying the letter *r*, the first letter of the word *radix* (root). However, research into the 15th- and 16th-century German manuscripts suggests that the radical sign may have actually evolved from a dot. For details see *A History of Mathematical Notations*, by Florian Cajori, pp. 366–368 (La Salle, Illinois: The Open Court Publishing Company, 1928.)

notation $\sqrt[n]{b}$ for the *n*th root. The definition and examples in the following box summarize our discussion up to this point.

| Definition 2:  Principal *n*th root | |
|---|---|
| | EXAMPLES |
| 1. Let *n* be a natural number. If *a* and *b* are nonnegative real numbers, then<br><br>$\sqrt[n]{b} = a$  if and only if  $b = a^n$.<br><br>The number *a* is the **principal *n*th root** of *b*.<br><br>2. If *a* and *b* are negative and *n* is an odd natural number, then<br><br>$\sqrt[n]{b} = a$  if and only if  $b = a^n$. | $\sqrt{25} = 5;\quad \sqrt{25} \neq -5$<br>$-\sqrt{25} = -5$<br><br>$\sqrt[4]{\dfrac{1}{16}} = \dfrac{1}{2};\quad \sqrt[3]{8} = 2$<br><br>$\sqrt[5]{-\dfrac{1}{32}} = -\dfrac{1}{2}$ |

There are five properties of *n*th roots that are frequently used in simplifying certain expressions. The first four are similar to the properties of square roots that are developed in elementary algebra. For reference, we list these properties side by side in the following box. Property 5 is listed here only for the sake of completeness; we'll postpone discussing it until Section 1.7.

| PROPERTY SUMMARY | Properties of *n*th Roots | Corresponding Properties for Square Roots |
|---|---|---|
| | Suppose that *x* and *y* are real numbers and that *m* and *n* are natural numbers. Then each of the following properties holds, provided only that the expressions on both sides of the equation are defined (and so represent real numbers). | |
| | 1. $(\sqrt[n]{x})^n = x$ | $(\sqrt{x})^2 = x$ |
| | 2. $\sqrt[n]{xy} = \sqrt[n]{x}\,\sqrt[n]{y}$ | $\sqrt{xy} = \sqrt{x}\,\sqrt{y}$ |
| | 3. $\sqrt[n]{\dfrac{x}{y}} = \dfrac{\sqrt[n]{x}}{\sqrt[n]{y}}$ | $\sqrt{\dfrac{x}{y}} = \dfrac{\sqrt{x}}{\sqrt{y}}$ |
| | 4. *n* even:  $\sqrt[n]{x^n} = \lvert x \rvert$<br><br>   *n* odd:  $\sqrt[n]{x^n} = x$ | $\sqrt{x^2} = \lvert x \rvert$ |
| | 5. $\sqrt[m]{\sqrt[n]{x}} = \sqrt[mn]{x}$ | |

Our immediate use for these properties will be in simplifying expressions involving *n*th roots. In general, we try to factor the expression under the radical so that one factor is the largest perfect *n*th power that we can find. Then we apply Property 2 or 3. For instance, the expression $\sqrt{72}$ is simplified as follows:

$$\sqrt{72} = \sqrt{(36)(2)} = \sqrt{36}\sqrt{2} = 6\sqrt{2}$$

In this procedure, we began by factoring 72 as $(36)(2)$. Note that 36 is the largest factor of 72 that is a perfect square. If we had begun instead with a different factorization, say $72 = (9)(8)$, we could still arrive at the same answer, but it would take longer. (Check this for yourself.) As another example, let us simplify $\sqrt[3]{40}$. First, what (if any) is the largest perfect cube factor of 40? Since the first few perfect cubes are

$$1^3 = 1 \qquad 2^3 = 8 \qquad 3^3 = 27 \qquad 4^3 = 64$$

we see that 8 is a perfect cube factor of 40, and we write

$$\sqrt[3]{40} = \sqrt[3]{(8)(5)} = \sqrt[3]{8}\sqrt[3]{5} = 2\sqrt[3]{5}$$

**EXAMPLE 1**    Simplify:   (a) $\sqrt{12} + \sqrt{75}$    (b) $\sqrt{\dfrac{162}{49}}$

Solution    (a) $\sqrt{12} + \sqrt{75} = \sqrt{(4)(3)} + \sqrt{(25)(3)}$
$$= \sqrt{4}\sqrt{3} + \sqrt{25}\sqrt{3}$$
$$= 2\sqrt{3} + 5\sqrt{3} = (2 + 5)\sqrt{3}$$
$$= 7\sqrt{3}$$

(b) $\sqrt{\dfrac{162}{49}} = \dfrac{\sqrt{162}}{\sqrt{49}} = \dfrac{\sqrt{81}\sqrt{2}}{7} = \dfrac{9\sqrt{2}}{7}$

**EXAMPLE 2**    Simplify $\sqrt[3]{16} + \sqrt[3]{250} - \sqrt[3]{128}$.

Solution    $\sqrt[3]{16} + \sqrt[3]{250} - \sqrt[3]{128} = \sqrt[3]{8}\sqrt[3]{2} + \sqrt[3]{125}\sqrt[3]{2} - \sqrt[3]{64}\sqrt[3]{2}$
$$= 2\sqrt[3]{2} + 5\sqrt[3]{2} - 4\sqrt[3]{2}$$
$$= 3\sqrt[3]{2}$$

**EXAMPLE 3**    Simplify each of the following expressions by removing the largest possible perfect square factor from within the radical.

(a) $\sqrt{8x^2}$

(b) $\sqrt{8x^2}$    assuming $x \geq 0$

(c) $\sqrt{18a^7}$    assuming $a \geq 0$

(d) $\sqrt[3]{16y^5}$

Solution    (a) $\sqrt{8x^2} = \sqrt{(4)(2)(x^2)} = \sqrt{4}\sqrt{2}\sqrt{x^2}$
$$= 2\sqrt{2}|x|$$

(b) $\sqrt{8x^2} = \sqrt{4}\sqrt{2}\sqrt{x^2}$
$$= 2\sqrt{2}x \qquad \sqrt{x^2} = x \text{ because } x \geq 0$$

(c) $\sqrt{18a^7} = \sqrt{(9a^6)(2a)} = \sqrt{9a^6}\sqrt{2a}$
$$= 3a^3\sqrt{2a}$$

(d) $\sqrt[3]{16y^5} = \sqrt[3]{8y^3}\sqrt[3]{2y^2} = 2y\sqrt[3]{2y^2}$

**EXAMPLE 4**    Simplify each of the following, assuming that $a$, $b$, and $c$ are positive.

(a)  $\sqrt{8ab^2c^5}$    (b)  $\sqrt[4]{\dfrac{32a^6b^5}{c^8}}$

**Solution**    (a)  $\sqrt{8ab^2c^5} = \sqrt{(4b^2c^4)(2ac)}$
$= \sqrt{4b^2c^4}\sqrt{2ac}$
$= 2bc^2\sqrt{2ac}$

(b)  $\sqrt[4]{\dfrac{32a^6b^5}{c^8}} = \dfrac{\sqrt[4]{32a^6b^5}}{\sqrt[4]{c^8}}$

$= \dfrac{\sqrt[4]{16a^4b^4}\sqrt[4]{2a^2b}}{c^2}$

$= \dfrac{2ab\sqrt[4]{2a^2b}}{c^2}$

There are times when it is convenient to rewrite fractions involving radicals in alternate forms. Suppose, for example, that we want to rewrite the fraction $5/\sqrt{3}$ in an equivalent form not involving a radical in the denominator. This is called **rationalizing the denominator**. The procedure here is to multiply by 1 in this way:

$$\frac{5}{\sqrt{3}} = \frac{5}{\sqrt{3}} \cdot 1 = \frac{5}{\sqrt{3}} \cdot \frac{\sqrt{3}}{\sqrt{3}} = \frac{5\sqrt{3}}{3}$$

That is, $\dfrac{5}{\sqrt{3}} = \dfrac{5\sqrt{3}}{3}$, as required.

To rationalize a denominator of the form $a + \sqrt{b}$, we need to multiply the fraction not by $\dfrac{\sqrt{b}}{\sqrt{b}}$, but rather by $\dfrac{a - \sqrt{b}}{a - \sqrt{b}}\ (= 1)$. To see why this is necessary, notice that

$$(a + \sqrt{b})\sqrt{b} = a\sqrt{b} + b$$

which still contains a radical, whereas

$$(a + \sqrt{b})(a - \sqrt{b}) = a^2 - a\sqrt{b} + a\sqrt{b} - b$$
$$= a^2 - b$$

which is free of radicals. Similarly, to rationalize a denominator of the form $a - \sqrt{b}$, we will need to multiply the fraction by $\dfrac{a + \sqrt{b}}{a + \sqrt{b}}$. (The quantities $a + \sqrt{b}$ and $a - \sqrt{b}$ are said to be **conjugates** of each other.) The next two examples make use of these ideas.

**EXAMPLE 5**  Simplify $\dfrac{1}{\sqrt{2}} - 3\sqrt{50}$.

Solution  First, rationalize the denominator in the fraction $\dfrac{1}{\sqrt{2}}$:

$$\frac{1}{\sqrt{2}} = \frac{1}{\sqrt{2}} \cdot 1 = \frac{1}{\sqrt{2}} \cdot \frac{\sqrt{2}}{\sqrt{2}} = \frac{\sqrt{2}}{2}$$

Next, simplify the expression $3\sqrt{50}$:

$$3\sqrt{50} = 3\sqrt{(25)(2)} = 3\sqrt{25}\sqrt{2} = (3)(5)\sqrt{2} = 15\sqrt{2}$$

Now, putting things together, we have

$$\frac{1}{\sqrt{2}} - 3\sqrt{50} = \frac{\sqrt{2}}{2} - 15\sqrt{2} = \frac{\sqrt{2}}{2} - \frac{30\sqrt{2}}{2}$$

$$= \frac{\sqrt{2} - 30\sqrt{2}}{2} = \frac{(1 - 30)\sqrt{2}}{2}$$

$$= \frac{-29\sqrt{2}}{2}$$

**EXAMPLE 6**  Rationalize the denominator in the expression $\dfrac{4}{2 + \sqrt{3}}$.

Solution  We multiply by 1, writing 1 as $\dfrac{2 - \sqrt{3}}{2 - \sqrt{3}}$.

$$\frac{4}{2 + \sqrt{3}} \cdot 1 = \frac{4}{2 + \sqrt{3}} \cdot \frac{2 - \sqrt{3}}{2 - \sqrt{3}}$$

$$= \frac{4(2 - \sqrt{3})}{4 - (\sqrt{3})^2} = \frac{4(2 - \sqrt{3})}{4 - 3} = \frac{8 - 4\sqrt{3}}{1}$$

$$= 8 - 4\sqrt{3}$$

NOTE:  Check for yourself that multiplying the original fraction by $\dfrac{2 + \sqrt{3}}{2 + \sqrt{3}}$ does *not* eliminate radicals in the denominator.

In the next example, we are asked to rationalize the numerator rather than the denominator. This is useful at times in calculus.

**EXAMPLE 7**  Rationalize the *numerator*:  $\dfrac{\sqrt{x} - \sqrt{3}}{x - 3}$   $(x \geq 0, x \neq 3)$

Solution  $\dfrac{\sqrt{x} - \sqrt{3}}{x - 3} \cdot 1 = \dfrac{\sqrt{x} - \sqrt{3}}{x - 3} \cdot \dfrac{\sqrt{x} + \sqrt{3}}{\sqrt{x} + \sqrt{3}}$

$$= \frac{(\sqrt{x})^2 - (\sqrt{3})^2}{(x - 3)(\sqrt{x} + \sqrt{3})} = \frac{x - 3}{(x - 3)(\sqrt{x} + \sqrt{3})}$$

$$= \frac{1}{\sqrt{x} + \sqrt{3}}$$

The strategy for rationalizing numerators or denominators involving *n*th roots is similar to that used for square roots. To rationalize a numerator or a denominator involving an *n*th root, we multiply that numerator or denominator by a factor that yields a product that itself is a perfect *n*th power. The next example displays two instances of this.

**EXAMPLE 8**    (a) Rationalize the denominator: $\dfrac{6}{\sqrt[3]{7}}$.

(b) Rationalize the denominator: $\dfrac{ab}{\sqrt[4]{a^2b^3}}$, where $a > 0$ and $b > 0$.

**Solution**    (a) $\dfrac{6}{\sqrt[3]{7}} \cdot 1 = \dfrac{6}{\sqrt[3]{7}} \cdot \dfrac{\sqrt[3]{7^2}}{\sqrt[3]{7^2}}$

$= \dfrac{6\sqrt[3]{7^2}}{\sqrt[3]{7^3}}$

$= \dfrac{6\sqrt[3]{49}}{7}$    as required

(b) $\dfrac{ab}{\sqrt[4]{a^2b^3}} \cdot 1 = \dfrac{ab}{\sqrt[4]{a^2b^3}} \cdot \dfrac{\sqrt[4]{a^2b}}{\sqrt[4]{a^2b}}$

$= \dfrac{ab\sqrt[4]{a^2b}}{\sqrt[4]{a^4b^4}} = \dfrac{ab\sqrt[4]{a^2b}}{ab} = \sqrt[4]{a^2b}$

## EXERCISE SET 1.6

### A

*In Exercises 1–10, determine whether each statement is true or false.*

1. $\sqrt{81} = -9$

2. $\sqrt{256} = 16$

3. $-\sqrt{100} = -10$

4. $\sqrt{49} = -7$

5. $\sqrt{4/3} = 2/\sqrt{3}$

6. $\sqrt{10 + 6} = \sqrt{10} + \sqrt{6}$

7. $(\sqrt[3]{10})^3 = 10$

8. $\sqrt{(-5)^2} = -5$

9. $\sqrt[5]{32a^5} = 2a$

10. $\sqrt[4]{x^4} = x$   $(x < 0)$

*For Exercises 11–20, evaluate each expression. If the expression is undefined (i.e., does not represent a real number), say so.*

11. (a) $\sqrt[3]{-64}$    (b) $\sqrt[4]{-64}$

12. (a) $\sqrt[5]{32}$    (b) $\sqrt[5]{-32}$

13. (a) $\sqrt[3]{8/125}$    (b) $\sqrt[3]{-8/125}$

14. (a) $\sqrt[3]{-1/1000}$    (b) $\sqrt[6]{-1/1000}$

15. (a) $\sqrt{-16}$    (b) $\sqrt[4]{-16}$

16. (a) $-\sqrt[4]{16}$    (b) $-\sqrt[4]{-16}$

17. (a) $\sqrt[4]{256/81}$    (b) $\sqrt[3]{-27/125}$

18. (a) $\sqrt[6]{64}$    (h) $\sqrt[6]{-64}$

19. (a) $\sqrt[5]{-32}$    (b) $-\sqrt[5]{-32}$

20. (a) $\sqrt[4]{(-10)^4}$    (b) $\sqrt[3]{(-10)^3}$

*In Exercises 21–50, simplify each expression. (Unless otherwise specified, assume that all letters in Exercises 37–50 represent positive numbers.)*

21. (a) $\sqrt{18}$    (b) $\sqrt[3]{54}$

22. (a) $\sqrt{150}$    (b) $\sqrt[3]{375}$

23. (a) $\sqrt{98}$    (b) $\sqrt[5]{-64}$

24. (a) $\sqrt{27}$    (b) $\sqrt[3]{-108}$

25. (a) $\sqrt{25/4}$    (b) $\sqrt[4]{16/625}$

26. (a) $\sqrt{225/49}$    (b) $\sqrt[3]{-256/243}$

27. (a) $\sqrt{2} + \sqrt{8}$    (b) $\sqrt[3]{2} + \sqrt[3]{16}$

28. (a) $4\sqrt{3} - 2\sqrt{27}$    (b) $2\sqrt[3]{81} + 3\sqrt[3]{24}$

29. (a) $4\sqrt{50} - 3\sqrt{128}$    (b) $\sqrt[4]{32} + \sqrt[4]{162}$

30. (a) $\sqrt{3} - \sqrt{12} + \sqrt{48}$

(b) $\sqrt[3]{-2} + \sqrt[3]{-64} - \sqrt[3]{486}$

31. (a) $\sqrt{0.09}$        (b) $\sqrt[3]{0.008}$

32. (a) $\sqrt[3]{-2} + \sqrt[3]{2}$        (b) $\sqrt{81/121} - \sqrt[3]{-8/1331}$

33. $4\sqrt{24} - 8\sqrt{54} + 2\sqrt{6}$

34. $\sqrt[3]{192} + \sqrt[3]{-81} + \sqrt{\sqrt[3]{9}}$

35. $\sqrt{\sqrt{64}}$        36. $\sqrt[3]{\sqrt{4096}}$

37. (a) $\sqrt{36x^2}$        (b) $\sqrt{36y^2}, \quad y < 0$

38. (a) $\sqrt{225x^4y^3}$        (b) $\sqrt[4]{16a^4}, \quad a < 0$

39. (a) $\sqrt{ab^2}\sqrt{a^2b}$        (b) $\sqrt{ab^3}\sqrt{a^3b}$

40. (a) $\sqrt[3]{125x^6}$        (b) $\sqrt[4]{64y^4}, \quad y < 0$

41. $\sqrt{72a^3b^4c^5}$        42. $\sqrt{\dfrac{(a+b)^5}{16a^2b^2}}$

43. $\sqrt[4]{16a^4b^5}$        44. $\sqrt[3]{8a^4b^6}$

45. $\sqrt{18a^3b^2}$        46. $\sqrt[5]{64a^6b^{12}}$

47. $\sqrt[3]{\dfrac{16a^{12}b^2}{c^9}}$        48. $\sqrt[4]{ab^3}\sqrt[4]{a^3b}$

49. $\sqrt[6]{\dfrac{5a^7}{a^{-5}b^6}}$        50. $\sqrt[3]{a^2b}\sqrt[3]{ab}\sqrt[3]{b^4}$

*In Exercises 51–66, rationalize the denominators and simplify where possible. (Assume that all letters represent positive quantities.)*

51. $\dfrac{4}{\sqrt{7}}$        52. $\dfrac{3}{\sqrt{3}}$

53. $\dfrac{1}{\sqrt{8}}$        54. $\dfrac{\sqrt{2}}{\sqrt{3}}$

55. $\dfrac{1}{1+\sqrt{5}}$        56. $\dfrac{\sqrt{2}}{1-\sqrt{2}}$

57. $\dfrac{1+\sqrt{3}}{1-\sqrt{3}}$        58. $\dfrac{\sqrt{a}+\sqrt{b}}{\sqrt{a}-\sqrt{b}}$

59. $\dfrac{1}{\sqrt{5}} + 4\sqrt{45}$        60. $\dfrac{3}{\sqrt{8}} - \sqrt{450}$

61. $\dfrac{1}{\sqrt[3]{25}}$        62. $\dfrac{4}{\sqrt[3]{16}}$

63. $\dfrac{3}{\sqrt[4]{3}}$        64. $\dfrac{\sqrt[3]{5}}{\sqrt[3]{6}}$

65. $\dfrac{1}{\sqrt[4]{2ab^5}}$        66. $\dfrac{1}{\sqrt[3]{4a^2b^8}}$

*In Exercises 67–70, rationalize the numerators.*

67. $\dfrac{\sqrt{x}-\sqrt{5}}{x-5}$        68. $\dfrac{\sqrt{a}-\sqrt{b}}{a-b}$

69. $\dfrac{\sqrt{2+h}-\sqrt{2}}{h}$        70. $\dfrac{\sqrt{x+h}-\sqrt{x}}{h}$

**B**

71. ⓒ (a) Evaluate $\sqrt{8-2\sqrt{7}}$ and $\sqrt{7}-1$. What do you observe?

(b) Prove that $\sqrt{8-2\sqrt{7}} = \sqrt{7}-1$. *Hint:* In view of the definition of a principal square root, you need to check that $(\sqrt{7}-1)^2 = 8 - 2\sqrt{7}$.

72. Verify that the principal square root of $16 - 2\sqrt{55}$ is $\sqrt{11} - \sqrt{5}$. *Hint:* In other words, check that $(\sqrt{11}-\sqrt{5})^2 = 16 - 2\sqrt{55}$.

73. Simplify $\dfrac{\sqrt{a}}{\sqrt{a}+\sqrt{b}} + \dfrac{\sqrt{b}}{\sqrt{a}-\sqrt{b}}$.

*Hint:* First rationalize the denominators.

ⓒ 74. Use a calculator to provide empirical evidence indicating that both of the following equations may be correct:

$$\frac{2-\sqrt{3}}{\sqrt{2}-\sqrt{2-\sqrt{3}}} + \frac{2+\sqrt{3}}{\sqrt{2}+\sqrt{2+\sqrt{3}}} = \sqrt{2}$$

$$\sqrt{\sqrt{97.5 - \frac{1}{11}}} = \pi$$

The point of this exercise is to remind you that as useful as calculators may be, there is still the need for proofs in mathematics. In fact, it can be shown that the first equation is indeed correct, but the second is not.

75. Rationalize the denominator: $\dfrac{1}{1+\sqrt{2}+\sqrt{3}}$.

*Hint:* First multiply by $\dfrac{1+\sqrt{2}-\sqrt{3}}{1+\sqrt{2}-\sqrt{3}}$.

## RATIONAL EXPONENTS

**1.7**

*. . . Nicole Oresme, a bishop in Normandy (about 1323–1382), first conceived the notion of fractional powers, afterwards rediscovered by [Simon] Stevin [1548–1620]. . . .*

FLORIAN CAJORI in *A History of Elementary Mathematics* (New York: Macmillan Company, 1917)

*If Descartes . . . had discarded the radical sign altogether and had introduced the notation for fractional as well as integral exponents, . . . it is conceivable that generations upon generations of pupils would have been saved the necessity of mastering the operations with two . . . notations when one alone (the exponential) would have answered all purposes.*

FLORIAN CAJORI, *A History of Mathematical Notations*, Vol. 1, (La Salle, Illinois: The Open Court Publishing Company, 1928)

We can use the concept of an $n$th root to give a meaning to fractional exponents that is useful and, at the same time, consistent with our earlier work. First, by way of motivation, suppose that we want to assign a value to $5^{1/3}$. Assuming that the usual properties of exponents continue to apply here, we can write

$$(5^{1/3})^3 = 5^1 = 5$$

That is,

$$(5^{1/3})^3 = 5$$

or

$$5^{1/3} = \sqrt[3]{5}$$

By replacing 5 and 3 with $b$ and $n$, respectively, we can see that we want to define $b^{1/n}$ to mean $\sqrt[n]{b}$. Also, by thinking of $b^{m/n}$ as $(b^{1/n})^m$, we see that the definition for $b^{m/n}$ ought to be $(\sqrt[n]{b})^m$. These definitions are formalized in the box that follows.

---

**Definition:   Rational Exponents**

EXAMPLES

1. Let $b$ denote a real number and $n$ a natural number. We define $b^{1/n}$ by

$$b^{1/n} = \sqrt[n]{b}$$

(If $n$ is even, we require that $b \geq 0$.)

$$4^{1/2} = \sqrt{4} = 2$$
$$(-8)^{1/3} = \sqrt[3]{-8} = -2$$

2. Let $m/n$ be a rational number reduced to lowest terms. Assume that $n$ is positive and that $\sqrt[n]{b}$ exists. Then,

$$b^{m/n} = (\sqrt[n]{b})^m$$

or, equivalently,

$$b^{m/n} = \sqrt[n]{b^m}$$

$$8^{2/3} = (\sqrt[3]{8})^2 = 2^2 = 4$$

or, equivalently,

$$8^{2/3} = \sqrt[3]{8^2} = \sqrt[3]{64} = 4$$

---

It can be shown that the four properties of exponents that we listed in Section 1.5 (on page 22) continue to hold for rational exponents in general. In fact, we'll take this for granted rather than follow the lengthy argument needed for its verification. We will also assume that these properties apply to irrational exponents. So, for instance, we have

$$(2^{\sqrt{5}})^{\sqrt{5}} = 2^5 = 32$$

(The definition of irrational exponents is discussed in Section 6.1.) In the next three examples, we display the basic techniques for working with rational exponents.

**EXAMPLE 1**　Simplify each of the following. Express the answers using positive exponents. If an expression does not represent a real number, say so.

(a) $49^{1/2}$　　(b) $-49^{1/2}$　　(c) $(-49)^{1/2}$　　(d) $49^{-1/2}$

**Solution**　(a) $49^{1/2} = \sqrt{49} = 7$

(b) $-49^{1/2} = -(49^{1/2}) = -\sqrt{49} = -7$

(c) Does not represent a real number because there is no real number $x$ such that $x^2 = -49$.

(d) $49^{-1/2} = \sqrt{49^{-1}} = \sqrt{\dfrac{1}{49}} = \dfrac{\sqrt{1}}{\sqrt{49}} = \dfrac{1}{7}$

Alternatively, we have

$$49^{-1/2} = (49^{1/2})^{-1} = 7^{-1} = \frac{1}{7}$$

**EXAMPLE 2**　Simplify each of the following. Write the answers using positive exponents. (Assume that $a > 0$.)

(a) $(5a^{2/3})(4a^{3/4})$

(b) $\sqrt[5]{\dfrac{16a^{1/3}}{a^{1/4}}}$

(c) $(x^2 + 1)^{1/5}(x^2 + 1)^{4/5}$

**Solution**　(a) $(5a^{2/3})(4a^{3/4}) = 20a^{(2/3)+(3/4)}$

$$= 20a^{17/12} \qquad \text{because } \frac{2}{3} + \frac{3}{4} = \frac{17}{12}$$

(b) $\sqrt[5]{\dfrac{16a^{1/3}}{a^{1/4}}} = \left(\dfrac{16a^{1/3}}{a^{1/4}}\right)^{1/5}$

$$= (16a^{1/12})^{1/5} \qquad \text{because } \frac{1}{3} - \frac{1}{4} = \frac{1}{12}$$

$$= 16^{1/5}a^{1/60}$$

(c) $(x^2 + 1)^{1/5}(x^2 + 1)^{4/5} = (x^2 + 1)^1 = x^2 + 1$

**EXAMPLE 3**　Simplify:　(a) $32^{-2/5}$　　(b) $(-8)^{4/3}$

**Solution**　(a) $32^{-2/5} = (\sqrt[5]{32})^{-2} = 2^{-2} = \dfrac{1}{2^2} = \dfrac{1}{4}$

Alternatively, we have

$$32^{-2/5} = (2^5)^{-2/5} = 2^{-2} = \frac{1}{2^2} = \frac{1}{4}$$

(b) $(-8)^{4/3} = (\sqrt[3]{-8})^4 = (-2)^4 = 16$

Alternatively, we can write

$$(-8)^{4/3} = [(-2)^3]^{4/3} = (-2)^4 = 16$$

One of the properties of $n$th roots that was listed (but not discussed) in the previous section was $\sqrt[m]{\sqrt[n]{x}} = \sqrt[mn]{x}$. Using exponents, it is easy to verify this property. We have

$$\sqrt[m]{\sqrt[n]{x}} = (x^{1/n})^{1/m}$$
$$= x^{1/mn} = \sqrt[mn]{x}, \qquad \text{as we wished to show.}$$

Along the same lines, the next two examples indicate how rational exponents can be used to simplify expressions containing radicals.

**EXAMPLE 4**   Consider the expression $\sqrt{x}\sqrt[3]{y^2}$, where $x$ and $y$ are positive.

(a) Rewrite the expression using rational exponents.
(b) Rewrite the expression using only one radical sign.

Solution   (a) $\sqrt{x}\sqrt[3]{y^2} = x^{1/2}y^{2/3}$

(b) $\sqrt{x}\sqrt[3]{y^2} = x^{1/2}y^{2/3}$
$$= x^{3/6}y^{4/6} \qquad \text{We've rewritten the fractions using a common denominator.}$$
$$= \sqrt[6]{x^3}\sqrt[6]{y^4}$$
$$= \sqrt[6]{x^3y^4}$$

**EXAMPLE 5**   Rewrite the following expression using rational exponents. (Assume that $x$, $y$, and $z$ are positive.)

$$\sqrt{\frac{\sqrt[3]{x}\sqrt[4]{y^3}}{\sqrt[5]{z^4}}}$$

Solution   $$\sqrt{\frac{\sqrt[3]{x}\sqrt[4]{y^3}}{\sqrt[5]{z^4}}} = \left(\frac{\sqrt[3]{x}\sqrt[4]{y^3}}{\sqrt[5]{z^4}}\right)^{1/2} = \left(\frac{x^{1/3}y^{3/4}}{z^{4/5}}\right)^{1/2}$$
$$= \frac{x^{1/6}y^{3/8}}{z^{2/5}} = x^{1/6}y^{3/8}z^{-2/5}$$

# EXERCISE SET 1.7

## A

*In Exercises 1–36, evaluate or simplify each expression. Express the answers using positive exponents. If an expression is undefined (i.e., does not represent a real number), say so. (Assume that all letters represent positive numbers.)*

1. $16^{1/2}$
2. $100^{1/2}$
3. $(\frac{1}{36})^{1/2}$
4. $0.09^{1/2}$
5. $(-16)^{1/2}$
6. $(-1)^{1/2}$
7. $625^{1/4}$
8. $(\frac{1}{81})^{1/4}$
9. $8^{1/3}$
10. $0.001^{1/3}$
11. $8^{2/3}$
12. $64^{2/3}$
13. $(-32)^{1/5}$
14. $(-\frac{1}{125})^{1/3}$
15. $(-1000)^{1/3}$
16. $(-243)^{1/5}$
17. $49^{-1/2}$
18. $121^{-1/2}$
19. $(-49)^{-1/2}$
20. $(-64)^{-1/3}$
21. $36^{-3/2}$

22. $(-0.001)^{-2/3}$
23. $125^{2/3}$
24. $125^{-2/3}$
25. $(-1)^{3/5}$
26. $27^{4/3} + 27^{-4/3} + 27^0$
27. $32^{4/5} - 32^{-4/5}$
28. $64^{1/2} + 64^{-1/2} - 64^{4/3}$
29. $(\frac{9}{16})^{-5/2} - (\frac{1000}{27})^{4/3}$
30. $(256^{-3/4})^{4/3}$
31. $(2a^{1/3})(3a^{1/4})$
32. $\sqrt[5]{3a^4}$
33. $\sqrt[4]{\dfrac{64a^{2/3}}{a^{1/3}}}$
34. $(x^2 + 1)^{2/3}(x^2 + 1)^{4/3}$
35. $\dfrac{(x^2 + 1)^{3/4}}{(x^2 + 1)^{-1/4}}$
36. $\dfrac{(2x^2 + 1)^{-6/5}(2x^2 + 1)^{6/5}(x^2 + 1)^{-1/5}}{(x^2 + 1)^{9/5}}$

*For Exercises 37–44, follow Example 4 in the text to rewrite each expression in two ways: (a) using rational exponents; (b) using only one radical sign. (Assume x, y, and z are positive.)*

37. $\sqrt{3}\,\sqrt[3]{6}$

38. $\sqrt{5}\,\sqrt[3]{7}$

39. $\sqrt[3]{6}\,\sqrt[4]{2}$

40. $\sqrt[3]{2}\,\sqrt[5]{2}$

41. $\sqrt[3]{x^2}\,\sqrt[5]{y^4}$

42. $\sqrt{x}\,\sqrt[3]{y}\,\sqrt[4]{z}$

43. $\sqrt[4]{x^a}\,\sqrt[3]{x^b}\,\sqrt{x^{a/6}}$

44. $\sqrt[3]{27\sqrt{64x}}$

*In Exercises 45–52, rewrite each expression using rational exponents rather than radicals. (Assume that x, y, and z are positive.)*

45. $\sqrt[3]{(x+1)^2}$

46. $\dfrac{1}{\sqrt{x}}+\sqrt{x}$

47. $(\sqrt[5]{x}+y)^2$

48. $\sqrt{\sqrt{x}}$

49. $\sqrt[3]{\sqrt{x}}+\sqrt{\sqrt[3]{x}}$

50. $\sqrt[3]{\sqrt{2}}$

51. $\sqrt{\sqrt[3]{x}\,\sqrt[4]{y}}$

52. $\sqrt[5]{\dfrac{\sqrt{x}\,\sqrt[3]{y}}{\sqrt[4]{z^2}}}$

C 53. Which is larger, $9^{10/9}$ or $10^{9/10}$?

C 54. Which number is closer to 3: $13^{3/7}$ or $6560^{1/8}$?

C 55. Consider the expression $n^{1/n}$, where $n$ is a natural number. In calculus, it is shown that as $n$ takes on larger and larger values, the resulting value of the expression approaches 1. Confirm this empirically by completing the table. (Round off your results to four decimal places.)

| $n$ | 2 | 5 | 10 | 100 | $10^3$ | $10^4$ | $10^5$ | $10^6$ |
|-----|---|---|----|-----|--------|--------|--------|--------|
| $n^{1/n}$ | | | | | | | | |

**B**

C 56. In working with roots and radicals, it is not always apparent from the outset when two numbers are equal. (We've already seen examples of this in Exercise Set 1.3, Exercise 41.) In each of the following cases, use a calculator to verify that the quantity on the left side of the equation agrees with the quantity on the right side through the first six decimal places. (In each case, it can be shown that the numbers are indeed equal.)

(a) $(176+80\sqrt{5})^{1/5}=1+\sqrt{5}$

(b) $(7\sqrt[3]{20}-19)^{1/6}=\sqrt[3]{\tfrac{5}{3}}-\sqrt[3]{\tfrac{2}{3}}$

(c) $(2+\sqrt{5})^{1/3}+(2-\sqrt{5})^{1/3}=1$

57. Without using a calculator, decide which number is larger in each case.

(a) $2^{2/3}$ or $2^{3/2}$  (b) $5^{1/2}$ or $5^{-2}$

(c) $2^{1/2}$ or $2^{1/3}$  (d) $(\tfrac{1}{2})^{1/2}$ or $(\tfrac{1}{2})^{1/3}$

(e) $10^{1/10}$ or $(\tfrac{1}{10})^{10}$

58. Give an example showing that each statement is false.

(a) $x^{1/m}=\dfrac{1}{x^m}$  (b) $x^{-1/2}=\dfrac{1}{x^2}$

(c) $x^{m/n}=\sqrt[m]{x^n}$  (d) $\sqrt{x+y}=\sqrt{x}+\sqrt{y}$

(e) $\sqrt[3]{x+y}=\sqrt[3]{x}+\sqrt[3]{y}$

59. (a) Give an example in which a rational number raised to a rational power is irrational.

(b) Give an example in which an irrational number raised to a rational power is rational.

60. Without using a calculator, decide which number is closer to zero: $(0.5)^{1/3}$ or $(0.5)^{1/4}$. Then use a calculator to check your answer.

61. Without using a calculator, decide which number is smaller: $(-0.5)^{1/3}$ or $(-0.4)^{1/3}$. Then use a calculator to check your answer.

**C**

62. Simplify: $\sqrt[x]{x^{(x^2-x)}}$  $(x>0)$

63. Show that $\dfrac{a-b}{a+b}\sqrt{\dfrac{a+b}{a-b}}=\left(\dfrac{a-b}{a+b}\right)^{1/2}$
(Assume that $a>b>0$.)

64. Given $a+b=2c$, evaluate the following expression.
$$\left[\frac{(2^{a-b})^b(2^{b-c})^{c-a}}{(2^{c+b})^{c-b}}\right]^{1/c}$$

## POLYNOMIALS

**1.8**

As background for our work on polynomials, we first review the terms *constant* and *variable*. By way of example, consider the familiar expression for the area of a circle of radius $r$, namely, $\pi r^2$. Here $\pi$ is a constant; its value never changes throughout the discussion. On the other hand, $r$ is a variable; we can substitute any positive number for $r$ to obtain the area of a particular circle. More generally, by a **constant** we mean either a particular number (such as $\pi$, or $-17$, or $\sqrt{2}$) or a letter whose value remains fixed (although perhaps unspecified) throughout a given discussion. In contrast, a **variable** is a letter for which we can substitute any number selected from a given set of numbers. The given set of numbers is called the **domain** of the variable.

Some expressions will make sense only for certain values of the variable. For instance, $\dfrac{1}{x-3}$ will be undefined when $x$ is 3 (for then the denominator is zero). So in this case we would agree that the domain of the variable $x$ consists of all real numbers except $x = 3$. Similarly, throughout this chapter (with the exception of Section 1.11 on complex numbers), we adopt the following convention:

> **The Domain Convention**
>
> The domain of a variable in a given expression is the set of all real number values of the variable for which the expression is defined.

In algebra it's customary (but not mandatory) to use letters near the end of the alphabet for variables; letters from the beginning of the alphabet are generally used for constants. (This convention was begun by Descartes.) So, for example, in the expression $ax + b$, the letter $x$ is the variable and $a$ and $b$ are constants.

**EXAMPLE 1**   Specify the variable, the constants, and the domain of the variable for each expression.

(a) $3x + 4$   (b) $\dfrac{1}{(t+1)(t-1)}$   (c) $ay^2 + by + c$   (d) $4x + 3x^{-1}$

**Solution**

| | VARIABLE | CONSTANTS | DOMAIN |
|---|---|---|---|
| (a) $3x + 4$ | $x$ | $3, 4$ | The set of all real numbers. |
| (b) $\dfrac{1}{(t+1)(t-1)}$ | $t$ | $1, -1$ | The set of all real numbers except $t = 1$ and $t = -1$. (The denominator is zero when $t = \pm 1$.) |
| (c) $ay^2 + by + c$ | $y$ | $a, b, c$ | The set of all real numbers. |
| (d) $4x + 3x^{-1}$ | $x$ | $4, 3$ | The set of all real numbers except $x = 0$. (Remember, $x^{-1} = 1/x$.) |

The expressions in parts (a) and (c) of Example 1 are *polynomials*. By a **polynomial in $x$**, we mean an expression of the form

$$a_n x^n + a_{n-1} x^{n-1} + \cdots + a_1 x + a_0$$

where $n$ is a nonnegative integer and $a_n \neq 0$. The individual expressions $a_k x^k$ making up the polynomial are called **terms**. In this chapter, the **coefficients** $a_k$ will always be real numbers. For example, the terms of the polynomial $x^2 - 7x + 3$ are $x^2$, $-7x$, and 3; the coefficients are 1, $-7$, and 3. In writing a polynomial, it's customary (but not mandatory) to write the terms in order of decreasing powers of $x$. For instance, we'd usually write $x^2 - 7x + 3$ rather than $x^2 + 3 - 7x$. The highest power of $x$ in a polynomial is called the **degree** of the polynomial. For example, the degree of the polynomial $x^2 - 7x + 3$ is 2. In the case of a polynomial consisting only of a nonzero constant $a_0$, we say that the degree is zero (because $a_0 = a_0 x^0$). No degree is defined for the polynomial whose only term is zero.

There is some additional terminology that is useful in describing polynomials that have only a few terms. A polynomial with only one term (such as $3x^2$) is a **monomial**; a polynomial with only two terms (such as $2x + 3$) is a **binomial**; and a polynomial with only three terms (such as $5x^2 - 6x + 3$) is a **trinomial**. The table that follows provides examples of the terminology.

| EXPRESSION | POLYNOMIAL? | DEGREE (IF POLYNOMIAL) | TERMS (IF POLYNOMIAL) | COEFFICIENTS (IF POLYNOMIAL) |
|---|---|---|---|---|
| $2x^3 - 3x^2 + 4x - 1$ | yes | 3 | $2x^3, -3x^2, 4x, -1$ | $2, -3, 4, -1$ |
| $t^2 + 1$ | yes | 2 | $t^2, 1$ | $1, 1$ |
| $-12$ | yes | 0 | $-12$ | $-12$ |
| $2x^{-4} + 5$ | no | | | |
| $\dfrac{1}{2x - 3}$ | no | | | |
| $\sqrt{4x^2 + 1}$ | no | | | |

Just as we can combine two numbers through addition, subtraction, multiplication, or division, so can we combine polynomials. To add or subtract two polynomials, we follow the familiar process of combining like terms. For example, to add the polynomials $3x - 1$ and $x^2 + 8x - 4$, we have

$$(3x - 1) + (x^2 + 8x - 4) = \underbrace{x^2}_{0 + x^2} + \underbrace{11x}_{3x + 8x} \underbrace{- 5}_{-1 + (-4)}$$

Similarly, to subtract we write

$$(3x - 1) - (x^2 + 8x - 4) = \underbrace{-x^2}_{0 - x^2} \underbrace{- 5x}_{3x - 8x} + \underbrace{3}_{-1 - (-4)}$$

Since all the terms in a polynomial ultimately stand for real numbers, we can rely on the properties of real numbers in simplifying polynomial expressions. For instance, the following rules are useful in dealing with sums and differences of polynomials.

---

**Rules for Sums and Differences of Polynomials**

| RULE | EXAMPLE |
|---|---|
| (a) If the parentheses are preceded by a plus sign, or no sign appears, then the parentheses can be omitted. | $(3x^2 + 4x) + (2x^2 - 5x)$ <br> $= 3x^2 + 4x + 2x^2 - 5x$ <br> $= 5x^2 - x$ |
| (b) If the parentheses are preceded by a minus sign, then the parentheses can be omitted if the sign of every term within is reversed. | $(x^2 - 3x + 1) - (3x^2 - 8x + 4)$ <br> $= x^2 - 3x + 1 - 3x^2 + 8x - 4$ <br> $= -2x^2 + 5x - 3$ |

As we saw in Section 1.2, multiplication of polynomials can be carried out by means of the distributive properties. Example 2 displays three instances of this.

**EXAMPLE 2**  Multiply:

(a) $3x(4x^2 - 8x + 1)$
(b) $(x^2 - 4x + 2)(x + 3)$
(c) $(2x - 3)(4x + 5)$

**Solution**

(a) $3x(4x^2 - 8x + 1) = 3x(4x^2) + 3x(-8x) + 3x(1)$
$$= 12x^3 - 24x^2 + 3x$$

(b) $(x^2 - 4x + 2)(x + 3) = (x^2 - 4x + 2)x + (x^2 - 4x + 2)3$
$$= x^3 - 4x^2 + 2x + 3x^2 - 12x + 6$$
$$= x^3 - x^2 - 10x + 6$$

(c) $(2x - 3)(4x + 5) = (2x - 3)4x + (2x - 3)5$
$$= 8x^2 - 12x + 10x - 15$$
$$= 8x^2 - 2x - 15$$

In part (c) of the example just concluded, we showed the multiplication of two binomials. Because this type of product occurs so frequently, it's very useful to be able to carry out the work mentally. The so-called FOIL method allows us to accomplish this. "FOIL" is an acronym standing for First, Outer, Inner, Last. Here's how the procedure works in computing the product $(2x - 3)(4x + 5)$.

STEP 1  Multiply the first two terms: $(2x)(4x) = 8x^2$.

$$(2x - 3)(4x + 5)$$

first

STEP 2  Add the products of the outer and inner terms:
$(2x)(5) + (-3)(4x) = -2x$.

outer

$$(2x - 3)(4x + 5)$$

inner

STEP 3  Multiply the last two terms: $(-3)(5) = -15$.

$$(2x - 3)(4x + 5)$$

last

STEP 4  Add the results: $8x^2 - 2x - 15$ (as obtained previously).

**EXAMPLE 3**   Compute each product:

(a) $(2x + 3y)(5x - y)$        (b) $(A - B)(A + B)$

Solution   (a) $(2x + 3y)(5x - y) = 10x^2 + 13xy - 3y^2$        (FOIL method)
(b) $(A - B)(A + B) = A^2 - B^2$        (The middle term, $AB - AB$, is zero.)

There are several products that occur so frequently that it's convenient to memorize the results. The product $(A - B)(A + B) = A^2 - B^2$ in the example just concluded is one of these. The list that follows displays these special products. (Exercise 68 asks you to verify these results by carrying out the multiplication.) In memorizing these formulas, remember it's the *form* or *pattern* that's important, not the specific choice of letters.

| PROPERTY SUMMARY | Special Products |
|---|---|
| | 1. $(A - B)(A + B) = A^2 - B^2$ |
| | 2. (a) $(A + B)^2 = A^2 + 2AB + B^2$<br>(b) $(A - B)^2 = A^2 - 2AB + B^2$ |
| | 3. (a) $(A + B)^3 = A^3 + 3A^2B + 3AB^2 + B^3$<br>(b) $(A - B)^3 = A^3 - 3A^2B + 3AB^2 - B^3$ |
| | 4. (a) $(A + B)(A^2 - AB + B^2) = A^3 + B^3$<br>(b) $(A - B)(A^2 + AB + B^2) = A^3 - B^3$ |

**EXAMPLE 4**   Use the Special Product formulas to compute:

(a) $(3\sqrt{xy} - z)(3\sqrt{xy} + z)$        (b) $(4x^3 - 3)^2$        (c) $(2x + 5)^3$

Solution   (a) Using Special Product 1, we have
$$(3\sqrt{xy} - z)(3\sqrt{xy} + z) = (3\sqrt{xy})^2 - z^2$$
$$= 9xy - z^2$$

(b) Using Special Product 2(b) we have
$$(4x^3 - 3)^2 = (4x^3)^2 - 2(4x^3)(3) + 3^2$$
$$= 16x^6 - 24x^3 + 9$$

(c) Using Special Product 3(a) we have
$$(2x + 5)^3 = (2x)^3 + 3(2x)^2(5) + 3(2x)(5)^2 + (5)^3$$
$$= 8x^3 + 60x^2 + 150x + 125$$

We conclude this section by describing the process of long division for polynomials. The terms *quotient*, *remainder*, *divisor*, and *dividend* are used here in the same way they're used in ordinary division of numbers. For instance, when 7 is divided by 2, the quotient is 3 and the remainder is 1. We write this

$$\frac{7}{2} = 3 + \frac{1}{2}$$

or, equivalently,

$$7 = 2 \times 3 + 1$$

with arrows pointing up from each term labeled: dividend, divisor, quotient, remainder

The process of long division for polynomials follows the same four-step cycle used in ordinary long division of numbers: divide, multiply, subtract, bring down. As a first example, we divide $2x^2 - 7x + 8$ by $x - 2$. Notice that in setting up the division, we write both the dividend and the divisor in decreasing powers of $x$.

$$
\begin{array}{r}
2x - 3 \\
x - 2 \overline{)\, 2x^2 - 7x + 8} \\
2x^2 - 4x \\
\hline
-3x + 8 \\
-3x + 6 \\
\hline
2
\end{array}
$$

1. Divide the first term of the dividend by the first term of the divisor: $\dfrac{2x^2}{x} = 2x$. The result becomes the first term of the quotient, as shown.
2. Multiply the divisor $x - 2$ by the term $2x$ obtained in the previous step. This yields the quantity $2x^2 - 4x$, which is written below the dividend, as shown.
3. From the quantity $2x^2 - 7x$ in the dividend, subtract the quantity $2x^2 - 4x$. This yields $-3x$.
4. Bring down the $+8$ in the dividend, as shown. The resulting quantity $-3x + 8$ is now treated as the dividend and the entire process is repeated.

We've now found that when $2x^2 - 7x + 8$ is divided by $x - 2$, the quotient is $2x - 3$ and the remainder is 2. This is summarized by writing either

$$\frac{2x^2 - 7x + 8}{x - 2} = 2x - 3 + \frac{2}{x - 2}$$

or

$$2x^2 - 7x + 8 = (x - 2)(2x - 3) + 2$$

As a second example, we divide $3x^4 - 2x^3 + 2$ by $x^2 - 1$. Notice in what follows that we have inserted the terms whose coefficients are zero in the divisor and in the dividend. These terms serve as place holders.

$$
\begin{array}{r}
3x^2 - 2x + 3 \\
x^2 + 0x - 1 \overline{)\, 3x^4 - 2x^3 + 0x^2 + 0x + 2} \\
3x^4 + 0x^3 - 3x^2 \\
\hline
-2x^3 + 3x^2 + 0x \\
-2x^3 + 0x^2 + 2x \\
\hline
3x^2 - 2x + 2 \\
3x^2 + 0x - 3 \\
\hline
-2x + 5
\end{array}
$$

Thus the quotient is $3x^2 - 2x + 3$, the remainder is $-2x + 5$, and we can write

$$\frac{3x^4 - 2x^3 + 2}{x^2 - 1} = 3x^2 - 2x + 3 + \frac{-2x + 5}{x^2 - 1}$$

or

$$3x^4 - 2x^3 + 2 = (x^2 - 1)(3x^2 - 2x + 3) + (-2x + 5)$$

In the example just concluded, note that the degree of the remainder is 1, whereas the degree of the divisor is 2. So in this case we have

(degree of remainder) < (degree of divisor)

In fact, it can be shown that this inequality remains valid whenever we obtain a nonzero remainder in dividing two nonconstant polynomials. (We'll return to this idea in a later chapter.)

## EXERCISE SET 1.8

### A

*In Exercises 1–6, specify the domain of each variable.*

1. (a) $2x^2 - 3x + 5$  (b) $2x^{1/2} - 3x + 5$

2. (a) $y - 1$  (b) $\dfrac{2y}{y - 1}$

3. (a) $ax + b$  (b) $ax^{1/3} + b$

4. (a) $x + x^{-1}$  (b) $t^{-1} + 2t^{-2}$

5. (a) $4\sqrt{t} + t^2$  (b) $\dfrac{4}{(t - 1)\sqrt{t}}$

6. (a) $\dfrac{1}{x}$  (b) $\dfrac{1}{(x - 1)(x + 2)(x - 3)}$

*For Exercises 7–10, specify the degree and the (nonzero) coefficients of each polynomial.*

7. (a) $4x^3 - 2x^2 - 6x - 1$
   (b) $ax^2 + bx + c$  ($a$, $b$, and $c$ nonzero)
8. (a) $-4t^3 + 3t^4 + \sqrt{2}$  (b) $3\sqrt{2}x^2$
9. (a) $6x$  (b) $6x + x^2 - x^3$
10. (a) $18$  (b) $18x^{10}$

*In Exercises 11–40, carry out the indicated operations.*

11. $(12x^2 - 4x + 2) + (8x^2 + 6x - 1)$
12. $(4x^2 - 1) + (4x^2 + 1)$
13. $(x^2 - x - 1) - (x^2 + x + 1)$
14. $(x^3 + 2x^2 + 1) - (x^2 - 1)$
15. $(2x^2 - 4x - 4) - (6x^2 + 5) - (8x - 1)$
16. $(x^3 - 1) - (x^3 - 1)$
17. $(2x^2 - 6x - 1) + (5x^2 - 5x - 1) - (3x^2 + 8x + 12)$
18. $(ax + b) - (2ax - b)$

19. $(ax^2 + bx + c) - (2ax^2 - 3bx + c)$
20. $x^3 - 4x^2 + x - 2(x^3 + 2x^2 + 1)$
21. $2x(x^2 - 4x - 5)$  22. $4ab(a^2 + 2ab + b^2)$
23. $(x - 1)(x - 2)$  24. $(y^2 + 1)(y^2 - 3)$
25. $(2x + 4)(x + 1)$  26. $(a + 2b)(a - 3b)$
27. $(x^2 + 3x)(x^2 + x)$  28. $(\sqrt{x} + 1)(\sqrt{x} + 2)$
29. $(2xy - 3)(2xy - 1)$  30. $(5xy^2 - 2)(xy^2 - 3)$
31. $\left(x + \dfrac{1}{x}\right)\left(x + \dfrac{1}{x}\right)$  32. $(2pq + r)(pq - 2r)$
33. $(\sqrt{a + b} + 1)(\sqrt{a + b} + 3)$
34. $(x^{1/2} + y^{1/2})(x^{1/2} + 2y^{1/2})$
35. $(x^{1/2} - 1)(x^{1/2} - 2)$
36. $(x - 1)(x^2 + 4x + 1)$
37. $(y + 2)(y^2 - 3y - 5)$
38. $(a + b + c)(a + b - c)$
39. $(x - 2y + z)(x + 2y - z)$
40. $a(b - c) + b(c - a) + c(a - b)$

*For Exercises 41–52, use the Special Product formulas to compute the products.*

41. (a) $(x - y)(x + y)$  (b) $(x^2 - 5)(x^2 + 5)$
42. (a) $(3y - a^2)(3y + a^2)$
    (b) $(3\sqrt{y} - a^2)(3\sqrt{y} + a^2)$
43. (a) $(A - 4)(A + 4)$
    (b) $[(a + b) - 4][(a + b) + 4]$
44. (a) $(\sqrt{ab} + \sqrt{c})(\sqrt{ab} - \sqrt{c})$
    (b) $(a^{1/2} + b^{1/2})(a^{1/2} - b^{1/2})$
45. (a) $(x - 8)^2$  (b) $(2x^2 - 5)^2$

46. (a) $(2^m + 1)^2$     (b) $(a^m - a^{-m})^2$
47. (a) $(\sqrt{x} + \sqrt{y})^2$     (b) $(\sqrt{x+y} - \sqrt{x})^2$
48. (a) $(2x + y)^3$     (b) $(x - 2y)^3$
49. (a) $(a + 1)^3$     (b) $(3x^2 - 2a^2)^3$
50. (a) $(x - y)(x^2 + xy + y^2)$
     (b) $(x^2 - y^2)(x^4 + x^2y^2 + y^4)$
51. (a) $(x + 1)(x^2 - x + 1)$
     (b) $(x + 2)(x^2 - 2x + 4)$
52. (a) $(x^{1/3} - y^{1/3})(x^{2/3} + x^{1/3}y^{1/3} + y^{2/3})$
     (b) $(a^{1/3} + 5^{1/3})(a^{2/3} - a^{1/3} \cdot 5^{1/3} + 5^{2/3})$

*In Exercises 53–60, use long division to compute the quotients and remainders.*

53. (a) $\dfrac{x^2 + 6x - 1}{x - 1}$     (b) $\dfrac{x^3 + 6x - 1}{x - 1}$
     (c) $\dfrac{x^2 + 6x - 1}{x^2 - 1}$

54. (a) $\dfrac{2x^2 + x + 5}{x + 2}$     (b) $\dfrac{2x^2 + 5}{x + 2}$
     (c) $\dfrac{2x^3 + x + 5}{x + 2}$

55. $\dfrac{2x^3 + 4x^2 - 6x + 2}{2x + 4}$     56. $\dfrac{x^2 + 8x - 12}{x^2 + 1}$

57. $\dfrac{x^6 - 64}{x - 2}$     58. $\dfrac{x + 4}{x - 3}$

59. $\dfrac{ax + b}{x + c}$     60. $\dfrac{8x^6 - 36x^4 + 54x^2 - 27}{2x^2 - 3}$

[c] 61. Notice that the expression $(x^2 - 16)/(x - 4)$ is undefined when $x = 4$. In this exercise we investigate the values of the expression when $x$ is very close to 4.
     (a) Complete the tables.

| $x$ | $\dfrac{x^2 - 16}{x - 4}$ | $x$ | $\dfrac{x^2 - 16}{x - 4}$ |
|---|---|---|---|
| 3.9 | | 4.1 | |
| 3.99 | | 4.01 | |
| 3.999 | | 4.001 | |
| 3.9999 | | 4.0001 | |
| 3.99999 | | 4.00001 | |

     (b) What value does the expression $\dfrac{x^2 - 16}{x - 4}$ seem to be approaching as $x$ gets closer and closer to 4? This "target value" is referred to as the *limit* of $\dfrac{x^2 - 16}{x - 4}$ as $x$ approaches 4. (The notion of a limit is made more precise in calculus.)

[c] 62. Notice that the expression $\dfrac{x^3 - 8}{x - 2}$ is undefined when $x = 2$. In this exercise we investigate the values of the expression when $x$ is very close to 2.
     (a) Complete the tables.

| $x$ | $\dfrac{x^3 - 8}{x - 2}$ | $x$ | $\dfrac{x^3 - 8}{x - 2}$ |
|---|---|---|---|
| 1.9 | | 2.1 | |
| 1.99 | | 2.01 | |
| 1.999 | | 2.001 | |
| 1.9999 | | 2.0001 | |
| 1.99999 | | 2.00001 | |

     (b) On the basis of the tables, what "target value" does the expression $\dfrac{x^3 - 8}{x - 2}$ seem to be approaching as $x$ gets closer and closer to 2?

**B**

*For Exercises 63–66, verify each statement by carrying out the operations on the left-hand side of the equation.*

63. $(b - c)(b + c - a) + (c - a)(c + a - b) + (a - b)(a + b - c) = 0$
64. $(b^{-1} + c^{-1})(b + c - a) + (c^{-1} + a^{-1})(c + a - b) + (a^{-1} + b^{-1})(a + b - c) = 6$
65. $(x^2 + 8 - 4x)(x^2 + 8 + 4x) = x^4 + 64$
66. $(\sqrt{a+x} + \sqrt{a-x})(\sqrt{a+x} - \sqrt{a-x}) = 2x$

67. (a) Use long division to determine the remainder when $ax^2 + bx + c$ is divided by $x - r$.
     (b) Use long division to determine the remainder when $ax^3 + bx^2 + cx + d$ is divided by $x - r$.
     (c) On the basis of your answers in parts (a) and (b), guess the remainder when $ax^4 + bx^3 + cx^2 + dx + e$ is divided by $x - r$.

68. Carry out the multiplication to verify each of the Special Products listed on page 42.

69. (a) Complete the table.

| $n$ | 1 | 2 | 3 | 4 | 5 | 6 |
|---|---|---|---|---|---|---|
| $\dfrac{n^2}{n + 1}$ | | | | | | |

     (b) Are there any natural numbers $n$ for which $n^2/(n + 1)$ is itself a natural number? *Hint:* Use long division.

# FACTORING

**1.9**

There are many cases in algebra in which the process of *factoring* simplifies the work at hand. To **factor** a polynomial means to write it as a product of two or more nonconstant polynomials. For instance, a factorization of $x^2 - 9$ is given by

$$x^2 - 9 = (x - 3)(x + 3)$$

We'll consider several techniques for factoring in this section. These techniques will be applied in Section 1.10 (Fractional Expressions), in Chapter 2 (Equations and Inequalities), and throughout the text.

There is one convention that we need to agree on at the outset. If the polynomial or expression that we wish to factor contains only integer coefficients, then the factors (if any) should involve only integer coefficients. For example, according to this convention, we will not consider the following type of factorization in this section:

$$x^2 - 2 = (x - \sqrt{2})(x + \sqrt{2})$$

because it involves coefficients that are irrational numbers. (We should point out, however, that factorizations such as this are useful at times, particularly in calculus.) As it happens, $x^2 - 2$ is an example of a polynomial that cannot be factored using integer coefficients. We say in such a case that the polynomial is **irreducible over the integers**.

There are five elementary techniques for factoring. These are summarized in the box that follows. Notice that three of the formulas in the box are just restatements of the corresponding Special Product formulas considered in the previous section. In these cases, remember it is the form or pattern in the formula that is important, not the specific choice of letters.

| TECHNIQUE | EXAMPLE OR FORMULA | REMARK |
|---|---|---|
| Common factor | $3x^4 + 6x^3 - 12x^2 = 3x^2(x^2 + 2x - 4)$<br>$4(x^2 + 1) - x(x^2 + 1) = (x^2 + 1)(4 - x)$ | In any factoring problem, the first step always is to look for the common factor of highest degree. |
| Difference of squares | $x^2 - a^2 = (x - a)(x + a)$ | There is no corresponding formula for a sum of squares; $x^2 + a^2$ is irreducible over the integers. |
| Trial and error | $x^2 + 2x - 3 = (x + 3)(x - 1)$ | In this example, the only distinct possibilities or trials are:<br>**(a)** $(x - 3)(x - 1)$ **(b)** $(x - 3)(x + 1)$<br>**(c)** $(x + 3)(x - 1)$ **(d)** $(x + 3)(x + 1)$<br>By inspection or by carrying out the indicated multiplications, we find that only case **(c)** checks. |
| Difference of cubes | $x^3 - a^3 = (x - a)(x^2 + ax + a^2)$ | Verify these formulas for yourself by carrying out the multiplications. Then memorize the formulas. |
| Sum of cubes | $x^3 + a^3 = (x + a)(x^2 - ax + a^2)$ | |
| Grouping | $x^3 - x^2 + x - 1 = (x^3 - x^2) + (x - 1)$<br>$\quad = x^2(x - 1) + (x - 1) \cdot 1$<br>$\quad = (x - 1)(x^2 + 1)$ | This is actually an application of the common factor technique. |

The idea in factoring is to use one or more of these techniques until each of the factors obtained is irreducible. The examples that follow show how this works in practice.

**EXAMPLE 1**   Factor:   (a) $x^2 - 49$     (b) $2x^3 - 50x$

Solution   (a) $x^2 - 49 = x^2 - 7^2$

$$= (x - 7)(x + 7) \qquad \text{difference of squares}$$

(b) $2x^3 - 50x = 2x(x^2 - 25) \qquad \text{common factor}$

$$= 2x(x - 5)(x + 5) \qquad \text{difference of squares}$$

**EXAMPLE 2**   Factor:   $3x^5 - 3x$

Solution   $3x^5 - 3x = 3x(x^4 - 1) \qquad\qquad\quad \text{common factor}$

$$= 3x[(x^2)^2 - 1^2]$$

$$= 3x(x^2 - 1)(x^2 + 1) \qquad\quad \text{difference of squares}$$

$$= 3x(x - 1)(x + 1)(x^2 + 1) \qquad \text{difference of squares, again}$$

**EXAMPLE 3**   Factor:   (a) $x^2 - 4x - 5$     (b) $x^2 - 4x + 5$     (c) $8x^4 - 24x^3 + 18x^2$

Solution   (a) $x^2 - 4x - 5 = (x - 5)(x + 1) \qquad \text{trial and error}$

(b) Irreducible     trial and error

(c) The first step is to check for a common factor. It is $2x^2$, and we write

$$8x^4 - 24x^3 + 18x^2 = 2x^2(4x^2 - 12x + 9)$$

Next we consider the expression in parentheses in this last equation. It can be factored by trial and error; alternately, it can be factored by recalling the Special Product formula $(A - B)^2 = A^2 - 2AB + B^2$. In either case, we find that $4x^2 - 12x + 9 = (2x - 3)^2$. Thus, our final factorization is

$$8x^4 - 24x^3 + 18x^2 = 2x^2(2x - 3)^2$$

**EXAMPLE 4**   Factor:   $(t - b)^2 - (t^2 + b)^2$

Solution   Although we could begin by carrying out the indicated operations, it's much more efficient here to recognize that this is a difference of squares, for which the basic pattern is

$$x^2 - a^2 = (x - a)(x + a)$$

In this last equation we replace $x$ with the quantity $(t - b)$ and $a$ with the quantity $(t^2 + b)$ to obtain

$$(t - b)^2 - (t^2 + b)^2 = [(t - b) - (t^2 + b)][(t - b) + (t^2 + b)]$$

$$= (t - b - t^2 - b)(t - b + t^2 + b)$$

$$= (t - t^2 - 2b)(t + t^2)$$

Finally now, notice that on the right-hand side of this last equation, the common factor $t$ can be removed from the expression $t + t^2$. Thus, the required factorization is

$$(t - b)^2 - (t^2 + b)^2 = (t - t^2 - 2b)(1 + t)t$$

**EXAMPLE 5**   Factor:   $x^2 - y^2 + 10x + 25$

Solution   Familiarity with the Special Products in Section 1.8 suggests that we try grouping the terms this way:

$$(x^2 + 10x + 25) - y^2$$

Then we have

$$
\begin{aligned}
x^2 - y^2 + 10x + 25 &= (x^2 + 10x + 25) - y^2 \\
&= (x + 5)^2 - y^2 \\
&= [(x + 5) - y][(x + 5) + y] \quad \text{difference of squares} \\
&= (x - y + 5)(x + y + 5)
\end{aligned}
$$

**EXAMPLE 6**   Factor:   $ax + ay^2 + bx + by^2$

Solution   Factor $a$ from the first two terms and $b$ from the second two, to obtain

$$ax + ay^2 + bx + by^2 = a(x + y^2) + b(x + y^2)$$

Now recognize the quantity $(x + y^2)$ as a common expression that can be factored out. We have then

$$a(x + y^2) + b(x + y^2) = (x + y^2)(a + b)$$

The required factorization is therefore

$$ax + ay^2 + bx + by^2 = (x + y^2)(a + b)$$

As you may wish to check for yourself, this factorization can also be obtained by trial and error.

**EXAMPLE 7**   Use factoring to evaluate:   (a) $100^2 - 99^2$      (b) $8^3 - 6^3$

Solution   (a)
$$
\begin{aligned}
100^2 - 99^2 &= (100 - 99)(100 + 99) \quad \text{difference of squares} \\
&= (1)(199) \\
&= 199
\end{aligned}
$$

(b)
$$
\begin{aligned}
8^3 - 6^3 &= (8 - 6)(8^2 + 8 \cdot 6 + 6^2) \quad \text{difference of cubes} \\
&= 2(64 + 48 + 36) \\
&= 2[(64 + 36) + 48] = 2(148) \\
&= 296
\end{aligned}
$$

**EXAMPLE 8**   Factor:   $x^3 - 6ax^2 + 12a^2x - 8a^3$

Solution   Since this expression involves four terms, our first inclination is to apply the method shown in Example 6. We factor $x^2$ from the first two terms and $4a^2$ from the second two. This yields

$$x^3 - 6ax^2 + 12a^2x - 8a^3 = x^2(x - 6a) + 4a^2(3x - 2a)$$

But now (unlike the corresponding place in Example 6) there is no common expression on the right-hand side. So at this point we need to start over. Again we will use the method of Example 6; but first we rearrange our terms. We have

$$x^3 - 6ax^2 + 12a^2x - 8a^3 = \underbrace{x^3 - 8a^3}_{\substack{\text{differences} \\ \text{of cubes is} \\ \text{applicable}}} - \underbrace{6ax^2 + 12a^2x}_{\substack{\text{common term} \\ \text{is } -6ax}}$$

Thus

$$x^3 - 6ax^2 + 12a^2x - 8a^3 = (x - 2a)(x^2 + 2ax + 4a^2) - 6ax(x - 2a)$$

Now on the right-hand side of this last equation we factor out the common expression $(x - 2a)$ to obtain

$$x^3 - 6ax^2 + 12a^2x - 8a^3 = (x - 2a)[(x^2 + 2ax + 4a^2) - 6ax]$$
$$= (x - 2a)(x^2 - 4ax + 4a^2)$$
$$= (x - 2a)[(x - 2a)(x - 2a)]$$
$$= (x - 2a)^3$$

The required factorization is therefore

$$x^3 - 6ax^2 + 12a^2x - 8a^3 = (x - 2a)(x - 2a)(x - 2a)$$

or, more concisely,

$$x^3 - 6ax^2 + 12a^2x - 8a^3 = (x - 2a)^3$$

NOTE   You can check this answer by recalling the Special Product Formula for $(A - B)^3$. Indeed, if you're sufficiently familiar with that formula, the required factorization in this example can be obtained simply by inspection.

**EXAMPLE 9**   Factor:   $x^6 - y^6$

**Solution**   We will show two methods; the first is reasonably straightforward, the second involves a clever manipulation.

FIRST METHOD   $x^6 - y^6 = (x^3)^2 - (y^3)^2$
$$= (x^3 - y^3)(x^3 + y^3)$$
$$= [(x - y)(x^2 + xy + y^2)][(x + y)(x^2 - xy + y^2)]$$

The required factorization is therefore

$$x^6 - y^6 = (x - y)(x + y)(x^2 + xy + y^2)(x^2 - xy + y^2)$$

ALTERNATE METHOD   $x^6 - y^6 = (x^2)^3 - (y^2)^3$
$$= (x^2 - y^2)[(x^2)^2 + x^2y^2 + (y^2)^2]$$
$$= (x - y)(x + y)[x^4 + x^2y^2 + y^4]$$

The problem now is to factor the quantity in the brackets in this last equation. (Surely it must factor, judging from the answer obtained with the first method.) The technique here is to add zero, written in the form $x^2y^2 - x^2y^2$. We have

$$x^4 + x^2y^2 + y^4 = x^4 + x^2y^2 + y^4 + 0$$
$$= x^4 + x^2y^2 + y^4 + x^2y^2 - x^2y^2$$
$$= (x^4 + 2x^2y^2 + y^4) - x^2y^2$$
$$= (x^2 + y^2)^2 - (xy)^2$$

We now have a difference of squares, which can be factored:

$$(x^2 + y^2)^2 - (xy)^2 = [(x^2 + y^2) - xy][(x^2 + y^2) + xy]$$

or

$$(x^2 + y^2)^2 - (xy)^2 = (x^2 - xy + y^2)(x^2 + xy + y^2)$$

Putting things together, we finally obtain

$$x^6 - y^6 = (x - y)(x + y)(x^2 - xy + y^2)(x^2 + xy + y^2)$$

Except for the order of the last two factors, this agrees with the result obtained using the first method. ▓

## EXERCISE SET 1.9

### A

In Exercises 1–48, factor each polynomial or expression. If a polynomial is irreducible, state this. (In Exercises 1–6, the factoring techniques are specified.)

1. (Common factor and difference of squares)
   (a) $x^2 - 64$
   (b) $7x^4 + 14x^2$
   (c) $121z - z^3$
   (d) $a^2b^2 - c^2$

2. (Common factor and difference of squares)
   (a) $1 - t^4$
   (b) $x^6 + x^5 + x^4$
   (c) $u^2v^2 - 225$
   (d) $81x^4 - x^2$

3. (Trial and error)
   (a) $x^2 + 2x - 3$
   (b) $x^2 - 2x - 3$
   (c) $x^2 - 2x + 3$
   (d) $-x^2 + 2x + 3$

4. (Trial and error)
   (a) $2x^2 - 7x - 4$
   (b) $2x^2 + 7x - 4$
   (c) $2x^2 + 7x + 4$
   (d) $-2x^2 - 7x + 4$

5. (Sum and difference of cubes)
   (a) $x^3 + 1$
   (b) $x^3 + 216$
   (c) $1000 - 8x^6$
   (d) $64a^3x^3 - 125$

6. (Grouping)
   (a) $x^4 - 2x^3 + 3x - 6$
   (b) $a^2x + bx - a^2z - bz$

7. $2x - 2x^3$

8. $3x^4 - 48x^2$

9. $100x^3 - x^5$

10. $2x^2 + 5x - 12$

11. $2x^4 + 3x^3 - 9x^2$

12. $x^2 - 6x + 1$

13. $4x^3 - 20x^2 + 25x$

14. $ab - bc + a^2 - ac$

15. $x^2z^2 + xzt + xyz + yt$

16. $x^2 + 32x + 256$

17. $a^2t^2 + b^2t^2 - cb^2 - ca^2$

18. $x^4 - 6x^2 + 9$

19. $x^3 - 13x^2 - 90x$

20. $a^4 - 4a^2b^2c^2 + 4b^4c^4$

21. $4x^2 - 29xy - 24y^2$

22. $A^2 - B^2 + 16A + 64$

23. $x^2 + 2x + 16$

24. $(x - y)^2 - z^2$

25. $1 - (x + y)^2$

26. $(a + b + 1)^2 - (a - b + 1)^2$

27. $x^8 - 1$

28. $x^2 - y^2 + x - y$

29. $x^3 + 3x^2 + 3x + 1$

30. $-1 + 6x - 12x^2 + 8x^3$

31. $27x^3 + 108x^2 + 144x + 64$

32. $x^6 - 1$

33. $64a^6 - b^6$

34. $(a + b)^3 - 1$

35. $x^4 - 25x^2 + 144$

36. $x^4 + 25x^2 + 144$

37. $x^2 + 16y^2$

38. $81x^4 - 16y^4$

39. $x^3 + 2x^2 - 255x$

40. $4(x - a)^3 - (x - a)$

41. $x^3 + a^3 + x + a$

42. $8(x + a)^3 - (2x - a)^3$

43. $x^2 - a^2 + y^2 - 2xy$

44. $a^4 - (b + c)^4$

45. $21x^3 + 82x^2 - 39x$

46. $x^3a^2 - 8y^3a^2 - 4x^3b^2 + 32y^3b^2$

47. $12xy + 25 - 4x^2 - 9y^2$

48. $64x^7 - 729x$

In Exercises 49 and 50, evaluate the expressions using factoring techniques. (The point here is to do as little actual arithmetic as possible.)

49. (a) $100^2 - 99^2$   (b) $8^3 - 6^3$   (c) $1000^2 - 999^2$

50. (a) $10^3 - 9^3$   (b) $50^2 - 49^2$   (c) $\dfrac{15^3 - 10^3}{15^2 - 10^2}$

### B

In Exercises 51–54, factor each expression.

51. $ax^2 + (1 + ab)xy + by^2$

52. $ax^2 + (a + b)x + b$

53. $(5a^2 - 11a + 10)^2 - (4a^2 - 15a + 6)^2$

54. $(x + a)^4 - 2(x + a)^2(x + b)^2 + (x + b)^4$

*Exercises 55–60 involve factoring expressions with fractional exponents. For an expression such as $(x + 1)^{1/2} - (x + 1)^{3/2}$, where the exponents are positive, the common expression to factor out is the one with the smallest exponent, in this case $(x + 1)^{1/2}$. In the case of expressions with negative exponents, such as $(x + 1)^{-1/2} - (x + 1)^{-3/2}$, the common expression to be factored out is the one whose exponent has the largest absolute value, in this case $(x + 1)^{-3/2}$. Use these guidelines to factor the expressions.*

55. $(x + 1)^{1/2} - (x + 1)^{3/2}$

56. $(x^2 + 1)^{3/2} + (x^2 + 1)^{7/2}$

57. $(x + 1)^{-1/2} - (x + 1)^{-3/2}$

58. $(x^2 + 1)^{-2/3} + (x^2 + 1)^{-5/3}$

59. Write the final answer in a form that doesn't involve negative exponents: $x^2(x - 2)^{-4} + x(x - 2)^{-3}$

60. Write the final answer in a form not involving negative exponents: $x^2(a^2 - x^2)^{-1/2} + \sqrt{a^2 - x^2}$
*Hint:* Use $(a^2 - x^2)^{-1/2}$ for the common term.

**C**

*In Exercises 61–66, factor the expressions.*

61. $(x + y)^2 + (x + z)^2 - (z + t)^2 - (y + t)^2$

62. $(a - a^2)^3 + (a^2 - 1)^3 + (1 - a)^3$

63. $(b - c)^3 + (c - a)^3 + (a - b)^3$

64. $(a + b + c)^3 - a^3 - b^3 - c^3$

65. (a) $x^4 - 6x^2 + 9$    (b) $x^4 - 15x^2 + 9$
*Hint:* In Example 9, study the alternate solution.

66. $x^4 + 64$

## FRACTIONAL EXPRESSIONS

**1.10**

*But if you do use a rule involving mechanical calculation, be patient, accurate, and systematically neat in the working. It is well known to mathematical teachers that quite half the failures in algebraic exercises arise from arithmetical inaccuracy and slovenly arrangement.*

GEORGE CHRYSTAL (1893)

The rules of arithmetic for fractions were listed in Section 1.2. These same rules are used for fractions involving algebraic expressions. In algebra, as in arithmetic, we say that a fraction is *reduced to lowest terms* or *simplified* when the numerator and denominator contain no common factors (other than 1 and $-1$). The factoring techniques developed in the previous section are used to reduce fractions. For example, to reduce the fraction $\dfrac{x^2 - 9}{x^2 + 3x}$ we write

$$\frac{x^2 - 9}{x^2 + 3x} = \frac{(x - 3)(x + 3)}{x(x + 3)} = \frac{x - 3}{x}$$

In Example 1 we display two more instances in which factoring is used to reduce a fraction. After that, Example 2 indicates how these skills are used to multiply and divide fractional expressions.

**EXAMPLE 1**    Simplify:    (a) $\dfrac{x^3 - 8}{x^2 - 2x}$    (b) $\dfrac{x^2 - 6x + 8}{a(x - 2) + b(x - 2)}$

**Solution**    (a) $\dfrac{x^3 - 8}{x^2 - 2x} = \dfrac{(x - 2)(x^2 + 2x + 4)}{x(x - 2)} = \dfrac{x^2 + 2x + 4}{x}$

(b) $\dfrac{x^2 - 6x + 8}{a(x - 2) + b(x - 2)} = \dfrac{(x - 2)(x - 4)}{(x - 2)(a + b)} = \dfrac{x - 4}{a + b}$

**EXAMPLE 2**   Carry out the indicated operations and simplify:

(a) $\dfrac{x^3}{2x^2 + 3x} \cdot \dfrac{12x + 18}{4x^2 - 6x}$

(b) $\dfrac{2x^2 - x - 6}{x^2 + x + 1} \cdot \dfrac{x^3 - 1}{4 - x^2}$

(c) $\dfrac{x^2 - 144}{x^2 - 4} \div \dfrac{x + 12}{x + 2}$

**Solution**   (a) $\dfrac{x^3}{2x^2 + 3x} \cdot \dfrac{12x + 18}{4x^2 - 6x} = \dfrac{x^3}{x(2x + 3)} \cdot \dfrac{6(2x + 3)}{2x(2x - 3)}$

$$= \dfrac{6x^3}{2x^2(2x - 3)} = \dfrac{3x}{2x - 3}$$

(b) $\dfrac{2x^2 - x - 6}{x^2 + x + 1} \cdot \dfrac{x^3 - 1}{4 - x^2} = \dfrac{(2x + 3)(x - 2)}{(x^2 + x + 1)} \cdot \dfrac{(x - 1)(x^2 + x + 1)}{(2 - x)(2 + x)}$

$$= \dfrac{-(2x + 3)(x - 1)}{2 + x} \qquad \text{We've used the fact that } \dfrac{x - 2}{2 - x} = -1.$$

$$= \dfrac{-2x^2 - x + 3}{2 + x}$$

(c) $\dfrac{x^2 - 144}{x^2 - 4} \div \dfrac{x + 12}{x + 2} = \dfrac{x^2 - 144}{x^2 - 4} \cdot \dfrac{x + 2}{x + 12}$

$$= \dfrac{(x - 12)(x + 12)}{(x - 2)(x + 2)} \cdot \dfrac{x + 2}{x + 12} = \dfrac{x - 12}{x - 2}$$

As in arithmetic, to add or subtract two fractions, the denominators must be the same. The rules in this case are

$$\dfrac{a}{b} + \dfrac{c}{b} = \dfrac{a + c}{b} \qquad \dfrac{a}{b} - \dfrac{c}{b} = \dfrac{a - c}{b}$$

For example, we have

$$\dfrac{4x - 1}{x + 1} - \dfrac{2x - 1}{x + 1} = \dfrac{4x - 1 - (2x - 1)}{x + 1} = \dfrac{4x - 1 - 2x + 1}{x + 1} = \dfrac{2x}{x + 1}$$

Fractions with unlike denominators are added or subtracted by first converting to a common denominator. For instance, to add $9/a$ and $10/a^2$ we write

$$\dfrac{9}{a} + \dfrac{10}{a^2} = \dfrac{9}{a} \cdot \dfrac{a}{a} + \dfrac{10}{a^2}$$

$$= \dfrac{9a}{a^2} + \dfrac{10}{a^2} = \dfrac{9a + 10}{a^2}$$

Notice that the common denominator used was $a^2$. This is the **least common denominator**. In fact, other common denominators (such as $a^3$ or $a^4$) could be used here, but that would be less efficient. In general, the least common denominator for a given group of fractions is chosen as follows. Write down a product involving the irreducible factors from each denominator. The power of each factor should be equal to (but not greater than) the highest power of that factor appearing in any of the individual denominators. For example, the least common denominator for the two fractions $\dfrac{1}{(x+1)^2}$ and $\dfrac{1}{(x+1)(x+2)}$ is $(x+1)^2(x+2)$. In the example that follows, notice that the denominators must be in factored form before the least common denominator can be chosen.

**EXAMPLE 3**  Combine into a single fraction and simplify:

(a) $\dfrac{3}{4x}+\dfrac{7x}{10y^2}-2$   (b) $\dfrac{x}{x^2-9}-\dfrac{1}{x+3}$   (c) $\dfrac{15}{x^2+x-6}+\dfrac{x+1}{2-x}$

Solution  (a) Denominators:  $2^2\cdot x;\quad 2\cdot 5\cdot y^2;\quad 1$
Least common denominator:  $2^2\cdot 5xy^2 = 20xy^2$

$$\frac{3}{4x}+\frac{7x}{10y^2}-\frac{2}{1}=\frac{3}{4x}\cdot\frac{5y^2}{5y^2}+\frac{7x}{10y^2}\cdot\frac{2x}{2x}-\frac{2}{1}\cdot\frac{20xy^2}{20xy^2}$$

$$=\frac{15y^2+14x^2-40xy^2}{20xy^2}$$

(b) Denominators:  $x^2-9=(x-3)(x+3);\quad x+3$
Least common denominator:  $(x-3)(x+3)$

$$\frac{x}{x^2-9}-\frac{1}{x+3}=\frac{x}{(x-3)(x+3)}-\frac{1}{x+3}$$

$$=\frac{x}{(x-3)(x+3)}-\frac{1}{x+3}\cdot\frac{x-3}{x-3}$$

$$=\frac{x-(x-3)}{(x-3)(x+3)}=\frac{3}{(x-3)(x+3)}$$

(c) $\dfrac{15}{x^2+x-6}+\dfrac{x+1}{2-x}=\dfrac{15}{(x-2)(x+3)}-\dfrac{x+1}{x-2}$    We used the fact that $2-x=-(x-2)$.

$$=\frac{15}{(x-2)(x+3)}-\frac{x+1}{x-2}\cdot\frac{x+3}{x+3}$$

$$=\frac{15-(x^2+4x+3)}{(x-2)(x+3)}=\frac{-x^2-4x+12}{(x-2)(x+3)}$$

$$=\frac{(x-2)(-x-6)}{(x-2)(x+3)}=\frac{-x-6}{x+3}$$

Notice in the last line that we were able to simplify the answer by factoring the numerator and reducing the fraction. (This is why we prefer to leave the least common denominator in factored form, rather than multiply it out, in this type of problem.)

**EXAMPLE 4**   Simplify: $\dfrac{\dfrac{1}{3a} - \dfrac{1}{4b}}{\dfrac{5}{6a^2} + \dfrac{1}{b}}$

Solution   The least common denominator for the four individual fractions is $12a^2b$. Multiplying the given expression by $\dfrac{12a^2b}{12a^2b}$, which equals 1, yields

$$\frac{12a^2b}{12a^2b} \cdot \frac{\dfrac{1}{3a} - \dfrac{1}{4b}}{\dfrac{5}{6a^2} + \dfrac{1}{b}} = \frac{4ab - 3a^2}{10b + 12a^2}$$

**EXAMPLE 5**   Simplify $(x^{-1} + y^{-1})^{-1}$. (The answer is *not* $x + y$.)

Solution   After applying the definition of negative exponents to rewrite the given expression, we'll use the method shown in Example 4.

$$(x^{-1} + y^{-1})^{-1} = \left(\frac{1}{x} + \frac{1}{y}\right)^{-1}$$

$$= \frac{1}{\dfrac{1}{x} + \dfrac{1}{y}}$$

$$= \frac{xy}{xy} \cdot \frac{1}{\dfrac{1}{x} + \dfrac{1}{y}} = \frac{xy}{y + x}$$

**EXAMPLE 6**   Simplify $\dfrac{\dfrac{1}{x + h} - \dfrac{1}{x}}{h}$. (This type of expression occurs in calculus.)

Solution   $$\frac{\dfrac{1}{x + h} - \dfrac{1}{x}}{h} = \frac{(x + h)x}{(x + h)x} \cdot \frac{\dfrac{1}{x + h} - \dfrac{1}{x}}{h}$$

$$= \frac{x - (x + h)}{(x + h)xh} = \frac{-h}{(x + h)xh} = -\frac{1}{(x + h)x}$$

**EXAMPLE 7**   Simplify: $\dfrac{x - \dfrac{1}{x^2}}{\dfrac{1}{x^2} - 1}$

Solution   $$\frac{x - \dfrac{1}{x^2}}{\dfrac{1}{x^2} - 1} = \frac{x^2}{x^2} \cdot \frac{x - \dfrac{1}{x^2}}{\dfrac{1}{x^2} - 1}$$

$$= \frac{x^3 - 1}{1 - x^2} = -\frac{x^3 - 1}{x^2 - 1}$$

$$= -\frac{(x - 1)(x^2 + x + 1)}{(x - 1)(x + 1)} = -\frac{x^2 + x + 1}{x + 1}$$

## EXERCISE SET 1.10

### A

In Exercises 1–12, reduce the fractions to lowest terms.

1. $\dfrac{x^2 - 9}{x + 3}$

2. $\dfrac{25 - x^2}{x - 5}$

3. $\dfrac{x + 2}{x^4 - 16}$

4. $\dfrac{x^2 - x - 20}{2x^2 + 7x - 4}$

5. $\dfrac{x^2 + 2x + 4}{x^3 - 8}$

6. $\dfrac{a + b}{ax^2 + bx^2}$

7. $\dfrac{9ab - 12b^2}{6a^2 - 8ab}$

8. $\dfrac{a^3b^2 - 27b^5}{(ab - 3b^2)^2}$

9. $\dfrac{a^3 + a^2 + a + 1}{a^2 - 1}$

10. $\dfrac{(x - y)^2(a + b)}{(x^2 - y^2)(a^2 + 2ab + b^2)}$

11. $\dfrac{x^3 - y^3}{(x - y)^3}$

12. $\dfrac{x^4 - y^4}{(x^4y + x^2y^3 + x^3y^2 + xy^4)(x - y)^2}$

In Exercises 13–63, carry out the indicated operations and simplify where possible.

13. $\dfrac{2}{x - 2} \cdot \dfrac{x^2 - 4}{x + 2}$

14. $\dfrac{ax + 3}{2a + 1} \div \dfrac{a^2x^2 + 3ax}{4a^2 - 1}$

15. $\dfrac{x^2 - x - 2}{x^2 + x - 12} \cdot \dfrac{x^2 - 3x}{x^2 - 4x + 4}$

16. $(3t^2 + 4tx + x^2) \div \dfrac{3t^2 - 2tx - x^2}{t^2 - x^2}$

17. $\dfrac{x^3 + y^3}{x^2 - 4xy + 3y^2} \div \dfrac{(x + y)^3}{x^2 - 2xy - 3y^2}$

18. $\dfrac{a^2 - a - 42}{a^4 + 216a} \div \dfrac{a^2 - 49}{a^3 - 6a^2 + 36a}$

19. $\dfrac{x^2 + xy - 2y^2}{x^2 - 5xy + 4y^2} \cdot \dfrac{x^2 - 7xy + 12y^2}{x^2 + 5xy + 6y^2}$

20. $\dfrac{x^4y^2 - xy^5}{x^4 - 2x^2y^2 + y^4} \cdot \dfrac{x^2 + 2xy + y^2}{x^3y^2 + x^2y^3 + xy^4}$

21. $\dfrac{4}{x} - \dfrac{2}{x^2}$

22. $\dfrac{1}{3x} + \dfrac{1}{5x^2} - \dfrac{1}{30x^3}$

23. $\dfrac{6}{a} - \dfrac{a}{6}$

24. $\dfrac{1}{a} + \dfrac{1}{b} + \dfrac{1}{c}$

25. $\dfrac{1}{x + 3} + \dfrac{3}{x + 2}$

26. $\dfrac{4}{x - 4} - \dfrac{4}{x + 1}$

27. $\dfrac{3x}{x - 2} - \dfrac{6}{x^2 - 4}$

28. $1 + \dfrac{1}{x} - \dfrac{1}{x^2}$

29. $\dfrac{a}{x - 1} + \dfrac{2ax}{(x - 1)^2} + \dfrac{3ax^2}{(x - 1)^3}$

30. $\dfrac{a^2 + 5a - 4}{a^2 - 16} - \dfrac{2a}{2a^2 + 8a}$

31. $\dfrac{x}{x^2 - 9} + \dfrac{x - 1}{x^2 - 5x + 6}$

32. $\dfrac{1}{x - 1} + \dfrac{1}{1 - x}$

33. $\dfrac{4}{x - 5} - \dfrac{4}{5 - x}$

34. $\dfrac{x}{x + a} + \dfrac{a}{a - x}$

35. $\dfrac{a^2 + b^2}{a^2 - b^2} + \dfrac{a}{a + b} + \dfrac{b}{b - a}$

36. $\dfrac{3}{2x + 2} - \dfrac{5}{x^2 - 1} + \dfrac{1}{x + 1}$

37. $\dfrac{1}{x^2 + x - 20} - \dfrac{1}{x^2 - 8x + 16}$

38. $\dfrac{4}{6x^2 + 5x - 4} + \dfrac{1}{3x^2 + 4x} - \dfrac{1}{2x - 1}$

39. $\dfrac{2q + p}{2p^2 - 9pq - 5q^2} - \dfrac{p + q}{p^2 - 5pq}$

40. $\dfrac{1}{x - 1} + \dfrac{1}{x^2 - 1} + \dfrac{1}{x^3 - 1}$

41. $\dfrac{x}{(x - y)(x - z)} + \dfrac{y}{(y - z)(y - x)} + \dfrac{z}{(z - x)(z - y)}$

42. $\dfrac{3x}{x - 2} + \dfrac{4x^2}{x^2 + 2x + 1} - \dfrac{x^2}{x^2 - x - 2}$

43. $\dfrac{y + z}{x^2 - xy - xz + yz} - \dfrac{x + z}{xy - xz - y^2 + yz}$
    $+ \dfrac{x + y}{xy - yz - xz + z^2}$

44. $\left(x + \dfrac{1}{x}\right) \cdot \left(1 - \dfrac{1}{x}\right)^2 \div \left(x - \dfrac{1}{x}\right)$

45. $\dfrac{x^2 + x - a - 1}{a^2 - x^2} + \dfrac{x + 1}{x + a} \div \dfrac{x - a}{x - 1}$

46. $\dfrac{1}{2ax - 2a^2} - \dfrac{1}{2ax + 2a^2}$

47. $\dfrac{x^2 - 2ax + a^2}{px + q}\left(\dfrac{p}{x - a} + \dfrac{ap + q}{(x - a)^2}\right)$

48. $\dfrac{\dfrac{1}{a} + \dfrac{1}{b}}{\dfrac{1}{a} - \dfrac{1}{b}}$

49. $\dfrac{1 + \dfrac{4}{x}}{\dfrac{3}{x} - 2}$

50. $\dfrac{\dfrac{1}{a}}{1 + \dfrac{b}{c}}$

51. $\dfrac{a - \dfrac{1}{a}}{1 + \dfrac{1}{a}}$

52. $\dfrac{\dfrac{1}{x^2} - \dfrac{1}{y^2}}{\dfrac{1}{x} + \dfrac{1}{y}}$

53. $\dfrac{\dfrac{1}{2+h} - \dfrac{1}{2}}{h}$

54. $\dfrac{\dfrac{3}{x^2+h} - \dfrac{3}{x^2}}{h}$

55. $\dfrac{\dfrac{a}{x^2} + \dfrac{x}{a^2}}{a^2 - ax + x^2}$

56. $\dfrac{x + \dfrac{xy}{y-x}}{\dfrac{y^2}{x^2-y^2} + 1}$

**B**

57. $\dfrac{x+y}{2(x^2+y^2)} - \dfrac{1}{2(x+y)} + \dfrac{x-y}{x^2-y^2} - \dfrac{x^3-y^3}{x^4-y^4}$

58. $\dfrac{4y}{x^2-y^2} + \dfrac{1}{x+y} + \dfrac{1}{y-x} - \dfrac{2y}{x^2+y^2}$

59. $\dfrac{\dfrac{a+b}{a-b} + \dfrac{a-b}{a+b}}{\dfrac{a-b}{a+b} - \dfrac{a+b}{a-b}} \cdot \dfrac{ab^3 - a^3b}{a^2+b^2}$

60. $\dfrac{\sqrt{x+a}}{\sqrt{x-a}} - \dfrac{\sqrt{x-a}}{\sqrt{x+a}}$

**C**

61. $\dfrac{ap+q}{ax - bx - a^2 + ab} + \dfrac{bp+q}{bx - ax - b^2 + ab}$

62. $\dfrac{b-c}{a^2 - ab - ac + bc} + \dfrac{a-c}{b^2 - bc - ab + ac}$
$$+ \dfrac{a-b}{c^2 - ac - bc + ab}$$

63. $\dfrac{\dfrac{1}{a} - \dfrac{a-x}{a^2+x^2}}{\dfrac{1}{x} - \dfrac{x-a}{x^2+a^2}} + \dfrac{\dfrac{1}{a} - \dfrac{a+x}{a^2+x^2}}{\dfrac{1}{x} - \dfrac{x+a}{x^2+a^2}}$

64. Simplify: $\dfrac{\left(a + \dfrac{1}{b}\right)^a \left(a - \dfrac{1}{b}\right)^b}{\left(b + \dfrac{1}{a}\right)^a \left(b - \dfrac{1}{a}\right)^b}$

65. Simplify:
$$\dfrac{x^2 - qr}{(p-q)(p-r)} + \dfrac{x^2 - rp}{(q-r)(q-p)} + \dfrac{x^2 - pq}{(r-p)(r-q)}$$

66. Simplify: $\left(\dfrac{x^{-2} + y^{-2}}{x^{-2} - y^{-2}} - \dfrac{x^{-2} - y^{-2}}{x^{-2} + y^{-2}}\right)$
$$\div \left[\left(\dfrac{x+y}{x-y} + \dfrac{x-y}{x+y}\right)\left(\dfrac{x^2}{y^2} + \dfrac{y^2}{x^2} - 2\right)\right]^{-1}$$

67. Consider the three fractions
$$\dfrac{b-c}{1+bc}, \qquad \dfrac{c-a}{1+ca}, \qquad \text{and} \qquad \dfrac{a-b}{1+ab}$$

   (a) If $a = 1$, $b = 2$, and $c = 3$, find the sum of the three fractions. Also compute their product. What do you observe?
   (b) Show that the sum and the product of the three given fractions are, in fact, always equal.

68. Evaluate the expression
$$\dfrac{y-a}{x-a} - \dfrac{y+a}{x+a} + \dfrac{y-b}{x-b} - \dfrac{y+b}{x+b}$$

   when $x = \sqrt{ab}$.   *Hint:*   Keep the first and second terms separate from the third and fourth as long as possible.

69. Show that $(1 + a^{x-y})^{-1} + (1 + a^{y-x})^{-1} = 1$.

70. Simplify:
$$\dfrac{x^3 y^{-3} - y^3 x^{-3}}{(xy^{-1} - yx^{-1})(xy^{-1} + yx^{-1} - 1)} \cdot \dfrac{y^{-1} - x^{-1}}{x^{-2} + y^{-2} + x^{-1}y^{-1}}$$

# THE COMPLEX NUMBER SYSTEM

**1.11**

*The very first square root of negative number on record, $\sqrt{81 - 144}$, is in the Stereometrica of Hero of Alexandria [second half of the first century A.D.].*

DAVID WELLS, *The Penguin Dictionary of Curious and Interesting Numbers* (Harmondsworth, Middlesex, England: Penguin Books Ltd., 1986)

*In elementary algebra, complex numbers appear as expressions $a + bi$, where a and b are ordinary real numbers and $i^2 = -1$. . . . Complex numbers are manipulated by the usual rules of algebra, with the convention that $i^2$ is to be replaced by $-1$ whenever it occurs.*

RALPH BOAS, *Invitation to Complex Analysis* (New York: Random House, 1987)

The preceding quotation from Professor Boas summarizes the basic approach we will follow in this section. Near the end of the section, after you've become accustomed to working with complex numbers, we'll present a formal list of some of the basic definitions and properties that can be used to develop the subject more rigorously, as is required in more advanced courses.

When we solve equations in Chapter 2, you'll see instances in which the real number system proves to be inadequate. In particular, since the square of a real number is never negative, there is no real number $x$ such that $x^2 = -1$. To overcome this inconvenience, mathematicians define the symbol $i$ by the equation

$$i^2 = -1$$

For reasons that are more historical than mathematical, $i$ is referred to as the **imaginary unit**. This name is unfortunate in a sense, because to an engineer or a mathematician $i$ is neither less "real" nor less tangible than any real number. Having said this, however, we do have to admit that $i$ does not belong to the real number system.

Algebraically, we operate with the symbol $i$ as if it were any letter in a polynomial expression. However, when we see $i^2$, we must remember to replace it by $-1$. Here are four sample calculations involving $i$.

1. $3i + 2i = 5i$
2. $-2i^2 + 6i = -2(-1) + 6i = 2 + 6i$
3. $(-i)^2 = i^2 = -1$
4. $0i = 0$

An expression of the form $a + bi$, where $a$ and $b$ are real numbers, is called a **complex number**.* Thus, four examples of complex numbers are:

$$2 + 3i$$
$$4 + (-5)i \qquad \text{(usually written } 4 - 5i\text{)}$$
$$1 - \sqrt{2}i \qquad \text{(also written } 1 - i\sqrt{2}\text{)}$$
$$\frac{1}{2} + \frac{3}{2}i \qquad \left(\text{also written } \frac{1 + 3i}{2}\right)$$

---

* The phrase *complex number* is attributed to Carl Friedrich Gauss (1777–1855), as is the use of $i$ to denote $\sqrt{-1}$. The term *imaginary number* originated with René Descartes (1596–1650).

Given a complex number $a + bi$, we say that $a$ is the **real part** of $a + bi$, and $b$ is the **imaginary part** of $a + bi$. For example, the real part of $3 - 4i$ is 3, and the imaginary part is $-4$. Observe that both the real part and the imaginary part of a complex number are themselves real numbers.

We define the notion of *equality* for complex numbers in terms of their real and imaginary parts. Two complex numbers are said to be **equal** if their corresponding real and imaginary parts are equal. We can write this definition symbolically as follows:

$$a + bi = c + di \qquad \text{if and only if} \qquad a = c \quad \text{and} \quad b = d$$

**EXAMPLE 1**   Determine real numbers $c$ and $d$ such that $10 + 4i = 2c + di$.

Solution   Equating the real parts of the two complex numbers gives us $2c = 10$, and therefore $c = 5$. Similarly, equating the imaginary parts yields $d = 4$. These are the required values for $c$ and $d$.

As the next example indicates, addition, subtraction, and multiplication of complex numbers are carried out using the usual rules of algebra, with the understanding (as mentioned before) that $i^2$ is always to be replaced with $-1$. (We'll discuss division of complex numbers subsequently.)

**EXAMPLE 2**   Let $z = 2 + 5i$ and $w = 3 - 4i$. Compute each of the following.

(a) $w + z$   (b) $3z$   (c) $w - 3z$   (d) $zw$   (e) $wz$

Solution   (a) $w + z = (3 - 4i) + (2 + 5i) = (3 + 2) + (-4i + 5i) = 5 + i$

(b) $3z = 3(2 + 5i) = 6 + 15i$

(c) $w - 3z = (3 - 4i) - (6 + 15i) = (3 - 6) + (-4i - 15i) = -3 - 19i$

(d) $zw = (2 + 5i)(3 - 4i) = 6 - 8i + 15i - 20i^2$
$$= 6 + 7i - 20(-1) = 26 + 7i$$

(e) $wz = (3 - 4i)(2 + 5i) = 6 + 15i - 8i - 20i^2$
$$= 6 + 7i - 20(-1) = 26 + 7i$$

If you look over the result of part (a) in Example 2, you can see that the sum is obtained simply by adding the corresponding real and imaginary parts of the given numbers. Likewise in part (c), the difference is obtained by subtracting the corresponding real and imaginary parts. In part (d), however, notice that the product is *not* obtained in a similar fashion. (Exercise 54 provides some perspective on this.) Finally, notice that the results in parts (d) and (e) are identical. In fact, it can be shown that for any two complex numbers $z$ and $w$, we always have $zw = wz$. In other words, just as with real numbers, multiplication of complex numbers is commutative. Furthermore, along the same lines, it can be shown that all of the properties of real numbers listed in Section 1.2 continue to hold for complex numbers. We'll return to this point at the end of this section and in the exercises.

As background for the discussion of division of complex numbers, we introduce the notion of a *complex conjugate*, or simply a *conjugate*.

### Definition of Complex Conjugate

|  | EXAMPLES |
|---|---|
| Let $z = a + bi$. The **complex conjugate** of $z$, denoted by $\bar{z}$, is defined by $$\bar{z} = a - bi$$ | If $z = 3 + 4i$, then $\bar{z} = 3 - 4i$. <br> If $w = 9 - 2i$, then $\bar{w} = 9 + 2i$. |

**EXAMPLE 3**   (a) If $z = 6 - 3i$, compute $z\bar{z}$.   (b) If $w = a + bi$, compute $w\bar{w}$.

**Solution**   (a) $z\bar{z} = (6 - 3i)(6 + 3i)$
$\qquad = 36 + 18i - 18i - 9i^2$
$\qquad = 36 + 9 = 45$

(b) $w\bar{w} = (a + bi)(a - bi)$
$\qquad = a^2 - abi + abi - b^2i^2$
$\qquad = a^2 + b^2$

NOTE   The result in part (b) shows that the product of a complex number and its conjugate is always a real number.

Quotients of complex numbers are easy to compute using conjugates. Suppose, for example, that we wish to compute the quotient

$$\frac{5 - 2i}{3 + 4i}$$

To do this, take the conjugate of the denominator, namely, $3 - 4i$, and then multiply the given fraction by $\dfrac{3 - 4i}{3 - 4i}$, which equals 1 (assuming that the usual rules of algebra are in force). This yields

$$\frac{5 - 2i}{3 + 4i} = \frac{5 - 2i}{3 + 4i} \cdot \frac{3 - 4i}{3 - 4i}$$
$$= \frac{15 - 26i + 8i^2}{9 - 16i^2}$$
$$= \frac{7 - 26i}{25} \qquad \text{since } 8i^2 = -8 \text{ and } -16i^2 = 16.$$
$$= \frac{7}{25} - \frac{26}{25}i, \qquad \text{as required}$$

In the box that follows, we summarize our procedure for computing quotients. The condition $w \neq 0$ means that $w$ is any complex number other than $0 + 0i$.

### Procedure for Computing Quotients

Let $z$ and $w$ be two complex numbers, and $w \neq 0$. Then $z/w$ is computed as follows:

$$\frac{z}{w} = \frac{z}{w} \cdot \frac{\bar{w}}{\bar{w}}$$

**EXAMPLE 4**   Let $z = 3 + 4i$ and $w = 1 - 2i$. Compute each of the following.

(a) $\dfrac{1}{z}$      (b) $\dfrac{z}{w}$

**Solution**   (a) $\dfrac{1}{z} = \dfrac{1}{z} \cdot \dfrac{\bar{z}}{\bar{z}} = \dfrac{1}{3 + 4i} \cdot \dfrac{3 - 4i}{3 - 4i}$

$$= \dfrac{3 - 4i}{9 - 16i^2} = \dfrac{3 - 4i}{25}$$

$$= \dfrac{3}{25} - \dfrac{4}{25}i$$

(b) $\dfrac{z}{w} = \dfrac{z}{w} \cdot \dfrac{\bar{w}}{\bar{w}} = \dfrac{3 + 4i}{1 - 2i} \cdot \dfrac{1 + 2i}{1 + 2i}$

$$= \dfrac{3 + 10i + 8i^2}{1 - 4i^2} = \dfrac{-5 + 10i}{5}$$

$$= -1 + 2i$$

We began this section by defining $i$ by the equation $i^2 = -1$. This can be rewritten

$$i = \sqrt{-1}$$

provided that we agree to certain conventions regarding principal square roots and negative numbers. In dealing with the principal square root of a negative real number, say $\sqrt{-5}$, we shall write

$$\sqrt{-5} = \sqrt{(-1)(5)} = \sqrt{-1}\sqrt{5} = i\sqrt{5}$$

In other words, we are allowing the use of the rule $\sqrt{ab} = \sqrt{a}\sqrt{b}$ when $a$ is $-1$ and $b$ is a positive real number. However, the rule $\sqrt{ab} = \sqrt{a}\sqrt{b}$ *cannot* be used when both $a$ and $b$ are negative. If that were allowed, we could write

$$1 = (-1)(-1)$$

and then

$$\sqrt{1} = \sqrt{(-1)(-1)} = \sqrt{-1}\sqrt{-1} = (i)(i)$$

Consequently,

$$1 = \sqrt{1} = i^2 = -1$$

or

$$1 = -1$$

which is a contradiction. Again, the point here is that the rule $\sqrt{ab} = \sqrt{a}\sqrt{b}$ cannot be applied when both $a$ and $b$ are negative.

**EXAMPLE 5**   Simplify:

(a) $i^4$      (b) $i^{101}$      (c) $\sqrt{-12} + \sqrt{-27}$      (d) $\sqrt{-9}\sqrt{-4}$

**Solution** **(a)** We make use of the defining equation for $i$, which is $i^2 = -1$. Thus, we have

$$i^4 = (i^2)^2 = (-1)^2 = 1$$

(The result, $i^4 = 1$, is worth remembering.)

**(b)** $i^{101} = i^{100}i = (i^4)^{25}i = 1^{25}i = i$

**(c)** $\sqrt{-12} + \sqrt{-27} = \sqrt{12}\sqrt{-1} + \sqrt{27}\sqrt{-1}$
$$= \sqrt{4}\sqrt{3}i + \sqrt{9}\sqrt{3}i$$
$$= 2\sqrt{3}i + 3\sqrt{3}i = 5\sqrt{3}i$$

**(d)** $\sqrt{-9}\sqrt{-4} = (3i)(2i) = 6i^2 = -6$

NOTE $\quad \sqrt{-9}\sqrt{-4} \neq \sqrt{36}$; why?

Near the beginning of this section, we mentioned that we'd eventually present a formal list of some of the basic definitions and properties that can be used to develop the subject more rigorously. These are given in the two boxes that follow. As you'll see in some of the exercises, these definitions and properties are indeed consistent with our work in this section and with the properties of real numbers that were listed in Section 1.2.

---

**Definition of Addition, Subtraction, Multiplication, and Division for Complex Numbers**

Let $z = a + bi$ and $w = c + di$. Then $z + w$, $z - w$, and $zw$ are defined as follows.

1. $z + w = (a + bi) + (c + di) = (a + c) + (b + d)i$
2. $z - w = (a + bi) - (c + di) = (a - c) + (b - d)i$
3. $zw = (a + bi)(c + di) = (ac - bd) + (ad + bc)i$

Furthermore, if $w \neq 0$, then $z/w$ is defined as follows.

4. $\dfrac{z}{w} = \dfrac{a + bi}{c + di} \cdot \dfrac{c - di}{c - di} = \left(\dfrac{ac + bd}{c^2 + d^2}\right) + \left(\dfrac{bc - ad}{c^2 + d^2}\right)i$

---

**Definition and Properties of Complex Conjugates**

The complex conjugate of $z = a + bi$ is $\bar{z} = a - bi$. The complex conjugate has the following properties.

1. $\bar{\bar{z}} = z$
2. $z = \bar{z}$ if and only if $z$ is a real number
3. $\overline{z + w} = \bar{z} + \bar{w}$; $\quad \overline{z - w} = \bar{z} - \bar{w}$
4. $\overline{zw} = \bar{z}\bar{w}$; $\quad \dfrac{\bar{z}}{\bar{w}} = \overline{\left(\dfrac{z}{w}\right)}$
5. $(\bar{z})^n = \overline{(z^n)}$ $\quad$ for each natural number $n$

## EXERCISE SET 1.11

### A

1. Complete the table.

| $i^2$ | $i^3$ | $i^4$ | $i^5$ | $i^6$ | $i^7$ | $i^8$ |
|-------|-------|-------|-------|-------|-------|-------|
| $-1$  |       |       |       |       |       |       |

2. Simplify the following expression, and write the answer in the form $a + bi$.

$$1 + 3i - 5i^2 + 4 - 2i - i^3$$

*For Exercises 3 and 4, specify the real and imaginary parts of each complex number.*

3. (a) $4 + 5i$      (b) $4 - 5i$
   (c) $\frac{1}{2} - i$      (d) $16i$
4. (a) $-2 + \sqrt{7}i$      (b) $1 + 5^{1/3}i$
   (c) $-3i$      (d) $0$

5. Determine the real numbers $c$ and $d$ such that

$$8 - 3i = 2c + di$$

6. Determine the real numbers $a$ and $b$ such that

$$27 - 64i = a^3 - b^3 i$$

7. Simplify each of the following.
   (a) $(5 - 6i) + (9 + 2i)$      (b) $(5 - 6i) - (9 + 2i)$
8. If $z = 1 + 4i$, compute $z - 10i$.
9. Compute each of the following.
   (a) $(3 - 4i)(5 + i)$      (b) $(5 + i)(3 - 4i)$
   (c) $\dfrac{3 - 4i}{5 + i}$      (d) $\dfrac{5 + i}{3 - 4i}$

10. Compute each of the following.
    (a) $(2 + 7i)(2 - 7i)$      (b) $\dfrac{-1 + 3i}{2 + 7i}$
    (c) $\dfrac{1}{2 + 7i}$      (d) $\dfrac{1}{2 + 7i} \cdot (-1 + 3i)$

*In Exercises 11–36, evaluate each expression using the values $z = 2 + 3i$, $w = 9 - 4i$, and $w_1 = -7 - i$.*

11. (a) $z + w$    (b) $\bar{z} + w$    (c) $z + \bar{z}$
12. (a) $\bar{z} + \bar{w}$    (b) $\overline{(z + w)}$    (c) $w - \bar{w}$
13. $(z + w) + w_1$      14. $z + (w + w_1)$
15. $zw$      16. $wz$
17. $z\bar{z}$      18. $w\bar{w}$
19. $z(ww_1)$      20. $(zw)w_1$
21. $z(w + w_1)$      22. $zw + zw_1$
23. $z^2 - w^2$      24. $(z - w)(z + w)$
25. $(zw)^2$      26. $z^2 w^2$
27. $z^3$      28. $z^4$

29. $\dfrac{z}{w}$      30. $\dfrac{w}{z}$
31. $\dfrac{\bar{z}}{\bar{w}}$      32. $\overline{\left(\dfrac{z}{w}\right)}$
33. $\dfrac{z}{\bar{z}}$      34. $\dfrac{\bar{z}}{z}$
35. $\dfrac{w - \bar{w}}{2i}$      36. $\dfrac{w + \bar{w}}{2}$

*For Exercises 37–40, compute each quotient.*

37. $\dfrac{i}{5 + i}$      38. $\dfrac{1 - i\sqrt{3}}{1 + i\sqrt{3}}$
39. $\dfrac{1}{i}$      40. $\dfrac{i + i^2}{i^3 + i^4}$

*In Exercises 41–48, simplify each expression.*

41. $\sqrt{-49} + \sqrt{-9} + \sqrt{-4}$      42. $\sqrt{-25} + i$
43. $\sqrt{-20} - 3\sqrt{-45} + \sqrt{-80}$
44. $\sqrt{-4}\sqrt{-4}$
45. $1 + \sqrt{-36}\sqrt{-36}$      46. $i - \sqrt{-100}$
47. $3\sqrt{-128} - 4\sqrt{-18}$      48. $64 + \sqrt{-64}\sqrt{-64}$
49. Let $z = a + bi$ and $w = c + di$. Compute each of the following, and then check that your results agree with the definitions in the box on page 61.

    (a) $z + w$    (b) $z - w$    (c) $zw$    (d) $\dfrac{z}{w}$

### B

50. Show that $\left(\dfrac{-1 + i\sqrt{3}}{2}\right)^2 + \left(\dfrac{-1 - i\sqrt{3}}{2}\right)^2 = -1$.

51. Let $z = \dfrac{-1 + i\sqrt{3}}{2}$ and $w = \dfrac{-1 - i\sqrt{3}}{2}$. Verify the following statements.
    (a) $z^3 = 1$ and $w^3 = 1$      (b) $zw = 1$
    (c) $z = w^2$ and $w = z^2$
    (d) $(1 - z + z^2)(1 + z - z^2) = 4$

52. Let $z = a + bi$ and $w = c + di$.
    (a) Show that $\bar{\bar{z}} = z$.
    (b) Show that $\overline{(z + w)} = \bar{z} + \bar{w}$.

53. Show that the complex number $0 (= 0 + 0i)$ has the following properties.
    (a) $0 + z = z$ and $z + 0 = z$, for all complex numbers $z$. *Hint:* Let $z = a + bi$.
    (b) $0 \cdot z = 0$ and $z \cdot 0 = 0$, for all complex numbers $z$.

**54.** This exercise indicates one of the reasons why multiplication of complex numbers is not carried out by simply multiplying the corresponding real and imaginary parts of the numbers. (Recall that addition and subtraction *are* carried out in this manner) Suppose for the moment that we were to define multiplication in this seemingly less complicated way:

$$(a + bi)(c + di) = ac + (bd)i \qquad (*)$$

(a) Compute $(2 + 3i)(5 + 4i)$, assuming that multiplication is defined by (*).

(b) Still assuming that multiplication is defined by (*), find two complex numbers $z$ and $w$ such that $z \neq 0$, $w \neq 0$, but $zw = 0$ (where 0 denotes the complex number $0 + 0i$).

Now notice that the result in part (b) is contrary to our expectation or desire that the product of two nonzero numbers should be nonzero, as is the case for real numbers. On the other hand, it can be shown that when multiplication is carried out as described in the text, then the product of two complex numbers is nonzero if and only if both factors are nonzero.

**55. (a)** Show that addition of complex numbers is commutative. That is, show that $z + w = w + z$ for all complex numbers $z$ and $w$.  *Hint:*  Let $z = a + bi$ and $w = c + di$.

(b) Show that multiplication of complex numbers is commutative. That is, show that $zw = wz$ for all complex numbers $z$ and $w$.

**56.** Let $z = a + bi$.

(a) Show that $(\bar{z})^2 = \overline{(z^2)}$.

(b) Show that $(\bar{z})^3 = \overline{(z^3)}$.

## C

**57.** Let $a$ and $b$ be real numbers. Find the real and imaginary parts of the quantity

$$\frac{a + bi}{a - bi} + \frac{a - bi}{a + bi}$$

**58.** Find the real and imaginary parts of the quantity

$$\left(\frac{a + bi}{a - bi}\right)^2 - \left(\frac{a - bi}{a + bi}\right)^2$$

**59.** Find the real part of $\dfrac{(a + bi)^2}{a - bi} - \dfrac{(a - bi)^2}{a + bi}$

**60. (a)** Let $\alpha = \left(\dfrac{\sqrt{a^2 + b^2} + a}{2}\right)^{1/2}$ and $\beta = \left(\dfrac{\sqrt{a^2 + b^2} - a}{2}\right)^{1/2}$ Show that the square of the complex number $\alpha + \beta i$ is $a + bi$.

(b) Use the result in part (a) to find a complex number $z$ such that $z^2 = i$.

(c) Use the result in part (a) to find a complex number $z$ such that $z^2 = -7 + 24i$.

| | SUMMARY OF PRINCIPAL TERMS AND NOTATION | |
|---|---|---|

| TERMS OR NOTATIONS | PAGE REFERENCE | COMMENTS |
|---|---|---|
| 1. Coefficient, base, exponent | 2 | In the expression $ax^n$, the number $a$ is the coefficient, $x$ is the base, and $n$ is the exponent. |
| 2. Commutative, associative, identity, inverse, and distributive properties | 6 | These are some of the most basic properties of the real number system. All of these properties are also valid for the complex number system. |
| 3. Natural numbers, integers, rational numbers, and irrational numbers | 12 | The box on page 12 provides both definitions and examples. Also note the theorem in the box that explains how to distinguish between rationals and irrationals in terms of their decimal representations. |
| 4. $a < b$ $\quad$ $b > a$ | 14 | $a$ is less than $b$. $b$ is greater than $a$. |
| 5. $(a, b)$ | 14 | The open interval $(a, b)$ consists of all real numbers between $a$ and $b$, excluding $a$ and $b$. |
| $[a, b]$ | | The closed interval $[a, b]$ consists of all real numbers between $a$ and $b$, including $a$ and $b$. |

| TERMS OR NOTATIONS | PAGE REFERENCE | COMMENTS |
|---|---|---|
| **6.** $\lvert x \rvert$ | 16, 17 | The absolute value of a real number $x$ is the distance of $x$ from the origin. This is equivalent to the following algebraic definition: $$\lvert x \rvert = \begin{cases} x & \text{if } x \geq 0 \\ -x & \text{if } x < 0 \end{cases}$$ |
| **7.** $a^0 = 1$ <br> $a^{-n} = 1/a^n$ <br> $a^{m/n} = \sqrt[n]{a^m} = (\sqrt[n]{a})^m$ | 22 <br> 23 <br> 35 | These are the defining equations for zero as an exponent, for negative exponents, and for rational exponents. The definition $a^{m/n}$ presupposes that $m/n$ is in lowest terms and that $\sqrt[n]{a}$ exists. |
| **8.** Scientific notation | 24 | This is a convenient and standard form that is used to express very large and very small numbers. The box on page 25 gives directions for converting to scientific notation. See also the comments in Appendix A1. |
| **9.** Constant, variable, and the domain convention | 38, 39 | By a constant we mean either a particular number (such as $-8$ or $\pi$) or a letter whose value remains fixed (although perhaps unspecified) throughout a given discussion. In contrast, a variable is a letter for which we can substitute any number from a given set of numbers. The given set is called the domain of the variable. According to the domain convention, the domain of a variable in a given expression consists of all real numbers for which the expression is defined. |
| **10.** Polynomial | 39 | A polynomial is an expression of the form $a_n x^n + a_{n-1} x^{n-1} + \cdots + a_1 x + a_0$. The individual expressions $a_k x^k$ making up the polynomial are called *terms*. The numbers $a_k$ are called *coefficients*. |
| **11.** Degree of a polynomial | 39 | For a polynomial that is not simply a constant, the degree is the highest power (that is, exponent) of $x$ that appears. The degree of a nonzero constant polynomial is zero. Degree is undefined for the zero polynomial. |
| **12.** Special products | 42 | The box on page 42 lists four basic types of products that you should memorize because they occur so frequently. |
| **13.** Factoring techniques | 46 | The techniques for factoring listed in the box on page 46 are fundamental for simplifying expressions (in Section 1.10) and for solving equations (in Chapter 2). |
| **14.** $i^2 = -1$ | 57 | This is the defining equation for the symbol $i$. The quantity $i$ is not a real number. |
| **15.** Complex numbers | 57 | An expression of the form $a + bi$, where $a$ and $b$ are real numbers, is a complex number. With the exception of division, complex numbers are combined using the usual rules of algebra, along with the convention $i^2 = -1$. Division is carried out using complex conjugates (defined below), as indicated in the box on page 59. All of the properties of real numbers listed on page 6 also apply to complex numbers. |
| **16.** If $z = c + di$, then $\bar{z} = c - di$. | 59 | $\bar{z}$ is called the *conjugate* of $z$. |

# CHAPTER 1   REVIEW EXERCISES

NOTE   The fifteen exercises in this set with colored problem numbers constitute a chapter test.

*Factor each expression in Exercises 1–20.*

1. $a^2 - 16b^2$

2. $x^2 + ax + yx + ay$

3. $8 - (a + 1)^3$

4. $x^2 - 18x + 81$

5. $a^2x^3 + 2ax^2b + b^2x$

6. $8a^2x^2 + 16a^3$

7. $8x^2 + 6x + 1$

8. $12x^2 - 2x - 4$

9. $a^4x^4 - x^8a^8$

10. $2x^2 - 2bx + ax - ab$

11. $8 + 12a + 6a^2 + a^3$

12. $(x^2 + 2x - 8)^2 - (2x + 1)^2$

13. $4x^2y^2z^3 - 3xyz^3 - z^3$

14. $(x + y - 1)^2 - (x - y + 1)^2$

15. $1 - x^6$

16. $12x^3 + 44x^2 - 16x$

17. $a^2x^2 + 2abx + b^2 - 4a^2b^2x^2$

18. $a^2 + 2ab + b^2 + a + b$

19. $a^2 - b^2 + ac - bc + a^2b - b^2a$

20. $5(a + 1)^2 + 29(a + 1) - 144$

*Simplify each expression in Exercises 21–36. (Assume that the letters represent positive quantities.)*

21. $4^{3/2}$

22. $49^{-1/2}$

23. $[(3025)^{1/2}]^0$

24. $(-125)^{2/3}$

25. $8^{-4/3}$

26. $10^{-3} + 10^3$

27. $(-243)^{-2/5}$

28. $\left(\frac{625}{16}\right)^{3/4}$

29. $(a^2b^6c^8)^{1/2}$

30. $\sqrt{25a^4b^{10}}$

31. $\sqrt{a^3b^5}\sqrt{4ab^3}$

32. $\sqrt{28a^3} + a\sqrt{63a}$

33. $\sqrt[3]{16} - \sqrt[3]{-54}$

34. $\sqrt[5]{\dfrac{-32a^{15}b^{10}}{c^5}}$

35. $\sqrt{24a^2b^3} + ba\sqrt{54b}$

36. $\sqrt{\sqrt{256x^8}}$

*Simplify each expression in Exercises 37–40. (Do not assume that the letters represent positive quantities.)*

37. $\sqrt{t^2}$

38. $\sqrt[3]{t^3}$

39. $\sqrt[4]{16x^4}$

40. $\sqrt[3]{8x^4}$

*In Exercises 41–44, rewrite each expression in two ways:*
(a)  using rational exponents;
(b)  using a single radical.
*Assume that all of the letters represent nonnegative quantities.*

41. $\sqrt[3]{x}\sqrt[4]{x^3}$

42. $\sqrt[4]{\sqrt[3]{x}}$

43. $\sqrt{\sqrt[3]{t}\sqrt[5]{t^4}}$

44. $\sqrt[3]{\dfrac{\sqrt{a}\sqrt[4]{b^2}}{\sqrt[5]{c^3}}}$

*Rationalize the denominators in Exercises 45–51.*

45. $\dfrac{6}{\sqrt{3}}$

46. $\dfrac{\sqrt{3}}{\sqrt{2}}$

47. $\dfrac{1}{\sqrt{6} - \sqrt{3}}$

48. $\dfrac{1}{\sqrt{x + 2} - \sqrt{x}}$

49. $\dfrac{\sqrt{a^2 + x^2} + \sqrt{a^2 - x^2}}{\sqrt{a^2 + x^2} - \sqrt{a^2 - x^2}}$

50. $\dfrac{1}{\sqrt{a} + \sqrt{b} + \sqrt{c}}$

   *Hint:*   First multiply numerator and denominator by $\sqrt{a} + \sqrt{b} - \sqrt{c}$. Then multiply numerator and denominator by $(a + b - c) - 2\sqrt{ab}$.

51. $\dfrac{1}{\sqrt{2} + \sqrt{3} + \sqrt{5}}$   *Hint:*   See Exercise 50.

*In Exercises 52–55, rationalize the numerators.*

52. $\dfrac{\sqrt{x} - 5}{x - 25}$

53. $\dfrac{\sqrt{t} - a}{t - a^2}$

54. $\dfrac{\sqrt{a^2 + x^2} - \sqrt{a^2 - x^2}}{\sqrt{a^2 + x^2} + \sqrt{a^2 - x^2}}$

55. $\dfrac{\sqrt[3]{x} - 2}{x - 8}$   *Hint:*   The denominator can be written $(x^{1/3})^3 - 2^3$.

*Use the Special Products in Section 1.8 to compute the products in Exercises 56–70.*

56. $(9x - y)(9x + y)$

57. $(3x + 1)^2$

58. $(2ab - 3)^2$

59. $(3x^2 + y^2)^2$

60. $(ax + b)^2(ax - b)^2$

61. $(2a + 3)^3$

62. $(x^2 - 2)^3$

63. $(1 - 3a)(1 + 3a + 9a^2)$

64. $(x^n + y^n)(x^{2n} - x^ny^n + y^{2n})$

65. $(x^{1/2} - y^{1/2})(x^{1/2} + y^{1/2})$

66. $(\sqrt{x} - \sqrt{3})(\sqrt{x} + \sqrt{3})$

67. $(x^{1/3} + y^{1/3})^3$

68. $(x^{1/3} + y^{1/3})(x^{2/3} - x^{1/3}y^{1/3} + y^{2/3})$

69. $(x^2 - 3x + 1)^2$   *Hint:*   Rewrite the expression as $[x^2 - (3x - 1)]^2$.

70. $(x^2 + x - 1)(x^2 + x + 1)$

*For Exercises 71–78, carry out the indicated operations and express your answer in the standard form $a + bi$.*

71. $(3 - 2i)(3 + 2i) + (1 + 3i)^2$

72. $2i(1 + i)^2$

73. $(1 + i\sqrt{2})(1 - i\sqrt{2}) + (\sqrt{2} + i)(\sqrt{2} - i)$

74. $\dfrac{2 + 3i}{1 + i}$        75. $\dfrac{3 - i\sqrt{3}}{3 + i\sqrt{3}}$

76. $\dfrac{1 + i}{1 - i} + \dfrac{1 - i}{1 + i}$

77. $-\sqrt{-2}\sqrt{-9} + \sqrt{-8} - \sqrt{-72}$

78. $\dfrac{\sqrt{-4} - \sqrt{-3}\sqrt{-3}}{\sqrt{-100}}$

79. The **real part** of a complex number $z$ is denoted by $\text{Re}(z)$. For instance, $\text{Re}(2 + 5i) = 2$. Show that for any complex number $z$, we have $\text{Re}(z) = \frac{1}{2}(z + \bar{z})$. *Hint:* Let $z = a + bi$.

80. The **imaginary part** of a complex number $z$ is denoted by $\text{Im}(z)$. For instance, $\text{Im}(2 + 5i) = 5$. Show that for any complex number $z$, we have $\text{Im}(z) = \dfrac{1}{2i}(z - \bar{z})$.

81. The **absolute value** of the complex number $a + bi$ is defined by $|a + bi| = \sqrt{a^2 + b^2}$.
   (a) Compute $|6 + 2i|$ and $|6 - 2i|$.
   (b) As you know, the absolute value of the real number $-3$ is 3. Now write $-3$ in the form $-3 + 0i$ and compute its absolute value using the new definition. (The point here is to observe that the two results agree.)
   (c) Let $z = a + bi$. Show that $z\bar{z} = |z|^2$.

*In Exercises 82–86, identify the properties of the real numbers that justify each statement. (The properties are listed in Section 1.2.)*

82. $(x + y)^2 + z^2 = z^2 + (x + y)^2$

83. $(x + 1)(x + 2) = (x + 1)x + (x + 1)2 = (x^2 + x) + (2x + 2)$

84. $(ab)c = a(bc) = a(cb) = (ac)b$

85. $(x + a)y + (x + a)z = (x + a)(y + z)$

86. Since 3 and $\sqrt{\pi}$ are real numbers, $3\sqrt{\pi}$ is also a real number.

*For Exercises 87–91, rewrite the statements using absolute values and inequalities or equalities.*

87. The distance between $x$ and 6 is 2.

88. The distance between $x$ and $a$ is less than $\frac{1}{2}$.

89. The distance between $a$ and $b$ is 3.

90. The distance between $x$ and $-1$ is 5.

91. The distance between $x$ and 0 exceeds 10.

92. What can you say about $x$ if $|x - 5| = 0$?

*Rewrite each of the expressions in Exercises 93–98 in a form not containing absolute values.*

93. $|\sqrt{6} - 2|$

94. $|2 - \sqrt{6}|$

95. $|x^4 + x^2 + 1|$

96. (a) $|x - 3|$ if $x < 3$
   (b) $|x - 3|$ if $x > 3$

97. $|x - 2| + |x - 3|$ if:
   (a) $x < 2$     (b) $2 < x < 3$     (c) $x > 3$

98. $|x + 2| + |x - 1|$ if:
   (a) $x < -2$     (b) $-2 < x < 1$     (c) $x > 1$

99. Determine whether each of the following is true or false.
   (a) $-1 \le 0$     (b) $\sqrt{\frac{3}{2}} \ge 0$
   (c) $\frac{11}{10} \ge \frac{12}{11}$     (d) $0.9 \ge 1$

*In Exercises 100–104, sketch each interval on a number line.*

100. $(3, 5)$             101. $(3, 5]$

102. The set of all negative real numbers that are in the interval $[-5, 2]$.

103. The set of all real numbers greater than or equal to 2.

104. The set of real numbers that are in both of the intervals $(2, 6)$ and $(5, 7)$.

*In Exercises 105–108, write the numbers using scientific notation.*

105. 0.0014        106. 186,000

107. 12.001        108. $81 \times 10^{12}$

*In Exercises 109–116, carry out the indicated operations.*

109. $1 - 2[3 - 4(1 - 5)]$    110. $x^2 - 4x - (1 - 2x^2)$

111. $1 + 3(x^2 - 5x - 4) - [1 - (15x - 3x^2)]$

112. $(ax^2 - 1)(ax^2 + 1)$

113. $(x - 1)(x + 1)(x + 2)$

114. $(x + a)(x^2 + ax + 1)$

115. $x^2 + 4 - (x - 1)(x - 2)$

116. $(x^2 - x + 1)(x^2 + x - 1) + [(x - 1)^2 - x^4]$

*For Exercises 117–120, use long division to determine the quotient and remainder.*

117. $(x^3 - 2x^2 + x + 4) \div (x - 1)$

118. $(x^3 + 4x^2 - 2x - 5) \div (x + 1)$

119. $(x^4 - x^2 + 1) \div (x^2 + 1)$

120. $(2x^4 + x^3 - 4x^2 - 8x - 1) \div (x + 2)$

*In Exercises 121–125, sketch the intervals described by the given inequalities.*

121. $|x - 6| < 3$

122. $|x - \frac{1}{2}| < 1$

123. $|x + 1| \geq 1$

124. $|x| \geq 5$

125. (a) $0 < |x - 4| < 5$   (b) $|x - 4| < 5$

*In Exercises 126–128, express the number in the form p/q, where p and q are integers.*

126. $12.\overline{7}$   127. $12.\overline{71}$   128. $12.3\overline{91}$

c 129. Polynomials can be used to approximate more complicated expressions. For example, in calculus it is shown that when $x$ is close to zero, the expression $\sqrt{1 + x}$ can be approximated by the polynomial $1 + \frac{1}{2}x$. Complete the table to see evidence of this. (Give the values of $\sqrt{1 + x}$ to six decimal places without rounding off.)

| $x$ | $1 + \dfrac{x}{2}$ | $\sqrt{1 + x}$ |
|-----|-----|-----|
| 0.1 | | |
| 0.01 | | |
| 0.001 | | |

c 130. Using calculus, it can be shown that when $x$ is close to zero, the expression $1/\sqrt{1 - x^2}$ can be approximated by the polynomial $1 + \frac{1}{2}x^2 + \frac{3}{8}x^4$. Complete the table to see evidence of this. (Round off your answers to six significant digits.)

| $x$ | $1 + \dfrac{x^2}{2} + \dfrac{3x^4}{8}$ | $1/\sqrt{1 - x^2}$ |
|-----|-----|-----|
| 0.1 | | |
| 0.01 | | |

c 131. (a) Evaluate the expression $\dfrac{\sqrt{3 - \sqrt{5}}}{\sqrt{2} + \sqrt{7 - 3\sqrt{5}}}$.

(b) Evaluate the expression $\sqrt{5}/5$. If you do the calculator work carefully, your results in parts (a) and (b) will suggest (but not prove!) that the two expressions are equal. Follow steps (c), (d), and (e) to prove that the two expressions are indeed equal.

(c) Multiply the expression in part (a) by $\dfrac{\sqrt{2} - \sqrt{7 - 3\sqrt{5}}}{\sqrt{2} - \sqrt{7 - 3\sqrt{5}}}$, which equals 1. After carrying out the indicated operations, you should obtain

$$\frac{\sqrt{6 - 2\sqrt{5}} - 2\sqrt{9 - 4\sqrt{5}}}{-5 + 3\sqrt{5}}$$

(d) Verify that $\sqrt{6 - 2\sqrt{5}} = \sqrt{5} - 1$. Also verify that $\sqrt{9 - 4\sqrt{5}} = \sqrt{5} - 2$.

(e) Use the results in part (d) to simplify the expression in part (c). The final result, after rationalizing the denominator, should be $\sqrt{5}/5$.

*In Exercises 132–136, verify that the formulas are correct by carrying out the operations indicated on the left-hand side of the equations. (This list of formulas appears in A Treatise on Algebra, by George Peacock, published in 1845.)*

132. $\dfrac{1}{a + bi} + \dfrac{1}{a - bi} = \dfrac{2a}{a^2 + b^2}$

133. $\dfrac{1}{a - bi} - \dfrac{1}{a + bi} = \dfrac{2bi}{a^2 + b^2}$

134. $\dfrac{a + bi}{a - bi} + \dfrac{a - bi}{a + bi} = \dfrac{2(a^2 - b^2)}{a^2 + b^2}$

135. $\dfrac{a + bi}{a - bi} - \dfrac{a - bi}{a + bi} = \dfrac{4abi}{a^2 + b^2}$

136. $\dfrac{a + bi}{c + di} + \dfrac{a - bi}{c - di} = \dfrac{2(ac + bd)}{c^2 + d^2}$

*Exercises 137 and 138 are taken (verbatim) from Algebra, by Isaac Todhunter, first published in 1870.*

137. Simplify

$$\frac{bc(x - a)^2}{(a - b)(a - c)} + \frac{ca(x - b)^2}{(b - c)(b - a)} + \frac{ab(x - c)^2}{(c - a)(c - b)}$$

138. If $x = \sqrt[3]{a + \sqrt{a^2 + b^3}} + \sqrt[3]{a - \sqrt{a^2 + b^3}}$, shew that $x^3 + 3bx - 2a = 0$.

# 2 EQUATIONS AND INEQUALITIES

*A considerable part of the work in solving equations is mechanical. . . . However, there is a reason for doing each step that is taken, and one should be able to give this reason.*

From the text *Practical Mathematics* by C. I. PALMER and S. F. BIBB, first published in 1912.

**INTRODUCTION**   The first two sections of this chapter present the basic techniques for solving linear equations and quadratic equations. Those techniques are then used in Section 2.3 to solve a variety of applied problems. In Section 2.4, several methods are developed for solving equations that are neither linear nor quadratic. The remaining two sections of the chapter, Sections 2.5 and 2.6, deal with the solution of inequalities rather than equalities.

## LINEAR EQUATIONS

### 2.1

In this section and the next we review the terminology and skills used in solving certain basic kinds of equations. Perhaps the simplest type of equation is the *linear* or *first-degree* equation in one variable:

| Definition:   Linear Equation in One Variable |
| --- |
| A **linear** or **first-degree equation in one variable** is an equation that can be written in the form<br><br>$ax + b = 0$      with $a$ and $b$ real numbers and $a \neq 0$ |

As examples of linear equations in one variable, we cite the following:

$$2x - 10 = 0 \qquad 3m + 1 = 2 \qquad \frac{y}{2} = \frac{y}{3} + 1$$

As with any equation involving a variable, each of these equations is neither true nor false *until* we replace the variable with a number. By a **solution** or a **root** of an equation in one variable, we mean a value for $x$ (or $m$ or $y$ in the foregoing examples) that makes the equation a true statement. For example, the value $x = 5$ is a solution of the equation $2x = 10$ since, with $x = 5$, the equation becomes $2(5) = 10$, which is certainly true. We also say in this case that the value $x = 5$ **satisfies** the equation.

Equations that become true statements for all values in the domain of the variable are called **identities**. Two examples of identities are

$$x^2 - 9 = (x - 3)(x + 3) \qquad \text{and} \qquad \frac{4x^2}{x} = 4x$$

In contrast to this, a **conditional equation** is true only for some (or perhaps even none) of the values of the variable. Two examples of conditional equations are $2x = 10$ and $x = x + 1$. The first of these is true only when $x = 5$. The second equation has no solution (because, intuitively at least, no number can be one more than itself).

We say that two equations are **equivalent** when they have exactly the same solutions. In this section, and throughout the text, the basic method for solving an equation in one variable involves writing a sequence of equivalent equations until we finally reach an equation of the form

variable = a number

In generating equivalent equations, we rely on the following three principles. (These can be justified using the properties of real numbers discussed in Section 1.2.)

---

**Procedures That Yield Equivalent Equations**

1. Adding or subtracting the same quantity on both sides of an equation produces an equivalent equation.

2. Multiplying or dividing both sides of an equation by the same nonzero quantity produces an equivalent equation.

3. Simplifying an expression on either side of an equation (using the techniques from Chapter 1) produces an equivalent equation.

---

The examples that follow show how these principles are applied in solving various equations.

**EXAMPLE 1**  (a) Solve $4x + 5 = 33$.
(b) Solve $ax + b = c$   ($a$, $b$, and $c$ are constants and $a \neq 0$).

**Solution**  (a)  $4x + 5 = 33$

$$4x = 28 \qquad \text{subtracting 5 from both sides}$$
$$x = 7 \qquad \text{dividing both sides by 4}$$

CHECK   Replacing $x$ with 7 in the original equation yields

$$4(7) + 5 = 33$$
$$28 + 5 = 33 \qquad \text{True}$$

(b)  $ax + b = c$

$$ax = c - b \qquad \text{subtracting } b \text{ from both sides}$$
$$x = \frac{c - b}{a} \qquad \text{dividing both sides by } a$$

CHECK   Replacing $x$ with $\dfrac{c-b}{a}$ in the original equation yields

$$a\left(\frac{c-b}{a}\right) + b = c$$
$$c - b + b = c$$
$$c = c \qquad \text{True}$$

**EXAMPLE 2**   Solve $3[1 - 2(x+1)] = 2 - x$.

Solution

$$3[1 - 2(x+1)] = 2 - x$$
$$3[1 - 2x - 2] = 2 - x \qquad \text{simplifying the left-hand side}$$
$$3(-1 - 2x) = 2 - x$$
$$-3 - 6x = 2 - x$$
$$-3 - 5x = 2 \qquad \text{adding } x \text{ to both sides}$$
$$-5x = 5 \qquad \text{adding 3 to both sides}$$
$$x = -1 \qquad \text{dividing both sides by } -5$$

CHECK   Replacing $x$ with $-1$ in the original equation yields

$$3[1 - 2(0)] = 2 - (-1)$$
$$3(1) = 2 + 1 \qquad \text{True}$$

In the next two examples the equations involve fractions. The strategy in such cases is to multiply through by the least common denominator. That eliminates the need to work with fractions.

**EXAMPLE 3**   Solve:   (a) $\dfrac{x}{2} - \dfrac{3x}{4} = 1$   (b) $\dfrac{1}{x-2} - \dfrac{4}{x+2} = \dfrac{1}{x^2 - 4}$

Solution   (a)

$$\frac{x}{2} - \frac{3x}{4} = 1$$
$$4\left(\frac{x}{2}\right) - 4\left(\frac{3x}{4}\right) = 4(1) \qquad \begin{array}{l}\text{multiplying both sides by the}\\ \text{least common denominator}\end{array}$$
$$2x - 3x = 4 \qquad \text{simplifying}$$
$$-x = 4 \qquad \text{simplifying}$$
$$x = -4 \qquad \text{multiplying both sides by } -1$$

CHECK   Replacing $x$ with $-4$ in the original equation yields

$$\frac{-4}{2} - \frac{3(-4)}{4} = 1$$
$$-2 + 3 = 1 \qquad \text{True}$$

(b) By factoring the denominator on the right-hand side, we obtain

$$\frac{1}{x-2} - \frac{4}{x+2} = \frac{1}{(x-2)(x+2)}$$

From this we see that the least common denominator for the three fractions is $(x - 2)(x + 2)$. Now, multiplying both sides by this least common denominator, we have

$$\frac{(x - 2)(x + 2)}{x - 2} - \frac{4(x - 2)(x + 2)}{x + 2} = \frac{(x - 2)(x + 2)}{(x - 2)(x + 2)}$$

$$x + 2 - 4(x - 2) = 1 \qquad \text{simplifying}$$

$$-3x + 10 = 1 \qquad \text{simplifying}$$

$$-3x = -9 \qquad \text{subtracting 10 from both sides}$$

$$x = 3 \qquad \text{dividing both sides by } -3$$

CHECK   Replacing $x$ with 3 in the original equation yields

$$\frac{1}{3 - 2} - \frac{4}{3 + 2} = \frac{1}{(3 - 2)(3 + 2)}$$

$$1 - \frac{4}{5} = \frac{1}{5} \qquad \text{True}$$

In part (a) of Example 3, we multiplied both sides of the original equation by 4 to produce an equivalent equation. In part (b), however, we multiplied both sides of the equation by $(x - 2)(x + 2)$. Since, we didn't know at that point whether the quantity $(x - 2)(x + 2)$ was nonzero, we could not be certain that the resulting equation was actually an equivalent equation. For this reason, it is always necessary to check any solutions you obtain as a result of multiplying or dividing both sides of an equation by an expression involving the variable. Part (a) of the next example serves to emphasize this point.

**EXAMPLE 4**   Solve:   (a) $\dfrac{1}{x + 5} = \dfrac{2}{x - 3} + \dfrac{2x + 2}{(x + 5)(x - 3)}$

(b) $\dfrac{1}{x} = \dfrac{1}{a} + \dfrac{1}{b}$   ($a$ and $b$ are constants and $a + b \neq 0$)

Solution   (a)   The least common denominator for the fractions is $(x + 5)(x - 3)$. After multiplying both sides of the given equation by this least common denominator, we have

$$x - 3 = 2(x + 5) + 2x + 2$$

$$x - 3 = 2x + 10 + 2x + 2 \qquad \text{simplifying}$$

$$x - 3 = 4x + 12 \qquad \text{simplifying}$$

$$-3x - 3 = 12 \qquad \text{subtracting } 4x \text{ from both sides}$$

$$-3x = 15 \qquad \text{adding 3 to both sides}$$

$$x = -5 \qquad \text{dividing both sides by } -3$$

CHECK   The preceding steps show that *if* the equation has a solution, then the solution is $x = -5$. With $x = -5$, however, the left-hand side of the original equation becomes $\dfrac{1}{-5+5}$, or $\dfrac{1}{0}$, which is undefined. We conclude therefore that the given equation has no solution.

(b)         $\dfrac{1}{x} = \dfrac{1}{a} + \dfrac{1}{b}$

$(xab) \cdot \dfrac{1}{x} = (xab) \cdot \dfrac{1}{a} + (xab) \cdot \dfrac{1}{b}$     multiplying both sides by the least common denominator, $xab$

$ab = xb + xa$     simplifying

$ab = x(b + a)$     factoring

$\dfrac{ab}{b + a} = x$     dividing both sides by $b + a$

CHECK   As Exercise 65 asks you to verify, the value $x = \dfrac{ab}{a + b}$ indeed satisfies the original equation.   ▪▪

**EXAMPLE 5**   Solve for $x$ in terms of the other letters:   $y = \dfrac{ax + b}{cx + d}$. (Assume that $cx + d \neq 0$ and $yc - a \neq 0$.)

Solution   Multiplying both sides of the given equation by the quantity $cx + d$ yields

$y(cx + d) = ax + b$

$ycx + yd = ax + b$

$ycx - ax = b - yd$

$x(yc - a) = b - yd$

$x = \dfrac{b - yd}{yc - a}$

You should check for yourself now that the expression for $x$ on the right-hand side of this last equation indeed satisfies the original equation.   ▪▪

The techniques used in this section can also be applied in solving equations in which absolute values appear. The next example shows how this is done.

**EXAMPLE 6**   Solve $|2x - 1| = 7$.

Solution   There are two cases to consider.

If $2x - 1 \geq 0$, the equation becomes

$2x - 1 = 7$

$2x = 8$

$x = 4$

If $2x - 1 < 0$, the equation becomes

$-(2x - 1) = 7$

$-2x + 1 = 7$

$-2x = 6$

$x = -3$

We've now obtained the values $x = 4$ and $x = -3$. You should check for your-self that both of these numbers indeed satisfy the original equation.  ⬚

## EXERCISE SET 2.1

### A

*Solve each equation in Exercises 1–40.*

1. $2x - 3 = -5$
2. $x + 4 = 0$
3. $8x - 4 = 20$
4. $-6x + 1 = 49$
5. $1 - y = 12$
6. $3 - 4y = 8y + 3$
7. $2m - 1 + 3m + 5 = 6m - 8$
8. $1 - (2m + 5) = -3m$
9. $(x + 2)(x + 1) = x^2 + 11$
10. $t - \{4 - [t - (4 + t)]\} - 6$
11. $4(y - 1) + 5(y - 3) = 2(y + 1) - 28$
12. $2t - 3[t - 4(t + 5)] = -2 - 20t$
13. $1 - (x - 2)^2 = -x^2 + 5x + 3$
14. $(x + 2)^2 = x^2 + 4$
15. $(x - 1)^2 - (x^2 - 1) = 16$
16. $(x - 1)^2 - (x^2 - 1) = 2 - 2x$
17. $\dfrac{x}{3} + \dfrac{2x}{5} = \dfrac{-11}{5}$
18. $\dfrac{x}{2} + 1 = 0$
19. $1 - \dfrac{y}{3} = 6$
20. $y - \dfrac{y}{2} - \dfrac{y}{3} = 2$
21. $3 - \dfrac{x - 1}{4} = \dfrac{x}{9}$
22. $1 + \dfrac{x}{3} - \dfrac{x}{4} = x - \dfrac{5x}{6}$
23. $\dfrac{x - 1}{4} + \dfrac{2x + 3}{-1} = 0$
24. $\dfrac{x - 1}{2} - \dfrac{x - 2}{4} = 1 + \dfrac{x}{3}$
25. $\dfrac{1}{x - 3} - \dfrac{2}{x + 3} = \dfrac{1}{x^2 - 9}$
26. $\dfrac{1}{x - 5} + \dfrac{1}{x + 5} = \dfrac{6}{x^2 - 25}$
27. $\dfrac{1}{x - 5} + \dfrac{1}{x + 5} = \dfrac{2x}{x^2 - 25}$
28. $\dfrac{1}{x - 5} + \dfrac{1}{x + 5} = \dfrac{2x + 1}{x^2 - 25}$
29. $\dfrac{4}{x + 2} + \dfrac{1}{x - 2} = \dfrac{4}{x^2 - 4}$
30. $\dfrac{1}{2x + 1} + \dfrac{1}{x + 1} = \dfrac{5}{2x^2 + 3x + 1}$
31. $\dfrac{2(x + 1)}{x - 1} - 3 = \dfrac{5x - 1}{x - 1}$
32. $\dfrac{x - 1}{x + 1} = \dfrac{x + 1}{x - 1}$
33. $\dfrac{1}{x} = \dfrac{4}{x} - 1$
34. $\dfrac{1}{y} + 1 = \dfrac{3}{y} - \dfrac{1}{2y}$
35. $|x - 5| = 1$
36. $|x + 1| = 9$
37. $|6x - 5| = 25$
38. $|5 - 6x| = 0$

ⓒ 39. $4.50 - 9.11x = 1.72$  (Round off the answer to two decimal places.)

ⓒ 40. $\dfrac{x - 0.18}{x + 0.23} = \dfrac{x + 1.16}{x - 2.24}$  (Round off the answer to two decimal places.)

*For Exercises 41–50, solve each equation for x in terms of the other letters.*

41. $3ax - 2b = b + 3$
42. $ax + b = bx - a$
43. $ax + b = bx + a$
44. $\dfrac{x}{a} + \dfrac{x}{b} = 1$
45. $\dfrac{1}{x} = a + b$
46. $\dfrac{1}{ax} = \dfrac{1}{bx} - \dfrac{1}{c}$
47. $(x + b)(x + c) - a(a + b) = (x - a)(x + a) + bc$
48. $\dfrac{1}{a} - \dfrac{1}{x} = \dfrac{1}{x} - \dfrac{1}{b}$
49. $(a - x)^2 = x^2 + b^2$
50. (a) $y = mx + b$  $(m \neq 0)$
    (b) $y - y_1 = m(x - x_1)$  $(m \neq 0)$
    (c) $\dfrac{x}{a} + \dfrac{y}{b} = 1$
    (d) $Ax + By + C = 0$  $(A \neq 0)$

### B

51. Solve for x:  $\dfrac{|x - 3|}{2} + \dfrac{|x - 3|}{3} = \dfrac{-5x}{3}$

    *Hint:*  Consider separately the cases $x \geq 3$ and $x < 3$.

52. Solve for x:  $\dfrac{|x + 4|}{4} - |x + 1| = -\dfrac{9}{2}$

In Exercises 53–58, solve each equation for x in terms of the other letters.

**53.** $a^2(a - x) = b^2(b + x) - 2abx$   (Assume that $a \neq b$.)

**54.** $\dfrac{b}{ax - 1} - \dfrac{a}{bx - 1} = 0$   (Assume that $a \neq b$.)

**55.** $\dfrac{a - x}{a - b} - 2 = \dfrac{c - x}{b - c}$

**56.** $\dfrac{x + 2p}{2q - x} + \dfrac{x - 2p}{2q + x} - \dfrac{4pq}{4q^2 - x^2} = 0$

**57.** $\dfrac{x - a}{x - b} = \dfrac{b - x}{a - x}$

**58.** $1 - \dfrac{a}{b}\left(1 - \dfrac{a}{x}\right) - \dfrac{b}{a}\left(1 - \dfrac{b}{x}\right) = 0$

In Exercises 59–64, solve each equation for the indicated variable.

**59.** $d = \dfrac{r}{1 + rt}$; for $r$

**60.** $\dfrac{x_1 x}{a^2} + \dfrac{y_1 y}{b^2} = 1$; for $y$

**61.** $S = 2\pi r^2 + 2\pi rh$; for $h$

**62.** $s = \dfrac{n}{2}(a + l)$; for $a$

**63.** $S = \dfrac{rl - a}{r - 1}$; for $r$

**64.** $3y^2 y' + 5x^2 yy' - x^2 = 0$; for $y'$

**65.** Verify that $x = \dfrac{ab}{a + b}$ is a solution of the equation $\dfrac{1}{x} = \dfrac{1}{a} + \dfrac{1}{b}$. [This is to check the answer obtained in Example 4(b).]

## C

For Exercises 66–67, solve for x in terms of the other letters.

**66.** $\dfrac{q}{x + q} - \dfrac{p}{x + p} = \dfrac{q - p}{x - 2pq}$

**67.** $\dfrac{1}{x - p} - \dfrac{1}{x - q} = \dfrac{p - q}{x^2 - pq}$

**68.** Solve for x and simplify the answer.

$$\frac{x - a}{b} + \frac{x - b}{a} + \frac{x - 1}{ab} = \frac{2}{a} + \frac{2}{b} + 2$$

(Assume both $a$ and $b$ are nonzero and $a + b \neq -1$.)

## QUADRATIC EQUATIONS

### 2.2

*The formula for the quadratic equation seems first to have been discovered by the Moslems around 900 A.D., although quadratic equations had been solved by the Babylonians 3000 years earlier.*

CHARLES ROBERT HADLOCK, *Field Theory and Its Classical Problems* (Washington, D.C.: Mathematical Association of America, 1978)

An equation that can be written in the form

$$ax^2 + bx + c = 0 \qquad \text{with } a, b, \text{ and } c \text{ real numbers and } a \neq 0$$

is called a **quadratic equation**. Examples of quadratic equations are

$$x^2 - 8x - 9 = 0 \qquad 2y^2 - 5 = 0 \qquad 3m^2 = 1 - m$$

Recall that a **solution**, or **root**, of an equation is a value for the unknown that makes the equation a true statement. For example, the value $x = -1$ is a solution of the quadratic equation $x^2 - 8x - 9 = 0$, because when $x$ is replaced by $-1$ in this equation, we have

$$(-1)^2 - 8(-1) - 9 = 0$$
$$1 + 8 - 9 = 0$$
$$0 = 0 \qquad \text{True}$$

When applicable, one of the simplest techniques for solving quadratic equations involves factoring. This method relies on the following familiar and important property of the real number system.

| PROPERTY SUMMARY | Zero-Product Property of Real Numbers |
| --- | --- |
| | $pq = 0$     if and only if     $p = 0$ or $q = 0$ (or both) |

For example, to solve the equation $x^2 - 2x - 3 = 0$ by factoring, we have

$$x^2 - 2x - 3 = 0$$
$$(x - 3)(x + 1) = 0$$
$$x - 3 = 0 \quad | \quad x + 1 = 0$$
$$x = 3 \quad | \quad x = -1$$

As you can easily check, the values $x = 3$ and $x = -1$ both satisfy the given equation.

Unfortunately, not all quadratic expressions can be factored so easily. Consider, for example, the equation $x^2 - 2x - 4 = 0$. There are only three possible factorizations with integer coefficients, but none yields the appropriate middle term, $-2x$, when multiplied out:

$$(x - 4)(x + 1) \qquad (x + 4)(x - 1) \qquad (x - 2)(x + 2)$$

Moreover, it is not always obvious whether a particular equation, such as $x^2 + 156x + 5963 = 0$, can be solved by factoring. (We will come back to this equation in Example 3.) Clearly, we need a systematic approach to solving those quadratic equations that are not readily solvable through factoring. The technique of **completing the square** provides this approach. We'll demonstrate this technique by solving the equation

$$x^2 - 2x - 4 = 0$$

First, we rewrite the equation in the form

$$x^2 - 2x = 4 \qquad\qquad\qquad (1)$$

with the $x$-terms isolated on the left-hand side of the equation. To complete the square, we follow these two steps:

STEP 1   Take half of the coefficient of $x$ and square it.
STEP 2   Add the number obtained in step 1 to both sides of the equation.

For equation (1), the coefficient of $x$ is $-2$. Taking half of $-2$ and then squaring it gives us $(-1)^2$, or 1. Now, as directed in step 2, we add 1 to both sides of equation (1). This yields

$$x^2 - 2x + 1 = 4 + 1$$
$$(x - 1)^2 = 5$$
$$x - 1 = \pm\sqrt{5}$$
$$x = 1 \pm \sqrt{5}$$

We have now obtained the two solutions, $1 + \sqrt{5}$ and $1 - \sqrt{5}$. (Exercise 6 will ask you to verify that these values indeed satisfy the original equation.) In the box that follows, we summarize the technique of completing the square, and we indicate how the process got its name.

ALGEBRAIC PROCEDURE FOR COMPLETING THE SQUARE IN THE EXPRESSION $x^2 + bx$

Add the square of half of the $x$-coefficient:

$$(x^2 + bx) + \left(\frac{b}{2}\right)^2 = \left(x + \frac{b}{2}\right)^2$$

GEOMETRIC INTERPRETATION OF COMPLETING THE SQUARE FOR $x^2 + bx$

The colored region in the figure represents the quantity $x^2 + bx$ $\left(\text{since the area is } x^2 + \frac{b}{2}x + \frac{b}{2}x\right)$. By adding the white square to the colored region, we fill out or "complete" the larger square. The area of the white region that completes the square is $\frac{b}{2} \cdot \frac{b}{2} = \frac{b^2}{4}$.

The description in the box explains how to complete the square for expressions of the form $x^2 + bx$. To complete the square for a quadratic equation in which the coefficient of $x^2$ is not 1, we can first divide both sides of the equation by the coefficient of $x^2$. This is done in the next example.

**EXAMPLE 1** Solve by completing the square: $9x^2 + 36x - 1 = 0$

Solution
$$9x^2 + 36x - 1 = 0$$

$$x^2 + 4x - \frac{1}{9} = 0 \qquad \text{dividing both sides by 9}$$

$$x^2 + 4x = \frac{1}{9}$$

$$x^2 + 4x + 4 = \frac{1}{9} + 4 \qquad \begin{array}{l}\text{completing the square by adding}\\ (\frac{1}{2} \cdot 4)^2 = 4 \text{ to both sides}\end{array}$$

$$(x + 2)^2 = \frac{37}{9}$$

$$x + 2 = \pm\sqrt{\frac{37}{9}} = \pm\frac{\sqrt{37}}{3}$$

$$x = -2 \pm \frac{\sqrt{37}}{3} = \frac{-6 \pm \sqrt{37}}{3}$$

The two solutions are therefore

$$\frac{-6+\sqrt{37}}{3} \quad \text{and} \quad \frac{-6-\sqrt{37}}{3}$$

The technique of completing the square can be used to derive the **quadratic formula**, a useful formula that gives the solutions to any quadratic equation. The derivation of this formula runs as follows. We start with the general quadratic equation $ax^2 + bx + c = 0 \quad (a \neq 0)$ and divide both sides by $a$ to obtain

$$x^2 + \frac{b}{a}x + \frac{c}{a} = 0$$

Subtracting $c/a$ from both sides yields

$$x^2 + \frac{b}{a}x = -\frac{c}{a}$$

Now, to complete the square, we add $[\frac{1}{2}(b/a)]^2$, or $b^2/4a^2$, to both sides. That gives us

$$x^2 + \frac{b}{a}x + \frac{b^2}{4a^2} = \frac{b^2}{4a^2} - \frac{c}{a}$$

$$\left(x + \frac{b}{2a}\right)^2 = \frac{b^2 - 4ac}{4a^2}$$

$$x + \frac{b}{2a} = \pm\sqrt{\frac{b^2 - 4ac}{4a^2}} = \pm\frac{\sqrt{b^2 - 4ac}}{2\,|a|}$$

$$= \pm\frac{\sqrt{b^2 - 4ac}}{2a}$$

This last equality follows from the fact that for any real number $a \neq 0$, the expressions $\pm 2\,|a|$ and $\pm 2a$ both represent the same two numbers. We now conclude that the solutions are

$$x = -\frac{b}{2a} + \frac{\sqrt{b^2 - 4ac}}{2a} \quad \text{and} \quad x = -\frac{b}{2a} - \frac{\sqrt{b^2 - 4ac}}{2a}$$

These solutions are usually written in the more compact form displayed in the following box.

---

**The Quadratic Formula**

The solutions of the equation $ax^2 + bx + c = 0 \quad (a \neq 0)$ are given by

$$x = \frac{-b \pm \sqrt{b^2 - 4ac}}{2a}$$

**EXAMPLE 2**    Use the quadratic formula to solve $2x^2 = 4 - x$.

Solution    We first rewrite the equation $2x^2 + x - 4 = 0$, so that it has the form $ax^2 + bx + c = 0$. By comparing these last two equations, we see that $a = 2$, $b = 1$, and $c = -4$. Therefore,

$$x = \frac{-b \pm \sqrt{b^2 - 4ac}}{2a} = \frac{-1 \pm \sqrt{1^2 - 4(2)(-4)}}{2(2)} = \frac{-1 \pm \sqrt{33}}{4}$$

Thus, the two solutions are $\dfrac{-1 + \sqrt{33}}{4}$ and $\dfrac{-1 - \sqrt{33}}{4}$.

**EXAMPLE 3**    Solve $x^2 + 156x + 5963 = 0$.

Solution    We use the quadratic formula and a hand-held calculator. In this case,

$$a = 1 \qquad b = 156 \qquad c = 5963$$

and

$$x = \frac{-b \pm \sqrt{b^2 - 4ac}}{2a} = \frac{-156 \pm \sqrt{156^2 - 4(1)(5963)}}{2(1)}$$

$$= \frac{-156 \pm 22}{2} \qquad \text{using a calculator}$$

$$= -78 \pm 11$$

The two solutions are therefore $-89$ and $-67$. Although it's conceivable that the initial equation could be solved by discovering the factorization $(x + 89)(x + 67)$, in practice it is usually more efficient to proceed directly to the formula when numbers of this magnitude are involved.

In Examples 2 and 3 we found that each quadratic equation had two real roots. As the next example indicates, however, it's also possible for quadratic equations to have only one real root, or even no real roots.

**EXAMPLE 4**    Use the quadratic formula to solve the following.

(a)  $4x^2 - 12x + 9 = 0$       (b)  $x^2 + x + 2 = 0$

Solution    (a)  Here $a = 4$, $b = -12$, $c = 9$; so

$$x = \frac{12 \pm \sqrt{(-12)^2 - 4(4)(9)}}{2(4)} = \frac{12 \pm \sqrt{144 - 144}}{8}$$

$$= \frac{12 \pm 0}{8} = \frac{12}{8} = \frac{3}{2}$$

In this case our only solution is $\frac{3}{2}$. We refer to $\frac{3}{2}$ as a **double root**.

NOTE    We've used the quadratic formula here only for purposes of illustration; in this case you can solve the original equation more efficiently by factoring.

**(b)** Since $a = 1$, $b = 1$, $c = 2$, we have

$$x = \frac{-1 \pm \sqrt{(1)^2 - 4(1)(2)}}{2(1)} = \frac{-1 \pm \sqrt{-7}}{2}$$

$$= \frac{-1 \pm i\sqrt{7}}{2}$$

In this case our roots are the two complex numbers

$$\frac{-1 + i\sqrt{7}}{2} \quad \text{and} \quad \frac{-1 - i\sqrt{7}}{2}$$

The quantity $b^2 - 4ac$ that appears under the radical sign in the quadratic formula is called the **discriminant**. In Examples 2 and 3, the discriminant was positive and consequently each equation had two real solutions. In Example 4(a), in which the discriminant was zero, we obtained but one (real) solution. In Example 4(b) the discriminant was negative, and consequently we obtained two solutions involving the imaginary unit $i$. These observations are generalized in the box that follows.

---

### The Discriminant $b^2 - 4ac$

Consider the quadratic equation $ax^2 + bx + c = 0$, where $a$, $b$, and $c$ are real numbers and $a \neq 0$. The expression $b^2 - 4ac$ is called the **discriminant**.

1. If $b^2 - 4ac > 0$, then the equation has two distinct real roots.
2. If $b^2 - 4ac = 0$, then the equation has exactly one real root.
3. If $b^2 - 4ac < 0$, then the equation has no real roots.

---

**EXAMPLE 5** **(a)** Compute the discriminant to determine how many real solutions there are for the equation $x^2 + x - 1 = 0$.
**(b)** Find a value for $k$ such that the quadratic equation $x^2 + \sqrt{2}x + k = 0$ has exactly one real solution.

**Solution** **(a)** Here $a = 1$, $b = 1$, $c = -1$, and therefore

$$b^2 - 4ac = 1^2 - 4(1)(-1) = 5$$

Since the discriminant is positive, the equation has two distinct real solutions.

**(b)** For the equation to have exactly one real solution, the discriminant must be zero. That is,

$$b^2 - 4ac = 0$$
$$(\sqrt{2})^2 - 4(1)(k) = 0$$
$$2 - 4k = 0$$
$$k = \frac{1}{2}$$

The required value for $k$ is $\frac{1}{2}$.

## EXERCISE SET 2.2

### A

In Exercises 1–5, determine if the given value is a solution of the equation.

1. $2x^2 - 6x - 36 = 0; x = -3$
2. $(y - 4)(y - 5) = 0; y = -1$
3. $4x^2 - 1 = 0; x = -\frac{1}{4}$
4. $m^2 + m - \frac{5}{16} = 0; m = \frac{1}{4}$
5. $x^2 - 2x - 6 = 0; x = -1 - \sqrt{7}$
6. Verify that the numbers $1 + \sqrt{5}$ and $1 - \sqrt{5}$ both satisfy the equation $x^2 - 2x - 4 = 0$.

In Exercises 7–26, solve each equation by factoring.

7. $x^2 - 5x - 6 = 0$      8. $x^2 - 5x = -6$
9. $x^2 - 100 = 0$      10. $4y^2 + 4y + 1 = 0$
11. $25x^2 - 60x + 36 = 0$      12. $144 - t^2 = 0$
13. $10z^2 - 13z - 3 = 0$      14. $3t^2 - t - 4 = 0$
15. $(x + 1)^2 - 4 = 0$      16. $x^2 + 3x - 40 = 0$
17. $x(2x - 13) = -6$      18. $x(3x - 23) = 8$
19. $x(x + 1) = 156$      20. $x^2 + (2\sqrt{5})x + 5 = 0$
21. $x^2 - 49 = 0$      22. $x^2 - \frac{16}{9} = 0$
23. $\frac{1}{4} - x^2 = 0$      24. $2x^2 - 50 = 0$
25. $3x^2 - 192 = 0$
26. (a) $a^2x^2 - c^2 = 0$   $(a > 0$ and $c > 0)$
    (b) $a^2x^2 - c^2 = 0$   $(a > 0$ and $c < 0)$

In Exercises 27–38, solve each equation in two ways: (a) by completing the square; (b) by using the quadratic formula.

27. $x^2 + 10x + 9 = 0$      28. $x^2 - 10x + 25 = 0$
29. $x^2 - x - 5 = 0$      30. $x^2 - x + 5 = 0$
31. $2x^2 + 3x - 4 = 0$      32. $4x^2 - 3x - 9 = 0$
33. $12x^2 + 32x + 5 = 0$      34. $10x^2 - x - 1 = 0$
35. $2x^2 = x + 5$      36. $3 - 2x = -4x^2$
37. $-6x^2 + 12x = -1$      38. $-\sqrt{2}x^2 + x = -\sqrt{2}$

In Exercises 39 and 40, use the quadratic formula and a calculator to solve for x. Round off each answer to two decimal places. (Note:  After completing Exercises 39 and 40, compare the two equations and their solutions. The point here is that a slight change in one of the coefficients can sometimes radically alter the nature of the solutions.)

39. $x^2 + 3x + 2.249 = 0$
40. $x^2 + 3x + 2.251 = 0$

In Exercises 41–48, use the discriminant to determine how many real roots each equation has.

41. $x^2 - 12x + 16 = 0$      42. $2x^2 - 6x + 5 = 0$
43. $4x^2 - 5x - \frac{1}{2} = 0$      44. $4x^2 - 28x + 49 = 0$
45. $x^2 + \sqrt{3}x + \frac{3}{4} = 0$      46. $\sqrt{2}x^2 + \sqrt{3}x + 1 = 0$
47. $y^2 - \sqrt{5}y = -1$      48. $\frac{m^2}{4} - \frac{4m}{3} + \frac{16}{9} = 0$

In each of Exercises 49–52, find a value for k such that the equation has exactly one real root.

49. $x^2 + 12x + k = 0$      50. $3x^2 + (\sqrt{2k})x + 6 = 0$
51. $x^2 + kx + 5 = 0$      52. $kx^2 + kx + 1 = 0$

### B

Solve the equations in Exercises 53–60. [In these exercises, you'll need to multiply both sides of the equations by expressions involving the variable. Remember (as indicated in Section 2.1) to check your answers in these cases.]

53. $\dfrac{3}{x + 5} + \dfrac{4}{x} = 2$      54. $\dfrac{5}{x + 2} - \dfrac{2x - 1}{5} = 0$
55. $1 - x - \dfrac{2}{6x + 1} = 0$      56. $\dfrac{x^2 - 3x}{x + 1} = \dfrac{4}{x + 1}$
57. $\dfrac{3x^2 - 6x - 3}{(x + 1)(x - 2)(x - 3)} + \dfrac{5 - 2x}{x^2 - 5x + 6} = 0$
58. $\dfrac{6}{x^2 - 1} + \dfrac{x}{x + 1} = \dfrac{3}{2}$      59. $\dfrac{2x}{x^2 - 1} - \dfrac{1}{x + 3} = 0$
60. $\dfrac{x}{x - 2} + \dfrac{x}{x + 2} = \dfrac{8}{x^2 - 4}$

For Exercises 61 and 62, solve for x in terms of the other letters.

61. $(ax + b)^2 - (bx + a)^2 = 0$
62. $(x - p)^2 + (x - q)^2 = p^2 + q^2$
63. For which values of A and B will the roots of $x^2 + Ax + B = 0$ be A and B?
64. For which values of k will the equation $x^2 = 2x(3k + 1) - 7(2k + 3)$ have equal roots?
65. Let $\alpha$ and $\beta$ be the roots of the quadratic equation $ax^2 + bx + c = 0$. Verify each of the following statements.
    (a) $\alpha + \beta = -b/a$      (b) $\alpha\beta = c/a$
    (c) $\alpha^2 + \beta^2 = \dfrac{b^2 - 2ac}{a^2}$   Suggestion:  Use (a) and (b) along with the fact that $\alpha^2 + \beta^2 = (\alpha + \beta)^2 - 2\alpha\beta$.

(d) $\dfrac{1}{\alpha^2} + \dfrac{1}{\beta^2} = \dfrac{b^2 - 2ac}{c^2}$  *Suggestion:* Add the fractions $1/\alpha^2$ and $1/\beta^2$.

66. Here is an outline for a slightly different derivation of the quadratic formula. The advantage to this method is that fractions are avoided until the very last step. Fill in the details.
    (a) Beginning with $ax^2 + bx = -c$, multiply both sides by $4a$. Then add $b^2$ to both sides.
    (b) Now factor the resulting left-hand side and take square roots.

## C

67. Consider the quadratic equation $x^2 + px + q = 0$, where both $p$ and $q$ are nonzero.

(a) If one root is three times the other, show that $p^2/q = 16/3$.
(b) If one root is $n$ times the other, show that $p^2/q = (n + 1)^2/n$. (Assume $n \neq \pm 1$.)

*In Exercises 68–70, solve each equation for x in terms of the other letters.*

68. $(a + b + c)x^2 - 2(a + b)x + (a + b - c) = 0$
    $(c > 0,\ a + b + c \neq 0)$  *Hint:*  Let $d = a + b$.

69. $ab(a + b)x^2 - (a^2 + b^2)x - 2(a + b) = 0$
    (Assume that $a$ and $b$ are nonzero and $a \neq -b$.)

70. $x^2 - \left( \sqrt{\dfrac{a}{b}} + \sqrt{\dfrac{b}{a}} \right)x + 1 = 0$

# APPLICATIONS

## 2.3

*Solving problems is a practical art, like swimming, or skiing, or playing the piano: you can learn it only by imitation and practice. This book cannot offer you a magic key that opens all doors and solves all the problems, but it offers you good examples for imitation and many opportunities for practice. . . .*

GEORGE POLYA in his book *Mathematical Discovery* (New York: John Wiley & Sons, 1981)

The techniques developed in the previous two sections for solving linear and quadratic equations can be used in solving a wide variety of applied problems. As you approach each example and exercise in this section for the first time, you may find it useful to refer to the following suggestions for organizing your thoughts and "getting started." Eventually, of course, you'll probably want to modify this list to fit your own style.

> **Steps for Problem Solving**
>
> 1. Read the problem carefully. When appropriate, draw a picture or a chart to summarize the situation.
>
> 2. State in your own words, as specifically as you can, what the problem is asking for. (This usually requires rereading the problem.) Assign variables or expressions to denote the required quantities. Then, as necessary, assign auxiliary variables to any other quantities that appear relevant. Are there equations relating these auxiliary variables?
>
> 3. Set up an equation (or equations) involving the required variables. Where necessary, use the results from step 2 to obtain an equation involving but one required unknown.
>
> 4. Solve the equation that you obtained in step 3.
>
> 5. Check your answer. Relate it to the original problem. Does it satisfy the conditions of the *original* problem?

In Example 1 we solve two problems that appear to be very similar. As you'll see, however, the first problem involves a linear equation, the second a quadratic.

**EXAMPLE 1**   The sum of two numbers is 20.
(a) Find the two numbers, assuming that their difference is 96.
(b) Find the two numbers, assuming that their product is 96.

Solution   (a) Let

$$x = \text{the first number}$$

$$20 - x = \text{the second number}$$

(Notice that the sum of $x$ and $20 - x$ is indeed 20.) Now, since the difference of these two numbers is 96, we have

$$x - (20 - x) = 96$$
$$2x = 116$$
$$x = 58$$

With $x = 58$, the second number, $20 - x$, is $20 - 58$, or $-38$. As you can now check for yourself, 58 and $-38$ are indeed the required numbers: their sum is 20 and their difference is 96.

(b) Again we denote the two numbers by $x$ and $20 - x$. But this time we are assuming that their product (rather than their difference) is 96. Thus we have

$$x(20 - x) = 96$$
$$20x - x^2 = 96$$
$$x^2 - 20x + 96 = 0$$
$$(x - 8)(x - 12) = 0$$
$$x - 8 = 0 \quad \bigg| \quad x - 12 = 0$$
$$x = 8 \quad \bigg| \quad x = 12$$

We've now found two possibilities for the first number, 8 or 12. If $x = 8$, then the second number is $20 - x = 20 - 8 = 12$; if $x = 12$, then the second number is $20 - 12 = 8$. So in either case we obtain the two numbers 8 and 12. As you can see, these are the required numbers: their sum is 20 and their product is 96.

**EXAMPLE 2**   If you have three exam scores of 91, 76, and 84, what score do you need on the fourth exam to raise your average to 85?

Solution   Let

$$x = \text{the score on the fourth exam}$$

The average of the four scores 91, 76, 84, and $x$ is $(91 + 76 + 84 + x) \div 4$. Since this average is to be 85, we write

$$\frac{91 + 76 + 84 + x}{4} = 85$$

$$91 + 76 + 84 + x = 340 \qquad \text{multiplying by 4}$$
$$251 + x = 340$$
$$x = 340 - 251 = 89$$

We've now found that $x = 89$. This is the required score because, as you can check, the average of the four numbers 91, 76, 84, and 89 is in fact 85, as required.

If you look back over Examples 1 and 2, you'll see that the equations we solved were obtained by directly translating the words in the problems into algebraic notation. Example 3 is one step removed from this in that we obtain the necessary equation by means of a basic formula from geometry, the Pythagorean Theorem. For reference, we state this theorem in the box that follows. (For a proof of this theorem, see Exercise Set 3.1, Exercise 39.)

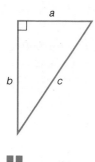

> ### Pythagorean Theorem
>
> In any right triangle, the lengths of the three sides are related by the equation
>
> $$a^2 + b^2 = c^2$$
>
> where $a$ and $b$ are the lengths of the sides forming the right angle and $c$ is the length of the hypotenuse (the side opposite the right angle).

**EXAMPLE 3**  Are there right triangles other than the 3-4-5 right triangle in which the lengths of the sides are three consecutive integers?

Solution  Any three consecutive integers can be denoted by $x$, $x + 1$, and $x + 2$ (see Figure 1). If these three quantities are to be the lengths of the sides of a right triangle, then according to the Pythagorean Theorem we must have

$$x^2 + (x + 1)^2 = (x + 2)^2$$
$$x^2 + x^2 + 2x + 1 = x^2 + 4x + 4$$
$$x^2 - 2x - 3 = 0 \qquad \text{simplifying}$$
$$(x - 3)(x + 1) = 0$$
$$x - 3 = 0 \quad \bigm| \quad x + 1 = 0$$
$$x = 3 \quad \bigm| \qquad x = -1$$

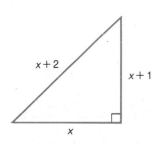

**FIGURE 1**

Thus, $x = 3$ or $x = -1$. We can immediately discard the value $x = -1$, for $x$ is supposed to represent a length. On the other hand, with $x = 3$ we find that

$$x + 1 = 3 + 1 = 4 \qquad \text{and} \qquad x + 2 = 3 + 2 = 5$$

From this we conclude that if the lengths of the three sides in the right triangle are consecutive integers, then those integers must be 3, 4, and 5. Or, to answer the original question, there are no other right triangles in which the lengths of the sides are three consecutive integers.

The next example involves *simple interest* earned on money in a savings account. As an illustration of simple interest, suppose that you place $100 in a savings account for one year at 10% simple interest. This means that at the end of the year, the bank contributes 10% of $100, or $10, to your account. Furthermore, assuming that you make no subsequent withdrawals or deposits, the bank continues to contribute $10 to your account at the end of each year. The following general formula can be derived for computing simple interest.

> **$I = Prt$**
>
> where $I$ is the simple interest earned on an initial deposit $P$ (the *principal*) after $t$ years at an annual rate $r$.

To use this formula, we need to convert the annual rate $r$ from a percent to a decimal. For instance, to compute the simple interest earned on a principal of $1000 invested for six years at 10%, we write

$$I = Prt = (\$1000)(0.1)(6) = \$600$$

**EXAMPLE 4**  Suppose that you open two savings accounts. In the first you deposit $5000 at an annual simple interest rate of 8%. In the second you deposit $10,000 at 12% (also simple interest). After how many years will the total interest amount to $16,000?

**Solution**  Let

$t =$ the required number of years

Then in $t$ years the interest on the first account is $(5000)(0.08)t$, while that on the second is $(10,000)(0.12)t$. Since the total interest is to be $16,000, we have

$$(5000)(0.08)t + (10,000)(0.12)t = 16,000$$
$$400t + 1200t = 16,000$$
$$1600t = 16,000$$
$$t = 10$$

Thus, after 10 years the total interest will amount to $16,000.

CHECK  After 10 years the interest from the first account will be $(\$5000)(0.08)(10)$, which is $4000. The interest from the second account after 10 years will be $(\$10,000)(0.12)(10)$, which is $12,000. So the total interest will be the sum of $4000 and $12,000, that is, $16,000, as required.

In some problems it's helpful to set up a chart or table to see the given data at a glance. This is done in the next example.

**EXAMPLE 5**  Suppose that a chemistry student can obtain two acid solutions from the stockroom. The first solution is 20% acid and the second is 45% acid. (The percents are by volume.) How many cubic centimeters ($cm^3$) of each solution should the student mix together to obtain $100 \, cm^3$ of a 30% acid solution?

**Solution**  We begin by assigning letters or expressions to denote the required quantities. Let $x$ denote the number of $cm^3$ of the 20% solution to be used. Then

$100 - x$

denotes the number of $cm^3$ of the 45% solution to be used. All the data can now be summarized in Table 1.

Original circle

$$\text{Area} = \pi r^2$$
$$= \pi (10^2) = 100\pi$$

New circle

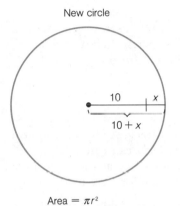

$$\text{Area} = \pi r^2$$
$$= \pi (10 + x)^2$$

**FIGURE 2**

**TABLE 1**

| TYPE OF SOLUTION | NUMBER OF CM$^3$ | PERCENT ACID | TOTAL ACID (CM$^3$) |
|---|---|---|---|
| First solution (20% acid) | $x$ | 20 | $(0.20)x$ |
| Second solution (45% acid) | $100 - x$ | 45 | $(0.45)(100 - x)$ |
| Mixture | 100 | 30 | $(0.30)(100)$ |

Looking at the data in the right-hand column of the table, we can write

$$\underbrace{0.20x}_{} + \underbrace{0.45(100 - x)}_{} = \underbrace{(0.30)(100)}_{}$$

Amount of acid in $x$ cm$^3$ of the 20% solution   Amount of acid in $(100 - x)$ cm$^3$ of the 45% solution   Amount of acid in the final mixture

So we have

$$0.20x + 0.45(100 - x) = 30$$
$$20x + 4500 - 45x = 3000 \qquad \text{multiplying by 100}$$
$$-25x = -1500$$
$$x = 60$$

Thus $x = 60$, and therefore $100 - x = 100 - 60 = 40$. Consequently, the student should use 60 cm$^3$ of the first solution and 40 cm$^3$ of the second solution.

CHECK   The amount of acid in 60 cm$^3$ of the first solution is $(60\,\text{cm}^3)(0.20)$, which is 12 cm$^3$. The amount of acid in 40 cm$^3$ of the second solution is $(40\,\text{cm}^3)(0.45)$, which is 18 cm$^3$. So the total amount of acid in the mixture is the sum of 12 cm$^3$ and 18 cm$^3$, which is 30 cm$^3$. Since the volume of the final mixture is 100 cm$^3$, the amount of acid there is $\frac{30}{100}(100)$; this is 30%, as required.

**EXAMPLE 6**   The radius of a circle is 10 cm. By how many centimeters should the radius be increased so that the area increases by $5\pi$ cm$^2$? Round off the answer to two decimal places.

Solution   Let

$x =$ the number of cm by which the radius should be increased

Then, as the calculations accompanying Figure 2 indicate, the area of the original circle is $100\pi$ cm$^2$, while the area of the new circle is $\pi(10 + x)^2$. Now, according to the statement of the problem, we want the two areas to differ by $5\pi$ cm$^2$. Consequently, we have

$$\pi(10 + x)^2 - 100\pi = 5\pi$$
$$(10 + x)^2 - 100 = 5 \qquad \text{dividing by } \pi$$
$$(10 + x)^2 = 105$$
$$10 + x = \sqrt{105} \qquad \text{disregarding the root } -\sqrt{105}. \text{ (Why is this?)}$$
$$x = \sqrt{105} - 10$$

Thus, the radius of the original circle should be increased by $(\sqrt{105} - 10)$ cm. Using a calculator and rounding off to two decimal places, we obtain $x \approx 0.25$ cm. As with all word problems, you should check for yourself here that the answer obtained satisfies the conditions of the original problem. (In checking, of course, use the exact value, $\sqrt{105} - 10$, not the calculator approximation, 0.25.)

Our final example is a motion problem. For travel at a constant speed or rate, the basic relationship is

distance = rate × time

or, more concisely,

$$d = rt$$

For example, if you drive 55 mph for 3 hours, then the distance you cover is $d = (55)(3) = 165$ miles. If a trip involves several parts, each at a different constant speed, then the *average speed* for the trip is defined by

$$\text{average speed} = \frac{\text{total distance}}{\text{total time}}$$

**EXAMPLE 7** Andy and Jerry are going to run on a quarter-mile track. Andy can run at a pace of 8 minutes per mile $(= \frac{1}{8}\text{ mile/min})$, while Jerry can run at a pace of 7 minutes per mile $(= \frac{1}{7}\text{ mile/min})$. If they start together at the same time (running in the same direction), how long will it take for Jerry to lap Andy?

Solution As in Example 5, we set up a table to organize the data. Table 2(a) indicates the format. (Notice that the top row is just taken from the formula $d = rt$.) In the $r$ column of Table 2(a), we've filled in the given rates for Andy and Jerry.

**TABLE 2(a)**

| | $d$ (MILES) | $r$ (MILES/MIN) | $t$ (MIN) |
|---|---|---|---|
| Andy | | $\frac{1}{8}$ | |
| Jerry | | $\frac{1}{7}$ | |

**TABLE 2(b)**

| | $d$ (MILES) | $r$ (MILES/MIN) | $t$ (MIN) |
|---|---|---|---|
| Andy | $\frac{1}{8}T$ | $\frac{1}{8}$ | $T$ |
| Jerry | $\frac{1}{7}T$ | $\frac{1}{7}$ | $T$ |

Now we complete the table as shown in Table 2(b). In the $t$ column, we've used $T$ to represent the (unknown) number of minutes for Jerry to lap Andy; the $d$ column was filled in using the fact that $d = rt$. Thus, Andy's and Jerry's distances are represented by $\frac{1}{8}T$ and $\frac{1}{7}T$, respectively. Consequently, the condition that Jerry be exactly $\frac{1}{4}$ mile ahead of Andy is

$$\frac{1}{7}T = \frac{1}{8}T + \frac{1}{4}$$
$$8T = 7T + 14 \qquad \text{multiplying by 56}$$
$$T = 14$$

Thus, after 14 minutes Jerry will have lapped Andy. (Check for yourself that this answer indeed satisfies the conditions of the problem.)

## EXERCISE SET 2.3

### A

1. Find two numbers whose sum is 7 and whose difference is 17.

2. The larger of two numbers is one less than twice the smaller. The sum of the two numbers is 29. Find the two numbers.

3. Find two numbers whose sum is 17 and whose product is 52.

4. The sum of two numbers is 6 and the sum of their squares is 90. What are the two numbers?

5. The sum of a number and its double is 63. What is the number?

6. Find three consecutive integers such that twice the first plus half of the second is nine more than twice the third.

7. The product of a certain positive number and four more than that number is 96. What is the number?

8. The sum of the squares of three consecutive positive integers is 365. Find the integers.

9. What number must be added to both the numerator and the denominator of the fraction $\frac{3}{7}$ to yield a fraction that is equal to $\frac{2}{3}$?

10. Find two consecutive even integers such that when they are divided by 8 and 9, respectively, the sum of the quotients is 4.

11. The difference of two positive numbers is one-third of their sum. The sum of the squares of the two numbers is 180. Find the two numbers.

12. (a) Is there a positive real number such that the sum of the number and its square is 1? If so, find the number.

    (b) Is there a positive real number such that the difference between the number and its square is 1? If so, find the number.

13. If you have exam scores of 70, 77, and 75, what score do you need on the fourth exam to raise your average to 75?

14. The average of three numbers is 42. The first two numbers differ by 1, and the average of the second and third numbers is 47. What are the three numbers?

15. The average of four numbers is 52. Find the fourth number, given that the first three are 80, 22, and 62.

16. The average of the reciprocals of two consecutive integers is $\frac{9}{40}$. Find the smaller of the two integers.

17. In a certain right triangle, the hypotenuse is 8 cm longer than the first leg and 1 cm longer than the second. Find the lengths of the sides of the triangle.

18. In a certain right triangle, the lengths of the two shorter sides are consecutive integers (measured in cm). Find the area of the triangle given that the sum of the lengths of the hypotenuse and longer leg is 50 cm.

19. When the sides of a square are each increased by 2 cm, the area increases by 14 cm$^2$. Find the length of a side in the original square.

20. In a certain rectangle, the length exceeds the width by 2 m. If the length is increased by 4 m and the width is decreased by 2 m, the area of the new rectangle is the same as that of the original rectangle. Find the dimensions of the original rectangle.

21. The perimeter and area of a rectangle are 104 cm and 640 cm$^2$, respectively. Find the length of the shorter side of the rectangle.

22. A rectangular picture with dimensions 5 in. by 8 in. is surrounded by a frame of uniform width. If the area of the frame (not including the space for the picture) is 114 in.$^2$ greater than the area of the picture, find the width of the frame.

23. A piece of wire 12 in. long is bent to form a right triangle in which the shortest side is 2 in. Find the lengths of the other two sides.

24. A piece of wire 24 cm long is cut into two pieces. Each piece is then formed into a square. If the sum of the areas of the two squares is 20 cm$^2$, find the lengths of the two pieces of wire.

25. In a certain isosceles triangle, each of the two equal legs is three times as long as the base. If the perimeter of the triangle is 70 cm, find the lengths of the sides.

26. In a certain triangle, the second angle is four times the first, while the third angle is 60° more than the first angle. Find the three angles. (Use the fact that the sum of the angles in any triangle is 180°.)

27. Suppose that you open two savings accounts. In the first you deposit $4000 at an annual simple interest rate of 7%. In the second you deposit $5000 at 9% (also simple interest). After how many years will the total interest amount to $5840?

28. You invest $600 in a stock paying a 6% annual dividend and $800 in a stock paying an 8% annual dividend. How many years will it take for the total earnings to equal the total initial investment?

29. If you deposit $10,000 at 11% (simple interest), how long will it be until the total amount in the account is $21,000?

30. Mr. X has \$5000 to deposit. He deposits part of this in an account earning 6%. The remainder is deposited in an account earning 8.5%. (Both rates are simple interest.) If the total interest earned each year is \$387.50, how much was initially deposited in each account?

31. You deposit \$7500 at 7.5% and \$4000 at 5%, simple interest. What is the **effective** interest rate? (In other words, what interest rate on \$11,500 would yield the same total interest at year's end?) Express your answer as a percent, rounded off to the nearest tenth of one percent.

32. You deposit \$2000 at 8% and \$3000 at 9%, simple interest. What additional amount would you need to invest at 12% to yield an overall return of 10% after one year?

33. Ms. X deposits \$8000 in one account and \$9000 in another. The interest rate on the \$8000 account is one percent less than that on the \$9000 account. If the \$9000 account earns \$165 more per year than the \$8000 account, find the interest rates on each account.

34. Jenny inherits \$42,900 and invests it for a year in two stocks. Stock $A$ earns 12.5% annually and stock $B$ earns 8.5% annually. How much was invested in each stock if stock $B$ earned three-quarters as much as stock $A$?

35. A student in a chemistry laboratory has available to her two acid solutions. The first is 10% acid and the second is 35% acid. (The percentages are by volume.) How many cm$^3$ of each solution should she mix to obtain 200 cm$^3$ of a 25% acid solution?

36. One salt solution is 15% salt and another is 20% salt. How many cm$^3$ of each solution must be mixed to obtain 50 cm$^3$ of a 16% salt solution?

37. A shopkeeper has two types of coffee beans. The first type sells for \$5.20/lb, the second for \$5.80/lb. How many pounds of the first type must be mixed with 5 lb of the second to produce a blend selling for \$5.35/lb?

38. How much pure copper must be melted with 300 lb of an alloy that is 20% copper to yield an alloy that is 37.5% copper? (The percentages are by weight.)

39. Two sources of iron ore are available to a steel company. The first source of ore contains 20% iron (by weight). The second source contains 35% iron. How many tons of the second ore must be added to 16 tons of the first ore to produce a mixture that is 25% iron?

40. How much water must be evaporated from 400 lb of a 10% salt solution to yield a solution that is 16% salt?

41. The two shorter sides of a right triangle differ by 7 cm. If the area is 60 cm$^2$, find the hypotenuse.

42. If the sides of a square are doubled, the area increases by 147 cm$^2$. Find the sides of the original square.

43. The area of a square with sides of length $1 - x$ is equal to the area of a rectangle with dimensions 1 by $x$. Find $x$.

44. In the accompanying figure, $\overline{PT}$ is tangent to the circle at $T$ and $PT = AB = 1$. Find $PB$. (Use the theorem from geometry stating that $PT^2 = PA \cdot PB$.)

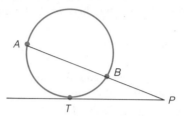

45. In the following figure, $AB = 1$ and $B$ divides $AC$ in such a way that the ratio of the whole segment to its larger part equals the ratio of the larger part to the smaller part. Find this ratio.

46. For a certain right circular cylinder, the height is 3 m and the total surface area is $8\pi$ m$^2$. Find the radius of the cylinder. (The formula for the total surface area $S$ is $S = 2\pi r^2 + 2\pi rh$, where $r$ is the radius and $h$ is the height.)

47. A ball is thrown straight up. Suppose that the height of the ball at time $t$ is given by $h = -16t^2 + 96t$, where $h$ is in feet and $t$ is in seconds, with $t = 0$ corresponding to the instant that the ball is first tossed.
    (a) How long does it take before the ball lands?
    (b) At what time is the height 80 ft? Why are there two answers here?

48. During a flu epidemic in a small town, a public health official finds that the total number $P$ of people who have caught the flu after $t$ days is closely approximated by the formula

$$P = -t^2 + 26t + 106 \qquad (1 \leq t \leq 13)$$

   (a) How many have caught the flu after 10 days?
   (b) After approximately how many days will 250 people have caught the flu?

49. Emily and Ernie are going to jog on a quarter-mile track. Emily can jog at a pace of 9 minutes per mile ($= \frac{1}{9}$ mile/min), while Ernie can jog at a pace of 11 minutes per mile ($= \frac{1}{11}$ mile/min). If they start together at the same time (running in the same direction), how long will it take for Emily to lap Ernie?

50. Consider the following motion problem. Suppose that you drive from town $A$ to town $B$ at 40 mph. At what

(constant) speed should you make the return trip so as to average 50 mph for the entire round trip? (The answer is not 60 mph.) Solve this problem by completing the steps that follow.

(a) Let $D$ denote the distance (in miles) from $A$ to $B$, and let $R$ denote the speed (in mph) on the return trip. Complete the table.

|  | d (MILES) | r (MPH) | t (H) |
|---|---|---|---|
| Going | $D$ |  |  |
| Returning |  | $R$ |  |

(b) Explain why the following equation is valid.

$$\frac{2D}{(D/40) + (D/R)} = 50$$

(c) Solve the equation in part (b) for $R$, and round off your result to the nearest mph.

**51.** The speed of an eastbound train is 5 mph greater than the speed of a westbound train. If the trains leave at the same time from stations 630 miles apart and pass each other after 6 hours, find the speed of the eastbound train.

**52.** Jim and Juan are going to drive their motorcycles from San Francisco to Los Angeles. Jim leaves at 8:00 A.M. but Juan can't leave until 8:30 A.M. If Jim drives at 50 mph and Juan at 55 mph, at what time will Juan catch up with Jim?

**53.** Two airplanes take off from the airport at the same time, heading in opposite directions. After 2 hours the planes are 1500 miles apart. If the speeds of the planes differ by 50 mph, find the speed of the slower plane.

**54.** Annie drives from town $A$ to town $B$ in 3 hours (at a fixed speed). The traffic is heavier on the return trip. As a result, Annie's speed is 16 mph less, and the return trip takes 4 hours. What were the speeds on both parts of the trip?

**55.** A charter boat captain has 2 hours in which to leave the dock, head out to sea as far as possible, and then return to the dock. The boat's speeds on the outward and return journeys are 20 mph and 12 mph, respectively. How far can the boat travel out to sea?

**C  56.** You invest $5200 in a stock paying a 6.75% annual dividend and $8400 in a stock paying a 7.85% annual dividend. How many years will it take for the total earnings to be $2500? (Round off your answer to the nearest one-half year.)

**C  57.** You deposit $2650 at 8.1% and $3300 at 9.2% simple interest. What additional amount would you need to deposit at 11.5% to yield an overall annual return of 10%? (Round off your answer to the nearest dollar.)

**C  58.** In a certain right triangle, the hypotenuse is 205 cm and the two legs differ by 23 cm. Find the length of the shorter leg.

**B**

**59.** At noon Larry and Linda start walking in the same direction. Larry's speed is $\frac{1}{15}$ mile/minute; Linda's is less than that. However, when Larry has walked 2 miles, Linda doubles her pace and catches up with Larry at 1:15 P.M. How fast was Linda walking initially?

**60.** The length of a rectangular sheet of metal is 4 in. more than the width. As indicated in the figure, 3-in. squares

are cut out from each corner of the rectangular sheet, and the sides are then folded up to form a box with no top. If the volume of the box is 180 in$^3$, determine the dimensions of the original rectangular sheet.

**61.** Suppose that you travel from town $A$ to town $B$ at a speed of $v_1$ mph and then return at a speed of $v_2$ mph. What is your average speed for the round trip (in terms of $v_1$ and $v_2$)?

**62.** If you travel for $a$ hours at $b$ mph and then $c$ hours at $d$ mph, what is your average speed for the trip?

**63.** In one savings account you deposit $d$ dollars at an annual simple interest rate of $r$%. At the same time you also deposit $D$ dollars at $R$% in a second account. After how many years will the total interest amount to $d + D$ dollars?

**64.** A chemistry student has two acid solutions available to him. The first is $a$% acid and the second is $b$% acid. (The percentages are by volume.) How many cm$^3$ of each solution should be mixed to obtain $d$ cm$^3$ of a solution that is $c$% acid? (Assume that $a$, $b$, $c$, and $d$ are positive and that $a < c < b$.)

65. A piece of wire $a$ inches long is bent to form a right triangle in which the shortest side is $b$ inches. Find the length of the hypotenuse.

66. How much water must be evaporated from $b$ pounds of an $a\%$ salt solution to obtain a solution that is $c\%$ salt? (Assume that the percentages are by weight and that $c > a$.)

67. If the length of each side of a square is multiplied by $a$, the area increases by $b^2$ square units. Find the sides of the original square.

68. A piece of wire $L$ inches long is cut into two pieces. Each piece is then bent to form a square. If the sum of the areas of the two squares is $5L^2/128$, how long are the two pieces of wire?

69. A piece of wire 16 cm long is cut into two pieces, and each piece is then bent to form a circle. If the sum of the areas of the two circles is $12\,\text{cm}^2$, how long is the shorter piece of wire? (Round off your answer to two decimal places.)

ⓒ 70. The height of a right circular cyclinder is 1 ft. Find the radius, given that the total surface area is $660\,\text{ft}^2$. (Round off your answer to two decimal places.)

*Exercises 71 and 72 are restatements of problems that were posed and solved in the ancient Chinese text "Nine Chapters on the Mathematical Art" (Chiu Chang Suan Shu), written sometime during the period 206 B.C.–221 A.D.*

71. Find the side of a square inscribed in a right triangle with legs $a$ and $b$. (See the accompanying figure.)

C

72. Find the diameter of a circle inscribed in a right triangle with legs $a$ and $b$.

## OTHER TYPES OF EQUATIONS

**2.4**

*If error is corrected wherever it is recognized as such, the path of error is the path of truth.*

HANS REICHENBACH (b. 1932)

The previous sections of this chapter dealt with linear and quadratic equations. In this section we consider several techniques that are useful in solving other types of equations. Throughout this section we'll concern ourselves only with solutions that are real numbers.

When $n$ is a natural number, equations of the form $x^n = a$ are solved simply by rewriting the equation in terms of the appropriate $n$th root or roots. For example, if $x^5 = 34$, then $x = \sqrt[5]{34}$. If the exponent $n$ is even, however, we have to remember that there are two real $n$th roots for each positive number. For instance, if

$$x^4 = 81$$

then

$$x = \pm\sqrt[4]{81} = \pm 3$$

**EXAMPLE 1** Solve: (a) $(x - 1)^4 = 15$    (b) $3x^4 = -48$

**Solution** (a) $(x - 1)^4 = 15$

$$x - 1 = \pm\sqrt[4]{15}$$
$$x = 1 \pm \sqrt[4]{15}$$

As you can check for yourself now, both of the values, $1 + \sqrt[4]{15}$ and $1 - \sqrt[4]{15}$, satisfy the given equation.

NOTE  You can estimate these values without a calculator. For instance, for $x = 1 + \sqrt[4]{15}$ we have

$$1 + \sqrt[4]{15} \approx 1 + \sqrt[4]{16} = 1 + 2 = 3$$

Therefore, $1 + \sqrt[4]{15}$ is a little less than 3. (In fact, a calculator shows that $1 + \sqrt[4]{15} \approx 2.97$.)

**(b)** For any real number $x$, the quantity $x^4$ is nonnegative. Therefore, the left-hand side of the equation is nonnegative, whereas the right-hand side is negative. Consequently, there are no real numbers satisfying the given equation.  ⚏

In Example 2 we solve two equations by factoring. As with quadratic equations, the factoring method is justified by the zero-product property of real numbers discussed in Section 2.2.

**EXAMPLE 2**  Solve:  **(a)** $3x^3 - 12x^2 - 15x = 0$    **(b)** $x^4 + x^2 - 6 = 0$

**Solution**  **(a)** $3x^3 - 12x^2 - 15x = 0$

$$3x(x^2 - 4x - 5) = 0$$
$$3x(x - 5)(x + 1) = 0$$

Now, by setting each factor equal to zero, we obtain the three values $x = 0$, $x = 5$, and $x = -1$. As you can check, each of these numbers indeed satisfies the original equation. So we have three solutions: 0, 5, and $-1$.

NOTE  If initially we'd divided both sides of the given equation by $3x$ to obtain $x^2 - 4x - 5 = 0$, then the solution $x = 0$ would have been overlooked.

**(b)**                 $x^4 + x^2 - 6 = 0$
$$(x^2 - 2)(x^2 + 3) = 0$$

| | |
|---|---|
| $x^2 - 2 = 0$ | $x^2 + 3 = 0$ |
| $x^2 = 2$ | $x^2 = -3$ |
| $x = \pm\sqrt{2}$ | |

By setting the first factor equal to zero, we've obtained the two values $x = \pm\sqrt{2}$. As you can check, these values do satisfy the given equation. When we set the second factor equal to zero, however, we obtained $x^2 = -3$, which has no real solutions (because the square of a real number must be nonnegative). We conclude therefore that the given equation in this case has just two real solutions, $\sqrt{2}$ and $-\sqrt{2}$.  ⚏

The equation $x^4 + x^2 - 6 = 0$ that we considered in Example 2(b) is called **an equation of quadratic type**. This means that with a suitable substitution, the resulting equation becomes quadratic. In particular here, suppose that we let

$$x^2 = t$$

and therefore

$$x^4 = t^2$$

Then the equation $x^4 + x^2 - 6 = 0$ becomes $t^2 + t - 6 = 0$, which is quadratic. This idea is exploited in the next three examples.

**EXAMPLE 3**   Solve:   $x^4 - 6x^2 + 4 = 0$

**Solution**   This equation cannot readily be solved by factoring. However, it is of quadratic type. We let $x^2 = t$. Then $x^4 = t^2$, and the equation becomes

$$t^2 - 6t + 4 = 0$$

The quadratic formula then yields

$$t = \frac{6 \pm \sqrt{36 - 4(4)}}{2} = \frac{6 \pm \sqrt{20}}{2}$$

$$= \frac{6 \pm 2\sqrt{5}}{2} = 3 \pm \sqrt{5}$$

Consequently, we have (in view of the definition of $t$)

$$x^2 = 3 \pm \sqrt{5}$$

and therefore

$$x = \pm\sqrt{3 \pm \sqrt{5}}$$

We've now found four values for $x$:

$$\sqrt{3 + \sqrt{5}} \qquad -\sqrt{3 + \sqrt{5}} \qquad \sqrt{3 - \sqrt{5}} \qquad -\sqrt{3 - \sqrt{5}}$$

As Exercise 70(a) asks you to check, each of these values indeed satisfies the original equation.

**EXAMPLE 4**   Solve:   $6x^{-2} - x^{-1} - 2 = 0$

**Solution**   We'll show two methods. The first method depends on recognizing the equation as one of quadratic type.

FIRST METHOD

Let $x^{-1} = t$. Then $x^{-2} = t^2$ and the equation becomes

$$6t^2 - t - 2 = 0$$
$$(3t - 2)(2t + 1) = 0$$

$$3t - 2 = 0 \quad \bigg| \quad 2t + 1 = 0$$
$$t = \frac{2}{3} \qquad \bigg| \qquad t = -\frac{1}{2}$$

If $t = \frac{2}{3}$, then $x^{-1} = \frac{2}{3}$. Therefore, $\frac{1}{x} = \frac{2}{3}$ and, consequently, $x = \frac{3}{2}$. On the other hand, if $t = -\frac{1}{2}$, then $\frac{1}{x} = -\frac{1}{2}$ and, consequently, $x = -2$. In summary, the two solutions are $\frac{3}{2}$ and $-2$.

ALTERNATE METHOD

Multiplying both sides of the given equation by $x^2$ yields

$$6 - x - 2x^2 = 0$$
$$2x^2 + x - 6 = 0$$
$$(2x - 3)(x + 2) = 0$$

$$2x - 3 = 0 \quad \bigg| \quad x + 2 = 0$$
$$x = \frac{3}{2} \qquad \bigg| \qquad x = -2$$

Thus we have $x = \frac{3}{2}$ or $x = -2$, as obtained using the first method.

**EXAMPLE 5**   Solve:   $4x^{4/3} + 15x^{2/3} - 4 = 0$

**Solution**   Let $x^{2/3} = t$. Then $x^{4/3} = t^2$, and the equation becomes

$$4t^2 + 15t - 4 = 0$$
$$(4t - 1)(t + 4) = 0$$

$$4t - 1 = 0 \qquad \bigg| \qquad t + 4 = 0$$
$$t = \frac{1}{4} \qquad \bigg| \qquad t = -4$$

We now have two cases to consider: $t = \frac{1}{4}$ and $t = -4$. With $t = \frac{1}{4}$ we have (in view of the definition of $t$)

$$x^{2/3} = \frac{1}{4}$$

$$x^2 = \frac{1}{64} \qquad \text{Two real numbers are equal if and only if their cubes are equal.}$$

$$x = \pm\frac{1}{8}$$

As Exercise 70(b) asks you to verify, the values $\frac{1}{8}$ and $-\frac{1}{8}$ satisfy the original equation. On the other hand, with $t = -4$, we have

$$x^{2/3} = -4$$

or

$$x^2 = -64 \qquad \text{cubing both sides}$$

This last equation has no real solutions, since the square of a real number is never negative. In summary, then, the only (real-number) solutions of the original equation are $\pm\frac{1}{8}$.

Some equations can be solved by raising both sides to the same power. In fact, if you review the solution given for Example 5, you'll see that we've already made use of this idea in cubing both sides of one of the equations there. As another example, consider the equation

$$\sqrt{x - 3} = 5$$

By squaring both sides of this equation we obtain $x - 3 = 25$ and, consequently, $x = 28$. As you can easily check, the value $x = 28$ does satisfy the original equation.

There is a complication, however, that may arise in raising both sides of an equation to the same power. Consider, for example, the equation

$$x - 2 = \sqrt{x}$$

By squaring both sides, we obtain

$$x^2 - 4x + 4 = x$$
$$x^2 - 5x + 4 = 0$$
$$(x - 4)(x - 1) = 0$$

Therefore

$$x = 4 \qquad \text{or} \qquad x = 1$$

The value $x = 4$ checks in the original equation, but the value $x = 1$ does not. (Verify this.) We say in this case that the value $x = 1$ is an **extraneous root** or **extraneous solution** of the original equation. [This extraneous root arises from the fact that if $a^2 = b^2$, then it need not be true that $a = b$. For instance, $(-3)^2 = 3^2$, but certainly $-3 \neq 3$.] In summary, any time that we square both sides of an equation (or raise both sides to an even integral power) there is the possibility of introducing extraneous roots. For this reason it is essential to check all candidates for solutions obtained in this manner.

Our final example for this section involves an equation containing several square root expressions. Before squaring both sides in such cases, it's usually a good idea to see if you can first isolate one of the radical expressions on one side of the equation.

**EXAMPLE 6** Solve: $\sqrt{9 + x} + \sqrt{1 + x} - \sqrt{x + 16} = 0$

**Solution**

$$\sqrt{9 + x} + \sqrt{1 + x} = \sqrt{x + 16} \qquad \text{adding } \sqrt{x + 16} \text{ to both sides in preparation for squaring}$$

$$(\sqrt{9 + x} + \sqrt{1 + x})^2 = (\sqrt{x + 16})^2$$

$$9 + x + 2\sqrt{9 + x}\sqrt{1 + x} + 1 + x = x + 16$$

$$2\sqrt{9 + x}\sqrt{1 + x} = 6 - x$$

$$(2\sqrt{9 + x}\sqrt{1 + x})^2 = (6 - x)^2$$

$$4(9 + 10x + x^2) = 36 - 12x + x^2 \qquad \text{(Why?)}$$

$$3x^2 + 52x = 0$$

$$x(3x + 52) = 0$$

$$x = 0 \qquad \Big| \qquad 3x + 52 = 0$$

$$x = -\tfrac{52}{3}$$

We've now found two possible solutions: 0 and $-\tfrac{52}{3}$. [Since these were obtained by squaring both sides of an equation, we need to check to see if either (or both) satisfy the original equation.] With $x = 0$, the original equation becomes $\sqrt{9} + \sqrt{1} - \sqrt{16} = 0$, or $3 + 1 - 4 = 0$, which is certainly true. On the other hand, with $x = -\tfrac{52}{3}$, the quantities underneath the radicals in the original equation are negative, and consequently the (real) square roots are undefined. So $-\tfrac{52}{3}$ is an extraneous solution, and $x = 0$ is the only solution of the given equation.

# EXERCISE SET 2.4

## A

*In Exercises 1–44, find all the real solutions of each equation.*

1. $3x^2 - 48x = 0$
2. $3x^3 - 48x = 0$
3. $t^3 - 125 = 0$
4. $t - t^3 = 0$
5. $7x^4 - 28x^2 = 0$
6. $y^4 - 81 = 0$
7. $225(x - 1) - x^2(x - 1) = 0$
8. $x^2(x + 4)^2 + 6x(x + 4)^2 + 9(x + 4)^2 = 0$
9. $4y^3 - 20y^2 + 25y = 0$
10. $y^4 + 27y = 0$
11. $t^4 + 2t^3 - 3t^2 = 0$
12. $2t^5 + 5t^4 - 12t^3 = 0$
13. $6x - 23x^2 - 4x^3 = 0$
14. $x^5 - 36x = 0$
15. $x^4 - x^2 - 6 = 0$
16. $x^4 - 5x^2 + 6 = 0$

17. $y^4 + 4y^2 - 5 = 0$

18. $y^4 + 6y^2 + 5 = 0$

19. $9t^4 - 3t^2 - 2 = 0$

20. $4t^4 - 5t^2 + 1 = 0$

21. $x^6 - 10x^4 + 24x^2 = 0$

22. $2x^5 - 15x^3 - 27x = 0$

23. $x^4 + x^2 - 1 = 0$

24. $x^4 - x^2 + 1 = 0$

25. $x^4 + 3x^2 - 2 = 0$

26. $2x^4 + x^2 + 2 = 0$

27. $x^6 + 7x^3 = 8$

28. $x^6 + 28x^3 + 27 = 0$

29. $t^{-2} - 7t^{-1} + 12 = 0$

30. $8t^{-2} - 17t^{-1} + 2 = 0$

31. $12y^{-2} - 23y^{-1} = -5$

32. $y^{-2} - y^{-1} = 0$

33. $4x^{-4} - 33x^{-2} - 27 = 0$

34. $x^{-4} + x^{-2} + 1 = 0$

35. $t^{2/3} = 9$

36. $t^{3/2} = 8$

37. $x^4 = 81$

38. $x^4 = -81$

39. $(y - 1)^3 = 7$

40. $(2y + 3)^4 = 5$

41. $(t + 1)^5 = -243$

42. $(t + 3)^4 = 625$

43. $9x^{4/3} - 10x^{2/3} + 1 = 0$

44. $x^{4/3} + 3x^{2/3} - 28 = 0$

*In Exercises 45–56, determine all the real number solutions for each equation.*

45. $\sqrt{1 - 3x} = 2$

46. $\sqrt{x^2 + 5x - 2} = 2$

47. $\sqrt{x^4 - 13x^2 + 37} = 1$

  *Hint:*  Let $x^2 = t$ and $x^4 = t^2$.

48. $\sqrt{y + 2} = y - 4$

49. $\sqrt{1 - 2x} + \sqrt{x + 5} = 4$

50. $\sqrt{x - 5} - \sqrt{x + 4} + 1 = 0$

51. $\sqrt{3 + 2t} + \sqrt{-1 + 4t} = 1$

52. $\sqrt{2t + 5} - \sqrt{8t + 25} + \sqrt{2t + 8} = 0$

53. $\sqrt{2y - 3} - \sqrt{3y + 3} + \sqrt{3y - 2} = 0$

54. $\sqrt{2x + 3} + \sqrt{x + 2} = 2$

55. $\sqrt{2x + 1} + \sqrt{x + 4} = 1$

56. $\sqrt{a - x} + \sqrt{b - x} = \sqrt{a + b - 2x}$   $(b > a > 0)$

*For Exercises 57–60, solve the equations and round off your answers to two decimal places.*

[c] 57. $x^4 - 2x^2 - 4 = 0$

[c] 58. $(3x - 5)^7 = 405$

[c] 59. $2x\sqrt{x^2 + 4} + 2x\sqrt{x^2 + 1} = 3$

[c] 60. The radii of three lead spheres are 10.01 cm, 15.43 cm, and 20.14 cm, respectively. Suppose that these three spheres are melted down and recast to form one large sphere. Find the radius of that sphere.  *Hint:*  Use the formula $V = \frac{4}{3}\pi r^3$ for the volume of a sphere.

## B

*In Exercises 61–69, find all the real solutions of each equation.*

61. $\sqrt{\sqrt{x} + \sqrt{a}} + \sqrt{\sqrt{x} - \sqrt{a}} = \sqrt{2\sqrt{x} + 2\sqrt{b}}$

62. $x = \sqrt{3x + x^2} - 3\sqrt{3x + x^2}$

63. $\dfrac{\sqrt{x} - a}{\sqrt{x}} - \dfrac{\sqrt{x} + a}{\sqrt{x} - b} = 0$   $(a > 0, b > 0)$

  *Suggestion:*  Let $\sqrt{x} = t$.

64. $x - \sqrt{x^2 - x} = \sqrt{x}$   $(x > 0)$

  *Suggestion:*  If $x \neq 0$, then you can divide through by $\sqrt{x}$ to obtain a simpler equation.

65. $\sqrt{x^2 - x - 1} - \dfrac{2}{\sqrt{x^2 - x - 1}} = 1$

  *Hint:*  Let $t = x^2 - x - 1$.

66. $\sqrt{x^2 + 3x - 4} - \sqrt{x^2 - 5x + 4} = x - 1$   $(x > 4)$

  *Hint:*  Factor the expressions beneath the radicals. Then note that $\sqrt{x - 1}$ is a factor of both sides of the equation.

67. $\sqrt{\dfrac{x - a}{x}} + 4\sqrt{\dfrac{x}{x - a}} = 5$   $(a \neq 0)$

  *Hint:*  Let $t = \dfrac{x - a}{x}$. Then $\dfrac{1}{t} = \dfrac{x}{x - a}$.

68. $\sqrt{p + 4q - 5t} + \sqrt{4p + q - 5t} = 3\sqrt{p + q - 2t}$

  (Assume that $p$ and $q$ are constants and $q > p$.)

69. $2\sqrt{\dfrac{x}{a}} + 3\sqrt{\dfrac{a}{x}} = \dfrac{b}{a} + \dfrac{6a}{b}$

70. (a) Verify that the four solutions obtained in Example 3 indeed satisfy the original equation.

  (b) Verify that the values $x = \frac{1}{8}$ and $x = -\frac{1}{8}$ satisfy the equation $4x^{4/3} + 15x^{2/3} - 4 = 0$.

## C

*For Exercises 71–75, find all real solutions of each equation. Exercises 74 and 75 are taken from* Higher Algebra *by H. S. Hall and S. R. Knight, first published in 1887.*

71. $\sqrt{8 + 2t} + \sqrt{5 + t} = \sqrt{15 + 3t}$

72. $-x = \sqrt{1 - \sqrt{1 + x}}$   *Hint:*  After squaring once and rearranging, look for a common factor appearing on both sides of the equation.

[c] 73. $\sqrt{x^2 + x} + \dfrac{1}{\sqrt{x^2 + x}} = \dfrac{5}{2}$   (Round off the answer to two decimal places.)

74. $\dfrac{\sqrt{x + 1} + \sqrt{x - 1}}{\sqrt{x + 1} - \sqrt{x - 1}} = \dfrac{4x - 1}{2}$

75. $\sqrt[n]{(a + x)^2} + 2\sqrt[n]{(a - x)^2} = 3\sqrt[n]{a^2 - x^2}$

## INEQUALITIES. MORE ABOUT ABSOLUTE VALUE

**2.5**

*The fundamental results of mathematics are often inequalities rather than equalities.*

E. BECKENBACH and R. BELLMAN in *An Introduction to Inequalities* (New York: Random House, 1961)

If we replace the equal sign in an equation with any one of the four symbols $<, \leq, >,$ or $\geq$, we obtain an **inequality**. As with equations in one variable, a real number is a **solution** of an inequality if we obtain a true statement when the variable is replaced by that real number. For example, the value $x = 5$ is a solution of the inequality $2x - 3 < 8$, because when $x = 5$ we have

$$2(5) - 3 < 8$$
$$7 < 8 \qquad \text{True}$$

We also say in this case that the value $x = 5$ **satisfies** the inequality. To **solve** an inequality means to find all of the solutions. The set of all solutions of an inequality is called (naturally enough) the **solution set**.

As we solve the inequalities in this section and the next, you'll see that the solution sets here are always one or more intervals of real numbers. In Section 1.3 we defined four types of bounded intervals:

$$(a, b) \qquad [a, b] \qquad (a, b] \qquad [a, b)$$

For our present purposes it is also convenient to introduce a notation for unbounded intervals of real numbers. As an example, we shall indicate the set of all real numbers that are greater than 2 with the notation $(2, \infty)$.

COMMENT AND CAUTION   The symbol $\infty$ is read *infinity*.* It is not a real number, and its use in the context $(2, \infty)$ is only to indicate that the interval has no right-hand boundary. In the box that follows we define the five types of unbounded intervals. Notice that the last interval in the table is actually the entire real number line.

| NOTATION | DEFINING INEQUALITY | GRAPH |
|---|---|---|
| $(a, \infty)$ | $x > a$ | |
| $[a, \infty)$ | $x \geq a$ | |
| $(-\infty, a)$ | $x < a$ | |
| $(-\infty, a]$ | $x \leq a$ | |
| $(-\infty, \infty)$ | | |

---

* The mathematician John Wallis (1616–1703) introduced the symbol $\infty$ in 1655. Some historians, knowing that Wallis was also a classical scholar, have suggested that he derived the infinity symbol from CIƆ, which was one of the Roman notations for 1000.

**EXAMPLE 1**  Indicate each set of real numbers on a number line.
(a)  $(-\infty, 4]$  (b)  $(-3, \infty)$

Solution  (a)  The interval $(-\infty, 4]$ consists of all real numbers that are less than or equal to 4. See Figure 1.
(b)  The interval $(-3, \infty)$ consists of all real numbers that are greater than $-3$. See Figure 2.

**FIGURE 1**

**FIGURE 2**

Recall that two equations are said to be equivalent if they have exactly the same solutions. Similarly, two inequalities are **equivalent** if they have the same solution set. Most of the procedures used in solving inequalities are quite similar to those for equalities. For example, adding or subtracting the same number on both sides of an inequality produces an equivalent inequality. We need to be careful, however, in multiplying or dividing both sides of an inequality by the same nonzero number. For instance, suppose that we start with the inequality $2 < 3$ and multiply both sides by 5. That yields $10 < 15$, which is certainly true. But if we multiply both sides of the inequality $2 < 3$ by $-5$, we obtain $-10 < -15$, which is false. In summary, multiplying both sides of an inequality by the same *positive* number preserves the inequality, whereas multiplying by a *negative* number reverses the inequality. In the following box, we list some of the principal properties of inequalities. In general, whenever we use properties 1 or 2 in solving an inequality, we obtain an equivalent inequality. Also, note that each property can be rewritten to reflect the fact that $a < b$ is equivalent to $b > a$. For example, Property 3 can just as well be written this way: If $b > a$ and $c > b$, then $c > a$.

| PROPERTY SUMMARY | Properties of Inequalities | |
|---|---|---|
| | PROPERTY | EXAMPLE |
| | 1. If $a < b$, then $a + c < b + c$ and $a - c < b - c$. | If $x - 3 < 0$, then $(x - 3) + 3 < 0 + 3$, and consequently $x < 3$. |
| | 2. (a) If $a < b$ and $c$ is positive, then $ac < bc$ and $a/c < b/c$. | If $\frac{1}{2}x < 4$, then $2(\frac{1}{2}x) < 2(4)$, and consequently $x < 8$. |
| | (b) If $a < b$ and $c$ is negative, then $ac > bc$ and $a/c > b/c$. | If $-\dfrac{x}{5} < 6$, then $(-5)\left(-\dfrac{x}{5}\right) > (-5)(6)$, and consequently $x > -30$. |
| | 3. The transitive property:  If $a < b$ and $b < c$, then $a < c$. | If $a < x$ and $x < 2$, then $a < 2$. |

**EXAMPLE 2**   Solve:   $2x - 3 < 5$

Solution   The work follows the same pattern we would use to solve the equation $2x - 3 = 5$. We have

$$2x - 3 < 5$$
$$2x < 8 \qquad \text{adding 3 to both sides}$$
$$x < 4 \qquad \text{dividing by 2}$$

FIGURE 3

From this sequence of steps, we see that the original inequality is equivalent to $x < 4$. The solution set is therefore $(-\infty, 4)$. Figure 3 shows this solution set on a number line.

**EXAMPLE 3**   Solve:   $4t + 8 \leq 7(1 + t)$

Solution   
$$4t + 8 \leq 7(1 + t)$$
$$4t + 8 \leq 7 + 7t$$
$$-3t \leq -1 \qquad \text{subtracting } 7t \text{ and 8 from both sides}$$
$$t \geq \frac{1}{3} \qquad \text{dividing by } -3 \text{ (this reverses the inequality)}$$

FIGURE 4

The solution set is therefore $[\frac{1}{3}, \infty)$. See Figure 4.

In the next example, we solve the inequality

$$-\frac{1}{2} < \frac{3 - x}{-4} \leq \frac{1}{2}$$

By definition, this is equivalent to the pair of inequalities

$$-\frac{1}{2} < \frac{3 - x}{-4} \qquad \text{and} \qquad \frac{3 - x}{-4} \leq \frac{1}{2}$$

One way to proceed here would be first to determine the solution set for each inequality. Then the set of real numbers common to both solution sets would be the solution set for the original inequality. However, the method shown in Example 4 is more efficient.

**EXAMPLE 4**   Solve:   $-\frac{1}{2} < \frac{3 - x}{-4} \leq \frac{1}{2}$

Solution   We begin by multiplying through by $-4$. Remember, this will reverse the inequalities:

$$2 > 3 - x \geq -2$$

Next, with a view toward isolating $x$, we subtract 3 to obtain

$$-1 > -x \geq -5$$

Finally, multiplying through by $-1$, we have

$$1 < x \leq 5$$

The solution set is therefore the interval $(1, 5]$.

**EXAMPLE 5**  Suppose that you inherit $5000 under the stipulation that it be invested in two stocks, A and B. Stock A pays an 8% annual dividend, and stock B pays a 9% annual dividend.* What is the largest amount that you can invest in stock A if the total dividends each year are to be at least $435?

**Solution**  First, just to get a feeling for the problem, let's see what would happen if the entire $5000 were invested in one stock. If you invested all $5000 in stock A, the annual dividend would be ($5000)(0.08)(1) = $400. That falls short of the required $435. On the other hand, if the $5000 were invested in stock B, the annual dividend would be ($5000)(0.09)(1), which is $450. That exceeds the required minimum of $435. The question the problem asks is, What is the most that can be invested in stock A if the total annual dividends are to be at least $435? Let $x$ denote the amount that can be invested in stock A. Then $5000 - x$ represents the amount (in dollars) to be invested in stock B. Since the combined dividends each year are to be at least $435, we have

$$x(0.08) + (5000 - x)(0.09) \geq 435$$
$$8x + (5000 - x)9 \geq 43500$$
$$8x + 45000 - 9x \geq 43500$$
$$-x \geq -1500$$
$$x \leq 1500$$

Thus, at most $1500 can be invested in stock A at 8%.

In the next example, reference is made to the Celsius and Fahrenheit scales for measuring temperature.[†] The formula relating the temperature readings on the two scales is

$$F = \frac{9}{5}C + 32$$

**EXAMPLE 6**  Over the temperature range $32° \leq F \leq 39.2°$ on the Fahrenheit scale, water contracts (rather than expands) with increasing temperature. What is the corresponding temperature range on the Celsius scale?

**Solution**

| | |
|---|---|
| $32 \leq F \leq 39.2$ | given |
| $32 \leq \frac{9}{5}C + 32 \leq 39.2$ | substituting $\frac{9}{5}C + 32$ for $F$ |
| $0 \leq \frac{9}{5}C \leq 7.2$ | subtracting 32 |
| $0 \leq C \leq \frac{5}{9}(7.2)$ | multiplying by $\frac{5}{9}$ |
| $0 \leq C \leq 4$ | |

---

* We compute dividends the same way we compute simple interest, by means of the formula $I = Prt$, where $I$ is the interest (or dividend), $P$ is the principal, $r$ is the rate, and $t$ is the time.

† The Celsius scale was devised in 1742 by the Swedish astronomer Anders Celsius. The Fahrenheit scale was first used by the German physicist Gabriel Fahrenheit in 1724.

Thus, a range of 32°F–39.2°F on the Fahrenheit scale corresponds to 0°C–4°C on the Celsius scale.   ▪▪

In preparation for the remaining examples in this section, we state the following useful theorem relating absolute values and inequalities.

| Theorem |
|---|
| If $a > 0$, then <br><br> $\quad \|u\| < a \qquad$ if and only if $\qquad -a < u < a$ <br> and <br> $\quad \|u\| > a \qquad$ if and only if $\qquad u < -a \quad$ or $\quad u > a$ |

FIGURE 5

You can see why this theorem is valid if you think in terms of distance and position on a number line. The condition $-a < u < a$ means that $u$ lies between $-a$ and $a$, as indicated in Figure 5. But this is the same as saying that the distance from $u$ to zero is less than $a$, which in turn can be written $|u| < a$. (The second part of the theorem can be justified in a similar manner.)

▪▪

**EXAMPLE 7**   Solve:   (a) $|x| < 1$   (b) $|x| \geq 1$

Solution   (a) According to the theorem just discussed, the condition $|x| < 1$ is equivalent to the inequality

$$-1 < x < 1$$

The solution set is therefore the open interval $(-1, 1)$.

(b) In view of the second part of the theorem, the inequality $|x| \geq 1$ is satisfied when $x$ satisfies either of the inequalities

$$x \leq -1 \qquad \text{or} \qquad x \geq 1$$

The solution sets for these last two inequalities are $(-\infty, -1]$ and $[1, \infty)$, respectively. Consequently, the solution set for the given inequality $|x| \geq 1$ consists of the two intervals $(-\infty, -1]$ and $[1, \infty)$.   ▪▪

In Example 7(b), we found that the solution set consisted of two intervals on the number line. There is a convenient notation for describing such sets. Given any two sets $A$ and $B$, we define the set $A \cup B$ (read $A$ **union** $B$) to be the set of all elements that are in $A$ or in $B$ (or in both). For example, if $A = \{1, 2, 3\}$ and $B = \{4, 5\}$, then $A \cup B = \{1, 2, 3, 4, 5\}$. As another example, the union of the two closed intervals $[3, 5]$ and $[4, 7]$ is given by

$$[3, 5] \cup [4, 7] = [3, 7]$$

because the numbers in the interval $[3, 7]$ are precisely those numbers that are in $[3, 5]$ or $[4, 7]$ (or in both). Using this notation, we can write the solution set for Example 7(b) as

$$(-\infty, -1] \cup [1, \infty)$$

**EXAMPLE 8**   Solve:  $|x - 3| < 1$

Solution   We'll show two methods.

FIRST METHOD   We use the theorem preceding Example 7. With $u = x - 3$ and $a = 1$, the theorem tells us that the given inequality is equivalent to

$$-1 < x - 3 < 1$$

or (by adding 3)

$$2 < x < 4$$

The solution set is therefore the open interval $(2, 4)$.

ALTERNATE METHOD   The given inequality tells us that $x$ must be less than one unit away from 3 on the number line. Looking one unit to either side of 3, then, we see that $x$ must lie strictly between 2 and 4. The solution set is therefore $(2, 4)$, as obtained using the first method.

**EXAMPLE 9**   Solve:  $\left|1 - \dfrac{t}{2}\right| \leq 5$

Solution   Referring again to the theorem, we use the fact that $|u| \leq a$ is equivalent to $-a \leq u \leq a$. In the present example, the given inequality is equivalent to

$$-5 \leq 1 - \frac{t}{2} \leq 5$$

$$-6 \leq -\frac{t}{2} \leq 4 \qquad \text{subtracting 1}$$

$$12 \geq t \geq -8 \qquad \text{multiplying by } -2 \text{ (remember, this reverses the inequalities)}$$

or (still equivalently)

$$-8 \leq t \leq 12$$

The solution set is therefore the closed interval $[-8, 12]$.

**EXAMPLE 10**   Solve:  $\left|1 - \dfrac{t}{2}\right| > 5$

Solution   FIRST METHOD   Applying the theorem, we have

$$1 - \frac{t}{2} < -5 \qquad \text{or} \qquad 1 - \frac{t}{2} > 5$$

$$-\frac{t}{2} < -6 \qquad\qquad\qquad -\frac{t}{2} > 4$$

$$t > 12 \qquad\qquad\qquad\qquad t < -8$$

This tells us that the given inequality is satisfied precisely when $t$ is in either of the intervals $(12, \infty)$ or $(-\infty, -8)$. In other words, the solution set is $(-\infty, -8) \cup (12, \infty)$.

ALTERNATE METHOD   The numbers satisfying the given inequality are precisely those numbers that *do not* satisfy $\left|1 - \dfrac{t}{2}\right| \leq 5$. But, as we've just seen in Example 9, the solution set for $\left|1 - \dfrac{t}{2}\right| \leq 5$ is the closed interval $[-8, 12]$. Therefore, the solution set for $\left|1 - \dfrac{t}{2}\right| > 5$ consists of those real numbers not in $[-8, 12]$. Thus, as indicated in Figure 6, the required solution set is $(-\infty, -8) \cup (12, \infty)$.

**FIGURE 6**

# EXERCISE SET 2.5

## A

*In Exercises 1–6, indicate each set of numbers on a number line.*

1. (a) $(0, \infty)$      (b) $[-2, 1] \cup (0, \infty)$
2. (a) $(-\infty, -5)$      (b) $(-\infty, -5) \cup (-6, 0]$
3. (a) $[1, \infty)$      (b) $[0, 1] \cup [1, \infty)$
4. $(-\infty, 1) \cup (-1, \infty)$
5. The set of all real numbers that are not in the open interval $(0, 1)$
6. The set of all real numbers that are not in the interval $(-\infty, 3)$

*Solve the inequalities in Exercises 7–38. (Specify the solution sets using interval notation.)*

7. $x + 5 < 4$
8. $2x - 7 < 11$
9. $1 - 3x \leq 0$
10. $6 - 4x \leq 22$
11. $4x - 3 > 33$
12. $5 - 8x > -19$
13. $\frac{3}{2}x + 4 \geq -1$
14. $1 - x \geq \pi$
15. $3x - 1 < 1 + x$
16. $4x + 6 < 3(x - 1) - x$
17. $2(t - 1) - 3(t + 1) \leq -5$
18. $1 - 2(t + 3) - t \leq 1 - 2t$
19. $t - 4[1 - (t - 1)] > 7 + 10t$
20. $(t - 1)(t - 2) \geq (t - 2)(t - 3)$
21. $(t + 4)(t - 4) \geq t(t + 3) - 1$
22. $(t + a)(t + b) < t^2$    $(a < 0$ and $b < 0)$
23. $\dfrac{3x}{5} - \dfrac{x - 1}{3} < 1$
24. $\dfrac{2x + 1}{2} + \dfrac{x - 1}{3} < x + \dfrac{1}{2}$
25. $\dfrac{x - 1}{4} - \dfrac{2x + 3}{5} \leq x$

26. $\dfrac{x}{2} - \dfrac{8x}{3} + \dfrac{x}{4} > \dfrac{23}{6}$
27. $3\left(\dfrac{x}{2} + 1\right) - 2\left(\dfrac{x}{3} - 1\right) \geq 12$
28. $2(1 - x) - \dfrac{x + 3}{4} \geq 1 + \dfrac{x}{2}$
29. $-2 \leq x - 6 \leq 0$
30. $-3 \leq 2x + 1 \leq 5$
31. $-6 < \dfrac{x + 4}{3} < -2$
32. $4 < 1 - t < 5$
33. $-1 \leq \dfrac{1 - 4t}{3} \leq 1$
34. $\dfrac{2}{3} \leq \dfrac{5 - 3t}{-2} \leq \dfrac{3}{4}$
35. $0.99 < \dfrac{x}{2} - 1 < 0.999$
36. $\dfrac{9}{10} < \dfrac{3x - 1}{-2} < \dfrac{91}{100}$
37. $1.86 < 4.11x - 0.15 < 1.95$ (Round off your answer to two decimal places.)   [c]
38. $2.713 < 1 - 3.332x < 3.001$ (Round off your answer to three decimal places.)   [c]
39. Suppose that you inherit $6000 under the stipulation that it be invested in two particular stocks, one paying a 7% dividend annually and the other paying 9%. What is the largest amount that you can invest in the stock paying the 7% dividend if you want the total yearly dividends to be at least $500?
40. Data from the Apollo 11 moon mission in July 1969 showed that temperature readings on the lunar surface vary over the interval $-183° \leq C \leq 112°$ on the Celsius scale. What is the corresponding interval on the Fahrenheit scale? (Round off the numbers you obtain to the nearest integers.)
41. From the cloud tops of Venus to the planet's surface, temperature readings range over the interval

$-25° \leq C \leq 475°$ on the Celsius scale. What is the corresponding range on the Fahrenheit scale?

42. Measurements sent back to Earth from the Viking I spacecraft in June 1976 indicated temperature readings on the surface of Mars in the range $-139° \leq C \leq -28°$. What is the corresponding range on the Fahrenheit scale? (Round off the numbers you obtain to the nearest integers.)

43. If an object is projected vertically upward with an initial velocity of $v_0$ ft/sec, then its velocity $v$ (in ft/sec) after $t$ sec is given by

$$v = -32t + v_0$$

If the initial velocity is 60 ft/sec, during what time interval will the velocity be in the range $50 \geq v \geq 40$? (Leave your answer in fractional form rather than in decimals.)

44. The width and length of a rectangle are 5 units and $x$ units, respectively.
    (a) Show that the perimeter $P$ is given by $P = 2x + 10$.
    (b) For which interval of $x$-values will the perimeter be in the range $100 \leq P \leq 200$?

57. $\left|\dfrac{x-2}{3}\right| < 4$

58. $\left|\dfrac{x+1}{2} - \dfrac{x-1}{3}\right| < 1$

59. $\left|\dfrac{3(x-2)}{4} + \dfrac{4(x-1)}{3}\right| \leq 2$

60. $|(x+1)^2 - x^2| < 0.01$

61. (a) $|(x+h)^2 - x^2| < 3h^2$ $(h > 0)$
    (b) $|(x+h)^2 - x^2| < 3h^2$ $(h < 0)$

62. (a) $|3(x+2)^2 - 3x^2| < \frac{1}{10}$
    (b) $|3(x+2)^2 - 3x^2| < \epsilon$ $(\epsilon > 0)$

**B**

[c] 63. Given two positive numbers $a$ and $b$, we define the **geometric mean**, the **arithmetic mean**, and the **root mean square** as follows:

$$\text{G.M.} = \sqrt{ab} \qquad \text{A.M.} = \frac{a+b}{2} \qquad \text{R.M.} = \sqrt{\frac{a^2+b^2}{2}}$$

Complete the table.

| $a$ | $b$ | $\sqrt{ab}$ (G.M.) | $\dfrac{a+b}{2}$ (A.M.) | $\sqrt{\dfrac{a^2+b^2}{2}}$ (R.M.) | WHICH IS LARGEST, G.M., A.M., OR R.M.? | WHICH IS SMALLEST, G.M., A.M., OR R.M.? |
|---|---|---|---|---|---|---|
| 1 | 2 | | | | | |
| 1 | 3 | | | | | |
| 1 | 4 | | | | | |
| 2 | 3 | | | | | |
| 3 | 4 | | | | | |
| 9 | 10 | | | | | |
| 99 | 100 | | | | | |
| 999 | 1000 | | | | | |

*Solve the inequalities in Exercises 45–62. (Specify the solution sets using interval notation.)*

45. $|x| < 6$

46. $|x| \leq \frac{1}{2}$

47. $|x| \geq 2$

48. $|x| > 0$

49. $|x| \geq 0$

50. $|x - 2| < 1$

51. $|x - 4| < 4$

52. $|1 - x| \leq 5$

53. $|3x + 5| < 17$

54. $|3x + 5| \geq 17$

55. $|4 - 5t| < \frac{1}{10}$

56. (a) $|x - a| < c$  (b) $|x - a| + b < c$

64. Prove that each of the following inequalities is valid for all positive numbers $a$ and $b$.  *Hint:* Use the following property of inequalities. If $x$ and $y$ are positive, then the inequality $x \leq y$ is equivalent to $x^2 \leq y^2$.

(a) $\sqrt{ab} \leq \dfrac{a+b}{2}$  (The geometric mean is less than or equal to the arithmetic mean.)

(b) $\dfrac{a+b}{2} \leq \sqrt{\dfrac{a^2+b^2}{2}}$  (The arithmetic mean is less than or equal to the root mean square.)

**65.** [c] **(a)** Complete the table, rounding off to four decimal places.

| $x$ | $y$ | $\dfrac{x}{y}+\dfrac{y}{x}$ | TRUE OR FALSE. $\dfrac{x}{y}+\dfrac{y}{x}\geq 2$ |
|---|---|---|---|
| 1 | 1 | | |
| 2 | 3 | | |
| 3 | 5 | | |
| 4 | 7 | | |
| 5 | 9 | | |
| 9 | 10 | | |
| 49 | 50 | | |
| 99 | 100 | | |

**(b)** Prove that for all positive numbers $x$ and $y$, we have $\dfrac{x}{y}+\dfrac{y}{x}\geq 2$. *Hint:* Use the inequality given in Exercise 64(a). Take $a=\dfrac{x}{y}$ and $b=\dfrac{y}{x}$.

**66.** Solve the inequality $|x-1|+|x-2|<3$. *Hint:* Begin by considering three cases: $x<1$; $1\leq x<2$; $x\geq 2$.

**67.** Suppose that $|x-2|<0.1$ and $|y-3|<0.01$. Show that $|xy-6|<0.321$. *Hint:* First check that $xy-6=x(y-3)+3(x-2)$. Then use the triangle inequality (page 18).

**C**

**68.** Let $a$ and $b$ be positive numbers. In Exercise 64 you were asked to show that

$$\sqrt{ab}\leq\frac{a+b}{2}\leq\sqrt{\frac{a^2+b^2}{2}}$$

This exercise shows how to establish these important inequalities using geometric, rather than algebraic,

methods. In the following figure, $\overline{CF}$ is the diameter of a semicircle with center $E$. Let $CD=a$ and $DF=b$.

**(a)** Explain why $CE=EH=\dfrac{a+b}{2}$.

**(b)** Show that $DG=\sqrt{ab}$ and $DH=\sqrt{\dfrac{a^2+b^2}{2}}$.

**(c)** Explain why $DG\leq EH\leq DH$, and conclude from this that $\sqrt{ab}\leq(a+b)/2\leq\sqrt{(a^2+b^2)/2}$.

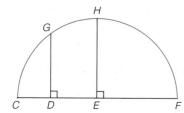

**69.** Let $a$, $b$, $c$, and $d$ be positive real numbers. Use the inequality in Exercise 64(a) to prove each of the following inequalities.

**(a)** $\sqrt{abcd}\leq\dfrac{ab+cd}{2}$

**(b)** $\sqrt{ab}+\sqrt{cd}\leq\sqrt{(a+c)(b+d)}$

**70.** Let $x$ denote the width of a rectangle whose perimeter is 30 ft.

**(a)** Show that the area $A$ (in square feet) of the rectangle is given by $A=x(15-x)$.

**(b)** Use the inequality in Exercise 64(a) to show that $A\leq 225/4$.

**(c)** For which value of $x$ is $A=225/4$? What are the dimensions of the rectangle in this case?

**71.** Let $x$ denote the width of a rectangle whose area is 25 ft$^2$.

**(a)** Show that the perimeter $P$ (in feet) of the rectangle is given by $P=2x+(50/x)$.

**(b)** Use the inequality in Exercise 64(a) to show that $P\geq 20$.

**(c)** For which value of $x$ is $P=20$? What are the dimensions of the rectangle in this case?

## MORE ON INEQUALITIES

**2.6**

$\left(-\infty,\frac{5}{2}\right)\quad\frac{5}{2}\quad\left(\frac{5}{2},\infty\right)$

**FIGURE 1**

In this section we are going to solve inequalities involving polynomials and quotients of polynomials. Some observations about linear polynomials and inequalities will be helpful in introducing the main ideas. Consider, for example, the linear inequality $2x-5<0$, along with its solution set $\left(-\infty,\frac{5}{2}\right)$. Let us call a solution of the corresponding equality $2x-5=0$ a **key number** for the given inequality. So, in this case, the only key number is $x=\frac{5}{2}$. As indicated in Figure 1, the key number $\frac{5}{2}$ divides the number line into two intervals, namely, $\left(-\infty,\frac{5}{2}\right)$ and $\left(\frac{5}{2},\infty\right)$. Notice that on each of the intervals determined by the key number, the algebraic sign of $2x - 5$ is constant. (That is, for $x < \frac{5}{2}$ the value of $2x - 5$ is always negative; for $x > \frac{5}{2}$ the value of $2x - 5$ is always positive.) More generally, it can be shown that this same type of behavior regarding *persistence of sign* occurs with all polynomials, and indeed with quotients of polynomials as well. This important fact, along with the definition of a key number, is presented in the box that follows.

---

### Key Numbers and Persistence of Sign

Let $P$ and $Q$ be polynomials with no common factors (other than constants), and consider the following four inequalities:

$$\frac{P}{Q} < 0 \qquad \frac{P}{Q} \leq 0 \qquad \frac{P}{Q} > 0 \qquad \frac{P}{Q} \geq 0$$

The **key numbers** for each of these inequalities are the real numbers for which $P = 0$ or $Q = 0$. It can be proved that the algebraic sign of $P/Q$ is constant on each of the intervals determined by these key numbers.

---

We'll show how this result is applied by solving the inequality

$$x^3 - 2x^2 - 3x > 0$$

First, using factoring techniques from Section 1.9, we rewrite the inequality in the equivalent form

$$x(x + 1)(x - 3) > 0$$

The key numbers then are the solutions of the equation $x(x + 1)(x - 3) = 0$. That is, the key numbers are $x = 0, x = -1$, and $x = 3$. Next, we locate these numbers on a coordinate line. As indicated in Figure 2, this divides the number line into four distinct intervals.

**FIGURE 2**

Now, according to the result stated in the box just prior to this example, no matter what $x$-value we choose in the interval $(-\infty, -1)$, the resulting sign of $x^3 - 2x^2 - 3x \; [= x(x + 1)(x - 3)]$ will always be the same. Thus, to see what that sign is, we first choose any convenient *test number* in the interval $(-\infty, -1)$, say, $x = -2$. Then, using $x = -2$, we determine the sign of $x(x + 1)(x - 3)$ simply by considering the sign of each factor, as indicated in Table 1(a).

---

TABLE 1(a)

On the interval $(-\infty, -1)$, the sign of $x^3 - 2x^2 - 3x \; [= x(x + 1)(x - 3)]$ is negative because it is the product of three negative factors.

| INTERVAL | TEST NUMBER | $x$ | $x + 1$ | $x - 3$ | $x(x + 1)(x - 3)$ |
|---|---|---|---|---|---|
| $(-\infty, -1)$ | $-2$ | neg. | neg. | neg. | neg. |

From Table 1(a), we conclude that the values of $x^3 - 2x^2 - 3x$ are negative *throughout* the interval $(-\infty, -1)$, and consequently none of the numbers in this interval satisfy the given inequality. Next, we carry out similar analyses for the remaining three intervals, as shown in Table 1(b). (You should verify for yourself that the entries in the table are correct.)

TABLE 1(b)

On the interval $(-1, 0)$, the product $x(x + 1)(x - 3)$ is positive because there are two negative factors and one positive factor. On $(0, 3)$, the product is negative since there are two positive factors and one negative factor. And on $(3, \infty)$, the product is positive because all of the factors are positive.

| INTERVAL | TEST NUMBER | $x$ | $x + 1$ | $x - 3$ | $x(x + 1)(x - 3)$ |
|----------|-------------|-----|---------|---------|-------------------|
| $(-1, 0)$ | $-1/2$ | neg. | pos. | neg. | pos. |
| $(0, 3)$ | $1$ | pos. | pos. | neg. | neg. |
| $(3, \infty)$ | $4$ | pos. | pos. | pos. | pos. |

Looking at the two tables now, we conclude that the sign of $x(x + 1)(x - 3)$ is positive throughout both of the intervals $(-1, 0)$ and $(3, \infty)$, and consequently all of the numbers in these intervals satisfy the given inequality. Moreover, our work also shows that the other two intervals that we considered are not part of the solution set. As you can readily check, the key numbers themselves do not satisfy the given inequality for this example. In summary, then, the solution set for the inequality $x^3 - 2x^2 - 3x > 0$ is $(-1, 0) \cup (3, \infty)$. Furthermore, it is important to notice that the work just carried out also provides us with three additional pieces of information:

The solution set for $x^3 - 2x^2 - 3x \geq 0$ is $[-1, 0] \cup [3, \infty)$;
The solution set for $x^3 - 2x^2 - 3x < 0$ is $(-\infty, -1) \cup (0, 3)$;
The solution set for $x^3 - 2x^2 - 3x \leq 0$ is $(-\infty, -1] \cup [0, 3]$.

In the box that follows, we summarize the steps for solving polynomial inequalities.

---

**Steps for Solving Polynomial Inequalities**

1. If necessary, rewrite the inequality so that the polynomial is on the left-hand side and zero is on the right-hand side.

2. Find the key numbers for the inequality and locate them on a number line.

3. List the intervals determined by the key numbers.

4. From each interval, choose a convenient test number. Then use the test number to determine the sign of the polynomial throughout the interval.

5. Use the information obtained in the previous step to specify the required solution set. [Don't forget to take into account whether the original inequality is strict ($<$ or $>$) or nonstrict ($\leq$ or $\geq$).]

**EXAMPLE 1**   Solve:   $x^4 \le 14x^3 - 48x^2$

Solution   First, we rewrite the inequality so that 0 is on the right-hand side. Then we factor the left-hand side as follows:

$$x^4 - 14x^3 + 48x^2 \le 0$$
$$x^2(x^2 - 14x + 48) \le 0$$
$$x^2(x - 6)(x - 8) \le 0$$

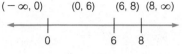

FIGURE 3

From this last line, we see that the key numbers are $x = 0$, 6, and 8. As indicated in Figure 3, these numbers divide the number line into four distinct intervals. We need to choose a test number from each interval and see whether the polynomial is positive or negative on the interval. This work is carried out in the following table.

| INTERVAL | TEST NUMBER | $x^2$ | $x - 6$ | $x - 8$ | $x^2(x - 6)(x - 8)$ |
|----------|-------------|-------|---------|---------|---------------------|
| $(-\infty, 0)$ | $-1$ | pos. | neg. | neg. | pos. |
| $(0, 6)$ | $1$ | pos. | neg. | neg. | pos. |
| $(6, 8)$ | $7$ | pos. | pos. | neg. | neg. |
| $(8, \infty)$ | $9$ | pos. | pos. | pos. | pos. |

The table shows that the quantity $x^2(x - 6)(x - 8)$ is negative only for $x$-values in the interval $(6, 8)$. Also, as we noted at the start, the quantity is equal to zero when $x = 0$, 6, or 8. Thus, the solution set consists of the numbers in the closed interval $[6, 8]$, along with the number 0. We can write this set

$$[6, 8] \cup \{0\}$$

where $\{0\}$ denotes the set whose only member is zero.

**EXAMPLE 2**   Solve:   **(a)** $x^2 - x + 15 > 0$     **(b)** $x^2 - x + 15 < 0$

Solution   **(a)** As you can readily check, the equation $x^2 - x + 15 = 0$ has no real solutions. So there are no key numbers, and consequently the polynomial $x^2 - x + 15$ never changes sign. To see what that sign is, choose the most convenient test number, namely, $x = 0$, and evaluate the polynomial: $0^2 - 0 + 15 > 0$. So the polynomial is positive for every value of $x$, and the solution set is $(-\infty, \infty)$, the set of all real numbers.

**(b)** Our work in part (a) shows that there are no real numbers satisfying the inequality $x^2 - x + 15 < 0$.

The technique used in the previous examples can also be used to solve inequalities involving quotients of polynomials. For these cases recall that the definition of a key number also includes the $x$-values for which the denominator is zero. For example, the key numbers for the inequality $\dfrac{x + 3}{x - 4} \ge 0$ are $-3$ and 4.

**EXAMPLE 3** Solve: $\dfrac{x+3}{x-4} \geq 0$

Solution The key numbers are $-3$ and $4$. As indicated in Figure 4, these numbers divide the number line into three intervals. In the table that follows, we've chosen a test number from each interval and determined the sign of the quotient $\dfrac{x+3}{x-4}$ for each interval.

$(-\infty, -3)$ $\quad$ $(-3, 4)$ $\quad$ $(4, \infty)$

$-3 \qquad 4$

FIGURE 4

| INTERVAL | TEST NUMBER | $x+3$ | $x-4$ | $\dfrac{x+3}{x-4}$ |
|---|---|---|---|---|
| $(-\infty, -3)$ | $-4$ | neg. | neg. | pos. |
| $(-3, 4)$ | $0$ | pos. | neg. | neg. |
| $(4, \infty)$ | $5$ | pos. | pos. | pos. |

From these results, we conclude that the solution set for $\dfrac{x+3}{x-4} \geq 0$ contains the two intervals $(-\infty, -3)$ and $(4, \infty)$. However, we still need to consider the two endpoints $-3$ and $4$. As you can easily check, the value $x = -3$ does satisfy the given inequality, but $x = 4$ does not. In summary, then, the solution is $(-\infty, -3] \cup (4, \infty)$.

**EXAMPLE 4** Solve: $\dfrac{2x-1}{x-1} - \dfrac{7}{x+1} < 1$

Solution Your first inclination here might be to multiply through by $(x-1)(x+1)$ to eliminate fractions. The problem with that is that we don't know whether the quantity $(x-1)(x+1)$ is positive or negative. Thus, we begin by rewriting the inequality in an equivalent form, with zero on the right-hand side and a single fraction on the left-hand side:

$$\frac{2x-1}{x-1} - \frac{7}{x+1} - 1 < 0$$

$$\frac{2x-1}{x-1} \cdot \frac{x+1}{x+1} - \frac{7}{x+1} \cdot \frac{x-1}{x-1} - 1 \cdot \frac{(x-1)(x+1)}{(x-1)(x+1)} < 0$$

$$\frac{2x^2 + x - 1 - 7(x-1) - (x^2 - 1)}{(x-1)(x+1)} < 0$$

$$\frac{x^2 - 6x + 7}{(x-1)(x+1)} < 0$$

The key numbers are those $x$-values for which the denominator or the numerator is zero. By inspection, the denominator is zero when $x = 1$ or $-1$. On the other hand, if the numerator is zero, we have $x^2 - 6x + 7 = 0$, and consequently (using the quadratic formula),

$$x = \frac{6 \pm \sqrt{36 - 28}}{2} = \frac{6 \pm \sqrt{8}}{2}$$

$$= \frac{6 \pm 2\sqrt{2}}{2} = 3 \pm \sqrt{2}$$

The four key numbers are therefore $\pm 1$ and $3 \pm \sqrt{2}$. As Figure 5 indicates, these numbers divide the number line into five distinct intervals. Now, just as in the previous examples, we choose a test number from each interval and determine the sign of the quotient for that interval. (You should check each entry in the following table for yourself.)

$$(-\infty, -1) \quad (-1, 1) \quad (1, 3 - \sqrt{2}) \quad (3 - \sqrt{2}, 3 + \sqrt{2}) \quad (3 + \sqrt{2}, \infty)$$

**FIGURE 5**

| INTERVAL | TEST NUMBER | $x^2 - 6x + 7$ | $x - 1$ | $x + 1$ | $\dfrac{x^2 - 6x + 7}{(x - 1)(x + 1)}$ |
|---|---|---|---|---|---|
| $(-\infty, -1)$ | $-2$ | pos. | neg. | neg. | pos. |
| $(-1, 1)$ | $0$ | pos. | neg. | pos. | neg. |
| $(1, 3 - \sqrt{2})$ | $\frac{3}{2}$ | pos. | pos. | pos. | pos. |
| $(3 - \sqrt{2}, 3 + \sqrt{2})$ | $3$ | neg. | pos. | pos. | neg. |
| $(3 - \sqrt{2}, \infty)$ | $5$ | pos. | pos. | pos. | pos. |

From these results we can see that the quotient $\dfrac{x^2 - 6x + 7}{(x - 1)(x + 1)}$ is negative on the two intervals $(-1, 1)$ and $(3 - \sqrt{2}, 3 + \sqrt{2})$. Now we need to check the endpoints of these two intervals. When $x = 3 \pm \sqrt{2}$, the quotient is zero, and so (in view of the original inequality) we exclude those two $x$-values from the solution set. Furthermore, the quotient is undefined when $x = \pm 1$, so we must also exclude those two values from the solution set. In summary, then, the solution set is $(-1, 1) \cup (3 - \sqrt{2}, 3 + \sqrt{2})$. ▪

# EXERCISE SET 2.6

## A

*Solve the inequalities in Exercises 1–54. Suggestion: A calculator is useful for approximating the key numbers in Exercises 19, 20, 33, 42, and 44.*

1. $x^2 + x - 6 < 0$
2. $x^2 + 4x - 32 < 0$
3. $x^2 - 11x + 18 > 0$
4. $2x^2 + 7x + 5 > 0$
5. $9x - x^2 \leq 20$
6. $3x^2 + x \leq 4$
7. $x^2 - 16 \geq 0$
8. $24 - x^2 \geq 0$
9. $16x^2 + 24x < -9$
10. $x^4 - 16 < 0$
11. $x^3 + 13x^2 + 42x > 0$
12. $2x^3 - 9x^2 + 4x \geq 0$
13. $225x \leq x^3$
14. $81x^2 \leq x^6$
15. $2x^2 + 1 \geq 0$
16. $1 + x^2 < 0$
17. $12x^3 + 17x^2 + 6x < 0$
18. $8x^4 < x^2 - 2x^3$
19. $x^2 + x - 1 > 0$
20. $2x^2 + 9x - 1 > 0$

21. $x^2 - 8x + 2 \leq 0$
22. $3x^2 - x + 5 \leq 0$
23. $(x - 1)(x + 3)(x + 4) \geq 0$
24. $x^4(x - 2)(x - 16) \geq 0$
25. $(x + 4)(x + 5)(x + 6) < 0$
26. $(x - \frac{1}{2})(x + \frac{1}{2})(x + \frac{3}{2}) < 0$
27. $(x - 2)^2(3x + 1)^3(3x - 1) > 0$
28. $(2x - 1)^3(2x - 3)^5(2x - 5) > 0$
29. $(x - 3)^2(x + 1)^4(2x + 1)^4(3x + 2) \leq 0$
30. $x^4 - 25x^2 + 144 \leq 0$
31. $x^4 - 9x^2 + 20 \geq 0$
32. $x(1 - x^2)^4 + (x + 3)(1 - x^2)^4 \geq 0$
33. $(x - 4)(2x^2 - 6x - 1) < 0$
34. $(x + 2)^3(x^2 - 4x - 2) < 0$
35. $x^3 + 2x^2 - x - 2 > 0$

36. $2x^4 + x^3 - 16x - 8 > 0$

37. $\dfrac{x-1}{x+1} \le 0$

38. $\dfrac{x+4}{2x-5} \le 0$

39. $\dfrac{2-x}{3-2x} \ge 0$

40. $\dfrac{x^2-1}{x^2+8x+15} \ge 0$

41. $\dfrac{x^2-8x-9}{x} < 0$

42. $\dfrac{x^2-3x+1}{1-x} < 0$

43. $\dfrac{2x^3+5x^2-7x}{3x^2+7x+4} > 0$

44. $\dfrac{x^2-x-1}{x^2+x-1} > 0$

45. $\dfrac{x+2}{x+5} \le 1$

46. $\dfrac{2}{x-4} - \dfrac{1}{x-3} \ge \dfrac{2}{3}$

47. $\dfrac{1}{x-2} - \dfrac{1}{x-1} \ge \dfrac{1}{6}$

48. $\dfrac{2x}{x+5} + \dfrac{x-1}{x-5} < \dfrac{1}{5}$

49. $\dfrac{1+x}{1-x} - \dfrac{1-x}{1+x} < -1$

50. $\dfrac{x+1}{x+2} > \dfrac{x-3}{x+4}$

51. $\dfrac{3-2x}{3+2x} > \dfrac{1}{x}$

52. $\dfrac{x}{x-2} - \dfrac{3}{x+1} \ge 2$

53. $1 + \dfrac{1}{x} \ge \dfrac{1}{1+x}$

54. $x - \dfrac{10}{x-1} \ge 4$

*For Exercises 55 and 56, determine the domain of the variable in each expression. (Recall the domain convention from Section 1.8: The domain is the set of all real number values for the variable for which the expression is defined.)*

55. (a) $\sqrt{x^2-4x-5}$    (b) $\sqrt{1/(x^2-4x-5)}$

56. (a) $\sqrt{\dfrac{x+2}{x-4}}$    (b) $\sqrt[3]{\dfrac{x+2}{x-4}}$

**B**

57. For which values of $b$ will the equation $x^2 + bx + 1 = 0$ have real solutions?

58. The sum of the first $n$ natural numbers is given by

$$1 + 2 + 3 + \cdots + n = \dfrac{n(n+1)}{2}$$

For which values of $n$ will the sum be less than 1225?

59. For which values of $a$ is $x = 1$ a solution of the inequality $\dfrac{2a+x}{x-2a} < 1$?

60. Solve: $\dfrac{ax+b}{\sqrt{x}} > 2\sqrt{ab}$    $(a > 0, b > 0)$

61. The two shorter sides in a right triangle have lengths $x$ and $1 - x$ $(x > 0)$. For which values of $x$ will the hypotenuse be less than $\sqrt{17}/5$?

62. A piece of wire 12 cm long is cut into two pieces. Denote the lengths of the two pieces by $x$ and $12 - x$. Both pieces are then bent into squares. For which values of $x$ will the combined areas of the squares exceed $5 \text{ cm}^2$?

63. Let $V$ and $S$ denote the volume and total surface area, respectively, for a right circular cylinder of radius $r$ and height 1. For which $r$-values will the ratio $V/S$ be less than $\frac{1}{3}$?

64. Let $V$ and $S$ denote the volume and total surface area, respectively, for a right circular cone of radius $r$ and height 1. For which $r$-values will the ratio $V/S$ be less than $\frac{4}{27}$?

65. (a) [c] Complete the table. (Round off the entries in the right-hand column to two decimal places.)

| $x$ | $x^2 + 1000$ | $2x^2 + x$ | $\dfrac{x^2+1000}{2x^2+x}$ |
|---|---|---|---|
| 1 | | | |
| 2 | | | |
| 5 | | | |
| 10 | | | |
| 100 | | | |
| 200 | | | |
| 10,000 | | | |

(b) For which values of $x$ will we have

$$\dfrac{x^2+1000}{2x^2+x} < 0.5?$$

(c) What is the smallest *natural number* $n$ for which

$$\dfrac{n^2+1000}{2n^2+n} < 0.5?$$

**C**

66. Find a nonzero value for $c$ so that the solution set for the inequality

$$x^2 + 2cx - 6c < 0$$

is the open interval $(-3c, c)$.

67. Solve $(x-a)^2 - (x-b)^2 > \dfrac{(a-b)^2}{4}$. Assume that $a > b$.

68. Solve $x^3 + \dfrac{1}{x^3} \ge 3$. (Use a calculator to approximate the key numbers.)

## CHAPTER 2 SUMMARY OF PRINCIPAL TERMS AND NOTATION

| TERMS OR NOTATIONS | PAGE REFERENCE | COMMENTS |
|---|---|---|
| 1. Linear (or first-degree) equation in one variable | 68 | These are equations that can be written in the form $ax + b = 0$, with $a \neq 0$. |
| 2. Solution (or root) of an equation | 68 | A solution or root is a number that, when substituted for the variable in an equation, yields a true statement. |
| 3. Equivalent equations | 69 | Two equations are equivalent if they have the same set of solutions. |
| 4. Quadratic equation | 74 | A quadratic equation is an equation that can be written in the form $$ax^2 + bx + c = 0$$ where $a \neq 0$. |
| 5. Zero-product property | 75 | This property of real numbers (and of complex numbers in general) can be stated as follows. If $pq = 0$ then $p = 0$ or $q = 0$; conversely, if $p = 0$ or $q = 0$, then $pq = 0$. (We used this property in solving quadratic equations by factoring.) |
| 6. Quadratic formula $$x = \frac{-b \pm \sqrt{b^2 - 4ac}}{2a}$$ | 77 | This is the quadratic formula; it provides the solutions of the quadratic equation $ax^2 + bx + c = 0$. The formula is derived on page 77 by means of the useful technique of completing the square. |
| 7. Discriminant | 79 | The discriminant of the quadratic equation $ax^2 + bx + c = 0$ is the number $b^2 - 4ac$. As indicated in the box on page 79, the discriminant provides information about the roots of the equation. |
| 8. Extraneous root | 94 | In solving equations, certain processes (such as squaring both sides) can lead to answers that do not check in the original equation. Such answers are called extraneous roots; they are not solutions of the original equation. |
| 9. Solution set of an inequality | 96 | This is the set of numbers that satisfy the inequality. |
| 10. $(a, \infty)$ | 96 | The unbounded interval $(a, \infty)$ consists of all real numbers $x$ such that $x > a$. The infinity symbol, $\infty$, does not denote a real number. It is used in the context $(a, \infty)$ to indicate that the interval has no right-hand boundary. For the definitions of the other types of unbounded intervals, see the box on page 96. |
| 11. $A \cup B$ | 100 | The set $A \cup B$ consists of all elements that are in at least one of the two sets $A$ and $B$. |
| 12. Key numbers of an inequality | 104 | Let $P$ and $Q$ denote polynomials and consider the following four inequalities: $$\frac{P}{Q} < 0 \qquad \frac{P}{Q} \leq 0 \qquad \frac{P}{Q} > 0 \qquad \frac{P}{Q} \geq 0$$ The key numbers for each of these inequalities are the real numbers for which $P = 0$ or $Q = 0$. It can be proved that the algebraic sign of $P/Q$ is constant on each of the intervals determined by these key numbers. |

## CHAPTER 2   REVIEW EXERCISES

NOTE   The fifteen exercises in this set with colored problem numbers constitute a chapter test.

*In Exercises 1–12, answer T if the statement is true without exception. Otherwise, answer F.*

**1.** If $x < 3$, then $x + 7 < 10$.

**2.** If $x < y$, then $x^2 < y^2$.

**3.** If $x \geq -4$, then $x \leq 4$.

**4.** If $-x > y$, then $x < -y$.

**5.** If $x < 2$, then $\dfrac{1}{x} < \dfrac{1}{2}$        **6.** $x \leq x^2$

**7.** If $\sqrt{x + 1} = 3$, then $x + 1 = 9$.

**8.** If $0 < x < 1$, then $\dfrac{1}{x} > x$.

**9.** If $x < 2$ and $y < 3$, then $x < y$.

**10.** If $x < 2$ and $y > 3$, then $x < y$.

**11.** If $a - b \leq a^2 - b^2$, then $1 \leq a + b$.

**12.** If $0 < a - b \leq a^2 - b^2$, then $1 \leq a + b$.

*In Exercises 13–64, find all the real solutions of each equation.*

**13.** $5 - 9x = 2$

**14.** $\dfrac{x}{3} - \dfrac{3x}{5} = \dfrac{x}{6} - 13$

**15.** $(t - 4)(t + 3) = (t + 5)^2$

**16.** $\dfrac{1 - x}{1 + x} = 1$

**17.** $\dfrac{2t - 1}{t + 2} = 5$

**18.** $\dfrac{x - 2}{3} - \dfrac{4(x + 1)}{2} + \dfrac{1}{6} = 0$

**19.** $\dfrac{2}{x + 4} - \dfrac{1}{x - 4} = \dfrac{-7}{x^2 - 16}$

**20.** $\dfrac{1}{1 + \dfrac{1}{x + 1}} = \dfrac{5}{6}$

**21.** $\dfrac{2y - 5}{4y + 1} = \dfrac{y - 1}{2y + 5}$

**22.** $\dfrac{2x - 3}{x - 2} = \dfrac{1}{x - 2}$

**23.** $|y + 4| = 2$

**24.** $|3t - 6| = 0$

**25.** $|3x - 1| = 2$

**26.** $|x + 3| - |3x - 5| = -16$

**27.** $12x^2 + 2x - 2 = 0$

**28.** $4y^2 - 21y = 18$

**29.** $\frac{1}{2}x^2 + x - 12 = 0$

**30.** $x^2 + \frac{13}{2}x + 10 = 0$

**31.** $\dfrac{1}{1 - x} + \dfrac{4}{2 - x} = \dfrac{11}{6}$

**32.** $\dfrac{2x + 1}{1 - 2x} + \dfrac{1 - 2x}{2x + 1} = 2$        **33.** $\dfrac{x}{5 - x} = \dfrac{-2}{11 - x}$

**34.** $\dfrac{x^2}{(x - 1)(x + 1)} = \dfrac{4}{x + 1} + \dfrac{4}{(x - 1)(x + 1)}$

**35.** $\dfrac{1}{3x - 7} - \dfrac{2}{5x - 5} - \dfrac{3}{3x + 1} = 0$

**36.** $x = \dfrac{5}{x + 1}$

**37.** $x^2 + 4x - 1 = 0$        **38.** $4x^2 + x - 2 = 0$

**39.** $t^2 + t - \frac{1}{2} = 0$        **40.** $\dfrac{1}{x - 1} = x\sqrt{2}$

**41.** $y^4 = 9$        **42.** $(y - 1)^4 = 81$

**43.** $x^4 - 7x^2 + 12 = 0$        **44.** $y^4 - 8y^2 + 11 = 0$

**45.** $x^5 - 2x^3 - 2x = 0$        **46.** $x^3 - 6x^2 + 7x = 0$

**47.** $1 + 14x^{-1} + 48x^{-2} = 0$

**48.** $x^{-2} - x^{-1} - 1 = 0$

**49.** $4x^{4/3} - 13x^{2/3} + 9 = 0$

**50.** $512x^3 - 152x^{3/2} - 27 = 0$

**51.** $x^{1/2} - 13x^{1/4} + 36 = 0$

**52.** $y^{2/3} - 21y^{1/3} + 80 = 0$

**53.** $\sqrt{4 - 3x} = 5$        **54.** $\sqrt{x} = \sqrt{x + 27} - 1$

**55.** $\sqrt{4x + 3} = \sqrt{11 - 8x} - 1$

**56.** $\sqrt{\sqrt{x - 4} + x} = 2$

**57.** $\sqrt{5 - 2x} - \sqrt{2 - x} - \sqrt{3 - x} = 0$

**58.** $2 - \sqrt{3\sqrt{2x - 1}} + x = 0$

**59.** $\sqrt{x + 48} - \sqrt{x} = 4$

**60.** $1 + \sqrt{2x^2 + 5x - 9} = \sqrt{2x^2 + 5x - 2}$

**61.** $\dfrac{2}{\sqrt{x^2 - 36}} + \dfrac{1}{\sqrt{x + 6}} - \dfrac{1}{\sqrt{x - 6}} = 0$

**62.** $\sqrt{x - \sqrt{1 - x}} + \sqrt{x - 1} = 0$

**63.** $\sqrt{x + 7} - \sqrt{x + 2} = \sqrt{x - 1} - \sqrt{x - 2}$

**64.** $\dfrac{\sqrt{x} - 4}{\sqrt{x} - 18} = 3$

*In Exercises 65–78, solve each equation for x in terms of the other letters.*

**65.** $\dfrac{ax + b}{cx + d} = e$   $(a \neq ce)$

**66.** $ax = bx + a^2 - b^2$   $(a \neq b)$

67. $(a^2 + b^2)x = a^3 + b^3 + abx$

68. $\dfrac{2}{x} = \dfrac{1}{a} + \dfrac{1}{b}$   $(a + b \neq 0)$    69. $\dfrac{x}{a} + \dfrac{a}{x} = 2$

70. $\dfrac{1}{a + x} + \dfrac{1}{b + x} = \dfrac{a + b}{ab}$   $(a + b \neq 0)$

71. $4x^2y^2 - 4xy = -1$   $(y \neq 0)$

72. $4x^4y^4 + 2x^2y^2 + \frac{1}{4} = 0$   $(y \neq 0)$

73. $x + \dfrac{1}{a} - \dfrac{1}{b} = \dfrac{2}{a^2x} + \dfrac{2}{abx}$

74. $\dfrac{a}{x - b} + \dfrac{b}{x - a} = \dfrac{x - a}{b} + \dfrac{x - b}{a}$   $(a \neq -b)$

   *Suggestion:*  Before clearing fractions, carry out the indicated additions on each side of the equation.

75. $\dfrac{1}{x + a + b} = \dfrac{1}{x} + \dfrac{1}{a} + \dfrac{1}{b}$   $(a + b \neq 0)$

76. $\dfrac{1}{x} + \dfrac{1}{a - x} + \dfrac{1}{x + 3a} = 0$   $(a \neq 0)$

77. $\dfrac{a^2 - b^2}{x} - \dfrac{2(a^2 + b^2)}{\sqrt{x}} = b^2 - a^2$   $(a > b > 0)$

   *Hint:*   Let $t = \dfrac{1}{\sqrt{x}}$ and $t^2 = \dfrac{1}{x}$.

78. $x^2 + \dfrac{1}{x^2} = a^2 + \dfrac{1}{a^2}$

*Solve the inequalities in Exercises 79–107.*

79. $4(1 + x) - 3(2x - 1) \geq 1$

80. $\dfrac{x}{2} - \dfrac{x - 1}{6} < 0$

81. $-1 < \dfrac{1 - 2(1 + x)}{3} < 1$

82. $3 < \dfrac{x - 1}{-2} < 4$    83. $\dfrac{3}{5} < \dfrac{3 - 2x}{-4} < \dfrac{4}{5}$

84. $|x| \geq 6$    85. $|x| \leq \frac{1}{2}$

86. $|t - 2| < 1$    87. $|x + 4| < \frac{1}{10}$

88. $|3 - 5x| < 2$    89. $|2x - 1| \geq 5$

90. $x^2 - 21x + 108 \leq 0$

91. $x^2 + 3x - 40 < 0$

92. $x^2 \geq 15x$    93. $x^2 - 6x - 1 < 0$

94. $(x + 12)(x - 1)(x - 8) < 0$

95. $(x - 4)^2(x + 8)^3 \geq 0$

96. $625 - x^4 \leq 0$    97. $x^4 - 34x^2 + 225 < 0$

98. $\dfrac{x + 12}{x - 5} > 0$    99. $\dfrac{(x - 7)^2}{(x + 2)^3} \geq 0$

100. $\dfrac{(x - 6)^2(x - 8)(x + 3)}{(x - 3)^2} \leq 0$

101. $\dfrac{x^2 - 10x + 9}{x^3 + 1} \leq 0$

102. $\dfrac{3x + 1}{x - 4} < 1$    103. $\dfrac{1 - 2x}{1 + 2x} \leq \dfrac{1}{2}$

104. $\dfrac{x}{x - 2} + \dfrac{1}{x - 1} \leq \dfrac{23}{12}$

105. $\dfrac{1}{x} + \dfrac{1}{x + 1} + \dfrac{1}{x + 2} \geq 0$

106. $x^2 + \dfrac{1}{x^2} > 3$   *Suggestion:*   Use a calculator to evaluate the key numbers.

107. $\sqrt{x} - \dfrac{5}{\sqrt{x}} \leq 4$

*For Exercises 108–111, find the values of k for which the roots of the equations are real numbers.*

108. $kx^2 - 6x + 5 = 0$    109. $x^2 + x + k^2 = 0$

110. $kx(x + 2) = -1$

111. $x^2 + (k + 1)x + 2k = 0$

112. How many pounds of an alloy that is 12% copper must be added to 1000 lb of an alloy that is 9% copper to yield a new alloy that is 11.2% copper?

113. Kona coffee beans are sold for \$6/lb and Colombian beans for \$5/lb. How many pounds of each bean must be mixed to yield 280 lb of a blend selling for \$5.60/lb?

114. Find four consecutive odd integers whose sum is 2104.

115. The sum of the digits in a certain two-digit number is 11. If the order of the digits is reversed, the number is increased by 27. Find the original number.

116. Robin deposits \$6000, part at 10% and the remainder at 12%. If the net interest rate for the entire \$6000 is $11\frac{1}{3}\%$, how much was deposited at each rate?

117. At 6 A.M. a train traveling at 50 mph leaves Los Angeles for San Francisco. Three hours later a train traveling 40 mph leaves San Francisco for Los Angeles. What time will the trains pass each other? Assume that the distance between the two cities is 450 miles.

118. Jennifer gives Lee a head start of $d$ miles. If Jennifer runs at $A$ mph and Lee runs at $B$ mph (where $A > B$), how long will it take for Jennifer to catch Lee?

119. The four corners of a square $ABCD$ have been cut off to form a regular octagon, as shown on the next page. If each side of the square is 1 cm long, how long is each side of the octagon?

**120.** Determine $p$ if the larger root of the equation $x^2 + px - 2 = 0$ is $p$.

**121.** A piece of wire $x$ cm long is bent into a square. For which values of $x$ will the area be (numerically) greater than the perimeter?

**122.** A piece of wire $6x$ cm long is bent into an equilateral triangle. For which values of $x$ will the area be (numerically) less than the perimeter?

**123.** Find three consecutive positive integers such that the sum of their squares is 1454.

**124.** The sum of two numbers is 1 and the sum of their cubes is 91. What are the numbers?

**125.** In a 10-mile race, Joan covers the first 4 miles at a constant rate. Then she speeds up a bit and runs the last 6 miles at a rate that is $\frac{1}{2}$ mph faster. If the entire race had been run at this faster pace, Joan's overall time would have been 2 minutes better. What was Joan's speed for the first 4 miles, and what was her time for the race?

**126.** The length of a rectangular piece of tin exceeds the width by 8 cm. A 1-cm square is cut from each corner of the piece of tin, and then the resulting flaps are turned up to form a box with no top. What are the dimensions of the box if its volume is 48 cm³?

**127.** The length and width of a rectangular flower garden are $a$ and $b$, respectively. The garden is bordered on all four sides by a gravel path of uniform width. Find the width of the path, given that the area of the garden equals the area of the path.

*If an object is thrown vertically upward from a height of $h_0$ feet with an initial speed of $v_0$ ft/sec, then its height $h$ (in feet) after $t$ seconds is given by*

$$h = -16t^2 + v_0 t + h_0$$

*Make use of this formula in working Exercises 128–130.*

**128.** A ball is thrown vertically upward from ground level with an initial speed of 64 ft/sec.
  (a) At what time will the height of the ball be 15 ft? (Two answers.)
  (b) For how long an interval of time will the height exceed 63 ft?

**129.** One ball is thrown vertically upward from a height of 50 ft with an initial speed of 40 ft/sec. At the same instant, another ball is thrown vertically upward from a height of 100 ft with an initial speed of 5 ft/sec. Which ball hits the ground first?

**130.** An object is projected vertically upward. Suppose that its height is $H$ ft at $t_1$ sec and again at $t_2$ sec. Express the initial speed in terms of $t_1$ and $t_2$.
*Answer:* $16(t_1 + t_2)$

**131.** A rectangle is inscribed in a semicircle of radius 1 cm, as shown. For which value of $x$ is the area of the rectangle 1 cm²? *Note:* $x$ is defined in the figure.

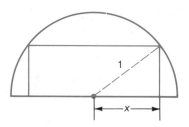

**132.** The height of an isosceles triangle is $a + b$ units $(a > b > 0)$ and the area does not exceed $a^2 - b^2$ square units. What is the range of possible values for the base of the triangle?

*Exercises 133 and 134 appear in the text* Introduction to Algebra *by George Chrystal, first published in 1898.*

**133.** Show that it is impossible to find a positive integer such that the sum of its square and its cube is an integral multiple of the square of the next highest integer. [*Hint:* After setting up and simplifying an equation, make use of the result in Exercise 69 of Exercise Set 1.8.]

**134.** A circle is inscribed in a quadrant of a circle of radius $r$ [as shown in the figure]. Find the radius of the inscribed circle.

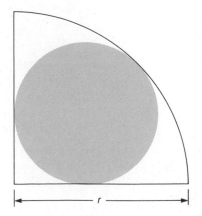

*Exercise 135 (including the clever solution that is outlined) appears in the book* The USSR Olympiad Problem Book *by D. O. Shklarsky et al. (San Francisco: W. H. Freeman and Co., 1962).*

135. Follow steps (a)–(d) to solve the equation

$$\sqrt{a - \sqrt{a + x}} = x \qquad (a \geq 1)$$

(a) By squaring as usual, obtain the equation

$$x^4 - 2ax^2 - x + a^2 - a = 0.$$

(b) Although the equation obtained in part (a) is a fourth-degree equation in $x$, it is only a quadratic in $a$. Solve for $a$ in terms of $x$ to obtain $a = x^2 + x + 1$ or $a = x^2 - x$.

(c) Solve each of the equations in part (b) for $x$ in terms of $a$.

(d) Check your results in part (c) to eliminate any extraneous roots.

# 3 COORDINATES AND GRAPHS

*The mathematics I was taught in my freshman year was algebra (old stuff), trigonometry (too much of it), and analytical geometry (a revelation).*

PAUL R. HALMOS in his book *I Want To Be a Mathematician* (N.Y.: Springer-Verlag, 1985)

**INTRODUCTION**    In Chapter 1 we saw an important connection between algebra and geometry: the points on a line can be labeled using the real numbers. In Chapter 2 we exploited this connection in solving inequalities and in graphing their solution sets. Now, just as each point on a line can be labeled with a unique real number, so each point in a *plane* can be labeled with a unique *pair* of real numbers. The details of this are explained in Section 3.1, Rectangular Coordinates. These ideas are then used in the next three sections to graph equations. The basic theme running through the work on graphing equations is this: We start with a given equation relating two variables, say $x$ and $y$. By choosing values for $x$, corresponding $y$-values are then determined. The resulting $(x,y)$ number pairs (*ordered pairs*, they're called) are then interpreted as points in the plane. This provides another basic connection between algebra and geometry: with an equation in two variables we associate its geometric picture, its graph. Sometimes it is easier to work directly from a graph, other times from the equation. But in any event, having the latitude to think either algebraically or geometrically (or both) will clarify many problems later on.

## RECTANGULAR COORDINATES

### 3.1

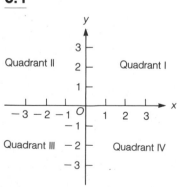

**FIGURE 1**

We begin by recalling the basic idea of a coordinate system in the plane. Consider two number lines, perpendicular to each other, as shown in Figure 1. The point of intersection of the two number lines is called the **origin** and is denoted by the letter $O$. The horizontal number line is called the **$x$-axis**; the vertical number line is called the **$y$-axis**. Unless otherwise indicated, we assume that the same unit of length is used on both axes. Referring to Figure 1, notice that the axes divide the plane into four regions or **quadrants**, labeled with Roman numerals I through IV, as shown.

Now look at the point $P$ in Figure 2(a). Starting from the origin $O$, one way to reach $P$ is to move three units in the positive $x$-direction and then two units in the positive $y$-direction. That is, the location of $P$ relative to the origin and the axes is "right 3, up 2." We say that the **coordinates** of $P$ are $(3,2)$. The first number within the parentheses conveys the information "right 3," and the second number conveys the information "up 2." We say that the **$x$-coordinate**

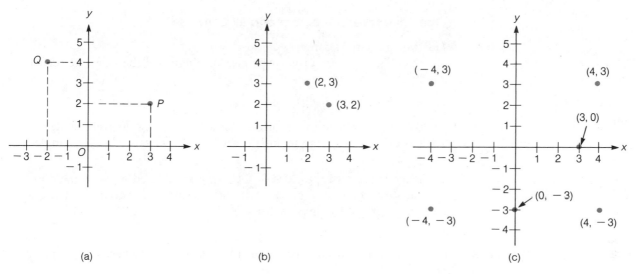

(a)　　　　　　　　(b)　　　　　　　　(c)

**FIGURE 2**

of $P$ is 3 and the **$y$-coordinate** of $P$ is 2. Likewise, the coordinates of point $Q$ in Figure 2(a) are $(-2, 4)$. With this coordinate notation in mind, observe in Figure 2(b) that $(3, 2)$ and $(2, 3)$ represent different points; that is, the order in which the two numbers appear within the parentheses affects the location of the point. Figure 2(c) displays various points with given coordinates; you should check for yourself that the coordinates are correct.

Some terminology and notation: the $x$-$y$ coordinate system that we have described is often called a **rectangular** or **Cartesian coordinate system**. The word *rectangular* refers to the fact that the $x$- and $y$-axes are straight lines meeting at right angles. The term *Cartesian* is used in honor of René Descartes, the seventeenth-century French philosopher and mathematician.* The coordinates $(x, y)$ of a point $P$ are referred to as an **ordered pair**. Recall, for example, that $(3, 2)$ and $(2, 3)$ represent different points; that is, the order of the numbers matters. The $x$-coordinate of a point is sometimes referred to as the **abscissa** of the point; the $y$-coordinate is the **ordinate**. The notation $P(x, y)$ means that $P$ is a point whose coordinates are $(x, y)$. At times, we abbreviate the phrase "the point whose coordinates are $(x, y)$" to simply "the point $(x, y)$."

In Example 1 we are going to use the Pythagorean Theorem to calculate the distance between two given points. Although we've already worked with this theorem (in Section 2.3), for emphasis and reference we restate it here, along with its converse, in the box that follows. (For a proof of the theorem, see Exercise 39.)

---

* Professor George F. Simmons in his book *Calculus with Analytic Geometry* (New York: McGraw-Hill, 1985) points out that a careful reading and comparison of the original sources clearly indicates that it is Pierre de Fermat (1601–1665), not Descartes, who should be credited with the invention of analytic geometry.

FIGURE 3

### The Pythagorean Theorem and Its Converse

1. Pythagorean Theorem
   (See Figure 3) In a right triangle, the lengths of the sides are related by the equation

   $$a^2 + b^2 = c^2$$

   where $a$ and $b$ are the lengths of the sides forming the right angle and $c$ is the length of the hypotenuse (the side opposite the right angle).

2. Converse
   If the lengths of the sides of a triangle are related by an equation of the form $a^2 + b^2 = c^2$, then the triangle is a right triangle, and $c$ is the length of the hypotenuse.

**EXAMPLE 1**  Use the Pythagorean Theorem to calculate the distance $d$ between points $(2, 1)$ and $(6, 3)$.

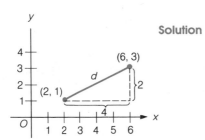

FIGURE 4

Solution  We plot the two given points and draw a line connecting them, as shown in Figure 4. Then we draw in the broken lines as shown, parallel to the axes, and apply the Pythagorean Theorem to the right triangle that is formed. The base of the triangle is four units long. You can see this by simply counting spaces or by using absolute value, as discussed in Section 1.4: $|6 - 2| = 4$. The height of the triangle is found to be two units, either by counting spaces or by computing the absolute value: $|3 - 1| = 2$. Thus, we have

$$d^2 = 4^2 + 2^2 = 20$$
$$d = \sqrt{20} = \sqrt{4}\,\sqrt{5}$$
$$d = 2\sqrt{5}$$

The method used in Example 1 can be applied to derive a general formula for the distance $d$ between any two points $(x_1, y_1)$ and $(x_2, y_2)$ (see Figure 5). Just as before, we draw in the right triangle and apply the Pythagorean Theorem. We have

$$d^2 = |x_2 - x_1|^2 + |y_2 - y_1|^2$$
$$= (x_2 - x_1)^2 + (y_2 - y_1)^2 \qquad \text{(Why?)}$$

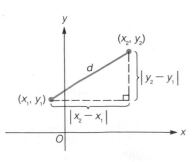

FIGURE 5

and therefore

$$d = \sqrt{(x_2 - x_1)^2 + (y_2 - y_1)^2}$$

This last equation is referred to as the **distance formula**. For reference, we restate it in the box that follows.

### The Distance Formula

The distance $d$ between the points $(x_1, y_1)$ and $(x_2, y_2)$ is given by

$$d = \sqrt{(x_2 - x_1)^2 + (y_2 - y_1)^2}$$

Examples 2 and 3 demonstrate some simple calculations involving the distance formula. *Note:* In computing the distance between two given points, it does not matter which one you treat as $(x_1, y_1)$ and which as $(x_2, y_2)$. This is because quantities such as $(x_2 - x_1)$ and $(x_1 - x_2)$ are negatives of each other, and so their squares are equal.

**EXAMPLE 2**   Calculate the distance between the points $(2, -6)$ and $(5, 3)$.

Solution   Substituting $(2, -6)$ for $(x_1, y_1)$ and $(5, 3)$ for $(x_2, y_2)$ in the distance formula, we have

$$\begin{aligned} d &= \sqrt{(5-2)^2 + [3-(-6)]^2} \\ &= \sqrt{3^2 + 9^2} = \sqrt{9 + 81} \\ &= \sqrt{90} = \sqrt{9}\sqrt{10} \\ &= 3\sqrt{10} \end{aligned}$$

You should check for yourself that the same answer is obtained using $(2, -6)$ as $(x_2, y_2)$ and $(5, 3)$ as $(x_1, y_1)$.

**EXAMPLE 3**   Is the triangle with vertices $D(-2, -1)$, $E(4, 1)$, $F(3, 4)$ a right triangle?

Solution   First, we sketch the triangle in question; see Figure 6. From the sketch it "appears" that angle $E$ could be a right angle, but certainly this is not a proof. The idea then is to use the distance formula to calculate the lengths of the three sides, then check whether any relation of the form $a^2 + b^2 = c^2$ holds. The calculations are as follows.

$$\begin{aligned} DE &= \sqrt{[4-(-2)]^2 + [1-(-1)]^2} = \sqrt{36 + 4} = \sqrt{40} \\ EF &= \sqrt{(4-3)^2 + (1-4)^2} = \sqrt{1 + 9} = \sqrt{10} \\ DF &= \sqrt{[3-(-2)]^2 + [4-(-1)]^2} = \sqrt{25 + 25} = \sqrt{50} \end{aligned}$$

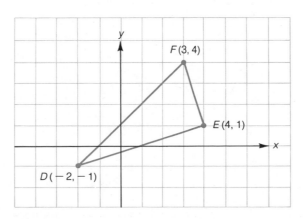

FIGURE 6

Because $(\sqrt{40})^2 + (\sqrt{10})^2 = (\sqrt{50})^2$, we are indeed guaranteed that $\triangle DEF$ is a right triangle. (In Section 3.3, you'll see that this result can be obtained more efficiently using the concept of "slope.")

As we've seen in this section, the basic idea behind coordinate geometry is to associate a unique ordered pair of numbers $(x,y)$ with each point in the plane. This allows us to deal with geometric relationships in a purely numerical or algebraic manner. At times this can be a real advantage. For instance, by using the methods of coordinate geometry, proofs of theorems from elementary geometry can be constructed almost at will, without the need for arguments involving congruent triangles, angles, or the like. Example 4 illustrates this idea. *Note:* You needn't memorize the result in Example 4. Of primary importance for our purposes are the *methods* used to obtain that result.

In Example 4 we make use of a simple result that you may recall from previous courses: the midpoint formula. This result is summarized in the box that follows. (For a proof of the formula, see Exercise 36.)

---

**The Midpoint Formula**

|  | EXAMPLE |
|---|---|
| The midpoint of the line segment joining the points $P(x_1, y_1)$ and $Q(x_2, y_2)$ is $$\left(\frac{x_1 + x_2}{2}, \frac{y_1 + y_2}{2}\right)$$ | The midpoint of the line segment joining $(2, -15)$ and $(4, 5)$ is $$\left(\frac{2+4}{2}, \frac{-15+5}{2}\right) = (3, -5)$$ |

---

**EXAMPLE 4** Use Figure 7 to prove the following theorem from geometry: The midpoint of the hypotenuse in a right triangle is equidistant from the vertices. (In Figure 7, $\triangle OBA$ is the right triangle and $M$ is the midpoint of the hypotenuse.)

**Solution** With reference to Figure 7, we have to show that $OM = MA = MB$. Now, by hypothesis, we already have that $MA = MB$, so we need only establish that $OM = MA$. The calculations are as follows.

Coordinates of $M$: $\left(\dfrac{a}{2}, \dfrac{b}{2}\right)$ using the midpoint formula

$$OM = \sqrt{\left(\frac{a}{2} - 0\right)^2 + \left(\frac{b}{2} - 0\right)^2} = \sqrt{\frac{a^2 + b^2}{4}} = \frac{\sqrt{a^2 + b^2}}{2}$$

$$MA = \sqrt{\left(a - \frac{a}{2}\right)^2 + \left(0 - \frac{b}{2}\right)^2} = \sqrt{\frac{a^2}{4} + \frac{b^2}{4}} = \sqrt{\frac{a^2 + b^2}{4}} = \frac{\sqrt{a^2 + b^2}}{2}$$

From these calculations we conclude that $OM = MA$ (since both quantities are equal to $\frac{1}{2}\sqrt{a^2 + b^2}$). Thus $OM = MA = MB$, as required.

NOTE   We could have avoided working with fractions had we initially made a more clever choice for the coordinates of $A$ and $B$. Since $a$ and $b$ were arbitrary, we could have used $(2a, 0)$ for $A$ and $(0, 2b)$ for $B$.

**FIGURE 7**

## EXERCISE SET 3.1

### A

1. Plot the points $(5,2)$, $(-4,5)$, $(-4,0)$, $(-1,-1)$, and $(5,-2)$.

2. Draw the square $ABCD$ whose vertices (corners) are $A(1,0)$, $B(0,1)$, $C(-1,0)$, and $D(0,-1)$.

3. (a) Draw the right triangle $PQR$ with vertices $P(1,0)$, $Q(5,0)$, and $R(5,3)$.
   (b) Use the formula for the area of a triangle, $A = \frac{1}{2}bh$, to find the area of triangle $PQR$ in part (a).

4. (a) Draw the trapezoid $ABCD$ with vertices $A(0,0)$, $B(7,0)$, $C(6,4)$, and $D(4,4)$.
   (b) Compute the area of the trapezoid. (See the inside front cover of this book for the appropriate formula.)

*In Exercises 5–10, calculate the distance between the given points.*

5. (a) $(0,0)$ and $(-3,4)$    (b) $(2,1)$ and $(7,13)$

6. (a) $(-1,-3)$ and $(-5,4)$
   (b) $(6,-2)$ and $(-1,1)$

7. (a) $(-5,0)$ and $(5,0)$    (b) $(0,-8)$ and $(0,1)$

8. (a) $(-5,-3)$ and $(-9,-6)$
   (b) $(\frac{9}{2},3)$ and $(-2\frac{1}{2},-1)$

9. $(1,\sqrt{3})$ and $(-1,-\sqrt{3})$

c 10. $(-3,1)$ and $(374,-335)$

11. Which point is farther from the origin?
    (a) $(3,-2)$ or $(4,\frac{1}{2})$?    (b) $(-6,7)$ or $(9,0)$?

12. Use the distance formula to show that the triangles with given vertices are isosceles triangles.
    (a) $(0,2)$, $(7,4)$, $(2,-5)$
    (b) $(-1,-8)$, $(0,-1)$, $(-4,-4)$
    (c) $(-7,4)$, $(-3,10)$, $(1,3)$

13. Which of the triangles with the given vertices are right triangles? *Hint:* Find the lengths of the sides and then use the *converse* of the Pythagorean Theorem.
    (a) $(7,-1)$, $(-3,5)$, $(-12,-10)$
    (b) $(4,5)$, $(-3,9)$, $(1,3)$
    (c) $(-8,-2)$, $(1,-1)$, $(10,19)$

14. (a) Two of the three triangles specified in Exercise 13 are right triangles. Find their areas.
    (b) Calculate the area of the remaining triangle in Exercise 13 by using the following formula for the area $A$ of a triangle with vertices $(x_1,y_1)$, $(x_2,y_2)$, and $(x_3,y_3)$:

    $$A = \frac{1}{2}|x_1y_2 - x_2y_1 + x_2y_3 - x_3y_2 + x_3y_1 - x_1y_3|$$

    The derivation of this formula is given in Exercise 38.

15. (c) Use the formula given in part (b) to check your answers in part (a).

15. Use the formula given in Exercise 14(b) to calculate the area of the triangle with vertices $(1,-4)$, $(5,3)$, and $(13,17)$. Conclusion?

16. The coordinates of points $A$, $B$, and $C$ are $A(-4,6)$, $B(-1,2)$, and $C(2,-2)$.
    (a) Show that $AB = BC$, using the distance formula.
    (b) Show that $AB + BC = AC$, using the distance formula.
    (c) What can you conclude from parts (a) and (b)?

*In Exercises 17 and 18, find the midpoint of the line segment joining the given points P and Q.*

17. (a) $P(3,2)$ and $Q(9,8)$
    (b) $P(-4,0)$ and $Q(5,-3)$
    (c) $P(3,-6)$ and $Q(-1,-2)$

18. (a) $P(12,0)$ and $Q(12,8)$
    (b) $P(\frac{3}{5},-\frac{2}{3})$ and $Q(0,0)$
    (c) $P(1,\pi)$ and $Q(3,3\pi)$

19. The coordinates of $A$ and $B$ are $(-1,2)$ and $(5,-3)$, respectively. If $B$ is the midpoint of line segment $\overline{AC}$, what are the coordinates of $C$?

20. The coordinates of the points $S$ and $T$ are $S(4,6)$ and $T(10,2)$. If $M$ is the midpoint of $\overline{ST}$, find the midpoint of $\overline{SM}$.

21. (a) Sketch the parallelogram with vertices $A(-7,-1)$, $B(4,3)$, $C(7,8)$, and $D(-4,4)$.
    (b) Compute the midpoints of the diagonals $\overline{AC}$ and $\overline{BD}$.
    (c) What conclusion can you draw from part (b)?

22. The vertices of $\triangle ABC$ are $A(1,1)$, $B(9,3)$, and $C(3,5)$.
    (a) Find the perimeter of $\triangle ABC$.
    (b) Find the perimeter of the triangle that is formed by joining the midpoints of the three sides of $\triangle ABC$.
    (c) Compute the ratio of the perimeter in part (a) to the perimeter in part (b).
    (d) What theorem from geometry provides the answer for part (c) without the need for the results in (a) and (b)?

23. This exercise refers to $\triangle ABC$ whose vertices are specified in Exercise 22.
    (a) Compute the sum of the squares of the lengths of the sides of $\triangle ABC$.
    (b) Compute the sum of the squares of the lengths of the three medians. (A **median** is a line segment drawn from a vertex to the midpoint of the opposite side.)

*(continued on next page)*

(c) Check that the ratio of the answer in part (a) to that in part (b) is 4/3. *Remark:* As Exercise 29 asks you to show, this ratio is 4/3 for any triangle.

24. The diagonals of a parallelogram bisect each other. Steps (a), (b), and (c) outline a proof of this theorem.
    (a) In the parallelogram $OABC$ in the accompanying figure, check that the coordinates of $B$ must be $(a + b, c)$.
    (b) Use the midpoint formula to calculate the midpoints of diagonals $\overline{OB}$ and $\overline{AC}$.

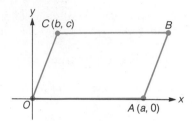

(c) Note that the two answers in part (b) are identical. This shows that the two diagonals do indeed bisect each other, as was to be proved.

25. Use the Pythagorean Theorem to find the length $a$ in the accompanying figure. Then find $b$, $c$, $d$, $e$, $f$, and $g$.

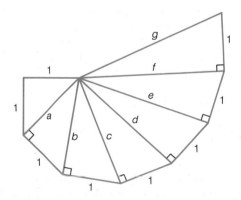

**B**

ⓒ 26. (A numerologist's delight) Using the Pythagorean Theorem and your calculator, compute the area of a right triangle in which the lengths of the hypotenuse and one leg are 2045 and 693, respectively.

27. Prove that in a parallelogram, the sum of the squares of the lengths of the diagonals equals the sum of the squares of the lengths of the four sides. (Use the figure in Exercise 24.)

28. Use the figure in Exercise 24 to prove the following theorem: If the lengths of the diagonals of a parallelo-gram are equal, then the parallelogram is a rectangle. *Hint:* Equate the two expressions for the lengths of the diagonals and conclude that $ab = 0$. The case in which $a = 0$ can be discarded.

29. The sum of the squares of the sides of any triangle equals 4/3 of the sum of the squares of the medians. Prove this, taking the vertices to be $(0,0)$, $(2,0)$, and $(2a, 2b)$. *Note:* The expression *squares of the sides* mean *squares of the lengths of the sides.*

30. In $\triangle ABC$, let $M$ be the midpoint of side $\overline{BC}$. Prove that

$$AB^2 + AC^2 = 2(BM^2 + AM^2)$$

*Hint:* Let the coordinates be $A(0,0)$, $B(2,0)$, and $C(2a, 2b)$.

31. If the point $(x, y)$ is equidistant from the points $(-3, -3)$ and $(5, 5)$, show that $x$ and $y$ satisfy the equation $x + y = 2$. *Hint:* Use the distance formula.

32. The point $(x, y)$ is four units away from $(0,0)$. Show that $x$ and $y$ are related by the equation $x^2 + y^2 = 16$.

33. The point $(x, y)$ is twice as far from $(0,0)$ as it is from $(4, 0)$. Show that $x$ and $y$ are related by the equation $3x^2 + 3y^2 - 32x + 64 = 0$.

34. Find a value for $t$ such that points $(0, 2)$ and $(12, t)$ are 13 units apart. *Hint:* By the distance formula, $13 = \sqrt{(12 - 0)^2 + (t - 2)^2}$. Now square both sides and solve for $t$.

35. (a) Find values for $t$ such that the points $(-2, 3)$ and $(t, 1)$ are six units apart.
    (b) Can you find a real number $t$ such that the points $(-2, 3)$ and $(t, 1)$ are one unit apart? Why or why not?

36. Suppose that the coordinates of points $P$, $Q$, and $M$ are as follows:

$$P(x_1, y_1) \qquad Q(x_2, y_2) \qquad M\left(\frac{x_1 + x_2}{2}, \frac{y_1 + y_2}{2}\right)$$

Follow steps (a) and (b) to prove that $M$ is the midpoint of the line segment from $P$ to $Q$.
    (a) By computing both of the distances $PM$ and $MQ$, show that $PM = MQ$. (This shows that $M$ lies on the perpendicular bisector of line segment $\overline{PQ}$, but it does not show that $M$ actually lies *on* $\overline{PQ}$.)
    (b) Show that $PM + MQ = PQ$. (This shows that $M$ does lie on $\overline{PQ}$.)

**C**

37. The point $(x, y)$ is located in such a way that the sum of its distance from $(-1, 0)$ and from $(1, 0)$ is 4. Show that $x$ and $y$ are related by the equation $3x^2 + 4y^2 = 12$.

**38.** This problem indicates a method for calculating the area of a triangle when the coordinates of the three vertices are given.

(a) Calculate the area of $\triangle ABC$ in the figure. *Hint:* First calculate the area of the rectangle enclosing $\triangle ABC$ and then subtract the areas of the three right triangles.

(b) Calculate the area of the triangle with vertices $(1,3)$, $(4,1)$, and $(10,4)$. *Hint:* Work with an enclosing rectangle and three right triangles, as in part (a).

(c) Using the same technique that you used in parts (a) and (b), show that the area of the triangle in the figure at the top of the next column is given by

$$A = \tfrac{1}{2}(x_1y_2 - x_2y_1 + x_2y_3 - x_3y_2 + x_3y_1 - x_1y_3)$$

*Remark:* If we use absolute value signs instead of the parentheses, then the formula will hold regardless of the relative positions or quadrants of the three vertices. Thus, the area of a triangle with vertices $(x_1,y_1)$, $(x_2,y_2)$, $(x_3,y_3)$ is given by

$$A = \tfrac{1}{2}|x_1y_2 - x_2y_1 + x_2y_3 - x_3y_2 + x_3y_1 - x_1y_3|$$

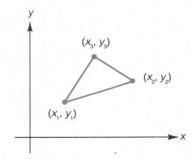

**39.** This problem outlines one of the shortest proofs of the Pythagorean Theorem. In the accompanying figure, we are given a right triangle $ACB$, with the right angle at $C$, and we want to prove that $a^2 + b^2 = c^2$. In the figure, $CD$ is drawn perpendicular to $AB$.

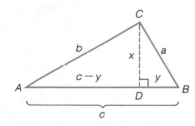

(a) Check that $\angle CAD = \angle DCB$ and that $\triangle BCD$ and $\triangle BAC$ are similar.

(b) Use the result in part (a) to obtain the equation $a/y = c/a$ and conclude that $a^2 = cy$.

(c) Show that $\triangle ACD$ is similar to $\triangle ABC$ and use this to deduce that $b^2 = c^2 - cy$.

(d) Combine the two equations deduced in parts (b) and (c) to arrive at $a^2 + b^2 = c^2$.

## GRAPHS AND SYMMETRY

**3.2**

*Fermat and Descartes discovered that it is possible to combine the methods and concepts of geometry and algebra in a way that is useful in both branches of mathematics. Their basic idea was to associate algebraic equations in two unknowns with geometric curves in the plane. . . . Using this correspondence, problems in geometry can be translated into problems solvable by algebraic manipulation, and insight into the nature of solutions of equations can be gained by consideration of the geometric interpretations.*

RICHARD S. HALL in *About Mathematics* (Englewood Cliffs, NJ: Prentice-Hall, 1973)

*Symmetry is a working concept. If all the object is symmetrical, then the parts must be halves (or some other rational fraction) and the amount of information necessary to describe the object is halved (etc.).*

ALAN L. MACAY, Department of Crystallography, University of London

One of the most useful ways to obtain information about an equation relating two variables is through a graph. The key definition is as follows.

**TABLE 1**

$y = 3x - 2$

| x | y |
|---|---|
| 0 | −2 |
| 1 | 1 |
| 2 | 4 |
| 3 | 7 |
| −1 | −5 |
| −2 | −8 |

> **Definition: The Graph of an Equation**
>
> The **graph** of an equation in two variables is the set of all points whose coordinates satisfy the equation.

Suppose, for example, that we want to graph the equation $y = 3x - 2$. We begin by noting that the domain of the variable $x$ in the expression $3x - 2$ is the set of all real numbers. Now we choose (at will) various values for $x$ and in each case compute the corresponding $y$-value from the equation $y = 3x - 2$. For example, if $x$ is zero, then $y = 3(0) - 2 = -2$. Table 1 summarizes the results of some of these calculations. The first line in Table 1 tells us that the point with coordinates $(0, -2)$ is on the graph of $y = 3x - 2$. Reading down the table, we see that some other points on the graph are $(1, 1)$, $(2, 4)$, $(3, 7)$, $(-1, -5)$, and $(-2, -8)$.

We now plot (locate) the points we have determined and note, in this example, that they all appear to lie on a straight line. We draw the line indicated; this is the graph of $y = 3x - 2$. (see Figure 1). In Example 1, we ask a simple question that will test your understanding of the process we've just described and of graphing in general.

**EXAMPLE 1**  Does the point $(1\frac{2}{3}, 2\frac{2}{3})$ lie on the graph of $y = 3x - 2$?

**Solution**  Looking at the graph in Figure 1, we certainly see that the point $(1\frac{2}{3}, 2\frac{2}{3})$ is very close to the line. But on these visual grounds alone, it is not clear whether this point actually lies on the line or not. To settle the question, then, we check to see if the values $x = 1\frac{2}{3} = \frac{5}{3}$ and $y = 2\frac{2}{3} = \frac{8}{3}$ together satisfy the equation $y = 3x - 2$. We have

$$\frac{8}{3} \overset{?}{=} 3\left(\frac{5}{3}\right) - 2$$

$$\frac{8}{3} \overset{?}{=} 3 \qquad \text{No!}$$

Since the coordinates $(1\frac{2}{3}, 2\frac{2}{3})$ evidently do not satisfy the equation $y = 3x - 2$, we conclude that the point $(1\frac{2}{3}, 2\frac{2}{3})$ is not on the graph. [The calculation further tells us that the point $(1\frac{2}{3}, 3)$ *is* on the graph.]

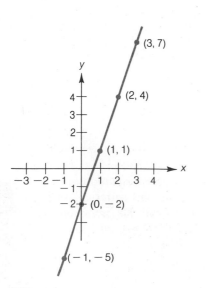

**FIGURE 1**  $y = 3x - 2$

In graphing an equation, it's helpful to know where the curve or line crosses the $x$- or $y$-axis. By a $y$-**intercept** of a graph, we mean the $y$-coordinate of a point where the graph crosses the $y$-axis. For instance, returning to Figure 1, the $y$-intercept is $-2$. Notice that this $y$-intercept can be obtained algebraically just by setting $x$ equal to zero in the given equation $y = 3x - 2$. In a similar fashion, an $x$-**intercept** of a graph is the $x$-coordinate of a point where the graph crosses the $x$-axis. In Figure 1 we can see that the $x$-intercept is a number between 0 and 1. To determine this $x$-intercept, set $y$ equal to zero in the equation $y = 3x - 2$. That yields $0 = 3x - 2$, and consequently $x = \frac{2}{3}$. So the $x$-intercept of the graph in Figure 1 is $\frac{2}{3}$.

In elementary graphing there is always the question of how many points must be plotted before the essential features of a graph are clear. As you'll see throughout this text, there are a number of techniques and concepts that make it unnecessary to plot a large number of points. In this section we introduce three types of *symmetry* that are useful in graphing equations.

## Three Types of Symmetry

| TYPE OF SYMMETRY | EXAMPLE | DEFINITION |
|---|---|---|
| 1. Symmetry about the *x*-axis | 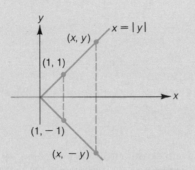<br>FIGURE 2  Symmetry about the *x*-axis | For each point $(x,y)$ on the graph, the point $(x, -y)$ is also on the graph. The points $(x,y)$ and $(x, -y)$ are *reflections of one another about the x-axis*. In Figure 2, the portions of the graph above and below the *x*-axis also are said to be reflections of one another about the *x*-axis. |
| 2. Symmetry about the *y*-axis | <br>FIGURE 3  Symmetry about the *y*-axis | For each point $(x,y)$ on the graph, the point $(-x,y)$ is also on the graph. The points $(x,y)$ and $(-x,y)$ are *reflections of one another about the y-axis*. In Figure 3, the portions of the graph to the right and left of the *y*-axis also are said to be reflections of one another about the *y*-axis. |
| 3. Symmetry about the origin | <br>FIGURE 4  Symmetry about the origin | For each point $(x,y)$ on the graph, the point $(-x, -y)$ is also on the graph. The points $(x,y)$ and $(-x, -y)$ are *reflections of one another about the origin*. In Figure 4, the first- and third-quadrant portions of the curve also are said to be reflections of one another about the origin. [In terms of the two previous symmetries, the point $(-x, -y)$ can be obtained from $(x,y)$ as follows: first reflect $(x,y)$ about the *y*-axis to obtain $(-x,y)$, then reflect $(-x,y)$ about the *x*-axis to obtain $(-x, -y)$.] |

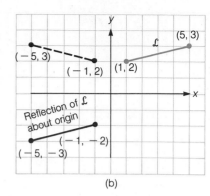

(a)                                (b)

**FIGURE 5**

**EXAMPLE 2**   A line segment $\mathscr{L}$ has endpoints $(1,2)$ and $(5,3)$. Sketch the reflection of $\mathscr{L}$ about:

(a) the $x$-axis;   (b) the $y$-axis;   (c) the origin.

**Solution**   (a) First reflect the endpoints $(1,2)$ and $(5,3)$ about the $x$-axis to obtain the new endpoints $(1,-2)$ and $(5,-3)$, respectively. Then join these two points as indicated in Figure 5(a), in quadrant IV.

(b) Reflect the given endpoints about the $y$-axis to obtain the new endpoints $(-1,2)$ and $(-5,3)$; then join these two points as shown in Figure 5(a), in quadrant II.

(c) As described in the box preceding this example, reflection about the origin can be carried out in two steps: first reflect about the $y$-axis, then reflect about the $x$-axis. In part (b) we obtained the reflection of $\mathscr{L}$ in the $y$-axis. So now we need only to reflect the line segment obtained in part (b) about the $x$-axis. See Figure 5(b).

In the box that follows we list three rules for testing whether the graph of an equation possesses any of the types of symmetry we've been discussing. (The validity of each rule follows directly from the definitions of symmetry.)

---

**Three Tests for Symmetry**

1. The graph of an equation is symmetric about the $y$-axis if replacing $x$ with $-x$ yields an equivalent equation.

2. The graph of an equation is symmetric about the $x$-axis if replacing $y$ with $-y$ yields an equivalent equation.

3. The graph of an equation is symmetric about the origin if replacing $x$ and $y$ with $-x$ and $-y$, respectively, yields an equivalent equation.

---

**EXAMPLE 3**   (a) Test for symmetry about the $y$-axis:   $y = x^4 - 3x^2 + 1$.

(b) Test for symmetry about the $x$-axis:   $x = \dfrac{y^2}{y-1}$.

**Solution**    (a)  Replacing $x$ with $-x$ in the original equation gives us

$$y = (-x)^4 - 3(-x)^2 + 1$$

or

$$y = x^4 - 3x^2 + 1$$

Since this last equation is the same as the original equation, we conclude that the graph is symmetric about the $y$-axis.

(b)  Replacing $y$ with $-y$ in the original equation gives us

$$x = \frac{(-y)^2}{(-y) - 1}$$

or

$$x = \frac{y^2}{-y - 1}$$

Since this last equation is not equivalent to the original equation, we conclude that the graph is not symmetric about the $x$-axis.  ▪

**EXAMPLE 4**    Is the graph of $y = 2x^3 - x$ symmetric about the origin?

**Solution**    Replacing $x$ and $y$ with $-x$ and $-y$, respectively, gives us

$$-y = 2(-x)^3 - (-x) = -2x^3 + x$$
$$y = 2x^3 - x \qquad \text{multiplying through by } -1$$

This last equation is identical to the given equation. Therefore the graph is symmetric about the origin.  ▪

In the next two examples we use the notions of symmetry and intercepts as guides for drawing the required graphs.

**EXAMPLE 5**    Graph the equation $y = -x^2 + 5$.

**Solution**    The domain of the variable $x$ in the expression $-x^2 + 5$ is the set of all real numbers. However, since the graph must be symmetric about the $y$-axis (why?), we need only to be able to sketch the graph to the right of the $y$-axis; the portion to the left will then be the mirror image of this (with the $y$-axis as the mirror). The $x$ and $y$-intercepts (if any) are computed as follows.

| $x$-INTERCEPTS | $y$-INTERCEPTS |
|---|---|
| $-x^2 + 5 = 0$ | $y = -(0)^2 + 5$ |
| $x^2 = 5$ | $y = 5$ |
| $x = \pm\sqrt{5} \ (\approx \pm 2.2)$ | |

In Figure 6(a) we've set up a short table of values and sketched the graph for $x \geq 0$. The complete graph is then obtained by reflection about the $y$-axis, as shown in Figure 6(b). [The curve in Figure 6(b) is a *parabola*. This type of curve will be considered in detail in later chapters.]  ▪

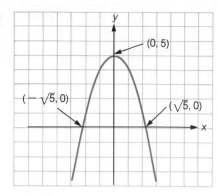

(a) $y = -x^2 + 5, \quad x \geq 0$

(b) The graph of $y = -x^2 + 5$ is symmetric about the $y$-axis.

**FIGURE 6**

**EXAMPLE 6**   Graph $y = 1/x$.

**Solution**   Before doing any calculations, note that the domain of the variable $x$ consists of all real numbers other than zero. Thus, however the resulting graph may look, it cannot contain a point whose $x$-coordinate is zero. In other words, the graph cannot cross the $y$-axis. (Also, check for yourself that the graph does not cross the $x$-axis.) Now, since the graph is symmetric about the origin (why?), we need only to be able to sketch the graph for $x > 0$; the portion corresponding to $x < 0$ can then be obtained by reflection about the origin. In Figure 7(a) we've set up a table of values and sketched the graph for $x > 0$. In Figure 7(b) we have reflected the first-quadrant portion of the graph, first

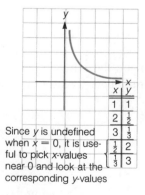

Since $y$ is undefined when $x = 0$, it is useful to pick $x$-values near 0 and look at the corresponding $y$-values

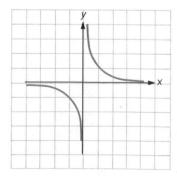

(a) $y = \dfrac{1}{x}, \quad x > 0$

(b) Reflections about the $y$-axis and then the $x$-axis yield a reflection about the origin.

(c) The graph of $y = 1/x$ is symmetric about the origin.

**FIGURE 7**

**FIGURE 8**

about the $y$-axis, then about the $x$-axis, to obtain the required reflection about the origin. Figure 7(c) shows our final graph of $y = 1/x$.

We can use the graph in Figure 7(a) to introduce the idea of an *asymptote*. A line is an **asymptote** for a curve if the distance between the line and the curve approaches zero as we move out farther and farther along the line. So, for the curve in Figure 7(a), both the $x$- and the $y$-axes are asymptotes. (We'll return to this idea in Chapters 4 and 5.)

Since we've been discussing symmetry in this section, it seems appropriate to conclude with a discussion of the circle, since in some sense, this is the most symmetric curve. We can use the distance formula to obtain the equation of a circle. Figure 8 shows a circle with center $(h, k)$ and radius $r$.

By definition, a point $(x, y)$ is on this circle if and only if the distance from $(x, y)$ to $(h, k)$ is $r$. Thus, we have

$$\sqrt{(x - h)^2 + (y - k)^2} = r$$

or, after squaring both sides,

$$(x - h)^2 + (y - k)^2 = r^2$$

This last equation is called the **standard form for the equation of a circle**.

---

**The Equation of a Circle in Standard Form**

The equation of a circle with center $(h, k)$ and radius $r$ is

$$(x - h)^2 + (y - k)^2 = r^2$$

---

**EXAMPLE 7**  Write the equation of the circle with center $(-2, 5)$ and radius 3.

**Solution**  In the equation $(x - h)^2 + (y - k)^2 = r^2$, we substitute the given values $h = -2$, $k = 5$, and $r = 3$. This yields

$$[x - (-2)]^2 + (y - 5)^2 = 3^2$$
$$(x + 2)^2 + (y - 5)^2 = 9$$

This is the standard form for the equation of the given circle. An alternative form of this answer is found by carrying out the indicated algebra. We have

$$(x + 2)^2 + (y - 5)^2 = 9$$

Therefore,

$$x^2 + 4x + 4 + y^2 - 10y + 25 = 9$$

and consequently,

$$x^2 + 4x + y^2 - 10y + 20 = 0$$

The disadvantage to this alternative form is that information regarding the center and radius is no longer so readily visible.

When the equation of a circle is not in standard form, the technique of completing the square (discussed in Section 2.2) can be used to convert the equation to standard form. Example 8 shows how this is done.

**EXAMPLE 8**   Determine the center and radius of the circle

$$4x^2 - 24x + 4y^2 + 16y + 51 = 0$$

Solution   First, divide through by 4 so that the coefficients of $x^2$ and $y^2$ are both 1:

$$x^2 - 6x + y^2 + 4y + \frac{51}{4} = 0$$

Now, in preparation for completing the squares, we write

$$(x^2 - 6x + \underline{\qquad}) + (y^2 + 4y + \underline{\qquad}) = -\frac{51}{4}$$

To complete the square in $x$, we need to add $(-\frac{6}{2})^2$, or 9. To complete the square in $y$, we need to add $(\frac{4}{2})^2$, or 4. Thus, we have

$$(x^2 - 6x + 9) + (y^2 + 4y + 4) = -\frac{51}{4} + 9 + 4$$

or

$$(x - 3)^2 + (y + 2)^2 = -\frac{51}{4} + \frac{52}{4} = \frac{1}{4}$$

$$(x - 3)^2 + (y + 2)^2 = \left(\frac{1}{2}\right)^2$$

This is the equation in standard form. The center of the circle is $(3, -2)$; the radius is $\frac{1}{2}$.

## EXERCISE SET 3.2

### A

*In Exercises 1–6, the graph of each equation is a straight line. Graph each equation after finding the x- and y-intercepts.*

1. $3x + 4y = 12$
2. $3x - 4y = 12$
3. $y = 2x - 4$
4. $x = 2y - 4$
5. $x + y = 1$
6. $2x - 3y = 6$

*In Exercises 7–16, determine any x- or y-intercepts for the graph of each equation. Note: You're not asked to draw the graph.*

7. $y = x^2 - 5x - 6$
8. $y = x^2 - 4x - 12$
9. $y = x^2 + x - 1$
10. $y = 2x^2 + x + 1$
11. $\frac{x}{2} + \frac{y}{3} = 1$
12. $y = |x - 1|$
13. $3x - 5y = 10$
14. $xy = 1$
15. $y = x^3 - 8$
16. $y = (x - 8)^3 + 1$

*In Exercises 17–20, you are given the endpoints of a line segment $\overline{AB}$. In each case, sketch the reflection of $\overline{AB}$ about (a) the x-axis; (b) the y-axis; (c) the origin.*

17. $A(1, 4)$ and $B(3, 1)$
18. $A(-1, -2)$ and $B(-5, -2)$
19. $A(-2, -3)$ and $B(2, -1)$
20. $A(-3, -3)$ and $B(-3, -1)$

*In Exercises 21–30, test each equation for symmetry about the x-axis, the y-axis, and the origin.*

21. $3x^2 + y = 16$
22. $x^2y + xy^2 = 0$
23. $y = x^4 - 3$
24. $y = x^5 - 4x^3 + x$
25. $x^2 + y^2 = 16$
26. $y = |x - 2|$
27. $y = x^2 + x^3$
28. $x = y^2 - 4y^6$

29. $x + y = 1$

30. $|x| + |y| = 1$

*In Exercises 31–48, graph each equation. Specify any x- or y-intercepts or symmetry about the x-axis, the y-axis, or the origin.*

31. $y = x^2$

32. $y = -x^3$

33. $y = 1/x$

34. $x = y^2 - 1$

35. $y = -x^2$

36. $y = 1/x^2$

37. $y = -1/x^2$

38. $y = |x|$

39. $y = \sqrt{x^2}$

40. $y = x + 1$

41. $y = x^2 - 2x + 1$

42. $x = y^3 - 1$

43. $y = 2x - 4$

44. $y = |2x - 4|$

45. $y = 2x^2 + x - 4$

46. (a) $y = x^2 - 4$　(b) $y = |x^2 - 4|$

47. (a) $y = 3x - 6$　(b) $y = |3x - 6|$

48. (a) $x + y = 2$　(b) $|x| + |y| = 2$

*In Exercises 49–56, determine the center, the radius, and the x- and y-intercepts (if any) for each circle.*

49. $(x - 3)^2 + (y - 1)^2 = 25$

50. $x^2 + (y + 1)^2 = 20$

51. $x^2 + y^2 = \sqrt{2}$

52. $x^2 + y^2 - 10x + 2y + 17 = 0$

53. $x^2 + y^2 + 8x - 6y = -21$

54. $4x^2 - 4x + 4y^2 - 63 = 0$

55. $9x^2 + 54x + 9y^2 - 6y + 64 = 0$

56. $3x^2 + 3y^2 + 5x - 4y = 1$

**B**

57. Find the area and the circumference of the circle $x^2 - 2ax + y^2 - 2by = c^2 - a^2 - b^2$. (Assume that $a$, $b$, and $c$ are constants and $c > 0$.)

58. (a) Find the equation of the circle tangent to the x-axis and with center $(3, 5)$.
    (b) Find the equation of the circle tangent to the y-axis and with center $(3, 5)$.
    (c) Find the equation of the circle passing through the origin and with center $(3, 5)$.

*In Exercises 59–62, determine the equation of the circle satisfying the given conditions. Write the equation in standard form.*

59. Line segment $\overline{PQ}$ is a diameter and the coordinates of $P$ and $Q$ are $P(-3, -4)$ and $Q(2, 8)$.

60. The circle passes through the origin and it is concentric with the circle $x^2 - 6x + y^2 - 4y + 4 = 0$.

61. The center of the circle is the point $(-3, 2)$, and the radius is equal to twice the distance between the two y-intercepts of the circle $x^2 - 8x + y^2 - 4 = 0$.

62. The circle passes through $(-4, 1)$, and its center is the midpoint of the line segment joining the centers of the two circles $x^2 + y^2 - 6x - 4y + 12 = 0$ and $x^2 + y^2 - 14x + 47 = 0$.

63. (a) Graph the equation $s = 16t^2$, which gives the distance $s$ (in feet) traveled by a freely falling object in $t$ seconds. (The time $t = 0$ corresponds to the instant the object begins its fall; thus, your graph should lie only in the first quadrant, where $t \geq 0$.)
    (b) Indicate the portion of the s-axis that shows the distance covered during the first second, from $t = 0$ to $t = 1$.
    (c) Indicate the portion of the s-axis that shows the distance covered during the next second, from $t = 1$ to $t = 2$.

64. Suppose that the formula $C = 5 + x$ gives the manufacturer's cost in dollars to produce $x$ units of a certain commodity.
    (a) Graph this equation.
    (b) Indicate on the C-axis the cost corresponding to a production level of $x = 2$ units.
    (c) Indicate on the x-axis the number of units corresponding to a cost of $9.

65. Draw the following graphs on the same set of axes and in each case include only the portion of the graph between $x = 0$ and $x = 1$. Draw the graphs carefully enough to ensure that you can make accurate comparisons among them.
    (a) $y = x$　(b) $y = x^2$　(c) $y = x^3$　(d) $y = x^4$
    What is the pattern here? What would you say that $y = x^{100}$ must look like in this interval?

66. The accompanying figure shows the graph of $y = \sqrt{x}$. Use this graph to estimate the following quantities (to one decimal place).
    (a) $\sqrt{2}$　(b) $\sqrt{3}$
    (c) $\sqrt{6}$　*Hint:* $\sqrt{ab} = \sqrt{a}\sqrt{b}$

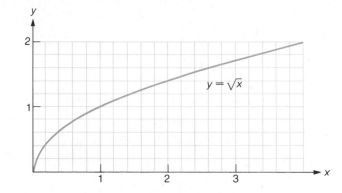

67. In a certain biology experiment, the number $N$ of bacteria increases with time $t$ as indicated in the following figure.
    (a) How many bacteria are initially present when $t = 0$?
    (b) Approximately how long does it take for the original colony to double in size?
    (c) For which value of $t$ is the population approximately 2500?
    (d) During which time interval does the population increase more rapidly, between $t = 0$ and $t = 1$ or between $t = 3$ and $t = 4$?

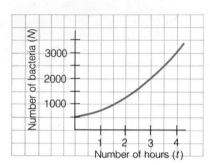

68. Determine the intercepts and symmetry for the equation $y = x^{2/3}$, and then graph the equation.

69. Find the intercepts and symmetry for $y = x^3 - 27x$. Then use a calculator (as necessary) to set up a table

of $x$- and $y$-values, with $x$ running from 0 to 6 in increments of $\frac{1}{2}$. Finally, graph the equation.

70. Set up a table of $x$- and $y$-values for the equation $y = 4x^2/(1 + x^2)$. Use $x$-values running from 0 to 5 in increments of $\frac{1}{2}$. Now use the results in your table, along with symmetry, to graph the equation.

## C

71. Graph the equation $(x^2 + y^2 - 1)(x^2 + y^2 - 4) = 0$.

72. Graph the equation
$$[(x - 1)^2 + y^2 - 1][(x - 3)^2 + y^2 - 1] \times [(x - 5)^2 + y^2 - 1] = 0.$$

73. Graph the equation $y = x + \dfrac{1}{\sqrt{x + 1}}$.

74. Suppose that the circle $x^2 + 2Ax + y^2 + 2By = C$ has two $x$-intercepts, $x_1$ and $x_2$, and two $y$-intercepts, $y_1$ and $y_2$. Prove each of the following statements.
    (a) $\dfrac{x_1 + x_2}{y_1 + y_2} = \dfrac{A}{B}$
    (b) $x_1 x_2 - y_1 y_2 = 0$
    (c) $x_1 x_2 + y_1 y_2 = -2C$

75. Find the equation of the circle that passes through the three points $(2, 5)$, $(0, 7)$, and $(-6, 1)$. Write the equation in standard form.

## SLOPE

### 3.3

Given a nonvertical line or line segment, we measure its direction or inclination using a number called the *slope*, which is defined as follows.

---

**Definition of Slope**

The **slope** of a nonvertical straight line passing through the two points $(x_1, y_1)$ and $(x_2, y_2)$ is the number $m$ defined by

$$m = \frac{y_2 - y_1}{x_2 - x_1}$$

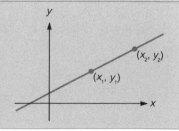

---

Note that the quantity $x_2 - x_1$ appearing in the definition of slope is the amount by which $x$ changes as we move from $(x_1, y_1)$ to $(x_2, y_2)$ along the line. We denote this change in $x$ by the symbol $\Delta x$ (read *delta x*). Thus $\Delta x = x_2 - x_1$ (see Figure 1). Similarly, the symbol $\Delta y$ is defined to mean the change in $y$:  $\Delta y = y_2 - y_1$. Using these ideas, we can rewrite our definition of slope as $m = \Delta y / \Delta x$.

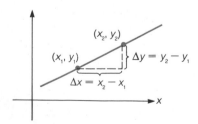

**FIGURE 1**

**EXAMPLE 1** Compute the slope of the line in Figure 2.

Solution We use the formula $m = \dfrac{y_2 - y_1}{x_2 - x_1}$. Which point will serve as $(x_1, y_1)$ and which as $(x_2, y_2)$? It will turn out not to matter how we label our points. Using $(-2, 1)$ as $(x_1, y_1)$ and $(4, 2)$ as $(x_2, y_2)$, we find $m = \dfrac{2 - 1}{4 - (-2)} = \dfrac{1}{6}$. If instead we use $(4, 2)$ as $(x_1, y_1)$ and $(-2, 1)$ as $(x_2, y_2)$, we find $m = \dfrac{1 - 2}{-2 - 4} = \dfrac{1}{6}$, the same result. This is not accidental, because in general, $\dfrac{y_2 - y_1}{x_2 - x_1} = \dfrac{y_1 - y_2}{x_1 - x_2}$.

REASON $\dfrac{y_2 - y_1}{x_2 - x_1} = \dfrac{-1(y_2 - y_1)}{-1(x_2 - x_1)} = \dfrac{-y_2 + y_1}{-x_2 + x_1} = \dfrac{y_1 - y_2}{x_1 - x_2}.$

Let us also compute $\Delta x$ and $\Delta y$ for the two points given in Figure 2. If we begin at the point $(-2, 1)$ and move up the line to $(4, 2)$, then we have

$$\Delta x = 4 - (-2) = 6 \qquad \text{and} \qquad \Delta y = 2 - 1 = 1$$

Thus $m = \Delta y / \Delta x = -1/(-6) = 1/6$, the same result as before, even though we start from the point $(4, 2)$ and move *down* the line to $(-2, 1)$, then

$$\Delta x = -2 - 4 = -6 \qquad \text{and} \qquad \Delta y = 1 - 2 = -1$$

Thus $m = \Delta y / \Delta x = -1/(-6) = 1/6$, the same result as before, even though $\Delta y$ and $\Delta x$ each differ by a sign from their previous values. In summary, the individual values of $\Delta x$ and $\Delta y$ depend on which direction you move along the given line; however, the value obtained for the slope $\Delta y / \Delta x$ is the same in both cases.

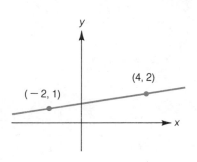

**FIGURE 2**

**EXAMPLE 2** Calculate the slope of the line in Figure 3, first using the points $(2, 1)$ and $(3, 3)$ and then using $(4, 5)$ and $(7, 11)$.

Solution Using the points $(2, 1)$ and $(3, 3)$, we have $m = \dfrac{3 - 1}{3 - 2} = \dfrac{2}{1} = 2$. Using the points $(4, 5)$ and $(7, 11)$, we have $m = \dfrac{11 - 5}{7 - 4} = \dfrac{6}{3} = 2$. Note that the two results are the same.

The point of Example 2 is to show you that the value obtained for the slope of a given line does not depend on which two points on the line are used to calculate the slope. To demonstrate this is true in general, consider Figure 4.

The two right triangles in the figure are similar (because the corresponding angles are equal). This implies that the corresponding sides of the two triangles are proportional, and so we have

$$\frac{a}{b} = \frac{c}{d}$$

Now just notice that the left-hand side of this equation represents the slope $\Delta y / \Delta x$ calculated using the points $A$ and $B$, whereas the right-hand side represents the slope calculated using the points $C$ and $D$. Thus the values we obtain

**FIGURE 3**

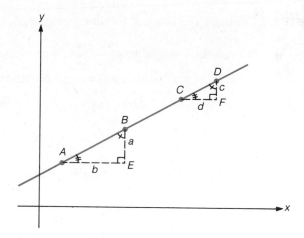

**FIGURE 4**

for the slope are indeed equal. (See Exercise 30 at the end of this section for another perspective on this.)

**EXAMPLE 3** Compute and compare the slopes of the three lines shown in Figure 5.

**Solution** The slopes are all calculated using the formula $m = \dfrac{y_2 - y_1}{x_2 - x_1}$. Writing $m_1, m_2, m_3$ for the respective slopes of lines 1, 2, and 3, we have

$$m_1 = \frac{4 - (-1)}{-2 - (-3)} = \frac{5}{1} = 5$$

$$m_2 = \frac{0 - (-1)}{2 - (-1)} = \frac{1}{3}$$

$$m_3 = \frac{3 - (-1)}{7 - 9} = \frac{4}{-2} = -2$$

Lines 1 and 2 both have positive slopes and slant upward to the right. Note that line 1 is steeper than line 2 and correspondingly has the larger slope. Line 3 has a negative slope and slants downward to the right.

**FIGURE 5**

FIGURE 7

FIGURE 6

The observations made in Example 3 are true in general. Lines with a positive slope slant upward to the right, the steeper line having the larger slope. Likewise, lines with a negative slope slant downward to the right. See Figure 6.

We have yet to mention slopes for horizontal or vertical lines. In Figure 7, line 1 is horizontal; it passes through the points $(a, b)$ and $(c, b)$. Note that the two $y$-coordinates must be the same in order for the line to be horizontal. Line 2 in Figure 7 is vertical; it passes through $(d, e)$ and $(d, f)$. Note that the two $x$ coordinates must be the same in order for the line to be vertical. For the slope of line 1 we have $m = \dfrac{b - b}{c - a} = \dfrac{0}{c - a} = 0$ (provided $a \neq c$). Thus the slope of line 1, a horizontal line, is zero. For the vertical line in Figure 7, the calculation of slope begins with writing $\dfrac{e - f}{d - d}$. But then the denominator is zero, and since division by zero is undefined, we conclude that slope is undefined for vertical lines. We summarize these results in the box that follows. Figure 8 shows for comparison some values of $m$ for various lines.

---

### Horizontal and Vertical Lines

1. The slope of a horizontal line is zero.
2. Slope is not defined for vertical lines.

---

FIGURE 8

**FIGURE 9**

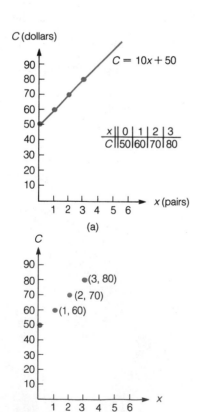

(a)

(b)

**FIGURE 10**

We conclude this section by looking at two ways in which slope is used in applications. As a first example, suppose that you are driving a car at a steady speed of 50 mph. So in 1 hr you travel 50 miles, in 2 hr you travel 100 miles, and so on. Letting $d$ denote the distance in miles that you would cover in $t$ hr at this speed, we can set up the following table.

| $t$ | 0 | 1 | 2 | 3 | 4 |
|---|---|---|---|---|---|
| $d$ | 0 | 50 | 100 | 150 | 200 |

Now we draw a graph, using the data from this table, as shown in Figure 9. We can calculate the slope of the line in Figure 9 using the points $(1, 50)$ and $(2, 100)$. We have

$$m = \frac{100 - 50}{2 - 1} = \frac{50}{1} = 50$$

OBSERVATION The value we have obtained here for the slope is the rate at which the car is traveling. Indeed, if we keep track of the units, our slope calculation will look like this:

$$m = \frac{100 \text{ miles} - 50 \text{ miles}}{2 \text{ hr} - 1 \text{ hr}} = \frac{50 \text{ miles}}{1 \text{ hr}} = 50 \text{ mph}$$

Thus *slope is a rate of change*. In this example, slope is the **velocity** or rate of change of distance with respect to time.

As a second example of how slope is used in applications, let us suppose that a small manufacturer of handmade running shoes knows that her total cost $C$, in dollars, for producing $x$ pairs of shoes each day is given by the formula

$$C = 10x + 50$$

Figure 10(a) displays a table and graph for this equation. Actually, since $x$ represents the number of pairs of shoes, $x$ can assume only whole-number values. So technically the graph should be that given in Figure 10(b). However, it turns out to be useful in practice to make use of the graph in Figure 10(a), and we shall follow this convention.

Using the points $(0, 50)$ and $(1, 60)$, we calculate the slope of the line in Figure 10(a):

$$m = \frac{\$60 - \$50}{1 \text{ pair} - 0 \text{ pair}} = \frac{\$10}{1 \text{ pair}} = \$10 \text{ per pair}$$

Again the slope is a rate. In this case, the slope represents the rate of increase of cost; each additional pair of shoes produced costs the manufacturer $10.

In the study of economics, an equation that gives the cost $C$ for producing $x$ unit of a commodity is called a **cost equation** or **cost function**. When the graph of the cost equation is a straight line, we define the **marginal cost** as the additional cost to produce one more unit. Thus, in the preceding example, the marginal cost is $10 per pair, and we see that the slope of the line in Figure 10(a) represents this marginal cost.

# EXERCISE SET 3.3

## A

In Exercises 1–3, compute the slope of the line passing through the two given points. In Exercise 3, include a sketch with your answers.

1. (a) $(-3, 2), (1, -6)$    (b) $(2, -5), (4, 1)$
   (c) $(-2, 7), (1, 0)$    (d) $(4, 5), (5, 8)$
2. (a) $(-3, 0), (4, 9)$    (b) $(-1, 2), (3, 0)$
   (c) $(\frac{1}{2}, -\frac{3}{5}), (\frac{3}{2}, \frac{3}{4})$    (d) $(\frac{17}{3}, -\frac{1}{2}), (-\frac{1}{2}, \frac{17}{3})$
3. (a) $(1, 1), (-1, -1)$    (b) $(0, 5), (-8, 5)$
   (c) $(-1, 1), (1, -1)$    (d) $(a, b), (b, a)$
                             (assume $a \neq b$)

4. Compute the slope of the line in the accompanying figure using the two points indicated.
   (a) $A$ and $B$    (b) $B$ and $C$    (c) $A$ and $C$
   The idea here is that no matter which pair of points you choose, the slope is the same.

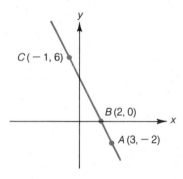

5. The slopes of four lines are indicated in the accompanying figure. List the numbers $m_1, m_2, m_3, m_4$ in order of increasing value.

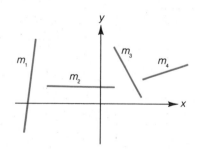

6. Compute the slopes of lines $L_1$, $L_2$, and $L_3$ in the following figure. (Assume that the broken lines are parallel to the axes.)

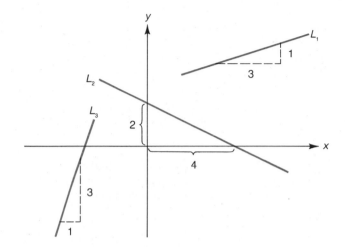

In Exercises 7–9, three points A, B, and C are specified. Find out if A, B, and C are collinear (lie on the same line) by checking to see whether the slope of $AB$ equals the slope of $BC$.

7. $A(8, 2), B(2, \frac{1}{2}), C(11, 1)$
8. $A(4, -3), B(-1, 0), C(-4, 2)$
9. $A(0, -5), B(3, 4), C(-1, -8)$

10. If the area of the "triangle" formed by three points is zero, then the points must in fact be collinear. Use this observation, along with the formula in Exercise 14(b) of Exercise Set 3.1, to rework Exercise 9.

11. (a) In the accompanying figure, find $\Delta x$, $\Delta y$, and $m = \Delta y/\Delta x$, starting at $(3, 1)$ and moving along the line up to $(5, 3)$.
    (b) Do the same as for part (a), but this time start at $(5, 3)$ and move down the line to $(3, 1)$.

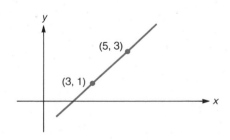

12. (a) In the figure, find $\Delta x$, $\Delta y$, and $m = \Delta y / \Delta x$, starting at $(6,0)$ and moving to $(4,3)$.
    (b) Do the same as for part (a), but start at $(4,3)$ and move down to $(6,0)$.

In Exercises 13 and 14, the graphs of the equations are straight lines. In each case, set up a table and draw the graph. Then pick any two points on the line and compute the slope.

13. (a) $y = 2x - 3$    (b) $y = 2x$    (c) $y = 2x + 1$
14. (a) $y = 3x - 4$    (b) $y = 3x$    (c) $y = 3x + 2$

15. Graph the equation $d = 5t$, which relates the distance $d$, in feet, covered by an object moving for $t$ sec at a speed of 5 ft/sec. Use $t$ for the horizontal axis. Your graph should be a straight line. Pick two points on the line and compute the slope. What do you observe?

16. Graph the equation $d = 30t$, which relates the distance $d$, in miles, covered by an object moving for $t$ hours at a rate of 30 mph. Your graph should be a straight line. Pick two points on the line and compute the slope. What do you observe?

17. The following graphs relate the distance and time for a moving object. Determine the velocity in each case.

(a)

(b)

(c)

18. Suppose that the cost $C$, in dollars, of producing $x$ units of a certain kind of stereo receiver is given by $C = 400 + 50x$.
    (a) Graph this equation.
    (b) Using the coordinates of two points on this line, compute the slope.
    (c) What is the marginal cost?

19. Suppose the cost $C$, in dollars, of producing $x$ record albums is given by $C = 0.5x + 500$. Graph this equation, determine the slope, and give the marginal cost.

**B**

20. Show that the slope of the line passing through the two points $(3,9)$ and $(3 + h, (3 + h)^2)$ is $6 + h$.

21. Show that the slope of the line passing through the two points $(x, x^2)$ and $(x + h, (x + h)^2)$ is $2x + h$.

22. Show that the slope of the line passing through the two points $(x, x^3)$ and $(x + h, (x + h)^3)$ is $3x^2 + 3xh + h^2$.

23. (a) Show that the slope of the line passing through the points $(x, \sqrt{x})$ and $(x + h, \sqrt{x + h})$ is $\dfrac{\sqrt{x + h} - \sqrt{x}}{h}$.
    (b) Show that the answer in part (a) can be written $\dfrac{1}{\sqrt{x + h} + \sqrt{x}}$.

24. Show that the slope of the line passing through the two points $\left(x, \dfrac{1}{x}\right)$ and $\left(x + h, \dfrac{1}{x + h}\right)$ is $\dfrac{-1}{x(x + h)}$.

25. Show that the slope of the line passing through the two points $(4,2)$ and $(4 + h, \sqrt{4 + h})$ can be written $\dfrac{1}{\sqrt{4 + h} + 2}$.

26. Find the slope of the line $L$ in the accompanying figure when the angle $\theta$ is as given.
    (a) $\theta = 30°$    *Hint:* In a 30°-60° right triangle, the side opposite the 30° angle is half the hypotenuse.
    (b) $\theta = 60°$
    (c) $\theta = 45°$

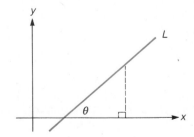

27. Why is the letter $m$ used to denote slope? Look up the verb *monter* in a French-to-English dictionary.

c 28. Sketch the curve $y = x^2$ and indicate the point $T(3,9)$ on the graph. In each case that follows, you're given the $x$-coordinate of a point $P$ on the curve. Find the corresponding $y$-coordinate, and then calculate the slope of the line through $P$ and $T$. Display your results in a table.

    (a) $x = 2.5$    (b) $x = 2.9$    (c) $x = 2.99$
    (d) $x = 2.999$    (e) $x = 2.9999$

    As $P$ gets closer to $T$, what numerical value does the slope of $\overline{PT}$ seem to approach? What would you estimate to be the slope of the line that is tangent to the curve $y = x^2$ at point $T$?

c 29. Follow the instructions for Exercise 28 using the curve $y = \sqrt{x}$, the point $T(1,1)$, and the following sequence of $x$-values.

    (a) $x = 1.1$    (b) $x = 1.01$    (c) $x = 1.001$
    (d) $x = 1.0001$    (e) $x = 1.00001$

    (Round off your answers to six decimal places.)

C

30. In this section, we saw in Example 2, that the slope of a given line does not depend on which two points on

the line are used in its calculation. We summarize this fact by saying that slope is *well-defined*. However, if the original definition of slope had instead been given by $m = \dfrac{y_2 + y_1}{x_2 + x_1}$, then the concept would not be well-defined in the sense described. Verify this as follows: Use the expression $\dfrac{y_2 + y_1}{x_2 + x_1}$ to calculate the "slope" of the line in Figure 3 (page 133), using the points $(2,1)$ and $(3,3)$ and then using the points $(4,5)$ and $(7,11)$. Note that the answers are *not* the same.

31. Find a point $P$ on the curve $y = x^3$ such that the slope of the line passing through $P$ and $(1,1)$ is $\frac{3}{4}$.

32. Find a point $P$ on the curve $y = \sqrt{x}$ such that the slope of the line through $P$ and $(1,1)$ is $\frac{1}{4}$.

33. Find a point $P$ on the curve $y = 1/x$ such that the slope of the line through $P$ and $(2,\frac{1}{2})$ is $-\frac{1}{16}$.

34. Find a point $P$ on the circle $x^2 + y^2 = 20$ such that the slope of the radius drawn from $(0,0)$ to $P$ is 2. (You will get two answers.)

## EQUATIONS OF LINES

### 3.4

*He [Pierre de Fermat (1601–1665)] introduced perpendicular axes and found the general equations of straight lines and circles and the simplest equations of parabolas, ellipses, and hyperbolas; and he further showed in a fairly complete and systematic way that every first- or second-degree equation can be reduced to one of these types.*

GEORGE F. SIMMONS in his book, *Calculus with Analytic Geometry* (New York: McGraw-Hill Book Company, 1985)

In this section we take a systematic look at equations of lines and their graphs. We begin by finding an equation for the line with slope $m$, passing through a given point $(x_1, y_1)$. See Figure 1(a) on the next page. Let $(x,y)$ be any other point on the line, as in Figure 1(b). Then the slope of the line is given by $m = \dfrac{y - y_1}{x - x_1}$. Thus $y - y_1 = m(x - x_1)$. Note that the given point $(x_1, y_1)$ also satisfies this last equation, because in that case we have $y_1 - y_1 = m(x_1 - x_1)$, or $0 = 0$. The equation $y - y_1 = m(x - x_1)$ is called the **point–slope formula**. We have shown that any point on the line satisfies this equation. (Conversely, it can be shown that if a point satisfies this equation, then the point does lie on the given line.)

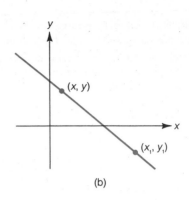

(a)                    (b)

**FIGURE 1**

| The Point–Slope Formula |
| --- |
| The equation of a line with slope $m$ passing through the point $(x_1, y_1)$ is <br><br> $y - y_1 = m(x - x_1)$ |

**EXAMPLE 1**   Write the equation of the line passing through $(-3, 1)$ with a slope of $-2$. Sketch a graph of the line.

**Solution**   Since the slope and a point are given, we use the point–slope formula:

$$y - y_1 = m(x - x_1)$$
$$y - 1 = -2[x - (-3)]$$
$$y - 1 = -2x - 6$$
$$y = -2x - 5$$

This is the required equation. One way to graph this line is to pick values for $x$ and then compute the corresponding $y$-values, as done in Section 3.2. The table and graph are displayed in Figure 2. Notice that if we know ahead of time that the graph is a straight line, a table as brief as that in Figure 2 is sufficient.

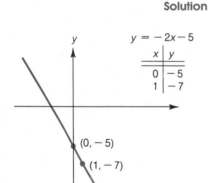

$y = -2x - 5$

| $x$ | $y$ |
| --- | --- |
| 0 | $-5$ |
| 1 | $-7$ |

**FIGURE 2**   $y = -2x - 5$

There is another way to go about graphing the line $y = -2x - 5$ in Example 1. Since slope is (change in $y$)/(change in $x$), a slope of $-2$ (i.e., $-2/1$) can be interpreted as telling us that if we start at $(-3, 1)$ and let $x$ increase by one unit, then $y$ must decrease by two units to bring us back to the line. Following this path, in Figure 3, takes us from $(-3, 1)$ to $(-2, -1)$. We now draw the line through these two points, as shown in Figure 3.

**EXAMPLE 2**   Find the equation of the line passing through the points $(-2, -3)$ and $(2, 5)$.

**Solution**   The slope of the line is

$$m = \frac{y_2 - y_1}{x_2 - x_1} = \frac{5 - (-3)}{2 - (-2)} = \frac{8}{4} = 2$$

**FIGURE 3**

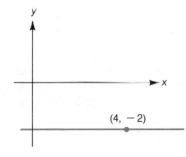

**FIGURE 4**

Knowing the slope, we can apply the point–slope formula, making use of either $(-2, -3)$ or $(2, 5)$. Using the point $(2, 5)$ as $(x_1, y_1)$, we have

$$y - y_1 = m(x - x_1)$$
$$y - 5 = 2(x - 2)$$
$$y - 5 = 2x - 4$$
$$y = 2x + 1$$

Thus the required equation is $y = 2x + 1$. You should check for yourself that the same answer is obtained using the point $(-2, -3)$ instead of $(2, 5)$ in the last set of calculations.

**EXAMPLE 3**   Find the equation of the horizontal line passing through the point $(4, -2)$. See Figure 4.

**Solution**   As we saw in Section 3.3, the slope of a horizontal line is zero. So we have

$$y - y_1 = m(x - x_1)$$
$$y - (-2) = 0(x - 4)$$
$$y = -2$$

Thus the equation of the horizontal line passing through $(4, -2)$ is $y = -2$.

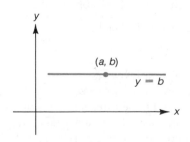

**FIGURE 5**

By using the point–slope formula exactly as we did in Example 3, we can show more generally that the equation of the horizontal line in Figure 5 passing through the point $(a, b)$ is $y = b$. The calculations look as follows:

$$y - y_1 = m(x - x_1)$$
$$y - b = 0(x - a)$$
$$y = b$$

FIGURE 6

What about the equation of the vertical line in Figure 6 passing through the point $(a,b)$? Because slope is not defined for vertical lines, the point–slope formula is not applicable this time. However, looking at Figure 6, note that as you move along the vertical line, your $x$-coordinate is always $a$; it is only the $y$-coordinate that varies. The equation $x = a$ expresses exactly these two facts; it says that $x$ must always be $a$ and it places no restrictions on $y$.

In the box that follows we summarize our results concerning horizontal and vertical lines.

### Equations of Vertical and Horizontal Lines

1. The equation of a horizontal line through the point $(a,b)$ is $y = b$. (See Figure 5.)

2. The equation of a vertical line through the point $(a,b)$ is $x = a$. (See Figure 6.)

**EXAMPLE 4**  What are the equations of the horizontal and vertical lines passing through the point $(-5,7)$?

Solution  The equation of the horizontal line through $(-5,7)$ is $y = 7$. The equation of the vertical line is $x = -5$.

Next let us find an equation for a straight line with slope $m$ and $y$-intercept $b$. See Figure 7. To say that the line has a $y$-intercept of $b$ is the same as saying that the line passes through $(0,b)$. The point–slope formula is applicable now, using the slope $m$ and the point $(0,b)$. We have

$$y - y_1 = m(x - x_1)$$
$$y - b = m(x - 0)$$
$$y = mx + b$$

FIGURE 7

This last equation is called the **slope–intercept formula**.

### The Slope–Intercept Formula

The equation of a line with slope $m$ and $y$-intercept $b$ is

$$y = mx + b$$

**EXAMPLE 5**  Write the equation of a line with slope 4 and $y$-intercept 1. Graph the line.

Solution  Substituting $m = 4$ and $b = 1$ in the equation $y = mx + b$ yields

$$y = 4x + 1$$

This is the required equation. We could draw the graph by first setting up a simple table, but for purposes of emphasis and review we proceed as we did at the end of Example 1. Starting from the point $(0,1)$, we interpret the slope of 4 as saying that if $x$ increases by 1, then $y$ increases by 4. This takes us from $(0,1)$ to the point $(1,5)$ and the line can now be sketched as in Figure 8.

**EXAMPLE 6** Find the slope and $y$-intercept of the line $3x - 5y = 15$.

Solution First solve for $y$ so that the equation is in the form $y = mx + b$:

$$3x - 5y = 15$$
$$-5y = -3x + 15$$
$$y = \frac{3}{5}x - 3$$

The slope $m$ and $y$-intercept $b$ can now be read directly from the equation: $m = \frac{3}{5}$; $b = -3$.

As a consequence of our work up to this point, we can say that the graph of any *linear equation*

$$Ax + By + C = 0, \qquad \text{where } A \text{ and } B \text{ are not both zero}$$

is a line. The reason for this is simply that an equation of this type can always be rewritten in one of the following three forms: $y = mx + b$, or $x = a$, or $y = b$. We summarize and expand upon this idea in the box that follows.

FIGURE 8  $y = 4x + 1$

---

| PROPERTY SUMMARY | The Graph of $Ax + By + C = 0$ |
|---|---|

The graph of a linear equation $Ax + By + C = 0$, where $A$ and $B$ are not both zero, is a line.

1. If $A \neq 0$ and $B \neq 0$, then the equation can be written

$$y = -\frac{A}{B}x - \frac{C}{B}$$

and the graph is a line with slope $-A/B$ and $y$-intercept $-C/B$.

2. If $A \neq 0$ and $B = 0$, then the equation can be written

$$x = -\frac{C}{A}$$

and the graph is a vertical line with $x$-intercept $-C/A$.

3. If $A = 0$ and $B \neq 0$, then the equation can be written

$$y = -\frac{C}{B}$$

and the graph is a horizontal line with $y$-intercept $-C/B$.

---

**EXAMPLE 7** Graph the (nonlinear) equation $(3x - 4y - 12)(3x + 4y - 12) = 0$

Solution From the given equation we obtain two linear equations:

$$3x - 4y - 12 = 0 \qquad \text{or} \qquad 3x + 4y - 12 = 0$$

The graphs of these two equations are lines. Perhaps the simplest method for graphing these lines is to obtain the intercepts.

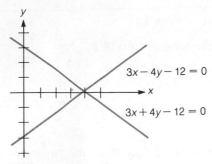

**FIGURE 9** The graph of
$(3x - 4y - 12)(3x + 4y - 12) = 0$
consists of the two lines
$3x - 4y - 12 = 0$ and
$3x + 4y - 12 = 0$.

$x$-intercept for $3x - 4y - 12 = 0$:     $3x - 12 = 0$
$$x = 4$$

$y$-intercept for $3x - 4y - 12 = 0$:     $-4y - 12 = 0$
$$y = -3$$

Since two points determine a line, the graph of $3x - 4y - 12 = 0$ is obtained now just by drawing the line through the two points $(4,0)$ and $(0, -3)$. In a similar way, you can check that the graph of the other line, $3x + 4y - 12 = 0$, passes through $(4,0)$ and $(0,3)$. The required graph then consists of the two lines shown in Figure 9.

We conclude this section by discussing two useful relationships regarding the slopes of parallel lines and the slopes of perpendicular lines. First, nonvertical parallel lines have the same slope. This should seem reasonable if you recall that slope is a number indicating the direction or slant of a line. The relationship concerning slopes of perpendicular lines is not so obvious. The slopes of two nonvertical perpendicular lines are negative reciprocals of each other. That is, if $m_1$ and $m_2$ denote the slopes of the two perpendicular lines, then $m_1 = -1/m_2$, or, equivalently, $m_1 m_2 = -1$. For example, if a line has a slope of $\frac{2}{3}$, then any line perpendicular to it must have a slope of $-\frac{3}{2}$. For reference we summarize these facts in the box that follows. (Proofs of these facts are outlined in detail in Exercises 52 and 53 at the end of this section.)

**Parallel and Perpendicular Lines**

Let $m_1$ and $m_2$ denote the slopes of two nonvertical lines. Then:

1. The lines are parallel if and only if $m_1 = m_2$;
2. The lines are perpendicular if and only if $m_1 = -1/m_2$.

**EXAMPLE 8**   Are the two lines $3x - 6y - 8 = 0$ and $2y = x + 1$ parallel?

**Solution**   By solving each equation for $y$, we can see what the slopes are:

$$3x - 6y - 8 = 0 \qquad\qquad 2y = x + 1$$
$$-6y = -3x + 8 \qquad\qquad y = \frac{1}{2}x + \frac{1}{2}$$
$$y = \frac{-3}{-6}x + \frac{8}{-6}$$
$$y = \frac{1}{2}x - \frac{4}{3}$$

From this we see that both lines have the same slope, namely, $m = \frac{1}{2}$. It follows therefore that the lines are parallel.

**EXAMPLE 9**   Find the equation of the line parallel to $5x + 6y = 30$ and passing through the origin.

FIGURE 10

**Solution**   First, find the slope of the line $5x + 6y = 30$:

$$5x + 6y = 30$$
$$6y = -5x + 30$$
$$y = -\frac{5}{6}x + 5$$

The slope is the $x$-coefficient in the last equation: $m = -\frac{5}{6}$. A parallel line will have the same slope. Since the required line is to pass through $(0,0)$, we have

$$y - y_1 = m(x - x_1)$$
$$y - 0 = -\frac{5}{6}(x - 0)$$
$$y = -\frac{5}{6}x$$

This is the equation of a line parallel to $5x + 6y = 30$ and passing through the origin, as required. See Figure 10.

**EXAMPLE 10**   (a)  Sketch the circle $x^2 + y^2 = 25$.
(b)  Verify that the point $(3, -4)$ lies on the circle.
(c)  Find the equation of the line tangent to this circle at the point $(3, -4)$. Write the equation in the form $y = mx + b$.

**Solution**   (a)  and (b) From our work in Section 3.2, we know that the equation $x^2 + y^2 = 25$ represents a circle with center $(0,0)$ and radius 5. To show that the point $(3, -4)$ lies on the circle, we need only check that the values $x = 3, y = -4$ satisfy the given equation. We have

$$x^2 + y^2 = 25$$
$$(3)^2 + (-4)^2 = 25$$
$$9 + 16 = 25 \qquad \text{True}$$

This shows that the point $(3, -4)$ does lie on the circle. See Figure 11(a).

(c)  To find the equation of the required tangent line, first draw the radius from $(0,0)$ to $(3, -4)$ as in Figure 11(b). The slope of this radius is

$$\frac{-4 - 0}{3 - 0} = -\frac{4}{3}$$

FIGURE 11

From elementary geometry, it is known that the tangent line is perpendicular to this radius. Since the slope of the radius was found to be $-\frac{4}{3}$, it follows that the slope of the tangent line is $\frac{3}{4}$. The point–slope formula is now applicable:

$$y - y_1 = m(x - x_1)$$

$$y - (-4) = \frac{3}{4}(x - 3)$$

Check now for yourself that this can be simplified to $y = \frac{3}{4}x - \frac{25}{4}$, which is the required equation of the tangent line. ▪

## EXERCISE SET 3.4

### A

*In Exercises 1 and 2, find the equation for the line having the given slope and passing through the given point. Write your answers in the form $y = mx + b$.*

1. (a) $m = -5$; through $(-2, 1)$
   (b) $m = 4$; through $(4, -4)$
   (c) $m = \frac{1}{3}$; through $(-6, -\frac{2}{3})$
   (d) $m = -1$; through $(0, 1)$

2. (a) $m = 22$; through $(0, 0)$
   (b) $m = -222$; through $(0, 0)$
   (c) $m = \sqrt{2}$; through $(0, 0)$

*In Exercises 3 and 4, find the equation for the line passing through the two given points. Write your answer in the form $y = mx + b$.*

3. (a) $(4, 8)$ and $(-3, -6)$
   (b) $(-2, 0)$ and $(3, -10)$
   (c) $(-3, -2)$ and $(4, -1)$

4. (a) $(7, 9)$ and $(-11, 9)$
   (b) $(\frac{5}{4}, 2)$ and $(\frac{3}{4}, 3)$
   (c) $(12, 13)$ and $(13, 12)$

*In Exercises 5 and 6, write the equation of a vertical line passing through the given point. In Exercises 7 and 8, write the equation of a horizontal line passing through the given point.*

5. $(-3, 4)$                6. $(8, 5)$

7. $(-3, 4)$                8. $(8, 5)$

9. Is the graph of the line $x = 0$ the $x$-axis or the $y$-axis?

10. Is the graph of the line $y = 0$ the $x$-axis or the $y$-axis?

*In Exercises 11 and 12, find the equation of the line with the given slope and $y$-intercept.*

11. (a) slope $-4$; $y$-intercept 7
    (b) slope 2; $y$-intercept $\frac{3}{2}$
    (c) slope $-\frac{4}{3}$, $y$-intercept 14

12. (a) slope 0; $y$-intercept 14
    (b) slope 14; $y$-intercept 0

*In Exercises 13–20, find the slope and $y$-intercept for each line and draw the graph.*

13. $y = -7x + 4$

14. $3x - 4y = 12$

15. $x - y - 1 = 0$

16. $2x - 3 - 2y = 4y + 6$

17. $\dfrac{x}{2} + \dfrac{y}{3} = 1$

18. $y - \dfrac{5}{3} = 0$

19. $y - 4 = \frac{1}{2}(x - 2)$

20. $y = 2(x + 1)$

*In Exercises 21–26, find an equation for the line that is described, and sketch the graph. For Exercises 21–23, write the final answer in the form $y = mx + b$; for Exercises 24–26, write the answer in the form $Ax + By + C = 0$.*

21. (a) Passing through $(-3, -1)$, with slope 4
    (b) Passing through $(\frac{5}{2}, 0)$, with slope $\frac{1}{2}$
    (c) With $x$-intercept 6, $y$-intercept 5
    (d) With $x$-intercept $-2$, slope $\frac{3}{4}$
    (e) Passing through $(1, 2)$ and $(2, 6)$

22. (a) Passing through $(-7, -2)$ and $(0, 0)$
    (b) Passing through $(6, -3)$, with $y$-intercept 8
    (c) Passing through $(0, -1)$ and with the same slope as the line $3x + 4y = 12$
    (d) Passing through $(6, 2)$ and with the same $x$-intercept as the line $-2x + y = 1$
    (e) With $x$-intercept $-6$, $y$-intercept $\sqrt{2}$

23. With the same $x$- and $y$-intercepts as the circle $x^2 + y^2 + 4x - 4y + 4 = 0$

24. Passing through $(3, -5)$ and through the center of the circle $4x^2 + 8x + 4y^2 - 24y + 15 = 0$

25. Passing through $(-3, 4)$ and parallel to the $x$-axis

26. Passing through $(-3, 4)$ and parallel to the $y$-axis

*In Exercises 27–28, graph the equations.*

27. (a) $(x + y - 1)(x - y - 1) = 0$
    (b) $(2x + 3y + 6)(2x - 3y + 6) = 0$

28. (a) $(3x + 5y + 15)(3x - 5y + 15) = 0$
    (b) $(x - 2y - 4)^2 = 0$

29. State whether the two lines are parallel, perpendicular, or neither.
    (a) $3x - 4y = 12$; $4x - 3y = 12$
    (b) $y = 5x - 16$; $y = 5x + 2$
    (c) $5x - 6y = 25$; $6x + 5y = 0$
    (d) $y = -\frac{2}{3}x - 1$; $y = \frac{3}{2}x - 1$
    (e) $-2x - 5y = 1$; $y - \frac{2}{5}x - 4 = 0$
    (f) $x = 8y + 3$; $4y - \frac{1}{2}x = 32$

30. Are the lines $y = x + 1$ and $y = 1 - x$ parallel, perpendicular, or neither?

*In Exercises 31–36, find an equation for the line that is described. Write each answer in the two forms $y = mx + b$ and $Ax + By + C = 0$.*

31. Parallel to $2x - 5y = 10$ and passing through $(-1, 2)$

32. Parallel to $4x + 5y = 20$ and passing through $(0, 0)$

33. Perpendicular to $4y - 3x = 1$ and passing through $(4, 0)$

34. Perpendicular to $x - y + 2 = 0$ and passing through $(3, 1)$

35. Parallel to $3x - 5y = 25$ and with the same $y$-intercept as $6x - y + 11 = 0$

36. Perpendicular to $9y - 2x = 3$ and with the same $x$-intercept as the circle $x^2 + y^2 - 12x - 2y = -36$

37. (a) Sketch the circle $x^2 + y^2 = 25$.
    (b) Find the equation of the line tangent to this circle at the point $(-4, -3)$. Include a sketch.

38. (a) Tangents are drawn to the circle $x^2 + y^2 = 169$ at the points $(5, 12)$ and $(5, -12)$. Find the equations of the tangents.
    (b) Find the coordinates of the point where these two tangents intersect.

*As background for Exercises 39–42, refer to the discussions of marginal cost and velocity in Section 3.3.*

39. Suppose that the cost $C$ in dollars of producing $x$ electric fans is given by $C = 450 + 8x$.
    (a) Find the cost to produce 10 fans.
    (b) Find the cost to produce 11 fans.
    (c) Use your answers in parts (a) and (b) to find the marginal cost. (Then check that your answer is the slope of the line $C = 450 + 8x$.)

40. Suppose that the cost to a manufacturer of producing $x$ units of a certain motorcycle is given by $C = 220x + 4000$, where $C$ is in dollars.
    (a) Find the marginal cost.
    (b) Find the cost of producing 500 motorcycles.
    (c) Use your answers in parts (a) and (b) to find the cost of producing 501 motorcycles.

41. The distance $d$, in feet, covered by a particular object in $t$ sec is given by $d = 5t$.
    (a) Find the velocity of the object.
    (b) Find the distance covered in 15 sec.
    (c) Use your answers in parts (a) and (b) to find the distance covered in 16 sec.

42. Two points $A$ and $B$ move along the $x$-axis. After $t$ sec, their positions are given by the equations

    $A$: $\ x = 3t + 100$

    $B$: $\ x = 20t - 36$

(a) Which point is traveling faster, $A$ or $B$?
(b) Which point is farther to the right when $t = 0$?
(c) At what time $t$ do $A$ and $B$ have the same $x$-coordinate?

**B**

43. A line with a slope of $-5$ passes through the point $(3,6)$. Find the area of the triangle in the first quadrant formed by this line and the coordinate axes.

44. The $y$-intercept of the line in the figure is 6. Find the slope of the line if the area of the shaded triangle is 72 square units.

45. (a) Show that the equation of a line with $x$- and $y$-intercepts of $a$ and $b$, respectively, is $\dfrac{x}{a} + \dfrac{y}{b} = 1$. (Assume that both $a$ and $b$ are nonzero.)
   (b) Use the result in part (a) to find the equation of a line whose $x$- and $y$-intercepts are 6 and 7, respectively. Write your answer in the form $Ax + By + C = 0$.

46. Suppose that the two equations

   $$ax + by + c = 0 \qquad \text{and} \qquad Ax + By + C = 0$$

   represent two nonvertical lines.
   (a) Show that if the lines are parallel, then $aB - Ab = 0$.
   (b) Show that if $aB - Ab = 0$, then the lines are parallel.
   (c) Show that if the lines are perpendicular, then $Aa + Bb = 0$.
   (d) Show that if $Aa + Bb = 0$, then the lines are perpendicular.

47. (a) Sketch the line $y = \frac{1}{2}x - 5$ and the point $P(1,3)$. Follow parts (b)–(d) to calculate the perpendicular distance from point $P(1,3)$ to the line.
   (b) Find the equation of the line that passes through $P(1,3)$ and that is perpendicular to the line $y = \frac{1}{2}x - 5$.
   (c) Find the coordinates of the point where these two lines intersect.
   (d) Use the distance formula to find the perpendicular distance from $P(1,3)$ to the line $y = \frac{1}{2}x - 5$.

48. (a) Show that the equation of the line tangent to the circle $x^2 + y^2 = a^2$ at the point $(x_1, y_1)$ on the circle is
   $$x_1 x + y_1 y = a^2$$
   (b) Use the result in part (a) to rework Exercises 37(b) and 38(a).

49. Follow the instructions in parts (a)–(e) to determine the slope $m$ of the line in the figure.
   (a) Use the point–slope formula to show that the equation of the line is
   $$y - 1 = m(x - 2)$$
   (b) Show that the $y$-intercept for the line is $-2m + 1$. Show that the $x$-intercept is $\dfrac{2m - 1}{m}$.
   (c) Show that the area of the triangle, being one-half base times height, is $\dfrac{1}{2}\left(\dfrac{2m - 1}{m}\right)(-2m + 1)$.
   (d) Set the expression for area in part (c) equal to 4, as given, and simplify the resulting equation to obtain $4m^2 + 4m + 1 = 0$.
   (e) Solve the equation for $m$, either by factoring or by using the quadratic formula. You should obtain $m = -\frac{1}{2}$.

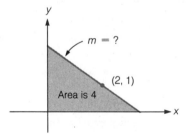

50. A line with a negative slope passes through the point $(3,1)$. The area of the triangle bounded by this line and the axes is 6 square units. Find the possible slopes of the line. *Hint:* See Exercise 49.

51. By analyzing sales figures, the accountant for College Stereo Company knows that 280 units of a top-of-the-line turntable can be sold each month when the price is $P = \$195$ per unit. The figures also show that for each \$15 hike in the price, 10 fewer units are sold monthly.
   (a) Let $x$ denote the number of units sold per month and $P$ the price per unit. Find an equation that expresses $x$ in terms of $P$, assuming that this relationship is linear. *Hint:* $\Delta x/\Delta P = -10/15 = -2/3$.

(b) Use the equation that you found in part (a) to determine how many units can be sold in a month when the price is $270 per unit.

(c) What should the price be to sell 205 units per month?

52. This exercise outlines a proof of the fact that two non-vertical lines are parallel if and only if their slopes are equal. The proof relies on the following observation for the given figure: The lines $y = m_1x + b_1$ and $y = m_2x + b_2$ will be parallel if and only if the two vertical distances $AB$ and $CD$ are equal. (In the figure, the points $C$ and $D$ both have an $x$-coordinate of 1.)

(a) Verify that the coordinates of $A$, $B$, $C$, and $D$ are:

$$A(0, b_1) \quad B(0, b_2) \quad C(1, m_1 + b_1) \quad D(1, m_2 + b_2)$$

(b) Using part (a), check that $AB = b_1 - b_2$ and $CD = (m_1 + b_1) - (m_2 + b_2)$.

(c) Use part (b) to show that the equation $AB = CD$ is equivalent to $m_1 = m_2$.

53. This exercise outlines a proof of the fact that two non-vertical lines with slopes $m_1$ and $m_2$ are perpendicular if and only if $m_1m_2 = -1$. In the accompanying figure, we've assumed that our two nonvertical lines $y = m_1x$ and $y = m_2x$ intersect at the origin. [If they did not

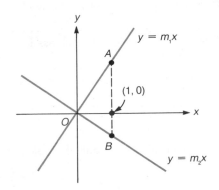

intersect there, we could just as well work with lines parallel to these that do intersect at $(0, 0)$, recalling that parallel lines have the same slope.] The proof relies on the following geometric fact:

$$OA \perp OB \quad \text{if and only if} \quad OA^2 + OB^2 = AB^2$$

(a) Verify that the coordinates of $A$ and $B$ are $A(1, m_1)$ and $B(1, m_2)$.

(b) Show that

$$OA^2 = 1 + m_1^2$$
$$OB^2 = 1 + m_2^2$$
$$AB^2 = m_1^2 - 2m_1m_2 + m_2^2$$

(c) Use part (b) to show that the equation $OA^2 + OB^2 = AB^2$ is equivalent to $m_1m_2 = -1$.

54. Find the equation of the line that is obtained when $y = mx + b$ is reflected about (a) the $x$-axis; (b) the $y$-axis; (c) the origin.

## C

55. The vertices of a triangle are $A(-4, 0)$, $B(2, 0)$, and $C(0, 6)$. Let $M_1$, $M_2$, and $M_3$ be the midpoints of $\overline{AB}$, $\overline{BC}$, and $\overline{AC}$, respectively. Let $H_1$, $H_2$, and $H_3$ be the feet of the altitudes on sides $\overline{AB}$, $\overline{BC}$, and $\overline{AC}$, respectively.

(a) Find the equation of the circle passing through $M_1$, $M_2$, and $M_3$. *Answer:* $3x^2 + 3y^2 + 3x - 11y = 0$

(b) Find the equation of the circle passing through $H_1$, $H_2$, and $H_3$. *Answer:* $3x^2 + 3y^2 + 3x - 11y = 0$

(c) Find the point $P$ at which the three altitudes intersect. *Answer:* $(0, \frac{4}{3})$

(d) Let $N_1$, $N_2$, and $N_3$ be the midpoints of $\overline{AP}$, $\overline{BP}$, and $\overline{CP}$, respectively. Show that the circle obtained in parts (a) and (b) passes through $N_1$, $N_2$, and $N_3$. *Note:* This circle is called the *nine-point circle* of triangle $ABC$.

(e) The *circumcircle* of triangle $ABC$ is the circle passing through the three points $A$, $B$, and $C$. Show that the coordinates of the center $Q$ of the circumcircle are $Q(-1, \frac{7}{3})$. *Hint:* Use the result from geometry that the perpendicular bisectors of the sides of the triangle intersect at the center of the circumcircle.

(f) For triangle $ABC$, show that the radius of the nine-point circle is one-half the radius of the circumcircle.

(g) Show that the midpoint of line segment $\overline{QP}$ is the center of the nine-point circle.

(h) Find the point $G$ where the three medians of triangle $ABC$ intersect.

(i) Show that $G$ lies on line segment $\overline{QP}$ and that $PG = 2(GQ)$.

## CHAPTER 3   SUMMARY OF PRINCIPAL FORMULAS AND DEFINITIONS

| FORMULAS OR NOTATIONS | PAGE REFERENCE | COMMENTS |
|---|---|---|
| 1. Pythagorean Theorem $a^2 + b^2 = c^2$ | 118 | In a right triangle, the lengths of the sides are related by this equation, where $c$ is the length of the hypotenuse. Conversely, if the lengths of the sides of a triangle are related by an equation of the form $a^2 + b^2 = c^2$, then the triangle is a right triangle, and $c$ is the length of the hypotenuse. |
| 2. $d = \sqrt{(x_2 - x_1)^2 + (y_2 - y_1)^2}$ | 118 | $d$ is the distance between $(x_1, y_1)$ and $(x_2, y_2)$. |
| 3. $\left(\dfrac{x_1 + x_2}{2}, \dfrac{y_1 + y_2}{2}\right)$ | 120 | These are the coordinates of the midpoint of the line segment joining $(x_1, y_1)$ and $(x_2, y_2)$. |
| 4. Symmetry about the $x$-axis | 125 | A graph is symmetric about the $x$-axis if, for each point $(x, y)$ on the graph, the point $(x, -y)$ is also on the graph. The points $(x, y)$ and $(x, -y)$ are *reflections* of each other about the $x$-axis. |
| 5. Symmetry about the $y$-axis | 125 | A graph is symmetric about the $y$-axis if, for each point $(x, y)$ on the graph, the point $(-x, y)$ is also on the graph. The points $(x, y)$ and $(-x, y)$ are *reflections* of each other about the $y$-axis. |
| 6. Symmetry about the origin | 125 | A graph is symmetric about the origin if, for each point $(x, y)$ on the graph, the point $(-x, -y)$ is also on the graph. The points $(x, y)$ and $(-x, -y)$ are *reflections* of each other about the origin. |
| 7. Symmetry tests | 126 | (i) The graph of an equation is symmetric about the $y$-axis if replacing $x$ with $-x$ yields an equivalent equation.<br>(ii) The graph of an equation is symmetric about the $x$-axis if replacing $y$ with $-y$ yields an equivalent equation.<br>(iii) The graph of an equation is symmetric about the origin if replacing $x$ and $y$ with $-x$ and $-y$, respectively, yields an equivalent equation. |
| 8. $(x - h)^2 + (y - k)^2 = r^2$ | 129 | This is the equation of a circle with center at $(h, k)$ and radius $r$. |
| 9. $m = \dfrac{y_2 - y_1}{x_2 - x_1}$ | 132 | $m$ is the slope of a nonvertical line passing through the points $(x_1, y_1)$ and $(x_2, y_2)$. |
| 10. $y - y_1 = m(x - x_1)$ | 140 | This is the point–slope form for the equation of a line passing through the point $(x_1, y_1)$ with slope $m$. |
| 11. $y = mx + b$ | 142 | This is the slope–intercept form for the equation of a line with slope $m$ and $y$-intercept $b$. |
| 12. $m_1 = m_2$ | 144 | Nonvertical parallel lines have the same slope. |
| 13. $m_1 = \dfrac{-1}{m_2}$ | 144 | Two nonvertical lines are perpendicular if and only if the slopes are negative reciprocals of one another. |

## CHAPTER 3   REVIEW EXERCISES

NOTE   The fifteen exercises in this set with colored problem numbers constitute a chapter test.

*In Exercises 1–18, find an equation for the line satisfying the given conditions. Write your answer in the form $y = mx + b$.*

1. Passing through $(-4, 2)$ and $(-6, 6)$
2. $m = -2$; $y$-intercept 5
3. $m = \frac{1}{4}$; passing through $(-2, -3)$
4. $m = \frac{1}{3}$; $x$-intercept $-1$
5. $x$-intercept $-4$; $y$-intercept 8
6. $m = -10$; $y$-intercept 0
7. $y$-intercept $-2$; parallel to the $x$-axis
8. Passing through $(0, 0)$ and parallel to $6x - 3y = 5$
9. Passing through $(1, 2)$ and perpendicular to the line $x + y + 1 = 0$
10. Passing through $(1, 1)$ and through the center of the circle $x^2 - 4x + y^2 - 8y + 16 = 0$
11. Passing through the centers of the circles $x^2 + 4x + y^2 + 2y = 0$ and $x^2 - 4x + y^2 - 16y = 0$
12. $m = 3$; the same $x$-intercept as the line $3x - 0y = 12$
13. Passing through the origin and the midpoint of the line segment joining the points $(-2, -3)$ and $(6, -5)$
14. Tangent to the circle $x^2 + y^2 = 20$ at the point $(-2, 4)$ on the circle
15. Tangent to the circle $x^2 - 6x + y^2 + 8y = 0$ at the point $(0, 0)$
16. Passing through $(2, 4)$; the $y$-intercept is twice the $x$-intercept
17. Passing through $(2, -1)$; the sum of the $x$- and $y$-intercepts is 2 (There are two answers.)
18. Passing through $(4, 5)$; no $x$-intercept
19. Find the distance between the points $(-1, 2)$ and $(4, -10)$.
20. (a) Find the perimeter of the triangle with vertices $A(3, 1)$, $B(7, 4)$, and $C(-2, 13)$.
    (b) Find the perimeter of the triangle formed by joining the midpoints of the sides of the triangle in part (a).
21. Which point is farther from the origin, $(15, 6)$ or $(16, 2)$?
22. Find the midpoint of the line segment joining
    (a) $(4, 9)$ and $(10, -2)$   (b) $(\frac{2}{3}, -1)$ and $(-\frac{3}{4}, -\frac{1}{3})$

*In Exercises 23–32, test each equation for symmetry about the x-axis, the y-axis, and the origin.*

23. $y = x^4 - 2x^2$
24. $y = x^4 - 2x^2 + 1$
25. $y = x^3 + 5x$
26. $y = x^3 + 5x + 1$
27. $y^2 = (x + y)^4$
28. $y = \dfrac{2}{1 - x^2}$
29. $y = 3x - \dfrac{1}{x}$
30. $y = 2^x + 2^{-x}$
31. $y = 2^x - 2^{-x}$
32. $x^2 + y^2 + 2x = 2\sqrt{x^2 + y^2}$

*For Exercises 33–42, tell whether each graph is symmetric about the x-axis, the y-axis, or the origin.*

33.

34.

35.

36.

37.

38.

39.

40.

**41.**

**42.**

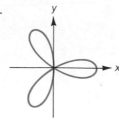

*Graph the equations in Exercises 43–62.*

**43.** $2x - 5y = 10$

**44.** $y = x^2 + 4x$

**45.** $x = 9 - y^2$

**46.** $y = -2x + 6$

**47.** $x^2 + y^2 = 1$

**48.** $(x + 2)^2 + y^2 = 1$

**49.** $y = |x| + 1$

**50.** $y = 1 - |x|$

**51.** $y = x^2 - 4x + 4$

**52.** $y = -x$

**53.** $y = |x - 2| + 2$

**54.** $y = |x - 2| - 2$

**55.** $3x + y = 0$

**56.** $|x| + |y| = 1$

**57.** $x^2 - 4x + y^2 + 6y = 0$

**58.** $x^2 + 2x + y^2 + 8y = 8$

**59.** $y = \dfrac{x^2 - 16}{x - 4}$

**60.** $y = 1 + \dfrac{1}{x}$

**61.** $(4x - y + 4)(4x + y - 4) = 0$

**62.** $(y - x)(y - x + 2)(y - x - 2) = 0$

**63.** Find a value for $t$ such that the slope of the line passing through $(2, 1)$ and $(5, t)$ is 6.

**64.** Express in terms of $x$ the slope of the line passing through $(x, x^2)$ and $(-3, 9)$.

**65.** Express in terms of $x$ the slope of the line passing through $(-2, -8)$ and $(x, x^3)$.

**66.** The vertices of a parallelogram are $(0, 0)$, $(5, 2)$, $(8, 7)$, and $(3, 5)$. Find the midpoint of each diagonal. Observation?

**67.** The vertices of a right triangle are $A(0, 0)$, $B(0, 2b)$, and $C(2c, 0)$. Let $M$ be the midpoint of the hypotenuse. Compute the three distances $MA$, $MB$, and $MC$. What do you observe?

**68.** The vertices of parallelogram $ABCD$ are $A(-4, -1)$, $B(2, 1)$, $C(3, 3)$, and $D(-3, 1)$.
   **(a)** Compute the sum of the squares of the two diagonals.
   **(b)** Compute the sum of the squares of the four sides. What do you observe?

*In Exercises 69–72, two points are given. In each case compute* **(a)** *the distance between the two points,* **(b)** *the slope of the line segment joining the two points, and* **(c)** *the midpoint of the line segment joining the two points.*

**69.** $(2, 5)$ and $(5, -6)$

**70.** $(\frac{3}{2}, 1)$ and $(\frac{7}{2}, 9)$

**71.** $(\sqrt{3}/2, 1/2)$ and $(-\sqrt{3}/2, -1/2)$

**72.** $(\sqrt{3}, \sqrt{2})$ and $(\sqrt{2}, \sqrt{3})$

**73.** A line passes through the points $(1, 2)$ and $(4, 1)$. Find the area of the triangle bounded by this line and the coordinate axes.

**74.** Suppose that the cost $C$ in dollars of producing $x$ typewriters is $C = 2200 + 50x$.
   **(a)** Find the cost of producing 100 typewriters.
   **(b)** Find the cost of producing 101 typewriters.
   **(c)** Use your answers in parts (a) and (b) to compute the marginal cost.

**75.** Suppose that the cost $C$ in dollars of producing $x$ tires is $C = 8500 + 12x$.
   **(a)** What is the marginal cost?
   **(b)** Find the cost of producing 1000 tires.
   **(c)** Use your answers in parts (a) and (b) to find the cost of producing 1001 tires. Check your answer by letting $x = 1001$ in the given equation.

**76.** A point moves along the $x$-axis, and its $x$-coordinate after $t$ seconds is $x = 4t + 10$. (Assume that $x$ is in centimeters.)
   **(a)** What is the velocity?
   **(b)** What is the $x$-coordinate when $t = 2$ sec?
   **(c)** Use your answers in parts (a) and (b) to find the $x$-coordinate when $t = 3$ sec. Check your answer by letting $t = 3$ in the given equation.

**77.** Let $P_1(x_1, y_1)$ and $P_2(x_2, y_2)$ be two given points. Let $Q$ be the point

$$(\tfrac{1}{3}x_1 + \tfrac{2}{3}x_2, \tfrac{1}{3}y_1 + \tfrac{2}{3}y_2)$$

   **(a)** Show that the points $P_1$, $Q$, and $P_2$ are collinear. *Hint:* Compute the slope of $\overline{P_1Q}$ and the slope of $\overline{QP_2}$.
   **(b)** Show that $P_1Q = \frac{2}{3}P_1P_2$. (In other words, $Q$ is on the line segment $\overline{P_1P_2}$ and two-thirds of the way from $P_1$ to $P_2$.)

**78.** **(a)** Let the vertices of $\triangle ABC$ be $A(-5, 3)$, $B(7, 7)$, and $C(3, 1)$. Find the point on each median that is two-thirds of the way from the vertex to the midpoint of the opposite side. (Recall that a median of a triangle is a line segment drawn from a vertex to the midpoint of the opposite side.) What do you observe? *Hint:* Use the result in Exercise 77.
   **(b)** Follow part (a) but take the vertices to be $A(0, 0)$, $B(2a, 0)$, and $C(2b, 2c)$. What do you observe? What does this prove?

**79.** Suppose that the circle $(x - h)^2 + (y - k)^2 = r^2$ has two $x$-intercepts, $x_1$ and $x_2$, and two $y$-intercepts, $y_1$ and $y_2$. Compute the quantity $x_1x_2 - y_1y_2$.

**80.** The point $(1, -2)$ is the midpoint of a chord of the circle $x^2 - 4x + y^2 + 2y = 15$. Find the length of the chord.

81. In the accompanying figure, points $P$ and $Q$ trisect the hypotenuse in $\triangle ABC$. Prove that the square of the hypotenuse is equal to $\frac{9}{5}$ the sum of the squares of the distances from the trisection points to the vertex $A$ of the right angle. *Hint:* Let the coordinates of $B$ and $C$ be $(0, 3b)$ and $(3c, 0)$, respectively. Then the coordinates of $P$ and $Q$ are $(c, 2b)$ and $(2c, b)$, respectively.

82. Figure (a) shows a triangle with sides of lengths $s$, $t$, and $u$ and a median of length $m$. (A median of a triangle is a line segment drawn from a vertex to the midpoint of the opposite side.) Prove that

$$m^2 = \frac{1}{2}(s^2 + t^2) - \frac{1}{4}u^2$$

*Hint:* Set up a coordinate system as indicated in Figure (b). Then each of the quantities $m^2$, $s^2$, $t^2$, and $u^2$ can be computed in terms of $a$, $b$, and $c$.

(a)

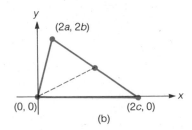

(b)

83. The accompanying figure shows the two tangent lines drawn from the point $(a, b)$ to the circle $x^2 + y^2 = R^2$.

Follow steps (a) through (d) to show that the equation of the line passing through the two points of tangency is

$$ax + by = R^2$$

(This problem, along with the following clever solution, appears in the classic text by Isaac Todhunter, *A Treatise on Plane Co-ordinate Geometry*, first published in 1855.)

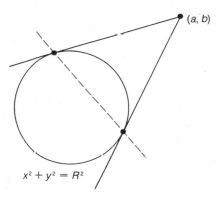

(a) Let $(x_1, y_1)$ be one of the points of tangency. Show that the equation of the tangent line through this point is $x_1 x + y_1 y = R^2$.
(b) Using the result in part (a), explain why $x_1 a + y_1 b = R^2$.
(c) In a similar fashion, explain why $x_2 a + y_2 b = R^2$, where $(x_2, y_2)$ is the other point of tangency.
(d) The equation $ax + by = R^2$ represents a line. Explain why this line must pass through $(x_1, y_1)$ and $(x_2, y_2)$.

84. In this exercise you'll derive a useful formula for the (perpendicular) distance $d$ from the point $(x_0, y_0)$ to the line $y = mx + b$. The formula is

$$d = \frac{|y_0 - mx_0 - b|}{\sqrt{1 + m^2}}$$

(a) Refer to the figure (on the next page). Use similar triangles to show that

$$\frac{d}{AB} = \frac{1}{\sqrt{1 + m^2}}.$$

Therefore, $d = AB/\sqrt{1 + m^2}$.
(b) Check that $AB = AC - BC = y_0 - mx_0 - b$.
(c) Conclude from parts (a) and (b) that

$$d = \frac{y_0 - mx_0 - b}{\sqrt{1 + m^2}}.$$

For the general case (in which the point and line may not be situated as in our figure), we need to use the absolute value of the quantity in the numerator to assure that $AB$ and $d$ are nonnegative.

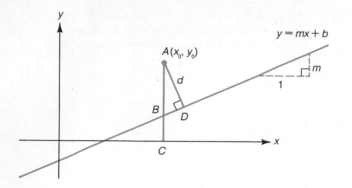

85. Use the formula given in Exercise 84 to find the distance from the point $(1, 2)$ to the line $y = \frac{1}{2}x - 5$.

86. Use the formula given in Exercise 84 to show that the distance from the point $(x_0, y_0)$ to the line $Ax + By + C = 0$ is

$$d = \frac{|Ax_0 + By_0 + C|}{\sqrt{A^2 + B^2}}$$

87. Use the formula given in Exercise 86 to find the distance from the point $(-1, -3)$ to the line $2x + 3y - 6 = 0$.

88. Find the equation of the circle with center $(2, 3)$ that is tangent to the line $x + y - 1 = 0$. *Hint:* Use the formula given in Exercise 86 to find the radius.

# 4 FUNCTIONS

*From the beginning of modern mathematics in the 17th century the concept of function has been at the very center of mathematical thought.*

RICHARD COURANT and FRITZ JOHN in *Introduction to Calculus and Analysis* (New York: Wiley-Interscience, 1965)

*The word "function" was introduced into mathematics by Leibniz, who used the term primarily to refer to certain kinds of mathematical formulas. It was later realized that Leibniz's idea of function was much too limited in scope, and the meaning of the word has since undergone many steps of generalization.*

TOM M. APOSTOL in *Calculus*, second edition (New York: John Wiley & Sons, 1967)

**INTRODUCTION**  We begin the study of functions in this chapter. The first section deals with matters of definition and notation. In Sections 4.2 and 4.3 we discuss graphs of functions; several techniques are developed that allow us to graph functions with a minimum of calculation. In Sections 4.4 and 4.5 we find that two functions can be combined in various ways to produce new functions. One of these ways of combining two functions is known as *composition of functions*. As you'll see, this is the unifying theme between Sections 4.4 and 4.5. Finally, in Section 4.6 we study certain types of functions and formulas that occur so frequently in the sciences that a special terminology exists to describe them.

## THE DEFINITION OF A FUNCTION

### 4.1

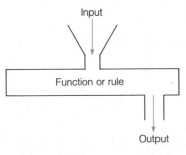

Input

Function or rule

Output

FIGURE 1

There are numerous instances in mathematics and its applications in which one quantity corresponds to or depends on another according to some definite rule. Consider, for example, the equation $y = 3x - 2$. Each time that you select an $x$-value, a corresponding $y$-value is determined, in this case according to the rule *multiply by 3, then subtract 2*. In this sense, the equation $y = 3x - 2$ is an example of a *function*. It is a *rule* specifying a $y$-value corresponding to each $x$-value. It is useful to think of the $x$-values as inputs and the corresponding $y$-values as outputs. The function or rule then tells you what output results from a given input. This is indicated schematically in Figure 1. As another example, the area $A$ of a circle depends on the radius $r$ according to the rule or function $A = \pi r^2$. For each value of $r$ there is a corresponding value for $A$ obtained by the rule *square $r$ and multiply the result by $\pi$*. In this case the inputs are the values of $r$ and the outputs are the corresponding values of $A$.

For most of the functions studied in this text (and in beginning calculus), the inputs and outputs are real numbers and the function or rule is specified by means of an equation. This was the case with the two examples given above;

however, this need not always be the case. Consider, for example, the correspondences set up in Table 1, which indicate the number of moons each planet in the solar system was known to have as of 1989. If we think of the inputs as the planets listed in the left-hand column of Table 1 and the outputs as the numbers in the right-hand column, then these correspondences constitute a function. The rule for this function may be stated, *Assign to each planet in the solar system the number of moons it was known to have in 1989.*

The following is one definition of the term *function*. As you will see, the definition is broad enough to encompass all the examples we have just looked at.

**TABLE 1**

| | |
|---|---|
| Mercury | 0 |
| Venus | 0 |
| Earth | 1 |
| Mars | 2 |
| Jupiter | 17 |
| Saturn | 22 |
| Uranus | 15 |
| Neptune | 3 |
| Pluto | 1 |

### Definition of a Function

Let $A$ and $B$ be two nonempty sets. A **function** from $A$ to $B$ is a rule of correspondence that assigns to each element in $A$ exactly one element in $B$.

The set $A$ in the definition we've just given is called the **domain** of the function. Think of the domain as the set of all possible inputs. The set of all outputs, on the other hand, is called the **range** of the function. When a function is defined by means of an equation, the letter representing elements from the domain (that is, the inputs) is called the **independent variable**. For example, in the equation $y = 3x - 2$, the independent variable is $x$. The letter representing elements from the range (that is, the outputs) is called the **dependent variable**. In the equation $y = 3x - 2$, $y$ is the dependent variable; its value *depends* on $x$. This is also expressed by saying that $y$ is a function of $x$.

For functions defined by equations, we'll agree to the following convention regarding the domain: Unless otherwise indicated, the domain is assumed to be the set of all real numbers that lead to unique real-number outputs. Thus, the domain of the function defined by $y = 3x - 2$ is the set of all real numbers, whereas the domain of the function defined by $y = \dfrac{1}{x - 5}$ is the set of all real numbers except 5. (The expression $\dfrac{1}{x - 5}$ is undefined when $x = 5$, because the denominator is then zero.)

**EXAMPLE 1** Find the domain of the function defined by each equation:

(a) $y = \sqrt{2x + 6}$     (b) $s = \dfrac{1}{t^2 - 6t - 7}$

**Solution**
(a) The quantity under the radical sign must be nonnegative, so we have

$$2x + 6 \geq 0$$
$$2x \geq -6$$
$$x \geq -3$$

The domain is therefore the interval $[-3, \infty)$.

(b) Since division by zero is undefined, the domain of this function consists of all real numbers $t$ except those for which the denominator is zero.

Thus to find out which values of $t$ to exclude, we solve the equation $t^2 - 6t - 7 = 0$. We have

$$t^2 - 6t - 7 = 0$$
$$(t - 7)(t + 1) = 0$$

$$t - 7 = 0 \quad | \quad t + 1 = 0$$
$$t = 7 \quad | \quad t = -1$$

It follows now that the domain of the function defined by $s = \dfrac{1}{t^2 - 6t - 7}$ is the set of all real numbers except $t = 7$ and $t = -1$.

**EXAMPLE 2**  Find the range of the function defined by $y = \dfrac{x + 2}{x - 3}$.

**Solution**  The range of this function is the set of all outputs $y$. One way to see what restrictions the given equation imposes on $y$ is to solve the equation for $x$ as follows:

$$y(x - 3) = x + 2 \qquad \text{multiplying by } (x - 3)$$
$$xy - 3y = x + 2$$
$$xy - x = 3y + 2$$
$$x(y - 1) = 3y + 2$$
$$x = \frac{3y + 2}{y - 1}$$

From this last equation we see that the value of $y$ cannot be 1. (The denominator is zero when $y = 1$.) The range therefore consists of all real numbers except $y = 1$.

We often use single letters in order to name functions. If $f$ is a function and $x$ is an input for the function, then the resulting output is denoted by $f(x)$. This is read $f$ *of* $x$ or *the value of* $f$ *at* $x$. As an example of this notation, suppose that $f$ is the function defined by

$$f(x) = x^2 - 3x + 1 \tag{1}$$

Then $f(-2)$ denotes the output that results when the input is $-2$. To calculate this output, just replace $x$ with $-2$ throughout equation (1). This yields

$$f(-2) = (-2)^2 - 3(-2) + 1$$
$$= 4 + 6 + 1 = 11$$

That is, $f(-2) = 11$. Figure 2 summarizes this result and the notation.

$f(-2) = 11$

Output

Input

Name of
function

**FIGURE 2**

**EXAMPLE 3**  Let $f(x) = \dfrac{1}{x - 1}$. Compute:  **(a)** $f(0)$  **(b)** $f(t)$  **(c)** $f(x - 1)$

**Solution**  **(a)** $f(0) = \dfrac{1}{0 - 1} = -1$

**(b)** $f(t) = \dfrac{1}{t - 1}$

**(c)** Replace $x$ with the quantity $x - 1$ throughout the given equation. This yields

$$f(x - 1) = \frac{1}{(x - 1) - 1} = \frac{1}{x - 2}$$

The next example points out a faulty assumption that students sometimes make in using function notation. Don't automatically assume that $f(a + b) = f(a) + f(b)$; it is not, in general, true.

**EXAMPLE 4**    Let $f(x) = x^2$. Show that $f(2 + 3) \neq f(2) + f(3)$.

**Solution**    $f(2 + 3) = f(5) = (5)^2 = 25$; but $f(2) = (2)^2 = 4$ and $f(3) = (3)^2 = 9$. Since $25 \neq 4 + 9$, we have shown that $f(2 + 3) \neq f(2) + f(3)$.

**EXAMPLE 5**    Let $g(x) = 1 - x^2$. Compute $g(x - 1)$.

**Solution**    In the equation $g(x) = 1 - x^2$, we substitute the quantity $x - 1$ in place of each occurrence of $x$. This gives us

$$\begin{aligned} g(x - 1) &= 1 - (x - 1)^2 \\ &= 1 - (x^2 - 2x + 1) \\ &= -x^2 + 2x \end{aligned}$$

Thus, $g(x - 1) = -x^2 + 2x$.

Here is a slightly different perspective on function notation that we can apply in Example 5 and similar problems. Instead of writing $g(x) = 1 - x^2$, we can write $g(\ ) = 1 - (\ )^2$, with the understanding that whatever quantity goes in the parentheses on the left side of the equation must also be placed in the parentheses on the right side. In particular, if we want $g(x - 1)$, we simply write $x - 1$ inside each set of parentheses:

$$g(\ ) = 1 - (\ )^2$$

Therefore,

$$g(x - 1) = 1 - (x - 1)^2$$

From here on, the algebra is the same as in Example 5. We again obtain $g(x - 1) = -x^2 + 2x$.

The next two examples using function notation involve calculations with **difference quotients**. These are expressions of the form

$$\frac{f(x + h) - f(x)}{h}$$

or

$$\frac{f(x) - f(a)}{x - a}$$

For now we'll concentrate on the algebraic techniques used in calculating these quantities. (Later you'll see some applications, in Exercise Set 4.2, Exercises 47–49.)

**EXAMPLE 6**    Let $f(x) = x^2 + 3x$. Compute $\dfrac{f(x) - f(2)}{x - 2}$.

**Solution**
$$\frac{f(x) - f(2)}{x - 2} = \frac{(x^2 + 3x) - [2^2 + 3(2)]}{x - 2}$$
$$= \frac{x^2 + 3x - 10}{x - 2}$$
$$= \frac{(x - 2)(x + 5)}{x - 2} = x + 5$$

The difference quotient is $x + 5$.

**EXAMPLE 7**   Let $G(x) = 2/x$. Find $\dfrac{G(x + h) - G(x)}{h}$.

**Solution**
$$\frac{G(x + h) - G(x)}{h} = \frac{2/(x + h) - 2/x}{h}$$

An easy way to simplify this expression is to multiply it by $\dfrac{(x + h)x}{(x + h)x}$, which equals 1. This yields

$$\frac{G(x + h) - G(x)}{h} = \frac{(x + h)x}{(x + h)x} \cdot \frac{\dfrac{2}{x + h} - \dfrac{2}{x}}{h}$$
$$= \frac{2x - 2(x + h)}{h(x + h)x}$$
$$= \frac{2x - 2x - 2h}{h(x + h)x} = \frac{-2h}{h(x + h)x}$$
$$= \frac{-2}{(x + h)x}$$

The difference quotient is therefore $-2/(x + h)x$.

In Examples 1 through 7 we've considered functions that were defined by means of equations. It is important to understand, however, that there are equations and rules that do not define functions. For example, consider the equation $y^2 = x$ and the input $x = 4$. Then we have $y^2 = 4$ and consequently $y = \pm 2$. So we have *two* outputs in this case, whereas the definition of a function requires that there be *exactly one* output. Example 8 provides some additional perspective on this.

**EXAMPLE 8**   Let $A = \{b, g\}$ and $B = \{s, t, u, z\}$. Which of the four correspondences in Figure 3 represent functions from $A$ to $B$? For those correspondences that do represent functions, specify the range in each case.

**Solution**   FIGURE 3(a):   Not a function. The definition requires that *each* element in $A$ be assigned an element in $B$. The element $g$ in this case has no assignment.

FIGURE 3(b):   Not a function. The definition requires *exactly one* output for a given input. In this case there are two outputs for the input $g$.

FIGURES 3(c) AND (d):   Both of these rules qualify as functions from $A$ to $B$. For each input there is exactly one output. (Regarding the function in Figure 3(d) in particular, notice that there is nothing in the definition of a function that prohibits two different inputs from producing the same output.) For the

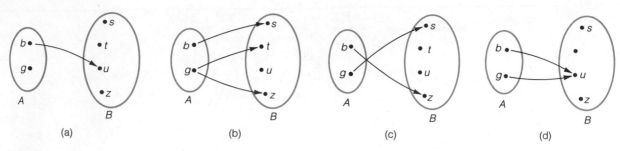

**FIGURE 3**

function in Figure 3(c), the outputs are *s* and *z*, and so the range is the set $\{s, z\}$. For the function in Figure 3(d), the only output is *u*, and consequently the range is the set $\{u\}$.

Because the function concept is so central to the rest of the material in this text, for reference, we conclude here with a summary of the terminology introduced in this section.

| PROPERTY SUMMARY | Terminology for Functions | |
|---|---|---|
| TERM | DEFINITION AND COMMENTS | EXAMPLE |
| Function | Given two nonempty sets *A* and *B*, a function *f* (from *A* to *B*) is a rule that assigns to each element *x* in *A* exactly one element *f(x)* in *B*. We think of each element *x* in *A* as an input and the corresponding element *f(x)* in *B* as an output. | The equation $f(x) = x^2$ defines a function. The rule in this case is: *For each real number x, compute its square.* Given the input $x = 3$, for example, the output is $f(3) = 9$. |
| Domain and range | The domain of a function is the set of all inputs; this is the set *A* in the definition. The range is the set of all outputs; see Figure 4. | The domain and the range of the function defined by $f(x) = x^2$ are the sets $(-\infty, \infty)$ and $[0, \infty)$, respectively. |
| Independent and dependent variables | The letter used to represent the elements in the domain of a function (the inputs) is the independent variable. The letter used for elements in the range (the outputs) is the dependent variable. | For the function $y = x^2$, the independent variable is *x* and the dependent variable is *y*. The value of *y depends* on *x*. |

**FIGURE 4**   A function *f* from *A* to *B*.

## EXERCISE SET 4.1

## A

*In Exercises 1–6, find the domain of each function.*

1. (a) $y = -5x + 1$
   (b) $y = \sqrt{-5x + 1}$
   (c) $y = \dfrac{x}{\sqrt{|-5x + 1|}}$

2. (a) $y = x^2 - 3x + 4$
   (b) $y = \sqrt{x^2 - 3x + 4}$
   (c) $y = \sqrt{x^2 - 3x - 4}$

3. (a) $y = \dfrac{x + 4}{x - 4}$
   (b) $y = \dfrac{x^2 + 4}{x^2 - 4}$
   (c) $y = \sqrt{\dfrac{x^2 + 4}{x^2 - 4}}$

4. (a) $y = \dfrac{1}{x^3 - 16x}$
   (b) $y = \dfrac{1}{x^3 - 16x^2}$
   (c) $y = \sqrt[3]{x^3 - 16x}$

5. (a) $y = \sqrt{x^2 - 4x - 5}$
   (b) $y = \dfrac{2x}{\sqrt{x^2 - 4x - 5}}$
   (c) $y = \sqrt[3]{x^2 - 4x - 5}$

6. (a) $y = x^3 - 14x^2 - 6x - 1$
   (b) $y = \dfrac{1}{x^3 + x^2 - 2x - 2}$
   (c) $y = \dfrac{1}{x^3 + x^2 - 2x}$

*In Exercises 7–10, find the range of each function.*

7. (a) $y = \dfrac{x + 3}{x - 5}$    (b) $y = \dfrac{x - 5}{x + 3}$

8. (a) $y = \dfrac{3x - 2}{x + 4}$    (b) $y = \dfrac{1}{x + 4}$

9. (a) $y = x^2 + 4$    (b) $y = x^3 + 4$

10. (a) $y = \dfrac{ax + b}{x}$    (*a* and *b* are real numbers and $b \neq 0$)
    (b) $y = cx^3 + d$    $(c \neq 0)$

11. Let $A = \{x, y, z\}$ and $B = \{1, 2, 3\}$. Which of the rules displayed in the figure represent functions from $A$ to $B$?

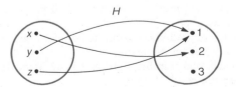

12. Let $D = \{a, b\}$ and $C = \{i, j, k\}$. Which of the rules displayed in the figure represent functions from $D$ to $C$?

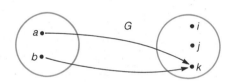

13. (a) Specify the range for each rule that represents a function in Exercise 11.
(b) Specify the range for each rule that represents a function in Exercise 12.

14. (a) Suppose that in Exercise 11 all the arrows were reversed. Which, if any, of the new rules would be functions from $B$ to $A$?
(b) Suppose that in Exercise 12 all the arrows were reversed. Which, if any, of the new rules would be functions from $C$ to $D$?

15. Each of the following rules defines a function whose domain is the set of all real numbers. Express each rule by means of an equation.

*Example* The rule *For each real number, compute its square* can be written $y = x^2$.

(a) For each real number, subtract 3 and then square the result.
(b) For each real number, compute its square and then subtract 3 from the result.
(c) For each real number, multiply it by 3 and then square the result.

(d) For each real number, compute its square and then multiply the result by 3.

16. Each of the following rules defines a function whose domain is the set of all real numbers. Express each rule in words.
(a) $y = 2x^3 + 1$      (b) $y = 2(x + 1)^3$
(c) $y = (2x + 1)^3$      (d) $y = (2x)^3 + 1$

17. Let $f(x) = x^2 - 3x + 1$. Compute the following.
(a) $f(1)$    (b) $f(0)$    (c) $f(-1)$
(d) $f(\frac{3}{2})$    (e) $f(z)$    (f) $f(x + 1)$
(g) $f(a + 1)$    (h) $f(-x)$    (i) $|f(1)|$
(j) $f(\sqrt{3})$    (k) $f(1 + \sqrt{2})$    (l) $|1 - f(2)|$

18. Let $H(x) = 1 - x + x^2 - x^3$.
(a) Which number is larger, $H(0)$ or $H(1)$?
(b) Find $H(\frac{1}{2})$. Does $H(\frac{1}{2}) + H(\frac{1}{2}) = H(1)$?

19. Let $f(x) = 3x^2$. Find the following.
(a) $f(2x)$    (b) $2f(x)$    (c) $f(x^2)$
(d) $[f(x)]^2$    (e) $f(x/2)$    (f) $f(x)/2$
*For checking:* No two answers are the same.

20. Let $f(x) = 4 - 3x$. Find the following.
(a) $f(2)$    (b) $f(3)$    (c) $f(2) + f(3)$
(d) $f(2 + 3)$    (e) $f(2x)$    (f) $2f(x)$
(g) $f(x^2)$    (h) $f(1/x)$    (i) $f[f(x)]$
(j) $x^2 f(x)$    (k) $1/f(x)$    (l) $f(-x)$
(m) $-f(x)$    (n) $-f(-x)$

21. Let $H(x) = 1 - 2x^2$. Find the following.
(a) $H(0)$      (b) $H(2)$
(c) $H(\sqrt{2})$      (d) $H(\frac{5}{6})$
(e) $H(x + 1)$      (f) $H(x + h)$
(g) $H(x + h) - H(x)$      (h) $\dfrac{H(x + h) - H(x)}{h}$

22. (a) If $f(x) = 2x + 1$, does $f(3 + 1) = f(3) + f(1)$?
(b) If $f(x) = 2x$, does $f(3 + 1) = f(3) + f(1)$?
(c) If $f(x) = \sqrt{x}$, does $f(3 + 1) = f(3) + f(1)$?

23. Let $R(x) = \dfrac{2x - 1}{x - 2}$. Find the following.
(a) The domain and range of $R$    (b) $R(0)$
(c) $R(\frac{1}{2})$      (d) $R(-1)$
(e) $R(x^2)$      (f) $R(1/x)$
(g) $R(a)$      (h) $R(x - 1)$

24. Let $g(x) = 2$, for all $x$. Find the following.
(a) $g(0)$    (b) $g(5)$    (c) $g(x + h)$

25. Let $d(t) = -16t^2 + 96t$.
(a) Compute $d(1)$, $d(\frac{3}{2})$, $d(2)$, $d(t_0)$.
(b) For which values of $t$ is $d(t) = 0$?
(c) For which values of $t$ is $d(t) = 1$?

26. Let $A(x) = |x^2 - 1|$. Compute $A(2)$, $A(1)$, and $A(0)$.

27. Let $g(t) = |t - 4|$. Find $g(3)$. Find $g(x + 4)$.

28. Let $f(x) = x^2/|x|$.
    (a) What is the domain of $f$?
    (b) Find $f(2)$, $f(-2)$, $f(20)$, and $f(-20)$.
    (c) What is the range of $f$? *Hint:* Look over your results in part (b).

29. Compute $\dfrac{f(x+h) - f(x)}{h}$ for the function $f$ specified in each case.
    (a) $f(x) = x^2$    (b) $f(x) = 2x^2 - 3x + 1$
    (c) $f(x) = x^3$

30. Compute $\dfrac{f(x) - f(a)}{x - a}$ for the three functions given in Exercise 29.

31. Let $f(x) = \dfrac{x}{x-1}$. Compute the following difference quotients.
    (a) $\dfrac{f(x) - f(a)}{x - a}$    (b) $\dfrac{f(x) - f(3)}{x - 3}$
    (c) $\dfrac{f(x+h) - f(x)}{h}$    (d) $\dfrac{f(3+h) - f(3)}{h}$

32. Let $g(t) = \dfrac{1}{3t - 5}$. Compute the following difference quotients
    (a) $\dfrac{g(t) - g(a)}{t - a}$    (b) $\dfrac{g(t) - g(2)}{t - 2}$
    (c) $\dfrac{g(t+h) - g(t)}{h}$    (d) $\dfrac{g(2+h) - g(2)}{h}$

33. In each case a pair of functions $f$ and $g$ are given. Find all real numbers $x_0$ for which $f(x_0) = g(x_0)$.
    (a) $f(x) = 4x - 3$,   $g(x) = 8 - x$
    (b) $f(x) = x^2 - 4$,   $g(x) = 4 - x^2$
    (c) $f(x) = x^2$,   $g(x) = x^3$
    (d) $f(x) = 2x^2 - x$,   $g(x) = 3$

34. Let $h(x) = \sqrt{x}$.
    (a) Compute $\dfrac{h(5) - h(1)}{5 - 1}$ and $\dfrac{1}{h(5) + 1}$. Round off both answers to three decimal places.
    (b) Why are the two answers in part (a) the same?

35. When \$1000 is deposited in a savings account at an annual rate of 12%, compounded quarterly, the amount in the account after $t$ years is given by

$$A(t) = 1000\left(1 + \frac{0.12}{4}\right)^{4t}$$

(a) Compute $A(1) - A(0)$. This is the amount by which the account will grow in the first year.
(b) Compute $A(10) - A(9)$. This is the amount by which the account will grow in the tenth year.

36. Let $f(n) = (1 + 1/n)^n$.
    (a) Complete the table. (Round off to three decimal places.)

| $n$ | 1 | 2 | 5 | 10 | 15 | 20 |
|---|---|---|---|---|---|---|
| $f(n)$ | | | | | | |

(b) By trial and error, find the smallest natural number $n$ such that $f(n) > 2.7$.
(c) Using your calculator, can you find a number $n$ such that $f(n) \geq 2.8$?

37. Let $g(n) = n^{1/n}$.
    (a) Complete the table. (Round off to four decimal places.)

| $n$ | 2 | 3 | 4 | 5 | 6 | 7 | 8 |
|---|---|---|---|---|---|---|---|
| $g(n)$ | | | | | | | |

(b) By trial and error, find the smallest natural number $n$ such that $g(n) < 1.2$.

38. Consider the function $f$ defined by

$$f(x) = x^2 + \frac{2}{x^2}, \qquad x > 0.$$

(a) Complete the table. (Round off to four decimal places.)

| $x$ | 1 | 1.05 | 1.10 | 1.15 | 1.20 | 1.25 |
|---|---|---|---|---|---|---|
| $f(x)$ | | | | | | |

(b) Which $x$-value in the table yields the smallest value for $f(x)$?
(c) It can be shown using calculus that the input $x$ yielding the smallest possible output $f(x)$ for this function is $x = 2^{1/4}$. Which $x$-value in the table is closest to this?
(d) Compute $f(2^{1/4})$. Which value of $f(x)$ in the table is closest to $f(2^{1/4})$?

B

39. Let $f(x) = \dfrac{x - a}{x + a}$.
    (a) Find $f(a)$, $f(2a)$, and $f(3a)$. Is it true that $f(3a) = f(a) + f(2a)$?
    (b) Show that $f(5a) = 2f(2a)$.

40. Let $M(x) = \dfrac{x - a}{x + a}$. Compute $M\left(\dfrac{1}{x}\right)$.

41. Let $\phi(y) = 2y - 3$. Show that $\phi(y^2) \neq [\phi(y)]^2$

42. Let $k(x) = 5x^3 + \dfrac{5}{x^3} - x - \dfrac{1}{x}$. Show that $k(x) = k(1/x)$.

43. Let $f(x) = 2x + 3$. Find values for $a$ and $b$ such that the equation $f(ax + b) = x$ is true for all values of $x$.

44. If $p(x) = 2^x$, verify the following identities.
    (a) $2p(x) = p(x + 1)$
    (b) $p(a + b) = p(a) \cdot p(b)$
    (c) $p(x) \cdot p(-x) - 1 = 0$

45. Let $f(t) = \dfrac{t - x}{t + y}$. Show that

$$f(x + y) + f(x - y) = \dfrac{-2y^2}{x^2 + 2xy}$$

46. Let $f(z) = \dfrac{3z - 4}{5z - 3}$. Find $f\left(\dfrac{3z - 4}{5z - 3}\right)$.

47. Let $F(x) = \dfrac{ax + b}{cx - a}$. Show that $F\left(\dfrac{ax + b}{cx - a}\right) = x$. (Assume that $a^2 + bc \neq 0$.)

48. If $f(x) = -2x^2 + 6x + k$ and $f(0) = -1$, find $k$.

49. If $g(x) = x^2 - 3xk - 4$ and $g(1) = -2$, find $k$.

50. If $f(x) = 1/x^2$, show that

$$\dfrac{f(x + h) - f(x)}{h} = \dfrac{-2x - h}{(x + h)^2 x^2}$$

51. A function doesn't always have to be given by an algebraic formula. For example, let the function $L$ be defined by the following rule: $L(x)$ *is the power to which 2 must be raised to yield $x$.* (For the moment, we won't concern ourselves with the domain and range.) Then $L(8) = 3$, for example, since the power to which 2 must be raised to yield 8 is 3 ($8 = 2^3$). Find the following.
    (a) $L(1)$  (b) $L(2)$  (c) $L(4)$  (d) $L(64)$
    (e) $L(\frac{1}{2})$  (f) $L(\frac{1}{4})$  (g) $L(\frac{1}{64})$  (h) $L(\sqrt{2})$
    The function $L$ is called a *logarithm function*. The usual notation for $L(x)$ in this example is $\log_2 x$. Logarithm functions will be studied in Chapter 6.

52. Let $f(x) = ax^2 + bx + c$. Show that

$$\dfrac{f(x + h) - f(x)}{h} = 2ax + ah + b$$

53. Let $q(x) = ax^2 + bx + c$. Evaluate

$$q\left(\dfrac{-b + \sqrt{b^2 - 4ac}}{2a}\right)$$

54. By definition, a **fixed point** for the function $f$ is a number $x_0$ such that $f(x_0) = x_0$. For instance, to find any fixed points for the function $f(x) = 3x - 2$, we write $3x_0 - 2 = x_0$. On solving this last equation, we find that $x_0 = 1$. Thus, 1 is a fixed point for $f$. Calculate the fixed points (if any) for the following functions.

    (a) $f(x) = 6x + 10$   (b) $g(x) = x^2 - 2x - 4$
    (c) $S(t) = t^2$   (d) $R(z) = \dfrac{z + 1}{z - 1}$

55. Let $f(x) = \dfrac{3x - 4}{x - 3}$.
    (a) Find $f[f(x)]$.
    (b) Find $f[f(\frac{22}{7})]$. Try not to do it the hard way!

56. Consider the following two rules, $F$ and $G$, where $F$ is the rule that assigns to each person his or her mother, and $G$ is the rule that assigns to each person his or her aunt. Explain why $F$ is a function but $G$ is not.

57. A **prime number** is a positive whole number with no factors other than itself and 1. For example, 2, 13, and 37 are primes, but 24 and 39 are not. By convention, 1 is not considered prime, so the list of the first few primes looks like this:

    $2, 3, 5, 7, 11, 13, 17, 19, 23, 29, \ldots$

    Let $G$ be the rule that assigns to each positive integer the nearest prime. For example, $G(8) = 7$, since 7 is the prime nearest 8. Explain why $G$ is not a function. How could you alter the definition of $G$ to make it a function?

58. Let $f$ be the function that assigns to each natural number $x$ the number of primes that are less than or equal to $x$. For example, $f(12) = 5$, because, as you can easily check, there are five primes that are less than or equal to 12. Similarly, $f(3) = 2$, because there are two primes that are less than or equal to 3. Find $f(8)$, $f(10)$, and $f(50)$.

**C**

59. If $P(x) = x^2 - x + 41$, find $P(1)$, $P(2)$, $P(3)$, and $P(4)$. Can you find a natural number $x$ for which $P(x)$ is not prime?

60. Let $S(x) = \dfrac{3^x - 3^{-x}}{2}$ and $C(x) = \dfrac{3^x + 3^{-x}}{2}$. Show that the functions $S$ and $C$ possess the following properties.
    (a) $S(0) = 0$; $C(0) = 1$; $S(1) = \frac{4}{3}$; $C(1) = \frac{5}{3}$
    (b) $[C(x)]^2 - [S(x)]^2 = 1$
    (c) $S(-x) = -S(x)$ and $C(-x) = C(x)$
    (d) $S(x + y) = S(x)C(y) + C(x)S(y)$
    (e) $C(x + y) = C(x)C(y) + S(x)S(y)$
    (f) $S(2x) = 2S(x)C(x)$ and $C(2x) = [C(x)]^2 + [S(x)]^2$
    (g) $S(3x) = 3S(x) + 4[S(x)]^3$
    (h) $[S(x) + C(x)]^2 = S(2x) + C(2x)$
    (i) $S\left(\dfrac{x}{2}\right) = \pm\sqrt{\dfrac{C(x) - 1}{2}}$
    (j) $C\left(\dfrac{x}{2}\right) = \sqrt{\dfrac{C(x) + 1}{2}}$

## THE GRAPH OF A FUNCTION

**4.2**

*In my own case, I got along fine without knowing the name of the distributive law until my sophomore year in college; meanwhile I had drawn lots of graphs.*

PROFESSOR DONALD E. KNUTH (Professor Knuth is judged by many to be the world's preeminent computer scientist.)

When the domain and range of a function are sets of real numbers, the function can be graphed in the same way in which equations were graphed in Chapter 3. In graphing functions, the usual practice is to reserve the horizontal axis for the independent variable and the vertical axis for the dependent variable. The function or rule then tells you how you must pick your $y$-coordinate, once you have selected an $x$-coordinate.

> **Definition of the Graph of a Function**
>
> The **graph** of a function $f$ in the $x$-$y$ plane consists of those points $(x, y)$ such that $x$ is in the domain of $f$ and $y = f(x)$. See Figure 1.

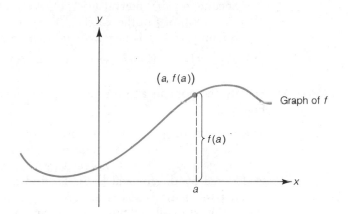

**FIGURE 1**

**EXAMPLE 1**   Graph the functions $f$ and $g$ defined as follows.

(a) $f(x) = x^2 - 1$     (b) $g(x) = |x^2 - 1|$

Solution   (a) The graph of this function is by definition the graph of the equation $y = x^2 - 1$. (Whether we label the vertical axis $y$ or $f(x)$ is immaterial.) As preparation for drawing the graph, we first determine the domain of $f$, the $x$- and $y$-intercepts of the graph, and any symmetries the graph may have. As you can readily check, the results are as follows.

$$\begin{aligned}
\text{domain:} \quad & (-\infty, \infty) \\
x\text{-intercepts:} \quad & 1, -1 \\
y\text{-intercept:} \quad & -1 \\
\text{symmetry:} \quad & \text{about the } y\text{-axis}
\end{aligned}$$

The required graph is then sketched in Figure 2(a) (on the next page).

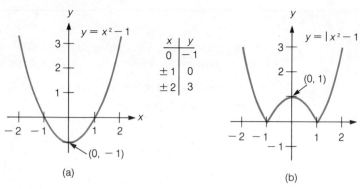

FIGURE 2

**(b)** The graph of $g$ is shown in Figure 2(b). It is obtained from the graph of $f$ in Figure 2(a) as follows. First, for $x \geq 1$ or $x \leq -1$, the graph of $g$ is identical to that of $f$ because in this case, we have

$$g(x) = |x^2 - 1| = x^2 - 1 = f(x)$$

Second, on the interval $(-1, 1)$, the graph of $g$ is obtained by reflecting the graph of $f$ in the $x$-axis. This is because on the interval $(-1, 1)$, the quantity $x^2 - 1$ is negative and therefore

$$g(x) = |x^2 - 1| = -(x^2 - 1) = -f(x)$$

**EXAMPLE 2**  Specify the domain and the range of the function $g$ whose graph appears in Figure 3(a).

**Solution**  The domain of $g$ is just that portion of the $x$-axis (the inputs) utilized in graphing $g$. As Figure 3(b) indicates, this amounts to all real numbers $x$ from 1 to 5, inclusive: $1 \leq x \leq 5$. Recall from Section 1.3 that this set of numbers is denoted by the interval $[1, 5]$. To find the range of $g$, you need to check which part of the $y$-axis is utilized in graphing $g$. As Figure 3(b) indicates, this is the set of all real numbers $y$ between 2 and 4, inclusive: $2 \leq y \leq 4$. Our shorthand notation for this interval of numbers is $[2, 4]$.

FIGURE 3

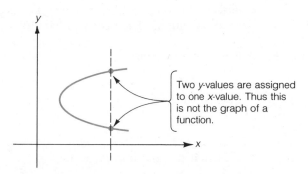

**FIGURE 4**

**EXAMPLE 3**  A portion of the graph of a function $h$ is shown in Figure 4. The open circle in the figure is used to indicate that the point $(3,3)$ does not belong to the graph of $h$. The two filled-in circles, on the other hand, are used to indicate that the points $(-4,-2)$ and $(3,4)$ do belong to the graph of $h$. Find the following:

(a) $h(-2)$     (b) $h(3)$     (c) $\left|h(-4)\right|$

**Solution**  (a) The function notation $h(-2)$ stands for the $y$-coordinate of that point on the graph of $h$ whose $x$-coordinate is $-2$. Since the point $(-2,-1)$ is on the graph of $h$, we conclude $h(-2) = -1$.

(b) We conclude that $h(3) = 4$ because the point $(3,4)$ lies on the graph of $h$. Note that $h$ would not be considered a function if the point $(3,3)$ were also part of the graph. (Why?)

(c) Since the point $(-4,-2)$ lies on the graph of $h$, we write $h(-4) = -2$. Thus $\left|h(-4)\right| = \left|-2\right| = 2$.

Many of the graphs that we looked at in Chapter 3 are graphs of functions. However, it's important to understand that there are graphs that do not represent functions. Consider, for example, the graph in Figure 5.

Two $y$-values are assigned to one $x$-value. Thus this is not the graph of a function.

**FIGURE 5**

Figure 5 also shows a vertical line intersecting the graph in two distinct points. The specific coordinates of the two points are unimportant. What the vertical line helps us to see is that two different $y$-values have been assigned to the same $x$-value, and therefore the graph cannot be the graph of a function $y = f(x)$.

The preceding remarks can be summarized as follows.

> **Vertical Line Test**
>
> A graph in the $x$-$y$ plane represents a function $y = f(x)$ provided that any vertical line intersects the graph in at most one point.

**EXAMPLE 4** The vertical line test implies that the graph in Figure 6(a) represents a function, while that in Figure 6(b) does not.

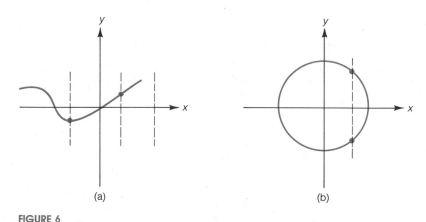

(a)                    (b)

**FIGURE 6**

There are six basic functions that arise frequently enough to make it worth your while to memorize the basic shapes and features of their graphs. Figure 7 displays these graphs. Exercise 6 at the end of this section asks you to set up tables and verify for yourself that the graphs in Figure 7 are indeed correct. From now on (with the exception of Exercise 6), if you need to sketch a graph of one of these basic functions, you should do so from memory.

All the graphs in Figure 7 arise from functions defined by rather simple equations. In some instances, however, functions may be defined by combinations of equations. The next two examples display instances of this.

**EXAMPLE 5** A function $g$ is defined by

$$g(x) = \begin{cases} x^2 & \text{if } x < 2 \\ \dfrac{1}{x} & \text{if } x \geq 2 \end{cases}$$

(a) Find $g(1)$, $g(2)$, and $g(3)$.
(b) Sketch the graph of $g$.

**Solution** (a) To find $g(1)$, do we substitute the value $x = 1$ in the expression $x^2$ or in the expression $1/x$? According to the instructions contained in the given

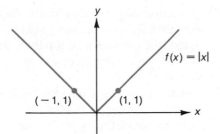

The absolute value
function $f(x) = |x|$
Domain: $(-\infty, \infty)$
Range: $[0, \infty)$

(a)

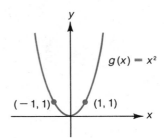

The squaring function
$g(x) = x^2$
Domain: $(-\infty, \infty)$
Range: $[0, \infty)$

(b)

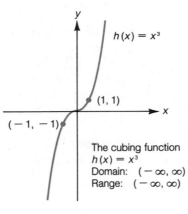

The cubing function
$h(x) = x^3$
Domain: $(-\infty, \infty)$
Range: $(-\infty, \infty)$

(c)

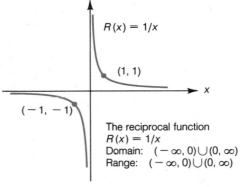

The reciprocal function
$R(x) = 1/x$
Domain: $(-\infty, 0) \cup (0, \infty)$
Range: $(-\infty, 0) \cup (0, \infty)$

(d)

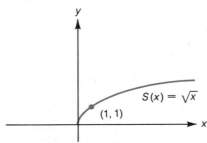

The square root function
$S(x) = \sqrt{x}$
Domain: $[0, \infty)$
Range: $[0, \infty)$

(e)

The function $C(x) = \sqrt{1 - x^2}$
(The graph is the upper
half of the circle $x^2 + y^2 = 1$.)
Domain: $[-1, 1]$
Range: $[0, 1]$

(f)

**FIGURE 7**

definition of $g$, we should use the expression $x^2$ whenever the inputs are less than 2. Thus we have $g(1) = 1^2 = 1$; that is, $g(1) = 1$. On the other hand, the definition of $g$ tells us to use the expression $1/x$ whenever the inputs are greater than or equal to 2. So in this case we have $g(2) = \frac{1}{2}$ and $g(3) = \frac{1}{3}$.

**(b)** For the graph of $g$, we look back at Figures 7(b) and (d) and we choose the appropriate portion of each. The result is displayed in Figure 8. The open circle in the figure is used to indicate that the point $(2, 4)$ does not belong to the graph of $g$. The filled-in circle, on the other hand, is used to indicate that the point $(2, \frac{1}{2})$ does belong to the graph of $g$.

**EXAMPLE 6**   Graph the function $f$ defined by

$$f(x) = \begin{cases} x^2 - 1 & \text{if } x \le 2 \\ 5 - x & \text{if } x > 2 \end{cases}$$

**Solution**   In view of the definition of $f$, we'll set up two tables. For the first table, the inputs $x$ will all be less than or equal to 2, and the outputs will be computed using the expression $x^2 - 1$. In the second table, the inputs $x$ will be greater than 2, and the outputs will be computed from the expression $5 - x$. These tables and the graph derived from them are shown in Figure 9. (Actually, we could save a few steps by referring to the graph of $y = x^2 - 1$ shown in Figure 2(a).)

We make two observations here. First, notice that, in setting up a table for $y = 5 - x$ with $x > 2$, only two points are needed, since the graph is a portion

$y = x^2 - 1, \quad x \le 2$

| $x$ | $\pm 2$ | $\pm 1$ | 0 |
|---|---|---|---|
| $y$ | 3 | 0 | $-1$ |

FIGURE 8

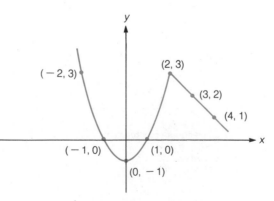

FIGURE 9   $f(x) = \begin{cases} x^2 - 1, & x \le 2 \\ 5 - x, & x > 2 \end{cases}$

$y = 5 - x, \quad x > 2$

| $x$ | 3 | 4 |
|---|---|---|
| $y$ | 2 | 1 |

of a straight line. Second, notice that in this example, as opposed to the case in Example 5, the two portions of the graph together form a graph with no breaks or gaps in it. We say that the function $f$ (whose graph appears in Figure 9) is *continuous* at $x = 2$ but that the function $g$ (whose graph appears in Figure 8) is *discontinuous* at $x = 2$. A rigorous definition of continuity is properly a subject for calculus. However, even at the intuitive level at which we've presented the idea here, you'll find that the concept is useful in helping you to organize your thoughts about the graph of a function.

In Example 7 we are going to graph the *greatest integer function*. In order to define this function, we introduce the notation $[x]$ to denote the greatest (largest) integer that is less than or equal to $x$. For instance,

$[3] = 3$, since 3 is the greatest integer that is less than or equal to 3;

$[4\frac{1}{2}] = 4$, since 4 is the greatest integer that is less than or equal to $4\frac{1}{2}$;

$[-\frac{4}{3}] = -2$, since $-2$ is the largest integer that is less than or equal to $-\frac{4}{3}$.

The **greatest integer function** then is the function defined by the equation $y = [x]$. The domain of this function is the set of all real numbers.

**EXAMPLE 7**   Graph the greatest integer function $y = [x]$.

Solution   To help you to see what is happening, first consider the behavior of this function for values of $x$ between 0 and 2. See Table 1.

TABLE 1

| $x$ | 0 | 0.1 | 0.5 | 0.9 | 1.0 | 1.5 | 1.9 | 2.0 |
|---|---|---|---|---|---|---|---|---|
| $y = [x]$ | 0 | 0 | 0 | 0 | 1 | 1 | 1 | 2 |

As indicated by Table 1, the value of $y$ is 0 throughout the interval $[0, 1)$. Then at $x = 1$, the value of $y$ jumps up to 1. The value of $y$ remains 1 throughout the interval $[1, 2)$. Then at $x = 2$, the value of $y$ jumps up to 2. We can summarize this behavior as follows:

on the interval $[0, 1)$:   $y = 0$

on the interval $[1, 2)$:   $y = 1$

and, more generally,

on the interval $[p, p + 1)$:   $y = p$      (for each integer $p$)

Now, using this information, we can sketch the graph of $y = [x]$ as shown in Figure 10. You should check the coordinates of a few points on the graph for yourself to be sure that you understand why the graph looks as it does. Although Figure 10 displays only that portion of the graph for which $-2 \leq x < 3$, it is understood that the pattern continues indefinitely in both directions. Finally, notice that the function is discontinuous whenever $x$ is an integer.

FIGURE 10   $y = [x]$

## EXERCISE SET 4.2

### A

1. The accompanying figure displays the graph of a function $f$.
   (a) Is $f(0)$ positive or negative?
   (b) Find $f(-2), f(1), f(2)$, and $f(3)$.
   (c) Which is larger, $f(2)$ or $f(4)$?
   (d) Find $f(4) - f(1)$.
   (e) Find $|f(4) - f(1)|$.
   (f) Write the domain and range of $f$ using the interval notation $[a, b]$.

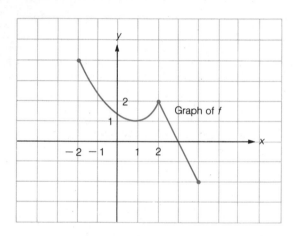

2. The following figure shows the graph of a function $h$.
   (a) Find $h(a), h(b), h(c)$, and $h(d)$.
   (b) Is $h(0)$ positive or negative?
   (c) For which values of $x$ does $h(x) = 0$?
   (d) Which is larger, $h(b)$ or $h(0)$?
   (e) As $x$ increases from $c$ to $d$, do the corresponding values of $h(x)$ increase or decrease?
   (f) As $x$ increases from $a$ to $b$, do the corresponding values of $h(x)$ increase or decrease?

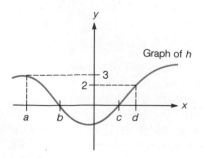

*In Exercises 3 and 4, refer to the graphs of the functions $f$ and $g$ in the following figure. Assume that the domain of each function is $[-3, 3]$ and that the axes are marked off in one-unit intervals.*

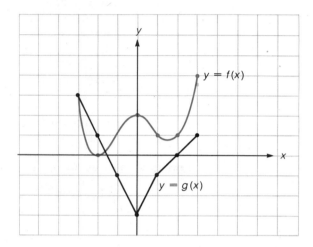

3. (a) Which is larger, $f(-2)$ or $g(-2)$?
   (b) Compute $f(0) - g(0)$.
   (c) Which among the following three quantities is the smallest?

   $$f(1) - g(1), \qquad f(2) - g(2), \qquad f(3) - g(3)$$

   (d) For which value(s) of $x$ does $g(x) = f(1)$?
   (e) Is the number 4 in the range of $f$ or in the range of $g$?

4. (a) For the interval $[0, 3]$, is the quantity $g(x) - f(x)$ positive or negative?
   (b) For the interval $(-3, -2)$, is the quantity $g(x) - f(x)$ positive or negative?
   (c) Compute $\dfrac{f(x) - f(2)}{x - 2}$ when $x = 3$.
   (d) Compute $\dfrac{g(x) - g(-2)}{x + 2}$ when $x = -3$.
   (e) Specify the range of $f$ and the range of $g$.

5. Let $f(x) = [x]$.
   (a) Compute $f(3), f(\pi), f(\frac{39}{10})$, and $f(-\frac{3}{2})$.
   (b) What is the range of the greatest integer function?

6. Set up tables and graph the given functions. (The symmetry tests from Section 3.2 are helpful here.)
   (a) $f(x) = |x|$   (b) $g(x) = x^2$
   (c) $h(x) = x^3$   (d) $R(x) = 1/x$
   (e) $S(x) = \sqrt{x}$   (f) $C(x) = \sqrt{1 - x^2}$

*In Exercises 7 and 8, graph the given functions.*

7. (a) $f(x) = x^3$        (b) $g(x) = |x^3|$
   (c) $h(x) = x^3 + 1$    (d) $H(x) = |x^3 + 1|$

8. (a) $f(x) = \dfrac{1}{x}$        (b) $F(x) = \left|\dfrac{1}{x}\right|$
   (c) $g(x) = 4 - x^2$    (d) $G(x) = |4 - x^2|$

9. The graph of a function $G$ is shown in the following figure. The domain of $G$ is $[0, 9]$.
   (a) What is the range of $G$?
   (b) List the following numbers in order of increasing size: $G(0), G(2), G(3), G(7), G(9)$.
   (c) What are the coordinates of the highest point on the graph of $G$? The $y$-coordinate of this point is called the **maximum value** of $G$.
   (d) What are the coordinates of the lowest point on the graph of $G$? The $y$-coordinate of this point is called the **minimum value** of $G$.

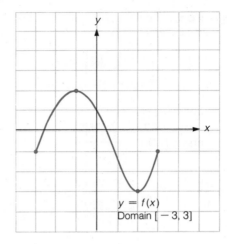

$y = f(x)$
Domain $[-3, 3]$

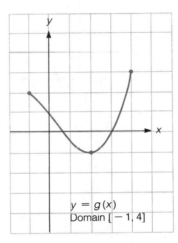

$y = g(x)$
Domain $[-1, 4]$

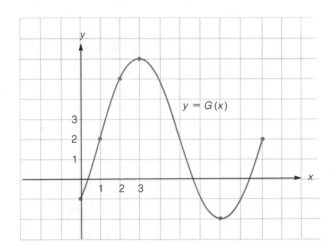

$y = G(x)$

10. First do Exercise 9. Find the maximum and minimum values for each of the functions shown; assume that the axes are marked off in one-unit intervals.

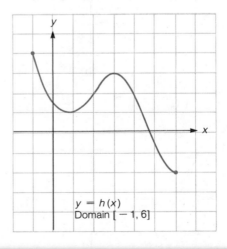

$y = h(x)$
Domain $[-1, 6]$

*In Exercises 11 and 12, use the vertical line test to determine which graphs represent functions $y = f(x)$.*

**11.**

(a)                                   (b)

(c)                                   (d)

**12.**

(a)                                   (b)

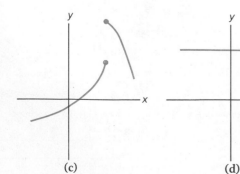

(c)                                   (d)

**13.** Let $f(x) = x^2$. Find the slope of the straight line that passes through the points $(3, f(3))$ and $(4, f(4))$. Include a sketch.

**14.** Let $G(x) = x^3$. Find the slope of the straight line that passes through the points $(0, G(0))$ and $(-2, G(-2))$. Include a sketch.

**15.** Let $T(x) = \sqrt{x}$. Which line has the larger slope, the line that passes through $(1, T(1))$ and $(4, T(4))$ or the line that passes through $(4, T(4))$ and $(9, T(9))$?

**16.** In each case, sketch a graph of the function for the given domain.
(a) $p(x) = x^2$ with domain $[-1, 2]$
(b) $q(x) = x^3$ with domain $[0, 1]$
(c) $r(x) = 1/x$ with domain $(0, \infty)$
(d) $s(x) = \sqrt{x}$ with domain $[1, 4]$

**17.** In each case, sketch a graph of the function for the given domain.
(a) $k(x) = \sqrt{1 - x^2}$ with domain $[0, 1]$
(b) $m(x) = \sqrt{1 - x^2}$ with domain $(0, 1)$
(c) $n(x) = |x|$ with domain consisting only of the numbers 0, 1, 2, 3, and 4
(d) $z(x) = x^2$ with domain consisting of the numbers $-2, -1, 0, 1$, and 4 as well as those numbers in the open interval $(2, 3)$

**18.** What are the domain and range of the function whose graph is the horizontal line $y = 3$?

*In Exercises 19–32, graph the functions defined by the given rules.*

**19.** $f(x) = \begin{cases} |x| & \text{if } x \le 0 \\ x^2 & \text{if } x > 0 \end{cases}$

**20.** (a) $g(x) = \begin{cases} |x| & \text{if } x \le 0 \\ x + 1 & \text{if } x > 0 \end{cases}$

(b) $F(x) = \begin{cases} |x| & \text{if } x < 0 \\ x + 1 & \text{if } x \ge 0 \end{cases}$

**21.** $A(x) = \begin{cases} x^3 & \text{if } -2 \le x \le -1 \\ x^2 & \text{if } x > -1 \end{cases}$

**22.** $B(x) = \begin{cases} \sqrt{1 - x^2} & \text{if } -1 \le x < 1 \\ 1/x & \text{if } x \ge 1 \end{cases}$

**23.** $C(x) = \begin{cases} x^3 & \text{if } x < 1 \\ \sqrt{x} & \text{if } x > 1 \end{cases}$

**24.** (a) $f(x) = \begin{cases} x^2/|x| & \text{if } x \ne 0 \\ 0 & \text{if } x = 0 \end{cases}$

(b) $F(x) = \begin{cases} x^2/|x| & \text{if } x \ne 0 \\ 1 & \text{if } x = 0 \end{cases}$

**25.** (a) $g(x) = \begin{cases} x/|x| & \text{if } x \ne 0 \\ 0 & \text{if } x = 0 \end{cases}$

(b) $G(x) = \begin{cases} x/|x| & \text{if } x \ne 0 \\ 1 & \text{if } x = 0 \end{cases}$

26. $U(x) = \begin{cases} 1 & \text{if } x \le -2 \\ -1 & \text{if } x > -2 \end{cases}$

27. $V(x) = \begin{cases} x & \text{if } x \le -2 \\ 1 & \text{if } x > -2 \end{cases}$

28. (a) $W(x) = \begin{cases} x & \text{if } x \le -2 \\ -2 & \text{if } x > -2 \end{cases}$

    (b) $D(x) = \begin{cases} x & \text{if } x < -2 \\ -2 & \text{if } x > -2 \end{cases}$

29. $f(x) = \begin{cases} 1/x & \text{if } x < -1 \\ x & \text{if } -1 \le x \le 1 \\ 1/x & \text{if } x > 1 \end{cases}$

30. $g(x) = \begin{cases} 1/x & \text{if } x < -\frac{1}{2} \\ 1 & \text{if } -\frac{1}{2} \le x \le 1 \\ x^3 & \text{if } x > 1 \end{cases}$

31. $y = \begin{cases} x & \text{if } 0 \le x < 1 \\ x-1 & \text{if } 1 \le x < 2 \\ x-2 & \text{if } 2 \le x < 3 \end{cases}$

32. $T(x) = \begin{cases} 2x+2 & \text{if } -1 \le x < 0 \\ -2x+2 & \text{if } 0 \le x < 1 \\ 2x-2 & \text{if } 1 \le x < 2 \\ -2x+6 & \text{if } 2 \le x \le 3 \end{cases}$

**B**

33. (a) Sketch the graph of $f(x) = \dfrac{x^2 - 25}{x + 5}$. What is the domain of $f$?

    (b) Graph the function $F$ defined by the rule

    $$F(x) = \begin{cases} \dfrac{x^2 - 25}{x + 5} & \text{if } x \ne -5 \\ -10 & \text{if } x = -5 \end{cases}$$

34. In the accompanying figure, find the coordinates of the point $T$ if the slope of the straight line passing through the two points $P$ and $T$ is  (a) 5  *Hint:*  let the co-ordinates of $T$ be $(x, x^2)$; (b) 1000; (c) $k$.

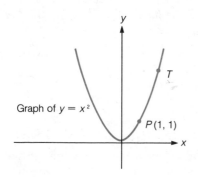

Graph of $y = x^2$

$T$

$P(1, 1)$

35. Let $f(x) = 1/x$. Find a number $t$ such that the slope of the straight line passing through the two points $(1, 1)$ and $(t, f(t))$ on the graph of $f$ is $-\frac{1}{5}$. Include a sketch.

36. Find the coordinates of a point $P$ on the graph of $f(x) = \sqrt{x}$ if the slope of the straight line passing through $P$ and the point $(1, 1)$ is $\frac{1}{7}$. Include a sketch.

*Graph the functions in Exercises 37–42.*

37. $p(x) = [2x]$                38. $q(x) = [x/2]$

39. $r(x) = x - [x]$            40. $R(x) = [x] - x$

41. $f(x) = \{[x]\}^2$          42. $F(x) = [x^2]$

43. (a) Graph the function defined by $y = -[-x]$.

    (b) For each real number $x$, let $f(x)$ be the smallest integer that is greater than or equal to $x$. Graph the function $f$. What do you observe?

44. Let $f(x) = 2x + 1$.

    (a) Graph the functions $g$ and $h$ that are defined by

    $$g(x) = \frac{f(x) - f(-x)}{2} \qquad h(x) = \frac{f(x) + f(-x)}{2}$$

    (b) Verify that the equation $f(x) = g(x) + h(x)$ holds for all values of $x$.

45. The graph of a function $f$ is a straight line. As the figure shows, $A$ and $B$ are two points on the line and the $x$-coordinates of $A$ and $B$ are $h$ units apart. The coordinates of $A$ are $(x, f(x))$. What are the coordinates of $B$? Show that the slope of the line is $\dfrac{f(x + h) - f(x)}{h}$.

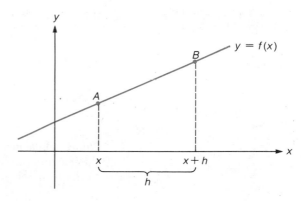

46. Let $F(x) = 2x + 3$. Using the values $x = 1$ and $h = \frac{1}{2}$, compute the value of $\dfrac{F(x + h) - F(x)}{h}$. Include a sketch of the type shown in Exercise 45.

47. The **average rate of change** of a function $f$ between $x = a$ and $x = b$ is defined to be the quantity $\dfrac{f(b) - f(a)}{b - a}$.

(a) By means of a sketch, indicate why the average rate of change represents the slope of the line joining the two points on the graph of $f$ whose $x$-coordinates are $a$ and $b$.

(b) Which is larger, the average rate of change of $f(x) = x^2$ between 0 and 1 or between 10 and 11?

(c) Which of the following functions has the largest average rate of change between 3 and 4?

$$f(x) = x^2 \qquad g(x) = x^3 \qquad h(x) = \frac{1}{x}$$

(d) Find the average rate of change for the function $f(x) = x^2$ over the following intervals:
   (i) $a = 1$ to $b = 1.1$
   (ii) $a = 1$ to $b = 1.01$
   (iii) $a = 1$ to $b = 1.001$
   As $b$ gets closer and closer to 1, what value does your average rate of change seem to be approaching? This "target value" is sometimes called the **instantaneous rate of change**.

48. Suppose that during the first 4 hours of a laboratory experiment, the temperature of a certain substance is given by the formula $f(t) = t^3 - 6t^2 + 9t$, where $t$ is measured in hours, with $t = 0$ corresponding to the time the experiment begins, and where $f(t)$ is the temperature (degrees Fahrenheit) of the substance at time $t$. Calculate the average rate of change of temperature between the following times:
   (a) $t = 0$ and $t = 1$     (b) $t = 1$ and $t = 2$
   (c) $t = 0$ and $t = 3$     (d) $t = 0$ and $t = 4$
   What are the units associated with your answers?

49. Assume that the distance covered by a falling object is given by the formula $f(t) = 16t^2$, where $t$ is measured

in seconds, $f(t)$ is in feet, and $t = 0$ corresponds to the instant the object first begins to fall. Thus, after 1 sec, the object has fallen $16(1)^2$, or 16 ft; after 2 sec, the object has fallen a total of $16(2)^2$, or 64 ft; and so on.

(a) Calculate $f(2) - f(1)$, the distance the object has fallen during the time interval from $t = 1$ to $t = 2$ (i.e., during the second second). Include a sketch of the graph of $f$. On the $t$-axis, indicate the time interval from $t = 1$ to $t = 2$; on the $y$- or $f(t)$-axis, indicate the interval corresponding to the distance $f(2) - f(1)$.

(b) Calculate $f(3) - f(2)$, the distance the object has fallen during the time interval from $t = 2$ to $t = 3$ (i.e., during the third second). Include a sketch as you did in part (a).

(c) The **average velocity** during the time interval from $t = a$ to $t = b$ is defined to be the quantity $\dfrac{f(b) - f(a)}{b - a}$. Note that this is just distance divided by elapsed time and that the units are feet per second. Calculate the average velocity from $t = 1$ to $t = 3$. On a graph of $f$, locate the two points $(1, f(1))$ and $(3, f(3))$. What is the slope of the line joining these two points?

(d) Calculate the average velocity over each of the following intervals:
   (i) $a = 1$ to $b = 1.1$
   (ii) $a = 1$ to $b = 1.001$
   (iii) $a = 1$ to $b = 1.00001$
   As $b$ comes closer and closer to 1, what value does your average velocity seem to be approaching? This "target value" is called the **instantaneous velocity** (or just **velocity**) at $t = 1$.

## TECHNIQUES IN GRAPHING

**4.3**

*One good picture is worth a thousand words. For us, in our study of functions, this means draw graphs. Even more, cultivate the habit of thinking graphically, to the point where it becomes almost second nature.*

GEORGE F. SIMMONS in his book, *Calculus with Analytic Geometry* (New York: McGraw Hill, 1985)

*. . . geometrical figures are graphic formulas.*

DAVID HILBERT (1862–1943)

Given the graph of a function $f$, it is useful to know how to sketch efficiently the graphs of the following closely related functions ($c$ denotes a constant throughout this discussion):

$$y = f(x) + c$$
$$y = f(x + c)$$
$$y = -f(x)$$
$$y = f(-x)$$

We shall approach this by looking at examples.

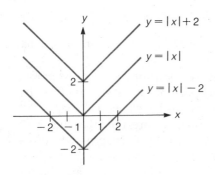

**FIGURE 1**

Figure 1 displays portions of the graphs of $y = |x|$, $y = |x| + 2$, and $y = |x| - 2$. All three graphs have identical shapes. We can view the graph of $y = |x| + 2$ as that obtained by moving the graph of $y = |x|$ up two units in the $y$-direction. Such a movement or shift in position is called a **translation**. Similarly, we can view the graph of $y = |x| - 2$ as that obtained by moving or translating the graph of $y = |x|$ down two units in the $y$-direction. We summarize these observations as follows.

---

**How to Graph $y = f(x) + c$**

If $c > 0$, translate the graph of $y = f(x)$ a distance of $c$ units in the positive $y$-direction.

If $c < 0$, translate the graph of $y = f(x)$ a distance of $|c|$ units in the negative $y$-direction.

---

**EXAMPLE 1**   Graph $y = x^2 - 4$.

**Solution**   Begin with the graph of $y = x^2$ and move it four units in the negative $y$-direction. See Figure 2.

**EXAMPLE 2**   Figure 3(a) shows the graph of a function $s$. Sketch the graph of $y = s(x) + 2$.

**Solution**   To graph $y = s(x) + 2$, move the graph of $s$ two units in the positive $y$-direction. See Figure 3(b).

**FIGURE 2**

(a)

(b)

**FIGURE 3**

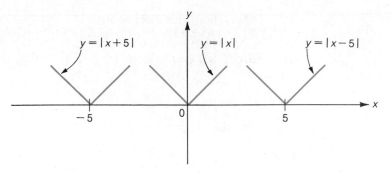

Next we look at examples that indicate the relationship between the graphs of $y = f(x)$ and $y = f(x + c)$. Figure 4 displays portions of the graphs of $y = |x|$, $y = |x - 5|$, and $y = |x + 5|$. Again all three graphs have identical shapes, but this time the movement is in the $x$-direction. Furthermore, note that the graph of $y = |x - 5|$ is five units to the *right*, not left, of $y = |x|$. To help see why this is so, consider the output $y = 0$. To produce the output $y = 0$ using the function $y = |x|$, you clearly need an input of $x = 0$. But to produce that same output of $y = 0$ using $y = |x - 5|$, you need an input of $x = 5$, which is five units to the right of $x = 0$.

---

**How to Graph $y = f(x + c)$**

If $c > 0$, move the graph of $y = f(x)$ a distance of $c$ units in the negative $x$-direction.

If $c < 0$, move the graph of $y = f(x)$ a distance of $|c|$ units in the positive $x$-direction.

---

**EXAMPLE 3**   Graph the functions defined by $y = \dfrac{1}{x - 1}$ and $y = \dfrac{1}{x - 1} + 1$.

Solution   Begin with the graph of $y = 1/x$ in Figure 5(a). As explained in Section 3.2, the $x$- and $y$-axes are asymptotes for this graph. Moving this graph one unit to the right yields the graph of $y = \dfrac{1}{x - 1}$, shown in Figure 5(b). (Note that the vertical asymptote moves one unit to the right also, but the horizontal asymptote is unchanged.) Next, move the graph in Figure 5(b) up one unit

(a)                          (b)                          (c)

$y = x^2$

$y = (x - 2)^2$

$y = (x - 2)^2 + 1$

(a)

(b)

(c)

**FIGURE 6**

to obtain the graph of $y = \dfrac{1}{x - 1} + 1$, as shown in Figure 5(c). (You should verify for yourself that the $y$-intercepts in Figures 5(b) and 5(c) are $-1$ and $0$, respectively.)

**EXAMPLE 4**    Graph $y = (x - 2)^2 + 1$.

Solution    Begin with the graph of $y = x^2$ in Figure 6(a). Move the curve two units to the right to obtain the graph of $y = (x - 2)^2$; see Figure 6(b). Then move the curve in Figure 6(b) up one unit to obtain the graph of $y = (x - 2)^2 + 1$; see Figure 6(c).

In addition to using translation as an aid in graphing functions, we can also use the idea of *reflection* about the $x$- and $y$-axes, as discussed in Section 3.2. We'll explain this through two examples.

First consider the graphs of $y = \sqrt{x}$ and $y = \sqrt{-x}$ in Figure 7. Notice that each point on the graph of $y = \sqrt{-x}$ is the reflection about the $y$-axis of a corresponding point on $y = \sqrt{x}$. For example, the point $(-1, 1)$ on $y = \sqrt{-x}$ is the reflection of $(1, 1)$ on $y = \sqrt{x}$. More generally, each point on $y = \sqrt{-x}$ is of the form $(-x, y)$, where $(x, y)$ is a point on $y = \sqrt{x}$. These observations lead to the following rule.

| **How to Graph $y = f(-x)$** |
| --- |
| Reflect (each point of) the graph of $y = f(x)$ about the $y$-axis to obtain the graph of $y = f(-x)$. |

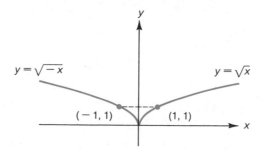

$y = \sqrt{-x}$      $y = \sqrt{x}$

$(-1, 1)$      $(1, 1)$

**FIGURE 7**

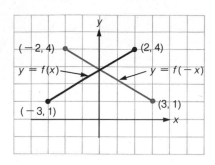

FIGURE 8    FIGURE 9

**EXAMPLE 5**   The graph of $y = f(x)$ is a line segment joining the points $(-3, 1)$ and $(2, 4)$, as indicated in Figure 8. Sketch the graph of $y = f(-x)$.

Solution   Reflect the points $(-3, 1)$ and $(2, 4)$ about the $y$-axis to obtain $(3, 1)$ and $(-2, 4)$, respectively. Then join these two reflected points with a line segment to obtain the graph of $y = f(-x)$, as indicated in Figure 9. (For reference, Figure 9 also shows the graph of $y = f(x)$.)

**EXAMPLE 6**   Figure 10 shows the graph of a function $g$. Sketch the following:
(a) $y = g(-x)$      (b) $y = g(-x) + 2$

Solution   (a) Reflect the graph of $g$ about the $y$-axis to obtain the graph of $y = g(-x)$. See Figure 11(a).
(b) To graph $y = g(-x) + 2$, move the graph obtained in part (a) up two units. See Figure 11(b).

In just the same way that the graphs of $y = \sqrt{x}$ and $y = \sqrt{-x}$ are reflections of each other about the $y$-axis, the graphs of $y = \sqrt{x}$ and $y = -\sqrt{x}$ are reflections of one another about the $x$-axis; see Figures 12(a) and (b). For each point $(x, y)$ on the graph of $y = \sqrt{x}$ in Figure 12(b), there is a corresponding point $(x, -y)$ on the graph of $y = -\sqrt{x}$. For instance, $(1, 1)$ is on the graph of $y = \sqrt{x}$ and $(1, -1)$ is on the graph of $y = -\sqrt{x}$. We can summarize these observations as follows.

FIGURE 10

(a)

FIGURE 11

(b)

(a)

FIGURE 12

(b)

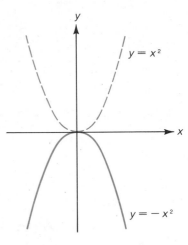

FIGURE 13

---

### How to Graph $y = -f(x)$

Reflect the graph of $y = f(x)$ about the $x$-axis.

---

**EXAMPLE 7**   Graph $y = -x^2$.

Solution   See Figure 13. The dotted lines show the graph of $y = x^2$. By reflecting this graph about the $x$-axis, we obtain the graph of $y = -x^2$.

**EXAMPLE 8**   The graph of a function $F$ is shown in Figure 14. Sketch the graph of $y = -F(x)$.

Solution   The graph of $y = -F(x)$ is obtained by reflecting the graph of $y = F(x)$ about the $x$-axis. Begin by reflecting the points whose coordinates are given. See Figure 15(a). After these points are obtained, the rest of the graph of $y = -F(x)$ can be drawn in as in Figure 15(b).

FIGURE 14

(a)

(b)

FIGURE 15

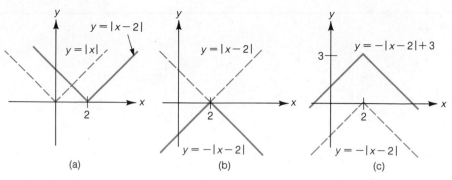

**FIGURE 16**

**EXAMPLE 9**   Graph $y = -|x - 2| + 3$.

**Solution**   Begin with the graph of $y = |x|$. Move the graph two units to the right to obtain the graph of $y = |x - 2|$. See Figure 16(a). Reflect this graph about the $x$-axis to obtain $y = -|x - 2|$. See Figure 16(b). And, finally, move this last graph up three units, as in Figure 16(c).

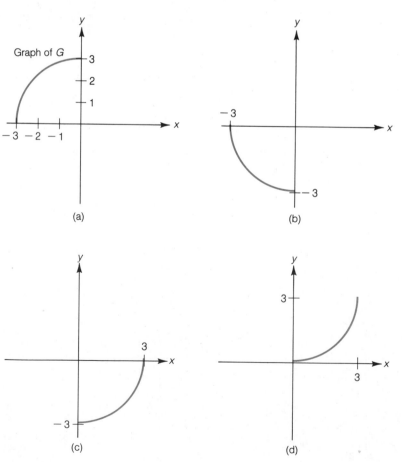

**FIGURE 17**

**EXAMPLE 10**  Figure 17(a) on the previous page shows the graph of $y = G(x)$. What equations describe the graphs in Figures 17(b), (c), and (d)?

Solution

| GRAPH IN GIVEN FIGURE | HOW OBTAINED | EQUATION |
|---|---|---|
| 17(b) | Reflection about the $x$-axis of the graph of $G$ | $y = -G(x)$ |
| 17(c) | Reflection about the $y$-axis of the graph of $y = -G(x)$ | $y = -G(-x)$ |
| 17(d) | Translation of the graph of $y = -G(-x)$ by three units in the positive $y$-direction | $y = -G(-x) + 3$ or $y = 3 - G(-x)$ |

We conclude this section with some additional remarks on translation. Consider, as an example, the two equations

$$y = (x - 1)^2 \qquad \text{and} \qquad y = x^2 + 1$$

As we've seen, the first equation involves a translation in the positive $x$-direction, and the second involves a translation in the positive $y$-direction. To see an underlying pattern here, rewrite the second equation as $y - 1 = x^2$. Then the situation is this:

| EQUATION | HOW GRAPH OBTAINED |
|---|---|
| $y = (x - 1)^2$ | translation in positive $x$-direction |
| $y - 1 = x^2$ | translation in positive $y$-direction |

Observation:  *In the equation $y = x^2$, the effect of replacing $x$ with $x - 1$ is to translate the graph one unit in the positive $x$-direction. Similarly, replacing $y$ with $y - 1$ translates the graph one unit in the positive $y$-direction.*

As a second example, before we generalize, consider the equations

$$y = (x + 1)^2 \qquad \text{and} \qquad y = x^2 - 1$$

The first equation involves a translation in the negative $x$-direction; the second involves a translation in the negative $y$-direction. As before, to see the underlying pattern, we rewrite the second equation as $y + 1 = x^2$. Now we have the following situation.

| EQUATION | HOW GRAPH OBTAINED |
|---|---|
| $y = (x + 1)^2$ | translation in negative $x$-direction |
| $y + 1 = x^2$ | translation in negative $y$-direction |

Observation:  *In the equation $y = x^2$, the effect of replacing $x$ with $x + 1$ is to translate the graph one unit in the negative $x$-direction. Similarly, replacing $y$ with $y + 1$ translates the graph one unit in the negative $y$-direction.*

Both of the examples that we have just considered are specific instances of the following basic result. (The result is valid whether or not the given equation and graph represent a function.)

| PROPERTY SUMMARY | Translation and Coordinates |
| --- | --- |

Suppose that we have an equation that determines a graph in the $x$-$y$ plane, and let $h$ and $k$ denote positive numbers. Then, replacing $x$ with $x - h$ or $x + h$, or replacing $y$ with $y - k$ or $y + k$, has the following effects on the graph of the original equation.

| REPLACEMENT | RESULTING TRANSLATION |
| --- | --- |
| 1. $x$ replaced with $x - h$ | $h$ units in the positive $x$-direction |
| 2. $y$ replaced with $y - k$ | $k$ units in the positive $y$-direction |
| 3. $x$ replaced with $x + h$ | $h$ units in the negative $x$-direction |
| 4. $y$ replaced with $y + k$ | $k$ units in the negative $y$-direction |

**EXAMPLE 11**   The graph of the equation $|x| + |y| = 1$ is shown in Figure 18. Sketch the graph of each of the following equations. (Question for review:   Is the graph in Figure 18 the graph of a function?)

(a) $|x - 4| + |y| = 1$     (b) $|x + 4| + |y| = 1$
(c) $|x| + |y - 4| = 1$     (d) $|x| + |y + 4| = 1$
(e) $|x - 4| + |y + 4| = 1$

**Solution**

| EQUATION | HOW TO GRAPH |
| --- | --- |
| (a) $|x - 4| + |y| = 1$ | Translate the graph of $|x| + |y| = 1$ four units to the right. See Figure 19(a). |
| (b) $|x + 4| + |y| = 1$ | Translate the graph of $|x| + |y| = 1$ four units to the left. See Figure 19(a). |
| (c) $|x| + |y - 4| = 1$ | Translate the graph of $|x| + |y| = 1$ up four units (i.e., four units in the positive $y$-direction). See Figure 19(b). |
| (d) $|x| + |y + 4| = 1$ | Translate the graph of $|x| + |y| = 1$ down four units (i.e., four units in the negative $y$-direction). See Figure 19(b). |
| (e) $|x - 4| + |y + 4| = 1$ | Translate the graph of $|x| + |y| = 1$ four units to the right and four units down. See Figure 19(c). |

FIGURE 18

FIGURE 19          (a)                    (b)                    (c)

# EXERCISE SET 4.3

## A

*In Exercises 1–24, sketch the graph of each function.*

1. $y = x^2 - 3$
2. $y = x^2 + 3$
3. $y = (x + 4)^2$
4. $y = (x + 4)^2 - 3$
5. $y = (x - 4)^2$
6. $y = (x - 4)^2 + 1$
7. $y = -x^2$
8. $y = -x^2 - 3$
9. $y = -(x - 3)^2$
10. $y = -(x - 3)^2 - 3$
11. $y = \sqrt{x - 3}$
12. $y = \sqrt{x - 3} + 1$
13. $y = -\sqrt{x + 1}$
14. $y = -\sqrt{x + 1} + 1$
15. $y = \dfrac{1}{x + 2} + 2$
16. $y = \dfrac{1}{x - 3} - 1$
17. $y = (x - 2)^3$
18. $y = (x - 2)^3 + 1$
19. $y = -x^3 + 4$
20. $y = -(x - 1)^3 + 4$
21. $y = [-x]$
22. $y = -[x]$
23. $y = -[-x]$
24. $y = 1 - [-x]$

*In Exercises 25–36, sketch the graph of each function, given that f, F, and g are defined as follows:*

$$f(x) = |x| \qquad F(x) = \frac{1}{x} \qquad g(x) = \sqrt{1 - x^2}$$

25. $y = f(x - 5)$
26. $y = -f(x - 5)$
27. $y = 1 - f(x - 5)$
28. $y = f(-x)$
29. $y = F(x + 3)$
30. $y = F(x) + 3$
31. $y = -F(x + 3)$
32. $y = F(-x) + 3$
33. $y = g(x - 2)$
34. $y = -g(x - 2)$
35. $y = 1 - g(x - 2)$
36. $y = g(-x)$

*In Exercises 37 and 38, refer to the graph of the function f shown in the figure.*

37. Graph the following functions.
    (a) $y = -f(x)$   (b) $y = f(-x)$
    (c) $y = -f(-x)$

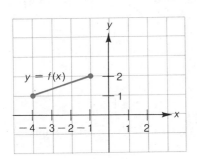

38. Graph the following functions.
    (a) $y = f(x) - 2$   (b) $y = f(x - 2)$
    (c) $y = f(x - 2) - 2$

*In Exercises 39 and 40, refer to the graph of the function g shown in the figure.*

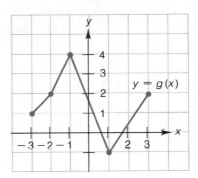

39. Graph the following functions.
    (a) $y = g(-x)$   (b) $y = -g(x)$
    (c) $y = -g(-x)$

40. Graph the function $y = -g(x - 3) - 1$.

41. The next figure shows the graph of the function $y = 10^x$. Sketch the graphs of the following functions.
    (a) $y = 10^{-x}$   (b) $y = -10^x$
    (c) $y = -10^{-x}$   (d) $y = 10^{x-1}$
    (e) $y = 10^x + 1$   (f) $y = -10^{x-1} - 1$

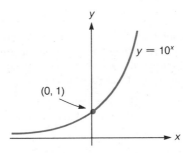

42. Let $f(x) = x^2$. In each of the following cases, sketch the graph of the given function and determine its $x$- and $y$-intercepts, if any.
    (a) $y = f(x) + 4$   (b) $y = f(x) - 3$
    (c) $y = f(x + 1)$   (d) $y = f(x - 4) + 3$
    (e) $y = -f(x - 4) + 2$   (f) $y = f(x - 100)$

In Exercises 43–46, graph the equations, using the following two graphs for reference. (The graph of the equation $4x^2 + 9y^2 = 36$ is an ellipse; this curve will be studied in a later chapter).

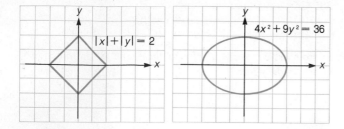

43. (a) $|x - 2| + |y| = 2$    (b) $|x + 2| + |y| = 2$
    (c) $|x| + |y - 2| = 2$    (d) $|x| + |y + 2| = 2$
    (e) $|x - 2| + |y + 2| = 2$
    (f) $|x + 2| + |y + 2| = 2$

44. (a) $|x - 3| + |y - 4| = 2$
    (b) $|x + 3| + |y - 4| = 2$
    (c) $|x + 3| + |y + 4| = 2$
    (d) $|x - 3| + |y + 4| = 2$
    (e) $|x - 3| + |y| = 2$
    (f) $|x| + |y + 4| = 2$

45. (a) $4(x - 1)^2 + 9y^2 = 36$
    (b) $4x^2 + 9(y - 1)^2 = 36$
    (c) $4(x + 1)^2 + 9(y + 1)^2 = 36$
    (d) $4(x - 1)^2 + 9(y - 1)^2 = 36$

46. (a) $4(x - 3)^2 + 9y^2 = 36$
    (b) $4x^2 + 9(y + 2)^2 = 36$
    (c) $4(x + 3)^2 + 9(y - 1)^2 = 36$
    (d) $4(x + 6)^2 + 9y^2 = 36$

## B

For Exercises 47–48, assume that $(a, b)$ is a point on the graph of $y = f(x)$, and specify the corresponding point on the graph of each equation. [For example, the point that corresponds to $(a, b)$ on the graph of $y = f(x - 1)$ is $(a + 1, b)$.]

47. (a) $y = f(x - 3)$    (b) $y = f(x) - 3$
    (c) $y = f(x - 3) - 3$    (d) $y = -f(x)$
    (e) $y = f(-x)$    (f) $y = -f(-x)$

48. (a) $y = f(-x) + 2$    (b) $y = -f(-x) + 2$
    (c) $y = -f(x - 3)$    (d) $y = 1 - f(x + 1)$

49. Show that $\dfrac{x}{x - 1} = \dfrac{1}{x - 1} + 1$, provided $x \neq 1$. Then use this fact to graph the function $g$ defined by $g(x) = \dfrac{x}{x - 1}$.

50. Verify that $-x^2 + 2x + 2 = -(x - 1)^2 + 3$. Then use this fact to graph the function $h$ defined by $h(x) = -x^2 + 2x + 2$.

51. Show that $x^3 - 6x^2 + 12x - 9 = (x - 2)^3 - 1$. Then use this fact to sketch the graph of the equation $y = x^3 - 6x^2 + 12x - 9$.

52. Show that $\dfrac{2 - x}{1 - \sqrt{x - 1}} = 1 + \sqrt{x - 1}$, provided $x \geq 1$ and $x \neq 2$. Then use this information to graph the function $f$ defined by $f(x) = \dfrac{2 - x}{1 - \sqrt{x - 1}}$.

53. What is the range of the function $f$ defined by $f(x) = \dfrac{x^2}{1 - \sqrt{1 - x^2}}$? *Hint:* The function is easy to graph after rationalizing the denominator.

54. (a) A function $f$ is said to be **even** if the equation $f(-x) = f(x)$ is satisfied by all values of $x$ in the domain of $f$. Explain why the graph of an even function must be symmetric about the y-axis.
    (b) Show that the following functions are even by computing $f(-x)$ and then noting that $f(x)$ and $f(-x)$ are equal.
       (i) $f(x) = x^2$    (ii) $f(x) = 2x^4 - 6$
       (iii) $f(x) = 3x^6 - \dfrac{4}{x^2} + 1$

55. (a) A function $f$ is said to be **odd** if the equation $f(-x) = -f(x)$ is satisfied by all values of $x$ in the domain of $f$. Show that if $(x, y)$ is a point on the graph of an odd function $f$, then the point $(-x, -y)$ is also on the graph. (This implies that the graph of an odd function must be symmetric about the origin.)
    (b) Show that the following are odd functions by computing $f(-x)$ as well as $-f(x)$ and then noting that the two expressions obtained are equal.
       (i) $f(x) = x^3$    (ii) $f(x) = -2x^5 + 4x^3 - x$
       (iii) $f(x) = \dfrac{|x|}{x + x^7}$

56. Are the following functions odd, even, or neither? (See Exercises 54 and 55 for definitions.)
    (a) $f(x) = \dfrac{1 - x^2}{2 + x^2}$    (b) $g(x) = \dfrac{x - x^3}{2x + x^3}$
    (c) $h(x) = x^2 + x$    (d) $F(x) = (x^2 + x)^2$
    (e) $G(x) = \begin{cases} 1 & \text{if } x > 0 \\ 0 & \text{if } x = 0 \\ -1 & \text{if } x < 0 \end{cases}$

*Suggestion for part (e):* **Look at the graph.**

## METHODS OF COMBINING FUNCTIONS

**4.4**

Given two numbers $a$ and $b$, there are various ways they can be combined to produce a third number. For instance, we can form the sum $a + b$ or the difference $a - b$ or the product $ab$. Also, if $b \neq 0$, we can form the quotient $a/b$. Similarly, two functions can be combined in various ways to produce a third function. Suppose, for example, that we start with the two functions $y = x^2$ and $y = x^3$. Then it would seem natural to define their sum, difference, product, and quotient as follows:

$$\text{sum:} \quad y = x^2 + x^3$$
$$\text{difference:} \quad y = x^2 - x^3$$
$$\text{product:} \quad y = x^2 x^3 = x^5$$
$$\text{quotient:} \quad y = \frac{x^2}{x^3} = \frac{1}{x}$$

Indeed, this is the idea behind the formal definitions we now give.

---

**Definition**

Let $f$ and $g$ be two functions. Then the **sum** $f + g$, the **difference** $f - g$, the **product** $fg$, and the **quotient** $f/g$ are functions defined by the following equations:

$$(f + g)(x) = f(x) + g(x) \tag{1}$$
$$(f - g)(x) = f(x) - g(x) \tag{2}$$
$$(fg)(x) = f(x) \cdot g(x) \tag{3}$$
$$(f/g)(x) = f(x)/g(x) \qquad \text{provided } g(x) \neq 0 \tag{4}$$

For the functions defined by equations (1), (2), and (3), the domain is the set of all inputs $x$ belonging to both the domain of $f$ and the domain of $g$. For the quotient function in equation (4), we impose the additional restriction that the domain exclude all inputs $x$ for which $g(x) = 0$.

---

**EXAMPLE 1**  Let $f(x) = 3x + 1$ and $g(x) = x - 1$. Compute the following.

**(a)** $(f + g)(x)$  **(b)** $(f - g)(x)$  **(c)** $(fg)(x)$  **(d)** $(f/g)(x)$

**Solution**  **(a)** $(f + g)(x) = f(x) + g(x) = (3x + 1) + (x - 1) = 4x$
**(b)** $(f - g)(x) = f(x) - g(x) = (3x + 1) - (x - 1) = 2x + 2$
**(c)** $(fg)(x) = [f(x)][g(x)] = (3x + 1)(x - 1) = 3x^2 - 2x - 1$

**(d)** $(f/g)(x) = \dfrac{f(x)}{g(x)} = \dfrac{3x + 1}{x - 1} \qquad (x \neq 1)$

For the remainder of this section, we are going to discuss a method for combining functions known as **composition of functions**. As you will see, this method is based on the familiar algebraic process of substitution. Suppose, for example, that $f$ and $g$ are two functions defined by

$$f(x) = x^2 \qquad g(x) = 3x + 1$$

Choose any number in the domain of $g$, say $x = -2$. We can compute $g(-2)$:

$$g(-2) = 3(-2) + 1 = -5$$

Now let's use the output $-5$ that $g$ has produced as an *input* for $f$. We obtain

$$f(-5) = (-5)^2 = 25$$

and consequently,

$$f[g(-2)] = 25$$

So, beginning with the input $-2$, we've successively applied $g$ and then $f$ to obtain the output 25. Similarly, we could carry out this same procedure for any other number in the domain of $g$. Here is a summary of the procedure:

1. Start with an input $x$ and calculate $g(x)$.
2. Use $g(x)$ as an input for $f$; that is, calculate $f[g(x)]$.

We use the notation $f \circ g$ to denote the function, or rule, that tells us to assign the output $f[g(x)]$ to the initial input $x$. In other words, $f \circ g$ denotes the following rule consisting of two steps: *First apply $g$; then apply $f$.* We read the notation $f \circ g$ as "$f$ circle $g$" or "$f$ composed with $g$." In Figure 1, we summarize these ideas.

**FIGURE 1**   Diagram for the function $f \circ g$.

When we write $g(x)$, we assume that $x$ is in the domain of the function $g$. Likewise, for the notation $f[g(x)]$ to make sense, the outputs $g(x)$ must themselves be acceptable inputs for the function $f$. Our formal definition, then, for the composite function $f \circ g$ is as follows.

---

**Composition of Functions: Definition of $f \circ g$**

Given two functions $f$ and $g$, the function $f \circ g$ is defined by

$$(f \circ g)(x) = f[g(x)]$$

The domain of $f \circ g$ consists of those inputs $x$ (in the domain of $g$) for which $g(x)$ is in the domain of $f$.

---

**EXAMPLE 2**   Let $f(x) = x^2$ and $g(x) = 3x + 1$. Compute $(f \circ g)(x)$ and also $(g \circ f)(x)$.

**Solution**
$$\begin{aligned}
(f \circ g)(x) &= f[g(x)] &&\text{definition of } f \circ g \\
&= f(3x + 1) &&\text{definition of } g \\
&= (3x + 1)^2 &&\text{definition of } f \\
&= 9x^2 + 6x + 1
\end{aligned}$$

$$(g \circ f)(x) = g[f(x)] \qquad \text{definition of } g \circ f$$
$$= g(x^2) \qquad \text{definition of } f$$
$$= 3(x^2) + 1 \qquad \text{definition of } g$$
$$= 3x^2 + 1$$

Notice that the two results obtained in Example 2 are not the same. This shows that, in general, $f \circ g$ and $g \circ f$ represent different functions.

**EXAMPLE 3**  Let $f$ and $g$ be defined as in Example 2:  $f(x) = x^2$ and $g(x) = 3x + 1$. Compute $(f \circ g)(-2)$.

**Solution**  We will show two methods.

FIRST METHOD  Using the formula for $(f \circ g)(x)$ developed in Example 2, we have

$$(f \circ g)(x) = 9x^2 + 6x + 1$$

and therefore,

$$(f \circ g)(-2) = 9(-2)^2 + 6(-2) + 1$$
$$= 36 - 12 + 1$$
$$= 25$$

ALTERNATIVE METHOD  Working directly from the definition of $f \circ g$, we have

$$(f \circ g)(-2) = f[g(-2)]$$
$$= f[3(-2) + 1]$$
$$= f(-5)$$
$$= (-5)^2 = 25$$

In Examples 2 and 3, the domain of both $f$ and $g$ is the set of all real numbers. And as you can easily check, the domain of both $f \circ g$ and $g \circ f$ is also the set of all real numbers. In Example 4, however, some care needs to be taken in describing the domain of the composite function.

**EXAMPLE 4**  Let $f$ and $g$ be defined as follows:

$$f(x) = x^2 + 1 \qquad g(x) = \sqrt{x}$$

Compute $(f \circ g)(x)$. Find the domain of $f \circ g$ and sketch its graph.

**Solution**
$$(f \circ g)(x) = f[g(x)]$$
$$= f(\sqrt{x}) = (\sqrt{x})^2 + 1 = x + 1$$

FIGURE 2  Graph of the function $f \circ g$ in Example 4.

So we have $(f \circ g)(x) = x + 1$. Now what about the domain of $f \circ g$? Our first inclination may be to say (incorrectly!) that the domain is the set of all real numbers, since any real number can be used as an input in the expression $x + 1$. However, the definition of $f \circ g$ on page 188 tells us that the inputs for $f \circ g$ must first of all be acceptable inputs for $g$. In view of the definition of $g$, then, we must require that $x$ be nonnegative. On the other hand, for any nonnegative input $x$, the number $g(x)$ will be an acceptable input for $f$. (Why?) In summary, then, the domain of $f \circ g$ is the interval $[0, \infty)$. The graph of $f \circ g$ is shown in Figure 2.

One reason for studying the composition of functions is that it lets us express a given function in terms of simpler functions. This is often useful in calculus. Suppose, for example, that we wish to express the function $C$ defined by

$$C(x) = (2x^3 - 5)^2$$

as a composition of simpler functions. That is, we want to come up with two functions $f$ and $g$ so that the equation

$$C(x) = (f \circ g)(x)$$

holds for every $x$ in the domain of $C$.

We begin by thinking what we would do to compute $(2x^3 - 5)^2$ for a given value of $x$. First we would compute the quantity $2x^3 - 5$, then we would square the result. Therefore, recalling that the rule $f \circ g$ tells us to do $g$ *first*, we let $g(x) = 2x^3 - 5$. And then, since the next step is squaring, we let $f(x) = x^2$. Now let us see if these choices for $f$ and $g$ are correct; that is, let us calculate $(f \circ g)(x)$ and see if it really is the same as $C(x)$.

Using $f(x) = x^2$ and $g(x) = 2x^3 - 5$, we have

$$(f \circ g)(x) = f[g(x)] = f(2x^3 - 5)$$
$$= (2x^3 - 5)^2 = C(x)$$

This shows that our choices for $f$ and $g$ were indeed correct, and we have expressed $C$ as a composition of two simpler functions.*

Before leaving this example, note that in expressing $C$ as $f \circ g$, we chose $g$ to be the "inner" function, that is, the quantity inside parentheses: $g(x) = 2x^3 - 5$. This observation is used in Example 5.

**EXAMPLE 5**  Let $s(x) = \sqrt{1 + x^4}$. Express the function $s$ as a composition of two simpler functions $f$ and $g$.

**Solution**  Let $g$ be the "inner" function; that is, let $g(x)$ be the quantity inside the radical:

$$g(x) = 1 + x^4$$

And let us take $f$ to be the square root function:

$$f(x) = \sqrt{x}$$

Now we need to verify that these are appropriate choices for $f$ and $g$; that is, we need to check that the equation $(f \circ g)(x) = s(x)$ is true for every $x$ in the domain of $s$. We have

$$(f \circ g)(x) = f[g(x)] = f(1 + x^4)$$
$$= \sqrt{1 + x^4} = s(x)$$

Thus, $(f \circ g)(x) = s(x)$, as required.

## EXERCISE SET 4.4

### A

*In Exercises 1–10, compute the indicated expressions, given that the functions $f$, $g$, $h$, $k$, and $m$ are defined as follows:*

$$f(x) = 2x - 1 \qquad k(x) = 2, \quad \text{for all } x$$
$$g(x) = x^2 - 3x - 6 \qquad m(x) = x^2 - 9$$
$$h(x) = x^3$$

1. (a) $(f + g)(x)$   (b) $(f - g)(x)$   (c) $(f - g)(0)$
2. (a) $(fh)(x)$   (b) $(h/f)(x)$   (c) $(f/h)(1)$
3. (a) $(m - f)(x)$   (b) $(f - m)(x)$
4. (a) $(fg)(x)$   (b) $(fg)(\frac{1}{2})$
5. (a) $(fk)(x)$   (b) $(kf)(x)$   (c) $(fk)(1) - (kf)(2)$

---

* Other answers are possible, too. For instance, if $F(x) = (x - 5)^2$ and $G(x) = 2x^3$, then (as you should verify for yourself) $C(x) = F[G(x)]$.

6. (a) $(g + m)(x)$     (b) $(g + m)(x) - (g - m)(x)$
7. (a) $(f/m)(x) - (m/f)(x)$     (b) $(f/m)(0) - (m/f)(0)$
8. (a) $[h \cdot (f + m)](x)$  *Note:* $h$ and $(f + m)$ are two functions; the notation $h \cdot (f + m)$ denotes the product function.
   (b) $(hf)(x) + (hm)(x)$
9. (a) $[m \cdot (k - h)](x)$     (b) $(mk)(x) - (mh)(x)$
   (c) $(mk)(-1) - (mh)(-1)$
10. (a) $(g + g)(x)$     (b) $(g - g)(x)$
    (c) $(kg)(x)$     (d) $(g + g)(-3) - (kg)(-3)$
11. Let $f(x) = 3x + 1$ and $g(x) = -2x - 5$. Compute the following.
    (a) $(f \circ g)(x)$     (b) $(f \circ g)(10)$
    (c) $(g \circ f)(x)$     (d) $(g \circ f)(10)$
12. Let $f(x) = 1 - 2x^2$ and $g(x) = x + 1$. Compute the following.
    (a) $(f \circ g)(x)$     (b) $(f \circ g)(-1)$
    (c) $(g \circ f)(x)$     (d) $(g \circ f)(-1)$
    (e) $(f \circ f)(x)$     (f) $(g \circ g)(-1)$
13. Compute $(f \circ g)(x)$, $(f \circ g)(-2)$, $(g \circ f)(x)$, and $(g \circ f)(-2)$ for each pair of functions.
    (a) $f(x) = 1 - x$,  $g(x) = 1 + x$
    (b) $f(x) = x^2 - 3x - 4$,  $g(x) = 2 - 3x$
    (c) $f(x) = x/3$,  $g(x) = 1 - x^4$
    (d) $f(x) = 2^x$,  $g(x) = x^2 + 1$
    (e) $f(x) = x$,  $g(x) = 3x^5 - 4x^2$
    (f) $f(x) = 3x - 4$,  $g(x) = \frac{1}{3}(x + 4)$
14. Let $h(x) = 4x^2 - 5x + 1$, $k(x) = x$, and $m(x) = 7$ for all $x$. Compute the following.
    (a) $h[k(x)]$     (b) $k[h(x)]$     (c) $h[m(x)]$
    (d) $m[h(x)]$     (e) $k[m(x)]$     (f) $m[k(x)]$
15. Let $F(x) = \dfrac{3x - 4}{3x + 3}$ and $G(x) = \dfrac{x + 1}{x - 1}$. Compute the following.
    (a) $(F \circ G)(x)$     (b) $F[G(t)]$     (c) $(F \circ G)(2)$
    (d) $(G \circ F)(x)$     (e) $G[F(y)]$     (f) $(G \circ F)(2)$
16. Let $f(x) = \dfrac{1}{x^2} + 1$ and $g(x) = \dfrac{1}{x - 1}$.
    (a) Compute $(f \circ g)(x)$.
    (b) What is the domain of $f \circ g$?
    (c) Graph the function $f \circ g$.
17. Let $M(x) = \dfrac{2x - 1}{x - 2}$.
    (a) Compute $M(7)$ and then $M[M(7)]$.
    (b) Compute $(M \circ M)(x)$.
    (c) Compute $(M \circ M)(7)$, using the formula you obtained in part (b). Check that your answer agrees with that obtained in part (a).
18. Let $F(x) = (x + 1)^5$, $f(x) = x^5$, and $g(x) = x + 1$. Which of the following is true for all $x$?
    $(f \circ g)(x) = F(x)$     or     $(g \circ f)(x) = F(x)$

19. Refer to the graphs of the functions $f$, $g$, and $h$ to compute the required quantities. Assume that all the axes are marked off in one-unit intervals.
    (a) $f[g(3)]$     (b) $g[f(3)]$     (c) $f[h(3)]$
    (d) $(h \circ g)(2)$     (e) $h\{f[g(3)]\}$
    (f) $(g \circ f \circ h \circ f)(2)$  *Note:* This means first do $f$, then $h$, then $f$, then $g$.

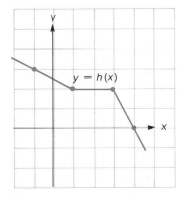

20. Let $f(x) = 2x - 1$. Find $f[f(x)]$. If $z = f[f(x)]$, find $f(z)$.
21. (a) Let $T(x) = 4x^3 - 3x^2 + 6x - 1$ and $I(x) = x$. Find $(T \circ I)(x)$ and $(I \circ T)(x)$.
    (b) Let $G(x) = ax^2 + bx + c$ and $I(x) = x$. Find $(G \circ I)(x)$ and $(I \circ G)(x)$.
    (c) What general conclusion do you arrive at from the results of parts (a) and (b)?
22. The figure (on the next page) shows the graphs of two functions, $F$ and $G$. Use the graphs to compute the following quantities.
    (a) $(G \circ F)(1)$     (b) $(F \circ G)(1)$
    (c) $(F \circ F)(1)$     (d) $(G \circ G)(-3)$
    (e) $(G \circ F)(5)$     (f) $(F \circ F)(5)$
    (g) $(F \circ G)(-1)$     (h) $(G \circ F)(2)$

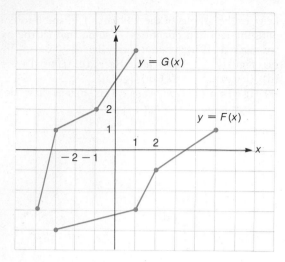

**FIGURE FOR EXERCISE 22**

| $y$ | 0 | $\dfrac{1}{4}$ | $\dfrac{\sqrt{2}}{4}$ | $\dfrac{1}{2}$ | $\dfrac{\sqrt{2}}{2}$ | $\dfrac{3}{4}$ | $\dfrac{\sqrt{3}}{2}$ | 1 |
|---|---|---|---|---|---|---|---|---|
| $g(y)$ | $\dfrac{\pi}{2}$ | $\pi$ | 0 | $\dfrac{\pi}{3}$ | $\dfrac{\pi}{4}$ | 0 | $\dfrac{\pi}{6}$ | 0 |

Use this information to complete the following table of values for $(g \circ f)(x)$.

| $x$ | 0 | $\dfrac{\pi}{6}$ | $\dfrac{\pi}{4}$ | $\dfrac{\pi}{3}$ | $\dfrac{\pi}{2}$ |
|---|---|---|---|---|---|
| $(g \circ f)(x)$ | | | | | |

**23.** The domain of a function $f$ consists of the numbers $-1$, 0, 1, 2, and 3. The following table shows the output $f$ assigns to each input.

| $x$ | $-1$ | 0 | 1 | 2 | 3 |
|---|---|---|---|---|---|
| $f(x)$ | 2 | 2 | 0 | 3 | 1 |

The domain of a function $g$ consists of the numbers 0, 1, 2, 3, and 4. The following table shows the output $g$ assigns to each input.

| $x$ | 0 | 1 | 2 | 3 | 4 |
|---|---|---|---|---|---|
| $g(x)$ | 3 | 2 | 0 | 4 | $-1$ |

Use this information to complete the tables for $f \circ g$ and $g \circ f$. *Note:* Two of the entries will be undefined.

| $x$ | 0 | 1 | 2 | 3 | 4 |
|---|---|---|---|---|---|
| $(f \circ g)(x)$ | | | | | |

| $x$ | $-1$ | 0 | 1 | 2 | 3 | 4 |
|---|---|---|---|---|---|---|
| $(g \circ f)(x)$ | | | | | | |

**24.** The following two tables show certain pairs of inputs and outputs for functions $f$ and $g$.

| $x$ | 0 | $\dfrac{\pi}{6}$ | $\dfrac{\pi}{4}$ | $\dfrac{\pi}{3}$ | $\dfrac{\pi}{2}$ |
|---|---|---|---|---|---|
| $f(x)$ | 0 | $\dfrac{1}{2}$ | $\dfrac{\sqrt{2}}{2}$ | $\dfrac{\sqrt{3}}{2}$ | 1 |

**25.** Let $f(x) = 2x + 1$ and $g(x) = 3x - 4$.
(a) Find $(f \circ g)(x)$ and graph the function $f \circ g$.
(b) Find $(g \circ f)(x)$ and graph the function $g \circ f$.

**26.** Let $F(x) = x^2$ and $G(x) = x - 4$.
(a) Find $F[G(x)]$ and graph the function $F \circ G$.
(b) Find $G[F(x)]$ and graph the function $G \circ F$.

**27.** Let $g(x) = \sqrt{x} - 3$ and $f(x) = x - 1$.
(a) Sketch a graph of $g$. Specify the domain and range.
(b) Sketch a graph of $f$. Specify the domain and range.
(c) Compute $(f \circ g)(x)$. Graph the function $f \circ g$ and specify its domain and range.
(d) Find a formula for $g[f(x)]$. Which values of $x$ are acceptable inputs here? That is, what is the domain of $g \circ f$?
(e) Use the results of part (d) to sketch a graph of the function $g \circ f$.

**28.** Let $F(x) = -x^2$ and $G(x) = \sqrt{x}$. Determine the domains of $F \circ G$ and $G \circ F$.

**29.** Let $C(x) = (3x - 1)^4$. Express $C$ as a composition of two simpler functions.

**30.** Let $C(x) = (1 + x^2)^3$. Find functions $f$ and $g$ so that $C(x) = (f \circ g)(x)$ is true for all values of $x$.

**31.** Express each function as a composition of two functions.
(a) $F(x) = \sqrt[3]{3x + 4}$  (b) $G(x) = |2x - 3|$
(c) $H(x) = (ax + b)^5$  (d) $T(x) = 1/\sqrt{x}$

**32.** Let $a(x) = x^2$, $b(x) = |x|$, $c(x) = 3x - 1$. Express each of the following functions as a composition of two of the given functions.
(a) $f(x) = (3x - 1)^2$  (b) $g(x) = |3x - 1|$
(c) $h(x) = 3x^2 - 1$

**33.** Let $a(x) = 1/x$, $b(x) = \sqrt[3]{x}$, $c(x) = 2x + 1$, and $d(x) = x^2$. Express each of the following functions as a composition of two of the given functions.
(a) $f(x) = \sqrt[3]{2x + 1}$  (b) $g(x) = 1/x^2$
(c) $h(x) = 2x^2 + 1$  (d) $K(x) = 2\sqrt[3]{x} + 1$
(e) $l(x) = \dfrac{2}{x} + 1$  (f) $m(x) = \dfrac{1}{2x + 1}$
(g) $n(x) = x^{2/3}$

**34.** Express $G(x) = 1/(1 + x^4)$ as a composition of two simpler functions, one of which is $f(x) = 1/x$.

**B**

**35.** The circumference of a circle with radius $r$ is given by $C(r) = 2\pi r$. Suppose that the circle is shrinking in size and that the radius at time $t$ is given by

$$r = f(t) = \frac{1}{t^2 + 1}$$

Assume that $t$ is measured in minutes and that $r$, or $f(t)$, is in feet. Find $(C \circ f)(t)$ and use this to find the circumference when $t = 3$ min.

**36.** A spherical weather balloon is being inflated in such a way that the radius is given by

$$r = g(t) = \tfrac{1}{2}t + 2$$

Assume that $r$ is in meters and $t$ is in seconds, with $t = 0$ corresponding to the time that inflation begins. If the volume of a sphere of radius $r$ is given by

$$V(r) = \tfrac{4}{3}\pi r^3$$

compute $V[g(t)]$ and use this to find the time at which the volume of the balloon is $36\pi \, \mathrm{m}^3$.

**37.** Suppose that a manufacturer knows that the daily production cost to build $x$ bicycles is given by the function $C$, where

$$C(x) = 100 + 90x - x^2 \qquad (0 \le x \le 40)$$

That is, $C(x)$ represents the cost in dollars of building $x$ bicycles. Furthermore, suppose that the number of bicycles that can be built in $t$ hr is given by the function $f$, where

$$x = f(t) = 5t \qquad (0 \le t \le 8)$$

(a) Compute $(C \circ f)(t)$.
(b) Compute the production cost on a day that the factory operates for $t = 3$ hr.
(c) If the factory runs for 6 hr instead of 3 hr, is the cost twice as much?

**38.** Suppose that in a certain biology lab experiment, the number of bacteria is related to the temperature of the environment by the function

$$N(T) = -2T^2 + 240T - 5400 \qquad (40 \le T \le 90)$$

Here, $N(T)$ represents the number of bacteria present when the temperature is $T$ degrees Fahrenheit.
Also, suppose that $t$ hr after the experiment begins, the temperature is given by

$$T(t) = 10t + 40 \qquad (0 \le t \le 5)$$

(a) Compute $N[T(t)]$.

(b) How many bacteria are present when $t = 0$? When $t = 2$ hr? When $t = 5$ hr?

**39.** Let $g(x) = 4x - 1$. Find $f(x)$, given that the equation $(g \circ f)(x) = x + 5$ is true for all values of $x$.

**40.** Let $g(x) = 2x + 1$. Find $f(x)$, given that $(g \circ f)(x) = 10x - 7$.

**41.** Let $f(x) = -2x + 1$ and $g(x) = ax + b$. Find values for $a$ and $b$ so the equation

$$f[g(x)] = x$$

holds for all values of $x$.

**42.** Let $f(x) = \dfrac{3x - 4}{x - 3}$.
(a) Compute $(f \circ f)(x)$.
(b) Find $f[f(\frac{113}{355})]$. (Try not to do it the hard way.)

**43.** Let $f(x) = x^2$ and $g(x) = 2x - 1$.
(a) Compute $\dfrac{f[g(x)] - f[g(a)]}{g(x) - g(a)}$.
(b) Compute $\dfrac{f[g(x)] - f[g(a)]}{x - a}$.

**44.** Suppose that $y = f(x)$ and $g(y) = x$. Show that $(g \circ f)(x) - (f \circ g)(y) = x - y$.

**45.** Given three functions $f$, $g$, and $h$, we define the function $f \circ g \circ h$ by the formula

$$(f \circ g \circ h)(x) = f\{g[h(x)]\}$$

For example, if $f(x) = x^2$, $g(x) = x + 1$, and $h(x) = x/2$, we evaluate $(f \circ g \circ h)(x)$ as

$$(f \circ g \circ h)(x) = f\{g[h(x)]\}$$
$$= f\left[g\left(\frac{x}{2}\right)\right]$$
$$= f\left(\frac{x}{2} + 1\right)$$
$$= \left(\frac{x}{2} + 1\right)^2 = \frac{x^2}{4} + x + 1$$

Using $f(x) = x^2$, $g(x) = x + 1$, and $h(x) = x/2$, compute the following.
(a) $(g \circ h \circ f)(x)$    (b) $(h \circ f \circ g)(x)$
(c) $(g \circ f \circ h)(x)$    (d) $(f \circ h \circ g)(x)$
(e) $(h \circ g \circ f)(x)$
*Hint for checking:* No two answers are the same.

**46.** Let $F(x) = 2x - 1$, $G(x) = 4x$, and $H(x) = 1 + x^2$. Compute the following.
(a) $(F \circ G \circ H)(x)$    (b) $(G \circ F \circ H)(x)$
(c) $(F \circ H \circ G)(x)$    (d) $(H \circ F \circ G)(x)$
(e) $(H \circ G \circ F)(x)$    (f) $(G \circ H \circ F)(x)$

**47.** Let $f(x) = x^2$, $g(x) = 1 - x$, and $h(x) = 3x$. Express the following functions as compositions of $f$, $g$, and $h$.

(a) $p(x) = 1 - 9x^2$  (b) $q(x) = 3 - 3x^2$
(c) $r(x) = 1 - 6x + 9x^2$  (d) $s(x) = 3 - 6x + 3x^2$

48. Suppose that three functions $f'$, $g$, and $g'$ are defined as follows ($f'$ is read "f prime"):

$$f'(x) = 3x^2 \qquad g(x) = 2x^2 + 5 \qquad g'(x) = 4x$$

(a) Find $f'[g(x)] \cdot g'(x)$. (The dot denotes multiplication, not composition.)
(b) Evaluate $f'[g(-1)] \cdot g'(-1)$.
(c) Find $f'[g(t)] \cdot g'(t)$.

49. Suppose that three functions $F'$, $G$, and $G'$ are defined by

$$F'(x) = \frac{1}{2\sqrt{x}}$$

$$G(x) = x^2 + 2x + 2$$

$$G'(x) = 2x + 2$$

(a) Find $F'[G(x)] \cdot G'(x)$. (The dot denotes multiplication.)
(b) Find $F'[G(9)] \cdot G'(9)$.

C

50. Let $p$ and $q$ be two numbers whose sum is 1. Define the function $F$ by

$$F(x) = p - \frac{1}{x + q}$$

Show that $F\{F[F(x)]\} = x$.

51. Let $i(x) = x$, $a(x) = -x$, $b(x) = 1/x$, and $c(x) = -1/x$. Assume that the domain of each function is $(0, \infty)$.

(a) Complete the following table, where the operation is composition of functions. For example, $c$ is the proper entry in the second row, third column, since if you compute $(a \circ b)(x)$, you'll get $-1/x$, which is $c(x)$.

| $\circ$ | $i$ | $a$ | $b$ | $c$ |
|---|---|---|---|---|
| $i$ | | | | |
| $a$ | | | $c$ | |
| $b$ | | | | |
| $c$ | | | | |

(b) According to your table, is the operation commutative (e.g., does $a \circ b = b \circ a$)?
(c) Use your table to find $a^2$ (meaning $a \circ a$), $b^2$, $c^2$, and $c^3$.
(d) Use your table to find out if $(a \circ b) \circ c = a \circ (b \circ c)$.
(e) Let $G$ denote the set consisting of the four functions you've worked with in this exercise. Now define a function $f$, with domain $G$, as follows: $f(t) = t^2$ for each element $t$ in the set $G$. [For example, $f(a) = a^2 = a \circ a = i$.] What is the range of $f$?
(f) Verify that $f(a \circ b) = [f(a)] \circ [f(b)]$.
(g) Find out if $f(a \circ b \circ c) = [f(a)] \circ [f(b)] \circ [f(c)]$.

---

## INVERSE FUNCTIONS

**4.5**

PROBLEM
SOLUTION

Illustration by Scott Kim. © 1981 Scott Kim.

*If anybody ever told me why the graph of $y = x^{1/2}$ is the reflection of the graph of $y = x^2$ in a 45° line, it didn't sink in. To this day, there are textbooks that expect students to think that it is so obvious as to need no explanation. This is a pity, if only because [in calculus] it is such a common practice to define the natural logarithm first and then define the exponential function as its inverse.*

PROFESSOR RALPH P. BOAS, recalling his student days in the article "Inverse Functions" in *The College Mathematics Journal* **16** (1985): 42.

We shall introduce the idea of inverse functions through an easy example. Consider the function $f$ defined by

$$f(x) = 2x$$

A short table of values for $f$ and the graph of $f$ are displayed in Figure 1.

Now we ask this question: What happens if we take the table for $f$ and interchange the entries in the $x$ and $y$ columns to obtain a new table that looks like the following?

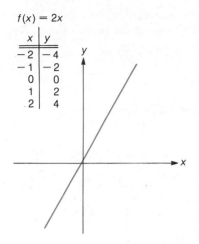

$f(x) = 2x$

| x | y |
|----|----|
| −2 | −4 |
| −1 | −2 |
| 0 | 0 |
| 1 | 2 |
| 2 | 4 |

**FIGURE 1**

| x | y |
|----|----|
| −4 | −2 |
| −2 | −1 |
| 0 | 0 |
| 2 | 1 |
| 4 | 2 |

← New table formed by interchanging the inputs and outputs for $f$

Several observations are in order. Our new table does itself represent a function (in this case); for each input $x$ there is exactly one output $y$. Let us call this new function $g$. Note that each output is half the corresponding input. So the function $g$ can be described by the formula $g(x) = x/2$. In summary, we began with the "doubling function," $f(x) = 2x$; then, by interchanging the inputs with the outputs, we obtained the "halving function," $g(x) = x/2$.

The doubling function $f$ and the halving function $g$ are in a certain sense opposites; each reverses or undoes the effect of the other. That is, if you begin with $x$ and then calculate $g[f(x)]$, you get $x$ again; similarly, $f[g(x)]$ is also equal to $x$. (Verify these last two statements for yourself.)

The two functions $f$ and $g$ that we've been discussing are an example of a pair of *inverse functions*. The general definition for inverse functions is given in the box that follows.

---

**Definition of Inverse Functions**

Two functions $f$ and $g$ are said to be **inverses** of one another provided

$$f[g(x)] = x \qquad \text{for each } x \text{ in the domain of } g$$

and

$$g[f(x)] = x \qquad \text{for each } x \text{ in the domain of } f$$

---

**EXAMPLE 1**    Verify that the functions $f$ and $g$ are inverses, where

$$f(x) = \frac{1}{3}x + 2 \qquad \text{and} \qquad g(x) = 3x - 6$$

**Solution**    In view of the definition, we must check that $f[g(x)] = x$ and $g[f(x)] = x$. We have

$$f[g(x)] = f(3x - 6)$$

$$= \frac{1}{3}(3x - 6) + 2 = x \qquad \text{(Check the algebra.)}$$

Thus, $f[g(x)] = x$. Now we still need to check that $g[f(x)] = x$. We have

$$g[f(x)] = g\left(\frac{1}{3}x + 2\right)$$

$$= 3\left(\frac{1}{3}x + 2\right) - 6 = x \qquad \text{(Again, check the algebra.)}$$

Having shown that $f[g(x)] = x$ and $g[f(x)] = x$, we conclude that $f$ and $g$ are indeed inverse functions.

**EXAMPLE 2**    Suppose that $f$ and $g$ are a pair of inverse functions. If $f(2) = 3$, what is $g(3)$?

**Solution**    We can use either of two methods.

FIRST METHOD    The quantities 3 and $f(2)$ are declared to be equal. Thus, whether we use 3 or $f(2)$ as an input for $g$, the result must be the same. We then have

$$3 = f(2)$$
$$g(3) = g[f(2)]$$
$$g(3) = 2 \qquad \text{using the fact that } g[f(x)] = x \text{ for any } x \text{ in the domain of } f$$

Thus $g(3) = 2$.

ALTERNATIVE METHOD    Think of a table of values for the function $f$. Since $f(2) = 3$, one entry in the table must look like this:

| $x$ | | 2 | |
|---|---|---|---|
| $y$ | | 3 | |

Now, $g$ reverses the roles of $x$ and $y$, so in the table for $g$, one entry must look like this:

| $x$ | | 3 | |
|---|---|---|---|
| $y$ | | 2 | |

Thus, $g(3) = 2$, as obtained previously.    ▪

It is customary to use the notation $f^{-1}$ (read "$f$ inverse") for the function that is the inverse of $f$. So in Example 1, for instance, we have

$$f(x) = \frac{1}{3}x + 2 \qquad \text{and} \qquad f^{-1}(x) = 3x - 6$$

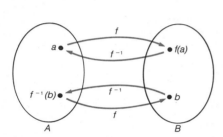

**FIGURE 2**   Diagram illustrating the action of inverse functions. The set $A$ in the figure is both the domain of $f$ and the range of $f^{-1}$. The set $B$ is both the domain of $f^{-1}$ and the range of $f$.

NOTE    In the context of functions, $f^{-1}$ does *not* in general mean $1/f$. (For an example of this, see Exercise 4 at the end of this section.) In the following box, we rewrite the defining equations for inverse functions, using the $f^{-1}$ notation rather than $g$. Figure 2 provides a graphic summary of these equations.

$$f[f^{-1}(x)] = x \qquad \text{for each } x \text{ in the domain of } f^{-1}$$
$$\text{and}$$
$$f^{-1}[f(x)] = x \qquad \text{for each } x \text{ in the domain of } f$$

**EXAMPLE 3**    Suppose $f(x) = 4x^3 + 7$. Find $f[f^{-1}(5)]$. (Assume that $f^{-1}$ exists and that 5 is in its domain.)

Solution    By definition, $f[f^{-1}(x)] = x$. Substituting $x = 5$ on both sides of this identity directly gives us

$$f[f^{-1}(5)] = 5$$

which is the required solution. Note that we didn't need to make use of the formula $f(x) = 4x^3 + 7$, nor did we need to find a formula for $f^{-1}(x)$.    ▪

We'll postpone until the end of this section a discussion of which functions have inverses and which do not. For functions that do have inverses, however, there's a simple method that can often be used to determine those inverses. This method is illustrated in Examples 4 and 5.

**EXAMPLE 4**   Let $f(x) = 3x - 4$. Find $f^{-1}(x)$.

Solution   Begin by rewriting the given equation as

$$y = 3x - 4$$

We know that $f^{-1}$ interchanges the inputs and outputs of $f$. So to determine $f^{-1}$, we only need to switch the $x$'s and $y$'s in the equation $y = 3x - 4$. That gives us $x = 3y - 4$. Now we solve for $y$ as follows:

$$x = 3y - 4$$
$$x + 4 = 3y$$
$$\frac{x + 4}{3} = y$$

Thus, the inverse function is $y = \dfrac{x + 4}{3}$. We can also write this as

$$f^{-1}(x) = \frac{x + 4}{3}$$

Actually, we should call this result just a *candidate* for $f^{-1}$, since certain technical matters remain to be discussed. However, if you compute $f[f^{-1}(x)]$ and $f^{-1}[f(x)]$ and find that they both do equal $x$, then the matter is settled, by definition. Exercise 13 at the end of this section asks you to carry out those calculations.

Here is a summary of our procedure for calculating $f^{-1}(x)$.

---

**To Find $f^{-1}(x)$ for the Function $y = f(x)$:**

1. Interchange $x$ and $y$ in the equation $y = f(x)$.
2. Solve the resulting equation for $y$.

---

**EXAMPLE 5**   Let $f(x) = \dfrac{2x - 1}{3x + 5}$. Find $f^{-1}(x)$.

Solution   STEP 1   Write the given function as $y = \dfrac{2x - 1}{3x + 5}$. Interchange $x$ and $y$ to get

$$x = \frac{2y - 1}{3y + 5}$$

STEP 2   Solve the resulting equation for $y$.

$$x = \frac{2y - 1}{3y + 5}$$
$$3xy + 5x = 2y - 1 \qquad \text{multiplying by } 3y + 5$$
$$3xy - 2y = -5x - 1 \qquad \text{rearranging}$$
$$y(3x - 2) = -5x - 1$$
$$y = \frac{-5x - 1}{3x - 2}$$

Thus, the inverse function is given by

$$f^{-1}(x) = \frac{-5x - 1}{3x - 2}$$

There is a certain symmetry that always exists between the graph of a function and the graph of its inverse. As background for this discussion, we need the following definition of **symmetry about a line**.

---

**Definition:** **Symmetry about a Line**

Refer to Figure 3. Two points $P$ and $Q$ are said to be **symmetric about the line** $\mathscr{L}$ provided that

$\overline{PQ}$ is perpendicular to $\mathscr{L}$

and

points $P$ and $Q$ are equidistant from $\mathscr{L}$

FIGURE 3

---

FIGURE 4

In other words, $P$ and $Q$ are symmetric about the line $\mathscr{L}$ if $\mathscr{L}$ is the perpendicular bisector of line segment $\overline{PQ}$. We say that $\mathscr{L}$ is the **axis of symmetry** and that $P$ and $Q$ are mirror images, or **reflections**, of each other through $\mathscr{L}$. (Actually, you already encountered two instances of this type of symmetry in Section 3.2, where we discussed symmetry about the $x$- and $y$-axes.) In addition, we say that two curves are symmetric about a line if each point on one curve is the reflection of a corresponding point on the other curve, and vice versa (see Figure 4).

**EXAMPLE 6**   Verify that the points $P(4, 1)$ and $Q(1, 4)$ are symmetric about the line $y = x$. See Figure 5.

**Solution**   We have to show that the line $y = x$ is the perpendicular bisector of the line segment $\overline{PQ}$. Now, the slope of $\overline{PQ}$ is

$$m = \frac{4 - 1}{1 - 4} = -1$$

On the other hand, the slope of the line $y = x$ is 1. Since these two slopes are negative reciprocals, we conclude that the line $y = x$ is perpendicular to $\overline{PQ}$. Next, we must show that the line $y = x$ passes through the midpoint of $\overline{PQ}$. But (as you should check for yourself), the midpoint of $\overline{PQ}$ is $(\frac{5}{2}, \frac{5}{2})$, which does lie on the line $y = x$. In summary, then, we've shown that the line $y = x$ passes through the midpoint of $\overline{PQ}$ and is perpendicular to $\overline{PQ}$. Thus $P$ and $Q$ are symmetric about the line $y = x$, as we wished to show.

Just as we've shown that $(4, 1)$ and $(1, 4)$ are symmetric about the line $y = x$, we can also show in general that $(a, b)$ and $(b, a)$ are always symmetric about

FIGURE 5

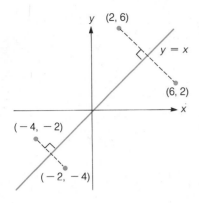

**FIGURE 6**

the line $y = x$. Figure 6 displays two examples of this, and Exercise 38 at the end of this section asks you to supply a proof for the general situation.

Now let us see how these ideas about symmetry relate to inverse functions. By way of example, consider the functions $f$ and $f^{-1}$ from Example 4:

$$f(x) = 3x - 4 \qquad f^{-1}(x) = \frac{1}{3}x + \frac{4}{3}$$

Figure 7 shows the graphs of $f$, $f^{-1}$, and the line $y = x$. Note that the graphs of $f$ and $f^{-1}$ are symmetric about the line $y = x$. This kind of symmetry, in fact, always occurs for the graphs of any pair of inverse functions.

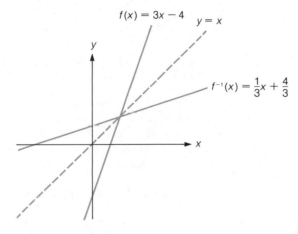

**FIGURE 7**

Why is it that the graphs of $f$ and $f^{-1}$ are always mirror images of each other about the line $y = x$? First, recall that the function $f^{-1}$ switches the inputs and outputs of $f$. Thus, $(a, b)$ is on the graph of $f$ if and only if $(b, a)$ is on the graph of $f^{-1}$. But as we have stated, the points $(a, b)$ and $(b, a)$ are mirror images in the line $y = x$. It follows that the graphs of $f$ and $f^{-1}$ are reflections of each other about the line $y = x$. For reference, we restate this useful fact in the box that follows.

| PROPERTY SUMMARY | The Graphs of $f$ and $f^{-1}$ |
|---|---|

The graphs of $f$ and $f^{-1}$ are symmetric about the line $y = x$.

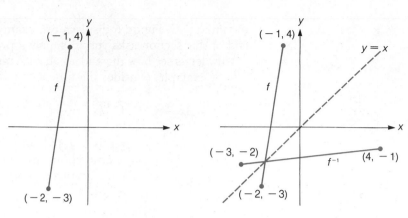

FIGURE 8                    FIGURE 9

**EXAMPLE 7**   The graph of a function $f$ consists of the straight line segment joining the points $(-2, -3)$ and $(-1, 4)$, as shown in Figure 8. Sketch a graph of $f^{-1}$.

Solution   The graph of $f^{-1}$ is obtained by reflecting the graph of $f$ about the line $y = x$. The reflection of $(-2, -3)$ is $(-3, -2)$, and the reflection of $(-1, 4)$ is $(4, -1)$. To graph $f^{-1}$, then, we plot the reflected points $(-3, -2)$ and $(4, -1)$ and connect them with a straight line segment, as shown in Figure 9. For reference, Figure 9 also shows the graphs of $f$ and $y = x$.

**EXAMPLE 8**   Let $g(x) = x^3$. Find $g^{-1}(x)$ and then, on the same set of axes, sketch the graphs of $g$, $g^{-1}$, and $y = x$.

Solution   Begin by writing $y = x^3$. Then switch $x$ and $y$ and solve for $y$. We first have

$$x = y^3$$

To solve this equation for $y$, we take the cube root of both sides to obtain

$$\sqrt[3]{x} = y$$

So the inverse function is $g^{-1}(x) = \sqrt[3]{x}$. We could graph the function $g^{-1}$ by plotting points. But the easier way is to reflect the graph of $g(x) = x^3$ about the line $y = x$. See Figure 10.

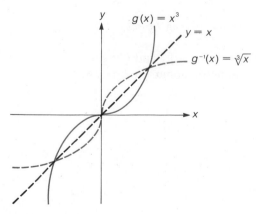

FIGURE 10

TABLE 1

| x | y |
|---|---|
| 1 | 1 |
| 2 | 4 |
| 3 | 9 |
| −1 | 1 |
| −2 | 4 |
| −3 | 9 |

Reverse inputs and outputs $\longrightarrow$

| x | y |
|---|---|
| 1 | 1 |
| 4 | 2 |
| 9 | 3 |
| 1 | −1 |
| 4 | −2 |
| 9 | −3 |

(a) $F(x) = x^2$

(b) The table formed by reversing the inputs and outputs for $F(x) = x^2$

Earlier in this section, we mentioned that not every function has an inverse function. For instance, Tables 1(a) and (b) indicate what happens if we begin with the function defined by $F(x) = x^2$ and interchange the inputs and outputs. The resulting table, Table 1(b), does not define a function. For instance, the input 4 in Table 1(b) produces two distinct outputs, 2 and −2, whereas the definition of a function requires exactly one output for a given input.

Why didn't this difficulty arise when we looked for the inverse of $f(x) = 2x$ at the beginning of this section? The answer is that there is an essential difference between the functions defined by $f(x) = 2x$ and $F(x) = x^2$. With $f(x) = 2x$, distinct inputs never yield the same output. That is, if $x_1 \neq x_2$, then $f(x_1) \neq f(x_2)$. This condition guarantees that interchanging the inputs and outputs of $f$ will yield a function. With $F(x) = x^2$, however, distinct inputs can yield the same output. For instance,

$$2 \neq -2$$

but

$$F(2) = F(-2) \qquad \text{because } 2^2 = 4 = (-2)^2$$

It's useful to have a name for those functions in which distinct inputs always yield distinct outputs. We call these functions **one-to-one functions**. So in view of the discussion in the previous paragraph, the function defined by $f(x) = 2x$ is one-to-one, but $F(x) = x^2$ is not. Using graphs, there is an easy way to tell which functions are one-to-one:

### Horizontal Line Test

A function $f$ is one-to-one if and only if each horizontal line intersects the graph of $y = f(x)$ in at most one point.

**EXAMPLE 9** Which of the following functions are one-to-one?

**(a)** $F(x) = 2x$ **(b)** $H(x) = x^2$ **(c)** $h(x) = (\sqrt{x})^4$

Solution We sketch the graphs and apply the horizontal line test, as in Figure 11. The horizontal line test tells us that $F$ and $h$ are one-to-one, but $H$ is not.

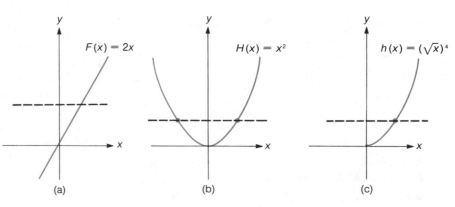

FIGURE 11

The following theorem tells us that the functions with inverses are precisely the one-to-one functions. (The proof of the theorem isn't difficult, but we'll omit giving it here.)

> **Theorem**
>
> A function $f$ has an inverse function if and only if $f$ is one-to-one.

**EXAMPLE 10**  Which of the three functions in Example 9 have inverse functions?

**Solution**  According to our theorem, the functions with inverses are those that are one-to-one. In Example 9, we found that $F$ and $h$ were one-to-one, but $H$ was not. So both $F$ and $h$ each have inverses, but $H$ does not.

## EXERCISE SET 4.5

### A

1. Verify that the given pairs of functions are inverse functions.
   (a) $f(x) = 3x$;  $g(x) = x/3$
   (b) $f(x) = 4x - 1$;  $g(x) = \frac{1}{4}x + \frac{1}{4}$
   (c) $g(x) = \sqrt{x}$;  $h(x) = x^2$  Assume that the domain of both $g$ and $h$ is $[0, \infty)$.

2. Which pairs of functions are inverses?
   (a) $f(x) = -3x + 2$;  $g(x) = \frac{2}{3} - \frac{1}{3}x$
   (b) $F(x) = 2x + 1$;  $G(x) = \frac{1}{2}x - 1$
   (c) $G(x) = x^3$;  $H(x) = 1 - x^3$
   (d) $f(t) = t^3$;  $g(t) = \sqrt[3]{t}$

3. Let $f(x) = x^3 + 2x + 1$ and assume that the domain of $f^{-1}$ is $(-\infty, \infty)$.
   (a) Find $f[f^{-1}(4)]$.
   (b) Find $f^{-1}[f(-1)]$.
   (c) Find $(f \circ f^{-1})(\sqrt{2})$.
   (d) Find $f[f^{-1}(t + 1)]$.
   (e) Find $f(0)$ and then use your answer to find $f^{-1}(1)$.
   (f) Find $f(-1)$ and then use your answer to find $f^{-1}(-2)$.

4. Let $f(x) = 2x + 1$.
   (a) Find $f^{-1}(x)$.
   (b) Calculate $f^{-1}(5)$ and $1/f(5)$. Are your answers the same?

5. Let $f(x) = 3x - 1$.
   (a) Compute $f^{-1}(x)$.
   (b) Verify that $f[f^{-1}(x)] = x$ and that $f^{-1}[f(x)] = x$.
   (c) On the same set of axes, sketch the graphs of $f$, $f^{-1}$, and the line $y = x$. Note that the graphs of $f$ and $f^{-1}$ are symmetric about the line $y = x$.

6. Follow Exercise 5, but use $f(x) = \frac{1}{3}x - 2$.

7. Follow Exercise 5, but use $f(x) = \sqrt{x - 1}$. [The domain of $f^{-1}$ will be $[0, \infty)$.]

8. Follow Exercise 5, but use $f(x) = 1/x$.

9. Let $f(x) = \dfrac{x + 2}{x - 3}$.
   (a) Find the domain and range of the function $f$.
   (b) Find $f^{-1}(x)$.
   (c) Find the domain and range of the function $f^{-1}$. What do you observe?

10. Let $f(x) = \dfrac{2x - 3}{x + 4}$. Find $f^{-1}(x)$. Find the domain and range for $f$ and $f^{-1}$. What do you observe?

11. Let $f(x) = 2x^3 + 1$. Find $f^{-1}(x)$.

12. In our preliminary discussion at the beginning of this section, we considered the functions $f(x) = 2x$ and $g(x) = x/2$. Verify that $f[g(x)] = x$ and that $g[f(x)] = x$. On the same set of axes, sketch the graphs of $f$, $g$, and $y = x$.

13. This exercise refers to the comments made at the end of Example 4. Compute $f[f^{-1}(x)]$ and $f^{-1}[f(x)]$ using the functions $f(x) = 3x - 4$ and $f^{-1}(x) = \frac{1}{3}x + \frac{4}{3}$. The point here is actually to carry out the calculations and to see in each case that the answer is indeed $x$. Then, on the same set of axes, sketch the graphs of $f$, $f^{-1}$, and the line $y = x$.

14. Let $f(x) = (x - 3)^2$ and take the domain to be $[3, \infty)$.
    (a) Find $f^{-1}(x)$. Give the domain of $f^{-1}$.
    (b) On the same set of axes, sketch the graphs of $f$ and $f^{-1}$.

15. Let $f(x) = (x - 3)^3 - 1$.
    (a) Compute $f^{-1}(x)$.
    (b) On the same set of axes, sketch the graphs of $f$, $f^{-1}$, and the line $y = x$.  *Hint:*  Sketch $f$ using the ideas of Section 4.3. Then reflect $f$ to get $f^{-1}$.

16. The following figure shows the graph of a function $f$. (The axes are marked off in one-unit intervals.) Graph the following functions.
    (a) $y = f^{-1}(x)$      (b) $y = f(x - 2)$
    (c) $y = f(x) - 2$      (d) $y = f(-x)$
    (e) $y = -f(x)$      (f) $y = f^{-1}(x - 2)$
    (g) $y = f^{-1}(x) - 2$      (h) $y = f^{-1}(x - 2) - 2$
    (i) $y = f^{-1}(-x)$      (j) $y = -f^{-1}(x)$

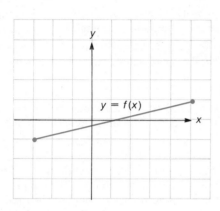

17. The following figure shows the graph of a function $g$. (The axes are marked off in one-unit intervals.) Graph the following functions.
    (a) $y = g^{-1}(x)$      (b) $y = g^{-1}(x) - 1$
    (c) $y = g^{-1}(x - 1)$      (d) $y = g^{-1}(-x)$
    (e) $y = -g^{-1}(x)$      (f) $y = -g^{-1}(-x)$

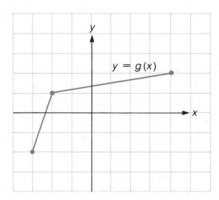

18. The following figure shows the graph of a function $y = h(x)$.
    (a) Sketch a graph that is the reflection of $h$ about the $y$-axis. What is the equation of the graph you obtain?

(b) On the same set of axes, sketch graphs of $h$, $h^{-1}$, and the line $y = x$.
(c) Let $k$ be the function whose graph is the reflection of $h$ about the $x$-axis. Sketch a graph of the function $k^{-1}$

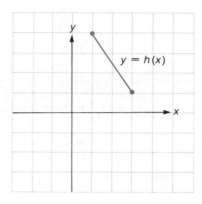

In Exercises 19–30, use the horizontal line test to determine which functions are one-to-one (and therefore have inverses).

19. $y = x^2 + 1$      20. $y = \sqrt{x}$
21. $f(x) = 1/x$      22. $g(x) = |x|$
23. $y = x^3$      24. $y = 1 - x^3$
25. $y = \sqrt{1 - x^2}$      26. $y = 3x$
27. $g(x) = 5$ (for all $x$)      28. $y = mx + b$  $(m \neq 0)$
29. $f(x) = \begin{cases} x^2 & \text{if } -1 \leq x \leq 0 \\ x^2 + 1 & \text{if } x > 0 \end{cases}$

30. $g(x) = \begin{cases} x^2 & \text{if } -1 \leq x < 0 \\ x^2 + 1 & \text{if } x \geq 0 \end{cases}$

31. Let $f(x) = \dfrac{3x - 2}{5x - 3}$. Show that $f[f(x)] = x$. This says that $f$ is its own inverse.

32. Let $f(x) = \dfrac{ax + b}{cx - a}$. Show that $f$ is its own inverse. (Assume that $a^2 + bc \neq 0$.)

33. Let $M(x) = \dfrac{ax + b}{cx + d}$.
    (a) Find $M^{-1}(x)$.      (b) Find $M[M(x)]$.
    (c) Find $M[M(0)]$.      (d) Find $M^{-1}[M(x)]$.
    (e) Find $M^{-1}[M(0)]$.

**B**

34. Let $f(x) = \sqrt{x}$.
    (a) Find $f^{-1}(x)$. What is the domain of $f^{-1}$? [The domain is not $(-\infty, \infty)$.]

**(b)** In each case, determine whether the given point lies on the graph of $f$ or $f^{-1}$.
- **(i)** $(4,2)$
- **(ii)** $(2,4)$
- **(iii)** $(5,\sqrt{5})$
- **(iv)** $(\sqrt{5},5)$
- **(v)** $(a,f(a))$  (Assume $a \geq 0$.)
- **(vi)** $(f(a),a)$
- **(vii)** $(b,f^{-1}(b))$  (Assume $b \geq 0$.)
- **(viii)** $(f^{-1}(b),b)$

**35.** Let $F(x) = x^3 + 7x - 5$. Find $F^{-1}(-5)$. [Assume that the domain of $F^{-1}$ is $(-\infty,\infty)$.]

**36.** Reread the definition of symmetry about a line given on page 198. Then use the method shown in Example 6 to show that the points $(7,-1)$ and $(-1,7)$ are symmetric about the line $y = x$.

**37.** Use the method shown in Example 6 to show that $(5,2)$ and $(2,5)$ are symmetric about $y = x$.

**38.** Use the method of Example 6 to show that the two points $(a,b)$ and $(b,a)$ are symmetric about the line $y = x$.

**39.** Points $P$ and $Q$ are both reflected in the line $y = x$ to obtain points $P'$ and $Q'$, respectively. Does the distance from $P$ to $Q$ equal the distance from $P'$ to $Q'$?

**40.** Pick any three points on the line $y = \frac{1}{2}x + 3$. Reflect each point about the line $y = x$. Show that the three reflected points all lie on one line. What is the equation of that line?

**C**

**41.** Determine values for $m$ and $b$ so that the points $(8,2)$ and $(4,8)$ are symmetric about the line $y = mx + b$.

**42.** Let $f(x) = 2x^3 - 5$. Define a function $F$ by $F(x) = f(x-1)$. Find $F[f^{-1}(x) + 1]$.

**43.** Let $f$ be a function whose domain is $(-\infty,\infty)$ and suppose that $f^{-1}$ exists. Let $F$ be the function defined by $F(x) = f(x-1)$. Show that $F^{-1}(x) = f^{-1}(x) + 1$.

**44.** Reflect the point $(2,1)$ about the line $y = 3x$. What are the coordinates of the point you obtain?

**45.** **(a)** If the point $(a,b)$ is reflected about the line $y = 3x$, show that the coordinates of the reflected point are $\left(\dfrac{3b-4a}{5}, \dfrac{3a+4b}{5}\right)$.
**(b)** If the point $(a,b)$ is reflected about the line $y = mx$, what are the coordinates of the reflected point?

**46.** **(a)** Find the reflection of the point $(a,b)$ about the line $y = -x$.  *Answer:* $(-b,-a)$
**(b)** Begin with the point $(a,b)$. Reflect this point about the $x$-axis to obtain a point $Q$. Then reflect $Q$ about the $y$-axis to obtain a point $R$. Then reflect $R$ about the line $y = x$ to obtain a point $S$. Show that $S$ is the reflection of $(a,b)$ in the line $y = -x$.

## VARIATION

**4.6**

*Perhaps the most dramatic moment in the development of physics during the 19th century occurred to J. C. Maxwell one day in the 1860's . . . Maxwell noted that the equations or the laws [for electricity, magnetism, and light] . . . were mutually inconsistent, . . . and in order for the whole system to be consistent, he had to add another term to his equations. With this new term there came an amazing prediction, which was that a part of the electric and magnetic fields would fall off much more slowly with the distance than the inverse square, namely inversely as the first power of the distance! And so he realized that electric currents in one place can affect other charges far away, and he predicted the basic effects with which we are familiar today—radio transmission, radar, and so on.*

RICHARD FEYNMAN, *The Feynman Lectures on Physics* (Reading, Mass.: Addison-Wesley, 1963)

Certain formulas or relationships between variables occur so frequently in the applications of mathematics that there is a special terminology used to describe them. For example, if $y = 9x$, we say that $y$ **varies directly as** $x$ or that $y$ is **directly proportional** to $x$. The number 9 in this case is called the **constant of proportionality**. As another example of this special terminology, suppose that the variables $A$ and $B$ are related by the equation $A = 3/B$. Then we say that $A$ **varies inversely as** $B$ or that $A$ is **inversely proportional** to $B$. The constant of proportionality in this case is the number 3. Before going on to list other forms of this terminology, let's look at examples of what we have just described.

**EXAMPLE 1**  Suppose that $F$ varies directly as $a$. If $F = 60$ when $a = 5$, find $F$ when $a = 6$.

Solution  Since $F$ varies directly as $a$, we write

$$F = ka$$

To find $k$, we substitute the given pair of values $F = 60$ and $a = 5$ in this equation. This yields

$$60 = k \cdot 5$$
$$k = 12$$

Now that we know $k = 12$, the equation $F = ka$ becomes

$$F = 12a$$

From this equation we can calculate $F$ whenever $a$ is given. In particular, then, if $a = 6$, we obtain $F = 12(6)$. Thus $F = 72$, as required. The formula $F = 12a$ is called the **law of variation** for this problem.

**EXAMPLE 2**  At a constant speed, the distance traveled varies directly as the time. Suppose a runner covers 1 mile in 7 min. At this rate, what distance would be covered in 30 min?

Solution  Let $d$ and $t$ denote distance and time, respectively. The statement that distance varies directly as time is written algebraically as

$$d = kt$$

As in Example 1, we first use the given data to calculate $k$. Since we are given that $d = 1$ when $t = 7$, we obtain

$$1 = k \cdot 7$$
$$k = \frac{1}{7}$$

Now that we know $k = \frac{1}{7}$, the equation $d = kt$ becomes

$$d = \frac{1}{7}t$$

Using this law of variation, we can calculate $d$ whenever $t$ is given. In particular then, if $t = 30$, we obtain $d = \frac{1}{7}(30)$. Thus in 30 min, the runner covers $\frac{30}{7} \approx 4.3$ miles.

**EXAMPLE 3**  Given that $y$ varies inversely as $x$ and that $y$ is 6 when $x$ is $-2$, find $y$ when $x$ is 4.

Solution  The statement that $y$ varies inversely as $x$ translates algebraically to

$$y = \frac{k}{x}$$

To first find $k$, we substitute the given pair of values $x = -2$ and $y = 6$ in this equation. This yields

$$6 = \frac{k}{-2}$$
$$k = -12$$

Now that we know $k = -12$, the equation $y = k/x$ becomes

$$y = -12/x$$

Using this law of variation, we can calculate $y$ whenever $x$ is given. In particular then, if $x = 4$, we obtain

$$y = -12/4 = -3$$

Thus when $x$ is 4, we have $y$ is $-3$; this is the required solution.

The list that follows shows some common formulas that can be expressed in the language of variation. In this list, $k$ denotes a constant in each case, the so-called **constant of proportionality**.

| FORMULA OR LAW OF VARIATION | TERMINOLOGY |
|---|---|
| 1. $y = kx^n$ | $y$ varies directly as $x^n$;  or $y$ is directly proportional to $x^n$. |
| 2. $y = \dfrac{k}{x^n}$ | $y$ varies inversely as $x^n$;  or $y$ is inversely proportional to $x^n$. |
| 3. $y = kxz$ | $y$ varies jointly as $x$ and $z$;  or $y$ varies directly as the product of $x$ and $z$. |
| 4. $y = \dfrac{kx}{z}$ | $y$ varies directly as $x$ and inversely as $z$. |

The rest of this section and the exercises that follow deal with this terminology. Notice that we have already discussed formulas 1 and 2 for the case $n = 1$.

**EXAMPLE 4**  The volume of a sphere varies directly as the cube of the radius. When the radius is 2 cm, the volume is $32\pi/3$ cm$^3$. Find the constant of proportionality $k$. Write the law of variation.

Solution  Let $V$ denote the volume and $r$ the radius. The statement that the volume varies directly as the cube of the radius translates algebraically to

$$V = kr^3$$

To find the constant of proportionality $k$, we substitute the given values for $r$ and $V$ in this equation to obtain

$$\frac{32\pi}{3} = k \cdot 2^3 = 8k$$

$$k = \frac{32\pi}{3 \cdot 8} = \frac{4\pi}{3}$$

Now that we have found the constant $k$, we can write the law of variation simply by substituting this value of $k$ in the equation $V = kr^3$. Thus the law of variation is $V = (4\pi/3)r^3$.

**EXAMPLE 5**   The quantity $B$ varies inversely as the square of $C$. What happens to the value of $B$ if $C$ is tripled?

Solution   The statement that $B$ varies inversely as the square of $C$ is written algebraically as

$$B = \frac{k}{C^2}$$

To see what happens when $C$ is tripled, replace $C$ by the quantity $(3C)$ in our equation.

$$B = \frac{k}{(3C)^2} = \frac{k}{9C^2}$$

Now compare the original expression for $B$ with the new expression.

original expression for $B$:   $\dfrac{k}{C^2}$

new expression for $B$ after $C$ is tripled:   $\dfrac{1}{9} \cdot \dfrac{k}{C^2}$

From this we can see that if $C$ is tripled, $B$ is one ninth of its original value. This is the required solution.

Example 5 differs from those preceding it in this section in that we did not need to evaluate the constant of proportionality $k$ specifically. Indeed, the problem did not supply enough information to calculate a value for $k$. Example 6 is similar in this respect.

**EXAMPLE 6**   The force of attraction between two objects varies directly as the product of their masses and inversely as the square of the distance between them. (This is Newton's Law of Gravitation.) What happens to the force if both masses are tripled and the distance is doubled?

Solution   We begin by assigning letters to denote the relevant quantities and then writing the law of variation. Let $F$, $m$, $M$, and $d$ denote, respectively, the force, the two masses, and the distance. Then the given statement translates algebraically to

$$F = \frac{kmM}{d^2}$$

We want to find out what happens to $F$ when both masses are tripled and the distance is doubled. To do this, we make the following substitutions in our law of variation:

replace $m$ by $(3m)$

replace $M$ by $(3M)$

replace $d$ by $(2d)$

We then have

$$F = \frac{k(3m)(3M)}{(2d)^2}$$

$$= \frac{9kmM}{4d^2}$$

Now compare the original expression for $F$ with the expression obtained after tripling the masses and doubling the distance.

original expression for $F$:  $\dfrac{kmM}{d^2}$

new expression:  $\dfrac{9}{4}\left(\dfrac{kmM}{d^2}\right)$

From this we can see that the new force is nine-fourths of the original force. In other words, the new force is $2\frac{1}{4}$ times as strong as the original. This is the required solution.

**EXAMPLE 7**  Suppose that $A$ varies jointly as $X$ and $Y$. Also, suppose that $X$ varies inversely as the square of $Y$. Show that $A$ varies inversely as $Y$.

Solution  Translating the two given statements into algebraic equations gives us

$$A = kXY \tag{1}$$

and

$$X = \frac{c}{Y^2} \tag{2}$$

Both $k$ and $c$ are constants here. We do need the two different letters since there is no reason to suppose that $k = c$. Now we are asked to show that $A$ varies inversely as $Y$. To do this, we just substitute for $X$ in equation (1) using the value given in equation (2). We have

$$A = kXY = k\left(\frac{c}{Y^2}\right)Y$$

$$= \frac{kc}{Y}$$

This last equation says that $A$ equals a constant divided by $Y$; that is, $A$ varies inversely as $Y$. If we want, we can denote $kc$ by the single letter $K$. Then we have $A = K/Y$, which again says that $A$ varies inversely as $Y$.

# EXERCISE SET 4.6

## A

*In Exercises 1–5, translate the given statement into an equation.*

1. (a) $y$ varies directly as $x$.
   (b) $A$ varies inversely as $B$.

2. (a) $y$ varies directly as the cube of $z$.
   (b) $s$ varies inversely as $t^2$.

3. (a) $x$ varies jointly as $u$ and $v^2$.
   (b) $z$ varies jointly as $A^2$ and $B^3$.

4. (a) $y$ varies directly as $u$ and inversely as $v$.
   (b) $y$ varies directly as the square root of $x$ and inversely as the product of $w$ and $z$.

5. (a) $F$ is inversely proportional to $r^2$.
   (b) $V^2$ is directly proportional to the sum of $U^2$ and $T^2$.

6. If $y$ varies directly as $x$, and $y$ is 20 when $x = 4$, find $y$ when $x = -6$.

7. Suppose $A$ varies inversely as $B$, and $A = -1$ when $B = 2$.
   (a) Find the constant of proportionality.
   (b) Write the law of variation.
   (c) Find $A$ when $B = \frac{5}{4}$.

8. Suppose $T$ varies directly as $V^2$, and $T = 8$ when $V = 4$.
   (a) Find the constant of proportionality and write the law of variation.
   (b) Find $T$ when $V = 10$.
   (c) Find $T$ when $V = \sqrt{10}$.

9. Assume $x$ varies jointly as $y$ and $z$. Also $x = 9$ when $y = 2$ and $z = -3$. Find $x$ when both $y$ and $z$ are 4.

10. Suppose $F$ varies directly as the product of $q$ and $Q$ and inversely as the square of $d$. When $q = 2$, $Q = 4$, and $d = \frac{1}{2}$, the value of $F$ is 8.
    (a) Find the constant of proportionality and write the law of variation.
    (b) Find $F$ when $q = 3$, $Q = 6$, and $d = 2$.

11. $A$ varies jointly as $B$ and $C$. If $B$ is tripled and $C$ is doubled, what happens to $A$?

12. If $S$ varies inversely as the square of $T$, what happens to $S$ if $T$ is cut in half?

13. If $x$ varies jointly as $B$ and $C$ and inversely as the square root of $A$, what happens to $x$ if $A$, $B$, and $C$ are all quadrupled?

14. Suppose that $xy = \frac{3}{4}$. Does $y$ vary directly or inversely with $x$? What is the constant of proportionality in this case?

15. The surface area of any sphere varies directly as the square of the radius. When the radius is 2 cm, the surface area is $16\pi$ cm$^2$.
    (a) Find the constant of proportionality and write the law of variation.
    (b) Find the surface area when the radius is $\sqrt{3}$ cm.

16. For motion at a constant speed, the distance varies directly as the time. Suppose a car covers 10 miles in 15 min. At this rate, what distance would be covered in 50 min?

17. The force of attraction between two objects varies jointly as their masses and inversely as the square of the distance between them. What happens to the force if one mass is tripled, the other mass is quadrupled, and the distance is cut in half?

18. The magnitude of the force between two electrical charges $Q_1$ and $Q_2$ varies directly as the product of $Q_1$ and $Q_2$ and inversely as the square of the distance between the charges. (This is Coulomb's Law, named after the French physicist Charles Coulomb, who verified the law experimentally in 1785.) What happens to the magnitude of the force if $Q_1$ is doubled, $Q_2$ is quadrupled, and the distance is tripled?

19. Neglecting air resistance, the distance any object falls starting from rest is directly proportional to the square of the time. (This law was discovered experimentally by Galileo, sometime before the year 1590. In his first experiments, Galileo used his own pulse to measure time.) Given that an object starting from rest will fall 490 m in 10 sec, how far will an object fall in 5 sec, starting from rest?

20. The volume of a right circular cone varies jointly as the height and the square of the radius of the base. When the radius and height are both 3 cm, the volume is $9\pi$ cm$^3$. Find the volume of a cone having a height of 4 cm and a radius of $\frac{3}{2}$ cm.

21. (Boyle's Law) In 1662, Robert Boyle observed that for a sample of gas at a constant temperature, the volume is inversely proportional to the pressure.
    (a) Suppose that at a certain fixed temperature, the pressure of a 2-liter sample of gas is 1.025 atm. (atm is the abbreviation for atmospheres, a unit of pressure, where 1 atm = 14.7 lb/in.$^2$, the air pressure at sea level.) Calculate the constant of proportionality and write the law of variation.
    (b) Suppose that, with the temperature still fixed, the pressure on our sample of gas is reduced to 1 atm. Calculate the new volume of the gas.

22. (Charles's Law) In 1787, Jacques Charles stated that for a sample of gas held at a constant pressure, the volume is directly proportional to the Kelvin temperature. Suppose that initially you have 300 ml of a gas at 300°K. Then, holding the pressure constant, you heat the gas to 310°K. What is the new volume?

23. (a) For a satellite in a circular orbit about the earth, the kinetic energy varies directly as the mass of the satellite and inversely as the radius of the orbit. What happens to the kinetic energy if we cut the radius in half and triple the mass?
    (b) The velocity of an earth satellite in a circular orbit varies inversely as the square root of the radius of the orbit. What happens to the velocity if we cut the radius in half?

24. (Hooke's Law) In 1676, the English physicist Robert Hooke found that the distance a spring is stretched varies directly as the force pulling on the spring. Suppose that, for a certain spring, a pull of 5 lb results in a stretch of 3 in.
    (a) Find the constant of proportionality and write the law of variation.
    (b) How far will a pull of 12 lb stretch the spring?

25. The weight of an object above the earth's surface varies inversely as the square of its distance from the center of the earth. How much would an astronaut weigh 500

miles above the surface of the earth, assuming her weight at the surface is 140 lb? (Assume that the radius of the earth is 4000 miles.)

26. One of the ways in which the human body adjusts blood flow is through changes in the radii of small arteries called *arterioles*. According to Poiseuille's Law [named after the French physiologist J. L. Poiseuille (1779–1869)], the blood flow in a given arteriole (measured, for example, in cm$^3$/sec) varies directly as the fourth power of the radius. By what fraction is the blood flow through an arteriole multiplied if the radius is decreased by 20%?

**B**

27. The graphs of six functions are displayed here. Which graph best represents the given statement?
    (a) $y$ varies directly as $x$, where the constant of proportionality is negative.
    (b) $y$ varies inversely as $x$.
    (c) $y$ varies directly as $x^2$, with a positive constant of proportionality.
    (d) $y$ varies directly as $x$, with a positive constant of proportionality.
    (e) $y$ varies directly as $x^2$, with a negative constant of proportionality.

28. Three solid lead spheres of radii $a$, $b$, and $c$, respectively, are melted down and recast into one large sphere of radius $R$. Show that $R = \sqrt[3]{a^3 + b^3 + c^3}$. Use the fact that the volume of a sphere varies directly as the cube of the radius.

29. The *period* of a pendulum is the time required for one complete swing, forward and back. If the period of a pendulum varies directly as the square root of the length of the pendulum, by how much should a pendulum of length $L$ be shortened in order to cut the period in half?

30. At highway speeds, the wind resistance against a car varies directly as the square of the speed.
    (a) If the speed is increased from 40 mph to 60 mph, by what factor is the wind resistance multiplied?
    (b) At what speed is the wind resistance twice as much as that at 50 mph?

31. The intensity of illumination due to a small light source varies inversely as the square of the distance from the light. Suppose a book is 6 ft from a lamp. At what distance from the lamp will the illumination be twice as great?

32. The volume of a coin varies jointly as the thickness and the square of the radius. Suppose that two silver dollars are melted down (illegal!) and cast into one coin of the same thickness as a silver dollar. Show that the diameter of the new coin is $\sqrt{2}$ times the diameter of a silver dollar.

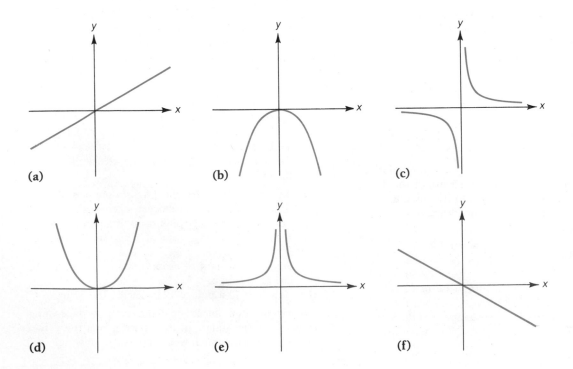

(a)        (b)        (c)

(d)        (e)        (f)

33. Show that the area of the shaded triangle in the figure varies jointly as $m$ and the square of $x$.

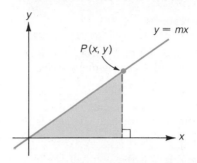

34. The weight of an object on or above the earth's surface varies inversely as the square of the object's distance from the center of the earth. At what height $x$ above the earth's surface will an object weigh half as much as it does on the surface? Assume the radius of the earth is 4000 miles. *Answer:* $4000(\sqrt{2} - 1) \approx 1656.9$ miles

c 35. In the year 1619, Johannes Kepler announced his Third Law of Planetary Motion. This law states that the square of the time it takes for a planet to complete one revolution about the sun varies directly as the cube of the planet's average distance from the sun. A table giving the average distance of each planet from the sun is shown below. Use a calculator to find the time it takes for each planet to complete one revolution about the sun. For Earth, assume the time is 365.2564 days.

| PLANET | AVERAGE DISTANCE FROM SUN MEASURED IN ASTRONOMICAL UNITS |
|--------|---------------------------------------|
| Mercury | 0.387099 |
| Venus | 0.723332 |
| Earth | 1.000000 |
| Mars | 1.523691 |
| Jupiter | 5.204377 |
| Saturn | 9.577971 |
| Uranus | 19.26020 |
| Neptune | 30.09421 |
| Pluto | 39.82984 |

c 36. Most astronomers accept the theory that the universe is expanding and that the galaxies are receding from one another at velocities that vary directly as their distances. Given that a galaxy in Virgo, which is 78,000,000 light years from us, is receding at 745 miles/sec, estimate the distance from us of the Hydra galaxy, which is receding from us at 37,881 miles/sec.

c

37. Suppose that $x$ varies directly as $z$ and that $y$ also varies directly as $z$.

(a) Does $x + y$ vary directly as $z$?
(b) Does $xy$ vary directly as $z$?
(c) Does $\sqrt{xy}$ vary directly as $z$? (Assume $x, y$, and $z$ are positive.)

38. Suppose that $A$ varies directly as $B$. Show that $AB$ varies directly as $A^2 + B^2$.

39. Suppose that $x + y$ varies directly as $x - y$. Also suppose that the constant of proportionality is not $\pm 1$.
(a) Show that $y$ varies directly as $x$.
(b) Show that $x^2 + y^2$ varies jointly as $x$ and $y$.
(c) Show that $x^3 + y^3$ varies jointly as $x, y$, and $x + y$.

40. Suppose that $A^3 + \dfrac{1}{B^3}$ varies directly as $A^3 - \dfrac{1}{B^3}$. Also suppose that the constant of proportionality is not $\pm 1$. Show that $B$ varies inversely as $A$.

41. Show that the area of the shaded triangle in the figure varies directly as $a^2$.

42. For this exercise, refer to the following figure.
(a) Show that the area of triangle $OAB$ varies inversely as $b$.
(b) Show that the area of triangle I varies directly as $a$ and inversely as the product of $b$ and $a - b$.
(c) Show that the area of triangle II varies inversely as $a - b$.

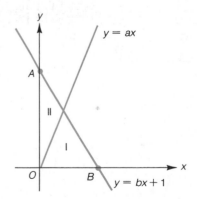

## CHAPTER 4   SUMMARY OF PRINCIPAL TERMS AND FORMULAS

| TERMS OR FORMULAS | PAGE REFERENCE | COMMENTS |
|---|---|---|
| 1. Function | 156 | Given two nonempty sets $A$ and $B$, a function from $A$ to $B$ is a rule that assigns to each element of $A$ exactly one element from $B$. |
| 2. Domain | 156 | The domain of a function is the set of all inputs for that function. If $f$ is a function from $A$ to $B$, then the domain is the set $A$. |
| 3. Range | 156 | The range of a function from $A$ to $B$ is the set of all elements in $B$ that are actually used as outputs. |
| 4. $f(x)$ | 157 | Given a function $f$, the notation $f(x)$ denotes the output that results from the input $x$. |
| 5. Graph of a function | 165 | The graph of a function $f$ consists of those points $(x,y)$ for which $x$ is in the domain of $f$ and $y = f(x)$. |
| 6. $[x]$ | 171 | $[x]$ denotes the greatest integer not exceeding $x$. |
| 7. $f \circ g$ | 188 | $(f \circ g)(x) = f[g(x)]$ |
| 8. Inverse functions | 195 | Two functions $f$ and $g$ are inverses of one another provided the following two conditions are met. First, $f[g(x)] = x$ for each $x$ in the domain of $g$. Second, $g[f(x)] = x$ for each $x$ in the domain of $f$. |
| 9. $f^{-1}$ | 196 | $f^{-1}$ denotes the inverse function for $f$.   *Note:*   In this context, $f^{-1}$ does not mean $1/f$. For an example, see Exercise 4 on page 202. |
| 10. Symmetry about a line | 198 | Two points $P$ and $Q$ are symmetric about a line if that line is the perpendicular bisector of $\overline{PQ}$. |
| 11. One-to-one function | 201 | A function $f$ is said to be one-to-one provided distinct inputs always yield distinct outputs. That is, if $x_1$ and $x_2$ are in the domain of $f$ and $x_1 \neq x_2$, then $f(x_1) \neq f(x_2)$. The relationship between inverse functions and the notion of *one-to-one* is given by the following theorem: A function $f$ has an inverse if and only if $f$ is one-to-one. |
| 12. Horizontal line test | 201 | A function $f$ is one-to-one provided that each horizontal line intersects the graph of $y = f(x)$ in at most one point. |
| 13. $y$ varies directly as $x$ | 204 | $y = kx$, where $k$ is a constant |
| 14. $y$ varies inversely as $x$ | 204 | $y = k/x$, where $k$ is a constant |
| 15. $y$ varies jointly as $x$ and $z$ | 206 | $y = kxz$, where $k$ is a constant |

## CHAPTER 4   REVIEW EXERCISES

NOTE   The fifteen exercises in this set with colored problem numbers constitute a chapter test.

*In Exercises 1–23, sketch the graphs and specify any x- or y-intercepts.*

1. $y = 4 - x^2$

2. $y = -(x - 1)^2 + 2$

3. $y = \dfrac{1}{x} + 1$

4. $f(x) = \dfrac{1}{x + 1}$

5. $y = \dfrac{1}{x + 1} + 1$

6. $y = |x + 3|$

7. $g(x) = -\sqrt{x - 4}$

8. $h(x) = \sqrt{1 - x^2}$

9. $f(x) = \sqrt{1 - x^2} + 1$

10. $y = 1 - (x + 1)^3$

11. $y = 4 - \sqrt{-x}$

12. $y = (\sqrt{x})^2$

13. $f \circ g$, where $g(x) = x + 3$ and $f(x) = x^2$

14. $f \circ g$, where $g(x) = \sqrt{x - 1}$ and $f(x) = -x^2$

15. $f(x) = \begin{cases} \sqrt{1 - x^2} & \text{if } -1 \le x \le 0 \\ \sqrt{x} + 1 & \text{if } x > 0 \end{cases}$

16. $F(x) = \begin{cases} -\sqrt{1 - x^2} & \text{if } -1 \le x < 0 \\ \sqrt{x} & \text{if } x \ge 0 \end{cases}$

17. $y = \begin{cases} |x - 1| & \text{if } 0 \le x \le 2 \\ |x - 3| & \text{if } 2 < x \le 4 \end{cases}$

18. $y = \begin{cases} 1/x & \text{if } 0 < x \le 1 \\ 1/(x - 1) & \text{if } 1 < x \le 2 \end{cases}$

19. $(y + |x| - 1)(y - |x| + 1) = 0$

20. $f^{-1}$, where $f(x) = \frac{1}{2}(x + 1)$

21. $g^{-1}$, where $g(x) = \sqrt[3]{x + 2}$

22. $f \circ f^{-1}$, where $f(x) = \sqrt{x - 2}$

23. $f^{-1} \circ f$, where $f(x) = \sqrt{x - 2}$

*For Exercises 24–27, assume that the graph of the function f is the straight line segment joining the points $(-4, -1)$ and $(0, 1)$.*

24. Find the slope of the line segment described by $y = -f(x)$.

25. Find the slope of the line segment described by $y = f(-x)$.

26. Find the midpoint of the line segment described by $y = -f(x + 2) + 5$.

27. Find the midpoint of the line segment described by $y = f^{-1}(x) + 1$.

*Find the domains of the given functions in Exercises 28–37.*

28. $y = \dfrac{1}{x^2 - 9}$

29. $y = x^3 - x^2$

30. $y = \sqrt{8 - 2x}$

31. $y = \dfrac{x}{6x^2 + 7x - 3}$

32. $y = \sqrt{|2 - 5x|}$

33. $y = \dfrac{25 - x^2}{\sqrt{x^2 + 1}}$

34. $y = \sqrt{x^2 - 2x - 3}$

35. $y = \sqrt{5 - x^2}$

36. $y = x + \dfrac{1}{x}$

37. $y = \dfrac{1}{x - [x]}$   (Recall that $[x]$ denotes the greatest integer not exceeding $x$, as defined in Section 4.2.)

*In Exercises 38–43, determine the range of each of the given functions.*

38. $y = \dfrac{x + 4}{3x - 1}$

39. $y = \dfrac{2x - 3}{x - 2}$

40. $f \circ g$, where $f(x) = 1/x$ and $g(x) = 3x + 4$

41. $g \circ f$, where $f(x) = \dfrac{x + 2}{x - 1}$ and $g(x) = \dfrac{x + 1}{x + 4}$

42. $f^{-1}$, where $f(x) = \dfrac{x}{3x - 6}$

43. $f^{-1}$, where $f(x) = \dfrac{5 - x}{1 + x}$

*In Exercises 44–51, express the functions as a composition of two or more of the following functions:*

$$f(x) = \frac{1}{x} \qquad g(x) = x - 1$$
$$F(x) = |x| \qquad G(x) = \sqrt{x}$$

44. $a(x) = \dfrac{1}{x - 1}$

45. $b(x) = \dfrac{1}{x} - 1$

46. $c(x) = \sqrt{x - 1}$

47. $d(x) = \sqrt{x} - 1$

48. $A(x) = \dfrac{1}{\sqrt{x}} - 1$

49. $B(x) = |x - 2|$

50. $C(x) = \sqrt[4]{x} - 1$

51. $D(x) = \dfrac{1}{\sqrt{x - 3}}$

*For Exercises 52–85, compute each of the indicated quantities using the functions f, g, and F defined as follows:*

$$f(x) = x^2 - x \qquad g(x) = 1 - 2x \qquad F(x) = \frac{x - 3}{x + 4}$$

52. $f(-3)$

53. $f(1 + \sqrt{2})$

54. $F(\frac{3}{4})$

55. $f(t)$

56. $f(-t)$

57. $g(2x)$

58. $f(x - 2)$

59. $g(x + h)$

60. $g(2) - g(0)$

61. $f(x) - g(x)$

62. $|f(1) - f(3)|$

63. $f(x + h) - f(x)$

64. $f(x^2)$

65. $f(x)/x \quad (x \neq 0)$

66. $[f(x)][g(x)]$

67. $f[f(x)]$

68. $f[g(x)]$

69. $g[f(3)]$

70. $(g \circ f)(x)$

71. $(g \circ f)(x) - (f \circ g)(x)$

72. $(F \circ g)(x)$

73. $\dfrac{g(x + h) - g(x)}{h}$

74. $\dfrac{f(x + h) - f(x)}{h}$

75. $\dfrac{F(x) - F(a)}{x - a}$

76. $F^{-1}(x)$

77. $F[F^{-1}(x)]$

78. $F^{-1}[F(x)]$

79. $F^{-1}(0) - \dfrac{1}{F(0)}$

80. $(g \circ g^{-1})(x)$

81. $g^{-1}(x)$

82. $g^{-1}(-x)$

83. $\dfrac{g^{-1}(x + h) - g^{-1}(x)}{h}$

84. $F^{-1}[F(\frac{22}{7})]$

85. $T^{-1}(x)$, where $T(x) = f(x)/x \quad (x \neq 0)$

In Exercises 86–97, refer to the graph of the function $f$ in the figure. (The axes are marked off in one-unit intervals.)

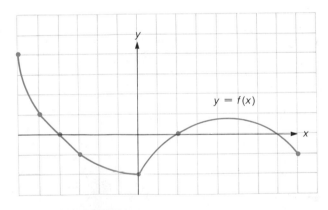

86. Is $f(0)$ positive or negative?

87. Specify the domain and range of $f$.

88. Find $f(-3)$.

89. Which is larger, $f(-\frac{5}{2})$ or $f(-\frac{1}{2})$?

90. Compute $f(0) - f(8)$.

91. Compute $|f(0) - f(8)|$.

92. As $x$ increases from $-4$ to $0$, do the corresponding $y$-values increase or decrease?

93. For which $x$-values is it true that $1 \leq f(x) \leq 4$? (Your answer should be a certain closed interval.)

94. What is the largest value of $f(x)$ when $|x| \leq 2$?

95. Is $f$ a one-to-one function?

96. Does $f$ possess an inverse function?

97. Compute $f[f(-4)]$.

For Exercises 98–109, refer to the graphs of the functions $f$ and $g$ in the figure. Assume that the domain of each function is $[0, 10]$.

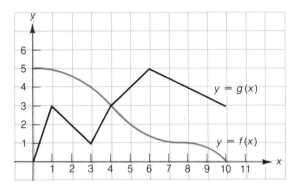

98. For which $x$-value is $f(x) = g(x)$?

99. For which $x$-values is it true that $g(x) \leq f(x)$?

100. (a) For which $x$-value is $f(x) = 0$?
     (b) For which $x$-value is $g(x) = 0$?

101. Compute $f(0) + g(0)$.

102. Compute each of the following.
     (a) $(f + g)(8)$  (b) $(f - g)(8)$
     (c) $(fg)(8)$  (d) $(f/g)(8)$

103. Compute each of the following.
     (a) $g[f(5)]$  (b) $f[g(5)]$
     (c) $(g \circ f)(5)$  (d) $(f \circ g)(5)$

104. Which is larger, $(f \circ f)(10)$ or $(g \circ g)(10)$?

105. Compute $g[f(10)] - f[g(10)]$.

106. For which $x$-values is it true that $f(x) \geq 3$?

107. For which $x$-values is it true that $|f(x) - 3| \leq 1$?

108. What is the largest number in the range of $g$?

109. Specify the coordinates of the highest point on the graph of each of the following equations.
     (a) $y = g(-x)$  (b) $y = -g(x)$
     (c) $y = g(x - 1)$  (d) $y = f(-x)$
     (e) $y = -f(x)$  (f) $y = -f(-x)$

110. Suppose $A$ varies jointly as $x$ and $y^2$. What happens to the value of $A$ when $x$ is tripled and $y$ is doubled?

111. Suppose $B$ varies directly as $x$ and inversely as $y^3$. When $x = 6$ and $y = 2$, the value of $B$ is 9. Find $B$ when $x = 3$ and $y = 4$.

112. The intensity of illumination from a small light source varies inversely as the square of the distance from the light. Suppose that a book is 10 ft from a lamp. At what distance from the lamp will the illumination be five times as great?

113. The velocity of an earth satellite in a circular orbit varies inversely as the square root of the radius of the orbit. What happens to the velocity if the radius is decreased by 10 percent?

114. The length of a side in an equilateral triangle varies directly as the square root of the area of the triangle. When the area is $16\sqrt{3}$ cm$^2$, the length of a side is 8 cm. Find the length of a side when the area is 20 cm$^2$.

115. If $f^{-1}(3) = 4$, what is $f(4)$?

116. Let $f(x) = 3x + 1$ and $g(x) = 2x + b$. Find $b$, given that the equation $f[g(x)] = g[f(x)]$ holds for all values of $x$.

117. Let $F(x) = x^2$ and $G(x) = 1/x$. Assume that the domain of both $F$ and $G$ is the set of all nonzero real numbers.
    (a) Compute $F[G(x)]$ and $G[F(x)]$.
    (b) Are $F$ and $G$ inverse functions?

118. Let $F$ be the function that assigns to each person their shoe size. Let $G$ be the function that assigns to each person their set of fingerprints. Which function has an inverse function?

119. Suppose that $(a, b)$ is a point on the graph of $y = f(x)$. Match the functions defined in the left-hand column with the points in the right-hand column. For example, the appropriate match for (a) in the left-hand column is determined as follows. The graph of $y = f(x) + 1$ is obtained by translating the graph of $y = f(x)$ up one unit. Thus, the point $(a, b)$ moves up to $(a, b + 1)$, and consequently (E) is the appropriate match for (a).

| | | | |
|---|---|---|---|
| (a) | $y = f(x) + 1$ | (A) | $(-a, b)$ |
| (b) | $y = f(x + 1)$ | (B) | $(b, a)$ |
| (c) | $y = f(x - 1) + 1$ | (C) | $(a - 1, b)$ |
| (d) | $y = f(-x)$ | (D) | $(-b, a + 1)$ |
| (e) | $y = -f(x)$ | (E) | $(a, b + 1)$ |
| (f) | $y - f(-x)$ | (F) | $(b, 1 - a)$ |
| (g) | $y = f^{-1}(x)$ | (G) | $(-a, -b)$ |
| (h) | $y = f^{-1}(x) + 1$ | (H) | $(-b, 1 - a)$ |
| (i) | $y = f^{-1}(x - 1)$ | (I) | $(b, -a)$ |
| (j) | $y = f^{-1}(-x) + 1$ | (J) | $(a, -b)$ |
| (k) | $y = -f^{-1}(x)$ | (K) | $(b + 1, a)$ |
| (l) | $y = -f^{-1}(-x) + 1$ | (L) | $(a + 1, b + 1)$ |
| (m) | $y = 1 - f^{-1}(x)$ | (M) | $(b, a + 1)$ |

# POLYNOMIAL FUNCTIONS AND RATIONAL FUNCTIONS

**INTRODUCTION** In the previous chapter, we studied some rather general rules for operating with functions and their graphs. Now, in Chapters 5 and 6, we focus our attention on several particular types of functions. Section 5.1 begins with a brief consideration of linear functions; the remainder of the section is devoted to quadratic functions. Section 5.2 could aptly be subtitled, "Translating English into Algebra." In this section, we develop the background skills that you will need in solving the maximum and minimum problems of Section 5.3. (If a friend of yours happens to be taking calculus, you will find that you can solve some of your friend's maximum and minimum problems without having to use calculus.) Sections 5.4 and 5.5 present a number of techniques for graphing polynomial and rational functions.

## LINEAR AND QUADRATIC FUNCTIONS

**5.1**

We are going to define the terms *linear function* and *quadratic function*. As you will soon see, many of the methods for dealing with these functions have already been developed in previous chapters. By a **linear function**, we mean a function defined by an equation of the form

$$f(x) = Ax + B$$

where $A$ and $B$ are constants. In this chapter the constants $A$ and $B$ will always be real numbers. From our work in Chapter 3 we know that the graph of $y = Ax + B$ is a straight line. So, aside from terminology and the use of function notation, there is nothing really new here.

**EXAMPLE 1** Suppose $f$ is a linear function. If $f(1) = 0$ and $f(2) = 3$, find an equation defining $f$.

**Solution** From the statement of the problem, we know that the graph of $f$ is a straight line passing through the points $(1, 0)$ and $(2, 3)$. Thus the slope of the line is

$$m = \frac{y_2 - y_1}{x_2 - x_1} = \frac{3 - 0}{2 - 1} = 3$$

Now we can use the point–slope formula to find the required equation. We have

$$y - y_1 = m(x - x_1)$$
$$y - 0 = 3(x - 1)$$
$$y = 3x - 3$$

This is the equation defining $f$. If we wish, we can rewrite it using function notation: $f(x) = 3x - 3$.

An important application of linear functions occurs in business and economics with *linear* or *straight-line depreciation*. In this situation we assume that the value $V$ of an asset (such as a machine or an apartment building) decreases linearly over time $t$; that is, $V = mt + b$, where the slope $m$ is negative.

**EXAMPLE 2**   A factory owner buys a new machine for $8000. After 10 years the machine has a salvage value of $500.

(a) Assuming linear depreciation (as indicated in Figure 1), find a formula for the value $V$ of the machine after $t$ years, where $0 \le t \le 10$.
(b) Use the formula derived in part (a) to find the value of the machine after 5 years.

Solution   (a) We need to determine $m$ and $b$ in the equation $V = mt + b$. From Figure 1 we see that the $V$-intercept of the line segment is $b = 8000$. Furthermore, since the line segment passes through the two points $(10, 500)$ and $(0, 8000)$, the slope $m$ is $\dfrac{8000 - 500}{0 - 10}$, or $-750$. So we have $m = -750$ and $b = 8000$, and the required equation is

$$V = -750t + 8000 \qquad (0 \le t \le 10)$$

(b) Substituting $t = 5$ in the equation $V = -750t + 8000$ yields

$$V = -750(5) + 8000$$
$$= -3750 + 8000 = 4250$$

Thus the value of the machine after 5 years is $4250.

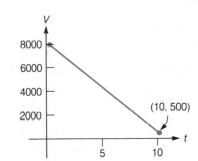

FIGURE 1

Next we turn to **quadratic functions**. These are functions defined by equations of the form

$$f(x) = ax^2 + bx + c \qquad (a \ne 0)$$

where $a$, $b$, and $c$ are constants and (as already indicated to the right of the equation) $a$ is not zero. In this chapter the constants $a$, $b$, and $c$ will always be real numbers. What we are going to see is that the graph of any quadratic function is a curve called a **parabola**, which is similar or identical in shape to the graph of $y = x^2$. Figure 2 displays the graphs of two typical parabolas.

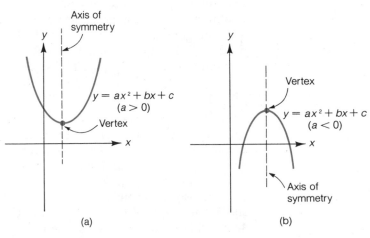

FIGURE 2

Subsequent examples will indicate that the parabola $y = ax^2 + bx + c$ opens upward when $a > 0$ and downward when $a < 0$. As Figure 2 indicates, the low or high point on the parabola is called the **vertex**. The **axis of symmetry** of the parabola $y = ax^2 + bx + c$ is the vertical line passing through the vertex.

At the beginning of this section we mentioned that many of the methods for dealing with quadratic functions were covered in previous chapters. In particular, the following two topics are prerequisites for understanding the rest of the examples in this section:

1. Completing the square; for a review, see Section 2.2.
2. Translations and reflections; for a review, see Section 4.3.

**EXAMPLE 3** Graph the function $y = x^2 - 2x + 3$.

Solution The idea here is to use the technique of completing the square; this will enable us to obtain the required graph simply by shifting the basic $y = x^2$ graph. We begin by writing the given equation

$$y = x^2 - 2x \qquad + 3$$

To complete the square for the $x$-terms we want to add 1. (Check this.) Of course, to keep the equation in balance we have to account for this. This is accomplished by writing

$$y = (x^2 - 2x + 1) + 3 - 1 \qquad \text{adding zero to the right side}$$

or

$$y = (x - 1)^2 + 2$$

Now, as we know from Section 4.3, the graph of this last equation is obtained by moving the parabola $y = x^2$ one unit in the positive $x$-direction and two units in the positive $y$-direction. This shifts the vertex from the origin to the point $(1, 2)$. See Figure 3.

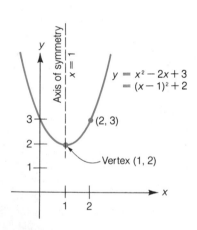

FIGURE 3

NOTE As a guide before sketching the graph yourself, you'll want to know the $y$-intercept. To find the $y$-intercept, substitute $x = 0$ in the given equation to obtain $y = 3$. Then, given the vertex $(1, 2)$ and the point $(0, 3)$, a reasonably accurate graph can be quickly sketched. [Actually, once you find that $(0, 3)$ is on the graph, you also know that the reflection of this point about the axis of symmetry is on the graph. This is why the point $(2, 3)$ is shown in Figure 3.]

As Figure 3 indicates, the lowest point on the graph of the function $y = x^2 - 2x + 3$ is the point $(1, 2)$. We say that 2 is the *minimum value* of this function, and we say that this minimum *occurs* at $x = 1$. More generally, given a function $f$, we say that $f$ has a **minimum** at $x = x_0$ if $f(x_0) \le f(x)$ for all $x$ in the domain of $f$. The number $f(x_0)$ in this case is called a **minimum value** of the function. We can also talk about a maximum value for a function. Figure 4, for example, shows the graph of $f(x) = -x^2 + 4$. The *maximum value* of $f$ in this case is 4, and this maximum occurs when $x = 0$. More generally, given a function $f$, we say that $f$ has a **maximum** at $x = x_0$ if $f(x_0) \ge f(x)$ for all $x$ in the domain

FIGURE 4

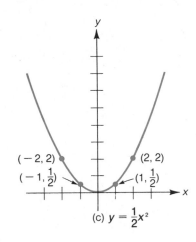

(a) $y = x^2$.

(b) $y = 2x^2$

(c) $y = \frac{1}{2}x^2$

FIGURE 5

of $f$. The number $f(x_0)$ in this case is called a **maximum value** of the function.

Now we want to compare the graphs of $y = x^2$, $y = 2x^2$, and $y = \frac{1}{2}x^2$. The last two equations were not discussed in Chapter 4, so for the moment you can think about graphing them by first setting up tables. Figure 5 displays the graphs. All three graphs are parabolas that open upward; but notice that the shapes are not identical. The parabola $y = 2x^2$ is narrower than $y = x^2$, while $y = \frac{1}{2}x^2$ is wider than $y = x^2$. These same observations about shape would also apply to $y = -x^2$, $y = -2x^2$, and $y = -\frac{1}{2}x^2$. In these cases, though, the parabolas open downward rather than upward. The observations that we have just made are generalized in the box that follows.

| PROPERTY SUMMARY | The Graph of $y = ax^2$ |
|---|---|
| | 1. The graph of $y = ax^2$ is a parabola with vertex at the origin. It is similar in shape to $y = x^2$. |
| | 2. The parabola $y = ax^2$ opens upward if $a > 0$, downward if $a < 0$. |
| | 3. The parabola $y = ax^2$ is narrower than $y = x^2$ if $|a| > 1$, wider than $y = x^2$ if $|a| < 1$. |

**EXAMPLE 4**   Sketch the graph of $y = 3(x - 1)^2$.

Solution   Because of the $x - 1$, we shift the basic parabola $y = x^2$ one unit to the right. The factor of 3 in the given equation tells us that we want to draw a parabola that is narrower than $y = x^2$ but that has the same vertex, $(1, 0)$. To see exactly how narrow to draw $y = 3(x - 1)^2$, we would like to know another point on the graph other than the vertex $(1, 0)$. An easy point to obtain is the $y$-intercept. Setting $x = 0$ in the equation yields

$$y = 3(0 - 1)^2 = 3$$

Now that we know the vertex $(1, 0)$ and the $y$-intercept 3, a reasonably accurate graph can be sketched. See Figure 6.

**FIGURE 6**  **FIGURE 7**

**EXAMPLE 5** Sketch the graph of $y = -3(x - 1)^2$.

Solution In Example 4 we sketched the graph of $y = 3(x - 1)^2$. By reflecting that graph in the $x$-axis, we obtain the graph of $y = -3(x - 1)^2$. See Figure 7.

**EXAMPLE 6** Graph the function $f(x) = -2x^2 + 12x - 16$ and specify the vertex, axis of symmetry, maximum or minimum value of $f$, and $x$- and $y$-intercepts.

Solution The idea is to complete the square, as in Example 3. We have

$$y = -2x^2 + 12x \qquad - 16$$
$$= -2(x^2 - 6x \quad) - 16$$
$$= -2(x^2 - 6x + 9) - 16 + 18 \qquad \text{adding } (2)(9) = 18 \text{ to}$$
$$\qquad\qquad\qquad\qquad\qquad\qquad\qquad \text{keep the equation in balance}$$
$$= -2(x - 3)^2 + 2$$

The graph of this last equation is obtained simply by shifting the graph of $y = -2x^2$ "right 3, up 2," so that the vertex is $(3, 2)$. As a guide for actually sketching the graph, we want to compute the intercepts. The $y$-intercept is $-16$. (Why?) For the $x$-intercepts, we replace $y$ with 0 in the original equation $y = -2x^2 + 12x - 16$ to obtain

$$-2x^2 + 12x - 16 = 0$$
$$x^2 - 6x + 8 = 0$$
$$(x - 4)(x - 2) = 0$$
$$x - 4 = 0 \quad \bigg| \quad x - 2 = 0$$
$$x = 4 \quad \bigg| \quad x = 2$$

Thus the $x$-intercepts are $x = 4$ and $x = 2$. Knowing these intercepts and the vertex, we can sketch the graph as in Figure 8. You should check for yourself that the information accompanying Figure 8 is correct.

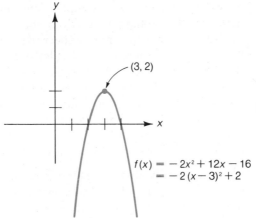

$$f(x) = -2x^2 + 12x - 16$$

| | |
|---|---|
| Vertex: | (3, 2) |
| Axis of symmetry: | $x = 3$ |
| Maximum value of $f$: | 2 |
| $y$-intercept: | $-16$ |
| $x$-intercepts: | 2 and 4 |

(3, 2)

$f(x) = -2x^2 + 12x - 16$
$\quad\quad = -2(x - 3)^2 + 2$

FIGURE 8

We are going to conclude this section with several examples involving maximum and minimum values of functions. As background for this, we first summarize our basic technique for graphing parabolas.

| PROPERTY SUMMARY | The Graph of the Parabola $y = ax^2 + bx + c$ |
|---|---|

By completing the square, the equation of the parabola $y = ax^2 + bx + c$ can always be rewritten in the form

$$y = a(x - h)^2 + k$$

In this form, the vertex of the parabola is $(h, k)$ and the axis of symmetry is the line $x = h$. The parabola opens upward if $a > 0$ and downward if $a < 0$.

**EXAMPLE 7**    Among all possible inputs for the function

$$g(x) = 3x^2 - 2x - 6$$

which yields the smallest output? What is this minimum output?

Solution    Think in terms of a graph. Since the graph of this function is a parabola opening upward, the input we want is just the $x$-coordinate of the vertex. By completing the square (as shown in Example 6), we find that the given equation can be rewritten

$$g(x) = 3\left(x - \frac{1}{3}\right)^2 - \frac{19}{3}$$

(You should verify this for yourself, using Example 6 as a model.) Thus, the vertex of the parabola is $(\frac{1}{3}, -\frac{19}{3})$. From this, and from the fact that the parabola opens upward, we conclude that the input $x = \frac{1}{3}$ produces the smallest output, and this minimum output is $g(\frac{1}{3}) = -\frac{19}{3}$.

For the previous two examples, keep in mind that we were able to find the maximum or minimum easily only because the functions were quadratics. By way of contrast, you cannot expect to find the minimum of $y = x^4 - 8x$ using the method of Examples 6 and 7, for this is not a quadratic function. In general, the techniques of calculus are required to find maxima and minima for functions other than quadratics. There are some cases, however, in which our present method can be adapted to functions that are closely related to quadratics. The next two examples show instances of this.

**EXAMPLE 8**  Let $D = \sqrt[3]{x^2 - x + 1}$. For which value of $x$ is $D$ a minimum?

Solution  The same $x$-value that makes $D$ as small as it can be will also cause the quantity $D^3 = x^2 - x + 1$ to be its smallest. That is, we only need to find the $x$-value that minimizes the quantity $x^2 - x + 1$. Call this quantity $y$, and complete the square as follows,

$$y = x^2 - x \qquad + 1$$
$$= x^2 - x + \frac{1}{4} + 1 - \frac{1}{4}$$
$$= \left(x - \frac{1}{2}\right)^2 + \frac{3}{4}$$

This shows that the vertex of the parabola $y = x^2 - x + 1$ is $(\frac{1}{2}, \frac{3}{4})$. Since this parabola opens upward, we conclude that the input $x = \frac{1}{2}$ will produce the minimum value for the quantity $x^2 - x + 1$. And according to our initial remark, this same value, $x = \frac{1}{2}$, also minimizes the quantity $D = \sqrt[3]{x^2 - x + 1}$. (Our work also shows that the minimum value of $D$ is $\sqrt[3]{\frac{3}{4}}$.)

**EXAMPLE 9**  Find the maximum value of the function $y = 16t^2 - 4t^4$.

Solution  Although this is not a quadratic function, we can complete the square as follows:

$$y = -4(t^4 - 4t^2 \quad )$$
$$= -4(t^4 - 4t^2 + 4) + 16$$
$$= -4(t^2 - 2)^2 + 16$$
$$= 16 - 4(t^2 - 2)^2$$

From this last equation we see that $y$ never exceeds 16 (because the quantity being subtracted from 16 is nonnegative). Furthermore, $y$ does attain the value 16 when $t^2 = 2$ or $t = \pm\sqrt{2}$. It now follows that the maximum value of the given function is 16.

# EXERCISE SET 5.1

## A

In Exercises 1–8, find the linear functions satisfying the given conditions.

1. $f(-1) = 0$ and $f(5) = 4$

2. $f(3) = 2$ and $f(-3) = -4$
3. $g(0) = 0$ and $g(1) = \sqrt{2}$
4. The graph passes through the points $(2, 4)$ and $(3, 9)$.

5. $f(\frac{1}{2}) = -3$ and the graph of $f$ is a line parallel to the line $x - y = 1$.

6. $g(2) = 1$ and the graph of $g$ is perpendicular to the line $6x - 3y = 2$.

7. The graph of $f$ is a horizontal line that passes through the larger of the two $y$-intercepts of the circle $x^2 - 2x + y^2 - 3 = 0$.

8. The $x$- and $y$-intercepts of the graph of $g$ are 1 and 4, respectively.

9. Let $f(x) = 3x - 4$ and $g(x) = 1 - 2x$. Find out if the function $f \circ g$ is linear.

10. Explain why there is no linear function whose graph passes through all three of the points $(-3, 2)$, $(1, 1)$, and $(5, 2)$.

11. A factory owner buys a new machine for $20,000. After 8 years, the machine has a salvage value of $1000. Find a formula for the value of the machine after $t$ years, where $0 \le t \le 8$.

12. A manufacturer buys a new machine costing $120,000. It is estimated that the machine has a useful lifetime of 10 years and a salvage value of $4000 at that time.
    (a) Find a formula for the value of the machine after $t$ years, where $0 \le t \le 10$.
    (b) Find the value of the machine after 8 years.

13. A factory owner installs a new machine costing $60,000. Its expected lifetime is 5 years, and at the end of that time the machine has no salvage value.
    (a) Find a formula for the value of the machine after $t$ years, where $0 \le t \le 5$.
    (b) Complete the following depreciation schedule.

| END OF YEAR | YEARLY DEPRECIATION | ACCUMULATED DEPRECIATION | VALUE $V$ |
|---|---|---|---|
| 0 | 0 | 0 | 60,000 |
| 1 | | | |
| 2 | | | |
| 3 | | | |
| 4 | | | |
| 5 | | 60,000 | 0 |

14. Let $x$ denote the temperature on the Celsius scale, and let $y$ denote the corresponding temperature on the Fahrenheit scale.
    (a) Find a linear function relating $x$ and $y$; use the facts that 32°F corresponds to 0°C, and 212°F corresponds to 100°C. Write your answer in the form $y = Ax + B$.
    (b) What Celsius temperature corresponds to 98.6°F?
    (c) Find a number $z$ for which $z$°F = $z$°C.

*In Exercises 15–28, graph the quadratic functions. In each case, specify the vertex, axis of symmetry, maximum or minimum value, and x- and y-intercepts.*

15. $y = (x + 2)^2$

16. $y = -(x + 2)^2$

17. $y = 2(x + 2)^2$

18. $y = 2(x + 2)^2 + 4$

19. $y = -2(x + 2)^2 + 4$

20. $y = x^2 + 6x - 1$

21. $f(x) = x^2 - 4x$

22. $F(x) = x^2 - 3x + 4$

23. $g(x) = 1 - x^2$

24. $y = 2x^2 + \sqrt{2}x$

25. $y = x^2 - 2x - 3$

26. $y = 2x^2 + 3x - 2$

27. $y = -x^2 + 6x + 2$

28. $y = -3x^2 + 12x$

29. Among all possible inputs for the given function, which yields the smallest or largest output (whichever is appropriate)? In each case, state whether the output is largest or smallest.
    (a) $y = 2x^2 - 4x + 11$
    (b) $f(x) = 8x^2 + x - 5$
    (c) $g(x) = -8x^2 + x - 5$
    (d) $s = -16t^2 + 196t + 80$
    (e) $y = x^2 - 10$
    (f) $y = -x^2 + \frac{3}{2}x + 1$

30. For the given function, find the maximum or minimum value (whichever is appropriate). In each case, state whether the value is a maximum or a minimum.
    (a) $y = x^2 - 8x + 3$
    (b) $f(x) = -2x^2 - 3x + 1$
    (c) $g(x) = -12x^2 + 1000$

31. How far from the origin is the vertex of the parabola $y = x^2 - 6x + 13$?

32. Let $f(x) = 2x - 3$, $g(x) = x^2 + 4x - 1$, and $h(x) = 1 - 2x^2$. Classify each of the following functions as linear, quadratic, or neither.
    (a) $f$
    (b) $g$
    (c) $h$
    (d) $f \circ g$
    (e) $g \circ f$
    (f) $g \circ h$

*In Exercises 33 and 34, determine the inputs that yield the minimum values for the functions. Compute the minimum value in each case.*

33. (a) $f(x) = \sqrt{x^2 - 6x + 73}$
    (b) $g(x) = \sqrt[3]{x^2 - 6x + 73}$
    (c) $h(x) = x^4 - 6x^2 + 73$

34. (a) $F(x) = (4x^2 - 4x + 109)^{1/2}$
    (b) $G(x) = (4x^2 - 4x + 109)^{1/3}$
    (c) $H(x) = 4x^4 - 4x^2 + 109$

*In Exercises 35 and 36, determine the maximum value for each function.*

35. (a) $f(x) = \sqrt{-x^2 + 4x + 12}$
    (b) $g(x) = \sqrt[3]{-x^2 + 4x + 12}$
    (c) $h(x) = -x^4 + 4x^2 + 12$

36. (a) $F(x) = (27x - x^2)^{1/2}$
    (b) $G(x) = (27x - x^2)^{1/3}$
    (c) $H(x) = 27x^2 - x^4$

c 37. Determine the linear function whose graph passes through the points $(1.27, -0.83)$ and $(2.06, 5.42)$. Round off the numbers in your answer to two decimal places.

c 38. Determine a value for $c$ such that the graph of $g(x) = x^2 + x + c$ passes through the point $(4.51, -3.02)$. Round off your answer to two decimal places.

c 39. A factory owner installs a new machine costing $26,450. After 7 years, the machine has no salvage value; in fact, it will cost the owner $1900 to have the machine disassembled and removed from its second-floor location. Assuming linear depreciation, find a formula for the value of the machine after $t$ years, where $0 \le t \le 7$.

**B**

In Exercises 40–43, find quadratic functions satisfying the given conditions.

40. The graph passes through the origin and the vertex is $(2, 2)$.

41. The graph is obtained by translating $y = x^2$ four units in the negative $x$-direction and three units in the positive $y$-direction.

42. The vertex is $(3, -1)$ and one $x$-intercept is $x = 1$.

43. The axis of symmetry is the line $x = 1$. The $y$-intercept is 1. There is only one $x$-intercept.

44. By completing the square, show that the coordinates of the vertex of the parabola $y = ax^2 + bx + c$ are

$$\left( \frac{-b}{2a}, \frac{4ac - b^2}{4a} \right)$$

45. Find the equation of the circle that passes through the origin and whose center is located at the vertex of the parabola $y = 2x^2 + 12x + 14$.

46. For which value of $c$ will the minimum value of the function $f(x) = x^2 + 2x + c$ be $\sqrt{2}$?

47. Find the $x$-coordinate of the vertex of the parabola $y = (x - a)(x - b)$, where $a$ and $b$ are constants.

48. Explain why the graph of $y = -x^2 - 10x - \frac{51}{2}$ has no $x$-intercepts by finding the vertex and noting whether the parabola opens up or down.

49. By analyzing sales figures, the economist for a stereo manufacturer knows that 150 units of a top-of-the-line turntable can be sold each month when the price is set at $p = \$200$ per unit. The figures also show that for each $10 hike in price, five fewer units are sold each month.
    (a) Let $x$ denote the number of units sold per month, and let $p$ denote the price per unit. Find a linear function relating $p$ and $x$.
    Hint: $\Delta p / \Delta x = 10/(-5) = -2$.
    (b) The revenue $R$ is given by $R = xp$. Find a quadratic function expressing $R$ in terms of $x$.

c 50. The minimum value of the function $f(x) = x^2 - bx$ is $-0.168$. What are the possible values for $b$? Round off your answer(s) to three decimal places.

c 51. In the twentieth century, the first runner to break the record for the one-mile run was John Paul Jones, in 1911. His time of $4:15.4$ bettered the old record (set in 1895) by two-tenths of a second. The table shows the evolution of the record for the one-mile run during the years 1911–1966. In the accompanying figure, we've plotted the $(x, y)$ pairs given in the table. The resulting plot is called a *scatter diagram*. A striking feature of this scatter diagram is that the records do not appear to be leveling off; rather, they seem to be decreasing in an approximately linear fashion. Using techniques from the field of statistics, it can be shown that the linear function that best fits this particular set of data is given by the equation

$$f(x) = -0.4164x + 1052.0433 \tag{1}$$

The graph of this line is shown in color in the figure. The line is referred to as the *regression line* (or the *least-squares line*).

Evolution of the Record for the Mile Run, 1911–1966.

| x (YEAR) | y (TIME) | |
|----------|----------|---|
| 1911 | 4:15.4 | (John Paul Jones, U.S.) |
| 1913 | 4:14.6 | (John Paul Jones, U.S.) |
| 1915 | 4:12.6 | (Norman Taber, U.S.) |
| 1923 | 4:10.4 | (Paavo Nurmi, Finland) |
| 1931 | 4:09.2 | (Jules Ladoumegue, France) |
| 1933 | 4:07.6 | (Jack Lovelock, New Zealand) |
| 1934 | 4:06.8 | (Glen Cunningham, U.S.) |
| 1937 | 4:06.4 | (Sidney Wooderson, Great Britain) |
| 1942 | 4:06.2 | (Gunder Haegg, Sweden) |
| 1942 | 4:06.2 | (Arne Andersson, Sweden) |
| 1942 | 4:04.6 | (Gunder Haegg, Sweden) |
| 1943 | 4:02.6 | (Arne Andersson, Sweden) |
| 1944 | 4:01.6 | (Arne Andersson, Sweden) |
| 1945 | 4:01.4 | (Gunder Haegg, Sweden) |
| 1954 | 3:59.4 | (Roger Bannister, Great Britain) |
| 1954 | 3:58.0 | (John Landy, Australia) |
| 1957 | 3:57.2 | (Derek Ibbotson, Great Britain) |
| 1958 | 3:54.5 | (Herb Elliott, Australia) |
| 1962 | 3:54.4 | (Peter Snell, New Zealand) |
| 1964 | 3:54.1 | (Peter Snell, New Zealand) |
| 1965 | 3:53.6 | (Michel Jazy, France) |
| 1966 | 3:51.3 | (Jim Ryun, U.S.) |

y (sec)

Scatter plot and regression line for Exercise 51

(a) Use the regression line to estimate (to the nearest tenth of a second) the mile record for the year 1967. That is, compute $f(1967)$. Next, compute the percent error in your estimate, given that the actual record in 1967, set by Jim Ryun, was 3:51.1 (or 231.1 sec). Percent error is calculated as follows:

$$\text{percent error} = \frac{|\text{actual value} - \text{predicted value}|}{\text{actual value}} \times 100$$

(b) Follow part (a) for the year 1980. The actual record for 1980 is 3:48.8 (or 228.8 sec), set by Steve Ovett of Great Britain.

(c) Follow part (a) for the year 1985. The actual record for 1985 is 3:46.31 (or 226.31 sec), set by Steve Cram of Great Britain.

(d) The June 1976 issue of *Scientific American* contained an article entitled "Future Performance in Footracing," by H. W. Ryder, H. J. Carr, and P. Herget. According to the authors of the article, "It appears likely that within 50 years the record [for the mile] will be down to 3:30." Use the regression line defined in equation (1) to make a projection for the mile time in the year 2026 (which will be 50 years after the article appeared). Is your projection close to the one given in the *Scientific American* article?

**C**

**52.** Consider the quadratic function $y = px^2 + px + r$, where $p \neq 0$.
(a) Show that if the vertex lies on the $x$-axis, then $p = 4r$.
(b) Show that if $p = 4r$, then the vertex lies on the $x$-axis.

**53.** Let $g(x) = x^2 + 2(a + b)x + 2(a^2 + b^2)$, where $a$ and $b$ are constants.
(a) Show that the coordinates of the vertex are

$$(-(a + b), (a - b)^2)$$

(b) Use the result in part (a) to explain why the graph of $g$ has no $x$-intercepts unless $a = b$, in which case there is only one $x$-intercept.

**54.** As you have seen in this section, if the quadratic equation $f(x) = x^2 + Bx + C = 0$ has two real roots, there is a geometric interpretation: these roots are the $x$-intercepts for the graph of $f$. Less well known is the fact that there is a geometric interpretation when the quadratic equation $f(x) = x^2 + Bx + C = 0$ has nonreal complex roots. If these roots are $a + bi$ and $a - bi$, where $a$ and $b$ are real numbers, show that the coordinates of the vertex for the graph of $f$ are $(a, b^2)$. *Hint:* If $r_1$ and $r_2$ are the roots of the equation $x^2 + Bx + C = 0$, then $x^2 + Bx + C$ can be written $(x - r_1)(x - r_2)$.

## APPLIED FUNCTIONS: SETTING UP EQUATIONS

**5.2**

*I hope that I shall shock a few people in asserting that the most important single task of mathematical instruction in the secondary schools is to teach the setting up of equations to solve word problems.*

GEORGE POLYA (1887–1985)

*Each problem that I solved became a rule which afterwards served to solve other problems.*

RENÉ DESCARTES (1596–1650)

One of the first steps in problem solving often involves defining a function. The function then serves to describe or summarize a given situation in a way that is both concise and (one hopes) revealing. In this section, we want to practice setting up equations that define such functions. Then, in Section 5.3, we will use this skill in solving a certain class of applied problems involving quadratic functions. For many of the examples we look at, we'll rely on the following four-step procedure to set up the required equation. You may eventually want to modify this procedure to fit your own style. The point, however, is that it is possible to approach these problems in a systematic manner. A word of advice: You're accustomed to working mathematics problems in which the answers are numbers. In this section, the answers are functions (or, more precisely, equations defining functions); you'll need to get used to this.

---

### Steps for Setting Up Equations

STEP 1   After reading the problem carefully, draw a picture that conveys the given information.

STEP 2   State in your own words, as specifically as you can, what the problem is asking for. (This usually requires rereading the problem.) Now, assuming that the problem asks you to find a particular quantity (or to find a formula for a particular quantity), assign a variable to denote that key quantity.

STEP 3   Label any other quantities in your figure that appear relevant. Are there equations relating these quantities?

STEP 4   Find an equation involving the key variable that you identified in step 2. (Some people prefer to do this right after step 2.) Now, as necessary, substitute in this equation using the auxiliary equations from step 3 to obtain an equation involving only the required variables.

---

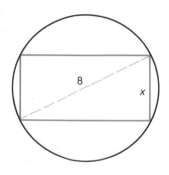

**FIGURE 1**

**EXAMPLE 1**   A rectangle is inscribed in a circle of diameter 8 cm. Express the perimeter of the rectangle as a function of its width $x$.

Solution   Let's follow our four-step procedure.

STEP 1   See Figure 1. Note that the diagonal of the rectangle is a diameter of the circle.

STEP 2   The problem asks us to come up with a formula or function that gives the perimeter of the rectangle in terms of $x$, the width. Let $P$ denote the perimeter.

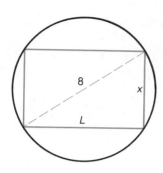

**FIGURE 2**

STEP 3 Let $L$ denote the length of the rectangle, as shown in Figure 2. Then by the Pythagorean Theorem, we have

$$L^2 + x^2 = 8^2$$
$$= 64$$

This equation relates the length $L$ and the width $x$. Rather than leaving the equation in this form, however, we'll solve for $L$ in terms of $x$ (because the instructions in the problem mention $x$, not $L$):

$$L^2 = 64 - x^2$$
$$L = \sqrt{64 - x^2} \tag{1}$$

STEP 4 The perimeter of the rectangle is the sum of the lengths of the four sides. Thus,

$$P = x + x + L + L$$
$$= 2x + 2L \tag{2}$$

This expresses $P$ in terms of $x$ and $L$. However, the problem asks for $P$ in terms of just $x$. Using equation (1) to substitute for $L$ in equation (2), we have

$$P = 2x + 2\sqrt{64 - x^2} \tag{3}$$

This is the required equation. It expresses the perimeter $P$ as a function of the width $x$, so if you know $x$, you can calculate $P$. To emphasize this dependence of $P$ on $x$, we can employ function notation to rewrite equation (3) as

$$P(x) = 2x + 2\sqrt{64 - x^2} \tag{4}$$

Before leaving this example, we need to specify the domain of the perimeter function in equation (4). An easy way to do this is to look back at Figure 1. Since $x$ represents a width, we certainly want $x > 0$. Furthermore, Figure 1 tells us that $x < 8$, because in any right triangle, a leg is always shorter than the hypotenuse. Putting these observations together, we conclude that the domain of the perimeter function in equation (4) is the open interval $(0,8)$.

NOTE Although it would make sense *algebraically* to use an input such as $x = -1$ in equation (4), it does not make sense in our *geometric* context, where $x$ denotes the width.

Perimeter = 100 cm

$x$

**FIGURE 3**

**EXAMPLE 2** The perimeter of a rectangle is 100 cm. Express the area of the rectangle in terms of the width $x$.

Solution STEP 1 See Figure 3.

STEP 2 We want to express the area of the rectangle in terms of $x$, the width. Let $A$ stand for the area of the rectangle.

Perimeter = 100 cm

$x$

$L$

**FIGURE 4**

STEP 3 Call the length of the rectangle $L$, as indicated in Figure 4. Then, since the perimeter is given as 100 cm, we have

$$2x + 2L = 100$$
$$x + L = 50$$
$$L = 50 - x \tag{5}$$

STEP 4   Area of a rectangle equals width times length:

$$A = x \cdot L$$
$$= x(50 - x) \qquad \text{substituting for } L \text{ using equation (5)}$$
$$= 50x - x^2 \tag{6}$$

This is the required equation expressing the area of the rectangle in terms of the width $x$. To emphasize this dependence of $A$ on $x$, we can use function notation to rewrite equation (6) as

$$A(x) = 50x - x^2$$

The domain of this area function is the open interval $(0, 50)$. To see why this is so, first note that $x > 0$, because $x$ denotes a width. Furthermore, in view of equation (5), we must have $x < 50$ (otherwise $L$, the length, would be zero or negative).

**EXAMPLE 3**   Let $P(x, y)$ be a point on the curve $y = \sqrt{x}$. Express the distance from $P$ to the point $(1, 0)$ as a function of $x$.

**Solution**   STEP 1   See Figure 5.

STEP 2   We want to express the length of the broken line in Figure 5 in terms of $x$. Call this length $D$.

STEP 3   There are no other quantities in Figure 5 that need labeling. But don't forget that we are given

$$y = \sqrt{x} \tag{7}$$

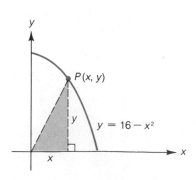

**FIGURE 5**

STEP 4   By the distance formula, we have

$$D = \sqrt{(x - 1)^2 + (y - 0)^2}$$
$$= \sqrt{x^2 - 2x + 1 + y^2} \tag{8}$$

Now use equation (7) to eliminate $y$ in equation (8):

$$D(x) = \sqrt{x^2 - 2x + 1 + (\sqrt{x})^2}$$
$$= \sqrt{x^2 - 2x + 1 + x}$$
$$= \sqrt{x^2 - x + 1} \tag{9}$$

Equation (9) expresses the distance as a function of $x$, as required. What about the domain of this distance function? Since the $x$-coordinate of a point on the curve $y = \sqrt{x}$ can be any nonnegative number, the domain of the distance function is $[0, \infty)$.

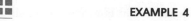

**FIGURE 6**

**EXAMPLE 4**   A point $P(x, y)$ lies in the first quadrant on the parabola $y = 16 - x^2$, as indicated in Figure 6. Express the area of the triangular region in Figure 6 as a function of $x$.

**Solution**   STEP 1   See Figure 6.

STEP 2    We want to express the area of the shaded triangle in terms of $x$. Let $A$ denote the area of this triangle.

STEP 3    Since the coordinates of $P$ are $(x,y)$, the base of our triangle is $x$ and the height of our triangle is $y$. Also, $x$ and $y$ are related by the given equation

$$y = 16 - x^2 \tag{10}$$

STEP 4    Area of a triangle equals $\frac{1}{2}$(base)(height):

$$A = \frac{1}{2}(x)(y)$$

$$A(x) = \frac{1}{2}(x)(16 - x^2) \qquad \text{substituting for } y \text{ using equation (10)}$$

$$= 8x - \frac{1}{2}x^3$$

This last equation expresses the area of the triangle as a function of $x$, as required. (Exercise 46 at the end of this section asks you to specify the domain of this area function.) ▪

**EXAMPLE 5**    A piece of wire $x$ in. long is bent into the shape of a circle. Express the area of the circle in terms of $x$.

Solution    STEP 1    See Figure 7.

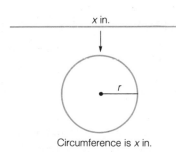

$x$ in.

Circumference is $x$ in.

FIGURE 7

STEP 2    We are supposed to express the area of the circle in terms of $x$, the circumference. Let $A$ denote the area.

STEP 3    The general formula for circumference in terms of radius is

$$C = 2\pi r$$

Since in our case the circumference is given as $x$, our equation becomes

$$x = 2\pi r$$

or

$$r = \frac{x}{2\pi} \tag{11}$$

STEP 4    The general formula for the area of a circle in terms of the radius is

$$A = \pi r^2 \tag{12}$$

This expresses $A$ in terms of $r$; but we want $A$ in terms of $x$. So we replace $r$ in equation (12) by the value given in equation (11). This yields

$$A(x) = \pi\left(\frac{x}{2\pi}\right)^2 = \frac{\pi x^2}{4\pi^2} = \frac{x^2}{4\pi}$$

Thus, we have

$$A(x) = \frac{x^2}{4\pi}$$

This is the required equation. It expresses the area of the circle in terms of the circumference $x$. In other words, if we know the length of the piece of wire that is to be bent into a circle, from that length we can calculate what the area of the circle will be.

**EXAMPLE 6**   Figure 8 displays a right circular cylinder, along with the formulas for the volume $V$ and the total surface area $S$. Given that the volume is $10 \text{ cm}^3$, express the surface area $S$ as a function of $r$, the radius of the base.

Solution   STEP 1   See Figure 8.

STEP 2   We are given a formula expressing the surface area $S$ in terms of both $r$ and $h$. We want to express $S$ in terms of just $r$.

STEP 3   We are given that $V = 10$ and also that $V = \pi r^2 h$. Thus,

$$\pi r^2 h = 10$$

and consequently, expressing $h$ in terms of $r$,

$$h = \frac{10}{\pi r^2} \qquad\qquad (13)$$

STEP 4   We take the given formula for $S$, namely,

$$S = 2\pi r^2 + 2\pi rh$$

and replace $h$ using the value given in equation (13). We get

$$S(r) = 2\pi r^2 + 2\pi r\left(\frac{10}{\pi r^2}\right)$$

$$= 2\pi r^2 + \frac{20}{r}$$

This is the required equation. It expresses the total surface area in terms of the radius $r$. Since the only restriction on $r$ is that it be positive, the domain of the area function is $(0, \infty)$.

We have followed the same four-step procedure in Examples 1 through 6. Of course, no single method can cover all possible cases. As usual, common sense and experience are often necessary. Also, you should not feel compelled to follow this procedure at any cost. Keep this in mind as you study the last three examples in this section.

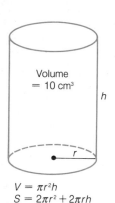

Volume
= 10 cm³

$h$

$r$

$V = \pi r^2 h$
$S = 2\pi r^2 + 2\pi rh$

**FIGURE 8**

**EXAMPLE 7**   Two numbers add up to 8. Express the product $P$ of these two numbers in terms of a single variable.

Solution   If we call the two numbers $x$ and $8 - x$, then their product $P$ is given by

$$P(x) = x(8 - x)$$
$$= 8x - x^2$$

That's it. This last equation expresses the product as a function of the variable $x$. Since there are no restrictions on $x$ (other than it being a real number), the domain of this function is $(-\infty, \infty)$.

**EXAMPLE 8**  Express the radius of a circle as a function of the area of the circle.

Solution  We already know a formula relating these two quantities:

$$A = \pi r^2$$

In essence, the problem asks us to solve this equation for $r$ instead of $A$. We then have

$$\pi r^2 = A$$
$$r^2 = \frac{A}{\pi}$$
$$r(A) = \sqrt{A/\pi}$$

This is the required equation. It tells us how to calculate the radius of the circle if the area is known.  *Question:*  Are there any restrictions on $A$ in this last equation?

**EXAMPLE 9**  In economics, the revenue $R$ generated by selling $x$ units at a price of $p$ dollars per unit is given by

$$R = x \cdot p$$

— price per unit

— number of units

In Figure 9, we are given a hypothetical function relating the selling price of a certain item to the number of units sold. Such a function is called a **demand function**. Express the revenue as a function of $x$.

Solution  More than anything else, this problem is an exercise in reading. After reading the problem several times, we find that it comes down to this:

given:  $R = x \cdot p$  and  $p = -\dfrac{1}{3}x + 4$  $(0 \le x \le 11)$

find:  An equation expressing $R$ in terms of $x$

In view of this, we write

$$R = x \cdot p$$
$$R(x) = x\left(-\frac{1}{3}x + 4\right)$$

Thus, we have

$$R(x) = -\frac{1}{3}x^2 + 4x \qquad (0 \le x \le 11)$$

$p = -\dfrac{1}{3}x + 4$  $(0 \le x \le 11)$

price per unit

number of units sold

**FIGURE 9**

This is the required function. It allows us to calculate the revenue when we know the number of units sold. Note that this revenue function is a quadratic function. *Question for review and also preview:*   How would you find the maximum revenue in this case?                                                                 ▪▪

## EXERCISE SET 5.2

### A

1. A rectangle is inscribed in a circle of diameter 12 in.
   (a) Express the perimeter of the rectangle as a function of its width $x$. *Suggestion:* First reread Example 1.
   (b) Express the area of the rectangle as a function of its width $x$.

2. (a) The perimeter of a rectangle is 16 cm. Express the area of the rectangle in terms of the width $x$. *Suggestion:*  Reread Example 2.
   (b) The area of a rectangle is 85 cm². Express the perimeter as a function of the width $x$.

3. A point $P(x,y)$ is on the curve $y = x^2 + 1$.
   (a) Express the distance from $P$ to the origin as a function of $x$. *Suggestion:*  First reread Example 3.
   (b) In part (a), you expressed the length of a certain line segment as a function of $x$. Now express the slope of that line segment in terms of $x$.

4. A point $P(x,y)$ lies on the curve $y = \sqrt{x}$, as shown in the following figure.
   (a) Express the area of the shaded triangle in terms of $x$. *Suggestion:*  First reread Example 4.
   (b) Express the perimeter of the shaded triangle in terms of $x$.

5. A piece of wire $\pi y$ in. long is bent into a circle.
   (a) Express the area of the circle as a function of $y$. *Suggestion:*  First reread Example 5.
   (b) If the original piece of wire were bent into a square instead of a circle, how would you express the area in terms of $y$?

6. The volume of a certain right circular cylinder is 20 in³. Express the total surface area of the cylinder as a function of $r$, the radius of the base. *Suggestion:*  First reread Example 6.

7. Two numbers add to 16.
   (a) Express the product of the two numbers in terms of a single variable. *Suggestion:*  First reread Example 7.
   (b) Express the sum of the squares of the two numbers in terms of a single variable.
   (c) Express the difference of the cubes of the two numbers in terms of a single variable. (There are two answers.)
   (d) What happens when you try to express the average of the two numbers in terms of one variable?

8. (a) Express the radius of a circle as a function of the circumference. *Suggestion:*  Reread Example 8.
   (b) Express the diameter of a circle as a function of the circumference.

9. Given a demand function $p = -\tfrac{1}{4}x + 8$, express the revenue as a function of $x$. (See Example 9 for terminology and definitions.)

10. In Example 2, we considered a rectangle with perimeter 100 cm. We found that the area of such a rectangle is given by $A(x) = 50x - x^2$, where $x$ is the width of the rectangle. Compute the numbers $A(1)$, $A(10)$, $A(20)$, $A(25)$, and $A(35)$. Which width $x$ seems to yield the largest area $A(x)$?

[c] 11. In Example 1, we considered a rectangle inscribed in a circle of diameter 8 cm. We found that the perimeter of such a rectangle is given by

$$P(x) = 2x + 2\sqrt{64 - x^2}$$

where $x$ is the width of the rectangle.
   (a) Use a calculator to complete the following table. Round off each answer to two decimal places.

| $x$    | 1 | 2 | 3 | 4 | 5 | 6 | 7 |
|--------|---|---|---|---|---|---|---|
| $P(x)$ |   |   |   |   |   |   |   |

   (b) In your table, what is the largest value for $P(x)$? What is the width $x$ in this case?

(c) Using calculus, it can be shown that among all possible widths $x$, the width $x = 4\sqrt{2}$ cm yields the largest possible perimeter. Use a calculator to compute the perimeter in this case, and check to see that the value you obtain is indeed larger than all of the values obtained in part (a).

12. In economics, the demand function for a given commodity tells us how the unit price $p$ is related to the number of units $x$ that are sold. Suppose we are given the demand function

$$p = 5 - \frac{x}{4} \qquad (p \text{ in dollars})$$

(a) Graph this demand function.
(b) How many units can be sold when the unit price is $3? Locate the point on the graph of the demand function that conveys this information.
(c) To sell 12 items, how should the unit price be set? Locate the point on the graph of the demand function that conveys this information.
(d) Find the revenue function corresponding to the given demand function. (Use the formula $R = x \cdot p$.) Graph the revenue function.
(e) Find the revenue when $x = 2$, when $x = 8$, and when $x = 14$.
(f) According to your graph in part (d), which $x$-value yields the greatest revenue? What is that revenue? What is the corresponding unit price?

13. Let $2s$ denote the length of the side of an equilateral triangle.
(a) Express the height of the triangle as a function of $s$.
(b) Express the area of the triangle as a function of $s$.
(c) Use the function you found in part (a) to determine the height of an equilateral triangle, each side of which is 8 cm long.
(d) Use the function you found in part (b) to determine the area of an equilateral triangle, each side of which is 5 in. long.

14. If $x$ denotes the length of a side of an equilateral triangle, express the area of the triangle as a function of $x$.

15. In a certain right circular cylinder, the height is twice the radius. Express the volume as a function of the radius.

16. Using the information given in Exercise 15, express the radius as a function of the volume.

17. The volume of a right circular cylinder is $12\pi \text{ in}^3$.
(a) Express the height as a function of the radius.
(b) Express the total surface area as a function of the radius.

18. In a certain right circular cylinder, the total surface area is $14 \text{ in}^2$. Express the volume as a function of the radius.

19. The volume $V$ and the surface area $S$ of a sphere of radius $r$ are given by the formulas $V = \frac{4}{3}\pi r^3$ and $S = 4\pi r^2$. Express $V$ as a function of $S$.

20. The base of a rectangle lies on the $x$-axis, while the upper two vertices lie on the parabola $y = 10 - x^2$. Suppose that the coordinates of the upper right vertex of the rectangle are $(x, y)$. Express the area of the rectangle as a function of $x$.

21. The hypotenuse of a right triangle is 20 cm. Express the area of the triangle as a function of the length $x$ of one of the legs.

22. (a) Express the area of the shaded triangle in the accompanying figure as a function of $x$.
(b) Express the perimeter of the shaded triangle as a function of $x$.

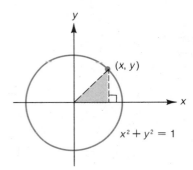

23. For the following figure, express the length of $\overline{AB}$ as a function of $x$. *Hint:* Note the similar triangles.

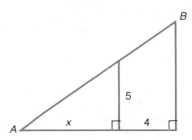

24. Five hundred feet of fencing are available to enclose a rectangular pasture alongside a river, the river serving as one side of the rectangle (so only three sides require fencing). Express the area of the rectangular pasture as a function of $x$. See the figure.

© 25. After rereading Example 2, complete the given table. Which $x$-value in the table yields the largest area $A$? What is the corresponding value of the length $L$ in that case?

| $x$ | 5 | 10 | 20 | 24 | 24.8 | 24.9 | 25 | 25.1 | 25.2 | 45 |
|---|---|---|---|---|---|---|---|---|---|---|
| $A(x)$ | | | | | | | | | | |

© 26. After rereading Example 3, complete the following three tables. For each table, specify the $x$-value that yields the smallest distance $D$. (For Table 2, round off the values of $D$ to two significant digits. For Table 3, round off the values of $D$ to seven significant digits.)

TABLE 1

| $x$ | 1 | 2 | 3 | 4 | 5 |
|---|---|---|---|---|---|
| $D$ | | | | | |

TABLE 2

| $x$ | 0.25 | 0.50 | 0.75 |
|---|---|---|---|
| $D$ | | | |

TABLE 3

| $x$ | 0.498 | 0.499 | 0.500 | 0.501 | 0.502 |
|---|---|---|---|---|---|
| $D$ | | | | | |

© 27. (a) After rereading Example 4, complete the following three tables. For each table, specify the $x$-value that yields the largest area $A$. [For Tables 2 and 3, round off the values of $A$ to six significant digits. In part (b), you'll see why we are asking for that many digits.]

(b) Using calculus, it can be shown that the $x$-value yielding the largest area $A$ is $x = 4\sqrt{3}/3$. Which $x$-value in the tables is closest to this $x$-value? To six significant digits, what is the area $A$ when $x = 4\sqrt{3}/3$?

TABLE 1

| $x$ | 1 | 2 | 3 | 4 |
|---|---|---|---|---|
| $A$ | | | | |

TABLE 2

| $x$ | 1.75 | 2.00 | 2.25 | 2.50 | 2.75 |
|---|---|---|---|---|---|
| $A$ | | | | | |

TABLE 3

| $x$ | 2.15 | 2.20 | 2.25 | 2.30 | 2.35 |
|---|---|---|---|---|---|
| $A$ | | | | | |

## B

*In Exercises 28–31, refer to the following figure, which displays a right circular cone along with the formulas for the volume V and the lateral surface area S.*

$$V = \frac{1}{3}\pi r^2 h$$
$$S = \pi r \sqrt{r^2 + h^2}$$

28. The volume of a right circular cone is $12\pi \text{ cm}^3$.
    (a) Express the height as a function of the radius.
    (b) Express the radius as a function of the height.

29. Suppose that, for a right circular cone, the height and radius are related by the equation $h = \sqrt{3}r$.
    (a) Express the volume as a function of $r$.
    (b) Express the lateral surface area as a function of $r$.

30. The volume of a right circular cone is $2 \text{ ft}^3$. Show that the lateral surface area as a function of $r$ is given by
$$S = \frac{\sqrt{\pi^2 r^6 + 36}}{r}$$

31. In a certain right circular cone, the volume is numerically equal to the lateral surface area.
    (a) Express the radius as a function of the height.
    (b) Express the height as a function of the radius.

32. A line is drawn from the origin $O$ to a point $P(x,y)$ in the first quadrant on the graph of $y = 1/x$. From point $P$, a line is drawn perpendicular to the $x$-axis, meeting the $x$-axis at $B$.
    (a) Draw a figure of the situation described.
    (b) Express the perimeter of triangle $OPB$ as a function of $x$.
    (c) Try to express the area of triangle $OPB$ as a function of $x$. What happens?

33. A piece of wire 14 in. long is cut into two pieces. The first piece is bent into a circle, the second into a square. Express the combined total area of the circle and square as a function of $x$, where $x$ denotes the length of the wire that is used for the circle.

34. A wire of length $L$ is cut into two pieces. The first piece is bent into a square, the second into an equilateral tri-

angle. Express the combined total area of the square and triangle as a function of $x$, where $x$ denotes the length of wire used for the triangle. ($L$ is to be considered a constant here, not another variable.)

35. An athletic field with a perimeter of $\frac{1}{4}$ mile consists of a rectangle with a semicircle at each end, as shown in the figure. Express the area of the field as a function of $r$, the radius of the semicircle.

36. A square of side $x$ is inscribed in a circle. Express the area of the circle as a function of $x$.

37. An equilateral triangle of side $x$ is inscribed in a circle. Express the area of the circle as a function of $x$.

38. Consider an object of mass $m$ moving at a velocity $v$. The kinetic energy $K$ is defined as $K = \frac{1}{2}mv^2$. The momentum $p$ is defined as $p = mv$. Express the kinetic energy in terms of mass and momentum (but not velocity).

39. An isosceles triangle is inscribed in a circle of radius $R$, where $R$ is a constant. Express the area within the circle but outside the triangle as a function of $h$, where $h$ denotes the height of the triangle.

40. Refer to the accompanying figure. An offshore oil rig is located at point $A$, which is 10 miles out to sea. An oil pipeline is to be constructed from $A$ to a point $C$ on the shore and then to an oil refinery at point $D$, farther up the coast. If it costs \$8000 per mile to lay the pipeline in the sea and \$2000 per mile on land, express the cost of laying the pipeline in terms of $x$, where $x$ is the distance from $B$ to $C$.

41. An open-top box is constructed from a 6-by-8 in. rectangular sheet of tin by cutting out equal squares at each corner and then folding up the flaps, as shown in the figure. Express the volume of the box as a function of $x$, the length of the side of each cut-out square.

42. Follow Exercise 41, but assume that the original piece of tin is a square, 12 in. on each side.

43. A Norman window is in the shape of a rectangle surmounted by a semicircle, as shown in the next figure. Assume that the perimeter of the window is 32 ft.
   (a) Express the area of the window as a function of $r$, the radius of the semicircle.
   (b) The function you were asked to find in part (a) is a quadratic function, so its graph is a parabola. Does the parabola open upward or downward? Does it pass through the origin? Show that the vertex of the parabola is
   $$\left( \frac{32}{\pi + 4}, \frac{512}{\pi + 4} \right).$$

44. Refer to the following figure. Express the lengths of $\overline{CB}$, $\overline{CD}$, $\overline{BD}$, and $\overline{AB}$ in terms of $x$. *Hint:* Recall the theorem from geometry stating that in a 30°–60°–90°

right triangle, the side opposite the 30° angle is half the hypotenuse.

**45.** Refer to the following figure. Let $s$ denote the ratio of $y$ to $z$.
   **(a)** Express $y$ as a function of $s$.
   **(b)** Express $s$ as a function of $y$.
   **(c)** Express $z$ as a function of $s$.
   **(d)** Express $s$ as a function of $z$.

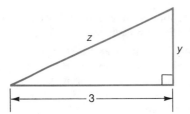

**46.** Refer to Example 4.
   **(a)** What is the $x$-intercept of the curve $y = 16 - x^2$ in Figure 6?
   **(b)** What is the domain of the area function in Example 4, assuming that the point $P$ does not lie on the $x$- or $y$-axis?

**47.** The following figure shows the parabola $y = x^2$ and a line segment $\overline{AP}$ drawn from the point $A(0, -1)$ to the point $P(a, a^2)$ on the parabola.

**(a)** Express the slope of $\overline{AP}$ in terms of $a$.
**(b)** Show that the area of the shaded triangle in the figure is given by

$$\text{area} = \frac{a^5}{2(a^2 + 1)}$$

**48.** A rancher who wishes to fence off a rectangular area finds that the fencing in the east-west direction will require extra reinforcement owing to strong prevailing winds. Because of this, fencing in the east-west direction will be $12 per (linear) yard, as opposed to a cost of $8 per yard for fencing in the north-south direction. Given that the rancher wants to spend $4800 on fencing, express the area of the rectangle as a function of $x$, its width. The required function is in fact a quadratic, so its graph is a parabola. Does the parabola open upward or downward? By considering this graph, find which width $x$ yields the rectangle of largest area. What is this maximum area?

**C**

**49.** The following figure shows two concentric squares. Express the area of the shaded triangle as a function of $x$.

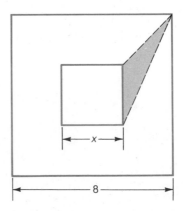

**50.** The following figure shows two concentric circles of radii $r$ and $R$. Let $A$ denote the area within the larger circle but outside the smaller one. Express $A$ as a function of $x$ (where $x$ is defined in the figure).

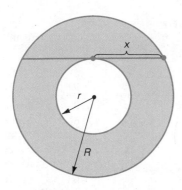

**51.** A straight line with slope $m$ $(m < 0)$ passes through the point $(1, 2)$ and intersects the line $y = 4x$ at a point in the first quadrant. Let $A$ denote the area of the triangle bounded by $y = 4x$, the $x$-axis, and the given line of slope $m$. Express $A$ as a function of $m$.

**52.** A line with slope $m$ $(m < 0)$ passes through the point $(a, b)$ in the first quadrant and intersects the line $y = Mx$ $(M > 0)$ at another point in the first quadrant. Let $A$ denote the area of the triangle bounded by $y = Mx$, the $x$-axis, and the given line with slope $m$. Express $A$ in terms of $m$, $M$, $a$, and $b$.   *Answer:*

$$A = \frac{M(am - b)^2}{2m(m - M)}.$$

**53.** A line with slope $m$ $(m < 0)$ passes through the point $(a, b)$ in the first quadrant. Express the area of the triangle bounded by this line and the axes in terms of $m$.

**54.** One corner of a page of width $a$ is folded over and just reaches the opposite side, as indicated in the following figure. Express $L$, the length of the crease, in terms of $x$ and $a$.

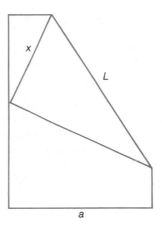

## MAXIMUM AND MINIMUM PROBLEMS

**5.3**

*. . . problems on maxima and minima, although new features in an English text-book, stand so little in need of apology with the scientific public that I offer none.*

G. CHRYSTAL, in his preface to *Textbook of Algebra*, 1886.

You have already met several examples of maximum and minimum problems in the first section of this chapter. In particular, before reading further in the present section, you should first look back and review Examples 7 through 9 on 221–222.

Actually, we begin this section's discussion of maximum and minimum problems at a more intuitive level. Consider, for example, the following question: If two numbers add to 9, what is the largest possible value of their product? To gain some insight here, we carry out a few preliminary calculations (see Table 1 on the next page).

We have circled the 20 in the right-hand column of Table 1, for that "appears" to be the largest product. We say *appears* because our table is incomplete. For instance, what if we allowed $x$- and $y$-values that are not whole numbers? Might we get a product exceeding 20? Table 2 shows the results of some additional calculations along these lines.

**TABLE 1**

| $x$ | $y$ | $x + y$ | $xy$ |
|---|---|---|---|
| $-2$ | 11 | 9 | $-22$ |
| $-1$ | 10 | 9 | $-10$ |
| 0 | 9 | 9 | 0 |
| 1 | 8 | 9 | 8 |
| 2 | 7 | 9 | 14 |
| 3 | 6 | 9 | 18 |
| 4 | 5 | 9 | ⑳ |

**TABLE 2**

| $x$ | $y$ | $x + y$ | $xy$ |
|---|---|---|---|
| 1 | 8 | 9 | 8 |
| 1.5 | 7.5 | 9 | 11.25 |
| 2 | 7 | 9 | 14 |
| 2.5 | 6.5 | 9 | 16.25 |
| 3 | 6 | 9 | 18 |
| 3.5 | 5.5 | 9 | 19.25 |
| 4 | 5 | 9 | 20 |
| 4.5 | 4.5 | 9 | 20.25 |

As you can see from Table 2, there is a product exceeding 20, namely, 20.25. Now the question is, if we further expand our tables, can we find yet another product, this time exceeding 20.25? And here we have come about as far as we want to go using this approach involving arithmetic and tables. For no matter what candidate we come up with for the largest product, there will always be the question of whether we might do still better using a larger table.

Nevertheless, this approach was useful, for it showed us what is really at the heart of a typical maximum or minimum problem. Essentially, we are trying to sort through an infinite number of possible cases and then pick out the required extreme case. For instance, in the example at hand, there are infinitely many pairs of numbers $x$ and $y$ adding to 9. Among all of these pairs, we are asked to look at the products and see which (if any!) is the largest. Example 1 shows how to apply our knowledge of quadratic functions to solve this problem in a definitive manner.

**EXAMPLE 1**  Two numbers add to 9. What is the largest possible value for their product?

**Solution**  Call the two numbers $x$ and $9 - x$. Then their product $P$ is given by

$$P = x(9 - x)$$
$$= 9x - x^2$$

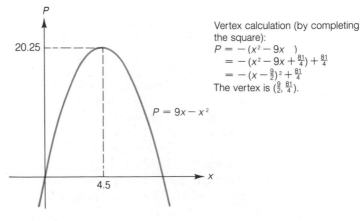

Vertex calculation (by completing the square):
$$P = -(x^2 - 9x \quad )$$
$$= -(x^2 - 9x + \tfrac{81}{4}) + \tfrac{81}{4}$$
$$= -(x - \tfrac{9}{2})^2 + \tfrac{81}{4}$$
The vertex is $(\tfrac{9}{2}, \tfrac{81}{4})$.

**FIGURE 1**

The graph of this quadratic function is the parabola in Figure 1. Note the accompanying calculation for the vertex. As Figure 1 and the accompanying calculations show, the largest value of the product $P$ is 20.25. This is the required solution. (Note from the graph that there is no smallest value of $P$.)

Example 1 illustrates the general strategy we will follow for solving the maximum and minimum problems in this section.

---

**Strategy for Solving the Maximum and Minimum Problems in This Section**

1. Express the quantity to be maximized or minimized in terms of a single variable. For instance, in Example 1, we found $P = 9x - x^2$. In trying to set up such functions, you'll want to keep in mind the four-step procedure used in Section 5.2.

2. Assuming that the function you have determined is a quadratic, note whether its graph, a parabola, opens upward or downward. Check whether this is consistent with what the problem asks for. For instance, the parabola in Figure 1 opens downward; so it makes sense to look for a largest value of $P$, not a smallest. Now complete the square to locate the vertex. (If the function is not a quadratic but is closely related to a quadratic, these ideas may still apply. See, for instance, Examples 8 and 9 in Section 5.1.)

3. After you have determined the vertex, you must relate that information to the original question. In Example 1, for instance, we were asked for the product $P$, not for $x$.

---

**EXAMPLE 2**    Among all rectangles having a perimeter of 10 ft, find the dimensions (length and width) of the one with the greatest area.

**Solution**    First we want to set up a function that expresses the area of the rectangle in terms of a single variable. In doing this, we'll be guided by the four-step procedure that we used in Section 5.2. Figure 2(a) displays the given information. Our problem is to determine the dimensions of the rectangle that has the greatest area. As Figure 2(b) indicates, we can label the dimensions $x$ and $y$. Since the perimeter is given as 10 ft, we have

$$2x + 2y = 10$$
$$x + y = 5$$
$$y = 5 - x \qquad (1)$$

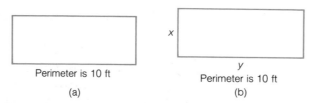

Perimeter is 10 ft

(a)

Perimeter is 10 ft

(b)

FIGURE 2

Letting $A$ denote the area of the rectangle, we can write

$$A = xy$$
$$A(x) = x(5 - x) \qquad \text{substituting for } y \text{ using equation (1)}$$
$$= 5x - x^2$$

This expresses the area of the rectangle in terms of the width $x$. Since the graph of this quadratic function is a parabola that opens downward, it does make sense to talk about the maximum. We can find the vertex of the parabola by completing the square:

$$A(x) = -(x^2 - 5x \qquad )$$
$$= -\left(x^2 - 5x + \frac{25}{4}\right) + \frac{25}{4}$$
$$= -\left(x - \frac{5}{2}\right)^2 + \frac{25}{4}$$

So the vertex is $(\frac{5}{2}, \frac{25}{4})$ and, in particular, the width $x = \frac{5}{2}$ ft yields the maximum area. Now, the problem asks us for the length and width of this rectangle, not its area. To compute the length $y$, we have

$$y = 5 - x$$
$$= 5 - \frac{5}{2} = \frac{5}{2}$$

Thus, among all rectangles having a perimeter of 10 ft, the one with the greatest area is actually the square, with dimensions of $2\frac{1}{2}$ ft by $2\frac{1}{2}$ ft.

**EXAMPLE 3**  Suppose that a baseball is tossed straight up and that its height as a function of time is given by the formula

$$h = -16t^2 + 64t + 6$$

In this formula, $h$ is measured in feet and $t$ in seconds, with $t = 0$ corresponding to the instant that the ball is released. What is the maximum height of the ball? When does the ball reach this height?

Solution  The given function tells us how the height $h$ of the ball depends on the time $t$. We want to know the largest possible value for $h$. Since the graph of the given function is a parabola opening downward, we can determine the largest value of $h$ just by finding the vertex of the parabola. After completing the square, we find that the original equation can be rewritten as

$$h = -16(t - 2)^2 + 70 \tag{2}$$

[You should verify this for yourself. (If you get stuck, follow the model for completing the square in Example 6 of Section 5.1.)] From equation (2) we see that the vertex of the parabola $h = -16t^2 + 64t + 6$ is $(2, 70)$. Therefore the maximum height of the ball is 70 ft, and the ball reaches this height at $t = 2$ sec.

**EXAMPLE 4**   Which point on the curve $y = \sqrt{x}$ is closest to the point $(1,0)$?

Solution   In Figure 3, we let $D$ denote the distance from a point $(x,y)$ on the curve to the point $(1,0)$. We are asked to find out exactly which point $(x,y)$ will make the distance $D$ as small as possible. Using the distance formula, we have

$$D = \sqrt{(x-1)^2 + (y-0)^2}$$
$$= \sqrt{x^2 - 2x + 1 + y^2} \qquad\qquad (3)$$

This expresses $D$ in terms of both $x$ and $y$. To express $D$ in terms of $x$ alone, we use the given equation $y = \sqrt{x}$ to substitute for $y$ in equation (3). This yields

$$D(x) = \sqrt{x^2 - 2x + 1 + (\sqrt{x})^2}$$
$$= \sqrt{x^2 - x + 1}$$

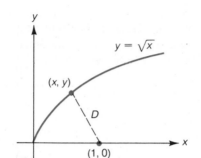

**FIGURE 3**

Since the same $x$-value that makes $D(x)$ as small as it can be will also cause the quantity $[D(x)]^2 = x^2 - x + 1$ to be its smallest, we need only to find the $x$-value that minimizes the quantity $x^2 - x + 1$. Denoting this quantity by $f(x)$ for the moment (so that we have an equation whose graph is a parabola), we have

$$f(x) = x^2 - x + 1$$
$$= \left(x - \frac{1}{2}\right)^2 + \frac{3}{4} \qquad \text{completing the square}$$

From this last equation we conclude that the quantity $x^2 - x + 1$ will be smallest when $x = \frac{1}{2}$.

Now that we have $x$, we want to calculate $y$. (The problem did not ask for $D$; we needn't calculate it.) We have

$$y = \sqrt{x} \qquad \text{and} \qquad x = \frac{1}{2}$$

Therefore,

$$y = \sqrt{\frac{1}{2}} = \frac{\sqrt{1}}{\sqrt{2}} = \frac{1}{\sqrt{2}}$$
$$= \frac{1}{\sqrt{2}} \cdot \frac{\sqrt{2}}{\sqrt{2}} = \frac{\sqrt{2}}{2}$$

Thus, among all points $(x,y)$ on the curve $y = \sqrt{x}$, the point closest to $(1,0)$ is $\left(\frac{1}{2}, \frac{\sqrt{2}}{2}\right)$. This is the required solution.

QUESTION   Why would it not have made sense to ask instead for the point farthest from $(1, 0)$?

**EXAMPLE 5**   Suppose that you have 600 m of fencing with which to build two adjacent rectangular corrals. The two corrals are to share a common fence on one side, as shown in Figure 4. Find the dimensions $x$ and $y$ so that the total enclosed area is as large as possible.

Solution   You have 600 m of fencing to be set up as shown in Figure 4. The question is how you should choose $x$ and $y$ so that the total area is a maximum. Since the total length of fencing is 600 m, we can relate $x$ and $y$ by writing

$$x + x + x + y + y = 600$$
$$3x + 2y = 600$$
$$2y = 600 - 3x$$
$$y = 300 - \frac{3}{2}x \qquad (4)$$

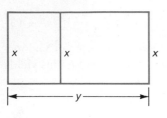

**FIGURE 4**

Letting $A$ denote the total area, we have

$$A = xy$$
$$A(x) = x\left(300 - \frac{3}{2}x\right) \qquad \text{using equation (4) to substitute for } y$$
$$= 300x - \frac{3}{2}x^2$$

This last equation expresses the area as a function of $x$. Note that the graph of this function is a parabola opening downward, so it does make sense to talk about a maximum. We complete the square, as usual, to locate the vertex. As you can check, the result is

$$A(x) = -\frac{3}{2}(x - 100)^2 + 15000$$

Thus the $x$-coordinate of the vertex is 100, and this is the $x$-value that maximizes the area. The corresponding $y$-value can now be calculated using equation (4):

$$y = 300 - \frac{3}{2}x$$
$$= 300 - \frac{3}{2}(100) = 300 - 3(50)$$
$$= 150$$

Thus, by choosing $x$ to be 100 m and $y$ to be 150 m, the total area in Figure 4 will be as large as possible. Incidentally, note that the exact location of the fence dividing the two corrals does not influence our work or final answer.

**EXAMPLE 6**  Suppose that the following function relates the selling price $p$ of an item to the quantity sold, $x$:

$$p = -\frac{1}{3}x + 40 \qquad (p \text{ in dollars})$$

For which value of $x$ will the corresponding revenue be a maximum? What are the maximum revenue and the unit price in this case?

Solution  First, let us recall the formula for revenue given on page 231:

$$R = \text{number of units} \times \text{price per unit}$$

In view of this, we have

$$R = x \cdot p$$
$$R(x) = x\left(-\frac{1}{3}x + 40\right)$$
$$= -\frac{1}{3}x^2 + 40x \tag{5}$$

We want to know which value of $x$ yields the largest revenue $R$. Since the graph of the revenue function in equation (5) is a parabola opening downward, the required $x$-value is the $x$-coordinate of the vertex of the parabola. We complete the square to locate the vertex:

$$R(x) = -\frac{1}{3}(x^2 - 120x \qquad )$$

$$= -\frac{1}{3}(x^2 - 120x + 3600) + 1200$$

$$= -\frac{1}{3}(x - 60)^2 + 1200$$

The vertex is therefore $(60, 1200)$. Consequently, the revenue is a maximum when $x = 60$ items, and this maximum revenue is $\$1200$. For the corresponding unit price, we have

$$p(x) = -\frac{1}{3}x + 40$$

$$p(60) = -\frac{1}{3}(60) + 40 = 20$$

Thus the unit price corresponding to the maximum revenue is $\$20$.

**FIGURE 5**

**EXAMPLE 7**  Figure 5 shows a computer-generated graph of the function

$$f(x) = x^4 - 3x^2$$

From the graph, it is clear that the minimum of the function is slightly less than $-2$. Find the exact value of this minimum and the corresponding $x$-values at which it occurs.

Solution  Although this is not a quadratic function, we can nevertheless use the technique of completing the square in this particular case. (You have already seen one instance of this in Example 9 of Section 5.1.) We have

$$f(x) = x^4 - 3x^2 + \frac{9}{4} - \frac{9}{4}$$

$$= \left(x^2 - \frac{3}{2}\right)^2 - \frac{9}{4}$$

$$= -\frac{9}{4} + \left(x^2 - \frac{3}{2}\right)^2$$

From this last equation we see that $f(x)$ is never less than $-\frac{9}{4}$ (because the quantity being added to $-\frac{9}{4}$ is nonnegative). Furthermore, $f(x)$ does attain the value $-\frac{9}{4}$ when $x^2 = \frac{3}{2}$; that is, when $x = \pm\sqrt{\frac{3}{2}}$. Thus the minimum value of $f$ is $-\frac{9}{4}$ and the corresponding $x$-values are

$$x = \pm\frac{\sqrt{3}}{\sqrt{2}} = \pm\frac{\sqrt{6}}{2} \qquad \text{rationalizing the denominator}$$

NOTE   Use your calculator to approximate these $x$-values, and then check that they are consistent with the graph in Figure 5. ▪

## EXERCISE SET 5.3

### A

1. Two numbers add to 5. What is the largest possible value of their product?

2. Find two numbers adding to 20 such that the sum of their squares is as small as possible.

3. The difference of two numbers is 1. What is the smallest possible value for the sum of their squares?

4. For each quadratic function, state whether it would make sense to look for a highest or a lowest point on the graph. Then determine the coordinates of that point.
   (a) $y = 2x^2 - 8x + 1$
   (b) $y = -3x^2 - 4x - 9$
   (c) $h = -16t^2 + 256t$
   (d) $f(x) = 1 - (x + 1)^2$
   (e) $g(t) = t^2 + 1$
   (f) $f(x) = 1000x^2 - x + 100$

5. Among all rectangles having a perimeter of 25 m, find the dimensions of the one with the largest area.

6. What is the largest possible area for a rectangle whose perimeter is 80 cm?

7. What is the largest possible area for a right triangle in which the sum of the lengths of the two shorter sides is 100 in.?

8. The perimeter of a rectangle is 12 m. Find the dimensions for which the diagonal is as short as possible.

9. Two numbers add to 6.
   (a) Let $T$ denote the sum of the squares of the two numbers. What is the smallest possible value for $T$?
   (b) Let $S$ denote the sum of the first number and the square of the second. What is the smallest possible value for $S$?
   (c) Let $U$ denote the sum of the first number and twice the square of the second number. What is the smallest possible value for $U$?
   (d) Let $V$ denote the sum of the first number and the square of twice the second number. What is the smallest possible value for $V$?

10. Suppose that the height of an object shot straight up is given by $h = 512t - 16t^2$ (here $h$ is in feet and $t$ is in seconds). Find the maximum height and the time at which the object hits the ground.

11. A baseball is thrown straight up, and its height as a function of time is given by the formula $h = -16t^2 + 32t$ (where $h$ is in feet and $t$ is in seconds).
    (a) Find the height of the ball when $t = 1$ and when $t = \frac{3}{2}$.

(b) Find the maximum height of the ball and the time at which that height is attained.

(c) At what times is the height 7 ft?

12. Find the point on the curve $y = \sqrt{x}$ that is nearest to the point $(3,0)$.

13. Which point on the curve $y = \sqrt{x-2} + 1$ is closest to the point $(4,1)$? What is this minimum distance?

14. Find the coordinates of the point on the line $y = 3x + 1$ closest to $(4,0)$.

15. (a) What number exceeds its square by the greatest amount?

(b) What number exceeds twice its square by the greatest amount?

16. Suppose that you have 1800 m of fencing with which to build three adjacent rectangular corrals, as shown in the figure. Find the dimensions so that the total enclosed area is as large as possible.

17. Five hundred feet of fencing is available for a rectangular pasture alongside a river, the river serving as one side of the rectangle (so only three sides require fencing). Find the dimensions yielding the greatest area.

18. Let $A = 3x^2 + 4x - 5$ and $B = x^2 - 4x - 1$. Find the minimum value of $A - B$.

19. Let $R = 0.4x^2 + 10x + 5$ and $C = 0.5x^2 + 2x + 101$. For which value of $x$ is $R - C$ a maximum?

20. Suppose that the revenue generated by selling $x$ units of a certain commodity is given by $R = -\frac{1}{5}x^2 + 200x$. Assume that $R$ is in dollars. What is the maximum revenue possible in this situation?

21. Suppose that the function $p = -\frac{1}{4}x + 30$ relates the selling price $p$ of an item to the number of units $x$ that are sold. Assume that $p$ is in dollars. For which value of $x$ will the corresponding revenue be a maximum? What is this maximum revenue and what is the unit price?

C  22. The action of sunlight on automobile exhaust produces air pollutants known as *photochemical oxidants*. In a study of cross-country runners in Los Angeles, it was shown that running performances can be adversely affected when the oxidant level reaches 0.03 part per million. Suppose that on a given day, the oxidant level $L$ is approximated by the formula

$$L = 0.059t^2 - 0.354t + 0.557 \qquad (0 \le t \le 7)$$

where $t$ is measured in hours, with $t = 0$ correspond-

ing to 12 noon, and $L$ is in parts per million. At what time is the oxidant level $L$ a minimum? At this time, is the oxidant level high enough to affect a runner's performance?

23. (a) Find the smallest possible value of the quantity $x^2 + y^2$ under the restriction that $2x + 3y = 6$.

(b) Find the radius of the circle whose center is at the origin and that is tangent to the line $2x + 3y = 6$. How does this answer relate to your answer in part (a)?

**B**

24. Through a type of chemical reaction known as *autocatalysis*, the human body produces the enzyme trypsin from the enzyme trypsinogen. (Trypsin then breaks down proteins into amino acids, which the body needs for growth.) Let $r$ denote the rate of this chemical reaction in which trypsin is formed from trypsinogen. It has been shown experimentally that $r = kx(a - x)$, where $k$ is a positive constant, $a$ is the initial amount of trypsinogen, and $x$ is the amount of trypsin produced (so $x$ increases as the reaction proceeds). Show that the reaction rate $r$ is a maximum when $x = a/2$. In other words, the speed of the reaction is greatest when the amount of trypsin formed is half the original amount of trypsinogen.

25. (a) Let $x + y = 15$. Find the minimum value of the quantity $x^2 + y^2$.

(b) Let $C$ be a constant and $x + y = C$. Show that the minimum value of $x^2 + y^2$ is $C^2/2$. Then use this result to check your answer in part (a).

26. Suppose that $A$, $B$, and $C$ are positive constants and that $x + y = C$. Show that the minimum value of $Ax^2 + By^2$ occurs when $x = \dfrac{BC}{A + B}$ and $y = \dfrac{AC}{A + B}$.

27. The following figure shows a square inscribed within a unit square. For which value of $x$ is the area of the inscribed square a minimum? What is the minimum area? *Hint:* Denote the lengths of the two segments that make up the base of the unit square by $t$ and $1 - t$. Now use the Pythagorean Theorem and congruent triangles to express $x$ in terms of $t$.

28. (a) Find the coordinates of the point on the line $y = mx + b$ that is closest to the origin. *Answer:* $\left( \dfrac{-mb}{1 + m^2}, \dfrac{b}{1 + m^2} \right)$.

    (b) Show that the perpendicular distance from the origin to the line $y = mx + b$ is $|b|/\sqrt{1 + m^2}$. *Suggestion:* Use the result in part (a).

    (c) Use part (b) to show that the perpendicular distance from the origin to the line
    $$Ax + By + C = 0 \quad \text{is} \quad \frac{|C|}{\sqrt{A^2 + B^2}}$$

29. The point $P$ lies in the first quadrant on the graph of the line $y = 7 - 3x$. From the point $P$, perpendiculars are drawn to both the $x$-axis and the $y$-axis. What is the largest possible area for the rectangle thus formed?

30. Show that the largest possible area for the shaded rectangle shown in the figure is $-b^2/4m$. Then use this to check your answer in Exercise 29.

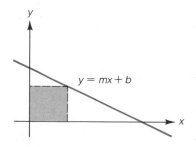

31. Show that the maximum possible area for a rectangle inscribed in a circle of radius $R$ is $2R^2$. *Hint:* Maximize the square of the area.

32. An athletic field with a perimeter of $\frac{1}{4}$ mile consists of a rectangle with a semicircle at each end, as shown in the figure. Find the dimensions $x$ and $r$ that yield the greatest possible area for the rectangular region.

33. A rancher who wishes to fence off a rectangular area finds that the fencing in the east-west direction will require extra reinforcement owing to strong prevailing winds. Because of this, the cost of fencing in the east-west direction will be $12 per (linear) yard, as opposed to a cost of $8 per yard for fencing in the north-south direction. Find the dimensions of the largest possible rectangular area that can be fenced for $4800.

34. Let $f(x) = (x - a)^2 + (x - b)^2 + (x - c)^2$, where $a$, $b$, and $c$ are constants. Show that $f(x)$ will be a minimum when $x$ is the average of $a$, $b$, and $c$.

35. Let $y = a_1(x - x_1)^2 + a_2(x - x_2)^2$, where $a_1$, $a_2$, $x_1$, and $x_2$ are all constants. In addition, suppose that $a_1$ and $a_2$ are both positive. Show that the minimum of this function occurs when
    $$x = \frac{a_1 x_1 + a_2 x_2}{a_1 + a_2}$$

36. Among all rectangles with a given perimeter $P$, find the dimensions of the one with the shortest diagonal.

37. By analyzing sales figures, the economist for a stereo manufacturer knows that 150 units of a top-of-the-line turntable can be sold each month when the price is set at $p = \$200$ per unit. The figures also show that for each $10 hike in price, five fewer units are sold each month.

    (a) Let $x$ denote the number of units sold per month and let $p$ denote the price per unit. Find a linear function relating $p$ and $x$. *Hint:* $\Delta p/\Delta x = 10/(-5) = -2$.

    (b) The revenue $R$ is given by $R = xp$. What is the maximum revenue? At what level should the price be set to achieve this maximum revenue?

38. Let $f(x) = x^2 + px + q$, and suppose that the minimum value of this function is 0. Show that $q = p^2/4$.

39. For which numbers $t$ will the value of the function $f(t) = t^2 - t^4$ be as large as possible?

40. Find the minimum value of the function $f(t) = t^4 - 8t^2$.

41. Among all possible inputs for the function $f(t) = -t^4 + 6t^2 - 6$, which ones yield the largest output?

C

42. A piece of wire 16 in. long is to be cut into two pieces. Let $x$ denote the length of the first piece and $16 - x$ the length of the second. The first piece is to be bent into a circle and the second piece into a square.

(a) Express the total combined area $A$ of the circle and the square as a function of $x$.

(b) For which value of $x$ is the area $A$ a minimum?

(c) Using the $x$-value that you found in part (b), find the ratio of the lengths of the shorter to the longer piece of wire.  *Answer:*  $\pi/4$.

43. Let $f(x) = x^2 + bx + 1$. Find a positive value for $b$ such that the distance from the origin to the vertex of the parabola is as small as possible.

44. The next figure shows a rectangle inscribed in a given triangle of base $b$ and height $h$. Find the ratio of the area of the triangle to the area of the rectangle when the area of the rectangle is a maximum.

45. A Norman window is in the shape of a rectangle surmounted by a semicircle, as shown in the figure. Assume that the perimeter of the window is $P$, a constant. Show that the area of the window is a maximum when both $x$ and $r$ are equal to $\dfrac{P}{\pi + 4}$. Show that this maximum area is $\dfrac{P^2}{2(\pi + 4)}$.

46. A triangle is inscribed in a semicircle of diameter $2R$, as shown in the figure. Show that the smallest possible value for the area of the shaded region is $(\frac{1}{2})(\pi - 2)R^2$. *Hint:* The area of the shaded region is a minimum when the area of the triangle is a maximum. Find the value of $x$ that maximizes the *square* of the area of the triangle. This will be the same $x$ that maximizes the area of the triangle.

47. (a) Complete the following table. Which $x$-$y$ pair in the table yields the smallest sum $x + y$?

| $x$ | 0.5 | 1 | 1.5 | 2 | 2.5 | 3 | 3.5 |
|---|---|---|---|---|---|---|---|
| $y$ | | | | | | | |
| $xy$ | 12 | 12 | 12 | 12 | 12 | 12 | 12 |
| $x + y$ | | | | | | | |

(b) Find two positive numbers whose product is 12 and whose sum is as small as possible.  *Hint:* The quantity that you need to minimize is $x + \dfrac{12}{x}$, where $x > 0$. But

$$x + \frac{12}{x} = \left(\sqrt{x} - \sqrt{\frac{12}{x}}\right)^2 + 2\sqrt{12}.$$

This last expression is minimized when the quantity within parentheses is zero. Why?

☐ (c) Use a calculator to verify that the two numbers obtained in part (b) produce a sum that is smaller than any of the sums obtained in part (a).

48. What is the smallest possible value for the sum of a positive number and its reciprocal?  *Hint:* After setting up the appropriate function, adapt the hint given in Exercise 47(b).

49. Let $f(x) = \dfrac{2x^2 - 4x + 1}{2x}$, and assume that $x > 0$.

☐ (a) Complete the following two tables. In both cases, specify the $x$-value that yields the smallest output. (Round off the outputs in the tables to six decimal places.)

| $x$ | 0.4 | 0.5 | 0.6 | 0.7 | 0.8 | 0.9 |
|---|---|---|---|---|---|---|
| $f(x)$ | | | | | | |

| $x$ | 0.68 | 0.69 | 0.70 | 0.71 | 0.72 | 0.73 |
|---|---|---|---|---|---|---|
| $f(x)$ | | | | | | |

(b) Among all positive inputs for the given function $f$, find the one for which $f(x)$ is a minimum.  *Hint:* Write $f(x)$ as $\left(x + \dfrac{1}{2x}\right) - 2$, and then adapt the hint given in Exercise 47(b).

C (c) Use a calculator to verify that the *x*-value obtained in part (b) produces an output that is smaller than any of the outputs obtained in part (a).

50. A line with a positive slope *m* passes through the point $(-2, 1)$.

(a) Show that the area of the triangle bounded by this line and the axes is $\left(2m + \dfrac{1}{2m}\right) + 2$.

(b) For which value of *m* will the area be a minimum? *Hint*: Adapt the hint given in Exercise 47(b).

## POLYNOMIAL FUNCTIONS

**5.4**

We can rephrase the definitions of linear and quadratic functions using the terminology for polynomials given in Section 1.8.

A function $f$ is **linear** if $f(x)$ is a **polynomial of degree 1**:

$$f(x) = a_1x + a_0 \tag{1}$$

A function $f$ is **quadratic** if $f(x)$ is a **polynomial of degree 2**:

$$f(x) = a_2x^2 + a_1x + a_0 \tag{2}$$

In view of equation (1), linear functions are sometimes called **polynomial functions of degree 1**. Similarly, in view of equation (2), quadratic functions are **polynomial functions of degree 2**. More generally, by a **polynomial function of degree n**, we mean a function defined by an equation of the form

$$f(x) = a_nx^n + a_{n-1}x^{n-1} + \cdots + a_1x + a_0 \tag{3}$$

where $n$ is a nonnegative integer and $a_n \neq 0$. Throughout the remainder of this chapter, the coefficients $a_k$ will always be real numbers.

In principle, we can obtain the graph of any polynomial function by setting up a table and plotting a sufficient number of points. Indeed, this is just the way a computer equipped with a curve plotter operates. However, to *understand* why the graphs look as they do, we want to discuss some additional methods for graphing polynomial functions.

There are three facts that we shall need. By way of example, look at the graphs of the polynomial functions in Figure 1. First, notice that both graphs are unbroken, smooth curves, with no "corners." As is shown in calculus, this

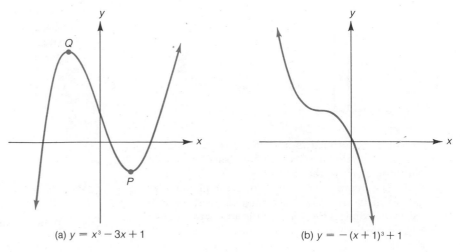

(a) $y = x^3 - 3x + 1$    (b) $y = -(x + 1)^3 + 1$

**FIGURE 1**

is true for the graph of every polynomial function. By way of contrast, the graphs in Figures 2 and 3 cannot represent polynomial functions. The graph in Figure 2 has a break in it, and the graph in Figure 3 has a corner, or **cusp**.

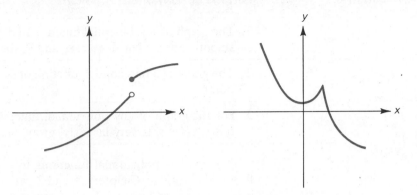

**FIGURE 2**  Since the graph has a break, it cannot represent a polynomial function.

**FIGURE 3**  Since the graph has a corner, it cannot represent a polynomial function.

Next, look again at the graph in Figure 1(a). The points $P$ and $Q$ are called **turning points**. These are points where the graph changes from rising to falling, or vice versa. It is a fact (proved in calculus) that the graph of a polynomial function of degree $n$ has *at most* $n - 1$ turning points. For instance, in Figure 1(a), there are two turning points, while the degree of the polynomial is 3. However, as Figure 1(b) indicates, we needn't have any turning points at all.

A third property concerning the graphs of polynomial functions is also evident in Figure 1. As $|x|$ gets very large, $|y|$ grows very large. It's easy to see from the equation $y = x^3 - 3x + 1$ why this is so. First, rewrite this equation

$$y = x^3 \left( 1 - \frac{3}{x^2} + \frac{1}{x^3} \right)$$

Now, when $|x|$ is very large, the quantity in parentheses is close to 1 (because the terms $3/x^2$ and $1/x^3$ are then close to zero). So in this case, we have the approximation $y \approx x^3$, which clearly shows that when $|x|$ is very large, $|y|$ is larger still. In view of this property, neither of the graphs in Figure 4 can

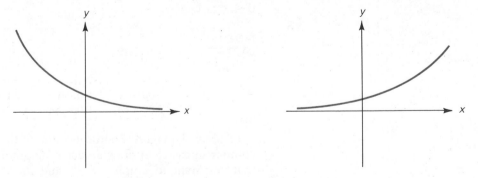

**FIGURE 4**  These two graphs cannot represent polynomial functions.

represent polynomial functions. For reference now, let us summarize our remarks up to this point in the box that follows.

| PROPERTY SUMMARY | Graphs of Polynomial Functions |
|---|---|
| | 1. The graph of a polynomial function of degree 2 or greater is an unbroken smooth curve. (For degrees 1 and 0, the graph is a line.) |
| | 2. The graph of a polynomial function of degree $n$ has at most $n - 1$ turning points. |
| | 3. For the graph of any polynomial function (other than a constant function), as $|x|$ gets very large, $|y|$ grows very large. |

The simplest polynomial functions to graph are those of the form $y = x^n$. From our work in Chapters 3 and 4 we already know about the graphs of $y = x$, $y = x^2$, and $y = x^3$. For reference, these are shown in Figure 5.

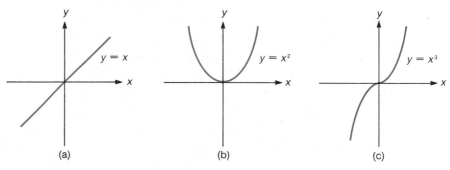

(a)                    (b)                    (c)

**FIGURE 5**

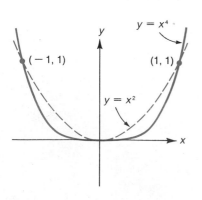

**FIGURE 6**

The graph of $y = x^n$, when $n$ is greater than 3, resembles the graph of $y = x^2$ or $y = x^3$, depending on whether $n$ is even or odd. Consider, for instance, the graph of $y = x^4$, shown in Figure 6 along with the graph of $y = x^2$. Just as with $y = x^2$, the graph of $y = x^4$ is a symmetric, U-shaped curve passing through the three points $(0,0)$, $(1,1)$, and $(-1,1)$. However, in the interval $(-1,1)$, the graph of $y = x^4$ is flatter than that of $y = x^2$. Similarly, the graph of $y = x^6$ in this interval would be flatter still. The data in Table 1 show why this is true. Figure 7 displays the graphs of $y = x^2$, $y = x^4$, and $y = x^6$ for the interval $0 \leq x \leq 1$.

**TABLE 1**

| $x$ | 0.2 | 0.4 | 0.6 | 0.8 | 1.0 |
|---|---|---|---|---|---|
| $x^2$ | 0.04 | 0.16 | 0.36 | 0.64 | 1.0 |
| $x^4$ | 0.0016 | 0.0256 | 0.1296 | 0.4096 | 1.0 |
| $x^6$ | 0.000064 | 0.004096 | 0.046656 | 0.262144 | 1.0 |

Incidentally, Figure 7 indicates one of the practical difficulties you may encounter in trying to draw an accurate graph of $y = x^n$. Suppose, for instance, that you want to graph $y = x^6$, and the thickness of the lines you draw is

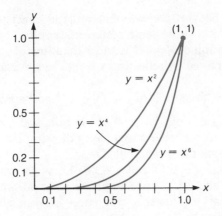

**FIGURE 7**

0.01 cm. Also, suppose that you use the same scale on both axes, taking the common unit to be 1 cm. Then in the first quadrant, your graph of $y = x^6$ will be indistinguishable from the $x$-axis when $x^6 < 0.01$, or $x < \sqrt[6]{0.01} \approx 0.46$ cm (using a calculator). This explains why sections of the graphs in Figure 7 appear horizontal.

For $|x| > 1$, the graph of $y = x^4$ rises more rapidly than that of $y = x^2$. Similarly, the graph of $y = x^6$ rises still more rapidly. This is shown in Figures 8(a) and (b). (Note the different scales used on the $y$-axes in the two figures.)

(a)

(b)

**FIGURE 8**

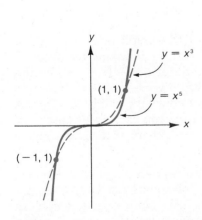

**FIGURE 9**

As mentioned earlier, when $n$ is odd, the graph of $y = x^n$ resembles that of $y = x^3$. In Figure 9 we compare the graphs of $y = x^3$ and $y = x^5$. Notice that both curves pass through $(0,0)$, $(1,1)$, and $(-1,-1)$. For reasons similar to those explained for even $n$, the graph of $y = x^5$ is flatter than that of $y = x^3$ in the interval $-1 < x < 1$, and the graph of $y = x^7$ would be flatter still. For $|x| > 1$, the graph of $y = x^5$ is steeper than that of $y = x^3$, and $y = x^7$ would be steeper still.

In Section 5.1, we observed the effect of the constant $a$ on the graph of $y = ax^2$. Those same comments apply to the graph of $y = ax^n$. For instance, the graph of $y = \frac{1}{2}x^4$ is wider than that of $y = x^4$, while the graph of $y = -\frac{1}{2}x^4$ is obtained by reflecting $y = \frac{1}{2}x^4$ in the $x$-axis.

**EXAMPLE 1**    Sketch the graph of $y = (x + 2)^5$ and specify the $y$-intercept.

**Solution**    The graph of $y = (x + 2)^5$ is obtained by moving the graph of $y = x^5$ two units to the left. As a guide to drawing the curve, we recall that $y = x^5$ passes through the points $(1, 1)$ and $(-1, -1)$. Thus, $y = (x + 2)^5$ must pass through $(-1, 1)$ and $(-3, -1)$, as shown in Figure 10. Although the curve rises and falls very sharply, it is important to realize that it is never really vertical. For instance, the curve eventually crosses the $y$-axis. To find the $y$-intercept, we set $x = 0$ to obtain $y = 2^5 = 32$. Thus, the $y$-intercept is 32.

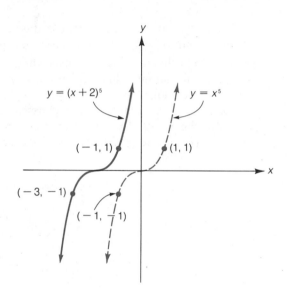

**FIGURE 10**

**EXAMPLE 2**    Graph the function $y = -2(x - 3)^4$.

**Solution**    Begin with the graph of $y = x^4$. By translating this graph three units to the right, we obtain the graph of $y = (x - 3)^4$. Next, by reflecting the graph of $y = (x - 3)^4$ in the $x$-axis, we obtain the graph of $y = -(x - 3)^4$. Finally, the graph of $y = -2(x - 3)^4$ will resemble that of $y = -(x - 3)^4$, but it will be narrower. As a guide as to just how narrow to draw the graph, we compute the $y$-values corresponding to $x = 4$ and $x = 2$. As you can readily check, the calculations show that the points $(4, -2)$ and $(2, -2)$ are on the graph. Figure 11 shows the required graph.

In general, the most efficient ways to graph a given polynomial function involve methods of calculus. However, if the polynomial can be written as a product of linear factors with real coefficients, then a fairly accurate graph can be obtained without the use of calculus. For example, let us graph the function

$$f(x) = (x + 2)(x - 1)(x - 3)$$

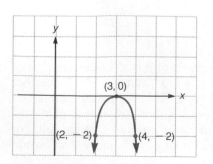

**FIGURE 11**   $y = -2(x - 3)^4$

**FIGURE 12** Near $x = -2$, the graph of $f(x) = (x + 2)(x - 1)(x - 3)$ is nearly vertical; it resembles the line $y = 15x + 30$.

**FIGURE 13** The graph of $f$ near its $x$-intercepts.

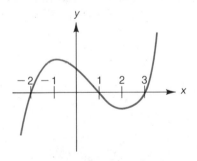

**FIGURE 14** Rough sketch of $f(x) = (x + 2)(x - 1)(x - 3)$.

By inspection, we see that $f(x) = 0$ when $x = -2, x = 1$, and $x = 3$. These are the $x$-intercepts of the graph. Also, note that the $y$-intercept is 6. (Why?) Next, we want to find out what the graph looks like in the immediate vicinity of each $x$-intercept. First, suppose that $x$ is very close to $-2$. We have

$$f(x) = (x + 2)\underbrace{(x - 1)}\underbrace{(x - 3)}$$

If $x$ is close to $-2$, this is close to $-3$.

If $x$ is close to $-2$, this is close to $-5$.

Thus, if $x$ is very close to $-2$, we have

$$\begin{aligned} f(x) &\approx (x + 2)(-3)(-5) \\ &\approx 15(x + 2) \\ &\approx 15x + 30 \end{aligned}$$

(Notice the technique used to obtain this approximation: We retained the factor corresponding to the intercept $-2$ and we estimated the remaining factors.) We conclude from this that when $x$ is very close to $-2$, the graph of $f$ resembles the line $y = 15x + 30$. In particular, the graph is quite steep and rises to the right (see Figure 12).

Next, we carry out similar analyses to see how the graph of $f$ looks near its other $x$-intercepts:

$x$ NEAR 1

$$\begin{aligned} f(x) &= (x + 2)(x - 1)(x - 3) \\ &\approx (1 + 2)(x - 1)(1 - 3) \\ &\approx (3)(x - 1)(-2) \\ &\approx -6x + 6 \end{aligned}$$

$x$ NEAR 3

$$\begin{aligned} f(x) &= (x + 2)(x - 1)(x - 3) \\ &\approx (3 + 2)(3 - 1)(x - 3) \\ &\approx (5)(2)(x - 3) \\ &\approx 10x - 30 \end{aligned}$$

Thus, for $x$ very close to 1, the graph of $f$ resembles the line $y = -6x + 6$, while for $x$ very close to 3, the graph resembles the line $y = 10x - 30$. Figure 13 summarizes our information about the behavior of the graph of $f$ near its $x$-intercepts. With this as a guide, we can then draw a rough sketch of the graph of $f$, as shown in Figure 14.

Notice that to draw a smooth curve satisfying the conditions of Figure 13, we need at least two turning points; one between $x = -2$ and $x = 1$, and one between $x = 1$ and $x = 3$. On the other hand, since the degree of $f(x)$ is 3, there can be no more than two turning points. This is why Figure 14 shows exactly two turning points. While the precise location of the turning points is a matter for calculus, we can nevertheless improve upon the sketch in Figure 14 by computing $f(x)$ for several specific values of $x$. Some reasonable choices here are $x = -1$ and $x = 2$. (We've already noted that when $x = 0$, we have $y = 6$.) As you can check, the resulting calculations show that $(-1, 8)$ and $(2, -4)$ are on the graph. We can now sketch the graph of $f(x) = (x + 2)(x - 1)(x - 3)$ as shown in Figure 15 on page 254. ▪▪

The next example further illustrates our technique for graphing those polynomial functions that can be expressed as a product of linear factors with real coefficients.

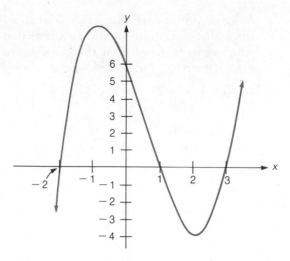

FIGURE 15  $f(x) = (x + 2)(x - 1)(x - 3)$

**EXAMPLE 3**  Sketch the graph of $y = x^3(x + 1)(x - 2)$.

**Solution**  From the given equation, we see that the y-intercept is 0, and the x-intercepts are $x = 0, x = -1$, and $x = 2$. Now, just as in the previous example, we want to see how the graph looks in the immediate vicinity of the x-intercepts.

| x NEAR 0 | x NEAR −1 | x NEAR 2 |
|---|---|---|
| $y \approx x^3(0 + 1)(0 - 2)$ | $y \approx (-1)^3(x + 1)(-1 - 2)$ | $y \approx 2^3(2 + 1)(x - 2)$ |
| $\approx -2x^3$ | $\approx 3(x + 1)$ | $\approx 24(x - 2)$ |
| | $\approx 3x + 3$ | $\approx 24x - 48$ |

These results are displayed in Figure 16. With this as a guide, we can draw a rough sketch of the required graph, as shown in Figure 17. To obtain a slightly

FIGURE 16  The graph of $y = x^3(x + 1)(x - 2)$ near its x-intercepts.

FIGURE 17  Rough sketch of $y = x^3(x + 1)(x - 2)$.

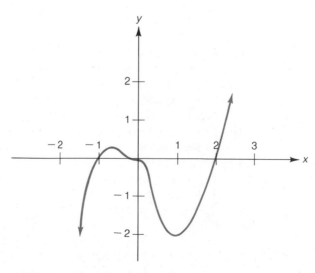

FIGURE 18  $y = x^3(x + 1)(x - 2)$

more accurate sketch, we compute the $y$-values corresponding to $x = -\frac{1}{2}$ and $x = 1$. As you can check, the curve passes through the points $(-\frac{1}{2}, \frac{5}{32})$ and $(1, -2)$. This additional information is used to draw the graph shown in Figure 18.

# EXERCISE SET 5.4

## A

*In Exercises 1–6, indicate whether the graph could possibly represent a polynomial function.*

1.

2.

3.

4.

5.

6.

*In Exercises 7–18, sketch the graph of each function and specify all x- and y-intercepts.*

7. $y = (x - 2)^2 + 1$

8. $y = -3x^4$

9. $y = -(x-1)^4$

10. $y = -(x+2)^3$

11. $y = (x-4)^3 - 2$

12. $y = -(x-4)^3 - 2$

13. $y = -2(x+5)^4$

14. $y = -2x^4 + 5$

15. $y = \frac{1}{2}(x+1)^5$

16. $y = \frac{1}{2}x^5 + 1$

17. $y = -(x-1)^3 - 1$

18. $y = x^8$

*In Exercises 19–32, first sketch the graph in the immediate vicinity of each of the x-intercepts. Then sketch the graph.*

19. $y = (x-2)(x-1)(x+1)$

20. $y = (x-3)(x+2)(x+1)$

21. $y = 2x(x+1)(x+3)$

22. $y = -x^2(x+2)$

23. $y = x^3(x+2)$

24. $y = (x-1)(x-4)^2$

25. $y = 2(x-1)(x-4)^3$

26. $y = (x-1)^2(x-4)^2$

27. $y = (x+1)^2(x-1)(x-3)$

28. $y = x^2(x-4)(x+2)$

29. $y = -x^3(x-4)(x+2)$

30. $y = 4(x-2)^2(x+2)^3$

31. $y = -4x(x-2)^2(x+2)^3$

32. $y = -3x^3(x+1)^4$

**B**

*In Exercises 33–38, six functions are defined as follows:*

$$f(x) = x \qquad g(x) = x^2 \qquad h(x) = x^3$$
$$F(x) = x^4 \qquad G(x) = x^5 \qquad H(x) = x^6$$

*Refer also to the following figure.*

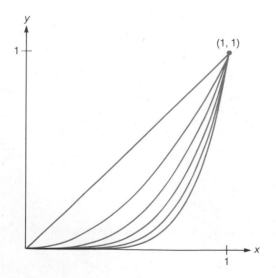

33. The six graphs in the figure are the graphs of the six given functions for the interval $[0, 1]$, but the graphs are not labeled. Which is which?

[c] 34. For which x-values in $[0, 1]$ will the graph of $g$ lie strictly below the horizontal line $y = 0.1$? Use a calculator to evaluate your answer. Round off the result to two significant figures.

[c] 35. Follow Exercise 34, using the function $H$ instead of $g$.

36. Find a number $t$ in $[0, 1]$ such that the vertical distance between $f(t)$ and $g(t)$ is $\frac{1}{4}$.

37. Is there a number $t$ in $[0, 1]$ such that the vertical distance between $g(t)$ and $F(t)$ is 0.26?

[c] 38. Find all numbers $t$ in $[0, 1]$ such that $F(t) = G(t) + H(t)$.

39. Do the graphs of $y = x$ and $y = \frac{1}{100}x^2$ intersect anywhere other than at the origin?

*In Exercises 40–45, graph each function. (Use the factoring techniques from Section 1.9 to factor the right-hand side of each equation.)*

40. $y = x^3 - 2x^2$

41. $y = 2x^3 - x^4$

42. $y = x^3 - 9x$

43. $y = x^3 + x^2 - 2x$

44. $y = x^3 + 3x^2 + 3x + 1$

45. $y = 4x^2 - x^4$

# GRAPHS OF RATIONAL FUNCTIONS

**5.5**

*We shall not attempt to explain the numerous situations in life sciences where the study of such graphs is important nor the chemical reactions which give rise to the rational functions $r(x)$.... Let it suffice to mention that, to the experimental biochemist, theoretical results concerning the shapes of rational functions are of considerable interest.*

W. G. BARDSLEY and R. M. W. WOOD in the article "Critical Points and Sigmoidicity of Positive Rational Functions," in *The American Mathematical Monthly*, **92** (1985) 37–42.

After the polynomial functions, the next simplest functions are the **rational functions**. These are functions defined by equations of the form

$$y = \frac{f(x)}{g(x)}$$

where $f(x)$ and $g(x)$ are polynomials. In general, throughout this section, when we write a function such as $y = f(x)/g(x)$, we assume that $f(x)$ and $g(x)$ contain no common factors (other than constants). [Exercises 34 and 35 ask you to consider several cases in which $f(x)$ and $g(x)$ do contain common factors.] Also, for each of the examples that we discuss, the degree of $f(x)$ is less than or equal to the degree of $g(x)$. [A case in which the degree of $f(x)$ exceeds the degree of $g(x)$ is developed in the exercises.]

**EXAMPLE 1**   Specify the domain of the rational function defined by

$$y = \frac{3x - 2}{x^2 - 1}$$

Also, find the $x$- and $y$-intercepts (if any) for the graph of this function.

**Solution**   By factoring the denominator, we can rewrite the given equation as

$$y = \frac{3x - 2}{(x - 1)(x + 1)}$$

Since the denominator is zero when $x = 1$ and when $x = -1$, it follows that the domain of the given function consists of all real numbers except 1 and $-1$. To determine the $x$-intercepts of the graph, we set $y = 0$ (as usual) in the given equation to obtain

$$\frac{3x - 2}{x^2 - 1} = 0 \qquad (x \neq 1, \ x \neq -1)$$

$$3x - 2 = 0 \qquad \text{multiplying both sides of the equation by the nonzero quantity } x^2 - 1, \text{ since } x \neq \pm 1$$

$$x = \frac{2}{3}$$

Thus, the only $x$-intercept is $x = \frac{2}{3}$. For the $y$-intercept, we set $x = 0$ (as usual) in the given equation. As you can readily check, this yields $y = 2$. In summary then, the $x$- and $y$-intercepts are $\frac{2}{3}$ and 2, respectively.

In the box that follows, we summarize the ideas used in Example 1 for determining the domain and $x$-intercepts of a rational function.

Let $R$ be a rational function defined by

$$R(x) = \frac{f(x)}{g(x)}$$

where $f(x)$ and $g(x)$ are polynomials with no common factors (other than constants). Then the domain of $R$ consists of all real numbers for which $g(x) \neq 0$; the $x$-intercepts (if any) are the real solutions of the equation $f(x) = 0$.

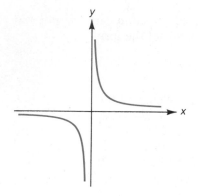

FIGURE 1   The graph of $y = \dfrac{1}{x}$

You already know about the graph of one rational function, because in Chapter 3 we graphed $y = 1/x$. For convenience, the graph is shown again in Figure 1. The graph in Figure 1 differs from the graph of every polynomial function in two important aspects. First, the graph has a break in it; it is composed of two distinct pieces or **branches**. (Recall from Section 5.4 that the graph of a polynomial function never has a break in it.) In general, the graph of a rational function has one more branch than the number of distinct real values for which the denominator is zero. For instance, referring to the function in Example 1, we can expect the graph of $y = \dfrac{3x - 2}{x^2 - 1}$ to have three branches.

The second way that the graph in Figure 1 differs from that of a polynomial function has to do with the *asymptotes*. Recall that a line is an **asymptote** for a curve if the distance between the line and the curve approaches zero as we move out farther and farther along the line. Thus, the $x$-axis is a horizontal asymptote for the graph in Figure 1, while the $y$-axis is a vertical asymptote. It can be shown that the graph of a polynomial function never has an asymptote. Figure 2 displays additional examples of curves with asymptotes.

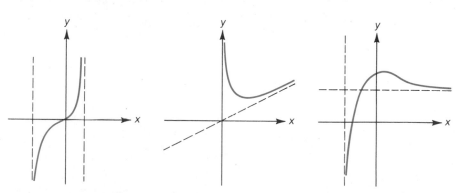

FIGURE 2   Curves with asymptotes

FIGURE 3

If $k$ is a positive constant, the graph of $y = k/x$ resembles that of $y = 1/x$. Figure 3 shows the graphs of $y = 1/x$ and $y = 4/x$. Once we know about the graph of $y = k/x$, we can graph any rational function of the form

$$y = \frac{ax + b}{cx + d}$$

The next three examples show how this is done. These examples make use of the techniques of translation and reflection developed in Chapter 4.

**EXAMPLE 2**   Graph $y = \dfrac{1}{x-1}$.

Solution   By translating the graph of $y = 1/x$ one unit to the right, we obtain the graph of $y = 1/(x-1)$ shown in Figure 4. Notice that the horizontal asymptote is unchanged by the translation. However, the vertical asymptote moves one unit to the right. (The $y$-intercept is $-1$; why?)

FIGURE 4   $y = \dfrac{1}{x-1}$

**EXAMPLE 3**   Graph $y = \dfrac{2}{x-1}$ and $y = \dfrac{-2}{x-1}$.

Solution   The graph of $y = 2/(x-1)$ has the same basic shape and location as the graph of $y = 1/(x-1)$ shown in Figure 4. As a further guide to sketching $y = 2/(x-1)$, we can pick several convenient $x$-values on either side of the asymptote $x = 1$ and then compute the corresponding $y$-values. After doing this, we obtain the graph shown in Figure 5(a). By reflecting this graph in the $x$-axis, we obtain the graph of $y = -2/(x-1)$, which is shown in Figure 5(b).

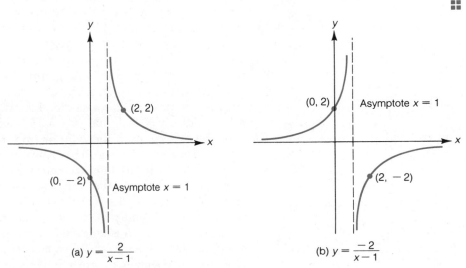

(a) $y = \dfrac{2}{x-1}$   (b) $y = \dfrac{-2}{x-1}$

FIGURE 5

**EXAMPLE 4**   Graph $y = \dfrac{4x - 2}{x - 1}$.

**Solution**   First, as you can readily check, the $x$- and $y$-intercepts are $\frac{1}{2}$ and 2, respectively. Next, using long division, we find that

$$\frac{4x - 2}{x - 1} = 4 + \frac{2}{x - 1}$$

We conclude that the required graph can be obtained by moving the graph of $y = 2/(x - 1)$ up four units in the $y$-direction (see Figure 6). Notice that the vertical asymptote is still $x = 1$, but the horizontal asymptote is now $y = 4$ instead of the $x$-axis.

**FIGURE 6**   $y = \dfrac{4x - 2}{x - 1}$

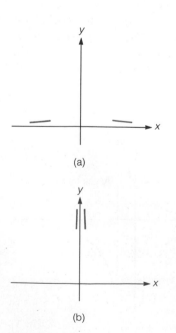

**FIGURE 7**

Now let us look at rational functions of the form $y = 1/x^n$. First consider $y = 1/x^2$. As with $y = 1/x$, the domain consists of all real numbers except $x = 0$. For $x \neq 0$, the quantity $1/x^2$ is always positive. This means that the graph will always lie above the $x$-axis. As $|x|$ becomes very large, the quantity $1/x^2$ approaches zero; this is true whether $x$ itself is positive or negative. So when $|x|$ is very large, we expect the graph of $y = 1/x^2$ to look as shown in Figure 7(a). On the other hand, when $x$ is a very small fraction close to zero, either negative or positive, the quantity $1/x^2$ is very large. For instance, if $x = \frac{1}{10}$, we find that $y = 100$, and if $x = \frac{1}{100}$, we find that $y = 10,000$. Thus, as $x$ approaches zero, from the right or from the left, the graph of $y = 1/x^2$ must look as shown in Figure 7(b). Now, by plotting several points and taking Figures 7(a) and (b) into account, we obtain the graph of $y = 1/x^2$, shown in Figure 8. Also, by following a similar line of reasoning, we find that the graph of $y = 1/x^3$ looks as shown in Figure 9. Note that $y = 1/x^2$ is symmetric about the $y$-axis, and $y = 1/x^3$ is symmetric about the origin.

In general, when $n$ is an even integer greater than 2, the graph of $y = 1/x^n$ resembles that of $y = 1/x^2$. When $n$ is an odd integer greater than 3, the graph of $y = 1/x^n$ resembles that of $y = 1/x^3$.

FIGURE 8   $y = \dfrac{1}{x^2}$          FIGURE 9   $y = \dfrac{1}{x^3}$

**EXAMPLE 5**   Graph $y = \dfrac{-1}{(x + 3)^2}$.

Solution   Refer to Figure 10. Begin with the graph of $y = 1/x^2$. By moving the graph three units to the left, we obtain the graph of $y = 1/(x + 3)^2$. Then by reflecting the graph of $y = 1/(x + 3)^2$ in the $x$-axis, we get the graph of $y = -1/(x + 3)^2$. *Question:*   What are the $y$-intercepts in Figures 10(b) and (c)?

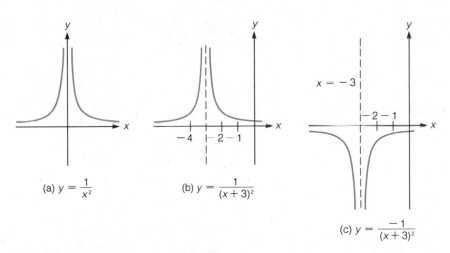

(a) $y = \dfrac{1}{x^2}$          (b) $y = \dfrac{1}{(x + 3)^2}$

(c) $y = \dfrac{-1}{(x + 3)^2}$

FIGURE 10

**EXAMPLE 6**   Graph $y = \dfrac{4}{(x - 2)^3}$.

Solution   Moving the graph of $y = 1/x^3$ two units to the right gives us the graph of $y = 1/(x - 2)^3$. The graph of $y = 4/(x - 2)^3$ will have the same basic shape and location. As a further guide to sketching the required graph, we can pick several convenient $x$-values near the asymptote $x = 2$ and compute the corresponding $y$-values. Using $x = 0$, $x = 1$, and $x = 3$, we find that the points

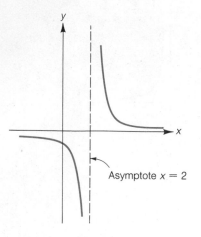

**FIGURE 11** $y = \dfrac{4}{(x-2)^3}$

$(0, -\frac{1}{2})$, $(1, -4)$, and $(3, 4)$ are on the graph. With this information, the graph can be sketched as in Figure 11.

In the previous section, you learned a technique for graphing those polynomial functions that could be expressed as a product of linear factors. That technique can also be used to graph rational functions in which both the numerator and denominator can be expressed as products of linear factors. In addition to investigating the behavior of the function near its $x$-intercepts, now we will also want to see how the function looks near its asymptotes. As the previous examples indicate, the vertical asymptotes of $y = f(x)/g(x)$ are known once we find the values of $x$ for which $g(x) = 0$. For instance, the vertical asymptotes of $y = 3x^2/(x^2 - 1)$ are the lines $x = 1$ and $x = -1$. To find the horizontal asymptotes, we divide the numerator and denominator by the highest power of $x$ that appears. (You'll see the reason for doing this in a moment.) In the case of $y = 3x^2/(x^2 - 1)$, we divide numerator and denominator by $x^2$ to obtain

$$y = \frac{3}{1 - \dfrac{1}{x^2}}$$

Now, when $|x|$ grows very large, $1/x^2$ approaches zero. We therefore have

$$y \approx \frac{3}{1 - 0} \qquad \text{when } |x| \text{ is very large}$$

This tells us that the line $y = 3$ is a horizontal asymptote for the graph of $y = 3x^2/(x^2 - 1)$. You'll see another example of this type of calculation in Example 7.

**EXAMPLE 7**  Graph the function $y = \dfrac{(x - 3)(x + 2)}{(x + 1)(x - 2)}$.

**Solution**  By inspection, we see that the $x$-intercepts are $x = 3$ and $x = -2$, the $y$-intercept is 3, and the vertical asymptotes are the lines $x = -1$ and $x = 2$. To find the horizontal asymptote, we write

$$y = \frac{x^2 - x - 6}{x^2 - x - 2}$$

$$= \frac{1 - \dfrac{1}{x} - \dfrac{6}{x^2}}{1 - \dfrac{1}{x} - \dfrac{2}{x^2}} \approx \frac{1 - 0 - 0}{1 - 0 - 0} \qquad \text{when } |x| \text{ is large}$$

The horizontal asymptote is therefore the line $y = 1$.

Now we want to see how the graph looks in the immediate vicinity of the $x$-intercepts and the vertical asymptotes. To do this, we use the approximation technique explained in the previous section for polynomial functions. Let's start with the $x$-intercept $x = 3$. We have, for $x$ near 3,

$$y = \frac{(x - 3)(x + 2)}{(x + 1)(x - 2)} \approx \frac{(x - 3)(3 + 2)}{(3 + 1)(3 - 2)} = \frac{5}{4}(x - 3)$$

So, in the immediate vicinity of the $x$-intercept $x = 3$, the required graph will closely resemble the graph of the line $y = \frac{5}{4}x - \frac{15}{4}$. The remaining calculations for approximating the graph near the other $x$-intercept and near the two vertical asymptotes are carried out in a similar manner. As Exercise 33 asks you to verify, the results are as follows:

$$x \text{ near } 3: \qquad y \approx \frac{5}{4}x - \frac{15}{4}$$

$$x \text{ near } -2: \qquad y \approx -\frac{5}{4}x - \frac{5}{2}$$

$$x \text{ near } -1: \qquad y \approx \frac{\frac{4}{3}}{x + 1}$$

$$x \text{ near } 2: \qquad y \approx \frac{-\frac{4}{3}}{x - 2}$$

We can summarize these results as follows. As the graph passes through the points $(3, 0)$ and $(-2, 0)$, it resembles the lines $y = \frac{5}{4}x - \frac{15}{4}$ and $y = -\frac{5}{4}x - \frac{5}{2}$, respectively. Near the vertical asymptote $x = -1$, the graph has the same basic shape as $y = 1/(x + 1)$. Near the vertical asymptote $x = 2$, the graph has the same basic shape as $y = -1/(x - 2)$ (see Figure 12). Next we need to see how the graph approaches the horizontal asymptote $y = 1$. When $|x|$ is large and $x > 0$, we find that $y$ is less than 1. (Check this for yourself using $x = 10$ or $x = 100$.) This means that to the right of the origin, the graph approaches the asymptote $y = 1$ from below. Similarly, when $|x|$ is large and $x < 0$, we find that $y$ is again less than 1. This means that to the left of the origin, the graph approaches the asymptote $y = 1$ from below. In Figure 13, we summarize what we have discovered up to this point about the graph. Then, using Figure 13 as a guide, we can sketch the required

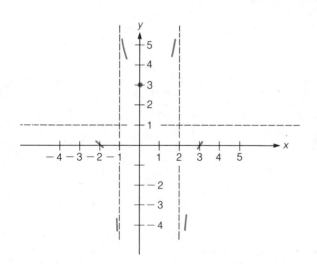

**FIGURE 12**  The graph of $y = \dfrac{(x - 3)(x + 2)}{(x + 1)(x - 2)}$
near its $x$-intercepts and vertical asymptotes.

**FIGURE 13**  The graph of $y = \dfrac{(x - 3)(x + 2)}{(x + 1)(x - 2)}$
near its $x$-intercepts and horizontal and vertical asymptotes.

graph as shown in Figure 14. *Note:* A general method for finding the coordinates of the lowest point on the middle branch of the graph is studied in calculus. Those coordinates can also be found, however, by applying algebraic techniques. (See Exercise 36.)

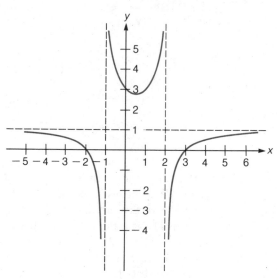

FIGURE 14 $\quad y = \dfrac{(x - 3)(x + 2)}{(x + 1)(x - 2)}$

## EXERCISE SET 5.5

### A

In Exercises 1–6, find the domain and the x- and y-intercepts for each rational function.

1. $y = \dfrac{3x + 15}{4x - 12}$

2. $y = \dfrac{(x + 6)(x + 4)}{(x - 1)^2}$

3. $y = \dfrac{x^2 - 8x - 9}{x^2 - x - 6}$

4. $y = \dfrac{2x^2 + x - 5}{x^2 + 1}$

5. $y = \dfrac{(x^2 - 4)(x^3 - 1)}{x^6}$

6. $y = \dfrac{x^5 - 2x^4 - 9x + 18}{8x^3 + 2x^2 - 3x}$

In Exercises 7–28, sketch the graph of each rational function. Specify the intercepts and the asymptotes.

7. $y = \dfrac{1}{x + 4}$

8. $y = \dfrac{-1}{x + 4}$

9. $y = \dfrac{3}{x + 2}$

10. $y = \dfrac{-3}{x + 2}$

11. $y = \dfrac{-2}{x - 3}$

12. $y = \dfrac{x - 1}{x + 1}$

13. $y = \dfrac{x-3}{x-1}$

14. $y = \dfrac{2x}{x+3}$

15. $y = \dfrac{4x-2}{2x+1}$

16. $y = \dfrac{3x+2}{x-3}$

17. $y = \dfrac{1}{(x-2)^2}$

18. $y = \dfrac{-1}{(x-2)^2}$

19. $y = \dfrac{3}{(x+1)^2}$

20. $y = \dfrac{-3}{(x+1)^2}$

21. $y = \dfrac{1}{(x+2)^3}$

22. $y = \dfrac{-1}{(x+2)^3}$

23. $y = \dfrac{-4}{(x+5)^3}$

24. $y = \dfrac{x}{(x+1)(x-3)}$

25. $y = \dfrac{-x}{(x+2)(x-2)}$

26. $y = \dfrac{2x}{(x+1)^2}$

27. $y = \dfrac{x}{(x-1)(x+3)}$

28. $y = \dfrac{x^2+1}{x^2-1}$

**B**

In Exercises 29–32, graph the functions. Note: Each graph intersects its horizontal asymptote once. Find the intersection point before sketching your final graph.

29. $y = \dfrac{(x-4)(x+2)}{(x-1)(x-3)}$

30. $y = \dfrac{(x-1)(x-3)}{(x+1)^2}$

31. $y = \dfrac{(x+1)^2}{(x-1)(x-3)}$

32. $y = \dfrac{2x^2-3x-2}{x^2-3x-4}$

33. (This exercise refers to Example 7.) Let

$$y = \dfrac{(x-3)(x+2)}{(x+1)(x-2)}$$

Verify each of the following approximations.
(a) When $x$ is close to $-2$, then $y \approx -\frac{5}{4}x - \frac{5}{2}$.

(b) When $x$ is close to $-1$, then $y \approx \dfrac{4/3}{x+1}$.

(c) When $x$ is close to 2, then $y \approx \dfrac{-4/3}{x-2}$.

In Exercises 34 and 35, graph the functions. Notice in each case that the numerator and denominator contain at least one common factor. Thus, you can simplify each quotient; but don't lose track of the domain of the function as it was initially defined.

34. (a) $y = \dfrac{x+2}{x+2}$    (b) $y = \dfrac{x^2-4}{x-2}$

(c) $y = \dfrac{x-1}{(x-1)(x-2)}$

35. (a) $y = \dfrac{x^2-9}{x+3}$    (b) $y = \dfrac{x^2-5x+6}{x^2-2x-3}$

(c) $y = \dfrac{(x-1)(x-2)(x-3)}{(x-1)(x-2)(x-3)(x-4)}$

36. This exercise shows you how to determine the coordinates of the lowest point on the middle branch of the curve in Figure 14. The basic idea is as follows. Suppose that the required coordinates are $(h, k)$. Then the horizontal line $y = k$ is the unique horizontal line intersecting the curve in one and only one point. (Any other horizontal line intersects the curve either in two points or not at all.) In steps (a) through (c) that follow, we use these observations to determine the point $(h, k)$.

(a) Given any horizontal line $y = k$, its intersection with the curve in Figure 14 is determined by solving the following pair of simultaneous equations:

$$\begin{cases} y = k \\ y = \dfrac{x^2-x-6}{x^2-x-2} \end{cases}$$

In the second equation of the system, replace $y$ with $k$ and show that the resulting equation can be written

$$(k-1)x^2 - (k-1)x + (6-2k) = 0 \qquad (1)$$

(b) If $k$ is indeed the required $y$-coordinate, then equation (1) must have exactly one real solution. Set the discriminant of the quadratic equation equal to zero to obtain

$$(k-1)^2 - 4(k-1)(6-2k) = 0$$

and deduce from this that $k = 1$ or $k = \frac{25}{9}$. The solution $k = 1$ can be discarded. (To see why, look at Figure 14.)

(c) Using the value $y = \frac{25}{9}$, show that the corresponding $x$-coordinate is $\frac{1}{2}$. Thus, the required point is $(\frac{1}{2}, \frac{25}{9})$.

37. Graph the function $y = \dfrac{x}{(x-3)^2}$. Use the technique explained in Exercise 36 to find the coordinates of any turning points on the graph.

38. Graph the function $y = \dfrac{2}{x - x^2}$. Use the technique explained in Exercise 36 to find the coordinates of any turning points on the graph.

*An asymptote that is neither horizontal nor vertical is called a **slant**, or **oblique**, asymptote. For example, as indicated in the following figure, the line $y = x$ is a slant asymptote for the graph of $y = \dfrac{x^2 + 1}{x}$. To understand why the line $y = x$ is an asymptote, we carry out the indicated division and write the function in the form*

$$y = x + \frac{1}{x} \qquad\qquad (2)$$

*From equation (2), we see that if $|x|$ is very large, then $y \approx x + 0$; that is, $y \approx x$, as we wished to show. Equation (2) actually tells us more than this. When $|x|$ is very close to zero, equation (2) yields $y \approx 0 + \dfrac{1}{x}$. In other words, as we approach the y-axis, the curve looks more and more like the graph of $y = 1/x$.*

The curve $y = \dfrac{x^2 + 1}{x}$ with slant asymptote $y = x$

*In general, if we have a rational function $f(x)/g(x)$ in which the degree of $f(x)$ is 1 greater than the degree of $g(x)$, then the graph has a slant asymptote that is obtained as follows.*

*Divide $f(x)$ by $g(x)$ to obtain*

$$\frac{f(x)}{g(x)} = (mx + b) + \frac{h(x)}{g(x)}$$

*where the degree of $h(x)$ is less than the degree of $g(x)$. Then the equation of the slant asymptote is $y = mx + b$. [For instance, using the example in the previous paragraph, we have $mx + b = x$ and $h(x)/g(x) = 1/x$. In Exercises 39–41, you are asked to graph functions that have slant asymptotes.*

39. Let $y = F(x) = \dfrac{x^2 + x - 6}{x - 3}$.

(a) Use long division to show that

$$\frac{x^2 + x - 6}{x - 3} = (x + 4) + \frac{6}{x - 3}$$

[c] (b) The result in part (a) shows that the line $y = x + 4$ is a slant asymptote for the graph of the function $F$. Verify this fact empirically by completing the following two tables.

| $x$ | $x + 4$ | $\dfrac{x^2 + x - 6}{x - 3}$ |
| --- | --- | --- |
| 10 | | |
| 100 | | |
| 1000 | | |

| $x$ | $x + 4$ | $\dfrac{x^2 + x - 6}{x - 3}$ |
| --- | --- | --- |
| $-10$ | | |
| $-100$ | | |
| $-1000$ | | |

(c) Determine the vertical asymptote and the $x$- and $y$-intercepts for the graph of $F$.

(d) Graph the function $F$. (Make use of the techniques in this section along with the fact that $y = x + 4$ is a slant asymptote.)

(e) Use the technique explained in Exercise 36 to find the coordinates of the two turning points on the graph of $F$.

40. (a) Show that the line $y = x - 2$ is a slant asymptote for the graph of $F(x) = x^2/(x + 2)$.

(b) Sketch the graph of $F$.

41. Show that the line $y = -x$ is a slant asymptote for the graph of $y = (1 - x^2)/x$. Then sketch the graph of this function.

| | CHAPTER 5   SUMMARY OF PRINCIPAL TERMS | |
|---|---|---|

| TERMS | PAGE REFERENCE | COMMENTS |
|---|---|---|
| **1.** Linear function | 216 | A linear function is a function of the form $f(x) = Ax + B$. The graph of a linear function is a straight line. |
| **2.** Quadratic function | 217 | A quadratic function is a function of the form $f(x) = ax^2 + bx + c$. The graph of a quadratic function is a parabola. See the boxes on pages 219 and 221. |
| **3.** Minimum value of a function | 218 | An output $f(x_0)$ is said to be a minimum value for the function $f$ provided the inequality $f(x_0) \leq f(x)$ holds for all $x$ in the domain of $f$. In the case of a parabola that opens upward, the minimum value of the function is just the $y$-coordinate of the vertex. |
| **4.** Maximum value of a function | 219 | An output $f(x_0)$ is said to be a maximum value for the function $f$ provided the inequality $f(x_0) \geq f(x)$ holds for all $x$ in the domain of $f$. In the case of a parabola that opens downward, the maximum value of the function is just the $y$-coordinate of the vertex. |
| **5.** Polynomial function | 248 | A polynomial function of degree $n$ is a function of the form $f(x) = a_n x^n + a_{n-1} x^{n-1} + \cdots + a_1 x + a_0$ where $n$ is a nonnegative integer and $a_n \neq 0$. |
| **6.** Turning point | 249 | A turning point on a graph is a point at which the graph changes from rising to falling, or falling to rising. For example, the origin is a turning point for the parabola $y = x^2$, whereas the graph of $y = x^3$ has no turning points. |
| **7.** Rational function | 257 | A rational function is a function of the form $y = \dfrac{f(x)}{g(x)}$, where $f(x)$ and $g(x)$ are polynomials. |
| **8.** Asymptote | 258 | A line is said to be an asymptote for a curve if the distance between the line and the curve approaches zero as we move out farther and farther along the line. See, for example, Figure 2 on page 258, in which the dashed lines are asymptotes. |

## CHAPTER 5   REVIEW EXERCISES

NOTE   The ten exercises in this set with colored problem numbers constitute a chapter test.

*In Exercises 1–6, find equations for the linear functions satisfying the given conditions. Write each answer in the form $f(x) = mx + b$.*

1. $f(3) = 5$ and $f(-2) = 0$

2. $f(1) = 5$, and the graph of $f$ passes through the origin

3. $f(4) = -1$, and the graph of $f$ is parallel to the line $3x - 8y = 16$

4. The graph passes through $(6, 1)$, and the $x$-intercept is twice the $y$-intercept

5. $f(-3) = 5$, and the graph of the inverse function passes through $(2, 1)$

6. The graph of $f$ passes through the vertices of the two parabolas $y = x^2 + 4x + 1$ and $x = 2y^2 - 8y - 1$

*In Exercises 7–12, graph the quadratic functions. In each case, specify the vertex and the x- and y-intercepts.*

7. $y = x^2 + 2x - 3$

8. $f(x) = x^2 - 2x - 15$

**9.** $y = -x^2 + 2\sqrt{3}x + 3$    **10.** $f(x) = 2x^2 - 2x + 1$

**11.** $y = -3x^2 + 12x$    **12.** $f(x) = -4x^2 + 16x$

**13.** Find the distance between the vertices of the two parabolas $y = x^2 - 4x + 6$ and $y = -x^2 - 4x - 5$.

**14.** Find the value of $a$, given that the maximum value of the function $f(x) = ax^2 + 3x - 4$ is 5.

**15.** The sum of two numbers is $\sqrt{3}$. Find the largest possible value for their product.

**16.** The sum of two numbers is $\frac{2}{3}$. What is the smallest possible value for the sum of their squares?

**17.** Suppose that an object is thrown vertically upward (from ground level) with an initial velocity of $v_0$ ft/sec. It can be shown that the height $h$ (in feet) after $t$ sec is given by the formula

$$h = v_0 t - 16t^2$$

   **(a)** At what time does the object reach its maximum height? What is that maximum height?

   **(b)** At what time does the object strike the ground?

**18.** Let $f(x) = 4x^2 - x + 1$ and $g(x) = (x - 3)/2$.

   **(a)** For which input will the value of the function $f \circ g$ be a minimum?

   **(b)** For which input will the value of $g \circ f$ be a minimum?

**19.** **(a)** Let $P$ be a point on the parabola $y = x^2$. Express the distance from $P$ to the point $(0, 2)$ in terms of $x$.

   **(b)** Which point in the second quadrant on the parabola $y = x^2$ is closest to the point $(0, 2)$?

**20.** Find the coordinates of the point on the line $y = 2x - 1$ closest to $(-5, 0)$.

**21.** Find all the values of $b$ such that the minimum distance from the point $(2, 0)$ to the line $y = \frac{4}{3}x + b$ is 5.

**22.** What number exceeds one-half its square by the greatest amount?

**23.** Suppose that $x + y = \sqrt{2}$. Find the minimum value of the quantity $x^2 + y^2$.

**24.** For which numbers $t$ will the value of $9t^2 - t^4$ be as large as possible?

**25.** Find the maximum area possible for a right triangle with a hypotenuse of 15 cm.    *Hint:*    Let $x$ denote the length of one leg. Show that the area is $A = \frac{1}{2}x\sqrt{225 - x^2}$. Now work with $A^2$.

**26.** For which point $(x, y)$ on the curve $y = 1 - x^2$ is the sum $x + y$ a maximum?

**27.** Let $f(x) = x^2 - (a^2 + 2a)x + 2a^3$, where $0 < a < 2$. For which value of $a$ will the distance between the $x$-intercepts of $f$ be a maximum?

**28.** Suppose that the revenue $R$ (in dollars) generated by selling $x$ units of a certain product is given by $R =$

$300x - \frac{1}{4}x^2$ $(0 \le x \le 1200)$. How many units should be sold to maximize the revenue? What is that maximum revenue?

**29.** Suppose that the function $p = 160 - \frac{1}{5}x$ relates the selling price $p$ of an item to the quantity $x$ that is sold. Assume that $p$ is in dollars. For which value of $x$ will the revenue $R$ $( = xp)$ be a maximum? What is the selling price $p$ in this case?

**30.** A piece of wire 16 cm long is cut into two pieces. Let $x$ denote the length of the first piece and $16 - x$ the length of the second. The first piece is formed into a rectangle in which the length is twice the width. The second piece of wire is also formed into a rectangle, but with the length three times the width. For which value of $x$ is the total area of the two rectangles a minimum?

*In Exercises 31–46, graph each function and specify the x- and y-intercepts and asymptotes, if any.*

**31.** $y = (x + 4)(x - 2)$    **32.** $y = (x + 4)(x - 2)^2$

**33.** $y = -(x + 5)^3$    **34.** $y = -x(x + 1)$

**35.** $y = -x^2(x + 1)$    **36.** $y = -x^3(x + 1)$

**37.** $y = x(x - 2)(x + 2)$

**38.** $y = (x - 3)(x + 1)(x + 5)$

**39.** $y = \dfrac{3x + 1}{x}$    **40.** $y = \dfrac{1 - 2x}{x}$

**41.** $y = \dfrac{-1}{(x - 1)^2}$    **42.** $y = \dfrac{x + 1}{x + 2}$

**43.** $y = \dfrac{x - 2}{x - 3}$    **44.** $y = \dfrac{x}{(x - 2)(x + 4)}$

**45.** $y = \dfrac{x^2 - 2x + 1}{x^2 - 4x + 4}$    **46.** $y = \dfrac{x(x - 2)}{(x - 4)(x + 4)}$

**47.** Let $f(x) = x^2 + 2bx + 1$.

   **(a)** If $b = 1$, find the distance from the vertex of the parabola to the origin.

   **(b)** If $b = 2$, find the distance from the vertex of the parabola to the origin.

   **(c)** For which real numbers $b$ will the distance from the vertex to the origin be as small as possible?

**48.** Find the range of the function $f(x) = -2x^2 + 12x - 5$. *Hint:*    Look at the graph.

**49.** The range of the function $y = x^2 - 2x + k$ is the interval $[5, \infty)$. Find the value of $k$.

**50.** Find a value for $b$ such that the range of the function $f(x) = x^2 + bx + b$ is the interval $[-15, \infty)$.

**51.** Find the range of the function $y = \dfrac{(x - 1)(x - 3)}{x - 4}$.

   *Hint:*    Solve the equation for $x$ in terms of $y$ using the quadratic formula. If you're careful with the algebra, you will find that the expression under the resulting radical sign is $y^2 - 8y + 4$. The range of the given function can then be found by solving the inequality $y^2 - 8y + 4 \geq 0$.

52. In the figure, triangle $OAB$ is equilateral and $AB$ is parallel to the $x$-axis. Find the length of a side and the area of the triangle $OAB$.

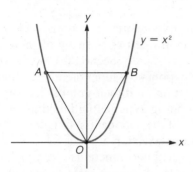

53. Let $A$ denote the area of the right triangle in the first quadrant that is formed by the $y$-axis and the lines $y = mx$ and $y = m$. (Assume $m > 0$.) Express the area of the triangle as a function of $m$.

54. (a) Express the distance from the origin to the vertex of the parabola $y = x^2 - 2bx$ $(b > 0)$ as a function of $b$.

(b) Let $V$ denote the vertex of the parabola in part (a), and let $A$ and $B$ denote the points where the curve meets the $x$-axis. Express the area of $\triangle VAB$ as a function of $b$.

(c) A circle is drawn with center $V$ and passing through $A$ and $B$. Then, through the smaller $y$-intercept of the circle, a tangent line is drawn. Express the $x$-intercept of this tangent line as a function of $b$.

55. In the accompanying figure, the radius of the circle is $OC = 1$. Express the area of $\triangle ABC$ as a function of $x$.

# 6 EXPONENTIAL AND LOGARITHMIC FUNCTIONS

**INTRODUCTION**  In this chapter we study exponential functions and their inverses. Figure 1 introduces one of the key properties of many of these exponential functions; they grow or increase very rapidly. Section 6.1 begins with an example involving perhaps the simplest exponential function, $y = 2^x$. In this section we discuss in detail exponential functions and their graphs. Section 6.2 introduces the constant $e$ and the corresponding exponential function $y = e^x$. In Sections 6.3 and 6.4 we study the inverses of the exponential functions; these inverse functions are called *logarithmic functions*. Section 6.5 is devoted to some of the applications of exponential and logarithmic functions. As you'll see, these applications are quite diverse, ranging from interest rates in banking to population growth and to nuclear energy.

**FIGURE 1***

## EXPONENTIAL FUNCTIONS

### 6.1

We begin with an example. Suppose that your mathematics instructor, in an effort to improve classroom attendance, offers to pay you each day for attending class! Say you are to receive 2¢ on the first day you attend class, 4¢ the second

---

* Drawing by Professor Ann Jones, University of Colorado, Boulder. From the cover of *The Physics Teacher*, Vol. 14, No. 7, Oct. '76.

**TABLE 1**

| $x$ (DAY NUMBER) | $y$ (AMOUNT EARNED THAT DAY) |
|---|---|
| 1 | 2¢ $(= 2^1)$ |
| 2 | 4¢ $(= 2^2)$ |
| 3 | 8¢ $(= 2^3)$ |
| 4 | 16¢ $(= 2^4)$ |
| 5 | 32¢ $(= 2^5)$ |
| ⋮ | ⋮ |
| $x$ | $2^x$ |

day, 8¢ the third day, and so on, as indicated in Table 1. How much money will you receive for attending class on the 30th day?

As you can see by looking at Table 1, the amount $y$ earned on day $x$ is given by the rule or *exponential function*

$$y = 2^x$$

Thus on the 30th day (when $x = 30$), you will receive

$$y = 2^{30} \text{ cents}$$

If you use a calculator, you will find this amount to be well over 10 million dollars. The point here is simply this. Although we begin with a small amount, $y = 2$¢, repeated doubling quickly leads to a very large amount. Put in other terms, the exponential function grows very rapidly.

Before leaving this example, we point out a simple method for quickly estimating numbers such as $2^{30}$ (or any power of two) in terms of the more familiar powers of ten. Begin by observing that

$$2^{10} \approx 10^3 \qquad \text{(a useful coincidence, worth remembering)}$$

Now just cube both sides to obtain

$$(2^{10})^3 \approx (10^3)^3$$

or

$$2^{30} \approx 10^9$$

Thus $2^{30}$ is about one billion. To convert this number of cents to dollars, you divide by 100 or $10^2$ to obtain

$$\frac{10^9}{10^2} = 10^7 \text{ dollars}$$

which is 10 million dollars, as mentioned before.

**EXAMPLE 1**  Estimate $2^{40}$ in terms of a power of 10.

Solution  Take the basic approximation $2^{10} \approx 10^3$ and raise both sides to the fourth power. This yields

$$(2^{10})^4 \approx (10^3)^4$$

or

$$2^{40} \approx 10^{12} \qquad \text{as required}$$

**EXAMPLE 2**  Estimate the power to which 10 must be raised to yield 2.

Solution  We begin with our approximation

$$10^3 \approx 2^{10}$$

Raising both sides to the power $\frac{1}{10}$ yields

$$(10^3)^{1/10} \approx (2^{10})^{1/10}$$

and consequently

$$10^{3/10} \approx 2$$

Thus the power to which 10 must be raised to yield 2 is approximately $\frac{3}{10}$. ▪

In Section 1.7 we defined the expression $b^x$, where $x$ is a rational number. We also stated that if $x$ is irrational, then $b^x$ can be defined so that the usual properties of exponents continue to apply. Although a rigorous definition of irrational exponents requires concepts from calculus, we can nevertheless indicate the basic idea by means of an example. (We need to do this before we give the general definition for exponential functions.)

How shall we assign a meaning to $2^{\sqrt{2}}$, for example? The basic idea is to evaluate successively the expression $2^x$ using rational numbers $x$ that are closer and closer to $\sqrt{2}$. The following table displays the results of some calculations along these lines.

---

**TABLE 2**

Values of $2^x$ for rational numbers $x$ approaching $\sqrt{2}$ $(= 1.41421356\ldots)$

| $x$ | 1.4 | 1.41 | 1.414 | 1.4142 | 1.41421 | 1.414213 |
|-----|-----|------|-------|--------|---------|----------|
| $2^x$ | $2.6\ldots$ | $2.65\ldots$ | $2.664\ldots$ | $2.6651\ldots$ | $2.66514\ldots$ | $2.665143\ldots$ |

The data in the table suggest that as $x$ approaches $\sqrt{2}$, the corresponding values of $2^x$ approach a unique real number, call it $t$, whose decimal expansion begins as 2.665. Furthermore, by continuing this process we could obtain (in theory, at least) as many places in the decimal expansion of $t$ as we wished. The value of the expression $2^{\sqrt{2}}$ is then defined to be this number $t$. The following theorem (stated here without proof) serves to summarize this discussion and also to pave the way for the definition we will give for exponential functions.

---

**Theorem**

Let $b$ denote an arbitrary positive real number. Then:

1. For each real number $x$, the quantity $b^x$ is a unique real number.
2. When $x$ is irrational, we can approximate $b^x$ as closely as we wish by evaluating $b^r$, where $r$ is a rational number sufficiently close to the number $x$.
3. The properties of rational exponents continue to hold for irrational exponents.
4. If $b^x = b^y$ and $b \neq 1$, then $x = y$.

---

For the remainder of this section, $b$ denotes an arbitrary positive constant other than 1. In the box that follows, we define the exponential function with base $b$.

### Definition:    The Exponential Function with Base *b*

EXAMPLES

Let *b* denote an arbitrary positive constant other than 1. The **exponential function with base *b*** is defined by the equation

$$y = b^x$$

1. The equations $y = 2^x$ and $y = 3^x$ define the exponential functions with bases 2 and 3, respectively.

2. The equation $y = (1/2)^x$ defines the exponential function with base 1/2.

3. The equations $y = x^2$ and $y = x^3$ do not define exponential functions.

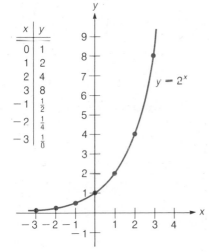

| x | y |
|---|---|
| 0 | 1 |
| 1 | 2 |
| 2 | 4 |
| 3 | 8 |
| −1 | $\frac{1}{2}$ |
| −2 | $\frac{1}{4}$ |
| −3 | $\frac{1}{8}$ |

**FIGURE 2**

To help with our analysis of exponential functions, let's set up a table and use it to graph the exponential function $y = 2^x$. This is done in Figure 2. (In drawing a smooth and unbroken curve, we are actually relying on the theorem that we just stated.) The key features of the exponential function $y = 2^x$ and its graph are these:

1. The domain of $y = 2^x$ is the set of all real numbers. The range is the set of all *positive* real numbers.
2. The *y*-intercept of the graph is 1. There is no *x*-intercept.
3. For $x > 0$, the function increases or grows very rapidly. For $x < 0$, the graph rapidly approaches the *x*-axis; the *x*-axis is a horizontal asymptote for the graph. (Recall from Section 5.5 that a line is an **asymptote** for a curve if the separation distance between the curve and the line approaches zero as we move out farther and farther along the line.)

You should memorize the basic shape and features of the graph of $y = 2^x$ so that you can sketch it as needed without first setting up a table. The next example shows why this is useful.

**EXAMPLE 3**    Graph the following functions. In each case specify the domain, range, intercept(s), and asymptote.

(a) $y = -2^x$    (b) $y = 2^{-x}$    (c) $y = (\frac{1}{2})^x$    (d) $y = 2^{-x} - 2$

**Solution**    (a) Recall that $-2^x$ means $-(2^x)$, not $(-2)^x$. The graph of $y = -2^x$ is obtained by reflecting the graph of $y = 2^x$ in the *x*-axis. See Figure 3(a).

(b) Similarly, the graph of $y = 2^{-x}$ is obtained from the graph of $y = 2^x$ by reflection in the *y*-axis. See Figure 3(b).

(c) Next, regarding $y = (\frac{1}{2})^x$, observe that

$$\left(\frac{1}{2}\right)^x = \frac{1^x}{2^x} = \frac{1}{2^x} = 2^{-x}$$

In other words, $y = (\frac{1}{2})^x$ is really the same function as $y = 2^{-x}$, and this we already graphed in Figure 3(b).

(d) Finally, to graph $y = 2^{-x} - 2$, take the graph of $y = 2^{-x}$ in Figure 3(b) and move it two units in the negative *y*-direction. Note that the asymptote and *y*-intercept will also move down two units. To find the *x*-intercept,

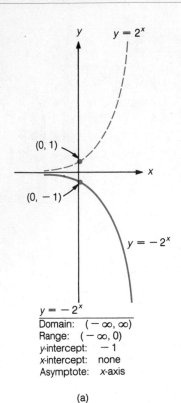

$y = -2^x$
Domain: $(-\infty, \infty)$
Range: $(-\infty, 0)$
$y$-intercept: $-1$
$x$-intercept: none
Asymptote: $x$-axis

(a)

$y = 2^{-x}$ (also $y = (\frac{1}{2})^x$)
Domain: $(-\infty, \infty)$
Range: $(0, \infty)$
$y$-intercept: $1$
$x$-intercept: none
Asymptote: $x$-axis

(b)

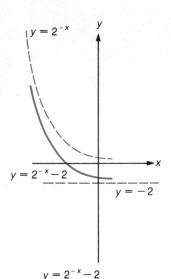

$y = 2^{-x} - 2$
Domain: $(-\infty, \infty)$
Range: $(-2, \infty)$
$y$-intercept: $-1$
$x$-intercept: $-1$
Asymptote: $y = -2$

(c)

FIGURE 3

set $y = 0$ in the given equation to obtain

$$2^{-x} - 2 = 0$$
$$2^{-x} = 2^1$$
$$-x = 1 \qquad \text{using property 4 in the theorem on page 272}$$
$$x = -1$$

Thus, the $x$-intercept is $-1$. See Figure 3(c).

In the next example, we apply our knowledge about the graph of $y = 2^x$ to solve an equation. In particular, we use the fact that the graph of $y = 2^x$ always lies above the $x$-axis; for no value of $x$ is $2^x$ ever zero.

**EXAMPLE 4**  Solve the equation $x^2 2^x - 2^x = 0$.

Solution  First, factor the left-hand side of the equation; the common term is $2^x$. This gives us

$$2^x(x^2 - 1) = 0$$

Since $2^x$ is never zero, we may now divide both sides of this last equation by $2^x$ to obtain

$$x^2 - 1 = 0$$
$$(x - 1)(x + 1) = 0$$

and consequently

$$x = 1 \qquad \text{or} \qquad x = -1$$

Thus the required solutions are $x = 1$, $x = -1$. (You should check for yourself that both of these numbers satisfy the original equation.) ▦

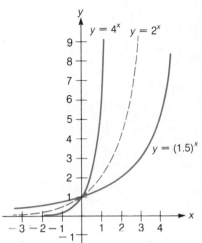

FIGURE 4

Now what about exponential functions with bases other than 2? As Figure 4 indicates, there is nothing essentially new here. As you would expect, the graph of $y = 4^x$ rises more rapidly than $y = 2^x$ when $x$ is positive. For negative $x$-values, the graph of $y = 4^x$ is below that of $y = 2^x$. You can see why this happens by taking $x = -1$, for example, and comparing the values of $4^x$ and $2^x$. If $x = -1$, then

$$2^x = 2^{-1} = \frac{1}{2} \qquad \text{but} \qquad 4^x = 4^{-1} = \frac{1}{4}$$

Therefore $4^x < 2^x$ when $x = -1$. Notice also in Figure 4 that all three graphs have the same $y$-intercept of 1. This follows from the fact that $b^0 = 1$ for any positive number $b$.

The exponential functions in Figure 4 all have bases larger than 1. To see examples in which the bases are less than 1, you need only reflect those graphs in Figure 4 in the $y$-axis. For instance, the reflection of $y = 4^x$ in the $y$-axis gives us the graph of $y = (\frac{1}{4})^x$. (We discussed the idea behind this in Example 3; see Figure 3(b), for instance.)

In the box that follows, we summarize what we've learned up to this point regarding the exponential function $y = b^x$.

| PROPERTY SUMMARY | The Exponential Function $y = b^x$ |
|---|---|

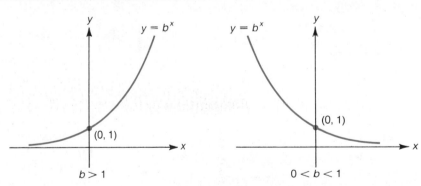

Domain: $(-\infty, \infty)$
Range: $(0, \infty)$
$y$-intercept: 1
$x$-intercept: none
Asymptote: $x$-axis

**EXAMPLE 5**  Graph the function $y = -3^{-x} + 1$.

Solution  The required graph is obtained by reflecting and translating the graph of $y = 3^x$, as shown in Figure 5. ▦

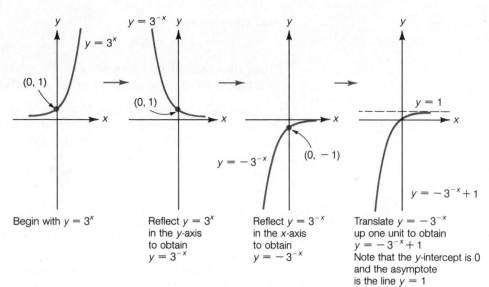

**FIGURE 5**

## EXERCISE SET 6.1

### A

In Exercises 1–4, estimate each quantity in terms of powers of 10, as in Example 1.

1. $2^{30}$    2. $2^{50}$    3. $2^{90}$    4. $4^{20}$

For Exercises 5–8, specify the domain of each function.

5. $y = 2^x$

6. $y = \dfrac{1}{2^x}$

7. $y = \dfrac{1}{2^{x-1}}$

8. $y = \dfrac{1}{2^x - 1}$

In Exercises 9–16, graph each pair of functions on the same set of axes.

9. $y = 2^x, y = 2^{-x}$

10. $y = 3^x, y = 3^{-x}$

11. $y = 3^x, y = -3^x$

12. $y = 4^x, y = -4^x$

13. $y = 2^x, y = 3^x$

14. $y = (\frac{1}{3})^x, y = 3^x$

15. $y = (\frac{1}{2})^x, y = (\frac{1}{3})^x$

16. $y = (\frac{1}{2})^{-x}, y = (\frac{1}{3})^{-x}$

For Exercises 17–24, graph each function and specify the domain, range, intercept(s), and asymptote.

17. $y = -2^x + 1$

18. $y = -3^x + 3$

19. $y = 3^{-x} + 1$

20. $y = 3^{-x} - 3$

21. $y = 2^{x-1}$

22. $y = 2^{x-1} - 1$

23. $y = 3^{x+1} + 1$

24. $y = 1 - 3^{x-1}$

In Exercises 25–28, solve each equation.

25. $3x(10^x) + 10^x = 0$

26. $4x^2(2^x) - 9(2^x) = 0$

27. $3(3^x) - 5x(3^x) + 2x^2(3^x) = 0$

28. $\dfrac{(x+4)10^x}{x-3} = 2x(10^x)$

### B

29. Let $f(x) = 2^x$. Show that
$$\frac{f(x+h) - f(x)}{h} = 2^x\left(\frac{2^h - 1}{h}\right)$$

30. Set up a table to graph each of the following functions. Indicate asymptotes and intercepts, if any.
    (a) $y = 2^{-x^2}$    (b) $y = 2^{|x|}$

31. Let $f(x) = 2^x$ and let $g$ denote the function that is the inverse of $f$.
    (a) On the same set of axes, sketch the graphs of $f$, $g$, and the line $y = x$.
    (b) On the basis of your graph in part (a), specify the domain, range, intercept, and asymptote for the function $g$.

32. Let $S(x) = \dfrac{2^x - 2^{-x}}{2}$ and $C(x) = \dfrac{2^x + 2^{-x}}{2}$. Compute $[C(x)]^2 - [S(x)]^2$.

33. Use the figure to estimate each of the following quantities.
    (a) $\sqrt{2}$    (b) $\sqrt[5]{2}$    (c) $\sqrt[5]{8}$

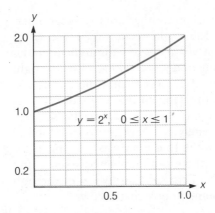

**FIGURE FOR EXERCISE 33**

**C**

**34.** Let $\phi(t) = 1 + a^t$. Show that

$$\frac{1}{\phi(t)} + \frac{1}{\phi(-t)} = 1$$

**35.** This exercise serves as a preview for the work on logarithms in Section 6.3. Follow steps (a)–(f) to complete the table. (Notice that one entry in the table is already filled in. Reread Example 2 in the text to see how that entry was obtained.)

**(a)** Fill in the entries corresponding to $x = 1$ and $x = 10$.

**(b)** Note that 4 and 8 are powers of 2. Use this information along with the approximation $10^{0.3} \approx 2$ to find the entries in the table corresponding to $x = 4$ and $x = 8$.

**(c)** Find the entry corresponding to $x = 5$. *Hint:* $5 = \dfrac{10}{2} \approx \dfrac{10}{10^{0.3}}$.

**(d)** Find the entry corresponding to $x = 7$. *Hint:* $7^2 \approx 50 = (5)(10)$. Now make use of your answer in part (c).

**(e)** Find the entry corresponding to $x = 3$. *Hint:* $3^4 \approx 80 = 8 \times 10$.

**(f)** Find the entries corresponding to $x = 6$ and $x = 9$. *Hint:* $6 = 3 \times 2$ and $9 = 3^2$.

REMARK   This table of powers of 10 is called a *table of logarithms to the base* 10. We say, for example, that the logarithm of 2 to the base 10 is (about) 0.3. We write this symbolically as $\log_{10} 2 \approx 0.3$.

| $x$ | POWER TO WHICH 10 MUST BE RAISED TO YIELD $x$ |
|---|---|
| 1 | |
| 2 | $\approx 0.3$ |
| 3 | |
| 4 | |
| 5 | |
| 6 | |
| 7 | |
| 8 | |
| 9 | |
| 10 | |

## THE EXPONENTIAL FUNCTION $y = e^x$

**6.2**

*This is undoubtedly the most important function in mathematics.*

WALTER RUDIN in his text, *Real and Complex Analysis* (New York: McGraw-Hill, 1966)

From the standpoint of calculus and scientific applications, there is one base for exponential functions that is by far the most useful. This base is a certain irrational number that lies between 2 and 3 and is denoted by the letter $e$.* For purposes of approximation, you'll need to know that

$$e \approx 2.7$$

At the precalculus level, it's hard to escape the feeling that $y = 2^x$ or $y = 10^x$ is by far more simple and more natural than $y = e^x$. So as we work with $e$ and $e^x$ now and again in this chapter, you will need to take it on faith that, in the long run, the number $e$ and the function $y = e^x$ make life simpler, not more complex.

---

* The Swiss mathematician Leonhard Euler introduced the letter $e$ to denote this number in the year 1741. To six decimal places, the value of $e$ is 2.718281....

Here is one way to define the number $e$. (For another approach see Exercise 31 at the end of this section; also see Table 3 in Section 6.5.) You know that the graph of each exponential function $y = b^x$ passes through the point $(0, 1)$. Figure 1(a) shows the exponential function $y = 2^x$ along with a straight line that is tangent to the curve at the point $(0, 1)$. By carefully measuring rise and run, it can be shown that the slope of this tangent line is about 0.7. Figure 1(b) shows a similar situation with the curve $y = 3^x$. There the slope of the tangent line through $(0, 1)$ is approximately 1.1. Now, since the slope of the tangent to $y = 2^x$ is a bit less than 1, while that for $y = 3^x$ is a bit more than 1, it seems reasonable to suppose that there is a number between 2 and 3, call it $e$, with the property that the slope of the tangent to $y = e^x$ through $(0, 1)$ is exactly 1. See Figure 2 and the summary box that follows.

(a)  (b)

**FIGURE 1**

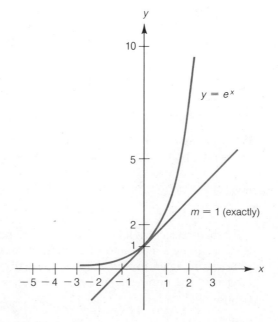

**FIGURE 2** The slope of the tangent to the curve $y = e^x$ at the point $(0, 1)$ is $m = 1$.

**PROPERTY SUMMARY**    **The Exponential Function $y = e^x$**

domain: $(-\infty, \infty)$
range: $(0, \infty)$
$y$-intercept: 1
asymptote: $x$-axis

**EXAMPLE 1**   Graph the following functions; in each case specify the domain, range, intercepts, and asymptotes.

(a) $y = e^{x-1}$    (b) $y = -e^{x-1}$    (c) $y = -e^{x-1} + 1$

Solution   (a)  Begin with the graph of $y = e^x$ in the figure at the top of this page. Moving the graph one unit to the right yields the graph of $y = e^{x-1}$ shown in Figure 3(a). The $x$-axis is still an asymptote for this translated graph, but the $y$-intercept will no longer be 1. For the $y$-intercept, we replace $x$ with 0 in the given equation to obtain

$$y = e^{-1} = 1/e \approx 0.37$$

$y = e^{x-1}$

domain: $(-\infty, \infty)$
range: $(0, \infty)$
$y$-intercept: $1/e$
asymptote: $x$-axis

(a)

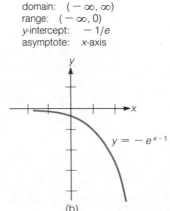

domain: $(-\infty, \infty)$
range: $(-\infty, 0)$
$y$-intercept: $-1/e$
asymptote: $x$-axis

$y = -e^{x-1}$

(b)

domain: $(-\infty, \infty)$
range: $(-\infty, 1)$
$x$-intercept: 1
$y$-intercept: $-1/e + 1$
asymptote: $y = 1$

$y = 1$

$y = -e^{x-1} + 1$

(c)

**FIGURE 3**

**(b)** Reflecting the graph from part (a) in the $x$-axis yields the graph of $y = -e^{x-1}$ [see Figure 3(b)]. Note that under this reflection, the $y$-intercept moves from $1/e$ to $-1/e$.

**(c)** Translating the graph in Figure 3(b) up one unit produces the graph of $y = -e^{x-1} + 1$, shown in Figure 3(c). Under this translation, the asymptote moves from $y = 0$ (the $x$-axis) to $y = 1$. Also, the $y$-intercept moves from $\dfrac{-1}{e}$ to $\dfrac{-1}{e} + 1$ ($\approx 0.63$). The $x$-intercept in Figure 3(c) is obtained by setting $y = 0$ in the equation $y = -e^{x-1} + 1$. This yields

$$0 = -e^{x-1} + 1$$
$$e^{x-1} = 1 = e^0$$
$$x - 1 = 0 \qquad \text{(Why?)}$$
$$x = 1$$

In some of the applications that arise in this section and in Section 6.5, we will need to work with equations of the form

$$e^{ax} = b \tag{1}$$

where $a$ and $b$ are constants. In Section 6.3 we'll see how to solve this equation. (It will require logarithms.) For now, however, it will be sufficient to isolate the quantity $e^x$ in equation (1). To do this, we just raise both sides of equation (1) to the power $1/a$:

$$(e^{ax})^{1/a} = b^{1/a}$$
$$e^x = b^{1/a}$$

**EXAMPLE 2**  Solve the equation $e^{4k} = 3$ for the quantity $e^k$.

**Solution**  To isolate $e^k$, raise both sides of the given equation to the power $1/4$. This yields

$$(e^{4k})^{1/4} = 3^{1/4}$$
$$e^k = 3^{1/4}$$

Thus, the value of $e^k$ is $3^{1/4}$. As you can check using a calculator, this is approximately 1.3.

We now turn to some applications involving the number $e$. In Section 6.5 we'll look at these and other applications in greater detail. (Thus your immediate goals in studying the remainder of this section are these: Begin to get used to seeing formulas involving $e^x$; master the algebra that you will be seeing in Examples 3 and 4.)

Under ideal conditions involving unlimited food and space, a colony of bacteria increases according to the *growth law*:

$$N = N_0 e^{kt}$$

In this formula, $N$ is the population at time $t$ and $k$ is a positive constant related to the growth rate of the population. The number $N_0$ is also a constant; it represents the size of the population at time $t = 0$. Sometimes, to emphasize the fact that $N$ depends on (is a function of) $t$, we employ function notation to rewrite the growth law as

$$N(t) = N_0 e^{kt}$$

**EXAMPLE 3**   Suppose that at the start of an experiment in a biology laboratory, there are 1200 bacteria in a colony. Four hours later, the population is found to be 3600. How many bacteria were there three hours after the experiment began?

**Solution**   We begin with the growth law $N = N_0 e^{kt}$. Our strategy will be to evaluate the quantity $e^k$; then we can determine the required value of $N$. (*Note:*   In Section 6.3 we'll see how to find $k$ itself, rather than $e^k$.) We are given that $N_0 = 1200$. Thus we can write

$$N = 1200 e^{kt} \qquad (2)$$

We are also given that $N = 3600$ when $t = 4$. Substituting these values in equation (2) gives us

$$3600 = 1200 e^{4k}$$

Now we divide both sides of this last equation by 1200 to obtain

$$3 = e^{4k} \qquad (3)$$

To isolate $e^k$, we raise both sides of the equation to the power $\frac{1}{4}$ (as in Example 2).

$$3^{1/4} = (e^{4k})^{1/4}$$

Therefore

$$3^{1/4} = e^k \qquad (4)$$

Now we can use this value of $e^k$ to determine the size of the colony after 3 hours, as required. We take equation (2) and replace $t$ by 3:

$$\begin{aligned} N &= 1200 e^{3k} \\ &= 1200 (e^k)^3 \\ &= 1200 (3^{1/4})^3 \qquad \text{using equation (4) to substitute for } e^k \\ &= 1200 (3^{3/4}) \end{aligned}$$

Using a calculator, we find that $1200(3^{3/4}) \approx 2735$. Rather than claim that there are precisely 2735 bacteria after 3 hours, we follow common sense and round off our answer to the nearest hundred. Thus we say that after 3 hours there are about 2700 bacteria in the colony.

In the example just concluded, we used the function $N = N_0 e^{kt}$ to describe population growth. It is a remarkable fact, shown in calculus, that the same general equation also governs radioactive decay, but in that case the constant $k$ is negative, not positive. Thus we will assume here the following *decay law* for radioactive substances:

$$N = N_0 e^{kt}$$

where $N_0$ is the original amount present at time $t = 0$, $N$ is the amount present at time $t$, and $k$ is a negative constant having to do with the rate of decay of the substance. In discussing radioactive decay, it is convenient to introduce the term half-life.

| Definition: Half-life | |
|---|---|
| | EXAMPLE |
| The **half-life** of a radioactive substance is the time required for half of a given sample to disintegrate. The half-life is an intrinsic property of the substance; it does not depend on the given sample size. | Iodine-131 is a radioactive substance with a half-life of 8 days. Suppose that 2 g are initially present. Then:<br><br>at $t = 0$, 2 g are present;<br>at $t = 8$ days, 1 g is left;<br>at $t = 16$ days, $\frac{1}{2}$ g is left;<br>at $t = 24$ days, $\frac{1}{4}$ g is left;<br>at $t = 32$ days, $\frac{1}{8}$ g is left. |

**EXAMPLE 4** Hospitals utilize the radioactive substance iodine-131 in the diagnosis of the thyroid gland. If a hospital acquires 2 g of iodine-131, how much of this sample will remain after 30 days? As stated earlier, the half-life of iodine-131 is 8 days.

Solution First, let's estimate the answer to get a feeling for the situation. We noted just prior to this example that, after 24 days, $\frac{1}{4}$ g will remain, while after 32 days, $\frac{1}{8}$ g will remain. Since 30 is between 24 and 32, it follows that, after 30 days, the amount remaining will be something between one-quarter and one-eighth of a gram.

Our actual calculations for the answer begin with the decay law

$$N = N_0 e^{kt} \tag{5}$$

We are asked to find $N$ when $t = 30$. As in Example 3, we first determine the quantity $e^k$. To do this, we use the half-life information. This says that when $t = 8$, the value of $N$ is $\frac{1}{2}N_0$. Using these values in equation (5), we find that

$$\frac{1}{2}N_0 = N_0 e^{8k}$$
$$\frac{1}{2} = e^{8k}$$
$$\left(\tfrac{1}{2}\right)^{1/8} = (e^{8k})^{1/8} = e^k \tag{6}$$

Returning to equation (5) now, we substitute $N_0 = 2$ and $t = 30$ to obtain

$$N = 2e^{30k}$$
$$= 2(e^k)^{30}$$
$$= 2[(\tfrac{1}{2})^{1/8}]^{30} \qquad \text{using equation (6) to substitute for } e^k$$

Using the properties for exponents, this can be simplified to

$$N = 2^{-11/4}$$

(Exercise 32 at the end of this section asks you to verify this.) Using a calculator now, we obtain

$$N = 0.148\ldots$$

Thus, after 30 days, approximately 0.15 g of the iodine-131 remains. As you can check, this figure is indeed between one-quarter and one-eighth, as we first estimated.

## EXERCISE SET 6.2

### A

*In Exercises 1–12, graph the functions and in each case specify the domain, range, intercept(s), and asymptote.*

1. $y = e^x$
2. $y = e^{-x}$
3. $y = -e^x$
4. $y = -e^{-x}$
5. $y = e^x + 1$
6. $y = e^{x+1}$
7. $y = e^{x+1} + 1$
8. $y = e^{x-1} - 1$
9. $y = -e^{x-2}$
10. $y = -e^{x-2} - 2$
11. $y = e - e^x$
12. $y = e^{-x} - e$

13. On the same set of axes, graph the functions $y = 2^x$ and $y = e^x$.

14. On the same set of axes, graph the functions $y = e^x$ and $y = 3^x$.

*In Exercises 15–18, solve the given equation for the quantity indicated.*

15. $e^{5k} = 32$; $e^k$
16. $e^{2k} = 225$; $e^k$
17. $e^{3t} = 4$; $e^t$
18. $e^{0.2t} = 10.3$; $e^t$

19. At the start of an experiment, there are 2000 bacteria in a colony. In 2 hours, the population has tripled. Assume that the growth law $N = N_0 e^{kt}$ applies.
    (a) Determine $e^k$.
    (b) Determine the population 10 hours after the start of the experiment.

20. At the start of an experiment there are $2 \times 10^4$ bacteria in a colony. After 8 hours the population is $3 \times 10^4$. Assume that the growth law $N = N_0 e^{kt}$ applies.
    (a) Determine the population 1 hour after it is $3 \times 10^4$.
    (b) Determine the population 24 hours after the start of the experiment.

21. A colony of bacteria grows according to the law $N = N_0 e^{kt}$. It is known that the value of $e^k$ is $\sqrt[5]{2}$. If the initial population is 3200, what will the population be 5 hours later?

22. The half-life of a certain radioactive substance is 5 days. Initially there are 8 g present. How many grams are present after (a) 5 days? (b) 10 days? (c) 15 days? (d) 20 days? (e) 50 days?

23. The half-life of iodine-131 is 8 days. How much of a 1-g sample will remain after 7 days?

24. The half-life of strontium-90 is 28 years. How much of a 10-g sample will remain after (a) 1 year? (b) 10 years?

25. Plutonium-239 is a product of nuclear reactors. The half-life of plutonium-239 is about 24,000 years. What percentage of a given sample will remain after 1000 years?

26. Krypton-91 has a half-life of 10 sec. What percentage of a given sample will remain after 5 min?

### B

27. If $E(x) = e^x$, show that
$$\frac{E(x + h) - E(x)}{h} = e^x \left( \frac{e^h - 1}{h} \right)$$

28. Set up a table and graph the function
$$S(x) = \frac{e^x - e^{-x}}{2}$$

29. Set up a table and graph the function
$$C(x) = \frac{e^x + e^{-x}}{2}$$

30. Look back at the graph of $y = e^x$ in Figure 2.
    (a) What is the equation of the tangent line shown there?
    (b) Provided that $x$ is close to zero, $e^x$ may be approximated by the quantity $x + 1$. Use part (a) to explain why.
    (c) Complete the table. What do you observe?

| $x$ | $x + 1$ | $e^x$ |
|---|---|---|
| 1.0 | | |
| 0.5 | | |
| 0.1 | | |
| 0.01 | | |
| 0.001 | | |
| 0.0001 | | |

© 31. Complete the following table. It can be shown that if the table is continued indefinitely, the numbers in the right-hand column approach the value $e$. (In fact, this is one way of defining $e$ in calculus.)

| $n$ | $\left(1 + \dfrac{1}{n}\right)$ | $\left(1 + \dfrac{1}{n}\right)^n$ |
|---|---|---|
| 1 | | |
| 10 | | |
| 100 | | |
| 1000 | | |
| 10000 | | |
| 100000 | | |

32. Carry out the simplification referred to in Example 4. That is, show that

$$2[(\tfrac{1}{2})^{1/8}]^{30} = 2^{-11/4}$$

*Hint:* First replace $\tfrac{1}{2}$ with $2^{-1}$.

© 33. Assume that the growth law $N = N_0 e^{kt}$, with which we described bacterial growth, also approximates the Earth's human population.
   (a) Use the following data to estimate the population for the year 1985. In 1850 (call this $t = 0$), the population was estimated to be one billion. In 1930, it was 2 billion.
   (b) Statistics from the United Nations say that the actual 1985 population was about 4.8 billion. How does your estimate from part (a) compare with this? Does the formula $N = N_0 e^{kt}$ predict too high or too low a figure?

34. Let $f(x) = e^x$ and $g(x) = x - 1$. Graph the functions $F$ and $G$ defined by

$$F(x) = (f \circ g)(x)$$
$$G(x) = (g \circ f)(x)$$

Specify any intercepts or asymptotes.

35. Let $f(x) = e^x$. Let $L$ denote the function that is the inverse of $f$.
   (a) On the same set of axes, sketch the graphs of $f$ and $L$.
   (b) Specify the domain, range, intercept, and asymptote for the function $L$ and its graph.
   (c) Graph each of the following. Specify the intercept and asymptote in each case.
      (i) $y = -L(x)$
      (ii) $y = L(-x)$
      (iii) $y = L(x - 1)$

## C

36. Let $S$ and $C$ denote the functions that were defined in Exercises 28 and 29, respectively. Prove the following identities.
   (a) $[C(x)]^2 - [S(x)]^2 = 1$
   (b) $S(-x) = -S(x)$
   (c) $C(-x) = C(x)$
   (d) $S(x + y) = S(x)C(y) + C(x)S(y)$
   (e) $C(x + y) = C(x)C(y) + S(x)S(y)$
   (f) $S(2x) = 2S(x)C(x)$
   (g) $C(2x) = [C(x)]^2 + [S(x)]^2$

## LOGARITHMIC FUNCTIONS

### 6.3

*[John Napier] hath set my head and hands a work with his new and admirable logarithms. I hope to see him this summer, if it please God, for I never saw book that pleased me better, or made me more wonder.*

HENRY BRIGGS, March 10, 1615

In the previous two sections we studied exponential functions. Now we consider functions that are inverses of exponential functions. These inverse functions are called *logarithmic functions.*

Having said this, let's back up for a moment to review briefly some of the basic ideas behind inverse functions (as discussed in Section 4.5). Start with a given function $F$, say $F(x) = 3x$ for example, that is one-to-one. (That is, for each output there is but one input.) Then, by interchanging the inputs and outputs, we obtain a new function, the so-called inverse function. In the case of $F(x) = 3x$, it's easy to find an equation defining the inverse function.

$f(x) = b^x$ $(b > 1)$

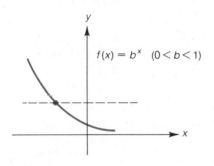

$f(x) = b^x$ $(0 < b < 1)$

**FIGURE 1** The exponential function $y = b^x$ is one-to-one.

Just interchange $x$ and $y$ in the equation $y = 3x$ to obtain $x = 3y$. Solving for $y$ in this last equation then gives us $y = \frac{1}{3}x$, which defines the inverse function. Using function notation, we can summarize the situation by writing $F(x) = 3x$ and $F^{-1}(x) = \frac{1}{3}x$.

In the preceding paragraph we saw that a particular linear function had an inverse, and we found a formula for that inverse. Now let's repeat that same reasoning beginning with an exponential function. First, we must make sure the exponential function $f$ defined by $f(x) = b^x$ is one-to-one. We can see this by applying the horizontal line test, as indicated in Figure 1. Next, since $f(x) = b^x$ is one-to-one, it has an inverse function. Let's study this inverse.

We begin by writing the exponential function $f(x) = b^x$ in the form

$$y = b^x \tag{1}$$

Then, to obtain an equation for $f^{-1}$, we interchange $x$ and $y$ in equation (1). This gives us

$$x = b^y \tag{2}$$

The crucial step now is to express equation (2) in words:

$$y \text{ is the power to which } b \text{ must be raised to yield } x \tag{3}$$

Statement (3) defines the function that is the inverse of $y = b^x$. Now we introduce a notation that will allow us to write this statement in a more compact form.

---

**Definition of $\log_b x$**

We define the expression $\log_b x$ to mean "the power to which $b$ must be raised to yield $x$." ($\log_b x$ is read "log base $b$ of $x$" or "the logarithm of $x$ to the base $b$.")

EXAMPLES

(a) $\log_2 8 = 3$, since 3 is the power to which 2 must be raised to yield 8.

(b) $\log_{10} \frac{1}{10} = -1$, since $-1$ is the power to which 10 must be raised to yield $\frac{1}{10}$.

(c) $\log_5 1 = 0$, since 0 is the power to which 5 must be raised to yield 1.

---

Using this notation, statement (3) becomes

$$y = \log_b x$$

Since equations (2) and (3) are equivalent, we have the following important relationship.

---

$$y = \log_b x \quad \text{is equivalent to} \quad x = b^y.$$

**TABLE 1**

| EXPONENTIAL FORM OF EQUATION | LOGARITHMIC FORM OF EQUATION |
|---|---|
| $8 = 2^3$ | $\log_2 8 = 3$ |
| $\dfrac{1}{9} = 3^{-2}$ | $\log_3 \dfrac{1}{9} = -2$ |
| $1 = e^0$ | $\log_e 1 = 0$ |
| $a = b^c$ | $\log_b a = c$ |

We say that the equation $y = \log_b x$ is in **logarithmic form** and that the equivalent equation $x = b^y$ is in **exponential form**. Table 1 displays some examples. Let us now summarize our discussion up to this point.

1.  According to the horizontal line test, the function $f(x) = b^x$ is one-to-one and therefore it possesses an inverse. This inverse function is written

    $$f^{-1}(x) = \log_b x$$

2.  $\log_b a = c$ means that $a = b^c$.

To graph the function $y = \log_b x$, we recall from Chapter 4 that the graph of a function and its inverse are reflections of one another through the line $y = x$. Thus, to graph $y = \log_b x$, we need only reflect the curve $y = b^x$ through the line $y = x$. This is shown in Figure 2.

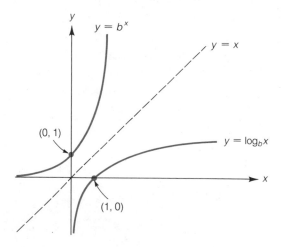

**FIGURE 2**

NOTE   For the rest of this chapter we assume that the base $b$ is greater than 1 when we use the expression $\log_b x$.

With the aid of Figure 2, we can make the following observations about the function $y = \log_b x$.

**PROPERTY SUMMARY**      **The Logarithmic Function $y = \log_b x$   ($b > 1$)**

Domain:  $(0, \infty)$
Range:  $(-\infty, \infty)$
$y$-intercept:  none
$x$-intercept:  1
Asymptote:  $y$-axis

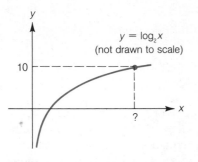

$y = \log_2 x$
(not drawn to scale)

10

?

x

**FIGURE 3**

One aspect of the function $y = \log_b x$ may not be immediately apparent to you from Figure 2. The function grows or increases *very* slowly. Consider $y = \log_2 x$, for example. Let us ask how large $x$ must be before the curve reaches the height $y = 10$. See Figure 3.

To answer this question, we substitute $y = 10$ in the equation $y = \log_2 x$. This gives us

$$10 = \log_2 x$$

Writing this equation in exponential form yields

$$x = 2^{10} = 1024$$

In other words, we must go out beyond 1000 on the $x$-axis before the curve $y = \log_2 x$ reaches a height of 10 units. Exercise 47 at the end of this section asks you to show that the graph of $y = \log_2 x$ doesn't reach a height of 100 until $x$ is greater than $10^{30}$. (Numbers as large as $10^{30}$ rarely occur in any of the sciences. For instance, the distance in inches to the Andromeda galaxy is less than $10^{24}$.) The point we are emphasizing here is this: The graph of $y = \log_2 x$ (or $\log_b x$) is always rising, but very slowly.

We conclude this section with a number of examples involving logarithms and logarithmic functions. In one way or another, each example makes use of the key fact that the equation $\log_b a = c$ is equivalent to $b^c = a$.

**EXAMPLE 1**  Which quantity is larger: $\log_3 10$ or $\log_7 40$?

**Solution**  First estimate the quantity $\log_3 10$. $\text{Log}_3 10$ represents the power to which 3 must be raised to yield 10. Since $3^2 = 9$ (less than 10), but $3^3 = 27$ (more than 10), we conclude that the quantity $\log_3 10$ lies between 2 and 3. In a similar way we can estimate the quantity $\log_7 40$. $\text{Log}_7 40$ represents the power to which 7 must be raised to yield 40. Since $7^1 = 7$ (less than 40), while $7^2 = 49$ (more than 40), we conclude that the quantity $\log_7 40$ lies between 1 and 2. It now follows from these two estimates that $\log_3 10$ is larger than $\log_7 40$.

**EXAMPLE 2**  Evaluate $\log_4 32$.

**Solution**  Let $y = \log_4 32$. Writing this equation in its equivalent exponential form gives us

$$4^y = 32$$
$$(2^2)^y = 2^5$$
$$2^{2y} = 2^5$$
$$2y = 5 \qquad \text{using property 4 in the box on page 272}$$
$$y = \frac{5}{2} \qquad \text{as required}$$

**EXAMPLE 3**  Solve $\log_{10}(x^2 + 3x + 12) = 1$ for $x$.

**Solution**  Writing the given equation in exponential form yields

$$x^2 + 3x + 12 = 10^1$$
$$x^2 + 3x + 2 = 0$$
$$(x + 2)(x + 1) = 0$$

Consequently, we have

$$x + 2 = 0 \quad \text{or} \quad x + 1 = 0$$

and therefore

$$x = -2 \quad \text{or} \quad x = -1$$

CHECK With $x = -2$ the given equation reads $\log_{10}(4 - 6 + 12) = 1$, or $\log_{10} 10 = 1$. Since this last equation is equivalent to $10^1 = 10$, we conclude that the value $x = -2$ does satisfy the given equation. In a similar manner, we find that $x = -1$ also satisfies the equation. In summary, then, the equation has two solutions, $x = -2$ and $x = -1$.

**EXAMPLE 4** Graph the following.

(a) $y = \log_{10} x$     (b) $y = -\log_{10} x$

Solution    (a) The function $y = \log_{10} x$ is the inverse function for the exponential function $y = 10^x$. Thus we obtain the graph of $y = \log_{10} x$ by reflecting the graph of $y = 10^x$ in the line $y = x$. See Figure 4(a).

(b) To graph $y = -\log_{10} x$, we reflect $y = \log_{10} x$ in the $x$-axis. See Figure 4(b).

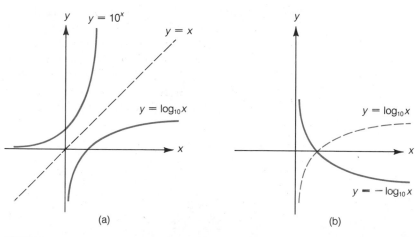

(a)                  (b)

**FIGURE 4**

**EXAMPLE 5** Find the domain of the function defined by $y = \log_2(12 - 4x)$.

Solution    As you can see by looking back at Figure 2 on page 286, the inputs for the logarithmic function must be positive. So, in the case at hand, we require that the quantity $12 - 4x$ be positive. Consequently, we have

$$12 - 4x > 0$$
$$-4x > -12$$
$$x < 3$$

Therefore the domain of the function defined by $y = \log_2(12 - 4x)$ is the interval $(-\infty, 3)$.

The next example concerns the exponential function $y = e^x$ and its inverse function $y = \log_e x$. Many books, as well as calculators, abbreviate the expression $\log_e x$ by $\ln x$, read "natural log of $x$." For reference and emphasis, we repeat this in the box that follows. (Incidentally, on calculators the expression "log" is an abbreviation for $\log_{10}$.)

---

**Definition**

$\ln x$  means  $\log_e x.$

---

**EXAMPLE 6**  Graph the following.

(a) $y = \ln x$    (b) $y = \ln(x - 1)$

**Solution**  (a)  The function $y = \ln x$ ($= \log_e x$) is the inverse of $y = e^x$. Thus its graph is obtained by reflecting $y = e^x$ in the line $y = x$, as in Figure 5(a).

(b)  To graph $y = \ln(x - 1)$, take the graph of $y = \ln x$ and move it one unit in the positive $x$-direction. See Figure 5(b).

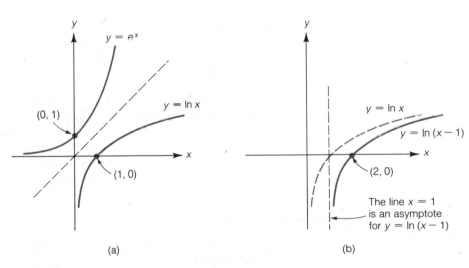

(a)                                                     (b)

**FIGURE 5**

**EXAMPLE 7**  Simplify each of the following.

(a) $\ln e$    (b) $\ln 1$

**Solution**  (a)  The expression $\ln e$ denotes the power to which $e$ must be raised to yield $e$. Clearly this power is 1. Thus $\ln e = 1$.

(b)  Similarly, $\ln 1$ denotes the power to which $e$ must be raised to yield 1. Since $e^0 = 1$, zero is the required power. Thus $\ln 1 = 0$.

**EXAMPLE 8**   Solve the following equations.

(a) $10^{2x} = 200$     (b) $e^{3t-1} = 2$

Solution    (a) Write the equation $10^{2x} = 200$ in logarithmic form to obtain

$$2x = \log_{10} 200$$

$$x = \frac{\log_{10} 200}{2}$$

Without a calculator or tables, we leave the answer in this form. On the other hand, using a calculator we obtain

$$x \approx 1.15$$

(b) To solve $e^{3t-1} = 2$, we write the equation in logarithmic form:

$$3t - 1 = \log_e 2$$

That is,

$$3t - 1 = \ln 2$$

$$3t = 1 + \ln 2$$

$$t = \frac{1 + \ln 2}{3}$$

Without a calculator or tables, this is the final answer. On the other hand, using a calculator we obtain

$$t \approx 0.56$$

In the previous section, we used the equation $N = N_0 e^{kt}$ to describe radio-active decay as well as population growth. If you look back at the examples and problems there, you'll see that every question was of the same type: Given $t$, find $N$. In other words, we were always asked "how much" or "how many," but never "when." Now that we have defined logarithms, we can solve a much wider range of problems, as the next example indicates.

**EXAMPLE 9**   Suppose that the half-life of a certain radioactive substance is 4 days.

(a) Compute the so-called *decay constant* $k$ in the formula $N = N_0 e^{kt}$.
(b) If you begin with a 2-g sample, how long will it be until only 0.01 g remains?

Solution    (a) The half-life is 4 days. This means that when $t$ is 4, the value of $N$ is $\frac{1}{2} N_0$. Using these values for $t$ and $N$ in the formula $N = N_0 e^{kt}$, we obtain

$$\frac{1}{2} N_0 = N_0 e^{k4}$$

and therefore

$$\frac{1}{2} = e^{4k}$$

We solve for $k$ by writing this last equation in logarithmic form:

$$4k = \ln \frac{1}{2}$$

$$k = \frac{\ln \frac{1}{2}}{4} = \frac{\ln 0.5}{4}$$

If a numerical value for $k$ is required, we can use tables or a calculator to obtain

$$k \approx -0.17$$

(b) We are given that $N_0 = 2$ g and we want to find the time $t$ at which $N$ is 0.01 g. Substituting the values $N_0 = 2$ and $N = 0.01$ into the decay law $N = N_0 e^{kt}$ yields

$$0.01 = 2e^{kt} \qquad \text{where } k \text{ has the value determined in part (a)}$$

or, dividing by 2,

$$0.005 = e^{kt}$$

Converting this last equation into its equivalent logarithmic form gives us

$$kt = \ln(0.005)$$

Therefore

$$t = \frac{\ln(0.005)}{k}$$

Finally, upon replacing $k$ by the expression $\frac{1}{4} \ln 0.5$, we obtain

$$t = \frac{\ln(0.005)}{\frac{1}{4} \ln 0.5}$$

$$= \frac{4 \ln(0.005)}{\ln 0.5}$$

$$\approx 30.58 \qquad \text{(using a calculator)}$$

Thus after about $30\frac{1}{2}$ days, only 0.01 g remains of the original 2-g sample.

## EXERCISE SET 6.3

### A

*Exercises 1–6 are review exercises dealing with inverse functions.*

1. Which of the following functions are one-to-one and therefore have inverses?
   (a) $y = x^2 + 1$      (b) $y = 3x$      (c) $y = (x + 1)^3$

2. Do either of the following functions have an inverse?
   (a) $f(x) = \begin{cases} x^2 & \text{if } -1 \leq x \leq 0 \\ x^2 + 1 & \text{if } x > 0 \end{cases}$

   (b) $g(x) = \begin{cases} x^2 & \text{if } -1 \leq x < 0 \\ x^2 + 1 & \text{if } x \geq 0 \end{cases}$

3. Let $f(x) = \dfrac{2x - 1}{3x + 4}$.
   (a) Find $f^{-1}(x)$.
   (b) Find $1/f(x)$.
   (c) Find $f^{-1}(0)$.
   (d) Find $1/f(0)$.

4. Let $f(x) = x^3 + 2x + 1$. Evaluate $f[f^{-1}(5)]$.
   [Assume that the domain of $f^{-1}$ is $(-\infty, \infty)$.]

5. The graph of $y = f(x)$ is a straight line segment joining the two points $(3, -2)$ and $(-1, 5)$. What are the corresponding endpoints for the graph of $y = f^{-1}(x - 1)$?

6. Which (if either) of the following two conditions tells us that a function is one-to-one?
   (a) For each input there is exactly one output.
   (b) For each output there is exactly one input.

7. Write each of the following equations in logarithmic form.
   (a) $9 = 3^2$      (b) $1000 = 10^3$
   (c) $7^3 = 343$      (d) $\sqrt{2} = 2^{1/2}$

8. Write each equation in logarithmic form.
   (a) $\frac{1}{125} = 5^{-3}$      (b) $e^0 = 1$
   (c) $5^x = 6$      (d) $e^{3t} = 8$

9. Write each equation in exponential form.
   (a) $\log_2 32 = 5$      (b) $\log_{10} 1 = 0$
   (c) $\log_e \sqrt{e} = \frac{1}{2}$      (d) $\log_3 \frac{1}{81} = -4$
   (e) $\log_t u = v$

10. Complete the tables.

(a)

| $x$ | 1 | 10 | $10^2$ | $10^3$ | $10^4$ | $10^{-1}$ | $10^{-2}$ | $10^{-3}$ |
|-----|---|----|--------|--------|--------|-----------|-----------|-----------|
| $\log_{10} x$ | | | | | | | | |

(b)

| $x$ | 1 | $e$ | $e^2$ | $e^3$ | $e^4$ | $e^{-1}$ | $e^{-2}$ | $e^{-3}$ |
|-----|---|-----|-------|-------|-------|----------|----------|----------|
| $\ln x$ | | | | | | | | |

11. Which quantity is larger, $\log_5 30$ or $\log_8 60$?

12. Which quantity is larger?
    (a) $\log_{10} 90$ or $\log_e e^5$      (b) $\log_2 3$ or $\log_3 2$

13. Evaluate each expression.
    (a) $\log_9 27$      (b) $\log_4 \frac{1}{32}$      (c) $\log_5 5\sqrt{5}$

14. Evaluate each expression.
    (a) $\log_{25} \frac{1}{625}$      (b) $\log_{16} \frac{1}{64}$
    (c) $\log_{10} 10$      (d) $\log_2 8\sqrt{2}$

15. Solve each equation for $x$.
    (a) $\log_x 256 = 8$      (b) $\log_5 x = -1$

16. Solve each equation for $x$.
    (a) $\log_{10}(x^2 + 36) = 2$
    (b) $\log_2(x^2 - 8x + 1) = 0$
    (c) $\log_{10}(x^2 - 5x + 14) = 1$

17. Find the domain of each function.
    (a) $y = \log_4 5x$      (b) $y = \log_{10}(3 - 4x)$
    (c) $y = \ln(x^2)$      (d) $y = (\ln x)^2$
    (e) $y = \ln(x^2 - 25)$

18. Find the domain of each function.
    (a) $y = \ln(2 - x - x^2)$

(b) $y = \log_{10} \dfrac{2x + 3}{x - 5}$

19. Find the domain and the range of the function defined by $y = 1/(1 - \ln x)$.

*In Exercises 20–24, graph the given function and specify the domain, range, intercept(s), and asymptote.*

20. $y = \log_2(x - 3)$

21. (a) $y = \log_{10}(x + 1)$      (b) $y = -\log_{10}(x + 1)$

22. $y = 1 + \log_2 x$

23. (a) $y = \ln x$      (b) $y = \ln(-x)$
    (c) $y = -1 + \ln(-x)$

24. $y = 1 - \log_2(x - 1)$

25. Simplify each expression.
    (a) $\ln e^4$      (b) $\ln(1/e)$      (c) $\ln \sqrt{e}$

26. Simplify each expression.
    (a) $\ln e$      (b) $\ln e^{-2}$      (c) $(\ln e)^{-2}$

*In Exercises 27–34, find all of the real-number solutions for each equation. In each case, give both the exact value of the answer and a calculator approximation rounded off to two decimal places.*

27. $10^x = 25$      28. $10^x = 145$

29. $10^{x^2} = 40$      30. $(10^x)^2 = 40$

31. $e^{2t+3} = 10$      32. $e^{t-1} = 16$

33. $e^{1-4t} = 12.405$      34. $e^{3x^2} = 112$

C 35. The half-life of uranium-238 is $4.5 \times 10^9$ years.
    (a) Find the decay constant $k$.
    (b) What percentage of a given amount $N_0$ remains after 1000 years?

C 36. Compute the half-life of carbon-14 if the value of the decay constant is $k = -0.00012$. (Assume that $t$ is in years.)

C 37. The half-life of a certain radioactive substance is 1 year.
    (a) Find the decay constant $k$.
    (b) Find the time required for 90% of a given 4-g sample to decay.

C 38. The initial population of a bacteria culture is $1.5 \times 10^5$. Three hours later, the population is found to be $2 \times 10^5$. Assume that the growth law $N = N_0 e^{kt}$ applies.
    (a) Find the growth constant $k$.
    (b) How long will it take for the initial population to double?

C 39. Initially a bacteria culture has a population of $2 \times 10^7$. Two hours later, the population is $3 \times 10^8$.
    (a) How long does it take for the culture to double its original size?
    (b) When does the population reach one billion?

40. The half-life of radium-226 is 1620 years.
    (a) Find the decay constant $k$.
    (b) How much of a 0.1-g sample remains after 100 years?

**B**

41. Let $f(x) = e^{x+1}$. Find $f^{-1}(x)$ and sketch its graph. Specify any intercepts or asymptotes.
42. Let $g(t) = \ln(t - 1)$. Find $g^{-1}(t)$ and draw its graph. Specify any intercepts or asymptotes.
43. Sketch the region bounded by $y = e^x$, $y = e^{-x}$, the $x$-axis, and the vertical lines $x = \pm 1$. Why must the area of this region be less than two square units?
44. Solve $2^{2x} - 2^{x+1} - 15 = 0$ for $x$.  *Hint:*  First show that the equation is equivalent to
    $$(2^x)^2 - 2(2^x) - 15 = 0.$$
45. Solve $e^{2x} - 5e^x - 6 = 0$ for $x$.
46. Solve $e^x - 3e^{-x} = 2$ for $x$.  *Hint:*  Multiply both sides by $e^x$.
47. Estimate a value for $x$ such that $\log_2 x = 100$. Use the approximation $10^3 \approx 2^{10}$ to express your answer as a power of 10.  *Answer:*  $10^{30}$
48. (a) How large must $x$ be before the graph of $y = \ln x$ reaches a height of $y = 100$?
    (b) How large must $x$ be before the graph of $y = e^x$ reaches a height of (i) $y = 100$? (ii) $y = 10^6$?
49. Suppose that $A = A_0 e^{k_1 t}$ and $B = B_0 e^{k_2 t}$. Find the value of $t$ for which $A = B$.  *Answer:*  $t = \dfrac{\ln(A_0/B_0)}{k_2 - k_1}$

    How would you interpret this question and answer in terms of radioactive decay?

50. Chemists define pH by the formula
    $$pH = -\log_{10}[H^+]$$
    where $[H^+]$ is the hydrogen ion concentration, measured in moles per liter. For example, if $[H^+] = 10^{-5}$, then pH = 5. Solutions with a pH of 7 are said to be *neutral*; a pH below 7 indicates an *acid* and a pH above 7 indicates a *base*.
    (a) For some fruit juices, $[H^+] = 3 \times 10^{-4}$. Determine the pH and classify as acid or base.
    (b) For sulfuric acid, $[H^+] = 1$. Find the pH.
    (c) An unknown substance has a hydrogen ion concentration of $3.5 \times 10^{-9}$. Classify the substance as acid or base.

51. This exercise indicates one of the ways the natural logarithm function is used in the study of prime numbers. Recall that a prime number is a natural number greater than 1 with no factors other than itself and 1. For example, the first ten prime numbers are 2, 3, 5, 7, 11, 13, 17, 19, 23, and 29.
    (a) Let $P(x)$ denote the number of prime numbers that do not exceed $x$. For instance, $P(6) = 3$, since there are three prime numbers (2, 3, and 5) that do not exceed 6. Compute $P(10)$, $P(18)$, and $P(19)$.
    (b) According to the *prime number theorem*, $P(x)$ can be approximated by $x/\ln x$ when $x$ is large, and in fact, the ratio $\dfrac{P(x)}{x/\ln x}$ approaches 1 as $x$ grows larger and larger. Verify this empirically by completing the following table. Round off your results to three decimal places. (These facts were discovered by Carl Friedrich Gauss in 1792, when he was 15 years old. It was more than 100 years later, however, in 1896, before the prime number theorem was formally proved by the French mathematician J. Hadamard and also by the Belgian mathematician C. J. de la Vallée-Poussin.)

| $x$ | $P(x)$ | $\dfrac{x}{\ln x}$ | $\dfrac{P(x)}{x/\ln x}$ |
|---|---|---|---|
| $10^2$ | 25 | | |
| $10^4$ | 1229 | | |
| $10^6$ | 78498 | | |
| $10^8$ | 5761455 | | |
| $10^9$ | 50847534 | | |
| $10^{10}$ | 455052512 | | |

52. The following figure shows a portion of the graph of $y = e^x$ for $1.5 \le x \le 1.7$. Using the figure, estimate the quantity $\ln 5$ to two decimal places. Then use a calculator to determine the percent error in your approximation.  *Hint:*  Write the equation $x = \ln 5$ in exponential form.

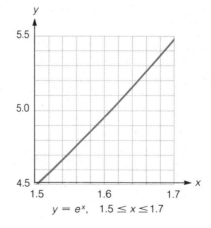

$y = e^x$,  $1.5 \le x \le 1.7$

## PROPERTIES OF LOGARITHMS

**6.4**

There are a few basic properties of logarithms that are used repeatedly. Our
procedure in this section will be to state these properties, discuss their proofs,
and then look at examples.

---

**Properties of Logarithms**

1. (a) $\log_b b = 1$    (b) $\log_b 1 = 0$
2. $\log_b PQ = \log_b P + \log_b Q$
   *The log of a product is the sum of the logs of the factors.*
3. $\log_b (P/Q) = \log_b P - \log_b Q$
   *The log of a quotient is the log of the numerator minus the log of the de-
   nominator.* As a useful particular case, we have $\log_b(1/Q) = -\log_b Q$.
4. $\log_b P^n = n \log_b P$
5. $b^{\log_b P} = P$

NOTE    $P$ and $Q$ are assumed to be positive in properties 2–5.

---

In essence, each of the properties follows from the equivalence of the two
equations $y = \log_b x$ and $b^y = x$. For instance, the equivalent exponential forms
for properties 1(a) and 1(b) are $b^1 = b$ and $b^0 = 1$, respectively, both of which
are certainly valid.

To prove property 2, we begin by letting $x = \log_b P$. The equivalent exponential form of this equation is

$$P = b^x \tag{1}$$

Similarly, we let $y = \log_b Q$. The exponential form of this equation is

$$Q = b^y \tag{2}$$

If we multiply equation (1) by equation (2), we get

$$PQ = b^x b^y$$

and therefore

$$PQ = b^{x+y} \tag{3}$$

Next we write equation (3) in its equivalent logarithmic form. This yields

$$\log_b PQ = x + y$$

But in view of the definitions of $x$ and $y$, this last equation is equivalent to

$$\log_b PQ = \log_b P + \log_b Q$$

That completes the proof of property 2.

The proof of property 3 is quite similar to the proof given for property 2. Exercise 76 at the end of this section asks you to carry out the proof of property 3.

We turn now to the proof of property 4. We begin by letting $x = \log_b P$. In exponential form, this last equation becomes

$$b^x = P \tag{4}$$

Now we raise both sides of equation (4) to the power $n$. This yields

$$(b^x)^n = b^{nx} = P^n$$

The logarithmic form of this last equation is

$$\log_b P^n = nx$$

or (in view of the definition of $x$)

$$\log_b P^n = n \log_b P$$

The proof of property 4 is now complete.

Property 5 is again just a restatement of the meaning of the expression $\log_b P$. To derive this property, let $x = \log_b P$. Therefore,

$$b^x = P$$

Now, in this last equation, just replace $x$ with $\log_b P$. The result is

$$b^{\log_b P} = P$$

as required.

By way of review, we point out that the proof just given for property 5 can be recast in terms of function notation and inverse functions. To do this, we define two functions $f$ and $g$ as follows:

$$f(P) = b^P \qquad \text{and} \qquad g(P) = \log_b P$$

Then, as explained in Section 6.3, $f$ and $g$ are inverses. We therefore have

$$f[g(P)] = P \qquad \text{using the definition of inverse functions}$$

and, consequently,

$$f(\log_b P) = P \quad \text{using the definition of } g$$

In view of the definition of $f$, this last equation can be written

$$b^{\log_b P} = P$$

as obtained previously.

Now let's see how these properties are used. To begin with, we display some simple numerical examples in the box that follows.

| PROPERTY SUMMARY | Properties of Logarithms |
|---|---|

| PROPERTY | EXAMPLE |
|---|---|
| $\log_b P + \log_b Q = \log_b PQ$ | Simplify: $\log_{10} 50 + \log_{10} 2$ |
| | Solution: $\log_{10} 50 + \log_{10} 2 = \log_{10}(50 \cdot 2)$ |
| | $\qquad\qquad\qquad\qquad\quad = \log_{10} 100$ |
| | $\qquad\qquad\qquad\qquad\quad = 2$ |
| $\log_b P - \log_b Q = \log_b \dfrac{P}{Q}$ | Simplify: $\log_8 56 - \log_8 7$ |
| | Solution: $\log_8 56 - \log_8 7 = \log_8 \dfrac{56}{7}$ |
| | $\qquad\qquad\qquad\qquad\quad = \log_8 8$ |
| | $\qquad\qquad\qquad\qquad\quad = 1$ |
| $\log_b P^n = n \log_b P$ | Simplify: $\log_2 \sqrt[5]{16}$ |
| | Solution: $\log_2 \sqrt[5]{16} = \log_2(16^{1/5})$ |
| | $\qquad\qquad\qquad\quad = \dfrac{1}{5} \log_2 16$ |
| | $\qquad\qquad\qquad\quad = \dfrac{1}{5} \cdot 4 = \dfrac{4}{5}$ |
| $b^{\log_b P} = P$ | Simplify: $3^{\log_3 7}$ |
| | Solution: $3^{\log_3 7} = 7$ |

The preceding examples showed how to simplify or shorten certain expressions involving logarithms. The next example is also of this type.

**EXAMPLE 1**  Express as a single logarithm with a coefficient of 1:

$$\tfrac{1}{2} \log_b x - \log_b(1 + x^2)$$

**Solution**  $\tfrac{1}{2} \log_b x - \log_b(1 + x^2) = \log_b x^{1/2} - \log_b(1 + x^2)$   using property 4

$$= \log_b \frac{x^{1/2}}{1 + x^2} \qquad\qquad \text{using property 3}$$

This last expression is the required answer.

Property 2 says that the logarithm of a product of two factors is equal to the sum of the logarithms of the two factors. This can be generalized to any number of factors. For instance, with three factors, we have

$$\log_b(ABC) = \log_b[A(BC)]$$
$$= \log_b A + \log_b BC \qquad \text{using property 2}$$
$$= \log_b A + \log_b B + \log_b C \qquad \text{using property 2 again}$$

The next example makes use of this idea.

**EXAMPLE 2** Express as a single logarithm with a coefficient of 1:

$$\ln(x^2 - 9) + 2\ln\frac{1}{x+3} + 4\ln x \qquad (x > 3)$$

**Solution**

$$\ln(x^2 - 9) + 2\ln\frac{1}{x+3} + 4\ln x = \ln(x^2 - 9) + \ln\left[\left(\frac{1}{x+3}\right)^2\right] + \ln x^4$$

$$= \ln(x^2 - 9) + \ln\frac{1}{(x+3)^2} + \ln x^4$$

$$= \ln\left[(x^2 - 9) \cdot \frac{1}{(x+3)^2} \cdot x^4\right]$$

$$= \ln\frac{(x^2 - 9)x^4}{(x+3)^2}$$

This last expression can be simplified still further by writing $x^2 - 9$ as $(x - 3)(x + 3)$. Then a factor of $x + 3$ can be divided out of the numerator and denominator of the fraction. The result is

$$\ln(x^2 - 9) + 2\ln\frac{1}{x+3} + 4\ln x = \ln\frac{(x-3)x^4}{x+3}$$

$$= \ln\frac{x^5 - 3x^4}{x+3}$$

In the examples considered so far, we've used the properties of logarithms to shorten given expressions. There are times, however, when we use these properties to expand an expression.

**EXAMPLE 3** Write each of the following quantities as sums and differences of simpler logarithmic expressions. Express the answers in such a way that no logarithms of products, quotients, or powers appear.

(a) $\log_{10}\sqrt{3x}$     (b) $\log_{10}\sqrt[3]{\dfrac{2x}{3x^2 + 1}}$     (c) $\ln\dfrac{x^2\sqrt{2x - 1}}{(2x + 1)^{3/2}}$

**Solution** (a) $\log_{10}\sqrt{3x} = \log_{10}(3x)^{1/2} = \frac{1}{2}\log_{10}(3x)$

$$= \frac{1}{2}(\log_{10} 3 + \log_{10} x)$$

(b) $\log_{10}\sqrt[3]{\dfrac{2x}{3x^2 + 1}} = \log_{10}\left[\left(\dfrac{2x}{3x^2 + 1}\right)^{1/3}\right]$

$$= \frac{1}{3}\log_{10}\frac{2x}{3x^2 + 1}$$

$$= \frac{1}{3}[\log_{10} 2x - \log_{10}(3x^2 + 1)]$$

$$= \frac{1}{3}[\log_{10} 2 + \log_{10} x - \log_{10}(3x^2 + 1)]$$

(c) $\ln\dfrac{x^2\sqrt{2x - 1}}{(2x + 1)^{3/2}} = \ln\dfrac{x^2(2x - 1)^{1/2}}{(2x + 1)^{3/2}}$

$$= \ln x^2 + \ln(2x - 1)^{1/2} - \ln(2x + 1)^{3/2}$$

$$= 2\ln x + \frac{1}{2}\ln(2x - 1) - \frac{3}{2}\ln(2x + 1)$$

**EXAMPLE 4**   Given that $\log_{10} A = a$, $\log_{10} B = b$, and $\log_{10} C = c$, express $\log_{10} \dfrac{A^3}{B^4 \sqrt{C}}$ in terms of $a$, $b$, and $c$.

Solution   $\log_{10} \dfrac{A^3}{B^4 \sqrt{C}} = \log_{10} \dfrac{A^3}{B^4 C^{1/2}}$

$= \log_{10} A^3 - \log_{10} B^4 C^{1/2}$

$= \log_{10} A^3 - (\log_{10} B^4 + \log_{10} C^{1/2})$

$= 3 \log_{10} A - (4 \log_{10} B + \tfrac{1}{2} \log_{10} C)$

$= 3 \log_{10} A - 4 \log_{10} B - \tfrac{1}{2} \log_{10} C$

$= 3a - 4b - \tfrac{1}{2}c$

The properties of logarithms are also useful in solving equations, as the next two examples demonstrate.

**EXAMPLE 5**   Use logarithms to the base $e$ to solve the equation $10 = 3e^{1-2x}$ for $x$.

Solution   We take the natural (base $e$) logarithm of both sides of the given equation to obtain

$\ln 10 = \ln(3e^{1-2x})$

$= \ln 3 + \ln e^{1-2x}$   using property 2

$= \ln 3 + (1 - 2x)$   using the definition of ln

$2x = \ln 3 - \ln 10 + 1$

$x = \dfrac{\ln 3 - \ln 10 + 1}{2}$

This is the required solution. If we wish, we can use property 3 to rewrite this solution as

$x = \dfrac{\ln \frac{3}{10} + 1}{2}$

(There are yet other ways to write this answer. See Exercise 77.)

**EXAMPLE 6**   Solve for $x$: $\log_3 x + \log_3 (x + 2) = 1$

Solution   Using property 2, we can write the given equation as

$\log_3 [x(x + 2)] = 1$

or

$\log_3 (x^2 + 2x) = 1$

Writing this last equation in exponential form yields

$x^2 + 2x = 3^1$

$x^2 + 2x - 3 = 0$

$(x + 3)(x - 1) = 0$

Thus, we have

$$x + 3 = 0 \qquad \text{or} \qquad x - 1 = 0$$

and consequently,

$$x = -3 \qquad \text{or} \qquad x = 1$$

Now let us check these values in the original equation to see if they are indeed solutions.

If $x = -3$, the equation becomes

$$\underbrace{\log_3(-3)} + \underbrace{\log_3(-1)} \overset{?}{=} 1$$

Neither expression is defined, since the domain of the logarithm function does not contain negative numbers.

If $x = 1$, the equation becomes

$$\log_3 1 + \log_3 3 \overset{?}{=} 1$$
$$0 + 1 \overset{?}{=} 1 \qquad \text{True}$$

Thus, the value $x = 1$ is a solution of the original equation, but $x = -3$ is not. Why was an extraneous solution ($x = -3$) generated along with the correct solution $x = 1$? The difficulty arose in the second line of our solution, when we used the property $\log_b PQ = \log_b P + \log_b Q$. This property is valid only when $P$ and $Q$ are both positive.

It is sometimes necessary to convert logarithms in one base to logarithms in another base. After the next example, we will state a formula for this. However, as the next example indicates, it is easy to work this type of problem from the basics, without relying on a formula.

**EXAMPLE 7**   Express the quantity $\log_2 5$ in terms of base 10 logarithms.

Solution   Let $z = \log_2 5$. The exponential form of this equation is

$$2^z = 5$$

We now take the base 10 logarithm of each side of this equation to obtain

$$\log_{10} 2^z = \log_{10} 5$$
$$z \log_{10} 2 = \log_{10} 5 \qquad \text{using property 4}$$
$$z = \frac{\log_{10} 5}{\log_{10} 2}$$

In view of the definition of $z$, this last equation can be written

$$\log_2 5 = \frac{\log_{10} 5}{\log_{10} 2}$$

This is the required answer.

The method shown in Example 7 can be used to convert between any two bases. Exercise 76(b) at the end of this section asks you to follow this method, using letters rather than numbers, to arrive at the following general formula.

| Change of Base Formula |
|---|
| $$\log_a x = \frac{\log_b x}{\log_b a}$$ |

Up to now, all the examples in this section have dealt with applications of the five properties of logarithms. However, you also need to understand what the properties *don't* say. For instance, property 3 does not apply to an expression such as $\dfrac{\log_{10} 5}{\log_{10} 2}$. (Property 3 *would* apply if the expression were $\log_{10} \frac{5}{2}$.)

**EXAMPLE 8** How does the statement $(\log_b P)^n = n \log_b P$ differ from property 4? Give an example showing that this equation is not an identity.

Solution In property 4, the exponent $n$ on the left-hand side of the equation applies only to $P$, not to the quantity $\log_b P$. To provide the required example, we choose (almost any) convenient values for $b$, $P$, and $n$ for which both sides of the equation can be easily evaluated. Using $b = 10$, $P = 10$, and $n = 2$, the equation reads

$$(\log_{10} 10)^2 \overset{?}{=} 2 \log_{10} 10$$
$$1^2 \overset{?}{=} (2)(1)$$
$$1 \overset{?}{=} 2 \qquad \text{No!}$$

Thus, the given statement is not always true. The point here is that you must learn not only what the properties say, but also what they don't say.

## EXERCISE SET 6.4

### A

*In Exercises 1–10, simplify the given expressions by using the definition and properties for logarithms.*

1. $\log_{10} 70 - \log_{10} 7$
2. $\log_{10} 40 + \log_{10} \frac{5}{2}$
3. $\log_7 \sqrt{7}$
4. $\log_9 25 - \log_9 75$
5. $\log_3 108 + \log_3 \frac{3}{4}$
6. $\ln e^3 - \ln e$
7. $-\frac{1}{2} + \ln \sqrt{e}$
8. $e^{\ln 3} + e^{\ln 2} - e^{\ln e}$
9. $2^{\log_2 5} - 3 \log_5 \sqrt[3]{5}$
10. $\log_b b^b$

*In Exercises 11–19, write the given expression as a single logarithm with a coefficient of 1.*

11. $\log_{10} 30 + \log_{10} 2$
12. $2 \log_{10} x - 3 \log_{10} y$
13. $\log_5 6 + \log_5 \frac{1}{3} + \log_5 10$
14. $p \log_b A - q \log_b B + r \log_b C$
15. (a) $\ln 3 - 2 \ln 4 + \ln 32$
    (b) $\ln 3 - 2(\ln 4 + \ln 32)$

16. (a) $\log_{10}(x^2 - 16) - 3 \log_{10}(x + 4) + 2 \log_{10} x$
    (b) $\log_{10}(x^2 - 16) - 3[\log_{10}(x + 4) + 2 \log_{10} x]$
17. $\log_b 4 + 3[\log_b(1 + x) - \frac{1}{2} \log_b(1 - x)]$
18. $\ln(x^3 - 1) - \ln(x^2 + x + 1)$
19. $4 \log_{10} 3 - 6 \log_{10}(x^2 + 1)$
    $\qquad + \frac{1}{2}[\log_{10}(x + 1) - 2 \log_{10} 3]$

*In Exercises 20–26, write each quantity using sums and differences of simpler logarithmic expressions. Express each answer so that logarithms of products, quotients, and powers do not appear.*

20. (a) $\log_{10} \sqrt{(x + 1)(x + 2)}$
    (b) $\ln \sqrt{\dfrac{(x + 1)(x + 2)}{(x - 1)(x - 2)}}$
21. (a) $\log_{10} \dfrac{x^2}{1 + x^2}$ (b) $\ln \dfrac{x^2}{\sqrt{1 + x^2}}$

22. (a) $\log_b \dfrac{\sqrt{1-x^2}}{x}$  (b) $\ln \dfrac{x\sqrt[3]{4x+1}}{\sqrt{2x-1}}$

23. (a) $\log_{10}\sqrt{9-x^2}$  (b) $\ln \dfrac{\sqrt{4-x^2}}{(x-1)(x+1)^{3/2}}$

24. (a) $\log_b \sqrt[3]{\dfrac{x+3}{x}}$  (b) $\ln \dfrac{1}{\sqrt{x^2+x+1}}$

25. (a) $\log_b \sqrt{x/b}$
    (b) $2\ln\sqrt{(1+x^2)(1+x^4)(1+x^6)}$

26. (a) $\log_b \sqrt[3]{\dfrac{(x-1)^2(x-2)}{(x+2)^2(x+1)}}$  (b) $\ln\left(\dfrac{e-1}{e+1}\right)^{3/2}$

27. Suppose that $\log_{10} A = a$, $\log_{10} B = b$, and $\log_{10} C = c$. Express each of the following in terms of $a$, $b$, and $c$.
    (a) $\log_{10} AB^2 C^3$  (b) $\log_{10}\sqrt{10ABC}$
    (c) $\log_{10}(10A/\sqrt{BC})$  (d) $\log_{10}(100A^2/B^4\sqrt[3]{C})$
    (e) $\log_{10}[(AB)^5/C]$

28. If $\ln x = t$ and $\ln y = u$, write each of the following in terms of $t$ and $u$.
    (a) $\ln xy - \ln x^2$  (b) $\ln\sqrt{xy}+\ln(x/e)$
    (c) $\ln(e^2 x^3\sqrt{y})$

In Exercises 29–35, solve the given equations. Express your answers in terms of base $e$ logarithms.

29. $5 = 2e^{2x-1}$  30. $100 = 3e^x$  31. $3e^{1+t}-2$
32. $4e^{1-2t}=7$  33. $2^x = 9$  34. $5^{3x-1}=27$
35. $10\cdot 2^x = 5^x$

In Exercises 36–45, solve each equation for x; check your answers to remove any extraneous roots.

36. $\log_6 x + \log_6(x+1) = 1$
37. $\log_9(x+1) = \tfrac12 + \log_9 x$
38. $\log_2(x+4) = 2 - \log_2(x+1)$
39. $\log_{10}(2x+4) + \log_{10}(x-2) = 1$
40. $\ln x + \ln(x+1) = \ln 12$
41. $\log_{10}(x+3) - \log_{10}(x-2) = 2$
42. $\ln(x+1) = 2 + \ln(x-1)$
43. $\log_b(x+1) = 2\log_b(x-1)$
44. $\log_2(2x^2+4) = 5$
45. $\log_{10}(x-6) + \log_{10}(x+3) = 1$
46. Solve for $x$ in terms of $a$:
    $\log_2(x+a) - \log_2(x-a) = 1$
47. Solve for $x$ in terms of $y$.
    (a) $\log_{10} x - y = \log_{10}(3x-1)$
    (b) $\log_{10}(x-y) = \log_{10}(3x-1)$
48. Solve for $x$ in terms of $b$:  $\log_b(1-3x) = 3 + \log_b x$.

In Exercises 49–54, express each of the quantities in terms of base 10 logarithms.

49. $\log_2 5$  50. $\log_5 10$  51. $\ln 3$
52. $\ln 10$  53. $\log_b 2$  54. $\log_2 b$

In Exercises 55–59, express each of the quantities in terms of natural logarithms,

55. $\log_{10} 6$  56. $\log_2 10$  57. $\log_{10} e$
58. $\log_b 2$, where $b = e^2$  59. $\log_{10}(\log_{10} x)$

60. Give specific examples showing that the following rules are *false*.
    (a) $\log(x+y) = \log x + \log y$
    (b) $(\log x)/(\log y) = \log x - \log y$
    (c) $(\log x)(\log y) = \log x + \log y$
    (d) $(\log x)^k = k\log x$

61. True or false?
    (a) $\log_{10} A + \log_{10} B - \tfrac12\log_{10} C = \log_{10}(AB/\sqrt{C})$
    (b) $\log_e \sqrt{e} = \tfrac12$
    (c) $\ln\sqrt{e} = \tfrac12$
    (d) $\ln x^3 = \ln 3x$
    (e) $\ln x^3 = 3\ln x$
    (f) $\ln 2x^3 = 3\ln 2x$
    (g) $\log_a c = b$ means $a^b = c$.
    (h) $\log_5 24$ is between $5^1$ and $5^2$.
    (i) $\log_5 24$ is between 1 and 2.
    (j) $\log_5 24$ is closer to 1 than to 2.
    (k) The domain of $g(x) = \ln x$ is the set of all real numbers.
    (l) The range of $g(x) = \ln x$ is the set of all real numbers.
    (m) The function $g(x) = \ln x$ is one-to-one.

C  Use a calculator for Exercises 62–75.

62. Check property 2 using the values $b = 10$, $P = \pi$, and $Q = \sqrt{2}$.
63. Let $P = 3$ and $Q = 4$. Show that $\ln(P+Q) \ne \ln P + \ln Q$.
64. Check property 3 using the values $b = 10$, $P = 2$, and $Q = 3$.
65. Let $P = 10$ and $Q = 20$. Show that $\ln(PQ) \ne (\ln P)(\ln Q)$.
66. Check property 3 using natural logarithms and the values $P = 19$ and $Q = 89$.
67. Show that $(\log_{10} 19)/(\log_{10} 89) \ne \log_{10} 19 - \log_{10} 89$.
68. Show that $(\ln 19)/(\ln 89) \ne \ln 19 - \ln 89$.
69. Check property 4 using the values $b = 10$, $P = \pi$, and $n = 7$.
70. Using the values given for $b$, $P$, and $n$ in Exercise 69, show that $\log_b P^n \ne (\log_b P)^n$.
71. Verify property 5 using the values $b = 10$ and $P = 1776$.

**72.** Verify that $\ln 2 + \ln 3 + \ln 4 = \ln 24$.

**73.** Verify that $\log_{10} A + \log_{10} B + \log_{10} C = \log_{10}(ABC)$, using the values $A = 11$, $B = 12$, and $C = 13$.

**74.** Let $f(x) = e^x$ and $g(x) = \ln x$. Compute $f[g(2345.6)]$.

**75.** Let $f(x) = 10^x$ and $g(x) = \log_{10} x$. Compute $g[f(0.123456)]$.

**B**

**76.** (a) Prove that $\log_b(P/Q) = \log_b P - \log_b Q$.

Hint: Study the proof of property 2 in the text.

(b) Prove the **change of base formula**:

$$\log_a x = \frac{\log_b x}{\log_b a}$$

Hint: Use the method of Example 7 in the text.

**77.** In Example 5 we solved the equation $10 = 3e^{1-2x}$ and found that $x = (\ln 0.3 + 1)/2$.

(a) Show that this $x$-value indeed satisfies the given equation.

(b) Show that an equivalent form for $x$ is $\frac{1}{2} + \ln\sqrt{0.3}$.

(c) Show that another equivalent form for $x$ is $\ln\sqrt{3e/10}$.

**78.** Solve $\log_b(2x + 1) = 2 + \log_b x$ for $x$ (in terms of $b$). What restrictions does your solution impose on $b$?

**79.** Show that $\log_b \dfrac{\sqrt{3} + \sqrt{2}}{\sqrt{3} - \sqrt{2}} = 2\log_b(\sqrt{3} + \sqrt{2})$.

**80.** (a) Show that $\log_b(P/Q) + \log_b(Q/P) = 0$.

(b) Simplify: $\log_a x + \log_{1/a} x$.

**81.** Simplify: $b^{3\log_b x}$.

**82.** Let $\log_{10} 2 = a$ and $\log_{10} 3 = b$. Express each of the following in terms of $a$ and/or $b$.

(a) $\log_{10} 4$    (b) $\log_{10} 8$    (c) $\log_{10} 5$

(d) $\log_{10} 6$    (e) $\log_{10} \frac{5}{64}$    (f) $\log_{10} 108$

(g) $\log_{10} \sqrt[3]{12}$    (h) $\log_{10} 0.0027$

**83.** Simplify: $\log_2 \sqrt[5]{4\sqrt{2}}$.

**84.** Solve for $x$ in terms of $\alpha$ and $\beta$: $\alpha \ln x + \ln \beta = 0$.

**85.** Solve for $x$ in terms of $\alpha$ and $\beta$: $3 \ln x = \alpha + 3 \ln \beta$.

**86.** Is there a constant $k$ such that the equation $e^x = 2^{kx}$ holds for all values of $x$?

**87.** Prove that $\log_b a = 1/(\log_a b)$.

**88.** Simplify: $(\log_2 3)(\log_3 4)(\log_4 5)$.

**C**

**89.** Prove that $(\log_a x)/(\log_{ab} x) = 1 + \log_a b$.

**90.** Simplify $a^x$ when $x = [\log_b(\log_b a)]/\log_b a$.

**91.** If $a^2 + b^2 = 7ab$, and $a$ and $b$ are positive, show that

$$\log[\tfrac{1}{3}(a + b)] = \tfrac{1}{2}(\log a + \log b)$$

no matter which base is used for the logarithms. (But it is understood that the same base is used throughout.)

**92.** Solve for $x$ (assuming $a > b > 0$):
$(a^4 - 2a^2b^2 + b^4)^{x-1} = (a - b)^{2x}(a + b)^{-2}$

Answer: $x = \dfrac{\ln(a - b)}{\ln(a + b)}$

**93.** Let $f(x) = \ln(x + \sqrt{x^2 + 1})$. Find $f^{-1}(x)$.

**94.** Suppose that $\log_{10} 2 = a$ and $\log_{10} 3 = b$. Solve the following equation for $x$ in terms of $a$ and $b$:

$$6^x = \frac{10}{3} - 6^{-x}$$

$\boxed{\text{c}}$ **95.** Let $y = \ln[\ln(\ln x)]$. First, complete the table using a calculator. After doing that, disregard the evidence in your table and prove (without a calculator, of course) that the range of the given function is actually the set of all real numbers.

| $x$ | 100 | 1000 | $10^6$ | $10^{20}$ | $10^{50}$ | $10^{99}$ |
|---|---|---|---|---|---|---|
| $y$ | | | | | | |

## APPLICATIONS

**6.5**

*$S = Pe^{rt}$. This result is remarkable both because of its simplicity and the occurrence of e. (Who would expect that number to pop up in finance theory?)*

PHILIP GILLETT in *Calculus and Analytic Geometry*, 3rd ed. (Lexington, Mass.: D. C. Heath & Company, 1988)

We begin this section by considering how money accumulates in a savings account. Eventually, this will lead us back to the number $e$ and the growth law $N = N_0 e^{kt}$, both of which arose in different contexts in Section 6.2.

The following idea from arithmetic is a prerequisite for our discussion. To increase a given quantity by, say, 15%, we multiply the quantity by 1.15. For instance, suppose we want to increase $100 by 15%. The calculations can be written

$$\$100 + 15\% \text{ of } \$100 = 100 + 0.15(100)$$
$$= 100(1 + 0.15) \qquad \text{factoring}$$
$$= 100(1.15)$$

Similarly, to increase a quantity by 30%, we would multiply by 1.30; and so on. The next example displays some calculations involving percentage increase. The results may surprise you unless you're already familiar with this topic.

**EXAMPLE 1**  An amount of $100 is increased by 15% and then the new amount is increased by 15%. Is this the same as an overall increase of 30%?

**Solution**  To increase $100 by 15%, we multiply by 1.15 to obtain $100(1.15)$. Now to increase this new amount by 15%, we multiply it by 1.15 to obtain

$$[(\$100)(1.15)](1.15) = \$100(1.15)^2$$
$$= \$132.25$$

On the other hand, if we increase the original $100 by 30%, we obtain

$$\$100(1.30) = \$130$$

Comparing, we see that the result of two successive 15% increases is greater than the result of a single 30% increase.

Now let's look at another example and use it to introduce some terminology. Suppose that you place $1000 in a savings account at 10% interest *compounded annually*. This means that at the end of each year, the bank contributes to your account 10% of the amount that is in the account at that time. Interest compounded in this manner is called **compound interest**. The original deposit of $1000 is called the **principal**, denoted $P$. The interest rate, expressed as a decimal, is denoted by $r$. Thus, $r = 0.10$ in this example. The variable $A$ is used to denote the **amount** in the account at any given time. The calculations displayed in Table 1 show how the account grows.

**TABLE 1**

| TIME PERIOD | ALGEBRA | ARITHMETIC |
|---|---|---|
| After 1 year | $A = P(1 + r)$ | $A = 1000(1.10)$ |
| | | $= \$1100$ |
| After 2 years | $A = [P(1 + r)](1 + r)$ | $A = 1000(1.10)^2$ |
| | $= P(1 + r)^2$ | $= \$1210$ |
| After 3 years | $A = [P(1 + r)^2](1 + r)$ | $A = 1000(1.10)^3$ |
| | $= P(1 + r)^3$ | $= \$1331$ |

There are several things we can learn from Table 1. First, consider how much interest is paid each year.

Interest paid for first year:    $\$1100 - \$1000 = \$100$

Interest paid for second year:   $\$1210 - \$1100 = \$110$

Interest paid for third year:    $\$1331 - \$1210 = \$121$

Thus the interest earned each year is not a constant; it increases each year. However, notice that the *ratio* of interest earned in successive years is constant:

$$\frac{110}{100} = 1.1$$

$$\frac{121}{110} = 1.1$$

This is because the given growth rate of 10% per year is constant.

If you look at the algebra in Table 1, you can see what the general formula should be for the amount after $t$ years.

---

**Compound Interest Formula**   (interest compounded annually)

Suppose that a principal of $P$ dollars is invested at an annual rate $r$, compounded annually. Then the amount $A$ after $t$ years is given by

$$A = P(1 + r)^t$$

---

**EXAMPLE 2**   Suppose that $2000 is invested at $7\frac{1}{2}\%$ interest compounded annually. How many years will it take for the money to double?

**Solution**   In the formula $A = P(1 + r)^t$, we use the given values $P = \$2000$ and $r = 0.075$. We want to find how long it will take for the money to double; that is, we want to find $t$ when $A = \$4000$. Making these substitutions in the formula, we obtain

$$4000 = 2000(1 + 0.075)^t$$

and therefore,

$$2 = 1.075^t \qquad \text{dividing by 2000}$$

We can solve this exponential equation by taking the logarithm of both sides. We will use base $e$ logarithms. (Base 10 would be just as convenient here.) This yields

$$\ln 2 = \ln 1.075^t$$

and consequently,

$$\ln 2 = t \ln 1.075 \qquad \text{(Why?)}$$

To isolate $t$, we divide both sides of this last equation by $\ln 1.075$. This yields

$$t = \frac{\ln 2}{\ln 1.075}$$

or, using a calculator

$$t \approx 9.6 \text{ years}$$

Now, assuming that the bank computes the compound interest only at the end of the year, we must round off the preliminary answer of 9.6 years and

**TABLE 2**

| $A = 2000(1.075)^t$ | |
| --- | --- |
| $t$ (YEARS) | $A$ (DOLLARS) |
| 9 | 3834.48 |
| 10 | 4122.06 |

say that when $t = 10$ years, the initial $2000 will have *more than* doubled. Table 2 adds some perspective to this. The table shows that after 9 years, something less than $4000 is in the account, whereas after 10 years, the amount exceeds $4000.

In Example 2, the interest was compounded annually. In practice, though, the interest is usually computed more often. For instance, a bank may advertise a rate of 10% per year compounded semiannually. This means that after half a year, the interest is compounded at 5%, and then after another half year, the interest is again compounded at 5%. If you review Example 1, you'll see that one compounding at a rate of $r$ is not the same as two compoundings, each at a rate of $r/2$. The formula $A = P(1 + r)^t$ can be generalized to cover such cases where interest is compounded more than once each year.

---

**Compound Interest Formula**
(interest compounded $n$ times per year)

Suppose that a principal of $P$ dollars is invested at an annual rate $r$ compounded $n$ times per year. Then the amount $A$ after $t$ years is given by

$$A = P\left(1 + \frac{r}{n}\right)^{nt}$$

---

**EXAMPLE 3** $1000 is placed in a savings account at 10% per annum. How much is in the account at the end of 1 year if the interest is **(a)** compounded once each year $(n = 1)$? **(b)** compounded quarterly $(n = 4)$?

**Solution** We use the formula $A = P(1 + r/n)^{nt}$.

**(a)** For $n = 1$, we obtain

$$A = 1000\left(1 + \frac{0.10}{1}\right)^{(1)(1)}$$
$$= 1000(1.1)$$
$$= \$1100$$

**(b)** For $n = 4$, we obtain

$$A = 1000\left(1 + \frac{0.10}{4}\right)^{4(1)}$$
$$= 1000(1.025)^4$$
$$= \$1103.81 \quad \text{using a calculator}$$

Notice here that compounding the interest quarterly rather than annually yields the greater amount. This is in agreement with our observations in Example 1.

The results in Example 3 will serve to illustrate some additional terminology used by financial institutions. In that example, the interest for the year under quarterly compounding was

$$\$1103.81 - \$1000 = \$103.81$$

Now, $103.81 is 10.381% of $1000. We say in this case that the **effective rate** of interest is 10.381%. The given rate of 10% per annum compounded once a year is called the **nominal rate**. The next example further illustrates these ideas.

**EXAMPLE 4** A bank offers an interest rate of 12% per annum for certain accounts. (This is the nominal rate.) Compute the effective rate if interest is compounded monthly.

Solution Let $P$ denote the principal earning 12% ($r = 0.12$), compounded monthly ($n = 12$). Then with $t = 1$, our formula yields

$$A = P\left(1 + \frac{0.12}{12}\right)^{12(1)}$$
$$= P(1.01)^{12}$$

Using a calculator to approximate the quantity $(1.01)^{12}$, we obtain

$$A \approx P(1.12683)$$

This shows that the effective interest rate is about 12.68%.

There are two rather natural questions to ask when upon first encountering compound interest calculations:

QUESTION 1 For a fixed period of time (say 1 year), does more and more frequent compounding of interest continue to yield greater and greater amounts?

QUESTION 2 Is there a limit on how much money can accumulate in a year when interest is compounded more and more frequently?

The answer to both of these questions is yes. If you look back over Example 1, you'll see evidence for the affirmative answer to question 1. For additional evidence, and for the answer to question 2, let's do some calculations. To keep things as simple as possible, suppose a principal of $1 is invested for 1 year at the nominal rate of 100% per annum. (More realistic figures could be used here, but the algebra becomes more cluttered.) With these data, our formula becomes

$$A = 1\left(1 + \frac{1}{n}\right)^{n(1)}$$

or

$$A = \left(1 + \frac{1}{n}\right)^{n}$$

Table 3 shows the results of compounding the interest more and more frequently.

Table 3 shows that the amount does increase with the number of compoundings. But assuming that the bank rounds off to the nearest penny, Table 3 also shows that there is no difference between compounding hourly, com-

**TABLE 3**

| NUMBER OF COMPOUNDINGS, $n$ | AMOUNT, $\left(1 + \frac{1}{n}\right)^n$ |
|---|---|
| $n = 1$ (annually) | $\left(1 + \frac{1}{1}\right)^1 = 2$ |
| $n = 2$ (semiannually) | $\left(1 + \frac{1}{2}\right)^2 = 2.25$ |
| $n = 4$ (quarterly) | $\left(1 + \frac{1}{4}\right)^4 \approx 2.44$ |
| $n = 12$ (monthly) | $\left(1 + \frac{1}{12}\right)^{12} \approx 2.61$ |
| $n = 365$ (daily) | $\left(1 + \frac{1}{365}\right)^{365} \approx 2.7146$ |
| $n = 8760$ (hourly) | $\left(1 + \frac{1}{8760}\right)^{8760} \approx 2.7181$ |
| $n = 525{,}600$ (each minute) | $\left(1 + \frac{1}{525{,}600}\right)^{525{,}600} \approx 2.71827$ |
| $n = 31{,}536{,}000$ (each second) | $\left(1 + \frac{1}{31{,}536{,}000}\right)^{31{,}536{,}000} \approx 2.71828$ |

pounding each minute, and compounding each second. In each case, the amount, when rounded off, is \$2.72.

The data in Table 3 suggest that the quantity $(1 + 1/n)^n$ gets closer and closer to the number $e$ as $n$ becomes larger and larger. In symbols,

$$\left(1 + \frac{1}{n}\right)^n \approx e \qquad \text{when } n \text{ is large}$$

Indeed, in many calculus books, the number $e$ is defined as the "limiting value" or "limit" of the quantity $(1 + 1/n)^n$ as $n$ grows ever larger. Admittedly, we have not defined here the meaning of "limiting value" or "limit." That is a topic for calculus. Nevertheless, Table 3 should give you a reasonable, if intuitive, appreciation of the idea.

Some banks advertise interest compounded not monthly, daily, or even hourly, but *continuously*—that is, at each instant. The formula for the amount earned under continuous compounding of interest is as follows.

**Compound Interest Formula**   (interest compounded continuously)

Suppose that a principal of $P$ dollars is invested at an annual rate $r$, compounded continuously. Then the amount $A$ after $t$ years is given by

$$A = Pe^{rt}$$

**EXAMPLE 5**  A sum of $100 is placed in a savings account at 5% per annum compounded continuously. Assuming no subsequent withdrawals or deposits, when will the balance reach $150?

**Solution**  Substitute the values $A = 150$, $P = 100$, and $r = 0.05$ in the formula $A = Pe^{rt}$ to obtain

$$150 = 100e^{0.05t}$$

and therefore,

$$1.5 = e^{0.05t}$$

To solve this last equation for $t$, we rewrite it in its equivalent exponential form:

$$0.05t = \ln 1.5$$
$$t = \frac{\ln 1.5}{0.05} \text{ years}$$

Using a calculator, we find that $t \approx 8.1$ years. In other words, it will take slightly more than 8 years and 1 month for the balance to reach $150.

In the next example, we compare the nominal rate with the effective rate under continuous compounding of interest.

**EXAMPLE 6**  (a)  Given a nominal rate of 8% per annum compounded continuously, compute the effective interest rate.
(b)  Given an effective rate of 8% per annum, compute the nominal rate.

**Solution**  (a)  With the values $r = 0.08$ and $t = 1$, the formula $A = Pe^{rt}$ yields

$$A = Pe^{0.08(1)}$$
$$A \approx P(1.08329) \qquad \text{using a calculator}$$

This shows that the effective interest rate is approximately 8.33% per year.
(b)  We now wish to compute the nominal rate $r$, given an effective rate of 8% per year. An effective rate of 8% means that the initial principal $P$ grows to $P(1.08)$ by the end of the year. Thus, in the formula $A = Pe^{rt}$, we make the substitutions $A = P(1.08)$ and $t = 1$. This yields

$$P(1.08) = Pe^{r(1)}$$

Dividing both sides of this last equation by $P$, we have

$$1.08 = e^r$$

To solve this equation for $r$, we rewrite it in its equivalent logarithmic form:

$$r = \ln(1.08)$$
$$r \approx 0.07696 \qquad \text{using a calculator}$$

Thus, a nominal rate of about 7.70% per annum yields an effective rate of 8%. Table 4 summarizes these results.

Now we come to one of the remarkable and characteristic features of growth governed by the formula $A = Pe^{rt}$. By the **doubling time** we mean, as the

**TABLE 4**

Comparison of nominal and effective rates in Example 6

| NOMINAL RATE (PER ANNUM) | EFFECTIVE RATE (PER ANNUM) |
|---|---|
| 8% | 8.33% |
| 7.70% | 8% |

name implies, the amount of time required for a given principal to double. The surprising fact here is that the doubling time does not depend on the principal $P$. To see why this is so, we begin with the formula

$$A = Pe^{rt}$$

We are interested in the time $t$ at which $A = 2P$. Replacing $A$ by $2P$ in the formula yields

$$2P = Pe^{rt}$$
$$2 = e^{rt}$$
$$rt = \ln 2$$
$$t = \frac{\ln 2}{r}$$

Denoting the doubling time by $T_2$, we have the following formula.

$$\text{Doubling time} = T_2 = \frac{\ln 2}{r}$$

As you can see, the formula for the doubling time $T_2$ does not involve $P$, but only $r$. Thus, at a given rate under continuous compounding, $2 and $2000 would both take the same amount of time to double. (This idea takes some getting used to.)

**EXAMPLE 7**  Compute the doubling time $T_2$ when a sum is invested at an interest rate of 4% per annum compounded continuously.

Solution    $T_2 = \dfrac{\ln 2}{r} = \dfrac{\ln 2}{0.04} \approx 17.3$ years     using a calculator

There is a convenient approximation that allows us easily to estimate doubling times. Using a calculator, we see that

$$\ln 2 \approx 0.7$$

Using this approximation, we have the following rule of thumb for estimating doubling times.

$$T_2 \approx \frac{0.7}{r}$$

Let's use this rule to rework Example 7. With $r = 0.04$, we obtain

$$T_2 \approx \frac{0.7}{0.04} = \frac{70}{4} = 17.5 \text{ years}$$

Notice that this estimation is quite close to the actual doubling time obtained in Example 7.

The idea of doubling time turns out to be useful in graphing the function $A = Pe^{rt}$. In this discussion, we assume that $P$ and $r$ are constants, so that the amount $A$ is a function of the time $t$. Suppose, for example, that a principal of $1000 is invested at 10% per annum compounded continuously. Then the function we wish to graph is

$$A = 1000e^{0.1t}$$

Now, the doubling time in this situation is

$$T_2 \approx \frac{0.7}{r} = \frac{0.7}{0.1} = 7 \text{ years}$$

Table 5 shows the results of doubling a principal of $1000 every 7 years.

**TABLE 5**

| $t$ (years) | 0 | 7 | 14 | 21 | 28 |
|---|---|---|---|---|---|
| $A$ (dollars) | 1000 | 2000 | 4000 | 8000 | 16000 |

In view of the data in Table 5, we'll mark off units on the $t$-axis in multiples of 7; on the $A$-axis, we'll use multiples of 2000. Figure 1 shows the result of plotting the points from Table 5 and then joining them with a smooth curve. Notice that the domain in this context is $[0, \infty)$.

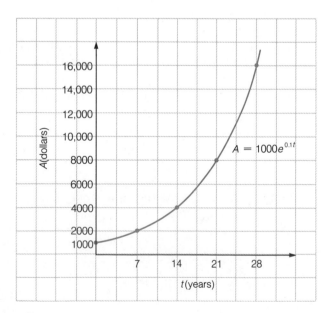

**FIGURE 1**

You may have already noticed that aside from the arbitrary choice of letters, there is no difference between the functions $A = Pe^{rt}$ and $N = N_0 e^{kt}$. Both formulas have the same form. Thus, this single function serves as a model for phenomena as diverse as continuous compounding of interest, population growth, and radioactive decay.

In newspapers and in everyday speech, the term *exponential growth* is used rather loosely to describe any situation involving rapid growth. In the sciences, however, **exponential growth** refers specifically to growth governed by functions of the form $N = N_0 e^{kt}$, with $k > 0$. And since the function $A = Pe^{rt}$ has this form, we say that money grows exponentially under continuous compounding of interest. Similarly, in the sciences, **exponential decay** refers specifically to decay governed by functions of the form $N = N_0 e^{kt}$, with $k < 0$.

In the next example we predict the population of the world in the year 2000, assuming that the population grows exponentially. In this context, it can be shown that the constant $k$ in the formula $N = N_0 e^{kt}$ represents the **relative growth rate** of the population $N$.* Algebraically, we will work with $k$ just as we did in the calculations with compound interest, where $k$ represented an interest rate.

**EXAMPLE 8**  Statistical projections from the United Nations indicate that the relative growth rate for the world's population during the years 1985–2000 will be approximately 1.6% per year. (This is down from the all-time high of 2% per year in 1970.) Use the formula $N = N_0 e^{kt}$ to predict the world population in the year 2000, given that the population in 1985 was 4.837 billion.

**Solution**  Let $t = 0$ correspond to the year 1985. Then the year 2000 corresponds to the value $t = 15$. Our given data therefore are $k = 0.016$, $t = 15$, and $N_0 = 4.837$ (in units of one billion). Using these values in the formula $N = N_0 e^{kt}$, we obtain

$$N = 4.837 e^{(0.016)15}$$

$$N \approx 6.149$$

Thus, our prediction for the world's population in the year 2000 is 6.149 billion.

In previous sections, we used the function $N = N_0 e^{kt}$, with $k < 0$, to describe radioactive decay. Let's return to that idea now. Just as we used the idea of a doubling time to graph an exponential growth function, we can use the half-life concept, introduced in Section 6.2, to graph an exponential decay function. Consider, for example, the radioactive element iodine-131, which has a half-life of about 8 days. Table 6 shows what fraction of an initial amount remains at 8-day intervals. Using the data in this table, we can draw the graph for the decay function $N = N_0 e^{kt}$ for iodine-131 (see Figure 2). Note that we are able to construct this graph without specifically evaluating the decay constant $k$.

TABLE 6

| $t$ (days) | 0 | 8 | 16 | 24 | 32 |
|---|---|---|---|---|---|
| $N$ (amount) | $N_0$ | $\frac{1}{2}N_0$ | $\frac{1}{4}N_0$ | $\frac{1}{8}N_0$ | $\frac{1}{16}N_0$ |

---

* The relative growth rate is $(\Delta N/N)/(\Delta t)$; it is the fractional change in $N$ per unit time. This is distinct from (but sometimes confused with) the average growth rate, $\Delta N/\Delta t$.

$$N = N_0 e^{kt}$$

**FIGURE 2**

The next example deals with the subject of nuclear energy. The purpose of the example is not to present an argument for or against the use of nuclear power. Rather, the purpose is to show you that with an understanding of exponential decay, you will be better equipped to read about and to evaluate the issues on your own.

**EXAMPLE 9**   An article on nuclear energy appeared in the January 1976 issue of *Scientific American*. The author of the article is physicist Hans Bethe, a Nobel Prize winner. At one point in the article, Professor Bethe discusses the disposal (through burial) of radioactive waste material from a nuclear reactor. The particular waste product under discussion is plutonium-239.

> ... Plutonium-239 has a half-life of nearly 25,000 years, and 10 half-lives are required to cut the radioactivity by a factor of 1000. Thus, the buried wastes must be kept out of the biosphere for 250,000 years.

(a) Supply the detailed calculations behind the statement that ten half-lives are required before the radioactivity is reduced by a factor of 1000.

(b) Show how the figure of ten half-lives can be obtained by estimation, as opposed to detailed calculation.

**Solution**   (a) Let $N_0$ denote the initial amount of plutonium-239 at time $t = 0$. Then the amount $N$ present at time $t$ is given by $N = N_0 e^{kt}$. We wish to determine $t$ when $N = \frac{1}{1000} N_0$. First, we determine the value of $k$ by using the half-life information. To say that the half-life is 25,000 years is to say that $N = \frac{1}{2}N_0$ when $t = 25,000$. Substituting these values into the formula $N = N_0 e^{kt}$ yields

$$\tfrac{1}{2}N_0 = N_0 e^{k(25,000)}$$
$$\tfrac{1}{2} = e^{25,000k} \qquad \text{dividing by } N_0$$

Writing this last equation in its equivalent logarithmic form, we obtain

$$25,000k = \ln\tfrac{1}{2}$$

and therefore,

$$k = \frac{\ln\tfrac{1}{2}}{25,000}$$

Now we can find $t$ when $N = \frac{1}{1000}N_0$, as required. Substituting $\frac{1}{1000}N_0$ for $N$ in the decay law gives us

$$\frac{1}{1000}N_0 = N_0 e^{kt} \qquad \left(\text{where } k \text{ is } \frac{\ln\tfrac{1}{2}}{25,000}\right)$$

$$\frac{1}{1000} = e^{kt}$$

$$kt = \ln\frac{1}{1000}$$

$$t = \frac{\ln\frac{1}{1000}}{k} = \frac{\ln\frac{1}{1000}}{(\ln\tfrac{1}{2})/25,000}$$

$$t = \frac{(\ln\frac{1}{1000})(25,000)}{\ln\tfrac{1}{2}}$$

Using a calculator, we obtain the value 249,144.6 years; however, given the time scale involved, it would be ludicrous to announce the answer in this form. Instead, we round off to the nearest thousand years and say that after 249,000 years, the radioactivity will have decreased by a factor of 1000. Notice that this result confirms Professor Bethe's ballpark estimate of ten half-lives, or 250,000 years.

**(b)** AFTER ONE HALF-LIFE: $\qquad N = \dfrac{N_0}{2}$

AFTER TWO HALF-LIVES: $\qquad N = \dfrac{1}{2}\left(\dfrac{N_0}{2}\right) = \dfrac{N_0}{2^2}$

AFTER THREE HALF-LIVES: $\qquad N = \dfrac{1}{2}\left(\dfrac{N_0}{2^2}\right) = \dfrac{N_0}{2^3}$

Following this pattern, we see that after ten half-lives, we should have

$$N = \frac{N_0}{2^{10}}$$

But as we noted in the first section of this chapter, $2^{10}$ is approximately 1000. Therefore, we have

$$N \approx \frac{N_0}{1000} \qquad \text{after ten half-lives}$$

This is in agreement with Professor Bethe's statement.

## EXERCISE SET 6.5

1. You invest $800 at 6% interest, compounded annually. How much is in the account after 4 years, assuming that you make no subsequent withdrawals or deposits?

2. A sum of $1000 is invested at an interest rate of $5\frac{1}{2}\%$ compounded annually. How many years will it take before the sum exceeds $2500? (First find out when the amount equals $2500; then round off as in Example 2.)

3. At what interest rate compounded annually will a sum of $4000 grow to $6000 in 5 years?

4. A bank pays 7% interest compounded annually. What principal will grow to $10,000 in 10 years?

5. You place $500 in a savings account at 5% compounded annually. After 4 years you withdraw all of your money and take it to a different bank, which advertises a rate of 6% compounded annually. What is the balance in this new account after 4 more years? (As usual, assume that no subsequent withdrawals or deposits are made.)

6. A sum of $3000 is placed in a savings account at 6% per annum. How much is in the account after 1 year if the interest is (a) compounded annually? (b) compounded semiannually? (c) compounded daily?

7. A sum of $1000 is placed in a savings account at 7% per annum. How much is in the account after 20 years if the interest is (a) compounded annually? (b) compounded quarterly?

8. Your friend invests $2000 at $5\frac{1}{4}\%$ per annum compounded semiannually. You invest an equal amount at the same yearly rate, but compounded daily. How much larger is your account than your friend's after 8 years?

9. You invest $100 at 6% per annum compounded quarterly. How long will it take for your balance to exceed $120? (Round off your answer to the nearest next quarter.)

10. A bank offers an interest rate of 7% per annum compounded daily. What is the effective rate?

11. What principal should you deposit at $5\frac{1}{2}\%$ per annum compounded semiannually so as to have $6000 after 10 years?

12. You place a sum of $800 in a savings account at 6% per annum compounded continuously. Assuming that you make no subsequent withdrawals or deposits, how much is in the account after 1 year? Also, when will the balance reach $1000?

13. A bank offers an interest rate of $6\frac{1}{2}\%$ per annum compounded continuously. What principal will grow to $5000 in 10 years under these conditions?

14. Given a nominal rate of 6% per annum, compute the effective rate under continuous compounding of interest.

15. Suppose that under continuous compounding of interest, the effective rate is 6% per annum. Compute the nominal rate.

16. You have two savings accounts, each with an initial principal of $1000. The nominal rate on both accounts is $5\frac{1}{4}\%$ per annum. In the first account, interest is compounded semiannually. In the second account, interest is compounded continuously. How much more is in the second account after 12 years?

17. You want to invest $10,000 for 5 years and you have a choice between two accounts. The first pays 6% per annum compounded annually. The second pays 5% per annum compounded continuously. Which is the better investment?

18. Suppose that a certain principal is invested at 6% per annum compounded continuously.
    (a) Use the rule $T_2 \approx 0.7/r$ to estimate the doubling time.
    (b) Compute the doubling time using the formula $T_2 = (\ln 2)/r$.
    (c) Do your answers in (a) and (b) differ by more than 2 months?

19. A sum of $1500 is invested at 5% per annum compounded continuously.
    (a) Estimate the doubling time.
    (b) Compute the actual doubling time.
    (c) Let $d_1$ and $d_2$ denote the actual and estimated doubling times, respectively. Define $d$ by
    $$d = |d_1 - d_2|$$
    What percent is $d$ of the actual doubling time?

20. A sum of $5000 is invested at 10% per annum compounded continuously.
    (a) Estimate the doubling time.
    (b) Estimate the time required for the $5000 to grow to $40,000.

21. After carrying out the calculations in this problem, you'll see one of the reasons why the government imposes inheritance taxes and why there are laws that prohibit savings accounts from being passed from generation to generation without restrictions. Suppose that a family invests $1000 at 8% per annum compounded continuously. If this account were to remain intact, being passed from generation to generation, for 300 years, how much would be in the account at the end of those 300 years?

**22.** A principal of $500 is invested at 7% per annum compounded continuously.
  (a) Estimate the doubling time.
  (b) Sketch a graph similar to the one in Figure 1, showing how the amount increases with time.

**23.** A principal of $7000 is invested at 5% per annum, compounded continuously.
  (a) Estimate the doubling time.
  (b) Sketch a graph showing how the amount increases with time.

**24.** In one savings account, a principal of $1000 is deposited at 5% per annum. In a second account, a principal of $500 is deposited at 10% per annum. Both accounts compound interest continuously.
  (a) Estimate the doubling time for each account.
  (b) On the same set of axes, sketch graphs showing the amount of money in each account over time. Give the (approximate) coordinates of the point where the two curves meet. In financial terms, what is the significance of this point? (In working this problem, assume that the initial deposits in each account were made at the same time.)
  (c) During what period of time does the first account have the larger balance?

The statistics presented in Exercises 25–27 are taken from the Global 2000 Report to the President. *The following statement by then President Jimmy Carter provides some background on the report.*

*I am directing the Council on Environmental Quality and the Department of State, working in cooperation with . . . other appropriate agencies, to make a one-year study of the probable changes in the world's population, natural resources, and environment through the end of the century.*

President Jimmy Carter May 23, 1977

**25.** Complete the following table on world population projections, assuming that the populations grow exponentially and that the indicated relative growth rates are valid through the year 2000.

| REGION | 1975 POPULATION (BILLIONS) | PERCENT OF POPULATION IN 1975 | RELATIVE GROWTH RATE (PERCENT PER YEAR) | YEAR 2000 POPULATION (BILLIONS) | PERCENT OF WORLD POPULATION IN 2000 |
|---|---|---|---|---|---|
| World | 4.090 | 100 | 1.8 | ? | ? |
| More developed regions | 1.131 | ? | 0.6 | ? | ? |
| Less developed regions | 2.959 | ? | 2.1 | ? | ? |

**26.** Complete the following table. Assume that the growth is exponential and that the given relative growth rates are in effect through the year 2000.

| COUNTRY | 1975 POPULATION (BILLIONS) | RELATIVE GROWTH RATE (PERCENT PER YEAR) | YEAR 2000 POPULATION (BILLIONS) | PERCENT INCREASE IN POPULATION |
|---|---|---|---|---|
| United States | 0.214 | 0.6 | ? | ? |
| People's Republic of China | 0.935 | 1.4 | ? | ? |
| Mexico | 0.060 | 3.1 | ? | ? |

**27.** The *Global 2000 Report* actually contains three sets of world population estimates and projections, as shown in the following table. Complete the table.

| PROJECTION | 1975 POPULATION (BILLIONS) | RELATIVE GROWTH RATE (PERCENT PER YEAR) | YEAR 2000 POPULATION (BILLIONS) | PERCENT INCREASE IN POPULATION |
|---|---|---|---|---|
| Low | 4.043 | 1.5 | ? | ? |
| Medium | 4.090 | 1.8 | ? | ? |
| High | 4.134 | 2.0 | ? | ? |

28. According to the United States Bureau of the Census, the population of the United States grew most rapidly during the period 1800–1810, and least rapidly during the period 1930–1940. Use the following data to compute the relative growth rate (percent per year) for each of these two periods. In 1800, the population was 5,308,483; in 1810, it was 7,239,881. In 1930, the population was 123,202,624; in 1940, it was 132,164,569.

29. According to the U.S. Bureau of the Census, the population of the United States in the year 1850 was 23,191,876; in 1900, the population was 62,947,714.
    (a) Assuming that the population grew exponentially during this period, compute the growth constant $k$.
    (b) Assuming continued growth at the same rate, predict the 1950 population.
    (c) The actual population for 1950, according to the Bureau of the Census, was 150,697,361. How does this compare to the prediction in part (b)? Was the actual growth over that period (1900–1950) faster or slower than exponential growth with the growth constant $k$ determined in part (a)?

*In Exercises 30–32, use the population statistics given in the following table, provided by the U.S. Bureau of the Census.*

| STATE | 1930 | 1940 | 1950 |
|---|---|---|---|
| California | 5,677,251 | 6,907,387 | 10,586,223 |
| New York | 12,588,066 | 13,479,142 | 14,830,192 |
| North Dakota | 680,845 | 641,935 | 619,636 |

30. (a) Assume that the population of California grew exponentially over the period 1930–1940. Compute the growth constant $k$ and express it as a percent (per year).
    (b) Assume that the population of California grew exponentially over the period 1940–1950. Compute the growth constant $k$ and express it as a percent.
    (c) What would the 1980 California population have been if the relative growth rate obtained in part (b) had remained in effect throughout the period 1950–1980? *Hint:* Let $t = 0$ correspond to 1950. Then you want to find $N$ when $t = 30$.
    (d) How does your prediction in part (c) compare to the actual 1980 population of 23,668,562?

31. Repeat Exercise 30 for New York. *Note:* The 1980 population of New York was 17,557,288.

32. Repeat Exercise 30 for North Dakota. *Note:* The 1980 population of North Dakota was 652,695.

33. (a) The half-life of radium-226 is 1620 years. Draw a graph of the decay function for radium-226, similar to that shown in Figure 2.

(b) The half-life of radium A is 3 min. Draw a graph of the decay function for radium A.

34. (a) The half-life of thorium-232 is $1.4 \times 10^{10}$ years. Draw a graph of the decay function.
    (b) The half-life of thorium A is 0.16 sec. Draw a graph of the decay function.

35. The half-life of plutonium-241 is 13 years.
    (a) Compute the decay constant $k$.
    (b) What fraction of an initial sample $N_0$ would remain after 10 years? after 100 years?

36. The half-life of thorium-229 is 7340 years.
    (a) Compute the time required for a given sample to be reduced by a factor of 1000. Show detailed calculations, as in Example 9(a).
    (b) Express your answer in part (a) in terms of half-lives.
    (c) As in Example 9(b), estimate the time required for a given sample of thorium-229 to be reduced by a factor of 1000. Compare your answer with that obtained in part (b).

37. Strontium-90, with a half-life of 28 years, is one of the waste products from nuclear fission reactors. One of the reasons great care is taken in the storage and disposal of this substance stems from the fact that strontium-90 is, in some chemical respects, similar to ordinary calcium. Thus, strontium-90 in the biosphere, entering the food chain via plants or animals, would eventually be absorbed into our bones. (In fact, everyone already has a measurable amount of strontium-90 in their bones as a result of fallout from atmospheric nuclear tests.)
    (a) Compute the decay constant $k$ for strontium-90.
    (b) Compute the time required if a given quantity of strontium-90 is to be stored until the radioactivity is reduced by a factor of 1000.
    (c) Using half-lives, estimate the time required for a given sample to be reduced by a factor of 1000. Compare your answer with that in (b).

38. (a) Suppose that country $X$ violates the ban against above-ground nuclear testing and, as a result, an island is contaminated with debris containing the radioactive substance iodine-131. A team of scientists from the United Nations wants to visit the island to look for clues in determining which country was involved. However, the level of radioactivity from the iodine-131 is estimated to be 30,000 times the safe level. Approximately how long must the team wait before it is safe to visit the island? The half-life of iodine-131 is 8 days.
    (b) Rework part (a), assuming instead that the radioactive substance is strontium-90 rather than iodine-131. The half-life of strontium-90 is 28 years. Assume, as before, that the initial level of radioactivity

is 30,000 times the safe level. (The purpose of this problem is to underscore the difference between a half-life of 8 days and one of 28 years.)

39. In 1969, the United States National Academy of Sciences issued a report entitled *Resources and Man*. One conclusion in the report is that a world population of 10 billion "is close to (if not above) the maximum that an intensively managed world might hope to support with some degree of comfort and individual choice." (The figure "10 billion" is sometimes referred to as the "carrying capacity" of the Earth.)
   (a) When the report was issued in 1969, the world population was about 3.6 billion, with a relative growth rate of 2% per year. Assuming continued exponential growth at this rate, estimate the year in which the Earth's carrying capacity of 10 billion might be reached.
   (b) The data in the following table, from the *Global 2000 Report*, show the world population estimates for 1975. Use the "low" figures to estimate the year in which the Earth's carrying capacity of 10 billion might be reached.
   (c) Use the "high" figures to estimate the year in which the Earth's carrying capacity of 10 billion might be reached.

| PROJECTION | POPULATION (BILLIONS) | RELATIVE GROWTH RATE (PERCENT PER YEAR) |
|---|---|---|
| Low | 4.043 | 1.5 |
| High | 4.134 | 2.0 |

40. *Depletion of Nonrenewable Energy Resources* Suppose that the world population grows exponentially. Then, as a first approximation, it is reasonable to assume that the use of energy resources such as petroleum and coal also grows exponentially. Under these conditions, the following formula can be derived (using calculus):

$$A = \frac{A_0}{k} (e^{kT} - 1)$$

where $A$ is the amount of the resource consumed from time $t = 0$ to $t = T$, the quantity $A_0$ is the amount of the

resource consumed during the year $t = 0$, and $k$ is the relative growth rate of annual consumption.
   (a) Show that solving the formula for $T$ yields

$$T = \frac{\ln[(Ak/A_0) + 1)]}{k}$$

This formula gives the "life expectancy" $T$ for a given resource. In the formula, $A_0$ and $k$ are as previously defined, and $A$ represents the total amount of the resource available.
   (b) In the *Global 2000 Report to the President*, it was estimated that the 1976 worldwide consumption of oil was 21.7 billion barrels and that the total remaining oil resources ultimately available were 1661 billion barrels. Compute the "life expectancy" for oil (i) if the relative growth rate $k$ is 1% per year; (ii) if $k = 2$% per year; (iii) if $k = 3$% per year. (This last is, in fact, the predicted rate for the next decade.)
   (c) In the *Global 2000 Report*, it was estimated that the 1976 worldwide consumption of natural gas was approximately 50 tcf (trillions of cubic feet), while the total remaining gas ultimately available was 8493 tcf. Compute the "life expectancy" for natural gas if $k = 2$% per year.

41. In this exercise, the term *nonrenewable energy resources* refers collectively to petroleum, natural gas, coal, shale oil, and uranium. In the *Global 2000 Report*, it was estimated that in 1976, the total worldwide consumption of these nonrenewable resources amounted to 250 quadrillion Btu. It was also estimated that in 1976, the total nonrenewable resources ultimately available were 161,241 quadrillion Btu. Use the formula for $T$ in Exercise 40 to compute the "life expectancy" for these nonrenewable energy resources
   (a) if $k = 1$% per year    (b) if $k = 2$% per year

42. The data in the following table, from the *Global 2000 Report*, show the world reserves for selected minerals. The term *world reserves* refers to known resources available with current technology. Use the formula for $T$ derived in Exercise 40 to estimate the "depletion date" for each mineral under these conditions (depletion date = $T + 1976$).

| MINERAL | 1976 WORLD RESERVES | 1976 CONSUMPTION | RELATIVE GROWTH RATE (PERCENT PER YEAR) |
|---|---|---|---|
| Fluorine (million short tons) | 37 | 2.1 | 4.58 |
| Silver (million troy ounces) | 6,100 | 305 | 2.33 |
| Tin (thousand metric tons) | 10,000 | 241 | 2.05 |
| Copper (million short tons) | 503 | 8.0 | 2.94 |
| Phosphate rock (million metric tons) | 25,732 | 107 | 5.17 |
| Aluminum in bauxite (million short tons) | 5,610 | 18 | 4.29 |

**43.** In 1956, scientists B. Gutenberg and C. F. developed a formula to estimate the amount of energy $E$ released in an earthquake. The formula is

$$\log_{10} E = 11.4 + 1.5M$$

where $E$ is the energy in units of ergs and $M$ is the Richter magnitude.

**(a)** Solve the formula for $E$.

**(b)** The table shows the Richter magnitudes for two large earthquakes that occurred in 1985. If $E_1$ denotes the energy of the quake in Indonesia, and $E_2$ denotes the energy of the quake in Mexico, compute the ratio $E_2/E_1$ to compare the energies of the two earthquakes.

| DATE | LOCATION | M (MAGNITUDE) |
|------|----------|---------------|
| April 13, 1985 | Indonesia | 6.8 |
| September 19, 1985 | Mexico | 7.8 |

**44.** The age of some rocks can be estimated by measuring the ratio of the amounts of certain chemical elements within the rock. The method known as the *rubidium-strontium method* will be discussed here. This method has been used in dating the moon rocks brought back on the Apollo missions.

Rubidium-87 is a radioactive substance with a half-life of $4.7 \times 10^{10}$ years. Rubidium-87 decays into strontium-87, which is stable (nonradioactive). We are going to derive the following formula for the age of a rock:

$$T = \frac{\ln(N_s/N_r + 1)}{-k}$$

where $T$ is the age of the rock, $k$ is the decay constant for rubidium-87, $N_s$ is the number of atoms of strontium-87 now present in the rock, and $N_r$ is the number of atoms of rubidium-87 now present in the rock.

**(a)** Assume that initially, when the rock was formed, there were $N_0$ atoms of rubidium-87 and none of strontium-87. Then as time goes by, some of the rubidium atoms decay into strontium atoms, but the total number of atoms must still be $N_0$. Thus, after $T$ years, we have

$$N_0 = N_r + N_s$$

or, equivalently,

$$N_s = N_0 - N_r \tag{1}$$

However, according to the law of exponential decay for the rubidium-87, we must have $N_r = N_0 e^{kT}$. Solve this equation for $N_0$ and then use the result to eliminate $N_0$ from equation (1). Show that the

result can be written

$$N_s = N_r e^{-kT} - N_r \tag{2}$$

**(b)** Solve equation (2) for $T$ to obtain the formula given at the beginning of this exercise.

**45.** (Continuation of Exercise 44) The half-life of rubidium-87 is $4.7 \times 10^{10}$ years. Compute the decay constant $k$.

**46.** (Continuation of Exercise 44) Analysis of lunar rock samples taken on the Apollo 11 mission showed the strontium-rubidium ratio to be

$$\frac{N_s}{N_r} = 0.0588$$

Estimate the age of these lunar rocks.

**47.** (Continuation of Exercise 44) Analysis of the so-called genesis rock sample taken on the Apollo 15 mission revealed a strontium-rubidium ratio of 0.0636. Estimate the age of this rock.

**48.** *Radiocarbon Dating* Because rubidium-87 decays so slowly, the technique of rubidium-strontium dating discussed in the previous exercises is generally considered effective only for objects older than 10 million years. By way of contrast, archeologists and geologists rely on the method of *radiocarbon dating* in assigning ages ranging from 500 to 50,000 years.

There are two types of carbon that occur naturally in our environment. The first is carbon-12, which is nonradioactive; the second is carbon-14, which has a half-life of 5730 years. All living plant and animal tissue contains both types of carbon, always in the same ratio. (The ratio is one part carbon-14 to $10^{12}$ parts carbon-12.) As long as the plant or animal is living, this ratio is maintained. When the organism dies, however, no new carbon-14 is absorbed, and the amount of carbon-14 begins to decrease exponentially. Since the amount of carbon-14 decreases exponentially, it follows that the level of radioactivity also must decrease exponentially. The formula describing this situation is

$$N = N_0 e^{kT}$$

where $T$ is the age of the sample, $N$ is the present level of radioactivity (in units of disintegrations per hour per gram of carbon), and $N_0$ is the level of radioactivity $T$ years ago, when the organism was alive. Given that the half-life of carbon-14 is 5730 years and that $N_0 = 920$ disintegrations per hour per gram, show that the age $T$ of a sample is given by

$$T = \frac{5730 \ln(N/920)}{\ln \frac{1}{2}}$$

In Exercises 49–52, use the formula derived in Exercise 48 to estimate the age of each sample. *Note: There are some technical complications that arise in interpreting such results. Studies have shown that the ratio of carbon-12 to carbon-14 in the air, and therefore in living matter, has not in fact been constant over time. For instance, air pollution from factory smokestacks tends to increase the level of carbon-12. In the other direction, nuclear bomb testing increases the level of carbon-14.*

49. Prehistoric cave paintings were discovered in the Lascaux cave in France. Charcoal from the site was analyzed and the level of radioactivity was found to be $N = 141$ disintegrations per hour per gram. Estimate the age of the paintings.

50. Before radiocarbon dating was used, historians estimated that the tomb of Vizier Hemaka, in Egypt, was constructed about 4900 years ago. After radiocarbon dating became available, wood samples from the tomb were analyzed, and it was determined that the radioactivity level was 510 disintegrations per hour per gram. Estimate the age of the tomb on the basis of this reading and compare your answer to the figure already mentioned.

51. Analyses of the oldest campsites in the Western Hemisphere reveal a carbon-14 radioactivity level of $N = 226$ disintegrations per hour per gram. Show that this implies an age of 11,500 years, to the nearest 500 years. (This would correspond to the last Ice Age. At that time, the sea level was significantly lower than today. It is believed that at the time of the last Ice Age, there was land across the Bering Strait and humans first entered the Western Hemisphere by this route.)

52. The Dead Sea Scrolls are a collection of ancient manuscripts discovered in caves along the west bank of the Dead Sea. (The discovery occurred by accident when an Arab herdsman of the Taamireh tribe was searching for a stray goat.) When the linen wrappings on the scrolls were analyzed, the carbon-14 radioactivity level was found to be 723 disintegrations per hour per gram. Estimate the age of the scrolls using this information. Historical evidence suggests that some of the scrolls date back somewhere between 150 B.C. and A.D. 40. How do these dates compare with the estimate derived using radiocarbon dating?

## CHAPTER 6 SUMMARY OF PRINCIPAL TERMS AND FORMULAS

| TERMS OR FORMULAS | PAGE REFERENCE | COMMENTS |
|---|---|---|
| 1. Exponential function with base $b$ | 273 | The exponential function with base $b$ is defined by the equation $y = b^x$. It is understood here that the base $b$ is a positive number other than 1. |
| 2. Asymptote | 273 | A line is said to be an asymptote for a curve if the separation distance between the curve and the line approaches zero as we move out farther and farther along the line. |
| 3. The number $e$ | 277 | The irrational number $e$ is one of the basic constants in mathematics, as is the irrational number $\pi$. To five decimal places, the value of $e$ is 2.71828. In calculus $e$ is the base most commonly used for exponential functions. The graph of the exponential function $y = e^x$ is shown in the box on page 279. |
| 4. $N = N_0 e^{kt}$     $(k > 0)$ | 280 | In many instances throughout this chapter, we studied situations in which a quantity $N$ grows or increases according to this formula, called the *growth law*. In this formula, the quantity $N$ is a function of the time $t$, and $N_0$ denotes the value of $N$ at time $t = 0$. The number $k$ is a constant, called the *growth constant*, which is related to the rate at which $N$ increases. |
| 5. $N = N_0 e^{kt}$     $(k < 0)$ | 281 | This is called the *decay law* for radioactive substances. In the formula, $N$ denotes the amount of the substance present at time $t$, and $N_0$ is the amount of the substance present at time $t = 0$. The negative number $k$ is a constant, called the *decay constant*, which is related to the rate at which the decay takes place. |

| TERMS OR FORMULAS | PAGE REFERENCE | COMMENTS |
|---|---|---|
| 6. Half-life | 282 | The half-life of a radioactive substance is the time required for half of a given sample to disintegrate. Notice that the half-life is an amount of time, not an amount of the substance. |
| 7. $\log_b x$ | 285 | The expression $\log_b x$ denotes the power to which $b$ must be raised to yield $x$. The equation $\log_b x = y$ is equivalent to $b^y = x$. |
| 8. $\ln x$ | 289 | The expression $\ln x$ means $\log_e x$. Logarithms to the base $e$ are known as *natural logarithms*. For the graph of $y = \ln x$, see Figure 5 on page 289. |
| 9. $\log_a x = \dfrac{\log_b x}{\log_b a}$ | 300 | This is the change of base formula for converting logarithms from one base to another. |
| 10. $A = P\left(1 + \dfrac{r}{n}\right)^{nt}$ | 305 | This formula gives the amount $A$ that accumulates after $t$ years when a principal of $P$ dollars is invested at an annual rate $r$, compounded $n$ times per year. |
| 11. $A = Pe^{rt}$ | 307 | This formula gives the amount $A$ that accumulates after $t$ years when a principal of $P$ dollars is invested at an annual rate $r$, compounded continuously. |

## CHAPTER 6   REVIEW EXERCISES

NOTE   The fifteen exercises in this set with colored problem numbers constitute a chapter test.

*In Exercises 1–12, graph each function and specify any asymptotes or intercepts.*

1. $y = e^x$
2. $y = -e^{-x}$
3. $y = \ln x$
4. $y = \ln(x + 2)$
5. $y = 2^{x+1} + 1$
6. $y = \log_{10}(-x)$
7. $y = (1/e)^x$
8. $y = (\frac{1}{2})^{-x}$
9. $y = e^{x+1} + 1$
10. $y = -\log_2(x + 1)$
11. $y = \ln(e^x)$
12. $y = e^{\ln x}$

*In Exercises 13–29, solve each equation for x. (When logarithms appear in your answer, leave the answer in that form, rather than using a calculator.)*

13. $\log_4 x + \log_4(x - 3) = 1$
14. $\log_3 x + \log_3(2x + 5) = 1$
15. $\ln x + \ln(x + 2) = \ln 15$
16. $\log_6 \dfrac{x + 4}{x - 1} = 1$
17. $\log_2 x + \log_2(3x + 10) - 3 = 0$
18. $2 \ln x - 1 = 0$
19. $3 \log_9 x = \frac{1}{2}$
20. $e^{2x} = 6$
21. $e^{1 - 5x} = 3\sqrt{e}$
22. $2^x = 100$
23. $\log_{10} x - 2 = \log_{10}(x - 2)$
24. $\log_{10}(x^2 - x - 10) = 1$
25. $\ln(x + 2) = \ln x + \ln 2$
26. $\ln(2x) = \ln 2 + \ln x$
27. $\ln x^4 = 4 \ln x$
28. $(\ln x)^4 = 4 \ln x$
29. $\log_{10} x = \ln x$
30. Solve for $x$:   $(\ln x)/(\ln 3) = \ln x - \ln 3$. Use a calculator to evaluate your result; round off to the nearest integer.   *Answer:* 206765

*In Exercises 31–46, simplify each expression (without using a calculator).*

31. $\log_{10} \sqrt{10}$
32. $\log_7 1$
33. $\ln \sqrt[5]{e}$
34. $\log_3 54 - \log_3 2$
35. $\log_{10} \pi - \log_{10} 10\pi$
36. $\log_2 2$
37. $10^t$ where $t = \log_{10} 16$
38. $e^{\ln 5}$
39. $\ln(e^4)$
40. $\log_{10}(10^{\sqrt{2}})$

**41.** $\log_{12} 2 + \log_{12} 18 + \log_{12} 4$

**42.** $(\log_{10} 8)/(\log_{10} 2)$    **43.** $(\ln 100)/(\ln 10)$

**44.** $\log_5 2 + \log_{1/5} 2$    **45.** $\log_2 \sqrt[7]{16\sqrt[3]{2\sqrt{2}}}$

**46.** $\log_{1/8}(\frac{1}{16}) + \log_5 0.02$

*In Exercises 47–50, express the given quantities in terms of a, b, and c, where $a = \log_{10} A$, $b = \log_{10} B$, and $c = \log_{10} C$.*

**47.** $\log_{10} A^2 B^3 \sqrt{C}$    **48.** $\log_{10} \sqrt[3]{AC/B}$

**49.** $16 \log_{10} \sqrt{A} \sqrt[4]{B}$    **50.** $6 \log_{10}[B^{1/3}/(A\sqrt{C})]$

*In Exercises 51–56, find consecutive integers n and n + 1 such that the given expression lies between n and n + 1. Do not use a calculator.*

**51.** $\log_{10} 209$    **52.** $\ln 2$

**53.** $\log_6 100$    **54.** $\log_{10} \frac{1}{12}$

**55.** $\log_{10} 0.003$    **56.** $\log_3 244$

**57.** **(a)** On the same set of axes, graph the curves $y = \ln(x + 2)$ and $y = \ln(-x) - 1$. According to your graph, in which quadrant do the two curves intersect?

   **(b)** Find the x-coordinate of the intersection point.

**58.** The curve $y = ae^{bt}$ passes through the point $(3, 4)$ and has a y-intercept of 2. Find $a$ and $b$.

**59.** A certain radioactive substance has a half-life of $T$ years. Find the decay constant (in terms of $T$).

**60.** Find the half-life of a certain radioactive substance if it takes $T$ years for one-third of a given sample to disintegrate. (Your answer will be in terms of $T$.)

**61.** A radioactive substance has a half-life of $M$ min. What percentage of a given sample will remain after $4M$ min?

**62.** At the start of an experiment, there are $a$ bacteria in a colony. After $b$ hours, there are $c$ bacteria. Determine the population $d$ hours after the start of the experiment.

**63.** The half-life of a radioactive substance is $d$ days. If you begin with a sample weighing $b$ g, how long will it be until $c$ g remain?

**64.** Find the half-life of a radioactive substance if it takes $D$ days for $P\%$ of a given sample to disintegrate.

*In Exercises 65–70, write the given expression as a single logarithm with a coefficient of 1.*

**65.** $\log_{10} 8 + \log_{10} 3 - \log_{10} 12$

**66.** $4 \log_{10} x - 2 \log_{10} y$

**67.** $\ln 5 - 3 \ln 2 + \ln 16$

**68.** $\ln(x^4 - 1) - \ln(x^2 - 1)$

**69.** $a \ln x + b \ln y$

**70.** $\ln(x^3 + 8) - \ln(x + 2) - 2\ln(x^2 - 2x + 4)$

*In Exercises 71–78, write each quantity using sums and differences of simpler logarithmic expressions. Express each answer so that logarithms of products, quotients, and powers do not appear.*

**71.** $\ln \sqrt{(x - 3)(x + 4)}$    **72.** $\log_{10} \dfrac{x^2 - 4}{x + 3}$

**73.** $\log_{10} \dfrac{x^3}{\sqrt{1 + x}}$    **74.** $\ln \dfrac{x^2 \sqrt{2x + 1}}{2x - 1}$

**75.** $\log_{10} \sqrt[3]{x/100}$    **76.** $\log_{10} \sqrt{\dfrac{x^2 + 8}{(x^2 + 9)(x + 1)}}$

**77.** $\ln\left(\dfrac{1 + 2e}{1 - 2e}\right)^3$    **78.** $\log_{10} \sqrt[3]{x\sqrt{1 + y^2}}$

**79.** Suppose that $A$ dollars are invested at $R\%$ compounded annually. How many years will it take for the money to double?

**80.** A sum of \$2800 is placed in a savings account at 9% per annum. How much is in the account after 2 years if the interest is compounded quarterly?

**81.** A bank offers an interest rate of 9.5% per annum compounded monthly. Compute the effective interest rate.

**82.** A sum of $D$ dollars is placed in an account at $R\%$ per annum compounded continuously. When will the balance reach $E$ dollars?

**83.** Compute the doubling time for a sum of $D$ dollars invested at $R\%$ per annum compounded continuously.

**84.** Your friend invests $D$ dollars at $R\%$ per annum compounded semiannually. You invest an equal amount at the same yearly rate, but compounded daily. How much larger is your account than your friend's after $T$ years?

**85.** ▣ **(a)** You invest \$660 at 5.5% per annum, compounded quarterly. How long will it take for your balance to reach \$1000? Round off your answer (upward) to the nearest quarter of a year.

   **(b)** You invest $D$ dollars at $R\%$ per annum, compounded quarterly. How long will it take for your balance to reach $nD$ dollars? (Assume that $n > 1$.)

*In Exercises 86–90, find the domain of each function.*

**86.** **(a)** $y = \ln x$    **87.** **(a)** $y = \log_{10} \sqrt{x}$

   **(b)** $y = e^x$    **(b)** $y = \sqrt{\log_{10} x}$

**88.** **(a)** $f \circ g$   where $f(x) = \ln x$ and $g(x) = e^x$

   **(b)** $g \circ f$   where $f$ and $g$ are as in part (a)

**89.** **(a)** $y = \log_{10}|x^2 - 2x - 15|$

   **(b)** $y = \log_{10}(x^2 - 2x - 15)$

**90.** $y = \dfrac{2 + \ln x}{2 - \ln x}$

**91.** Find the range of the function defined by $y = \dfrac{e^x + 1}{e^x - 1}$.

*In Exercises 92–102, use the accompanying graph and the properties of logarithms to estimate each quantity to the nearest tenth.*

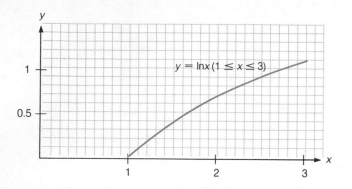

**92.** $\ln 2$    **93.** $\ln 0.5$    **94.** $\ln 3$

**95.** $\ln \frac{1}{9}$    **96.** $\ln 6$    **97.** $\ln 72$

**98.** $\ln 40.5$   *Hint:*   $40.5 = \frac{81}{2}$

**99.** $e$   *Hint:*   If $x = e$, then $\ln x = 1$.

**100.** $e^{1.1}$   *Hint:*   If $x = e^{1.1}$, then $\ln x = 1.1$.

**101.** $\log_2 3$    **102.** $\log_3 6$

**103.** Lester R. Brown, in his book *State of the World 1985* (New York: W. W. Norton and Company, 1985), makes the following statement:

*The projected [population] growth for North America, all of Europe, and the Soviet Union is less than the additions expected in either Bangladesh or Nigeria.*

In this exercise, you are asked to carry out the type of calculations that could be used to support Lester Brown's projection for the period 1990–2025. The source of the data in the following table is the Population Division of the United Nations.

**(a)** Complete the table, assuming that the populations grow exponentially and that the indicated growth rates are valid over the period 1990–2025.

**(b)** According to the projections in part (a), what will be the net increase in Nigeria's population over the period 1990–2025?

**(c)** According to the projections in part (a), what will be the net increase in the combined populations of North America, the Soviet Union, and Europe over the period 1990–2025?

**(d)** Compare your answers in parts (b) and (c). Do your results support or contradict Lester Brown's projection?

**104.** As of 1987, there were ten countries with populations exceeding 100 million. These countries are listed (in order of population, from largest to smallest) in the accompanying table. According to *World Population Profile: 1987*, published by the United States Census Bureau, "The latest projections suggest that India's population may surpass China's in less than 60 years, or before today's youngsters in both countries reach old age." Using the data in the table, estimate the populations of the ten countries in the year 2050. (Assume that the indicated growth rates are valid over the period 1987–2050.) Tabulate your final answers, listing the countries in order of population, from largest to smallest. Do your results for China and India support the projection by the Census Bureau?

| COUNTRY | 1987 POPULATION (MILLIONS) | RELATIVE GROWTH RATE (PERCENT PER YEAR) |
|---|---|---|
| 1. China | 1062.0 | 1.3 |
| 2. India | 800.3 | 2.1 |
| 3. Soviet Union | 284.0 | 0.9 |
| 4. United States | 243.8 | 0.7 |
| 5. Indonesia | 174.9 | 2.1 |
| 6. Brazil | 141.5 | 2.1 |
| 7. Japan | 122.2 | 0.6 |
| 8. Nigeria | 108.6 | 2.8 |
| 9. Bangladesh | 107.1 | 2.7 |
| 10. Pakistan | 104.6 | 2.9 |

Source: *1987 World Population Data Sheet*, published by Population Reference Bureau, Inc., Washington, D.C.

**105.** The figure on the following page shows a portion of the graph of $y = e^x$, for $1.0970 \leq x \leq 1.1000$.

**(a)** Use the graph to estimate $\ln 3$ to four decimal places.

| REGION | 1990 PROJECTED POPULATION (MILLIONS) | PROJECTED RELATIVE GROWTH RATE (PERCENT PER YEAR) | 2025 PROJECTED POPULATION (MILLIONS) |
|---|---|---|---|
| North America | 275.2 | 0.7 | ? |
| Soviet Union | 291.3 | 0.7 | ? |
| Europe | 499.5 | 0.2 | ? |
| Nigeria | 113.3 | 3.1 | ? |

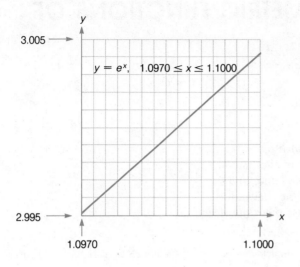

3.005

$y = e^x, \quad 1.0970 \le x \le 1.1000$

2.995

1.0970          1.1000

C **(b)** Using your calculator, compute the percent error in the approximation in part (a)

**(c)** Using the approximation in part (a), estimate the quantities $\ln \sqrt{3}$, $\ln 9$, and $\ln \frac{1}{3}$.

# 7 TRIGONOMETRIC FUNCTIONS OF ANGLES

*The subject of trigonometry is an excellent example of a branch of mathematics . . . which was motivated by both practical and intellectual interests—surveying, map-making, and navigation on the one hand, and curiosity about the size of the universe on the other. With it the Alexandrian mathematicians triangulated the universe and rendered precise their knowledge about the Earth and the heavens.*

MORRIS KLEIN in his text, *Mathematics in Western Culture* (New York: Oxford University Press, 1953)

**INTRODUCTION** There are two general approaches to the study of trigonometry at the precalculus level. The first approach centers around the study of triangles. Indeed, the word "trigonometry" is derived from two Greek words, *trigonon*, meaning triangle, and *metria*, meaning measurement. This is the approach that we shall follow in this chapter. The second approach, which we take up in Chapter 8, is in a sense a generalization of the first. Before getting down to specifics, we offer the following overview of Chapters 7 and 8, in terms of functions.

|  | INPUTS FOR THE FUNCTIONS | OUTPUTS | EMPHASIS |
|---|---|---|---|
| *Chapter 7* | Angles | Certain real numbers | The trigonometric functions are used to study triangles. |
| *Chapter 8* | Certain real numbers | Certain real numbers | The objects of study are the trigonometric functions themselves. |

## TRIGONOMETRIC FUNCTIONS OF ACUTE ANGLES

**7.1**

Most of the functions that we have considered up to this point were defined by means of equations that involved only the basic operations of algebra. Examples of such functions are

$$f(x) = x^2 \qquad g(x) = mx + b \qquad h(x) = \sqrt{x}$$

In this and the next two chapters, we will study the so-called trigonometric functions. The definitions of these functions involve some geometry, as well as algebra.

Let $\theta$ denote an acute angle in a right triangle, as indicated in Figure 1.* We define the six trigonometric functions of the acute angle $\theta$ in the box that follows. Notice that $\sin\theta$ and $\csc\theta$ are reciprocals of one another. Similarly, $\cos\theta$ and $\sec\theta$ are reciprocals, as are $\tan\theta$ and $\cot\theta$.

**FIGURE 1**

| NAME OF FUNCTION | ABBREVIATION | DEFINITION |
|---|---|---|
| sine | sin | $\sin\theta = \dfrac{\text{length of side opposite angle }\theta}{\text{length of hypotenuse}}$ |
| cosine | cos | $\cos\theta = \dfrac{\text{length of side adjacent angle }\theta}{\text{length of hypotenuse}}$ |
| tangent | tan | $\tan\theta = \dfrac{\text{length of side opposite angle }\theta}{\text{length of side adjacent angle }\theta}$ |
| cosecant | csc | $\csc\theta = \dfrac{\text{length of hypotenuse}}{\text{length of side opposite angle }\theta}$ |
| secant | sec | $\sec\theta = \dfrac{\text{length of hypotenuse}}{\text{length of side adjacent angle }\theta}$ |
| cotangent | cot | $\cot\theta = \dfrac{\text{length of side adjacent angle }\theta}{\text{length of side opposite angle }\theta}$ |

**EXAMPLE 1**    Figure 2 shows two right triangles. The first right triangle has sides 5, 12, and 13. The second right triangle is similar to the first (the angles are the same), but each side is 10 times longer than the corresponding side in the first triangle. Calculate and compare the values of $\sin\theta$, $\cos\theta$, and $\tan\theta$ for both triangles.

**Solution**

SMALL TRIANGLE

$$\sin\theta = \frac{\text{opposite}}{\text{hypotenuse}} = \frac{5}{13}$$

$$\cos\theta = \frac{\text{adjacent}}{\text{hypotenuse}} = \frac{12}{13}$$

$$\tan\theta = \frac{\text{opposite}}{\text{adjacent}} = \frac{5}{12}$$

LARGE TRIANGLE

$$\sin\theta = \frac{50}{130} = \frac{5}{13}$$

$$\cos\theta = \frac{120}{130} = \frac{12}{13}$$

$$\tan\theta = \frac{50}{120} = \frac{5}{12}$$

---

* The symbol "$\theta$" is the lowercase Greek letter *theta*. Greek letters are often used to represent angles. See the back endpapers for a listing of the complete Greek alphabet.

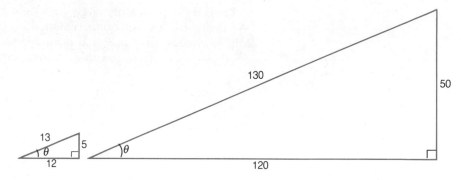

**FIGURE 2**

OBSERVATION   The corresponding values for $\sin \theta$, $\cos \theta$, and $\tan \theta$ are the same for both triangles.

The point of Example 1 is this. For a given acute angle $\theta$, the values of the trigonometric functions depend on the *ratios* of the lengths of the sides but not on the size of the particular right triangle in which $\theta$ resides.

**EXAMPLE 2**   Let $\theta$ be the acute angle indicated in Figure 3. Determine the six quantities $\sin \theta$, $\cos \theta$, $\tan \theta$, $\csc \theta$, $\sec \theta$, and $\cot \theta$.

**Solution**   Use the definitions.

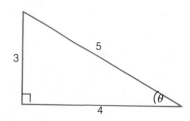

$$\sin \theta = \frac{\text{opposite}}{\text{hypotenuse}} = \frac{3}{5} \qquad \csc \theta = \frac{\text{hypotenuse}}{\text{opposite}} = \frac{5}{3}$$

$$\cos \theta = \frac{\text{adjacent}}{\text{hypotenuse}} = \frac{4}{5} \qquad \sec \theta = \frac{\text{hypotenuse}}{\text{adjacent}} = \frac{5}{4}$$

$$\tan \theta = \frac{\text{opposite}}{\text{adjacent}} = \frac{3}{4} \qquad \cot \theta = \frac{\text{adjacent}}{\text{opposite}} = \frac{4}{3}$$

**FIGURE 3**

Note the pairs of answers that are reciprocals. (This helps in memorizing the definitions.)

**EXAMPLE 3**   Let $\beta$ be the acute angle indicated in Figure 4. Find $\sin \beta$ and $\cos \beta$.

**Solution**   In view of the definitions, we need to know the length of the hypotenuse in Figure 4. If we call this length $x$, then by the Pythagorean Theorem we have

$$x^2 = 3^2 + 1^2 = 10$$
$$x = \sqrt{10}$$

Therefore

$$\sin \beta = \frac{\text{opposite}}{\text{hypotenuse}} = \frac{3}{\sqrt{10}} = \frac{3\sqrt{10}}{10}$$

and

$$\cos \beta = \frac{\text{adjacent}}{\text{hypotenuse}} = \frac{1}{\sqrt{10}} = \frac{\sqrt{10}}{10}$$

**FIGURE 4**

There are certain angles $\theta$ for which we can evaluate the trigonometric functions without the use of calculators or tables. Two such angles are 30° and 60°. To do this, we rely on the following theorem from geometry. (For a proof of this theorem, see Exercise 43 at the end of this section.)

---

**Theorem**

---

In a 30°–60° right triangle, the length of the side opposite the 30° angle is half the length of the hypotenuse.

**FIGURE 5**

---

If we let $y$ denote the length of the unmarked side of the triangle in Figure 5, then (according to the Pythagorean Theorem) we have

$$y^2 + x^2 = (2x)^2$$
$$y^2 = 4x^2 - x^2$$
$$y^2 = 3x^2$$
$$y = \sqrt{3}x$$

Our 30°–60° right triangle can now be labeled as in Figure 6. From this figure, the required trigonometric values can be written down by inspection. We will list only the values of the sine, cosine, and tangent, since the remaining three values are just the reciprocals of these.

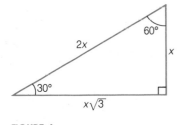

**FIGURE 6**

$$\sin 30° = \frac{x}{2x} = \frac{1}{2} \qquad\qquad \sin 60° = \frac{x\sqrt{3}}{2x} = \frac{\sqrt{3}}{2}$$

$$\cos 30° = \frac{x\sqrt{3}}{2x} = \frac{\sqrt{3}}{2} \qquad\qquad \cos 60° = \frac{x}{2x} = \frac{1}{2}$$

$$\tan 30° = \frac{x}{x\sqrt{3}} = \frac{1}{\sqrt{3}} = \frac{\sqrt{3}}{3} \qquad \tan 60° = \frac{x\sqrt{3}}{x} = \sqrt{3}$$

There are two observations to be made here. First, the final answers do not involve $x$. Once again this shows that for a given angle, it is the ratios of the sides that determine the trigonometric values, not the particular triangle. Second, notice that

$$\sin 30° = \cos 60°$$

and

$$\sin 60° = \cos 30°$$

**FIGURE 7**

It is easy to see that this is no coincidence. The terms "opposite" and "adjacent" are relative ones. As a glance at Figure 6 shows, the side opposite to one acute angle is automatically adjacent to the other acute angle.

Another angle for which we can compute the trigonometric values without the aid of a calculator or tables is 45°. Figure 7(a) and the calculations that accompany it show that if one acute angle in a right triangle is 45°, then the other acute angle is also 45°. It then follows from a theorem of geometry that the triangle is isosceles (that is, it has two equal legs), as indicated in Figure 7(b). If we use $y$ to denote the length of the unmarked side in Figure 7(b), then we have

$$y^2 = x^2 + x^2 = 2x^2$$
$$y = \sqrt{2}x$$

Our 45°–45° right triangle can then be labeled as in Figure 8. From this figure, the required trigonometric values can be obtained by inspection. As before, we list only the values for sine, cosine, and tangent, since the remaining three values are just the reciprocals of these.

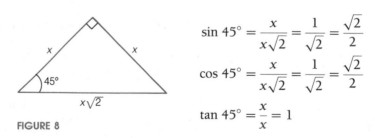

**FIGURE 8**

$$\sin 45° = \frac{x}{x\sqrt{2}} = \frac{1}{\sqrt{2}} = \frac{\sqrt{2}}{2}$$

$$\cos 45° = \frac{x}{x\sqrt{2}} = \frac{1}{\sqrt{2}} = \frac{\sqrt{2}}{2}$$

$$\tan 45° = \frac{x}{x} = 1$$

Table 1 summarizes the results we've now obtained for angles of 30, 45, and 60 degrees. These results should be memorized.

**TABLE 1**

| $\theta$ | $\sin \theta$ | $\cos \theta$ | $\tan \theta$ |
|---|---|---|---|
| 30° | $\dfrac{1}{2}$ | $\dfrac{\sqrt{3}}{2}$ | $\dfrac{\sqrt{3}}{3}$ |
| 45° | $\dfrac{\sqrt{2}}{2}$ | $\dfrac{\sqrt{2}}{2}$ | 1 |
| 60° | $\dfrac{\sqrt{3}}{2}$ | $\dfrac{1}{2}$ | $\sqrt{3}$ |

There are numerous identities involving the trigonometric functions. For now, in the context of right-triangle trigonometry, we'll consider only the three most basic ones. In Section 7.4 and in Chapter 8, we will take up this subject in greater detail and from a broader perspective. Recall that an **identity** is an equation that is satisfied by all relevant values of the variables concerned. Two examples of identities are $(x - y)(x + y) = x^2 - y^2$ and $x^3 = x^4/x$. The first of these is true no matter what real numbers are used for $x$ and $y$; the second is true for all real numbers except $x = 0$. In the box that follows, the notations

$$\sin^2 \theta \qquad \text{and} \qquad \cos^2 \theta$$

stand for

$$(\sin \theta)^2 \qquad \text{and} \qquad (\cos \theta)^2$$

respectively. (We will say more about this common notational convention in Section 7.4.)

| PROPERTY SUMMARY | Basic Identities for Sine and Cosine $\quad (0° < \theta < 90°)$ |
|---|---|
| IDENTITY | EXAMPLES |
| 1. $\sin^2 \theta + \cos^2 \theta = 1$ | 1. $\sin^2 30° + \cos^2 30° = 1$ <br> $\sin^2 19° + \cos^2 19° = 1$ |
| 2. $\dfrac{\sin \theta}{\cos \theta} = \tan \theta$ | 2. $\dfrac{\sin 60°}{\cos 60°} = \tan 60°$ |
| 3. $\sin(90° - \theta) = \cos \theta \quad$ and $\quad \cos(90° - \theta) = \sin \theta$ | 3. $\sin 70° = \cos 20° \qquad \sin 20° = \cos 70°$ |

**FIGURE 9**

All three of these identities can be proved by referring to Figure 9.

PROOF THAT $\sin^2 \theta + \cos^2 \theta = 1$
Looking at Figure 9, we have $\sin \theta = a/c$ and $\cos \theta = b/c$. Thus

$$\sin^2 \theta + \cos^2 \theta = \left(\frac{a}{c}\right)^2 + \left(\frac{b}{c}\right)^2 = \frac{a^2}{c^2} + \frac{b^2}{c^2} = \frac{a^2 + b^2}{c^2}$$

$$= \frac{c^2}{c^2} \qquad \text{using } a^2 + b^2 = c^2, \text{ the Pythagorean Theorem}$$

$$= 1 \qquad \text{as required}$$

PROOF THAT $(\sin \theta)/(\cos \theta) = \tan \theta$
Again, with reference to Figure 9, we have $\sin \theta = a/c$, $\cos \theta = b/c$, and $\tan \theta = a/b$. Therefore

$$\frac{\sin \theta}{\cos \theta} = \frac{a/c}{b/c}$$

$$= \frac{a}{c} \times \frac{c}{b} = \frac{a}{b}$$

$$= \tan \theta \qquad \text{as required}$$

PROOF THAT $\sin(90° - \theta) = \cos\theta$

First of all, since the sum of the angles in any triangle is $180°$, we have

$$\theta + \beta + 90° = 180°$$
$$\beta = 90° - \theta$$

Then $\sin(90° - \theta) = \sin\beta = b/c$. But also $\cos\theta = b/c$. Thus

$$\sin(90° - \theta) = \cos\theta$$

since both expressions equal $b/c$. This is what we wanted to prove.

The proof that $\cos(90° - \theta) = \sin\theta$ is entirely similar and we shall omit it. We can conveniently summarize these last two results by recalling the notion of complementary angles. Two acute angles are said to be **complementary** provided their sum is $90°$. Thus, the two angles $\theta$ and $90° - \theta$ are complementary. In view of this, we can restate the last two results as follows.

> If two angles are complementary, then the sine of (either) one equals the cosine of the other.

Incidentally, this result gives us an insight into the origin of the term *cosine*. *Cosine* is a shortened form of the phrase *complement's sine*.

**EXAMPLE 4**   Suppose that $B$ is an acute angle and $\cos B = \frac{2}{5}$. Find $\sin B$ and $\tan B$.

Solution   We'll show two methods. The first uses the identities $\sin^2 B + \cos^2 B = 1$ and $(\sin B)/(\cos B) = \tan B$. The second makes direct use of the Pythagorean Theorem.

FIRST METHOD   Replace $\cos B$ with $\frac{2}{5}$ in the identity $\sin^2 B + \cos^2 B = 1$. This yields

$$\sin^2 B + \left(\frac{2}{5}\right)^2 = 1$$
$$\sin^2 B = 1 - \frac{4}{25} = \frac{21}{25}$$
$$\sin B = \sqrt{\frac{21}{25}} = \frac{\sqrt{21}}{5}$$

Notice that we've chosen the positive square root here. This is because the values of the trigonometric functions of an acute angle are by definition positive. *Caution:* In Section 7.3 we'll extend the definitions of the trigonometric functions to include angles that are not acute. In those cases, some of the trigonometric values will not be positive; we'll then need to pay more attention to the matter of signs.

For $\tan B$, we have

$$\tan B = \frac{\sin B}{\cos B} = \frac{\sqrt{21}/5}{2/5} = \frac{\sqrt{21}}{2}$$

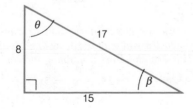

FIGURE 10

ALTERNATE METHOD    Since $\cos B = \dfrac{2}{5} = \dfrac{\text{adjacent}}{\text{hypotenuse}}$, we can work with a right triangle labeled as in Figure 10. Using the Pythagorean Theorem, we have $2^2 + x^2 = 5^2$, from which it follows that $x = \sqrt{21}$. Consequently,

$$\sin B = \frac{\text{opposite}}{\text{hypotenuse}} = \frac{x}{5} = \frac{\sqrt{21}}{5}$$

and

$$\tan B = \frac{\text{opposite}}{\text{adjacent}} = \frac{x}{2} = \frac{\sqrt{21}}{2} \qquad \text{as obtained previously.}$$

**EXAMPLE 5**    If $\theta$ is an acute angle and $\sin \theta = x$, express the other five trigonometric values as functions of $x$.

Solution    We can use either of the methods shown in Example 4. We'll demonstrate the first method here. Replacing $\sin \theta$ with $x$ in the identity $\sin^2 \theta + \cos^2 \theta = 1$ yields

$$x^2 + \cos^2 \theta = 1$$
$$\cos^2 \theta = 1 \quad x^2$$
$$\cos \theta = \sqrt{1 - x^2} \qquad \text{(Why is the positive root appropriate in this context?)}$$

For $\tan \theta$, now, we have

$$\tan \theta = \frac{\sin \theta}{\cos \theta} = \frac{x}{\sqrt{1 - x^2}}$$

The values of $\csc \theta$, $\sec \theta$, and $\cot \theta$ are, by definition, the reciprocals of $\sin \theta$, $\cos \theta$, and $\tan \theta$, respectively. So we have

$$\csc \theta = \frac{1}{x} \qquad \sec \theta = \frac{1}{\sqrt{1 - x^2}} \qquad \cot \theta = \frac{\sqrt{1 - x^2}}{x}$$

# EXERCISE SET 7.1

## A

1.  (a) Check that $8^2 + 15^2 = 17^2$. This shows that the triangle in the accompanying figure is indeed a right triangle.
    (b) Find $\sin \theta$ and $\cos \beta$.
    (c) Find $\cos \theta$ and $\sin \beta$.

    (d) Find $\tan \theta$, $\csc \theta$, $\sec \theta$, and $\cot \theta$.
    (e) Find $\tan \beta$, $\csc \beta$, $\sec \beta$, and $\cot \beta$.

2.  (a) Find $x$ in the accompanying figure.
    (b) Find the values of the six trigonometric functions of $\theta$.
    (c) Find the values of the six trigonometric functions of $\beta$.

*In Exercises 3–10, refer to the following figure. (Each problem, however, is independent of the others.)*

3. Suppose $AC = 3$ and $BC = 2$.
   (a) Find $\sin A$, $\cos A$, and $\tan A$.
   (b) Using the values in part (a), verify that $\sin^2 A + \cos^2 A = 1$ and $(\sin A)/(\cos A) = \tan A$.
   (c) Find $\cos B$. Then check that $\sin^2 A + \cos^2 B \neq 1$.

4. Suppose $AC = 6$ and $BC = 2$.
   (a) Compute $\sin B$, $\cos B$, and $\tan B$.
   (b) Using the values in part (a), verify that $\sin^2 B + \cos^2 B = 1$.   and   $(\sin B)/(\cos B) = \tan B$.
   (c) Find $\sin A$. Then check that $\sin^2 A + \cos^2 B \neq 1$.

5. If $AB = 13$ and $BC = 5$, compute the values of the six trigonometric functions of angle $A$.

6. If $AB = 29$ and $AC = 21$, compute the values of the six trigonometric functions of angle $B$.

7. Suppose $AC = 1$ and $BC = \frac{3}{4}$.
   (a) Compute $\sin B$ and $\cos A$.
   (b) Compute $\cos B$ and $\sin A$.
   (c) Compute the product $(\tan A)(\tan B)$.

8. Suppose $AC = BC = 1$.
   (a) Compute $\csc A$, $\sec A$, and $\cot A$.
   (b) Compute $\csc B$, $\sec B$, and $\cot B$.

9. Suppose $AC = \sqrt{11}$ and $BC = \sqrt{5}$.
   (a) Compute $\sin A$ and $\cos A$. Then verify that $\sin^2 A + \cos^2 A = 1$.
   (b) Compute $\tan B$, $\sin B$, and $\cos B$. Then verify that $\tan B = (\sin B)/(\cos B)$.

10. Suppose $AC = \sqrt{2} + 1$ and $BC = \sqrt{2} - 1$. Compute the values of the six trigonometric functions of angle $B$. For each answer, rationalize the denominator.

*In Exercises 11–22, verify that each equation is correct. (The purpose of Exercises 11–22 is twofold. First, doing the problems will help you to memorize the values in Table 1. Second, the exercises serve as an algebra review.)*

11. $\cos 60° = \cos^2 30° - \sin^2 30°$

12. $\cos 60° = 1 - 2\sin^2 30°$

13. $\sin^2 30° + \sin^2 45° + \sin^2 60° = \frac{3}{2}$

14. $\sin 30° \cos 60° + \cos 30° \sin 60° = 1$

15. $2 \sin 30° \cos 30° = \sin 60°$

16. $2 \sin 45° \cos 45° = 1$

17. $\sin 30° = \sqrt{\dfrac{1 - \cos 60°}{2}}$

18. $\cos 30° = \sqrt{\dfrac{1 + \cos 60°}{2}}$

19. $\tan 30° = \dfrac{\sin 60°}{1 + \cos 60°}$

20. $\tan 30° = \dfrac{1 - \cos 60°}{\sin 60°}$

21. $1 + \tan^2 45° = \sec^2 45°$

22. $1 + \cot^2 60° = \csc^2 60°$

*For Exercises 23–26, refer to the following figures. In each case, express the indicated trigonometric values as functions of $x$. (Rationalize any denominators containing radicals.)*

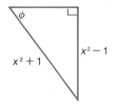

23. (a) $\sin \theta$, $\cos \theta$, $\tan \theta$
    (b) $\sin^2 \theta$, $\cos^2 \theta$, $\tan^2 \theta$
    (c) $\sin(90° - \theta)$, $\cos(90° - \theta)$, $\tan(90° - \theta)$

24. (a) $\csc \alpha$, $\sec \alpha$, $\cot \alpha$
    (b) $\sin^2 \alpha + \cos^2 \alpha + \tan^2 \alpha$

25. (a) $\sin \beta$, $\cos \beta$, $\tan \beta$
    (b) $\csc \beta$, $\sec \beta$, $\cot \beta$
    (c) $\sin(90° - \beta)$, $\cos(90° - \beta)$, $\tan(90° - \beta)$

26. (a) $\sin \phi$, $\cos \phi$, $\tan \phi$
    (b) $(\csc \phi)(\sec \phi)(\cot \phi)$
    (c) $\sin(90° - \phi)$, $\cos(90° - \phi)$, $\tan(90° - \phi)$

*In Exercises 27–32, use the given information to determine the values of the remaining five trigonometric functions. (The angles are assumed to be acute angles.)*

27. $\cos B = \frac{4}{7}$

28. $\cos B = \frac{3}{8}$

29. $\sin \theta = 2\sqrt{3}/5$

30. $\sin \theta = \frac{7}{25}$

31. $\tan A = \dfrac{\sqrt{2} - 1}{\sqrt{2} + 1}$

32. $\tan A = \dfrac{2 - \sqrt{3}}{2 + \sqrt{3}}$

*In Exercises 33–36, use the given information to express the remaining five trigonometric values of the acute angle $\theta$ in terms of $x$. (Rationalize any denominators containing radicals.)*

33. $\sin \theta = x/2$

34. $\sin \theta = 3x/5$

35. $\cos \theta = x^2$

36. $\cos \theta = (x - 1)/(x + 1)$

*Exercises 37–40 are calculator exercises. Their point is simply to familiarize you with the use of your calculator for computing trigonometric values. Round off each of the final answers to five decimal places. Note:* Check to see that your calculator is set in the "degree mode," rather than the "radian mode." (Radian measure for angles is discussed in the next chapter.)

[c] 37. (a) Evaluate $\cos 30°$, $\cos 45°$, and $\sin 60°$.
   (b) In Table 1, there are expressions (involving radicals) for $\cos 30°$, $\cos 45°$, and $\sin 60°$. Evaluate these expressions, and check to see that your results agree with those in part (a).

[c] 38. (a) Evaluate $\tan 30°$, and $\tan 60°$.
   (b) Evaluate the expressions for $\tan 30°$ and $\tan 60°$ given in Table 1. Check to see that the results agree with those in part (a).

[c] 39. Evaluate $\csc 25°$, $\sec 25°$, and $\cot 25°$. (To evaluate $\csc 25°$, for example, first evaluate $\sin 25°$, then use the $\boxed{1/x}$ key.)

[c] 40. (a) Evaluate $\dfrac{\sin 59°}{\cos 59°}$.

   (b) Evaluate $\tan 59°$, and check that your answer agrees with that in part (a).
   (c) Use your calculator to evaluate $\sin^2 59° + \cos^2 59°$.

**B**

41. This exercise shows how to obtain radical expressions for $\sin 15°$ and $\cos 15°$. In the figure, assume that $AB = BD = 2$.

   (a) In the right triangle $BCD$, note that $DC = 1$ because $DC$ is opposite the 30° angle. Use the Pythagorean Theorem to show that $BC = \sqrt{3}$.
   (b) Use the Pythagorean Theorem to show that $AD = 2\sqrt{2 + \sqrt{3}}$.
   (c) Show that the expression for $AD$ in part (b) is equal to $\sqrt{6} + \sqrt{2}$. *Hint:* Two nonnegative quantities are equal if and only if their squares are equal.
   (d) Explain why $\angle BAD = \angle BDA$.
   (e) According to a theorem from geometry, an exterior angle of a triangle is equal to the sum of the two nonadjacent interior angles. Apply this to $\triangle ABD$ with exterior angle $DBC = 30°$, and show that $\angle BAD = 15°$.
   (f) Using the figure and the values that you have obtained for the lengths, conclude that

$$\sin 15° = \frac{1}{\sqrt{6} + \sqrt{2}} \qquad \cos 15° = \frac{2 + \sqrt{3}}{\sqrt{6} + \sqrt{2}}$$

   (g) Rationalize the denominators in part (f) to obtain

$$\sin 15° = \frac{\sqrt{6} - \sqrt{2}}{4} \qquad \cos 15° = \frac{\sqrt{6} + \sqrt{2}}{4}$$

   (h) Use your calculator to check the results in part (g).

42. Suppose that $\beta$ is an acute angle and

$$\sin \beta = \frac{m^2 - n^2}{m^2 + n^2} \quad (m > n > 0)$$

Show that

$$\cos \beta = \frac{2mn}{m^2 + n^2} \quad \text{and} \quad \tan \beta = \frac{m^2 - n^2}{2mn}$$

43. The following figure shows a 30°–60° right triangle, $\triangle ABC$. Prove that $AC = 2AB$. *Suggestion:* Construct $\triangle DBC$ as shown, congruent to $\triangle ABC$. Then note that $\triangle ADC$ is equilateral.

**C**

44. This exercise shows how to obtain radical expressions for $\sin 18°$ and $\cos 18°$, using the following figure.
   (a) Find $\angle B$, $\angle BDC$, and $\angle ADC$.
   (b) Why does $AC = BC = BD$?
   For the rest of this problem, assume $AD = 1$.
   (c) Why does $CD = 1$?

**(d)** Let $x$ denote the common lengths of $\overline{AC}$, $\overline{BC}$, and $\overline{BD}$. Use similar triangles to deduce that $\dfrac{x}{1+x} = \dfrac{1}{x}$.

Then show that $x = \dfrac{1+\sqrt{5}}{2}$.

**(e)** In $\triangle BDC$, draw an altitude from $B$ to $\overline{DC}$, meeting $\overline{DC}$ at $F$. Use right triangle $BFC$ to conclude that

$$\sin 18° = \frac{1}{1+\sqrt{5}}.$$

**(f)** Rationalize the denominator in part (e) to obtain

$$\sin 18° = \tfrac{1}{4}(\sqrt{5} - 1)$$

**(g)** Use the result in part (f); along with the identity $\sin^2 \theta + \cos^2 \theta = 1$, to show that

$$\cos 18° = \tfrac{1}{4}\sqrt{10 + 2\sqrt{5}}$$

## RIGHT-TRIANGLE APPLICATIONS

**7.2**

We continue the work we began in Section 7.1 on right-triangle trigonometry. As prerequisites for this section, you'll need to have memorized the definitions of the trigonometric functions, given in the box on page 325, and the table of trigonometric values for 30°, 45°, and 60° on page 328.

**EXAMPLE 1** Use one of the trigonometric functions to find $x$ in Figure 1.

Solution  Relative to the given 30° angle, $x$ is the adjacent side. The length of the hypotenuse is 100 cm. Since the adjacent side and the hypotenuse are involved, we use the cosine function here:

$$\cos 30° = \frac{\text{adjacent}}{\text{hypotenuse}} = \frac{x}{100}$$

That is,

$$\cos 30° = \frac{x}{100}$$

and consequently

$$x = 100 \cos 30°$$
$$= (100)\left(\frac{\sqrt{3}}{2}\right) = 50\sqrt{3} \text{ cm}$$

This is the result we were looking for.

**FIGURE 1**

We used the cosine function in Example 1 because the adjacent side and the hypotenuse were involved. We could instead use the secant. In that case, with reference again to Figure 1, the calculations look like this:

$$\sec 30° = \frac{\text{hypotenuse}}{\text{adjacent}} = \frac{100}{x}$$

$$\frac{2}{\sqrt{3}} = \frac{100}{x}$$

$$2x = 100\sqrt{3}$$

$$x = \frac{100\sqrt{3}}{2} = 50\sqrt{3} \text{ cm} \qquad \text{as obtained previously}$$

**EXAMPLE 2**    Find $y$ in Figure 1 (on the previous page).

Solution    As you can see from Figure 1, the side of length $y$ is opposite the 30° angle. Furthermore, we are given the length of the hypotenuse. Since the opposite side and the hypotenuse are involved, we employ the sine function. This yields

$$\sin 30° = \frac{\text{opposite}}{\text{hypotenuse}} = \frac{y}{100}$$

That is,

$$\sin 30° = \frac{y}{100}$$

and therefore

$$y = 100 \sin 30°$$
$$= (100)\left(\frac{1}{2}\right) = 50 \text{ cm}$$

This is the required answer. Actually, we could have obtained this particular result much faster by recalling that in the 30°–60° right triangle, the side opposite the 30° angle, namely $y$, is half the hypotenuse. That is, $y = \frac{100}{2} = 50$ cm, as obtained previously.

**EXAMPLE 3**    A ladder, which is leaning against the side of a building, forms an angle of 50° with the ground. If the foot of the ladder is 12 ft from the base of the building, how far up the side of the building does the ladder reach? See Figure 2.

Solution    In Figure 2 we have used $y$ to denote the required distance. Notice that $y$ is opposite the 50° angle, while the given side is adjacent to that angle. Since the opposite and adjacent sides are involved, we employ the tangent function. (The cotangent function could also be used.) We have

$$\tan 50° = \frac{y}{12}$$

and therefore

$$y = 12 \tan 50°$$

Without the use of a calculator or tables, this is our final answer. On the other hand, using a calculator we find that $\tan 50°$ is approximately 1.19. Thus we have

$$y \approx (12)(1.19) = 14 \text{ ft} \qquad \text{to the nearest foot}$$

In the next example we derive a useful formula for the area of a triangle. The formula can be used when we are given two sides and the included angle, but not the height.

**FIGURE 2**

**EXAMPLE 4**   Show that the area of the triangle in Figure 3(a) is given by $A = \frac{1}{2}ab\sin\theta$.

Solution   Draw an altitude as shown in Figure 3(b). Call the length of this altitude $h$. Then we have

$$\sin\theta = \frac{h}{a}$$

and

$$h = a\sin\theta$$

This value for $h$ can now be used in the usual formula for the area of a triangle:

$$A = \frac{1}{2}bh$$

$$= \frac{1}{2}b(a\sin\theta)$$

$$= \frac{1}{2}ab\sin\theta$$

(a)

(b)

**FIGURE 3**

The formula that we just derived in Example 4 is worth remembering. In words, the formula states that *the area of a triangle is equal to half the product of the lengths of two of the sides times the sine of the included acute angle.* In Section 7.3, we will see that the formula remains valid even when the included angle is not an acute angle.

**EXAMPLE 5**   Find the area of the triangle in Figure 4.

Solution   In the formula $A = \frac{1}{2}ab\sin\theta$, we let $a = 3$ cm, $b = 12$ cm, and $\theta = 60°$. This gives us

$$A = \frac{1}{2}(3)(12)\sin 60°$$

$$= (18)\frac{\sqrt{3}}{2}$$

$$= 9\sqrt{3} \text{ cm}^2$$

This is the required area.

**FIGURE 4**

**EXAMPLE 6**   Figure 5 shows a regular pentagon inscribed in a circle of radius 2 in. (**Regular** means that all of the sides are equal and all of the angles are equal.) Find the area of the pentagon.

Solution   The idea here is to find first the area of triangle *BOA* using the area formula from Example 4. Then, since the pentagon is composed of five such identical triangles, the area of the pentagon will be five times the area of triangle *BOA*. We will make use of the result from geometry that in a regular *n*-sided polygon, the central angle is $360°/n$. In our case, we therefore have

$$\angle BOA = \frac{360°}{5} = 72°$$

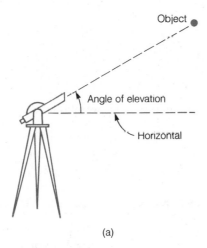

**FIGURE 5**

The area of triangle *BOA* can now be found. We have

$$\text{area} = \frac{1}{2} ab \sin \theta$$

$$= \frac{1}{2}(2)(2) \sin 72°$$

$$= 2 \sin 72° \text{ in.}^2$$

The area of the pentagon is five times this, or $10 \sin 72°$ in.$^2$ (Using a calculator, this is about 9.51 in.$^2$)

Now let us introduce some terminology that will be used in the next two examples. Suppose that a surveyor sights an object at a point above the horizontal, as indicated in Figure 6(a). Then the angle between the line of sight and the horizontal is called the **angle of elevation**. The **angle of depression** is similarly defined for an object below the horizontal, as shown in Figure 6(b).

(a)                                            (b)

**FIGURE 6**                                            **FIGURE 7**

**EXAMPLE 7**    A helicopter hovers 800 ft directly above a small island that is off the California coast. From the helicopter, the pilot takes a sighting to a point *P* directly ashore on the mainland, at the water's edge. If the angle of depression is 35°, how far off the coast is the island? See Figure 7.

**Solution**    Let *x* denote the distance from the island to the mainland. Then, as you can see from Figure 7, we have $\theta + 35° = 90°$, from which it follows that $\theta = 55°$. Now we can write

$$\tan 55° = \frac{x}{800}$$

or

$$x = 800 \tan 55° \approx 1150 \text{ ft} \qquad \text{using a calculator and rounding to the nearest 50 ft}$$

**EXAMPLE 8**   Two satellite tracking stations, located at points $A$ and $B$ in California's Mojave Desert, are 200 miles apart. At a prearranged time, both stations measure the angle of elevation of a satellite. If the angles of elevation from $A$ and from $B$ are $\alpha$ and $\beta$, respectively, express the altitude $h$ of the satellite in terms of $\alpha$ and $\beta$. See Figure 8.

**Solution**   We want to express the length $h = SC$ in Figure 8 in terms of the angles $\alpha$ and $\beta$. Note that $\overline{SC}$ is a side of both of the right triangles $SCA$ and $SCB$. Working first in right triangle $SCB$, we have

$$\cot \beta = \frac{CA + 200}{h}$$

or

$$h \cot \beta = CA + 200 \tag{1}$$

We can eliminate $CA$ from equation (1) as follows. Looking at right triangle $SCA$, we have

$$\cot \alpha = \frac{CA}{h}$$

or

$$CA = h \cot \alpha$$

Using this last equation to substitute for $CA$ in equation (1), we obtain

$$h \cot \beta = h \cot \alpha + 200$$
$$h \cot \beta - h \cot \alpha = 200$$
$$h(\cot \beta - \cot \alpha) = 200$$

$$h = \frac{200}{\cot \beta - \cot \alpha} \text{ miles} \qquad \text{as required}$$

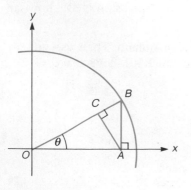

**FIGURE 8**

**EXAMPLE 9**   The arc shown in Figure 9 is a portion of the unit circle, $x^2 + y^2 = 1$. Express the following quantities in terms of $\theta$.

**(a)** $OA$    **(b)** $AB$    **(c)** Area of $\triangle OAB$    **(d)** $OC$

**Solution**   **(a)** In right triangle $OAB$:

$$\cos \theta = \frac{OA}{OB} = \frac{OA}{1}$$
$$OA = \cos \theta$$

**(b)** In right triangle $OAB$:

$$\sin \theta = \frac{AB}{OB} = \frac{AB}{1}$$
$$AB = \sin \theta$$

**(c)** Area $\triangle OAB = \frac{1}{2}(OA)(OB)\sin \theta$
$$= \frac{1}{2}(\cos \theta)(1)(\sin \theta)$$
$$= \frac{1}{2}\cos \theta \sin \theta$$

**FIGURE 9**

**(d)** In right triangle $OAC$:

$$\cos \theta = \frac{OC}{OA}$$

$$OC = OA \cos \theta$$

$$= (\cos \theta)(\cos \theta) = \cos^2 \theta$$

**EXAMPLE 10**  Determine the angle $\theta$ in Figure 10.

Solution  By inspection, $\tan \theta = \frac{3}{4} = 0.75$. From here, we have three choices.

CHOICE 1  We can take the easy way out and say, "$\theta$ is the acute angle whose tangent is $\frac{3}{4}$." True, this doesn't say what $\theta$ really is in degrees; nevertheless, this form of the answer will often suffice in calculus.

CHOICE 2  Using a table of the trigonometric functions, we can hunt until we find an angle whose tangent is as close to 0.75 as possible.

CHOICE 3  We can use a calculator to do all the work. For instance, with a Texas Instruments calculator, we push the buttons

0.75    |INV|    |TAN|

The result in this case is

$$\theta \approx 36.8°$$

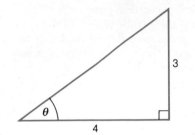

FIGURE 10

NOTE  When performing the calculations in Example 10, check to see that your calculator is set in the "degree mode," rather than the "radian mode." (Radian measure for angles is discussed in the next chapter, as are the inverse trigonometric functions.)

# EXERCISE SET 7.2

## A

*For Exercises 1–6, refer to the following figure. (However, each problem is independent of the others.)*

1. If $\angle A = 30°$ and $AB = 60$ cm, find $AC$ and $BC$.
2. If $\angle A = 60°$ and $AB = 12$ cm, find $AC$ and $BC$.
3. If $\angle B = 60°$ and $AC = 16$ cm, find $BC$ and $AB$.
4. If $\angle B = 45°$ and $AC = 9$ cm, find $BC$ and $AB$.

[c] 5. If $\angle B = 50°$ and $AB = 15$ cm, find $BC$ and $AC$. (Round off your answers to one decimal place.)

[c] 6. If $\angle A = 25°$ and $AC = 100$ cm, find $BC$ and $AB$. (Round off your answers to one decimal place.)

7. A ladder 18 ft long leans against a building. The ladder forms an angle of 60° with the ground.
   **(a)** How high up the side of the building does the ladder reach? [Here and in part (b), give two forms for your answers: one with radicals and one (using a calculator) with decimals, rounded off to two places.]
   **(b)** Find the horizontal distance from the foot of the ladder to the base of the building.

[c] 8. From a point level with and 1000 ft away from the base of the Washington Monument, the angle of elevation to the top of the monument is 29.05°. Determine the height of the monument to the nearest half foot.

9. Refer to the figure. At certain times, the planets Earth and Mercury line up in such a way that ∠ EMS is a right angle, as shown in the figure. At such times, ∠ SEM is found to be 21.16°. Use this information to estimate the distance MS of Mercury from the Sun. Assume that the distance from the Earth to the Sun is 93 million miles. (Round off your answer to the nearest million miles. Because Mercury's orbit is not exactly circular, the actual distance of Mercury from the Sun varies from about 28 million miles to 43 million miles.)

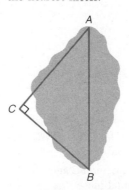

10. Determine the distance $AB$ across the lake shown in the figure, using the following data: $AC = 400$ m, $\angle C = 90°$, and $\angle CAB = 40°$. Round off the answer to the nearest meter.

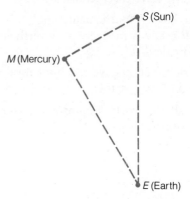

11. Find the area of each triangle. In part (b) use a calculator and round off the final answer to two decimal places.

(a)          (b)

12. Show that the perimeter of the pentagon in Example 6 is 20 sin 36°.   *Hint:*   In Figure 5, draw a perpendicular from O to $\overline{AB}$.

13. A regular decagon (ten sides) is inscribed in a circle of radius 20 cm. Find the area of the decagon. Round off the answer to one decimal place.

14. In triangle $OAB$, lengths $OA = OB = 6$ in. and $\angle AOB = 72°$. Find $AB$.   *Hint:*   Draw a perpendicular from O to, $\overline{AB}$. Round off the answer to one decimal place.

15. The accompanying figure shows two ships at points $P$ and $Q$, which are in the same vertical plane with an airplane at point $R$. When the height of the airplane is 3500 ft, the angle of depression to $P$ is 48° and that to $Q$ is 25°. Find the distance between the two ships. Round off the answer to the nearest 10 ft.

16. An observer in a lighthouse is 66 ft above the surface of the water. The observer sees a ship and finds the angle of depression to be 0.7°. Estimate the distance of the ship from the base of the lighthouse. Round off the answer to the nearest 5 ft.

17. From a point on ground level, you measure the angle of elevation to the top of a mountain to be 38°. Then you walk 200 meters farther away from the mountain and find that the angle of elevation is now 20°. Find the height of the mountain. Round off the answer to the nearest meter.

18. A surveyor stands 30 yd from the base of a building. On top of the building is a vertical radio antenna. Let $\alpha$ denote the angle of elevation when the surveyor sights to the top of the building. Let $\beta$ denote the angle of elevation when the surveyor sights to the top of the antenna. Express the length of the antenna in terms of the angles $\alpha$ and $\beta$.

19. In $\triangle ACD$, you are given $\angle C = 90°$, $\angle A = 60°$, and $AC = 18$ cm. If $B$ is a point on $\overline{CD}$ and $\angle BAC = 45°$, find $BD$. Express the answer in terms of a radical (rather than using a calculator).

20. The radius of the circle in the following figure is 1 unit. Express the lengths $OA$, $AB$, and $DC$ in terms of $\alpha$.

(a)

(b)

(c)

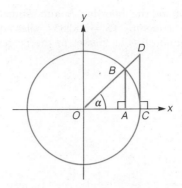

**21.** The arc in the next figure is a portion of the unit circle, $x^2 + y^2 = 1$.

(a) Express the following angles in terms of $\theta$: $\angle OBA$, $\angle OAB$, $\angle BAP$, $\angle BPA$.

(b) Express the following lengths in terms of $\sin \theta$ and $\cos \theta$: $AO$, $AP$, $OB$, $BP$.

**22.** In the accompanying figure, suppose that $AD = 1$. Find the length of each of the other line segments in the figure. When radicals appear in an answer, leave the answer in that form, rather than using a calculator.

**23.** (First look over Example 10.) Use a calculator to determine the angle $\theta$ in each figure. Round off each answer to one decimal place.

**24.** First, use the Pythagorean Theorem to find the lengths $OA$, $OB$, $OC$, $OD$, and $OE$ shown in the figure. Then use a calculator or tables to estimate the angles $\alpha, \beta, \gamma, \delta$, and $\epsilon$. (Actually, you don't need any aid for two of the angles.)

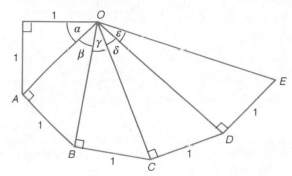

**B**

**25.** Express $AC$ as a function of $\theta$.

**26.** If $AB = 8$ in., express $x$ as a function of $\theta$.

27. In the figure, line segment $\overline{BA}$ is tangent to the circle at $A$. Also, $\overline{CF}$ is tangent to the circle at $F$. Express the following lengths in terms of $\theta$.
    (a) $DE$    (b) $OE$    (c) $CF$
    (d) $OC$    (e) $AB$    (f) $OB$

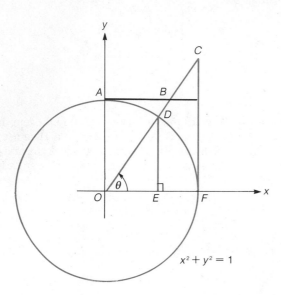

$$x^2 + y^2 = 1$$

28. At point $P$ on the earth's surface, the moon is observed to be directly overhead, while at the same time at point $T$, the moon is just visible. See the figure.
    (a) Show that $MP = \dfrac{OT}{\cos\theta} - OP$.
    c (b) Use a calculator and the following data to estimate the distance $MP$ from the earth to the moon: $\theta = 89.05°$ and $OT = OP = 4000$ miles. Round off your answer to the nearest thousand miles. (Because the moon's orbit is not really circular, the actual distance varies from about 216,400 miles to 247,000 miles.)

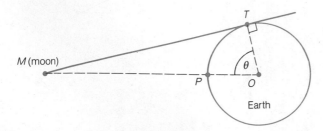

29. Refer to the figure. Let $r$ denote the radius of the moon.
    (a) Show that $r = \left(\dfrac{\sin\theta}{1 - \sin\theta}\right) PS$.

c (b) Use a calculator and the following data to estimate the radius $r$ of the moon:  $PS = 238{,}857$ miles and $\theta = 0.257°$. Round off your answer to the nearest 10 miles.

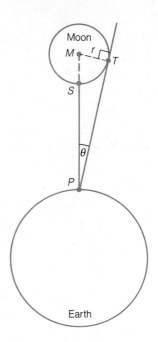

30. Figure $A$ shows a regular hexagon inscribed in a circle of radius 1. Figure $B$ shows a regular heptagon (seven-sided polygon) inscribed in a circle of radius 1. In Figure $A$, a line segment drawn from the center of the circle perpendicular to one of the sides is called an **apothem** of the polygon.
    (a) Show that the length of the apothem in Figure $A$ is $\sqrt{3}/2$.
    (b) Show that the length of one side of the heptagon in Figure $B$ is $2\sin(180°/7)$.

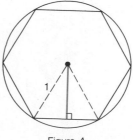

Figure $A$          Figure $B$

c (c) Use a calculator to evaluate the expressions in parts (a) and (b). Round off each answer to four decimal places, and note how close the two values are. Approximately two thousand years ago, Heron of Alexandria made use of this coincidence when he used the length of the apothem of the hexagon to approximate the length of the side of the heptagon. (The apothem of the hexagon can be constructed with ruler and compass; the side of the regular heptagon cannot.)

31. **(a)** Show that the area of a regular $n$-gon inscribed in a circle of radius 1 unit is given by

$$A_n = \frac{n}{2} \sin \frac{360°}{n}$$

C **(b)** Use a calculator and the formula in part (a) to complete the following table.

| $n$ | 5 | 10 | 50 | 100 | 1000 | 5000 | 10,000 |
|-----|---|----|----|-----|------|------|--------|
| $A_n$ | | | | | | | |

**(c)** Explain why the successive values of $A_n$ in your table get closer and closer to $\pi$.

## C

32. In the accompanying figure, the smaller circle is tangent to the larger circle. Ray $PQ$ is a common tangent and ray $PR$ passes through the centers of both circles. If the radius of the smaller circle is $a$ and the radius of the larger circle is $b$, show that

$$\sin \theta = \frac{b - a}{a + b}$$

Then, using the identity $\sin^2 \theta + \cos^2 \theta = 1$, show that

$$\cos \theta = \frac{2\sqrt{ab}}{a + b}$$

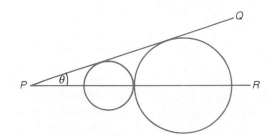

33. A vertical tower of height $h$ stands on level ground. From a point $P$ at ground level and due south of the tower, the angle of elevation to the top of the tower is $\theta$. From a point $Q$ at ground level and due west of the tower, the angle of elevation to the top of the tower is $\beta$. If $d$ is the distance between $P$ and $Q$, show that

$$h = \frac{d}{\sqrt{\cot^2 \theta + \cot^2 \beta}}$$

34. The following problem is taken from *A Treatise on Plane Trigonometry*, 7th ed., by E. W. Hobson (New York: Cambridge University Press, 1928): *ABCD* is the rectangular floor of a room whose length *AB* is $a$ feet. Find its height, which at $C$ subtends at $A$ an angle $\alpha$, and at $B$ an angle $\beta$.

## TRIGONOMETRIC FUNCTIONS OF GENERAL ANGLES

### 7.3

*There is nothing strange in the circle being the origin of any and every marvel.*

ARISTOTLE

Recall that the definitions of the trigonometric functions given in Section 7.1 apply only to acute angles. That is, as defined on page 325, the domains of trigonometric functions consist of all angles $\theta$ such that $0° < \theta < 90°$. In this section, we want to expand our definitions of the trigonometric functions to include angles of any size. To do this, we will need to be more explicit than we have been about angles and their size or measure.

An angle is formed by two **rays**, or half-lines, issuing from a common point called the **vertex**. For our purposes, it is useful to think of the two rays as originally coincident. Then, while one ray is held fixed, the other is rotated to create the given angle. As Figure 1 indicates, the fixed ray is called the **initial**

**FIGURE 1**

**FIGURE 2**

**side** of the angle, while the rotated ray is called the **terminal side**. By convention, we take the measure of an angle to be **positive** if the rotation is counterclockwise and **negative** if the rotation is clockwise. Figure 2 shows some examples of this.

We say that an angle is in **standard position** provided:

**(a)** the vertex of the angle coincides with the origin in a rectangular coordinate system;

**(b)** the initial side of the angle lies along the positive horizontal axis.

Figure 3 displays examples of angles in standard position.

**FIGURE 3**

Now, to extend our definitions of the trigonometric functions to accommodate any angle $\theta$, we begin by placing the angle $\theta$ in standard position; see Figure 4(a). Next we draw the *unit circle* $x^2 + y^2 = 1$, as shown in Figure 4(b).

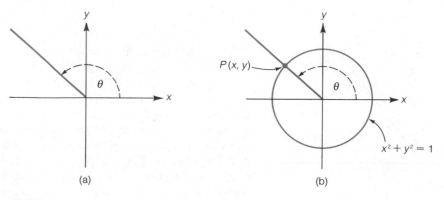

(a)                                                   (b)

**FIGURE 4**

(Recall from Chapter 3 that the equation $x^2 + y^2 = 1$ represents a circle of radius 1, with center at the origin.) As Figure 4(b) indicates, $P(x,y)$ denotes the point where the terminal side of angle $\theta$ intersects the unit circle. Then with this notation, we state the following definitions.

---

**Trigonometric Functions of General Angles**

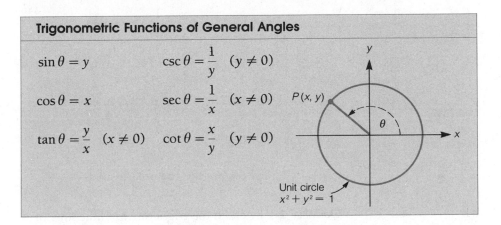

$$\sin\theta = y \qquad\qquad \csc\theta = \frac{1}{y} \quad (y \neq 0)$$

$$\cos\theta = x \qquad\qquad \sec\theta = \frac{1}{x} \quad (x \neq 0)$$

$$\tan\theta = \frac{y}{x} \quad (x \neq 0) \qquad \cot\theta = \frac{x}{y} \quad (y \neq 0)$$

Unit circle
$x^2 + y^2 = 1$

---

The preceding definitions of sine and cosine can be summarized this way:

*$\sin\theta$ is the second coordinate of that point where the terminal side of angle $\theta$ intersects the unit circle; $\cos\theta$ is the first coordinate of that point.*

For the cases in which the angle $\theta$ is acute, these definitions are really equivalent to the original right-triangle definitions. Consider, for instance, the acute angle $\theta$ in Figure 5. According to our "new" definition of sine, we have

$$\sin\theta = y$$

On the other hand, applying the original right-triangle definition in Figure 5 yields

$$\sin\theta = \frac{\text{opposite}}{\text{hypotenuse}} = \frac{y}{1} = y$$

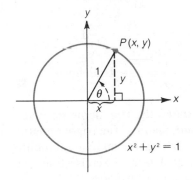

**FIGURE 5**

Thus, in both cases we obtain the same result. Using Figure 5, you should check for yourself that the same agreement also occurs for the other trigonometric functions.

In fact, these definitions agree in other ways, too, with our previous work. Notice that the definitions imply that $\tan\theta = (\sin\theta)/(\cos\theta)$. Additionally, from the definitions we see that $\sin\theta$ and $\csc\theta$ are reciprocals, as are $\cos\theta$ and $\sec\theta$ and $\tan\theta$ and $\cot\theta$.

**EXAMPLE 1**  Compute $\sin 90°$, $\cos 90°$, $\tan 90°$, $\csc 90°$, $\sec 90°$, and $\cot 90°$.

**Solution**  We place the angle $\theta = 90°$ in standard position. Then, as Figure 6 indicates, the terminal side of the angle meets the unit circle at the point $(0, 1)$. Now we apply the definitions.

$(0, 1)$

By definition, $\cos 90°$ is this number.  By definition, $\sin 90°$ is this number.

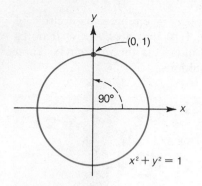

**FIGURE 6**

Thus,

$$\sin 90° = 1 \qquad \text{and} \qquad \cos 90° = 0$$

For the remaining trigonometric functions of 90°, we have

$$\tan 90° = \frac{y}{x} = \frac{1}{0}, \text{Undefined}$$

$$\csc 90° = \frac{1}{y} = \frac{1}{1} = 1$$

$$\sec 90° = \frac{1}{x} = \frac{1}{0}, \text{Undefined}$$

$$\cot 90° = \frac{x}{y} = \frac{0}{1} = 0$$

These are the required results.

**EXAMPLE 2** Evaluate the trigonometric functions of $-180°$.

**Solution** The results can be read off from Figure 7.

**FIGURE 7**

$$\sin(-180°) = y = 0 \qquad \csc(-180°) = \frac{1}{y} = \frac{1}{0}, \text{Undefined}$$

$$\cos(-180°) = x = -1 \qquad \sec(-180°) = \frac{1}{x} = \frac{1}{-1} = -1$$

$$\tan(-180°) = \frac{y}{x} = \frac{0}{-1} = 0 \qquad \cot(-180°) = \frac{x}{y} = \frac{-1}{0}, \text{Undefined}$$

In the two examples just concluded, we evaluated the trigonometric functions for $\theta = 90°$ and for $\theta = -180°$. In the same manner, the trigonometric functions can be evaluated just as easily for any angle that is an integral multiple of 90°. Table 1 shows the results of such calculations. Exercise 6 at the end of this section asks you to make these calculations for yourself.

**TABLE 1**

| $\theta$ | $\sin\theta$ | $\cos\theta$ | $\tan\theta$ | $\csc\theta$ | $\sec\theta$ | $\cot\theta$ |
|---|---|---|---|---|---|---|
| 0° | 0 | 1 | 0 | undefined | 1 | undefined |
| 90° | 1 | 0 | undefined | 1 | undefined | 0 |
| 180° | 0 | -1 | 0 | undefined | -1 | undefined |
| 270° | -1 | 0 | undefined | -1 | undefined | 0 |
| 360° | 0 | 1 | 0 | undefined | 1 | undefined |

To evaluate the trigonometric functions for angles that are not multiples of 90°, we introduce the notion of a *reference angle*. Given an angle $\theta$ that is not a multiple of 90°, the **reference angle** associated with $\theta$ is the acute angle

The reference angle
for 135° is 45°.
[180° − 135° = 45°]

The reference angle
for 210° is 30°.
[210° − 180° = 30°]

The reference angle
for 300° is 60°.
[360° − 300° = 60°]

**FIGURE 8**

formed by the *x*-axis and the terminal side of *θ*. Figure 8 displays examples of angles and their respective reference angles.

Now let's look at an example to see how reference angles are used to evaluate the trigonometric functions. Suppose we want to evaluate cos 150°. In Figure 9(a) we've placed the angle *θ* = 150° in standard position. As you can also see, the reference angle for 150° is 30°. By definition, the value of cos 150° is the *x*-coordinate of the point *P* in Figure 9(a). To determine this *x*-coordinate, we'll use Figure 9(b), in which the line segment $\overline{PQ}$ is parallel to the *x*-axis. There are two facts that follow from the symmetry of the situation in Figure 9(b). (Exercise 35 outlines a detailed verification of these two facts.)

FIRST FACT   The *x*-coordinate of *P* is just the negative of the *x*-coordinate of *Q*.

SECOND FACT   The reference angle *POB* is equal to angle *QOA*, and therefore angle *QOA* is 30°.

Now we apply these two facts. Since angle *QOA* is 30°, the *x*-coordinate of *Q* is by definition cos 30°, or $\sqrt{3}/2$. The *x*-coordinate of *P* is then the negative of this. That is, the *x*-coordinate of *P* is $-\sqrt{3}/2$. It follows now, again by definition, that the value of cos 150° is $-\sqrt{3}/2$.

The same method that we have just used to evaluate cos 150° can be used to evaluate any of the trigonometric functions when the angles are not multiples of 90°. The following three steps summarize this method.

STEP 1   Determine the reference angle associated with the given angle.

STEP 2   Evaluate the given trigonometric function using the reference angle for the input.

STEP 3   Affix the appropriate sign to the number found in Step 2.

The next three examples illustrate this procedure.

(a)

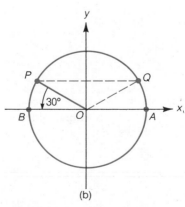

(b)

**FIGURE 9**

**EXAMPLE 3**   Evaluate the following.

(a) $\sin 135°$     (b) $\cos 135°$     (c) $\tan 135°$

**Solution**   As Figure 10 indicates, the reference angle associated with $135°$ is $45°$.

(a) STEP 1   The reference angle is $45°$.

STEP 2   $\sin 45° = \sqrt{2}/2$

STEP 3   $\operatorname{Sin}\theta$ is the $y$-coordinate. In the second quadrant $y$-coordinates are positive. Thus $\sin 135°$ is positive, since the terminal side of $\theta = 135°$ lies in the second quadrant. We have therefore

$$\sin 135° = \frac{\sqrt{2}}{2}$$

(b) STEP 1   The reference angle is $45°$.

STEP 2   $\cos 45° = \sqrt{2}/2$

STEP 3   $\operatorname{Cos}\theta$ is the $x$-coordinate. In the second quadrant $x$-coordinates are negative. Thus $\cos 135°$ is negative, since the terminal side of $\theta = 135°$ lies in the second quadrant. We have therefore

$$\cos 135° = -\frac{\sqrt{2}}{2}$$

(c) STEP 1   The reference angle is $45°$.

STEP 2   $\tan 45° = 1$

STEP 3   By definition, $\tan\theta = y/x$. Now, the terminal side of $\theta = 135°$ lies in the second quadrant, in which $y$ is positive and $x$ is negative. Thus $\tan 135°$ is negative. We have therefore

$$\tan 135° = -1$$

**FIGURE 10**

**EXAMPLE 4**   Evaluate each of the following.

(a) $\cos(-120°)$     (b) $\cot(-120°)$

**Solution**   (a) As Figure 11 shows, the reference angle for $-120°$ is $60°$.

STEP 1   The reference angle for $-120°$ is $60°$.

STEP 2   $\cos 60° = \frac{1}{2}$

STEP 3   $\operatorname{Cos}\theta$ is the $x$-coordinate. In the third quadrant $x$-coordinates are negative. Thus, $\cos(-120°)$ is negative, since the terminal side of $\theta = -120°$ lies in the third quadrant. It now follows that

$$\cos(-120°) = -\frac{1}{2}$$

(b) STEP 1   The reference angle for $-120°$ is $60°$.

STEP 2   $\cot 60° = \sqrt{3}/3$

**FIGURE 11**

STEP 3   By definition $\cot\theta = x/y$. Now, the terminal side of $\theta = -120°$ lies in the third quadrant, in which both $x$ and $y$ are negative. So $x/y$ is positive and we have

$$\cot(-120°) = \frac{\sqrt{3}}{3}$$

In the first section of this chapter, we showed that the identity $\sin^2\theta + \cos^2\theta = 1$ is valid for all acute angles $\theta$. Now we're in a position to show that this identity holds for *all* angles, not just acute angles. We are also going to prove two useful results concerning supplementary angles. Recall that two angles are **supplementary** provided their sum is $180°$. Our theorem is as follows.

---

**Theorem**

1. For any angle $\theta$, we have

   $$\sin^2\theta + \cos^2\theta = 1$$

2. If $0° < \theta < 180°$, then

   $$\cos(180° - \theta) = -\cos\theta$$

   and

   $$\sin(180° - \theta) = \sin\theta$$

---

To prove that $\sin^2\theta + \cos^2\theta = 1$, we need to use only the unit circle definitions of sine and cosine (see Figure 12). Since the point $(x, y)$ lies on the unit circle, we have

$$x^2 + y^2 = 1$$

Replacing $x$ by $\cos\theta$ and $y$ by $\sin\theta$ in this equation then yields

$$(\cos\theta)^2 + (\sin\theta)^2 = 1$$

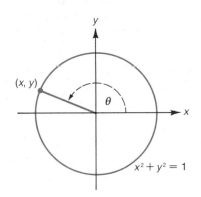

FIGURE 12

That is,

$$\sin^2\theta + \cos^2\theta = 1 \qquad \text{as required}$$

We will prove the second part of our theorem for the case in which $\theta$ is an acute angle. (See Exercises 36 and 37 at the end of this section for the cases in which $\theta$ is an angle between $90°$ and $180°$.) Now, referring to Figure 13, notice that the reference angle for $180° - \theta$ is $\theta$ itself. Thus, we have

reference angle for $180° - \theta$

$$\cos(180° - \theta) = \underline{-\cos\theta}$$

negative, since the terminal side of $180° - \theta$ lies in the second quadrant

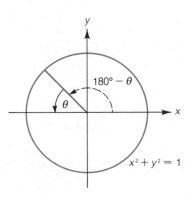

FIGURE 13

and

$$\sin(180° - \theta) = \underset{\underset{\text{positive, since the terminal side of}}{\uparrow}}{\underbrace{\sin\theta}}$$

reference angle for $180° - \theta$

positive, since the terminal side of $180° - \theta$ lies in the second quadrant

These observations complete the proof of the second part of our theorem.

**EXAMPLE 5** Given that $\sin\theta = \frac{2}{3}$ and $90° < \theta < 180°$, find $\cos\theta$ and $\tan\theta$.

**Solution** The substitution $\sin\theta = \frac{2}{3}$ in the identity $\sin^2\theta + \cos^2\theta = 1$ yields

$$\left(\frac{2}{3}\right)^2 + \cos^2\theta = 1$$

$$\cos^2\theta = 1 - \frac{4}{9} = \frac{5}{9}$$

$$\cos\theta = \pm\sqrt{\frac{5}{9}} = \frac{\pm\sqrt{5}}{3}$$

To decide whether to choose the positive or negative value, note that the given inequality, $90° < \theta < 180°$, tells us that the terminal side of $\theta$ lies in the second quadrant. Since in the second quadrant $x$-coordinates are negative, we choose the negative value here. Thus

$$\cos\theta = \frac{-\sqrt{5}}{3}$$

For $\tan\theta$, we have

$$\tan\theta = \frac{y}{x} = \frac{\sin\theta}{\cos\theta}$$

$$= \frac{2/3}{-\sqrt{5}/3} = -\frac{2}{\sqrt{5}}$$

$$= -\frac{2\sqrt{5}}{5} \qquad \text{rationalizing the denominator}$$

**EXAMPLE 6** Show that the area $A$ of the triangle in Figure 14(a) is given by

$$A = \frac{1}{2}ab\sin\theta$$

NOTE: In Section 7.2 we proved this formula for the case in which $\theta$ is an acute angle.

**Solution** The area $A$ of any triangle is given by $A = \frac{1}{2}(\text{base})(\text{height})$. Thus, referring to Figure 14(b), we have

$$A = \frac{1}{2}bh \tag{1}$$

Also from Figure 14(b), we have

$$\sin(180° - \theta) = \frac{\text{opposite}}{\text{hypotenuse}} = \frac{h}{a}$$
$$h = a\sin(180° - \theta)$$
$$h = a\sin\theta \qquad (2)$$

Now we can use equation (2) to substitute for $h$ in equation (1). This yields

$$A = \tfrac{1}{2}b(a\sin\theta)$$
$$= \tfrac{1}{2}ab\sin\theta$$

The formula for the area of a triangle that we derived in Example 6 is a useful one. We will use it in later work to prove the *law of sines*. For reference, we summarize the result from Example 6 in the box that follows.

**Formula for the Area of a Triangle**

If $a$ and $b$ are the lengths of two sides of a triangle and $\theta$ is the angle included between those two sides, then the area of the triangle is given by

$$\text{area} = \tfrac{1}{2}ab\sin\theta$$

IN WORDS:   The area of a triangle equals half the product of the lengths of two sides times the sine of the included angle.

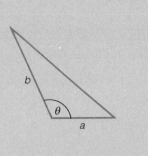

# EXERCISE SET 7.3

## A

*In Exercises 1 and 2, sketch each angle in standard position, and specify the reference angle.*

1. (a) 110°   (b) 240°   (c) 60°   (d) −60°
2. (a) 300°   (b) 1000°   (c) −15°   (d) 15°

*In Exercises 3–5, evaluate each expression. (If the value is not defined, say so.)*

3. (a) $\cos 0°$   (b) $\sin 450°$
   (c) $\sin 270°$   (d) $\sin(-630°)$

**4. (a)** $\sin(-270°)$     **(b)** $\tan 180°$
    **(c)** $\sec(-90°)$     **(d)** $\sec 90°$

**5. (a)** $\csc(-90°)$     **(b)** $\cot 720°$
    **(c)** $\cos(-540°)$     **(d)** $\sin 810°$

**6.** Verify each of the entries in Table 1 on page 346.

*In Exercises 7 and 8, refer to the following figure, in which the x-coordinate of P is $\frac{3}{4}$ and the y-coordinate of Q is $\frac{1}{5}$.*

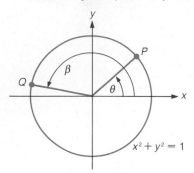

**7. (a)** Evaluate $\cos\theta$.
    **(b)** Evaluate $\sin\theta$ and $\tan\theta$.
    **(c)** Which is larger: $\cos(\beta - 90°)$ or $\cos(\theta + 90°)$?
       *Hint:* Sketch both angles in standard position.

**8. (a)** Evaluate $\sin\beta$.
    **(b)** Evaluate $\cos\beta$ and $\cot\beta$.
    **(c)** Which is larger: $\cos(\beta + 180°)$ or $\sin(\theta + 180°)$?
       *Hint:* Sketch both angles in standard position.

*In Exercises 9–12, use the method shown in Example 3 to evaluate each expression.*

**9. (a)** $\sin 315°$     **(b)** $\cos 300°$
    **(c)** $\tan 330°$     **(d)** $\sin(-315°)$

**10. (a)** $\cos 210°$     **(b)** $\sec 210°$
    **(c)** $\csc 225°$     **(d)** $\cos(-210°)$

**11. (a)** $\tan 135°$     **(b)** $\cot 120°$
    **(c)** $\cot 480°$     **(d)** $\tan(-135°)$

**12. (a)** $\cos 120°$     **(b)** $\csc 240°$
    **(c)** $\sec 600°$     **(d)** $\cos(-120°)$

**13.** Evaluate the six trigonometric functions of $\theta = -30°$.

**14.** Evaluate the six trigonometric functions of $\theta = -45°$.

**15.** Complete the following two tables.

**(a)**

| $\theta$ | 0° | 90° | 180° | 270° | 360° | 450° | 540° | 630° | 720° |
|---|---|---|---|---|---|---|---|---|---|
| $\sin\theta$ | | | | | | | | | |
| $\cos\theta$ | | | | | | | | | |

**(b)**

| $\theta$ | 30° | 60° | 90° | 120° | 150° | 180° | 210° | 240° | 270° | 300° | 330° | 360° |
|---|---|---|---|---|---|---|---|---|---|---|---|---|
| $\sin\theta$ | | | | | | | | | | | | |

**16.** Complete the following tables.

**(a)**

| $\theta$ | $\sin\theta$ | $\cos\theta$ | $\tan\theta$ |
|---|---|---|---|
| 0° | | | |
| 30° | | | |
| 45° | | | |
| 60° | | | |
| 90° | | | |
| 120° | | | |
| 135° | | | |
| 150° | | | |
| 180° | | | |

**(b)**

| $\theta$ | $\sin\theta$ | $\cos\theta$ | $\tan\theta$ |
|---|---|---|---|
| 180° | | | |
| 210° | | | |
| 225° | | | |
| 240° | | | |
| 270° | | | |
| 300° | | | |
| 315° | | | |
| 330° | | | |
| 360° | | | |

*In Exercises 17–20, use the given information to determine the remaining five trigonometric values.*

**17.** $\sin\theta = \frac{1}{5}$, $90° < \theta < 180°$

**18.** $\sin\theta = -\frac{24}{25}$, $180° < \theta < 270°$

**19.** $\cos\theta = -\frac{3}{5}$, $180° < \theta < 270°$

**20.** $\cos\theta = \frac{1}{4}$, $270° < \theta < 360°$

*In Exercises 21–24, use the given information to determine the area of each triangle.*

**21.** Two of the sides are 5 cm and 7 cm, and the angle between those sides is 120°.

22. Two of the sides are 3 m and 6 m, and the included angle is 150°.

C 23. Two of the sides are 21.4 cm, and 28.6 cm, and the included angle is 98.5°. (Round off the final answer to one decimal place.)

C 24. Two of the sides are 5.98 cm and 8.05 cm, and the included angle is 107.1°. (Round off the answer to one decimal place.)

25. List three angles for which the sine of each is equal to 1.

26. List three angles for which the cosine of each is $-\frac{1}{2}$.

27. Use the following figure to estimate the following quantities.
    (a) sin 10°; cos 10°        (b) sin 20°; cos 20°
    (c) sin 30°; cos 30°        (d) sin 40°; cos 40°
    (e) sin 50°; cos 50°
    *Hint:* $\sin \theta = \cos(90° - \theta)$.
    (f) sin 70°; cos 70°        (g) sin 80°; cos 80°
    (h) sin 100°; cos 100°      (i) sin 130°; cos 130°

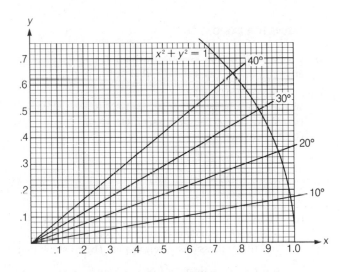

28. (a) Complete the following table, using the words "positive" or "negative" as appropriate.

| TERMINAL SIDE OF ANGLE $\theta$ LIES IN | | | |
| Quadrant I | Quadrant II | Quadrant III | Quadrant IV |
|---|---|---|---|
| $\sin \theta$ | positive | positive | | |
| $\cos \theta$ | | | | |
| $\tan \theta$ | | | | |

(b) The mnemonic (memory device) ASTC (*all students take calculus*) is sometimes used to recall

the signs of the trigonometric values in each quadrant:

A . . . . . All are positive in the first quadrant.
S . . . . . Sine is positive in the second quadrant.
T . . . . . Tangent is positive in the third quadrant.
C . . . . . Cosine is positive in the fourth quadrant.

Check the validity of this mnemonic against your chart in part (a).

**B**

29. Given that $\sin 15° = \frac{1}{4}(\sqrt{6} - \sqrt{2})$ and $\cos 18° = \frac{1}{4}\sqrt{10 + 2\sqrt{5}}$, evaluate the following.
    (a) sin 195°        (b) cos 162°        (c) tan 345°
    (d) sin(−15°)       (e) cos(−18°)       (f) cos 918°

30. Suppose that A, B, and C are the three angles of a triangle. Prove the following identities.
    (a) $\sin A = \sin(B + C)$
    (b) $\cos A = -\cos(B + C)$
    (c) $\cos \dfrac{B + C}{2} = \sin \dfrac{A}{2}$
    (d) $\sin \dfrac{B + C}{2} = \cos \dfrac{A}{2}$
    (e) $\tan A = -\tan(B + C)$

31. Refer to the accompanying figure.
    (a) What are the coordinates of P? What are the coordinates of Q?
    (b) Show that the length of line segment $\overline{PQ}$ is
    $$\sqrt{2}\sqrt{1 - \cos \theta \cos \phi - \sin \theta \sin \phi}$$

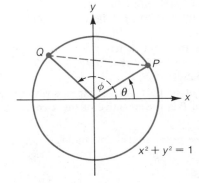

32. Use the figure (on the next page) to prove the following theorem: The area of a quadrilateral is equal to half the product of the diagonals times the sine of the included angle. *Hint:* Find the areas of each of the four triangles that make up the quadrilateral. Show that the sum of those areas is $\frac{1}{2}(\sin \theta)[qs + rq + rp + ps]$. Then factor the quantity within the brackets.

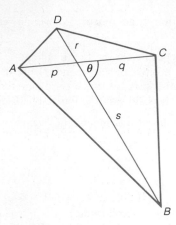

**FIGURE FOR EXERCISE 32**

**33.** Prove the following identity for $0° < \theta < 180°$:

$$2\log_{10}\sin\theta = \log_{10}(1 - \cos\theta) + \log_{10}(1 + \cos\theta)$$

Why is the restriction $0° < \theta < 180°$ necessary?

**34.** Use the identity $\sin^2\theta + \cos^2\theta = 1$ to explain why for any angle $\theta$ we have

$$|\sin\theta| \le 1 \quad \text{and} \quad |\cos\theta| \le 1$$

**35.** In this exercise we are going to verify two statements that were made in the text concerning Figure 9(b). In particular, with reference to that figure, we'll show that angle $QOA$ is $30°$, and also that the $x$-coordinate of $P$ is the negative of the $x$-coordinate of $Q$.

(a) Since $\overline{PQ}$ is parallel to $\overline{BA}$, the two angles $BOP$ and $QPO$ are equal. Thus angle $QPO$ is $30°$. Now explain why angle $PQO$ is $30°$. *Hint:* The base angles of an isosceles triangle are equal.

(b) Since angle $PQO$ is $30°$, it follows that angle $QOA$ is $30°$. Why?

(c) Let $C$ denote the point where the line segment $\overline{PQ}$ meets the $y$-axis. Show that triangle $OPC$ is congruent to triangle $OQC$.

(d) Use the result in part (c) to explain why the $x$-coordinate of $P$ is the negative of the $x$-coordinate of $Q$.

**36.** In this exercise we prove that the formulas

$$\cos(180° - \theta) = -\cos\theta$$

and

$$\sin(180° - \theta) = \sin\theta$$

are valid when $90° < \theta < 180°$. (Recall that the proof in the text assumes $\theta$ is an acute angle.)

The statements that follow refer to the given figure. Supply the reason or reasons for each statement.

(a) The coordinates of $P$ are $(\cos\theta, \sin\theta)$.

(b) Draw $\overline{PQ}$ parallel to the $x$-axis, as shown in the figure. Then the coordinates of $Q$ are $(-\cos\theta, \sin\theta)$.

(c) $\angle QOA = 180° - \theta$

(d) The coordinates of $Q$ are $(\cos(180° - \theta), \sin(180° - \theta))$.

(e) $\cos(180° - \theta) = -\cos\theta$ and $\sin(180° - \theta) = \sin\theta$.

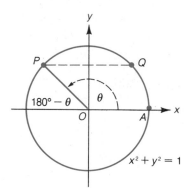

**FIGURE FOR EXERCISE 36**

**37.** Check that the two formulas in part (e) of Exercise 36 are valid in the cases $\theta = 0°$, $\theta = 90°$, and $\theta = 180°$.

**38.** *Formula for* $\sin(\alpha + \beta)$
In the figure below, $\overline{AD} \perp \overline{BC}$ and $AD = 1$.

(a) Show that $AC = \sec\alpha$ and $AB = \sec\beta$.

(b) Show that

$$\text{area } \triangle ADC = \tfrac{1}{2}\sec\alpha\sin\alpha$$
$$\text{area } \triangle ADB = \tfrac{1}{2}\sec\beta\sin\beta$$
$$\text{area } \triangle ABC = \tfrac{1}{2}\sec\alpha\sec\beta\sin(\alpha + \beta)$$

(c) The sum of the areas of the two smaller triangles in part (b) equals the area of $\triangle ABC$. Use this fact and the expressions given in part (b) to show that

$$\sin(\alpha + \beta) = \sin\alpha\cos\beta + \cos\alpha\sin\beta$$

(d) Find $\sin 75°$. *Hint:* $75° = 30° + 45°$

(e) Show that $\sin 75° \ne \sin 30° + \sin 45°$.

(f) Compute $\sin 105°$ and then check that $\sin 105° \ne \sin 45° + \sin 60°$.

## ALGEBRA AND THE TRIGONOMETRIC FUNCTIONS

**7.4**

*It was Robert of Chester's translation from the Arabic that resulted in our word "sine." The Hindus had given the name* jiva *to the half chord in trigonometry, and the Arabs had taken this over as* jiba. *In the Arabic language there is also a word* jaib *meaning "bay" or "inlet." When Robert of Chester came to translate the technical word* jiba, *he seems to have confused this with the word* jaib *(perhaps because vowels were omitted); hence he used the word* sinus, *the Latin word for "bay" or "inlet."*

CARL B. BOYER in his text, *A History of Mathematics* (New York: Wiley, 1968)

In this section, we are going to concentrate on the algebra involved in manipulating trigonometric expressions. This will help to pave the way for some of the more analytical parts of trigonometry that are developed in the next chapter. For reference, in the box that follows we summarize some common notational conventions that have been introduced in the previous sections.

---

### Some Notational Conventions

1. The quantity $(\sin\theta)^n$ is usually written $\sin^n\theta$. For example, $(\sin\theta)^2$ is written $\sin^2\theta$. The same convention also applies to the other five trigonometric functions.*

2. Parentheses are often omitted in multiplication. For example, the product $(\sin\theta)(\cos\theta)$ is usually written $\sin\theta\cos\theta$. Similarly, $2(\sin\theta)$ is written $2\sin\theta$.

3. An expression such as $\sin\theta$ really means $\sin(\theta)$, where sin or sine is the name of the function and $\theta$ is an input. It is for historical rather than mathematical reasons that the parentheses are suppressed. An exception to this, however, occurs in expressions such as $\sin(A+B)$; here the parentheses are necessary.

---

**EXAMPLE 1** Multiply $(1-\sin\theta)^2$.

**Solution** Do this the same way you would expand $(1-s)^2$. Since

$$(1-s)^2 = 1 - 2s + s^2$$

it follows that

$$(1-\sin\theta)^2 = 1 - 2\sin\theta + \sin^2\theta$$

---

* A single exception to this convention is the case $n = -1$. The meaning of expressions such as $\sin^{-1}x$ will be explained in Section 8 of Chapter 8.

**EXAMPLE 2**   Add $\sin \theta + \dfrac{1}{\cos \theta}$.

**Solution**   The least common denominator is the quantity $\cos \theta$. We have

$$\sin \theta + \frac{1}{\cos \theta} = \frac{\sin \theta}{1} + \frac{1}{\cos \theta}$$

$$= \frac{\sin \theta}{1} \cdot \frac{\cos \theta}{\cos \theta} + \frac{1}{\cos \theta}$$

$$= \frac{\sin \theta \cos \theta}{\cos \theta} + \frac{1}{\cos \theta}$$

$$= \frac{\sin \theta \cos \theta + 1}{\cos \theta}$$

This is the required result.

**EXAMPLE 3**   Show that the statement $\cos A + \cos B = \cos(A + B)$ is not true in general.

**Solution**   Back up for a minute. How would you convince a beginning algebra student that the statement $(x + y)^2 = x^2 + y^2$ is not true in general? One way would be simply to pick specific values for $x$ and $y$ and then show the equation fails in that case. For instance, using $x = 1$ and $y = 1$ would give

$$(1 + 1)^2 \overset{?}{=} 1^2 + 1^2$$
$$4 \overset{?}{=} 2 \qquad \text{No!}$$

We can do the same thing in the problem at hand. Let $A = 30°$ and $B = 30°$. Then we have

$$\cos A + \cos B \overset{?}{=} \cos(A + B)$$
$$\cos 30° + \cos 30° \overset{?}{=} \cos(30° + 30°)$$
$$\frac{\sqrt{3}}{2} + \frac{\sqrt{3}}{2} \overset{?}{=} \cos 60°$$
$$\sqrt{3} \overset{?}{=} \frac{1}{2} \qquad \text{No!}$$

We conclude that the statement $\cos A + \cos B = \cos(A + B)$ is not true in general. [To see what $\cos(A + B)$ does equal, see Exercise 48 at the end of this section.]

**EXAMPLE 4**   Factor:   $\tan^2 A + 5 \tan A + 6$

**Solution**   PRELIMINARY SOLUTION   To help you focus on the algebra that is actually involved, replace each occurrence of the quantity $\tan A$ by the letter $T$. Then

$$T^2 + 5T + 6 = (T + 3)(T + 2)$$

ACTUAL SOLUTION

$$\tan^2 A + 5\tan A + 6 = (\tan A + 3)(\tan A + 2)$$

NOTE   After you are accustomed to working with trigonometric expressions, you will be able to eliminate the preliminary step.  ▪▪

**EXAMPLE 5**   Simplify:  $\dfrac{\dfrac{\sin\theta + 1}{\sin\theta} + 1}{\dfrac{\sin\theta - 1}{\sin\theta} - 1}$

**Solution**   PRELIMINARY SOLUTION

$$\frac{\dfrac{S+1}{S} + 1}{\dfrac{S-1}{S} - 1} = \frac{S\left(\dfrac{S+1}{S} + 1\right)}{S\left(\dfrac{S-1}{S} - 1\right)}$$

$$= \frac{S + 1 + S}{S - 1 - S}$$

$$= \frac{2S + 1}{-1}$$

$$= -2S - 1$$

ACTUAL SOLUTION

$$\frac{\dfrac{\sin\theta + 1}{\sin\theta} + 1}{\dfrac{\sin\theta - 1}{\sin\theta} - 1} = \frac{\sin\theta\left(\dfrac{\sin\theta + 1}{\sin\theta} + 1\right)}{\sin\theta\left(\dfrac{\sin\theta - 1}{\sin\theta} - 1\right)}$$

$$= \frac{\sin\theta + 1 + \sin\theta}{\sin\theta - 1 - \sin\theta}$$

$$= \frac{2\sin\theta + 1}{-1}$$

$$= -2\sin\theta - 1$$  ▪▪

**EXAMPLE 6**   Combine and simplify:  $\dfrac{\sin A}{\cos A} + \dfrac{\cos A}{\sin A}$

**Solution**   The common denominator is $\cos A \sin A$. Therefore we have

$$\frac{\sin A}{\cos A} + \frac{\cos A}{\sin A} = \frac{\sin A}{\cos A} \cdot \frac{\sin A}{\sin A} + \frac{\cos A}{\sin A} \cdot \frac{\cos A}{\cos A}$$

$$= \frac{\sin^2 A}{\cos A \sin A} + \frac{\cos^2 A}{\sin A \cos A}$$

$$= \frac{\sin^2 A + \cos^2 A}{\cos A \sin A}$$

$$= \frac{1}{\cos A \sin A}$$

This is the required result. If we wish, we can write this answer in an alternative form that doesn't involve fractions:

$$\frac{1}{\cos A \sin A} = \frac{1}{\cos A} \cdot \frac{1}{\sin A}$$

$$= \sec A \csc A$$  ▪▪

One of the most useful techniques for simplifying a trigonometric expression is first to rewrite it in terms of sines and cosines and then to carry out the usual algebraic simplifications (as discussed in Section 1.10). The next example shows how this works.

**EXAMPLE 7**  Simplify the expression $\dfrac{\sec A + 1}{\sin A + \tan A}$.

**Solution**

$$\frac{\sec A + 1}{\sin A + \tan A} = \frac{\dfrac{1}{\cos A} + 1}{\sin A + \dfrac{\sin A}{\cos A}}$$

$$= \frac{\cos A\left(\dfrac{1}{\cos A} + 1\right)}{\cos A\left(\sin A + \dfrac{\sin A}{\cos A}\right)} \qquad \text{multiplying numerator and denominator by } \cos A$$

$$= \frac{1 + \cos A}{\cos A \sin A + \sin A}$$

$$= \frac{1 + \overset{1}{\cancel{\cos A}}}{\sin A(\underset{1}{\cancel{\cos A + 1}})}$$

$$= \frac{1}{\sin A}$$

This last expression can also be written $\csc A$.

In the next three examples, we are asked to show that certain trigonometric equations are, in fact, identities. This type of problem is similar to Example 7, which we just completed, except now we are given an answer to work toward. The identities in these examples should not be memorized; they are too specialized. Instead, you should concentrate on the proofs themselves, noting where the basic definitions are used, and where the fundamental identities (such as $\sin^2 \theta + \cos^2 \theta = 1$) come into play.

**EXAMPLE 8**  Prove that the equation $\csc A \tan A \cos A = 1$ is an identity.

**Solution**  Begin with the left-hand side and express each factor in terms of sines or cosines.

$$\csc A \tan A \cos A = \frac{1}{\underset{1}{\cancel{\sin A}}} \cdot \frac{\overset{1}{\cancel{\sin A}}}{\underset{1}{\cancel{\cos A}}} \cdot \overset{1}{\cancel{\cos A}}$$

$$= 1 \qquad \text{as required}$$

**EXAMPLE 9**  Prove that $\cos^2 B - \sin^2 B = \dfrac{1 - \tan^2 B}{1 + \tan^2 B}$.

Solution   We begin with the right-hand side this time; it is the more complicated expression. As in the previous examples, we express everything in terms of sines and cosines.

$$\frac{1 - \tan^2 B}{1 + \tan^2 B} = \frac{1 - \dfrac{\sin^2 B}{\cos^2 B}}{1 + \dfrac{\sin^2 B}{\cos^2 B}}$$

$$= \frac{\cos^2 B \left( 1 - \dfrac{\sin^2 B}{\cos^2 B} \right)}{\cos^2 B \left( 1 + \dfrac{\sin^2 B}{\cos^2 B} \right)}$$

$$= \frac{\cos^2 B - \sin^2 B}{\cos^2 B + \sin^2 B}$$

$$= \frac{\cos^2 B - \sin^2 B}{1}$$

$$= \cos^2 B - \sin^2 B \qquad \text{as required}$$

**EXAMPLE 10**   Prove that $\dfrac{\cos \theta}{1 - \sin \theta} = \dfrac{1 + \sin \theta}{\cos \theta}$.

Solution   The suggestions given in the previous examples are not applicable here. Everything is already in terms of sines and cosines. Furthermore, neither side appears more complicated than the other. A technique that does work here is to begin with the left-hand side and multiply numerator *and* denominator by the same quantity, namely, $1 + \sin \theta$. Doing so gives us

$$\frac{\cos \theta}{1 - \sin \theta} = \frac{\cos \theta}{1 - \sin \theta} \cdot \frac{1 + \sin \theta}{1 + \sin \theta}.$$

$$= \frac{\cos \theta (1 + \sin \theta)}{1 - \sin^2 \theta}$$

$$= \frac{\cos \theta (1 + \sin \theta)}{\cos^2 \theta}$$

$$= \frac{1 + \sin \theta}{\cos \theta} \qquad \text{as required}$$

The general strategy for each of the proofs in Examples 8 through 10 was the same. In each case we worked with one side of the given equation, and we transformed it (into equivalent expressions) until it was identical to the other side of the equation. This is not the only strategy that can be used. For instance, an alternate way to establish the identity in Example 10 is as follows. The given equation is equivalent to

$$\frac{\cos \theta}{1 - \sin \theta} - \frac{1 + \sin \theta}{\cos \theta} = 0 \tag{1}$$

Now we show that equation (1) is an identity. To do this, we combine the two fractions on the left-hand side, using the common denominator $\cos\theta(1 - \sin\theta)$. We have

$$\frac{\cos\theta}{1 - \sin\theta} - \frac{1 + \sin\theta}{\cos\theta} = \frac{\cos^2\theta - (1 + \sin\theta)(1 - \sin\theta)}{\cos\theta(1 - \sin\theta)}$$

$$= \frac{\cos^2\theta - (1 - \sin^2\theta)}{\cos\theta(1 - \sin\theta)}$$

$$= \frac{\cos^2\theta + \sin^2\theta - 1}{\cos\theta(1 - \sin\theta)} = \frac{1 - 1}{\cos\theta(1 - \sin\theta)}$$

$$= 0$$

This shows that equation (1) is an identity, and thus the equation given in Example 10 is an identity.

Another strategy that can be used in establishing trigonometric identities is to work independently with each side of the given equation until a common expression is obtained. This is the strategy used in Example 11.

**EXAMPLE 11**  Prove that $\dfrac{\cos(180° - A) + \sin(180° - A)}{\cos(180° - A) - \sin(180° - A)} = \dfrac{1 - \tan A}{1 + \tan A}$

**Solution**

$$\text{Left-hand side} = \frac{-\cos A + \sin A}{-\cos A - \sin A}$$

$$= \frac{\cos A - \sin A}{\cos A + \sin A}$$

$$\text{Right-hand side} = \frac{1 - \dfrac{\sin A}{\cos A}}{1 + \dfrac{\sin A}{\cos A}} \cdot \frac{\cos A}{\cos A}$$

$$= \frac{\cos A - \sin A}{\cos A + \sin A}$$

We've now established the required identity by showing that both sides are equal to the same quantity.

# EXERCISE SET 7.4

## A

*In Exercises 1 and 2, carry out the indicated operations and simplify where possible.*

1. (a) $(1 - \cos\theta)^2$   (b) $(\sin\theta + \cos\theta)^2$
   (c) $\cos\theta + \dfrac{1}{\sin\theta}$

2. (a) $(\sin A + \cos A)(\csc A + \sec A)$
   (b) $(\sin^2 B + 1)(\cos^2 B + 1)$
   (c) $(\sin\theta\csc\theta + \cos\theta\sec\theta)^2$

*In Exercises 3 and 4, factor each expression.*

3. (a) $\tan^2\theta - 5\tan\theta - 6$
   (b) $\sin^2 B - \cos^2 B$
   (c) $\cos^2 A + 2\cos A + 1$

4. (a) $3\sin\theta\cos^2\theta + 6\sin\theta$
   (b) $\csc^2\alpha - 2\csc\alpha - 3$
   (c) $6\sin^2\theta + 7\sin\theta + 2$

*In Exercises 5–24, simplify each expression.*

5. $\dfrac{\dfrac{\cos\theta + 1}{\cos\theta} + 1}{\dfrac{\cos\theta - 1}{\cos\theta} - 1}$

6. $\dfrac{\dfrac{\cot\beta + 1}{\cot\beta} + 1}{\dfrac{\cot\beta - 1}{\cot\beta} - 1}$

7. $\dfrac{\dfrac{1 - \tan\theta}{\sin\theta}}{\dfrac{\sin\theta}{\cos\theta} - 1}$

8. $\dfrac{2 + \dfrac{1}{\cos\theta}}{\dfrac{1}{\cos^2\theta}}$   *Hint:*  Multiply numerator and denominator by $\cos^2\theta$.

9. $\dfrac{\sin^2 A - \cos^2 A}{\sin A - \cos A}$   *Hint:*  Factor the numerator.

10. $\dfrac{\sin^4 A - \cos^4 A}{\cos A - \sin A}$

11. $\sin^2\theta \cos\theta \csc^3\theta \sec\theta$

12. $\sin\theta \csc\theta \tan\theta$

13. $\cot B \sin^2 B \cot B$

14. $\dfrac{3\sin\theta + 6}{\sin^2\theta - 4}$

15. $\dfrac{\cos^2 A + \cos A - 12}{\cos A - 3}$

16. (a) $\dfrac{x - y}{y - x}$   (b) $\dfrac{\sin\theta - \cos\theta}{\cos\theta - \sin\theta}$

17. $\dfrac{\cos A - 2\sin A \cos A}{\cos^2 A - \sin^2 A + \sin A - 1}$

18. $\dfrac{\tan\theta}{\sec\theta - 1} + \dfrac{\tan\theta}{\sec\theta + 1}$

19. $\sec A \csc A - \tan A - \cot A$

20. $\cot\theta + \dfrac{1 - 2\cos^2\theta}{\sin\theta \cos\theta}$

21. $\dfrac{\sec\theta - 1}{\sec\theta + 1} - \dfrac{\tan\theta - \sin\theta}{\tan\theta + \sin\theta}$

22. $(\sec A + \tan A)(\sec A - \tan A)$

23. $\dfrac{\cot^2\theta}{\csc^2\theta} + \dfrac{\tan^2\theta}{\sec^2\theta}$

24. $\dfrac{\tan\theta + \tan\theta \sin\theta - \cos\theta \sin\theta}{\sin\theta \tan\theta}$

*In Exercises 25–43, prove that the equations are identities.*

25. $\sin\theta \cos\theta \sec\theta \csc\theta = 1$

26. $\tan^2 A + 1 = \sec^2 A$

27. $(\sin\theta \sec\theta)/(\tan\theta) = 1$

28. $\tan\beta \sin\beta = \sec\beta - \cos\beta$

29. $\dfrac{1 - 5\sin x}{\cos x} = \sec x - 5\tan x$

30. $\dfrac{1}{\sin\theta} - \sin\theta = \cot\theta \cos\theta$

31. $\cos A(\sec A - \cos A) = \sin^2 A$

32. $\dfrac{\sin\theta}{\csc\theta} + \dfrac{\cos\theta}{\sec\theta} = 1$

33. $(1 - \sin\theta)(\sec\theta + \tan\theta) = \cos\theta$

34. $(\cos\theta - \sin\theta)^2 + 2\sin\theta \cos\theta = 1$

35. $(\sec\alpha - \tan\alpha)^2 = \dfrac{1 - \sin\alpha}{1 + \sin\alpha}$

36. $\dfrac{\sin B}{1 + \cos B} + \dfrac{1 + \cos B}{\sin B} = 2\csc B$

37. $\sin A + \cos A = \dfrac{\sin A}{1 - \cot A} - \dfrac{\cos A}{\tan A - 1}$

38. $(1 - \cos C)(1 + \sec C) = \tan C \sin C$

39. $\csc^2\theta + \sec^2\theta = \csc^2\theta \sec^2\theta$

40. $1 - \sin(90° - \theta)\cos\theta = \sin^2\theta$

41. $\dfrac{2\sin^3\beta}{1 - \cos\beta} = 2\sin\beta + 2\sin\beta \cos\beta$

*Hint:*  Write $\sin^3\beta$ as $(\sin\beta)(\sin^2\beta)$.

42. $\dfrac{\cos B \tan B}{\tan(90° - B)} = \sin^2 B \sec B$

43. $\dfrac{\sin^3\theta + \cos^3\theta}{\sin\theta + \cos\theta} = 1 - \sin\theta \cos\theta$

44. Prove the identity $\dfrac{\sin\theta}{1 - \cos\theta} = \dfrac{1 + \cos\theta}{\sin\theta}$ in two different ways.
   (a) Adapt the method of Example 10.
   (b) Begin with the left-hand side and multiply numerator and denominator by $\sin\theta$.

**B**

45. Prove the identity $\dfrac{\sec\theta - \csc\theta}{\sec\theta + \csc\theta} = \dfrac{\tan\theta - 1}{\tan\theta + 1}$.

46. (a) Factor the expression $\cos^3\theta - \sin^3\theta$.
   (b) Prove the identity.
   $$\dfrac{\cos\phi \cot\phi - \sin\phi \tan\phi}{\csc\phi - \sec\phi} = 1 + \sin\phi \cos\phi$$

47. Prove the identity.
   $$(r\sin\theta \cos\phi)^2 + (r\sin\theta \sin\phi)^2 + (r\cos\theta)^2 = r^2$$

48. *Formula for $\cos(\alpha + \beta)$*  The figure shown is constructed as follows. Begin with right triangle $ABC$ in which $AC = 1$ and $\alpha + \beta = \angle A$. Extend line segment $\overline{AP}$ and draw $\overline{CE}$ perpendicular to it. Draw $\overline{ED}$ perpendicular to $\overline{BC}$. Finally, from $E$, draw a perpendicular to $\overline{AB}$ extended, meeting this extension at $F$:

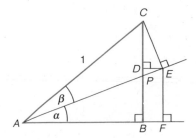

(a) Show that $CE = \sin \beta$. (Look at triangle $AEC$.)
(b) Show that $\angle DCE = \alpha$.
(c) Show that $DE = \sin \alpha \sin \beta$. (Look at triangle $CDE$.)
(d) Show that $AE = \cos \beta$.
(e) Show that $AF = \cos \alpha \cos \beta$.
(f) Show that $\cos(\alpha + \beta) = AF - DE$.
   *Hint:* $DE = BF$.
(g) On the basis of steps (f), (e), and (c), conclude that

$$\cos(\alpha + \beta) = \cos \alpha \cos \beta - \sin \alpha \sin \beta$$

(h) Find $\cos 75°$. *Hint:* $75° = 30° + 45°$
(i) Show that $\cos 75° \neq \cos 30° + \cos 45°$.

49. Suppose that

$$A \sin \theta + \cos \theta = 1$$

and

$$B \sin \theta - \cos \theta = 1$$

Show that $AB = 1$. *Hint:* Solve the first equation for $A$, the second for $B$, and then compute $AB$.

50. (a) Make the substitution $u = a \sin \theta$ in the expression $\sqrt{a^2 - u^2}$ and show that the expression becomes $a \cos \theta$. (Assume $a > 0$ and $0° < \theta < 90°$.)
   (b) Make the substitution $u = a \tan \theta$ in the expression $\sqrt{a^2 + u^2}$ and show that the result is $a \sec \theta$. (Assume $a > 0$ and $0° < \theta < 90°$.)
   (c) Make the substitution $x = 2 \sin \theta$ in the expression $x^2/(4 - x^2)^{3/2}$ and show that the result is $\frac{1}{2} \tan^2 \theta \sec \theta$.

51. Show that the equation of line $L$ in the following figure is $x \cos \theta + y \sin \theta = d$.

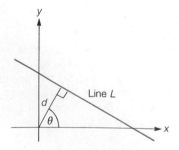

52. Suppose that $\tan \theta + \sin \theta = a$ and $\tan \theta - \sin \theta = b$. Show that $a^2 - b^2 = 4\sqrt{ab}$.

53. If $\sin \alpha + \cos \alpha = a$ and $\sin \alpha - \cos \alpha = b$, show that

$$\tan \alpha = \frac{a + b}{a - b}$$

## C

54. In the figure, triangle $ABC$ is a right triangle with the right angle at $C$. The remainder of the figure is constructed as follows. From $A$ a line is drawn perpendicular to $\overline{AB}$, meeting $\overline{BC}$ (extended) at $E$. Similarly, from $B$ a line is drawn perpendicular to $\overline{AB}$, meeting $\overline{AC}$ (extended) at $D$. Finally, segment $\overline{DE}$ is drawn.
   (a) Show that $\tan \theta = \tan^3 \beta$.
   (b) Find the ratio of the area of triangle $ABC$ to the area of triangle $CED$.

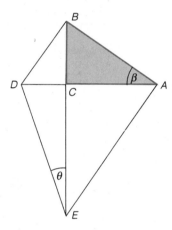

55. Prove the identity $\dfrac{\tan \theta + \sec \theta - 1}{\tan \theta - \sec \theta + 1} = \dfrac{1 + \sin \theta}{\cos \theta}$.

56. Given that $\dfrac{\sin \alpha}{\sin \beta} = p$ and $\dfrac{\cos \alpha}{\cos \beta} = q$, express $\tan \alpha$ and $\tan \beta$ in terms of $p$ and $q$. (Assume that $\alpha$ and $\beta$ are acute angles.)

## CHAPTER 7 SUMMARY OF PRINCIPAL FORMULAS AND NOTATION

| FORMULAS OR NOTATIONS | PAGE REFERENCE | COMMENT |
|---|---|---|
| 1. $\sin\theta$<br>$\cos\theta$<br>$\tan\theta$<br>$\csc\theta$<br>$\sec\theta$<br>$\cot\theta$ | 325 | When $\theta$ is an acute angle in a right triangle, the six trigonometric functions of $\theta$ can be defined as follows:<br><br>$\sin\theta = \dfrac{\text{opposite}}{\text{hypotenuse}}$  $\csc\theta = \dfrac{\text{hypotenuse}}{\text{opposite}}$<br><br>$\cos\theta = \dfrac{\text{adjacent}}{\text{hypotenuse}}$  $\sec\theta = \dfrac{\text{hypotenuse}}{\text{adjacent}}$<br><br>$\tan\theta = \dfrac{\text{opposite}}{\text{adjacent}}$  $\cot\theta = \dfrac{\text{adjacent}}{\text{opposite}}$ |
|  | 345 | More generally, if $\theta$ denotes any angle in standard position, and $P(x,y)$ is the point where the terminal side of the angle meets the unit circle, then the six trigonometric functions of $\theta$ can be defined as follows:<br><br>$\sin\theta = y$    $\csc\theta = \dfrac{1}{y}$  $(y \neq 0)$<br><br>$\cos\theta = x$    $\sec\theta = \dfrac{1}{x}$  $(x \neq 0)$<br><br>$\tan\theta = \dfrac{y}{x}$  $(x \neq 0)$    $\cot\theta = \dfrac{x}{y}$  $(y \neq 0)$ |
| 2. $\sin(90° - \theta) = \cos\theta$<br>$\cos(90° - \theta) = \sin\theta$ | 329 | If two angles are complementary, then the sine of (either) one equals the cosine of the other. |
| 3. $\tan\theta = \dfrac{\sin\theta}{\cos\theta}$ | 329, 345 | This equation is an identity. It holds for all angles $\theta$ except those for which $\cos\theta$ is zero. |
| 4. $\sin^2\theta + \cos^2\theta = 1$ | 329, 349 | This equation is an identity. It holds for all angles $\theta$. |
| 5. $\cos(180° - \theta) = -\cos\theta$<br>$\sin(180° - \theta) = \sin\theta$ | 349 | In Section 7.3, we saw that these two identities are valid for all angles $\theta$ such that $0° < \theta < 180°$. In the next chapter, we will see that these two identities actually hold for all angles $\theta$. |
| 6. $A = \frac{1}{2}ab\sin\theta$ | 351 | The area of a triangle equals half the product of the lengths of two sides times the sine of the included angle. |
| 7. $\sin^n\theta$ | 355 | $\sin^n\theta$ means $(\sin\theta)^n$. The same convention also applies to the other five trigonometric functions. |

# CHAPTER 7 REVIEW EXERCISES

NOTE  The fifteen exercises in this set with colored problem numbers constitute a chapter test.

*For Exercises 1–16, evaluate each expression. (Don't use a calculator.)*

1. $\sin 135°$
2. $\cos(-60°)$
3. $\tan(-240°)$
4. $\sin 450°$
5. $\csc 210°$
6. $\sec 225°$
7. $\sin 270°$
8. $\cot(-330°)$
9. $\cos(-315°)$
10. $\cos 180°$
11. $\cos 1800°$
12. $\sec 120°$
13. $\csc 240°$
14. $\csc(-45°)$
15. $\sec 780°$
16. $\sin^2 33° + \sin^2 57°$

17. A student who was asked to simplify the expression $\sin^2 33° + \sin^2 57°$ wrote $\sin^2 33° + \sin^2 57° = \sin^2(33° + 57°) = \sin^2 90° = 1^2 = 1$.
    (a) Where is the error?
    (b) What is the correct answer?

18. If $\theta = 45°$, find each of the following.
    (a) $\cos \theta$      (b) $\cos^3 \theta$      (c) $\cos 2\theta$
    (d) $\cos 3\theta$      (e) $\cos(2\theta/3)$      (f) $(\cos 3\theta)/3$
    (g) $\cos(-\theta)$      (h) $\cos^3 5\theta$

*In Exercises 19–31, the lengths of the three sides of a triangle are denoted a, b, and c; the angles opposite these sides are A, B, and C, respectively. In each exercise, use the given information to find the required quantities.*

|  | GIVEN | FIND |
|---|---|---|
| 19. | $B = 90°, A = 30°,$ $b = 1$ | $a$ and $c$ |
| 20. | $B = 90°, A = 60°,$ $a = 1$ | $c$ and $b$ |
| 21. | $B = 90°, \sin A = \frac{2}{5},$ $a = 7$ | $b$ |
| 22. | $B = 90°, \sec C = 4,$ $c = \sqrt{2}$ | $b$ |
| 23. | $B = 90°, \cos A = \frac{3}{8}$ | $\sin A$ and $\cot A$ |
| 24. | $B = 90°, b = 1,$ $\tan C = \sqrt{5}$ | $a$ |
| 25. | $b = 4, c = 5,$ $A = 150°$ | area of $\triangle ABC$ |
| 26. | $A = 120°, b = 8,$ area of $\triangle ABC = 12\sqrt{3}$ | $c$ |
| 27. | $c = 4, a = 2, B = 90°$ | $\sin^2 A + \cos^2 B$ |
| 28. | $B = 90°, 2a = b$ | $A$ |
| 29. | $A = 30°, B = 120°,$ $b = 16$ | $a$ |
| 30. | $a = 7, b = 8,$ $\sin C = \frac{1}{4}$ | area of $\triangle ABC$ |
| 31. | $a = b = 5,$ $\sin(\frac{1}{2}C) = \frac{9}{10}$ | $c$ and area of $\triangle ABC$ |

32. For the following figure, show that $y = x[\tan(\alpha + \beta) - \tan \beta]$.

33. For the following figure, show that
$$y = \frac{x}{\cot \alpha - \cot \beta}.$$

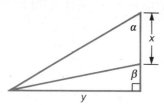

34. For the following figure, show that
$$\cot \theta = \frac{a}{b} + \cot \alpha.$$

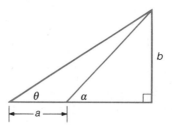

35. Suppose that $\theta$ is an acute angle in a right triangle and $\sin \theta = \dfrac{2p^2 q^2}{p^4 + q^4}$. Find $\cos \theta$ and $\tan \theta$.

36. This problem is adapted from the text, *An Elementary Treatise on Plane Trigonometry*, by R. D. Beasley, first published in 1884.
    (a) Prove the following identity, which will be used in part (b): $1 + \tan^2 \alpha = \sec^2 \alpha$.
    (b) Suppose that $\alpha$ and $\theta$ are acute angles and $\tan \theta = \dfrac{1 + \tan \alpha}{1 - \tan \alpha}$. Express $\sin \theta$ and $\cos \theta$ in terms of $\alpha$.
    *Hint:* Draw a right triangle, label one of the angles $\theta$, and let the lengths of the sides opposite and adjacent to $\theta$ be $1 + \tan \alpha$ and $1 - \tan \alpha$, respectively. *Answer:*

$$\sin \theta = \frac{1}{\sqrt{2}}(\cos \alpha + \sin \alpha)$$

$$\cos \theta = \frac{1}{\sqrt{2}}(\cos \alpha - \sin \alpha)$$

*In Exercises 37–46, convert each expression into one involving only sines and cosines and then simplify. (Leave your answers in terms of sines and/or cosines.)*

37. $\dfrac{\sin A + \cos A}{\sec A + \csc A}$      38. $\dfrac{\csc A \sec A}{\sec^2 A + \csc^2 A}$

39. $\dfrac{\sin A \sec A}{\tan A + \cot A}$      40. $\cos A + \tan A \sin A$

**41.** $\dfrac{\cos A}{1 - \tan A} + \dfrac{\sin A}{1 - \cot A}$

**42.** $\dfrac{1}{\sec A - 1} \div \dfrac{1}{\sec A + 1}$

**43.** $(\sec A + \csc A)^{-1}[(\sec A)^{-1} + (\csc A)^{-1}]$

**44.** $\dfrac{\dfrac{\tan^2 A - 1}{\tan^3 A + \tan A}}{\dfrac{\tan A + 1}{\tan^2 A + 1}}$

**45.** $\dfrac{\dfrac{\sin A + \cos A}{\sin A - \cos A} - \dfrac{\sin A - \cos A}{\sin A + \cos A}}{\dfrac{\sin A + \cos A}{\sin A - \cos A} + \dfrac{\sin A - \cos A}{\sin A + \cos A}}$

**46.** $\dfrac{\sin A \tan A - \cos A \cot A}{\sec A - \csc A}$

*In Exercises 47–56, evaluate each expression in terms of a, where $a = \cos 20°$.*

**47.** $\sin 20°$          **48.** $\tan 20°$

**49.** $\cos 70°$          **50.** $\sin 70°$

**51.** $\cos 160°$         **52.** $\cos 340°$

**53.** $\cos(-160°)$       **54.** $\cos 200°$

**55.** $\sin 200°$         **56.** $\cot 200°$

**57.** Simplify: $\dfrac{\sin^4 \theta - \cos^4 \theta}{\sin^2 \theta - \cos^2 \theta} \div \dfrac{1 + \sin \theta \cos \theta}{\sin^3 \theta - \cos^3 \theta}$

**58.** Simplify: $\dfrac{1 + \sin \theta + \sin^2 \theta}{1 - \sin^3 \theta}$

**59.** Suppose that $\alpha$ and $\beta$ are acute angles such that $\sin \alpha / \sin \beta = \sqrt{2}$ and $\tan \alpha / \tan \beta = \sqrt{3}$. Find $\alpha$ and $\beta$.

**60.** In the following figure, $\overline{AC}$ is tangent to the unit circle at $P$. Express the area of triangle $AOC$ as a function of $\theta$.

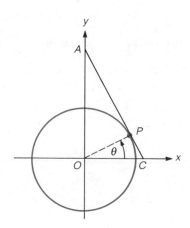

*In Exercises 61–70, use the given information to find the required quantities.*

| | GIVEN | FIND |
|---|---|---|
| **61.** | $\cos \theta = \frac{3}{5}$, $0° < \theta < 90°$ | $\sin \theta$ and $\tan \theta$ |
| **62.** | $\sin \theta = -\frac{12}{13}$, $180° < \theta < 270°$ | $\cos \theta$ |
| **63.** | $\sec \theta = \frac{25}{7}$, $270° < \theta < 360°$ | $\tan \theta$ |
| **64.** | $\cot \theta = \frac{4}{3}$, $0° < \theta < 90°$ | $\sin \theta$ |
| **65.** | $\csc \theta = \frac{13}{12}$, $0° < \theta < 90°$ | $\cot \theta$ |
| **66.** | $\sin \theta = \frac{1}{5}$, $0° < \theta < 90°$ | $\cos(90° - \theta)$ |
| **67.** | $\cos \theta = -\frac{3}{4}$, $90° < \theta < 180°$ | $\tan(90° - \theta)$ |
| **68.** | $\sin \theta = -\sqrt{5}/6$, $180° < \theta < 270°$ | $\sec \theta$ |
| **69.** | $\cos \theta = \frac{7}{9}$, $0° < \theta < 90°$ | $\cos(180° - \theta)$ |
| **70.** | $\cos \theta = -\frac{1}{4}$, $90° < \theta < 180°$ | $\cos(180° - \theta)$ |

**71.** Suppose $\theta$ is an acute angle and $\tan \theta + \cot \theta = 2$. Show that $\sin \theta + \cos \theta = \sqrt{2}$.

**72.** A 100-ft vertical antenna is on the roof of a building. From a point on the ground, the angles of elevation to the top and the bottom of the antenna are $51°$ and $37°$, respectively. Find the height of the building.

**73.** In an isosceles triangle, the two base angles are each $35°$, and the length of the base is 120 cm. Find the area of the triangle.

**74.** Find the perimeter and the area of a regular pentagon inscribed in a circle of radius 9 cm.

**75.** In triangle $ABC$, let $h$ denote the length of the altitude from $A$ to $\overline{BC}$. Show that $h = a/(\cot B + \cot C)$.

**76.** The length of each side of an equilateral triangle is $2a$. Show that the radius of the inscribed circle is $a/\sqrt{3}$ and the radius of the circumscribed circle is $2a/\sqrt{3}$.

**77.** From a helicopter $h$ ft above the sea, the angle of depression to the pilot's horizon is $\theta$. Show that $\cot \theta = R/\sqrt{2Rh + h^2}$, where $R$ is the radius of the Earth.

**78.** In triangle $ABC$, angle $C$ is a right angle. Show that
$$\frac{\sin A - \sin B}{\sin A + \sin B} = \frac{a - b}{a + b}$$

*In Exercises 79–108, show that each equation is an identity.*

**79.** $\sin^2(90° - \theta) \csc\theta - \tan^2(90° - \theta)\sin\theta = 0$

**80.** $\dfrac{1 - \sin\theta\cos\theta}{\cos\theta(\sec\theta - \csc\theta)} \times \dfrac{\sin^2\theta - \cos^2\theta}{\sin^3\theta + \cos^3\theta} = \sin\theta$

**81.** $\cos A \cot A = \csc A - \sin A$

**82.** $\sec A - 1 = \sec A(1 - \cos A)$

**83.** $\dfrac{\cot A - 1}{\cot A + 1} = \dfrac{\cot A - \sin A}{\cos A + \sin A}$

**84.** $\dfrac{\sin A}{\csc A - \cot A} = 1 + \cos A$

**85.** $\cos^2\theta - \sin^2\theta = 2\cos^2\theta - 1$

**86.** $\cos^2\theta - \sin^2\theta = 1 - 2\sin^2\theta$

**87.** $\sin A \tan A = \dfrac{1 - \cos^2 A}{\cos A}$

**88.** $\dfrac{\cot A - 1}{\cot A + 1} = \dfrac{1 - \tan A}{1 + \tan A}$

**89.** $\cot^2 A + \csc^2 A = -\cot^4 A + \csc^4 A$

**90.** $\sin^3 A + \cos^3 A = (\sin A + \cos A)(1 - \sin A \cos A)$

**91.** $\dfrac{\cot^2 A - \tan^2 A}{(\cot A + \tan A)^2} = 2\cos^2 A - 1$

**92.** $\dfrac{\sin A - \cos A}{\sin A} + \dfrac{\cos A - \sin A}{\cos A} = 2 - \sec A \csc A$

**93.** $\dfrac{\cos A - \sin A}{\cos A + \sin A} = \dfrac{1 - \tan A}{1 + \tan A}$

**94.** $\dfrac{\cos A - \sin A}{\sec A \cot A - \csc A \tan A} = \cos A \sin A$

**95.** $\dfrac{1 - \sin A}{\cos A} - \dfrac{1}{\sec A + \tan A} = 0$

**96.** $\dfrac{\cos A}{1 - \tan A} + \dfrac{\sin A}{1 - \cot A} = \cos A + \sin A$

**97.** $\tan A \tan B = \dfrac{\tan A + \tan B}{\cot A + \cot B}$

**98.** $\dfrac{\sin A}{1 + \cos A} + \dfrac{1 + \cos A}{\sin A} = 2\csc A$

**99.** $\tan A - \dfrac{\sec A \sin^3 A}{1 + \cos A} = \sin A$

**100.** $\dfrac{1}{1 - \cos A} + \dfrac{1}{1 + \cos A} = 2 + 2\cot^2 A$

**101.** $\dfrac{1}{\csc A - \cot A} - \dfrac{1}{\csc A + \cot A} = 2\cot A$

**102.** $\dfrac{\sin^3 A}{\cos A - \cos^3 A} = \tan A$

**103.** $\dfrac{1}{1 + \sin^2 A} + \dfrac{1}{1 + \cot^2 A} + \dfrac{1}{1 + \sec^2 A}$

$$+ \dfrac{1}{1 + \csc^2 A} = 2$$

**104.** $\dfrac{\sec A - \tan A}{\sec A + \tan A} = \left(\dfrac{\cos A}{1 + \sin A}\right)^2$

**105.** $\dfrac{\sec A - \csc A}{\sec A + \csc A} = \dfrac{\sin A - \cos A}{\sin A + \cos A}$

**106.** $\sin^6 A + \cos^6 A = 1 - 3\sin^2 A \cos^2 A$

**107.** $\sin A \cos A = \dfrac{\tan A}{1 + \tan^2 A}$

**108.** $\dfrac{2\tan A}{1 - \tan^2 A} + \dfrac{1}{\cos^2 A - \sin^2 A} = \dfrac{\cos A + \sin A}{\cos A - \sin A}$

**109.** The radius of the circle in the following figure is 1 unit. The line segment $\overline{PT}$ is tangent to the circle at $P$, and $\overline{PN}$ is perpendicular to $\overline{OT}$. Express each of the following quantities in terms of $\theta$.
  **(a)** $PN$ **(b)** $ON$ **(c)** $PT$
  **(d)** $OT$ **(e)** $NA$ **(f)** $NT$

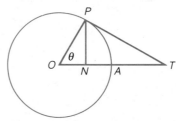

**110.** Refer to the figure accompanying Exercise 109.
  **(a)** Show that the ratio of the area of triangle $ONP$ to the area of triangle $NPT$ is $\cot^2\theta$.
  **(b)** Show that the ratio of the area of triangle $NPT$ to the area of triangle $OPT$ is $\sin^2\theta$.

*In Exercises 111–114, specify a value for each expression. (Do not use a calculator.)*

**111.** **(a)** $\cos 30°$ **(b)** $\tan 60°$ **(c)** $\sin^2 7° + \cos^2 7°$

**112.** **(a)** $\sin 30°$ **(b)** $\cot 30°$ **(c)** $\tan 40° - \dfrac{\sin 40°}{\cos 40°}$

**113.** **(a)** $\cos 330°$ **(b)** $\csc 315°$

**114.** **(a)** $\cos(-30°)$ **(b)** $\sin(-210°)$

**115.** Factor: $6\tan^2\theta + 11\tan\theta + 3$

**116.** Factor: $\sin^4\theta - \cos^4\theta$

**117.** Simplify:

$$\dfrac{\dfrac{\tan\theta + 1}{\tan\theta} + 1}{\dfrac{\tan\theta - 1}{\tan\theta} - 1}$$

**118.** Simplify: $\dfrac{\sin^2\theta + \sin\theta - 12}{3 - \sin\theta}$

**119.** Find the area of a triangle in which the lengths of two sides are 8 cm and 12 cm, and the angle included between these sides is 135°.

**120.** The area of $\triangle ABC$ is $1\,\text{cm}^2$, and $\angle B = 150°$. Find $AB$ and $BC$, given that $BC$ is twice $AB$.

**121.** In the following figure, $\angle DAB = 25°$, $\angle CAB = 55°$, and $AB = 50$ cm. Find $CD$. (Leave your answer in terms of the trigonometric functions, rather than decimals.)

*In Exercises 122–124, prove that each equation is an identity.*

**122.** $\dfrac{\cos\theta + 1}{\csc\theta + \cot\theta} = \sin\theta$

**123.** $\dfrac{1}{\sin\theta + \cot\theta\cos\theta} = \sin\theta$

**124.** $\cos^2\theta + \dfrac{1 - \cot^2\theta}{1 + \cot^2\theta} = \sin^2\theta$

**125.** The arc in the following figure is a portion of the unit circle. Express $AB$ and $BP$ in terms of $\theta$.

# 8 TRIGONOMETRIC FUNCTIONS OF REAL NUMBERS

**INTRODUCTION** In the previous chapter we measured angles using the familiar units of degrees. However, there is another unit for measuring angles, called the *radian*, that has proven to be more useful in calculus and in many of the sciences. In Section 8.1 we define radian measure. Then, in Section 8.2, we use radian measure to restate the definitions of the trigonometric functions in such a way that the domains are sets of real numbers. Among other advantages, this allows us to analyze the trigonometric functions using the graphing techniques developed in Chapter 4. With this in mind, we discuss the graphs of the trigonometric functions in Sections 8.5 and 8.6. In the two sections prior to this, Sections 8.3 and 8.4, we develop the basic trigonometric identities. Then in Section 8.7 we apply some of these identities in solving trigonometric equations. Finally, in Section 8.8 we consider the inverse trigonometric functions.

## RADIAN MEASURE

### 8.1

*. . . I wrote to him [i.e., to Alexander J. Ellis, in 1874], and he declared at once for the form "radian," on the ground that it could be viewed as a contraction for "radial angle. . . ."*

THOMAS MUIR, in a letter appearing in the April 7, 1910, issue of *Nature*

*I shall be very pleased to send Dr. Muir a copy of my father's examination questions of June, 1873, containing the word* radian*. . . . It thus appears that* radian *was thought of independently by Dr. Muir and my father, and, what is really more important than the exact form of the name, they both independently thought of the necessity of giving a name to the unit-angle.*

JAMES THOMSON, in a letter appearing in the June 16, 1910, issue of *Nature*

For the portion of trigonometry dealing with triangles, the units of degrees are quite suitable for measuring angles. For the more analytical portions of trigonometry, however, radian measure is used. The radian measure of an angle is defined as follows.

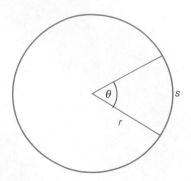

**FIGURE 1**

---

**Definition of the Radian Measure of an Angle**

Place the vertex of the angle $\theta$ at the center of a circle of radius $r$. Let $s$ denote the length of the arc intercepted by $\theta$, as indicated in Figure 1. Then the **radian measure** of $\theta$ is the ratio of the arc length $s$ to the radius $r$. In symbols,

$$\theta = \frac{s}{r}$$

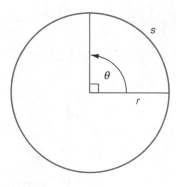

**FIGURE 2**

At first it may appear to you that the radian measure of $\theta$ depends on the radius of the particular circle that we use. But as you will see, this is not the case.

To gain some experience in working with the definition of radian measure, let us calculate the radian measure of the right angle in Figure 2. We begin with the formula

$$\theta = \frac{s}{r} \qquad (1)$$

Now, since $\theta$ is a right angle, the arc length $s$ is one-quarter of the entire circumference. Thus,

$$s = \frac{1}{4}(2\pi r)$$

$$= \frac{\pi r}{2} \qquad (2)$$

Using equation (2) to substitute for $s$ in equation (1), we get

$$\theta = \frac{\pi r/2}{r}$$

$$= \frac{\pi r}{2} \times \frac{1}{r} = \frac{\pi}{2}$$

In other words,

$$90° = \frac{\pi}{2} \text{ radians} \qquad (3)$$

Notice that the radius $r$ does not appear in our answer.

For practical reasons, we would like to be able to convert rapidly between degree and radian measure. Multiplying both sides of equation (3) by 2 yields

$$180° = \pi \text{ radians} \qquad (4)$$

Equation (4) is useful and should be memorized. For instance, dividing both sides of equation (4) by 6 yields

$$\frac{180°}{6} = \frac{\pi}{6}$$

or

$$30° = \frac{\pi}{6} \text{ radians}$$

Similarly, dividing both sides of equation (4) by 4 and 3, respectively, yields

$$45° = \frac{\pi}{4} \text{ radians} \qquad \text{and} \qquad 60° = \frac{\pi}{3} \text{ radians}$$

And, multiplying both sides of equation (4) by 2 gives us

$$360° = 2\pi \text{ radians}$$

Figure 3 summarizes some of these results.

FIGURE 3

**EXAMPLE 1**   (a)  Express 1° in radian measure.
(b)  Express 1 radian in terms of degrees.

Solution   (a)  Dividing both sides of equation (4) by 180 yields

$$1° = \frac{\pi}{180} \text{ radian}$$

(b)  Dividing both sides of equation (4) by $\pi$ yields

$$\frac{180°}{\pi} = 1 \text{ radian}$$

In other words, 1 radian is approximately 180°/3.14, or 57.3°.

From the results in Example 1, we have the following rules for converting between radians and degrees.

---

To convert from degrees to radians, multiply by $\frac{\pi}{180°}$. To convert from radians to degrees, multiply by $\frac{180°}{\pi}$.

---

**EXAMPLE 2**   Convert 150° to radians.

Solution   $$150°\left(\frac{\pi}{180°}\right) = \frac{5\pi}{6} \qquad \text{reducing the fraction}$$

Thus,

$$150° = \frac{5\pi}{6} \text{ radians}$$

**EXAMPLE 3**   Convert $11\pi/6$ radians to degrees.

Solution   $$\frac{11\pi}{6}\left(\frac{180°}{\pi}\right) = \frac{(11)(180°)}{6}$$
$$= 11(30°) = 330°$$

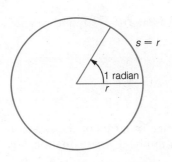

FIGURE 4 An angle of 1 radian intercepts an arc whose length is the same as the length of the radius.

Thus,

$$\frac{11\pi}{6} \text{ radians} = 330°$$

We saw in Example 1(b) that 1 radian is approximately 57°. It is also useful to be able to visualize an angle of 1 radian without thinking in terms of degree measure. This is done as follows. In the equation $\theta = s/r$, we let $\theta = 1$. This yields $1 = s/r$, and consequently, $r = s$. In other words, in a circle, 1 radian is the central angle that intercepts an arc equal in length to the radius of the circle (see Figure 4).

From now on, when we specify the measure of an angle, we will assume that the units are radians unless the degree symbol is explicitly used. (This convention is also used in calculus.) For instance, the equation $\theta = 2$ means that $\theta$ is 2 radians. Similarly, the expression $\sin \pi/6$ refers to the sine of $\pi/6$ radians.

**EXAMPLE 4** Evaluate the following.

(a) $\sin \dfrac{\pi}{6}$    (b) $\cos 2\pi$

Solution    (a) $\dfrac{\pi}{6}$ radians $= 30°$

Therefore,

$$\sin \frac{\pi}{6} = \sin 30°$$

$$= \frac{1}{2} \qquad \text{because } \sin 30° = \frac{1}{2}$$

(b) $2\pi$ radians $= 360°$

Therefore,

$$\cos 2\pi = \cos 360°$$

$$= 1 \qquad \text{because } \cos 360° = 1$$

**EXAMPLE 5** Without using a calculator, determine whether the following values are positive or negative.

(a) $\cos 3$    (b) $\sin 1$    (c) $\tan 6$

Solution    (a) Since $\pi$ radians is 180°, we estimate that 3 radians is slightly less than 180° (because 3 is slightly less than $\pi$). Thus, in standard position, the terminal side of an angle of 3 radians lies in the second quadrant. Consequently, $\cos 3$ is negative.

(b) Since 1 radian is approximately 60° [as we saw in Example 1(b) and in Figure 4], $\sin 1$ is positive.

(c) We know that 6 radians is slightly less than $2\pi$ radians, which is one revolution, or 360°. Thus, in standard position, the terminal side of an angle of 6 radians lies in the fourth quadrant. Consequently, $\tan 6$ is negative.

Let us now return to the defining equation for radians, $\theta = s/r$, and rewrite it as follows:

| **Formula for Arc Length** | |
| --- | --- |
| $s = r\theta$ | (5) |

This useful equation expresses arc length on a circle in terms of the radius $r$ and the central angle $\theta$ subtended by the arc. In applying this equation, we must be certain that $\theta$ is expressed in radians.

**EXAMPLE 6** Find the arc length $s$ in Figure 5.

**Solution** We have seen previously that $30° = \pi/6$ radians. Thus

$$s = r\theta = (10)\left(\frac{\pi}{6}\right)$$

$$= \frac{5\pi}{3} \text{ cm} \qquad \text{(This is approximately 5 cm. Why?)}$$

Note that the units for $s$ and for $r$ are the same.

The next example indicates how radian measure is used in the study of rotating objects. As a prerequisite for this example, we first define the terms *angular speed* and *linear speed*.

FIGURE 5

| **Definition: Angular Speed and Linear Speed** |
| --- |
| Suppose that a wheel rotates about its axis at a constant rate. |

1. [Refer to Figure 6(a)] If a radial line turns through an angle $\theta$ in time $t$, then the **angular speed** of the wheel (denoted by the Greek letter $\omega$) is defined to be

$$\text{angular speed} = \omega = \frac{\theta}{t}$$

2. [Refer to Figure 6(b)] If a point $P$ on the rotating wheel travels a distance $d$ in time $t$, then the **linear speed** of $P$, denoted by $v$, is defined to be

$$\text{linear speed} = v = \frac{d}{t}$$

FIGURE 6

(a)                              (b)

**EXAMPLE 7**  A circular gear in a motor rotates at the rate of 100 rpm (revolutions per minute).

(a) What is the angular speed of the gear in radians per minute?
(b) Find the linear speed of a point on the gear 4 cm from the center.

**Solution**   (a) Each revolution of the gear is $2\pi$ radians. So in 100 revolutions, there are

$$\theta = 100(2\pi) = 200\pi \text{ radians}$$

Consequently, we have

$$\omega = \frac{\theta}{t} = \frac{200\pi \text{ radians}}{1 \text{ min}}$$
$$= 200\pi \text{ radians/min}$$

(b) We can use the formula $s = r\theta$ to find the distance traveled by the point in 1 min. Using $r = 4$ cm and $\theta = 200\pi$, we obtain

$$s = r\theta$$
$$= 4(200\pi) = 800\pi \text{ cm}$$

The linear speed therefore is

$$v = \frac{d}{t} = \frac{800\pi \text{ cm}}{1 \text{ min}}$$
$$= 800\pi \text{ cm/min}$$

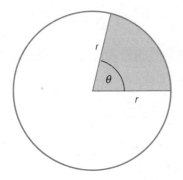

One of the advantages of using radian measure is that many formulas then take on particularly simple forms. The formula $s = r\theta$ is one example of this. As another example, let us derive a formula for the area of a *sector* of a circle. (In Figure 7, the shaded region is the sector.) From geometry, we know that the area $A$ of the sector is directly proportional to the measure of its central angle $\theta$. In symbols,

$$A = k\theta \tag{6}$$

To determine the constant $k$, we use the fact that when $\theta = 2\pi$, we have a complete circle. Thus, when $\theta = 2\pi$, the area must be $\pi r^2$. Substituting the values $\theta = 2\pi$ and $A = \pi r^2$ in equation (6) yields

$$\pi r^2 = k(2\pi)$$

and therefore,

$$\tfrac{1}{2}r^2 = k$$

**FIGURE 7**

Using this value for $k$ in equation (6) gives us the following result:

| Formula for the Area of a Sector |
|---|
| $A = \tfrac{1}{2}r^2\theta$ |

This is the formula for the area of a sector of a circle in terms of the central angle $\theta$ and the radius $r$. The formula is valid only when $\theta$ is expressed in radians.

**EXAMPLE 8**  Compute the area of the sector in Figure 8.

**Solution**  We first convert $120°$ to radians:

$$120° \left( \frac{\pi}{180°} \right) = \frac{2\pi}{3} \text{ radians}$$

We then have

$$A = \frac{1}{2} r^2 \theta$$

$$= \frac{1}{2} (5^2) \frac{2\pi}{3}$$

$$= \frac{25\pi}{3} \text{ cm}^2$$

**FIGURE 8**

**EXAMPLE 9**  Express the area of the shaded region in Figure 9 as a function of $\theta$.

**Solution**  Let $S$ denote the area of the shaded region in Figure 9. Then we have

$$S = (\text{area of sector } OAB) - (\text{area } \triangle OAB)$$

Use the formula area $= \frac{1}{2} r^2 \theta$ to find this.

Use the result from Chapter 7 stating that the area of a triangle equals half the product of two sides times the sine of the included angle.

Thus,

$$S = \tfrac{1}{2}(2^2)\theta - \tfrac{1}{2}(2)(2)\sin\theta$$

$$= 2\theta - 2\sin\theta$$

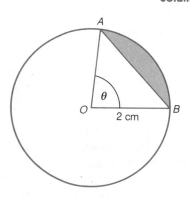

**FIGURE 9**

The required formula is therefore $S = 2\theta - 2\sin\theta$ or, equivalently, $S = 2(\theta - \sin\theta)$.

# EXERCISE SET 8.1

## A

*In Exercises 1 and 2, convert to radian measure.*

1. (a) $60°$  (b) $225°$  (c) $36°$
   (d) $450°$  (e) $0°$
2. (a) $35°$  (b) $22.5°$  (c) $2°$  (d) $100°$

*In Exercises 3 and 4, convert the radian measures to degrees.*

3. (a) $\pi/12$  (b) $3\pi/2$  (c) $6\pi$
   (d) $\pi/10$  (e) $\pi/2$  (f) $3$
4. (a) $11\pi/6$  (b) $3\pi/7$  (c) $2\pi$
   (d) $2$  (e) $1/\pi$  (f) $\pi^2$

5. Suppose that $\theta = \frac{3}{2}$. Without using a calculator or tables, decide whether $\theta$ is larger or smaller than a right angle.

*In Exercises 6–8, complete each table.*

6.

| $\theta$ | 0 | $\pi/2$ | $\pi$ | $3\pi/2$ | $2\pi$ |
|---|---|---|---|---|---|
| $\sin\theta$ | | | | | |

7.

| $\theta$ | 0 | $\pi/2$ | $\pi$ | $3\pi/2$ | $2\pi$ |
|---|---|---|---|---|---|
| $\cos\theta$ | | | | | |

8.

| $\theta$ | 0 | $\pi/6$ | $\pi/4$ | $\pi/3$ | $\pi/2$ |
|---|---|---|---|---|---|
| $\tan\theta$ | | | | | |

*In Exercises 9–12, determine whether the given quantities are positive or negative; do not use a calculator.*

9. (a) $\sin 2$      (b) $\sin 3$
10. (a) $\cos 2$      (b) $\tan 2$
11. (a) $\sin 3.16$      (b) $\tan 3.16$
12. (a) $\sec 6.3$      (b) $\sec(-6.3)$

*In Exercises 13–16, find the arc length s in each case.*

13.

14.

15.

16.

17. In a circle of radius 6 cm, an arc of length 1 cm subtends an angle $\theta$ at the center of the circle. What is the radian measure of $\theta$?

18. In a circle of radius 4 in., an arc of length 4 in. subtends an angle $\theta$ at the center of the circle. What is the radian measure of $\theta$?

*In Exercises 19–24, you are given the rate of rotation of a wheel and its radius. In each case, determine the following:*
(a) *the angular speed, in units of radians/sec;*
(b) *the linear speed, in units of cm/sec, of a point on the circumference of the wheel;*
(c) *the linear speed, in cm/sec, of a point halfway between the center of the wheel and the circumference.*

19. 6 revolutions/sec; $r = 12$ cm
20. 15 revolutions/sec; $r = 20$ cm
21. $1080°$/sec; $r = 25$ cm
22. $2160°$/sec; $r = 60$ cm
23. 500 rpm; $r = 45$ cm
24. 1250 rpm; $r = 10$ cm

25. Find the area of the shaded sectors.

(a)

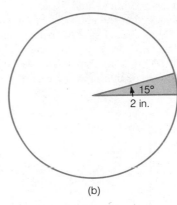

(b)

26. In a circle of radius 1 cm, the area of a certain sector is $\pi/5$ cm². Find the central angle of this sector.

27. Find the area of the shaded region in the following figure.

28. Express the area of the shaded region in the figure on the following page as a function of $\theta$. *Hint:* The region is composed of a sector and a triangle.

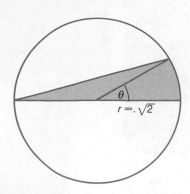

**FIGURE FOR EXERCISE 28**

## B

29. Suppose that you have two sticks and a piece of wire, each of length 1 ft, fastened at the ends to form an equilateral triangle. See Figure (a). If side $\overline{BC}$ is bent out to form an arc of a circle with center $A$, then the angle at $A$ will decrease from 60° to something less. See Figure (b). What is the measure of this new angle at $A$?

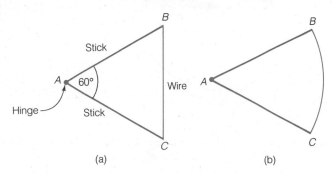

(a)                    (b)

30. **(a)** When a clock reads 4:00, what is the radian measure of the (smaller) angle between the hour hand and minute hand?
    **(b)** When a clock reads 5:30, what is the radian measure of the (smaller) angle between the hour hand and minute hand?

© 31. A wheel 3 ft in diameter makes $x$ revolutions. Find $x$, given that the distance traveled by a point on the circumference of the wheel is 22619 ft. (Round off your answer to the nearest whole number.)

32. A point $P$ is on a rotating wheel, 12 cm from the center. If the linear speed of $P$ is 204 cm/sec, find the angular speed of the wheel (in radians/sec).

© 33. For this problem, assume that the earth is a sphere with a radius of 3960 miles and a rotation rate of 1 revolution per 24 hours.
    **(a)** Find the angular speed. Express your answer in units of radians/sec, and round off to two significant digits.

**(b)** Find the linear speed of a point on the equator. Express the answer in units of miles per hour, and round off to the nearest 10 mph.

34. The accompanying figure shows a circular sector with radius $r$ cm and central angle $2\theta$ (radian measure). The perimeter of the sector is 12 cm.
    **(a)** Express $r$ as a function of $\theta$.
    **(b)** Express the area $A$ of the sector as a function of $\theta$.

35. The following figure shows a semicircle of radius 1 unit and two adjacent sectors, $AOC$ and $COB$.
    **(a)** Show that the product $P$ of the areas of the two sectors is given by

$$P = \frac{\pi\theta}{4} - \frac{\theta^2}{4}$$

   Is this a quadratic function?
    **(b)** For what value of $\theta$ is $P$ a maximum?

## C

36. Refer to the accompanying figure. From a point $R$, two tangent lines are drawn to a circle of radius 1. These tangents meet the circle at the points $P$ and $Q$. At the midpoint of the circular arc $\overset{\frown}{PQ}$, a third tangent line is drawn, meeting the other two tangents at $T$ and $U$, as shown. Express the area of triangle $RTU$ as a function of $\theta$ ($\theta$ is defined in the figure).

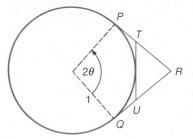

# TRIGONOMETRIC FUNCTIONS OF REAL NUMBERS

## 8.2

The definitions of the trigonometric functions in Section 7.3 are based on the unit circle. Let us look at radian measure in this context. By definition, the radian measure of $\theta$ is given by $\theta = s/r$. So when $r = 1$, we obtain

$$\theta = \frac{s}{1} = s$$

Thus, in the unit circle, the arc length *is* the radian measure of the angle $\theta$ (see Figure 1).

In view of this observation, we can now define the trigonometric functions in such a way that the domains are sets of real numbers, rather than angles. There are two reasons for doing this. First, for analytical work, the domains of the trigonometric functions are sets of real numbers. Second, if the domains are sets of real numbers, then the techniques for graphing that we developed in Chapter 4 can be used to analyze these functions.

In the definitions that follow, you may think of $\theta$ as either the measure of an arc length or the radian measure of an angle. But in both cases—and this is the point—$\theta$ denotes a real number. The conventions regarding the measurement of arc length on the unit circle are the same as those used previously for angles: We measure from the point $(1,0)$ and we assume that the positive direction is counterclockwise.

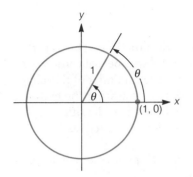

**FIGURE 1** In the unit circle, the length of the intercepted arc is the radian measure of the central angle.

---

### Trigonometric Functions of Real Numbers

(Refer to Figure 2.) Let $P(x,y)$ denote the point on the unit circle whose arc length from $(1,0)$ is $\theta$. Or, equivalently, let $P(x,y)$ denote the point where the terminal side of the angle with radian measure $\theta$ intersects the unit circle. Then the six trigonometric functions are defined as follows.

$$\sin\theta = y \qquad\qquad \csc\theta = \frac{1}{y} \quad (y \neq 0)$$

$$\cos\theta = x \qquad\qquad \sec\theta = \frac{1}{x} \quad (x \neq 0)$$

$$\tan\theta = \frac{y}{x} \quad (x \neq 0) \qquad \cot\theta = \frac{x}{y} \quad (y \neq 0)$$

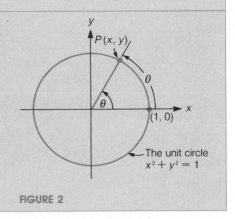

**FIGURE 2**

---

Let us immediately note that there is nothing new here as far as evaluating the trigonometric functions. For instance, $\sin \pi/6$ is still equal to $\frac{1}{2}$. [If this seems unfamiliar, review Example 4(a) on page 371.] Also, whether we are working in radians or degrees, our three-step procedure for evaluating the trigonometric functions is applicable. For instance, in Example 3(b) on page 348, we computed $\cos 135°$ ($\cos 3\pi/4$). Here is how that same procedure looks when radian measure is used. You should carefully compare each of the corresponding steps.

STEP 1   The reference angle for $3\pi/4$ is $\pi/4$.* $(\pi - 3\pi/4 = \pi/4)$

---

* Some texts refer to this as the reference *number*, instead of reference angle, to emphasize the fact that $3\pi/4$ is a real number.

STEP 2  $\cos \pi/4 = \sqrt{2}/2$

STEP 3  Cos $\theta$ is the $x$-coordinate. In the second quadrant, $x$-coordinates are negative. Thus, $\cos 3\pi/4$ is negative, since the terminal side of $\theta = 3\pi/4$ lies in the second quadrant. (Equivalently, the arc of length $3\pi/4$ terminates in the second quadrant.) Thus, we have

$$\cos \frac{3\pi}{4} = -\frac{\sqrt{2}}{2}$$

There are many identities involving trigonometric functions. (In fact, that is one reason why these functions are so useful.) We will be concerned with these identities, in one form or another, throughout the remainder of this chapter. Some of the identities will probably seem familiar to you from the text and exercises in Chapter 7. However, keep in mind that now the identities will be true for all real numbers $\theta$ for which the expressions are defined, rather than for a limited range of angles.

First of all, there are five trigonometric identities that are immediate consequences of the definitions. These are as follows.

$$\csc \theta = \frac{1}{\sin \theta} \qquad \tan \theta = \frac{\sin \theta}{\cos \theta}$$

$$\sec \theta = \frac{1}{\cos \theta} \qquad \cot \theta = \frac{\cos \theta}{\sin \theta}$$

$$\cot \theta = \frac{1}{\tan \theta}$$

The next identities that we consider are the three *Pythagorean identities*.

### The Pythagorean Identities

$$\sin^2 \theta + \cos^2 \theta = 1$$
$$\tan^2 \theta + 1 = \sec^2 \theta$$
$$\cot^2 \theta + 1 = \csc^2 \theta$$

To prove the identity $\sin^2 \theta + \cos^2 \theta = 1$, we refer back to Figure 2. Since $(x, y)$ is a point on the unit circle, we have

$$x^2 + y^2 = 1$$

Replacing $x$ by $\cos \theta$ and $y$ by $\sin \theta$ then gives us

$$\cos^2 \theta + \sin^2 \theta = 1$$

which is essentially what we wished to show. Incidentally, you should also become familiar with the equivalent forms of this identity:

$$\cos^2 \theta = 1 - \sin^2 \theta \qquad \text{and} \qquad \sin^2 \theta = 1 - \cos^2 \theta$$

To prove the second of the Pythagorean identities, we begin with $\sin^2\theta + \cos^2\theta = 1$ and divide both sides by the quantity $\cos^2\theta$ to obtain

$$\frac{\sin^2\theta}{\cos^2\theta} + \frac{\cos^2\theta}{\cos^2\theta} = \frac{1}{\cos^2\theta} \qquad (\text{assuming } \cos\theta \neq 0)$$

and consequently,

$$\tan^2\theta + 1 = \sec^2\theta \qquad \text{as required}$$

Since the proof of the third Pythagorean identity is similar to this, we omit giving it here.

**EXAMPLE 1**  If $\sin\theta = \frac{2}{3}$ and $\pi/2 < \theta < \pi$, compute $\cos\theta$ and $\tan\theta$.

Solution
$$\cos^2\theta = 1 - \sin^2\theta \qquad \text{using the first Pythagorean identity}$$

$$= 1 - \left(\frac{2}{3}\right)^2 \qquad \text{substituting}$$

$$= 1 - \frac{4}{9} = \frac{5}{9}$$

Consequently,

$$\cos\theta = \frac{\sqrt{5}}{3} \qquad \text{or} \qquad \cos\theta = \frac{-\sqrt{5}}{3}$$

Now, since $\pi/2 < \theta < \pi$, it follows that $\cos\theta$ is negative. (Why?) Thus,

$$\cos\theta = \frac{-\sqrt{5}}{3} \qquad \text{as required}$$

To compute $\tan\theta$, we use the identity $\tan\theta = \sin\theta/\cos\theta$ to obtain

$$\tan\theta = \frac{2/3}{-\sqrt{5}/3} = \frac{2}{3} \times \frac{3}{-\sqrt{5}}$$

$$= -\frac{2}{\sqrt{5}}$$

If required, we can rationalize the denominator (by multiplying by $\sqrt{5}/\sqrt{5}$) to obtain

$$\tan\theta = -\frac{2\sqrt{5}}{5}$$

**EXAMPLE 2**  If $\sec\theta = -\frac{5}{3}$ and $\pi < \theta < 3\pi/2$, compute $\cos\theta$ and $\tan\theta$.

Solution  Since $\cos\theta$ is the reciprocal of $\sec\theta$, we have $\cos\theta = -\frac{3}{5}$. We can compute $\tan\theta$ using the second Pythagorean identity, as follows.

$$\tan^2\theta = \sec^2\theta - 1$$

$$= \left(-\frac{5}{3}\right)^2 - 1$$

$$= \frac{25}{9} - \frac{9}{9} = \frac{16}{9}$$

Therefore,

$$\tan \theta = \frac{4}{3} \quad \text{or} \quad \tan \theta = -\frac{4}{3}$$

Since $\theta$ is between $\pi$ and $3\pi/2$, $\tan \theta$ is positive. Thus,

$$\tan \theta = \frac{4}{3}$$

The next example shows how certain radical expressions can be simplified through an appropriate trigonometric substitution. (This is often useful in calculus.)

**EXAMPLE 3**  In the expression $u/\sqrt{u^2 - 1}$, make the substitution $u = \sec \theta$ and show that the resulting expression is equal to $\csc \theta$. (Assume that $0 < \theta < \pi/2$.)

**Solution**  Replacing $u$ by $\sec \theta$ in the given expression yields

$$\frac{u}{\sqrt{u^2 - 1}} = \frac{\sec \theta}{\sqrt{\sec^2 \theta - 1}}$$

$$= \frac{\sec \theta}{\sqrt{\tan^2 \theta}} \qquad \text{using the second Pythagorean identity}$$

$$= \frac{\sec \theta}{\tan \theta}$$

$$= \frac{1/\cos \theta}{\sin \theta/\cos \theta} = \frac{1}{\cos \theta} \times \frac{\cos \theta}{\sin \theta}$$

$$= \frac{1}{\sin \theta}$$

$$= \csc \theta$$

QUESTION  Where did we use the condition $0 < \theta < \pi/2$?

The next three identities indicate the effects of replacing $\theta$ by $-\theta$ in the expressions $\sin \theta$, $\cos \theta$, and $\tan \theta$.

$$\cos(-\theta) = \cos \theta$$
$$\sin(-\theta) = -\sin \theta$$
$$\tan(-\theta) = -\tan \theta$$

To see why the first two of these identities are true, consider Figure 3. (Although Figure 3 shows an arc of length $\theta$ terminating in the first quadrant, the same kind of argument we use will work when $\theta$ terminates in any quadrant.) By definition, the coordinates of $P$ and $Q$ are as follows:

$P$:  $(\cos \theta, \sin \theta)$
$Q$:  $(\cos(-\theta), \sin(-\theta))$

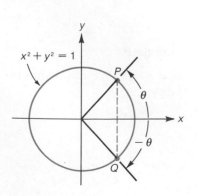

**FIGURE 3**

However, as you can see by looking at Figure 3, the $x$-coordinates of $P$ and $Q$ are the same, while the $y$-coordinates are negatives of each other. Thus,

$$\cos(-\theta) = \cos\theta$$

and

$$\sin(-\theta) = -\sin\theta \qquad \text{as we wished to show}$$

Now we can establish the third identity, involving $\tan(-\theta)$, as follows:

$$\begin{aligned}
\tan(-\theta) &= \frac{\sin(-\theta)}{\cos(-\theta)} \\
&= \frac{-\sin\theta}{\cos\theta} = -\frac{\sin\theta}{\cos\theta} \\
&= -\tan\theta
\end{aligned}$$

The last identities we are going to discuss in this section are simply consequences of the fact that the circumference $C$ of the unit circle is $2\pi$.* Thus, if we begin at any point $P$ on the unit circle and travel a distance of $2\pi$ units along the circle, we return to the same point $P$. In other words, arc lengths of $\theta$ and $\theta + 2\pi$ [measured from $(1,0)$, as usual] yield the same terminal point $P$ on the unit circle. Since the trigonometric functions are defined in terms of the coordinates of that point $P$, we conclude that

$$\sin(\theta + 2\pi) = \sin\theta$$
$$\cos(\theta + 2\pi) = \cos\theta$$

These two identities are true for all real numbers $\theta$. Similar identities hold for the other trigonometric functions in their respective domains:

$$\tan(\theta + 2\pi) = \tan\theta \qquad \csc(\theta + 2\pi) = \csc\theta$$
$$\cot(\theta + 2\pi) = \cot\theta \qquad \sec(\theta + 2\pi) = \sec\theta$$

**EXAMPLE 4**   Evaluate $\sin(5\pi/2)$.

**Solution**   First, simply as a matter of arithmetic, we observe that $5\pi/2 = \pi/2 + 2\pi$. Thus, in view of our earlier remarks, we have

$$\sin\frac{5\pi}{2} = \sin\frac{\pi}{2} = 1$$

The preceding set of identities can be generalized as follows. If we start at a point $P$ on the unit circle and make two complete revolutions, the arc length we travel is

$$2\pi + 2\pi = 4\pi$$

* This is because $C = 2\pi r$. Thus, when $r = 1$ we obtain $C = 2\pi$.

Similarly, the arc length covered in three complete revolutions is

$$2\pi + 2\pi + 2\pi = 6\pi$$

And the arc length for $k$ complete revolutions is $2k\pi$. (When $k$ is positive, the revolutions are counterclockwise; when $k$ is negative, the revolutions are clockwise.) It follows that the arc lengths $\theta$ and $\theta + 2\pi k$ yield the same terminal point $P$. We summarize this as follows.

> For any integer $k$, the following identities hold.
> $$\sin(\theta + 2\pi k) = \sin\theta$$
> $$\cos(\theta + 2\pi k) = \cos\theta$$

**EXAMPLE 5**   Evaluate $\cos(-17\pi)$.

**Solution**
$$\cos(-17\pi) = \cos(\pi - 18\pi)$$
$$= \cos\pi$$
$$= -1$$

## EXERCISE SET 8.2

### A

1. Complete the following table.

| $\theta$ | $\sin\theta$ | $\cos\theta$ | $\tan\theta$ | $\csc\theta$ | $\sec\theta$ | $\cot\theta$ |
|---|---|---|---|---|---|---|
| $0$ | | | | | | |
| $\dfrac{\pi}{6}$ | | | | | | |
| $\dfrac{\pi}{4}$ | | | | | | |
| $\dfrac{\pi}{3}$ | | | | | | |
| $\dfrac{\pi}{2}$ | | | | | | |
| $\dfrac{2\pi}{3}$ | | | | | | |
| $\dfrac{3\pi}{4}$ | | | | | | |
| $\dfrac{5\pi}{6}$ | | | | | | |
| $\pi$ | | | | | | |

2. Complete the following table.

| $\theta$ | $\sin\theta$ | $\cos\theta$ | $\tan\theta$ | $\csc\theta$ | $\sec\theta$ | $\cot\theta$ |
|---|---|---|---|---|---|---|
| $\pi$ | | | | | | |
| $\dfrac{7\pi}{6}$ | | | | | | |
| $\dfrac{5\pi}{4}$ | | | | | | |
| $\dfrac{4\pi}{3}$ | | | | | | |
| $\dfrac{3\pi}{2}$ | | | | | | |
| $\dfrac{5\pi}{3}$ | | | | | | |
| $\dfrac{7\pi}{4}$ | | | | | | |
| $\dfrac{11\pi}{6}$ | | | | | | |

3. If $\sin\theta = -\frac{3}{5}$ and $\pi < \theta < 3\pi/2$, compute $\cos\theta$ and $\tan\theta$.

4. If $\cos\theta = \frac{5}{13}$ and $3\pi/2 < \theta < 2\pi$, compute $\sin\theta$ and $\cot\theta$.

5. If $\sin t = \sqrt{3}/4$ and $\pi/2 < t < \pi$, compute $\tan t$.

6. If $\sec\theta = -\sqrt{13}/2$ and $\sin\theta > 0$, compute $\tan\theta$.

7. If $\tan\alpha = \frac{12}{5}$ and $\cos\alpha > 0$, compute $\sec\alpha$, $\cos\alpha$, and $\sin\alpha$.

8. If $\cot\theta = -1/\sqrt{3}$ and $\cos\theta < 0$, compute $\csc\theta$ and $\sin\theta$.

9. In the expression $\sqrt{9 - x^2}$, make the substitution $x = 3\sin\theta$ $(0 < \theta < \pi/2)$, and show that the result is $3\cos\theta$.

10. Make the substitution $u = 2\cos\theta$ in the expression $1/\sqrt{4 - u^2}$, and simplify the result. (Assume that $0 < \theta < \pi$.)

11. In the expression $1/(u^2 - 25)^{3/2}$, make the substitution $u = 5\sec\theta$ $(0 < \theta < \pi/2)$, and show that the result is $(\cot^3\theta)/125$.

12. In the expression $1/(x^2 + 5)^2$, replace $x$ by $\sqrt{5}\tan\theta$, and show that the result is $(\cos^4\theta)/25$.

13. In the expression $1/\sqrt{u^2 + 7}$, let $u = \sqrt{7}\tan\theta$ $(0 < \theta < \pi/2)$, and simplify the result.

14. In the expression $\sqrt{x^2 - a^2}/x$ $(a > 0)$, let $x = a\sec\theta$ $(0 < \theta < \pi/2)$, and simplify the result.

15. (a) If $\sin\theta = \frac{2}{3}$, find $\sin(-\theta)$.
    (b) If $\sin\phi = -\frac{1}{4}$, find $\sin(-\phi)$.
    (c) If $\cos\alpha = \frac{1}{5}$, find $\cos(-\alpha)$.
    (d) If $\cos\beta = -\frac{1}{5}$, find $\cos(-\beta)$.

16. (a) If $\sin\theta = 0.35$, find $\sin(-\theta)$.
    (b) If $\sin\phi = -0.47$, find $\sin(-\phi)$.
    (c) If $\cos\alpha = 0.21$, find $\cos(-\alpha)$.
    (d) If $\cos\beta = -0.56$, find $\cos(-\beta)$.

17. If $\cos\theta = -\frac{1}{3}$, $\pi/2 < \theta < \pi$, compute the following.
    (a) $\sin(-\theta) + \cos(-\theta)$    (b) $\sin^2(-\theta) + \cos^2(-\theta)$

18. If $\sin(-\theta) = \frac{3}{5}$ and $\pi < \theta < 3\pi/2$, compute the following.
    (a) $\sin\theta$    (b) $\cos(-\theta)$
    (c) $\cos\theta$    (d) $\tan\theta + \tan(-\theta)$

19. Evaluate each of the following using the identities $\cos(\theta + 2\pi) = \cos\theta$ and $\sin(\theta + 2\pi) = \sin\theta$.
    (a) $\cos\left(\dfrac{\pi}{4} + 2\pi\right)$    (b) $\sin\left(\dfrac{\pi}{3} + 2\pi\right)$
    (c) $\sin\left(\dfrac{\pi}{2} - 6\pi\right)$

20. Evaluate each of the following using the identities $\cos(\theta + 2\pi k) = \cos\theta$ and $\sin(\theta + 2\pi k) = \sin\theta$.
    (a) $\sin\dfrac{17\pi}{4}$    (b) $\sin\left(-\dfrac{17\pi}{4}\right)$

(c) $\cos 11\pi$    (d) $\cos\dfrac{53\pi}{4}$

(e) $\tan\left(\dfrac{-7\pi}{4}\right)$    (f) $\cos\dfrac{7\pi}{4}$

(g) $\sec\left(\dfrac{11\pi}{6} + 2\pi\right)$    (h) $\csc\left(2\pi - \dfrac{\pi}{3}\right)$

*In Exercises 21–24, use the Pythagorean identities to simplify the given expressions.*

21. $\dfrac{\sin^2\theta + \cos^2\theta}{\tan^2\theta + 1}$    22. $\dfrac{\sec^2\theta - 1}{\tan^2\theta}$

23. $\dfrac{\sec^2\theta - \tan^2\theta}{1 + \cot^2\theta}$    24. $\dfrac{\csc^4\theta - \cot^4\theta}{\csc^2\theta + \cot^2\theta}$

*In Exercises 25–32, prove that the equations are identities.*

25. $\csc\theta = \sin\theta + \cot\theta\cos\theta$

26. $\sin^2\theta - \cos^2\theta = \dfrac{1 - \cot^2\theta}{1 + \cot^2\theta}$

27. $\dfrac{1}{1 + \sec\theta} + \dfrac{1}{1 - \sec\theta} = -2\cot^2\theta$

28. $\dfrac{1 + \tan\theta}{1 - \tan\theta} = \dfrac{\sec^2\theta + 2\tan\theta}{2 - \sec^2\theta}$

29. $\cot\theta + \tan\theta + 1 = \dfrac{\cot\theta}{1 - \tan\theta} + \dfrac{\tan\theta}{1 - \cot\theta}$

30. $\dfrac{\sec s + \cot s\csc s}{\cos s} = \csc^2 s\sec^2 s$

31. $(\tan\theta)(1 - \cot^2\theta) + (\cot\theta)(1 - \tan^2\theta) = 0$

32. $(\cos\alpha\cos\beta - \sin\alpha\sin\beta) \times$
    $(\cos\alpha\cos\beta + \sin\alpha\sin\beta) = \cos^2\alpha - \sin^2\beta$

## B

33. Use the accompanying figure to explain why the following four identities are valid. (The identities can be used to provide an algebraic foundation for the "reference angle technique" that we've used to evaluate the trigonometric functions.)

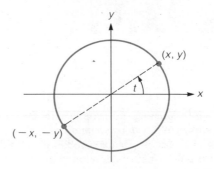

(i) $\sin(t + \pi) = -\sin t$
(ii) $\sin(t - \pi) = -\sin t$
(iii) $\cos(t + \pi) = -\cos t$
(iv) $\cos(t - \pi) = -\cos t$

**34.** Suppose that $\tan \theta = (b^2 - a^2)/2ab$, where $a$ and $b$ are positive and $\pi/2 < \theta < \pi$. Show that $\cos \theta = -2ab/(a^2 + b^2)$.

**35.** In the equation $x^4 + 6x^2y^2 + y^4 = 32$, make the substitutions

$$x = X\cos\frac{\pi}{4} - Y\sin\frac{\pi}{4}$$

and

$$y = X\sin\frac{\pi}{4} + Y\cos\frac{\pi}{4}$$

and show that the resulting equation simplifies to

$$X^4 + Y^4 = 16$$

**36.** Suppose that $\tan \theta = 2$ and $0 < \theta < \pi/2$.
   **(a)** Compute $\sin \theta$ and $\cos \theta$.
   **(b)** Using the values obtained in part (a), make the substitutions

   $$x = X\cos\theta - Y\sin\theta$$

   and

   $$y = X\sin\theta + Y\cos\theta$$

   in the expression $7x^2 - 8xy + y^2$, and simplify the result.

**37.** In Section 8.1, we pointed out that one of the advantages in using radian measure is that many formulas then take on particularly simple forms. Another reason for using radian measure is that the trigonometric functions can be closely approximated by very simple polynomial functions. To see an example of this, complete the following table. The values for $\cos \theta$ in the table were obtained using a calculator.

| $\theta$ | $1 - \dfrac{\theta^2}{2}$ | $\cos \theta$ |
|---|---|---|
| 0.1 | | 0.995004 ... |
| 0.2 | | 0.980066 ... |
| 0.3 | | 0.955336 ... |

NOTE: The approximating polynomials in Exercises 37 and 38 are known as *Taylor polynomials*, after the English mathematician Brook Taylor (1685–1731). The theory of Taylor polynomials is developed in calculus.

**38.** Complete the following table, as in Exercise 37.

| $\theta$ | $\theta - \dfrac{\theta^3}{6}$ | $\sin \theta$ |
|---|---|---|
| 0.1 | | 0.099833 ... |
| 0.2 | | 0.198669 ... |
| 0.3 | | 0.295520 ... |

**39.** Let $a = \sin^2 \theta + \csc^2 \theta$, $b = \cos^2 \theta + \sec^2 \theta$, and $c = \tan^2 \theta + \cot^2 \theta$. Show that $a + b - c = 3$.

**40.** [c] **(a)** Use a calculator to complete the following table. (Set your calculator in the radian mode.)

| $\theta$ | $\sin \theta$ | WHICH IS LARGER, $\theta$ OR $\sin \theta$? |
|---|---|---|
| 0.1 | | |
| 0.2 | | |
| 0.3 | | |
| 0.4 | | |
| 0.5 | | |

   **(b)** From the following figure, explain why

   $$PQ < PR < \theta$$

   Then use this to show that if $0 < \theta < \pi/2$, then

   $$\sin \theta < \theta$$

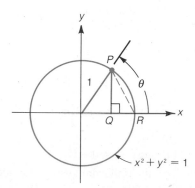

**41.** **(a)** List four distinct real numbers $\theta$ for which $\sin \theta = 0$.
   **(b)** What is the domain of the function $f(\theta) = 1/\sin \theta$?
   **(c)** What is the domain of the function $g(\theta) = 1/\sin \pi\theta$?

# THE ADDITION FORMULAS

## 8.3

*It has long been recognized that the addition formulas are the heart of trigonometry. Indeed, Professor Rademacher and others have shown that the entire body of trigonometry can be derived from the assumption that there exist functions S and C such that*

**(1)** $\quad S(x - y) = S(x)C(x) - C(x)S(y)$
**(2)** $\quad C(x - y) = C(x)C(y) + S(x)S(y)$

**(3)** $\quad \lim_{x \to 0^+} \dfrac{S(x)}{x} = 1$

PROFESSOR FREDERICK H. YOUNG in his article, "The Addition Formulas," in *The Mathematics Teacher*, Vol. L (1957), pp. 45–48

For any real numbers $r$, $s$, and $t$, it is always true that $r(s + t) = rs + rt$. This is the so-called **distributive law** for real numbers. If $f$ is a function, however, it is not true in general that $f(s + t) = f(s) + f(t)$. By way of example, consider the cosine function. It is not true in general that $\cos(s + t) = \cos s + \cos t$. For instance, with $s = \pi/6$ and $t = \pi/3$, we have

$$\cos\left(\frac{\pi}{6} + \frac{\pi}{3}\right) \overset{?}{=} \cos\frac{\pi}{6} + \cos\frac{\pi}{3}$$

$$\cos\frac{\pi}{2} \overset{?}{=} \cos\frac{\pi}{6} + \cos\frac{\pi}{3}$$

$$0 \overset{?}{=} \frac{\sqrt{3}}{2} + \frac{1}{2} \qquad \text{No!}$$

In this section, we will see just what $\cos(s + t)$ does equal. The correct formula for $\cos(s + t)$ is one of a group of four important trigonometric identities called the **addition formulas**.

| The Addition Formulas |
|---|
| $\sin(s + t) = \sin s \cos t + \cos s \sin t$ |
| $\sin(s - t) = \sin s \cos t - \cos s \sin t$ |
| $\cos(s + t) = \cos s \cos t - \sin s \sin t$ |
| $\cos(s - t) = \cos s \cos t + \sin s \sin t$ |

To prove the fourth formula in the box, we use Figure 1. The idea behind the proof is as follows.* We begin in Figure 1(a) with the unit circle and the angles (or arc lengths) $s$, $t$, and $s - t$. Then we rotate $\triangle OPQ$ about the origin until the point $P$ coincides with the point $(1, 0)$, as indicated in Figure 1(b). Although this rotation changes the coordinates for the points $P$ and $Q$, it certainly has no effect upon the length of the line segment $\overline{PQ}$. Thus, whether we

---

\* This idea for the proof can be traced back to the great French mathematician Augustin-Louis Cauchy (1789–1857).

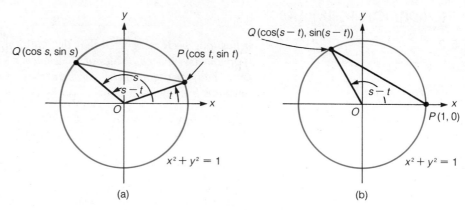

**FIGURE 1**

calculate $PQ$ using the coordinates in Figure 1(a) or those in Figure 1(b), the results must be the same. As you'll see, by equating the two expressions for $PQ$, we will obtain the required formula.

Using the distance formula in Figure 1(a), we have

$$PQ = \sqrt{(\cos s - \cos t)^2 + (\sin s - \sin t)^2}$$
$$= \sqrt{\cos^2 s - 2\cos s \cos t + \cos^2 t + \sin^2 s - 2\sin s \sin t + \sin^2 t}$$
$$= \sqrt{2 - 2\cos s \cos t - 2\sin s \sin t} \qquad \text{(Why?)} \qquad (1)$$

Next, using the distance formula in Figure 1(b), we have

$$PQ = \sqrt{[\cos(s - t) - 1]^2 + [\sin(s - t) - 0]^2}$$
$$= \sqrt{\cos^2(s - t) - 2\cos(s - t) + 1 + \sin^2(s - t)}$$

The right-hand side of this last equation can be simplified using the fact that $\cos^2(s - t) + \sin^2(s - t) = 1$. We then have

$$PQ = \sqrt{2 - 2\cos(s - t)} \qquad (2)$$

From equations (1) and (2), we conclude that

$$2 - 2\cos(s - t) = 2 - 2\cos s \cos t - 2\sin s \sin t$$
$$-2\cos(s - t) = -2\cos s \cos t - 2\sin s \sin t$$
$$\cos(s - t) = \cos s \cos t + \sin s \sin t$$

This completes the proof of the fourth addition formula. Before deriving the other three addition formulas, let us look at some applications of this result.

**EXAMPLE 1**   Simplify the expression $\cos 2\theta \cos \theta + \sin 2\theta \sin \theta$.

Solution   According to the identity that we just proved, we have

$$\cos 2\theta \cos \theta + \sin 2\theta \sin \theta = \cos(2\theta - \theta)$$
$$= \cos \theta$$

Thus, the given expression is equal to $\cos \theta$.

**EXAMPLE 2**   Simplify $\cos(\theta - \pi)$.

**Solution**   We use the formula for $\cos(s - t)$, with $s$ and $t$ replaced by $\theta$ and $\pi$, respectively. This yields

$$\cos(s - t) = \cos s \cos t + \sin s \sin t$$
$$\cos(\theta - \pi) = \cos\theta \cos\pi + \sin\theta \sin\pi$$
$$= (\cos\theta)(-1) + (\sin\theta)(0)$$
$$= -\cos\theta$$

Thus, the required simplification is $\cos(\theta - \pi) = -\cos\theta$.

In Example 2, we found that $\cos(\theta - \pi) = -\cos\theta$. This type of identity is often referred to as a **reduction formula**. The next example develops two basic reduction formulas that we will need to use later in this section.

**EXAMPLE 3**   Prove the following identities.

(a) $\cos\left(\dfrac{\pi}{2} - \alpha\right) = \sin\alpha$     (b) $\sin\left(\dfrac{\pi}{2} - \beta\right) = \cos\beta$

**Solution**   (a) $\cos\left(\dfrac{\pi}{2} - \alpha\right) = \cos\dfrac{\pi}{2}\cos\alpha + \sin\dfrac{\pi}{2}\sin\alpha$

$$= (0)\cos\alpha + (1)\sin\alpha$$
$$= \sin\alpha$$

This proves the identity. Incidentally, if we use degree measure instead of radian measure, this identity states that

$$\cos(90° - \alpha) = \sin\alpha$$

as we saw earlier in Chapter 7.

(b) Since the identity $\cos\left(\dfrac{\pi}{2} - \alpha\right) = \sin\alpha$ holds for all values of $\alpha$, we can

simply replace $\alpha$ by the quantity $\dfrac{\pi}{2} - \beta$ to obtain

$$\cos\left[\dfrac{\pi}{2} - \left(\dfrac{\pi}{2} - \beta\right)\right] = \sin\left(\dfrac{\pi}{2} - \beta\right)$$
$$\cos\left(\dfrac{\pi}{2} - \dfrac{\pi}{2} + \beta\right) = \sin\left(\dfrac{\pi}{2} - \beta\right)$$
$$\cos\beta = \sin\left(\dfrac{\pi}{2} - \beta\right)$$

This proves the identity.

The two identities in Example 3 are worth memorizing. For simplicity, we replace both $\alpha$ and $\beta$ by the single letter $\theta$ to get

$$\cos\left(\frac{\pi}{2} - \theta\right) = \sin\theta$$

$$\sin\left(\frac{\pi}{2} - \theta\right) = \cos\theta$$

We developed the fourth addition formula under the assumption that the angles were measured in radians. However, the formula is still valid when the angles are measured in degrees. In the next example, we apply the formula in just such a case.

**EXAMPLE 4** Use the formula for $\cos(s - t)$ to determine the exact value of $\cos 15°$.

**Solution** First observe that $15° = 45° - 30°$. Then we have

$$\cos 15° = \cos(45° - 30°)$$
$$= \cos 45° \cos 30° + \sin 45° \sin 30° \qquad \text{using the formula for } \cos(s-t)$$
$$\text{with } s = 45° \text{ and } t = 30°$$

$$= \left(\frac{\sqrt{2}}{2}\right)\left(\frac{\sqrt{3}}{2}\right) + \left(\frac{\sqrt{2}}{2}\right)\left(\frac{1}{2}\right)$$

$$= \frac{\sqrt{6}}{4} + \frac{\sqrt{2}}{4}$$

$$= \frac{\sqrt{6} + \sqrt{2}}{4}$$

Thus, the exact value of $\cos 15°$ is $\dfrac{\sqrt{6} + \sqrt{2}}{4}$.

Now let us return to our derivations of the addition formulas. Using the fourth addition formula, we can easily derive the third formula as follows. In the formula

$$\cos(s - t) = \cos s \cos t + \sin s \sin t$$

we replace $t$ by the quantity $-t$. This is permissible because the formula holds for all real numbers. We obtain

$$\cos[s - (-t)] = \cos s \cos(-t) + \sin s \sin(-t)$$

On the right-hand side of this equation, we can use the identities developed for $\cos(-t)$ and $\sin(-t)$ in the previous section. Doing this yields

$$\cos(s + t) = (\cos s)(\cos t) + (\sin s)(-\sin t)$$

which is equivalent to

$$\cos(s + t) = \cos s \cos t - \sin s \sin t$$

This is the third addition formula, as we wished to prove.

Next we derive the formula for $\sin(s + t)$. We have

$$\sin(s + t) = \cos\left[\frac{\pi}{2} - (s + t)\right] \qquad \text{replacing } \theta \text{ by } s + t \text{ in the identity } \sin\theta = \cos[(\pi/2) - \theta]$$

$$= \cos\left[\left(\frac{\pi}{2} - s\right) - t\right]$$

$$= \cos\left(\frac{\pi}{2} - s\right)\cos t + \sin\left(\frac{\pi}{2} - s\right)\sin t \qquad \text{(Why?)}$$

$$= \sin s \cos t + \cos s \sin t \qquad \text{(Why?)}$$

This proves the first addition formula.

Finally, we can use the first addition formula to prove the second, as follows:

$$\sin(s - t) = \sin[s + (-t)]$$

$$= \sin s \cos(-t) + \cos s \sin(-t)$$

$$= (\sin s)(\cos t) + (\cos s)(-\sin t)$$

$$= \sin s \cos t - \cos s \sin t$$

This completes the proofs of the four addition formulas.

**EXAMPLE 5** If $\sin s = \frac{3}{5}$, $0 < s < \pi/2$, and $\sin t = -\sqrt{3}/4$, $\pi < t < 3\pi/2$, compute $\sin(s - t)$.

Solution

$$\sin(s - t) = \underbrace{\sin s}_{\substack{\uparrow \\ \text{given as } \frac{3}{5}}} \cos t - \cos s \underbrace{\sin t}_{\substack{\uparrow \\ \text{given as } \frac{-\sqrt{3}}{4}}} \qquad (3)$$

In view of equation (3), we need to find only $\cos t$ and $\cos s$. These can be determined using the Pythagorean identity $\cos^2\theta = 1 - \sin^2\theta$. We have

$$\cos^2 t = 1 - \sin^2 t$$

$$= 1 - \left(\frac{-\sqrt{3}}{4}\right)^2 = 1 - \frac{3}{16} = \frac{13}{16}$$

Therefore,

$$\cos t = \frac{\sqrt{13}}{4} \qquad \text{or} \qquad \cos t = \frac{-\sqrt{13}}{4}$$

We choose the negative value here for cosine, since it is given that $\pi < t < 3\pi/2$. Thus,

$$\cos t = \frac{-\sqrt{13}}{4}$$

Similarly, to find $\cos s$, we have

$$\cos^2 s = 1 - \sin^2 s$$

$$= 1 - \left(\frac{3}{5}\right)^2$$

$$= 1 - \frac{9}{25} = \frac{16}{25}$$

Therefore,

$$\cos s = \frac{4}{5} \qquad \cos s \text{ is positive, since } 0 < s < \pi/2.$$

Finally, we substitute the values we've obtained for $\cos t$ and $\cos s$, along with the given data, back into equation (3). This yields

$$\sin(s - t) = \left(\frac{3}{5}\right)\left(\frac{-\sqrt{13}}{4}\right) - \left(\frac{4}{5}\right)\left(\frac{-\sqrt{3}}{4}\right)$$
$$= \frac{-3\sqrt{13}}{20} + \frac{4\sqrt{3}}{20}$$
$$= \frac{-3\sqrt{13} + 4\sqrt{3}}{20}$$

## EXERCISE SET 8.3

### A

*In Exercises 1–6, use the addition formulas to simplify the expressions.*

1.  (a) $\sin\theta\cos 2\theta + \cos\theta\sin 2\theta$
    (b) $\sin\dfrac{\pi}{6}\cos\dfrac{\pi}{3} + \cos\dfrac{\pi}{6}\sin\dfrac{\pi}{3}$

2.  (a) $\sin 3\theta\cos\theta - \cos 3\theta\sin\theta$
    (b) $\sin 110°\cos 20° - \cos 110°\sin 20°$

3.  (a) $\cos 2u\cos 3u - \sin 2u\sin 3u$
    (b) $\cos 2u\cos 3u + \sin 2u\sin 3u$

4.  (a) $\cos\dfrac{3\pi}{10}\cos\dfrac{\pi}{5} - \sin\dfrac{3\pi}{10}\sin\dfrac{\pi}{5}$
    (b) $\cos\dfrac{2\pi}{9}\cos\dfrac{\pi}{18} + \sin\dfrac{2\pi}{9}\sin\dfrac{\pi}{18}$

5.  $\sin(A + B)\cos A - \cos(A + B)\sin A$

6.  $\cos(s - t)\cos t - \sin(s - t)\sin t$

*In Exercises 7–10, simplify each term (as was done in Example 2).*

7.  $\sin\left(\theta - \dfrac{3\pi}{2}\right)$

8.  $\cos\left(\dfrac{3\pi}{2} + \theta\right)$

9.  $\cos(\theta + \pi)$

10. $\sin(\theta - \pi)$

11. Expand $\sin(\theta + 2\pi)$ using the appropriate addition formula, and check to see that your answer agrees with the formula for $\sin(\theta + 2\pi)$ given on page 381.

12. Follow the directions in Exercise 11, but use $\cos(\theta + 2\pi)$.

13. Use the formula for $\cos(s + t)$ to compute the exact value for $\cos 75°$.

14. Use the formula for $\sin(s - t)$ to compute the exact value of $\sin(\pi/12)$.

15. Use the formula for $\sin(s + t)$ to find $\sin(7\pi/12)$.

16. If $\sin s = \frac{4}{5}, 0 < s < \pi/2$, and $\sin t = -\frac{1}{4}, \pi < t < 3\pi/2$, compute $\sin(s - t)$.

*In Exercises 17–20, use the addition formulas to simplify each expression.*

17. $\sin\left(\dfrac{\pi}{4} + s\right) - \sin\left(\dfrac{\pi}{4} - s\right)$

18. $\sin\left(t + \dfrac{\pi}{6}\right) - \sin\left(t - \dfrac{\pi}{6}\right)$

19. $\cos\left(\dfrac{\pi}{3} - \theta\right) - \cos\left(\dfrac{\pi}{3} + \theta\right)$

20. $\cos\left(\theta - \dfrac{\pi}{4}\right) + \cos\left(\theta + \dfrac{\pi}{4}\right)$

21. If $\sin\alpha = \frac{12}{13}, \pi/2 < \alpha < \pi$, and $\cos\beta = \frac{3}{5}, 3\pi/2 < \beta < 2\pi$, compute the following.
    (a) $\sin(\alpha + \beta)$     (b) $\sin(\alpha - \beta)$
    (c) $\cos(\alpha + \beta)$     (d) $\cos(\alpha - \beta)$

22. Suppose that $\sin\theta = \frac{1}{5}$ and $0 < \theta < \pi/2$.
    (a) Compute $\cos\theta$.
    (b) Compute $\sin 2\theta$.   *Hint:* $\sin 2\theta = \sin(\theta + \theta)$

23. Suppose that $\cos\theta = \frac{12}{13}$ and $3\pi/2 < \theta < 2\pi$.
    (a) Compute $\sin\theta$.
    (b) Compute $\cos 2\theta$.   *Hint:* $\cos 2\theta = \cos(\theta + \theta)$

24. Given $\tan\theta = -\frac{2}{3}, \pi/2 < \theta < \pi$, and $\csc\beta = 2$, $0 < \beta < \pi/2$, find $\sin(\theta + \beta)$ and $\cos(\beta - \theta)$.

25. Given $\sec s = \frac{5}{4}$, $\sin s < 0$, and $\cot t = -1$, $\pi/2 < t < \pi$, find $\sin(s - t)$ and $\cos(s + t)$.

26. In the following figure, $AB = CD = 1$ and $BC = 2$. Find $\alpha + \beta$ and express your answer using radian measure.
*Hint:* Find $\cos(\alpha + \beta)$.

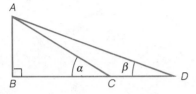

**B**

In Exercises 27–34, prove that each equation is an identity.

27. $\dfrac{\sin(s + t)}{\cos s \cos t} = \tan s + \tan t$

28. $\dfrac{\cos(s - t)}{\cos s \sin t} = \cot t + \tan s$

29. $\cos(A - B) - \cos(A + B) = 2 \sin A \sin B$

30. $\sin(A - B) + \sin(A + B) = 2 \sin A \cos B$

31. $\cos(A + B)\cos(A - B) = \cos^2 A - \sin^2 B$

32. $\sin(A + B)\sin(A - B) = \cos^2 B - \cos^2 A$

33. $\cos(\alpha + \beta)\cos \beta + \sin(\alpha + \beta)\sin \beta = \cos \alpha$

34. $\cos\left(\theta + \dfrac{\pi}{4}\right) + \sin\left(\theta - \dfrac{\pi}{4}\right) = 0$

35. Let $f(t) = \cos^2 t + \cos^2\left(t + \dfrac{2\pi}{3}\right) + \cos^2\left(t - \dfrac{2\pi}{3}\right)$.

   c (a) Complete the table.

   | $t$ | 1 | 2 | 3 | 4 |
   |-----|---|---|---|---|
   | $f(t)$ | | | | |

   (b) On the basis of your results in part (a), make a conjecture about the function $f$. Prove that your conjecture is correct.

36. Let $A$, $B$, and $C$ be the angles of a triangle.
   (a) Show that $\sin(A + B) = \sin C$.
   (b) Show that $\cos(A + B) = -\cos C$.
   (c) Show that $\tan(A + B) = -\tan C$.

37. Let $f$ be the function defined by $f(\theta) = \sin \theta$. Prove that
$$\frac{f(\theta + h) - f(\theta)}{h} = (\sin \theta)\left(\frac{\cos h - 1}{h}\right) + (\cos \theta)\left(\frac{\sin h}{h}\right)$$

38. Prove the following identities.
   (a) $\sin 2\theta = 2 \sin \theta \cos \theta$.  *Hint:* $\sin 2\theta = \sin(\theta + \theta)$
   (b) $\cos 2\theta = \cos^2 \theta - \sin^2 \theta$

39. Prove that
$$\frac{\sin(\alpha - \beta)}{\cos \alpha \cos \beta} + \frac{\sin(\beta - \gamma)}{\cos \beta \cos \gamma} + \frac{\sin(\gamma - \alpha)}{\cos \gamma \cos \alpha} = 0$$

40. Suppose that $A$, $B$, and $C$ are the angles of a triangle. Show that
$$\cos^2 A + \cos^2 B + \cos^2 C + 2 \cos A \cos B \cos C = 1$$

41. Suppose that $a^2 + b^2 = 1$ and $c^2 + d^2 = 1$. Prove that $|ac + bd| \le 1$.  *Hint:* Let $a = \cos \theta$, $b = \sin \theta$, $c = \cos \phi$, and $d = \sin \phi$.

*Exercises 42 and 43 outline simple geometric derivations of the formulas for $\sin(\alpha + \beta)$ and $\cos(\alpha + \beta)$ in the case where $\alpha$ and $\beta$ are acute angles, with $\alpha + \beta < 90°$. The exercises rely on the accompanying figures, which are constructed as follows. Begin, in Figure A, with $\alpha = \angle GAD$, $\beta = \angle HAG$, and $AH = 1$. Then, from H, draw perpendiculars to $\overline{AD}$ and to $\overline{AG}$, as shown in Figure B. Finally, draw $FE \perp BH$ and $\overline{FC} \perp \overline{AD}$.*

**FIGURE A**          **FIGURE B**

42. *Formula for $\sin(\alpha + \beta)$*  Supply the reasons or steps behind each statement.
   (a) $BH = \sin(\alpha + \beta)$   (b) $FH = \sin \beta$
   (c) $\angle BHF = \alpha$
   (d) $EH = \cos \alpha \sin \beta$  *Hint:* Use $\triangle EFH$ and the result in part (b).
   (e) $AF = \cos \beta$   (f) $CF = \sin \alpha \cos \beta$
   (g) $\sin(\alpha + \beta) = \sin \alpha \cos \beta + \cos \alpha \sin \beta$
      *Hint:* $\sin(\alpha + \beta) = BH = EH + CF$.

43. *Formula for $\cos(\alpha + \beta)$*  Supply the reasons or steps behind each statement.
   (a) $\cos(\alpha + \beta) = AB$
   (b) $AC = \cos \alpha \cos \beta$  *Hint:* Use $\triangle ACF$ and the result in Exercise 42(e).
   (c) $EF = \sin \alpha \sin \beta$
   (d) $\cos(\alpha + \beta) = \cos \alpha \cos \beta - \sin \alpha \sin \beta$
      *Hint:* $AB = AC - BC$

## FURTHER IDENTITIES

### 8.4

There are a number of basic identities that follow from the addition formulas for sine and cosine. In the box, we summarize some of the most useful of these.

---

**Addition Formulas for Tangent**

(a) $\tan(s + t) = \dfrac{\tan s + \tan t}{1 - \tan s \tan t}$

(b) $\tan(s - t) = \dfrac{\tan s - \tan t}{1 + \tan s \tan t}$

**Double-Angle Formulas**

(a) $\sin 2\theta = 2 \sin \theta \cos \theta$

(b) $\cos 2\theta = \cos^2 \theta - \sin^2 \theta$

(c) $\tan 2\theta = \dfrac{2 \tan \theta}{1 - \tan^2 \theta}$

**Half-Angle Formulas**

(a) $\sin \dfrac{s}{2} = \pm \sqrt{\dfrac{1 - \cos s}{2}}$

(b) $\cos \dfrac{s}{2} = \pm \sqrt{\dfrac{1 + \cos s}{2}}$

(c) $\tan \dfrac{s}{2} = \dfrac{\sin s}{1 + \cos s}$

---

To prove the formula for $\tan(s + t)$, we begin with

$$\tan(s + t) = \frac{\sin(s + t)}{\cos(s + t)}$$

$$= \frac{\sin s \cos t + \cos s \sin t}{\cos s \cos t - \sin s \sin t} \tag{1}$$

Now we divide both numerator and denominator on the right-hand side of equation (1) by the quantity $\cos s \cos t$. This yields

$$\tan(s + t) = \frac{\dfrac{\sin s \cos t}{\cos s \cos t} + \dfrac{\cos s \sin t}{\cos s \cos t}}{\dfrac{\cos s \cos t}{\cos s \cos t} - \dfrac{\sin s \sin t}{\cos s \cos t}}$$

$$= \frac{\tan s + \tan t}{1 - \tan s \tan t}$$

This proves the formula for $\tan(s + t)$. The formula for $\tan(s - t)$ can be deduced from this with the aid of the identity $\tan(-t) = -\tan t$, which was derived in Section 8.2. We have

$$\tan(s - t) = \tan[s + (-t)]$$

$$= \frac{\tan s + \tan(-t)}{1 - \tan s \tan(-t)} = \frac{\tan s + (-\tan t)}{1 - (\tan s)(-\tan t)}$$

$$= \frac{\tan s - \tan t}{1 + \tan s \tan t} \qquad \text{as required}$$

**EXAMPLE 1**  Compute $\tan \dfrac{7\pi}{12}$, using the fact that $\dfrac{7\pi}{12} = \dfrac{\pi}{4} + \dfrac{\pi}{3}$.

Solution  $\tan \dfrac{7\pi}{12} = \tan\left(\dfrac{\pi}{4} + \dfrac{\pi}{3}\right)$

$$= \frac{\tan(\pi/4) + \tan(\pi/3)}{1 - \tan(\pi/4)\tan(\pi/3)} \qquad \begin{array}{l}\text{using the formula for } \tan(s + t)\\ \text{with } s = \pi/4 \text{ and } t = \pi/3\end{array}$$

$$= \frac{1 + \sqrt{3}}{1 - (1)(\sqrt{3})}$$

Thus, $\tan \dfrac{7\pi}{12} = \dfrac{1 + \sqrt{3}}{1 - \sqrt{3}}$. We can write this answer in a more compact form by rationalizing the denominator. As you can check, the result is

$$\tan \frac{7\pi}{12} = -2 - \sqrt{3}$$

We turn now to the identities for $\sin 2\theta$, $\cos 2\theta$, and $\tan 2\theta$. All of these are derived in the same way: we replace $2\theta$ by $(\theta + \theta)$ and use the appropriate addition formula. For instance, for $\sin 2\theta$ we have

$$\sin 2\theta = \sin(\theta + \theta)$$

$$= \sin\theta \cos\theta + \cos\theta \sin\theta$$

$$= 2\sin\theta \cos\theta$$

This establishes the formula for $\sin 2\theta$. (Exercise 29 asks you to carry out the corresponding derivations for $\cos 2\theta$ and $\tan 2\theta$.)

**EXAMPLE 2**  If $\sin\theta = \frac{4}{5}$ and $\pi/2 < \theta < \pi$, find $\cos\theta$, $\sin 2\theta$, and $\cos 2\theta$.

Solution  We have

$$\cos^2\theta = 1 - \sin^2\theta$$

$$= 1 - \left(\frac{4}{5}\right)^2 = \frac{9}{25}$$

Consequently,

$$\cos\theta = \frac{3}{5} \qquad \text{or} \qquad \cos\theta = -\frac{3}{5}$$

We want the negative value for the cosine here, since $\pi/2 < \theta < \pi$. Thus,

$$\cos\theta = -\frac{3}{5}$$

Now that we know the values of $\cos\theta$ and $\sin\theta$, the double-angle formulas can be used to determine $\sin 2\theta$ and $\cos 2\theta$. We have

$$\sin 2\theta = 2\sin\theta\cos\theta$$

$$= 2\left(\frac{4}{5}\right)\left(-\frac{3}{5}\right) = -\frac{24}{25}$$

and

$$\cos 2\theta = \cos^2\theta - \sin^2\theta$$

$$= \left(-\frac{3}{5}\right)^2 - \left(\frac{4}{5}\right)^2 = \frac{9}{25} - \frac{16}{25}$$

$$= -\frac{7}{25}$$

The required values are therefore $\cos\theta = -\frac{3}{5}$, $\sin 2\theta = -\frac{24}{25}$, and $\cos 2\theta = -\frac{7}{25}$.

**EXAMPLE 3**  If $x = 4\sin\theta$, $0 < \theta < \pi/2$, express $\sin 2\theta$ in terms of $x$.

Solution  The given equation is equivalent to $\sin\theta = x/4$, so we have

$$\sin 2\theta = 2\sin\theta\cos\theta = 2\left(\frac{x}{4}\right)\cos\theta$$

$$= \frac{x}{2}\sqrt{1 - \sin^2\theta} \qquad \text{(Why is the positive root appropriate?)}$$

$$= \frac{x}{2}\sqrt{1 - \frac{x^2}{16}} = \frac{x}{2}\sqrt{\frac{16 - x^2}{16}}$$

$$= \frac{x\sqrt{16 - x^2}}{8}$$

**EXAMPLE 4**  Prove the following identity:

$$\cos 3\theta = 4\cos^3\theta - 3\cos\theta$$

Solution  $\cos 3\theta = \cos(2\theta + \theta)$

$$= \underbrace{\cos 2\theta}_{(\cos^2\theta - \sin^2\theta)}\cos\theta - \underbrace{\sin 2\theta}_{(2\sin\theta\cos\theta)}\sin\theta$$

$$= (\cos^2\theta - \sin^2\theta)\cos\theta - (2\sin\theta\cos\theta)\sin\theta$$

$$= \cos^3\theta - \sin^2\theta\cos\theta - 2\sin^2\theta\cos\theta$$

Collecting like terms now gives us

$$\cos 3\theta = \cos^3\theta - 3\sin^2\theta\cos\theta$$

Finally, we replace $\sin^2\theta$ by the quantity $1 - \cos^2\theta$. This yields

$$\cos 3\theta = \cos^3\theta - 3(1 - \cos^2\theta)\cos\theta$$

$$= \cos^3\theta - 3\cos\theta + 3\cos^3\theta$$

$$= 4\cos^3\theta - 3\cos\theta \qquad \text{as required}$$

In the box that follows, we list several alternate ways of writing the formula for $\cos 2\theta$. Formulas (c) and (d) are quite useful in calculus.

---

**Equivalent Forms of the Formula $\cos 2\theta = \cos^2 \theta - \sin^2 \theta$**

(a) $\cos 2\theta = 2\cos^2 \theta - 1$    (b) $\cos 2\theta = 1 - 2\sin^2 \theta$

(c) $\cos^2 \theta = \dfrac{1 + \cos 2\theta}{2}$    (d) $\sin^2 \theta = \dfrac{1 - \cos 2\theta}{2}$

---

One way to prove identity (a) is as follows:

$$\cos 2\theta = \cos^2 \theta - \sin^2 \theta$$
$$= \cos^2 \theta - (1 - \cos^2 \theta)$$
$$= \cos^2 \theta - 1 + \cos^2 \theta$$
$$= 2\cos^2 \theta - 1 \qquad \text{as required}$$

If, in this last equation, we add 1 to both sides and then divide by 2, the result is identity (c). (Verify this.) The proofs for (b) and (d) are similar; see Exercise 30.

**EXAMPLE 5**  Express $\cos^4 t$ in a form that does not involve powers of the trigonometric functions.

**Solution**

$$\cos^4 t = (\cos^2 t)^2 = \left(\frac{1 + \cos 2t}{2}\right)^2 \qquad \text{using the formula for } \cos^2 \theta$$

$$= \frac{1 + 2\cos 2t + \cos^2 2t}{4}$$

$$= \frac{1 + 2\cos 2t + \frac{1}{2}(1 + \cos 4t)}{4} \qquad \begin{array}{l}\text{using the formula} \\ \text{for } \cos^2 \theta \text{ with } \theta = 2t\end{array}$$

An easy way to simplify this last expression is to multiply both the numerator and the denominator by 2. As you should check for yourself, the final result is

$$\cos^4 t = \frac{3 + 4\cos 2t + \cos 4t}{8}$$

The last three formulas we are going to prove in this section are the **half-angle formulas**:

$$\cos \frac{s}{2} = \pm \sqrt{\frac{1 + \cos s}{2}}$$

$$\sin \frac{s}{2} = \pm \sqrt{\frac{1 - \cos s}{2}}$$

$$\tan \frac{s}{2} = \frac{\sin s}{1 + \cos s}$$

To derive the formula for $\cos(s/2)$, we begin with one of the alternate forms of the cosine double-angle formula:

$$\cos^2 \theta = \frac{1 + \cos 2\theta}{2}$$

or

$$\cos\theta = \pm\sqrt{\frac{1 + \cos 2\theta}{2}}$$

Since this identity holds for all values of $\theta$, we may replace $\theta$ by $s/2$ to obtain

$$\cos\frac{s}{2} = \pm\sqrt{\frac{1 + \cos 2(s/2)}{2}}$$

$$= \pm\sqrt{\frac{1 + \cos s}{2}}$$

This is the required formula for $\cos(s/2)$. To derive the formula for $\sin(s/2)$, we follow exactly the same procedure, except that we begin with the identity $\sin^2\theta = \dfrac{1 - \cos 2\theta}{2}$. [Exercise 30(c) asks you to complete the proof.] In both formulas, the sign before the radical is determined by the quadrant in which the arc (or angle) $s/2$ terminates.

**EXAMPLE 6** Evaluate $\cos 105°$ using the half-angle formula.

**Solution**
$$\cos 105° = \cos\frac{210°}{2} = \pm\sqrt{\frac{1 + \cos 210°}{2}} \qquad \text{using the formula for} \atop \cos(s/2), \text{with } s = 210°$$

$$= \pm\sqrt{\frac{1 + (-\sqrt{3}/2)}{2}}$$

$$= \pm\sqrt{\frac{1 - (\sqrt{3}/2)}{2} \cdot \frac{2}{2}} = \pm\sqrt{\frac{2 - \sqrt{3}}{4}}$$

$$= \frac{\pm\sqrt{2 - \sqrt{3}}}{2}$$

We want to choose the negative value here, since the terminal side of $105°$ lies in the second quadrant. Thus, we finally obtain

$$\cos 105° = \frac{-\sqrt{2 - \sqrt{3}}}{2}$$

Our last task is to establish the formula for $\tan(s/2)$. To do this, we first prove the equivalent identity $\tan\theta = \dfrac{\sin 2\theta}{1 + \cos 2\theta}$.

PROOF THAT $\tan\theta = \dfrac{\sin 2\theta}{1 + \cos 2\theta}$

$$\frac{\sin 2\theta}{1 + \cos 2\theta} = \frac{2\sin\theta\cos\theta}{2\cos^2\theta} \qquad \text{using the identity } \cos 2\theta = 2\cos^2\theta - 1 \atop \text{in the denominator}$$

$$= \frac{\sin\theta}{\cos\theta} = \tan\theta$$

If we now replace $\theta$ by $s/2$ in the identity $\tan\theta = \dfrac{\sin 2\theta}{1 + \cos 2\theta}$, the result is

$$\tan\frac{s}{2} = \frac{\sin s}{1 + \cos s}$$

This is the half-angle formula for the tangent.

We conclude this section with a summary of the principal trigonometric identities developed in this and the previous section. For completeness, the list also includes two sets of trigonometric identities that we did not discuss in the text but that are occasionally useful. These are the so-called **product-to-sum formulas** and **sum-to-product formulas**. Proofs and applications of these formulas are included in the exercises.

| PROPERTY SUMMARY | Principal Trigonometric Identities |
|---|---|

**I.** Consequences of the Definitions

(a) $\csc\theta = \dfrac{1}{\sin\theta}$      (b) $\sec\theta = \dfrac{1}{\cos\theta}$

(c) $\cot\theta = \dfrac{1}{\tan\theta}$      (d) $\tan\theta = \dfrac{\sin\theta}{\cos\theta}$

(e) $\cot\theta = \dfrac{\cos\theta}{\sin\theta}$

**II.** Pythagorean Identities

(a) $\sin^2\theta + \cos^2\theta = 1$      (b) $\tan^2\theta + 1 = \sec^2\theta$

(c) $\cot^2\theta + 1 = \csc^2\theta$

**III.** Opposite-Angle Formulas

(a) $\sin(-\theta) = -\sin\theta$      (b) $\cos(-\theta) = \cos\theta$

(c) $\tan(-\theta) = -\tan\theta$

**IV.** Reduction Formulas

(a) $\sin(\theta + 2\pi k) = \sin\theta$      (b) $\cos(\theta + 2\pi k) = \cos\theta$

(c) $\sin\left(\dfrac{\pi}{2} - \theta\right) = \cos\theta$      (d) $\cos\left(\dfrac{\pi}{2} - \theta\right) = \sin\theta$

**V.** Addition Formulas

(a) $\sin(s + t) = \sin s \cos t + \cos s \sin t$

(b) $\sin(s - t) = \sin s \cos t - \cos s \sin t$

(c) $\cos(s + t) = \cos s \cos t - \sin s \sin t$

(d) $\cos(s - t) = \cos s \cos t + \sin s \sin t$

(e) $\tan(s + t) = \dfrac{\tan s + \tan t}{1 - \tan s \tan t}$

(f) $\tan(s - t) = \dfrac{\tan s - \tan t}{1 + \tan s \tan t}$

| PROPERTY SUMMARY | Principal Trigonometric Identities |
|---|---|

**VI. Double-Angle Formulas**

(a) $\sin 2\theta = 2\sin\theta\cos\theta$  (b) $\cos 2\theta = \cos^2\theta - \sin^2\theta$

(c) $\tan 2\theta = \dfrac{2\tan\theta}{1 - \tan^2\theta}$

**VII. Half-Angle Formulas**

(a) $\sin\dfrac{\theta}{2} = \pm\sqrt{\dfrac{1 - \cos\theta}{2}}$  (b) $\cos\dfrac{\theta}{2} = \pm\sqrt{\dfrac{1 + \cos\theta}{2}}$

(c) $\tan\dfrac{\theta}{2} = \dfrac{\sin\theta}{1 + \cos\theta}$

**VIII. Product-to-Sum Formulas**

(a) $\sin A \sin B = \frac{1}{2}[\cos(A - B) - \cos(A + B)]$

(b) $\sin A \cos B = \frac{1}{2}[\sin(A + B) + \sin(A - B)]$

(c) $\cos A \cos B = \frac{1}{2}[\cos(A + B) + \cos(A - B)]$

**IX. Sum-to-Product Formulas**

(a) $\sin\alpha + \sin\beta = 2\sin\dfrac{\alpha + \beta}{2}\cos\dfrac{\alpha - \beta}{2}$

(b) $\sin\alpha - \sin\beta = 2\cos\dfrac{\alpha + \beta}{2}\sin\dfrac{\alpha - \beta}{2}$

(c) $\cos\alpha + \cos\beta = 2\cos\dfrac{\alpha + \beta}{2}\cos\dfrac{\alpha - \beta}{2}$

(d) $\cos\alpha - \cos\beta = -2\sin\dfrac{\alpha + \beta}{2}\sin\dfrac{\alpha - \beta}{2}$

# EXERCISE SET 8.4

## A

*In Exercises 1–12, refer to the two triangles and compute the quantities indicated.*

1. (a) $\tan(\theta + t)$
   (b) $\tan(\theta - t)$

2. (a) $\tan(s + \beta)$
   (b) $\tan(s - \beta)$

3. (a) $\tan\left(\theta + \dfrac{\pi}{4}\right)$
   (b) $\tan\left(\theta - \dfrac{\pi}{4}\right)$

4. (a) $\tan\left(\beta + \dfrac{\pi}{3}\right)$
   (b) $\tan\left(\beta - \dfrac{\pi}{3}\right)$

5. (a) $\sin 2\theta$
   (b) $\cos 2\theta$
   (c) $\tan 2\theta$

6. (a) $\sin 2t$
   (b) $\cos 2t$
   (c) $\tan 2t$

7. (a) $\sin 2\beta$
   (b) $\cos 2\beta$
   (c) $\tan 2\beta$

8. (a) $\sin 2s$
   (b) $\cos 2s$
   (c) $\tan 2s$

9. (a) $\sin(\theta/2)$
   (b) $\cos(\theta/2)$
   (c) $\tan(\theta/2)$

10. (a) $\sin(s/2)$
    (b) $\cos(s/2)$
    (c) $\tan(s/2)$

11. (a) $\sin(\beta/2)$
    (b) $\cos(\beta/2)$
    (c) $\tan(\beta/2)$

12. (a) $\sin(t/2)$
    (b) $\cos(t/2)$
    (c) $\tan(t/2)$

In Exercises 13–16, use the given information to compute each of the following: (a) $\sin 2\theta$; (b) $\cos 2\theta$; (c) $\sin(\theta/2)$; (d) $\cos(\theta/2)$.

13. $\sin\theta = \frac{3}{4}$ and $\pi/2 < \theta < \pi$

14. $\cos\theta = \frac{2}{5}$ and $3\pi/2 < \theta < 2\pi$

15. $\cos\theta = -\frac{1}{3}$ and $\pi < \theta < 3\pi/2$

16. $\sin\theta = -\frac{1}{10}$ and $3\pi/2 < \theta < 2\pi$

In Exercises 17–20, use an appropriate half-angle formula to evaluate each quantity,

17. (a) $\sin(\pi/12)$
    (b) $\cos(\pi/12)$
    (c) $\tan(\pi/12)$

18. (a) $\sin(\pi/8)$
    (b) $\cos(\pi/8)$
    (c) $\tan(\pi/8)$

19. (a) $\sin 105°$
    (b) $\cos 105°$
    (c) $\tan 105°$

20. (a) $\sin 165°$
    (b) $\cos 165°$
    (c) $\tan 165°$

In Exercises 21–24, use the given information to express $\sin 2\theta$ and $\cos 2\theta$ in terms of $x$.

21. $x = 5\sin\theta$, $0 < \theta < \pi/2$

22. $x = \sqrt{2}\cos\theta$, $0 < \theta < \pi/2$

23. $x - 1 = 2\sin\theta$, $0 < \theta < \pi/2$

24. $x + 1 = 3\sin\theta$, $\pi/2 < \theta < \pi$

In Exercises 25–28, express each quantity in a form that does not involve powers of the trigonometric functions (as in Example 5).

25. $\sin^4\theta$

26. $\cos^6\theta$

27. $\sin^4(\theta/2)$

28. $\sin^6(\theta/4)$

29. Prove each of the following double-angle formulas. *Hint:* As in the text, replace $2\theta$ with $\theta + \theta$, and use an appropriate addition formula.
    (a) $\cos 2\theta = \cos^2\theta - \sin^2\theta$
    (b) $\tan 2\theta = \dfrac{2\tan\theta}{1 - \tan^2\theta}$

30. (a) Beginning with the identity $\cos 2\theta = \cos^2\theta - \sin^2\theta$, prove that $\cos 2\theta = 1 - 2\sin^2\theta$.
    (b) Using the result in part (a), prove that $\sin^2\theta = \frac{1}{2}(1 - \cos 2\theta)$.
    (c) Derive the formula for $\sin(s/2)$ as follows. Using the identity in part (b), replace $\theta$ with $s/2$, and then take square roots.

**B**

In Exercises 31–46, prove that the given equations are identities.

31. $\cos 2s = \dfrac{1 - \tan^2 s}{1 + \tan^2 s}$

32. $1 + \cos 2t = \cot t \sin 2t$

33. $\cos\theta = 2\cos^2(\theta/2) - 1$

34. $\dfrac{\sin 2\theta}{\sin\theta} - \dfrac{\cos 2\theta}{\cos\theta} = \sec\theta$

35. $\sin^4\theta = \dfrac{3 - 4\cos 2\theta + \cos 4\theta}{8}$

36. $\sin 3\theta = 3\sin\theta - 4\sin^3\theta$

37. $\sin 2\theta = \dfrac{2\tan\theta}{1 + \tan^2\theta}$

38. $2\csc 2\theta = \dfrac{\csc^2\theta}{\cot\theta}$

39. $\sin 2\theta = 2\sin^3\theta\cos\theta + 2\sin\theta\cos^3\theta$

40. $\cot\theta = \dfrac{1 + \cos 2\theta}{\sin 2\theta}$

41. $\dfrac{1 + \tan(\theta/2)}{1 - \tan(\theta/2)} = \tan\theta + \sec\theta$

42. $\tan\theta + \cot\theta = 2\csc 2\theta$

43. $2\sin^2(45° - \theta) = 1 - \sin 2\theta$

44. $(\sin\theta - \cos\theta)^2 = 1 - \sin 2\theta$

45. $1 + \tan\theta\tan 2\theta = \tan 2\theta\cot\theta - 1$

46. $\tan\left(\dfrac{\pi}{4} + \theta\right) - \tan\left(\dfrac{\pi}{4} - \theta\right) = 2\tan 2\theta$

47. Let triangle $ABC$ be a right triangle with $\angle C = 90°$. Prove the following statements. [In part (c), the letters $a, b,$ and $c$ denote the lengths of the sides opposite angles $A, B,$ and $C,$ respectively.]
    (a) $\sin 2A = \sin 2B$   (b) $\cos 2A + \cos 2B = 0$
    (c) $\sin 3A = \dfrac{3ab^2 - a^3}{c^3}$

48. Prove that $\tan\dfrac{\theta}{2} = \dfrac{1 - \cos\theta}{\sin\theta}$.

49. If $\cos(\alpha + \beta) = 0$, show that $\sin(\alpha + 2\beta) = \sin\alpha$.

50. If $\tan\alpha = \frac{1}{11}$ and $\tan\beta = \frac{5}{6}$, find $\alpha + \beta$, given that $0 < \alpha < \pi/2$ and $0 < \beta < \pi/2$.

51. If $\sin\theta = \dfrac{a^2 - b^2}{a^2 + b^2}$, show that $\tan\dfrac{\theta}{2} = \dfrac{a - b}{a + b}$. (Assume that $\theta$ is an acute angle and $a$ and $b$ are positive.)

52. Let $z = \tan\theta$. Show that
$$\cos 2\theta = \dfrac{1 - z^2}{1 + z^2} \quad \text{and} \quad \sin 2\theta = \dfrac{2z}{1 + z^2}$$

53. Prove the following three formulas, known as the product-to-sum formulas.
    (a) $\sin A \sin B = \frac{1}{2}[\cos(A - B) - \cos(A + B)]$
    (b) $\sin A \cos B = \frac{1}{2}[\sin(A + B) + \sin(A - B)]$
    (c) $\cos A \cos B = \frac{1}{2}[\cos(A + B) + \cos(A - B)]$

54. Express each product as a sum or difference of sines or cosines.
    (a) $4\sin 6\theta \cos 2\theta$     (b) $2\sin A \sin 3A$
    (c) $\cos 4\theta \cos 2\theta$

55. (a) Show that $\sin(\pi/4)\cos(\pi/12) = \frac{1}{4}(\sqrt{3} + 1)$.
    (b) Evaluate $2(\sin 82.5°)(\cos 37.5°)$ without using a calculator.

56. Prove the following identities.
    (a) $\sin(A + B)\sin(A - B) = \sin^2 A - \sin^2 B$
    (b) $\cos(A + B)\cos(A - B) = \cos^2 A - \sin^2 B$

57. In the formula (from Exercise 53)

    $\sin A \cos B = \frac{1}{2}[\sin(A + B) + \sin(A - B)]$

    make the substitutions $A + B = \alpha$ and $A - B = \beta$ and deduce that

    $$\sin \alpha + \sin \beta = 2\sin \frac{\alpha + \beta}{2} \cos \frac{\alpha - \beta}{2}$$

    This is one of the so-called sum-to-product formulas mentioned in the text.

58. Prove the following sum-to-product identities.   *Hint:* See the previous exercise.

    (a) $\sin \alpha - \sin \beta = 2\cos \dfrac{\alpha + \beta}{2} \sin \dfrac{\alpha - \beta}{2}$

    (b) $\cos \alpha + \cos \beta = 2\cos \dfrac{\alpha + \beta}{2} \cos \dfrac{\alpha - \beta}{2}$

    (c) $\cos \alpha - \cos \beta = -2\sin \dfrac{\alpha + \beta}{2} \sin \dfrac{\alpha - \beta}{2}$

59. (a) Show that $\cos \dfrac{2\pi}{9} + \cos \dfrac{\pi}{9} = \sqrt{3}\, \cos \dfrac{\pi}{18}$.
    (b) Show that $\sin 105° + \sin 15° = \sqrt{6}/2$.

60. Prove that $\dfrac{\sin \theta + \sin 3\theta}{\cos \theta + \cos 3\theta} = \tan 2\theta$.   *Hint:* Use the sum-to-product formulas.

61. Prove that $\dfrac{\sin 7\theta - \sin 5\theta}{\cos 7\theta + \cos 5\theta} = \tan \theta$.

62. Prove that $\tan(x + y) = \dfrac{\sin 2x + \sin 2y}{\cos 2x + \cos 2y}$.

63. The following figure shows a semicircle with radius $AO = 1$.
    (a) Use the figure to derive the formula

    $$\tan \frac{\theta}{2} = \frac{\sin \theta}{1 + \cos \theta}$$

    *Hint:* Show that $CD = \sin \theta$ and $OD = \cos \theta$. Then look at right triangle $ADC$ to find $\tan(\theta/2)$.

    (b) Use the formula developed in part (a) to show that
    (i) $\tan 15° = \dfrac{1}{2 + \sqrt{3}} = 2 - \sqrt{3}$
    (ii) $\tan(\pi/8) = \sqrt{2} - 1$

64. This exercise outlines geometric derivations of the formulas for $\sin 2\theta$ and $\cos 2\theta$. Refer to the following figure, which shows a semicircle with diameter $\overline{SV}$ and radius $UV = 1$. In each case, supply the reasons for the given statement.
    (a) $TU = \cos 2\theta$       (b) $TV = 1 + \cos 2\theta$
    (c) $WT = \sin 2\theta$
    (d) $WV = 2\cos \theta$   *Hint:* Consider right triangle $SWV$.
    (e) $SW = 2\sin \theta$
    (f) $\sin \theta = \dfrac{\sin 2\theta}{2\cos \theta}$   *Hint:* Use $\triangle WTV$ and the results in parts (c) and (d).
    (g) $\sin 2\theta = 2\sin \theta \cos \theta$
    (h) $\cos \theta = \dfrac{1 + \cos 2\theta}{2\cos \theta}$   *Hint:* Use $\triangle WTV$
    (i) $\cos 2\theta = 2\cos^2 \theta - 1$

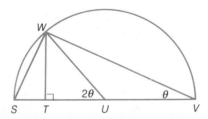

C

65. (a) Prove that

    $$\tan \theta \tan(60° - \theta)\tan(60° + \theta) = \tan 3\theta.$$

    (b) Prove that $\tan 20° \tan 40° \tan 80° = \sqrt{3}$.

66. *Calculation of* $\sin 18°$, $\cos 18°$, *and* $\sin 3°$
    (a) Prove that $\cos 3\theta = 4\cos^3 \theta - 3\cos \theta$.
    (b) Supply a reason for each statement.
       (i) $\sin 36° = \cos 54°$
       (ii) $2\sin 18° \cos 18° = 4\cos^3 18° - 3\cos 18°$
       (iii) $2\sin 18° = 4\cos^2 18° - 3$

(c) In equation (iii), replace $\cos^2 18°$ by $1 - \sin^2 18°$ and then solve the resulting equation for $\sin 18°$. Thus, show that

$$\sin 18° = \tfrac{1}{4}(\sqrt{5} - 1)$$

(d) Show that

$$\cos 18° = \tfrac{1}{4}\sqrt{10 + 2\sqrt{5}}$$

(e) Show that

$$\sin 3° = \tfrac{1}{16}[(\sqrt{5} - 1)(\sqrt{6} + \sqrt{2})$$
$$- 2(\sqrt{3} - 1)\sqrt{5 + \sqrt{5}}]$$

*Hint:* $3° = 18° - 15°$.

**c** (f) Use your calculator to check the results in parts (c), (d), and (e).

67. (a) The angles of a triangle are $A = 20°$, $B = 50°$, and $C = 110°$. Use your calculator to compute the two quantities $\tan A + \tan B + \tan C$ and $\tan A \tan B \tan C$. What do you observe?

(b) The angles of a triangle are $\alpha = \pi/10$, $\beta = 3\pi/10$, and $\gamma = 3\pi/5$. Use your calculator to compute $\tan \alpha + \tan \beta + \tan \gamma$ and $\tan \alpha \tan \beta \tan \gamma$.

(c) If triangle $ABC$ is not a right triangle, prove that

$$\tan A + \tan B + \tan C = \tan A \tan B \tan C$$

## GRAPHS OF THE SINE AND COSINE FUNCTIONS

### 8.5

In this section and the next, we consider the graphs of the trigonometric functions. Our focus in this section is on the sine and cosine functions. As preparation for the discussion, we want to understand what is meant by the term *periodic function.* By way of example, all of the functions in Figure 1 are **periodic**. That is, their graphs display patterns that repeat themselves at regular intervals.

(a)

(b)

(c)

**FIGURE 1**

In Figure 1(a), the graph of the function $f$ repeats itself every six units. We say that *the period of $f$ is* 6. Similarly, the period of $g$ in Figure 1(b) is 2, while the period of $h$ in Figure 1(c) is $2\pi$. Notice that in each case, the period represents the minimum number of units that we must travel along the horizontal axis before the graph begins to repeat itself. Because of this, we can state the definition of a periodic function as follows.

---

**Definition:   A Periodic Function and Its Period**

A function $f$ is said to be **periodic** provided there is a number $p > 0$ such that

$$f(x + p) = f(x)$$

for all $x$ in the domain of $f$. The smallest such number $p$ is called the **period** of $f$.

---

We also want to define the term **amplitude** as it applies to periodic functions. For a function such as $h$ in Figure 1(c), in which the graph is centered about the horizontal axis, the amplitude is simply the maximum height of the graph above that horizontal axis. Thus, the amplitude of $h$ is 4. More generally, we define the amplitude of any periodic function as follows.

---

**Definition:   Amplitude**

Let $f$ be a periodic function and let $m$ and $M$ denote, respectively, the smallest and largest values of the function. Then, by definition, the **amplitude** of $f$ is the number

$$\frac{M - m}{2}$$

---

For the function $h$ in Figure 1(c), this definition tells us that the amplitude is $\dfrac{4 - (-4)}{2} = 4$, which agrees with our previous value. Check for yourself that the amplitudes of the functions $f$ and $g$ in Figure 1(a) and (b) are, respectively, 1 and $\frac{3}{2}$.

Let us now graph the sine function $f(\theta) = \sin \theta$. We are assuming that $\theta$ is in radians, so that the domain of the sine function is the set of all real numbers. Even before making any calculations, we can gain a strong intuitive insight into how the graph must look by carrying out the following experiment. After drawing the unit circle, $x^2 + y^2 = 1$, place your fingertip at the point $(1, 0)$ and then move your finger counterclockwise around the circle. As you do this, keep track of what happens to the $y$-coordinate of your fingertip. (The $y$-coordinate is $\sin \theta$.) Figure 2 indicates the general results.

In summary, as $\theta$ increases from 0 to $2\pi$, the graph of $y = \sin \theta$ should display the general features shown in Figure 3. Furthermore, since at $\theta = 2\pi$ we've returned to our starting point, $(1, 0)$, additional counterclockwise trips around the unit circle will just result in repetitions of the pattern established in Figure 3. In other words, the period of $y = \sin \theta$ is $2\pi$.

Now let us set up a table so that we can sketch the graph of $y = \sin \theta$ more accurately. Since the sine function is periodic (as we've just observed), with

(a) As $\theta$ increases from 0 to
$\pi/2$, the $y$-coordinate (sin $\theta$)
increases from 0 to 1.

(b) As $\theta$ increases from $\pi/2$ to $\pi$,
the $y$-coordinate decreases from
1 back down to 0.

(c) As $\theta$ increases from $\pi$ to $3\pi/2$,
the $y$-coordinate decreases from
0 to $-1$.

(d) As $\theta$ increases from $3\pi/2$ to $2\pi$,
the $y$-coordinate increases from
$-1$ back up to 0.

**FIGURE 2**

$$y = \sin\theta, \quad 0 \le \theta \le 2\pi$$

**FIGURE 3**

period $2\pi$, our table needs only to contain values of $\theta$ between 0 and $2\pi$, as shown in Table 1. This will establish the basic pattern for the graph.

**TABLE 1**

| $\theta$ | 0 | $\dfrac{\pi}{6}$ | $\dfrac{\pi}{3}$ | $\dfrac{\pi}{2}$ | $\dfrac{2\pi}{3}$ | $\dfrac{5\pi}{6}$ | $\pi$ | $\dfrac{7\pi}{6}$ | $\dfrac{4\pi}{3}$ | $\dfrac{3\pi}{2}$ | $\dfrac{5\pi}{3}$ | $\dfrac{11\pi}{6}$ | $2\pi$ |
|---|---|---|---|---|---|---|---|---|---|---|---|---|---|
| $\sin\theta$ | 0 | $\dfrac{1}{2}$ | $\dfrac{\sqrt{3}}{2}$ | 1 | $\dfrac{\sqrt{3}}{2}$ | $\dfrac{1}{2}$ | 0 | $-\dfrac{1}{2}$ | $-\dfrac{\sqrt{3}}{2}$ | $-1$ | $-\dfrac{\sqrt{3}}{2}$ | $-\dfrac{1}{2}$ | 0 |

In plotting the points obtained from this table, we use the approximation $\sqrt{3}/2 \approx 0.87$. Rather than approximating $\pi$, however, we mark off units on the horizontal axis in terms of $\pi$. The resulting graph is shown in Figure 4. By continuing this same pattern to the left and right, we obtain the complete graph of $f(\theta) = \sin\theta$, as indicated in Figure 5.

Before going on to analyze the sine function or to study the graphs of the other trigonometric functions, we are going to make a slight change in the

**FIGURE 5**

notation we have been using. To conform with common usage, we will use $x$ instead of $\theta$ on the horizontal axis, and we will use $y$ for the vertical axis. The sine function is then written simply as $y = \sin x$, where the real number $x$ denotes the radian measure of an angle or, equivalently, the length of the corresponding arc on the unit circle. For reference, we redraw Figures 4 and 5 using this familiar $x$-$y$ notation, as shown in Figure 6.

You should memorize the graph of the sine curve in Figure 6(b) so that you can sketch it without first setting up a table. Of course, once you know the shape of the basic sine wave in Figure 6(a), you automatically know the graph of the full sine curve in Figure 6(b).

We can use the graphs in Figure 6 to help us list some of the key properties of the sine function.

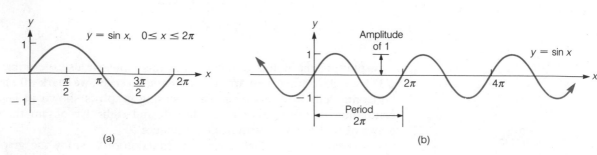

(a)                                                                        (b)

**FIGURE 6**

| PROPERTY SUMMARY | The Sine Function:  $y = \sin x$ |
|---|---|

1. The domain of the sine function is the set of all real numbers. The range of the sine function is the closed interval $[-1, 1]$. Another way to express this last fact is

$$-1 \leq \sin x \leq 1, \qquad \text{for all } x$$

2. The sine function is periodic, with period $2\pi$. The amplitude is 1.

3. (Refer to Figure 7.) The basic sine wave crosses the $x$-axis at the beginning, midpoint, and end of the period. The curve reaches its highest point after one-quarter of the period and its lowest point after three-quarters of the period.

**FIGURE 7**

The graph of the cosine function is readily obtained from that of the sine function through the use of the identity

$$\cos x = \sin\left(x + \frac{\pi}{2}\right) \tag{1}$$

First, to verify that this is indeed an identity, we have

$$\sin\left(x + \frac{\pi}{2}\right) = \sin x \cos \frac{\pi}{2} + \cos x \sin \frac{\pi}{2}$$
$$= (\sin x)(0) + (\cos x)(1)$$
$$= \cos x$$

From equation (1), it follows that the graph of $y = \cos x$ is obtained by translating (shifting) the sine curve $\pi/2$ units to the left. The result is shown in Figure 8(a). Figure 8(b) displays one complete cycle of the cosine curve, from $x = 0$ to $x = 2\pi$.

As with the sine function, you should memorize the graph and basic features of the cosine function. The key features of the cosine are summarized in the box that follows.

**FIGURE 8**

| PROPERTY SUMMARY | The Cosine Function:  $y = \cos x$ |
|---|---|

1. The domain of the cosine function is the set of all real numbers. The range of the cosine function is the closed interval $[-1, 1]$. Another way to express this last fact is

$$-1 \le \cos x \le 1, \qquad \text{for all } x$$

2. The cosine function is periodic, with period $2\pi$. The amplitude is 1.

3. (Refer to Figure 9.) At the beginning and at the end of the period, the basic cosine wave reaches its highest point. At the midpoint of the period, the curve reaches its lowest point. The $x$-intercepts occur one-quarter and three-quarters of the way through the period.

**FIGURE 9**

The graphs of $y = \sin x$ and $y = \cos x$ are the building blocks we need for graphing functions of the form

$$y = A\sin(Bx + C) \qquad \text{and} \qquad y = A\cos(Bx + C)$$

where $A$, $B$, and $C$ are constants and neither $A$ nor $B$ is zero. These functions are used throughout the sciences to analyze a variety of periodic phenomena ranging from the vibrations of an electron to the variations in the size of an animal population as it interacts with its environment.

As a first example, consider $y = 2\sin x$. To obtain the graph of $y = 2\sin x$ from that of $y = \sin x$, we multiply each $y$-coordinate on the graph of $y = \sin x$ by 2. As indicated in Figure 10, this changes the amplitude from 1 to 2, but it does not affect the period, which remains $2\pi$. In a similar way, Figure 11 shows the graphs of $y = \cos x$ and $y = \frac{1}{2}\cos x$. Note that the amplitude of $y = \frac{1}{2}\cos x$ is $\frac{1}{2}$ and the period is, again, $2\pi$. More generally, graphs of functions of the form $y = A\sin x$ and $y = A\cos x$ always have an amplitude of $|A|$ and a period of $2\pi$.

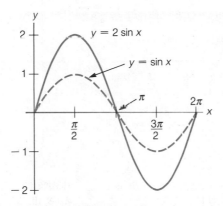

FIGURE 10    The amplitude of $y = 2\sin x$ is 2. Both $y = \sin x$ and $y = 2\sin x$ have a period of $2\pi$.

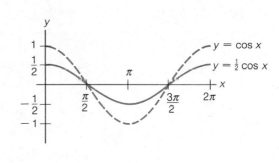

FIGURE 11    The amplitude of $y = \frac{1}{2}\cos x$ is $\frac{1}{2}$. Both $y = \cos x$ and $y = \frac{1}{2}\cos x$ have a period of $2\pi$.

**EXAMPLE 1**    Graph the function $y = -2\sin x$ over one period.

Solution    In Chapter 4 we saw that the graph of $y = -f(x)$ is obtained from that of $y = f(x)$ by reflection about the $x$-axis. Thus, we need only to take the graph of $y = 2\sin x$ from Figure 10 and reflect it about the $x$-axis. See Figure 12. Note that both functions have an amplitude of 2 and a period of $2\pi$.

Now let us look at functions of the form $y = \sin Bx$, with $B > 0$. We know that $y = \sin x$ begins its basic pattern when $x = 0$ and completes that pattern when $x = 2\pi$. Thus, $y = \sin Bx$ will begin its basic pattern when $Bx = 0$, and it will complete that pattern when $Bx = 2\pi$. From the equation $Bx = 0$, we conclude that $x = 0$, and from the equation $Bx = 2\pi$, we conclude that $x = 2\pi/B$. Thus, the function $y = \sin Bx$ begins its basic pattern at $x = 0$ and ends at $x = 2\pi/B$. In other words, the period of $y = \sin Bx$ is $2\pi/B$. Combining this information with our earlier observations about amplitude yields the graph shown in Figure 13.

**FIGURE 12**

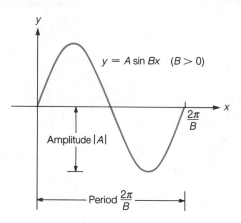

**FIGURE 13** The period of $y = A \sin Bx$ is $2\pi/B$; the amplitude is $|A|$.

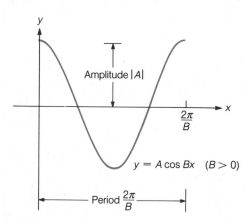

**FIGURE 14** The period of $y = A \cos Bx$ is $2\pi/B$; the amplitude is $|A|$.

Before turning to some examples, we point out that the same analysis we have used for $y = A \sin Bx$ applies equally well to $y = A \cos Bx$. See Figure 14.

**EXAMPLE 2** Determine the amplitude and period of the function $y = 3 \sin \frac{1}{2}x$. Graph the function over one period, and indicate the coordinates of the highest and lowest points on the graph.

**Solution** Comparing the equations

$$y = A \sin Bx$$

and

$$y = 3 \sin \tfrac{1}{2}x$$

we see that $A = 3$ and $B = \frac{1}{2}$. Thus, we have

$$\text{period} = \frac{2\pi}{B} = \frac{2\pi}{\left(\frac{1}{2}\right)} = 4\pi$$

Also, the amplitude is $|A| = |3| = 3$. The required graph is thus a sine wave with period $4\pi$ and amplitude 3, as displayed in Figure 15.

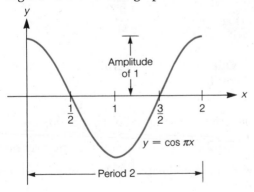

FIGURE 15

The highest point on the sine wave occurs when $x$ is one-quarter of the way through the period, that is, when

$$x = \frac{1}{4}(4\pi) = \pi$$

Since the largest $y$-value is (in this case) the amplitude, we conclude that the coordinates of the highest point on the graph in Figure 15 are $(\pi, 3)$. In a similar fashion, the lowest point occurs when $x$ is three-quarters of the way through the period, that is, when

$$x = \frac{3}{4}(4\pi) = 3\pi$$

Thus, the lowest point on the graph is $(3\pi, -3)$.

**EXAMPLE 3** Determine the amplitude and period of $y = \cos \pi x$. Sketch the graph of this function over one period and specify the $x$-intercepts.

**Solution** For the function $y = \cos \pi x$, we have $A = 1$ and $B = \pi$. Thus,

$$\text{amplitude} = |A| = 1$$

and,

$$\text{period} = \frac{2\pi}{B} = \frac{2\pi}{\pi} = 2$$

Figure 16 shows the graph of a cosine wave with these specifications. Since

FIGURE 16

the $x$-intercepts of the cosine wave occur one-quarter and three-quarters of the way through the period, we compute these intercepts by writing

$$\frac{1}{4}(2) = \frac{1}{2} \qquad \text{and} \qquad \frac{3}{4}(2) = \frac{3}{2}$$

These $x$-intercepts are shown in Figure 16.

We now have at our disposal all the tools needed to graph functions of the form

$$y = A \sin(Bx + C) \qquad \text{or} \qquad y = A \cos(Bx + C) \qquad (B > 0)$$

First, as a matter of algebra, observe that

$$A \sin(Bx + C) = A\left[\sin B\left(x + \frac{C}{B}\right)\right]$$

The conclusion we draw from this is that the graph of $y = A \sin(Bx + C)$ is obtained by taking the graph of $y = A \sin Bx$ and translating it horizontally a distance of $|C/B|$ units. If $C/B$ is positive, the movement is to the left; if $C/B$ is negative, the movement is to the right. The quantity $-C/B$ is called the **phase shift**. As Figure 17(a) indicates, the phase shift gives us the $x$-coordinate at which to begin drawing $y = A \sin(Bx + C)$ over one period. Note that the amplitude and period are the same as they were for $y = A \sin Bx$.

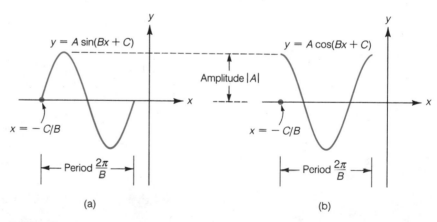

FIGURE 17

Our remarks concerning $y = A \sin(Bx + C)$ apply equally well to $y = A \cos(Bx + C)$. See Figure 17(b).

**EXAMPLE 4**  Determine the amplitude, period, and phase shift for the function $y = \frac{1}{2}\sin\left(2x - \frac{2\pi}{3}\right)$. Sketch the graph over one period, indicating the $x$-intercepts and the coordinates of the highest and lowest points on the graph over this period.

**Solution** Comparing the given equation with $y = A \sin(Bx + C)$, we find that $A = \frac{1}{2}$, $B = 2$, and $C = -2\pi/3$. Consequently, we have

Amplitude: $|A| = \dfrac{1}{2}$

Period: $\dfrac{2\pi}{B} = \dfrac{2\pi}{2} = \pi$

Phase shift: $\dfrac{-C}{B} = \dfrac{2\pi/3}{2} = \dfrac{\pi}{3}$

From these data, we conclude that this particular sine wave begins at $x = \pi/3$ and ends the period at

$$x = \frac{\pi}{3} + \pi = \frac{4\pi}{3}$$

Thus, two of the $x$-intercepts are $\pi/3$ and $4\pi/3$. Since the third $x$-intercept is midway between these two, we can compute it by taking the average:

$$\frac{\dfrac{\pi}{3} + \dfrac{4\pi}{3}}{2} - \frac{5\pi/3}{2} = \frac{5\pi}{6}$$

The three $x$-intercepts are therefore $\pi/3$, $5\pi/6$, and $4\pi/3$. Next, the $x$-coordinate of the highest point on the graph is midway between $\pi/3$ and $5\pi/6$. Computing the average of these two numbers, we have

$$\frac{\dfrac{\pi}{3} + \dfrac{5\pi}{6}}{2} = \frac{\dfrac{2\pi}{6} + \dfrac{5\pi}{6}}{2} = \frac{\dfrac{7\pi}{6}}{2} = \frac{7\pi}{12}$$

Thus, the coordinates of the highest point on the graph are $(7\pi/12, \frac{1}{2})$. Finally, in a similar manner, the $x$-coordinate of the lowest point on the graph can be obtained by averaging the numbers $5\pi/6$ and $4\pi/3$. As you can check, this average is $13\pi/12$. It follows that the lowest point on the graph is $(13\pi/12, -\frac{1}{2})$. Figure 18 displays the required graph. ∷

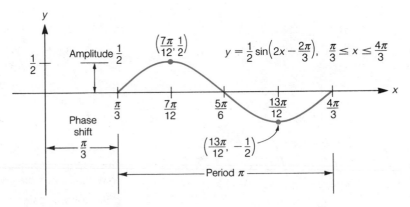

FIGURE 18

# EXERCISE SET 8.5

## A

In Exercises 1–8, specify the period and amplitude for each function.

**1.**

**2.**

**3.**

**4.**

**5.**

**6.**

**7.**

**8.**

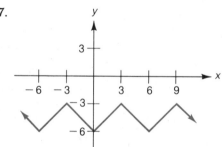

In Exercises 9–20, graph the functions for one period. In each case, specify the amplitude, period, x-intercepts, and highest and lowest points on the graph.

**9.** (a) $y = 2\sin x$
   (b) $y = -\sin 2x$

**10.** (a) $y = 3\sin x$
   (b) $y = \sin 3x$

11. (a) $y = \cos 2x$
    (b) $y = 2\cos 2x$

12. (a) $y = \cos(x/2)$
    (b) $y = -\frac{1}{2}\cos(x/2)$

13. (a) $y = 3\sin(\pi x/2)$
    (b) $y = -3\sin(\pi x/2)$

14. (a) $y = 2\sin \pi x$
    (b) $y = -2\sin \pi x$

15. (a) $y = \cos 2\pi x$
    (b) $y = -4\cos 2\pi x$

16. (a) $y = -2\cos(x/4)$
    (b) $y = -2\cos(\pi x/4)$

17. $y = 1 + \sin 2x$

18. $y = \sin(x/2) - 2$

19. $y = 1 - \cos(\pi x/3)$

20. $y = -2 - 2\cos 3\pi x$

*In Exercises 21–32, determine the amplitude, period, and phase shift for the given function. Graph the function over one period. Indicate the x-intercepts and the coordinates of the highest and lowest points on the graph.*

21. $y = \sin\left(2x - \dfrac{\pi}{2}\right)$

22. $y = \sin\left(3x + \dfrac{\pi}{2}\right)$

23. $y = \cos(2x - \pi)$

24. $y = \cos\left(x - \dfrac{\pi}{2}\right)$

25. $y = 3\sin\left(\dfrac{1}{2}x + \dfrac{\pi}{6}\right)$

26. $y = 2\sin(\pi x + \pi)$

27. $y = 4\cos\left(3x - \dfrac{\pi}{4}\right)$

28. $y = \cos(x + 1)$

29. $y = \dfrac{1}{2}\sin\left(\dfrac{\pi x}{2} - \pi^2\right)$

30. $y = \cos\left(2x - \dfrac{\pi}{3}\right) + 1$

31. $y = 1 - \cos\left(2x - \dfrac{\pi}{3}\right)$

32. $y = 3\cos\left(\dfrac{2x}{3} + \dfrac{\pi}{6}\right)$

**B**

33. (a) Beginning with the identity $\sin\dfrac{s}{2} = \pm\sqrt{\dfrac{1 - \cos s}{2}}$, show that $\sin^2 x = \frac{1}{2} - \frac{1}{2}\cos 2x$.
    (b) Graph the function $y = \sin^2 x$ for one period.

34. (a) Beginning with the identity $\cos\dfrac{s}{2} = \pm\sqrt{\dfrac{1 + \cos s}{2}}$, show that $\cos^2 x = \frac{1}{2} + \frac{1}{2}\cos 2x$.
    (b) Graph $y = \cos^2 x$ for one period.

35. Graph the function $y = \sin x \cos x$ for one period. *Hint:* $2\sin x \cos x = \sin 2x$.

36. Graph the function $y = \cos^2 x - \sin^2 x$ for one period.

37. A weight is attached to a spring hung from the ceiling, as shown in the figure. The coordinate system has been chosen so that the equilibrium position of the weight corresponds to $s = 0$. Now suppose that the weight is pulled down to the position where $s = -3$, and then it is released, so that it oscillates up and down in a periodic fashion. Assume that the position of the weight at time $t$ is given by the function

$$s = -3\cos\frac{\pi t}{3}$$

where $s$ is in feet and $t$ is in seconds.

*s*-axis

(a) Find the amplitude and period of this function, and sketch its graph over the interval $0 \le t \le 12$.

(b) Use your graph to determine the times in this interval at which the weight is farthest from the origin.

(c) When during this interval of time is the weight passing through the equilibrium position? (Specify the values of $t$.)

(d) Now assume that under these conditions, the velocity $v$ of the weight is given by the function

$$v = \pi \sin\frac{\pi t}{3}$$

where $t$ is in seconds and $v$ is in feet per second. Graph this velocity function over the interval $0 \le t \le 12$. Specify the amplitude and period.

(e) Use your graph to find the times during this interval when the velocity is zero. At these times, what is the position ($s$-coordinate) of the weight?

(f) At what times during this interval is the weight moving downward?

(g) Use your graph to find the times when the velocity is greatest and least. What is the numerical value of the velocity in those cases? (Use $\pi \approx 3.14$.) Where is the weight located at those times when the velocity is greatest and least?

(h) On the same set of axes, graph the velocity function and the position function over the interval $0 \le t \le 12$.

38. The following figure shows a simple pendulum consisting of a string with a weight attached at one end, while the other end is suspended from a fixed point. It can be shown that if the pendulum is displaced 0.1 radian

from the equilibrium (vertical) position and then released, the size of the angle $\theta$ at time $t$ is very closely approximated by $\theta = 0.1 \cos(\sqrt{g/L})t$. In this formula, $t$ is measured in seconds, $L$ is the length of the pendulum, and $g$ is a constant (the acceleration due to gravity). Show that under these conditions, the time $T$ required for one complete oscillation of the pendulum is given by

$$T = 2\pi \sqrt{\frac{L}{g}}$$

[There are two conclusions that follow from this formula: (1) We can adjust a pendulum clock by altering the length $L$ of the pendulum. (2) On the other hand, altering the amplitude slightly will have no effect on the period, since the value of $T$ does not depend on the amplitude.]

39. **(a)** Show that $3 \sin x + 4 \cos x = 5 \sin(x + \delta)$, where $\delta$ is the acute angle shown in the following triangle.

   **(b)** Graph the function $y = 3 \sin x + 4 \cos x$ by using the identity given in part (a). *Hint:* See Example 10 on page 339 for methods of determining $\delta$. If you use a calculator, be sure to set it in the radian mode.

40. **(a)** Determine the amplitude and period of the function $y = \sin 2x + \sqrt{3} \cos 2x$.
   *Hint:* Write $\sin 2x + \sqrt{3} \cos 2x = A \sin(2x + \delta)$ and solve for $A$.
   **(b)** What is the smallest possible value of the expression $\sin 2x + \sqrt{3} \cos 2x$?

## GRAPHS OF THE TANGENT AND THE RECIPROCAL FUNCTIONS

**8.6**

*A third . . . function, the tangent of $\theta$, or $\tan \theta$, is of secondary importance, in that it is not associated with wave phenomena. Nevertheless, it enters into the body of analysis so prominently that we cannot ignore it.*

From the text *Elements of Mathematical Analysis* by SAMUEL E. URNER and WILLIAM B. ORANGE (Boston: Ginn and Company, 1950)

We have seen in the previous section that the sine and cosine functions are periodic. The remaining four trigonometric functions also are periodic, but their graphs differ significantly from those of sine and cosine. The graphs of $y = \tan x$, $\cot x$, $\sec x$, and $\csc x$ all possess asymptotes.

We begin with the tangent function, $y = \tan x$. Table 1 displays a list of values for this function using $x$-values in the interval $[0, \pi/2)$.

**TABLE 1**

As $x$ increases from 0 toward $\pi/2$, the values of $\tan x$ increase slowly at first, then more and more rapidly.

| $x$ | 0 | $\dfrac{\pi}{6}$ | $\dfrac{\pi}{4}$ | $\dfrac{\pi}{3}$ | $\dfrac{5\pi}{12}(=75°)$ | $\dfrac{17\pi}{36}(=85°)$ | $\dfrac{89\pi}{180}(=89°)$ | $\dfrac{\pi}{2}$ |
|---|---|---|---|---|---|---|---|---|
| $\tan x$ | 0 | $\dfrac{\sqrt{3}}{3} \approx 0.58$ | 1 | $\sqrt{3} \approx 1.73$ | 3.73 | 11.43 | 57.29 | undefined |

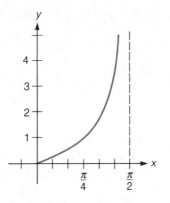

**FIGURE 1** $y = \tan x$, $0 \leq x < \pi/2$

As indicated in Table 1, $\tan x$ is undefined when $x = \pi/2$. This follows from the identity

$$\tan x = \frac{\sin x}{\cos x} \qquad (1)$$

When $x = \pi/2$, the denominator in this identity is zero. Indeed, when $x$ is equal to any odd integral multiple of $\pi/2$ (e.g., $\pm 3\pi/2$, $\pm 5\pi/2$), the denominator in equation (1) will be zero, and consequently, $\tan x$ will be undefined.

Because $\tan x$ is undefined when $x = \pi/2$, we want to see how the graph behaves as $x$ gets closer and closer to $\pi/2$. This is why the $x$-values $5\pi/12$, $17\pi/36$, and $89\pi/180$ are used in Table 1. In Figure 1, we've used the data in Table 1 to draw the graph of $y = \tan x$ for $0 \leq x < \pi/2$. As the figure indicates, the vertical line $x = \pi/2$ is an asymptote for the graph.

The graph of $y = \tan x$ can now be completed without further need for tables or a calculator. First, the identity $\tan(-x) = -\tan x$ (from Section 8.2) tells us that the graph of $y = \tan x$ is symmetric about the origin. So, after reflecting the graph in Figure 1 about the origin, we can draw the graph of $y = \tan x$ on the interval $(-\pi/2, \pi/2)$, as shown in Figure 2.

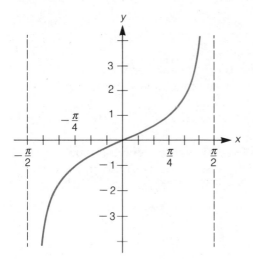

**FIGURE 2** $y = \tan x$, $-\pi/2 < x < \pi/2$
The graph is symmetric about the origin.
The lines $x = \pm \pi/2$ are asymptotes.

To complete the graph of $y = \tan x$, we'll need to rely on the identity

$$\tan(x + \pi) = \tan x \qquad (2)$$

Using the addition formula for $\tan(s + t)$, we can verify identity (2) as follows:

$$\tan(x + \pi) = \frac{\tan x + \tan \pi}{1 - \tan x \tan \pi} = \frac{\tan x + 0}{1 - (\tan x)(0)}$$

$$= \tan x \qquad \text{as required}$$

Identity (2) tells us that the graph of $y = \tan x$ must repeat itself at intervals of length $\pi$. Taking this fact into account, along with Figure 2, we conclude that the period of $y = \tan x$ is exactly $\pi$. Our final graph of $y = \tan x$ is shown in the summary box that follows.

| PROPERTY SUMMARY | The Tangent Function:   $y = \tan x$ |
| --- | --- |

Domain:   The set of all real numbers other than
$\quad\quad\quad \pm\pi/2,\ \pm 3\pi/2,\ \pm 5\pi/2, \ldots$

Range:   The set of all real numbers

Period:   $\pi$ (See Figure 3.)

Asymptotes:   $x = \pm\pi/2,\ x = \pm 3\pi/2,\ x = \pm 5\pi/2, \ldots$

$x$-intercepts:   $0,\ \pm\pi,\ \pm 2\pi,\ \pm 3\pi, \ldots$

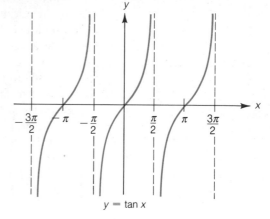

$y = \tan x$

**FIGURE 3**

**EXAMPLE 1**   Graph the following functions for one period. In each case, specify the period, the asymptotes, and the intercepts.

$$\text{(a)}\ \ y = \tan\!\left(x - \frac{\pi}{4}\right) \qquad \text{(b)}\ \ y = -\tan\!\left(x - \frac{\pi}{4}\right)$$

**Solution**   We begin with the graph of one period of $y = \tan x$, as shown in Figure 4(a). By translating this graph $\pi/4$ units to the right, we obtain the graph of

$y = \tan x$

Period:   $\pi$
Asymptotes:   $x = \pm\pi/2$
$x$-intercept:   0
$y$-intercept:   0

(a)

$y = \tan(x - \frac{\pi}{4})$

Period:   $\pi$
Asymptotes:   $x = -\pi/4,\ x = 3\pi/4$
$x$-intercept:   $\pi/4$
$y$-intercept:   $-1$

(b)

$y = -\tan(x - \frac{\pi}{4})$

Period:   $\pi$
Asymptotes:   $x = -\pi/4,\ x = 3\pi/4$
$x$-intercept:   $\pi/4$
$y$-intercept:   1

(c)

**FIGURE 4**

$y = \tan\left(x - \dfrac{\pi}{4}\right)$, in Figure 4(b). With this translation, note that the left asymp-

tote shifts from $x = -\dfrac{\pi}{2}$ to $x = -\dfrac{\pi}{2} + \dfrac{\pi}{4} = -\dfrac{\pi}{4}$; the right asymptote shifts

from $x = \dfrac{\pi}{2}$ to $x = \dfrac{\pi}{2} + \dfrac{\pi}{4} = \dfrac{3\pi}{4}$; and the x-intercept shifts from 0 to $\pi/4$. For

the y-intercept of $y = \tan\left(x - \dfrac{\pi}{4}\right)$, replace x with 0 in the equation. This yields

$$y = \tan\left(0 - \dfrac{\pi}{4}\right) = \tan\left(-\dfrac{\pi}{4}\right)$$

$$= -\tan\dfrac{\pi}{4} = -1$$

Finally, for the graph of $y = -\tan\left(x - \dfrac{\pi}{4}\right)$, we need only to reflect the graph in Figure 4(b) about the x-axis. See Figure 4(c).

**EXAMPLE 2**  Graph the function $y = \tan(x/2)$ for one period.

**Solution**  First we refer back to Figure 2, which shows the basic pattern for one period of $y = \tan x$. In this basic pattern, the asymptotes occur when the angle (or arc length) equals $-\pi/2$ or $\pi/2$. Consequently, for $y = \tan(x/2)$, the asymptotes occur when $x/2 = -\pi/2$ and when $x/2 = \pi/2$. From the equation $x/2 = -\pi/2$, we conclude that $x = -\pi$, and from the equation $x/2 = \pi/2$, we conclude that $x = \pi$. Thus, the asymptotes for $y = \tan(x/2)$ are $x = -\pi$ and $x = \pi$. The distance between these asymptotes, namely, $2\pi$, is the period of $y = \tan(x/2)$. This is twice the period of $y = \tan x$. So basically, we want to draw a curve with the same general shape as $y = \tan x$ but twice as wide. See Figure 5. Note that the graph in Figure 5 passes through the origin, for when $x = 0$, we have

$$y = \tan\dfrac{0}{2} = \tan 0 = 0$$

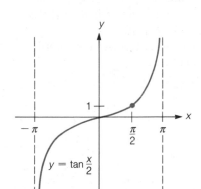

**FIGURE 5**

The graph of the cotangent function can be obtained from that of the tangent function by means of the identity

$$\cot x = -\tan\left(x - \dfrac{\pi}{2}\right) \qquad (3)$$

[Exercise 33 provides a suggestion for deriving this identity. (Why is the formula for $\tan(s - t)$ not helpful here?)] According to identity (3), the graph of $y = \cot x$ can be obtained by first translating the graph of $y = \tan x$ (in Figure 3) $\pi/2$ units to the right and then reflecting the translated graph about the x-axis. When this is done, we obtain the graph shown in the box that follows.

| PROPERTY SUMMARY | The Cotangent Function: $y = \cot x$ |
| --- | --- |

Domain:   The set of all real numbers other than $0, \pm\pi, \pm 2\pi, \ldots$

Range:   The set of all real numbers

Period:   $\pi$ (See Figure 6.)

Asymptotes:   $x = 0, x = \pm\pi, x = \pm 2\pi, \ldots$

$x$-intercepts:   $\pm\dfrac{\pi}{2}, \pm\dfrac{3\pi}{2}, \pm\dfrac{5\pi}{2}, \ldots$

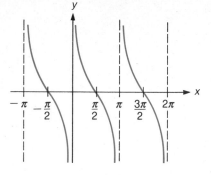

FIGURE 6   $y = \cot x$

**EXAMPLE 3**   Graph each of the following functions for one period.

(a) $y = \cot \pi x$      (b) $y = \frac{1}{2}\cot \pi x$      (c) $y = -\frac{1}{2}\cot \pi x$

Solution   (a) Looking at the graph of $y = \cot x$ in Figure 6, we see that one complete pattern or cycle of the graph is bounded by the asymptotes $x = 0$ and $x = \pi$. Now, for the function we are given, $x$ has been replaced by $\pi x$. Thus, the corresponding asymptotes occur when $\pi x = 0$ and when $\pi x = \pi$; in other words, when $x = 0$ and when $x = 1$. See Figure 7(a).

(b) The graph of $y = \frac{1}{2}\cot \pi x$ will have the same general shape as that of $y = \cot \pi x$, but each $y$-coordinate on $y = \frac{1}{2}\cot \pi x$ will be half of the corresponding coordinate on $y = \cot \pi x$. See Figure 7(b).

(c) The graph of $y = -\frac{1}{2}\cot \pi x$ is obtained by reflecting the graph of $y = \frac{1}{2}\cot \pi x$ about the $x$-axis. See Figure 7(c).

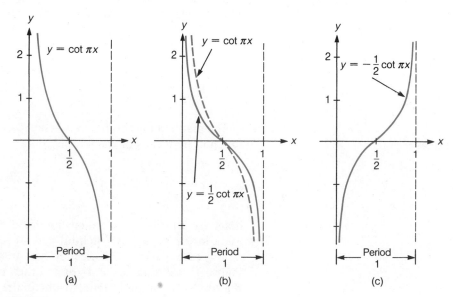

FIGURE 7

We conclude this section with a discussion of the graphs of $y = \csc x$ and $y = \sec x$. We will obtain these graphs in a series of easy steps, relying on the ideas of symmetry and translation. First consider the function $y = \csc x$. In Figure 8(a), we've set up a table and used it to sketch the graph of $y = \csc x$ on the interval $[\pi/2, \pi)$. Note that $\csc x$ is undefined when $x = \pi$. (Why?) As Figure 8(a) indicates, the vertical line $x = \pi$ is an asymptote for the graph. The graph in Figure 8(a) can be extended to the interval $(0, \pi)$ by means of the identity

$$\csc\left(\frac{\pi}{2} + s\right) = \csc\left(\frac{\pi}{2} - s\right)$$

(Exercise 34 asks you to verify this identity.) This identity tells us that, starting at $x = \pi/2$, whether we travel a distance $s$ to the right or a distance $s$ to the left, the value of $y = \csc x$ is the same. In other words, the graph of $y = \csc x$ is symmetric about the line $x = \pi/2$. In view of this symmetry, we can sketch the graph of $y = \csc x$ on the interval $(0, \pi)$ as shown in Figure 8(b).

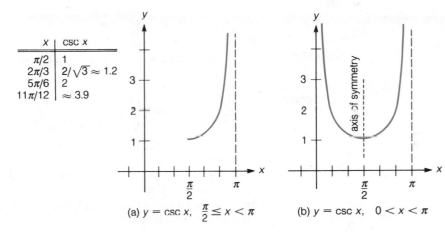

| $x$ | $\csc x$ |
|-----|----------|
| $\pi/2$ | 1 |
| $2\pi/3$ | $2/\sqrt{3} \approx 1.2$ |
| $5\pi/6$ | 2 |
| $11\pi/12$ | $\approx 3.9$ |

(a) $y = \csc x$, $\frac{\pi}{2} \leq x < \pi$

(b) $y = \csc x$, $0 < x < \pi$

FIGURE 8

The next step in obtaining the graph of $y = \csc x$ is to use the fact that the graph is symmetric about the origin. To verify this, we need to check that $\csc(-x) = -\csc x$. We have

$$\csc(-x) = \frac{1}{\sin(-x)}$$

$$= \frac{1}{-\sin x} = -\csc x \qquad \text{as required}$$

Now, taking into account this symmetry about the origin and the portion of the graph that we've already obtained in Figure 8(b), we can sketch the graph of $y = \csc x$ over the interval $(-\pi, \pi)$, as shown in Figure 9.

To complete the graph of $y = \csc x$, we observe that the values of $\csc x$ must repeat themselves at intervals of $2\pi$. This is because $\csc x = 1/(\sin x)$, and the sine function has a period of $2\pi$. In view of this, we can draw the graph of $y = \csc x$ as shown in the box that follows.

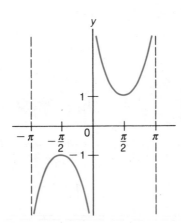

FIGURE 9 $y = \csc x$, $-\pi < x < \pi$
The graph is symmetric about the origin.

---

**PROPERTY SUMMARY**  **The Cosecant Function:**  $y = \csc x$

Domain:  All real numbers other than 0,
  $\pm \pi, \pm 2\pi, \ldots$

Range:  $(-\infty, -1] \cup [1, \infty)$

Period:  $2\pi$ (See Figure 10.)

Asymptotes:  $x = 0, x = \pm \pi, x = \pm 2\pi, \ldots$

Intercepts:  None

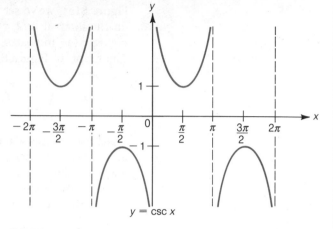

$y = \csc x$

**FIGURE 10**

---

**EXAMPLE 4**  Graph the function $y = \csc \dfrac{x}{3}$ for one period.

**Solution**  Since $\csc \dfrac{x}{3} = \dfrac{1}{\sin(x/3)}$, it will be helpful first to graph one period of $y = \sin(x/3)$. This is done in Figure 11(a) (using the techniques of the previous section). Note that the period of $y = \sin(x/3)$ is $6\pi$. This is also the period of $y = \csc(x/3)$, because $\csc(x/3)$ and $\sin(x/3)$ are just reciprocals. The asymptotes for $y = \csc(x/3)$ occur when $\sin(x/3) = 0$. From Figure 11(a), we see that $\sin(x/3) = 0$ when $x = 0$, when $x = 3\pi$, and when $x = 6\pi$. These asymptotes are sketched in Figure 11(b). The colored points in Figure 11(b) indicate where the value of $\sin(x/3)$ is 1 or $-1$. The graph of $y = \csc(x/3)$ must pass through these points. (Why?) Finally, using the points and the asymptotes in Figure 11(b), we can sketch the graph of $y = \csc(x/3)$, as shown in Figure 11(c).

(a) $y = \sin \dfrac{x}{3}$

(b) The asymptotes
for $y = \csc(x/3)$ occur
when $\sin(x/3) = 0$.

(c) $y = \csc \dfrac{x}{3}$

**FIGURE 11**

The graph of $y = \sec x$ can be obtained simply by translating the graph of $y = \csc x$ a distance of $\pi/2$ units in the negative $x$-direction. This is a consequence of the identity

$$\csc\left(x + \frac{\pi}{2}\right) = \sec x$$

which can be derived as follows.

$$\csc\left(x + \frac{\pi}{2}\right) = \frac{1}{\sin\left(x + \frac{\pi}{2}\right)} = \frac{1}{\sin x \cos\frac{\pi}{2} + \cos x \sin\frac{\pi}{2}}$$

$$= \frac{1}{(\sin x)0 + (\cos x)1} = \sec x$$

After translating the graph of $y = \csc x$ (shown in Figure 10) a distance of $\pi/2$ units in the negative $x$-direction, we obtain the results for $y = \sec x$ that appear in the box that follows.

---

**PROPERTY SUMMARY**        **The Secant Function:** $y = \sec x$

Domain:  All real numbers other than $\pm\pi/2, \pm 3\pi/2$, $\pm 5\pi/2, \ldots$

Range:  $(-\infty, -1] \cup [1, \infty)$

Period:  $2\pi$ (See Figure 12.)

Asymptotes:  $x = \pm\pi/2, x = \pm 3\pi/2$, $x = \pm 5\pi/2, \ldots$

$y$-intercept:  1

$x$-intercept:  None

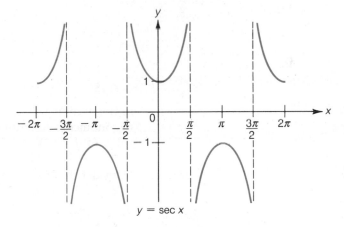

$y = \sec x$

FIGURE 12

---

# EXERCISE SET 8.6

## A

*In Exercises 1–29, graph each function for one period, and show (or specify) the intercepts and asymptotes.*

1.  (a) $y = \tan\left(x + \dfrac{\pi}{4}\right)$

    (b) $y = -\tan\left(x + \dfrac{\pi}{4}\right)$

2.  (a) $y = \tan\left(x - \dfrac{\pi}{3}\right)$

    (b) $y = -\tan\left(x - \dfrac{\pi}{3}\right)$

3.  (a) $y = \tan(x/3)$

    (b) $y = -\tan(x/3)$

4.  (a) $y = 2\tan \pi x$

    (b) $y = -2\tan \pi x$

5.  $y = \frac{1}{2}\tan(\pi x/2)$

6.  $y = -\frac{1}{2}\tan 2\pi x$

7.  $y = \cot(\pi x/2)$

8.  $y = \cot 2\pi x$

9.  $y = -\cot\left(x - \dfrac{\pi}{4}\right)$

10.  $y = \cot\left(x + \dfrac{\pi}{6}\right)$

11. $y = \frac{1}{2} \cot 2x$        12. $y = -\frac{1}{2} \cot(x/2)$

13. $y = \csc\left(x - \dfrac{\pi}{4}\right)$        14. $y = \csc\left(x - \dfrac{\pi}{6}\right)$

15. $y = -\csc(x/2)$        16. $y = 2 \csc x$

17. $y = \frac{1}{3} \csc \pi x$        18. $y = -\frac{1}{2} \csc 2\pi x$

19. $y = -\sec x$        20. $y = -2 \sec x$

21. $y = \sec(x - \pi)$        22. $y = \sec(x + 1)$

23. $y = 3 \sec(\pi x/2)$        24. $y = -2 \sec(\pi x/3)$

**B**

25. $y = 2 \csc(2x + \frac{1}{3}\pi)$   *Hint:*   First graph one period of
    $y = 2 \sin(2x + \frac{1}{3}\pi)$.

26. $y = -\frac{1}{2}\sec(\frac{1}{3}x + \pi)$   27. $y = |\tan x|$

28. $y = |\csc x|$   29. (a) $y = \sec^2 x$
    (b) $y = \tan^2 x + 1$

30. (a) Prove the identity $\sec 2x \csc 2x = 2 \csc 4x$.
    (b) Graph the function $y = \sec 2x \csc 2x$ for one period.

31. (a) Prove the identity
    $$\frac{\cos x - \cos 2x}{\sin x + \sin 2x} = \tan \frac{x}{2}$$

(b) Find the period of the function
$$y = \frac{\cos x - \cos 2x}{\sin x + \sin 2x}$$

32. Find the value of the constant $k$, given that the graph of $y = \tan kx$ passes through the point $(\frac{3}{2}, 1)$ and that $2 < k < 3$.

33. In the text, we obtained the graph of $y = \cot x$ from that of $y = \tan x$ by means of the identity $\cot x = -\tan\left(x - \dfrac{\pi}{2}\right)$. Derive this identity using the fact that
$$\tan\left(x - \frac{\pi}{2}\right) = \frac{\sin\left(x - \dfrac{\pi}{2}\right)}{\cos\left(x - \dfrac{\pi}{2}\right)}$$

34. In the text, we made use of the following identity in graphing $y = \csc x$:
$$\csc\left(\frac{\pi}{2} + s\right) = \csc\left(\frac{\pi}{2} - s\right)$$

(a) How was this identity used?
(b) Prove this identity.

## TRIGONOMETRIC EQUATIONS

**8.7**

In this section, we consider some techniques for solving equations involving the trigonometric functions. As usual, by a **solution** of an equation, we mean a value of the variable for which the equation becomes a true statement.

**EXAMPLE 1**   Consider the trigonometric equation $\sin x + \cos x = 1$. Is $x = \pi/4$ a solution? Is $x = \pi/2$ a solution?

**Solution**   To see if the value $x = \pi/4$ satisfies the given equation, we write

$$\sin\frac{\pi}{4} + \cos\frac{\pi}{4} \overset{?}{=} 1$$

$$\frac{\sqrt{2}}{2} + \frac{\sqrt{2}}{2} \overset{?}{=} 1$$

$$\sqrt{2} \overset{?}{=} 1 \qquad \text{No!}$$

Thus, $x = \pi/4$ is not a solution. In a similar fashion, we can check to see if $x = \pi/2$ is a solution:

$$\sin\frac{\pi}{2} + \cos\frac{\pi}{2} \overset{?}{=} 1$$

$$1 + 0 \overset{?}{=} 1 \qquad \text{Yes!}$$

Thus, $x = \pi/2$ is a solution

The example that we have just concluded points out the difference between trigonometric *identities* and trigonometric *equations*. **Trigonometric identities** are true for all values of the variable in the given domain. **Trigonometric equations**, on the other hand, are only true for certain values of the unknown. Indeed, for some trigonometric equations, there may be no values of the unknown that satisfy the equation. For instance, the equation $\sin \theta = 2$ has no solution, since the value of $\sin \theta$ never exceeds 1. In general, there is no single technique that can be used to solve every trigonometric equation. In the examples that follow, we illustrate some of the more common approaches to solving trigonometric equations.

**EXAMPLE 2**   Solve the equation $\sin \theta = \frac{1}{2}$.

**Solution**   First of all, we know that $\sin(\pi/6) = \frac{1}{2}$. Thus, one solution is certainly $\theta = \pi/6$. To find another solution, we note that $\sin \theta$ is positive in the second quadrant, as well as in the first. In the second quadrant, the angle with a reference angle of $\pi/6$ is $5\pi/6$ (150°). Thus, $\sin(5\pi/6) = \frac{1}{2}$, and we conclude that $\theta = 5\pi/6$ is also a solution to the given equation. (See Figure 1.)

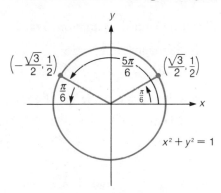

**FIGURE 1**   $\sin \dfrac{\pi}{6} = \sin \dfrac{5\pi}{6} = \dfrac{1}{2}$

Since $\sin \theta$ is negative in the third and fourth quadrants, we needn't look there for solutions. Nevertheless, there are other solutions, because the sine function is periodic, with period $2\pi$. Therefore, since $\theta = \pi/6$ is a solution, we know that

$$\theta = \frac{\pi}{6} + 2\pi$$

is another solution. So is $\theta = (\pi/6) + 4\pi$ a solution, and so on. Similarly, adding multiples of $2\pi$ to $\theta = 5\pi/6$ produces additional solutions. Therefore, the solutions of the equation $\sin \theta = \frac{1}{2}$ are

$$\frac{\pi}{6} + 2\pi k \qquad \text{and} \qquad \frac{5\pi}{6} + 2\pi k$$

where $k$ is any integer. If we wish, we may express these solutions in degrees:

$$30° + 360k° \qquad \text{and} \qquad 150° + 360k°$$

The next example shows how factoring can be used to solve a trigonometric equation.

**EXAMPLE 3**   Solve the equation $\cos^2 x + \cos x - 2 = 0$.

**Solution**   By factoring the expression on the left-hand side of the equation, we obtain

$$(\cos x + 2)(\cos x - 1) = 0$$

Therefore,

$$\cos x = -2 \qquad \text{or} \qquad \cos x = 1$$

We discard the result $\cos x = -2$, since the value of $\cos x$ is never less than $-1$. From the equation $\cos x = 1$, we conclude that $x = 0$ is one solution. After $x = 0$, the next time we have $\cos x = 1$ is when $x = 2\pi$. (You can see this by looking at the graph of $y = \cos x$ or by considering the unit circle definition of the cosine.) In general, then, the solutions of the equation $\cos^2 x + \cos x - 2 = 0$ are given by $x = 2\pi k$, where $k$ is an integer.

In some equations, more than one trigonometric function is present. A common approach here is to express the various functions in terms of a single one. The next example demonstrates this technique.

**EXAMPLE 4**   Find all solutions of the equation $3 \tan^2 x - \sec^2 x - 5 = 0$.

**Solution**   We use the Pythagorean identity

$$\tan^2 x + 1 = \sec^2 x$$

to substitute for $\sec^2 x$ in the given equation. This gives us

$$3 \tan^2 x - (\tan^2 x + 1) - 5 = 0$$
$$2 \tan^2 x - 6 = 0$$
$$2 \tan^2 x = 6$$
$$\tan^2 x = 3$$
$$\tan x = \pm\sqrt{3}$$

Since the period of the tangent function is $\pi$, we need only to find values of $x$ between $0$ and $\pi$ that satisfy $\tan x = \pm\sqrt{3}$. The other solutions will then be obtained by adding multiples of $\pi$ to these solutions. Now, in the first quadrant, the tangent is positive and we know that $\tan x = \sqrt{3}$ when $x = \pi/3$. In the second quadrant, $\tan x$ is negative and we know that $\tan 2\pi/3 = -\sqrt{3}$, since the reference angle for $2\pi/3$ is $\pi/3$. Thus, the solutions between $0$ and $\pi$ are $x = \pi/3$ and $x = 2\pi/3$. It follows that all of the solutions to the equation $3 \tan^2 x - \sec^2 x - 5 = 0$ are given by

$$x = \frac{\pi}{3} + \pi k \qquad \text{and} \qquad x = \frac{2\pi}{3} + \pi k$$

where $k$ is an interger.

The technique used in Example 4, of expressing the various functions in terms of a single function, is most useful when it does not involve introducing a radical expression. For instance, consider the equation $\sin s + \cos s = 1$. Although we could begin by replacing $\cos s$ by the expression $\pm\sqrt{1 - \sin^2 s}$, it turns out to be easier in this situation to begin by squaring both sides of the given equation. This is done in the next example.

**EXAMPLE 5**   Find all solutions of the equation $\sin s + \cos s = 1$ satisfying $0° \leq s < 360°$.

**Solution**   Squaring both sides of the equation yields

$$(\sin s + \cos s)^2 = 1^2$$

$$\underbrace{\sin^2 s + 2 \sin s \cos s + \cos^2 s}_{\text{These add to 1}} = 1$$

Consequently, we have

$$2 \sin s \cos s = 0$$

From this last equation, we conclude that $\sin s = 0$ or $\cos s = 0$. When $\sin s = 0$, we know that $s = 0°$ or $s = 180°$. And when $\cos s = 0$, we know that $s = 90°$ or $s = 270°$. Now we must go back and check which (if any) of these values is a solution to the *original* equation. This must be done whenever we square both sides in the process of solving an equation.

| | | |
|---|---|---|
| $s = 0°$: | $\sin 0° + \cos 0° \overset{?}{=} 1$ | |
| | $0 + 1 \overset{?}{=} 1$ | True |
| $s = 90°$: | $\sin 90° + \cos 90° \overset{?}{=} 1$ | |
| | $1 + 0 \overset{?}{=} 1$ | True |
| $s = 180°$: | $\sin 180° + \cos 180° \overset{?}{=} 1$ | |
| | $0 + (-1) \overset{?}{=} 1$ | False |
| $s = 270°$: | $\sin 270° + \cos 270° \overset{?}{=} 1$ | |
| | $-1 + 0 \overset{?}{=} 1$ | False |

We conclude that the only solutions of the equation $\sin s + \cos s = 1$ on the interval $0° \leq s < 360°$ are $s = 0°$ and $s = 90°$.

In the example that follows, we consider an equation that involves a multiple of the unknown angle. (Methods for solving other types of equations involving multiple angles are indicated in the exercises.)

**EXAMPLE 6**   Solve the equation $\sin 3x = 1$ on the interval $0 \leq x \leq 2\pi$.

**Solution**   We know that $\sin(\pi/2) = 1$. Thus, one solution can be found by writing

$$3x = \frac{\pi}{2}$$

from which we conclude

$$x = \frac{\pi}{6}$$

We can look for other solutions in the required interval by writing, more generally,

$$3x = \frac{\pi}{2} + 2\pi k$$

from which it follows that

$$x = \frac{\pi}{6} + \frac{2\pi k}{3}$$

Thus, when $k = 1$, we obtain

$$x = \frac{\pi}{6} + \frac{2\pi}{3} = \frac{\pi}{6} + \frac{4\pi}{6} = \frac{5\pi}{6}$$

When $k = 2$, we obtain

$$x = \frac{\pi}{6} + \frac{2\pi(2)}{3} = \frac{\pi}{6} + \frac{8\pi}{6}$$

$$= \frac{9\pi}{6} = \frac{3\pi}{2}$$

When $k = 3$, we obtain

$$x = \frac{\pi}{6} + \frac{2\pi(3)}{3}$$

$$= \frac{\pi}{6} + 2\pi \qquad \text{which is greater than } 2\pi$$

We conclude that the solutions of $\sin 3x = 1$ on the interval $0 \le x \le 2\pi$ are $\pi/6$, $5\pi/6$, and $3\pi/2$.

In Example 6, notice that we did not need to make use of a formula for $\sin 3x$, even though the expression $\sin 3x$ did appear in the given equation. In the next example, however, we do make use of the identity $\sin 2x = 2 \sin x \cos x$.

**EXAMPLE 7**   Solve the equation $\sin x \cos x = 1$.

Solution   We could begin by squaring both sides. (Exercise 51 will ask you to use this approach.) However, with the double-angle formula for sine in mind, we can proceed instead as follows. We multiply both sides of the given equation by 2. This yields

$$2 \sin x \cos x = 2$$

and consequently,

$$\sin 2x = 2 \qquad \text{using the double-angle formula}$$

This last equation has no solution, since the value of the sine function never exceeds 1. Thus, the equation $\sin x \cos x = 1$ has no solution.

For the last example in this section, we look at a case in which tables or a calculator are required. As background for this, review Example 10 in Section 7.2.

**EXAMPLE 8**   Find all angles $\theta$ between $0°$ and $360°$ satisfying the equation $\sin \theta = 2 \cos \theta$.

Solution   We first want to rewrite the given equation using a single function. The easiest way to do this is to divide both sides by $\cos \theta$. (Nothing is lost here in assuming that $\cos \theta \ne 0$. If $\cos \theta$ were 0, then $\theta$ would be $90°$ or $270°$; but neither of those angles is a solution of the given equation.) Dividing through by $\cos\theta$, we obtain

$$\frac{\sin\theta}{\cos\theta} = \frac{2\cos\theta}{\cos\theta} \quad \text{or} \quad \tan\theta = 2$$

From experience, we know that none of the angles with which we are familiar (the multiples of 30°and 45°) has a value of 2 for their tangent. Thus, a calculator or table is required at this point. Using a Texas Instruments calculator (set in the degree mode), we find an angle whose tangent is 2 by using the following keystrokes:

2 $\boxed{\text{INV}}$ $\boxed{\text{TAN}}$

By so doing, we find that

$$\theta \approx 63.4°$$

(Even if we were to retain all the decimals displayed by the calculator, we would still have only an approximate solution.) To find another value of $\theta$, we note that the tangent is also positive in the third quadrant. Thus, a second (approximate) solution is given by

$$\theta = 180° + 63.4°$$
$$= 243.4°$$

We conclude that the solutions between 0° and 360° to the equation $\sin\theta = 2\cos\theta$ are approximately 63.4° and 243.4°. Note that the calculator provided only one value for $\theta$; we still needed to work out the second value ourselves using the reference-angle concept.

# EXERCISE SET 8.7

## A

**1.** Is $\theta = \pi/2$ a solution of the following equation?

$$2\cos^2\theta - 3\cos\theta = 0$$

**2.** Is $x = 15°$ a solution of $(\sqrt{3}/3)\cos 2x + \sin 2x = 1$?

**3.** Is $x = 3\pi/4$ a solution of $\tan^2 x - 3\tan x + 2 = 0$?

**4.** Is $t = 2\pi/3$ a solution of $2\sin t + 2\cos t = \sqrt{3} - 1$?

*In Exercises 5–22, determine all solutions of the given equations. Express your answers using radian measure.*

**5.** $\sin\theta = \sqrt{3}/2$

**6.** $\sin\theta = \sqrt{2}/2$

**7.** $\sin\theta = -\frac{1}{2}$

**8.** $\sin\theta + \dfrac{\sqrt{2}}{2} = 0$

**9.** $\cos\theta = -1$

**10.** $\cos\theta = \frac{1}{2}$

**11.** $\tan\theta = \sqrt{3}$

**12.** $\tan\theta + \dfrac{\sqrt{3}}{3} = 0$

**13.** $\tan x = 0$

**14.** $2\sin^2 x - 3\sin x + 1 = 0$

**15.** $2\cos^2\theta + \cos\theta = 0$

**16.** $\sin^2 x - \sin x - 6 = 0$

**17.** $\cos^2 t\sin t - \sin t = 0$

**18.** $\cos\theta + 2\sec\theta = -3$

**19.** $2\cos^2 x - \sin x - 1 = 0$

**20.** $2\cot^2 x + \csc^2 x - 2 = 0$

**21.** $\sqrt{3}\sin t - \sqrt{1 + \sin^2 t} = 0$

**22.** $\sec\alpha + \tan\alpha = \sqrt{3}$

*In Exercises 23–30, determine all solutions in the interval $0° \le \theta < 360°$.*

**23.** $\cos 3\theta = 1$

**24.** $\tan 2\theta = -1$

**25.** $\sin 3\theta = -\sqrt{2}/2$

**26.** $\sin(\theta/2) = \frac{1}{2}$

**27.** $\sin\theta = \cos\dfrac{\theta}{2}$  *Hint:*  $\sin\theta = 2\sin\dfrac{\theta}{2}\cos\dfrac{\theta}{2}$

28. $2 \sin^2 \theta - \cos 2\theta = 0$

29. $\sin 2\theta = \sqrt{3} \cos 2\theta$   *Hint:*   Divide by $\cos 2\theta$.

30. $\sin 2\theta = -2 \cos \theta$

[c] *In Exercises 31–36, use a calculator where necessary to find all solutions in the interval $0° \leq \theta < 360°$.*

31. $\sin \theta = \frac{1}{4}$          32. $3 \cos \theta = 5$

33. $2 \tan \theta = -4$

34. $3 \sin^2 \theta - 2 \sin \theta - 1 = 0$

35. $\cos^2 x - \cos x - 1 = 0$     36. $2 \sin x = 3$

**B**

37. Find all solutions of the equation $\tan 3x - \tan x = 0$ in the interval $0 \leq x < 2\pi$.   *Hint:*   Write $\tan 3x = \tan(2x + x)$ and use the addition formula for tangent.

[c] 38. Find all solutions of the equation $2 \sin x = 1 - \cos x$ in the interval $0° \leq x < 360°$.

39. Find all solutions of the equation $\cos(x/2) = 1 + \cos x$ in the interval $0 \leq x < 2\pi$.

40. Find all solutions of the equation $\sin 3x \cos x + \cos 3x \sin x = \sqrt{3}/2$ in the interval $0 < x < 2\pi$.

41. Find all real numbers $\theta$ for which $\sec 4\theta + 2 \sin 4\theta = 0$.

42. Consider the equation $\sin^2 x - \cos^2 x = \frac{7}{25}$.
    (a) Solve the equation for $\cos x$.
    [c] (b) Find all solutions of the equation satisfying $0 < x < \pi$.

*Each of the equations in Exercises 43–48 can be solved with one of the sum-to-product identities listed on page 398. Find all solutions of the equations in Exercises 43–48 on the interval $0 \leq x \leq 2\pi$.*

43. $\sin 5x = \sin 3x$          44. $\cos 3x = \cos 2x$

45. $\sin 3x = \cos 2x$   *Hint:*   $\cos 2x = \sin\left(\dfrac{\pi}{2} - 2x\right)$

46. $\sin 5x - \sin 3x + \sin x = 0$   *Hint:*   First subtract $\sin x$ from both sides.

47. $\sin\left(x + \dfrac{\pi}{18}\right) = \cos\left(x - \dfrac{2\pi}{9}\right)$

48. $\cos 3x + \cos 2x + \cos x = 0$

[c] 49. Find a solution of the equation $4 \sin \theta - 3 \cos \theta = 2$ in the interval $0° < \theta < 90°$.   *Hint:*   Add $3 \cos \theta$ to both sides, and then square.

50. Find all solutions of the equation $\sin^3 \theta \cos \theta - \sin \theta \cos^3 \theta = -\frac{1}{4}$ in the interval $0 < \theta < \pi$. *Hint:*   Factor the left-hand side, and then use the double-angle formulas.

51. Consider the equation $\sin x \cos x = 1$.
    (a) Square both sides, and then replace $\cos^2 x$ by $1 - \sin^2 x$. Show that the resulting equation can be written $\sin^4 x - \sin^2 x + 1 = 0$.
    (b) Show that the equation $\sin^4 x - \sin^2 x + 1 = 0$ has no solutions. Conclude from this that the original equation has no real-number solutions.

**C**

52. In this exercise, you will see how certain cubic equations can be solved by using the following identity (which we proved in Example 3 in Section 8.4):

$$4 \cos^3 \theta - 3 \cos \theta = \cos 3\theta \qquad (1)$$

For example, suppose that we wish to solve the equation

$$8x^3 - 6x - 1 = 0 \qquad (2)$$

To transform this equation into a form where the stated identity is useful, we make the substitution $x = a \cos \theta$, where $a$ is a constant to be determined. With this substitution, equation (2) can be written

$$8a^3 \cos^3 \theta - 6a \cos \theta = 1 \qquad (3)$$

In equation (3), the coefficient of $\cos^3 \theta$ is $8a^3$. Since we want this coefficient to be 4 [as it is in equation (1)], we divide both sides of equation (3) by $2a^3$ to obtain

$$4 \cos^3 \theta - \frac{3}{a^2} \cos \theta = \frac{1}{2a^3} \qquad (4)$$

Next, a comparison of equations (4) and (1) leads us to require that

$$\frac{3}{a^2} = 3$$

Thus, $a = \pm 1$. For convenience, we choose $a = 1$; equation (4) then becomes

$$4 \cos^3 \theta - 3 \cos \theta = \frac{1}{2} \qquad (5)$$

Comparing equation (5) with the identity in (1) leads us to the equation

$$\cos 3\theta = \frac{1}{2}$$

As you can check, the solutions here are of the form

$$\theta = 20° + 120k° \qquad \text{and} \qquad \theta = 100° + 120k°)$$

Thus,

$$x = \cos(20° + 120k°) \quad \text{and} \quad x = \cos(100° + 120k°)$$

Now, however, as you can again check, only three of the angles yield distinct values for $\cos \theta$, namely, $\theta = 20°$, $\theta = 140°$, and $\theta = 260°$. Thus, the solutions of the equation $8x^3 - 6x - 1 = 0$ are given by $x = \cos 20°$, $x = \cos 140°$, and $x = \cos 260°$.

Use the method just described to solve the following equations.

**(a)** $x^3 - 3x + 1 = 0$

*Answers:* $2\cos 40°$, $-2\cos 20°$, $2\cos 80°$

**(b)** $x^3 - 36x - 72 = 0$

**(c)** $x^3 - 6x + 4 = 0$   *Answers:* $2, -1 \pm \sqrt{3}$

[c] **(d)** $x^3 - 7x - 7 = 0$   (Round off your answers to three decimal places.)

## THE INVERSE TRIGONOMETRIC FUNCTIONS

### 8.8

*This notation* $\cos^{-1} e$ *must not be understood to signify* $\dfrac{1}{\cos e}$.

JOHN HERSCHEL in *Philosophical Transactions of London*, 1813

According to the horizontal line test (page 201), the function $y = \sin x$ is not one-to-one (see Figure 1). Since the function is not one-to-one, there is no inverse function. However, let us consider the **restricted sine function**:

$$y = \sin x \qquad \left( -\frac{\pi}{2} \le x \le \frac{\pi}{2} \right)$$

**FIGURE 1**   $y = \sin x$ is not one-to-one.

As Figure 2 indicates, this restricted sine function is one-to-one according to the horizontal line test.

Since the restricted sine function is one-to-one, the inverse function does exist. There are two notations commonly used to denote this inverse:

$$y = \sin^{-1} x \qquad \text{and} \qquad y = \arcsin x$$

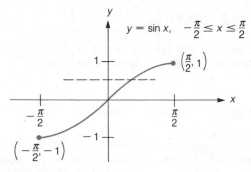

**FIGURE 2**   The restricted sine function is one-to-one.

Initially at least, we will use the notation $y = \sin^{-1}x$. The graph of $y = \sin^{-1}x$ is easily obtained using the fact that the graph of a function and its inverse are reflections of one another about the line $y = x$. Figure 3(a) shows the graph of the restricted sine function and its inverse, $y = \sin^{-1}x$. Figure 3(b) shows the graph of $y = \sin^{-1}x$ alone. From Figure 3(b), we see that the domain of $y = \sin^{-1}x$ is the interval $[-1, 1]$, while the range is $[-\pi/2, \pi/2]$.

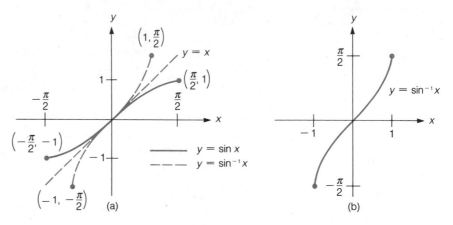

**FIGURE 3**

Values of $\sin^{-1}x$ are computed according to the following rule.

$\sin^{-1}x$ is that number in the interval $[-\pi/2, \pi/2]$ whose sine is $x$.

To see why this is so, we begin with the restricted sine function:

$$y = \sin x \qquad \left(-\frac{\pi}{2} \le x \le \frac{\pi}{2}\right)$$

As explained in Chapter 4, the inverse function is obtained by interchanging $x$ and $y$ (the inputs and the outputs). So, for the inverse function, we have

$$x = \sin y \qquad \left(-\frac{\pi}{2} \le y \le \frac{\pi}{2}\right) \tag{1}$$

Equation (1) tells us that $y$ is the number in the interval $[-\pi/2, \pi/2]$ whose sine is $x$. This is what we wished to show.

**EXAMPLE 1**    Evaluate each of the following.

(a) $\sin^{-1}\left(\frac{1}{2}\right)$      (b) $\sin^{-1}\left(-\frac{1}{2}\right)$

Solution    (a) $\sin^{-1}\left(\frac{1}{2}\right)$ is that number in the interval $[-\pi/2, \pi/2]$ whose sine is $\frac{1}{2}$. Since $\sin(\pi/6) = \frac{1}{2}$, we conclude that $\sin^{-1}\left(\frac{1}{2}\right) = \pi/6$.

(b) Since $\sin(-\pi/6) = -\frac{1}{2}$ and $-\pi/6$ belongs to the interval $[-\pi/2, \pi/2]$, we conclude that $\sin^{-1}\left(-\frac{1}{2}\right) = -\pi/6$.

Using the arcsin notation, we can write the result in Example 1(a) in the form

$$\arcsin \frac{1}{2} = \frac{\pi}{6}$$

Now, as Figure 4 indicates, the arc length with a sine of $\frac{1}{2}$ is $\pi/6$. This is the idea behind the arcsin notation.

**EXAMPLE 2**   Evaluate $\arcsin \frac{3}{4}$.

**Solution**   The quantity $\arcsin \frac{3}{4}$ is that number (or arc length, or angle in radians) in the interval $[-\pi/2, \pi/2]$ whose sine is $\frac{3}{4}$. Since we are not familiar with an angle with a sine of $\frac{3}{4}$, we use a calculator. For instance, using a Texas Instruments calculator, we set the calculator to the radian mode and then use the keystrokes

$$0.75 \quad \boxed{\text{INV}} \quad \boxed{\text{SIN}}$$

The result is

$$\arcsin \frac{3}{4} \approx 0.85$$

**EXAMPLE 3**   Show that the following two expressions are not equal:

$$\sin^{-1} 0 \qquad \text{and} \qquad \frac{1}{\sin 0}$$

**Solution**   The quantity $\sin^{-1} 0$ is that number in the interval $[-\pi/2, \pi/2]$ whose sine is 0. Since $\sin 0 = 0$, we conclude that

$$\sin^{-1} 0 = 0$$

On the other hand, since $\sin 0 = 0$, the expression $1/(\sin 0)$ is not even defined. Thus, the two given expressions certainly are not equal.

If $f$ and $f^{-1}$ are any pair of inverse functions, then by definition,

$$f[f^{-1}(x)] = x \qquad \text{for every } x \text{ in the domain of } f^{-1}$$

and

$$f^{-1}[f(x)] = x \qquad \text{for every } x \text{ in the domain of } f$$

Applying these facts to the restricted sine function, $y = \sin x$, and its inverse, $y = \sin^{-1} x$, we obtain the following two identities.

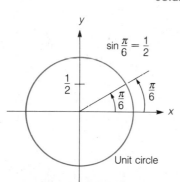

$sin \dfrac{\pi}{6} = \dfrac{1}{2}$

Unit circle

**FIGURE 4**

$$\sin(\sin^{-1} x) = x \qquad (-1 \le x \le 1)$$

$$\sin^{-1}(\sin x) = x \qquad \left( -\frac{\pi}{2} \le x \le \frac{\pi}{2} \right)$$

The following example indicates that the domain restrictions accompanying these two identities cannot be disregarded.

**EXAMPLE 4**    Compute each of the following that is defined.

(a) $\sin^{-1}\left(\sin\dfrac{\pi}{4}\right)$      (b) $\sin^{-1}(\sin\pi)$

(c) $\sin(\sin^{-1}2)$      (d) $\sin\left[\sin^{-1}\left(-\dfrac{1}{\sqrt{5}}\right)\right]$

**Solution**    (a) Since $\pi/4$ lies in the domain of the restricted sine function, the identity $\sin^{-1}(\sin x) = x$ is applicable here. Thus,

$$\sin^{-1}\left(\sin\frac{\pi}{4}\right) = \frac{\pi}{4}$$

CHECK    $\sin(\pi/4) = \sqrt{2}/2$. Therefore, $\sin^{-1}(\sin\pi/4) = \sin^{-1}(\sqrt{2}/2) = \pi/4$.

(b) The number $\pi$ is not in the domain of the restricted sine function, so the identity $\sin^{-1}(\sin x) = x$ does not apply in this case. However, since $\sin\pi = 0$, we have

$$\sin^{-1}(\sin\pi) = \sin^{-1}0 = 0$$

Thus, $\sin^{-1}(\sin\pi)$ is equal to 0, not $\pi$.

(c) The number 2 is not in the domain of the inverse sine function. Thus, the expression $\sin(\sin^{-1}2)$ is undefined.

(d) The identity $\sin(\sin^{-1}x) = x$ is applicable here. (Why?) Thus,

$$\sin\left[\sin^{-1}\left(-\frac{1}{\sqrt{5}}\right)\right] = -\frac{1}{\sqrt{5}}$$

To define the inverse cosine function, we begin by defining the **restricted cosine function**:

$$y = \cos x \qquad (0 \le x \le \pi)$$

As indicated by the horizontal line test in Figure 5, this function is one-to-one. Since the restricted cosine function is one-to-one, the inverse function exists. We denote this inverse by

$$y = \cos^{-1}x \qquad \text{or} \qquad y = \arccos x$$

The graph of $y = \cos^{-1}x$ is obtained by reflecting the restricted cosine function about the line $y = x$. Figure 6(a) displays the graph of the restricted cosine function along with $y = \cos^{-1}x$. Figure 6(b) shows the graph of $y = \cos^{-1}x$ alone. From Figure 6(b), we see that the domain of $y = \cos^{-1}x$ is the interval $[-1, 1]$, while the range is $[0, \pi]$. By reasoning the way we did for the inverse sine, we obtain the following results for the inverse cosine.

**FIGURE 5**

$\cos^{-1}x$ is that number in the interval $[0, \pi]$ whose cosine is $x$.

$$\cos(\cos^{-1}x) = x \qquad (-1 \le x \le 1)$$
$$\cos^{-1}(\cos x) = x \qquad (0 \le x \le \pi)$$

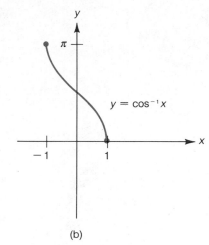

(a)

(b)

**FIGURE 6**

**EXAMPLE 5**   Compute each of the following.

(a) $\cos^{-1}(0)$      (b) $\arccos(\cos\frac{2}{5})$

Solution   (a) $\cos^{-1}(0)$ is that number in the interval $[0,\pi]$ whose cosine is 0. Since $\cos(\pi/2) = 0$, we have

$$\cos^{-1}(0) = \frac{\pi}{2}$$

(b) Since the number $\frac{2}{5}$ is in the domain of the restricted cosine function, the identity $\arccos(\cos x) = x$ is applicable. Thus,

$$\arccos\left(\cos\frac{2}{5}\right) = \frac{2}{5}$$

**EXAMPLE 6**   Show that $\sin(\cos^{-1} x) = \sqrt{1 - x^2}$ for $-1 \le x \le 1$.

Solution   We use the identity

$$\sin y = \sqrt{1 - \cos^2 y}, \qquad \text{which is valid for } 0 \le y \le \pi$$

Substituting $\cos^{-1} x$ for $y$ in this identity, we obtain

$$\sin(\cos^{-1} x) = \sqrt{1 - [\cos(\cos^{-1} x)]^2}$$

or

$$\sin(\cos^{-1} x) = \sqrt{1 - x^2} \qquad \text{as required}$$

Before leaving Example 6, we point out an alternate method of solution that is useful when the restriction on $x$ is $0 < x < 1$. In this case, we let $\theta = \cos^{-1} x$. Then $\theta$ is the radian measure of the acute angle whose cosine is $x$ and we can sketch $\theta$ as shown in Figure 7. The sides of the triangle in Figure 7 have been labeled in such a way that $\cos\theta = x$. Then by the Pythagorean Theorem

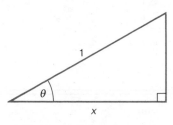

**FIGURE 7**

we find that the third side of the triangle is $\sqrt{1-x^2}$. We therefore have

$$\sin\theta = \frac{\text{opposite}}{\text{hypotenuse}} = \frac{\sqrt{1-x^2}}{1} = \sqrt{1-x^2}$$

Now let us turn to the definition of the inverse tangent function. We begin by defining the **restricted tangent function**:

$$y = \tan x \qquad \left(-\frac{\pi}{2} < x < \frac{\pi}{2}\right)$$

Figure 8 shows the graph of this function. As you can check by applying the horizontal line test, the restricted tangent function is one-to-one; therefore the inverse function exists. We denote the inverse tangent function by

$$y = \tan^{-1} x \qquad \text{or} \qquad y = \arctan x$$

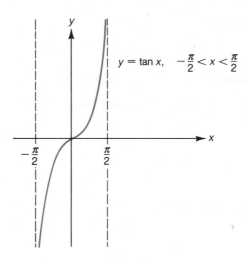

$$y = \tan x, \quad -\frac{\pi}{2} < x < \frac{\pi}{2}$$

**FIGURE 8**

The graph of $y = \tan^{-1} x$ is obtained by reflecting the restricted tangent function about the line $y = x$. Figure 9(a) shows the graphs of the restricted tangent

$$y = \tan^{-1} x$$

$$—— \quad y = \tan x$$

$$---- \quad y = \tan^{-1} x$$

(a)                    (b)

**FIGURE 9**

function and its inverse, while Figure 9(b) shows the graph of $y = \tan^{-1} x$ alone. From Figure 9(b), we see that the domain of $y = \tan^{-1} x$ is the set of all real numbers, while the range is the open interval $(-\pi/2, \pi/2)$. Applying the same type of reasoning used for the inverse sine, we obtain the results in the box that follows.

> $\tan^{-1} x$ is that number in the interval $(-\pi/2, \pi/2)$ whose tangent is $x$.
>
> $$\tan(\tan^{-1} x) = x \qquad \text{(for all } x\text{)}$$
> $$\tan^{-1}(\tan x) = x \qquad \left(-\frac{\pi}{2} < x < \frac{\pi}{2}\right)$$

**EXAMPLE 7**  Compute the following.

(a) $\tan^{-1}(-1)$    (b) $\tan(\tan^{-1}\sqrt{5})$

Solution    (a) The quantity $\tan^{-1}(-1)$ is that number in the interval $(-\pi/2, \pi/2)$ whose tangent is $-1$. Since $\tan(-\pi/4) = -1$, we have

$$\tan^{-1}(-1) = -\frac{\pi}{4}$$

(b) The identity $\tan(\tan^{-1} x) = x$ holds for all real numbers $x$. We therefore have

$$\tan(\tan^{-1}\sqrt{5}) = \sqrt{5}$$

When suitable restrictions are placed on the domains of the cosecant, secant, and cotangent functions, corresponding inverse functions can be defined. However, since these three inverses are rarely, if ever, encountered in calculus, we omit a discussion of them here. Instead, we look at three additional examples involving the inverse sine and the inverse tangent.

**EXAMPLE 8**  Evaluate $\cos\left(\sin^{-1}\frac{3}{5}\right)$.

Solution    Let $\theta = \sin^{-1}(\frac{3}{5})$. Then $\theta$ is the measure of the acute angle whose sine is $\frac{3}{5}$, as indicated in Figure 10. Using the Pythagorean Theorem, we find that the length of the third side of the triangle is 4. Thus,

$$\cos\theta = \frac{\text{adjacent}}{\text{hypotenuse}} = \frac{4}{5}$$

In view of the definition of $\theta$, this last equation states that

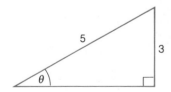

FIGURE 10

$$\cos\left(\sin^{-1}\frac{3}{5}\right) = \frac{4}{5}$$

This is the required answer.

**EXAMPLE 9** If $\sin\theta = x/3$ and $0 < \theta < \pi/2$, express the following quantity as a function of $x$:

$$\theta - \sin 2\theta$$

**Solution** The given conditions tell us that $\theta = \sin^{-1}(x/3)$. That expresses $\theta$ in terms of $x$. To express $\sin 2\theta$ in terms of $x$, we have

$$\sin 2\theta = 2\sin\theta\cos\theta$$

$$= 2\sin\theta\sqrt{1 - \sin^2\theta}$$

$$= 2\left(\frac{x}{3}\right)\sqrt{1 - \frac{x^2}{9}} = \frac{2x}{3}\sqrt{\frac{9 - x^2}{9}}$$

$$= \frac{2x\sqrt{9 - x^2}}{9}$$

Now, combining the results, we have

$$\theta - \sin 2\theta = \sin^{-1}\frac{x}{3} - \frac{2x\sqrt{9 - x^2}}{9}$$

**EXAMPLE 10** Show that $\pi = 4[\tan^{-1}(\frac{1}{2}) + \tan^{-1}(\frac{1}{5}) + \tan^{-1}(\frac{1}{8})]$.

**Solution** The given equation is equivalent to

$$\underbrace{\frac{\pi}{4} - \tan^{-1}\frac{1}{2}}_{A} = \underbrace{\tan^{-1}\frac{1}{5} + \tan^{-1}\frac{1}{8}}_{B}$$

Then, using the letters $A$ and $B$ as indicated, we wish to show that $A = B$. We will accomplish this if we can show that $\tan A = \tan B$, since both $A$ and $B$ lie between 0 and $\pi/2$. (Why?) Using the formula

$$\tan(x - y) = \frac{\tan x - \tan y}{1 + \tan x \tan y}$$

we have

$$\tan A = \tan\left(\frac{\pi}{4} - \tan^{-1}\frac{1}{2}\right)$$

$$= \frac{\tan\dfrac{\pi}{4} - \tan\left(\tan^{-1}\dfrac{1}{2}\right)}{1 + \tan\dfrac{\pi}{4}\tan\left(\tan^{-1}\dfrac{1}{2}\right)}$$

$$= \frac{1 - \frac{1}{2}}{1 + 1(\frac{1}{2})} = \frac{\frac{1}{2}}{\frac{3}{2}} = \frac{1}{3}$$

Similarly, using the formula

$$\tan(x + y) = \frac{\tan x + \tan y}{1 - \tan x \tan y}$$

we have

$$\tan B = \tan\left(\tan^{-1}\frac{1}{5} + \tan^{-1}\frac{1}{8}\right)$$

$$= \frac{\tan(\tan^{-1}\frac{1}{5}) + \tan(\tan^{-1}\frac{1}{8})}{1 - [\tan(\tan^{-1}\frac{1}{5})][\tan(\tan^{-1}\frac{1}{8})]}$$

$$= \frac{\frac{1}{5} + \frac{1}{8}}{1 - (\frac{1}{5})(\frac{1}{8})} = \frac{1}{3} \qquad \text{(Check the arithmetic.)}$$

We have now shown that $\tan A = \tan B$, since both quantities are equal to $\frac{1}{3}$. It now follows that $A = B$, as we wished to show. ▪

## EXERCISE SET 8.8

### A

*In Exercises 1–20, evaluate each of the quantities that is defined, without using a calculator or tables. If a quantity is undefined, say so.*

1. $\sin^{-1}(\sqrt{3}/2)$
2. $\cos^{-1}(-1)$
3. $\tan^{-1}\sqrt{3}$
4. $\arccos(-\sqrt{2}/2)$
5. $\arctan(-1/\sqrt{3})$
6. $\arcsin(-1)$
7. $\tan^{-1}1$
8. $\sin^{-1}0$
9. $\cos^{-1}2\pi$
10. $\arctan 0$
11. $\sin(\sin^{-1}\frac{1}{4})$
12. $\cos(\cos^{-1}\frac{4}{3})$
13. $\cos(\cos^{-1}\frac{3}{4})$
14. $\tan(\arctan 3\pi)$
15. $\arctan[\tan(-\pi/7)]$
16. $\sin(\arcsin 2)$
17. $\arcsin[\sin(\pi/2)]$
18. $\arccos[\cos(\pi/8)]$
19. $\arccos(\cos 2\pi)$
20. $\sin^{-1}[\sin(3\pi/2)]$

*In Exercises 21–30, evaluate the given quantities without using a calculator or tables.*

21. $\tan(\sin^{-1}\frac{4}{5})$
22. $\cos(\arcsin\frac{2}{7})$
23. $\sin(\tan^{-1}1)$
24. $\sin[\tan^{-1}(-1)]$
25. $\tan(\arccos\frac{5}{13})$
26. $\cos(\sin^{-1}\frac{2}{3})$
27. $\cos(\arctan\sqrt{3})$
28. $\sin(\cos^{-1}\frac{1}{3})$
29. $\sin[\arccos(-\frac{1}{3})]$
30. $\tan(\arcsin\frac{20}{29})$

c 31. Use a calculator to evaluate each of the following quantities. Express your answers to two decimal places; don't round off.
  (a) $\sin^{-1}(\frac{3}{4})$    (b) $\cos^{-1}(\frac{2}{3})$
  (c) $\tan^{-1}\pi$    (d) $\tan^{-1}(\tan^{-1}\pi)$

*In Exercises 32–34, evaluate the given expressions without using a calculator or tables.*

32. $\csc[\sin^{-1}(\frac{1}{2}) - \cos^{-1}(\frac{1}{2})]$
33. $\sec[\cos^{-1}(\sqrt{2}/2) + \sin^{-1}(-1)]$
34. $\cot[\cos^{-1}(-1/2) + \cos^{-1}(0) + \tan^{-1}(1/\sqrt{3})]$
35. Show that $\cos(\sin^{-1}x) = \sqrt{1-x^2}$ for $-1 \le x \le 1$. *Suggestion:* Use the method of Example 6 in the text.
36. If $\sin\theta = 2x$ and $0 < \theta < \pi/2$, express $\theta + \cos 2\theta$ as a function of $x$.
37. If $\sin\theta = 3x/2$ and $0 < \theta < \pi/2$, express $(\theta/4) - \sin 2\theta$ as a function of $x$.
38. If $\cos\theta = x - 1$ and $0 < \theta < \pi/2$, express $2\theta - \cos 2\theta$ as a function of $x$.
39. If $\tan\theta = \frac{1}{2}(x-1)$ and $0 < \theta < \pi/2$, express $\theta - \cos\theta$ as a function of $x$.
40. If $\tan\theta = \frac{1}{3}(x+1)$ and $0 < \theta < \pi/2$, express $2\theta + \tan 2\theta$ as a function of $x$.

### B

41. Evaluate $\sin(2\tan^{-1}4)$.  *Hint:* $\sin 2\theta = 2\sin\theta\cos\theta$
42. Evaluate $\cos[2\sin^{-1}(\frac{5}{13})]$.
43. Evaluate $\sin(\arccos\frac{3}{5} - \arctan\frac{7}{13})$. *Suggestion:* Use the formula for $\sin(x-y)$.
44. Show that $\sin\left(\sin^{-1}\frac{1}{3} + \sin^{-1}\frac{1}{4}\right) = \frac{\sqrt{15}+2\sqrt{2}}{12}$.
45. Evaluate $\cos(\tan^{-1}2 + \tan^{-1}3)$.
46. (a) Explain why $0 < \tan^{-1}(\frac{1}{3}) + \tan^{-1}(\frac{1}{2}) < \pi/2$.
    (b) Show that $\pi/4 = \tan^{-1}(\frac{1}{3}) + \tan^{-1}(\frac{1}{2})$.
47. Show that $\tan^{-1}(\frac{4}{3}) - \tan^{-1}(\frac{1}{7}) = \pi/4$.
48. Show that $\sin^{-1}x + \cos^{-1}x = \pi/2$. *Suggestion:* Let $\theta = \sin^{-1}x$ and $\beta = \cos^{-1}x$. First explain why $-\pi/2 \le \theta + \beta \le 3\pi/2$. Then show $\sin(\theta + \beta) = 1$.
49. If $-1 < x < 1$, show that $\sin^{-1}x = \tan^{-1}(x/\sqrt{1-x^2})$.

50. Show that $\arctan x + \arctan y = \arctan \dfrac{x+y}{1-xy}$ when $x$ and $y$ are positive and $xy < 1$.

51. Show that $\tan^{-1}(\frac{1}{7}) + 2\tan^{-1}(\frac{1}{3}) = \pi/4$.

52. Evaluate $\arcsin \frac{4}{5} + \arctan \frac{3}{4}$.

53. Show that $\sin(2\sin^{-1}x) = 2x\sqrt{1-x^2}$ for $-1 \le x \le 1$.

54. Let $f(x) = 2\tan^{-1}\left(\dfrac{1+x}{1-x}\right) + \sin^{-1}\left(\dfrac{1-x^2}{1+x^2}\right)$
$(0 < x < 1)$.

 ©  (a) Set your calculator in the radian mode and complete the table. (Record your answers using the same number of decimal places as is shown on your calculator.)

| $x$ | 0.1 | 0.2 | 0.3 | 0.4 | 0.5 |
|-----|-----|-----|-----|-----|-----|
| $f(x)$ | | | | | |

(b) On the basis of your results in part (a), make a conjecture about the function $f$. Prove that your conjecture is correct. *Hint:* For the proof, compute $\sin[f(x)]$.

## C

55. Show that $\tan^{-1}1 + \tan^{-1}2 + \tan^{-1}3 = \pi$.

56. Suppose that $\arccos A + \arccos B + \arccos C = \pi$. Show that $A^2 + B^2 + C^2 + 2ABC = 1$.

57. (a) Show that $2\tan^{-1}x = \tan^{-1}[2x/(1-x^2)]$ for $0 \le x < 1$.
(b) Show that $4\tan^{-1}(\frac{1}{5}) = \tan^{-1}(\frac{120}{119})$.
(c) Show that $4\tan^{-1}(\frac{1}{5}) - \tan^{-1}(\frac{1}{239}) = \pi/4$. (In 1706, John Machin made use of this formula in calculating the value of $\pi$ to 100 decimal places.)

---

## CHAPTER 8  SUMMARY OF PRINCIPAL FORMULAS AND TERMS

| FORMULAS OR TERMS | PAGE REFERENCE | COMMENT |
|---|---|---|
| 1. $\theta = \dfrac{s}{r}$ | 368 | This equation defines the *radian measure* $\theta$ of an angle. We assume here that the vertex of the angle is placed at the center of a circle of radius $r$ and that $s$ is the length of the intercepted arc. See Figure 1 in Section 8.1. |
| 2. $s = r\theta$ | 372 | This formula expresses the arc length $s$ on a circle in terms of the radius $r$ and the radian measure $\theta$ of the central angle subtended by the arc. |
| 3. $A = \frac{1}{2}r^2\theta$ | 373 | This formula expresses the area of a sector of a circle in terms of the radius $r$ and the radian measure $\theta$ of the central angle. |
| 4. $\sin\theta$ $\cos\theta$ $\tan\theta$ $\csc\theta$ $\sec\theta$ $\cot\theta$ | 377 | Let $\theta$ denote the radian measure of an angle in standard position, and let $P(x,y)$ denote the point where the terminal side of the angle intersects the unit circle. Or, equivalently, let $\theta$ denote the measure of the corresponding arc length on the unit circle. Then the values of $\theta$ are real numbers, and we define the six trigonometric functions of $\theta$ as follows: $\sin\theta = y$   $\csc\theta = \dfrac{1}{y}$ $(y \ne 0)$ $\cos\theta = x$   $\sec\theta = \dfrac{1}{x}$ $(x \ne 0)$ $\tan\theta = \dfrac{y}{x}$ $(x \ne 0)$   $\cot\theta = \dfrac{x}{y}$ $(y \ne 0)$ |

| FORMULAS OR TERMS | PAGE REFERENCE | COMMENT |
|---|---|---|
| 5. $\csc\theta = \dfrac{1}{\sin\theta}$ <br><br> $\sec\theta = \dfrac{1}{\cos\theta}$ <br><br> $\cot\theta = \dfrac{1}{\tan\theta}$ <br><br> $\tan\theta = \dfrac{\sin\theta}{\cos\theta}$ <br><br> $\cot\theta = \dfrac{\cos\theta}{\sin\theta}$ | 378 | These five identities are direct consequences of the definitions of the trigonometric functions. |
| 6. $\sin^2\theta + \cos^2\theta = 1$ <br> $\tan^2\theta + 1 = \sec^2\theta$ <br> $\cot^2\theta + 1 = \csc^2\theta$ | 378 | These are the three *Pythagorean identities*. |
| 7. $\cos(-\theta) = \cos\theta$ <br> $\sin(-\theta) = -\sin\theta$ <br> $\tan(-\theta) = -\tan\theta$ | 380 | These three identities are sometimes referred to as the *opposite angle identities*. In each case, a trigonometric function of $-\theta$ is expressed in terms of that same function of $\theta$. |
| 8. $\sin(s + t) = \sin s\cos t + \cos s\sin t$ <br> $\sin(s - t) = \sin s\cos t - \cos s\sin t$ <br> $\cos(s + t) = \cos s\cos t - \sin s\sin t$ <br> $\cos(s - t) = \cos s\cos t + \sin s\sin t$ | 385 | These identities are referred to as the *addition formulas* for sine and cosine. |
| 9. $\cos\left(\dfrac{\pi}{2} - \theta\right) = \sin\theta$ <br><br> $\sin\left(\dfrac{\pi}{2} - \theta\right) = \cos\theta$ | 388 | These two identities, called *reduction formulas*, hold for all real numbers $\theta$. In terms of angles, the identities state that the sine of an angle is equal to the cosine of its complement. |
| 10. $\tan(s + t) = \dfrac{\tan s + \tan t}{1 - \tan s\tan t}$ <br><br> $\tan(s - t) = \dfrac{\tan s - \tan t}{1 + \tan s\tan t}$ | 392 | These two identities are referred to as the *addition formulas* for tangent. |
| 11. $\sin 2\theta = 2\sin\theta\cos\theta$ <br> $\cos 2\theta = \cos^2\theta - \sin^2\theta$ <br><br> $\tan 2\theta = \dfrac{2\tan\theta}{1 - \tan^2\theta}$ | 392 | These identities are known as the *double-angle formulas*. There are four other forms of the double-angle formula for cosine. They appear in the box on page 395. |
| 12. $\sin\dfrac{s}{2} = \pm\sqrt{\dfrac{1 - \cos s}{2}}$ <br><br> $\cos\dfrac{s}{2} = \pm\sqrt{\dfrac{1 + \cos s}{2}}$ <br><br> $\tan\dfrac{s}{2} = \dfrac{\sin s}{1 + \cos s}$ | 395 | These three identities are referred to as the *half-angle formulas*. In the case of the half-angle formulas for sine and cosine, the sign before the radical is determined by the quadrant in which the angle or arc terminates. |

| FORMULAS OR TERMS | PAGE REFERENCE | COMMENT |
|---|---|---|
| **13.** Periodic function | 401–402 | A function $f$ is periodic if there is a positive number $p$ such that the equation $f(x + p) = f(x)$ holds for all $x$ in the domain of $f$. The smallest such number $p$ is called the *period* of $f$. Important examples: The period of $y = \sin x$ is $2\pi$; the period of $y = \cos x$ is $2\pi$; the period of $y = \tan x$ is $\pi$. |
| **14.** Amplitude | 402 | Let $m$ and $M$ denote the smallest and the largest values, respectively, of the periodic function $f$. Then the amplitude of $f$ is defined to be the number $\frac{1}{2}(M - m)$. Important examples: The amplitude for both $y = \sin x$ and $y = \cos x$ is 1; amplitude is not defined for the function $y = \tan x$. |
| **15.** Period $= \dfrac{2\pi}{B}$ | 407 | This formula gives the period for the functions $y = A \sin(Bx + C)$ and $y = A \cos(Bx + C)$. |
| **16.** Phase shift $= \dfrac{-C}{B}$ | 410 | The phase shift serves as a guide in graphing functions of the form $y = A \sin(Bx + C)$ or $y = A \cos(Bx + C)$. For instance, to graph one complete cycle of $y = A \sin(Bx + C)$, first sketch one complete cycle of $y = A \sin Bx$, beginning at $x = 0$. Then draw a curve with exactly the same shape, but beginning at $x = -C/B$ rather than $x = 0$. This will represent one cycle of $y = A \sin(Bx + C)$. |
| **17.** Restricted sine function | 429 | The domain of the function $y = \sin x$ is the set of all real numbers. By allowing inputs only from the closed interval $[-\pi/2, \pi/2]$, we obtain the restricted sine function. |
| **18.** $\sin^{-1} x$ | 430 | $\sin^{-1} x$ is that number in the interval $[-\pi/2, \pi/2]$ whose sine is $x$. An alternate, but entirely equivalent, form of this expression is $\arcsin x$. The function defined by $y = \sin^{-1} x$ is the inverse function for the restricted sine function. |
| **19.** Restricted cosine function | 432 | The domain of the function $y = \cos x$ is the set of all real numbers. By allowing only inputs from the closed interval $[0, \pi]$, we obtain the restricted cosine function. |
| **20.** $\cos^{-1} x$ | 432 | $\cos^{-1} x$ is that number in the closed interval $[0, \pi]$ whose cosine is $x$. An alternate, but entirely equivalent, expression is $\arccos x$. The function defined by $y = \cos^{-1} x$ is the inverse function for the restricted cosine function. |
| **21.** Restricted tangent function | 434 | The restricted tangent function is defined by the equation $y = \tan x$ for $-\pi/2 < x < \pi/2$. |
| **22.** $\tan^{-1} x$ | 435 | $\tan^{-1} x$ is that number in the open interval $(-\pi/2, \pi/2)$ whose tangent is $x$. An alternate, but entirely equivalent, form of this expression is $\arctan x$. The function defined by $y = \tan^{-1} x$ is the inverse function for the restricted tangent function. |

# CHAPTER 8   REVIEW EXERCISES

NOTE   The fifteen exercises in this set with colored problem numbers constitute a chapter test.

*In Exercises 1–8, convert the radian measures to degrees.*

1. $3\pi/4$
2. $\pi/15$
3. $5\pi$
4. $\pi/3$
5. $5\pi/6$
6. $1$
7. $2$
8. $180/\pi$

*In Exercises 9–16, convert the degree measures to radians.*

9. $360°$
10. $36°$
11. $1°$
12. $330°$
13. $7.5°$
14. $\pi°$
15. $(1/\pi)°$
16. $(\pi/180)°$

*In Exercises 17–28, evaluate each expression without using a calculator or tables.*

17. $\cos\pi$
18. $\sin(-3\pi/2)$
19. $\csc(2\pi/3)$
20. $\tan(\pi/3)$
21. $\cot(11\pi/6)$
22. $\cos 0$
23. $\sin(\pi/6)$
24. $\sec(3\pi/4)$
25. $\cot(5\pi/4)$
26. $\tan(-7\pi/4)$
27. $\csc(-5\pi/6)$
28. $\sin^2(\pi/7) + \cos^2(\pi/7)$

*In Exercises 29–36, use the arc length formula $s = r\theta$ and the sector area formula $A = \frac{1}{2}r^2\theta$ to find the required quantities.*

|  | GIVEN | FIND |
|---|---|---|
| 29. | $\theta = \pi/8, r = 16$ cm | $s$ and $A$ |
| 30. | $\theta = 120°, r = 12$ cm | $s$ and $A$ |
| 31. | $r = s = 1$ cm | $A$ |
| 32. | $r = s = \sqrt{3}$ cm | $\theta$ |
| 33. | $s = 4$ cm, $\theta = 36°$ | $r$ and $A$ |
| 34. | $\theta = \pi/10, A = 200\pi$ cm$^2$ | $r$ |
| 35. | $s = 12$ cm, $\theta = r + 1$ | $r$ and $\theta$ |
| 36. | $\theta = 50°, A = 20\pi$ cm$^2$ | $s$ |

37. Two angles in a triangle are $40°$ and $70°$. What is the radian measure of the third angle?

38. The radian measures of two angles in a triangle are $\frac{1}{6}$ and $\frac{5}{12}$. What is the radian measure of the third angle?

[c] *Exercises 39–48 are calculator exercises. (Set your calculator to the radian mode.) In Exercises 39–44, where numerical answers are required, round off your results to three decimal places.*

39. Evaluate $\sin 1$.
40. Evaluate $\cos 2$.

41. Evaluate $\sin(3\pi/2)$.
42. Evaluate $\sin(0.78)$.
43. Evaluate $\sin(\sin 0.0123)$.
44. Evaluate $\sin[\sin(\sin 0.0123)]$.
45. Verify that $\sin^2 1986 + \cos^2 1986 = 1$.
46. Verify that $\sin 14 = 2\sin 7\cos 7$.
47. Verify that $\cos(0.5) = \cos^2(0.25) - \sin^2(0.25)$.
48. Verify that $\cos(0.3) = [\frac{1}{2}(1 + \cos 0.6)]^{1/2}$.
49. In the expression $\sqrt{25 - x^2}$, make the substitution $x = 5\sin\theta$ $(0 < \theta < \pi/2)$ and simplify the result.
50. In the expression $(49 + x^2)^{1/2}$, make the substitution $x = 7\tan\theta$ $(0 < \theta < \pi/2)$ and simplify the result.
51. In the expression $(x^2 - 100)^{1/2}$, make the substitution $x = 10\sec\theta$ $(0 < \theta < \pi/2)$ and simplify the result.
52. In the expression $(x^2 - 4)^{-3/2}$, make the substitution $x = 2\sec\theta$ $(0 < \theta < \pi/2)$ and simplify the result.
53. In the expression $(x^2 + 5)^{-1/2}$, make the substitution $x = \sqrt{5}\tan\theta$ $(0 < \theta < \pi/2)$ and simplify the result.
54. If $\sin\theta = -\frac{5}{13}$ and $\pi < \theta < 3\pi/2$, compute $\cos\theta$.
55. If $\sin\theta = \frac{3}{5}$, $\tan\theta$ is positive, and $0 < \theta < 2\pi$, compute $\sin(\theta/2)$.
56. If $\cos\theta = \frac{8}{17}$, $\sin\theta$ is negative, and $0 < \theta < 2\pi$, compute $\tan\theta$.
57. If $\tan\theta = \frac{7}{24}$, $\cos\theta$ is positive, and $0 < \theta < 2\pi$, compute $\cos 2\theta$.
58. If $\sec\theta = -\frac{25}{7}$ and $\pi < \theta < 3\pi/2$, compute $\cot\theta$.
59. If $\csc\theta = \frac{29}{20}$, $\cos\theta$ is negative, and $0 < \theta < 2\pi$, compute $\cos(\theta/2)$.
60. If $\cos\theta = \frac{3}{5}$, $\sin\theta$ is positive, and $0 < \theta < 2\pi$, compute (a) $\sin 3\theta$;   (b) $\sin(\theta/4)$.
61. If $\tan 2\theta = \frac{7}{24}$ and $\pi < 2\theta < 3\pi/2$, compute $\sin\theta$.
62. Express the area of the shaded region in the following figure in terms of $r$ and $\theta$.

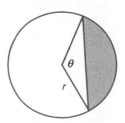

63. In the following figure, *ABCD* is a square, each side of which is 1 cm. The two arcs are portions of circles with

radii of 1 cm and with centers $A$ and $C$, respectively. Find the area of the shaded region. *Hint:* Draw $\overline{BD}$ and use your result from Exercise 62.

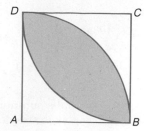

64. In the following figure, the three circles are mutually tangent, and each circle has a radius of 2 cm. Compute the area of the shaded region.

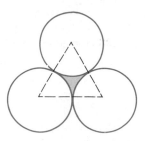

c 65. The following identity is known as *Euler's formula:*

$$\cos(36° + A) + \cos(36° - A)$$
$$= \cos A + \sin(18° + A) + \sin(18° - A)$$

Use a calculator to evaluate each side of the formula for the following values of $A$. In each case, round off your final results to three decimal places.
(a) 1°    (b) 4°    (c) 6°

66. In Euler's formula (given in Exercise 65), let $A = 0°$. Then use one of the double-angle formulas for cosine to transform the equation you've obtained into an equation that is quadratic in the quantity $\sin 18°$. Solve this equation to obtain the exact value of $\sin 18°$ in radical form.

*In Exercises 67–73, prove that the given equations are identities.*

67. $(\sin x + \cos x)^2 + (\sin x - \cos x)^2 = 2$

68. $(\sin x + \cos x)^3 - (\sin x - \cos x)^3$
$$= 6\cos x - 4\cos^3 x$$

69. $(\cos x + \sin x - 1)(\cos x + \sin x + 1) = 2\sin x \cos x$

70. $\dfrac{1}{1 - \sin x} + \dfrac{1}{1 + \sin x} + \dfrac{\sin x}{1 + \cos x} + \dfrac{1 + \cos x}{\sin x}$
$$= 2(\csc x + \sec^2 x)$$

71. $(9 - 4\sin x - \cos^2 x)(9 + 4\sin x - \cos^2 x)$
$$= \sin^4 x + 64$$

72. $(\sin x - \cos x)(1 + \tan x + \cot x)$
$$= (\sin^2 x \cos^2 x)(\sec^3 x - \csc^3 x)$$

73. $(a\sin x - b\cos x)^2 + (a\cos x + b\sin x)^2$
$$= a^2 + b^2$$

74. If $A$ and $B$ are acute angles such that $\tan A = \frac{12}{5}$ and $\tan B = \frac{3}{4}$, compute $\sin(A + B)$.

75. Using the data in Exercise 74, compute $\cos(A + B)$.

*In Exercises 76–85, use the addition formulas to simplify the expressions.*

76. $\sin\left(x + \dfrac{3\pi}{2}\right)$         77. $\sin(\pi - x)$

78. $\cos(\pi - x)$

79. $\sin 10° \cos 80° + \cos 10° \sin 80°$

80. $\cos 175° \cos 25° + \sin 175° \sin 25°$

81. $\cos\dfrac{2\pi}{5}\cos\dfrac{\pi}{10} - \sin\dfrac{2\pi}{5}\sin\dfrac{\pi}{10}$

82. $\cos\left(x - \dfrac{2\pi}{3}\right) - \cos\left(x + \dfrac{2\pi}{3}\right)$

83. $\sin\left(x + \dfrac{\pi}{6}\right) - \sin\left(x - \dfrac{\pi}{6}\right)$

84. $\tan(x + 45°)\tan(x - 45°)$

85. $\tan x \tan y + \tan y \tan z + \tan z \tan x$, where $y = x + 60°$ and $z = x - 60°$

86. (a) Show that $\tan(\pi/12) = 2 - \sqrt{3}$.
    *Hint:* $\pi/12 = (\pi/4) - (\pi/6)$
    (b) Show that $\cot(\pi/12) = 2 + \sqrt{3}$.

87. If $x, y,$ and $z$ are acute angles with sines $\frac{2}{5}, \frac{5}{13},$ and $\frac{8}{17},$ respectively, compute $\sin(x + y + z)$.

*In Exercises 88–92, a function of the form $y = A\sin Bx$ or $y = A\cos Bx$ is graphed for one period. Specify the equation in each case.*

88.

$(6, -3)$

**89.**

**90.**

**91.**

**92.**

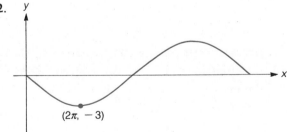

**93.** Find the amplitude of the function
$y = A \sin\left(\dfrac{3\pi x}{2} - \pi\right)$, given that the graph passes through the point $(1, \sqrt{2})$.

**94.** Find the amplitude of the function
$y = -3 \sin(2x - 3)\cos(2x - 3)$. *Hint:* Make use of the formula for $\sin 2\theta$.

*In Exercises 95–102, sketch the graph of each function for one complete cycle. In each case, specify the x-intercepts and the coordinates of the highest and lowest points on the graph.*

**95.** $y = -3 \cos 4x$

**96.** $y = 2 \sin(3\pi x/4)$

**97.** $y = 2 \sin\left(\dfrac{\pi x}{2} - \dfrac{\pi}{4}\right)$

**98.** $y = -\sin(x - 1)$

**99.** $y = 3 \cos\left(\dfrac{\pi x}{3} - \dfrac{\pi}{3}\right)$

**100.** $y = -2 \cos(x + \pi)$

**101.** $y = -\cos 2x \sin 2x$

**102.** $y = 1 - 2 \sin^2 x$

*In Exercises 103–108, sketch the graph of each function for one period.*

**103.** (a) $y = \tan(\pi x/4)$
(b) $y = \cot(\pi x/4)$

**104.** (a) $y = 2 \cot 2x$
(b) $y = 2 \tan 2x$

**105.** (a) $y = 3 \sec(x/4)$
(b) $y = 3 \csc(x/4)$

**106.** (a) $y = \sec \pi x$
(b) $y = \csc \pi x$

**107.** $y = \tan(x - 2)$

**108.** $y = \dfrac{\sin 2x}{1 + \cos 2x}$

*In Exercises 109–134, prove that the equations are identities.*

**109.** $\cot(x + y) = \dfrac{\cot x \cot y - 1}{\cot x + \cot y}$

**110.** $\cos 2x = \dfrac{1 - \tan^2 x}{1 + \tan^2 x}$

**111.** $\sin 2x = \dfrac{2 \tan x}{1 + \tan^2 x}$

**112.** $\sin^2 x - \sin^2 y = \sin(x + y)\sin(x - y)$

**113.** $\tan^2 x - \tan^2 y = \dfrac{\sin(x + y)\sin(x - y)}{\cos^2 x \cos^2 y}$

**114.** $2 \csc 2x = \sec x \csc x$

**115.** $(\sin x)\left(\tan\dfrac{x}{2} + \cot\dfrac{x}{2}\right) = 2$

**116.** $\tan\left(x + \dfrac{\pi}{4}\right) = \dfrac{1 + \tan x}{1 - \tan x}$

**117.** $\tan\left(\dfrac{\pi}{4} + x\right) - \tan\left(\dfrac{\pi}{4} - x\right) = 2 \tan 2x$

**118.** $\dfrac{\cot x - 1}{\cot x + 1} = \dfrac{1 - \sin 2x}{\cos 2x}$

**119.** $2 \sin\left(\dfrac{\pi}{4} - \dfrac{x}{2}\right)\cos\left(\dfrac{\pi}{4} - \dfrac{x}{2}\right) = \cos x$

**120.** $\dfrac{\tan(x + y) - \tan y}{1 + \tan(x + y)\tan y} = \tan x$

121. $\tan 2x + \sec 2x = \dfrac{\cos x + \sin x}{\cos x - \sin x}$

122. $\cos^4 x - \sin^4 x = \cos 2x$

123. $2\sin x + \sin 2x = \dfrac{2\sin^3 x}{1 - \cos x}$

124. $1 + \tan x \tan(x/2) = \sec x$

125. $\tan \dfrac{x}{2} = \dfrac{1 - \cos x + \sin x}{1 + \cos x + \sin x}$

126. $\dfrac{\sin 3x}{\sin x} - \dfrac{\cos 3x}{\cos x} = 2$

127. $\sin(x + y)\cos y - \cos(x + y)\sin y = \sin x$

128. $\dfrac{\sin x + \sin 2x}{\cos x - \cos 2x} = \cot \dfrac{x}{2}$

129. $\dfrac{1 - \tan^2(x/2)}{1 + \tan^2(x/2)} = \cos x$

130. $4\sin \dfrac{x}{4}\cos \dfrac{x}{4}\cos \dfrac{x}{2} = \sin x$

131. $\sin 4x = 4\sin x \cos x - 8\sin^3 x \cos x$

132. $\cos 4x = 8\cos^4 x - 8\cos^2 x + 1$

133. $\sin 5x = 16\sin^5 x - 20\sin^3 x + 5\sin x$

134. $\cos 5x = 16\cos^5 x - 20\cos^3 x + 5\cos x$

*In Exercises 135–143, establish the identities by applying the sum-to-product formulas.*

135. $\sin 80° - \sin 20° = \cos 50°$

136. $\sin 65° + \sin 25° = \sqrt{2}\cos 20°$

137. $\dfrac{\cos x - \cos 3x}{\sin x + \sin 3x} = \tan x$

138. $\dfrac{\sin 3° + \sin 33°}{\cos 3° + \cos 33°} = \tan 18°$

139. $\sin(5\pi/12) + \sin(\pi/12) = \sqrt{6}/2$

140. $\cos 10° - \sin 10° = \sqrt{2}\sin 35°$

141. $\dfrac{\cos 3y + \cos(2x - 3y)}{\sin 3y + \sin(2x - 3y)} = \cot x$

142. $\dfrac{\sin 10° - \sin 50°}{\cos 50° - \cos 10°} = \sqrt{3}$

143. $\dfrac{\sin 40° - \sin 20°}{\cos 20° - \cos 40°} = \dfrac{\sin 10° - \sin 50°}{\cos 50° - \cos 10°}$

144. Suppose that in triangle $ABC$, $\sin A + \sin B = \cos A + \cos B$. Show that angle $C$ is a right angle. *Suggestion:* Begin with the sum-to-product formulas.

*In Exercises 145–157, find all solutions of each equation in the range $0 \le x < 2\pi$.*

145. $\tan^2 x - 3 = 0$

146. $\cot^2 x - \cot x = 0$

147. $1 + \sin x = \cos x$

148. $2\sin 3x - \sqrt{3} = 0$

149. $\sin x - \cos 2x + 1 = 0$

150. $\sin x + \sin 2x = 0$

151. $3\csc x - 4\sin x = 0$

152. $2\sin^2 x + \sin x - 1 = 0$

153. $2\sin^4 x - 3\sin^2 x + 1 = 0$

154. $\sec^2 x - \sec x - 2 = 0$

155. $\sin^4 x + \cos^4 x = \dfrac{5}{8}$

156. $4\sin^2 2x + \cos 2x - 2\cos^2 x - 2 = 0$

157. $\cot x + \csc x + \sec x = \tan x$
  *Suggestion:* Using sines and cosines, the given equation becomes $\cos^2 x - \sin^2 x + \cos x + \sin x = 0$, which can be factored.

158. If $A$ and $B$ both are solutions of the equation $a\cos x + b\sin x = c$, show that $\tan \dfrac{A + B}{2} = \dfrac{b}{a}$. *Hint:* The given information yields two equations. After subtracting one of those equations from the other and rearranging, you will have $\dfrac{\cos A - \cos B}{\sin A - \sin B} = -(b/a)$. Now use the sum-to-product formulas.

159. If $a\sin x + b\cos x = a\csc x + b\sec x$, show that $\tan x = -\sqrt[3]{a/b}$.

160. Evaluate $\cos \tan^{-1}\sin \tan^{-1}(\sqrt{2}/2)$.

*In Exercises 161–182, evaluate each expression (without using a calculator or tables).*

161. $\cos^{-1}(-\sqrt{2}/2)$

162. $\arctan(\sqrt{3}/3)$

163. $\sin^{-1} 0$

164. $\arcsin \frac{1}{2}$

165. $\arctan \sqrt{3}$

166. $\cos^{-1}(\frac{1}{2})$

167. $\tan^{-1}(-1)$

168. $\cos^{-1}(-\frac{1}{2})$

169. $\sin(\sin^{-1} 1)$

170. $\cos[\cos^{-1}(\frac{2}{7})]$

171. $\sin[\arccos(-\frac{1}{2})]$

172. $\sin[\tan^{-1}(-1)]$

173. $\cot[\cos^{-1}(\frac{1}{2})]$

174. $\sec[\cos^{-1}(\sqrt{2}/3)]$

175. $\sin\left(\dfrac{3\pi}{2} + \arccos \dfrac{3}{5}\right)$

176. $\tan\left[\dfrac{\pi}{4} + \sin^{-1}\left(\dfrac{5}{13}\right)\right]$

177. $\sin[2\sin^{-1}(\frac{4}{5})]$

178. $\tan(2\tan^{-1} 2)$

179. $\sin^{-1}[\sin(\pi/7)]$

180. $\cos[\frac{1}{2}\cos^{-1}(\frac{4}{5})]$

181. $\sin(\arctan \frac{1}{2} + \arctan \frac{1}{3})$

182. $\tan[\frac{1}{2}(\arcsin \frac{1}{3} + \arccos \frac{1}{2})]$

*In Exercises 183–190, show that each equation is an identity.*

183. $\tan(\tan^{-1} x + \tan^{-1} y) = (x + y)/(1 - xy)$

184. $\tan^{-1}(x/\sqrt{1 - x^2}) = \sin^{-1} x$

185. $\sin(2\arctan x) = 2x/(1 + x^2)$

186. $\cos(2\cos^{-1} x) = 2x^2 - 1$

**187.** $\sin[\frac{1}{2}\sin^{-1}(x^2)] = \sqrt{\frac{1}{2} - \frac{1}{2}\sqrt{1 - x^4}}$

**188.** $\tan^{-1}(\frac{1}{3}) + \tan^{-1}(\frac{1}{5}) = \tan^{-1}(\frac{4}{7})$

**189.** $\arcsin(4\sqrt{41}/41) + \arcsin(\sqrt{82}/82) = \pi/4$

**190.** $\tan\left(\sin^{-1}\frac{1}{3} + \cos^{-1}\frac{1}{2}\right) = \dfrac{8\sqrt{2} + 9\sqrt{3}}{5}$

**191.** Ⓒ **(a)** Use a calculator to compute the quantity $\cos 20° \cos 40° \cos 60° \cos 80°$. Give your answer to as many decimal places as is shown on your calculator.

**(b)** Now use a product-to-sum formula to *prove* that the display on your calculator is the exact value of the given expression, not an approximation.

# 9 ADDITIONAL TOPICS IN TRIGONOMETRY

*As the text and the examples of the work have been tested by considerable experience in teaching, the hope may be entertained that they will be suitable for imparting a sound and comprehensive knowledge of Plane Trigonometry, together with readiness in the application of this knowledge to the solution of problems.*

ISAAC TODHUNTER in the preface to his text, *Plane Trigonometry*, 5th ed. (London: Macmillan and Company, 1874)

**INTRODUCTION**   In this chapter we develop five topics that require a background in basic trigonometry. The first two of these topics, discussed in Section 9.1, are the law of sines and the law of cosines. These two laws relate the angles and the lengths of the sides in a triangle. The next two sections, 9.2 and 9.3, introduce the important topic of vectors, first from a geometric standpoint, then from an algebraic standpoint. In Section 9.4 we present an alternative to the familiar *x-y* coordinate system; this is the system of polar coordinates. The fifth and last topic that we introduce in this chapter is the trigonometric form for complex numbers. As you'll see, when a complex number is written in trigonometric form, the processes of multiplication, division, and computation of roots are simplified.

## THE LAW OF SINES AND THE LAW OF COSINES

### 9.1

The law of sines and the law of cosines are, in essence, formulas relating the sides and angles in any triangle. These formulas can be used to determine an unknown side or angle using given information about the triangle. As you will see, which formula to apply in a particular case depends on what data are initially given. To help you keep track of this, we'll use the following notation in our examples.

| NOTATION | EXPLANATION |
|----------|-------------|
| SAA | One side and two angles are given. |
| SSA | Two sides and an angle opposite one of those sides are given. |
| SAS | Two sides and the included angle are given. |
| SSS | Three sides of the triangle are given. |

Also, we will often follow the convention of denoting the angles of a triangle by *A*, *B*, and *C* and the lengths of the corresponding opposite sides by *a*, *b*, and *c* (see Figure 1). With this notation, we are ready to state the law of sines.

> ### Law of Sines
>
> In any triangle, the sines of the angles are proportional to the lengths of the opposite sides:
>
> $$\frac{\sin A}{a} = \frac{\sin B}{b} = \frac{\sin C}{c}$$
>
>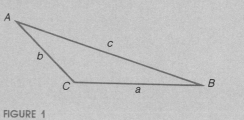
>
> FIGURE 1

The proof of the law of sines is easy. We use the following result from Chapter 7: The area of any triangle is equal to half the product of two sides times the sine of the included angle. Thus, with reference to Figure 1, we have

$$\tfrac{1}{2}bc\sin A = \tfrac{1}{2}ac\sin B = \tfrac{1}{2}ab\sin C$$

since each of these three expressions equals the area of triangle $ABC$. Now we just multiply through by the quantity $2/abc$ to get

$$\frac{\sin A}{a} = \frac{\sin B}{b} = \frac{\sin C}{c}$$

which completes the proof.

**EXAMPLE 1**   (SAA) Find the length $x$ in Figure 2.

**Solution**   We can determine $x$ directly by applying the law of sines. We have

$$\frac{\sin 30°}{20} = \frac{\sin 135°}{x}$$

length of side opposite the 30° angle ⟜      ⟜ length of side opposite the 135° angle

$$x\sin 30° = 20\sin 135°$$

$$x\left(\frac{1}{2}\right) = 20\left(\frac{\sqrt{2}}{2}\right)$$

$$x = 20\sqrt{2}\ \text{cm}$$

FIGURE 2

**EXAMPLE 2**   (SAA) Find the length $y$ in Figure 3 (on the next page).

**Solution**   To determine $y$ using the law of sines, it is necessary to know the opposite angle, denoted by $\theta$ in Figure 3. Since the sum of the angles in any triangle is 180°, we have

$$\theta + 75° + 35° = 180°$$

and consequently,

$$\theta = 180° - 75° - 35° = 70°$$

FIGURE 3

Then, by the law of sines, we obtain

$$\frac{\sin 75°}{10.0} = \frac{\sin 70°}{y}$$

$$y \sin 75° = (10.0) \sin 70°$$

$$y = \frac{(10.0)(\sin 70°)}{\sin 75°} \text{ in.}$$

Without the use of a calculator or tables, this is the final form of the answer. On the other hand (as you should check for yourself), a calculator yields

$$y \approx \frac{(10.0)(0.939)}{0.965} \approx 9.7 \text{ in.}$$

**EXAMPLE 3**     (SSA) In triangle $ABC$, we are given $\angle C = 45°$, $b = 4\sqrt{2}$ ft, and $c = 8$ ft. Determine the remaining side and angles.

Solution     First draw a preliminary sketch conveying the given data. See Figure 4. (The sketch must be considered tentative. At the outset, we don't know whether the other angles are acute or even whether the given data are compatible.) To find angle $B$, we have (according to the law of sines)

$$\frac{\sin B}{4\sqrt{2}} = \frac{\sin 45°}{8}$$

FIGURE 4

and therefore,

$$8 \sin B = 4\sqrt{2} \sin 45°$$

$$= 4\sqrt{2}\left(\frac{\sqrt{2}}{2}\right) = 4$$

$$\sin B = \frac{1}{2}$$

From our previous work, we know that one possibility for $B$ is 30°, since $\sin 30° = \frac{1}{2}$.* However, there is another possibility. Since the reference angle for 150° is 30°, we know that $\sin 150°$ is also equal to $\frac{1}{2}$. Which angle do we want? For the problem at hand, this is easy to answer. Since angle $C$ is given as 45°, angle $B$ cannot equal 150°, for the sum of 45° and 150° exceeds 180°. We conclude that

$$\angle B = 30°$$

Next, since $\angle B = 30°$ and $\angle C = 45°$, we have

$$\angle A = 180° - 30° - 45° = 105°$$

* If the value were not so familiar, say, $\sin B = \frac{2}{3}$, we could use a calculator or tables, as in Example 10 of Section 7.2.

Finally, since angle $A$ is opposite to side $\overline{BC}$, the length of $\overline{BC}$ can be determined using the law of sines:

$$\frac{\sin C}{c} = \frac{\sin A}{a}$$

$$\frac{\sin 45°}{8} = \frac{\sin 105°}{a}$$

$$a \sin 45° = 8 \sin 105°$$

$$a = \frac{8 \sin 105°}{\sin 45°} = \frac{8 \sin 105°}{\sqrt{2}/2}$$

$$= \frac{16}{\sqrt{2}} \sin 105° = \frac{16\sqrt{2}}{2} \sin 105°$$

$$= 8\sqrt{2} \sin 105° \text{ ft}$$

Using a calculator, this expression for $a$ can be evaluated directly. The result is $a = 11$ ft to the nearest foot. On the other hand, using tables, it would first be necessary to use the identity $\sin(180° - \theta) = \sin \theta$ to write

$$\sin 105° = \sin(180° - 105°) = \sin 75°$$

This is because trigonometric tables generally do not include values of angles beyond $90°$.

In the preceding example, two possibilities arose for angle $B$: both $30°$ and $150°$. However, it turned out that the value $150°$ was incompatible with the given information in the problem. In Exercise 11 at the end of this section, you will see a case in which both of two possibilities are compatible with the given data. This results in two distinct solutions to the problem. In contrast to this, Exercise 10 shows a case in which there is no triangle fulfilling the given conditions. For these reasons, the case SSA is sometimes referred to as the **ambiguous case**.

Now we turn to the law of cosines.

---

### Law of Cosines

In any triangle, the square of the length of any side equals the sum of the squares of the lengths of the other two sides minus twice the product of the lengths of those other two sides times the cosine of their included angle.

$$a^2 = b^2 + c^2 - 2bc \cos A$$
$$b^2 = c^2 + a^2 - 2ca \cos B$$
$$c^2 = a^2 + b^2 - 2ab \cos C$$

Before looking at a proof of this law, we make two preliminary comments. First, it is important to understand that the three equations in the box all follow the same pattern. For example, look at the first equation.

$$\underbrace{a^2 = b^2 + c^2 - 2bc\cos A}$$

side and opposite angle

sides that include this angle

Now check for yourself that the other two equations also follow this pattern. It is the pattern that is important here; after all, not every triangle will be labeled *ABC*.

The second observation is that the law of cosines is a generalization of the Pythagorean Theorem. In fact, look what happens to the equation

$$a^2 = b^2 + c^2 - 2bc\cos A$$

when angle *A* is a right angle:

$$a^2 = b^2 + c^2 - 2bc\underbrace{\cos 90°}_{0}$$

$$a^2 = b^2 + c^2 \qquad \text{which is the Pythagorean Theorem}$$

Now let us prove the law of cosines:

$$a^2 = b^2 + c^2 - 2bc\cos A$$

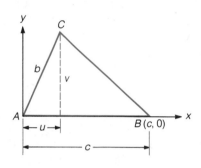

**FIGURE 5**

(The other two equations would be proved in the same way. Indeed, just re-labeling the figure would suffice.) The proof that we give uses coordinate geometry in a very nice way to complement the trigonometry. We begin by placing angle *A* in standard position, as indicated in Figure 5. (So, in the figure, angle *A* is then identified with angle *CAB*.) Then if *u* and *v* denote the lengths indicated in Figure 5, the coordinates of *C* are $(u, v)$ and we have

$$\cos A = \frac{\text{adjacent}}{\text{hypotenuse}} = \frac{u}{b}$$

and therefore,

$$u = b\cos A$$

Similarly, we have

$$\sin A = \frac{\text{opposite}}{\text{hypotenuse}} = \frac{v}{b}$$

and therefore,

$$v = b\sin A$$

Thus, the coordinates of *C* are

$$(b\cos A, b\sin A)$$

(Exercise 29 at the end of this section asks you to check that these represent the coordinates of *C* even when angle *A* is not acute.) Now we use the distance formula,

$$d = \sqrt{(x_2 - x_1)^2 + (y_2 - y_1)^2}$$

to compute the required distance $a$ between the points $C(b \cos A, b \sin A)$ and $B(c, 0)$. We have

$$a = \sqrt{(b \cos A - c)^2 + (b \sin A - 0)^2}$$
$$= \sqrt{b^2 \cos^2 A - 2bc \cos A + c^2 + b^2 \sin^2 A}$$
$$= \sqrt{b^2 \underbrace{(\cos^2 A + \sin^2 A)}_{1} - 2bc \cos A + c^2}$$
$$= \sqrt{b^2 + c^2 - 2bc \cos A}$$

Squaring both sides of this last equation gives us the law of cosines, as we set out to prove.

**EXAMPLE 4** (SAS) Compute the length $x$ in Figure 6.

Solution The law of cosines is directly applicable. We have

$$x^2 = 7^2 + 8^2 - 2(7)(8) \cos 120°$$
$$= 49 + 64 - 112(-\tfrac{1}{2}) = 169$$
$$x = \sqrt{169} = 13 \, \text{cm}$$

FIGURE 6

If the equation $a^2 = b^2 + c^2 - 2bc \cos A$ is solved for $\cos A$, the result is

$$\cos A = \frac{b^2 + c^2 - a^2}{2bc}$$

This expresses the cosine of an angle in a triangle in terms of the lengths of the sides. In a similar fashion, we obtain the corresponding formulas

$$\cos B = \frac{c^2 + a^2 - b^2}{2ca}$$

and

$$\cos C = \frac{a^2 + b^2 - c^2}{2ab}$$

These alternative forms for the law of cosines are used in the next example.

**EXAMPLE 5** (SSS) In triangle $ABC$, the sides are $a = 3$ units, $b = 5$ units, and $c = 7$ units. Find the angles.

Solution Figure 7 summarizes the given data. We now have

$$\cos A = \frac{b^2 + c^2 - a^2}{2bc}$$
$$= \frac{5^2 + 7^2 - 3^2}{2(5)(7)} = \frac{65}{70} = \frac{13}{14}$$

FIGURE 7

Using a calculator (set in the degree mode), we obtain $\angle A \approx 21.8°$. For example, the sequence of keystrokes on a Texas Instruments calculator is as follows:

$$\boxed{13} \;\; \boxed{\div} \;\; \boxed{14} \;\; \boxed{=} \;\; \boxed{\text{INV}} \;\; \boxed{\cos}$$

In a similar manner, we have

$$\cos B = \frac{c^2 + a^2 - b^2}{2ca}$$

$$= \frac{7^2 + 3^2 - 5^2}{2(7)(3)} = \frac{33}{42} = \frac{11}{14}$$

Thus, $\cos B = \frac{11}{14}$ and, using a calculator, we find that $\angle B \approx 38.2°$.

Finally, working in the same manner to compute angle $C$, we have

$$\cos C = \frac{a^2 + b^2 - c^2}{2ab}$$

$$= \frac{3^2 + 5^2 - 7^2}{2(3)(5)} = \frac{-15}{30} = -\frac{1}{2}$$

Thus, $\cos C = -\frac{1}{2}$. A calculator is not needed in this case. On the basis of our previous work, we conclude that $\angle C = 120°$. The required angles are thus

$$\angle A \approx 21.8°$$
$$\angle B \approx 38.2°$$
$$\angle C = 120°$$

## EXERCISE SET 9.1

### A

*In Exercises 1–8, assume that the vertices and the lengths of the sides of a triangle are labeled as in Figure 1 in the text. For Exercises 1–4, leave your answers in terms of radicals or the trigonometric functions; that is, don't use a calculator. In Exercises 5–8, use a calculator and round off your final answers to one decimal place.*

1. If $\angle A = 60°$, $\angle B = 45°$, and $BC = 12$ cm, find $AC$.

2. If $\angle A = 30°$, $\angle B = 135°$, and $BC = 4$ cm, find $AC$.

3. If $\angle B = 100°$, $\angle C = 30°$, and $AB = 10$ cm, find $BC$.

4. If $\angle A = \angle B = 35°$, and $AB = 16$ cm, find $AC$ and $BC$.

5. If $\angle A = 36°$, $\angle B = 50°$, and $b = 12.61$ cm, find $a$ and $c$. [c]

6. If $\angle B = 81°$, $\angle C = 55°$, and $b = 6.24$ cm, find $c$ and $a$. [c]

7. If $a = 29.45$ cm, $b = 30.12$ cm, and $\angle B = 66°$, find the remaining side and angles of the triangle. [c]

8. If $a = 52.15$ cm, $c = 42.90$ cm, and $\angle A = 125°$, find the remaining side and angles of the triangle. [c]

9. (a) In triangle $ABC$, $\sin B = \sqrt{2}/2$. What are the possible values for angle $B$?
   (b) In triangle $ABC$, if $\cos A = -\sqrt{3}/2$, what are the possible values for angle $A$?
   (c) In triangle $CDE$, $\sin D = \frac{1}{4}$. Using a calculator, determine the possible values for angle $D$. [c]

10. Show that there is no triangle for which $a = 23$, $b = 50$, and $\angle A = 30°$.   *Hint:*  Try computing $\sin B$ using the law of sines.

11. Let $b = 1$, $a = \sqrt{2}$, and $\angle B = 30°$.
    (a) Use the law of sines to show that $\sin A = \sqrt{2}/2$.
    (b) Conclude that $A = 45°$ or $A = 135°$.
    (c) Assuming that $A = 45°$, determine the remaining parts of triangle $ABC$.
    (d) Assuming that $A = 135°$, determine the remaining parts of triangle $ABC$.

12. Let $a = 30$, $b = 36$, and $\angle A = 20°$. Find the remaining parts of triangle $ABC$. (There are two distinct triangles possible.) [c]

13. Find the lengths $a$, $b$, $c$, and $d$ in the following figure. Leave your answers in terms of trigonometric functions (rather than decimals) as necessary.

14. Find the area of triangle $ABC$ if $c = 7$ in., $\angle A = 40°$, and $\angle B = 85°$.

15. Use the law of cosines to determine $x$ in each of the figures. For part (b), use a calculator or tables and round off to one decimal place.

(a)  (b)

16. In applying the law of cosines to the following figure, a student incorrectly writes

$$x^2 = 5^2 + 11^2 - 2(5)(11)\cos 95°$$

Why is this incorrect? What is the correct equation?

*In Exercises 17 and 18, use the given information to find the cosine of each angle in* $\triangle ABC$.

17. $a = 6$ cm, $b = 7$ cm, $c = 10$ cm

18. $a = 17$ cm, $b = 8$ cm, $c = 15$ cm (For this particular triangle, you can check your answers because there is an alternate method of solution that does not require the law of cosines.)

*In Exercises 19–22, round off each answer to one decimal place.*

[c] 19. A regular pentagon is inscribed in a circle of unit radius. Find the perimeter of the pentagon. *Hint:* First find the length of a side using the law of cosines.

[c] 20. In triangle $ABC$, $a = 2$, $b = 3$, and $c = 4$. Find each angle.

[c] 21. In triangle $ABC$, $\angle A = 40°$, $b = 6.1$ cm, and $c = 3.2$ cm.
(a) Find $a$ using the law of cosines.
(b) Find angle $C$ using the law of sines.
(c) Find angle $B$.

[c] 22. In parallelogram $ABCD$, you are given $AB = 6$ in., $AD = 4$ in., and $\angle A = 40°$. Find the length of each diagonal.

23. In square $ABCD$, each side has unit length. Point $P$ is on diagonal $AC$ such that $AP = 1$. Find the distance from $P$ to $D$. (Express your answer using radicals.)

[c] 24. Two points $P$ and $Q$ are on opposite sides of a river (see the sketch). From $P$ to another point $R$ on the same side is 300 ft. Angles $PRQ$ and $RPQ$ are found to be 20° and 120°, respectively. Compute the distance from $P$ to $Q$, across the river. (Round off your answer to the nearest foot.)

**B**

25. (a) Let $m$ and $n$ be positive numbers, with $m > n$. Furthermore, suppose that in triangle $ABC$ the lengths $a$, $b$, and $c$ are given by

$$a = 2mn + n^2 \qquad b = m^2 - n^2$$
$$c = m^2 + n^2 + mn$$

Show that $\cos C = -\frac{1}{2}$ and conclude that $\angle C = 120°$.

(b) Give an example of a triangle in which the lengths of the sides are whole numbers and one of the angles is 120°. (Specify the three sides; you needn't find the other angles.)

26. If the lengths of two adjacent sides of a parallelogram are $a$ and $b$, and if the acute angle formed by these two sides is $\theta$, show that the product of the lengths of the two diagonals is given by the expression

$$\sqrt{(a^2 + b^2)^2 - 4a^2b^2 \cos^2 \theta}$$

[c] 27. Compute the length $x$ in the following figure. *Suggestion:* Draw the line segment joining the vertices of the 80° and 50° angles.

28. In the following figure, $\overline{AD}$ bisects angle $A$ in triangle $ABC$. Show that

$$\frac{n}{m} = \frac{c}{b}$$

In words, this says that the bisector of an angle in a triangle divides the opposite side into segments that are proportional to the adjacent sides. *Hint:* Apply the law of sines in triangles $ABD$ and $ACD$.

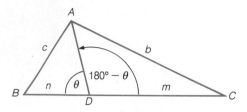

29. Let the positive numbers $u$ and $v$ denote the lengths indicated in the following figure, so that the coordinates of $C$ are $(-u, v)$. Show that $u = -b\cos A$ and $v = b\sin A$. Conclude from this that the coordinates of $C$ are

$$(b\cos A, b\sin A)$$

*Hint:* Use the right triangle definitions for cosine and sine along with the formulas for $\cos(180° - \theta)$ and $\sin(180° - \theta)$.

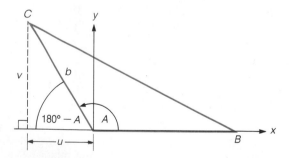

30. *Heron's formula*  Approximately 2000 years ago, Heron of Alexandria derived a formula for the area of a triangle in terms of the lengths of the sides. A more modern derivation of Heron's formula using the law of cosines is indicated in the steps that follow.

(a) Solve the equation $c^2 = a^2 + b^2 - 2ab\cos C$ for $\cos C$ to obtain

$$\cos C = \frac{a^2 + b^2 - c^2}{2ab}$$

(b) Show that $1 - \cos C = \dfrac{(c + a - b)(c - a + b)}{2ab}$.

*Suggestion:* Work with $1 - \cos C$. You can factor the expression $c^2 - (a - b)^2$ that arises using the difference of squares technique.

(c) Using the result in part (a), show that $1 + \cos C = \dfrac{(a + b + c)(a + b - c)}{2ab}$.

(d) Show that $\sin^2 C = \dfrac{(a + b + c)(-a + b + c)(a - b + c)(a + b - c)}{4a^2b^2}$

(e) Using the formula

$$\text{area } \triangle ABC = \tfrac{1}{2}ab\sin C$$

show that area $\triangle ABC =$
$$\tfrac{1}{4}\sqrt{(a + b + c)(-a + b + c)(a - b + c)(a + b - c)}$$

(f) The formula in part (e) expresses the area in terms of $a$, $b$, and $c$. We can simplify this formula somewhat by introducing the following notation.

Let $s$ denote one-half the perimeter of triangle $ABC$. That is, let

$$s = \tfrac{1}{2}(a + b + c)$$

Using this notation, check that
  (i) $a + b + c = 2s$
  (ii) $-a + b + c = 2(s - a)$
  (iii) $a - b + c = 2(s - b)$
  (iv) $a + b - c = 2(s - c)$
Then show that with this notation, we obtain

$$\text{area } \triangle ABC = \sqrt{s(s - a)(s - b)(s - c)}$$

This is *Heron's formula*. [For historical background and a purely geometric proof, see *An Introduction to the History of Mathematics*, 5th edition, by Howard Eves (Philadelphia: Saunders, 1983), pp. 133 and 147.]

31. The accompanying figure shows a quadrilateral, with sides $a$, $b$, $c$, and $d$, inscribed in a circle. If $\lambda$ denotes the length of the diagonal indicated in the figure, prove that

$$\lambda^2 = \frac{(ab + cd)(ac + bd)}{bc + ad}$$

This result is known as *Brahmagupta's Theorem*. It is named after its discoverer, a seventh-century Hindu mathematician. *Hint:* Assume as given the theorem from geometry stating that when a quadrilateral is inscribed in a circle, the opposite angles are supplementary. Apply the law of cosines in both of the triangles in the figure to obtain expressions for $\lambda^2$. Then eliminate $\cos\theta$ from one equation.

32. *Alternate derivation for the law of sines*
    The following figure shows triangle *ABC* inscribed in a circle of radius *r*.
    (a) Show that $\angle AOD = \angle C$. *Hint:* Use the following theorem from geometry: The measure of an angle inscribed in a circle is half the measure of the corresponding central angle. (In the figure, angle *C* is the inscribed angle and angle *AOB* is the central angle.)

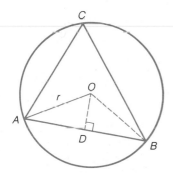

(b) Show that $r = c/(2\sin C)$.
(c) Using the reasoning in parts (a) and (b) conclude that

$$r = \frac{a}{2\sin A} \quad \text{and} \quad r = \frac{b}{2\sin B}$$

Then show that

$$\frac{\sin A}{a} = \frac{\sin B}{b} = \frac{\sin C}{c}$$

(d) In the figure we used, the center of the circle falls within triangle *ABC*. Draw a figure in which the center lies outside of triangle *ABC* and prove that the equation $r = c/(2\sin C)$ is true in this case, too.

C

33. Suppose that triangle *ABC* is a right triangle with the right angle at *C*. Use the law of cosines to prove the following statements.
    (a) The square of the distance from *C* to the midpoint of the hypotenuse is equal to one-fourth the square of the hypotenuse.
    (b) The sum of the squares of the distances from *C* to the two points that trisect the hypotenuse is equal to five-ninths the square of the hypotenuse.
    (c) Let *P*, *Q*, and *R* be points on the hypotenuse such that $AP = PQ = QR = RB$. Derive a result similar to your results in parts (a) and (b) for the sum of the squares of the distances from *C* to *P*, *Q*, and *R*.

## VECTORS IN THE PLANE, A GEOMETRIC APPROACH

### 9.2

*The idea of a parallelogram of velocities may be found in various ancient Greek authors, and the concept of a parallelogram of forces was not uncommon in the sixteenth and seventeenth centuries. By the early nineteenth century parallelograms of physical entities frequently appeared in treatises, and this usage indirectly led to vector analysis, for this area provided a striking example of how vectorial entities could be used for physical applications.*

MICHAEL J. CROWE in *A History of Vector Analysis* (Notre Dame, Ind.: University of Notre Dame Press, 1967)

Certain quantities, such as temperature, length, and mass, can be specified by means of a single number (assuming that a system of units has been agreed on). We call these quantities **scalars**. On the other hand, quantities such as force and velocity are characterized by both a *magnitude* (a number) and a *direction*. We call these quantities **vectors**.

Geometrically, a vector is a directed line segment or arrow. The vector in Figure 1, for instance, represents a wind velocity of 5 mph from the west. The

**FIGURE 1**  A vector representing a wind velocity of 5 mph from the west.

length of this vector represents the magnitude of the wind velocity, while the direction of the vector indicates the direction of the wind velocity. As another example, the vector in Figure 2 represents a force acting on an object: the magnitude of the force is 3 lb, and the force acts at an angle of 135° with the horizontal.

In a moment, we are going to discuss the important concept of vector addition, but first let us agree on some matters of notation. Suppose that we have a vector drawn from a point $P$ to point $Q$, as shown in Figure 3.

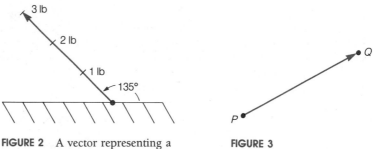

**FIGURE 2** A vector representing a force of 3 lb acting at an angle of 135° with the horizontal.

**FIGURE 3**

The point $P$ in Figure 3 is called the **initial point** of the vector, and $Q$ is the **terminal point**. We can denote this vector by

$$\overrightarrow{PQ}$$

The length of the vector $\overrightarrow{PQ}$ is denoted by $|\overrightarrow{PQ}|$. On the printed page, vectors are often indicated by boldface letters, such as **a**, **A**, and **v**.

If two vectors **a** and **b** have the same length and the same direction, we say that they are *equal* and we write **a** = **b** (see Figure 4). Notice that our definition for vector equality involves magnitude and direction, but not location. Thus, when it is convenient to do so, we are free to move a given vector to another location, provided we do not alter the magnitude or direction.

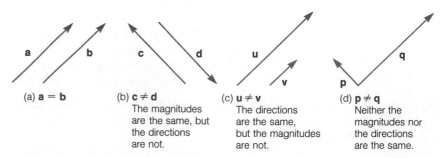

(a) **a** = **b**

(b) **c** ≠ **d**
The magnitudes are the same, but the directions are not.

(c) **u** ≠ **v**
The directions are the same, but the magnitudes are not.

(d) **p** ≠ **q**
Neither the magnitudes nor the directions are the same.

**FIGURE 4**

As motivation for the definition of vector addition, let us suppose that an object moves from a point $P$ to a point $Q$. Then we can represent this *displacement* by the vector $\overrightarrow{PQ}$. (Indeed, the word *vector* is derived from the Latin *vectus*, meaning "carried.") Now suppose that after moving from $P$ to $Q$, the

**FIGURE 5**

object moves from $Q$ to $R$. Then, as you can see in Figure 5, the net effect is a displacement from $P$ to $R$. We say in this case that the vector $\overrightarrow{PR}$ is the **sum** or **resultant** of the vectors $\overrightarrow{PQ}$ and $\overrightarrow{QR}$, and we write

$$\overrightarrow{PQ} + \overrightarrow{QR} = \overrightarrow{PR}$$

These ideas are formalized in the definition that follows.

**Definition of Vector Addition**

Let **u** and **v** be two vectors. Position **v** (without changing its magnitude or direction) so that its initial point coincides with the terminal point of **u**, as in Figure 6(a). Then, as indicated in Figure 6(b), the vector **u** + **v** is the directed line segment from the initial point of **u** to the terminal point of **v**. The vector **u** + **v** is called the **sum** or **resultant** of **u** and **v**.

(a)                               (b)

**FIGURE 6**

**EXAMPLE 1**  Refer to Figure 7.

(a) Determine the initial and terminal points of **u** + **v**.
(b) Compute $|\mathbf{u} + \mathbf{v}|$.

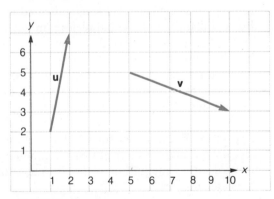

**FIGURE 7**

**Solution**  (a) According to the definition, we first need to move **v** (without changing its length or direction) so that its initial point coincides with the terminal point of **u**. From Figure 7, we see that this can be accomplished by moving each point of **v** three units in the negative $x$-direction and two units in the positive $y$-direction. Figure 8(a) shows the new location of **v**, and Figure 8(b) indicates the sum **u** + **v**. From Figure 8(b), we see that the initial and terminal points of **u** + **v** are $(1, 2)$ and $(7, 5)$, respectively.

(a)

(b)

**FIGURE 8**

**FIGURE 9**   The upper triangle shows the sum **u** + **v**, whereas the lower triangle shows the sum **v** + **u**. Since in both cases the resultant is $\overrightarrow{PQ}$, it follows that **u** + **v** = **v** + **u**.

**FIGURE 10**   The parallelogram law for vector addition.

**(b)** We can use the distance formula to determine $|\mathbf{u} + \mathbf{v}|$. Using the points $(1, 2)$ and $(7, 5)$ that were obtained in part (a), we have

$$|\mathbf{u} + \mathbf{v}| = \sqrt{(7 - 1)^2 + (5 - 2)^2} = \sqrt{45} = \sqrt{9 \cdot 5} = 3\sqrt{5}$$

One important consequence of our definition for vector addition is that this operation is commutative. That is, for any two vectors **u** and **v**, we have

$$\mathbf{u} + \mathbf{v} = \mathbf{v} + \mathbf{u}$$

Figure 9 indicates why this is so.

In view of Figure 9, vector addition can also be carried out by using the so-called **parallelogram law**: To determine **u** + **v**, position **u** and **v** so that their initial points coincide. Then, as indicated in Figure 10, the vector **u** + **v** is the directed diagonal of the parallelogram determined by **u** and **v**.

It is a fact—and it has been verified experimentally—that if two forces **F** and **G** act on an object, the net effect is the same as if just the resultant **F** + **G** acted on the object. In Example 2, we use the parallelogram law to compute the resultant of two forces. Note that the units of force used in this example are **newtons** (N) where $1\,\text{N} \approx 0.2248\,\text{lb}$.

**EXAMPLE 2**   Two forces **F** and **G** act on an object. As indicated in Figure 11, the force **G** acts horizontally to the right with a magnitude of 12 N, while **F** acts vertically upward with a magnitude of 16 N. Determine the magnitude and direction of the resultant force.

Solution   We complete the parallelogram, as shown in Figure 12. Now we need to calculate the length of **F** + **G** and the angle $\theta$. Applying the Pythagorean Theorem in Figure 12, we have

$$|\mathbf{F} + \mathbf{G}| = \sqrt{12^2 + 16^2} = \sqrt{144 + 256} = \sqrt{400} = 20$$

Also from Figure 12, we have

$$\tan \theta = \frac{16}{12} = \frac{4}{3}$$

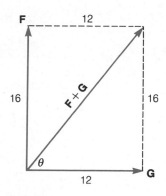

FIGURE 11                    FIGURE 12

Consequently,

$$\theta = \tan^{-1} \frac{4}{3} \approx 53.1°$$

Now we can summarize our results. The magnitude of **F** + **G** is 20 N, and the angle $\theta$ between **F** + **G** and the horizontal is (approximately) 53.1°.   ▪▪

In Example 2, we determined the resultant for two perpendicular forces. The next example shows how to compute the resultant when the forces are not perpendicular. Our calculations will make use of both the law of sines and the law of cosines.

**EXAMPLE 3**    Determine the resultant of the two forces in Figure 13. (Round off the answers to one decimal place.)

Solution    As in the previous example, we complete the parallelogram. In Figure 14, the angle in the lower right-hand corner of the parallelogram is 140°. This is because the sum of two adjacent angles in any parallelogram is always 180°. Letting $d$ denote the length of the diagonal in Figure 14, we can use the law of cosines to write

FIGURE 13

$$d^2 = 15^2 + 5^2 - 2(15)(5)\cos 140°$$
$$d = \sqrt{250 - 150\cos 140°} = \sqrt{250 + 150\cos 40°} \qquad \text{(Why?)}$$
$$= \sqrt{25(10 + 6\cos 40°)} = 5\sqrt{10 + 6\cos 40°}$$
$$\approx 19.1 \qquad \text{using a calculator}$$

FIGURE 14

So the magnitude of the resultant is 19.1 N (to one decimal place). To specify the direction of the resultant, we need to determine the angle $\theta$ in Figure 14.

Using the law of sines, we have

$$\frac{\sin \theta}{5} = \frac{\sin 140°}{d}$$

and consequently,

$$\sin \theta = \frac{5 \sin 40°}{d} \qquad \text{(Why?)}$$

$$= \frac{5 \sin 40°}{5 \sqrt{10 + 6 \cos 40°}} = \frac{\sin 40°}{\sqrt{10 + 6 \cos 40°}}$$

Using a calculator now, we obtain

$$\theta = \sin^{-1}\left(\frac{\sin 40°}{\sqrt{10 + 6 \cos 40°}}\right) \approx 9.7°$$

In summary, the magnitude of the resultant force is about 19.1 N, and the angle $\theta$ between the resultant and the 15-N force is approximately 9.7°.

As background for the next example, we introduce the notion of *components* of a vector. (You will see this concept again in the next section, but in a more algebraic context.) Suppose that the initial point of a vector **v** is located at the origin of a rectangular coordinate system, as shown in Figure 15. Now suppose we draw perpendiculars from the terminal point of **v** to the axes, as indicated in Figure 16. Then the coordinates $v_x$ and $v_y$ in Figure 16 are called the **components** of the vector **v**. For an example involving components, refer back to Figure 12. The horizontal component of the vector **F** + **G** is 12 N, and the vertical component is 16 N.

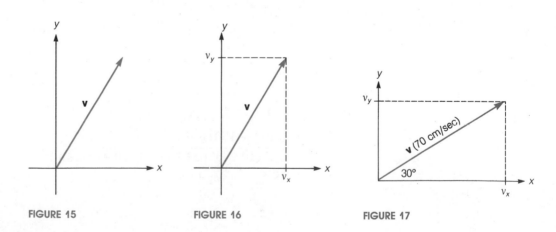

FIGURE 15          FIGURE 16          FIGURE 17

**EXAMPLE 4** Determine the horizontal and vertical components of the velocity vector **v** in Figure 17.

Solution From Figure 17, we can write

$$\cos 30° = \frac{\text{adjacent}}{\text{hypotenuse}} = \frac{v_x}{70}$$

and consequently,

$$v_x = (\cos 30°)(70) = \frac{\sqrt{3}}{2}(70)$$

$$= 35\sqrt{3} \approx 61 \, \text{cm/sec} \qquad \text{(to two significant digits)}$$

Similarly, we have

$$\sin 30° = \frac{v_y}{70}$$

and therefore,

$$v_y = (\sin 30°)(70) = 35 \, \text{cm/sec}$$

In summary, now, the $x$-component of the velocity is about 61 cm/sec, and the $y$-component is 35 cm/sec. ▪

Our last example in this section will indicate how vectors are used in navigation. First, however, let us introduce some terminology. Suppose that an airplane has a **heading** of due east. This means that the airplane is pointed due east, and if there were no wind effects, the plane would indeed travel due east with respect to the ground. The **air speed** is the speed of the airplane relative to the air, whereas the **ground speed** is the plane's speed relative to the ground. Again, if there were no wind effects, then the air speed and the ground speed would be equal. Now suppose that the heading and air speed of an airplane are represented by the velocity vector **V** in Figure 18. (The direction of **V** is the heading; the magnitude of **V** is the air speed.) Also suppose that the wind velocity is represented by the vector **W** in Figure 18. Then the vector sum **V + W** represents the actual velocity of the plane with respect to the ground. The direction of **V + W** is called the **course**; it is the direction in which the airplane is moving with respect to the ground. The magnitude of **V + W** is the ground speed (which was defined previously). The angle $\theta$ in Figure 18 is called, naturally enough, the **drift angle**.

In navigation, directions are given in terms of the angle measured clockwise from true north. For example, the direction of the velocity vector **v** in Figure 19 is 120°. We say in this case that the **bearing** of **v** is 120°.

**FIGURE 18**

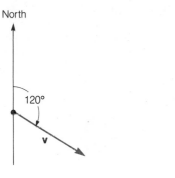

**FIGURE 19**

▪

**EXAMPLE 5** The heading and air speed of an airplane are 60° and 250 mph, respectively. If the wind is 40 mph from 150°, find the ground speed, the drift angle, and the course.

**Solution** In Figure 20(a), the vector $\overrightarrow{OA}$ represents the air speed of 250 mph and the heading of 60°. The vector $\overrightarrow{OW}$ represents a wind of 40 mph from 150°. Also, angle $WOA = 90°$. (Why?) In Figure 20(b), we have completed the parallelogram to obtain the vector sum $\overrightarrow{OA} + \overrightarrow{OW} = \overrightarrow{OB}$. The length of $\overrightarrow{OB}$ represents the ground speed, $\theta$ is the drift angle, and $\alpha$ is the bearing of the course. Because triangle $BOA$ is a right triangle, we have

$$\tan \theta = \frac{40}{250} = \frac{4}{25} \qquad \left| \overrightarrow{OB} \right| = \sqrt{250^2 + 40^2}$$

$$\theta \approx 9.1° \qquad\qquad\qquad \approx 253.2$$

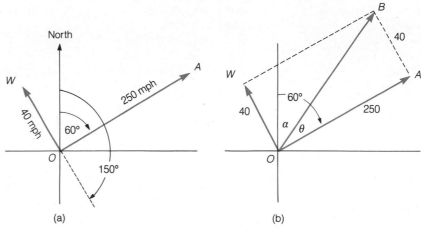

**FIGURE 20**

From these calculations, we conclude that the ground speed is approximately 253.2 mph and the drift angle is about 9.1°. We still need to compute the bearing of the course. From Figure 20(b), we have

$$\alpha = 60° - \theta \approx 60° - 9.1° = 50.9°$$

Thus, the bearing of the course is 50.9° (to one decimal place).

## EXERCISE SET 9.2

### A

*In Exercises 1–26, assume that the coordinates of the points P, Q, R, S, and O are as follows:*

$$P(-1,3) \qquad Q(4,6) \qquad R(4,3) \qquad S(5,9) \qquad O(0,0)$$

*For each exercise, draw the indicated vector (using graph paper) and compute its magnitude. In Exercises 7–20, compute the sums using the definition given on page 457. In Exercises 21–26, use the parallelogram law to compute the sums.*

1. $\vec{PQ}$
2. $\vec{QP}$
3. $\vec{SQ}$
4. $\vec{QS}$
5. $\vec{OP}$
6. $\vec{PO}$
7. $\vec{PQ} + \vec{QS}$
8. $\vec{SQ} + \vec{QP}$
9. $\vec{OP} + \vec{PQ}$
10. $\vec{OS} + \vec{SQ}$
11. $(\vec{OS} + \vec{SQ}) + \vec{QP}$
12. $(\vec{OS} + \vec{SP}) + \vec{PR}$
13. $\vec{OP} + \vec{QS}$
14. $\vec{QS} + \vec{PO}$
15. $\vec{SR} + \vec{PO}$
16. $\vec{OS} + \vec{QO}$
17. $\vec{OP} + \vec{RQ}$
18. $\vec{OP} + \vec{QR}$
19. $\vec{SQ} + \vec{RO}$
20. $\vec{SQ} + \vec{OR}$

21. $\vec{OP} + \vec{OR}$
22. $\vec{OP} + \vec{OQ}$
23. $\vec{RP} + \vec{RS}$
24. $\vec{QP} + \vec{QR}$
25. $\vec{SO} + \vec{SQ}$
26. $\vec{SQ} + \vec{SR}$

*In Exercises 27–32, the vectors $\mathbf{F}$ and $\mathbf{G}$ denote two forces that act on an object: $\mathbf{G}$ acts horizontally to the right, and $\mathbf{F}$ acts vertically upward. In each case, use the information that is given to compute $|\mathbf{F} + \mathbf{G}|$ and $\theta$, where $\theta$ is the angle between $\mathbf{G}$ and the resultant.*

27. $|\mathbf{F}| = 4\,\text{N}, \quad |\mathbf{G}| = 5\,\text{N}$
28. $|\mathbf{F}| = 15\,\text{N}, \quad |\mathbf{G}| = 6\,\text{N}$
29. $|\mathbf{F}| = |\mathbf{G}| = 9\,\text{N}$
30. $|\mathbf{F}| = 28\,\text{N}, \quad |\mathbf{G}| = 1\,\text{N}$
31. $|\mathbf{F}| = 3.22\,\text{N}, \quad |\mathbf{G}| = 7.21\,\text{N}$
32. $|\mathbf{F}| = 4.06\,\text{N}, \quad |\mathbf{G}| = 26.83\,\text{N}$

*In Exercises 33–38, the vectors $\mathbf{F}$ and $\mathbf{G}$ represent two forces acting on an object, as indicated in the following figure. In each case, use the given information to compute (to two decimal* *places) the magnitude and direction of the resultant. (Give the direction of the resultant by specifying the angle θ between* **F** *and the resultant.)*

33. $|\mathbf{F}| = 5\,\text{N}, \quad |\mathbf{G}| = 4\,\text{N}, \quad \alpha = 80°$
34. $|\mathbf{F}| = 8\,\text{N}, \quad |\mathbf{G}| = 10\,\text{N}, \quad \alpha = 60°$
35. $|\mathbf{F}| = 16\,\text{N}, \quad |\mathbf{G}| = 25\,\text{N}, \quad \alpha = 35°$
36. $|\mathbf{F}| = 4.24\,\text{N}, \quad |\mathbf{G}| = 9.01\,\text{N}, \quad \alpha = 45°$
37. $|\mathbf{F}| = 50\,\text{N}, \quad |\mathbf{G}| = 25\,\text{N}, \quad \alpha = 130°$
38. $|\mathbf{F}| = 1.26\,\text{N}, \quad |\mathbf{G}| = 2.31\,\text{N}, \quad \alpha = 160°$

*In Exercises 39–46, the initial point for each vector is the origin, and θ denotes the angle (measured counterclockwise) from the x-axis to the vector. In each case, compute the horizontal and vertical components of the given vector. (Round off your answers to two decimal places.)*

39. The magnitude of **V** is 16 cm/sec, and $\theta = 30°$.
40. The magnitude of **V** is 40 cm/sec, and $\theta = 60°$.
41. The magnitude of **F** is 14 N, and $\theta = 75°$.
42. The magnitude of **F** is 23.12 N, and $\theta = 52°$.
43. The magnitude of **V** is 1 cm/sec, and $\theta = 135°$.
44. The magnitude of **V** is 12 cm/sec, and $\theta = 120°$.
45. The magnitude of **F** is 1.25 N, and $\theta = 145°$.
46. The magnitude of **F** is 6.34 N, and $\theta = 175°$.

*In Exercises 47–50, use the given flight data to compute the ground speed, the drift angle, and the course. (Round off your answers to two decimal places.)*

47. The heading and air speed are 30° and 300 mph, respectively; the wind is 25 mph from 120°.

48. The heading and air speed are 45° and 275 mph, respectively; the wind is 50 mph from 135°.

49. The heading and air speed are 100° and 290 mph, respectively; the wind is 45 mph from 190°.

50. The heading and air speed are 90° and 220 mph, respectively; the wind is 80 mph from **(a)** 180°; **(b)** 90°.

**B**

51. A block weighing 12 lb rests on an inclined plane, as indicated in the following figure. Determine the components of the weight perpendicular to and parallel to the plane. Round off your answers to two decimal places. *Hint:* In Figure (b), the component of the weight perpendicular to the plane is $|\overrightarrow{OR}|$; the component parallel to the plane is $|\overrightarrow{OP}|$. Why does angle $QOR$ equal 35°?

(a)

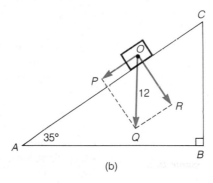

(b)

*In Exercises 52 and 53, you are given the weight of a block on an inclined plane, along with the angle θ that the inclined plane makes with the horizontal. In each case, determine the components of the weight perpendicular to and parallel to the plane. (Round off your answers to two decimal places where necessary.)*

52. 15 lb; $\theta = 30°$          53. 12 lb; $\theta = 10°$

54. A block rests on an inclined plane that makes an angle of 20° with the horizontal. The component of the weight parallel to the plane is 34.2 lb.
   **(a)** Determine the weight of the block. (Round off your answer to one decimal place.)
   **(b)** Determine the component of the weight perpendicular to the plane. (Round off your answer to one decimal place.)

55. In Section 9.3, we will see that vector addition is associative. That is, for any three vectors **A**, **B**, and **C**, we have $(\mathbf{A} + \mathbf{B}) + \mathbf{C} = \mathbf{A} + (\mathbf{B} + \mathbf{C})$. In this exercise, you

are going to check that this property holds in a particular case. Let A, B, and C be the vectors whose initial and terminal points are as follows:

| VECTOR | INITIAL POINT | TERMINAL POINT |
|--------|---------------|----------------|
| A | $(-1, 2)$ | $(2, 4)$ |
| B | $(1, 2)$ | $(3, 0)$ |
| C | $(6, 2)$ | $(4, -3)$ |

(a) Use the definition of vector addition on page 457 to determine the initial and terminal points of $(A + B) + C$. *Suggestion:* Use graph paper.

(b) Use the definition of vector addition to determine the initial and terminal points of $A + (B + C)$. [Your answers should agree with those in part (a).]

## VECTORS IN THE PLANE, AN ALGEBRAIC APPROACH

**9.3**

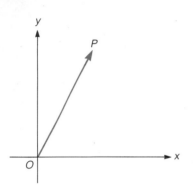

**FIGURE 1**

*A great many of the mathematical ideas that apply to physics and engineering are collected in the concept of vector spaces. This branch of mathematics has applications in such practical problems as calculating the vibrations of bridges and airplane wings. Logical extensions to spaces of infinitely many dimensions are widely used in modern theoretical physics as well as in many branches of mathematics itself.*

From *The Mathematical Sciences*, edited by the Committee on Support of Research in the Mathematical Sciences with the collaboration of George Boehm (Cambridge, Mass.: The MIT Press, 1969)

The geometric concept of a vector in the plane can be recast in an algebraic setting. This is useful both for computational purposes and (as our opening quotation implies) for more advanced work.

Consider an $x$-$y$ coordinate system and a vector $\overrightarrow{OP}$ whose initial point coincides with the origin, as shown in Figure 1. We call $\overrightarrow{OP}$ the **position vector** (or **radius vector**) of the point $P$. Most of our work in this section will involve such position vectors. There is no loss of generality in focusing on these types of vectors, for, as indicated in Figure 2, each vector **v** in the plane is equal to a unique position vector $\overrightarrow{OP}$.

If the coordinates of the point $P$ are $(a, b)$, we call $a$ and $b$ the **components** of the vector $\overrightarrow{OP}$, and we use the notation

$$\langle a, b \rangle$$

to denote this vector (see Figure 3).

**FIGURE 2**

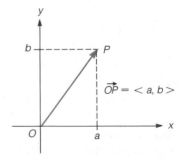

**FIGURE 3** The vector $\overrightarrow{OP}$ is denoted by $\langle a, b \rangle$. The coordinates $a$ and $b$ are the components of the vector.

In the previous section, we said that two vectors are equal provided they have the same length and the same direction. For vectors $\langle a, b \rangle$ and $\langle c, d \rangle$, this implies that

$$\langle a, b \rangle = \langle c, d \rangle \qquad \text{if and only if} \qquad a = c \quad \text{and} \quad b = d$$

(Notice the similarity here to equality of complex numbers. Two complex numbers are equal provided their corresponding real and imaginary parts are equal; two vectors are equal provided their corresponding components are equal.)

It's easy to calculate the length of a vector $\mathbf{v}$ when its components are given. Suppose that $\mathbf{v} = \langle v_1, v_2 \rangle$. Applying the Pythagorean Theorem in Figure 4, we have

$$|\mathbf{v}|^2 = v_1^2 + v_2^2$$

and consequently,

$$|\mathbf{v}| = \sqrt{v_1^2 + v_2^2}$$

Although Figure 4 shows the point $(v_1, v_2)$ in the first quadrant, you can check for yourself that the same formula results when $(v_1, v_2)$ is located in any of the other three quadrants. We therefore have the following general formula.

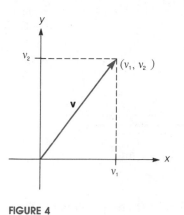

**FIGURE 4**

---

**The Length of a Vector**

If $\mathbf{v} = \langle v_1, v_2 \rangle$, then

$$|\mathbf{v}| = \sqrt{v_1^2 + v_2^2}$$

---

**EXAMPLE 1**  Compute the length of a vector $\mathbf{v} = \langle 2, -4 \rangle$.

**Solution**
$$\begin{aligned}
|\mathbf{v}| &= \sqrt{v_1^2 + v_2^2} \\
&= \sqrt{2^2 + (-4)^2} = \sqrt{20} = 2\sqrt{5}
\end{aligned}$$

Even if the initial point of a vector $\mathbf{v}$ is not the origin, we can still find the *components* of $\mathbf{v}$ by determining the components of the equivalent position vector. The formula that we are going to derive for this is as follows.

---

If the coordinates of the points $P$ and $Q$ are $P(x_1, y_1)$ and $Q(x_2, y_2)$, then

$$\overrightarrow{PQ} = \langle x_2 - x_1, y_2 - y_1 \rangle$$

---

To derive this formula, we first construct the right triangle shown in Figure 5(a). Now we let $R$ denote the point $(x_2 - x_1, y_2 - y_1)$ and draw the position vector $\overrightarrow{OR}$ shown in Figure 5(b). Since the right triangles in Figures 5(a) and 5(b) are congruent and have corresponding legs that are parallel, we have

$$\overrightarrow{PQ} = \overrightarrow{OR} = \langle x_2 - x_1, y_2 - y_1 \rangle$$

as required. We can summarize this result as follows. For any vector $\mathbf{v}$, we have

$x$-component of $\mathbf{v}$:

$x$-coordinate of terminal point $-$ $x$-coordinate of initial point

$y$-component of $\mathbf{v}$:

$y$-coordinate of terminal point $-$ $y$-coordinate of initial point

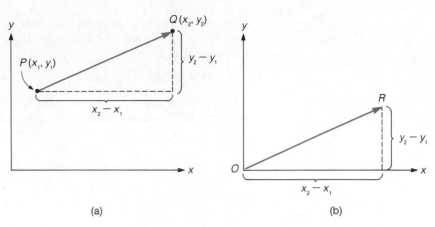

(a)                    (b)

FIGURE 5

**EXAMPLE 2**   Let $P$ and $Q$ be the points $P(3, 1)$ and $Q(7, 3)$. Find the components of $\overrightarrow{PQ}$.

**Solution**   $x$-component of $\overrightarrow{PQ}$ = $x$-coordinate of $Q$ − $x$-coordinate of $P$
$$= 7 - 3 = 4$$
$y$-component of $\overrightarrow{PQ}$ = $y$-coordinate of $Q$ − $y$-coordinate of $P$
$$= 3 - 1 = 2$$

Consequently,

$$\overrightarrow{PQ} = \langle 4, 2 \rangle$$

Vector addition is particularly simple to carry out when the vectors are in component form. Indeed, we can use the parallelogram law to verify the following result.

> ### Theorem
>
> If $\mathbf{u} = \langle u_1, u_2 \rangle$ and $\mathbf{v} = \langle v_1, v_2 \rangle$, then $\mathbf{u} + \mathbf{v} = \langle u_1 + v_1, u_2 + v_2 \rangle$.

This theorem tells us that vector addition can be carried out *componentwise*; in other words, to add two vectors, just add the corresponding components. For example,

$$\langle 1, 2 \rangle + \langle 3, 7 \rangle = \langle 4, 9 \rangle$$

To see why this theorem is valid, consider Figure 6. (*Caution:* In reading the derivation that follows, don't confuse notation such as $OA$ with $\overrightarrow{OA}$; recall that $OA$ denotes the length of the line segment $\overline{OA}$). Since the $x$-component of $\mathbf{u}$ is $u_1$, we have

$$OB = u_1$$

Also,

$$BC = v_1 \qquad \text{(Why?)}$$

Therefore,

$$OC = OB + BC$$
$$= u_1 + v_1$$

FIGURE 6

But $OC$ is the $x$-component of the vector $\overrightarrow{OP}$, and by the parallelogram law, $\overrightarrow{OP} = \mathbf{u} + \mathbf{v}$. In other words, the $x$-component of $\mathbf{u} + \mathbf{v}$ is $u_1 + v_1$, as we wished to show. The fact that the $y$-component of $\mathbf{u} + \mathbf{v}$ is $u_2 + v_2$ is proved in a similar fashion.

The vector $\langle 0,0 \rangle$ is called the **zero vector**, and it is denoted by $\mathbf{0}$. Notice that for any vector $\mathbf{v} = \langle v_1, v_2 \rangle$, we have

$\mathbf{v} + \mathbf{0} = \mathbf{v}$      because      $\langle v_1, v_2 \rangle + \langle 0,0 \rangle = \langle v_1, v_2 \rangle$

and

$\mathbf{0} + \mathbf{v} = \mathbf{v}$      because      $\langle 0,0 \rangle + \langle v_1, v_2 \rangle = \langle v_1, v_2 \rangle$

So for the operation of vector addition, the zero vector plays the same role as does the real number zero in addition of real numbers. There are, in fact, several other ways in which vector addition resembles ordinary addition of real numbers. We'll return to this point again near the end of this section.

In the box that follows, we define an operation called **scalar multiplication**, in which a vector is "multiplied" by a real number (a **scalar**) to obtain another vector.

| **Definition of Scalar Multiplication** | |
|---|---|
| | EXAMPLES |
| For each real number $k$ and each vector $\mathbf{v} = \langle x, y \rangle$, we define a vector $k\mathbf{v}$ by the equation $$k\mathbf{v} = k\langle x,y \rangle = \langle kx, ky \rangle$$ | If $\mathbf{v} = \langle 2,1 \rangle$, then $$2\mathbf{v} = \langle 4,2 \rangle$$ $$3\mathbf{v} = \langle 6,3 \rangle$$ $$-1\mathbf{v} = \langle -2,-1 \rangle$$ |

In geometric terms, the length of $k\mathbf{v}$ is $|k|$ times the length of $\mathbf{v}$. The vectors $\mathbf{v}$ and $k\mathbf{v}$ have the same direction if $k > 0$ and opposite directions if $k < 0$. For example, let $\mathbf{v} = \langle 2,1 \rangle$. In Figure 7(a) we show the vectors $\mathbf{v}$ and $3\mathbf{v}$, while in Figure 7(b) we show $\mathbf{v}$ and $-2\mathbf{v}$.

(a) The vectors **v** and 3**v** have the same direction. The length of 3**v** is three times that of **v**.

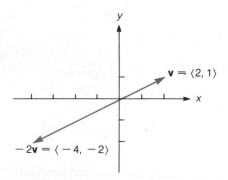

(b) The vectors **v** and $-2$**v** have opposite directions. The length of $-2$**v** is twice that of **v**.

FIGURE 7

**EXAMPLE 3** Let $\mathbf{v} = \langle 3,4 \rangle$ and $\mathbf{w} = \langle -1,2 \rangle$. Compute each of the following.

(a) $\mathbf{v} + \mathbf{w}$     (b) $-2\mathbf{v} + 3\mathbf{w}$     (c) $|-2\mathbf{v} + 3\mathbf{w}|$

**Solution**

(a) $\mathbf{v} + \mathbf{w} = \langle 3,4 \rangle + \langle -1,2 \rangle = \langle 3 - 1, 4 + 2 \rangle = \langle 2,6 \rangle$

(b) $-2\mathbf{v} + 3\mathbf{w} = -2\langle 3,4 \rangle + 3\langle -1,2 \rangle = \langle -6, -8 \rangle + \langle -3,6 \rangle$
$= \langle -9, -2 \rangle$

(c) $|-2\mathbf{v} + 3\mathbf{w}| = |\langle -9, -2 \rangle| = \sqrt{(-9)^2 + (-2)^2} = \sqrt{85}$

For each vector $\mathbf{v}$, we define a vector $-\mathbf{v}$, called the **negative** of $\mathbf{v}$, by the equation

$$-\mathbf{v} = -1\mathbf{v}$$

Thus, if $\mathbf{v} = \langle a,b \rangle$, then $-\mathbf{v} = \langle -a, -b \rangle$. As indicated in Figure 8, the vectors $\mathbf{v}$ and $-\mathbf{v}$ have the same length but opposite directions.

We can use the ideas in the preceding paragraph to define **vector subtraction**. Given two vectors $\mathbf{u}$ and $\mathbf{v}$, we define a vector $\mathbf{u} - \mathbf{v}$ by the equation

$$\mathbf{u} - \mathbf{v} = \mathbf{u} + (-\mathbf{v}) \tag{1}$$

First let's see what equation (1) is saying in terms of components. Then we will indicate a nice geometric interpretation of vector subtraction.

If $\mathbf{u} = \langle u_1, u_2 \rangle$ and $\mathbf{v} = \langle v_1, v_2 \rangle$, then equation (1) tells us that

$$\mathbf{u} - \mathbf{v} = \langle u_1, u_2 \rangle + \langle -v_1, -v_2 \rangle$$
$$= \langle u_1 + (-v_1), u_2 + (-v_2) \rangle$$

That is,

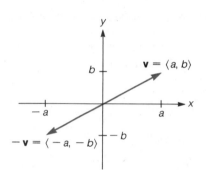

FIGURE 8

$$\mathbf{u} - \mathbf{v} = \langle u_1 - v_1, u_2 - v_2 \rangle \tag{2}$$

In other words, to subtract two vectors, just subtract the corresponding components.

Vector subtraction can be interpreted geometrically. According to the formula in the box on page 465, the right-hand side of equation (2) represents a vector drawn from the terminal point of $\mathbf{v}$ to the terminal point of $\mathbf{u}$. Figure 9 summarizes this fact, and Figure 10 provides a geometric comparison of vector addition and vector subtraction.

FIGURE 9

**EXAMPLE 4** Let $\mathbf{u} = \langle 5,3 \rangle$ and $\mathbf{v} = \langle -1,2 \rangle$. Compute $3\mathbf{u} - \mathbf{v}$.

**Solution**

$3\mathbf{u} - \mathbf{v} = 3\langle 5,3 \rangle - \langle -1,2 \rangle$
$= \langle 15,9 \rangle - \langle -1,2 \rangle = \langle 16,7 \rangle$

The next three examples deal with unit vectors. By definition, any vector with a length of 1 is called a **unit vector**. Two particularly useful unit vectors are

$$\mathbf{i} = \langle 1,0 \rangle \quad \text{and} \quad \mathbf{j} = \langle 0,1 \rangle$$

These are shown in Figure 11.

**FIGURE 10** $\mathbf{u} + \mathbf{v}$ and $\mathbf{u} - \mathbf{v}$ are the directed diagonals of parallelogram $OABC$: $\mathbf{u} + \mathbf{v} = \overrightarrow{OB}$ and $\mathbf{u} - \mathbf{v} = \overrightarrow{AC}$.

**FIGURE 11**

Any vector $\mathbf{v} = \langle x, y \rangle$ can be uniquely expressed in terms of the unit vectors $\mathbf{i}$ and $\mathbf{j}$ as follows:

$$\langle x, y \rangle = x\mathbf{i} + y\mathbf{j} \tag{3}$$

To verify equation (3), we have

$$
\begin{aligned}
x\mathbf{i} + y\mathbf{j} &= x\langle 1, 0 \rangle + y\langle 0, 1 \rangle \\
&= \langle x, 0 \rangle + \langle 0, y \rangle \\
&= \langle x, y \rangle
\end{aligned}
$$

**EXAMPLE 5** (a) Express the vector $\langle 3, -7 \rangle$ in terms of the unit vectors $\mathbf{i}$ and $\mathbf{j}$.
(b) Express the vector $\mathbf{v} = -4\mathbf{i} + 5\mathbf{j}$ in component form.

**Solution** (a) Using equation (3), we can write

$$\langle 3, -7 \rangle = 3\mathbf{i} + (-7)\mathbf{j} = 3\mathbf{i} - 7\mathbf{j}$$

(b) $\mathbf{v} = -4\mathbf{i} + 5\mathbf{j} = -4\langle 1, 0 \rangle + 5\langle 0, 1 \rangle$
$= \langle -4, 0 \rangle + \langle 0, 5 \rangle = \langle -4, 5 \rangle$

Thus, the component form of $\mathbf{v}$ is $\langle -4, 5 \rangle$.

**EXAMPLE 6** Find a unit vector $\mathbf{u}$ that has the same direction as the vector $\mathbf{v} = \langle 3, 4 \rangle$.

**Solution** First, let us determine the length of $\mathbf{v}$:

$$|\mathbf{v}| = \sqrt{3^2 + 4^2} = \sqrt{25} = 5$$

So we want a vector whose length is one-fifth that of $\mathbf{v}$ and whose direction is the same as that of $\mathbf{v}$. Such a vector is

$$\mathbf{u} = \frac{1}{5}\mathbf{v} = \frac{1}{5}\langle 3, 4 \rangle = \left\langle \frac{3}{5}, \frac{4}{5} \right\rangle$$

(You should check for yourself now that the length of this vector is 1.)

**EXAMPLE 7** The angle from the positive $x$-axis to the unit vector **u** is $\pi/3$, as indicated in Figure 12. Determine the components of **u**.

**Solution** Let $P$ denote the terminal point of **u**. Since $P$ lies on the unit circle, the coordinates of $P$ are by definition $(\cos(\pi/3), \sin(\pi/3))$. We therefore have

$$\mathbf{u} = \left\langle \frac{1}{2}, \frac{\sqrt{3}}{2} \right\rangle.$$

**FIGURE 12**

It was mentioned earlier that there are several ways in which vector addition resembles ordinary addition of real numbers. In the previous section, for example, we saw that vector addition is commutative. That is, for any two vectors **u** and **v**,

$$\mathbf{u} + \mathbf{v} = \mathbf{v} + \mathbf{u}$$

By using components, we can easily verify that this property holds. (In the previous section, we used a geometric argument to establish this property.) We begin by letting $\mathbf{u} = \langle u_1, u_2 \rangle$ and $\mathbf{v} = \langle v_1, v_2 \rangle$. Then we have

$$
\begin{aligned}
\mathbf{u} + \mathbf{v} &= \langle u_1, u_2 \rangle + \langle v_1, v_2 \rangle \\
&= \langle u_1 + v_1, u_2 + v_2 \rangle \\
&= \langle v_1 + u_1, v_2 + u_2 \rangle \quad \text{Addition of real numbers is commutative.} \\
&= \langle v_1, v_2 \rangle + \langle u_1, u_2 \rangle \\
&= \mathbf{v} + \mathbf{u}
\end{aligned}
$$

which is what we wanted to show.

There are a number of other properties of vector addition and scalar multiplication that can be proved in a similar fashion. In the following box, we list a particular collection of these properties, known as the **vector space properties**. (Exercises 55–58 ask that you verify these properties by using components, just as we did for the commutative property.)

---

**Properties of Vector Addition and Scalar Multiplication**

For all vectors **u**, **v**, and **w**, and for all scalars (real numbers) $a$ and $b$, the following properties hold.

1. $\mathbf{u} + (\mathbf{v} + \mathbf{w}) = (\mathbf{u} + \mathbf{v}) + \mathbf{w}$      5. $a(\mathbf{u} + \mathbf{v}) = a\mathbf{u} + a\mathbf{v}$
2. $\mathbf{0} + \mathbf{v} = \mathbf{v} + \mathbf{0} = \mathbf{v}$      6. $(a + b)\mathbf{v} = a\mathbf{v} + b\mathbf{v}$
3. $\mathbf{v} + (-\mathbf{v}) = \mathbf{0}$      7. $(ab)\mathbf{v} = a(b\mathbf{v})$
4. $\mathbf{u} + \mathbf{v} = \mathbf{v} + \mathbf{u}$      8. $1\mathbf{v} = \mathbf{v}$

---

## EXERCISE SET 9.3

### A

*In Exercises 1–6, sketch each vector in an x-y coordinate system, and compute the length of the vector.*

1. $\langle 4, 3 \rangle$      2. $\langle 5, 12 \rangle$      3. $\langle -4, 2 \rangle$

4. $\langle -6, -6 \rangle$      5. $\langle \frac{3}{4}, -\frac{1}{2} \rangle$      6. $\langle -3, 0 \rangle$

*In Exercises 7–12, the coordinates of two points P and Q are given. In each case, determine the components of the vector $\overrightarrow{PQ}$. Write your answers in the form $\langle a, b \rangle$.*

7. $P(2, 3)$ and $Q(3, 7)$

8. $P(5,1)$ and $Q(4,9)$

9. $P(-2,-3)$ and $Q(-3,-2)$

10. $P(0,-4)$ and $Q(0,-8)$

11. $P(-5,1)$ and $Q(3,-4)$

12. $P(1,0)$ and $Q(0,1)$

*In Exercises 13–32, assume that the vectors* **a**, **b**, **c**, *and* **d** *are defined as follows:*

$$\mathbf{a} = \langle 2,3 \rangle \quad \mathbf{b} = \langle 5,4 \rangle \quad \mathbf{c} = \langle 6,-1 \rangle \quad \mathbf{d} = \langle -2,0 \rangle$$

*Compute each of the indicated quantities.*

13. $\mathbf{a} + \mathbf{b}$

14. $\mathbf{c} + \mathbf{d}$

15. $2\mathbf{a} + 4\mathbf{b}$

16. $-2\mathbf{c} + 2\mathbf{d}$

17. $|\mathbf{b} + \mathbf{c}|$

18. $|5\mathbf{b} + 5\mathbf{c}|$

19. $|\mathbf{a} + \mathbf{c}| - |\mathbf{a}| - |\mathbf{c}|$

20. $1/|\mathbf{d}|$

21. $\mathbf{a} + (\mathbf{b} + \mathbf{c})$

22. $(\mathbf{a} + \mathbf{b}) + \mathbf{c}$

23. $3\mathbf{a} + 4\mathbf{a}$

24. $|4\mathbf{b} + 5\mathbf{b}|$

25. $\mathbf{a} - \mathbf{b}$

26. $\mathbf{b} - \mathbf{c}$

27. $3\mathbf{b} - 4\mathbf{d}$

28. $\dfrac{1}{|3\mathbf{b} - 4\mathbf{d}|} \cdot (3\mathbf{b} - 4\mathbf{a})$

29. $\mathbf{a} - (\mathbf{b} + \mathbf{c})$

30. $(\mathbf{a} - \mathbf{b}) - \mathbf{c}$

31. $|\mathbf{c} + \mathbf{d}|^2 - |\mathbf{c} - \mathbf{d}|^2$

32. $|\mathbf{a} + \mathbf{b}|^2 + |\mathbf{a} - \mathbf{b}|^2 - 2|\mathbf{a}|^2 - 2|\mathbf{b}|^2$

*In Exercises 33–38, express each vector in terms of the unit vectors* **i** *and* **j**.

33. $\langle 3,8 \rangle$

34. $\langle 4,-2 \rangle$

35. $\langle -8,-6 \rangle$

36. $\langle -9,0 \rangle$

37. $3\langle 5,3 \rangle + 2\langle 2,7 \rangle$

38. $|\langle 12,5 \rangle|\langle 3,4 \rangle + |\langle 3,4 \rangle|\langle 12,5 \rangle$

*In Exercises 39–42, express each vector in the form* $\langle a,b \rangle$.

39. $\mathbf{i} + \mathbf{j}$

40. $\mathbf{i} - 2\mathbf{j}$

41. $5\mathbf{i} - 4\mathbf{j}$

42. $\dfrac{1}{|\mathbf{i} + \mathbf{j}|} (\mathbf{i} + \mathbf{j})$

*In Exercises 43–48, find a unit vector having the same direction as the given vector.*

43. $\langle 4,8 \rangle$

44. $\langle -3,3 \rangle$

45. $\langle 6,-3 \rangle$

46. $\langle -12,5 \rangle$

47. $8\mathbf{i} - 9\mathbf{j}$

48. $\langle 7,3 \rangle - \mathbf{i} + \mathbf{j}$

*In Exercises 49–54, you are given an angle $\theta$ measured counterclockwise from the positive x-axis to a unit vector* $\mathbf{u} = \langle u_1,u_2 \rangle$. *In each case, determine the components $u_1$ and $u_2$.*

49. $\theta = \pi/6$

50. $\theta = \pi/4$

51. $\theta = 2\pi/3$

52. $\theta = 3\pi/4$

53. $\theta = 5\pi/6$

54. $\theta = 3\pi/2$

**B**

*In Exercises 55–58, let* $\mathbf{u} = \langle u_1,u_2 \rangle$, $\mathbf{v} = \langle v_1,v_2 \rangle$, *and* $\mathbf{w} = \langle w_1,w_2 \rangle$.

55. Verify properties 1 and 2 in the box on page 470.

56. Verify properties 3 and 4 in the box on page 470.

57. Verify properties 5 and 6 in the box on page 470.

58. Verify properties 7 and 8 in the box on page 470.

**C**

59. Refer to the accompanying figure. In this exercise you are going to derive the following formula for the cosine of the angle $\theta$ between two nonzero vectors $\langle x_1,y_1 \rangle$ and $\langle x_2,y_2 \rangle$:

$$\cos \theta = \frac{x_1 x_2 + y_1 y_2}{|\langle x_1,y_1 \rangle| \cdot |\langle x_2,y_2 \rangle|}$$

(a) Compute $|\overrightarrow{PQ}|$, using the fact that $\overrightarrow{PQ} = \langle x_2 - x_1, y_2 - y_1 \rangle$. Use your calculations to conclude that

$$|\overrightarrow{PQ}|^2 = |\overrightarrow{OP}|^2 + |\overrightarrow{OQ}|^2 - 2(x_1 x_2 + y_1 y_2)$$

(b) Independent of part (a), use the law of cosines to verify that

$$|\overrightarrow{PQ}|^2 = |\overrightarrow{OP}|^2 + |\overrightarrow{OQ}|^2 - 2|\overrightarrow{OP}| \cdot |\overrightarrow{OQ}| \cos \theta$$

(c) Equate the two expressions for $|\overrightarrow{PQ}|^2$ that you obtained in parts (a) and (b). Then solve for $\cos \theta$ to obtain the required formula.

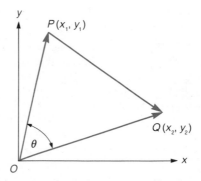

60. Use the formula for $\cos \theta$ given in Exercise 59 to determine the cosine of the angle between each pair of vectors.

(a) $\langle 2,1 \rangle$ and $\langle 1,4 \rangle$

(b) $\langle 2,6 \rangle$ and $\langle 5,-1 \rangle$

## INTRODUCTION TO POLAR COORDINATES

**9.4**

Up until now, we have always specified the location of a point in the plane by means of a rectangular coordinate system. In this section, we introduce another coordinate system that can be used to locate points in the plane. This is the system of **polar coordinates**. We begin by drawing a half-line or ray emanating from a fixed point $O$. The fixed point $O$ is called the **pole** or **origin**, and the half-line is called the **polar axis**. As a matter of convention, the polar axis is usually depicted as being horizontal and extending to the right, as indicated in Figure 2(a). Now let $P$ be any point in the plane. As indicated in Figure 2(b), we initially let $r$ denote the distance from $O$ to $P$ and we let $\theta$ denote the angle measured from the polar axis counterclockwise to $\overline{OP}$. Then the ordered pair $(r, \theta)$ serves to locate the point $P$ with respect to the pole and the polar axis. We refer to $r$ and $\theta$ as the polar coordinates of $P$, and we write $P(r, \theta)$ to indicate that $P$ is the point with polar coordinates $(r, \theta)$.

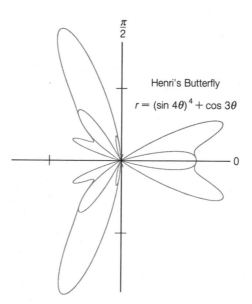

Henri's Butterfly

$r = (\sin 4\theta)^4 + \cos 3\theta$

**FIGURE 1**  This polar coordinate graph was discovered by Henri Berger, a student in one of the author's mathematics classes at U.C.L.A., Spring 1988.

Pole

$O$    Polar axis

(a)

$P(r, \theta)$

$r$

$\theta$

$O$    Polar axis

(b)

**FIGURE 2**

Plotting points in polar coordinates is facilitated by the use of polar coordinate graph paper, such as that shown in Figure 3(a). Figure 3(b) shows the points with polar coordinates $A(2, \pi/6)$, $B(3, 2\pi/3)$, and $C(4, 0)$.

There is a minor complication that arises in using polar rather than rectangular coordinates. Consider, for example, the point $C(4, 0)$ in Figure 3(b). This point could just as well have been labeled with the coordinates $(4, 2\pi)$ or $(4, 2k\pi)$ for any integral value of $k$. Similarly, the coordinates $(r, \theta)$ and $(r, \theta + 2k\pi)$ represent the same point for all integral values of $k$. This is in marked contrast to the situation with rectangular coordinates, where the coordinate representation of each point is unique. Also, what polar coordinates should we assign to the origin? For if $r = 0$ in Figure 2(b), we cannot really define an angle $\theta$. The convention agreed on to cover this case is that the coordinates $(0, \theta)$ denote the origin for all values of $\theta$. Finally, we point out that in working with polar coordinates, it is sometimes useful to let $r$ take on negative values. For example, consider the point $P(2, 5\pi/4)$ in Figure 4. The coordinates $(2, 5\pi/4)$ indicate that to reach $P$ from the origin, we go 2 units in the direction $5\pi/4$. Alternately, we can describe this as $-2$ units in the $\pi/4$ direction. (Refer again to Figure 4.) For reasons such as this, we will adhere to the convention that the polar coordinates $(r, \theta)$ and $(-r, \theta + \pi)$ represent the same point.

(a)

FIGURE 3

(b)

FIGURE 4

FIGURE 5

It is often useful to consider simultaneously both rectangular and polar coordinates. To do this, we draw the two coordinate systems so that the origins coincide and the positive $x$-axis coincides with the polar axis (see Figure 5). Suppose now that a point $P$, other than the origin, has rectangular coordinates $(x, y)$ and polar coordinates $(r, \theta)$, as indicated in Figure 5. We wish to find equations relating the two sets of coordinates. From Figure 5, we see that

$$x^2 + y^2 = r^2 \qquad \sin\theta = \frac{y}{r} \qquad \cos\theta = \frac{x}{r} \qquad \tan\theta = \frac{y}{x}$$

Although Figure 5 displays the point $P$ in the first quadrant, it can be shown that these same equations hold when $P$ is in any quadrant. For reference we summarize these equations as follows.

**Relations between Polar and Rectangular Coordinates**

$$x = r\cos\theta$$
$$y = r\sin\theta$$
$$x^2 + y^2 = r^2$$
$$\tan\theta = \frac{y}{x}$$

**EXAMPLE 1**  The polar coordinates of a point are $(5, \pi/6)$. What are the rectangular coordinates?

Solution  We are given that $r = 5$ and $\theta = \pi/6$. Thus,

$$x = r\cos\theta = 5\cos\frac{\pi}{6} = 5\left(\frac{\sqrt{3}}{2}\right)$$

and

$$y = r\sin\theta = 5\sin\frac{\pi}{6} = 5\left(\frac{1}{2}\right)$$

The rectangular coordinates are therefore $\left(\dfrac{5\sqrt{3}}{2}, \dfrac{5}{2}\right)$.

The definition of the graph of an equation in polar coordinates is similar to the corresponding definition for rectangular coordinates. The **graph** of an equation in polar coordinates is the set of all points $(r, \theta)$ whose coordinates satisfy the given equation. It is often the case that the equation of a curve is simpler in one coordinate system than in another. The next two examples show instances of this.

**EXAMPLE 2**  Convert the polar equation $r^2 = \sin 2\theta$ to rectangular form.

Solution  Replacing $r^2$ by $x^2 + y^2$ and using the double-angle formula for $\sin 2\theta$ yields

$$x^2 + y^2 = 2\sin\theta\cos\theta$$
$$= 2\left(\frac{y}{r}\right)\left(\frac{x}{r}\right) = \frac{2xy}{r^2}$$
$$= \frac{2xy}{x^2 + y^2}$$

Multiplying both sides of this last equation by the quantity $x^2 + y^2$ gives us

$$(x^2 + y^2)^2 = 2xy$$
$$x^4 + 2x^2y^2 + y^4 = 2xy$$
$$x^4 + 2x^2y^2 + y^4 - 2xy = 0$$

This is the rectangular form of the given equation. Notice how much simpler the equation is in its polar coordinate form.

**EXAMPLE 3** Convert the rectangular equation $x^2 + y^2 + ax = a\sqrt{x^2 + y^2}$ to polar form, expressing $r$ as a function of $\theta$. Assume that $a$ is a constant.

Solution Using the relations $x^2 + y^2 = r^2$ and $x = r\cos\theta$, we obtain

$$r^2 + ar\cos\theta = ar$$

Notice that this equation is satisfied by $r = 0$. In other words, the graph of this equation will pass through the origin. This is consistent with the fact that the original equation is satisfied when $x$ and $y$ are both zero. Now assume for the moment that $r \neq 0$. Then we can divide both sides of the last equation by $r$ to obtain

$$r + a\cos\theta = a$$

or

$$r = a - a\cos\theta$$

This expresses $r$ as a function of $\theta$, as required. Notice that when $\theta = 0$, we obtain

$$r = a - a(1) = 0$$

That is, nothing has been lost in dividing through by $r$; the graph will still pass through the origin.

Now let us turn to the problem of graphing curves defined by polar equations. As a first example, consider the equation

$$r = 2\cos\theta$$

We need to keep in mind at this stage that $r$ and $\theta$ are polar, not rectangular, coordinates. Thus, we should not expect the graph of $r = 2\cos\theta$ to be the familiar cosine wave.

First we set up a table using convenient values of $\theta$, as shown in Table 1.

TABLE 1

Values for $r = 2\cos\theta$

| $\theta$ | 0 | $\dfrac{\pi}{6}$ | $\dfrac{\pi}{4}$ | $\dfrac{\pi}{3}$ | $\dfrac{\pi}{2}$ | $\dfrac{2\pi}{3}$ | $\dfrac{3\pi}{4}$ | $\dfrac{5\pi}{4}$ | $\pi$ |
|---|---|---|---|---|---|---|---|---|---|
| $r = 2\cos\theta$ | 2 | $\sqrt{3} \approx 1.7$ | $\sqrt{2} \approx 1.4$ | 1 | 0 | $-1$ | $-\sqrt{2} \approx -1.4$ | $-\sqrt{3} \approx -1.7$ | $-2$ |

As you can check, values of $\theta$ beyond $\pi$ in this case merely lead to points already listed. For instance, with $\theta = 4\pi/3$, we have

$$r = 2\cos\frac{4\pi}{3} = 2\left(-\frac{1}{2}\right) = -1$$

However, the point $(-1, 4\pi/3)$ is the same point as $(1, \pi/3)$, according to our convention regarding negative values of $r$. In Figure 6, we have plotted the points obtained in Table 1 and connected them with a smooth curve. (The curve is in fact a circle, as we will show.)

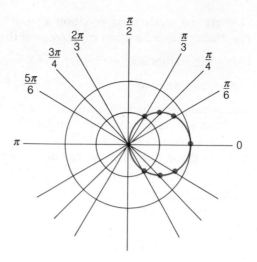

**FIGURE 6**   $r = 2\cos\theta$

As a check on our work in the example just completed, we can convert the equation $r = 2\cos\theta$ to rectangular form. We obtain

$$\sqrt{x^2 + y^2} = 2\left(\frac{x}{\sqrt{x^2 + y^2}}\right)$$

Multiplying both sides of this equation by $\sqrt{x^2 + y^2}$, we have

$$x^2 + y^2 = 2x$$

or

$$x^2 - 2x + y^2 = 0$$

Completing the square now, we find

$$x^2 - 2x + 1 + y^2 = 1$$

or

$$(x - 1)^2 + y^2 = 1$$

This represents a circle whose center is $(1, 0)$ and whose radius is 1, in agreement with Figure 6.

**EXAMPLE 4**   Graph the equation $r = 2 + 2\cos\theta$.

**Solution**   Again we need to keep in mind that we are using polar rather than rectangular coordinates. Thus, even though we have just seen that $r = 2\cos\theta$ represents a circle, we cannot expect the graph of $r = 2 + 2\cos\theta$ to be simply a translate of the circle. As before, we begin by setting up a table. This time, however, we find that values of $\theta$ beyond $\pi$ yield new points on the graph. Thus, we continue the table to $\theta = 2\pi$. In Figure 7, we have plotted the points obtained from Table 2 and connected them with a smooth curve. The resulting heart-shaped curve is known as a **cardioid**.

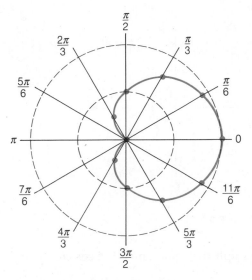

**FIGURE 7**  $r = 2 + 2\cos\theta$

**TABLE 2**

Values for $r = 2 + 2\cos\theta$

| $\theta$ | 0 | $\dfrac{\pi}{6}$ | $\dfrac{\pi}{3}$ | $\dfrac{\pi}{2}$ | $\dfrac{2\pi}{3}$ | $\dfrac{5\pi}{6}$ | $\pi$ | $\dfrac{7\pi}{6}$ | $\dfrac{4\pi}{3}$ | $\dfrac{3\pi}{2}$ | $\dfrac{5\pi}{3}$ | $\dfrac{11\pi}{6}$ | $2\pi$ |
|---|---|---|---|---|---|---|---|---|---|---|---|---|---|
| $r = 2 + 2\cos\theta$ | 4 | 3.7 | 3 | 2 | 1 | 0.3 | 0 | 0.3 | 1 | 2 | 3 | 3.7 | 4 |

We saw in Chapter 3 that symmetry considerations can often be used to lessen the amount of work involved in graphing equations. This is also true for polar equations. In the box that follows, we list four tests for symmetry in polar coordinates.

| **Symmetry Tests in Polar Coordinates** | |
|---|---|
| The graph is symmetric about: | If the following substitution yields an equivalent equation |
| 1. the x-axis | replacing $\theta$ with $-\theta$ |
| 2. (a) the y-axis | replacing $r$ and $\theta$ with $-r$ and $-\theta$, respectively |
| (b) the y-axis | replacing $\theta$ with $\pi - \theta$ |
| 3. the origin | replacing $r$ with $-r$ |

The validity of the first test follows from the fact that the points $(r, \theta)$ and $(r, -\theta)$ are reflections of each other about the x-axis [see Figure 8(a)]. (This first test could have been used to reduce the labor involved in graphing the cardioid $r = 2 + 2\cos\theta$ in Figure 7.) The validity of test 2(a) follows from the fact that the points $(r, \theta)$ and $(-r, -\theta)$ are reflections of each other about the y-axis, as indicated in Figure 8(b). The other tests can be justified in a similar manner.

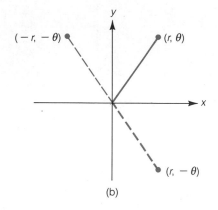

(a)

(b)

**FIGURE 8**

**EXAMPLE 5** Graph the equation $r^2 = 4\cos 2\theta$.

**Solution** As $\theta$ varies from 0 to $\pi$, the values of $2\theta$ run from 0 to $2\pi$, and hence $\cos 2\theta$ runs through one complete cycle of values. Thus, in setting up a table to graph this equation, we do not need to consider values of $\theta$ beyond $\pi$. Furthermore, we do not need to consider values of $\theta$ in the interval $\pi/4 < \theta < 3\pi/4$ because $\cos 2\theta$ is negative there, whereas $r^2$ is always nonnegative. We begin by setting up a table, using a calculator as necessary. See Table 3. If we plot the points in Table 3 and join them with a smooth curve, we obtain the graph in Figure 9(a).

**TABLE 3**

| $\theta$ | 0 | $\dfrac{\pi}{12}$ | $\dfrac{\pi}{8}$ | $\dfrac{\pi}{6}$ | $\dfrac{\pi}{4}$ |
|---|---|---|---|---|---|
| $r = \pm 2\sqrt{\cos 2\theta}$ | $\pm 2$ | $\pm 1.86$ | $\pm 1.68$ | $\pm 1.41$ | 0 |

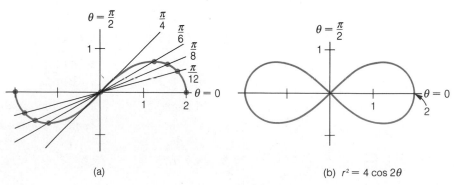

(a)

(b) $r^2 = 4\cos 2\theta$

**FIGURE 9**

Rather than set up another table with values of $\theta$ running from $3\pi/4$ to $\pi$, we can rely on symmetry to complete the graph. According to the first two symmetry tests, the curve is symmetric about both the $x$-axis and the $y$-axis. Thus, we obtain the graph shown in Figure 9(b). This curve is called a **lemniscate**.

All the polar equations we have graphed so far have involved the trigonometric functions. As the next example indicates, this need not always be the case.

**EXAMPLE 6**   Graph the equation $r = \theta/\pi$ for $\theta \geq 0$.

Solution   The equation shows that as $\theta$ increases, so does $r$. We set up a table of values as in the previous examples. Using the entries in Table 4, we construct the graph shown in Figure 10. The curve is known as the **spiral of Archimedes.**

**TABLE 4**

| $\theta$ | 0 | $\dfrac{\pi}{4}$ | $\dfrac{\pi}{2}$ | $\dfrac{3\pi}{4}$ | $\pi$ | $\dfrac{5\pi}{4}$ | $\dfrac{3\pi}{2}$ | $\dfrac{7\pi}{4}$ | $2\pi$ | $\dfrac{5\pi}{2}$ | $3\pi$ | $4\pi$ |
|---|---|---|---|---|---|---|---|---|---|---|---|---|
| $r$ | 0 | $\dfrac{1}{4}$ | $\dfrac{1}{2}$ | $\dfrac{3}{4}$ | 1 | $\dfrac{5}{4}$ | $\dfrac{3}{2}$ | $\dfrac{7}{4}$ | 2 | $\dfrac{5}{2}$ | 3 | 4 |

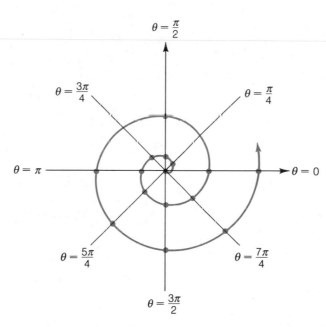

**FIGURE 10**   $r = \theta/\pi$   $(\theta \geq 0)$

# EXERCISE SET 9.4

## A

In Exercises 1–3, convert the given polar coordinates to rectangular coordinates.

1.  (a) $(3, 2\pi/3)$   (b) $(4, 11\pi/6)$   (c) $(4, -\pi/6)$
2.  (a) $(5, \pi/4)$   (b) $(-5, \pi/4)$   (c) $(-5, -\pi/4)$
3.  (a) $(1, \pi/2)$   (b) $(1, 5\pi/2)$   (c) $(-1, \pi/8)$

In Exercises 4 and 5, convert the given rectangular coordinates to polar coordinates. Express your answers in such a way that $r$ is nonnegative and $0 \leq \theta \leq 2\pi$.

4.  $(3, \sqrt{3})$   5.  $(-1, -1)$

*In Exercises 6–15, convert the polar equations to rectangular form.*

6. $r = 6 \sin \theta$

7. $r = 2 \cos \theta$

8. $2 \sin \theta - 3 \cos \theta = r$

9. $r = \tan \theta$

10. $r = 4$

11. $r = 3 \cos 2\theta$

12. $r = 4 \sin 2\theta$

13. $r^2 = \dfrac{8}{2 - \sin^2 \theta}$

14. $r^2 = \dfrac{1}{3 + \cos^2 \theta}$

15. $r(\cos \theta + 2 \sin \theta) = 1$

*In Exercises 16–23, convert the rectangular equations to polar form.*

16. $x^2 + y^2 = 25$

17. $3x - 4y = 2$

18. $y = x^2$

19. $y^2 = x^3$

20. $x^2 + 4x + y^2 + 4y = 0$

21. $2xy = 1$

22. $x^2(x^2 + y^2) = y^2$

23. $9x^2 + y^2 = 9$

*In Exercises 24–45, graph the given equations.*

24. $r = 2 \sin \theta$

25. $r = 3 \cos \theta$

26. $r = 1 + \cos \theta$

27. $r = 1 - \cos \theta$

28. $r^2 = 2 \cos 2\theta$

29. $r^2 = 2 \sin 2\theta$

30. $r = 2\theta/\pi, \quad \theta \geq 0$

31. $r = 1$

32. $\theta = \pi/6$

33. $\theta = 1$

34. $r(2 \sin \theta + \cos \theta) = 1$   *Hint:* First convert to rectangular form.

35. $r = 2 \sin 2\theta$   *(four-leafed rose)*

36. $r = \sin 3\theta$   *(three-leafed rose)*

37. $r = 2 + \sin \theta$   *(limacon)*

38. $r = 1 + 2 \sin \theta$   *(limacon with an inner loop)*

39. $r = 1 - 2 \cos \theta$   *(limacon with an inner loop)*

40. $r = 2 + 2 \sin \theta$   *(cardioid)*

41. $r^2 = 4 \sin 2\theta$   *(lemniscate)*

42. $r = 8 \tan \theta$   *(kappa curve)*

43. $r = \csc \theta + 2$   *(conchoid of Nicomedes)*

44. $r = e^\theta$   *(logarithmic spiral)*

45. $r = \sec^2(\theta/2)$   *(parabola)*

**B**

46. Show that the distance between the two points $P_1(r_1, \theta_1)$ and $P_2(r_2, \theta_2)$ is given by
$$d = \sqrt{r_1^2 + r_2^2 - 2r_1 r_2 \cos(\theta_1 - \theta_2)}$$

47. By converting the polar equation
$$r = a \cos \theta + b \sin \theta$$

to rectangular form, show that the graph is a circle, and find the center and the radius.

48. Show that the rectangular form of the equation $r = a \sin 3\theta$ is $(x^2 + y^2)^2 = ay(3x^2 - y^2)$.

49. Show that the rectangular form of the equation
$$r = \frac{ab}{1 - a \cos \theta} \qquad (a < 1)$$
is $(1 - a^2)x^2 + y^2 - 2a^2bx - a^2b^2 = 0$.

50. Show that the polar form of the equation $\dfrac{x^2}{a^2} - \dfrac{y^2}{b^2} = 1$
is $r^2 = \dfrac{a^2 b^2}{b^2 \cos^2 \theta - a^2 \sin^2 \theta}$.

51. Let $k$ denote a positive constant, and let $F_1$ and $F_2$ denote the points with rectangular coordinates $(-k, 0)$ and $(k, 0)$, respectively. A curve known as the *lemniscate of Bernoulli* is defined as the set of points $P(x, y)$ such that $(F_1 P) \times (F_2 P) = k^2$.
    (a) Show that the rectangular equation of the curve is $(x^2 + y^2)^2 = 2k^2(x^2 - y^2)$.
    (b) Show that the polar equation is $r^2 = 2k^2 \cos 2\theta$.
    (c) Graph the equation $r^2 = 2k^2 \cos 2\theta$.   *Hint:* See Example 5 in the text.

52. A rigid wire of length $2k$ is constrained to move in such a way that its ends are always on the coordinate axes. Let $P$ denote the point where the perpendicular from the origin to the wire meets the wire. Then as the wire moves, $P$ moves and traces out a certain curve.
    (a) If the polar coordinates of $P$ are $(r, \theta)$, show that $r = k \sin 2\theta$.
    (b) Graph the curve $r = k \sin 2\theta$, which represents the path of $P$.
    (c) Show that in rectangular form, the equation becomes $(x^2 + y^2)^3 = 4k^2 x^2 y^2$.

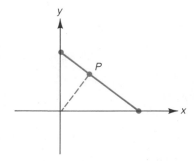

## TRIGONOMETRIC FORM FOR COMPLEX NUMBERS

**9.5**

*Go to Mr. DeMoivre; he knows these things better than I do.*

ISAAC NEWTON [according to Boyer's *A History of Mathematics* (New York: Wiley, 1968)]

In this section, we explore one of the many important connections between trigonometry and the complex number system. We begin with an observation that was first made in the year 1797 by the Norwegian surveyor and mathematician Caspar Wessel. Wessel realized, essentially, that the complex numbers could be visualized as points in the *x-y* plane, the complex number $a + bi$ being identified with the point $(a, b)$. In this context, we often refer to the *x-y* plane as the **complex plane**, and we refer to the complex numbers as points in this plane.

**EXAMPLE 1**  Plot the point $2 + 3i$ in the complex plane.

**Solution**  The complex number $2 + 3i$ is identified with the point $(2, 3)$. See Figure 1.

**FIGURE 1**

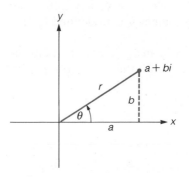

**FIGURE 2**

As indicated in Figure 2, the distance from the origin to the point $a + bi$ is denoted by $r$. We call the distance $r$ the **modulus** of the complex number $a + bi$. The angle $\theta$ in Figure 2 (measured counterclockwise from the positive *x*-axis) is referred to as the **argument** of the complex number $a + bi$. (Using the terminology of the previous section, $r$ and $\theta$ are the **polar coordinates** of the point $a + bi$.)

From Figure 2, we have the following three equations relating the quantities $a$, $b$, $r$, and $\theta$. (Although Figure 2 shows $a + bi$ in the first quadrant, the equations remain valid for the other quadrants as well.)

$$r = \sqrt{a^2 + b^2} \tag{1}$$
$$a = r\cos\theta \tag{2}$$
$$b = r\sin\theta \tag{3}$$

If we have a complex number $z = a + bi$, we can use equations (2) and (3) to write

$$z = (r\cos\theta) + (r\sin\theta)i = r(\cos\theta + i\sin\theta)$$

That is,

$$z = r(\cos\theta + i\sin\theta) \tag{4}$$

The expression that appears on the right-hand side of equation (4) is called the **trigonometric** (or **polar**) **form** of the complex number $z$. In contrast to this, the expression $a + bi$ is referred to as the **rectangular form** of the complex number $z$.

**EXAMPLE 2**   Express the complex number $z = 3\left(\cos\dfrac{\pi}{3} + i\sin\dfrac{\pi}{3}\right)$ in rectangular form.

**Solution**

$$z = 3\left(\cos\frac{\pi}{3} + i\sin\frac{\pi}{3}\right)$$

$$= 3\left(\frac{1}{2} + i\frac{\sqrt{3}}{2}\right) = \frac{3}{2} + \frac{3\sqrt{3}}{2}i$$

The rectangular form is therefore $\frac{3}{2} + \frac{3}{2}\sqrt{3}\,i$.

**EXAMPLE 3**   Find the trigonometric form for the complex number $-\sqrt{2} + i\sqrt{2}$.

**Solution**   We are asked to write the given number in the form $r(\cos\theta + i\sin\theta)$, so we need to find $r$ and $\theta$. Using equation (1) and the values $a = -\sqrt{2}$, $b = \sqrt{2}$, we have

$$r = \sqrt{(-\sqrt{2})^2 + (\sqrt{2})^2} = \sqrt{2+2} = \sqrt{4} = 2$$

Now that we know $r$, we can use equations (2) and (3) to determine $\theta$. From equation (2), we obtain

$$\cos\theta = \frac{a}{r} = \frac{-\sqrt{2}}{2} \tag{5}$$

Similarly, equation (3) gives us

$$\sin\theta = \frac{b}{r} = \frac{\sqrt{2}}{2} \tag{6}$$

One angle satisfying both of equations (5) and (6) is $\theta = 3\pi/4$. (There are other angles, and we'll return to this point in a moment.) In summary, then, we have $r = 2$ and $\theta = 3\pi/4$, so the required trigonometric form is

$$2\left(\cos\frac{3\pi}{4} + i\sin\frac{3\pi}{4}\right)$$

In the example just completed, we noted that $\theta = 3\pi/4$ was only one angle satisfying the conditions $\cos\theta = -\sqrt{2}/2$ and $\sin\theta = \sqrt{2}/2$. Another such angle is $(3\pi/4) + 2\pi$. Indeed, any angle of the form $(3\pi/4) + 2\pi k$, where $k$ is an integer, would do just as well. The upshot of this is that $\theta$, the argument of a complex number, is not uniquely determined. In Example 3, we followed a common convention in converting to trigonometric form: we picked $\theta$ in the interval $0 \le \theta < 2\pi$. Furthermore, although it won't cause us any difficulties in this section, you might also note that the argument $\theta$ is undefined for the complex number $0 + 0i$. (Why?)

We are now ready to derive a formula that will make it easy to multiply two complex numbers in trigonometric form. Suppose that the two complex numbers are

$$r(\cos\alpha + i\sin\alpha) \qquad \text{and} \qquad R(\cos\beta + i\sin\beta)$$

Then their product is

$$rR[(\cos\alpha + i\sin\alpha)(\cos\beta + i\sin\beta)]$$
$$= rR[(\cos\alpha\cos\beta - \sin\alpha\sin\beta) + i(\sin\alpha\cos\beta + \cos\alpha\sin\beta)]$$
$$= rR[\cos(\alpha + \beta) + i\sin(\alpha + \beta)] \qquad \text{using the addition formulas from Section 8.3}$$

Notice that the modulus of the product is $rR$, which is the product of the two original moduli. Also, the argument is $\alpha + \beta$, which is the *sum* of the two original arguments. So, to multiply two complex numbers, just multiply the moduli and add the arguments. There is a similar rule for obtaining the quotient of two complex numbers: Divide their moduli and subtract the arguments. These two rules are stated more precisely in the box that follows. (For a proof of the division rule, see Exercise 75.)

Let $z = r(\cos\alpha + i\sin\alpha)$ and $w = R(\cos\beta + i\sin\beta)$. Then

$$zw = rR[\cos(\alpha + \beta) + i\sin(\alpha + \beta)]$$

Also, if $R \neq 0$, then

$$\frac{z}{w} = \frac{r}{R}[\cos(\alpha - \beta) + i\sin(\alpha - \beta)]$$

**EXAMPLE 4** Let $z = 8\left(\cos\dfrac{5\pi}{3} + i\sin\dfrac{5\pi}{3}\right)$ and $w = 4\left(\cos\dfrac{2\pi}{3} + i\sin\dfrac{2\pi}{3}\right)$. Compute (a) $zw$; (b) $z/w$. Express each answer in both trigonometric and rectangular form.

**Solution** (a) $zw = (8)(4)\left[\cos\left(\dfrac{5\pi}{3} + \dfrac{2\pi}{3}\right) + i\sin\left(\dfrac{5\pi}{3} + \dfrac{2\pi}{3}\right)\right]$

$\qquad = 32\left(\cos\dfrac{7\pi}{3} + i\sin\dfrac{7\pi}{3}\right) = 32\left(\cos\dfrac{\pi}{3} + i\sin\dfrac{\pi}{3}\right)$    trigonometric form

$\qquad = 32\left(\dfrac{1}{2} + i\dfrac{\sqrt{3}}{2}\right) = 16 + 16\sqrt{3}\,i$    rectangular form

(b) $\dfrac{z}{w} = \dfrac{8}{4}\left[\cos\left(\dfrac{5\pi}{3} - \dfrac{2\pi}{3}\right) + i\sin\left(\dfrac{5\pi}{3} - \dfrac{2\pi}{3}\right)\right]$

$\qquad = 2(\cos\pi + i\sin\pi)$    trigonometric form

$\qquad = 2(-1 + i\cdot 0) = -2$    rectangular form

**EXAMPLE 5** Compute $z^2$, where $z = r(\cos\theta + i\sin\theta)$.

**Solution** $z^2 = [r(\cos\theta + i\sin\theta)][r(\cos\theta + i\sin\theta)]$

$\qquad = r^2[\cos(\theta + \theta) + i\sin(\theta + \theta)]$

$\qquad = r^2(\cos 2\theta + i\sin 2\theta)$

The result in Example 5 is a particular case of an important theorem attributed to Abraham DeMoivre (1667–1754). In the box that follows, we state DeMoivre's Theorem. (The theorem can be proved using *mathematical induction*, which is discussed in Chapter 13.)

### DeMoivre's Theorem

Let $n$ be a natural number. Then

$$[r(\cos\theta + i\sin\theta)]^n = r^n(\cos n\theta + i\sin n\theta)$$

**EXAMPLE 6** Use DeMoivre's Theorem to compute $(-\sqrt{2} + i\sqrt{2})^5$. Express your answer in rectangular form.

**Solution** In Example 3, we saw that the trigonometric form of $-\sqrt{2} + i\sqrt{2}$ is given by

$$-\sqrt{2} + i\sqrt{2} = 2\left(\cos\frac{3\pi}{4} + i\sin\frac{3\pi}{4}\right)$$

Therefore,

$$(-\sqrt{2} + i\sqrt{2})^5 = 2^5\left(\cos\frac{15\pi}{4} + i\sin\frac{15\pi}{4}\right)$$

$$= 32\left[\frac{\sqrt{2}}{2} + i\left(-\frac{\sqrt{2}}{2}\right)\right]$$

$$= 16\sqrt{2} - 16\sqrt{2}\,i$$

The next two examples show how DeMoivre's Theorem is used in computing roots. If $n$ is a natural number and $z^n = w$, then we say that $z$ is an **nth root** of $w$. The work in the examples also relies on the following observation about equality between nonzero complex numbers in trigonometric form: If $r(\cos\theta + i\sin\theta) = R(\cos A + i\sin A)$, then $r = R$ and $\theta = A + 2\pi k$, where $k$ is an integer.

**EXAMPLE 7** Find the cube roots of $8i$.

**Solution** First we express $8i$ in trigonometric form. As can be seen from Figure 3,

$$8i = 8\left(\cos\frac{\pi}{2} + i\sin\frac{\pi}{2}\right)$$

Now we let $z = r(\cos\theta + i\sin\theta)$ denote a cube root of $8i$. Then the equation $z^3 = 8i$ becomes

$$r^3(\cos 3\theta + i\sin 3\theta) = 8\left(\cos\frac{\pi}{2} + i\sin\frac{\pi}{2}\right) \qquad (7)$$

From equation (7), we conclude that $r^3 = 8$ and, consequently, that $r = 2$. Also from equation (7), we have

$$3\theta = \frac{\pi}{2} + 2\pi k$$

or

$$\theta = \frac{\pi}{6} + \frac{2\pi k}{3} \qquad \text{dividing by 3} \qquad (8)$$

If we let $k = 0$, equation (8) yields $\theta = \pi/6$. Thus, one of the cube roots of $8i$ is

$$2\left(\cos\frac{\pi}{6} + i\sin\frac{\pi}{6}\right) = 2\left(\frac{\sqrt{3}}{2} + i\cdot\frac{1}{2}\right) = \sqrt{3} + i$$

**FIGURE 3**

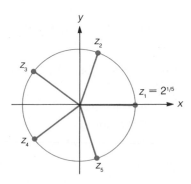

FIGURE 4

Next we let $k = 1$ in equation (8). As you can check, this yields $\theta = 5\pi/6$. So another cube root of $8i$ is

$$2\left(\cos\frac{5\pi}{6} + i\sin\frac{5\pi}{6}\right) = 2\left(\frac{-\sqrt{3}}{2} + i\cdot\frac{1}{2}\right) = -\sqrt{3} + i$$

Similarly, using $k = 2$ in equation (8) yields $\theta = 3\pi/2$. (Verify this.) Consequently, a third cube root is

$$2\left(\cos\frac{3\pi}{2} + i\sin\frac{3\pi}{2}\right) = 2[0 + i(-1)] = -2i$$

We have now found three distinct cube roots of $8i$. If we were to continue the process, using $k = 3$, for example, we would find that no additional roots are obtained in this manner. We therefore conclude that there are exactly three cube roots of $8i$. We have plotted these cube roots in Figure 4. Notice that the points lie equally spaced on a circle of radius 2 about the origin.

When we used DeMoivre's Theorem in Example 7, we found that the number $8i$ had three distinct cube roots. Along these same lines, it is true in general that any nonzero number $a + bi$ possesses exactly $n$ distinct $n$th roots. This is true even if $b = 0$. In the next example, for instance, we compute the five fifth roots of 2.

**EXAMPLE 8** Compute the five fifth roots of 2.

Solution We will follow the procedure we used in Example 7. In trigonometric form, the number 2 becomes $2(\cos 0 + i\sin 0)$. Now we let $z = r(\cos\theta + i\sin\theta)$ denote a fifth root of 2. Then the equation $z^5 = 2$ becomes

$$r^5(\cos 5\theta + i\sin 5\theta) = 2(\cos 0 + i\sin 0) \tag{9}$$

From equation (9), we see that $r^5 = 2$ and, consequently, $r = 2^{1/5}$. Also from equation (9), we have

$$5\theta = 0 + 2\pi k \qquad \text{or} \qquad \theta = \frac{2\pi k}{5}$$

Using the values $k = 0, 1, 2, 3$, and 4 in succession, we obtain the following results:

| $k$ | 0 | 1 | 2 | 3 | 4 |
|-----|---|---|---|---|---|
| $\theta$ | 0 | $2\pi/5$ | $4\pi/5$ | $6\pi/5$ | $8\pi/5$ |

The five fifth roots of 2 are therefore

$$z_1 = 2^{1/5}(\cos 0 + i\sin 0) = 2^{1/5} \qquad z_4 = 2^{1/5}\left(\cos\frac{6\pi}{5} + i\sin\frac{6\pi}{5}\right)$$

$$z_2 = 2^{1/5}\left(\cos\frac{2\pi}{5} + i\sin\frac{2\pi}{5}\right) \qquad z_5 = 2^{1/5}\left(\cos\frac{8\pi}{5} + i\sin\frac{8\pi}{5}\right)$$

$$z_3 = 2^{1/5}\left(\cos\frac{4\pi}{5} + i\sin\frac{4\pi}{5}\right)$$

In Figure 5, we have plotted these five fifth roots. Notice that the points are equally spaced on a circle of radius $2^{1/5}$ about the origin.

FIGURE 5

## EXERCISE SET 9.5

### A

*In Exercises 1–8, plot each point in the complex plane.*

1. $4 + 2i$
2. $4 - 2i$
3. $-5 + i$
4. $-3 - 5i$
5. $1 - 4i$
6. $i$
7. $-i$
8. $1 (= 1 + 0i)$

*In Exercises 9–18, convert each complex number to rectangular form.*

9. $2\left(\cos\dfrac{\pi}{4} + i\sin\dfrac{\pi}{4}\right)$

10. $6\left(\cos\dfrac{5\pi}{3} + i\sin\dfrac{5\pi}{3}\right)$

11. $4\left(\cos\dfrac{5\pi}{6} + i\sin\dfrac{5\pi}{6}\right)$

12. $3\left(\cos\dfrac{3\pi}{2} + i\sin\dfrac{3\pi}{2}\right)$

13. $\sqrt{2}\,(\cos 225° + i\sin 225°)$

14. $\frac{1}{2}(\cos 240° + i\sin 240°)$

15. $\sqrt{3}\left(\cos\dfrac{\pi}{2} + i\sin\dfrac{\pi}{2}\right)$

16. $5(\cos\pi + i\sin\pi)$

17. $4(\cos 75° + i\sin 75°)$   *Hint:* Use the addition formulas from Section 8.3 to evaluate $\cos 75°$ and $\sin 75°$.

18. $2\left(\cos\dfrac{\pi}{8} + i\sin\dfrac{\pi}{8}\right)$   *Hint:* Use the half-angle formulas from Section 8.4 to evaluate $\cos(\pi/8)$ and $\sin(\pi/8)$.

*In Exercises 19–28, convert from rectangular to trigonometric form. (In each case, choose an argument $\theta$ such that $0 \le \theta < 2\pi$.)*

19. $\frac{1}{2}\sqrt{3} + \frac{1}{2}i$

20. $\sqrt{2} + \sqrt{2}i$

21. $-1 + \sqrt{3}i$

22. $-4$

23. $-2\sqrt{3} - 2i$

24. $-3\sqrt{2} - 3\sqrt{2}i$

25. $-6i$

26. $-4 - 4\sqrt{3}i$

27. $\frac{1}{4}\sqrt{3} - \frac{1}{4}i$

28. $16$

*In Exercises 29–54, carry out the indicated operations. Express your results in rectangular form for those cases in which the trigonometric functions are readily evaluated without tables or a calculator.*

29. $2(\cos 22° + i\sin 22°) \times 3(\cos 38° + i\sin 38°)$

30. $4(\cos 5° + i\sin 5°) \times 6(\cos 130° + i\sin 130°)$

31. $\sqrt{2}\left(\cos\dfrac{\pi}{3} + i\sin\dfrac{\pi}{3}\right) \times \sqrt{2}\left(\cos\dfrac{4\pi}{3} + i\sin\dfrac{4\pi}{3}\right)$

32. $\left(\cos\dfrac{\pi}{5} + i\sin\dfrac{\pi}{5}\right) \times \left(\cos\dfrac{\pi}{20} + i\sin\dfrac{\pi}{20}\right)$

33. $3\left(\cos\dfrac{\pi}{7} + i\sin\dfrac{\pi}{7}\right) \times \sqrt{2}\left(\cos\dfrac{\pi}{7} + i\sin\dfrac{\pi}{7}\right)$

34. $\sqrt{3}\,(\cos 3° + i\sin 3°) \times \sqrt{3}\,(\cos 38° + i\sin 38°)$

35. $6(\cos 50° + i\sin 50°) \div 2(\cos 5° + i\sin 5°)$

36. $\sqrt{3}\,(\cos 140° + i\sin 140°) \div 3(\cos 5° + i\sin 5°)$

37. $2^{4/3}\left(\cos\dfrac{5\pi}{12} + i\sin\dfrac{5\pi}{12}\right) \div 2^{1/3}\left(\cos\dfrac{\pi}{4} + i\sin\dfrac{\pi}{4}\right)$

38. $\sqrt{6}\left(\cos\dfrac{16\pi}{9} + i\sin\dfrac{16\pi}{9}\right) \div \sqrt{2}\left(\cos\dfrac{\pi}{9} + i\sin\dfrac{\pi}{9}\right)$

39. $\left(\cos\dfrac{2\pi}{5} + i\sin\dfrac{2\pi}{5}\right) \div \left(\cos\dfrac{2\pi}{5} + i\sin\dfrac{2\pi}{5}\right)$

40. $\left(\cos\dfrac{2\pi}{5} + i\sin\dfrac{2\pi}{5}\right) \div \left(\cos\dfrac{\pi}{10} + i\sin\dfrac{\pi}{10}\right)$

41. $\left[3\left(\cos\dfrac{\pi}{3} + i\sin\dfrac{\pi}{3}\right)\right]^5$

42. $\left[\sqrt{2}\left(\cos\dfrac{5\pi}{6} + i\sin\dfrac{5\pi}{6}\right)\right]^4$

43. $\left[\dfrac{1}{2}\left(\cos\dfrac{\pi}{24} + i\sin\dfrac{\pi}{24}\right)\right]^6$

44. $[\sqrt{3}\,(\cos 70° + i\sin 70°)]^3$

45. $[2^{1/5}(\cos 63° + i\sin 63°)]^{10}$

46. $\left[2\left(\cos\dfrac{\pi}{5} + i\sin\dfrac{\pi}{5}\right)\right]^3$

47. $2(\cos 200° + i\sin 200°) \times \sqrt{2}\,(\cos 20° + i\sin 20°)$
    $\times \frac{1}{2}(\cos 5° + i\sin 5°)$

48. $\left[\dfrac{\cos(\pi/8) + i\sin(\pi/8)}{\cos(-\pi/8) + i\sin(-\pi/8)}\right]^5$

49. $\left(\dfrac{1}{2} - \dfrac{\sqrt{3}}{2}i\right)^5$   *Hint:* Convert to trigonometric form.

50. $(1 - i)^3$

51. $(-2 - 2i)^5$

52. $(-\frac{1}{2} + \frac{1}{2}\sqrt{3}i)^6$

53. $(-2\sqrt{3} - 2i)^4$

54. $(1 + i)^{16}$

*In Exercises 55–61, use DeMoivre's Theorem to find the indicated roots. Express the results in rectangular form.*

55. Cube roots of $-27i$
56. Cube roots of $2$
57. Eighth roots of $1$
58. Square roots of $i$
59. Cube roots of $64$
60. Square roots of $-\frac{1}{2} - \frac{1}{2}\sqrt{3}i$
61. Sixth roots of $729$

© *Use a calculator to complete Exercises 62–65.*

**62.** Compute $(9 + 9i)^6$.          **63.** Compute $(7 - 7i)^8$.

**64.** Compute the cube roots of $1 + 2i$. Express your answers in rectangular form, with the real and imaginary parts rounded off to two decimal places.

**65.** Compute the fifth roots of $i$. Express your answers in rectangular form, with the real and imaginary parts rounded off to two decimal places.

**B**

*In Exercises 66–68, find the indicated roots. Express the results in rectangular form.*

**66.** Find the fourth roots of $i$.   *Hint:*   Use the half-angle formulas from Section 8.4.

**67.** Find the fourth roots of $8 - 8\sqrt{3}i$.   *Hint:*   Use the addition formulas or the half-angle formulas.

**68.** Find the square roots of $7 + 24i$.   *Hint:*   You'll need to use the half-angle formulas from Section 8.4.

**69.** (a) Compute the three cube roots of 1.
(b) Let $z_1$, $z_2$, and $z_3$ denote the three cube roots of 1. Verify that $z_1 + z_2 + z_3 = 0$ and $z_1 z_2 + z_2 z_3 + z_3 z_1 = 0$.

**70.** (a) Compute the four fourth roots of 1.
(b) Verify that the sum of these four fourth roots is 0.

**71.** Evaluate $(-\frac{1}{2} + \frac{1}{2}\sqrt{3}i)^5 + (-\frac{1}{2} - \frac{1}{2}\sqrt{3}i)^5$.
*Hint:*   Use DeMoivre's Theorem.

**72.** Show that $(-\frac{1}{2} + \frac{1}{2}\sqrt{3}i)^6 + (-\frac{1}{2} - \frac{1}{2}\sqrt{3}i)^6 = 2$.

**73.** Compute $(\cos\theta + i\sin\theta)(\cos\theta - i\sin\theta)$.

**74.** In the identity $(\cos\theta + i\sin\theta)^2 = \cos 2\theta + i\sin 2\theta$, carry out the actual multiplication on the left-hand side of the equation. Then equate the corresponding real parts and the corresponding imaginary parts from each side of the equation that results. What do you obtain?

**75.** Show that
$$\frac{r(\cos\alpha + i\sin\alpha)}{R(\cos\beta + i\sin\beta)} = \frac{r}{R}[\cos(\alpha - \beta) + i\sin(\alpha - \beta)]$$

*Suggestion:*   Begin with the quantity on the left side and multiply it by $\dfrac{\cos\beta - i\sin\beta}{\cos\beta - i\sin\beta}$.

**76.** Show that
$$1 + \cos\theta + i\sin\theta = 2\cos\left(\frac{\theta}{2}\right)\left[\cos\left(\frac{\theta}{2}\right) + i\sin\left(\frac{\theta}{2}\right)\right]$$

**77.** If $z = r(\cos\theta + i\sin\theta)$, show that
$$\frac{1}{z} = \frac{1}{r}(\cos\theta - i\sin\theta)$$

*Hint:*   $1/z = [1(\cos 0 + i\sin 0)]/[r(\cos\theta + i\sin\theta)]$.

**78.** Show that $\dfrac{1 + \sin\theta + i\cos\theta}{1 + \sin\theta - i\cos\theta} = \sin\theta + i\cos\theta$.   Assume that $\theta \neq \dfrac{3\pi}{2} + 2\pi k$. where $k$ is an integer.

*Hint:*   Work with the left-hand side; first "rationalize" the denominator by multiplying by
$$\frac{(1 + \sin\theta) + i\cos\theta}{(1 + \sin\theta) + i\cos\theta}.$$

## CHAPTER 9   SUMMARY OF PRINCIPAL TERMS AND FORMULAS

| TERMS OR FORMULAS | PAGE REFERENCE | COMMENT |
|---|---|---|
| 1. $\dfrac{\sin A}{a} = \dfrac{\sin B}{b} = \dfrac{\sin C}{c}$ | 447 | This is the *law of sines*. In words, it states that in any triangle, the sines of the angles are proportional to the lengths of the opposite sides. |
| 2. $a^2 = b^2 + c^2 - 2bc\cos A$ | 449 | This is the *law of cosines*. We can use it to calculate the third side of a triangle when we are given two sides and the included angle. The formula can also be used to calculate the angles of a triangle when we know the three sides, as in Example 5 on page 451. |

| TERMS OR FORMULAS | PAGE REFERENCE | COMMENT |
|---|---|---|
| **3.** Vector | 455<br>464 | Geometrically, a *vector* in the plane is a directed line segment. Vectors can be used to represent quantities that have both magnitude and direction, such as force and velocity. For examples, see Figures 1 and 2 in Section 9.2. Algebraically, a vector in the plane is an ordered pair of real numbers, denoted by $\langle x,y \rangle$. (See Figure 3 in Section 9.3.) The numbers $x$ and $y$ are called the *components* of the vector $\langle x,y \rangle$. |
| **4.** Vector equality | 456<br>465 | Geometrically, two vectors are said to be equal provided they have the same length and the same direction. In terms of components, this means that two vectors are equal if and only if their corresponding components are equal. |
| **5.** Vector addition | 457<br>458<br>466 | Geometrically, two vectors can be added by using the parallelogram law, as in Figure 10 in Section 9.2. Algebraically, this is equivalent to the following componentwise formula for vector addition:<br>$$\langle a,b \rangle + \langle c,d \rangle = \langle a+c, b+d \rangle$$ |
| **6.** $\lvert \mathbf{v} \rvert = \sqrt{v_1^2 + v_2^2}$ | 456<br>465 | The expression on the left-hand side of the equation denotes the length or magnitude of the vector $\mathbf{v}$. The expression on the right-hand side of the equation tells us how to compute the length of $\mathbf{v}$ in terms of its respective $x$- and $y$-components, $v_1$ and $v_2$. |
| **7.** $\overrightarrow{PQ} = \langle x_2 - x_1, y_2 - y_1 \rangle$ | 465 | This formula gives the components of the vector $\overrightarrow{PQ}$, where $P$ and $Q$ are the points $P(x_1, y_1)$ and $Q(x_2, y_2)$. See Figure 5 in Section 9.3 for the derivation of this formula. |
| **8.** The zero vector | 467 | The *zero vector*, denoted by **0**, is the vector $\langle 0,0 \rangle$. The zero vector plays the same role in vector addition as does the real number zero in addition of real numbers. For any vector $\mathbf{v}$, we have<br>$$\mathbf{0} + \mathbf{v} = \mathbf{v} + \mathbf{0} = \mathbf{v}$$ |
| **9.** Scalar multiplication | 467 | For each real number $k$ and each vector $\mathbf{v} = \langle x,y \rangle$, the vector $k\mathbf{v}$ is defined by the equation<br>$$k\mathbf{v} = \langle kx, ky \rangle$$<br>Geometrically, the length of $k\mathbf{v}$ is $\lvert k \rvert$ times the length of $\mathbf{v}$. The operation whereby the vector $k\mathbf{v}$ is formed from the scalar $k$ and the vector $\mathbf{v}$ is called *scalar multiplication*. |
| **10.** $\mathbf{i} = \langle 1,0 \rangle$<br>$\mathbf{j} = \langle 0,1 \rangle$ | 468 | These two equations define the *unit vectors* $\mathbf{i}$ and $\mathbf{j}$. See Figure 11 in Section 9.3. Any vector $\langle x,y \rangle$ can be expressed in terms of the unit vectors $\mathbf{i}$ and $\mathbf{j}$ as follows:<br>$$\langle x,y \rangle = x\mathbf{i} + y\mathbf{j}$$ |
| **11.** $x = r\cos\theta$<br>$y = r\sin\theta$<br>$x^2 + y^2 = r^2$<br>$\tan\theta = y/x$ | 474 | These formulas relate the rectangular coordinates of a point $(x,y)$ to its polar coordinates $(r,\theta)$. See the boxed figure on page 474. |

| TERMS OF FORMULAS | PAGE REFERENCE | COMMENT |
|---|---|---|
| 12. The complex plane | 481 | Refer to the following figure. A complex number $a + bi$ can be identified with the point $(a, b)$ in the $x$-$y$ plane. In this context, the $x$-$y$ plane is referred to as the *complex plane*. The distance $r$ in the figure is the *modulus* of the complex number $a + bi$; the angle $\theta$ is an *argument* of $a + bi$. |

y

● (a, b) or a + bi

r

θ

x

| 13. $z = r(\cos\theta + i\sin\theta)$ | 481 | The expression on the right-hand side is the *trigonometric form* of the complex number $z = a + bi$. (The expression $a + bi$ is the *rectangular form* of the complex number $z$.) The trigonometric form and the rectangular form are related by the following equations: $$r = \sqrt{a^2 + b^2} \qquad x = r\cos\theta \qquad y = r\sin\theta$$ |
| 14. DeMoivre's Theorem | 483 | Let $n$ be a natural number and let $z = r(\cos\theta + i\sin\theta)$. Then DeMoivre's Theorem states that $$z^n = r^n(\cos n\theta + i\sin n\theta)$$ For applications of this theorem, see Examples 6–8 in Section 9.5. |

# CHAPTER 9   REVIEW EXERCISES

NOTE   The fifteen exercises in this set with colored problem numbers constitute a chapter test.

*In Exercises 1–8, the lengths of the three sides of a triangle are denoted by a, b, and c; the angles opposite these sides are A, B, and C, respectively. In each exercise, use the given data to find the remaining sides and angles. Use a calculator and round off the final answers to one decimal place.*

1.  $\angle A = 40°$, $\angle B = 85°$, $c = 16$ cm
2.  $\angle C = 84°$, $\angle B = 16°$, $a = 9$ cm
3.  (a)  $a = 8$ cm, $b = 9$ cm, $\angle A = 52°$, $\angle B < 90°$
    (b)  Use the data in part (a), but assume $\angle B > 90°$.
4.  (a)  $a = 6.25$ cm, $b = 9.44$ cm, $\angle A = 12°$, $\angle B < 90°$
    (b)  Use the data in part (a), but assume $\angle B > 90°$.
5.  $a = 18$ cm, $b = 14$ cm, $\angle C = 24°$
6.  $a = 32.16$ cm, $b = 50.12$ cm, $\angle C = 156°$

7.  $a = 4$ cm, $b = 7$ cm, $c = 9$ cm
8.  $a = 12.61$ cm, $b = 19.01$ cm, $c = 14.14$ cm

© *In Exercises 9–18, refer to the following figure. In each case, determine the indicated quantity. Use a calculator, and round off your result to two decimal places.*

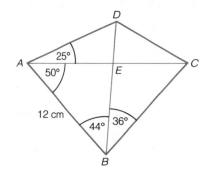

9. *BE*

10. *BC*

11. area of $\triangle BCE$

12. *BD*

13. area of $\triangle ABD$

14. *DE*

15. *CD*

16. *AD*

17. *AC*

18. area of $\triangle DEC$

[c] *In Exercises 19 and 20, refer to the following figure. In the figure, $\overline{BD}$ bisects $\angle ABC$.*

19. If $AB = 24$ cm, $BC = 40$ cm, and $AC = 56$ cm, find the length of the angle bisector $\overline{BD}$.

20. If $AB = 105$ cm, $BC = 120$ cm, and $AC = 195$ cm, find the length of the angle bisector $\overline{BD}$.

[c] *In Exercises 21 and 22, refer to the following figure, in which $AD = DC$.*

21. If $AB = 12$ cm, $BC = 26$ cm, and $AD = DC = 17$ cm, find $BD$.

22. If $AB = 16$ cm, $BC = 22$ cm, and $AD = DC = 17$ cm, find $BD$.

[c] 23. Two forces **F** and **G** act on an object, **G** horizontally to the right with a magnitude of 15 N, and **F** vertically upward with a magnitude of 20 N. Determine the magnitude and direction of the resultant force. (Use a calculator to determine the angle between the horizontal and the resultant; round off the result to one decimal place.)

[c] 24. Determine the resultant of the two forces in the accompanying figure. (Use a calculator, and round off the values you obtain for the magnitude and direction to one decimal place.)

[c] 25. Determine the horizontal and vertical components of the velocity vector **v** in the following figure. (Use a calculator, and round off your answers to one decimal place.)

[c] 26. The heading and air speed of an airplane are 50° and 220 mph, respectively. If the wind is 60 mph from 140°, find the ground speed, the drift angle, and the course. (Use a calculator, and round off your answers to one decimal place.)

[c] 27. A block rests on an inclined plane that makes an angle of 24° with the horizontal. The component of the weight parallel to the plane is 14.8 pounds. Determine the weight of the block and the component of the weight perpendicular to the plane. (Use a calculator, and round off your answers to one decimal place.)

28. Find the length of the vector $\langle 20, 99 \rangle$.

29. For which values of $b$ will the vectors $\langle 2, 6 \rangle$ and $\langle -5, b \rangle$ have the same length?

30. The coordinates of the points $A$ and $B$ are $(2, 6)$ and $(-7, 4)$, respectively. Find the components of the following vectors. Write your answers in the form $\langle x, y \rangle$.
(a) $\overrightarrow{AB}$  (b) $\overrightarrow{BA}$  (c) $3\overrightarrow{AB}$  (d) $\overrightarrow{AB}/|\overrightarrow{AB}|$

*In Exercises 31–42, compute each of the indicated quantities, given that the vectors **a**, **b**, **c**, and **d** are defined as follows:*

$$\mathbf{a} = \langle 3, 5 \rangle \qquad \mathbf{b} = \langle 7, 4 \rangle \qquad \mathbf{c} = \langle 2, -1 \rangle \qquad \mathbf{d} = \langle 0, 3 \rangle$$

31. $\mathbf{a} + \mathbf{b}$

32. $\mathbf{b} - \mathbf{d}$

33. $3\mathbf{c} + 2\mathbf{a}$

34. $|\mathbf{a}|$

35. $|\mathbf{b} + \mathbf{d}|^2 - |\mathbf{b} - \mathbf{d}|^2$

36. $\mathbf{a} + (\mathbf{b} + \mathbf{c})$

37. $(\mathbf{a} + \mathbf{b}) + \mathbf{c}$

38. $\mathbf{a} - (\mathbf{b} - \mathbf{c})$

39. $(\mathbf{a} - \mathbf{b}) - \mathbf{c}$

40. $|\mathbf{a} + \mathbf{b}|^2 + |\mathbf{a} - \mathbf{b}|^2 - 2|\mathbf{a}|^2 - 2|\mathbf{b}|^2$

41. $4\mathbf{c} + 2\mathbf{a} - 3\mathbf{b}$

42. $|4\mathbf{c} + 2\mathbf{a} - 3\mathbf{b}|$

43. Express the vector $\langle 7, -6 \rangle$ in terms of the unit vectors **i** and **j**.

44. Express the vector $4\mathbf{i} - 6\mathbf{j}$ in the form $\langle a, b \rangle$.

45. Find a unit vector having the same direction as $\langle 6, 4 \rangle$.

*In Exercises 46 and 47, you are given an angle $\theta$ measured counterclockwise from the positive x-axis to a unit vector $\langle u_1, u_2 \rangle$. In each case, determine the components $u_1$ and $u_2$.*

46. $\theta = 7\pi/4$

**47.** $\theta = \pi/12$   *Hint:*   Use the addition formulas from Section 8.3.

**48.** Prove properties 5 and 6 in the box on page 470.   *Hint:* For property 5, let $\mathbf{u} = \langle u_1, u_2 \rangle$ and $\mathbf{v} = \langle v_1, v_2 \rangle$. Compute the quantities on each side of the equation in property 5 and compare the results.

*In Exercises 49–56, graph the polar equations.*

**49.** (a) $r = 2 - 2\cos\theta$
    (b) $r = 2 - 2\sin\theta$

**50.** (a) $r^2 = 4\cos 2\theta$
    (b) $r^2 = 4\sin 2\theta$

**51.** (a) $r = 2\cos\theta - 1$
    (b) $r = 2\sin\theta - 1$

**52.** (a) $r = 2\cos\theta$
    (b) $r = 2\cos\theta + 1$

**53.** (a) $r = 4\sin 2\theta$
    (b) $r = 4\cos 2\theta$

**54.** (a) $r = 4\sin 3\theta$
    (b) $r = 4\cos 3\theta$

**55.** (a) $r = 1 + 2\sin(\theta/2)$
    (b) $r = 1 - 2\cos(\theta/2)$

**56.** (a) $r = (1.5)^\theta$   $(\theta \geq 0)$
    (b) $r = (1.5)^\theta$   $(\theta \leq 0)$

*In Exercises 57–60, convert each complex number to rectangular form.*

**57.** $3\left(\cos\dfrac{\pi}{3} + i\sin\dfrac{\pi}{3}\right)$

**58.** $\cos\dfrac{\pi}{6} + i\sin\dfrac{\pi}{6}$

**59.** $2^{1/4}[\cos(7\pi/4) + i\sin(7\pi/4)]$

**60.** $5[\cos(-\pi/4) + i\sin(-\pi/4)]$

*In Exercises 61–64, express the complex numbers in trigonometric form.*

**61.** $\frac{1}{2} + \frac{1}{2}\sqrt{3}i$

**62.** $3i$

**63.** $-3\sqrt{2} - 3\sqrt{2}i$

**64.** $2\sqrt{3} - 2i$

*In Exercises 65–74, carry out the indicated operations. Express your results in rectangular form for those cases in which the trigonometric functions are readily evaluated without tables or a calculator.*

**65.** $5\left(\cos\dfrac{\pi}{7} + i\sin\dfrac{\pi}{7}\right) \times 2\left(\cos\dfrac{3\pi}{28} + i\sin\dfrac{3\pi}{28}\right)$

**66.** $4\left(\cos\dfrac{\pi}{12} + i\sin\dfrac{\pi}{12}\right) \times 3\left(\cos\dfrac{\pi}{12} + i\sin\dfrac{\pi}{12}\right)$

**67.** $8\left(\cos\dfrac{\pi}{12} + i\sin\dfrac{\pi}{12}\right) \div 4\left(\cos\dfrac{\pi}{3} + i\sin\dfrac{\pi}{3}\right)$

**68.** $4(\cos 32° + i\sin 32°) \div 2^{1/2}(\cos 2° + i\sin 2°)$

**69.** $\left(\cos\dfrac{\pi}{9} + i\sin\dfrac{\pi}{9}\right) \times 3\left(\cos\dfrac{4\pi}{9} + i\sin\dfrac{4\pi}{9}\right)$

**70.** $5(\cos 3° + i\sin 3°) \div 4(\cos 5° + i\sin 5°)$

**71.** $\left[3^{1/4}\left(\cos\dfrac{\pi}{36} + i\sin\dfrac{\pi}{36}\right)\right]^{12}$

**72.** $\left[2\left(\cos\dfrac{2\pi}{15} + i\sin\dfrac{2\pi}{15}\right)\right]^5$

**73.** $(\sqrt{3} + i)^{10}$

**74.** $(\sqrt{2} - \sqrt{2}i)^{15}$

*In Exercises 75–78, use DeMoivre's Theorem to find the indicated roots. Express your results in rectangular form.*

**75.** Sixth roots of 1

**76.** Cube roots of $-64i$

**77.** Square roots of $\sqrt{2} - \sqrt{2}i$

**78.** Fourth roots of $1 + \sqrt{3}i$

**[c] 79.** Find the five fifth roots of $1 + i$. Express the roots in rectangular form. (Round off each decimal to two places in the final answer.)

**80.** If $z = r(\cos\theta + i\sin\theta)$, show that
$$r[\cos(\theta + \pi) + i\sin(\theta + \pi)] = -z$$

**81.** Trigonometric identities for $\cos n\theta$ and for $\sin n\theta$ can be derived using DeMoivre's Theorem. For example, according to DeMoivre's Theorem, we have $(\cos\theta + i\sin\theta)^3 = \cos 3\theta + i\sin 3\theta$. Now expand the expression on the left-hand side of this last equation, and then equate the corresponding real parts and imaginary parts in the equation that results. What identities do you obtain?

**82.** The following problem is taken verbatim from *An Elementary Treatise on Plane Trigonometry* by R. D. Beasley, first published in 1858.

If $a_1, a_2, a_3$ be respectively the sides of a regular pentagon, hexagon, and decagon inscribed in a circle, then [show]
$$a_1^2 = a_2^2 + a_3^2$$

[*Hint:* Use the law of cosines.]

# 10 SYSTEMS OF EQUATIONS

**INTRODUCTION**   In this chapter we consider systems of equations. Roughly speaking, a system of equations is just a collection of equations with a common set of unknowns. In solving such systems, we try to find values for the unknowns that simultaneously satisfy each of the equations in the system. In Section 10.1 we review two techniques that are often presented in intermediate algebra for solving systems involving two linear equations in two unknowns. An important technique for solving larger systems of equations is developed in Section 10.2: this technique is known as *Gaussian elimination*. After that, in Section 10.3, we introduce matrices as a tool for reducing the amount of bookkeeping involved in Gaussian elimination. Two additional techniques for solving systems of linear equations are explained in Sections 10.4 and 10.5. The technique in Section 10.4 uses the idea of an inverse matrix; the technique in Section 10.5 involves determinants and Cramer's rule. Section 10.6 contains a brief discussion of nonlinear systems of equations. In the last section of this chapter, Section 10.7, we consider systems of inequalities. As an application of this and of the earlier work in the chapter, we conclude Section 10.7 with an introduction to linear programming.

## SYSTEMS OF TWO LINEAR EQUATIONS IN TWO UNKNOWNS

### 10.1

Both in theory and in applications, it's often necessary to solve two equations in two unknowns. You may have been introduced to the idea of simultaneous equations in a previous course in algebra; however, to put matters on a firm foundation, we begin here with the basic definitions. By a **linear equation in two variables** we mean an equation of the form

$$ax + by = c$$

where the constants $a$ and $b$ are not both zero. The two variables needn't always be denoted by the letters $x$ and $y$, of course; it is the *form* of the equation that matters. Table 1 displays some examples.

An ordered pair of numbers $(x_0, y_0)$ is said to be a **solution of the linear equation** $ax + by = c$ provided we obtain a true statement when we replace $x$ and $y$ in the equation by $x_0$ and $y_0$, respectively. For example, the ordered pair $(3, 2)$ is a solution of the equation $x - y = 1$, since $3 - 2 = 1$. On the other hand, $(2, 3)$ is not a solution of $x - y = 1$, since $2 - 3 \neq 1$.

Now consider a **system** of two linear equations in two unknowns:

$$\begin{cases} ax + by = c \\ dx + ey = f \end{cases}$$

If we can find an ordered pair that is a solution to both equations, then we say that this ordered pair is a **solution of the system**. Sometimes, to emphasize the

**TABLE 1**

| EQUATION IN TWO VARIABLES | IS IT LINEAR? | |
|---|---|---|
| | Yes | No |
| $3x - 8y = 12$ | ✓ | |
| $-s + 4t = 0$ | ✓ | |
| $2x - 3y^2 = 1$ | | ✓ |
| $y = 4 - 2x$ | ✓ | |
| $\dfrac{4}{u} + \dfrac{5}{v} = 3$ | | ✓ |

fact that a solution must satisfy both equations, we refer to the system as a pair of **simultaneous equations**. A system that has at least one solution is said to be **consistent**. If there are no solutions, the system is **inconsistent**.

**EXAMPLE 1**     Consider the system

$$\begin{cases} x + y = 2 \\ 2x - 3y = 9 \end{cases}$$

**(a)** Is $(1, 1)$ a solution of the system?
**(b)** Is $(3, -1)$ a solution of the system?

**Solution**     **(a)** Although $(1, 1)$ is a solution of the first equation, it is not a solution of the system because it does not satisfy the second equation. (Check this for yourself.)

**(b)** $(3, -1)$ satisfies the first equation:

$$3 + (-1) \stackrel{?}{=} 2 \qquad \text{True}$$

$(3, -1)$ satisfies the second equation:

$$2(3) - 3(-1) \stackrel{?}{=} 9$$
$$6 + 3 \stackrel{?}{=} 9 \qquad \text{True}$$

Since $(3, -1)$ satisfies both equations, it is a solution of the system.

We can gain an important perspective on systems of linear equations by looking at Example 1 in graphical terms. Table 2 shows how each of the statements in that example can be rephrased using geometric ideas with which we are already familiar.

**TABLE 2**

| ALGEBRAIC IDEA | CORRESPONDING GEOMETRIC IDEA |
| --- | --- |
| 1. The ordered pair $(1, 1)$ is a solution of the equation $x + y = 2$. | 1. The point $(1, 1)$ lies on the line $x + y = 2$. See Figure 1. |
| 2. The ordered pair $(1, 1)$ is not a solution of the equation $2x - 3y = 9$. | 2. The point $(1, 1)$ does not lie on the line $2x - 3y = 9$. See Figure 1. |
| 3. The ordered pair $(3, -1)$ is a solution of the system $$\begin{cases} x + y = 2 \\ 2x - 3y = 9 \end{cases}$$ | 3. The point $(3, -1)$ lies on both of the lines $x + y = 2$ and $2x - 3y = 9$. See Figure 1. |

In Example 1, we verified that $(3, -1)$ is a solution of the system

$$\begin{cases} x + y = 2 \\ 2x - 3y = 9 \end{cases}$$

Are there any other solutions of this particular system? No; Figure 1 shows us that there are no other solutions, since $(3, -1)$ is clearly the only point common to both lines. In a moment, we'll look at two important methods for solving systems of linear equations in two unknowns. But even before we consider these methods, we can say something about the solutions of systems.

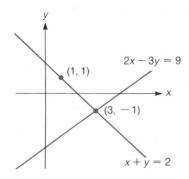

**FIGURE 1**

| PROPERTY SUMMARY | Possibilities for Solutions of Systems |
|---|---|

Given a system of two linear equations in two unknowns, exactly one of the following cases must occur.

CASE 1   The graphs of the two linear equations intersect in exactly one point. Thus, there is exactly one solution to the system. See Figure 2.

CASE 2   The graphs of the two linear equations are parallel lines. Therefore, the lines do not intersect, and the system has no solution. See Figure 3.

CASE 3   The two equations actually represent the same line. Thus, there are infinitely many points of intersection and correspondingly infinitely many solutions. See Figure 4.

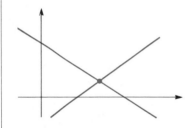

**FIGURE 2**   A consistent system with exactly one solution

**FIGURE 3**   An inconsistent system has no solution

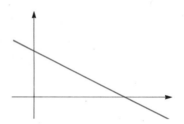

**FIGURE 4**   A consistent system with infinitely many solutions

We are going to review two methods from basic algebra for solving systems of two linear equations in two unknowns. These methods are the **substitution method** and the **addition–subtraction method**. We'll begin by demonstrating the substitution method. Consider the system

$$\begin{cases} 3x + 2y = 17 & \text{(1)} \\ 4x - 5y = -8 & \text{(2)} \end{cases}$$

We first choose one of the two equations and then use it to express one of the variables in terms of the other. In the case at hand, neither equation appears particularly simpler than the other, so let's just start with the first equation and solve for $x$ in terms of $y$. We have

$$3x = 17 - 2y$$
$$x = \tfrac{1}{3}(17 - 2y) \qquad \qquad \text{(3)}$$

Now we use equation (3) to substitute for $x$ in the equation that we have not used yet, namely, equation (2). This yields

$$4[\tfrac{1}{3}(17 - 2y)] - 5y = -8$$
$$4(17 - 2y) - 15y = -24 \qquad \text{multiplying by 3}$$
$$-23y = -92$$
$$y = 4$$

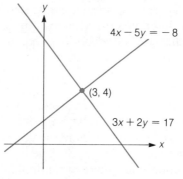

**FIGURE 5**

The value $y = 4$ that we have just obtained can be used in equation (3) to find $x$. Replacing $y$ with 4 in equation (3) yields

$$x = \tfrac{1}{3}[17 - 2(4)] = \tfrac{1}{3}(9) = 3$$

We have now found that $x = 3$ and $y = 4$. As you can easily check, this pair of values indeed satisfies both of the original equations. We write our solution as the ordered pair $(3,4)$. Figure 5 provides a graphical summary of the situation. The figure shows that the system is consistent and that $(3,4)$ is the only solution.

Generally speaking, it is not necessary to graph the equations in a given system in order to decide whether the system is consistent. Rather, this information will emerge as you attempt to follow an algebraic method of solution. Examples 2 and 3 will illustrate this.

**EXAMPLE 2**  Solve the system

$$\begin{cases} \tfrac{3}{2}x - 3y = -9 \\ x - 2y = 4 \end{cases}$$

**Solution**  We use the substitution method. Since it is easy to solve the second equation for $x$, we begin there:

$$x - 2y = 4$$
$$x = 4 + 2y$$

Now we substitute this result in the first equation of our system to obtain

$$\tfrac{3}{2}(4 + 2y) - 3y = -9$$
$$6 + 3y - 3y = -9$$
$$6 = -9 \qquad \text{False}$$

Since the substitution process leads us to this obviously false statement, we conclude that the given system has no solution; that is, the system is inconsistent.   *Question:*  What can you say about the graphs of the two given equations?

**EXAMPLE 3**  Solve the system

$$\begin{cases} 3x + 4y = 12 \\ 2y = 6 - \tfrac{3}{2}x \end{cases}$$

**Solution**  We use the method of substitution. Since it is easy to solve the second equation for $y$, we begin there:

$$2y = 6 - \tfrac{3}{2}x$$
$$y = 3 - \tfrac{3}{4}x$$

Now we use this result to substitute for $y$ in the first equation of the original system. The result is

$$3x + 4(3 - \tfrac{3}{4}x) = 12$$
$$3x + 12 - 3x = 12$$
$$3x - 3x = 12 - 12$$
$$0 = 0 \qquad \text{Always true}$$

This last identity imposes no restrictions on $x$. Graphically speaking, this says that our two lines intersect for every value of $x$. In other words, the two lines coincide. We could have foreseen this initially had we solved both equations for $y$. As you can verify, the result in both cases is

$$y = -\tfrac{3}{4}x + 3$$

Every point on this line yields a solution to our system of equations. In summary then, our system is consistent and the solutions to the system have the form $(x, -\tfrac{3}{4}x + 3)$, where $x$ can be any real number. For instance, when $x = 0$, we obtain the solution $(0, 3)$. When $x = 1$, we obtain the solution $(1, \tfrac{9}{4})$. The idea here is that for *each* value of $x$, we obtain a solution; thus, there are infinitely many solutions. ▪▪

Now let us turn to the addition–subtraction method of solving systems of equations. By way of example, consider the system

$$\begin{cases} 2x + 3y = 5 \\ 4x - 3y = -1 \end{cases}$$

Notice that if we add these two equations, the result will be an equation involving only the unknown $x$. Carrying this out gives us

$$6x = 4$$
$$x = \tfrac{4}{6} = \tfrac{2}{3}$$

There are now several ways in which the corresponding value of $y$ may be obtained. As you can easily check, substituting the value $x = \tfrac{2}{3}$ in either of the original equations leads to the result $y = \tfrac{11}{9}$.

Another way to find $y$ is by multiplying both sides of the first equation by $-2$. (You'll see why in a moment.) We display the work this way:

$$2x + 3y = 5 \quad \xrightarrow{\text{Multiply by } -2} \quad -4x - 6y = -10$$
$$4x - 3y = -1 \quad \xrightarrow{\text{No change}} \quad 4x - 3y = -1$$

Adding the last two equations then gives us

$$-9y = -11$$
$$y = \tfrac{11}{9}$$

The required solution is therefore $(\tfrac{2}{3}, \tfrac{11}{9})$.

In the previous example, we were able to find $x$ directly by adding the two equations. As the next example shows, it may be necessary first to multiply both sides of each equation by an appropriate constant.

**EXAMPLE 4**  Solve the system

$$\begin{cases} 5x - 3y = 4 \\ 2x + 4y = 1 \end{cases}$$

**Solution**  To eliminate $x$, we could multiply the second equation by $\tfrac{5}{2}$ and then subtract the resulting equation from the first equation. However, to avoid working with fractions, we proceed as follows.

$$5x - 3y = 4 \quad \xrightarrow{\text{Multiply by 2}} \quad 10x - 6y = 8 \tag{4}$$
$$2x + 4y = 1 \quad \xrightarrow{\text{Multiply by 5}} \quad 10x + 20y = 5 \tag{5}$$

Subtracting equation (5) from equation (4) then yields

$$-26y = 3$$
$$y = -\tfrac{3}{26}$$

To find $x$, we return to the original system and work in a similar manner:

$$5x - 3y = 4 \xrightarrow{\text{Multiply by 4}} 20x - 12y = 16$$
$$2x + 4y = 1 \xrightarrow{\text{Multiply by 3}} 6x + 12y = 3$$

Upon adding the last two equations, we obtain

$$26x = 19$$
$$x = \tfrac{19}{26}$$

The solution of the given system of equations is therefore $(\tfrac{19}{26}, -\tfrac{3}{26})$. ▪

We conclude this section with some problems that can be solved using simultaneous equations.

**EXAMPLE 5** Determine constants $b$ and $c$ so that the parabola $y = x^2 + bx + c$ passes through the points $(-3, 1)$ and $(1, -2)$.

**Solution** Since the point $(-3, 1)$ lies on the curve $y = x^2 + bx + c$, the coordinates must satisfy the equation. Thus, we have

$$1 = (-3)^2 + b(-3) + c$$
$$-8 = -3b + c \tag{6}$$

This gives us one equation in two unknowns. We need another equation involving $b$ and $c$. Since the point $(1, -2)$ also lies on the graph of $y = x^2 + bx + c$, we must have

$$-2 = 1^2 + b(1) + c$$
$$-3 = b + c \tag{7}$$

Rewriting equations (6) and (7), we have the system

$$\begin{cases} -3b + c = -8 & (8) \\ \phantom{-3}b + c = -3 & (9) \end{cases}$$

Subtracting equation (9) from (8) then yields

$$-4b = -5$$
$$b = \tfrac{5}{4}$$

One way to obtain the corresponding value of $c$ is to replace $b$ by $\tfrac{5}{4}$ in equation (9). This yields

$$\tfrac{5}{4} + c = -3$$
$$c = -3 - \tfrac{5}{4} = -\tfrac{17}{14}$$

The required values of $b$ and $c$ are therefore

$$b = \tfrac{5}{4}, \qquad c = -\tfrac{17}{4}$$

(Exercise 54 at the end of this section asks you to check that the parabola $y = x^2 + \tfrac{5}{4}x - \tfrac{17}{4}$ indeed passes through the given points.) ▪

**EXAMPLE 6**  In Figure 6, find the area of the triangular region bounded by the lines $y = 3x$, $y = -\frac{1}{2}x + 7$, and the $x$-axis.

Solution  To compute the area of the triangle, we will need to know the base and the height. To determine the base, we compute the $x$-intercept of the line $y = -\frac{1}{2}x + 7$:

$$0 = -\tfrac{1}{2}x + 7$$
$$\tfrac{1}{2}x = 7$$
$$x = 14$$

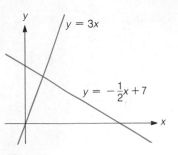

FIGURE 6

The height of the triangle is the $y$-coordinate of the intersection point of the lines $y = 3x$ and $y = -\frac{1}{2}x + 7$ (see Figure 6). The substitution method can be used to solve this pair of simultaneous equations for $y$. From the equation $y = 3x$, we obtain $x = \frac{1}{3}y$. Then substituting this in the equation $y = -\frac{1}{2}x + 7$ yields

$$y = -\tfrac{1}{2}(\tfrac{1}{3}y) + 7$$
$$6y = -y + 42$$
$$7y = 42$$
$$y = 6$$

Now that we know that the base of the triangle is 14 and the height is 6, we can compute the required area:

$$A = \tfrac{1}{2}bh$$
$$= \tfrac{1}{2}(14)(6)$$
$$= 42 \text{ square units}$$

In the next example, we solve a mixture problem using a system of two equations in two unknowns.

**EXAMPLE 7**  Suppose that a chemistry student can obtain two acid solutions from the stockroom. The first solution is 20% acid and the second solution is 45% acid. (The percentages are by volume.) How many cubic centimeters of each solution should the student mix together to obtain 100 cm³ of a 30% acid solution?

Solution  We begin by assigning letters to denote the required quantities.

Let $x$ denote the number of cubic centimeters of the 20% solution to be used.

Let $y$ denote the number of cubic centimeters of the 45% solution to be used.

Then the data can be summarized as in Table 3. Since the final mixture must total 100 cm³, we have the equation

$$x + y = 100 \tag{10}$$

This gives us one equation in two unknowns. However, we need a second equation. Looking at the data in the right-hand column of Table 3, we can write

$$\underbrace{0.20x}_{\substack{\text{Amount of acid} \\ \text{in } x\,\text{cm}^3 \text{ of the} \\ 20\% \text{ solution}}} + \underbrace{0.45y}_{\substack{\text{Amount of acid} \\ \text{in } y\,\text{cm}^3 \text{ of the} \\ 45\% \text{ solution}}} = \underbrace{(0.30)(x+y)}_{\substack{\text{Amount of acid in} \\ \text{the final mixture}}}$$

**TABLE 3**

| TYPE OF SOLUTION | NUMBER OF CM³ | PERCENT OF ACID | TOTAL ACID (CM³) |
|---|---|---|---|
| First solution (20% acid) | $x$ | 20 | $(0.20)x$ |
| Second solution (45% acid) | $y$ | 45 | $(0.45)y$ |
| Mixture | $x+y$ | 30 | $(0.30)(x+y)$ |

Thus,

$$0.20x + 0.45y = 0.30(x+y)$$
$$20x + 45y = 30(x+y)$$
$$4x + 9y = 6(x+y) = 6x + 6y$$
$$-2x + 3y = 0 \qquad\qquad (11)$$

Equations (10) and (11) can be solved by either the substitution method or the addition method. As Exercise 55 at the end of this section asks you to show, the results are

$$x = 60\,\text{cm}^3 \qquad \text{and} \qquad y = 40\,\text{cm}^3$$

# EXERCISE SET 10.1

## A

1. Which of the following are linear equations in two variables?
   (a) $3x + 3y = 10$  (b) $2x + 4xy + 3y = 1$
   (c) $u - v = 1$  (d) $x = 2y + 6$

2. Which of the following are linear equations in two variables?
   (a) $y = x$  (b) $y = x^2$
   (c) $\dfrac{4}{x} - \dfrac{3}{y} = -1$  (d) $2w + 8z = -4w + 3$

3. Is $(5, 1)$ a solution of the following system?
$$\begin{cases} 2x - 8y = 2 \\ 3x + 7y = 22 \end{cases}$$

4. Is $(14, -2)$ a solution of the following system?
$$\begin{cases} x + y = 12 \\ x - y = 4 \end{cases}$$

5. Is $(0, -4)$ a solution of the following system?
$$\begin{cases} \frac{1}{6}x + \frac{1}{2}y = -2 \\ \frac{2}{3}x + \frac{3}{4}y = 2 \end{cases}$$

6. Is $(12, -8)$ a solution of the system in Exercise 5?

7. Is $(3, -2)$ a solution of the following system?
$$\begin{cases} \frac{2}{7}x - \frac{1}{5}y = \frac{44}{35} \\ \frac{1}{3}x - \frac{5}{4}y = \frac{7}{2} \end{cases}$$

*In Exercises 8–18, use the substitution method to find all solutions of each system.*

8. $\begin{cases} 4x - y = 7 \\ -2x + 3y = 9 \end{cases}$   9. $\begin{cases} 3x - 2y = -19 \\ x + 4y = -4 \end{cases}$

10. $\begin{cases} 6x - 2y = -3 \\ 5x + 3y = 4 \end{cases}$   11. $\begin{cases} 4x + 2y = 3 \\ 10x + 4y = 1 \end{cases}$

12. $\begin{cases} \frac{3}{2}x - 5y = 1 \\ x + \frac{3}{4}y = -1 \end{cases}$   13. $\begin{cases} 13x - 8y = -3 \\ -7x + 2y = 0 \end{cases}$

14. $\begin{cases} 4x + 6y = 3 \\ -6x - 9y = -\frac{9}{2} \end{cases}$   15. $\begin{cases} -\frac{2}{5}x + \frac{1}{4}y = 3 \\ \frac{1}{4}x - \frac{2}{5}y = -3 \end{cases}$

16. $\begin{cases} 0.02x - 0.03y = 1.06 \\ 0.75x + 0.50y = -0.01 \end{cases}$

17. $\begin{cases} \sqrt{2}x - \sqrt{3}y = \sqrt{3} \\ \sqrt{3}x - \sqrt{8}y = \sqrt{2} \end{cases}$   18. $\begin{cases} 7x - 3y = -12 \\ \frac{14}{3}x - 2y = 2 \end{cases}$

*In Exercises 19–28, use the addition–subtraction method to find all solutions of each system of equations.*

19. $\begin{cases} 5x + 6y = 4 \\ 2x - 3y = -3 \end{cases}$   20. $\begin{cases} -8x + y = -2 \\ 4x - 3y = 1 \end{cases}$

21. $\begin{cases} 4x + 13y = -5 \\ 2x - 54y = -1 \end{cases}$   22. $\begin{cases} 16x - 3y = 100 \\ 16x + 10y = 10 \end{cases}$

23. $\begin{cases} \frac{1}{4}x - \frac{1}{3}y = 4 \\ \frac{2}{7}x - \frac{1}{7}y = \frac{1}{10} \end{cases}$

   *Suggestion:* First clear both equations of fractions.

24. $\begin{cases} 2.1x - 3.5y = 1.2 \\ 1.4x + 2.6y = 1.1 \end{cases}$   25. $\begin{cases} 8x + 16y = 5 \\ 2x + 5y = \frac{5}{4} \end{cases}$

26. $\begin{cases} 8x + 16y = 5 \\ 2x + 4y = 1 \end{cases}$

27. $\begin{cases} \sqrt{6}x - \sqrt{3}y = 3\sqrt{2} - \sqrt{3} \\ \sqrt{2}x - \sqrt{5}y = \sqrt{6} + \sqrt{5} \end{cases}$

28. $\begin{cases} 125x - 40y = 45 \\ \frac{1}{10}x + \frac{1}{10}y = \frac{3}{10} \end{cases}$

29. Find $b$ and $c$, given that the parabola $y = x^2 + bx + c$ passes through $(0, 4)$ and $(2, 14)$.

30. Determine the constants $a$ and $b$, given that the parabola $y = ax^2 + bx + 1$ passes through $(-1, 11)$ and $(3, 1)$.

31. Determine the constants $A$ and $B$, given that the line $Ax + By = 2$ passes through the points $(-4, 5)$ and $(7, -9)$.

32. Find the area of the triangular region in the first quadrant bounded by the lines $y = 5x$ and $y = -3x + 6$ and the $x$-axis.

33. Find the area of the triangular region in the first quadrant bounded by the $x$-axis and the lines $y = 2x - 5$ and $y = -\frac{1}{2}x + 3$.

34. Find the area of the triangular region in the first quadrant bounded by the $y$-axis and the lines $y = 2x + 2$ and $y = -\frac{3}{2}x + 9$.

35. A student in a chemistry laboratory has access to two acid solutions. The first solution is 10% acid and the second is 35% acid. (The percentages are by volume.) How many cubic centimeters of each should she mix together to obtain $200 \, \text{cm}^3$ of a 25% acid solution?

36. One salt solution is 15% salt and another is 20% salt. How many cubic centimeters of each solution must be mixed to obtain $50 \, \text{cm}^3$ of a 16% salt solution?

37. A shopkeeper has two types of coffee beans on hand. One type sells for \$5.20/lb, the other for \$5.80/lb. How many pounds of each type must be mixed to produce 16 lb of a blend that sells for \$5.50/lb?

38. A certain alloy contains 10% tin and 30% copper. (The percentages are by weight.) How many pounds of tin and how many pounds of copper must be melted with 1000 lb of the given alloy to yield a new alloy containing 20% tin and 35% copper?

**B**

39. Find $x$ and $y$ in terms of $a$ and $b$:

$$\begin{cases} \dfrac{x}{a} + \dfrac{y}{b} = 1 \\ \dfrac{x}{b} + \dfrac{y}{a} = 1 \end{cases}$$

   Does your solution impose any conditions on $a$ and $b$?

40. Solve the following system for $x$ and $y$ in terms of $a$ and $b$, where $a \neq b$:

$$\begin{cases} ax + by = \dfrac{1}{a} \\ b^2x + a^2y = 1 \end{cases}$$

41. Solve the following system for $x$ and $y$ in terms of $a$ and $b$, where $a \neq b$:

$$\begin{cases} ax + a^2y = 1 \\ bx + b^2y = 1 \end{cases}$$

   Does your solution impose any additional conditions on $a$ and $b$?

42. Solve the following system for $s$ and $t$:

$$\begin{cases} \dfrac{3}{s} - \dfrac{4}{t} = 2 \\ \dfrac{5}{s} + \dfrac{1}{t} = -3 \end{cases}$$

   *Hint:* Make the substitutions $1/s = x$ and $1/t = y$ in order to obtain a system of two linear equations.

43. Solve the following system for $s$ and $t$:

$$\begin{cases} \dfrac{1}{2s} - \dfrac{1}{2t} = -10 \\ \dfrac{2}{s} + \dfrac{3}{t} = 5 \end{cases}$$

   (Use the hint in Exercise 42.)

In Exercises 44–49, find all solutions of the given systems. For Exercises 47–49, round off the final answers to two decimal places.

**44.** $\begin{cases} 0.5x - 0.8y = 0.3 \\ 0.4x - 0.1y = 0.9 \end{cases}$

**45.** $\begin{cases} \dfrac{2w - 1}{3} + \dfrac{z + 2}{4} = 4 \\ \dfrac{w + 3}{2} - \dfrac{w - z}{3} = 3 \end{cases}$

**46.** $\dfrac{x - y}{2} = \dfrac{x + y}{3} = 1$

Ⓒ **47.** $\begin{cases} 1.03x - 2.54y = 5.47 \\ 3.85x + 4.29y = -1.84 \end{cases}$

Ⓒ **48.** $\begin{cases} 2.39x + 8.16y = -2.83 \\ 1.01x + 2.98y = 4.41 \end{cases}$

Ⓒ **49.** $\begin{cases} \sqrt{5}x - 4\sqrt{3}y = 6 \\ \sqrt{2}x + 3\sqrt{3}y = 8 \end{cases}$

**50.** The sum of two numbers is 64. Twice the larger number plus five times the smaller number is 20. Find the two numbers. (Let $x$ denote the larger number and let $y$ denote the smaller number.)

**51.** In a two-digit number, the sum of the digits is 14. Twice the tens digit exceeds the units digit by one. Find the number.

**52.** The sum of the digits in a two-digit number is 14. Furthermore, the number itself is 2 greater than 11 times the tens digit. Find the number.

**53.** The perimeter of a rectangle is 34 in. If the length is 2 in. more than twice the width, find the length and the width.

**54.** Verify that the parabola $y = x^2 + \frac{5}{4}x - \frac{17}{4}$ indeed passes through $(-3, 1)$ and $(1, -2)$.

**55.** Consider the following system:

$$\begin{cases} x + y = 100 \\ -2x + 3y = 0 \end{cases}$$

(a) Solve this system using the method of substitution. *Answer:* $(60, 40)$

(b) Solve the system using the addition–subtraction method.

**56.** Consider the following system:

$$\begin{cases} 2x - 5y = -2 \\ 3x + 4y = 5 \end{cases}$$

(a) Solve the system using the substitution method.

(b) Solve the system using the addition–subtraction method. *Answer:* $\left(\frac{17}{23}, \frac{16}{23}\right)$

(c) Verify that your solution indeed satisfies both equations of the system.

**57.** Solve the following system for $x$ and $y$:

$$\begin{cases} by = x + ab \\ cy = x + ac \end{cases}$$

(Assume $a$, $b$, and $c$ are constants and that $b \neq c$.)

**58.** Solve for $x$ and $y$ in terms of $a$, $b$, $c$, $d$, $e$, and $f$:

$$\begin{cases} ax + by = c \\ dx + ey = f \end{cases}$$

(Assume that $ae - bd \neq 0$.)

**59.** Solve the following system for $x$ and $y$ in terms of $a$ and $b$, where $a \neq b$:

$$\begin{cases} \dfrac{a}{bx} + \dfrac{b}{ay} = a + b \\ \dfrac{b}{x} + \dfrac{a}{y} = a^2 + b^2 \end{cases}$$

## C

**60.** Solve the following system for $x$ and $y$ in terms of $a$ and $b$, where $ab \neq -1$:

$$\begin{cases} \dfrac{x + y}{x - y + 1} = a \\ \dfrac{y - x + 1}{x - y + 1} = ab \end{cases}$$

*Answer:* $\left(\dfrac{a + 1}{ab + 1}, \dfrac{a(b + 1)}{ab + 1}\right)$

**61.** Given that the lines $7x + 5y = 4$, $x + ky = 3$, and $5x + y + k = 0$ are concurrent (pass through a common point), what are the possible values for $k$? *Answer:* 2 and $-\frac{29}{7}$

**62.** Solve the following system for $x$ and $y$ in terms of $a$ and $b$:

$$\begin{cases} (a + b)x + (a^2 + b^2)y = a^3 + b^3 \\ (a - b)x + (a^2 - b^2)y = a^3 - b^3 \end{cases}$$

(Assume that $a$ and $b$ are nonzero and that $a \neq \pm b$.)

**63.** The vertices of triangle $ABC$ are $A(2a, 0)$, $B(2b, 0)$, and $C(0, 2)$.

(a) Show that the equations of the sides are $x + by = 2b$, $x + ay = 2a$, and $y = 0$.

(b) Show that the equations of the medians are $x + (2a - b)y = 2a$, $x + (2b - a)y = 2b$, and $2x + (a + b)y = 2(a + b)$.

(c) Show that all three medians intersect at the point $\left(\frac{2}{3}(a + b), \frac{2}{3}\right)$.

(d) Show that the equations of the altitudes are $bx - y = 2ab$, $ax - y = 2ab$, and $x = 0$.

(e) Show that all three altitudes intersect at the point $(0, -2ab)$.

(f) Show that the equations of the perpendicular bisectors of the sides are $x = a + b$, $bx - y = b^2 - 1$, and $ax - y = a^2 - 1$.

(g) Show that all three perpendicular bisectors intersect at the point $(a + b, ab + 1)$.

(h) Show that the three points found in parts (c), (e), and (g) all lie on a line. This is the **Euler line** of the triangle.

## GAUSSIAN ELIMINATION

**10.2**

*A method of solution is perfect if we can foresee from the start, and even prove, that following that method we shall obtain our aim.*

GOTTFRIED WILHELM VON LEIBNIZ (1646–1716)

In the previous section, we solved systems of linear equations in two unknowns. In this section, we introduce the technique known as Gaussian elimination for solving systems of linear equations in which there are more than two unknowns.*

As a first example, consider the following system of three linear equations in the three unknowns $x$, $y$, and $z$:

$$\begin{cases} 3x + 2y - z = -3 \\ 5y - 2z = 2 \\ 5z = 20 \end{cases}$$

This system is easy to solve using the process of *back-substitution*. Dividing both sides of the third equation by 5 yields $z = 4$. Then substituting $z = 4$ back into the second equation gives us

$$5y - 2(4) = 2$$
$$5y = 10$$
$$y = 2$$

Finally, substituting the values $z = 4$ and $y = 2$ back into the first equation yields

$$3x + 2(2) - 4 = -3$$
$$3x = -3$$
$$x = -1$$

We have now found that $x = -1$, $y = 2$, and $z = 4$. If you go back and check, you will find that these values indeed satisfy all three equations in the given system. Furthermore, the algebra we've just carried out shows that these are the only possible values for $x, y$, and $z$ satisfying all three equations. We summarize by saying that the **ordered triple** $(-1, 2, 4)$ is the solution of the given system.

The system that we just considered was easy to solve (using back substitution) because of the special form in which it was written. This form is called

---

* The technique is named in honor of Carl Friedrich Gauss (1777–1885), however, the technique itself was in existence before Gauss's time. Indeed, the essentials of the method appear in a Chinese text, "Nine Chapters on the Mathematical Art" (*Chiu Chang Suan Shu*), written approximately 2000 years ago.

**upper-triangular form**. Although the following definition of upper-triangular form refers to systems with three unknowns, the same type of definition can be given for systems with any number of unknowns. Table 1 displays examples of systems in upper-triangular form.

---

**Upper-Triangular Form (Three Variables)**

A system of linear equations in $x, y$, and $z$ is said to be in **upper-triangular form** provided $x$ appears in no equation after the first and $y$ appears in no equation after the second. (It is possible that $y$ may not even appear in the second equation.)

---

TABLE 1

Examples of systems in upper-triangular form

| 2 UNKNOWNS: $x, y$ | 3 UNKNOWNS: $x, y, z$ | 4 UNKNOWNS: $x, y, z, t$ |
|---|---|---|
| $\begin{cases} 3x + 5y = 7 \\ \phantom{3x + {}} 8y = 5 \end{cases}$ | $\begin{cases} 4x - 3y + \phantom{7}2z = -5 \\ \phantom{4x - {}} 7y + \phantom{2}z = 9 \\ \phantom{4x - 7y + {}} -4z = 3 \end{cases}$ | $\begin{cases} x - y + \phantom{2}z - 4t = 1 \\ \phantom{x - {}} 3y - 2z + \phantom{5}t = -1 \\ \phantom{x - 3y + {}} 3z - 5t = 4 \\ \phantom{x - 3y + 3z + {}} 6t = 7 \end{cases}$ |
| | $\begin{cases} 15x - 2y + \phantom{3}z = 1 \\ \phantom{15x - 2y + {}} 3z = -8 \end{cases}$ | $\begin{cases} 2x + y + 2z - \phantom{3}t = -3 \\ \phantom{2x + y + {}} 4z + 3t = 1 \\ \phantom{2x + y + 4z + {}} 5t = 6 \end{cases}$ |
| | | $\begin{cases} 8x + 3y - z + \phantom{4}t = 2 \\ \phantom{8x + {}} 2y + z - 4t = 1 \end{cases}$ |

When we were solving linear systems of equations with two unknowns in the previous section, we observed that there were three possibilities: a unique solution; infinitely many solutions; and no solution. As the next three examples indicate, the situation is similar when dealing with larger systems.

**EXAMPLE 1** Find all solutions of the system

$$\begin{cases} x + y + 2z = 2 \\ \phantom{x + {}} 3y - 4z = -5 \\ \phantom{x + 3y + {}} 6z = 3 \end{cases}$$

Solution The system is in upper-triangular form, and we can use back-substitution. Dividing the third equation by 6 yields $z = \frac{1}{2}$. Substituting this value for $z$ back into the second equation then yields

$$3y - 4(\tfrac{1}{2}) = -5$$
$$3y = -3$$
$$y = -1$$

Now, substituting the values $z = \frac{1}{2}$ and $y = -1$ back in the first equation, we obtain

$$x + (-1) + 2(\tfrac{1}{2}) = 2$$
$$x = 2$$

As you can easily check, the values $x = 2$, $y = -1$, and $z = \frac{1}{2}$ indeed satisfy all three equations. Furthermore, the algebra we've just carried out shows that these are the only possible values for $x, y$, and $z$ satisfying all three equations. We summarize by saying that the unique solution to our system is the ordered triple $(2, -1, \frac{1}{2})$.

**EXAMPLE 2** Find all solutions of the system

$$\begin{cases} -2x + y + 3z = 6 \\ \phantom{-2x + y + {}} 2z = 10 \end{cases}$$

**Solution** Again, the system is in upper-triangular form, and we use back-substitution. Solving the second equation for $z$ yields $z = 5$. Then replacing $z$ by 5 in the first equation gives us

$$-2x + y + 3(5) = 6$$
$$-2x + y = -9$$
$$y = 2x - 9$$

At this point, we've made use of both equations in the given system. There is no third equation to provide additional restrictions on $x$, $y$, or $z$. We know from the previous section that the equation $y = 2x - 9$ has infinitely many solutions, all of the form

$(x, 2x - 9)$     where $x$ is a real number

It follows, then, that there are infinitely many solutions to the given system and they may be written

$(x, 2x - 9, 5)$     where $x$ is a real number

For instance, choosing in succession $x = 0$, $x = 1$, and $x = 2$ yields the solutions $(0, -9, 5)$, $(1, -7, 5)$, and $(2, -5, 5)$. (We remark in passing that any linear system in upper-triangular form in which the number of unknowns exceeds the number of equations will always have infinitely many solutions.)

**EXAMPLE 3** Find all solutions of the system

$$\begin{cases} 4x - 7y + 3z = 1 \\ 3x + \phantom{7}y - 2z = 4 \\ 4x - 7y + 3z = 6 \end{cases}$$

(Note that the system is not in upper-triangular form.)

**Solution** Look at the left-hand sides of the first and third equations: they are identical. Thus, if there were values for $x, y$, and $z$ that satisfied both equations, it would follow that $1 = 6$, which is clearly impossible. We conclude that the given system has no solutions.

As Examples 1 and 2 have demonstrated, systems in upper-triangular form can be readily solved. In view of this, it would be useful to have a technique for converting a given system into an equivalent system in upper-triangular form. (The expression **equivalent system** means a system with exactly the same set of solutions as the original system.) **Gaussian elimination** is one such technique. We will demonstrate this technique in Examples 4, 5, and 6. In using Gaussian elimination, we will rely on what are called the three **elementary operations**, listed in the box that follows. These are operations that can be performed on an equation in a system without altering the set of solutions of the system.

---

### The Elementary Operations

1. Multiply both sides of an equation by a nonzero constant.
2. Interchange the order in which two equations of a system are listed.
3. To one equation add a multiple of another equation in the system.

---

**EXAMPLE 4**  Find all solutions of the system

$$\begin{cases} x + 2y + z = 3 \\ 2x + y + z = 16 \\ x + y + 2z = 9 \end{cases}$$

**Solution**  First we want to eliminate $x$ from the second and third equations. To eliminate $x$ from the second equation, we add to it $-2$ times the first equation. The result is the equivalent system

$$\begin{cases} x + 2y + z = 3 \\ -3y - z = 10 \\ x + y + 2z = 9 \end{cases}$$

To eliminate $x$ from the third equation, we add to it $-1$ times the first equation. The result is the equivalent system

$$\begin{cases} x + 2y + z = 3 \\ -3y - z = 10 \\ -y + z = 6 \end{cases}$$

Now to bring the system into upper-triangular form, we need to eliminate $y$ from the third equation. We could do this by adding to the third equation $-\frac{1}{3}$ times the second equation. However, to avoid working with fractions as long as possible, we proceed instead to interchange the second and third equations to obtain the equivalent system

$$\begin{cases} x + 2y + z = 3 \\ -y + z = 6 \\ -3y - z = 10 \end{cases}$$

Now we add to the last equation $-3$ times the second equation to obtain the equivalent system

$$\begin{cases} x + 2y + z = 3 \\ -y + z = 6 \\ -4z = -8 \end{cases}$$

The system is now in upper-triangular form, and back-substitution yields, in turn, $z = 2, y = -4,$ and $x = 9$. (Check this for yourself.) The required solution is therefore $(9, -4, 2)$.

In Table 2 we list some convenient abbreviations that are used in describing the elementary operations. You should plan on using these abbreviations yourself: they'll make it simpler for you (and your instructor) to check your work. In Table 2, the notation $E_i$ stands for the $i$th equation in a system. For instance, for the initial system in Example 4, the symbol $E_1$ denotes the first equation: $x + 2y + z = 3$.

**TABLE 2**

Abbreviations for the elementary operations

| ABBREVIATION | EXPLANATION |
|---|---|
| 1. $cE_i$ | Multiply both sides of the $i$th equation by $c$. |
| 2. $E_i \leftrightarrow E_j$ | Interchange the $i$th and $j$th equations. |
| 3. $cE_i + E_j$ | To the $j$th equation, add $c$ times the $i$th equation. |

**EXAMPLE 5** Solve the system

$$\begin{cases} 4x - 3y + 2z = 40 \\ 5x + 9y - 7z = 47 \\ 9x + 8y - 3z = 97 \end{cases}$$

**Solution**

$$\begin{cases} 4x - 3y + 2z = 40 \\ 5x + 9y - 7z = 47 \\ 9x + 8y - 3z = 97 \end{cases} \xrightarrow{(-1)E_2 + E_1} \begin{cases} -x - 12y + 9z = -7 \\ 5x + 9y - 7z = 47 \\ 9x + 8y - 3z = 97 \end{cases}$$ adding $-1$ times the second equation *to* the first equation

$$\xrightarrow[9E_1 + E_3]{5E_1 + E_2} \begin{cases} -x - 12y + 9z = -7 \\ -51y + 38z = 12 \\ -100y + 78z = 34 \end{cases}$$

$$\xrightarrow{\frac{1}{2}E_3} \begin{cases} -x - 12y + 9z = -7 \\ -51y + 38z = 12 \\ -50y + 39z = 17 \end{cases}$$

$$\xrightarrow{(-1)E_3 + E_2} \begin{cases} -x - 12y + 9z = -7 \\ -y - z = -5 \\ -50y + 39z = 17 \end{cases}$$ to allow working with smaller but integral coefficients

$$\xrightarrow{-50E_2 + E_3} \begin{cases} -x - 12y + 9z = -7 \\ -y - z = -5 \\ 89z = 267 \end{cases}$$

The system is now in upper-triangular form. Solving the third equation, we obtain $z = 3$. Substituting this value back into the second equation yields $y = 2$. (Check this for yourself.) Finally, substituting $z = 3$ and $y = 2$ back into the first equation yields $x = 10$. (Again, check this for yourself.) The solution to the system is therefore $(10, 2, 3)$.

**EXAMPLE 6**  Solve the system

$$\begin{cases} x + 2y + 4z = 0 \\ x + 3y + 9z = 0 \end{cases}$$

Solution  This system is similar to the one in Example 2 in that there are fewer equations than there are unknowns. By subtracting the first equation from the second, we readily obtain an equivalent system in upper-triangular form:

$$\begin{cases} x + 2y + 4z = 0 \\ \phantom{x + 2}y + 5z = 0 \end{cases}$$

Although the system is now in upper-triangular form, notice that the second equation does not determine $y$ or $z$ uniquely; that is, there are infinitely many number pairs $(y, z)$ satisfying the second equation. We can solve the second equation for $y$ in terms of $z$; the result is $y = -5z$. Now we replace $y$ with $-5z$ in the first equation to obtain

$$x + 2(-5z) + 4z = 0 \qquad \text{or} \qquad x = 6z$$

At this point, we've used both of the equations in the system to express $x$ and $y$ in terms of $z$. Furthermore, there is no third equation in the system to provide additional restrictions on $x$, $y$, or $z$. We therefore conclude that the given system has infinitely many solutions. These solutions have the form

$$(6z, -5z, z) \qquad \text{where } z \text{ is any real number}$$

**EXAMPLE 7**  Solve the system

$$\begin{cases} x - 4y + \phantom{2}z = 3 \\ 3x + 5y - 2z = -1 \\ 7x + 6y - 3z = 2 \end{cases}$$

Solution  $\begin{cases} x - 4y + \phantom{2}z = 3 \\ 3x + 5y - 2z = -1 \\ 7x + 6y - 3z = 2 \end{cases}$  $\xrightarrow[-7E_1 + E_3]{-3E_1 + E_2}$  $\begin{cases} x - 4y + \phantom{2}z = 3 \\ \phantom{x}17y - \phantom{1}5z = -10 \\ \phantom{x}34y - 10z = -19 \end{cases}$

$\xrightarrow{-2E_2 + E_3}$  $\begin{cases} x - 4y + \phantom{2}z = 3 \\ \phantom{x}17y - 5z = -10 \\ \phantom{x17y - 5z}0 = 1 \end{cases}$

From the third equation in this last system, we conclude that this system, and consequently the original system, has no solution. (Reason: If there *were* values for $x$, $y$, and $z$ satisfying the original system, then it would follow that $0 = 1$, which is clearly impossible.)

In the next two examples, we introduce the subject of **partial fractions**. The basic goal here is to write a given fractional expression as a sum or difference of two or more *simpler* fractions. For instance, in Example 8, we will be given the fraction $\dfrac{1}{(x-1)(x+1)}$ and we will be asked to find constants $A$ and $B$ so that

$$\frac{1}{(x-1)(x+1)} = \frac{A}{x-1} + \frac{B}{x+1}$$

When $A$ and $B$ are determined, we refer to the right-hand side of the preceding equation as the **partial fraction decomposition** of the given fraction. Detailed techniques for working with partial fractions are developed as the need arises in calculus courses. The two examples and the exercises that follow will provide you with ample background for that later work.

**EXAMPLE 8**   Determine constants $A$ and $B$ so that the following equation is an identity (i.e., the equation is to hold for all values of $x$ for which the denominators are not zero):

$$\frac{1}{(x-1)(x+1)} = \frac{A}{x-1} + \frac{B}{x+1} \tag{1}$$

**Solution**   First, to clear equation (1) of fractions, we multiply both sides of the equation by the quantity $(x-1)(x+1)$. This yields

$$
\begin{aligned}
1 &= (x+1)A + (x-1)B \\
&= Ax + A + Bx - B \\
1 &= (A+B)x + (A-B)
\end{aligned}
\tag{2}
$$

Since this last equation is to be an identity, we now reason as follows:

$$\left[\begin{matrix}\text{coefficient of } x \text{ on the right-}\\ \text{hand side of equation (2)}\end{matrix}\right] = \left[\begin{matrix}\text{coefficient of } x \text{ on the left-}\\ \text{hand side of equation (2)}\end{matrix}\right]$$

Therefore,

$$A + B = 0 \tag{3}$$

Similarly, we have

$$\left[\begin{matrix}\text{constant term on the right-}\\ \text{hand side of equation (2)}\end{matrix}\right] = \left[\begin{matrix}\text{constant term on the left-}\\ \text{hand side of equation (2)}\end{matrix}\right]$$

Therefore,

$$A - B = 1 \tag{4}$$

Rewriting equations (3) and (4) gives us a system of two linear equations in the unknowns $A$ and $B$:

$$\begin{cases} A + B = 0 \\ A - B = 1 \end{cases}$$

Adding these two equations yields

$$2A = 1$$
$$A = \tfrac{1}{2}$$

Subtracting the two equations gives us

$$B - (-B) = 0 - 1$$
$$2B = -1$$
$$B = -\tfrac{1}{2}$$

We have now found that $A = \tfrac{1}{2}$ and $B = -\tfrac{1}{2}$, as required. The partial fraction decomposition of $\dfrac{1}{(x-1)(x+1)}$ is therefore

$$\frac{1}{(x-1)(x+1)} = \frac{\tfrac{1}{2}}{x-1} + \frac{-\tfrac{1}{2}}{x+1}$$
$$= \frac{1}{2(x-1)} - \frac{1}{2(x+1)}$$

You should check for yourself that the result of combining the two fractions on the right-hand side of this last equation is indeed $\dfrac{1}{(x-1)(x+1)}$.

In Example 8, we were able to find the partial fraction decomposition by solving a system of two linear equations. There is a shortcut that is sometimes useful in such problems. As before, we multiply both sides of equation (1) by $(x-1)(x+1)$ to obtain

$$1 = (x+1)A + (x-1)B \tag{5}$$

Now, equation (5) is to hold for all values of $x$; in particular, it must hold when $x$ is $-1$. Replacing $x$ with $-1$ in the equation then yields

$$1 = -2B$$
$$B = -\tfrac{1}{2} \qquad \text{as obtained previously}$$

The value of $A$ can be obtained in a similar way. We replace $x$ by 1 in equation (5). As you can check, this readily yields $A = \tfrac{1}{2}$, which again agrees with the result obtained previously.

**EXAMPLE 9**    Determine the constants $A, B,$ and $C$ so that the following equation is an identity:

$$\frac{2x+1}{(x+2)(x-3)^2} = \frac{A}{x+2} + \frac{B}{(x-3)} + \frac{C}{(x-3)^2}$$

**Solution**    First, to clear the given equation of fractions, we multiply both sides by the quantity $(x+2)(x-3)^2$. After dividing out the common factors, we have

$$2x + 1 = A(x-3)^2 + B(x+2)(x-3) + C(x+2)$$
$$= A(x^2 - 6x + 9) + B(x^2 - x - 6) + C(x+2)$$
$$= Ax^2 - 6Ax + 9A + Bx^2 - Bx - 6B + Cx + 2C$$
$$2x + 1 = (A+B)x^2 + (-6A - B + C)x + 9A - 6B + 2C$$

Equating the coefficients of $x^2$ on both sides of this identity yields

$$A + B = 0$$

Similarly, equating the $x$-coefficients on both sides yields

$$-6A - B + C = 2$$

Finally, equating the constant terms on both sides gives us

$$9A - 6B + 2C = 1$$

We now have a system of three equations in the three unknowns $A$, $B$, and $C$:

$$\begin{cases} A + B = 0 \\ -6A - B + C = 2 \\ 9A - 6B + 2C = 1 \end{cases}$$

Exercise 12 at the end of this section asks you to verify that the required values here are $A = -\frac{3}{25}$, $B = \frac{3}{25}$, and $C = \frac{7}{5}$. The partial fraction decomposition is therefore

$$\frac{2x + 1}{(x + 2)(x - 3)^2} = \frac{-3}{25(x + 2)} + \frac{3}{25(x - 3)} + \frac{7}{5(x - 3)^2}$$

Here is an alternate method for working Example 9, which relies on the short-cut mentioned earlier. Again, we first multiply the given equation by $(x + 2)(x - 3)^2$ to obtain

$$2x + 1 = A(x - 3)^2 + B(x + 2)(x - 3) + C(x + 2) \tag{6}$$

Now we just replace $x$ by 3 in this equation. This yields

$$7 = 5C$$
$$C = \tfrac{7}{5} \qquad \text{as obtained previously}$$

The value of $A$ can be obtained in a similar manner. We replace $x$ by $-2$ in equation (6). As you can check, this readily yields $A = -\frac{3}{25}$, which again agrees with the value obtained previously. We've now computed $A$ and $C$ using the short method. But as you'll see if you try, the value of $B$ cannot be obtained in the same way. One way to find $B$ here is by substituting the values found for $A$ and $C$ in equation (6) to obtain

$$2x + 1 = -\tfrac{3}{25}(x - 3)^2 + B(x + 2)(x - 3) + \tfrac{7}{5}(x + 2)$$

This equation is to hold for all values of $x$; in particular, it must hold for the convenient value $x = 0$. Substituting $x = 0$ in this last equation yields

$$1 = -\tfrac{3}{25}(-3)^2 + B(2)(-3) + \tfrac{7}{5}(2)$$
$$25 = -27 - 150B + 70 \qquad \text{multiplying by 25}$$
$$150B = 18 \qquad \text{collecting like terms}$$
$$B = \tfrac{18}{150} = \tfrac{3}{25} \qquad \text{dividing by 150 and simplifying}$$

We have now found that $B = \frac{3}{25}$, in agreement with the value obtained previously.

# EXERCISE SET 10.2

## A

In Exercises 1–10, the systems of linear equations are in upper-triangular form. Find all solutions of each system.

1. $\begin{cases} 2x + y + z = -9 \\ \quad\ 3y - 2z = -4 \\ \quad\qquad 8z = -8 \end{cases}$

2. $\begin{cases} -3x + 7y + 2z = -19 \\ \qquad\ y + z = 1 \\ \qquad\quad -2z = -2 \end{cases}$

3. $\begin{cases} 8x + 5y + 3z = 1 \\ \quad\ 3y + 4z = 2 \\ \qquad\ 5z = 3 \end{cases}$

4. $\begin{cases} 2x \qquad + 7z = -4 \\ \quad 5y \quad 3z = 6 \\ \qquad\ 6z = 18 \end{cases}$

5. $\begin{cases} -4x + 5y \qquad = 0 \\ \quad\ 3y + 2z = 1 \\ \qquad\ 3z = -1 \end{cases}$

6. $\begin{cases} 3x - 2y + z = 4 \\ \qquad\ 3z = 9 \end{cases}$

7. $\begin{cases} -x + 8y + 3z = 0 \\ \qquad\ 2z = 0 \end{cases}$

8. $\begin{cases} -x + y + z + w = 9 \\ \quad\ 2y - z - w = 9 \\ \qquad\ 3z + 2w = 1 \\ \qquad\qquad 11w = 22 \end{cases}$

9. $\begin{cases} 2x + 3y + z + w = -6 \\ \quad\ y + 3z - 4w = 23 \\ \qquad\ 6z - 5w = 31 \\ \qquad\qquad -2w = 10 \end{cases}$

10. $\begin{cases} 7x - y - z + w = 3 \\ \quad\ 2y - 3z - 4w = -2 \\ \qquad\qquad 3w = 6 \end{cases}$

In Exercises 11–30, find all solutions of each system.

11. $\begin{cases} x + y + z = 12 \\ 2x - y - z = -1 \\ 3x + 2y + z = 22 \end{cases}$

12. $\begin{cases} A + B \qquad = 0 \\ -6A - B + C = 2 \\ 9A - 6B + 2C = 1 \end{cases}$ Answer: $\left(-\frac{3}{25}, \frac{3}{25}, \frac{7}{5}\right)$

13. $\begin{cases} 2x - 3y + 2z = 4 \\ 4x + 2y + 3z = 7 \\ 5x + 4y + 2z = 7 \end{cases}$

14. $\begin{cases} x \qquad + 2z = 5 \\ \quad y - 30z = -16 \\ x - 2y + 4z = 8 \end{cases}$

15. $\begin{cases} 3x + 3y - 2z = 13 \\ 6x + 2y - 5z = 13 \\ 7x + 5y - 3z = 26 \end{cases}$

16. $\begin{cases} 2x + 5y - 3z = 4 \\ 4x - 3y + 2z = 9 \\ 5x + 6y - 2z = 18 \end{cases}$

17. $\begin{cases} x + y + z = 1 \\ -2x + y + z = -2 \\ 3x + 6y + 6z = 5 \end{cases}$

18. $\begin{cases} 7x + 5y - 7z = -10 \\ 2x + y + z = 7 \\ x + y - 3z = -8 \end{cases}$

19. $\begin{cases} 2x - y + z = -1 \\ x + 3y - 2z = 2 \\ -5x + 6y - 5z = 5 \end{cases}$

20. $\begin{cases} -2x + 2y - z = 0 \\ 3x - 4y + z = 1 \\ 5x - 8y + z = 4 \end{cases}$

21. $\begin{cases} 2x - y + z = 4 \\ x + 3y + 2z = -1 \\ 7x \qquad + 5z = 11 \end{cases}$

22. $\begin{cases} 3x + y - z = 10 \\ 8x - y - 6z = -3 \\ 5x - 2y - 5z = 1 \end{cases}$

23. $\begin{cases} x + y + z + w = 4 \\ x - 2y - z - w = 3 \\ 2x - y + z - w = 2 \\ x - y + 2z - 2w = -7 \end{cases}$

24. $\begin{cases} x + y - 3z + 2w = 0 \\ -2x - 2y + 6z + w = -5 \\ -x + 3y + 3z + 3w = 5 \\ 2x + y - 3z - w = 4 \end{cases}$

25. $\begin{cases} 2x + 3y + 2z = 5 \\ x + 4y - 3z = 1 \end{cases}$

26. $\begin{cases} 4x - y - 3z = 2 \\ 6x + 5y - z = 0 \end{cases}$

27. $\begin{cases} x - 2y - 2z + 2w = -10 \\ 3x + 4y - z - 3w = 11 \\ -1x \quad 3y \quad 3z + 8w = -21 \end{cases}$

28. $\begin{cases} 2x + y + z + w = 1 \\ x + 3y - 3z - 3w = 0 \\ -3x - 4y + 2z + 2w = -1 \end{cases}$

29. $\begin{cases} 4x - 2y + 3z = -2 \\ 6y - 4z = 6 \end{cases}$

30. $\begin{cases} 6x - 2y - 5z + w = 3 \\ 5x - y + z + 5w = 4 \\ -4y - 7z - 9w = -5 \end{cases}$

In Exercises 31–40, determine the constants (denoted by capital letters) so that each equation is an identity.

31. (a) $\dfrac{1}{(x-2)(x+2)} = \dfrac{A}{x-2} + \dfrac{B}{x+2}$

 (b) $\dfrac{x}{(x-2)(x+2)} = \dfrac{A}{x-2} + \dfrac{B}{x+2}$

32. (a) $\dfrac{2}{(x+3)(x+2)} = \dfrac{A}{x+3} + \dfrac{B}{x+2}$

 (b) $\dfrac{x-9}{(x+3)(x+2)} = \dfrac{A}{x+3} + \dfrac{B}{x+2}$

33. $\dfrac{3x}{(x+4)(x-5)} = \dfrac{A}{x+4} + \dfrac{B}{x-5}$

34. $\dfrac{4x+26}{(x-1)(x+5)} = \dfrac{A}{x-1} + \dfrac{B}{x+5}$

35. (a) $\dfrac{3}{(x-4)(x+4)} = \dfrac{A}{x-4} + \dfrac{B}{x+4}$

(b) $\dfrac{3}{(x-4)(x+4)^2} = \dfrac{A}{x-4} + \dfrac{B}{x+4} + \dfrac{C}{(x+4)^2}$

36. (a) $\dfrac{1}{(x+1)(x+6)} = \dfrac{A}{x+1} + \dfrac{B}{x+6}$

(b) $\dfrac{1}{(x+1)(x+6)^2} = \dfrac{A}{x+1} + \dfrac{B}{x+6} + \dfrac{C}{(x+6)^2}$

37. $\dfrac{1}{(x+1)(x^2-x+1)} = \dfrac{A}{x+1} + \dfrac{Bx+C}{x^2-x+1}$

38. $\dfrac{1}{(x+1)^2(x^2-x+1)} = \dfrac{A}{x+1} + \dfrac{B}{(x+1)^2} + \dfrac{Cx+D}{x^2-x+1}$

39. $\dfrac{4}{x(1-x)} = \dfrac{A}{x} + \dfrac{B}{1-x}$

40. $\dfrac{4}{x^2(1-x)} = \dfrac{A}{x} + \dfrac{B}{x^2} + \dfrac{C}{1-x}$

**B**

41. Determine the constants $A$, $B$, $C$, and $D$ so that the following equation is an identity.

$$\dfrac{1}{(x^2+x+1)(x^2-x+1)} = \dfrac{Ax+B}{x^2+x+1} + \dfrac{Cx+D}{x^2-x+1}$$

42. Suppose that the height of an object as a function of time is given by $f(t) = at^2 + bt + c$, where $t$ is time in seconds, $f(t)$ is the height in feet at time $t$, and $a$, $b$, and $c$ are certain constants. If after 1, 2, and 3 seconds the corresponding heights are 184 ft, 136 ft, and 56 ft, find the time at which the object is at ground level (height = 0 ft).

43. Find the equation of the circle passing through the points $(-1, -5)$, $(1, 2)$, and $(-2, 0)$. Write your answer in the form $Ax^2 + Ay^2 + Bx + Cy + D = 0$.

44. Find the equation of a circle that passes through the origin and the points $(8, -4)$ and $(7, -1)$. Specify the center and radius.

45. Solve the following system for $x$, $y$, and $z$:

$$\begin{cases} e^x + e^y - 2e^z = 2a \\ e^x + 2e^y - 4e^z = 3a \\ \tfrac{1}{2}e^x - 3e^y + e^z = -5a \end{cases}$$

(Assume that $a > 0$.) *Hint*: Let $A = e^x$, $B = e^y$, and $C = e^z$. Solve the resulting linear system for $A$, $B$, and $C$.

46. Solve the following system for $\alpha$, $\beta$, and $\gamma$, given that each of these numbers lies in the closed interval $[0, \pi]$.

$$\begin{cases} \cos\alpha - \cos\beta - \cos\gamma = 2 \\ 3\cos\alpha + 5\cos\beta - 2\cos\gamma = 1 \\ 2\cos\alpha - 4\cos\beta + \cos\gamma = 2 \end{cases}$$

*Hint*: First make the substitutions $x = \cos\alpha$, $y = \cos\beta$, and $z = \cos\gamma$; then solve the resulting linear system for $x$, $y$, and $z$.

47. The following figure displays three circles that are mutually tangent. The line segments joining the centers have lengths $a$, $b$, and $c$, as shown. Let $r_1$, $r_2$, and $r_3$ denote the radii of the circles, as indicated in the figure. Show that

$$r_1 = \dfrac{a+c-b}{2}$$

$$r_2 = \dfrac{a+b-c}{2}$$

$$r_3 = \dfrac{b+c-a}{2}$$

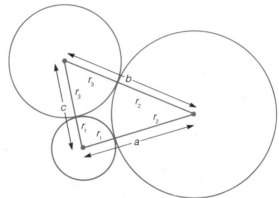

48. Consider the system

$$\begin{cases} \lambda x + y + z = a \\ x + \lambda y + z = b \\ x + y + \lambda z = c \end{cases}$$

(a) Assuming that the value of the constant $\lambda$ is neither 1 nor $-2$, find all solutions of the system.

(b) If $\lambda = -2$, how must $a$, $b$, and $c$ be related for the system to have solutions? Find these solutions.

(c) If $\lambda = 1$, how must $a$, $b$, and $c$ be related for the system to have solutions? Find these solutions.

*In Exercises 49 and 50, the lowercase letters $a$, $b$, $p$, and $q$ denote given constants. In each case, determine the values of $A$ and $B$ so that the equation is an identity.*

49. (a) $\dfrac{1}{(x-a)(x-b)} = \dfrac{A}{x-a} + \dfrac{B}{x-b}$ $\quad (a \neq b)$

(b) $\dfrac{px+q}{(x-a)(x-b)} = \dfrac{A}{x-a} + \dfrac{B}{x-b}$ $\quad (a \neq b)$

50. (a) $\dfrac{1}{(x-a)(x+a)} = \dfrac{A}{x-a} + \dfrac{B}{x+a}$ $\quad (a \neq 0)$

(b) $\dfrac{px+q}{(x-a)(x+a)} = \dfrac{A}{x-a} + \dfrac{B}{x+a}$ $\quad (a \neq 0)$

## C

**51.** Express $A$, $B$, and $C$ in terms of $a$, $b$, $c$, $p$, and $q$ so that the following is an identity. (Assume $a$, $b$, and $c$ are all unequal.)

$$\frac{x^2 + px + q}{(x-a)(x-b)(x-c)} = \frac{A}{x-a} + \frac{B}{x-b} + \frac{C}{x-c}$$

**52.** Express $A$, $B$, and $C$ in terms of $a$, $b$, and $c$ so that the following equation is an identity. (Assume $a$, $b$, and $c$ are all unequal.)

$$\frac{1}{(1-ax)(1-bx)(1-cx)} = \frac{A}{1-ax} + \frac{B}{1-bx} + \frac{C}{1-cx}$$

*The following exercise appears in* Algebra for Colleges and Schools *by H. S. Hall and S. R. Knight, revised by F. L. Sevenoak* (*New York: The Macmillan Company, 1906*).

**53.** $A$, $B$, and $C$ are three towns forming a triangle. A man has to walk from one to the next, ride thence to the next, and drive thence to his starting point. He can walk, ride, and drive a mile in $a$, $b$, and $c$ minutes, respectively. If he starts from $B$ he takes $a + c - b$ hours, if he starts from $C$ he takes $b + a - c$ hours, and if he starts from $A$ he takes $c + b - a$ hours. Find the length of the circuit. [Assume that the circuit from $A$ to $B$ to $C$ is counterclockwise.]

## MATRICES

### 10.3

As you know from the previous section, there can be a good deal of book-keeping involved in using Gaussian elimination to solve systems of equations. We can lighten the load somewhat by using matrices, where a **matrix** is simply a rectangular array of numbers, enclosed in parentheses or brackets. Here are three examples:

$$\begin{pmatrix} 2 & 3 \\ -5 & 4 \end{pmatrix} \qquad \begin{pmatrix} -6 & 0 & 1 & \frac{1}{4} \\ \frac{2}{3} & 1 & 5 & 8 \end{pmatrix} \qquad \begin{pmatrix} \pi & 0 & 0 \\ 0 & 1 & .9 \\ -1 & -2 & 3 \\ -4 & 8 & 6 \end{pmatrix}$$

The particular numbers constituting a matrix are called its **entries** or **elements**. For instance, the entries in the matrix $\begin{pmatrix} 2 & 3 \\ -5 & 4 \end{pmatrix}$ are the four numbers $2, 3, -5,$ and $4$. In this section, the entries will always be real numbers. However, it is also possible to consider matrices in which some or all of the entries are nonreal complex numbers.

It is convenient to agree on a standard system for labeling the rows and columns of a matrix. The rows are numbered from top to bottom and the columns from left to right, as indicated in the following example:

$$\begin{matrix} & \text{column 1} & \text{column 2} \\ \text{row 1} \longrightarrow & 5 & -3 \\ \text{row 2} \longrightarrow & 0 & 2 \\ \text{row 3} \longrightarrow & -1 & 16 \end{matrix}$$

We express the **size**, or **dimension**, of a matrix by specifying the number of rows and the number of columns, in that order. For instance, we would say that the matrix

$$\begin{pmatrix} 5 & -3 \\ 0 & 2 \\ -1 & 16 \end{pmatrix}$$

is a 3 × 2 (read "3 by 2") matrix, not 2 × 3. The following example will help fix in your mind the terminology that we have introduced.

**EXAMPLE 1**  Consider the matrix

$$\begin{pmatrix} 1 & 3 & 5 \\ 7 & 9 & 11 \end{pmatrix}$$

(a) List the entries.
(b) What is the size of the matrix?
(c) Which element is in the second row, third column?

**Solution**  (a) The entries are 1, 3, 5, 7, 9, and 11.
(b) Since there are two rows and three columns, this is a 2 × 3 matrix.
(c) To locate the element in the second row, third column, draw lines through the second row and the third column and see where they intersect:

$$\begin{pmatrix} 1 & 3 & 5 \\ 7 & 9 & 11 \end{pmatrix}$$

Thus, the entry in row 2, column 3 is 11.

There is a natural way to use matrices to describe and solve systems of linear equations. Consider, for example, the following system of linear equations in **standard form** (with the $x$, $y$, and $z$ terms lined up on the left-hand side and the constant terms on the right-hand side of each equation):

$$\begin{cases} x + 2y - 3z = 4 \\ 3x \qquad + z = 5 \\ -x - 3y + 4z = 0 \end{cases}$$

The **coefficient matrix** of this system is the matrix

$$\begin{pmatrix} 1 & 2 & -3 \\ 3 & 0 & 1 \\ -1 & -3 & 4 \end{pmatrix}$$

As the name implies, the coefficient matrix of the system is the matrix whose entries are the coefficients of $x$, $y$, and $z$, written in the same relative positions as they appear in the system. Notice the zero appearing in the second row, second column of the matrix. It is there because the coefficient of $y$ in the second equation is in fact zero. The **augmented matrix** of the system of equations considered here is

$$\begin{pmatrix} 1 & 2 & -3 & 4 \\ 3 & 0 & 1 & 5 \\ -1 & -3 & 4 & 0 \end{pmatrix}$$

As you can see, the augmented matrix is formed by *augmenting* the coefficient matrix with the column of constant terms taken from the right-hand side of the given system of equations. To help relate the augmented matrix to the original system of equations, we will write the augmented matrix this way:

$$\begin{pmatrix} 1 & 2 & -3 & | & 4 \\ 3 & 0 & 1 & | & 5 \\ -1 & -3 & 4 & | & 0 \end{pmatrix}$$

**EXAMPLE 2**   Write the coefficient matrix and the augmented matrix for the system

$$\begin{cases} 8x - 2y + z = 1 \\ 3x - 4z + y = 2 \\ 12y - 3z - 6 = 0 \end{cases}$$

**Solution**   First write the system in standard form, with the $x$, $y$, and $z$ terms lined up and the constant terms on the right. This yields

$$\begin{cases} 8x - 2y + \phantom{4}z = 1 \\ 3x + \phantom{1}y - 4z = 2 \\ \phantom{3x +} 12y - 3z = 6 \end{cases}$$

The coefficient matrix is then

$$\begin{pmatrix} 8 & -2 & 1 \\ 3 & 1 & -4 \\ 0 & 12 & -3 \end{pmatrix}$$

while the augmented matrix is

$$\begin{pmatrix} 8 & -2 & 1 & \vdots & 1 \\ 3 & 1 & -4 & \vdots & 2 \\ 0 & 12 & -3 & \vdots & 6 \end{pmatrix}$$

In the previous section we used the three elementary operations in solving systems of linear equations. In Table 1 we express these operations in the language of matrices. The matrix operations are called the **elementary row operations**.

TABLE 1

|  | ELEMENTARY OPERATIONS FOR A SYSTEM OF LINEAR EQUATIONS | CORRESPONDING ELEMENTARY ROW OPERATION FOR A MATRIX |
|---|---|---|
| 1. | Multiply both sides of an equation by a nonzero constant. | 1′. Multiply each entry in a given row by a nonzero constant. |
| 2. | Interchange two equations. | 2′. Interchange two rows. |
| 3. | To one equation, add a multiple of another equation. | 3′. To one row, add a multiple of another row. |

Table 2, on the next page, displays an example of each elementary row operation. In the table, notice that the notation for describing these operations is essentially the same as that introduced in the previous section. For example, in the previous section, $10E_1$ indicated that the first equation in a system was multiplied by 10. Now, $10R_1$ indicates that each entry in the first *row* of a matrix is multiplied by 10.

We are now ready to use matrices to solve systems of equations. In the example that follows, we'll use the same system of equations used in Example 5 of the previous section. *Suggestion:* After reading the next example, carefully compare each step with the corresponding one taken in Example 5 of the previous section.

**TABLE 2**

| EXAMPLES OF THE ELEMENTARY ROW OPERATIONS | COMMENTS |
|---|---|
| $\begin{pmatrix} 1 & 2 & 3 \\ 4 & 5 & 6 \\ 7 & 8 & 9 \end{pmatrix} \xrightarrow{\ 10R_1\ } \begin{pmatrix} 10 & 20 & 30 \\ 4 & 5 & 6 \\ 7 & 8 & 9 \end{pmatrix}$ | Multiply each entry in row 1 by 10. |
| $\begin{pmatrix} 1 & 2 & 3 \\ 4 & 5 & 6 \\ 7 & 8 & 9 \end{pmatrix} \xrightarrow{\ R_2 \leftrightarrow R_3\ } \begin{pmatrix} 1 & 2 & 3 \\ 7 & 8 & 9 \\ 4 & 5 & 6 \end{pmatrix}$ | Interchange rows 2 and 3. |
| $\begin{pmatrix} 1 & 2 & 3 \\ 4 & 5 & 6 \\ 7 & 8 & 9 \end{pmatrix} \xrightarrow{\ -4R_1 + R_2\ } \begin{pmatrix} 1 & 2 & 3 \\ 0 & -3 & -6 \\ 7 & 8 & 9 \end{pmatrix}$ | To each entry in row 2, add $-4$ times the corresponding entry in row 1. |

**EXAMPLE 3** Solve the system

$$\begin{cases} 4x - 3y + 2z = 40 \\ 5x + 9y - 7z = 47 \\ 9x + 8y - 3z = 97 \end{cases}$$

**Solution**

$$\begin{pmatrix} 4 & -3 & 2 & | & 40 \\ 5 & 9 & -7 & | & 47 \\ 9 & 8 & -3 & | & 97 \end{pmatrix} \xrightarrow{\ (-1)R_2 + R_1\ } \begin{pmatrix} -1 & -12 & 9 & | & -7 \\ 5 & 9 & -7 & | & 47 \\ 9 & 8 & -3 & | & 97 \end{pmatrix}$$

$$\xrightarrow[\ 9R_1 + R_3\ ]{\ 5R_1 + R_2\ } \begin{pmatrix} -1 & -12 & 9 & | & -7 \\ 0 & -51 & 38 & | & 12 \\ 0 & -100 & 78 & | & 34 \end{pmatrix}$$

$$\xrightarrow{\ \frac{1}{2}R_3\ } \begin{pmatrix} -1 & -12 & 9 & | & -7 \\ 0 & -51 & 38 & | & 12 \\ 0 & -50 & 39 & | & 17 \end{pmatrix}$$

$$\xrightarrow{\ (-1)R_3 + R_2\ } \begin{pmatrix} -1 & -12 & 9 & | & -7 \\ 0 & -1 & -1 & | & -5 \\ 0 & -50 & 39 & | & 17 \end{pmatrix}$$

$$\xrightarrow{\ -50R_2 + R_3\ } \begin{pmatrix} -1 & -12 & 9 & | & -7 \\ 0 & -1 & -1 & | & -5 \\ 0 & 0 & 89 & | & 267 \end{pmatrix}$$

This last augmented matrix represents a system of equations in upper-triangular form:

$$\begin{cases} -x - 12y + 9z = -7 \\ -y - z = -5 \\ 89z = 267 \end{cases}$$

As you should now check for yourself, this yields the values $z = 3$, then $y = 2$, then $x = 10$. The solution of the original system is therefore $(10, 2, 3)$.

For the remainder of this section (and in part of the next section) we are going to study matrices, without reference to systems of equations. As motivation for this, we point out that matrices are essential tools in many fields of study. For example, a knowledge of matrices and their properties is needed for work in computer graphics. To begin, we need to say what it means for two matrices to be equal.

### Definition of Equality for Matrices

EXAMPLES

Two matrices are equal provided that they have the same size (same number of rows, same number of columns) and that the corresponding entries are equal.

$$\begin{pmatrix} 2 & 3 \\ 4 & 5 \end{pmatrix} = \begin{pmatrix} 2 & 3 \\ 4 & 5 \end{pmatrix}$$

$$\begin{pmatrix} 2 & 3 & 0 \\ 4 & 5 & 0 \end{pmatrix} \neq \begin{pmatrix} 2 & 3 \\ 4 & 5 \end{pmatrix}$$

$$\begin{pmatrix} 2 & 3 \\ 4 & 5 \end{pmatrix} \neq \begin{pmatrix} 2 & 3 \\ 5 & 4 \end{pmatrix}$$

Now we can define matrix addition and subtraction. These operations are defined only between matrices of the same size.

### Definition of Matrix Addition

EXAMPLES

To add (or subtract) two matrices of the same size, add (or subtract) the corresponding entries.

$$\begin{pmatrix} 2 & 3 \\ -1 & 4 \end{pmatrix} + \begin{pmatrix} 6 & 1 \\ 0 & -4 \end{pmatrix} = \begin{pmatrix} 8 & 4 \\ -1 & 0 \end{pmatrix}$$

$$\begin{pmatrix} 2 & 3 \\ -1 & 4 \\ 9 & 10 \end{pmatrix} - \begin{pmatrix} 6 & 1 \\ 0 & -4 \\ 3 & 2 \end{pmatrix} = \begin{pmatrix} -4 & 2 \\ -1 & 8 \\ 6 & 8 \end{pmatrix}$$

Many properties of the real numbers also apply to matrices. For instance, matrix addition is *commutative*:

$$A + B = B + A \qquad \text{where } A \text{ and } B \text{ are matrices of the same size}$$

Matrix addition is also *associative*:

$$A + (B + C) = (A + B) + C \qquad \text{where } A, B, \text{ and } C \text{ are matrices of the same size}$$

In the next example, we verify these properties in two specific instances.

**EXAMPLE 4**  Let $A = \begin{pmatrix} 1 & 2 \\ 3 & 4 \end{pmatrix}$, $B = \begin{pmatrix} 0 & -5 \\ 8 & -1 \end{pmatrix}$, and $C = \begin{pmatrix} 6 & 7 \\ 8 & 9 \end{pmatrix}$.

(a) Show that $A + C = C + A$.
(b) Show that $A + (B + C) = (A + B) + C$.

Solution  (a) $A + C = \begin{pmatrix} 1 & 2 \\ 3 & 4 \end{pmatrix} + \begin{pmatrix} 6 & 7 \\ 8 & 9 \end{pmatrix} = \begin{pmatrix} 7 & 9 \\ 11 & 13 \end{pmatrix}$

$C + A = \begin{pmatrix} 6 & 7 \\ 8 & 9 \end{pmatrix} + \begin{pmatrix} 1 & 2 \\ 3 & 4 \end{pmatrix} = \begin{pmatrix} 7 & 9 \\ 11 & 13 \end{pmatrix}$

This shows that $A + C = C + A$, since both $A + C$ and $C + A$ represent the matrix $\begin{pmatrix} 7 & 9 \\ 11 & 13 \end{pmatrix}$.

**(b)** First we compute $A + (B + C)$:

$$A + (B + C) = \begin{pmatrix} 1 & 2 \\ 3 & 4 \end{pmatrix} + \left[ \begin{pmatrix} 0 & -5 \\ 8 & -1 \end{pmatrix} + \begin{pmatrix} 6 & 7 \\ 8 & 9 \end{pmatrix} \right]$$

$$= \begin{pmatrix} 1 & 2 \\ 3 & 4 \end{pmatrix} + \begin{pmatrix} 6 & 2 \\ 16 & 8 \end{pmatrix}$$

$$= \begin{pmatrix} 7 & 4 \\ 19 & 12 \end{pmatrix}$$

Next we compute $(A + B) + C$:

$$(A + B) + C = \left[ \begin{pmatrix} 1 & 2 \\ 3 & 4 \end{pmatrix} + \begin{pmatrix} 0 & -5 \\ 8 & -1 \end{pmatrix} \right] + \begin{pmatrix} 6 & 7 \\ 8 & 9 \end{pmatrix}$$

$$= \begin{pmatrix} 1 & -3 \\ 11 & 3 \end{pmatrix} + \begin{pmatrix} 6 & 7 \\ 8 & 9 \end{pmatrix}$$

$$= \begin{pmatrix} 7 & 4 \\ 19 & 12 \end{pmatrix}$$

We conclude from these calculations that $A + (B + C) = (A + B) + C$, since both sides of that equation represent the matrix $\begin{pmatrix} 7 & 4 \\ 19 & 12 \end{pmatrix}$. ▪

We will now define an operation on matrices called **scalar multiplication**. First of all, the word *scalar* here just means *real number*. So we are talking about multiplying a matrix by a real number. (In more advanced work, nonreal complex scalars are also considered.)

| Definition of Scalar Multiplication | |
|---|---|
| | EXAMPLES |
| To multiply a matrix by a scalar, multiply each entry in the matrix by that scalar. | $2 \begin{pmatrix} 5 & 9 & 0 \\ -1 & 2 & 3 \end{pmatrix} = \begin{pmatrix} 10 & 18 & 0 \\ -2 & 4 & 6 \end{pmatrix}$ |
| | $1 \begin{pmatrix} 1 & 2 \\ 3 & 4 \end{pmatrix} = \begin{pmatrix} 1 & 2 \\ 3 & 4 \end{pmatrix}$ |

There are two simple but useful properties of scalar multiplication that are worth noting at this point. We'll omit the proofs of these two properties; however, Example 5 does ask us to verify the properties for a particular case.

| PROPERTY SUMMARY | Properties of Scalar Multiplication |
|---|---|
| | **1.** $c(kM) = (ck)M$   for all scalars $c$ and $k$ and any matrix $M$ |
| | **2.** $c(M + N) = cM + cN$,   where $c$ is any scalar and $M$ and $N$ are any matrices of the same size. |

**EXAMPLE 5**  Let $c = 2$, $k = 3$, $M = \begin{pmatrix} 1 & 2 \\ 3 & 4 \end{pmatrix}$, and $N = \begin{pmatrix} 5 & 6 \\ 7 & 8 \end{pmatrix}$.

(a) Show that $c(kM) = (ck)M$.

(b) Show that $c(M + N) = cM + cN$.

**Solution**  (a) $c(kM) = 2\left[ 3\begin{pmatrix} 1 & 2 \\ 3 & 4 \end{pmatrix} \right] = 2\begin{pmatrix} 3 & 6 \\ 9 & 12 \end{pmatrix} = \begin{pmatrix} 6 & 12 \\ 18 & 24 \end{pmatrix}$

$(ck)M = (2 \cdot 3)\begin{pmatrix} 1 & 2 \\ 3 & 4 \end{pmatrix} = 6\begin{pmatrix} 1 & 2 \\ 3 & 4 \end{pmatrix} = \begin{pmatrix} 6 & 12 \\ 18 & 24 \end{pmatrix}$

Thus, $c(kM) = (ck)M$, since in both cases the result is $\begin{pmatrix} 6 & 12 \\ 18 & 24 \end{pmatrix}$.

(b) $c(M + N) = 2\left[ \begin{pmatrix} 1 & 2 \\ 3 & 4 \end{pmatrix} + \begin{pmatrix} 5 & 6 \\ 7 & 8 \end{pmatrix} \right]$

$= 2\begin{pmatrix} 6 & 8 \\ 10 & 12 \end{pmatrix} = \begin{pmatrix} 12 & 16 \\ 20 & 24 \end{pmatrix}$

$cM + cN = 2\begin{pmatrix} 1 & 2 \\ 3 & 4 \end{pmatrix} + 2\begin{pmatrix} 5 & 6 \\ 7 & 8 \end{pmatrix}$

$= \begin{pmatrix} 2 & 4 \\ 6 & 8 \end{pmatrix} + \begin{pmatrix} 10 & 12 \\ 14 & 16 \end{pmatrix}$

$= \begin{pmatrix} 12 & 16 \\ 20 & 24 \end{pmatrix}$

Thus, $c(M + N) = cM + cN$, since both sides equal $\begin{pmatrix} 12 & 16 \\ 20 & 24 \end{pmatrix}$.

A matrix with zeros for all of its entries plays the same role in matrix addition as does the number zero in ordinary addition of real numbers. For instance, in the case of $2 \times 2$ matrices, we have

$$\begin{pmatrix} a & b \\ c & d \end{pmatrix} + \begin{pmatrix} 0 & 0 \\ 0 & 0 \end{pmatrix} = \begin{pmatrix} a & b \\ c & d \end{pmatrix}$$

and

$$\begin{pmatrix} 0 & 0 \\ 0 & 0 \end{pmatrix} + \begin{pmatrix} a & b \\ c & d \end{pmatrix} = \begin{pmatrix} a & b \\ c & d \end{pmatrix}$$

for all real numbers $a$, $b$, $c$, and $d$. The matrix $\begin{pmatrix} 0 & 0 \\ 0 & 0 \end{pmatrix}$ is called the **additive identity** for $2 \times 2$ matrices. Similarly, any matrix with all zero entries is the additive identity for matrices of that size. It is sometimes convenient to denote an additive identity matrix by a boldface zero: **0**. With this notation, we can write

$$A + \mathbf{0} = \mathbf{0} + A = A \qquad \text{for any matrix } A$$

For the preceding matrix equation, it is understood that the size of the matrix **0** is the same as the size of $A$. With this notation, we also have

$$A - A = 0 \qquad \text{for any matrix } A$$

Our last topic in this section is matrix multiplication. We will begin with the simplest case and then work up to the more general situation. By convention, a matrix with only one row is called a **row vector**. Examples of row vectors are

$$(2 \quad 13), \qquad (-1 \quad 4 \quad 3), \qquad \text{and} \qquad (0 \quad 0 \quad 0 \quad 1)$$

Similarly, a matrix with only one column is called a **column vector**. Examples are

$$\begin{pmatrix} 2 \\ 13 \end{pmatrix}, \qquad \begin{pmatrix} -1 \\ 4 \\ 3 \end{pmatrix}, \qquad \text{and} \qquad \begin{pmatrix} 0 \\ 0 \\ 0 \\ 1 \end{pmatrix}$$

The following definition tells us how to multiply a row vector and a column vector when they have the same number of entries.

---

### Inner Product of a Row Vector and a Column Vector

EXAMPLES

Let $A$ be a row vector and $B$ a column vector, and assume that the number of columns in $A$ is the same as the number of rows in $B$. Then the **inner product** $A \cdot B$ is defined to be the number obtained by multiplying the corresponding entries and then adding the products.

$$(1 \quad 2 \quad 3) \cdot \begin{pmatrix} 4 \\ 5 \\ 6 \end{pmatrix} = 1 \cdot 4 + 2 \cdot 5 + 3 \cdot 6$$
$$= 32$$

$$(1 \quad 2) \cdot \begin{pmatrix} 4 \\ 5 \\ 6 \end{pmatrix} \quad \text{is not defined}$$

$$(1 \quad 2 \quad 3) \cdot \begin{pmatrix} 4 \\ 5 \end{pmatrix} \quad \text{is not defined}$$

---

An important observation here is that the end result of taking the inner product is always just a number. The definition of matrix product that we now give depends on this observation.

---

### The Product of Two Matrices

Let $A$ and $B$ be two matrices, and assume that the number of columns in $A$ is the same as the number of rows in $B$. Then the product matrix $AB$ is computed according to the following rule.

The entry in the $i$th row and the $j$th column of $AB$ is the inner product of the $i$th row of $A$ with the $j$th column of $B$.

The matrix $AB$ will have as many rows as $A$ and as many columns as $B$.

As an example of matrix multiplication, we will compute the product $AB$, where $A = \begin{pmatrix} 1 & 2 \\ 3 & 4 \end{pmatrix}$ and $B = \begin{pmatrix} 5 & 6 & 0 \\ 7 & 8 & 1 \end{pmatrix}$. In other words, we will compute $\begin{pmatrix} 1 & 2 \\ 3 & 4 \end{pmatrix}\begin{pmatrix} 5 & 6 & 0 \\ 7 & 8 & 1 \end{pmatrix}$. Before carrying out the calculations of any matrix multiplication, however, we should check on two points.

1. Is the product defined? That is, does the number of columns in $A$ equal the number of rows in $B$? In this case, yes; the common number is 2.
2. What is the size of the product? According to the definition, the product $AB$ will have as many rows as $A$ and as many columns as $B$. Thus, the size of $AB$ will be $2 \times 3$.

Schematically, then, the situation looks like this:

$$\begin{pmatrix} 1 & 2 \\ 3 & 4 \end{pmatrix}\begin{pmatrix} 5 & 6 & 0 \\ 7 & 8 & 1 \end{pmatrix} = \begin{pmatrix} ? & ? & ? \\ ? & ? & ? \end{pmatrix}$$

We have six positions to fill. The computations are presented in Table 3.

**TABLE 3**

| POSITION | HOW TO COMPUTE | COMPUTATION |
|---|---|---|
| row 1, column 1 | inner product of row 1 and column 1 $$\begin{pmatrix} 1 & 2 \\ 3 & 4 \end{pmatrix}\begin{pmatrix} 5 & 6 & 0 \\ 7 & 8 & 1 \end{pmatrix}$$ | $1 \cdot 5 + 2 \cdot 7 = 19$ |
| row 1, column 2 | inner product of row 1 and column 2 $$\begin{pmatrix} 1 & 2 \\ 3 & 4 \end{pmatrix}\begin{pmatrix} 5 & 6 & 0 \\ 7 & 8 & 1 \end{pmatrix}$$ | $1 \cdot 6 + 2 \cdot 8 = 22$ |
| row 1, column 3 | inner product of row 1 and column 3 $$\begin{pmatrix} 1 & 2 \\ 3 & 4 \end{pmatrix}\begin{pmatrix} 5 & 6 & 0 \\ 7 & 8 & 1 \end{pmatrix}$$ | $1 \cdot 0 + 2 \cdot 1 = 2$ |
| row 2, column 1 | inner product of row 2 and column 1 $$\begin{pmatrix} 1 & 2 \\ 3 & 4 \end{pmatrix}\begin{pmatrix} 5 & 6 & 0 \\ 7 & 8 & 1 \end{pmatrix}$$ | $3 \cdot 5 + 4 \cdot 7 = 43$ |
| row 2, column 2 | inner product of row 2 and column 2 $$\begin{pmatrix} 1 & 2 \\ 3 & 4 \end{pmatrix}\begin{pmatrix} 5 & 6 & 0 \\ 7 & 8 & 1 \end{pmatrix}$$ | $3 \cdot 6 + 4 \cdot 8 = 50$ |
| row 2, column 3 | inner product of row 2 and column 3 $$\begin{pmatrix} 1 & 2 \\ 3 & 4 \end{pmatrix}\begin{pmatrix} 5 & 6 & 0 \\ 7 & 8 & 1 \end{pmatrix}$$ | $3 \cdot 0 + 4 \cdot 1 = 4$ |

Reading from the table, we see that our result is

$$AB = \begin{pmatrix} 1 & 2 \\ 3 & 4 \end{pmatrix}\begin{pmatrix} 5 & 6 & 0 \\ 7 & 8 & 1 \end{pmatrix} = \begin{pmatrix} 19 & 22 & 2 \\ 43 & 50 & 4 \end{pmatrix}$$

**EXAMPLE 6** Let $A = \begin{pmatrix} 1 & 2 \\ 3 & 4 \end{pmatrix}$ and $B = \begin{pmatrix} 5 & 6 \\ 7 & 8 \end{pmatrix}$. By computing $AB$ and then $BA$, show that $AB \neq BA$. This shows that in general, matrix multiplication is not commutative.

Solution

$$AB = \begin{pmatrix} 1 & 2 \\ 3 & 4 \end{pmatrix}\begin{pmatrix} 5 & 6 \\ 7 & 8 \end{pmatrix} = \begin{pmatrix} 1\cdot 5 + 2\cdot 7 & 1\cdot 6 + 2\cdot 8 \\ 3\cdot 5 + 4\cdot 7 & 3\cdot 6 + 4\cdot 8 \end{pmatrix}$$

$$= \begin{pmatrix} 19 & 22 \\ 43 & 50 \end{pmatrix}$$

$$BA = \begin{pmatrix} 5 & 6 \\ 7 & 8 \end{pmatrix}\begin{pmatrix} 1 & 2 \\ 3 & 4 \end{pmatrix} = \begin{pmatrix} 5\cdot 1 + 6\cdot 3 & 5\cdot 2 + 6\cdot 4 \\ 7\cdot 1 + 8\cdot 3 & 7\cdot 2 + 8\cdot 4 \end{pmatrix}$$

$$= \begin{pmatrix} 23 & 34 \\ 31 & 46 \end{pmatrix}$$

Comparing the two matrices $AB$ and $BA$, we conclude that $AB \neq BA$.

## EXERCISE SET 10.3

### A

*In Exercises 1–4, specify the size of each matrix.*

1. (a) $\begin{pmatrix} -4 & 0 & 5 \\ 2 & 8 & -1 \end{pmatrix}$  (b) $\begin{pmatrix} 7 & 1 \\ 4 & -3 \\ 0 & 0 \end{pmatrix}$

2. (a) $\begin{pmatrix} 1 & 0 \\ 0 & -1 \end{pmatrix}$  (b) $\begin{pmatrix} 1 \\ 6 \\ 8 \\ 1 \end{pmatrix}$

3. $\begin{pmatrix} 1 & a & b & c \\ a & 1 & 0 & a \\ b & 0 & 1 & b \\ c & a & b & 1 \\ 0 & 0 & 0 & 1 \end{pmatrix}$  4. $(-3 \quad 1 \quad 6 \quad 0)$

*In Exercises 5–8, write the coefficient matrix and the augmented matrix for each system.*

5. $\begin{cases} 2x + 3y + 4z = 10 \\ 5x + 6y + 7z = 9 \\ 8x + 9y + 10z = 8 \end{cases}$  6. $\begin{cases} 5x - y + z = 0 \\ 4y + 2z = 1 \\ 3x + y + z = -1 \end{cases}$

7. $\begin{cases} x + z + w = -1 \\ x + y + 2w = 0 \\ y + z + w = 1 \\ 2x - y - z = 2 \end{cases}$

8. $\begin{cases} 8x - 8y = 5 \\ x - y + z = 1 \end{cases}$

*In Exercises 9–22, use matrices to solve the systems of equations.*

9. $\begin{cases} x - y + 2z = 7 \\ 3x + 2y - z = -10 \\ -x + 3y + z = -2 \end{cases}$

10. $\begin{cases} 2x - 3y + 4z = 14 \\ 3x - 2y + 2z = 12 \\ 4x + 5y - 5z = 16 \end{cases}$

11. $\begin{cases} x + z = -2 \\ -3x + 2y = 17 \\ x - y - z = -9 \end{cases}$

12. $\begin{cases} 5x + y + 10z = 23 \\ 4x + 2y - 10z = 76 \\ 3x - 4y = 18 \end{cases}$

13. $\begin{cases} x + y + z = -4 \\ 2x - 3y + z = -1 \\ 4x + 2y - 3z = 33 \end{cases}$

14. $\begin{cases} 2x + 3y - 4z = 7 \\ x - y + z = -\frac{3}{2} \\ 6x - 5y - 2z = -7 \end{cases}$

15. $\begin{cases} 3x - 2y + 6z = 0 \\ x + 3y + 20z = 15 \\ 10x - 11y - 10z = -9 \end{cases}$

16. $\begin{cases} 3A - 3B + C = 4 \\ 6A + 9B - 3C = -7 \\ A - 2B - 2C = -3 \end{cases}$

17. $\begin{cases} 4x - 3y + 3z = 2 \\ 5x + y - 4z = 1 \\ 9x - 2y - z = 3 \end{cases}$

**18.** $\begin{cases} 6x + y - z = -1 \\ -3x + 2y + 2z = 2 \\ 5y + 3z = 1 \end{cases}$

**19.** $\begin{cases} x - y + z + w = 6 \\ x + y - z + w = 4 \\ x + y + z - w = -2 \\ -x + y + z + w = 0 \end{cases}$

**20.** $\begin{cases} x + 2y - z - 2w = 5 \\ 2x + y + 2z + w = -7 \\ -2x - y - 3z - 2w = 10 \\ z + w = -3 \end{cases}$

**21.** $\begin{cases} 15A + 14B + 26C = 1 \\ 18A + 17B + 32C = -1 \\ 21A + 20B + 38C = 0 \end{cases}$

**22.** $\begin{cases} A + B + C + D + E = -1 \\ 3A - 2B - 2C + 3D + 2E = 13 \\ 3C + 4D - 4E = 7 \\ 5A - 4B + E = 30 \\ C - 2E = 3 \end{cases}$

*In Exercises 23–50, the matrices A, B, C, D, E, F, and G are defined as follows:*

$$A = \begin{pmatrix} 2 & 3 \\ -1 & 4 \end{pmatrix} \quad B = \begin{pmatrix} 1 & -1 \\ 3 & 0 \end{pmatrix} \quad C = \begin{pmatrix} 1 & 0 \\ 0 & 1 \end{pmatrix}$$

$$D = \begin{pmatrix} -1 & 2 & 3 \\ 4 & 0 & 5 \end{pmatrix} \quad E = \begin{pmatrix} 2 & 1 \\ 8 & -1 \\ 6 & 5 \end{pmatrix}$$

$$F = \begin{pmatrix} 5 & -1 \\ -4 & 0 \\ 2 & 3 \end{pmatrix} \quad G = \begin{pmatrix} 0 & 0 \\ 0 & 0 \\ 0 & 0 \end{pmatrix}$$

*In each exercise, carry out the indicated matrix operations if they are defined. If an operation is not defined, say so.*

**23.** $A + B$       **24.** $A - B$       **25.** $2A + 2B$

**26.** $2(A + B)$     **27.** $AB$          **28.** $BA$

**29.** $AC$          **30.** $CA$          **31.** $3D + E$

**32.** $E + F$        **33.** $2F - 3G$     **34.** $DE$

**35.** $ED$          **36.** $DF$          **37.** $FD$

**38.** $A + D$        **39.** $G + A$       **40.** $DG$

**41.** $GD$          **42.** $(A + B) + C$  **43.** $A + (B + C)$

**44.** $CD$          **45.** $DC$          **46.** $5E - 3F$

**47.** $A^2 (= AA)$   **48.** $A^2A$        **49.** $AA^2$

**50.** $C^2$

**51.** Let

$$A = \begin{pmatrix} -1 & 3 & 4 \\ 3 & 2 & -3 \\ 9 & 1 & 6 \end{pmatrix} \quad B = \begin{pmatrix} 7 & 0 & 1 \\ 0 & 0 & 3 \\ -1 & 2 & 4 \end{pmatrix}$$

$$C = \begin{pmatrix} 4 & 6 & 1 \\ 2 & 1 & 3 \\ -1 & -1 & 2 \end{pmatrix}$$

**(a)** Compute $A(B + C)$.     **(b)** Compute $AB + AC$.
**(c)** Compute $(AB)C$.       **(d)** Compute $A(BC)$.

**B**

**52.** Let $A = \begin{pmatrix} 1 & 2 \\ 3 & 4 \end{pmatrix}$ and $B = \begin{pmatrix} 5 & 6 \\ 7 & 8 \end{pmatrix}$. Let $A^2$ and $B^2$ denote the matrix products $AA$ and $BB$, respectively. Compute each of the following.
**(a)** $(A + B)(A + B)$       **(b)** $A^2 + 2AB + B^2$
**(c)** $A^2 + AB + BA + B^2$

**53.** Let $A = \begin{pmatrix} 3 & 5 \\ 7 & 9 \end{pmatrix}$ and $B = \begin{pmatrix} 2 & 4 \\ 6 & 8 \end{pmatrix}$. Compute each of the following.
**(a)** $A^2 - B^2$           **(b)** $(A - B)(A + B)$
**(c)** $(A + B)(A - B)$       **(d)** $A^2 + AB - BA - B^2$

**54.** Let

$$A = \begin{pmatrix} 1 & 0 \\ 0 & 1 \end{pmatrix} \quad B = \begin{pmatrix} 1 & 0 \\ 0 & -1 \end{pmatrix}$$

$$C = \begin{pmatrix} -1 & 0 \\ 0 & 1 \end{pmatrix} \quad D = \begin{pmatrix} -1 & 0 \\ 0 & -1 \end{pmatrix}$$

Complete the following multiplication table.

|   | A | B | C | D |
|---|---|---|---|---|
| A |   |   |   |   |
| B |   |   | D |   |
| C |   |   |   |   |
| D |   |   |   |   |

*Hint:* $D$ is the proper entry in the second row, third column because (as you can check) $BC = D$.

**55.** In this exercise, let us agree to write the coordinates $(x, y)$ of a point in the plane as the $2 \times 1$ matrix $\begin{pmatrix} x \\ y \end{pmatrix}$.

**(a)** Let $A = \begin{pmatrix} 1 & 0 \\ 0 & -1 \end{pmatrix}$ and $Z = \begin{pmatrix} x \\ y \end{pmatrix}$. Compute the matrix $AZ$. After computing $AZ$, observe that it represents the point obtained by reflecting $\begin{pmatrix} x \\ y \end{pmatrix}$ about the $x$-axis.

**(b)** Let $B = \begin{pmatrix} -1 & 0 \\ 0 & 1 \end{pmatrix}$ and $Z = \begin{pmatrix} x \\ y \end{pmatrix}$. Compute the matrix $BZ$. After computing $BZ$, observe that it represents the point obtained by reflecting $\begin{pmatrix} x \\ y \end{pmatrix}$ about the y-axis.

(c) Let A, B, and Z represent the matrices defined in parts (a) and (b). Compute the matrix $(AB)Z$, and then interpret it in terms of reflection about the axes.

56. In this exercise, we continue to explore some of the connections between matrices and geometry. As in Exercise 55, we will use $2 \times 1$ matrices to specify the coordinates of points in the plane. Let P, S, and T be the matrices defined as follows:

$$P = \begin{pmatrix} \cos x \\ \sin x \end{pmatrix} \qquad S = \begin{pmatrix} \cos \theta & -\sin \theta \\ \sin \theta & \cos \theta \end{pmatrix}$$

$$T = \begin{pmatrix} \cos \beta & -\sin \beta \\ \sin \beta & \cos \beta \end{pmatrix}$$

Notice that the point P lies on the unit circle.

(a) Compute the matrix $SP$. After computing $SP$, observe that it represents the point on the unit circle obtained by rotating P (about the origin) through an angle $\theta$.

(b) Show that

$$ST = TS = \begin{pmatrix} \cos(\theta + \beta) & -\sin(\theta + \beta) \\ \sin(\theta + \beta) & \cos(\theta + \beta) \end{pmatrix}.$$

(c) Compute $(ST)P$. What is the angle through which P is rotated?

57. A function $f$ is defined as follows. The domain of $f$ is the set of all $2 \times 2$ matrices (with real entries). If $A = \begin{pmatrix} a & b \\ c & d \end{pmatrix}$, then $f(A) = ad - bc$.

(a) Let $A = \begin{pmatrix} 1 & 2 \\ 3 & 4 \end{pmatrix}$ and $B = \begin{pmatrix} 3 & -1 \\ 5 & 8 \end{pmatrix}$. Compute $f(A)$, $f(B)$, and $f(AB)$. Is it true, in this case, that $f(A) \cdot f(B) = f(AB)$?

(b) Let $A = \begin{pmatrix} a & b \\ c & d \end{pmatrix}$ and $B = \begin{pmatrix} e & f \\ g & h \end{pmatrix}$. Show that $f(A) \cdot f(B) = f(AB)$.

58. The **trace** of a $2 \times 2$ matrix $\begin{pmatrix} a & b \\ c & d \end{pmatrix}$ is defined by

$$\mathrm{tr}\begin{pmatrix} a & b \\ c & d \end{pmatrix} = a + d$$

(a) If $A = \begin{pmatrix} 1 & 2 \\ 3 & 4 \end{pmatrix}$ and $B = \begin{pmatrix} 5 & 6 \\ 7 & 8 \end{pmatrix}$, verify that $\mathrm{tr}(A + B) = \mathrm{tr}\,A + \mathrm{tr}\,B$.

(b) If $A = \begin{pmatrix} a & b \\ c & d \end{pmatrix}$ and $B = \begin{pmatrix} e & f \\ g & h \end{pmatrix}$, show that $\mathrm{tr}(A + B) = \mathrm{tr}\,A + \mathrm{tr}\,B$.

59. Let $A = \begin{pmatrix} a & b \\ c & d \end{pmatrix}$. The **transpose** of A is the matrix denoted by $A^{\mathrm{T}}$ and defined by

$$A^{\mathrm{T}} = \begin{pmatrix} a & c \\ b & d \end{pmatrix}$$

In other words, $A^{\mathrm{T}}$ is obtained by switching the columns and rows of A. Show that the following equations hold for all $2 \times 2$ matrices A and B.

(a) $(A + B)^{\mathrm{T}} = A^{\mathrm{T}} + B^{\mathrm{T}}$  (b) $(A^{\mathrm{T}})^{\mathrm{T}} = A$

(c) $(AB)^{\mathrm{T}} = B^{\mathrm{T}}A^{\mathrm{T}}$

60. Find an example of two $2 \times 2$ matrices A and B for which $AB = 0$ but neither A nor B is 0.

## THE INVERSE OF A SQUARE MATRIX

**10.4**

A matrix in which there are as many rows as there are columns is called a **square matrix**. So, two examples of square matrices are

$$A = \begin{pmatrix} 1 & 2 \\ 3 & 4 \end{pmatrix} \qquad \text{and} \qquad B = \begin{pmatrix} -5 & 6 & 7 \\ \frac{1}{2} & 0 & 1 \\ 8 & 4 & -3 \end{pmatrix}$$

The matrix A is said to be a square matrix of **order two** (or, more simply, a $2 \times 2$ matrix); B is a square matrix of **order three** (that is, a $3 \times 3$ matrix). Initially in this section, we will present the concepts and techniques in terms of square matrices of order two. After that, we'll show how the ideas carry over to larger square matrices.

We begin by defining a special matrix $I_2$:

$$I_2 = \begin{pmatrix} 1 & 0 \\ 0 & 1 \end{pmatrix}$$

The matrix $I_2$ has the following property. For every $2 \times 2$ matrix $A = \begin{pmatrix} a & b \\ c & d \end{pmatrix}$, we have (as you can easily verify)

$$\begin{pmatrix} a & b \\ c & d \end{pmatrix}\begin{pmatrix} 1 & 0 \\ 0 & 1 \end{pmatrix} = \begin{pmatrix} a & b \\ c & d \end{pmatrix} \quad \text{and} \quad \begin{pmatrix} 1 & 0 \\ 0 & 1 \end{pmatrix}\begin{pmatrix} a & b \\ c & d \end{pmatrix} = \begin{pmatrix} a & b \\ c & d \end{pmatrix}$$

In other words,

$$AI_2 = A \quad \text{and} \quad I_2A = A$$

This shows that the matrix $I_2$ plays the same role in the multiplication of $2 \times 2$ matrices as does the number 1 in the multiplication of real numbers. For this reason, $I_2$ is called the **multiplicative identity** for square matrices of order two.

If we have two real numbers $a$ and $b$ such that $ab = 1$, then (as described in Chapter 1) we say that $a$ and $b$ are (multiplicative) inverses. In the box that follows we apply this same terminology to matrices.

---

**Definition:  The Inverse of a 2 × 2 Matrix**

If $A$ and $B$ are $2 \times 2$ matrices such that

$$AB = I_2 \quad \text{and} \quad BA = I_2$$

then $A$ and $B$ are said to be **inverses** of one another.

---

**EXAMPLE 1**  Find the inverse of $A = \begin{pmatrix} 1 & 2 \\ 3 & 4 \end{pmatrix}$

**Solution**  We need to find numbers $a$, $b$, $c$, and $d$ such that the following two matrix equations are valid.

$$\begin{pmatrix} 1 & 2 \\ 3 & 4 \end{pmatrix}\begin{pmatrix} a & b \\ c & d \end{pmatrix} = \begin{pmatrix} 1 & 0 \\ 0 & 1 \end{pmatrix} \tag{1}$$

$$\begin{pmatrix} a & b \\ c & d \end{pmatrix}\begin{pmatrix} 1 & 2 \\ 3 & 4 \end{pmatrix} = \begin{pmatrix} 1 & 0 \\ 0 & 1 \end{pmatrix} \tag{2}$$

From equation (1), we have

$$\begin{pmatrix} a + 2c & b + 2d \\ 3a + 4c & 3b + 4d \end{pmatrix} = \begin{pmatrix} 1 & 0 \\ 0 & 1 \end{pmatrix}$$

For this last equation to be valid, the corresponding entries in the two matrices must be equal. (Why?) Consequently, we obtain four equations, two involving $a$ and $c$, and two involving $b$ and $d$:

$$\begin{cases} a + 2c = 1 \\ 3a + 4c = 0 \end{cases} \qquad \begin{cases} b + 2d = 0 \\ 3b + 4d = 1 \end{cases}$$

As Exercise 32(a) asks you to show, the solution to the first system is $a = -2$ and $c = \frac{3}{2}$; and for the second system, $b = 1$ and $d = -\frac{1}{2}$. Furthermore (as you can check), these same values are obtained for $a$, $b$, $c$, and $d$ if we begin with matrix equation (2) rather than (1). Thus, the required inverse matrix is

$$\begin{pmatrix} -2 & 1 \\ \frac{3}{2} & -\frac{1}{2} \end{pmatrix}$$

Exercise 32(b) asks you to carry out the matrix multiplication to confirm that we indeed have

$$\begin{pmatrix} 1 & 2 \\ 3 & 4 \end{pmatrix}\begin{pmatrix} -2 & 1 \\ \frac{3}{2} & -\frac{1}{2} \end{pmatrix} = \begin{pmatrix} -2 & 1 \\ \frac{3}{2} & -\frac{1}{2} \end{pmatrix}\begin{pmatrix} 1 & 2 \\ 3 & 4 \end{pmatrix}$$

$$= \begin{pmatrix} 1 & 0 \\ 0 & 1 \end{pmatrix}$$

In Example 1 we found the inverse of a square matrix by solving systems of equations. However, as you've seen in previous sections, there are systems of equations that do not have solutions. In the present context, this implies that there are matrices that do not have inverses. For example (as Exercise 33 asks you to show), the matrix $\begin{pmatrix} 2 & 5 \\ 6 & 15 \end{pmatrix}$ does not have an inverse. A matrix that does not have an inverse is called a **singular matrix**. If a matrix does possess an inverse, we say that the matrix is **nonsingular**. So, according to the result in Example 1, the matrix $\begin{pmatrix} 1 & 2 \\ 3 & 4 \end{pmatrix}$ is nonsingular.

It can be proven that if a square matrix has an inverse, then that inverse is unique. (In other words, you can never find two different inverses for the same matrix.) The inverse of a nonsingular matrix $A$ is denoted by $A^{-1}$.

There are several methods that can be used to compute the inverse of a matrix (or to show that there is no inverse). One method involves working with systems of equations, as in Example 1. A more efficient method, however, is the following: (We'll demonstrate the method with an example, but we won't give a formal proof of its validity.)

Suppose, for example, that we want to compute the inverse (if it exists) for the matrix

$$A = \begin{pmatrix} 6 & 2 \\ 13 & 4 \end{pmatrix}$$

We begin by writing a larger matrix, formed by merging $A$ and $I_2$ as follows.

$$\begin{pmatrix} 6 & 2 & \vdots & 1 & 0 \\ 13 & 4 & \vdots & 0 & 1 \end{pmatrix}$$

$$A \underset{\underrightarrow{\hspace{2cm}}}{} \qquad \underset{\underrightarrow{\hspace{2cm}} I_2}{}$$

A convenient abbreviation for this matrix is $(A \mid I_2)$. Now we carry out elementary row operations on this matrix until one of the following two situations occurs:

**1.** Either the matrix takes on the form

$$\begin{pmatrix} 1 & 0 & \vdots & a & b \\ 0 & 1 & \vdots & c & d \end{pmatrix}$$

$$I_2 \underset{\underrightarrow{\hspace{2cm}}}{}$$

In this case, $A^{-1} = \begin{pmatrix} a & b \\ c & d \end{pmatrix}$.

**2.** Or, one of the rows to the left of the dashed line consists entirely of zeros. In this case, $A^{-1}$ does not exist.

The calculations for our example then run as follows.

$$\begin{pmatrix} 6 & 2 & \vdots & 1 & 0 \\ 13 & 4 & \vdots & 0 & 1 \end{pmatrix} \xrightarrow{\;-2R_1 + R_2\;} \begin{pmatrix} 6 & 2 & \vdots & 1 & 0 \\ 1 & 0 & \vdots & -2 & 1 \end{pmatrix}$$

$$\xrightarrow{\;R_1 \leftrightarrow R_2\;} \begin{pmatrix} 1 & 0 & \vdots & -2 & 1 \\ 6 & 2 & \vdots & 1 & 0 \end{pmatrix}$$

$$\xrightarrow{\;-6R_1 + R_2\;} \begin{pmatrix} 1 & 0 & \vdots & -2 & 1 \\ 0 & 2 & \vdots & 13 & -6 \end{pmatrix}$$

$$\xrightarrow{\;\frac{1}{2}R_2\;} \begin{pmatrix} 1 & 0 & \vdots & -2 & 1 \\ 0 & 1 & \vdots & \frac{13}{2} & -3 \end{pmatrix}$$

The inverse matrix can now be read off:

$$A^{-1} = \begin{pmatrix} -2 & 1 \\ \frac{13}{2} & -3 \end{pmatrix}$$

(You should verify for yourself that we indeed have $AA^{-1} = I_2$ and $A^{-1}A = I_2$.)

Inverse matrices can be used in solving certain systems of equations in which the number of unknowns is the same as the number of equations. Before explaining how this works, we describe how a system of equations can be written in matrix form. Consider the system

$$\begin{cases} x + 2y = 8 \\ 3x + 4y = 6 \end{cases} \tag{3}$$

As defined in the previous section, the coefficient matrix $A$ for this system is

$$A = \begin{pmatrix} 1 & 2 \\ 3 & 4 \end{pmatrix}$$

Now we define two matrices $X$ and $B$:

$$X = \begin{pmatrix} x \\ y \end{pmatrix} \qquad B = \begin{pmatrix} 8 \\ 6 \end{pmatrix}$$

Then system (3) can be written as a single matrix equation:

$$AX = B \tag{4}$$

To see why this is so, we expand equation (4) to obtain

$$\begin{pmatrix} 1 & 2 \\ 3 & 4 \end{pmatrix}\begin{pmatrix} x \\ y \end{pmatrix} = \begin{pmatrix} 8 \\ 6 \end{pmatrix} \qquad \text{using the definitions of } A, X, \text{ and } B$$

$$\begin{pmatrix} x + 2y \\ 3x + 4y \end{pmatrix} = \begin{pmatrix} 8 \\ 6 \end{pmatrix} \qquad \text{carrying out the matrix multiplication}$$

By equating the corresponding entries of the matrices in this last equation, we obtain $x + 2y = 8$ and $3x + 4y = 6$, as given initially in system (3).

**EXAMPLE 2** Use an inverse matrix to solve the system

$$\begin{cases} x + 2y = 8 \\ 3x + 4y = 6 \end{cases}$$

Solution   As explained just prior to this example, the matrix form for this system is

$$AX = B \qquad\qquad (5)$$

where

$$A = \begin{pmatrix} 1 & 2 \\ 3 & 4 \end{pmatrix} \qquad X = \begin{pmatrix} x \\ y \end{pmatrix} \qquad B = \begin{pmatrix} 8 \\ 6 \end{pmatrix}$$

From Example 1, we know that

$$A^{-1} = \begin{pmatrix} -2 & 1 \\ \frac{3}{2} & -\frac{1}{2} \end{pmatrix}$$

Now we multiply both sides of equation (5) by $A^{-1}$ to obtain

$$A^{-1}(AX) = A^{-1}B$$
$$(A^{-1}A)X = A^{-1}B \qquad \text{Matrix multiplication is associative.}$$
$$I_2 X = A^{-1}B$$
$$X = A^{-1}B \qquad \text{(Why?)}$$

Substituting the actual matrices into this last equation, we have

$$\begin{pmatrix} x \\ y \end{pmatrix} = \begin{pmatrix} -2 & 1 \\ \frac{3}{2} & -\frac{1}{2} \end{pmatrix}\begin{pmatrix} 8 \\ 6 \end{pmatrix}$$
$$= \begin{pmatrix} -10 \\ 9 \end{pmatrix}$$

Therefore $x = -10$ and $y = 9$, as required.

All of the ideas we have discussed for square matrices of order two can be carried over rather directly to larger square matrices. It is easy to check that the following matrices $I_3$ and $I_4$ are the multiplicative identities for $3 \times 3$ and $4 \times 4$ matrices, respectively.

$$I_3 = \begin{pmatrix} 1 & 0 & 0 \\ 0 & 1 & 0 \\ 0 & 0 & 1 \end{pmatrix} \qquad I_4 = \begin{pmatrix} 1 & 0 & 0 & 0 \\ 0 & 1 & 0 & 0 \\ 0 & 0 & 1 & 0 \\ 0 & 0 & 0 & 1 \end{pmatrix}$$

These identity matrices are described by saying that they have ones down the **main diagonal** and zeros everywhere else. (Larger identity matrices can be defined following this same pattern.) Sometimes, when it is clear from the context, or as a matter of convenience, we'll omit the subscript and denote the appropriate-sized identity matrix simply by $I$. (This is done in the box that follows.)

| PROPERTY SUMMARY | The Inverse of a Square Matrix |
|---|---|

1. Suppose that $A$ and $B$ are square matrices of the same size, and let $I$ denote the identity matrix of the same size. Then $A$ and $B$ are said to be **inverses** provided

$$AB = I \quad \text{and} \quad BA = I$$

2. It can be shown that every square matrix has at most one inverse.

3. A square matrix that has an inverse is said to be **nonsingular** (or **invertible**). A square matrix that does not have an inverse is called **singular**.

4. If the matrix $A$ is nonsingular, then the inverse of $A$ is denoted by $A^{-1}$. In this case, we have

$$AA^{-1} = I \quad \text{and} \quad A^{-1}A = I$$

**EXAMPLE 3**  Let $A = \begin{pmatrix} 5 & 0 & 2 \\ 2 & 2 & 1 \\ -3 & 1 & -1 \end{pmatrix}$. Use the elementary row operations to compute $A^{-1}$, if it exists.

**Solution**  Following the method that we described for the $2 \times 2$ case, we first write down the matrix $(A \mid I_3)$ formed by merging $A$ and $I_3$:

$$\begin{pmatrix} 5 & 0 & 2 & \mid & 1 & 0 & 0 \\ 2 & 2 & 1 & \mid & 0 & 1 & 0 \\ -3 & 1 & -1 & \mid & 0 & 0 & 1 \end{pmatrix}$$

Now we carry out the elementary row operations, trying to obtain $I_3$ to the left of the dashed line. We have

$$\begin{pmatrix} 5 & 0 & 2 & \mid & 1 & 0 & 0 \\ 2 & 2 & 1 & \mid & 0 & 1 & 0 \\ -3 & 1 & -1 & \mid & 0 & 0 & 1 \end{pmatrix} \xrightarrow{-2R_2 + R_1} \begin{pmatrix} 1 & -4 & 0 & \mid & 1 & -2 & 0 \\ 2 & 2 & 1 & \mid & 0 & 1 & 0 \\ -3 & 1 & -1 & \mid & 0 & 0 & 1 \end{pmatrix}$$

$$\xrightarrow[3R_1 + R_3]{-2R_1 + R_2} \begin{pmatrix} 1 & -4 & 0 & \mid & 1 & -2 & 0 \\ 0 & 10 & 1 & \mid & -2 & 5 & 0 \\ 0 & -11 & -1 & \mid & 3 & -6 & 1 \end{pmatrix}$$

$$\xrightarrow{1R_3 + R_2} \begin{pmatrix} 1 & -4 & 0 & \mid & 1 & -2 & 0 \\ 0 & -1 & 0 & \mid & 1 & -1 & 1 \\ 0 & -11 & -1 & \mid & 3 & -6 & 1 \end{pmatrix} \xrightarrow{(-1)R_2} \begin{pmatrix} 1 & -4 & 0 & \mid & 1 & -2 & 0 \\ 0 & 1 & 0 & \mid & -1 & 1 & -1 \\ 0 & -11 & -1 & \mid & 3 & -6 & 1 \end{pmatrix}$$

$$\xrightarrow[11R_2 + R_3]{4R_2 + R_1} \begin{pmatrix} 1 & 0 & 0 & \mid & -3 & 2 & -4 \\ 0 & 1 & 0 & \mid & -1 & 1 & -1 \\ 0 & 0 & -1 & \mid & -8 & 5 & -10 \end{pmatrix} \xrightarrow{(-1)R_3} \begin{pmatrix} 1 & 0 & 0 & \mid & -3 & 2 & -4 \\ 0 & 1 & 0 & \mid & -1 & 1 & -1 \\ 0 & 0 & 1 & \mid & 8 & -5 & 10 \end{pmatrix}$$

The required inverse is therefore

$$A^{-1} = \begin{pmatrix} -3 & 2 & -4 \\ -1 & 1 & -1 \\ 8 & -5 & 10 \end{pmatrix}$$

## EXERCISE SET 10.4

### A

In Exercises 1–4, the matrices A, B, C, and D are defined as follows.

$$A = \begin{pmatrix} 4 & -1 \\ -5 & 2 \end{pmatrix} \qquad B = \begin{pmatrix} \frac{1}{2} & 5 \\ 3 & 1 \end{pmatrix}$$

$$C = \begin{pmatrix} 3 & 0 & -2 \\ 0 & 5 & 6 \\ 1 & 4 & -7 \end{pmatrix} \qquad D = \begin{pmatrix} 1 & 2 & 3 \\ 4 & 5 & 6 \\ 7 & 8 & 9 \end{pmatrix}$$

1. Compute $AI_2$ and $I_2A$ to verify that $AI_2 = I_2A = A$.
2. Compute $BI_2$ and $I_2B$ to verify that $BI_2 = I_2B = B$.
3. Compute $CI_3$ and $I_3C$ to verify that $CI_3 = I_3C = C$.
4. Compute $DI_3$ and $I_3D$ to verify that $DI_3 = I_3D = D$.

In Exercises 5–12, compute $A^{-1}$, if it exists, using the method of Example 1.

5. $A = \begin{pmatrix} 7 & 9 \\ 4 & 5 \end{pmatrix}$

6. $A = \begin{pmatrix} 3 & -8 \\ 2 & -5 \end{pmatrix}$

7. $A = \begin{pmatrix} -3 & 1 \\ 5 & 6 \end{pmatrix}$

8. $A = \begin{pmatrix} -4 & 0 \\ 9 & 3 \end{pmatrix}$

9. $A = \begin{pmatrix} -2 & 3 \\ -4 & 6 \end{pmatrix}$

10. $A = \begin{pmatrix} \frac{5}{3} & -2 \\ -\frac{2}{3} & 1 \end{pmatrix}$

11. $A = \begin{pmatrix} \frac{1}{3} & \frac{1}{3} \\ -\frac{1}{9} & \frac{2}{9} \end{pmatrix}$

12. $A = \begin{pmatrix} -3 & 7 \\ 12 & -28 \end{pmatrix}$

In Exercises 13–26, compute the inverse matrix, if it exists, using elementary row operations (as shown in Example 3).

13. $\begin{pmatrix} 2 & 1 \\ 3 & 2 \end{pmatrix}$

14. $\begin{pmatrix} -6 & 5 \\ 18 & -15 \end{pmatrix}$

15. $\begin{pmatrix} 0 & -11 \\ 1 & 6 \end{pmatrix}$

16. $\begin{pmatrix} -2 & 13 \\ -4 & 25 \end{pmatrix}$

17. $\begin{pmatrix} \frac{2}{3} & -\frac{1}{4} \\ -8 & 3 \end{pmatrix}$

18. $\begin{pmatrix} -\frac{2}{5} & \frac{1}{3} \\ -6 & 5 \end{pmatrix}$

19. $\begin{pmatrix} -5 & 4 & -3 \\ 10 & -7 & 6 \\ 8 & -6 & 5 \end{pmatrix}$

20. $\begin{pmatrix} 1 & 0 & -2 \\ -3 & -1 & 6 \\ 2 & 1 & -5 \end{pmatrix}$

21. $\begin{pmatrix} 1 & 2 & -1 \\ 0 & 3 & 0 \\ -4 & 0 & 5 \end{pmatrix}$

22. $\begin{pmatrix} 1 & -4 & -8 \\ 1 & 2 & 5 \\ 1 & 1 & 3 \end{pmatrix}$

23. $\begin{pmatrix} -7 & 5 & 3 \\ 3 & -2 & -2 \\ 3 & -2 & -1 \end{pmatrix}$

24. $\begin{pmatrix} 2 & -1 & -1 \\ 1 & 0 & -1 \\ -2 & 1 & 2 \end{pmatrix}$

25. $\begin{pmatrix} 1 & 2 & 3 \\ 4 & 5 & 6 \\ 7 & 8 & 9 \end{pmatrix}$

26. $\begin{pmatrix} 2 & 1 & 3 \\ 4 & 5 & -7 \\ 2 & 1 & 3 \end{pmatrix}$

27. If $A = \begin{pmatrix} 3 & 8 \\ 4 & 11 \end{pmatrix}$, then $A^{-1} = \begin{pmatrix} 11 & -8 \\ -4 & 3 \end{pmatrix}$. Use this fact and the method of Example 2 to solve the following systems.

(a) $\begin{cases} 3x + 8y = 5 \\ 4x + 11y = 7 \end{cases}$ (b) $\begin{cases} 3x + 8y = -12 \\ 4x + 11y = 0 \end{cases}$

28. If $A = \begin{pmatrix} 3 & -7 \\ 4 & -9 \end{pmatrix}$, then $A^{-1} = \begin{pmatrix} -9 & 7 \\ -4 & 3 \end{pmatrix}$. Use this fact and the method of Example 2 to solve the following systems.

(a) $\begin{cases} 3x - 7y = 30 \\ 4x - 9y = 39 \end{cases}$ (b) $\begin{cases} 3x - 7y = -45 \\ 4x - 9y = -71 \end{cases}$

29. The inverse of the matrix $A = \begin{pmatrix} 3 & 2 & 6 \\ 1 & 1 & 2 \\ 2 & 2 & 5 \end{pmatrix}$ is

$A^{-1} = \begin{pmatrix} 1 & 2 & -2 \\ -1 & 3 & 0 \\ 0 & -2 & 1 \end{pmatrix}$. Use this fact to solve the following systems.

(a) $\begin{cases} 3x + 2y + 6z = 28 \\ x + y + 2z = 9 \\ 2x + 2y + 5z = 22 \end{cases}$

(b) $\begin{cases} 3x + 2y + 6z = -7 \\ x + y + 2z = -2 \\ 2x + 2y + 5z = -6 \end{cases}$

30. The inverse of the matrix $A = \begin{pmatrix} 1 & -1 & 1 \\ 2 & -3 & 2 \\ -4 & 6 & 1 \end{pmatrix}$ is

$A^{-1} = \begin{pmatrix} 3 & -\frac{7}{5} & -\frac{1}{5} \\ 2 & -1 & 0 \\ 0 & \frac{2}{5} & \frac{1}{5} \end{pmatrix}$. Use this fact to solve the following systems.

(a) $\begin{cases} x - y + z = 5 \\ 2x - 3y + 2z = -15 \\ -4x + 6y + z = 25 \end{cases}$

(b) $\begin{cases} x - y + z = 1 \\ 2x - 3y + 2z = -2 \\ -4x + 6y + z = 0 \end{cases}$

**B**

31. Let $A = \begin{pmatrix} 1 & -6 & 3 \\ 2 & -7 & 3 \\ 4 & -12 & 5 \end{pmatrix}$.

    (a) Compute the matrix product $AA$. What do you observe?
    (b) Use the result in part (a) to solve the following system.
    $$\begin{cases} x - 6y + 3z = \frac{19}{2} \\ 2x - 7y + 3z = 11 \\ 4x - 12y + 5z = 19 \end{cases}$$

32. (a) Solve the following two systems and then check to see that your results agree with those given in Example 1.
    $$\begin{cases} a + 2c = 1 \\ 3a + 4c = 0 \end{cases} \qquad \begin{cases} b + 2d = 0 \\ 3b + 4d = 1 \end{cases}$$
    (b) At the end of Example 1, it was asserted that
    $$\begin{pmatrix} 1 & 2 \\ 3 & 4 \end{pmatrix}\begin{pmatrix} -2 & 1 \\ \frac{3}{2} & -\frac{1}{2} \end{pmatrix} = \begin{pmatrix} -2 & 1 \\ \frac{3}{2} & -\frac{1}{2} \end{pmatrix}\begin{pmatrix} 1 & 2 \\ 3 & 4 \end{pmatrix}$$
    $$= \begin{pmatrix} 1 & 0 \\ 0 & 1 \end{pmatrix}$$

Carry out the indicated matrix multiplications to verify that these equations are valid.

33. Use the technique in Example 1 to show that the matrix $\begin{pmatrix} 2 & 5 \\ 6 & 15 \end{pmatrix}$ does not have an inverse.

34. Use the elementary row operations (as in Example 3) to find the inverse of the following matrix.
    $$\begin{pmatrix} 1 & 1 & 1 & 1 \\ 1 & 2 & 3 & 4 \\ 1 & 3 & 6 & 10 \\ 1 & 4 & 10 & 20 \end{pmatrix}$$

35. Let $A = \begin{pmatrix} 2 & 3 \\ 4 & 5 \end{pmatrix}$ and $B = \begin{pmatrix} 7 & 8 \\ 6 & 7 \end{pmatrix}$.
    (a) Compute $A^{-1}$, $B^{-1}$, and $B^{-1}A^{-1}$.
    (b) Compute $(AB)^{-1}$. What do you observe?

36. Let $A = \begin{pmatrix} a & b \\ c & d \end{pmatrix}$. Compute $A^{-1}$. (Assume that $ad - bc \neq 0$.)

## DETERMINANTS AND CRAMER'S RULE

**10.5**

As you saw in the previous section, square matrices and their inverses can be used to solve certain systems of equations. In this section, we are going to associate a number with each square matrix. This number is called the **determinant** of the matrix. As you'll see, this too has an application in solving systems of equations.

The determinant of a matrix $A$ may be denoted by $\det A$ or $|A|$. Determinants are also denoted simply by replacing the parentheses of matrix notation with vertical lines. Thus, three examples of determinants are

$$\begin{vmatrix} 1 & 2 \\ 3 & 4 \end{vmatrix} \qquad \begin{vmatrix} 1 & 2 & 3 \\ 4 & 5 & 6 \\ 7 & 8 & 9 \end{vmatrix} \qquad \begin{vmatrix} 3 & 7 & 8 & 9 \\ 5 & 6 & 4 & 3 \\ -9 & 9 & 0 & 1 \\ 1 & 3 & -2 & 1 \end{vmatrix}$$

A determinant with $n$ rows and $n$ columns is said to be an **nth-order determinant**. Therefore, the determinants we've just written are, respectively, second-, third-, and fourth-order determinants. As with matrices, we speak of the numbers in a determinant as its **entries**. We also number the rows and the columns of a determinant as we do with matrices. However, unlike matrices, each determinant has a definite value. The value of a second-order determinant is defined as follows.

$$\begin{vmatrix} a & b \\ c & d \end{vmatrix} = ad - bc$$

Table 1 illustrates how this definition is used to evaluate, or *expand*, a second-order determinant.

**TABLE 1**

| DETERMINANT | VALUE OF DETERMINANT |
|---|---|
| $\begin{vmatrix} 3 & 7 \\ 5 & 10 \end{vmatrix}$ | $\begin{vmatrix} 3 & 7 \\ 5 & 10 \end{vmatrix} = 3(10) - 7(5) = 30 - 35 = -5$ |
| $\begin{vmatrix} 3 & -7 \\ 5 & 10 \end{vmatrix}$ | $\begin{vmatrix} 3 & -7 \\ 5 & 10 \end{vmatrix} = 3(10) - (-7)(5) = 30 + 35 = 65$ |
| $\begin{vmatrix} a & a^3 \\ 1 & a^2 \end{vmatrix}$ | $\begin{vmatrix} a & a^3 \\ 1 & a^2 \end{vmatrix} = a(a^2) - a^3(1) = a^3 - a^3 = 0$ |

In general, the value of an $n$th-order determinant $(n > 2)$ is defined in terms of certain $(n - 1)$st-order determinants. For instance, the value of a third-order determinant is defined in terms of second-order determinants. We'll use the following example to introduce the necessary terminology here:

$$\begin{vmatrix} 8 & 3 & 5 \\ 2 & 4 & 6 \\ 9 & 1 & 7 \end{vmatrix}$$

Pick a given entry—say, 8—and imagine crossing out all entries occupying the same row or the same column as 8.

$$\begin{vmatrix} 8 & 3 & 5 \\ 2 & 4 & 6 \\ 9 & 1 & 7 \end{vmatrix}$$

Then we are left with the second-order determinant $\begin{vmatrix} 4 & 6 \\ 1 & 7 \end{vmatrix}$. The second-order determinant is called the **minor** of the entry 8. Similarly, to find the minor of the entry 6 in the original determinant, imagine crossing out all entries that occupy the same row or the same column as 6.

$$\begin{vmatrix} 8 & 3 & 5 \\ 2 & 4 & 6 \\ 9 & 1 & 7 \end{vmatrix}$$

We are left with the second-order determinant $\begin{vmatrix} 8 & 3 \\ 9 & 1 \end{vmatrix}$, which by definition is the minor of the entry 6. In the same manner, *the minor of any element is the determinant obtained by crossing out the entries occupying the same row or column as the given element.*

Closely related to the minor of an entry in a determinant is the cofactor of that entry. The **cofactor** of an entry is defined as the minor multiplied by $+1$ or $-1$, according to the scheme displayed in Figure 1.

$$\begin{vmatrix} + & - & + \\ - & + & - \\ + & - & + \end{vmatrix}$$

**FIGURE 1**

After looking at an example, we'll give a more formal rule for computing cofactors, one that will not rely on a figure and that will also apply to larger determinants.

**EXAMPLE 1**  Consider the determinant

$$\begin{vmatrix} 1 & 2 & 3 \\ 4 & 5 & 6 \\ 7 & 8 & 9 \end{vmatrix}$$

Compute the minor and the cofactor of the entry 4.

**Solution**  By definition, we have

$$\text{minor of } 4 = \begin{vmatrix} 2 & 3 \\ 8 & 9 \end{vmatrix} = 18 - 24 = -6$$

Thus, the minor of the entry 4 is $-6$. To compute the cofactor of 4, we first notice that 4 is located in the second row and first column of the given determinant. Upon checking the corresponding position in Figure 1, we see a negative sign. Therefore, we have by definition

$$\text{cofactor of } 4 = (-1)(\text{minor of } 4)$$
$$= (-1)(-6) = 6$$

The cofactor of 4 is therefore 6.

The following rule tells us how cofactors can be computed without relying on Figure 1. For reference, we also restate the definition of a minor. (You should verify for yourself that this rule yields results that are consistent with Figure 1.)

---

### Minors and Cofactors

The **minor** of an entry $b$ in a determinant is the determinant formed by suppressing the entries in the row and in the column in which $b$ appears.

Suppose that the entry $b$ is in the $i$th row and the $j$th column. Then the **cofactor** of $b$ is given by the expression

$$(-1)^{i+j}(\text{minor of } b)$$

---

We are now prepared to state the definition that tells us how to evaluate a third-order determinant. Actually, as you'll see later, the definition is quite general and may be applied to determinants of any size.

---

### Definition for the Value of a Determinant

Multiply each entry in the first row of the determinant by its cofactor and then add the results. The value of the determinant is defined to be this sum.

---

To see how this definition is used, let us evaluate the determinant

$$\begin{vmatrix} 1 & 2 & 3 \\ 4 & 5 & 6 \\ 7 & 8 & 9 \end{vmatrix}$$

The definition tells us to multiply each entry in the first row by its cofactor and then add the results. Carrying out this procedure, we have

$$\begin{vmatrix} 1 & 2 & 3 \\ 4 & 5 & 6 \\ 7 & 8 & 9 \end{vmatrix} = 1 \begin{vmatrix} 5 & 6 \\ 8 & 9 \end{vmatrix} - 2 \begin{vmatrix} 4 & 6 \\ 7 & 9 \end{vmatrix} + 3 \begin{vmatrix} 4 & 5 \\ 7 & 8 \end{vmatrix}$$

$$= 1(45 - 48) - 2(36 - 42) + 3(32 - 35)$$
$$= -3 - 2(-6) + 3(-3)$$
$$= -3 + 12 - 9 = 0$$

So the value of this particular determinant is zero. The procedure we've used here is referred to as **expanding the determinant along its first row**. The following theorem (stated here without proof) tells us that the value of a determinant can be obtained by expanding along any row or along any column; the results are the same in all cases.

---

**Theorem**

Select any row or any column in a determinant and multiply each element in that row or column by its cofactor. Then add the results. The number obtained will be the value of the determinant. (In other words, the number obtained will be the same as that obtained by expanding the determinant along its first row.)

---

According to this theorem, we could have evaluated the determinant

$$\begin{vmatrix} 1 & 2 & 3 \\ 4 & 5 & 6 \\ 7 & 8 & 9 \end{vmatrix}$$

by expanding it along any row or any column. Let's expand it along the second column and check to see that the result agrees with the value obtained earlier. Expanding along the second column, we have

$$\begin{vmatrix} 1 & 2 & 3 \\ 4 & 5 & 6 \\ 7 & 8 & 9 \end{vmatrix} = -2 \begin{vmatrix} 4 & 6 \\ 7 & 9 \end{vmatrix} + 5 \begin{vmatrix} 1 & 3 \\ 7 & 9 \end{vmatrix} - 8 \begin{vmatrix} 1 & 3 \\ 4 & 6 \end{vmatrix}$$

$$= -2(36 - 42) + 5(9 - 21) - 8(6 - 12)$$
$$= -2(-6) + 5(-12) - 8(-6)$$
$$= 12 - 60 + 48 = 0 \qquad \text{as obtained previously}$$

We would have obtained the same result had we chosen to begin with any other row or column. (Exercise 13 at the end of this section asks you to verify this.)

There are three basic rules that make it easier to evaluate determinants. These are summarized in the box that follows. (Suggestions for proving these can be found in the exercises.)

**Three Rules for Manipulating Determinants**

EXAMPLES

1. If each entry in a given row is multiplied by the constant $k$, then the value of the determinant is multiplied by $k$. This is true for columns, also.

$$10 \begin{vmatrix} 1 & 3 & 4 \\ 1 & 2 & 3 \\ 4 & 5 & 6 \end{vmatrix} = \begin{vmatrix} 10 & 30 & 40 \\ 1 & 2 & 3 \\ 4 & 5 & 6 \end{vmatrix}$$

$$k \begin{vmatrix} a & b & c \\ d & e & f \\ g & h & i \end{vmatrix} = \begin{vmatrix} a & kb & c \\ d & ke & f \\ g & kh & i \end{vmatrix}$$

2. If a multiple of one row is added to another row, the value of the determinant is not changed. This applies to columns, also.

$$\begin{vmatrix} a & b & c \\ d & e & f \\ g & h & i \end{vmatrix} = \begin{vmatrix} a & b & c \\ d + ka & e + kb & f + kc \\ g & h & i \end{vmatrix}$$

3. If two rows are interchanged, then the value of the determinant is multiplied by $-1$. This applies to columns, also.

$$\begin{vmatrix} 1 & 2 & 3 \\ 4 & 5 & 6 \\ a & b & c \end{vmatrix} = - \begin{vmatrix} 4 & 5 & 6 \\ 1 & 2 & 3 \\ a & b & c \end{vmatrix}$$

**EXAMPLE 2**  Evaluate the determinant

$$\begin{vmatrix} 15 & 14 & 26 \\ 18 & 17 & 32 \\ 21 & 20 & 42 \end{vmatrix}$$

**Solution**

$$\begin{vmatrix} 15 & 14 & 26 \\ 18 & 17 & 32 \\ 21 & 20 & 42 \end{vmatrix} = (3 \times 2) \begin{vmatrix} 5 & 14 & 13 \\ 6 & 17 & 16 \\ 7 & 20 & 21 \end{vmatrix}$$

using rule 1 to factor 3 from the first column and 2 from the third column

$$= 6 \begin{vmatrix} 5 & 1 & 13 \\ 6 & 1 & 16 \\ 7 & -1 & 21 \end{vmatrix}$$

using rule 2 to subtract the third column from the second column

$$= 6 \begin{vmatrix} 12 & 0 & 34 \\ 13 & 0 & 37 \\ 7 & -1 & 21 \end{vmatrix}$$

using rule 2 to add the third row to the first and second rows

$$= 6 \left[ -\left( -1 \begin{vmatrix} 12 & 34 \\ 13 & 37 \end{vmatrix} \right) \right]$$

expanding the determinant along the second column

$$= 6 \begin{vmatrix} 12 & 34 \\ 1 & 3 \end{vmatrix}$$

using rule 2 to subtract the first row from the second row

$$= 6(36 - 34) = 12$$

The value of the given determinant is therefore 12. Notice the general strategy. We use rules 1 and 2 until one column (or one row) contains two zeros. At that point, it is a simple matter to expand the determinant along that column (or row).

**EXAMPLE 3** Show that

$$\begin{vmatrix} 1 & 1 & 1 \\ a & b & c \\ a^2 & b^2 & c^2 \end{vmatrix} = (b-a)(c-a)(c-b)$$

**Solution**

$$\begin{vmatrix} 1 & 1 & 1 \\ a & b & c \\ a^2 & b^2 & c^2 \end{vmatrix} = \begin{vmatrix} 1 & 0 & 0 \\ a & b-a & c-a \\ a^2 & b^2-a^2 & c^2-a^2 \end{vmatrix}$$

using rule 2 to subtract the first column from the second and the third columns

$$= \begin{vmatrix} 1 & 0 & 0 \\ a & b-a & c-a \\ a^2 & (b-a)(b+a) & (c-a)(c+a) \end{vmatrix}$$

$$= (b-a)(c-a)\begin{vmatrix} 1 & 0 & 0 \\ a & 1 & 1 \\ a^2 & b+a & c+a \end{vmatrix}$$

using rule 1 to factor $(b-a)$ from the second column and $(c-a)$ from the third column

$$= (b-a)(c-a)[(c+a)-(b+a)]$$

expanding the determinant along the first row

$$= (b-a)(c-a)(c-b)$$

The definition that we gave for third-order determinants can also be applied to define fourth-order (or larger) determinants. Consider, for example, the fourth-order determinant $A$ given by

$$A = \begin{vmatrix} 25 & 40 & 5 & 10 \\ 9 & 0 & 3 & 6 \\ -2 & 3 & 11 & -17 \\ -3 & 4 & 7 & 2 \end{vmatrix}$$

By definition, then, we could evaluate this determinant by selecting the first row, multiplying each entry by its cofactor, and then adding the results. This yields

$$A = 25\begin{vmatrix} 0 & 3 & 6 \\ 3 & 11 & -17 \\ 4 & 7 & 2 \end{vmatrix} - 40\begin{vmatrix} 9 & 3 & 6 \\ -2 & 11 & -17 \\ -3 & 7 & 2 \end{vmatrix}$$

$$+ 5\begin{vmatrix} 9 & 0 & 6 \\ -2 & 3 & -17 \\ -3 & 4 & 2 \end{vmatrix} - 10\begin{vmatrix} 9 & 0 & 3 \\ -2 & 3 & 11 \\ -3 & 4 & 7 \end{vmatrix}$$

The problem is now reduced to evaluating four third-order determinants. If we had instead expanded down the second column, the ensuing work would be somewhat less because of the zero in that column. Nevertheless, there would still be three third-order determinants to evaluate.

Because of the amount of computation involved, we in fact rarely evaluate fourth-order determinants directly from the definition. Instead, we use the three rules (given just before Example 2) to simplify matters first. In the example that follows, we show how this is done. (We will accept the fact that the three rules are applicable in evaluating determinants of any size.)

**EXAMPLE 4**  Evaluate the determinant $A$ given by

$$A = \begin{vmatrix} 25 & 40 & 5 & 10 \\ 9 & 0 & 3 & 6 \\ -2 & 3 & 11 & -17 \\ -3 & 4 & 7 & 2 \end{vmatrix}$$

**Solution**

$$A = 5 \times 3 \begin{vmatrix} 5 & 8 & 1 & 2 \\ 3 & 0 & 1 & 2 \\ -2 & 3 & 11 & -17 \\ -3 & 4 & 7 & 2 \end{vmatrix} = 15 \begin{vmatrix} 2 & 8 & 0 & 0 \\ 3 & 0 & 1 & 2 \\ -35 & 3 & 0 & -39 \\ -24 & 4 & 0 & -12 \end{vmatrix}$$

Factor 5 from the first row and 3 from the second row.

Subtract the second row from the first. Subtract 11 times the second row from the third. Subtract 7 times the second row from the fourth.

$$= 15 \times 2 \times 4 \begin{vmatrix} 1 & 4 & 0 & 0 \\ 3 & 0 & 1 & 2 \\ -35 & 3 & 0 & -39 \\ -6 & 1 & 0 & -3 \end{vmatrix} = 120 \cdot (-1) \begin{vmatrix} 1 & 4 & 0 \\ -35 & 3 & -39 \\ -6 & 1 & -3 \end{vmatrix}$$

Factor 2 from the first row and 4 from the fourth row.

Expand the determinant along the third column.

$$= -120(-3) \begin{vmatrix} 1 & 4 & 0 \\ -35 & 3 & 13 \\ -6 & 1 & 1 \end{vmatrix} = 360 \begin{vmatrix} 1 & 0 & 0 \\ -35 & 143 & 13 \\ -6 & 25 & 1 \end{vmatrix}$$

Factor $(-3)$ from the third column.

Subtract 4 times the first column from the second.

$$= 360 \begin{vmatrix} 143 & 13 \\ 25 & 1 \end{vmatrix} = 360 \begin{vmatrix} -7 & 7 \\ 25 & 1 \end{vmatrix}$$

Expand the determinant along the first row.

Subtract 6 times the second row from the first row.

$$= 360(-7 - 175) = -65520$$

Do the arithmetic!

The value of the given fourth-order determinant is therefore $-65520$.

Determinants can be used to solve certain systems of linear equations in which there are as many unknowns as there are equations. In the box that follows, we state **Cramer's rule** for solving a system of three linear equations in three unknowns.* A more general, but entirely similar, version of Cramer's rule holds for $n$ equations in $n$ unknowns.

---

### Cramer's Rule

Consider the system

$$\begin{cases} a_1x + b_1y + c_1z = d_1 \\ a_2x + b_2y + c_2z = d_2 \\ a_3x + b_3y + c_3z = d_3 \end{cases}$$

Let the four determinants $D$, $D_x$, $D_y$, and $D_z$ be defined as follows:

$$D = \begin{vmatrix} a_1 & b_1 & c_1 \\ a_2 & b_2 & c_2 \\ a_3 & b_3 & c_3 \end{vmatrix} \qquad D_x = \begin{vmatrix} d_1 & b_1 & c_1 \\ d_2 & b_2 & c_2 \\ d_3 & b_3 & c_3 \end{vmatrix}$$

$$D_y = \begin{vmatrix} a_1 & d_1 & c_1 \\ a_2 & d_2 & c_2 \\ a_3 & d_3 & c_3 \end{vmatrix} \qquad D_z = \begin{vmatrix} a_1 & b_1 & d_1 \\ a_2 & b_2 & d_2 \\ a_3 & b_3 & d_3 \end{vmatrix}$$

Then if $D \neq 0$, the unique values of $x$, $y$, and $z$ satisfying the system are given by

$$x = \frac{D_x}{D} \qquad y = \frac{D_y}{D} \qquad z = \frac{D_z}{D}$$

---

Notice that the determinant $D$ in Cramer's rule is just the determinant of the coefficient matrix of the given system. If you replace the first column of $D$ with the column of numbers on the right side of the given system, you obtain $D_x$. The determinants $D_y$ and $D_z$ are obtained similarly.

Before actually proving Cramer's rule, let's take a look at how the rule is applied.

**EXAMPLE 5**   Use Cramer's rule to find all solutions of the following system of equations:

$$\begin{cases} 2x + 2y - 3z = -20 \\ x - 4y + z = 6 \\ 4x - y + 2z = -1 \end{cases}$$

**Solution**   First, we list the determinants $D$, $D_x$, $D_y$, and $D_z$:

$$D = \begin{vmatrix} 2 & 2 & -3 \\ 1 & -4 & 1 \\ 4 & -1 & 2 \end{vmatrix} \qquad D_x = \begin{vmatrix} -20 & 2 & -3 \\ 6 & -4 & 1 \\ -1 & -1 & 2 \end{vmatrix}$$

$$D_y = \begin{vmatrix} 2 & -20 & -3 \\ 1 & 6 & 1 \\ 4 & -1 & 2 \end{vmatrix} \qquad D_z = \begin{vmatrix} 2 & 2 & -20 \\ 1 & -4 & 6 \\ 4 & -1 & -1 \end{vmatrix}$$

---

* The rule is named after one of its discoverers, the Swiss mathematician Gabriel Cramer (1704–1752).

The calculations for evaluating $D$ begin as follows:

$$D = \begin{vmatrix} 2 & 2 & -3 \\ 1 & -4 & 1 \\ 4 & -1 & 2 \end{vmatrix} = \begin{vmatrix} 0 & 10 & -5 \\ 1 & -4 & 1 \\ 0 & 15 & -2 \end{vmatrix}$$

Subtract twice the second row from the first.
Subtract 4 times the second row from the third.

Since we now have two zeros in the first column, it is an easy matter to expand $D$ along that column to obtain

$$D = -1 \begin{vmatrix} 10 & -5 \\ 15 & -2 \end{vmatrix}$$

$$= -1[-20 - (-75)] = -1(55) = -55$$

The value of $D$ is therefore $-55$. (Since this value is nonzero, Cramer's rule does apply.) As Exercise 31 at the end of this section asks you to verify, the values of the other three determinants are

$$D_x = 144 \qquad D_y = 61 \qquad D_z = -230$$

By Cramer's rule, then, the unique values of $x$, $y$, and $z$ that satisfy the system are

$$x = \frac{D_x}{D} = \frac{144}{-55} = -\frac{144}{55}$$

$$y = \frac{D_y}{D} = \frac{61}{-55} = -\frac{61}{55}$$

$$z = \frac{D_z}{D} = \frac{-230}{-55} = \frac{46}{11}$$

One way we can prove Cramer's rule is to use Gaussian elimination to solve the system

$$\begin{cases} a_1x + b_1y + c_1z = d_1 \\ a_2x + b_2y + c_2z = d_2 \\ a_3x + b_3y + c_3z = d_3 \end{cases} \tag{1}$$

A much shorter and simpler proof, however, has been found by D. E. Whitford and M. S. Klamkin.* This is the proof we give here; it makes effective use of the rules employed in this section for manipulating determinants.

Consider the system of equations (1) and assume that $D \neq 0$. We will show that *if* $x$, $y$, and $z$ satisfy the system, then in fact $x = D_x/D$, with similar equations giving $y$ and $z$. (Exercise 62 at the end of this section then shows how

---

* The proof was published in the *American Mathematical Monthly*, **60** (1953):186–187.

to check that these values indeed satisfy the given system.) We have

$$D_x = \begin{vmatrix} d_1 & b_1 & c_1 \\ d_2 & b_2 & c_2 \\ d_3 & b_3 & c_3 \end{vmatrix} \qquad \text{by definition}$$

$$= \begin{vmatrix} (a_1x + b_1y + c_1z) & b_1 & c_1 \\ (a_2x + b_2y + c_2z) & b_2 & c_2 \\ (a_3x + b_3y + c_3z) & b_3 & c_3 \end{vmatrix} \qquad \begin{array}{l}\text{using the equations in (1)} \\ \text{to substitute for } d_1, d_2, \text{ and } d_3\end{array}$$

$$= \begin{vmatrix} a_1x & b_1 & c_1 \\ a_2x & b_2 & c_2 \\ a_3x & b_3 & c_3 \end{vmatrix} \qquad \begin{array}{l}\text{from the first column, subtracting} \\ y \text{ times the second column as well} \\ \text{as } z \text{ times the third column}\end{array}$$

$$= x\begin{vmatrix} a_1 & b_1 & c_1 \\ a_2 & b_2 & c_2 \\ a_3 & b_3 & c_3 \end{vmatrix} \qquad \begin{array}{l}\text{factoring } x \text{ out of} \\ \text{the first column}\end{array}$$

$$= xD \qquad \text{by definition}$$

We now have $D_x = xD$, which is equivalent to $x = D_x/D$, as required. The formulas for $y$ and $z$ are obtained similarly.

## EXERCISE SET 10.5

### A

In Exercises 1–6, evaluate the determinants.

1. (a) $\begin{vmatrix} 2 & -17 \\ 1 & 6 \end{vmatrix}$     (b) $\begin{vmatrix} 1 & 6 \\ 2 & -17 \end{vmatrix}$

2. (a) $\begin{vmatrix} 1 & 0 \\ 0 & 1 \end{vmatrix}$     (b) $\begin{vmatrix} 0 & 1 \\ 0 & 1 \end{vmatrix}$

3. (a) $\begin{vmatrix} 5 & 7 \\ 500 & 700 \end{vmatrix}$     (b) $\begin{vmatrix} 5 & 500 \\ 7 & 700 \end{vmatrix}$

4. (a) $\begin{vmatrix} -8 & -3 \\ 4 & -5 \end{vmatrix}$     (b) $\begin{vmatrix} -3 & -8 \\ -5 & 4 \end{vmatrix}$

5. $\begin{vmatrix} \sqrt{2}-1 & \sqrt{2} \\ \sqrt{2} & \sqrt{2}+1 \end{vmatrix}$

6. $\begin{vmatrix} \sqrt{3}+\sqrt{2} & 1+\sqrt{5} \\ 1-\sqrt{5} & \sqrt{3}-\sqrt{2} \end{vmatrix}$

In Exercises 7–12, refer to the following determinant:

$$\begin{vmatrix} -6 & 3 & 8 \\ 5 & -4 & 1 \\ 10 & 9 & -10 \end{vmatrix}$$

7. Evaluate the minor of the entry 3.

8. Evaluate the cofactor of the entry 3.

9. Evaluate the minor of $-10$.

10. Evaluate the cofactor of $-10$.

11. (a) Multiply each entry in the first row by its minor and find the sum of the results.
    (b) Multiply each entry in the first row by its cofactor and find the sum of the results.
    (c) Does the answer in part (a) or part (b) give you the value of the determinant?

12. (a) Multiply each entry in the first column by its cofactor and find the sum of the results.
    (b) Follow the same instructions as in part (a), but use the second column.
    (c) Follow the same instructions as in part (a), but use the third column.

13. Let $A = \begin{vmatrix} 1 & 2 & 3 \\ 4 & 5 & 6 \\ 7 & 8 & 9 \end{vmatrix}$. Evaluate $A$ by expanding it along the indicated row or column.
    (a) second row    (b) third row
    (c) first column    (d) third column

*In Exercises 14–21, evaluate the determinants.*

14. $\begin{vmatrix} 1 & 2 & -1 \\ 2 & -1 & 1 \\ 4 & 0 & 2 \end{vmatrix}$

15. $\begin{vmatrix} 5 & 10 & 15 \\ 1 & 2 & 3 \\ -9 & 11 & 7 \end{vmatrix}$

16. $\begin{vmatrix} 8 & 4 & 2 \\ 3 & 9 & 3 \\ -2 & 8 & 6 \end{vmatrix}$

17. $\begin{vmatrix} 1 & 2 & -3 \\ 4 & 5 & -9 \\ 0 & 0 & 1 \end{vmatrix}$

18. $\begin{vmatrix} 3 & 0 & 0 \\ 0 & 19 & 0 \\ 0 & 0 & 10 \end{vmatrix}$

19. $\begin{vmatrix} -6 & -8 & 18 \\ 25 & 12 & 15 \\ -9 & 4 & 13 \end{vmatrix}$

20. $\begin{vmatrix} 23 & 0 & 47 \\ -37 & 0 & 18 \\ 14 & 0 & 25 \end{vmatrix}$

21. $\begin{vmatrix} 16 & 0 & -64 \\ -8 & 15 & -12 \\ 30 & -20 & 10 \end{vmatrix}$

22. Use the method illustrated in Example 3 in the text to show that

$$\begin{vmatrix} 1 & 1 & 1 \\ a & b & c \\ a^3 & b^3 & c^3 \end{vmatrix} = (b - a)(c - a)(c^2 - b^2 + ac - ab)$$

$$= (b - a)(c - a)(c - b)(a + b + c)$$

23. Use the method shown in Example 3 to express the following determinant as a product:

$$\begin{vmatrix} 1 & x & x^2 \\ 1 & y & y^2 \\ 1 & z & z^2 \end{vmatrix}$$

24. Show that

$$\begin{vmatrix} 1 & 1 & 1 \\ a^2 & b^2 & c^2 \\ a^3 & b^3 & c^3 \end{vmatrix}$$

$$= (b - a)(c - a)(bc^2 - b^2c + ac^2 - ab^2)$$

$$= (b - a)(c - a)(c - b)(bc + ac + ab)$$

25. Simplify the following:

$$\begin{vmatrix} 1 & 1 & 1 \\ 1 & 1 + x & 1 \\ 1 & 1 & 1 + y \end{vmatrix}$$

26. Use the method shown in Example 3 to express the following determinant as a product of three factors:

$$\begin{vmatrix} 1 & 1 & 1 \\ a & b & c \\ bc & ca & ab \end{vmatrix}$$

*In Exercises 27–30, evaluate the determinants.*

27. $\begin{vmatrix} 1 & -1 & 0 & 2 \\ 0 & 1 & -1 & 0 \\ 2 & 1 & 0 & -1 \\ -2 & 2 & 1 & 1 \end{vmatrix}$

28. $\begin{vmatrix} 3 & -2 & 3 & 4 \\ 1 & 4 & -3 & 2 \\ 6 & 3 & -6 & -3 \\ -1 & 0 & 1 & 5 \end{vmatrix}$

29. $\begin{vmatrix} 2 & 0 & 0 & 0 \\ 0 & 3 & 0 & 0 \\ 0 & 0 & 4 & 0 \\ 0 & 0 & 0 & 5 \end{vmatrix}$

30. $\begin{vmatrix} 7 & -8 & 1 & 2 \\ 21 & 4 & 3 & -1 \\ -35 & 8 & 3 & -2 \\ 14 & 16 & 0 & 1 \end{vmatrix}$

31. Verify the following statements.

(a) $\begin{vmatrix} -20 & 2 & -3 \\ 6 & -4 & 1 \\ -1 & -1 & 2 \end{vmatrix} = 144$

(b) $\begin{vmatrix} 2 & -20 & -3 \\ 1 & 6 & 1 \\ 4 & -1 & 2 \end{vmatrix} = 61$

(c) $\begin{vmatrix} 2 & 2 & -20 \\ 1 & -4 & 6 \\ 4 & -1 & -1 \end{vmatrix} = -230$

32. Consider the following system:

$$\begin{cases} x + y + 3z = 0 \\ x + 2y + 5z = 0 \\ x - 4y - 8z = 0 \end{cases}$$

(a) Without doing any calculations, find one obvious solution of this system.

(b) Calculate the determinant $D$.

(c) List all solutions of this system.

*In Exercises 33–42, use Cramer's rule to solve those systems for which $D \neq 0$. In cases where $D = 0$, use Gaussian elimination or matrix methods.*

33. $\begin{cases} 3x + 4y - z = 5 \\ x - 3y + 2z = 2 \\ 5x - 6z = -7 \end{cases}$

34. $\begin{cases} 3A - B - 4C = 3 \\ A + 2B - 3C = 9 \\ 2A - B + 2C = -8 \end{cases}$

35. $\begin{cases} 3x + 2y - z = -6 \\ 2x - 3y - 4z = -11 \\ x + y + z = 5 \end{cases}$

36. $\begin{cases} 5x - 3y - z = 16 \\ 2x + y - 3z = 5 \\ 3x - 2y + 2z = 5 \end{cases}$

37. $\begin{cases} 2x + 5y + 2z = 0 \\ 3x - y - 4z = 0 \\ x + 2y - 3z = 0 \end{cases}$

38. $\begin{cases} 4u + 3v - 2w = 14 \\ u + 2v - 3w = 6 \\ 2u - v + 4w = 2 \end{cases}$

39. $\begin{cases} 12x - 11z = 13 \\ 6x + 6y - 4z = 26 \\ 6x + 2y - 5z = 13 \end{cases}$

40. $\begin{cases} 3x + 4y + 2z = 1 \\ 4x + 6y + 2z = 7 \\ 2x + 3y + z = 11 \end{cases}$

**41.** $\begin{cases} x + y + z + w = -7 \\ x - y + z - w = -11 \\ 2x - 2y - 3z - 3w = 26 \\ 3x + 2y + z - w = -9 \end{cases}$

**42.** $\begin{cases} 2A - B - 3C + 2D = -2 \\ A - 2B + C - 3D = 4 \\ 3A - 4B + 2C - 4D = 12 \\ 2A + 3B - C - 2D = -4 \end{cases}$

**43.** Find all values of $x$ for which

$$\begin{vmatrix} x - 4 & 0 & 0 \\ 0 & x + 4 & 0 \\ 0 & 0 & x + 1 \end{vmatrix} = 0$$

**44.** Find all values of $x$ for which

$$\begin{vmatrix} 1 & x & x^2 \\ 1 & 1 & 1 \\ 4 & 5 & 0 \end{vmatrix} = 0$$

**45.** By expanding the determinant $\begin{vmatrix} a & b & c \\ a & b & c \\ d & e & f \end{vmatrix}$ down the first column, show that its value is zero.

**46.** By expanding the determinant $\begin{vmatrix} ka & kb & kc \\ d & e & f \\ g & h & i \end{vmatrix}$ along its first row, show that it is equal to

$$k \begin{vmatrix} a & b & c \\ d & e & f \\ g & h & i \end{vmatrix}$$

**B**

**47.** Show that

$$\begin{vmatrix} a_1 + A_1 & b_1 & c_1 \\ a_2 + A_2 & b_2 & c_2 \\ a_3 + A_3 & b_3 & c_3 \end{vmatrix} = \begin{vmatrix} a_1 & b_1 & c_1 \\ a_2 & b_2 & c_2 \\ a_3 & b_3 & c_3 \end{vmatrix} + \begin{vmatrix} A_1 & b_1 & c_1 \\ A_2 & b_2 & c_2 \\ A_3 & b_3 & c_3 \end{vmatrix}$$

**48.** Show that

$$\begin{vmatrix} a_1 & b_1 & c_1 \\ a_2 & b_2 & c_2 \\ a_3 & b_3 & c_3 \end{vmatrix} = \begin{vmatrix} a_1 + kb_1 & b_1 & c_1 \\ a_2 + kb_2 & b_2 & c_2 \\ a_3 + kb_3 & b_3 & c_3 \end{vmatrix}$$

**49.** By expanding each determinant along a row or column, show that

$$\begin{vmatrix} a_1 & b_1 & c_1 \\ a_2 & b_2 & c_2 \\ a_3 & b_3 & c_3 \end{vmatrix} = - \begin{vmatrix} a_2 & b_2 & c_2 \\ a_1 & b_1 & c_1 \\ a_3 & b_3 & c_3 \end{vmatrix}$$

**50.** Solve for $x$ in terms of $a$, $b$, and $c$:

$$\begin{vmatrix} a & a & x \\ c & c & c \\ b & x & b \end{vmatrix} = 0 \qquad (c \neq 0)$$

**51.** Show that

$$\begin{vmatrix} 1 + a & 1 & 1 & 1 \\ 1 & 1 + b & 1 & 1 \\ 1 & 1 & 1 + c & 1 \\ 1 & 1 & 1 & 1 + d \end{vmatrix}$$

$$= abcd \left( \frac{1}{a} + \frac{1}{b} + \frac{1}{c} + \frac{1}{d} + 1 \right)$$

**52.** Show that

$$\begin{vmatrix} 1 & a & a^2 \\ a^2 & 1 & a \\ a & a^2 & 1 \end{vmatrix} = (a^3 - 1)^2$$

**53.** Consider the determinant

$$D = \begin{vmatrix} a_1 & b_1 & c_1 \\ a_2 & b_2 & c_2 \\ a_3 & b_3 & c_3 \end{vmatrix}$$

Let $A_1$ denote the cofactor of $a_1$, let $B_1$ denote the cofactor of $b_1$, and so on. Prove that

$$\begin{vmatrix} B_2 & C_2 \\ B_3 & C_3 \end{vmatrix} = a_1 D$$

**54.** Show that

$$\begin{vmatrix} 1 & bc & b + c \\ 1 & ca & c + a \\ 1 & ab & a + b \end{vmatrix} = (b - c)(c - a)(a - b)$$

**55.** Find the value of each determinant.

(a) $\begin{vmatrix} 1 & 0 & 0 \\ x & 1 & 0 \\ x & y & 1 \end{vmatrix}$ (b) $\begin{vmatrix} 1 & 0 & 0 & 0 \\ x & 1 & 0 & 0 \\ x & y & 1 & 0 \\ x & y & z & 1 \end{vmatrix}$

**56.** Evaluate the determinant.

$$\begin{vmatrix} a & b & c \\ a & a + b & a + b + c \\ a & 2a + b & 3a + 2b + c \end{vmatrix}$$

**57.** Show that

$$\begin{vmatrix} 1 & a & a & a \\ 1 & b & a & a \\ 1 & a & b & a \\ 1 & u & u & b \end{vmatrix} = (b - a)^3$$

**58.** Show that

$$\begin{vmatrix} a & 1 & 1 & 1 \\ 1 & a & 1 & 1 \\ 1 & 1 & a & 1 \\ 1 & 1 & 1 & a \end{vmatrix} = (a-1)^3(a+3)$$

**59.** Solve the following system for $x$, $y$, and $z$:

$$\begin{cases} ax + by + cz = k \\ a^2x + b^2y + c^2z = k^2 \\ a^3x + b^3y + c^3z = k^3 \end{cases}$$

(Assume that the values of $a$, $b$, and $c$ are all distinct and all nonzero.)

**60.** Show that

$$\begin{vmatrix} x+a & a & a & a \\ b & x+b & b & b \\ c & c & x+c & c \\ d & d & d & x+d \end{vmatrix} = x^3(a+b+c+d+x)$$

*Hint:* To the first row add each of the three other rows. For convenience, let $t = a+b+c+d+x$.

**C**

**61.** Show that the area of the triangle in Figure (a) is $\frac{1}{2}\begin{vmatrix} a & b \\ c & d \end{vmatrix}$    *Hint:* Figure (b) indicates how the required

area may be found using a rectangle and three right triangles.

**62.** This exercise completes the derivation of Cramer's rule given in the text. Using the same notation, and assuming $D \neq 0$, we need to show that the values $x = D_x/D$, $y = D_y/D$, and $z = D_z/D$ satisfy the equations in (1) on page 539. We will show that these values satisfy the first equation in (1), the verification for the other equations being entirely similar.

**(a)** Check that substituting the values $x = D_x/D$, $y = D_y/D$, and $z = D_z/D$ in the first equation of (1) yields an equation equivalent to

$$a_1D_x + b_1D_y + c_1D_z - d_1D = 0$$

**(b)** Show that the equation in part (a) may be written

$$a_1\begin{vmatrix} b_1 & c_1 & d_1 \\ b_2 & c_2 & d_2 \\ b_3 & c_3 & d_3 \end{vmatrix} - b_1\begin{vmatrix} a_1 & c_1 & d_1 \\ a_2 & c_2 & d_2 \\ a_3 & c_3 & d_3 \end{vmatrix}$$
$$+ c_1\begin{vmatrix} a_1 & b_1 & d_1 \\ a_2 & b_2 & d_2 \\ a_3 & b_3 & d_3 \end{vmatrix} - d_1\begin{vmatrix} a_1 & b_1 & c_1 \\ a_2 & b_2 & c_2 \\ a_3 & b_3 & c_3 \end{vmatrix} = 0$$

**(c)** Show that the equation in part (b) may be written

$$\begin{vmatrix} a_1 & b_1 & c_1 & d_1 \\ a_1 & b_1 & c_1 & d_1 \\ a_2 & b_2 & c_2 & d_2 \\ a_3 & b_3 & c_3 & d_3 \end{vmatrix} = 0$$

**(d)** Now explain why the equation in (c) indeed holds.

**63.** Let $D$ denote the determinant of the matrix $\begin{pmatrix} a & b \\ c & d \end{pmatrix}$.

**(a)** Show that the inverse of this matrix is $\frac{1}{D}\begin{pmatrix} d & -b \\ -c & a \end{pmatrix}$. Assume that $ad - bc \neq 0$.

**(b)** Use the result in part (a) to find the inverse of the matrix $\begin{pmatrix} -6 & 7 \\ 1 & 9 \end{pmatrix}$.

(a)

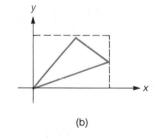

(b)

## NONLINEAR SYSTEMS OF EQUATIONS

**10.6**

In the previous sections of this chapter, we looked at several techniques for solving systems of linear equations. In particular, we saw that if solutions exist, they can always be found by Gaussian elimination. In the present section, we consider **nonlinear systems** of equations, that is, systems in which at least one of the equations is not linear. There is no single technique that serves to solve all nonlinear systems. However, simple substitution will often suffice. The work in this section focuses on examples showing some of the more common approaches. In all the examples and in the exercises, we will be concerned exclusively with solutions $(x,y)$ in which both $x$ and $y$ are real numbers.

In Example 1, the system consists of one linear equation and one quadratic equation. Such a system can always be solved by substitution.

**EXAMPLE 1**   Find all solutions $(x, y)$ of the following system, where $x$ and $y$ are real numbers:

$$\begin{cases} 2x + y = 1 \\ y = 4 - x^2 \end{cases}$$

Solution   We use the second equation to substitute for $y$ in the first equation. This will yield an equation with only one unknown:

$$2x + (4 - x^2) = 1$$
$$-x^2 + 2x + 3 = 0$$
$$x^2 - 2x - 3 = 0$$
$$(x - 3)(x + 1) = 0$$

$$x - 3 = 0 \quad \bigg| \quad x + 1 = 0$$
$$x = 3 \quad \bigg| \quad x = -1$$

The values $x = 3$ and $x = -1$ can now be substituted back into either of the original equations. Substituting $x = 3$ in the equation $y = 4 - x^2$ yields $y = -5$. Similarly, substituting $x = -1$ in the equation $y = 4 - x^2$ gives us $y = 3$. We have now determined the two ordered pairs $(3, -5)$ and $(-1, 3)$. As you can easily check, both of these are solutions of the given system. Figure 1 displays the graphical interpretation of this result. The line $2x + y = 1$ intersects the parabola $y = 4 - x^2$ at the points $(-1, 3)$ and $(3, -5)$.

**FIGURE 1**

**EXAMPLE 2**   Where do the graphs of the parabola $y = x^2$ and the circle $x^2 + y^2 = 1$ intersect? See Figure 2.

Solution   The system that we wish to solve is

$$\begin{cases} y = x^2 & (1) \\ x^2 + y^2 = 1 & (2) \end{cases}$$

In view of equation (1), we can replace the $x^2$-term of equation (2) by $y$. Doing this yields

$$y + y^2 = 1$$

or

$$y^2 + y - 1 = 0$$

This last equation can be solved by using the quadratic formula with $a = 1$, $b = 1$, and $c = -1$. As you can check, the results are

$$y = \frac{-1 + \sqrt{5}}{2} \quad \text{and} \quad y = \frac{-1 - \sqrt{5}}{2}$$

However, from Figure 2, it is clear that the $y$-coordinate at each intersection point is positive. Therefore, we discard the negative number $y = -\frac{1}{2}(1 + \sqrt{5})$ from further consideration in this context. Substituting the positive number

**FIGURE 2**

$y = \frac{1}{2}(-1 + \sqrt{5})$ back in the equation $y = x^2$ then gives us

$$x^2 = \frac{1}{2}(-1 + \sqrt{5})$$

Therefore,

$$x = \pm\sqrt{\frac{1}{2}(-1 + \sqrt{5})}$$

By choosing the positive square root, we obtain the x-coordinate for the intersection point in the first quadrant. That point is therefore

$$\left(\sqrt{\frac{1}{2}(-1 + \sqrt{5})}, \frac{1}{2}(-1 + \sqrt{5})\right) \approx (0.79, 0.62) \qquad \text{using a calculator}$$

Similarly, the negative square root yields the x-coordinate for the intersection point in the second quadrant. That point is

$$\left(-\sqrt{\frac{1}{2}(-1 + \sqrt{5})}, \frac{1}{2}(-1 + \sqrt{5})\right) \approx (-0.79, 0.62) \qquad \text{using a calculator}$$

We have now found the two intersection points, as required.

**EXAMPLE 3**   Find all solutions $(x, y)$ of the following system, where $x$ and $y$ are real numbers:

$$\begin{cases} xy = 1 & (3) \\ y = 3x + 1 & (4) \end{cases}$$

**Solution**   Since these equations are easy to graph, we do so, because that will tell us something about the required solutions. As Figure 3 indicates, there are two intersection points, one in the first quadrant, the other in the third quadrant. One way to begin now would be to solve equation (3) for one unknown in terms of the other. However, to avoid introducing fractions at the outset, let us use equation (4) to substitute for $y$ in equation (3). This yields

$$x(3x + 1) = 1$$

or

$$3x^2 + x - 1 = 0$$

This last equation can be solved by using the quadratic formula. As you should verify, the solutions are

$$x = \frac{-1 + \sqrt{13}}{6} \qquad \text{and} \qquad x = \frac{-1 - \sqrt{13}}{6}$$

The corresponding y-values can now be obtained by substituting for $x$ in either of the given equations. We will substitute in equation (4). (Exercise 23 at the end of this section asks you to substitute in equation (3) as well and then to show that the very different-looking answers obtained in that way are in fact equal to those found here.) Substituting $x = \frac{1}{6}(-1 + \sqrt{13})$ in the equation $y = 3x + 1$ gives us

$$y = 3\left(\frac{-1 + \sqrt{13}}{6}\right) + 1$$

$$= \frac{-1 + \sqrt{13}}{2} + \frac{2}{2} = \frac{-1 + \sqrt{13} + 2}{2}$$

$$= \frac{1 + \sqrt{13}}{2}$$

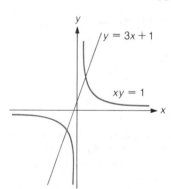

**FIGURE 3**

Thus, one of the intersection points is

$$(\tfrac{1}{6}(-1 + \sqrt{13}), \tfrac{1}{2}(1 + \sqrt{13}))$$

Notice that this must be the first-quadrant point of intersection, since both coordinates are positive. The intersection point in the third quadrant is obtained in exactly the same manner. As you can check, substituting the value $x = -\tfrac{1}{6}(1 + \sqrt{13})$ in the equation $y = 3x + 1$ yields $y = \tfrac{1}{2}(1 - \sqrt{13})$. Thus, the other intersection point is

$$(-\tfrac{1}{6}(1 + \sqrt{13}), \tfrac{1}{2}(1 - \sqrt{13}))$$

As Figure 3 indicates, there are no other solutions.

**EXAMPLE 4**  Find all real numbers $x$ and $y$ satisfying the system of equations

$$\begin{cases} y = \sqrt{x} \\ (x + 2)^2 + y^2 = 1 \end{cases}$$

**Solution**  We use the first equation to substitute for $y$ in the second equation. This yields

$$(x + 2)^2 + (\sqrt{x})^2 = 1$$
$$x^2 + 4x + 4 + x = 1$$
$$x^2 + 5x + 3 = 0$$

Using the quadratic formula to solve this last equation, we have

$$x = \frac{-5 \pm \sqrt{5^2 - 4(1)(3)}}{2(1)} = \frac{-5 \pm \sqrt{13}}{2}$$

The two values of $x$ are thus

$$-\tfrac{1}{2}(5 - \sqrt{13}) \qquad \text{and} \qquad -\tfrac{1}{2}(5 + \sqrt{13})$$

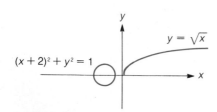

$(x + 2)^2 + y^2 = 1$

$y = \sqrt{x}$

**FIGURE 4**

However, notice that both of these quantities are negative. (The second is obviously negative; without using a calculator, can you explain why the first is negative?) Thus, neither of these quantities is an appropriate $x$-input in the equation $y = \sqrt{x}$, since we are looking for $y$-values that are real numbers. We conclude from this that there are no pairs of real numbers $x$ and $y$ satisfying the given system. In geometric terms, this means that the two graphs do not intersect. See Figure 4.

In the next example, we look at a system that can be reduced to a linear system through appropriate substitutions.

**EXAMPLE 5**  Solve the system

$$\begin{cases} \dfrac{2}{x^2} - \dfrac{3}{y^2} = -6 \\[2mm] \dfrac{3}{x^2} + \dfrac{4}{y^2} = 59 \end{cases}$$

**Solution**  Let $u = 1/x^2$ and $v = 1/y^2$, so that the system becomes

$$\begin{cases} 2u - 3v = -6 \\ 3u + 4v = 59 \end{cases}$$

This is now a linear system. As Exercise 46 asks you to show, the solution is $u = 9$, $v = 8$. In view of the definitions of $u$ and $v$, then, we have

$$\frac{1}{x^2} = 9 \qquad \qquad \frac{1}{y^2} = 8$$
$$x^2 = 1/9 \qquad \qquad y^2 = 1/8$$
$$x = \pm 1/3 \qquad \qquad y = \pm 1/\sqrt{8} = \pm 1/(2\sqrt{2})$$
$$= \pm\sqrt{2}/4$$

This gives us four possible solutions for the original system:

$$\left(\frac{1}{3}, \frac{\sqrt{2}}{4}\right) \qquad \left(\frac{1}{3}, -\frac{\sqrt{2}}{4}\right) \qquad \left(-\frac{1}{3}, \frac{\sqrt{2}}{4}\right) \qquad \left(-\frac{1}{3}, -\frac{\sqrt{2}}{4}\right)$$

As you can check, all four of these pairs satisfy the given system.

**EXAMPLE 6** Determine all solutions $(x, y)$ of the following system, where $x$ and $y$ are real numbers:

$$\begin{cases} y = 3^x & (5) \\ y = 3^{2x} - 2 & (6) \end{cases}$$

**Solution** We'll use the substitution method. First we rewrite equation (6) as

$$y = (3^x)^2 - 2 \qquad \qquad (7)$$

Now, in view of equation (5), we can replace $3^x$ with $y$ in equation (7) to obtain

$$y = y^2 - 2$$
$$0 = y^2 - y - 2$$
$$= (y + 1)(y - 2)$$

From this last equation, we see that $y = -1$ or $y = 2$. With $y = -1$, equation (5) becomes $-1 = 3^x$, contrary to the fact that $3^x$ is positive for all real numbers $x$. Thus, we discard the case where $y = -1$. On the other hand, if $y = 2$, equation (5) becomes

$$2 = 3^x$$

We can solve this exponential equation by taking the logarithm of both sides. Using base $e$ logarithms, we have

$$\ln 2 = \ln 3^x = x \ln 3$$

and consequently,

$$x = \frac{\ln 2}{\ln 3} \qquad \textit{Caution:} \quad \frac{\ln 2}{\ln 3} \neq \ln 2 - \ln 3.$$

We've now found that $x = (\ln 2)/(\ln 3)$ and $y = 2$. Figure 5 displays a graphical interpretation of this result. (Using a calculator for the $x$-coordinate, we find that $x \approx 0.6$, which is consistent with Figure 5.)

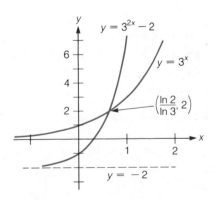

**FIGURE 5**

## EXERCISE SET 10.6

### A

In Exercises 1–22, find all solutions $(x, y)$ of the given systems, where $x$ and $y$ are real numbers.

1. $\begin{cases} y = 3x \\ y = x^2 \end{cases}$
2. $\begin{cases} y = x + 3 \\ y = 9 - x^2 \end{cases}$

3. $\begin{cases} x^2 + y^2 = 25 \\ 24y = x^2 \end{cases}$
4. $\begin{cases} 3x + 4y = 12 \\ x^2 - y + 1 = 0 \end{cases}$

5. $\begin{cases} xy = 1 \\ y = -x^2 \end{cases}$
6. $\begin{cases} x + 2y = 0 \\ xy = -2 \end{cases}$

7. $\begin{cases} 2x^2 + y^2 = 17 \\ x^2 + 2y^2 = 22 \end{cases}$
8. $\begin{cases} x - 2y = 1 \\ y^2 - x^2 = 3 \end{cases}$

9. $\begin{cases} y = 1 - x^2 \\ y = x^2 - 1 \end{cases}$
10. $\begin{cases} xy = 4 \\ y = 4x \end{cases}$

11. $\begin{cases} xy = 4 \\ y = 4x + 1 \end{cases}$
12. $\begin{cases} \dfrac{2}{x^2} + \dfrac{5}{y^2} = 3 \\ \dfrac{3}{x^2} - \dfrac{2}{y^2} = 1 \end{cases}$

13. $\begin{cases} \dfrac{1}{x^2} - \dfrac{3}{y^2} = 14 \\ \dfrac{2}{x^2} + \dfrac{1}{y^2} = 35 \end{cases}$
14. $\begin{cases} y = -\sqrt{x} \\ (x - 3)^2 + y^2 = 4 \end{cases}$

15. $\begin{cases} y = -\sqrt{x - 1} \\ (x - 3)^2 + y^2 = 4 \end{cases}$
16. $\begin{aligned} y &= -\sqrt{x - 6} \\ (x - 3)^2 + y^2 &= 4 \end{aligned}$

17. $\begin{cases} y = 2^x \\ y = 2^{2x} - 12 \end{cases}$
18. $\begin{cases} y = e^{4x} \\ y = e^{2x} + 6 \end{cases}$

19. $\begin{cases} 2(\log_{10} x)^2 - (\log_{10} y)^2 = -1 \\ 4(\log_{10} x)^2 - 3(\log_{10} y)^2 = -11 \end{cases}$

20. $\begin{cases} y = \log_2(x + 1) \\ y = 5 - \log_2(x - 3) \end{cases}$
21. $\begin{cases} 2^x \cdot 3^y = 4 \\ x + y = 5 \end{cases}$

22. $\begin{cases} a^{2x} + a^{2y} = 10 \\ a^{x+y} = 4 \end{cases} \qquad (a > 0)$

Hint: Use the substitutions $a^x = t$ and $a^y = u$.

23. Let $x = \frac{1}{6}(-1 + \sqrt{13})$. Using this $x$-value, show that the equations $y = 3x + 1$ and $y = 1/x$ yield the same $y$-value.

24. A sketch shows that the line $y = 100x$ intersects the parabola $y = x^2$ at the origin. Are there any other intersection points? If so, find them. If not, explain why not.

25. Solve the following system for $x$ and $y$:

$$\begin{cases} ax + by = 2 \\ \quad abxy = 1 \end{cases}$$

(Assume that neither $a$ nor $b$ is zero.)

### B

26. Let $a$, $b$, and $c$ be constants (with $a \neq 0$), and consider the system

$$\begin{cases} y = ax^2 + bx + c \\ y = k \end{cases}$$

For which value of $k$ (in terms of $a$, $b$, and $c$) will the system have exactly one solution? What is that solution? What is the relationship between the solution you've found and the graph of $y = ax^2 + bx + c$?

27. Find all solutions of the system

$$\begin{cases} x^3 + y^3 = 3473 \\ \quad x + y = 23 \end{cases}$$

28. Solve the following system for $x$, $y$, and $z$:

$$\begin{cases} yz = p^2 \\ zx = q^2 \\ xy = r^2 \end{cases}$$

(Assume that $p$, $q$, and $r$ are positive constants.)

29. If the diagonal of a rectangle has length $d$ and the perimeter of the rectangle is $2p$, express the lengths of the sides in terms of $d$ and $p$.

30. Solve the following system for $x$ and $y$:

$$\begin{cases} xy + pq = 2px \\ x^2y^2 + p^2q^2 = 2q^2y^2 \end{cases} \qquad (pq \neq 0)$$

Hint: Square the first equation, then substract the second. This results in an equation that can be written $(2px + qy)(px - qy) = 0$.

31. Solve the following system for $u$ and $v$:

$$\begin{cases} u^2 - v^2 = 9 \\ \sqrt{u + v} + \sqrt{u - v} = 4 \end{cases}$$

Hint: Square the second equation.

32. Solve the given system for $x$ and $y$ in terms of $p$ and $q$:

$$\begin{cases} q^{\ln x} = p^{\ln y} \\ (px)^{\ln a} = (qy)^{\ln b} \end{cases}$$

(Assume that all the constants and variables are positive.)

33. Solve the following system for $x$, $y$, and $z$ in terms of $p$, $q$, and $r$:

$$\begin{cases} x(x + y + z) = p^2 \\ y(x + y + z) = q^2 \\ z(x + y + z) = r^2 \end{cases}$$

(Assume that $p$, $q$, and $r$ are nonzero.)   *Hint:*   Denote $x + y + z$ by $w$; then add the three equations.

34. (a) Find the points where the line $y = -2x - 2$ intersects the parabola $y = \frac{1}{2}x^2$.
    (b) On the same set of axes, sketch the line $y = -2x - 2$ and the parabola $y = \frac{1}{2}x^2$. Be certain that your sketch is consistent with the results obtained in part (a).

35. The area of a right triangle is $180 \, \text{cm}^2$, and the hypotenuse is $41 \, \text{cm}$. Find the lengths of the two legs.

36. The sum of two numbers is 8, while their product is $-128$. What are the two numbers?

37. The perimeter of a rectangle is $16 \, \text{cm}$, and the area is $60 \, \text{cm}^2$. Find the length and the width.

38. Find all right triangles for which the perimeter is 24 units and the area is 24 square units.

39. Solve the following system for $x$ and $y$ using the substitution method:
$$\begin{cases} x^2 + y^2 = 5 & (1) \\ xy = 2 & (2) \end{cases}$$

40. The substitution method in Exercise 39 leads to a quadratic equation. Here is an alternative approach to solving that system; this approach leads to linear equations. Multiply equation (2) by 2 and add the resulting equation to equation (1). Now take square roots to conclude that $x + y = \pm 3$. Next multiply equation (2) by 2 and subtract the resulting equation from equation (1). Take square roots to conclude that $x - y = \pm 1$. You now have the following four linear systems, each of which can be solved (with almost no work) by the addition–subtraction method:

$$\begin{cases} x + y = 3 \\ x - y = 1 \end{cases} \qquad \begin{cases} x + y = 3 \\ x - y = -1 \end{cases}$$

$$\begin{cases} x + y = -3 \\ x - y = 1 \end{cases} \qquad \begin{cases} x + y = -3 \\ x - y = -1 \end{cases}$$

Solve these systems and compare your results with those obtained in Exercise 39.

41. Solve the following system using the method explained in Exercise 40:
$$\begin{cases} x^2 + y^2 = 7 \\ xy = 3 \end{cases}$$

42. Solve the following system using the substitution method:
$$\begin{cases} 3xy - 4x^2 = 2 \\ -5x^2 + 3y^2 = 7 \end{cases}$$

(Begin by solving the first equation for $y$.)

43. Here is an alternative approach for solving the system in Exercise 42. Let $y = mx$, where $m$ is a constant to be determined. Replace $y$ with $mx$ in both equations of the system to obtain the following pair of equations:

$$x^2(3m - 4) = 2 \qquad\qquad (1)$$
$$x^2(-5 + 3m^2) = 7 \qquad\qquad (2)$$

Now divide equation (2) by (1). After clearing fractions and simplifying, you can write the resulting equation as $2m^2 - 7m + 6 = 0$. Solve this last equation by factoring. The values of $m$ can then be used in equation (1) to determine values for $x$. In each case, the corresponding $y$-values are determined by the equation $y = mx$.

## C

44. Solve the following system for $x$ and $y$:
$$\begin{cases} \dfrac{1}{x^2} + \dfrac{1}{xy} = \dfrac{1}{a^2} \\ \dfrac{1}{y^2} + \dfrac{1}{xy} = \dfrac{1}{b^2} \end{cases}$$

(Assume that $a$ and $b$ are positive.)

45. Solve the following system for $x$ and $y$:
$$\begin{cases} x^4 = y^6 \\ \ln \dfrac{x}{y} = \dfrac{\ln x}{\ln y} \end{cases}$$

46. Solve the following system. (You should obtain $u = 9$ and $v = 8$, as stated in Example 5.)
$$\begin{cases} 2u - 3v = -6 \\ 3u + 4v = 59 \end{cases}$$

## SYSTEMS OF INEQUALITIES. LINEAR PROGRAMMING

### 10.7

In the first section of this chapter, we solved systems of equations in two unknowns. Now let us consider systems of inequalities in two unknowns. The techniques we develop are used in calculus when discussing functions of two variables, and in business and economics in the study of linear programming.

Let $a$, $b$, and $c$ denote real numbers, and assume that $a$ and $b$ are not both zero. Then all of the following are called **linear inequalities**:

$$ax + by + c < 0 \qquad ax + by + c > 0$$
$$ax + by + c \le 0 \qquad ax + by + c \ge 0$$

An ordered pair of numbers $(x_0, y_0)$ is said to be a **solution** of a given inequality (linear or not) provided we obtain a true statement upon substituting $x_0$ and $y_0$ for $x$ and $y$, respectively. For instance, $(-1, 1)$ is a solution of the inequality $2x + 3y - 6 < 0$, since substitution yields

$$2(-1) + 3(1) - 6 < 0$$
$$-5 < 0 \qquad \text{True}$$

As with a linear equation in two unknowns, a linear inequality has infinitely many solutions. For this reason, we often represent the solutions graphically. When we do this, we say that we are **graphing the inequality**.

For example, let us graph the inequality $y < 2x$. First we observe that the coordinates of the points on the line $y = 2x$ do not, by definition, satisfy this inequality. So it remains to consider points above the line and points below the line. We will in fact show that the required graph consists of all points that lie below the line. Take any point $P(x_0, y_0)$ not on the line $y = 2x$. Let $Q(x_0, 2x_0)$ be the point on $y = 2x$ with the same first coordinate as $P$. Then, as indicated in Figure 1, the point $P$ lies below the line if and only if the $y$-coordinate of $P$ is less than the $y$-coordinate of $Q$. In other words, $(x_0, y_0)$ lies below the line if and only if $y_0 < 2x_0$. This last statement is equivalent to saying that $(x_0, y_0)$ satisfies the inequality $y < 2x$. This shows that the graph of $y < 2x$ consists of all points below the line $y = 2x$ (see Figure 2). The broken line in Figure 2 indicates that the points on $y = 2x$ are not part of the required graph. If the original inequality had been $y \le 2x$, then we would use a solid line rather than a broken one, indicating that the line is included in the graph. And if the original inequality had been $y > 2x$, the graph would be the region above the line.

Just as the graph of $y < 2x$ is the region below the line $y = 2x$, it is true in general that the graph of $y < f(x)$ is the region below the graph of the function $f$. For example, Figures 3 and 4 display the graphs of $y < x^2$ and $y \ge x^2$.

**FIGURE 1**

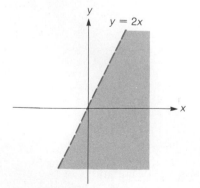

**FIGURE 2** The graph of $y < 2x$

**FIGURE 3**

**FIGURE 4**

Example 1 summarizes the technique developed so far for graphing an inequality. Following this example, we will point out a useful alternative method.

**EXAMPLE 1**   Graph the inequality $4x - 3y \leq 12$.

**Solution**   The graph will include the line $4x - 3y = 12$ and either the region above or the region below the line. To decide which region, we solve the inequality for $y$:

$$4x - 3y \leq 12$$
$$-3y \leq -4x + 12$$
$$y \geq \frac{4}{3}x - 4 \qquad \text{multiplying or dividing by a negative} \atop \text{number reverses an inequality}$$

This last inequality tells us that we want the region above the line, as well as the line itself. Figure 5 displays the required graph.

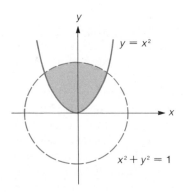

**FIGURE 5**   The graph of $4x - 3y \leq 12$.

There is another method that we can use in Example 1 to determine which side of the line we want. This method involves a **test point**. We pick any convenient point that is not on the line $4x - 3y = 12$. Then we test to see if this point satisfies the given inequality. For example, let us pick the point $(0,0)$. Substituting these coordinates in the inequality $4x - 3y \leq 12$ yields

$$4(0) - 3(0) \overset{?}{\leq} 12$$
$$0 \overset{?}{\leq} 12 \qquad \text{True}$$

We conclude from this that the required side of the line is that on which the point $(0,0)$ resides. In agreement with Figure 5, we see this is the region above the line.

Next, we discuss systems of inequalities in two unknowns. As with systems of equations, a **solution** of a system of inequalities is an ordered pair $(x_0, y_0)$ that satisfies all of the inequalities in the system. As a first example, let us graph the points that satisfy the following nonlinear system:

$$\begin{cases} y - x^2 \geq 0 \\ x^2 + y^2 < 1 \end{cases}$$

By writing the first inequality as $y \geq x^2$, we see that it describes the set of points on or above the parabola $y = x^2$. For the second inequality, we must decide whether it describes the points within or outside the circle $x^2 + y^2 = 1$. One way to do this is to choose $(0,0)$ as a test point. Substituting the values $x = 0$ and $y = 0$ in the inequality $x^2 + y^2 < 1$ yields the true statement $0^2 + 0^2 < 1$. Since $(0,0)$ lies within the circle and satisfies the inequality, we conclude that $x^2 + y^2 < 1$ describes the set of all points within the circle. Now we put our information together. We wish to graph the points that lie on or above the parabola $y = x^2$ but within the circle $x^2 + y^2 = 1$. This is the shaded region shown in Figure 6.

**FIGURE 6**

**EXAMPLE 2** Graph the system

$$\begin{cases} -x + 3y \le 12 \\ x + y \le 8 \\ x \ge 0 \\ y \ge 0 \end{cases}$$

**Solution** Solving the first inequality for $y$, we have

$$3y \le x + 12$$
$$y \le \tfrac{1}{3}x + 4$$

Thus, the first inequality is satisfied by the points on or below the line $-x + 3y = 12$. Similarly, by solving the second inequality for $y$, we see that it describes the set of points on or below the line $x + y = 8$. The third inequality, $x \ge 0$, describes the points on or to the right of the $y$-axis. Similarly, the fourth inequality, $y \ge 0$, describes the points on or above the $x$-axis. In Figure 7(a), we summarize these statements. The arrows indicate which side of each line we wish to consider. Finally, Figure 7(b) shows the graph of the given system. The coordinates of each point in the shaded region satisfy all four of the given inequalities. (Exercise 25 at the end of this section asks you to show that the coordinates listed in the figure are correct.)

Convex set

(a)

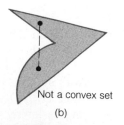

Not a convex set

(b)

**FIGURE 8**

(a)

(b)

**FIGURE 7**

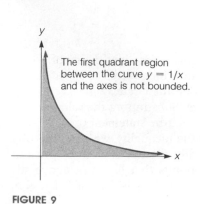

The first quadrant region between the curve $y = 1/x$ and the axes is not bounded.

**FIGURE 9**

We can use Figure 7(b) to introduce some terminology that is useful in describing sets of points in the plane. The **vertices** of a region are the corners, or points, where the adjacent bounding sides meet. Thus, the vertices of the shaded region in Figure 7(b) are the four points $(0,0)$, $(8,0)$, $(3,5)$, and $(0,4)$. The shaded region in Figure 7(b) is **convex**. This means that given any two points in that region, the straight line segment joining these two points lies wholly within the region. Figure 8(a) displays another example of a convex set, while Figure 8(b) shows a set that is not convex. The shaded region in Figure 8(b) is also an example of a **bounded region**. By this we mean that the region can be wholly contained within some (sufficiently large) circle. Perhaps the simplest example of a region that is not bounded is the entire $x$-$y$ plane itself. Figure 9 shows another example of an unbounded region.

We are prepared now to consider a typical problem in linear programming.

PROBLEM   Let $C = x + 3y$. Find the values of $x$ and $y$ that yield the largest possible value of $C$, given that $x$ and $y$ satisfy the following set of inequalities.

$$\begin{cases} -x + 3y \le 12 \\ x + y \le 8 \\ x \ge 0 \\ y \ge 0 \end{cases}$$

The function $C = x + 3y$ in this problem is known as the **objective function**. The given inequalities are called the **constraints**. In analyzing this problem we will take advantage of the fact that we've already graphed this system of inequalities in Example 2. For convenience, we repeat the graph here (Figure 10). Since the coordinates of each point in the shaded region of Figure 10 satisfy the constraints of the present problem, we refer to the region as the set of **feasible solutions**. The problem then is to find that point among the feasible solutions that yields the largest possible value for $C$. A feasible solution $(x_0, y_0)$ that yields the largest value for $C$ is called an **optimal solution**. Some linear programming problems ask for a smallest rather than a largest value of the objective function. A feasible solution yielding the smallest value for $C$ is also called an optimal solution.

We will see subsequently that there is a general method for solving this type of linear programming problem, but as a start, let us experiment. How large can the value of $C$ be, subject to the given constraints? For instance, can $C$ equal 30? With $C = 30$, we have

$$x + 3y = 30$$

Figure 11 shows the graph of this line. Since the line has no points in common with the set of feasible solutions, we conclude that $C$ cannot equal 30 under the conditions of the problem.

Still experimenting, in Figure 12 we have graphed the lines $x + 3y = C$ for various values of $C$. Observe that the lines are all parallel. To see why this is so, solve the equation $x + 3y = C$ for $y$. This yields

$$y = -\frac{1}{3}x + \frac{1}{3}C$$

FIGURE 10

FIGURE 11

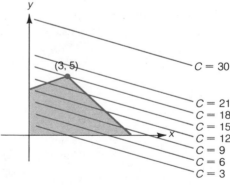

FIGURE 12

As you can see, the slope here is $m = -\frac{1}{3}$, independent of $C$. Since the lines all have the same slope, they are parallel. On the other hand, the equation $y = -\frac{1}{3}x + \frac{1}{3}C$ shows how the value of $C$ does influence the graph. The $y$-intercept of the line is $C/3$. So as $C$ increases, the line moves *up*, but the slope is always $-\frac{1}{3}$.

Now as you can see in Figure 12, with $C = 18$ the line $x + 3y = 18$ intersects the set of feasible solutions at the vertex $(3, 5)$. However, it is also evident from the figure and the remarks in the previous paragraph that when $C$ is greater than 18, the line $x + 3y = C$ will not intersect the set of feasible solutions. We conclude therefore that the vertex $(3, 5)$ is the optimal solution. That is, subject to the given constraints, the value of $C$ is largest when $x = 3$ and $y = 5$.

In the example just concluded, we found that the optimal solution to the linear programming problem occurred at a vertex. The following theorem (stated here without proof) tells us that the optimal solution, when it exists, always occurs at a vertex. (In the statement of the theorem, the word "nonstrict" refers to inequalities that involve the symbols $\leq$ or $\geq$ rather than $<$ or $>$.)

---

**Linear Programming Theorem**

Consider the objective function $C = ax + by + c$, where the nonnegative variables $x$ and $y$ are subject to constraints specified by nonstrict linear inequalities. Then if $C$ assumes a maximum (or minimum) value, this occurs at a vertex. Furthermore, if the set of feasible solutions is both convex and bounded, then a maximum and minimum always exist.

---

In view of this theorem, we can solve linear programming problems by first finding the vertices of the feasibility set and then computing and comparing the values of the objective function at those vertices. This is done in Example 3, which follows. Technically speaking, the theorem we've stated guarantees a maximum and a minimum only when the feasibility set is convex and bounded. However, this is not as great a restriction as it might at first seem. As it turns out, the graph of any set of linear inequalities forms a convex set. (For a look at a case in which the feasibility set is not bounded, see Exercise 44.)

**EXAMPLE 3** Let $C = 3x - 2y$. Find the maximum and the minimum value of $C$ subject to the following set of constraints.

$$\begin{cases} 6y - x \geq 5 \\ 4x + 3y \leq 34 \\ 2y - x \leq 8 \\ 4x - y \geq 3 \end{cases}$$

What are the optimal solutions in each case?

**Solution** In Figure 13, we have grapned the given system of inequalities using the method discussed in Example 2. Now we need to determine the coordinates of the four vertices. To find the coordinates of $E$, we solve the system

$$\begin{cases} 6y - x = 5 \\ 4x - y = 3 \end{cases} \tag{1}$$
$$\tag{2}$$

Solving equation (1) for $x$, we have

$$6y - 5 = x \tag{3}$$

Now we use equation (3) to substitute for $x$ in equation (2). This yields

$$4(6y - 5) - y = 3$$
$$24y - 20 - y = 3$$
$$23y = 23$$
$$y = 1$$

Finally, substituting $y = 1$ in equation (3) gives us $x = 6(1) - 5 = 1$.

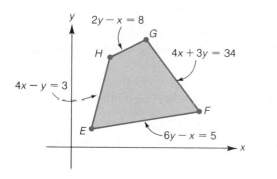

FIGURE 13                FIGURE 14

**TABLE 1**

| VERTEX | $C = 3x - 2y$ |
|--------|---------------|
| $E(1,1)$ | $3(1) - 2(1) = \phantom{0}1$ |
| $F(7,2)$ | $3(7) - 2(2) = 17$ |
| $G(4,6)$ | $3(4) - 2(6) = \phantom{0}0$ |
| $H(2,5)$ | $3(2) - 2(5) = -4$ |

We have now found that the coordinates of the vertex $E$ are $(1,1)$. The coordinates of $F$, $G$, and $H$ can be found in a similar fashion. The results are displayed in Figure 14. Notice that the set of feasible solutions forms a convex and bounded set. Thus the objective function attains a maximum and a minimum value at certain vertices. We compute the value of $C$ at each vertex and tabulate the results as shown in Table 1.

The largest value of $C$ in this table is 17. This is the required maximum. The corresponding optimal solution is $(7,2)$. Similarly, the smallest value of $C$ is $-4$. This is the required minimum. The corresponding optimal solution is $(2,5)$. ∎

The next example deals with a type of linear programming problem known as the **transportation problem**. In this type of problem we are required to minimize the cost of shipping goods from warehouses to retail outlets.

**EXAMPLE 4**   A motorcycle manufacturer must fill orders from two dealers. The first dealer, $D_1$, has ordered 20 motorcycles, while the second dealer, $D_2$, has ordered 30 motorcycles. The manufacturer has the motorcycles stored in two warehouses, $W_1$ and $W_2$. There are 40 motorcycles in $W_1$ and 15 in $W_2$. The shipping costs, per motorcycle, are as follows: \$15 from $W_1$ to $D_1$; \$13 from $W_1$ to $D_2$; \$14 from $W_2$ to $D_1$; \$16 from $W_2$ to $D_2$. Under these conditions, find the number of motorcycles to be shipped from each warehouse to each dealer if the total shipping cost is to be a minimum. What is this minimum cost?

**Solution**  The first step is to summarize the data. We do this as follows.

| SHIPPING COST PER MOTORCYCLE |
| --- |
| $W_1$ to $D_1$:   $15 |
| $W_1$ to $D_2$:   $13 |
| $W_2$ to $D_1$:   $14 |
| $W_2$ to $D_2$:   $16 |

| SUPPLY AND DEMAND | |
| --- | --- |
| $W_1$<br>(40 motorcycles<br>in stock) | $W_2$<br>(15 motorcycles<br>in stock) |
| $D_1$<br>(wants 20<br>motorcycles) | $D_2$<br>(wants 30<br>motorcycles) |

The problem asks how many motorcycles should be shipped from each warehouse to each dealer.

1. Let $x$ denote the number of motorcycles to be shipped from $W_1$ to $D_1$.
2. Then $20 - x$ denotes the number of motorcycles to be shipped from $W_2$ to $D_1$. (Why?)
3. Let $y$ denote the number of motorcycles to be shipped from $W_1$ to $D_2$.
4. Then $30 - y$ denotes the number of motorcycles to be shipped from $W_2$ to $D_2$.

If $C$ denotes the total shipping cost, we then have

As you can check, this equation simplifies to

$$C = x - 3y + 760$$

This is the objective function that we must minimize.

Now let us examine the constraints of the problem. The number of items shipped from the warehouse certainly must be a nonnegative number in each case. Thus we must have

$$x \geq 0$$
$$y \geq 0$$
$$20 - x \geq 0$$
$$30 - y \geq 0$$

Furthermore, since at most 40 motorcycles can be shipped from $W_1$, we must have

$$x + y \leq 40$$

Similarly, since at most 15 motorcycles can be shipped from $W_2$, we have

$$(20 - x) + (30 - y) \leq 15$$

or

$$50 - x - y \le 15$$
$$-x - y \le -35$$
$$x + y \ge 35$$

In summary, then, we have the following set of constraints:

$$\begin{cases} x \ge 0 \\ y \ge 0 \\ 20 - x \ge 0 \\ 30 - y \ge 0 \\ x + y \le 40 \\ x + y \ge 35 \end{cases}$$

Figure 15 shows the graph of the region described by this set of constraints. (Exercise 33 at the end of this section asks you to verify that the graph and the coordinates of the vertices are as indicated.) Since the region is both convex and bounded, we know that the objective function assumes its minimum value at one of the vertices. We compute the value of $C$ at each vertex.

| VERTEX | $C = x - 3y + 760$ |
|--------|--------------------|
| $(20, 15)$ | $C = 20 - 3(15) + 760 = 735$ |
| $(20, 20)$ | $C = 20 - 3(20) + 760 = 720$ |
| $(10, 30)$ | $C = 10 - 3(30) + 760 = 680$ |
| $(5, 30)$ | $C = 5 - 3(30) + 760 = 675$ |

As the table shows, the minimum shipping cost is \$675. The corresponding vertex is $(5, 30)$. Therefore, the shipping cost will be minimized if the deliveries are made as follows.

$x = 5$:      ship 5 motorcycles from $W_1$ to $D_1$

$20 - x = 15$:    ship 15 motorcycles from $W_2$ to $D_1$

$y = 30$:     ship 30 motorcycles from $W_1$ to $D_2$

$30 - y = 0$:     ship no motorcycles from $W_2$ to $D_2$

The linear programming problems that we have considered have involved only two unknowns, $x$ and $y$. Because there were only two unknowns, we were able to utilize graphical methods in solving the problems. As you might guess, linear programming problems actually arising in business and industry may involve dozens or even thousands of unknowns. For such cases, there is a purely algebraic method of solution called the **simplex method**, which makes use of matrices. Furthermore, this method can be programmed for computer solution. Several commercial firms now market computer packages for solving linear programming problems.*

The set of feasible solutions

**FIGURE 15**

---

## EXERCISE SET 10.7

### A

In Exercises 1 and 2, decide whether or not the ordered pairs are solutions of the given inequality.

1. $4x - 6y + 3 \geq 0$   (a) $(1,2)$   (b) $(0,\frac{1}{2})$
2. $5x + 2y < 1$   (a) $(-1,3)$   (b) $(0,0)$

In Exercises 3–16, graph the given inequalities.

3. $2x - 3y > 6$         4. $2x - 3y < 6$
5. $2x - 3y \geq 6$         6. $2x + 3y \leq 6$
7. $x - y < 0$           8. $y \leq \frac{1}{2}x - 1$
9. $x \geq 1$             10. $y < 0$
11. $x > 0$            12. $y \leq \sqrt{x}$
13. $y > x^3 + 1$       14. $y \leq |x - 2|$
15. $x^2 + y^2 \geq 25$     16. $y \geq e^x - 1$

In Exercises 17–22, graph the systems of inequalities.

17. $\begin{cases} y \leq x^2 \\ x^2 + y^2 \leq 1 \end{cases}$     18. $\begin{cases} y < x \\ x^2 + y^2 < 1 \end{cases}$

19. $\begin{cases} y \geq 1 \\ y \leq |x| \end{cases}$     20. $\begin{cases} x \geq 0 \\ y \geq 0 \\ y < \sqrt{x} \\ x \leq 4 \end{cases}$

21. $\begin{cases} x \geq 0 \\ y \geq 0 \\ y \leq 1 - x^2 \end{cases}$     22. $\begin{cases} y < 2x \\ y > \frac{1}{2}x \end{cases}$

In Exercises 23–34, graph the systems of linear inequalities. In each case specify the vertices. Is the region convex? Is the region bounded?

23. $\begin{cases} y \leq x + 5 \\ y \leq -2x + 14 \\ x \geq 0 \\ y \geq 0 \end{cases}$     24. $\begin{cases} y \geq x + 5 \\ y \geq -2x + 14 \\ x \geq 0 \\ y \geq 0 \end{cases}$

25. $\begin{cases} -x + 3y \leq 12 \\ x + y \leq 8 \\ x \geq 0 \\ y \geq 0 \end{cases}$     26. $\begin{cases} y \geq 2x \\ y \geq -x + 6 \end{cases}$

27. $\begin{cases} 0 \leq 2x - y + 3 \\ x + 3y \leq 23 \\ 5x + y \leq 45 \end{cases}$     28. $\begin{cases} 0 \leq 2x - y + 3 \\ x + 3y \leq 23 \\ 5x + y \leq 45 \\ x \geq 0 \\ y \geq 0 \end{cases}$

29. $\begin{cases} 5x + 6y < 30 \\ y > 0 \end{cases}$     30. $\begin{cases} 5x + 6y < 30 \\ x > 0 \end{cases}$

31. $\begin{cases} 5x + 6y < 30 \\ x > 0 \\ y > 0 \end{cases}$     32. $\begin{cases} 2x + 3y \geq 6 \\ 2x + 3y \leq 12 \end{cases}$

33. $\begin{cases} x \geq 0 \\ y \geq 0 \\ 20 - x \geq 0 \\ 30 - y \geq 0 \\ x + y \leq 40 \\ x + y \geq 35 \end{cases}$     34. $\begin{cases} x \geq 0 \\ y \geq 0 \\ 3x - y + 1 \geq 0 \\ 0 \leq x - y + 3 \\ y \leq 5 \\ x \leq \frac{1}{3}(17 - y) \\ x \leq \frac{1}{2}(y + 8) \end{cases}$

In Exercises 35–43, find the maximum and the minimum values of the objective function subject to the given constraints. Specify the corresponding optimal solutions in each case.

35. Objective function:   $C = 3y - x$

   Constraints: $\begin{cases} y \leq x + 1 \\ y \leq -2x + 10 \\ x \geq 0 \\ y \geq 0 \end{cases}$

36. Objective function:   $C = 2x - y + 8$

   Constraints: $\begin{cases} 4x + 3y \geq 12 \\ 6x + 5y \leq 30 \\ x \geq 0 \\ y \geq 0 \end{cases}$

37. Objective function:   $C = 2x + y$

   Constraints: $\begin{cases} x \geq 0 \\ x \leq 1 \\ y \geq 0 \\ 2y - x \leq 3 \end{cases}$

38. Objective function:   $C = 2x - y$
   Constraints:   Use the same constraints given in Exercise 37.

39. Objective function:   $C = 10y + 9x - 1$

   Constraints: $\begin{cases} x + y \geq 3 \\ y - x + 1 \geq 0 \\ y \leq -x + 12 \\ y \leq 6 \\ x \geq 1 \end{cases}$

40. Objective function:   $C = 10y + 11x - 1$
   Constraints:   Use the same constraints given in Exercise 39.

**41.** Objective function:   $C = 10y + 3x + 100$

Constraints: $\begin{cases} 0 \le 4x - y + 4 \\ x + 5y \le 41 \\ x + 3y \le 27 \\ 2x + y \le 24 \\ x \ge 0 \\ y \ge 0 \end{cases}$

**42.** Objective function:   $C = 20y + 7x + 100$
Constraints:   Use the same ones given in Exercise 41.

**43. (a)** Objective function:   $C = 19x + 100y$
Constraints:   Use the same constraints given in Exercise 41.
**(b)** Objective function:   $C = 21x + 100y$
Constraints:   Use the same constraints given in Exercise 41.
(This exercise shows that a slight change in the objective function may yield a relatively large change in the optimal solution.)

**44.** Consider the objective function $C = 4y - x$ subject to the constraints

$\begin{cases} y \ge -\frac{3}{2}x + 4 \\ 2y \ge x \\ y \ge 0 \\ x \ge 0 \end{cases}$

**(a)** Given that there is an optimal solution yielding a minimum value of $C$, find that solution and the corresponding value of $C$.
**(b)** Find a feasible solution for which $C = 100$.
**(c)** Find a feasible solution for which $C = 1000$.
**(d)** Find a feasible solution for which $C = 10^6$.

**45.** An automobile manufacturer must fill orders from two dealers. The first dealer, $D_1$, has ordered 40 cars, while the second dealer, $D_2$, has ordered 25 cars. The manufacturer has the cars stored in two locations, $W_1$ and $W_2$. There are 30 cars in $W_1$ and 50 cars in $W_2$. The shipping costs per car are as follows: \$180 from $W_1$ to $D_1$; \$150 from $W_1$ to $D_2$; \$160 from $W_2$ to $D_1$; \$170 from $W_2$ to $D_2$. Under these conditions, how many cars should be shipped from each storage location to each dealer so as to minimize the total shipping cost? What is this minimum cost?

**46.** A television manufacturer must fill orders from two retailers. The first retailer, $R_1$, has ordered 55 television sets, while the second retailer, $R_2$, has ordered 75 sets. The manufacturer has the television sets stored in two warehouses, $W_1$ and $W_2$. There are 100 sets in $W_1$ and 120 sets in $W_2$. The shipping costs, per television set, are as follows: \$8 from $W_1$ to $R_1$; \$12 from $W_1$ to $R_2$; \$13 from $W_2$ to $R_1$; \$7 from $W_2$ to $R_2$. Under these con-

ditions, find the number of television sets to be shipped from each warehouse to each retailer if the total shipping cost is to be a minimum. What is this minimum cost?

**47.** A factory produces two products, $A$ and $B$. The production time (in hours) for each unit is as follows.

| PRODUCT | ASSEMBLY | INSPECTION AND PACKING |
|---------|----------|------------------------|
| $A$ | 0.40 | 0.20 |
| $B$ | 0.50 | 0.40 |

Each day at most 500 man-hours of assembly time and 300 man-hours of inspection/packing time are available. The profit realized on each unit of $A$ is \$0.60; on each unit of $B$ it is \$0.80. Find the maximum daily profit under these conditions.   *Hints:*   Let $x$ denote the number of units of $A$ produced each day and let $y$ denote the number of units of $B$ produced each day. After graphing the constraints you'll find that the coordinates of one vertex are not whole numbers. Round these off to the nearest whole number. (Why?)

**48.** Due to increased energy costs, the owner of the factory described in Exercise 47 decides to limit production. In particular, the owner decides that no more than 800 units of $A$ and 500 units of $B$ should be produced daily. Under these additional constraints, how many units of $A$ and of $B$ should be produced daily to maximize profits? What is the maximum profit?

**49.** A farmer can allocate at most 600 acres for growing cherry tomatoes and regular tomatoes. The labor requirements and the profits are as follows.

|  | CHERRY TOMATOES | REGULAR TOMATOES |
|--|-----------------|------------------|
| **Labor** (hours per acre) | 3 | 2 |
| **Profit** (dollars per acre) | 50 | 36 |

What is the farmer's maximum profit if 1350 hr of labor are available? How many acres should be devoted to each crop to achieve the maximum profit?

**50.** Consider the situation described in Exercise 49. Now additionally suppose that due to market considerations, the farmer decides to devote at most 550 acres to regular tomatoes and at most 100 acres to cherry tomatoes. What is the farmer's maximum profit in this case?

**51.** Suppose that you inherit \$12,000 but there are stipulations attached. Some (or all) of the money must be invested in two stocks, $A$ and $B$. Stock $A$ is low-risk and $B$ is medium-risk. The stipulations require that at most

$6,000 be invested in $A$ and at least $2,000 be invested in $B$. Furthermore, the amount invested in $B$ is not to exceed twice the amount invested in $A$. The expected returns on the stocks are $0.06 per dollar for $A$ and $0.08 per dollar for $B$. Under these conditions, how much should you invest in each stock so as to maximize your expected returns?

52. You have $10,000 to invest in three stocks, $A$, $B$, and $C$. At least $1,000 must be invested in each stock. The combined investment in $A$ and $B$ must be at least $5,000, while the investment in $B$ cannot exceed five times that in $A$. The expected returns on the stocks are as follows: $0.04 per dollar for $A$; $0.05 per dollar for $B$; $0.06 per dollar for $C$. Under these conditions, how much should you invest in each stock so as to maximize the expected returns? *Hint:* Let $x$ and $y$ denote the dollar amounts invested in $A$ and $B$, respectively. Then $10,000 - x - y$ denotes the number of dollars invested in $C$.

53. In Exercise 52, suppose instead that the expected returns are $0.05 per dollar on $A$ and $0.04 per dollar on $B$. The return on $C$ remains at $0.06 per dollar. Now how much should you invest in each stock so as to maximize the expected returns?

54. In Exercise 52, suppose instead that the expected return on $B$ is $0.07 per dollar. The expected returns on $A$ and $C$ remain the same. Now how much should you invest to maximize the expected returns?

55. A veterinarian wishes to prepare a supply of dog food by mixing two commercially available foods, $A$ and $B$. The mixture is to contain at least 50 units of carbohydrates, at least 36 units of protein, and at least 40 units of fat. The specifications for the two commercially available products are listed in the following table.

| PRODUCT | UNITS OF CARBOHYDRATES (PER POUND) | UNITS OF PROTEIN (PER POUND) | UNITS OF FAT (PER POUND) |
|---|---|---|---|
| $A$ | 3 | 2 | 2 |
| $B$ | 5 | 4 | 8 |

The veterinarian can purchase food $A$ for $0.44 per pound and $B$ for $0.80 per pound. How many pounds of each food should be mixed to meet the dietary requirements at the least cost?

## CHAPTER 10   SUMMARY OF PRINCIPAL TERMS

| TERMS | PAGE REFERENCE | COMMENT |
|---|---|---|
| 1. Linear equation in two variables | 492 | A linear equation in two variables is an equation of the form $ax + by = c$, where $a$, $b$, and $c$ are constants and $x$ and $y$ are variables or unknowns. Similarly, a linear equation in three variables is an equation of the form $ax + by + cz = d$. |
| 2. Solution of a linear equation | 492 | A solution of the linear equation $ax + by = c$ is an ordered pair of numbers $(x_0, y_0)$ such that $ax_0 + by_0 = c$. Similarly, a solution of the linear equation $ax + by + cz = d$ is an ordered triple of numbers $(x_0, y_0, z_0)$ such that $ax_0 + by_0 + cz_0 = d$. |
| 3. Consistent system; inconsistent system | 493 | A system of equations is *consistent* if it has at least one solution; otherwise, the system is *inconsistent*. See Figures 2, 3, and 4 in Section 10.1 for a geometric interpretation of these terms. |
| 4. Upper-triangular form (three variables) | 503 | A system of linear equations in $x$, $y$, and $z$ is said to be in *upper-triangular form* if $x$ appears in no equation after the first, and $y$ appears in no equation after the second. This definition can be extended to include systems with any number of unknowns. See Table 1 in Section 10.2 for examples of systems that are in upper-triangular form. When a system is in upper-triangular form, it is a simple matter to obtain the solutions; see, for instance, Examples 1 and 2 in Section 10.2. |
| 5. Elementary operations | 505 | These are operations that can be performed on an equation in a system without altering the solution set. See the box on page 505 for a list of these operations. |

| TERMS | PAGE REFERENCE | COMMENT |
|---|---|---|
| 6. Gaussian elimination | 505 | This is a technique for converting a system of equations to upper-triangular form. See Examples 4 and 5 in Section 10.2 for demonstrations of this technique. |
| 7. Matrix | 513 | A *matrix* is a rectangular array of numbers, enclosed in parentheses or brackets. The numbers constituting the rectangular array are called the *entries* or the *elements* in the matrix. The *size* or *dimension* of a matrix is expressed by specifying the number of rows and the number of columns, in that order. For examples of this terminology, see Example 1 in Section 10.3. |
| 8. Matrix equality | 517 | Two matrices are said to be *equal* provided they are the same size and their corresponding entries are equal. |
| 9. Matrix addition and subtraction | 517 | To add two matrices of the same size, add the corresponding entries. Similarly, to subtract two matrices of the same size, subtract the corresponding entries. |
| 10. Matrix multiplication | 520 | Let $A$ and $B$ be two matrices. The matrix product $AB$ is defined only when the number of columns in $A$ is the same as the number of rows in $B$. In this case, the matrix $AB$ will have as many rows as $A$ and as many columns as $B$. The entry in the $i$th row and $j$th column of $AB$ is the number formed as follows. Multiply the corresponding entries in the $i$th row of $A$ and the $j$th column of $B$; then add the results. |
| 11. Square matrix | 524 | A matrix in which there are as many rows as there are columns is called a *square matrix*. An $n \times n$ square matrix is said to be an $n$th-order square matrix. |
| 12. Multiplicative identity matrix | 525, 528 | For the set of $n \times n$ matrices, the multiplicative identity matrix $I_n$ is the $n \times n$ matrix with ones down the main diagonal (upper left corner to bottom right corner) and zeros everywhere else. For example, $$I_2 = \begin{pmatrix} 1 & 0 \\ 0 & 1 \end{pmatrix} \quad \text{and} \quad I_3 = \begin{pmatrix} 1 & 0 & 0 \\ 0 & 1 & 0 \\ 0 & 0 & 1 \end{pmatrix}$$ For every $n \times n$ matrix $A$, we have $AI_n = A$ and $I_nA = A$. |
| 13. Inverse matrix | 525, 529 | Given an $n \times n$ matrix $A$, the inverse matrix (if it exists) is denoted by $A^{-1}$. The defining equations for $A^{-1}$ are $AA^{-1} = I$ and $A^{-1}A = I$. Here, $I$ stands for the multiplicative identity matrix that is the same size as $A$. If a square matrix has an inverse, it is said to be *nonsingular*; if there is no inverse, then the matrix is *singular*. |
| 14. Minor | 532, 533 | The *minor* of an entry $b$ in a determinant is the determinant formed by suppressing the entries in the row and column in which $b$ appears. |
| 15. Cofactor | 532, 533 | If an entry $b$ appears in the $i$th row and the $j$th column of a determinant, then the *cofactor* of $b$ is computed by multiplying the minor of $b$ times the number $(-1)^{i+j}$. |

| TERMS | PAGE REFERENCE | COMMENT |
|---|---|---|
| 16. Determinant | 533, 534 | The value of the $2 \times 2$ determinant $\begin{vmatrix} a & b \\ c & d \end{vmatrix}$ is defined to be $ad - bc$. For larger determinants, the value can be found as follows. Select any row or column and multiply each entry in that row or column by its cofactor. Then add the results. The resulting sum is the value of the determinant. It can be shown that this value is independent of the particular row or column that is chosen. |
| 17. Cramer's rule | 538 | This rule yields the solutions of certain systems of linear equations in terms of determinants. For a statement of the rule, see page 538. Example 5 on pages 538–539 shows how the rule is applied. The proof of Cramer's rule begins on page 539. |
| 18. Linear inequality in two variables | 550 | A linear equality in two variables is any one of the four types of inequalities that result when the equal sign in the equation $ax + by = c$ is replaced by one of the four symbols $<$, $\leq$, $>$, $\geq$. For any of these linear inequalities, a *solution* is an ordered pair of numbers $(x_0, y_0)$ with the property that a true statement is obtained when $x$ and $y$ (in the inequality) are replaced by $x_0$ and $y_0$, respectively. |
| 19. Vertices | 552 | The *vertices* of a region in the $x$-$y$ plane are the corners or points where the adjacent bounding sides meet. |
| 20. Convex region | 552 | A region in the $x$-$y$ plane is *convex* if, given any two points in the region, the line segment joining those points lies wholly within the region. |
| 21. Bounded region | 552 | A region in the $x$-$y$ plane is *bounded* if it can be wholly contained within some (sufficiently large) circle. |
| 22. Objective function and constraints | 553 | In linear programming problems, the function to be maximized or minimized is called the *objective function*. The inequalities restricting the variables in the objective function are referred to as the *constraints*. |
| 23. Feasible solutions | 553 | In a linear programming problem, the set of points obtained by graphing the constraints is called the *set of feasible solutions*. |
| 24. Optimal solution | 553 | In linear programming, a feasible solution that yields a largest or smallest value for the objective function is called an *optimal solution*. |
| 25. Linear programming theorem | 554 | Consider an objective function $C = ax + by + c$, where the nonnegative variables $x$ and $y$ are subject to constraints specified by nonstrict inequalities. Then if $C$ assumes a maximum (or a minimum) value, this occurs at a vertex. Furthermore, if the set of feasible solutions is both convex and bounded, then both a maximum and a minimum always exist. |

# CHAPTER 10   REVIEW EXERCISES

NOTE   The fifteen exercises in this set with colored problem numbers constitute a chapter test.

*In Exercises 1–38, solve each system of equations. If there are no solutions in a particular case, say so. In cases where there are literal (rather than numerical) coefficients, specify any restrictions that your solutions impose on those coefficients.*

1. $\begin{cases} x + y = -2 \\ x - y = 8 \end{cases}$

2. $\begin{cases} x - y = 1 \\ x + y = 5 \end{cases}$

3. $\begin{cases} 2x + y = 2 \\ x + 2y = 7 \end{cases}$

4. $\begin{cases} 3x + 2y = 6 \\ 5x + 4y = 4 \end{cases}$

5. $\begin{cases} 7x + 2y = 9 \\ 4x + 5y = 63 \end{cases}$

6. $\begin{cases} \dfrac{x}{2} + \dfrac{y}{3} = 9 \\ \dfrac{x}{5} - \dfrac{y}{2} = -4 \end{cases}$

7. $\begin{cases} 2x - \dfrac{y}{2} = -8 \\ \dfrac{x}{3} + \dfrac{y}{8} = -1 \end{cases}$

8. $\begin{cases} 3x - 14y - 1 = 0 \\ -6x + 28y - 3 = 0 \end{cases}$

9. $\begin{cases} 3x + 5y - 1 = 0 \\ 9x - 10y - 8 = 0 \end{cases}$

10. $\begin{cases} 9x + 15y - 1 = 0 \\ 6x + 10y + 1 = 0 \end{cases}$

11. $\begin{cases} \frac{2}{3}x = -\frac{1}{2}y - 12 \\ \dfrac{x}{2} = y + 2 \end{cases}$

12. $\begin{cases} 0.1x + 0.2y = -5 \\ -0.2x - 0.5y = 13 \end{cases}$

13. $\begin{cases} \dfrac{1}{x} + \dfrac{1}{y} = -1 \\ \dfrac{2}{x} + \dfrac{5}{y} = -14 \end{cases}$

14. $\begin{cases} \dfrac{2}{x} + \dfrac{15}{y} = -9 \\ \dfrac{1}{x} + \dfrac{10}{y} = -2 \end{cases}$

15. $\begin{cases} ax + (1 - a)y = 1 \\ (1 - a)x + y = 0 \end{cases}$

16. $\begin{cases} ax - by - 1 = 0 \\ (a - 1)x + by + 2 = 0 \end{cases}$

17. $\begin{cases} 2x - y = 3a^2 - 1 \\ 2y + x = 2 - a^2 \end{cases}$

18. $\begin{cases} 3ax + 2by = 3a^2 - ab + 2b^2 \\ 3bx + 2ay = 2a^2 + 5ab - 3b^2 \end{cases}$

19. $\begin{cases} 5x - y = 4a^2 - 6b^2 \\ 2x + 3y = 5a^2 + b^2 \end{cases}$

20. $\begin{cases} \dfrac{2b}{x} - \dfrac{3}{y} = 7ab \\ \dfrac{4a}{x} + \dfrac{5a}{by} = 3a^2 \end{cases}$

21. $\begin{cases} px - qy = q^2 \\ qx + py = p^2 \end{cases}$

22. $\begin{cases} x - y = \dfrac{a - b}{a + b} \\ x + y = 1 \end{cases}$

23. $\begin{cases} \dfrac{4a}{x} - \dfrac{3b}{y} = a - 7b \\ \dfrac{3a^2}{x} - \dfrac{2b^2}{y} = (3a + b)(a - 2b) \end{cases}$

24. $\begin{cases} 6b^2x^{-1/2} - 5a^2y^{-1/2} - a^2b^2 = 0 \\ 2bx^{-1/2} + a^2y^{-1/2} - 3a^2b = 0 \end{cases}$

   *Hint:* Let $u = x^{-1/2}$ and $v = y^{-1/2}$.

25. $\begin{cases} x + y + z = 9 \\ x - y - z = -5 \\ 2x + y - 2z = -1 \end{cases}$

26. $\begin{cases} x - 4y + 2z = 9 \\ 2x + y + z = 3 \\ 3x - 2y - 3z = -18 \end{cases}$

27. $\begin{cases} 4x - 4y + z = 4 \\ 2x + 3y + 3z = -8 \\ x + y + z = -3 \end{cases}$

28. $\begin{cases} x - 8y + z = 1 \\ 5x + 16y + 3z = 3 \\ 4x - 4y - 4z = -4 \end{cases}$

29. $\begin{cases} -2x + y + z = 1 \\ x - 2y + z = -2 \\ x + y - 2z = 4 \end{cases}$

30. $\begin{cases} -x + y + z = 1 \\ x - y + z = -1 \\ x + y - z = 1 \end{cases}$

31. $\begin{cases} 4x + 2y - 3z = 15 \\ 2x + y + 3z = 3 \end{cases}$

32. $\begin{cases} 3x + 2y + 17z = 1 \\ x + 2y + 3z = 3 \end{cases}$

33. $\begin{cases} x + 2y - 3z = -2 \\ 2x - y + z = 1 \\ 3x - 4y + 5z = 1 \end{cases}$

34. $\begin{cases} 9x + y + z = 0 \\ -3x + y - z = 0 \\ 3x - 5y + 3z = 0 \end{cases}$

35. $\begin{cases} x + y + z = a + b \\ 2x - y + 2z = -a + 5b \\ x - 2y + z = -2a + 4b \end{cases}$

36. $\begin{cases} ax + by - 2az = 4ab + 2b^2 \\ x + y + z = 4a + 2b \\ bx + ay + 4az = 5a^2 + b^2 \end{cases}$

37. $\begin{cases} x + y + z + w = 8 \\ 3x + 3y - z - w = 20 \\ 4x - y - z + 2w = 18 \\ 2x + 5y + 5z - 5w = 8 \end{cases}$

38. $\begin{cases} x - 2y + 3z + w = 1 \\ x + y + z + w = 5 \\ 2x + 3y + 2z - w = 3 \\ 3x + y - z + 2w = 4 \end{cases}$

*For Exercises 39–50, determine the constants (denoted by capital letters) so that each equation is an identity.*

39. $\dfrac{1}{(x-10)(x+10)} = \dfrac{A}{x-10} + \dfrac{B}{x+10}$

40. $\dfrac{x}{(x-5)(x+5)} = \dfrac{A}{x-5} + \dfrac{B}{x+5}$

41. $\dfrac{2x}{(x+1)^2} = \dfrac{A}{x+1} + \dfrac{B}{(x+1)^2}$

42. $\dfrac{x+2}{(x-1)^2} = \dfrac{A}{x-1} + \dfrac{B}{(x-1)^2}$

43. $\dfrac{5}{x(x-4)} = \dfrac{A}{x} + \dfrac{B}{x-4}$

44. $\dfrac{6x}{(x+2)(x+3)} = \dfrac{A}{x+2} + \dfrac{B}{x+3}$

45. $\dfrac{1}{(x-1)(x+3)^2} = \dfrac{A}{x-1} + \dfrac{B}{x+3} + \dfrac{C}{(x+3)^2}$

46. $\dfrac{x^2}{(x+2)(x-2)^2} = \dfrac{A}{x+2} + \dfrac{B}{x-2} + \dfrac{C}{(x-2)^2}$

47. $\dfrac{4x^2+2x+15}{(x-1)(x^2+x+5)} = \dfrac{A}{x-1} + \dfrac{Bx+C}{x^2+x+5}$

48. $\dfrac{1}{x(x^2+16)} = \dfrac{A}{x} + \dfrac{Bx+C}{x^2+16}$

49. $\dfrac{1}{x^3+64} = \dfrac{A}{x+4} + \dfrac{Bx+C}{x^2-4x+16}$

50. $\dfrac{1}{x^3+3x^2+3x+1} = \dfrac{A}{x+1} + \dfrac{B}{(x+1)^2} + \dfrac{C}{(x+1)^3}$

*In Exercises 51–54, the lowercase letters a and b denote nonzero constants, with $a \neq b$. In each case, determine the values of A and B (and C, if applicable) so that the equation is an identity.*

51. $\dfrac{x}{(x-a)^3} = \dfrac{A}{x-a} + \dfrac{B}{(x-a)^2} + \dfrac{C}{(x-a)^3}$

52. $\dfrac{ax(1-a)}{(x-a)(x+a^2)} = \dfrac{A}{x-a} + \dfrac{B}{x+a^2}$

53. $\dfrac{(a-b)(a+b-x)}{(x-a)(x-b)} = \dfrac{A}{x-a} + \dfrac{B}{x-b}$

54. $\dfrac{4a^2x+6a^3}{(x+a)(x+2a)(x+3a)} = \dfrac{A}{x+a} + \dfrac{B}{x+2a} + \dfrac{C}{x+3a}$

*In Exercises 55–62, evaluate each of the determinants.*

55. $\begin{vmatrix} 1 & 5 \\ -6 & 4 \end{vmatrix}$

56. $\begin{vmatrix} \frac{1}{6} & 1 \\ 2 & 12 \end{vmatrix}$

57. $\begin{vmatrix} 4 & 0 & 3 \\ -2 & 1 & 5 \\ 0 & 2 & -1 \end{vmatrix}$

58. $\begin{vmatrix} 2 & 6 & 4 \\ 6 & 18 & 24 \\ 15 & 5 & -10 \end{vmatrix}$

59. $\begin{vmatrix} 1 & 5 & 7 \\ 1 & 5 & 7 \\ 17 & 19 & 21 \end{vmatrix}$

60. $\begin{vmatrix} 0 & 2 & 4 & 0 \\ 4 & 0 & 6 & 2 \\ 0 & 0 & 1 & 1 \\ 14 & 7 & 1 & 0 \end{vmatrix}$

61. $\begin{vmatrix} 1 & 0 & 0 & 0 \\ 0 & 2 & 0 & 0 \\ 0 & 0 & 3 & 0 \\ 0 & 0 & 0 & 4 \end{vmatrix}$

62. $\begin{vmatrix} 1 & a & b & c \\ 0 & 2 & d & e \\ 0 & 0 & 3 & f \\ 0 & 0 & 0 & 4 \end{vmatrix}$

63. Show that $\begin{vmatrix} a & b & c \\ b & c & a \\ c & a & b \end{vmatrix} = 3abc - a^3 - b^3 - c^3$.

64. Show that $\begin{vmatrix} 1 & 1 & 1 \\ 1 & 1+x & 1 \\ 1 & 1 & 1+x^2 \end{vmatrix} = x^3$.

65. Show that
$$\begin{vmatrix} a^2+x & b & c & d \\ -b & 1 & 0 & 0 \\ -c & 0 & 1 & 0 \\ -d & 0 & 0 & 1 \end{vmatrix} = a^2 + b^2 + c^2 + d^2 + x$$

66. In a two-digit number, the sum of the digits is 11. Four times the units digit exceeds the tens digit by 4. Find the number.

67. Determine constants $a$ and $b$ so that the parabola $y = ax^2 + bx - 1$ passes through the points $(-2, 5)$ and $(2, 9)$.

68. Find two numbers whose sum and difference are 52 and 10, respectively.

69. The vertices of triangle $ABC$ are $A(-2, 0)$, $B(4, 0)$, and $C(0, 6)$. Let $A_1$, $B_1$, and $C_1$ denote the midpoints of sides $\overline{BC}$, $\overline{AC}$, and $\overline{AB}$, respectively.
    (a) Find the point where the line segments $\overline{AA_1}$ and $\overline{BB_1}$ intersect. *Answer:* $(\frac{2}{3}, 2)$
    (b) Follow the instructions given in part (a) using $\overline{BB_1}$ and $\overline{CC_1}$.
    (c) Follow the instructions given in part (a) using $\overline{AA_1}$ and $\overline{CC_1}$.
    (d) Let $P$ denote the point $(\frac{2}{3}, 2)$ that you found in part (a). Compute each of the following ratios: $AP/PA_1$, $BP/PB_1$, $CP/PC_1$. What do you observe?

70. An **altitude** of a triangle is a line segment drawn from a vertex perpendicular to the opposite side. Suppose that the vertices of triangle $ABC$ are as given in Exercise 69. Find the intersection point for each pair of altitudes. What do you observe about the three answers?

71. This exercise appears in *Plane and Solid Analytic Geometry,* by W. F. Osgood and W. G. Graustein (New York: Macmillan, 1920): Let $P$ be any point $(a, a)$ of the line $x - y = 0$, other than the origin. Through $P$ draw two lines, of arbitrary slopes $m_1$ and $m_2$, inter-

secting the $x$-axis in $A_1$ and $A_2$, and the $y$-axis in $B_1$ and $B_2$, respectively. Prove that the lines $A_1B_2$ and $A_2B_1$ will, in general, meet on the line $x + y = 0$.

72. Determine constants $h$, $k$, and $r$ so that the circle $(x - h)^2 + (y - k)^2 = r^2$ passes through the three points $(0,0)$, $(0,1)$, and $(1,0)$.

73. The vertices of a triangle are the points of intersection of the lines $y = x - 1$, $y = -x - 2$, and $y = 2x + 3$. Find the equation of the circle passing through these three intersection points.

74. The vertices of triangle $ABC$ are $A(0,0)$, $B(3,0)$, and $C(0,4)$.
  (a) Find the center and the radius of the circle that passes through $A$, $B$, and $C$. This circle is called the **circumcircle** for triangle $ABC$.
  (b) The figure below shows the **inscribed circle** for triangle $ABC$; this is the circle that is tangent to all three sides of the triangle. Find the center and the radius of this circle using the following two facts.
   (i) The center of the inscribed circle is the common intersection point of the three angle bisectors of the triangle.
   (ii) The slope of the line that bisects angle $B$ is $-\frac{1}{2}$.
  (c) Verify the following statement for triangle $ABC$. (The statement actually holds for any triangle; the result is known as *Euler's theorem.*) Let $R$ and $r$ denote the radii of the circles in parts (a) and (b), respectively. Then the distance $d$ between the centers of the circles satisfies the equation
   $$d^2 = R^2 - 2rR$$
  (d) Verify the following statement for triangle $ABC$. (The statement actually holds for any triangle.) The area of triangle $ABC$ is equal to one-half the product of the perimeter and the radius of the inscribed circle.
  (e) Verify the following statement for triangle $ABC$. (The statement actually holds for any triangle.) The sum of the reciprocals of the lengths of the altitudes is equal to the reciprocal of the radius of the inscribed circle.

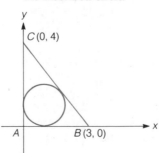

*In Exercises 75–98, the matrices A, B, C, D, E, and F are defined as follows.*

$$A = \begin{pmatrix} 3 & -2 \\ 1 & 5 \end{pmatrix} \qquad B = \begin{pmatrix} 2 & 1 \\ 1 & 8 \end{pmatrix} \qquad C = \begin{pmatrix} -1 & 0 \\ 0 & -1 \end{pmatrix}$$

$$D = \begin{pmatrix} -4 & 0 & 6 \\ 1 & 3 & 2 \end{pmatrix} \qquad E = \begin{pmatrix} 3 & -1 \\ 4 & 1 \\ -5 & 9 \end{pmatrix} \qquad F = \begin{pmatrix} 2 & 6 \\ 5 & 3 \\ 5 & 8 \end{pmatrix}$$

*In each exercise, carry out the indicated matrix operations if they are defined. If an operation is not defined, say so.*

75. $2A + 2B$  
76. $2(A + B)$  
77. $4B$  
78. $B + 4$  
79. $AB$  
80. $BA$  
81. $AB - BA$  
82. $B + E$  
83. $B + C$  
84. $A(B + C)$  
85. $AB + AC$  
86. $(B + C)A$  
87. $BA + CA$  
88. $D + E$  
89. $DE$  
90. $(EE)D$  
91. $E(ED)$  
92. $E + F$  
93. $EF$  
94. $3E - 2F$  
95. $(A + B) + C$  
96. $A + (B + C)$  
97. $(AB)C$  
98. $A(BC)$  

99. For a square matrix $A$, the notation $A^2$ means $AA$. Similarly, $A^3$ means $AAA$. If $A = \begin{pmatrix} 1 & 1 \\ 0 & 1 \end{pmatrix}$, verify that
$$A^2 = \begin{pmatrix} 1 & 2 \\ 0 & 1 \end{pmatrix} \quad \text{and} \quad A^3 = \begin{pmatrix} 1 & 3 \\ 0 & 1 \end{pmatrix}$$

100. Let $A$ be the matrix $\begin{pmatrix} 0 & 0 & 0 \\ a & 0 & 0 \\ b & c & 0 \end{pmatrix}$. Compute $A^2$ and $A^3$.

*For Exercises 101–104, in each case compute the inverse of the matrix in part (a), and then use that inverse to solve the system of equations in part (b).*

101. (a) $\begin{pmatrix} 1 & 5 \\ 2 & 9 \end{pmatrix}$  (b) $\begin{cases} x + 5y = 3 \\ 2x + 9y = -4 \end{cases}$

102. (a) $\begin{pmatrix} 5 & -4 \\ 14 & -11 \end{pmatrix}$  (b) $\begin{cases} 5x - 4y = 2 \\ 14x - 11y = 5 \end{cases}$

103. (a) $\begin{pmatrix} 1 & -2 & 3 \\ 2 & -5 & 10 \\ -1 & 2 & -2 \end{pmatrix}$  (b) $\begin{cases} x - 2y + 3z = -2 \\ 2x - 5y + 10z = -3 \\ -x + 2y - 2z = 6 \end{cases}$

104. (a) $\begin{pmatrix} -1 & 4 & 2 \\ -3 & 10 & 5 \\ 0 & 3 & 1 \end{pmatrix}$  (b) $\begin{cases} -x + 4y + 2z = 8 \\ -3x + 10y + 5z = 0 \\ 3y + z = -12 \end{cases}$

*In Exercises 105 and 106, find the inverse of each matrix.*

105. $\begin{pmatrix} 5 & 3 & 6 & -7 \\ 3 & -4 & 0 & -9 \\ 0 & 1 & -1 & -1 \\ 2 & 2 & 3 & -2 \end{pmatrix}$   106. $\begin{pmatrix} 1 & -1 & 0 & -3 \\ 5 & -2 & 3 & -11 \\ 2 & 3 & 2 & -3 \\ 4 & 5 & 5 & -5 \end{pmatrix}$

*In Exercises 107–114, compute $D$, $D_x$, $D_y$, $D_z$ (and $D_w$ where appropriate) for each system of equations. Use Cramer's rule to solve the systems in which $D \neq 0$. If $D = 0$, solve the system using Gaussian elimination or matrix methods.*

107. $\begin{cases} 2x - y + z = 1 \\ 3x + 2y + 2z = 0 \\ x - 5y - 3z = -2 \end{cases}$   108. $\begin{cases} x + 2y - z = -1 \\ 2x - 3y + 3z = 3 \\ 2x + 3y + z = 1 \end{cases}$

109. $\begin{cases} x + 2y + 3z = -1 \\ 4x + 5y + 6z = 2 \\ 7x + 8y + 9z = -3 \end{cases}$   110. $\begin{cases} 3x + 2y - 2z = 0 \\ 2x + 3y - z = 0 \\ 8x + 7y - 5z = 0 \end{cases}$

111. $\begin{cases} 3x + 2y - 2z = 1 \\ 2x + 3y - z = -2 \\ 8x + 7y - 5z = 0 \end{cases}$

112. $\begin{cases} x + y + z + w = 5 \\ x - y - z + w = 3 \\ 2x + 3y + 3z + 2w = 21 \\ 4z - 3w = -7 \end{cases}$

113. $\begin{cases} 2x - y + z + 3w = 15 \\ x + 2y + 2w = 12 \\ 3y + 3z + 4w = 12 \\ -4x + y - 4z = -11 \end{cases}$

114. $\begin{cases} x + y + z = (a + b)^2 \\ \dfrac{bx}{a} + \dfrac{ay}{b} - z = 0 \\ x + y - z = (a - b)^2 \end{cases}$

*In Exercises 115–128, find all solutions $(x,y)$ for each system, where $x$ and $y$ are real numbers.*

115. $\begin{cases} y = 6x \\ y = x^2 \end{cases}$   116. $\begin{cases} y = 4x \\ y = x^3 \end{cases}$

117. $\begin{cases} y = 9 - x^2 \\ y = x^2 - 9 \end{cases}$   118. $\begin{cases} x^3 - y = 0 \\ xy - 16 = 0 \end{cases}$

119. $\begin{cases} x^2 - y^2 = 9 \\ x^2 + y^2 = 16 \end{cases}$   120. $\begin{cases} 2x + 3y = 6 \\ y = \sqrt{x + 1} \end{cases}$

121. $\begin{cases} x^2 + y^2 = 1 \\ y = \sqrt{x} \end{cases}$   122. $\begin{cases} \dfrac{x}{11} + \dfrac{y}{12} = 2 \\ xy/132 - 1 \end{cases}$

123. $\begin{cases} x^2 + y^2 = 1 \\ y = 2x^2 \end{cases}$

124. $\begin{cases} x^2 - 3xy + y^2 = -11 \\ 2x^2 + xy - y^2 = 8 \end{cases}$

125. $\begin{cases} x^2 + 2xy + 3y^2 = 68 \\ 3x^2 - xy + y^2 = 18 \end{cases}$

126. $\begin{cases} \dfrac{x^2}{a^2} + \dfrac{y^2}{b^2} = 1 \\ \dfrac{x^2}{b^2} + \dfrac{y^2}{a^2} = 1 \end{cases}$   $(a > b > 0)$

127. $\begin{cases} 2(x - 3)^2 - (y + 1)^2 = -1 \\ -3(x - 3)^2 + 2(y + 1)^2 = 6 \end{cases}$
*Hint:* Let $u = x - 3$ and $v = y + 1$.

128. $\begin{cases} x^4 = y - 1 \\ y - 3x^2 + 1 = 0 \end{cases}$

*Exercises 129–134 appear (in German) in an algebra text by Leonhard Euler, first published in 1770. The English versions given here are taken from the translated version, Elements of Algebra, 5th ed., by Leonhard Euler (London: Longman, Orme, and Co., 1840). [This, in turn, has been reprinted by Springer-Verlag (New York, 1984).]*

129. Required two numbers, whose sum may be $s$, and their proportion as $a$ to $b$.   *Answer:*  $\dfrac{as}{a + b}$ and $\dfrac{bs}{a + b}$

130. The sum $2a$, and the sum of the squares $2b$, of two numbers being given; to find the numbers.   *Answer:* $a - \sqrt{b - a^2}$ and $a + \sqrt{b - a^2}$

131. To find three numbers, so that [the sum of] one-half of the first, one-third of the second, and one-quarter of the third, shall be equal to 62; one-third of the first, one-quarter of the second, and one-fifth of the third, equal to 47; and one-quarter of the first, one-fifth of the second, and one-sixth of the third, equal to 38.   *Answer:*  24, 60, 120

132. Required two numbers, whose product may be 105, and whose squares [when added] may together make 274.

133. Required two numbers, whose product may be $m$, and the sum of the squares $n$   $(n \geq 2m)$.

134. Required two numbers such that their sum, their product, and the difference of their squares may all be equal.

*In Exercises 135–140, graph each system of inequalities and specify whether the region is convex or bounded.*

135. $\begin{cases} x^2 + y^2 \geq 1 \\ y - 4x \leq 0 \\ y - x \geq 0 \\ x \geq 0 \\ y \geq 0 \end{cases}$   136. $\begin{cases} x^2 + y^2 \leq 1 \\ x \geq 0 \\ y \geq 0 \end{cases}$

137. $\begin{cases} y - \sqrt{x} \leq 0 \\ y \geq 0 \\ x \geq 1 \\ x - 4 \leq 0 \end{cases}$   138. $\begin{cases} y - |x| \leq 0 \\ x + 1 \geq 0 \\ x - 1 \leq 0 \\ y + 1 \geq 0 \end{cases}$

139. $\begin{cases} y \leq 1/x \\ y \geq 0 \\ x \geq 1 \end{cases}$   140. $\begin{cases} y - 100x \leq 0 \\ y - x^2 \geq 0 \\ x \geq 0 \end{cases}$

141. A factory assembles two types of motors, small and large. The cost of material is $15 for a small motor and $30 for a large motor. Two hours of labor are required to assemble a small motor and 6 hr are needed for a large motor. The cost of labor is $8/hr. The owner of the factory can allocate at most $1500 per day for material and $2000 per day for labor. Find the largest daily profit that the owner can make if the small motors are sold for $40 each and the large motors for $104 each. Assume that every motor produced is sold. *Hint:* Profit equals revenue minus cost.

142. Suppose that the factory owner in Exercise 141 finds it necessary to limit production to a total of at most 65 motors per day. Find the largest daily profit under this additional constraint.

# 11 ROOTS OF POLYNOMIAL EQUATIONS

**INTRODUCTION**  In this chapter we continue the work begun in Chapter 2 on solving polynomial equations. We begin in Section 11.1 with a second look at the long division process for polynomials. In this section you'll see how synthetic division is used to abbreviate the long division process. Section 11.2 presents two theorems about polynomials: the remainder theorem and the factor theorem. The remainder theorem and the factor theorem are then applied repeatedly in the next section (Section 11.3) to develop some fundamental results regarding polynomial equations and their solutions. You can view much of the material in this section as a kind of generalization of what you already know about quadratic equations. The last two sections of this chapter present additional results that are useful in actually solving polynomial equations, or in determining the nature of their solutions.

## MORE ON DIVISION OF POLYNOMIALS

**11.1**

In Section 1.8 we studied the long division process for polynomials. There is a theorem, commonly referred to as the *division algorithm*, that summarizes rather nicely the key results of that long division process. We state the theorem here without proof.

---

**The Division Algorithm**

Let $p(x)$ and $d(x)$ be polynomials and assume $d(x)$ is not the zero polynomial. Then there are unique polynomials $q(x)$ and $R(x)$ such that

$$p(x) = d(x) \cdot q(x) + R(x)$$

and where either $R(x)$ is the zero polynomial or the degree of $R(x)$ is less than the degree of $d(x)$.

The polynomials $p(x)$, $d(x)$, $q(x)$, and $R(x)$ are referred to, respectively, as the **dividend, divisor, quotient, and remainder**. In the case where $R(x) = 0$, we say that $d(x)$ and $q(x)$ are **factors of** $p(x)$.

---

**EXAMPLE 1**  Let $p(x) = x^3 + 2x^2 - 4$ and $d(x) = x - 3$. Use the long division process to find the polynomials $q(x)$ and $R(x)$ such that

$$p(x) = d(x) \cdot q(x) + R(x)$$

where either $R(x) = 0$ or the degree of $R(x)$ is less than the degree of $d(x)$.

**Solution**  After inserting the term $0x$ in the dividend $p(x)$, we use long division to divide $p(x)$ by $d(x)$.

$$
\begin{array}{r}
x^2 + 5x + 15 \\
x - 3 \overline{\smash{\big)}\, x^3 + 2x^2 + \phantom{1}0x - 4} \\
\underline{x^3 - 3x^2 \phantom{+ 0x - 4}} \\
5x^2 + \phantom{1}0x \phantom{-4} \\
\underline{5x^2 - 15x \phantom{-4}} \\
15x - \phantom{4}4 \\
\underline{15x - 45} \\
41
\end{array}
$$

We now have that

$$
\frac{x^3 + 2x^2 - 4}{x - 3} = x^2 + 5x + 15 + \frac{41}{x - 3} \tag{1}
$$

or, after multiplying through by $x - 3$,

$$
\underbrace{x^3 + 2x^2 - 4}_{p(x)} = \underbrace{(x - 3)}_{d(x)} \underbrace{(x^2 + 5x + 15)}_{q(x)} \underbrace{+ 41}_{R(x)} \tag{2}
$$

Thus $q(x) = x^2 + 5x + 15$ and $R(x) = 41$, as required. Notice that the degree of $R(x)$ is less than the degree of $d(x)$. [Also, notice that equation (2) is valid for all real numbers $x$, whereas equation (1) carries the implicit restriction that $x$ may not equal 3. For this reason, we often prefer to write our results in the form of equation (2).]

The long division procedure for polynomials can be streamlined when the divisor is of the form $x - a$. This shortened version, known as **synthetic division**, will be useful in subsequent sections when we are solving polynomial equations.

We can explain the idea behind synthetic division using the long division carried out in Example 1. The basic idea is that in the long division process, it is the *coefficients* of the various polynomials that carry all the necessary information. In our example, for instance, the quotient and remainder can be abbreviated by writing down a sequence of four numbers:

$$1 \qquad 5 \qquad 15 \qquad 41$$

By studying the long division process, you will find that these numbers are obtained through the following four steps:

STEP 1  Write down the first coefficient of the dividend. This will be the first coefficient of the quotient.

*Result*  $\boxed{1}$

STEP 2  Multiply the 1 obtained in the previous step by the $-3$ in the divisor. Then subtract this from the second coefficient of the dividend:

$$-3 \times 1 = -3$$
$$2 - (-3) = 5 \qquad\qquad\qquad \textit{Result} \quad \boxed{5}$$

STEP 3    Multiply the 5 obtained in the previous step by the $-3$ in the divisor. Then subtract this from the third coefficient of the dividend:

$$-3 \times 5 = -15$$
$$0 - (-15) = 15 \qquad\qquad\qquad\qquad\qquad \textit{Result} \quad \boxed{15}$$

STEP 4    Multiply the 15 obtained in the previous step by the $-3$ in the divisor. Then subtract this from the fourth coefficient of the dividend:

$$-3 \times 15 = -45$$
$$-4 - (-45) = 41 \qquad\qquad\qquad\qquad\qquad \textit{Result} \quad \boxed{41}$$

A convenient format for setting up this process involves writing the constant term of the divisor and the coefficients of the dividend as follows:

$$\underline{-3}\big| \quad 1 \quad\quad 2 \quad\quad 0 \quad\quad -4$$

Now, using this format, let us again go through the four steps we have just described:

STEP 1    Bring down the 1.

$$\underline{-3}\big| \quad 1 \quad\quad 2 \quad\quad 0 \quad\quad -4$$
$$\overline{\quad\quad 1}$$

STEP 2    $-3 \times 1 = -3$
             $2 - (-3) = 5$

$$\underline{-3}\big| \quad 1 \quad\quad 2 \quad\quad 0 \quad\quad -4$$
$$\quad\quad\quad\quad -3$$
$$\overline{\quad 1 \quad\quad 5}$$

STEP 3    $-3 \times 5 = -15$
             $0 - (-15) = 15$

$$\underline{-3}\big| \quad 1 \quad\quad 2 \quad\quad 0 \quad\quad -4$$
$$\quad\quad\quad\quad -3 \quad\quad -15$$
$$\overline{\quad 1 \quad\quad 5 \quad\quad 15}$$

STEP 4    $-3 \times 15 = -45$
             $-4 - (-45) = 41$

$$\underline{-3}\big| \quad 1 \quad\quad 2 \quad\quad 0 \quad\quad -4$$
$$\quad\quad\quad\quad -3 \quad\quad -15 \quad\quad -45$$
$$\overline{\quad 1 \quad\quad 5 \quad\quad 15 \quad\quad 41}$$

Although we have now obtained the required sequence of numbers, 1   5   15   41, there is one further simplification that can be made. In steps 2 through 4, we can add instead of subtract if we use 3 instead of $-3$ in the initial format. (You will see the motivation for this in the next section when we discuss the remainder theorem.) With this change, let us now summarize the technique of synthetic division using the example with which we've been working. The method is applicable for any polynomial division in which the form of the divisor is $x - a$.

---

**To Divide $x^3 + 2x^2 - 4$ by $x - 3$ Using Synthetic Division**

|  |  |  |  |  | COMMENTS |
|---|---|---|---|---|---|
| Format | $3\rfloor$ 1 2 0 $-4$ | | | | Since the divisor is $x - 3$, the format begins with 3. The coefficients from the dividend are written in the order corresponding to decreasing powers of $x$. A zero coefficient is inserted as a place holder. |

Procedure $\quad 3\rfloor \quad 1 \quad 2 \quad 0 \quad -4 \qquad$ **Step 1:** Bring down the 1

$\qquad\qquad\qquad\qquad\quad 3 \quad 15 \quad 45 \qquad$ **Step 2:** $3 \times 1 = 3; 2 + 3 = 5$

$\qquad\qquad\qquad\qquad \overline{1 \quad 5 \quad 15 \quad 41} \qquad$ **Step 3:** $3 \times 5 = 15; 0 + 15 = 15$

$\qquad\qquad\qquad\qquad\qquad\qquad\qquad\qquad\qquad$ **Step 4:** $3 \times 15 = 45; -4 + 45 = 41$

**Answer** $\quad$ Quotient: $\qquad\qquad\quad$ The degree of the first term in the
$\qquad\qquad x^2 + 5x + 15 \qquad\quad$ quotient is one less than the degree
$\qquad\qquad$ Remainder: 41 $\qquad\quad$ of the first term of the dividend.

---

**EXAMPLE 2** Use synthetic division to divide $x^3 - 6x + 4$ by $x - 2$.

Solution

$$2\rfloor \quad \begin{array}{cccc} 1 & 0 & -6 & 4 \\ & 2 & 4 & -4 \\ \hline 1 & 2 & -2 & 0 \end{array}$$

Looking at the third row of numbers in the synthetic division we've carried out, we see that the quotient is $x^2 + 2x - 2$ and the remainder is 0. In other words, both $x - 2$ and $x^2 + 2x - 2$ are factors of $x^3 - 6x + 4$, and we have

$$x^3 - 6x + 4 = (x - 2)(x^2 + 2x - 2)$$

As we have just seen, when the divisor is $x - 2$, we write $+2$ in the synthetic division format. In general, if the divisor is $x - a$, we write $a$. The next example shows what to do when the form of the divisor is $x + a$.

**EXAMPLE 3** Use synthetic division to divide $x^4 - 2x^3 + 5x^2 - 4x + 3$ by $x + 1$.

Solution We first need to write the divisor $x + 1$ in the form $x - a$. We have

$$x + 1 = x - (-1)$$

In other words, $a$ is $-1$, and this is the value we use to set up the synthetic division format. The format then is

$$-1\rfloor \quad 1 \quad -2 \quad 5 \quad -4 \quad 3$$

Now we carry out the synthetic division procedure:

$$-1\rfloor \quad \begin{array}{ccccc} 1 & -2 & 5 & -4 & 3 \\ & -1 & 3 & -8 & 12 \\ \hline 1 & -3 & 8 & -12 & 15 \end{array}$$

The quotient is therefore $x^3 - 3x^2 + 8x - 12$, and the remainder is 15. We can summarize this result by writing

$$x^4 - 2x^3 + 5x^2 - 4x + 3 = (x + 1)(x^3 - 3x^2 + 8x - 12) + 15$$

(Notice that the degree of the remainder is less than the degree of the divisor, in agreement with the division algorithm.) ▪▪

## EXERCISE SET 11.1

### A

*In Exercises 1–20, use synthetic division to find the quotients and remainders. Also, in each case, write the result of the division in the form $p(x) = d(x) \cdot q(x) + R(x)$, as in equation (2) in the text.*

1. $\dfrac{x^2 - 6x - 2}{x - 5}$

2. $\dfrac{3x^2 + 4x - 1}{x - 1}$

3. $\dfrac{4x^2 - x - 5}{x + 1}$

4. $\dfrac{x^2 - 1}{x + 2}$

5. $\dfrac{6x^3 - 5x^2 + 2x + 1}{x - 4}$

6. $\dfrac{x^4 - 4x^3 + 6x^2 - 4x + 1}{x - 1}$

7. $\dfrac{x^3 - 1}{x - 2}$

8. $\dfrac{x^3 - 8}{x - 2}$

9. $\dfrac{x^5 - 1}{x + 2}$

10. $\dfrac{x^3 - 8x^2 - 1}{x + 3}$

11. $\dfrac{x^4 - 6x^3 + 2}{x + 4}$

12. $\dfrac{3x^3 - 2x^2 + x + 1}{x - \frac{1}{2}}$

13. $\dfrac{x^3 - 4x^2 - 3x + 6}{x - 10}$

14. $\dfrac{1 + 3x + 3x^2 + x^3}{x + 1}$

15. $\dfrac{x^3 - x^2}{x + 5}$

16. $\dfrac{5x^4 - 4x^3 + 3x^2 - 2x + 1}{x + \frac{1}{2}}$

17. $\dfrac{14 - 27x - 27x^2 + 54x^3}{x - \frac{2}{3}}$

18. $\dfrac{14 - 27x - 27x^2 + 54x^3}{x + \frac{2}{3}}$

19. $\dfrac{x^4 + 3x^2 + 12}{x - 3}$

20. (a) $\dfrac{x^2 - a^2}{x - a}$ (b) $\dfrac{x^3 - a^3}{x - a}$

    (c) $\dfrac{x^4 - a^4}{x - a}$ (d) $\dfrac{x^5 - a^5}{x - a}$

### B

*In Exercises 21–26, use synthetic division to determine the quotient $q(x)$ and the remainder $R(x)$ in each case.*

21. $\dfrac{ax^2 + bx + c}{x - r}$

22. $\dfrac{ax^3 + bx^2 + cx + d}{x - r}$

23. $\dfrac{6x^2 - 8x + 1}{3x - 4}$ *Hint:* First divide both numerator and denominator by 3 so that the expression can be rewritten $\dfrac{2x^2 - \frac{8}{3}x + \frac{1}{3}}{x - \frac{4}{3}}$

24. $\dfrac{4x^3 + 6x^2 - 6x - 5}{2x - 3}$

25. $\dfrac{6x^3 + 1}{2x + 1}$

26. $\dfrac{5x^3 - 3x^2 + 1}{3x + 1}$

27. When $x^3 + kx + 1$ is divided by $x + 1$, the remainder is $-4$. Find $k$.

28. (a) Show that when $x^3 + kx + 6$ is divided by $x + 3$, the remainder is $-21 - 3k$.
    (b) Determine a value of $k$ such that $x + 3$ will be a factor of $x^3 + kx + 6$.

29. When $x^2 + 2px - 3q^2$ is divided by $x - p$, the remainder is zero. Show that $p^2 = q^2$.

30. Given that $x - 3$ is a factor of $x^3 - 2x^2 - 4x + 3$, solve the equation $x^3 - 2x^2 - 4x + 3 = 0$.

*The process of synthetic division applies equally well when some or all of the coefficients are nonreal complex numbers. In Exercises 31–34, use synthetic division to determine the quotient $q(x)$ and the remainder $R(x)$ in each case.*

31. $\dfrac{x^2 - 4x + 1}{x - i}$

32. $\dfrac{x^3 - 2x^2 - 4}{x - 3i}$

33. $\dfrac{x^2 - 2x + 2}{x - (1 + i)}$

34. $\dfrac{x^3 - x^2 + 4x - 4}{x + 2i}$

## ROOTS OF POLYNOMIAL EQUATIONS: THE REMAINDER THEOREM AND THE FACTOR THEOREM

**11.2**

The techniques for solving polynomial equations of degree 2 were discussed in Chapter 2. Now we want to extend those ideas. Our focus in this section and in the remainder of the chapter is on solving polynomial equations of any degree, that is, equations of the form

$$f(x) = a_n x^n + a_{n-1} x^{n-1} + \cdots + a_1 x + a_0 = 0 \tag{1}$$

Here, as in Chapter 2, a **root** or **solution** of equation (1) is a number $r$ that when substituted for $x$ leads to a true statement. Thus, $r$ is a root of equation (1) provided that $f(r) = 0$. We also refer to the number $r$ in this case as a **zero** of the function $f$.

**EXAMPLE 1**   (a)  Is $-3$ a zero of the function $f$ defined by $f(x) = x^4 + x^2 - 6$?
(b)  Is $\sqrt{2}$ a root of the equation $x^4 + x^2 - 6 = 0$?

**Solution**   (a)  By definition, $-3$ will be a zero of $f$ if $f(-3) = 0$. We have

$$f(-3) = (-3)^4 + (-3)^2 - 6 = 81 + 9 - 6 - 84 \neq 0$$

Thus, $-3$ is not a zero of the function $f$.
(b)  To check if $\sqrt{2}$ is a root of the given equation, we have

$$(\sqrt{2})^4 + (\sqrt{2})^2 - 6 \overset{?}{=} 0$$
$$4 + 2 - 6 \overset{?}{=} 0$$
$$0 \overset{?}{=} 0 \qquad \text{True}$$

Thus, $\sqrt{2}$ is a root of the equation $x^4 + x^2 - 6 = 0$.

There are cases in which a root of an equation is what we call a **repeated root**. Consider, for instance, the equation $x(x-1)(x-1) = 0$. We have

$$x(x-1)(x-1) = 0$$

| $x = 0$ | $x - 1 = 0$ | $x - 1 = 0$ |
|---------|-------------|-------------|
|         | $x = 1$     | $x = 1$     |

The roots of the equation are therefore 0, 1, and 1. The repeated root here is $x = 1$. We say in this case that 1 is a **double root** or, equivalently, that 1 is a **root of multiplicity** 2. More generally, if a root is repeated $k$ times, we call it a **root of multiplicity** $k$.

**EXAMPLE 2**   State the multiplicity of each root of the equation

$$(x - 4)^2 (x - 5)^3 = 0$$

**Solution**   We have $(x - 4)(x - 4)(x - 5)(x - 5)(x - 5) = 0$. By setting each factor equal to zero, we obtain the roots 4, 4, 5, 5, and 5. From this we see that the root 4 has multiplicity 2, while the root 5 has multiplicity 3. Notice that it is not really necessary to write out all the factors as we did here; the exponents of the factors in the original equation give us the required multiplicities.

There are two simple but important theorems that will form the basis for much of our subsequent work with polynomials. These are the *remainder theorem* and the *factor theorem*. We begin with a statement of the remainder theorem.

---

### The Remainder Theorem

When a polynomial $f(x)$ is divided by $x - r$, the remainder is $f(r)$.

---

Before turning to a proof of the remainder theorem, let us see what the theorem is saying in two particular cases. First, suppose that we divide the polynomial $f(x) = 2x^2 - 3x + 4$ by $x - 1$. Then according to the remainder theorem, the remainder in this case should be the number $f(1)$. Let's check:

$$
\begin{array}{r|rrr}
1\rfloor & 2 & -3 & 4 \\
         &   & 2  & -1 \\
\hline
         & 2 & -1 & \boxed{3}
\end{array}
\qquad
\begin{aligned}
f(x) &= 2x^2 - 3x + 4 \\
f(1) &= 2(1)^2 - 3(1) + 4 \\
f(1) &= \boxed{3}
\end{aligned}
$$

As the calculations show, the remainder is indeed equal to $f(1)$. As a second example, let us divide the polynomial $g(x) = ax^2 + bx + c$ by $x - r$. According to the remainder theorem, the remainder should be $g(r)$. Again, let us check:

$$
\begin{array}{r|rrr}
r\rfloor & a & b     & c \\
         &   & ar    & ar^2 + br \\
\hline
         & a & ar+b  & \boxed{ar^2 + br + c}
\end{array}
\qquad
\begin{aligned}
g(x) &= ax^2 + bx + c \\[1em]
g(r) &= \boxed{ar^2 + br + c}
\end{aligned}
$$

The calculations show that the remainder is equal to $g(r)$, as we wished to check.

In the example just concluded, we verified that the remainder theorem holds for any quadratic polynomial $g(x) = ax^2 + bx + c$. A general proof of the remainder theorem can easily be given along these same lines. The only drawback is that it becomes slightly cumbersome to carry out the synthetic division process when the dividend is

$$a_n x^n + a_{n-1} x^{n-1} + \cdots + a_1 x + a_0$$

For this reason, mathematicians often prefer to base the proof of the remainder theorem on the division algorithm given in the previous section. This is the path we will follow here.

To prove the remainder theorem, we must show that when the polynomial $f(x)$ is divided by $x - r$, the remainder is $f(r)$. Now according to the division algorithm, we can write

$$f(x) = (x - r) \cdot q(x) + R(x) \tag{2}$$

In this identity, either $R(x)$ is the zero polynomial or the degree of $R(x)$ is less than that of $x - r$. But since the degree of $x - r$ is 1, we must have in this case that the degree of $R(x)$ is zero. Thus, in *either* case, the remainder $R(x)$ is a constant. Denoting this constant by $c$, we can rewrite equation (2) as

$$f(x) = (x - r) \cdot q(x) + c$$

Now if we set $x = r$ in this identity, we obtain

$$f(r) = (r - r) \cdot q(r) + c = c$$

We have now shown that $f(r) = c$. But by definition, $c$ is the remainder $R(x)$. Thus, $f(r) = R(x)$. This proves the remainder theorem.

**EXAMPLE 3**  Let $f(x) = 2x^3 - 5x^2 + x - 6$. Use the remainder theorem to evaluate $f(3)$.

**Solution**  According to the remainder theorem, $f(3)$ is the remainder when $f(x)$ is divided by $x - 3$. Using synthetic division, we have

$$
\begin{array}{r|rrrr}
3 & 2 & -5 & 1 & -6 \\
  &   & 6  & 3 & 12 \\
\hline
  & 2 & 1  & 4 & 6
\end{array}
$$

The remainder is 6, and therefore $f(3) = 6$.

From our experience with quadratic equations, we know that there is a close connection between factoring a quadratic polynomial $f(x)$ and solving the polynomial equation $f(x) = 0$. The factor theorem states this relationship between roots and factors in a precise form. Furthermore, the factor theorem tells us that this relationship holds for polynomials of all degrees, not just quadratics.

---

### The Factor Theorem

Let $f(x)$ be a polynomial. If $f(r) = 0$, then $x - r$ is a factor of $f(x)$. Conversely, if $x - r$ is a factor of $f(x)$, then $f(r) = 0$.

---

In terms of roots, we can summarize the factor theorem by saying that $r$ is a root of the equation $f(x) = 0$ if and only if $x - r$ is a factor of $f(x)$. To prove the factor theorem, let us begin by assuming that $f(r) = 0$. We want to show that $x - r$ is a factor of $f(x)$. Now, according to the remainder theorem, if $f(x)$ is divided by $x - r$, the remainder is $f(r)$. So we can write

$$f(x) = (x - r) \cdot q(x) + f(r) \qquad \text{for some polynomial } q(x)$$

But since $f(r)$ is zero, this equation becomes

$$f(x) = (x - r) \cdot q(x)$$

This last equation tells us that $x - r$ is a factor of $f(x)$, as we wished to prove. Now, conversely, let us assume that $x - r$ is a factor of $f(x)$. We want to show that $f(r) = 0$. Since $x - r$ is a factor of $f(x)$, we can write

$$f(x) = (x - r) \cdot q(x) \qquad \text{for some polynomial } q(x)$$

If we now let $x = r$ in this last equation, we obtain

$$f(r) = (r - r) \cdot q(r) = 0$$

as we wished to show.

The example that follows indicates how the factor theorem can be used to solve equations.

**EXAMPLE 4**   Solve the equation $x^3 - 2x + 1 = 0$, given that one root is $x = 1$.

Solution   Since $x = 1$ is a root, the factor theorem tells us that $x - 1$ is a factor of $x^3 - 2x + 1$. In other words,

$$x^3 - 2x + 1 = (x - 1) \cdot q(x) \qquad \text{for some polynomial } q(x)$$

To determine this other factor $q(x)$, we divide $x^3 - 2x + 1$ by $x - 1$:

$$
\begin{array}{r|rrrr}
\underline{1} & 1 & 0 & -2 & 1 \\
  &   & 1 & 1 & -1 \\
\hline
  & 1 & 1 & -1 & 0
\end{array}
$$

Thus, $q(x) = x^2 + x - 1$ and we have the factorization

$$x^3 - 2x + 1 = (x - 1)(x^2 + x - 1)$$

Using this identity, the original equation becomes

$$(x - 1)(x^2 + x - 1) = 0$$

Now the problem is reduced to solving the linear equation $x - 1 = 0$ and the quadratic equation $x^2 + x - 1 = 0$. The linear equation yields $x = 1$, but we already knew that 1 was a root. So if we are to find any additional roots, they must come from the equation $x^2 + x - 1 = 0$. Using the quadratic formula, we obtain

$$x = \frac{-1 \pm \sqrt{(1)^2 - 4(1)(-1)}}{2(1)} = \frac{-1 \pm \sqrt{5}}{2}$$

We now have three roots of the cubic equation $x^3 - 2x + 1 = 0$. These are 1, $\frac{1}{2}(-1 + \sqrt{5})$, and $\frac{1}{2}(-1 - \sqrt{5})$. As you will see in the next section, a cubic equation can have at most three roots. So in this case, we have determined all the roots; that is, we have solved the equation.

Before going on to other examples, let's take a moment to summarize the technique we used in Example 4. We want to solve a polynomial equation $f(x) = 0$, given that one root is $x = r$. Since $r$ is a root, the factor theorem tells us that $x - r$ is a factor of $f(x)$. Then, with the aid of synthetic division, we obtain a factorization

$$f(x) = (x - r) \cdot q(x) \qquad \text{for some polynomial } q(x)$$

This gives rise to the two equations, $x - r = 0$ and $q(x) = 0$. Since the first of these only reasserts that $x = r$ is a root, we try to solve the second equation, $q(x) = 0$. We refer to the equation $q(x) = 0$ as the **reduced equation**. Example 4 showed you the idea behind this terminology; the degree of $q(x)$ is one less than that of $f(x)$. If, as in Example 4, the reduced equation happens to be a quadratic equation, then we can always determine the remaining roots by factoring or by the quadratic formula. In subsequent sections, we will look at techniques that are helpful in cases where $q(x)$ is not quadratic.

**EXAMPLE 5**   Solve the equation $x^4 + 2x^3 - 7x^2 - 20x - 12 = 0$, given that $x = 3$ and $x = -2$ are roots.

**Solution**   Since $x = 3$ is a root, the factor theorem tells us that $x - 3$ is a factor of the polynomial $x^4 + 2x^3 - 7x^2 - 20x - 12$. That is,

$$x^4 + 2x^3 - 7x^2 - 20x - 12 = (x - 3) \cdot q(x) \qquad \text{for some polynomial } q(x)$$

As in Example 4, we can find $q(x)$ by synthetic division. We have

$$
\begin{array}{r|rrrrr}
3 & 1 & 2 & -7 & -20 & -12 \\
  &   & 3 & 15 & 24 & 12 \\
\hline
  & 1 & 5 & 8 & 4 & 0
\end{array}
$$

Thus, $q(x) = x^3 + 5x^2 + 8x + 4$, and our original equation is equivalent to

$$(x - 3)(x^3 + 5x^2 + 8x + 4) = 0$$

Now, $x = -2$ is also a root of this equation. But $x = -2$ surely is not a root of the equation $x - 3 = 0$. Therefore, $x = -2$ must be a root of the reduced equation $x^3 + 5x^2 + 8x + 4 = 0$. Again the factor theorem is applicable. Since $x = -2$ is a root of the reduced equation, $x + 2$ must be a factor of $x^3 + 5x^2 + 8x + 4$. That is,

$$x^3 + 5x^2 + 8x + 4 = (x + 2) \cdot p(x) \qquad \text{for some polynomial } p(x)$$

We can use synthetic division to determine $p(x)$:

$$
\begin{array}{r|rrrr}
-2 & 1 & 5 & 8 & 4 \\
   &   & -2 & -6 & -4 \\
\hline
   & 1 & 3 & 2 & 0
\end{array}
$$

Thus, $p(x) = x^2 + 3x + 2$, and our reduced equation can be written

$$(x + 2)(x^2 + 3x + 2) = 0$$

This gives rise to a second reduced equation,

$$x^2 + 3x + 2 = 0$$

In this case, the roots are readily obtained by factoring. We have

$$x^2 + 3x + 2 = 0$$
$$(x + 2)(x + 1) = 0$$

$$
\begin{array}{c|c}
x + 2 = 0 & x + 1 = 0 \\
x = -2 & x = -1
\end{array}
$$

Now, of the two roots we've just found, the $-2$ happens to be one of the roots that we were initially given in the statement of the problem. The $-1$, on the other hand, is a distinct additional root. In summary, then, we have three distinct roots: $3$, $-2$, and $-1$. The root $-2$ has multiplicity 2. As you will see in the next section, a fourth-degree equation can have at most four (not necessarily distinct) roots. So in the case at hand, we have found all the roots of the given equation. ∎

**EXAMPLE 6**   Find a polynomial equation for which $-1$, $4$, and $5$ are roots.

**Solution**   If $f(x)$ is any polynomial containing the factors $(x + 1)$, $(x - 4)$, and $(x - 5)$, then the equation $f(x) = 0$ will certainly be satisfied when $x = -1$, $x = 4$, or

$x = 5$. The simplest polynomial equation in this case is therefore

$$(x + 1)(x - 4)(x - 5) = 0$$

After carrying out the indicated multiplication, we find that this last equation can be written $x^3 - 8x^2 + 11x + 20 = 0$. This is a polynomial equation that satisfies the required conditions.

## EXERCISE SET 11.2

### A

*In Exercises 1–6, determine whether the given value for the variable is a root of the equation.*

1. $12x - 8 = 112$; $x = 10$
2. $12x^2 - x - 20 = 0$; $x = \frac{5}{4}$
3. $x^2 - 2x - 4 = 0$; $x = 1 - \sqrt{5}$
4. $1 - x + x^2 - x^3 = 0$; $x = -1$
5. $2x^2 - 3x + 1 = 0$; $x = \frac{1}{2}$
6. $(x - 1)(x - 2)(x - 3) = 0$; $x = 4$

*In Exercises 7–14, determine whether the given value is a zero of the function.*

7. $f(x) = 3x - 2$; $x = \frac{2}{3}$
8. $g(x) = 1 + x^2$; $x = -1$
9. $h(x) = 5x^3 - x^2 + 2x + 8$; $x = -1$
10. $F(x) = -2x^5 + 3x^4 + 8x^3$; $x = 0$
11. $f(t) = 1 + 2t + t^3 - t^5$; $t = 2$
12. $f(t) = 1 + 2t + t^3 - t^5$; $t = \sqrt{2}$
13. $f(x) = 2x^3 - 3x + 1$;  (a) $x = \frac{1}{2}(\sqrt{3} - 1)$;
    (b) $x = \frac{1}{2}(\sqrt{3} + 1)$
14. $g(x) = x^4 + 8x^3 + 9x^2 - 8x - 10$;  (a) $x = 1$;
    (b) $x = \sqrt{6} - 4$;  (c) $x = \sqrt{6} + 4$
15. List the distinct roots of each of the following equations. In the case of a repeated root, give its multiplicity.
    (a) $(x - 1)(x - 2)^3(x - 3) = 0$
    (b) $(x - 1)(x - 1)(x - 1) = 0$
    (c) $(x - 5)^6(x + 1)^4 = 0$    (d) $x^5(x - 1) = 0$
16. In this exercise, we verify that the remainder theorem is valid for the cubic polynomial

    $g(x) = ax^3 + bx^2 + cx + d$

    (a) Compute $g(r)$.
    (b) Using synthetic division, divide $g(x)$ by $x - r$. Check that the remainder you obtain is the same as the answer in part (a).

*In Exercises 17–24, use the remainder theorem (as in Example 3) to evaluate $f(x)$ for the given value of $x$.*

17. $f(x) = 4x^3 - 6x^2 + x - 5$; $x = -3$

18. $f(x) = 2x^3 - x - 4$; $x = 4$
19. $f(x) = 6x^4 + 5x^3 - 8x^2 - 10x - 3$; $x = \frac{1}{2}$
20. $f(x) = x^5 - x^4 - x^3 - x^2 - x - 1$; $x = -2$
21. $f(x) = x^2 + 3x - 4$; $x = -\sqrt{2}$
22. $f(x) = -3x^3 + 8x^2 - 12$; $x = 5$
23. $f(x) = \frac{1}{2}x^3 - 5x^2 - 13x - 10$; $x = 12$
24. $f(x) = x^7 - 7x^6 + 5x^4 + 1$; $x = -3$

*In Exercises 25–38, you are given a polynomial equation and one or more roots. Solve each equation using the method shown in Examples 4 and 5. To help you decide if you have found all the roots in each case, you may rely on the following theorem, discussed in the next section: A polynomial equation of degree n has at most n (not necessarily distinct) roots.*

25. $x^3 - 4x^2 - 9x + 36 = 0$; $-3$ is a root.
26. $x^3 + 7x^2 + 11x + 5 = 0$; $-1$ is a root.
27. $x^3 + x^2 - 7x + 5 = 0$; $1$ is a root.
28. $x^3 + 8x^2 - 3x - 24 = 0$; $-8$ is a root.
29. $3x^3 - 5x^2 - 16x + 12 = 0$; $-2$ is a root.
30. $2x^3 - 5x^2 - 46x + 24 = 0$; $6$ is a root.
31. $2x^3 + x^2 - 5x - 3 = 0$; $-\frac{3}{2}$ is a root.
32. $6x^4 - 19x^3 - 25x^2 + 18x + 8 = 0$; $4$ and $-\frac{1}{3}$ are roots.
33. $x^4 - 15x^3 + 75x^2 - 125x = 0$; $5$ is a root.
34. $2x^3 + 5x^2 - 8x - 20 = 0$; $2$ is a root.
35. $x^4 + 2x^3 - 23x^2 - 24x + 144 = 0$; $-4$ and $3$ are roots.
36. $6x^5 + 5x^4 - 29x^3 - 25x^2 - 5x = 0$; $\sqrt{5}$ and $-\frac{1}{3}$ are roots.
37. $x^3 + 7x^2 - 19x - 9 = 0$; $-9$ is a root.
38. $4x^5 - 15x^4 + 8x^3 + 19x^2 - 12x - 4 = 0$; $1$, $2$, and $-1$ are roots.
39. Find a polynomial equation of degree 3 in which the coefficient of $x^3$ is 1 and in which $3$, $-4$, and $5$ are roots.
40. Find a polynomial equation of degree 3 in which the coefficients are integers and $\frac{1}{2}$, $\frac{2}{5}$, and $-\frac{3}{4}$ are roots.

41. Find a cubic equation of the form

    $$x^3 + bx^2 + cx + d = 0$$

    in which $-1$ and $-6$ are roots, $-1$ of multiplicity 2.

42. Find a fourth-degree equation in which the coefficient of $x^4$ is 1 and for which $0$, $-1$, and $1$ are roots, where the multiplicity of 0 is 2.

43. Find $c$ and $d$ such that 1 and 2 are roots of the equation $x^3 - 4x^2 + cx + d = 0$.

44. Let $g(x) = ax^2 + bx + c$. Show that when $g(x)$ is divided by $x + r$, the remainder is $g(-r)$.

⌐c⌐ In Exercises 45–48, use the remainder theorem (as in Example 3) to evaluate $f(x)$ for the given value of x. Use a calculator, and round off your answers to two decimal places.

45. $f(x) = x^3 - 3x^2 + 12x + 9$; $x = 1.16$

46. $f(x) = x^3 - 2x - 5$;   (a) $x = 2.09$;   (b) $x = 2.094$;
    (c) $x = 2.0945$

47. $f(x) = x^3 - 5x - 2$;   (a) $x = 2.41$;   (b) $x = 2.42$

48. $f(x) = x^4 - 2x^3 - 5x^2 + 10x - 3$;   (a) $x = -2.3$;
    (b) $x = -2.302$;   (c) $x = -2.30277$

**B**

In Exercises 49–52, determine whether the given value is a zero of the function.

49. $f(x) = \frac{1}{2}x^2 + bx + c$; $x = -b - \sqrt{b^2 - 2c}$

50. $Q(x) = ax^2 + bx + c$; $x = \dfrac{-b + \sqrt{b^2 - 4ac}}{2a}$

    Hint:   Look before you leap!

51. $F(x) = 2x^4 + 4x + 1$; $x = \frac{1}{2}(-\sqrt{2} + \sqrt{2\sqrt{2} - 2})$

52. $f(x) = x^3 - 3x^2 + 3x - 3$;   (a) $x = \sqrt[3]{2} - 1$;
    (b) $x = \sqrt[3]{2} + 1$

53. Determine values for $a$ and $b$ such that $x - 1$ is a factor of both $x^3 + x^2 + ax + b$ and $x^3 - x^2 - ax + b$.

54. Determine a quadratic equation with the given roots.

    (a) $\dfrac{a}{b}, -\dfrac{b}{a}$   (b) $-a + 2\sqrt{2b}, -a - 2\sqrt{2b}$

55. One root of the equation $x^2 + bx + 1 = 0$ is twice the other; find $b$. (There are two answers.)

56. Determine a value for $a$ such that one root of the equation $ax^2 + x - 1 = 0$ is five times the other.

## THE FUNDAMENTAL THEOREM OF ALGEBRA

**11.3**

*Every equation of algebra has as many solutions as the exponent of the highest term indicates.*

ALBERT GIRARD, 1629, as quoted in *The History of Mathematics, An Introduction*, by David M. Burton (Boston: Allyn and Bacon, 1985)

Does every polynomial equation (of degree at least 1) have a root? Or, to put the question another way, is it possible to write a polynomial equation that has no solution? Certainly, if we consider only real roots (i.e., roots that are real numbers), then it is easy to specify an equation with no real root:

$$x^2 = -1$$

This equation has no real root because the square of a real number is never negative. On the other hand, if we expand our base of operations from the real number system to the complex number system, then $x^2 = -1$ does indeed have a root. In fact, both $i$ and $-i$ are roots in this case.

As it turns out, the situation just described for the equation $x^2 = -1$ holds quite generally. That is, within the complex number system, every polynomial equation of degree at least 1 has at least one root. This is the substance of a remarkable theorem that was first proved by the great mathematician Carl Friedrich Gauss in 1799.* (Gauss was only 22 years old at the time.) Although

---

* In the opinion of George F. Simmons in his text, *Calculus with Analytic Geometry* (New York: McGraw-Hill, 1985), "Gauss was the greatest of all mathematicians and perhaps the most richly gifted genius of whom there is any record. This gigantic figure, towering at the beginning of the nineteenth century, separates the modern era in mathematics from all that went before."

there are many fundamental theorems in algebra, Gauss's result has come to be known as *the* **fundamental theorem of algebra**.

---

**The Fundamental Theorem of Algebra**

Every polynomial equation of the form

$$a_nx^n + a_{n-1}x^{n-1} + \cdots + a_1x + a_0 = 0 \qquad (n \geq 1, a_n \neq 0)$$

has at least one root among the complex numbers. (This root may be a real number.)

---

Although most of this section deals with polynomials with real coefficients, Gauss's theorem still applies in cases in which some or all of the coefficients are nonreal complex numbers. The proof of Gauss's theorem is usually given in the post-calculus course called complex variables.

The fundamental theorem of algebra asserts that every polynomial equation of degree at least 1 has a root. There are two initial observations to be made here. First, notice that the theorem says nothing about actually finding the root. Second, notice that the theorem deals only with polynomial equations. Indeed, it is easy to specify a nonpolynomial equation that does not have a root. Such an equation is $1/x = 0$. The expression on the left-hand side of this equation can never be zero because the numerator is 1.

**EXAMPLE 1**    Which of the following equations have at least one root?

(a) $x^3 - 17x^2 + 6x - 1 = 0$     (b) $\sqrt{2}x^{47} - \pi x^{25} + \sqrt{3} = 0$
(c) $x^2 - 2ix + (3 + i) = 0$

**Solution**    All three equations are polynomial equations. So, according to the fundamental theorem of algebra, each equation has at least one root.

In Chapter 2 we used factoring as a tool for solving quadratic equations. The next theorem, a consequence of the fundamental theorem of algebra, tells us that (in principle, at least) any polynomial of degree $n$ can be factored into a product of $n$ linear factors. In proving the **linear factors theorem**, we will need to use the factor theorem. If you reread the proof of that theorem in the previous section, you will see that it makes no difference whether the number $r$ appearing in the factor $x - r$ is a real number or a nonreal complex number. Thus, the factor theorem is valid in either case.

---

**The Linear Factors Theorem**

Let $f(x) = a_nx^n + a_{n-1}x^{n-1} + \cdots + a_1x + a_0$   $(n \geq 1, a_n \neq 0)$. Then $f(x)$ can be expressed as a product of $n$ linear factors:

$$f(x) = a_n(x - r_1)(x - r_2) \cdots (x - r_n)$$

(The complex numbers $r_k$ appearing in these factors are not necessarily all distinct, and some or all of the $r_k$ may be real numbers.)

---

PROOF OF THE LINEAR FACTORS THEOREM    According to the fundamental theorem of algebra, the equation $f(x) = 0$ has a root. Let us call this root $r_1$. By the factor theorem, $x - r_1$ is a factor of $f(x)$, and we can write

$$f(x) = (x - r_1) \cdot Q_1(x)$$

for some polynomial $Q_1(x)$ that has degree $n - 1$ and leading coefficient $a_n$. If the degree of $Q_1(x)$ happens to be zero, we are done. On the other hand, if the degree of $Q_1(x)$ is at least 1, another application of the fundamental theorem of algebra followed by the factor theorem gives us

$$Q_1(x) = (x - r_2) \cdot Q_2(x)$$

where the degree of $Q_2(x)$ is $n - 2$ and the leading coefficient of $Q_2(x)$ is $a_n$. We now have

$$f(x) = (x - r_1)(x - r_2) \cdot Q_2(x)$$

We continue this process until the quotient is $Q_n(x) = a_n$. As a result, we obtain

$$f(x) = (x - r_1)(x - r_2) \cdots (x - r_n)a_n$$
$$= a_n(x - r_1)(x - r_2) \cdots (x - r_n)$$

as we wished to show.

The linear factors theorem tells us that any polynomial can be expressed as a product of linear factors. The theorem gives us no information, however, as to how those factors can actually be obtained. The next example demonstrates a case in which the factors are readily obtainable; this is always the case with quadratic polynomials.

**EXAMPLE 2**    Express each of the following second-degree polynomials in the form $a_n(x - r_1)(x - r_2)$.

(a) $3x^2 - 5x - 2$    (b) $x^2 - 4x + 5$

Solution    (a) A factorization for $3x^2 - 5x - 2$ can be found by simple trial and error. We have

$$3x^2 - 5x - 2 = (3x + 1)(x - 2)$$

We now write the factor $3x + 1$ as $3(x + \frac{1}{3})$. This, in turn, can be written $3[x - (-\frac{1}{3})]$. The final factorization is then

$$3x^2 - 5x - 2 = 3[x - (-\frac{1}{3})](x - 2)$$

(b) From the factor theorem, or from our more elementary work with quadratic equations, we know that if $r_1$ and $r_2$ are the roots of the equation $x^2 - 4x + 5 = 0$, then $x - r_1$ and $x - r_2$ are the factors of $x^2 - 4x + 5$. That is, $x^2 - 4x + 5 = (x - r_1)(x - r_2)$. The values for $r_1$ and $r_2$ in this case are readily obtained by using the quadratic formula. As you can check, the results are

$$r_1 = 2 + i \quad \text{and} \quad r_2 = 2 - i$$

The required factorization is therefore

$$x^2 - 4x + 5 = [x - (2 + i)][x - (2 - i)]$$

Using the linear factors theorem, we can show that every polynomial equation of degree $n$  $(n \geq 1)$ has exactly $n$ roots. To help you follow the reasoning, we make two preliminary comments. First, we agree that a root of multiplicity $k$ will be counted as $k$ roots. For example, although the third-degree equation $(x - 1)(x - 4)^2 = 0$ has only two distinct roots, namely, 1 and 4, it has three roots *if* we agree to count the repeated root 4 two times. The second preliminary comment concerns the zero-product property, which states that $pq = 0$ if and only if $p = 0$ or $q = 0$. When we stated this in Section 2.2, we were working within the real number system. However, the property is also valid within the complex number system. Now let's state and prove our theorem.

---

### Theorem

Every polynomial equation of degree $n \geq 1$ has exactly $n$ roots, where a root of multiplicity $k$ is counted $k$ times.

---

PROOF OF THEOREM   Using the linear factors theorem, we can write the $n$th-degree polynomial equation $f(x) = a_n x^n + a_{n-1} x^{n-1} + \cdots + a_0$ as

$$f(x) = a_n(x - r_1)(x - r_2) \cdots (x - r_n) = 0 \qquad (a_n \neq 0) \tag{1}$$

By the factor theorem, each of the numbers $r_1, r_2, \ldots, r_n$ is a root. Some of these numbers may in fact be equal; in other words, we may have repeated roots in this list. In any case, if we agree to count a root of multiplicity $k$ as $k$ roots, then we obtain exactly $n$ roots from the list $r_1, r_2, \ldots, r_n$. Furthermore, the equation $f(x) = 0$ can have no other roots, as we now show. Suppose that $r$ is any number distinct from all the numbers $r_1, r_2, \ldots, r_n$. Replacing $x$ with $r$ in equation (1) yields

$$f(r) = a_n(r - r_1)(r - r_2) \cdots (r - r_n)$$

But the expression on the right-hand side of this last equation cannot be zero, because none of the factors is zero. Thus, $f(r)$ is not zero, and so $r$ is not a root. This completes the proof of the theorem.

**EXAMPLE 3**   Find a polynomial $f(x)$ with leading coefficient 1 such that the equation $f(x) = 0$ has only those roots specified in Table 1. What is the degree of this polynomial?

**Solution**   The expressions $(x - 3)^2$, $(x + 2)$, and $(x - 0)^2$ all must appear as factors of $f(x)$, and furthermore, no other linear factor can appear. The form of $f(x)$ is therefore

$$f(x) = a_n(x - 3)^2(x + 2)(x - 0)^2$$

Since the leading coefficient $a_n$ is to be 1, we can rewrite this last equation as

$$f(x) = x^2(x - 3)^2(x + 2)$$

This is the required polynomial. The degree here is 5. This can be seen either by multiplying out the factors or by simply adding the multiplicities of the roots in Table 1.

**TABLE 1**

| ROOT | MULTIPLICITY |
|------|--------------|
| 3 | 2 |
| −2 | 1 |
| 0 | 2 |

**EXAMPLE 4**  Find a quadratic function $f$ that has zeros of 3 and 5 and a graph that passes through the point $(2, -9)$.

**Solution**  The general form of a quadratic function with 3 and 5 as zeros is $f(x) = a_n(x - 3)(x - 5)$. Since the graph passes through $(2, -9)$, we have

$$-9 = a_n(2 - 3)(2 - 5) = a_n(3)$$
$$-3 = a_n$$

The required function is therefore

$$f(x) = -3(x - 3)(x - 5)$$

If we wish, we can carry out the multiplication and rewrite this as

$$f(x) = -3x^2 + 24x - 45$$

The next example shows how we can use the factored form of a polynomial $f(x)$ to determine the relationships between the coefficients of the polynomial and the roots of the equation $f(x) = 0$.

**EXAMPLE 5**  Let $r_1$ and $r_2$ be the roots of the equation $x^2 + bx + c = 0$. Show that

$$r_1 r_2 = c \qquad \text{and} \qquad r_1 + r_2 = -b$$

**Solution**  Since $r_1$ and $r_2$ are the roots of the equation $x^2 + bx + c = 0$, we have the identity

$$x^2 + bx + c = (x - r_1)(x - r_2)$$

After multiplying out the right-hand side, we can rewrite this identity as

$$x^2 + bx + c = x^2 - (r_1 + r_2)x + r_1 r_2$$

By equating coefficients, we readily obtain $r_1 r_2 = c$ and $r_1 + r_2 = -b$, as required.

The technique used in Example 5 can be used to obtain similar relationships between the roots and the coefficients of polynomial equations of any given degree. In Table 2, for instance, we show the relationships obtained in Example 5, along with the corresponding relationships that can be derived for a cubic equation. (Exercise 31 at the end of this section asks you to verify the results for the cubic equation.)

**TABLE 2**

| EQUATION | ROOTS | RELATIONSHIPS BETWEEN ROOTS AND COEFFICIENTS |
|---|---|---|
| $x^2 + bx + c = 0$ | $r_1, r_2$ | $r_1 + r_2 = -b$ <br> $r_1 r_2 = c$ |
| $x^3 + bx^2 + cx + d = 0$ | $r_1, r_2, r_3$ | $r_1 + r_2 + r_3 = -b$ <br> $r_1 r_2 + r_2 r_3 + r_3 r_1 = c$ <br> $r_1 r_2 r_3 = -d$ |

We conclude this section with some remarks concerning the solving of polynomial equations by formulas. You know that the roots of the quadratic equation $ax^2 + bx + c = 0$ are given by the formula

$$x = \frac{-b \pm \sqrt{b^2 - 4ac}}{2a}$$

The question is, are there similar formulas for the solutions of higher-degree equations? By "similar" we mean a formula involving the coefficients and radicals. To answer this question, we look at a bit of history. As early as 1700 B.C., Babylonian mathematicians were able to solve quadratic equations. This is clear from the study of the clay tablets with cuneiform numerals that archeologists have found. The ancient Greeks were also able to solve quadratic equations. Like the Babylonians, the Greeks worked without the aid of algebra as we know it. The mathematicians of ancient Greece used geometric constructions to solve equations. Of course, since all quantities were interpreted geometrically, negative roots were never considered.

The general quadratic formula was known to the Moslem mathematicians sometime before A.D. 1000. For the next 500 years, mathematicians searched for, but did not discover, a formula to solve the general cubic equation. Indeed, in 1494, Luca Pacioli stated in his text *Summa di Arithmetica* that the general cubic equation could not be solved by the algebra techniques then available. All of this was to change, however, within the next several decades.

Around 1515, the Italian mathematician Scipione del Ferro solved the cubic equation $x^3 + px + q = 0$ using algebraic techniques. This essentially constituted a solution of the seemingly more general equation $x^3 + bx^2 + cx + d = 0$. The reason for this is that if we make the substitution $x = y - b/3$ in the latter equation, the result is a cubic equation with no $y^2$-term. By 1540, the Italian mathematician Ludovico Ferrari had solved the general fourth-degree equation. Actually, at that time in Renaissance Italy, there was considerable controversy as to exactly who discovered the various formulas first. Details of the dispute can be found in any text on the history of mathematics. But for our purposes here, the point is simply that by the middle of the sixteenth century, all polynomial equations of degree 4 or less could be solved by the formulas that had been discovered. The common feature of these formulas was that they involved the coefficients, the four basic operations of arithmetic, and various radicals. For example, a formula for the solution of the equation $x^3 + px + q = 0$ is as follows:

$$x = \sqrt[3]{\frac{-q}{2} + \sqrt{\frac{q^2}{4} + \frac{p^3}{27}}} + \sqrt[3]{\frac{-q}{2} - \sqrt{\frac{q^2}{4} + \frac{p^3}{27}}}$$

To get some idea of the practical difficulties inherent in computing with this formula, try using it to show that $x = -2$ is a root of the equation $x^3 + 4x + 16 = 0$.

For more than 200 years after the cubic and quartic (fourth-degree) equations had been solved, mathematicians continued to search for a formula that would yield the solutions of the general fifth-degree equation. The first breakthrough, if it can be called that, occurred in 1770, when the French mathematician Joseph Louis Lagrange found a technique that served to unify and summarize all of the previous methods used for the equations of degrees 2, 3, and 4. However, Lagrange then showed that his technique could not work in the case of the general fifth-degree equation. While we will not describe the details of Lagrange's work here, it is worth pointing out that he relied on the types of relationships between roots and coefficients that we looked at in Table 2 of this section.

Finally, in 1828, the Norwegian mathematician Niels Henrik Abel proved that for the general polynomial equation of degree 5 or higher, there could be no formula yielding the solutions in terms of the coefficients and radicals. This is not to say that such equations do not possess solutions. In fact they must, as we saw earlier in this section. It is just that we cannot in every case express the solutions in terms of the coefficients and radicals. For example, it can be shown that the equation $x^5 - 6x + 3 = 0$ has a real root between 0 and 1, but this number cannot be expressed in terms of the coefficients and radicals. (We can, however, compute the root to as many decimal places as we wish, as you will see in the next section.) In 1830, the French mathematician Evariste Galois completed matters by giving conditions for determining exactly which polynomial equations can be solved in terms of coefficients and radicals.

## EXERCISE SET 11.3

### A

*According to the fundamental theorem of algebra, which of the equations in Exercises 1 and 2 have at least one root?*

1. (a) $x^5 - 14x^4 + 8x + 53 = 0$
   (b) $(4.17)x^3 + (2.06)x^2 + (0.01)x + 1.23 = 0$
   (c) $ix^2 + (2 + 3i)x - 17 = 0$
   (d) $x^{2.1} + 3x^{0.3} + 1 = 0$

2. (a) $\sqrt{3}x^{17} + \sqrt{2}x^{13} + \sqrt{5} = 0$
   (b) $17x^{\sqrt{3}} + 13x^{\sqrt{2}} + \sqrt{5} = 0$
   (c) $\dfrac{1}{x^2 + 1} = 0$   (d) $2^{3x} - 2^x - 1 = 0$

*In Exercises 3–10, express each polynomial in the form $a_n(x - r_1)(x - r_2) \cdots (x - r_n)$.*

3. $x^2 - 2x - 3$
4. $x^3 - 2x^2 - 3x$
5. $4x^2 + 23x - 6$
6. $6x^2 + x - 12$
7. $x^2 - 5$
8. $x^2 + 5$
9. $x^2 - 10x + 26$
10. $x^3 + 2x^2 - 3x - 6$

*In Exercises 11–16, find a polynomial $f(x)$ with leading coefficient 1 such that the equation $f(x) = 0$ has the given roots and no others. If the degree of $f(x)$ is 7 or more, express $f(x)$ in factored form; otherwise, express $f(x)$ in the form $a_n x^n + a_{n-1}x^{n-1} + \cdots + a_1 x + a_0$.*

11.

| Root | 1 | −3 |
|---|---|---|
| Multiplicity | 2 | 1 |

12.

| Root | 0 | 4 |
|---|---|---|
| Multiplicity | 2 | 1 |

13.

| Root | 2 | −2 | 2i | −2i |
|---|---|---|---|---|
| Multiplicity | 1 | 1 | 1 | 1 |

14.

| Root | 2 + i | 2 − i |
|---|---|---|
| Multiplicity | 1 | 1 |

15.

| Root | $\sqrt{3}$ | $-\sqrt{3}$ | 4i | −4i |
|---|---|---|---|---|
| Multiplicity | 2 | 2 | 1 | 1 |

16.

| Root | 5 | 1 | 1 − i | 1 + i |
|---|---|---|---|---|
| Multiplicity | 2 | 3 | 1 | 1 |

*In Exercises 17–20, express the polynomial $f(x)$ in the form $a_n x^n + a_{n-1}x^{n-1} + \cdots + a_1 x + a_0$.*

17. Find a quadratic function that has zeros −4 and 9 and a graph that passes through the point $(3, 5)$.

18. Find a quadratic function that has a maximum value of 2 and that has $-2$ and $4$ as zeros.

19. Find a third-degree polynomial function that has zeros $-5, 2$, and $3$ and a graph that passes through the point $(0, 1)$.

20. Find a fourth-degree polynomial function that has zeros $\sqrt{2}, -\sqrt{2}, 1$, and $-1$ and a graph that passes through $(2, -20)$.

*In Exercises 21–27, find a quadratic equation with the given roots and no others. Write your answers in the form $Ax^2 + Bx + C = 0$. Suggestion: Make use of Table 2.*

21. $r_1 = -i, r_2 = -\sqrt{3}$

22. $r_1 = 1 + i\sqrt{3}, r_2 = 1 - i\sqrt{3}$

23. $r_1 = 9, r_2 = -6$

24. $r_1 = 5, r_2 = \frac{3}{4}$

25. $r_1 = 1 + \sqrt{5}, r_2 = 1 - \sqrt{5}$

26. $r_1 = 6 - 5i, r_2 = 6 + 5i$

27. $r_1 = a + \sqrt{b}, r_2 = a - \sqrt{b}$ $(b > 0)$

28. True or false: mark **T** if the statement is true without exception; otherwise mark **F**.
    (a) Every equation has a root.
    (b) Every polynomial equation of degree at least 1 has a root.
    (c) Every polynomial equation of degree 4 has four distinct roots.
    (d) No cubic equation can have a root of multiplicity 4.
    (e) The degree of the polynomial
    $$x(x - 1)(x - 2)(x - 3)$$
    is 3.
    (f) The degree of the polynomial $6(x + 1)^2(x - 5)^4$ is 6.
    (g) Every polynomial of degree $n$, where $n \geq 1$, can be written in the form
    $$(x - r_1)(x - r_2) \cdots (x - r_n)$$
    (h) The sum of the roots of the polynomial equation $x^2 - px + q = 0$ is $p$.
    (i) The product of the roots of the polynomial equation $2x^3 - x^2 + 3x - 1 = 0$ is 1. *Hint:* Use Table 2.
    (j) Although a polynomial equation of degree $n$ may have $n$ distinct roots, the fundamental theorem of algebra tells us how to find only one of the roots.
    (k) Every polynomial equation of degree $n \geq 1$ has at least one real root.
    (l) If all the coefficients in a polynomial equation are real, then at least one of the roots must be real.

(m) Every cubic equation whose roots are $\sqrt{5}, \sqrt{6}$, and $\sqrt{7}$ can be written in the form
$$a_n(x - \sqrt{5})(x - \sqrt{6})(x - \sqrt{7}) = 0$$

**B**

29. Express the polynomial $x^4 + 64$ as a product of four linear factors. *Hint:* First add and subtract the quantity $16x^2$ so that you can use the difference-of-squares factoring formula.

30. Suppose that $p$ and $q$ are positive integers with $p > q$. Find a quadratic equation with integer coefficients whose roots are
$$\frac{\sqrt{p}}{\sqrt{p} \pm \sqrt{p - q}}$$

31. Let $r_1, r_2$, and $r_3$ be the roots of the equation $x^3 + bx^2 + cx + d = 0$. Use the method shown in Example 5 to verify the following relationships:
$$r_1 + r_2 + r_3 = -b$$
$$r_1r_2 + r_2r_3 + r_3r_1 = c$$
$$r_1r_2r_3 = -d$$

32. Let $r_1, r_2, r_3$, and $r_4$ be the roots of the equation $x^4 + bx^3 + cx^2 + dx + e = 0$. Use the method shown in Example 5 to prove the following facts:
$$r_1 + r_2 + r_3 + r_4 = -b$$
$$r_1r_2 + r_2r_3 + r_3r_4 + r_4r_1 + r_2r_4 + r_3r_1 = c$$
$$r_1r_2r_3 + r_2r_3r_4 + r_3r_4r_1 + r_4r_1r_2 = -d$$
$$r_1r_2r_3r_4 = e$$

33. Solve the equation $x^3 - 4x^2 - 9x + 36 = 0$, given that the sum of two of the roots is 0. *Suggestion:* Use Table 2.

34. Solve the equation $x^3 - 12x + 16 = 0$, given that two of the roots are equal. *Suggestion:* Use Table 2.

35. Let $\alpha$ and $\beta$ be the roots of the polynomial equation $x^2 + bx + c = 0$.
    (a) Show that $\alpha^2 + \beta^2 = b^2 - 2c$. *Hint:* Use Table 2 along with the identity $\alpha^2 + \beta^2 = (\alpha + \beta)^2 - 2\alpha\beta$.
    (b) Show that $\dfrac{1}{\alpha^2} + \dfrac{1}{\beta^2} = \dfrac{b^2 - 2c}{c^2}$.
    (c) Show that $\alpha^3 + \beta^3 = -b(b^2 - 3c)$.

**C**

36. (a) Let $r_1, r_2, r_3$, and $r_4$ be four real roots of the equation $x^4 + ax^2 + bx + c = 0$. Show that $r_1 + r_2 + r_3 + r_4 = 0$. *Hint:* Use the first formula in Exercise 32.

(b) Suppose a circle intersects the parabola $y = x^2$ in four points, $(x_1, y_1), \ldots, (x_4, y_4)$. Show that $x_1 + x_2 + x_3 + x_4 = 0$. *Hint:* Use the result in part (a).

37. (a) Let $r_1$, $r_2$, and $r_3$ be three distinct numbers that are roots of the equation $f(x) = Ax^2 + Bx + C = 0$. Show that $f(x) = 0$ for all values of $x$. *Hint:* You need to show that $A = B = C = 0$. First show that both $A$ and $B$ are zero as follows. If either $A$ or $B$ were nonzero, then the equation $f(x) = 0$ would be a polynomial equation of degree at most 2 with three distinct roots. Why is that impossible?

(b) Use the result in part (a) to prove the following identity:

$$\frac{a^2 - x^2}{(a - b)(a - c)} + \frac{b^2 - x^2}{(b - c)(b - a)}$$
$$+ \frac{c^2 - x^2}{(c - a)(c - b)} - 1 = 0$$

*Hint:* Let $f(x)$ denote the quadratic expression on the left-hand side of the equation. Compute $f(a)$, $f(b)$, and $f(c)$.

38. Prove that the following equation is an identity:

$$\frac{(x - a)(x - b)c^2}{(c - a)(c - b)} + \frac{(x - b)(x - c)a^2}{(a - b)(a - c)}$$
$$+ \frac{(x - c)(x - a)b^2}{(b - c)(b - a)} - x^2 = 0$$

*Hint:* Use the result in Exercise 37(a).

## RATIONAL AND IRRATIONAL ROOTS

### 11.4

As we saw in the previous section, not every polynomial equation has a real root. Furthermore, even if a polynomial equation does possess a real root, that root isn't necessarily a rational number. (The equation $x^2 = 2$ provides a simple example.) If a polynomial equation with integer coefficients does have a rational root, however, we can find that root by applying the **rational roots theorem**, which we now state.

> **The Rational Roots Theorem**
>
> Consider the polynomial equation
>
> $$a_n x^n + a_{n-1} x^{n-1} + \cdots + a_1 x + a_0 = 0 \qquad (n \geq 1, a_n \neq 0)$$
>
> and suppose that all the coefficients are integers. Let $p/q$ be a rational number, where $p$ and $q$ have no common factors other than $\pm 1$. If $p/q$ is a root of the equation, then $p$ is a factor of $a_0$ and $q$ is a factor of $a_n$.

A proof of the rational roots theorem is outlined in Exercise 39 at the end of this section. For the moment, though, let's just see why the theorem is plausible. Suppose that the two rational numbers $a/b$ and $c/d$ are the roots of a certain quadratic equation. Then, from our experience with quadratics (or by the linear factors theorem), we know that the equation can be written in the form

$$k\left(x - \frac{a}{b}\right)\left(x - \frac{c}{d}\right) = 0 \tag{1}$$

where $k$ is a constant. Now, as Exercise 28 asks you to check, if we carry out the multiplication and clear of fractions, equation (1) becomes

$$(kbd)x^2 - (kad + kbc)x + kac = 0 \tag{2}$$

Observe that $a$ and $c$ (the numerators of the two roots) are factors of the constant term $kac$ in equation (2), just as the rational roots theorem asserts. Furthermore, $b$ and $d$ (the denominators of the roots) are factors of the coefficient of the $x^2$-term in equation (2), again as the theorem asserts.

The following example shows how the rational roots theorem can be used to solve a polynomial equation.

**EXAMPLE 1**    Find the rational roots (if any) of the equation $2x^3 - x^2 - 9x - 4 = 0$. Then solve the equation.

Solution    First we list the factors of $a_0$, the factors of $a_n$, and the possibilities for rational roots:

$$\text{factors of } a_0 = 4: \quad \pm1, \pm2, \pm4$$
$$\text{factors of } a_3 = 2: \quad \pm1, \pm2$$
$$\text{possible rational roots:} \quad \pm\frac{1}{1}, \pm\frac{1}{2}, \pm\frac{2}{1}, \pm\frac{4}{1}$$

Now we can use synthetic division to test whether or not any of these possibilities is a root. (A zero remainder will tell us that we have a root.) As you can check, the first three possibilities ($1$, $-1$, and $\frac{1}{2}$) are not roots. However, using $-\frac{1}{2}$, we have

$$
\begin{array}{r|rrrr}
-1/2 & 2 & -1 & -9 & -4 \\
     &   & -1 &  1 &  4 \\
\hline
     & 2 & -2 & -8 &  0
\end{array}
$$

Thus, $x = -\frac{1}{2}$ is a root. We could now continue to check the remaining possibilities in this same manner. At this point, however, it is simpler to consider the reduced equation $2x^2 - 2x - 8 = 0$, or $x^2 - x - 4 = 0$. Since this is a quadratic equation, it can be solved directly. We have

$$x = \frac{-(-1) \pm \sqrt{(-1)^2 - 4(1)(-4)}}{2(1)} = \frac{1 \pm \sqrt{17}}{2}$$

We have now determined three distinct roots. Since the degree of the original equation is 3, there can be no other roots. We conclude that $x = -\frac{1}{2}$ is the only rational root. The three roots of the equation are $-\frac{1}{2}$, $(1 \pm \sqrt{17})/2$.

As Example 1 indicates, the number of possibilities for rational roots can be relatively large, even for rather simple equations. The next theorem we develop allows us to reduce the number of possibilities. We say that a real number $B$ is an **upper bound** for the roots of an equation if every real root is less than or equal to $B$. Similarly, a real number $b$ is a **lower bound** if every real root is greater than or equal to $b$. The following theorem tells us how synthetic division can be used in determining upper and lower bounds for roots.

> ### The Upper and Lower Bound Theorem for Real Roots
>
> Consider the polynomial equation
>
> $$f(x) = a_n x^n + a_{n-1} x^{n-1} + \cdots + a_1 x + a_0 = 0$$
>
> where all of the coefficients are real numbers and $a_n$ is positive.
>
> 1. If we use synthetic division to divide $f(x)$ by $x - B$, where $B > 0$, and we obtain a third row containing no negative numbers, then $B$ is an upper bound for the real roots of $f(x) = 0$.
> 2. If we use synthetic division to divide $f(x)$ by $x - b$, where $b < 0$, and we obtain a third row in which the numbers are alternately positive and negative, then $b$ is a lower bound for the real roots of $f(x) = 0$. (In determining whether the signs alternate in the third row, zeros are ignored.)

We will prove the first part of this theorem. A proof of the second part can be developed along similar lines. To prove the first part of the theorem, we use the division algorithm to write

$$f(x) = (x - B) \cdot Q(x) + R \tag{3}$$

The remainder $R$ here is a constant that may be zero. To show that $B$ is an upper bound, we must show that any number greater than $B$ is not a root. Toward this end, let $p$ be a number that is greater than $B$. Note that $p$ must be positive, since $B$ is positive. Then with $x = p$, equation (3) becomes

$$f(p) = (p - B) \cdot Q(p) + R \tag{4}$$

We are now going to show that the right-hand side of equation (4) is positive. This will tell us that $p$ is not a root. First, look at the factor $(p - B)$. This is positive, since $p$ is greater than $B$. Next consider $Q(p)$. By hypothesis, the coefficients of $Q(x)$ are all nonnegative. Furthermore, the leading coefficient of $Q(x)$ is a number $a_n$ that is positive. Since $p$ is also positive, it follows that $Q(p)$ must be positive. Finally, the number $R$ is nonnegative because, in the synthetic division of $f(x)$ by $x - B$, all the numbers in the third row are non-negative. It now follows that the right-hand side of equation (4) is positive. Consequently, $f(p)$ is not zero and $p$ is not a root of the equation $f(x) = 0$. This is what we wished to show.

**EXAMPLE 2**  Determine the rational roots, or show that none exist, for the equation

$$\frac{1}{4}x^4 - \frac{3}{4}x^3 + \frac{17}{4}x^2 + 4x + 5 = 0$$

**Solution**  We will use the rational roots theorem along with the upper and lower bound theorem. First of all, if we are to apply the rational roots theorem, then our equation must have integer coefficients. In view of this, we multiply both sides of the given equation by 4 to obtain

$$x^4 - 3x^3 + 17x^2 + 16x + 20 = 0$$

As in the previous example, we list the factors of $a_0$, the factors of $a_n$, and the possibilities for rational roots:

$$\text{factors of } a_0 = 20: \quad \pm 1, \pm 2, \pm 4, \pm 5, \pm 10, \pm 20$$
$$\text{factors of } a_4 = 1: \quad \pm 1$$
$$\text{possible rational roots:} \quad \pm 1, \pm 2, \pm 4, \pm 5, \pm 10, \pm 20$$

Our strategy here will be to first check for positive rational roots, beginning with 1 and working upward. The checks for $x = 1$, $x = 2$, and $x = 4$ are as follows:

$$
\begin{array}{r|rrrrr}
1 & 1 & -3 & 17 & 16 & 20 \\
  &   & 1 & -2 & 15 & 31 \\
\hline
  & 1 & -2 & 15 & 31 & 51
\end{array}
\qquad
\begin{array}{r|rrrrr}
2 & 1 & -3 & 17 & 16 & 20 \\
  &   & 2 & -2 & 30 & 92 \\
\hline
  & 1 & -1 & 15 & 46 & 112
\end{array}
\qquad
\begin{array}{r|rrrrr}
4 & 1 & -3 & 17 & 16 & 20 \\
  &   & 4 & 4 & 84 & 400 \\
\hline
  & 1 & 1 & 21 & 100 & 420
\end{array}
$$

As you can see, none of the remainders here is zero. However, notice that in the division corresponding to $x = 4$, all the numbers appearing in the third row are nonnegative. It therefore follows that 4 is an upper bound for the roots of the given equation. In view of this, we needn't bother to check the remaining values $x = 5$, $x = 10$, and $x = 20$, since none of those can be roots. At this point, we can conclude that the given equation has no positive rational roots.

Next we check for negative rational roots, beginning with $-1$ and working downward (if necessary). Checking $x = -1$, we have

$$
\begin{array}{r|rrrrr}
-1 & 1 & -3 & 17 & 16 & 20 \\
   &   & -1 & 4 & -21 & 5 \\
\hline
   & 1 & -4 & 21 & -5 & 25
\end{array}
$$

Two conclusions can be drawn from this synthetic division. First, $x = -1$ is not a root of the equation. Second, $-1$ is a lower bound for the roots because the signs in the third row of the synthetic division alternate. This means that we needn't bother to check whether any of the numbers $-2, -4, -5, -15$, or $-20$ are roots; none of them can be roots, since they are all less than $-1$.

Let us summarize our results. We have shown that the given equation has no positive rational roots and no negative rational roots. Furthermore, by inspection we see that zero is not a root of the equation. Thus, the given equation possesses no rational roots. ■

We conclude this section by demonstrating a method for approximating irrational roots. The method depends on the **location theorem**.

> ### The Location Theorem
>
> Let $f(x)$ be a polynomial, all of whose coefficients are real numbers. If $a$ and $b$ are real numbers such that $f(a)$ and $f(b)$ have opposite signs, then the equation $f(x) = 0$ has at least one real root between $a$ and $b$.

Figure 1 indicates why this theorem is plausible. If the point $(a, f(a))$ lies below the $x$-axis, and $(b, f(b))$ lies above the $x$-axis, then it certainly seems that the graph of $f$ must cross the $x$-axis at some point $x_0$ between $a$ and $b$. At this intercept, we have $f(x_0) = 0$; that is, $x_0$ is a root of the equation $f(x) = 0$. (The location theorem is a special case of the so-called intermediate value theorem; the proof is usually discussed in calculus courses.)

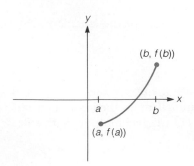

**FIGURE 1**

Our technique for approximating (or "locating") irrational roots uses the **method of successive approximations**. We will demonstrate this method in Example 3.

**EXAMPLE 3** The following equation has exactly one positive root. Locate this root between successive hundredths.

$$f(x) = x^3 + 2x - 4 = 0$$

Solution First we need to find two numbers $a$ and $b$ such that $f(a)$ and $f(b)$ have opposite signs. By inspection (or by trial and error), we find that $f(1) = -1$ and $f(2) = 8$. Thus, according to the location theorem, the equation $f(x) = 0$ has a real root in the interval $(1, 2)$.

Now that we have located the root between successive integers, we can locate it between successive tenths. We compute $f(1.0)$, $f(1.1)$, $f(1.2)$, and so on, up through $f(2.0)$ if necessary, until we find a sign change. As you can verify, using synthetic division and a calculator, the results are as follows:

$$f(1.0) = -1$$
$$f(1.1) = -0.469$$
$$f(1.2) = \phantom{-}0.128$$  ⟵ sign change

This shows that the root lies between 1.1 and 1.2.

Having located the root between successive tenths, we follow a similar process to locate the root between successive hundredths. Using synthetic division and a calculator, we obtain these results:

$$f(1.10) = -0.469 \qquad f(1.15) \approx -0.179$$
$$f(1.11) \approx -0.412 \qquad f(1.16) \approx -0.119$$
$$f(1.12) \approx -0.355 \qquad f(1.17) \approx -0.058$$
$$f(1.13) \approx -0.297 \qquad f(1.18) \approx \phantom{-}0.003$$  ⟵ sign change
$$f(1.14) \approx -0.238$$

This shows that the root lies between 1.17 and 1.18, so we've located the root between successive hundredths, as required.

The procedure described in Example 3 could be continued to yield closer and closer approximations for the required root. We note in passing that such calculations are easily handled with the aid of a programmable calculator or a computer. [See, for example, page 47 in the paperback text *Calculus by Calculator* by Maurice D. Weir (Englewood Cliffs, N.J.: Prentice-Hall, 1982).]

# EXERCISE SET 11.4

## A

*In Exercises 1–6, list the possibilities for rational roots.*

1. $4x^3 - 9x^2 - 15x + 3 = 0$

2. $x^4 - x^3 + 10x^2 - 24 = 0$

3. $8x^5 - x^2 + 9 = 0$

4. $18x^4 - 10x^3 + x^2 - 4 = 0$

5. $\frac{2}{3}x^3 - x^2 - 5x + 2 = 0$

6. $\frac{1}{2}x^4 - 5x^3 + \frac{4}{3}x^2 + 8x - \frac{1}{3} = 0$

*In Exercises 7–12, show that each equation has no rational roots.*

7. $x^3 - 3x + 1 = 0$

8. $x^3 + 8x^2 - 1 = 0$

9. $x^3 + x^2 - x + 1 = 0$

10. $x^4 + 4x^3 + 4x^2 - 16 = 0$

11. $12x^4 - x^2 - 6 = 0$

12. $4x^5 - x^4 - x^3 - x^2 + x - 8 = 0$

*For Exercises 13–25, find the rational roots of each equation and then solve the equation. (Use the rational roots theorem and the upper and lower bound theorem, as in Example 2.)*

13. $x^3 + 3x^2 - x - 3 = 0$

14. $2x^3 - 5x^2 - 3x + 9 = 0$

15. $4x^3 + x^2 - 20x - 5 = 0$

16. $3x^3 - 16x^2 + 17x - 4 = 0$

17. $9x^3 + 18x^2 + 11x + 2 = 0$

18. $4x^3 - 10x^2 - 25x + 4 = 0$

19. $x^4 + x^3 - 25x^2 - x + 24 = 0$

20. $10x^4 + 107x^3 + 301x^2 + 171x + 23 = 0$

21. $x^4 - 4x^3 + 6x^2 - 4x + 1 = 0$

22. $24x^3 - 46x^2 + 29x - 6 = 0$

23. $x^3 - \frac{5}{2}x^2 - 23x + 12 = 0$

24. $x^3 - \frac{17}{3}x^2 - \frac{10}{3}x + 8 = 0$

25. $2x^4 - \frac{9}{10}x^3 - \frac{29}{10}x^2 + \frac{27}{20}x - \frac{3}{20} = 0$

*In Exercises 26 and 27, determine integral upper and lower bounds for the real roots of the equations. (Follow the method used within the solution of Example 2.)*

26. (a) $x^3 + 2x^2 - 5x + 20 = 0$
    (b) $x^5 - 3x^2 + 100 = 0$

27. (a) $5x^4 - 10x - 12 = 0$
    (b) $3x^4 - 4x^3 + 5x^2 - 2x - 4 = 0$
    (c) $2x^4 - 7x^3 - 5x^2 + 28x - 12 = 0$

28. Referring to equation (1) in this section, multiply out the left-hand side and then clear the equation of fractions. Check that your result agrees with equation (2).

[c] *In Exercises 29–34, each equation has exactly one positive root. In each case, locate the root between successive hundredths. Use a calculator.*

29. $x^3 + x - 1 = 0$    30. $x^3 - 2x - 5 = 0$

31. $x^5 - 200 = 0$    32. $x^3 - 3x^2 + 3x - 26 = 0$

33. $x^3 - 8x^2 + 21x - 22 = 0$

34. $2x^4 - x^3 - 12x^2 - 16x - 8 = 0$

[c] *In Exercises 35–38, each equation has exactly one negative root. In each case, use a calculator to locate the root between successive hundredths.*

35. $x^3 + x^2 - 2x + 1 = 0$    36. $x^5 + 100 = 0$

37. $x^3 + 2x^2 + 2x + 101 = 0$

38. $x^4 + 4x^3 - 6x^2 - 8x - 3 = 0$

## B

39. This exercise outlines a proof of the rational roots theorem. At one point in the proof, we will need to rely on the following fact, which is proved in courses on number theory.

   FACT FROM NUMBER THEORY   Suppose that $A$, $B$, and $C$ are integers and that $A$ is a factor of the number $BC$. Then if $A$ has no factor in common with $C$ (other than $\pm 1$), $A$ must be a factor of $B$.

   (a) Let $A = 2$, $B = 8$, and $C = 5$. Verify that the fact from number theory is correct here.
   (b) Let $A = 20$, $B = 8$, and $C = 5$. Note that $A$ is a factor of $BC$, but $A$ is not a factor of $B$. Why doesn't this contradict the fact from number theory?
   (c) Now we're ready to prove the rational roots theorem. We begin with a polynomial equation with integer coefficients:

   $$a_n x^n + a_{n-1} x^{n-1} + \cdots + a_1 x + a_0 = 0$$
   $$(n \geq 1, a_n \neq 0)$$

   We assume that the rational number $p/q$ is a root of the equation and that $p$ and $q$ have no common factors other than 1. Why is the following equation now true?

   $$a_n \left(\frac{p}{q}\right)^n + a_{n-1} \left(\frac{p}{q}\right)^{n-1} + \cdots + a_1 \left(\frac{p}{q}\right) + a_0 = 0$$

   (d) Show that the last equation in part (c) can be written

   $$p(a_n p^{n-1} + a_{n-1} q p^{n-2} + \cdots + a_1 q^{n-1}) = -a_0 q^n$$

   Since $p$ is a factor of the left-hand side of this last equation, $p$ must also be a factor of the right-hand side. That is, $p$ must be a factor of $a_0 q^n$. But since $p$ and $q$ have no common factors, neither do $p$ and $q^n$. Our fact from number theory now tells us that $p$ must be a factor of $a_0$, as we wished to show. (The proof that $q$ is a factor of $a_n$ is carried out in a similar manner.)

40. The location theorem asserts that the polynomial equation $f(x) = 0$ has a root in the open interval $(a, b)$

whenever $f(a)$ and $f(b)$ have unlike signs. If $f(a)$ and $f(b)$ have the same sign, can the equation $f(x) = 0$ have a root between $a$ and $b$? *Hint:* Look at the graph of $f(x) = x^2 - 2x + 1$ with $a = 0$ and $b = 2$.

c  *Use a calculator for Exercises 41–43. Round off your answers to two decimal places.*

41. On the same set of axes, sketch the graphs of $y = x^3$ and $y = 1 - 3x$. Find the $x$-coordinate of the point where the graphs intersect.

42. On the same set of axes, sketch the graphs of $y = x^3$ and $y = x + 1$. Find the $x$-coordinate of the point where the graphs intersect.

43. Find the $x$-coordinate of the point where the curves $y = x^2 - 1$ and $y = \sqrt{x}$ meet.

## C

*In Exercises 44 and 45, you need to know that a **prime number** is a positive integer greater than 1 with no factors other than*

*itself and 1. Thus, the first seven prime numbers are 2, 3, 5, 7, 11, 13, and 17.*

44. Find all prime numbers $p$ for which the equation $x^2 + x - p = 0$ has a rational root.

45. Find all prime numbers $p$ for which the equation $x^3 + x^2 + x - p = 0$ has at least one rational root. For each value of $p$ that you find, determine the corresponding *real* roots of the equation.

46. Find all integral values of $b$ for which the equation $x^3 - b^2x^2 + 3bx - 4 = 0$ has a rational root.

47. Let $P(a, b)$ be a point on the first-quadrant portion of the curve $y = x^2$ such that the distance of $P$ from the origin is equal to $ab$. (Assume that $a \neq 0$.)

c  (a) By using the method demonstrated in this section, find the value of $a$; round off your answer to two decimal places.

(b) Find the exact value of $a$.

## CONJUGATE ROOTS AND DESCARTES'S RULE OF SIGNS

### 11.5

As you know from earlier work involving quadratic equations with real coefficients, when nonreal complex roots occur, they occur in conjugate pairs. For instance, as you can check by means of the quadratic formula, the roots of the equation $x^2 - 2x + 5 = 0$ are $1 + 2i$ and $1 - 2i$. The **conjugate roots theorem** tells us that the situation is the same for all polynomial equations with real coefficients.

> **The Conjugate Roots Theorem**
>
> Let $f(x)$ be a polynomial, all of whose coefficients are real numbers. Suppose that $a + bi$ is a root of the equation $f(x) = 0$, where $a$ and $b$ are real and $b \neq 0$. Then $a - bi$ is also a root of the equation.

To prove the conjugate roots theorem, we use four of the properties of complex conjugates listed in Chapter 1:

property 1:    $\bar{z}_1 \bar{z}_2 = \overline{z_1 z_2}$

property 2:    $(\bar{z})^m = \overline{z^m}$

property 3:    $\bar{r} = r$    for every real number $r$

property 4:    $\bar{z}_1 + \bar{z}_2 = \overline{z_1 + z_2}$

Using these properties, we can prove the theorem as follows. We begin with a polynomial with real coefficients:

$$f(x) = a_n x^n + a_{n-1}x^{n-1} + \cdots + a_1 x + a_0$$

We must show that if $z = a + bi$ is a root of $f(x) = 0$, then $\bar{z} = a - bi$ is also a root. We have

$$
\begin{aligned}
f(\bar{z}) &= a_n\bar{z}^n + a_{n-1}\bar{z}^{n-1} + \cdots + a_1\bar{z} + a_0 \\
&= \bar{a}_n\bar{z}^n + \overline{a_{n-1}}\bar{z}^{n-1} + \cdots + \bar{a}_1\bar{z} + \bar{a}_0 \qquad \text{properties 3 and 2} \\
&= \overline{a_n z^n} + \overline{a_{n-1}z^{n-1}} + \cdots + \overline{a_1 z} + \overline{a_0} \qquad \text{property 1} \\
&= \overline{a_n z^n + a_{n-1}z^{n-1} + \cdots + a_1 z + a_0} \qquad \text{property 4} \\
&= \overline{f(z)} = \bar{0} \qquad\qquad\qquad\qquad\quad f(z) = 0, \text{ since } z \text{ is a root} \\
&= 0 \qquad\qquad\qquad\qquad\qquad\qquad \text{property 3}
\end{aligned}
$$

We have now shown that $f(\bar{z}) = 0$, given that $f(z) = 0$. Thus, $\bar{z}$ is a root, as we wished to show.

Although the conjugate roots theorem concerns nonreal complex roots, it can nevertheless be used to obtain information about real roots, as the next two examples demonstrate.

**EXAMPLE 1**  Solve the equation $f(x) = 2x^4 - 3x^3 + 12x^2 + 22x - 60 = 0$, given that one root is $1 + 3i$.

**Solution**  Since all the coefficients of $f(x)$ are real numbers, we know that the conjugate of $1 + 3i$ must also be a root. Thus, $1 + 3i$ and $1 - 3i$ are roots, from which it follows that $[x - (1 + 3i)]$ and $[x - (1 - 3i)]$ are factors of $f(x)$. As you can check, the product of these two factors is $x^2 - 2x + 10$. Thus, we must have

$$f(x) = (x^2 - 2x + 10) \cdot Q(x)$$

for some polynomial $Q(x)$. We compute $Q(x)$ using long division:

$$
\begin{array}{r}
2x^2 + \phantom{0}x - \phantom{0}6 \\
x^2 - 2x + 10 \overline{\smash{\big)}\ 2x^4 - 3x^3 + 12x^2 + 22x - 60} \\
\underline{2x^4 - 4x^3 + 20x^2} \\
x^3 - \phantom{0}8x^2 + 22x \\
\underline{x^3 - \phantom{0}2x^2 + 10x} \\
-6x^2 + 12x - 60 \\
\underline{-6x^2 + 12x - 60} \\
0
\end{array}
$$

Thus, $Q(x) = 2x^2 + x - 6$, and the original equation becomes

$$f(x) = (x^2 - 2x + 10)(2x^2 + x - 6) = 0$$

We can now find any additional roots by solving the equation $2x^2 + x - 6 = 0$. We have

$$
\begin{aligned}
2x^2 + x - 6 &= 0 \\
(2x - 3)(x + 2) &= 0
\end{aligned}
$$

$$
\begin{array}{c|c}
2x - 3 = 0 & x + 2 = 0 \\
x = \dfrac{3}{2} & x = -2
\end{array}
$$

We now have four distinct roots of the original equation: $1 + 3i$, $1 - 3i$, $\frac{3}{2}$, and $-2$. Since the degree of the equation is 4, there can be no other roots.

**EXAMPLE 2**  Show that the equation $x^3 - 2x^2 + x - 1 = 0$ has at least one irrational root.

Solution  Allowing for multiple roots, the given equation has three roots. Therefore, since nonreal complex roots occur in pairs, the equation can have either two non-real complex roots or no nonreal complex roots. In either case, then, the remaining root or roots must be real. Furthermore, the equation has no rational roots; this is easily checked using the rational roots theorem. It follows now that the equation must have at least one irrational root. (The reasoning here relies on the fact that each real number is either rational or irrational, but not both.)

There is a theorem, similar to the conjugate roots theorem, that tells us about irrational roots of the form $a + b\sqrt{c}$. As background for this theorem, let us look at two preliminary examples. First, we consider the equation $x^2 - 2x - 5 = 0$. As you can check, the roots in this case are $1 + \sqrt{6}$ and $1 - \sqrt{6}$. However, it is not true in general that irrational roots such as these always occur in pairs. Consider as a second example the quadratic equation

$$(x + 2)(x - \sqrt{3}) = 0$$

or

$$x^2 + (2 - \sqrt{3})x - 2\sqrt{3} = 0$$

Here, one of the roots is $\sqrt{3}$, yet $-\sqrt{3}$ is not a root. This type of behavior can occur in polynomial equations where not all of the coefficients are rational. On the other hand, when the coefficients are all rational, we do have the following theorem. (See Exercise 11 at the end of this section for a proof.)

> **Theorem**
>
> Let $f(x)$ be a polynomial in which all the coefficients are rational. Suppose that $a + b\sqrt{c}$ is a root of the equation $f(x) = 0$, where $a$, $b$, and $c$ are rational and $\sqrt{c}$ is irrational. Then $a - b\sqrt{c}$ is also a root of the equation.

**EXAMPLE 3**  Find a quadratic equation with rational coefficients and a leading coefficient of 1 such that one of the roots is $r_1 = 4 + 5\sqrt{3}$.

Solution  If one root is $r_1 = 4 + 5\sqrt{3}$, then the other is $r_2 = 4 - 5\sqrt{3}$. We denote the required equation by $x^2 + bx + c = 0$. Then, according to Table 2 in Section 11.3, we have

$$b = -(r_1 + r_2) = -[(4 + 5\sqrt{3}) + (4 - 5\sqrt{3})] = -8$$

and

$$c = r_1 r_2 = (4 + 5\sqrt{3})(4 - 5\sqrt{3}) = 16 - 75 = -59$$

The required equation is therefore $x^2 - 8x - 59 = 0$. This answer can also be obtained without using the table. Since the roots are $4 \pm 5\sqrt{3}$, we can write the required equation as $[x - (4 + 5\sqrt{3})][x - (4 - 5\sqrt{3})] = 0$. As you can now check by multiplying the two factors, this equation is equivalent to $x^2 - 8x - 59 = 0$, as obtained previously.

We conclude this section with a discussion of **Descartes's rule of signs**. This rule, published by Descartes in 1637, provides us with information about the types of roots an equation can have, even before we attempt to solve the equation. In order to state Descartes's rule of signs, we first explain what is meant by a variation in sign in a polynomial with real coefficients. Suppose that $f(x)$ is a polynomial with real coefficients, written in descending (or ascending) powers of $x$. For example, let $f(x) = 2x^3 - 4x^2 - 3x + 1$. Then we say that there is a **variation in sign** if two successive coefficients have opposite signs. In the case of $f(x) = 2x^3 - 4x^2 - 3x + 1$, there are two variations in sign, the first occurring as we go from 2 to $-4$ and the second occurring as we go from $-3$ to 1. In looking for variations in sign, we ignore terms with zero coefficients. Table 1 shows a few more examples of how we count variations in sign.

We now state Descartes's rule of signs and look at some examples. Although the proof of this theorem is not difficult, it is rather lengthy, and we shall omit it here.

**TABLE 1**

| POLYNOMIAL | NUMBER OF VARIATIONS IN SIGN |
|---|---|
| $x^2 + 4x$ | 0 |
| $-3x^5 + x^2 + 1$ | 1 |
| $x^3 + 3x^2 - x + 6$ | 2 |

---

**Descartes's Rule of Signs**

Let $f(x)$ be a polynomial, all of whose coefficients are real numbers, and consider the equation $f(x) = 0$. Then:

a. The number of positive roots either is equal to the number of variations in sign of $f(x)$ or is less than that by an even integer.

b. The number of negative roots either is equal to the number of variations in sign of $f(-x)$ or is less than that by an even integer.

---

**EXAMPLE 4**    Use Descartes's rule of signs to obtain information regarding the roots of the equation $x^3 + 8x + 5 = 0$.

**Solution**    Let $f(x) = x^3 + 8x + 5$. Then, since there are no variations in sign for $f(x)$, we see from part a of Descartes's rule that the given equation has no positive roots. Next we compute $f(-x)$ to learn about the possibilities for negative roots: we have $f(-x) = -x^3 - 8x + 5$. So $f(-x)$ has one sign change, and consequently (by part b of Descartes's rule) the original equation has one negative root. Furthermore, notice that zero is not a root of the equation. Thus, the equation has only one real root, a negative root. Since the equation has a total of three roots, we can conclude that we have one negative root and two nonreal complex roots. The two nonreal roots will be complex conjugates. ▪

**EXAMPLE 5**    Use Descartes's rule to obtain information regarding the roots of the equation $x^4 + 3x^2 - 7x - 5 = 0$.

**Solution**    Let $f(x) = x^4 + 3x^2 - 7x - 5$. Then $f(x)$ has one variation in sign. So according to part a of Descartes's rule, the equation has one positive root. That leaves us three roots still to account for, since the degree of the equation is 4. We have $f(-x) = x^4 + 3x^2 + 7x - 5$. Since $f(-x)$ has one sign change, we know from part b of Descartes's rule that the equation has one negative root. Noting now that zero is not a root, we conclude that the two remaining roots must be nonreal complex roots. In summary, then, the equation has one positive root, one negative root, and two nonreal complex (conjugate) roots.

**EXAMPLE 6**  Use Descartes's rule to obtain information regarding the roots of the equation $f(x) = x^3 - x^2 + 3x + 2 = 0$.

**Solution**  Since $f(x)$ has two variations in sign, the given equation has either two positive roots or no positive roots. To see how many negative roots are possible, we compute $f(-x)$:

$$f(-x) = (-x)^3 - (-x)^2 + 3(-x) + 2$$
$$= -x^3 - x^2 - 3x + 2$$

Since $f(-x)$ has one variation in sign, we conclude from part b of Descartes's rule that the equation has exactly one negative root. In summary then, there are two possibilities.

Either:  one negative root and two positive roots

Or:  one negative root and two nonreal complex roots

By using Descartes's rule in Examples 4 and 5, we were able to determine the exact numbers of positive roots and negative roots for the given equations. As Example 6 indicates, however, there are cases in which a direct application of Descartes's rule provides several distinct possibilities for the types of roots, rather than a single definitive result. (Exercises 56 and 58 in the Review Exercises for this chapter illustrate a technique that is sometimes useful in gaining additional information from Descartes's rule. In particular, Exercise 58 will show you that the equation in Example 6 has no positive roots.)

## EXERCISE SET 11.5

### A

*In Exercises 1–16, an equation is given, followed by one or more roots of the equation. In each case, determine the remaining roots.*

1. $x^2 - 14x + 53 = 0$; $x = 7 - 2i$
2. $x^2 - x - \frac{1535}{4} = 0$; $x = \frac{1}{2} + 8\sqrt{6}$
3. $x^3 - 13x^2 + 59x - 87 = 0$; $x = 5 + 2i$
4. $x^4 - 10x^3 + 30x^2 - 10x - 51 = 0$; $x = 4 + i$
5. $x^4 + 10x^3 + 38x^2 + 66x + 45 = 0$; $x = -2 + i$
6. $2x^3 + 11x^2 + 30x - 18 = 0$; $x = -3 - 3i$
7. $4x^3 - 47x^2 + 232x + 61 = 0$; $x = 6 - 5i$
8. $9x^4 + 18x^3 + 20x^2 - 32x - 64 = 0$; $x = -1 + \sqrt{3}i$
9. $4x^4 - 32x^3 + 81x^2 - 72x + 162 = 0$; $x = 4 + \sqrt{2}i$
10. $2x^4 - 17x^3 + 137x^2 - 57x - 65 = 0$; $x = 4 - 7i$
11. $x^4 - 22x^3 + 140x^2 - 128x - 416 = 0$; $x = 10 + 2i$
12. $4x^4 - 8x^3 + 24x^2 - 20x + 25 = 0$; $x = \frac{1}{2}(1 + 3i)$

13. $15x^3 - 16x^2 + 9x - 2 = 0$; $x = \frac{1}{3}(1 + \sqrt{2}i)$
14. $x^5 - 5x^4 + 30x^3 + 18x^2 + 92x - 136 = 0$; $x = -1 + i\sqrt{3}$, $x = 3 - 5i$
15. $x^7 - 3x^6 - 4x^5 + 30x^4 + 27x^3 - 13x^2 - 64x + 26 = 0$; $x = 3 - 2i$, $x = -1 + i$, $x = 1$
16. $x^6 - 2x^5 - 2x^4 + 2x^3 + 2x + 1 = 0$; $x = 1 + \sqrt{2}$

*In Exercises 17–20, find a quadratic equation with rational coefficients, one of whose roots is the given number. Write your answer so that the coefficient of $x^2$ is 1. Use either of the methods shown in Example 3.*

17. $r_1 = 1 + \sqrt{6}$
18. $r_1 = 2 - \sqrt{3}$
19. $r_1 = \frac{1}{3}(2 + \sqrt{10})$
20. $r_1 = \frac{1}{2} + \frac{1}{4}\sqrt{5}$

*In Exercises 21–36, use Descartes's rule of signs to obtain information regarding the roots of the equations.*

21. $x^3 + 5 = 0$
22. $x^4 + x^2 + 1 = 0$

23. $2x^5 + 3x + 4 = 0$    24. $x^3 + 8x - 3 = 0$
25. $5x^4 + 2x - 7 = 0$    26. $x^3 - 4x^2 + x - 1 = 0$
27. $x^3 - 4x^2 - x - 1 = 0$
28. $x^8 + 4x^6 + 3x^4 + 2x^2 + 5 = 0$
29. $3x^8 + x^6 - 2x^2 - 4 = 0$
30. $12x^4 - 5x^3 - 7x^2 - 4 = 0$
31. $x^9 - 2 = 0$    32. $x^9 + 2 = 0$
33. $x^8 - 2 = 0$    34. $x^8 + 2 = 0$
35. $x^6 + x^2 - x - 1 = 0$
36. $x^7 + x^2 - x - 1 = 0$

## B

37. Consider the equation $x^4 + cx^2 + dx - e = 0$, where $c$, $d$, and $e$ are positive. Show that the equation has one positive root, one negative root, and two nonreal complex roots.

38. Consider the equation $x^n - 1 = 0$.
    (a) Show that the equation has $n - 2$ nonreal complex roots when $n$ is even.
    (b) How many nonreal complex roots are there when $n$ is odd?

39. Find the polynomial $f(x)$ of lowest degree with rational coefficients and with leading coefficient 1, such that $\sqrt{3} + 2i$ is a root of the equation $f(x) = 0$.

40. Find a quadratic polynomial $f(x)$ with integer coefficients, such that $\dfrac{2 + \sqrt{3}}{2 - \sqrt{3}}$ is a root of the equation $f(x) = 0$. *Hint:* First rationalize the given root.

## C

41. Let $f(x)$ be a polynomial, all of whose coefficients are rational. Suppose that $a + b\sqrt{c}$ is a root of $f(x) = 0$, where $a$, $b$, and $c$ are rational and $\sqrt{c}$ is irrational. Complete the following steps to prove that $a - b\sqrt{c}$ is also a root of the equation $f(x) = 0$.
    (a) If $b = 0$, we're done. Why?
    (b) (From now on we'll assume that $b \neq 0$.) Let $d(x) = [x - (a + b\sqrt{c})][x - (a - b\sqrt{c})]$. Explain why $d(a + b\sqrt{c}) = 0$.
    (c) Verify that $d(x) = (x - a)^2 - b^2 c$. Thus, $d(x)$ is a quadratic polynomial with rational coefficients.
    (d) Now suppose that we use the long division process to divide the polynomial $f(x)$ by the quadratic polynomial $d(x)$. We'll obtain a quotient $Q(x)$ and a remainder. Since the degree of $d(x)$ is 2, our remainder will be of degree 1 or less. In other words, the general form of this remainder will be $Cx + D$. Furthermore, $C$ and $D$ will have to be rational, because all of the coefficients in $f(x)$ and in $d(x)$ are rational. In summary, we have the identity

    $$f(x) = d(x) \cdot Q(x) + (Cx + D)$$

    Now in this identity, make the substitution $x = a + b\sqrt{c}$ and conclude that $C = D = 0$.
    (e) Using the result in part (d), we have

    $$f(x) = [x - (a + b\sqrt{c})][x - (a - b\sqrt{c})] \cdot Q(x)$$

    Let $x = a - b\sqrt{c}$ in this last identity and conclude from the result that $a - b\sqrt{c}$ is a root of $f(x) = 0$, as required.

42. Find a polynomial $f(x)$ with integer coefficients, such that one root of $f(x) = 0$ is $x = \sqrt{2} + \sqrt[3]{2}$.

## CHAPTER 11   SUMMARY OF PRINCIPAL TERMS AND THEOREMS

| TERMS OR THEOREM | PAGE REFERENCE | COMMENT |
|---|---|---|
| 1. Division algorithm | 568 | This theorem summarizes the results of the long division process for polynomials. Suppose that $p(x)$ and $d(x)$ are polynomials and $d(x)$ is not the zero polynomial. Then according to the *division algorithm*, there are unique polynomials $q(x)$ and $R(x)$ such that $$p(x) = d(x) \cdot q(x) + R(x)$$ where either $R(x)$ is the zero polynomial or the degree of $R(x)$ is less than the degree of $d(x)$. |
| 2. Root, solution, zero | 573 | A *root*, or *solution*, of a polynomial equation $f(x) = 0$ is a number $r$ such that $f(r) = 0$. The root $r$ is also called a *zero* of the function $f$. |
| 3. The remainder theorem | 574 | The *remainder theorem* asserts that when a polynomial $f(x)$ is divided by $x - r$, the remainder is $f(r)$. Example 3 in Section 11.2 shows how this theorem can be used with synthetic division to evaluate a polynomial. |
| 4. The factor theorem | 575 | The *factor theorem* makes two statements about a polynomial $f(x)$. First, if $f(r) = 0$, then $x - r$ is a factor of $f(x)$. And second, if $x - r$ is a factor of $f(x)$, then $f(r) = 0$. |
| 5. The fundamental theorem of algebra | 580 | Let $f(x)$ be a polynomial of degree 1 or greater. The *fundamental theorem of algebra* asserts that the equation $f(x) = 0$ has at least one root among the complex numbers. (See Example 1 in Section 11.3.) |
| 6. The linear factors theorem | 580 | Let $f(x) = a_n x^n + a_{n-1} x^{n-1} + \cdots + a_0$, where $n \geq 1$ and $a_n \neq 0$. Then this theorem asserts that $f(x)$ can be expressed as a product of $n$ linear factors: $$f(x) = a_n(x - r_1)(x - r_2) \cdots (x - r_n)$$ (The complex numbers $r_1 r_2, \ldots, r_n$ are not necessarily all distinct, and some or all of them may be real numbers.) On page 582, the linear factors theorem is used to prove that every polynomial equation of degree $n \geq 1$ has exactly $n$ roots, where a root of multiplicity $k$ is counted $k$ times. |
| 7. The rational roots theorem | 587 | Given a polynomial equation, this theorem tells us which rational numbers are candidates for roots of the equation. For a statement of the theorem, see page 587. A proof of the theorem is outlined in Exercise 39, Exercise Set 11.4. For an example of how the theorem is applied, see Example 1 in Section 11.4. |
| 8. Upper bound (for roots); lower bound (for roots) | 588 | A real number $B$ is an upper bound for the roots of an equation if every real root is less than or equal to $B$. Similarly, a real number $b$ is a lower bound for the roots if every real root is greater than or equal to $b$. |

| TERMS OR THEOREM | PAGE REFERENCE | COMMENT |
|---|---|---|
| 9. The upper and lower bound theorem for roots | 589 | This theorem tells how synthetic division can be used in determining upper and lower bounds for roots of equations. For the statement and proof of the theorem, see page 589 in Section 11.4. For a demonstration of how the theorem is applied, see Example 2 in Section 11.4. |
| 10. The location theorem | 590 | Let $f(x)$ be a polynomial, all of whose coefficients are real numbers. If $a$ and $b$ are real numbers such that $f(a)$ and $f(b)$ have opposite signs, then the equation $f(x) = 0$ has at least one root between $a$ and $b$. To see how this theorem is applied, see Example 3 in Section 11.4. |
| 11. The conjugate roots theorem | 593 | Let $f(x)$ be a polynomial, all of whose coefficients are real numbers, and suppose that $a + bi$ is a root of the equation $f(x) = 0$, where $a$ and $b$ are real numbers and $b \neq 0$. Then this theorem asserts that $a - bi$ is also a root of the equation. (In other words, for polynomial equations in which all of the coefficients are real numbers, when complex nonreal roots occur, they occur in conjugate pairs.) For illustrations of how this theorem is applied, see Examples 1 and 2 in Section 11.5. |
| 12. Variation in sign | 596 | Suppose that $f(x)$ is a polynomial with real coefficients, written in descending or ascending powers of $x$. Then a variation in sign occurs whenever two successive coefficients have opposite signs. For examples, see Table 1 in Section 11.5. |
| 13. Descartes's rule of signs | 596 | Let $f(x)$ be a polynomial, all of whose coefficients are real numbers, and consider the equation $f(x) = 0$. Then, according to Descartes's rule: <br><br> a. The number of positive roots either is equal to the number of variations in sign of $f(x)$ or is less than that by an even integer. <br> b. The number of negative roots either is equal to the number of variations in sign of $f(-x)$ or is less than that by an even integer. <br><br> Examples 4–6 in Section 11.5 show how Descartes's rule is applied. |

# CHAPTER 11    REVIEW EXERCISES

NOTE    The fifteen exercises in this set with colored problem numbers constitute a chapter test.

In Exercises 1–8, use synthetic division to find the quotients and the remainders. Also, in each case, write the result of the division in the form $p(x) = d(x) \cdot q(x) + R(x)$.

1. $\dfrac{x^4 - 2x^2 + 8}{x - 3}$

2. $\dfrac{x^3 - 1}{x - 2}$

3. $\dfrac{2x^3 - 5x^2 - 6x - 3}{x + 4}$

4. $\dfrac{x^3 + x - 3\sqrt{2}}{x - \sqrt{2}}$

5. $\dfrac{5x^2 - 19x - 4}{x + 0.2}$

6. $\dfrac{6x^3 - 3x^2 + 1}{3x - 4}$    Hint:  First divide both numerator and denominator by 3.

7. $\dfrac{x^2 - 6x + 13}{x - (3 + 2i)}$

8. $\dfrac{x^3 - 3a^2x^2 - 4a^4x + 9a^6}{x - a^2}$

9. State each of the following theorems.
   (a) The division algorithm
   (b) The remainder theorem

*In Exercises 10–16, use synthetic division and the remainder theorem to find the indicated values of the functions.*

10. $f(x) = x^5 - 10x + 4;\ f(10)$

11. $f(x) = x^4 + 2x^3 - x;\ f(-2)$

12. $f(x) = x^3 - 10x^2 + x - 1;\ f(\frac{1}{10})$

13. $f(x) = x^4 - 2a^2x^2 + 3a^3x - a^4;\ f(-a)$

14. $f(x) = x^3 + 3x^2 + 3x + 1;\ f(a - 1)$

[c] 15. $f(x) = x^3 - 1;\ f(1.1)$

[c] 16. $f(x) = x^4 + 4x^3 - 6x^2 - 8x - 2$
   (a) $f(-0.3)$ (Round off the result to two decimal places.)
   (b) $f(-0.39)$ (Round off the result to three decimal places.)
   (c) $f(-0.394)$ (Round off the result to five decimal places.)
   (d) $f(-4.907)$ (Round off the result to three decimal places.)

17. State each of the following theorems.
   (a) The factor theorem
   (b) The fundamental theorem of algebra

18. Find a value for $a$ such that 3 is a root of the equation $x^3 - 4x^2 - ax - 6 = 0$.

19. For which values of $b$ will $-1$ be a root of the equation $x^3 + 2b^2x^2 + x - 48 = 0$?

20. For which values of $a$ will $x - 1$ be a factor of $a^2x^3 + 3ax^2 + 2$?

21. Find a quadratic function that has zeros of 1 and $-8$ and a graph with a $y$-intercept of $-24$.

22. Find a cubic function that has zeros of $-5$, $i$, and $-i$, and a graph with a $y$-intercept of 20.

23. Find a polynomial $f(x)$ with leading coefficient 1 such that the equation $f(x) = 0$ has the roots listed in the following table and no others. Write your answer in the form

$$a_n(x - r_1)(x - r_2) \cdots (x - r_n)$$

| ROOT | MULTIPLICITY |
|------|--------------|
| 2 | 1 |
| 3i | 3 |
| $1 + \sqrt{2}$ | 2 |

24. Use synthetic division to verify that $\sqrt{2} - 1$ is a root of the equation $x^6 + 14x^3 - 1 = 0$.

25. Let $f(x) = ax^3 + bx^2 + cx + d$ and suppose that $r$ is a root of the equation $f(x) = 0$.
   (a) Show that $r - h$ is a root of the equation $f(x + h) = 0$.
   (b) Show that $-r$ is a root of the equation $f(-x) = 0$.
   (c) Show that $kr$ is a root of the equation $f(x/k) = 0$.

26. Suppose that $r$ is a root of the polynomial equation $a_2x^2 + a_1x + a_0 = 0$. Show that $mr$ is a root of the equation $a_2x^2 + ma_1x + m^2a_0 = 0$.

27. Consider the equation $x^4 - x^3 + 24 = 0$.
   (a) List the possibilities for rational roots.
   (b) Use the upper and lower bound theorem to show that 2 is an upper bound for the roots.
   (c) In view of parts (a) and (b), what possibilities now remain for positive rational roots?
   (d) Which (if any) of the possibilities in part (c) are actually roots?

*In Exercises 28–32, list the possibilities for the rational roots of the equations.*

28. $x^5 - 12x^3 + x - 17 = 0$

29. $\frac{1}{6}x^5 - 2x^3 + \frac{1}{6}x - 3 = 0$

30. $\frac{3}{5}x^3 - 8x^2 - \frac{1}{2}x + \frac{3}{2} = 0$

31. $x^3 + x - p = 0$ (where $p$ is a prime number)

32. $x^3 + x - pq = 0$ (where both $p$ and $q$ are prime numbers)

*In Exercises 33–40, each equation has at least one rational root. Solve the equations. Suggestion: Use the upper and lower bound theorem to eliminate some of the possibilities for rational roots.*

33. $2x^3 + x^2 - 7x - 6 = 0$

34. $x^3 + 6x^2 - 8x - 7 = 0$

35. $2x^3 - x^2 - 14x + 10 = 0$

36. $2x^3 + 12x^2 + 13x + 15 = 0$

37. $\frac{3}{2}x^3 + \frac{1}{2}x^2 + \frac{1}{2}x - 1 = 0$

38. $x^4 - 2x^3 - 13x^2 + 38x - 24 = 0$

39. $x^5 + x^4 - 14x^3 - 14x^2 + 49x + 49 = 0$

40. $8x^5 + 12x^4 + 14x^3 + 13x^2 + 6x + 1 = 0$

41. Solve the equation $x^3 - 9x^2 + 24x - 20 = 0$ using the fact that one of the roots has multiplicity 2.

42. One root of the equation $x^2 + kx + 2k = 0$ $(k \neq 0)$ is twice the other. Find $k$ and find the roots of the equation.

43. Solve the equation $x^3 + x^2 - 11x - 15 = 0$, given that one root is $-3$.

**44.** Find a quadratic equation with roots $a - \sqrt{a^2 - 1}$ and $a + \sqrt{a^2 - 1}$ $(a > 1)$.

In Exercises 45–48, · write each polynomial in the form $a_n(x - r_1)(x - r_2) \cdots (x - r_n)$.

**45.** $6x^2 + 7x - 20$

**46.** $x^2 + x - 1$

**47.** $x^4 - 4x^3 + 5x - 20$

**48.** $x^4 - 4x^2 - 5$

In Exercises 49–52, equations are given, followed by one or more roots. Solve the equations.

**49.** $x^3 - 7x^2 + 25x - 39 = 0$; $x = 2 - 3i$

**50.** $x^3 + 6x^2 - 24x + 160 = 0$; $x = 2 + 2i\sqrt{3}$

**51.** $x^4 - 2x^3 - 4x^2 + 14x - 21 = 0$; $x = 1 + i\sqrt{2}$

**52.** $x^5 + x^4 - x^3 + x^2 + x - 1 = 0$;
$x = \frac{1}{2}(1 + i\sqrt{3})$, $x = \frac{1}{2}(-1 - \sqrt{5})$

**53.** (a) Find the rational roots of the equation $2x^3 - x^2 - x - 3 = 0$.
(b) Find all solutions of the equation in part (a).

**54.** Consider the equation $x^3 - 36x - 84 = 0$.
(a) Use Descartes's rule to check that this equation has exactly one positive root.
(b) Use the upper and lower bound theorem to show that 7 is an upper bound for the positive root.
(c) Locate the positive root within successive hundredths

**55.** Consider the equation $x^3 - 3x + 1 = 0$.
(a) Use Descartes's rule to check that this equation has exactly one negative root.
(b) Use the upper and lower bound theorem to show that $-2$ is a lower bound for the negative root.
(c) Locate the negative root within successive hundredths.

**56.** Consider the equation $x^3 + x^2 + x + 1 = 0$.
(a) Use Descartes's rule to show that the equation has either one or three negative roots.
(b) Now show that the equation cannot have three negative roots. *Hint:* Multiply both sides of the equation by $x - 1$. Then simplify the left-hand side and reapply Descartes's rule to the new equation.
(c) Actually, the original equation can be solved using only the basic algebraic techniques discussed in Chapter 2. Solve the equation in this manner.

**57.** Use Descartes's rule of signs to obtain information regarding the roots of the equation
$$3x^4 + x^2 - 5x - 1 = 0$$

**58.** Use Descartes's rule to show that the equation $x^3 - x^2 + 3x + 2 = 0$ has no positive roots. *Hint:* Multiply both sides of the equation by $x + 1$ and apply Descartes's rule to the resulting equation.

In Exercises 59–64, use Descartes's rule of signs to obtain information regarding the roots of the equations.

**59.** $x^3 + 8x - 7 = 0$

**60.** $3x^4 + x^2 + 4x - 2 = 0$

**61.** $x^3 + 3x + 1 = 0$

**62.** $2x^6 + 3x^2 + 6 = 0$

**63.** $x^4 - 10 = 0$

**64.** $x^4 + 5x^2 - x + 2 = 0$

**65.** Solve the following equation, given that two of the roots are $-3i$ and $1 + \sqrt{3}$:
$$x^5 + 3x^3 - 4x^2 - 54x - 36 = 0$$

**66.** Let $P$ be the point in the first quadrant where the parabola $y = 4 - x^2$ intersects the curve $y = x^3$. Locate the $x$-coordinate of $P$ within successive hundredths.

**67.** Let $P$ be the point in the first quadrant where the curve $y = x^3$ intersects the circle $x^2 + y^2 = 1$. Locate the $x$-coordinate of $P$ within successive hundredths.

**68.** Find a quadratic polynomial $f(x)$ with integer coefficients, such that $\dfrac{1 + \sqrt{5}}{1 - \sqrt{5}}$ is a root of the equation $f(x) = 0$.

In Exercises 69–76, mark **T** if the statement is true without exception; otherwise mark **F**.

**69.** Every polynomial equation of degree at least 1 has at least one real root.

**70.** Every polynomial of degree $n$, where $n \geq 1$, can be written in the form
$$a_n(x - r_1)(x - r_2) \cdots (x - r_n)$$

**71.** According to the rational roots theorem, there are four possibilities for the rational roots of the equation $x^5 + 6x^2 - 2 = 0$.

**72.** According to the rational roots theorem, there are only two possibilities for the rational roots of the equation $\frac{1}{3}x^3 - 5x + 1 = 0$.

**73.** The sum of the roots of the equation $x^2 - 12x + 16 = 0$ is $-12$.

**74.** According to the location theorem, if $f(x)$ is a polynomial and $f(a)$ and $f(b)$ have the same signs, then the equation $f(x) = 0$ has at least one root between $a$ and $b$.

**75.** Let $f(x)$ be a polynomial with real coefficients, and let $a$ and $b$ be real numbers. If $f(a)$ and $f(b)$ have opposite signs, then the equation $f(x) = 0$ has a root between $f(a)$ and $f(b)$.

**76.** According to Descartes's rule, the equation $x^7 + 4x + 3 = 0$ has no positive real root.

In Exercises 77–80, find polynomial equations with integer coefficients and with the given values as roots.

**77.** $4 - \sqrt{5}$

78. $a + b$ and $a - b$ ($a$ and $b$ are integers)

79. $6 - 2i$ and $\sqrt{5}$

80. $\dfrac{5 + \sqrt{6}}{5 - \sqrt{6}}$   *Hint:*  First rationalize the denominator.

81. Find a cubic polynomial $f(x)$ with integer coefficients, such that $1 - 3i$ and $-2$ are roots of the equation $f(x) = 0$.

*In Exercises 82–86, first determine the zeros of each function, then sketch the graph.*

82. $y = x^3 - 2x^2 - 3x$

83. $y = x^4 + 3x^3 + 3x^2 + x$

84. $y = x^4 - 4x^2$

85. $y = x^3 + 6x^2 + 5x - 12$

86. $y = x^5 - 8x^4 + 25x^3 - 38x^2 + 28x - 8$

87. Find a fourth-degree polynomial equation with integer coefficients, such that $x = 1 + \sqrt{2} + \sqrt{3}$ is a root. *Hint:* Begin by writing the given relationship as $x - 1 = \sqrt{2} + \sqrt{3}$; then square both sides.

88. Find a cubic equation with integer coefficients, such that $x = 1 + \sqrt[3]{2}$ is a root. Is $1 - \sqrt[3]{2}$ also a root of the equation?

# 12 THE CONIC SECTIONS

*The Greeks knew the properties of the curves given by cutting a cone with a plane—the ellipse, the parabola and hyperbola. Kepler discovered by analysis of astronomical observations, and Newton proved mathematically . . . that the planets move in ellipses. The geometry of Ancient Greece thus became the cornerstone of modern astronomy.*

JOHN LIGHTON SYNGE (1897–1987)

**INTRODUCTION**  The **conic sections** are the curves formed when a plane intersects the surface of a right circular cone. As Figure 1 indicates, these curves are the *circle*, the *parabola*, the *ellipse*, and the *hyperbola*. For the parabola in Figure 1, the cutting plane is parallel to line $L$; for the hyperbola, the cutting plane is parallel to the axis of the cone. In this chapter, we'll discuss the equations describing these curves in an *x-y* plane. We won't, however, derive these equations from the geometry of Figure 1.

## THE PARABOLA

### 12.1

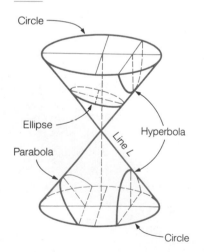

**FIGURE 1**

In Section 5.1, we saw that the graph of a quadratic function $y = ax^2 + bx + c$ is a symmetric U-shaped curve called a parabola. In this section, we give a more general definition of the parabola, a definition that emphasizes the geometric properties of the curve.

> **The Parabola**
>
> A **parabola** is the set of all points in the plane equally distant from a fixed line and a fixed point not on the line. The fixed line is called the **directrix**. The fixed point is called the **focus**.

Let us initially suppose that the focus of the parabola is the point $(0, p)$ and the directrix is the line $y = -p$. We will assume throughout this section that $p$ is positive. To understand the geometric content of the definition of the parabola, the special graph paper displayed in Figure 2(a) is useful. The common center of the concentric circles in Figure 2(a) is the focus $(0, p)$. Thus, all the points on a given circle are at a fixed distance from the focus. The radii of the circles increase in increments of $p$ units. Similarly, the broken horizontal lines in the figure are drawn at intervals that are multiples of $p$ units from the directrix $y = -p$. By considering the points where the circles intersect the horizontal lines, we can find a number of points equally distant from the focus $(0, p)$ and the directrix $y = -p$ (see Figure 2(b)).

FIGURE 2(a)

FIGURE 2(b)

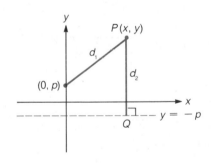

FIGURE 3

The graphical method just described lets us locate many points equally distant from the focus $(0, p)$ and the directrix $y = -p$. Figure 2(b) shows that the points on the parabola are symmetric about a line, in this case the $y$-axis. Also, by studying the figure, you should be able to convince yourself that in this case there can be no points below the $x$-axis that satisfy the stated condition. However, to completely describe the required set of points, and to show that our new definition is consistent with the old, Figure 2(b) is inadequate. We need to bring algebraic methods to bear on the problem. Thus, let $d_1$ denote the distance from the point $P(x, y)$ to the focus $(0, p)$ and let $d_2$ denote the distance from $P(x, y)$ to the directrix $y = -p$, as shown in Figure 3. The distance $d_1$ is then

$$d_1 = \sqrt{(x - 0)^2 + (y - p)^2}$$
$$= \sqrt{x^2 + y^2 - 2py + p^2}$$

The distance $d_2$ in Figure 3 is just the vertical distance between the points $P$ and $Q$. Thus,

$$d_2 = (y\text{-coordinate of } P) - (y\text{-coordinate of } Q)$$
$$= y - (-p) = y + p$$

(Absolute value signs are unnecessary here for, as noted earlier, $P$ cannot lie below the $x$-axis.) The condition that $P$ be equally distant from the focus $(0, p)$ and the directrix $y = -p$ can be expressed by the equation

$$d_1 = d_2$$

and, using the expressions we've found for $d_1$ and $d_2$, this last equation becomes

$$\sqrt{x^2 + y^2 - 2py + p^2} = y + p$$

We can obtain a simpler but equivalent equation by squaring both sides. (Two nonnegative quantities are equal if and only if their squares are equal.) Thus,

$$x^2 + y^2 - 2py + p^2 = y^2 + 2py + p^2$$

After combining like terms, this equation becomes

$$x^2 = 4py$$

This is the equation of a parabola with focus $(0, p)$ and directrix $y = -p$. In the box that follows, we summarize the properties of the parabola $x^2 = 4py$. As Figure 4(b) indicates, the terminology we've introduced applies equally well to an arbitrary parabola for which the axis of symmetry is not necessarily parallel to one of the coordinate axes and the vertex is not necessarily the origin.

| PROPERTY SUMMARY | The Parabola |
|---|---|

1. A **parabola** is the set of points equidistant from a fixed line called the **directrix** and a fixed point, not on the line, called the **focus**.

2. The **axis** of a parabola is the line drawn through the focus and perpendicular to the directrix.

3. The **vertex** of a parabola is the point where the parabola intersects its axis. The vertex is located halfway between the focus and the directrix. See Figure 4.

Axis   $x^2 = 4py$

Focus

$(0, p)$

Vertex   $y = -p$

Directrix

(a)

Axis

Focus

Vertex

Directrix

(b)

**FIGURE 4**

**EXAMPLE 1**   Determine the equation of the parabola in Figure 5, given that the curve passes through the point $(3, 5)$. Specify the focus and the directrix.

**Solution**   We have just seen that the general equation for a parabola in this position is $x^2 = 4py$. Since the point $(3, 5)$ lies on the curve, its coordinates must satisfy the equation $x^2 = 4py$. Thus,

$$3^2 = 4p(5)$$
$$9 = 20p$$
$$\frac{9}{20} = p$$

With $p = \frac{9}{20}$, the equation $x^2 = 4py$ becomes

$$x^2 = 4\left(\frac{9}{20}\right)y = \frac{9}{5}y \quad \text{or} \quad x^2 = \frac{9}{5}y \quad \text{as required}$$

Furthermore, since $p = \frac{9}{20}$, the focus is $(0, \frac{9}{20})$ and the directrix is $y = -\frac{9}{20}$.

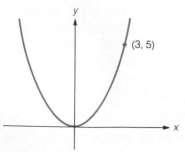

**FIGURE 5**

We have seen that the equation of a parabola with focus $(0, p)$ and directrix $y = -p$ is $x^2 = 4py$. By following the same method, we can obtain general equations for parabolas with other orientations. The basic results are summarized in Figure 6.

<table>
<tr><td>**PROPERTY SUMMARY**</td><td>**Basic Equations for the Parabola**</td></tr>
</table>

(a) $x^2 = 4py$

(b) $x^2 = -4py$

(c) $y^2 = 4px$

(d) $y^2 = -4px$

**FIGURE 6**

By a **chord** of a parabola, we mean a straight line segment joining any two points on the curve. If the chord passes through the focus, it is called a **focal chord**. For purposes of graphing, it is useful to know the length of the focal chord perpendicular to the axis of the parabola. This is the length of the horizontal line segment $\overline{AB}$ in Figure 7 and the vertical line segment $\overline{A'B'}$ in Figure 8. We will call this length the **focal width**.

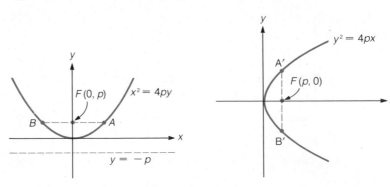

**FIGURE 7**  Focal width $= AB = 4p$      **FIGURE 8**  Focal width $= A'B' = 4p$

In Figure 7, the distance from $A$ to $F$ is the same as the distance from $A$ to the line $y = -p$. (Why?) But the distance from $A$ to the line $y = -p$ is $2p$. Therefore, $AF = 2p$ and $AB$ is twice this, or $4p$. We have shown that the focal width of the parabola is $4p$. In other words, given a parabola $x^2 = 4py$, the width at its focus is $4p$, the coefficient of $y$. In the same way, the length of the focal chord $\overline{A'B'}$ in Figure 8 is also $4p$, the coefficient of $x$ in that case.

**EXAMPLE 2** Find the focus and directrix of the parabola $y^2 = -4x$, and sketch the graph.

Solution Comparing the basic equation $y^2 = -4px$ [in Figure 6(d)] with the equation at hand, we see that $4p = 4$ and thus $p = 1$. The focus is therefore $(-1, 0)$, and the directrix is $x = 1$. The basic form of the graph will be as in Figure 6(d). For purposes of graphing, we note that the focal width is 4 (the absolute value of the coefficient of $x$). This, along with the fact that the vertex is $(0, 0)$, is enough information to draw the graph. See Figure 9.

**EXAMPLE 3** Graph the parabola $(y + 1)^2 = -4(x - 2)$. Specify the vertex, the focus, the directrix, and the axis of symmetry.

Solution The given equation is obtained from $y^2 = -4x$ (which we graphed in the previous example) by replacing $x$ and $y$ with $x - 2$ and $y + 1$, respectively. So (as we saw in Section 4.3 when we discussed translation) the required graph is obtained by translating the parabola in Figure 9 two units to the right and one unit down. In particular, this means that the vertex moves from $(0, 0)$ to $(2, -1)$; the focus moves from $(-1, 0)$ to $(1, -1)$; the directrix moves from $x = 1$ to $x = 3$; and the axis of symmetry moves from $y = 0$ (which is the $x$-axis) to $y = -1$. The required graph is shown in Figure 10.

FIGURE 9

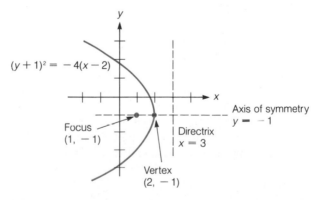

FIGURE 10

**EXAMPLE 4** Graph the parabola $2y^2 - 4y - x + 5 = 0$, and specify each of the following: vertex, focus, directrix, axis of symmetry, and focal width.

Solution Just as we did in Section 5.1, we use the technique of completing the square:

$$2(y^2 - 2y \quad\quad) = x - 5$$
$$2(y^2 - 2y + 1) = x - 5 + 2 \qquad \text{adding 2 to both sides}$$
$$(y - 1)^2 = \frac{1}{2}(x - 3)$$

The graph of this last equation is obtained by translating the graph of $y^2 = \frac{1}{2}x$ "right 3, up 1." This moves the vertex from $(0,0)$ to $(3,1)$. Now, for $y^2 = \frac{1}{2}x$, the focus and directrix are determined by setting $4p = \frac{1}{2}$. Therefore $p = \frac{1}{8}$, and consequently the focus and directrix of $y^2 = \frac{1}{2}x$ are $(\frac{1}{8},0)$ and $x = -\frac{1}{8}$, respectively. Thus, the focus and directrix of the translated curve are $(3\frac{1}{8},1)$ and $x = 2\frac{7}{8}$, respectively. Figure 11(a) shows the graph of $y^2 = \frac{1}{2}x$, and Figure 11(b) shows the translated graph. (You should check for yourself that the information accompanying Figure 11(b) is correct.) ▪▪

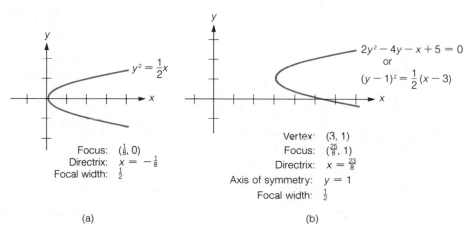

Focus: $(\frac{1}{8}, 0)$
Directrix: $x = -\frac{1}{8}$
Focal width: $\frac{1}{2}$

(a)

Vertex: $(3, 1)$
Focus: $(\frac{25}{8}, 1)$
Directrix: $x = \frac{23}{8}$
Axis of symmetry: $y = 1$
Focal width: $\frac{1}{2}$

(b)

**FIGURE 11**

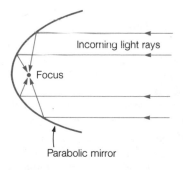

Incoming light rays

Focus

Parabolic mirror

**FIGURE 12**

Before concluding our discussion of the parabola, we point out that there are numerous applications of the parabola in the sciences. Parabolic reflectors are used in communications systems, in telescope mirrors, and in automobile headlights. In the case of the telescope mirror, an incoming ray of light that is parallel to the axis of the parabola will be reflected through the focus, as indicated in Figure 12. Indeed, the very word *focus* comes from a Latin word meaning "fireplace." [For a more complete description of the uses of the parabola, see the article "The Standup Conic Presents: The Parabola and Applications," by Lee Whitt, in *The Journal of Undergraduate Mathematics and its Applications*, Vol. III, No. 3 (1982).]

## EXERCISE SET 12.1

### A

*In Exercises 1–22, graph the parabolas. In each case, specify the focus, the directrix, and the focal width. For Exercises 13–22, also specify the vertex.*

1. $x^2 = 4y$
2. $x^2 = 16y$
3. $y^2 = -8x$
4. $y^2 = 12x$
5. $x^2 = -20y$
6. $x^2 - y = 0$
7. $y^2 + 28x = 0$
8. $4y^2 + x = 0$
9. $x^2 = 6y$
10. $y^2 = -10x$
11. $4x^2 = 7y$
12. $3y^2 = 4x$
13. $y^2 - 6y - 4x + 17 = 0$
14. $y^2 + 2y + 8x + 17 = 0$

15. $x^2 - 8x - y + 18 = 0$
16. $x^2 + 6y + 18 = 0$
17. $y^2 + 2y - x + 1 = 0$
18. $2y^2 - x + 1 = 0$
19. $2x^2 - 12x - y + 18 = 0$
20. $y + \sqrt{2} = (x - 2\sqrt{2})^2$
21. $2x^2 - 16x - y + 33 = 0$
22. $\frac{1}{4}y^2 - y - x + 1 = 0$

## B

*For Exercises 23 and 24, refer to the following figure.*

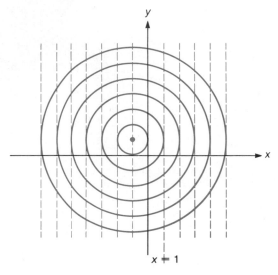

**FIGURE A**

23. Make a photocopy of Figure A. Then in your copy of Figure A, indicate (by drawing dots) eleven points that are equidistant from the point $(-1, 1)$ and the line $x = 1$. What is the equation of the line of symmetry for the set of dots?

24. The eleven dots that you located in Exercise 23 are part of a parabola. Find the equation of that parabola and sketch its graph. Specify the vertex, the focus and directrix, and the focal width.

*For Exercises 25–30, find the equation of the parabola satisfying the given conditions. In each case, assume that the vertex is at the origin.*

25. The focus is $(0, 3)$.

26. The directrix is $y - 8 = 0$.

27. The directrix is $x + 32 = 0$.

28. The focus lies on the $y$-axis, and the parabola passes through the point $(7, -10)$.

29. The parabola is symmetric about the $x$-axis, the $x$-coordinate of the focus is negative, and the length of the focal chord perpendicular to the $x$-axis is 9.

30. The focus is the smaller of the two $x$-intercepts of the circle $x^2 - 8x + y^2 - 6y + 9 = 0$.

31. Let $P$ denote the point $(8, 8)$ on the parabola $x^2 = 8y$, and let $\overline{PQ}$ be a focal chord.
    (a) Find the equation of the line through the point $(8, 8)$ and the focus.
    (b) Find the coordinates of $Q$.
    (c) Find the length of $\overline{PQ}$.
    (d) Find the equation of the circle with this focal chord as a diameter.
    (e) Show that the circle determined in part (d) intersects the directrix of the parabola in only one point. Conclude from this that the directrix is tangent to the circle. Draw a sketch of the situation.

32. The following figure shows a parabolic cross section of a reflecting mirror. Find the distance from the vertex of the parabola to the focus. *Hint:* Choose a convenient coordinate system.

33. An arch is in the shape of a parabola with a vertical axis. The arch is 15 ft high at the center and 40 ft wide at the base. At what height above the base is the width 20 ft? *Hint:* Choose a convenient coordinate system in which the vertex of the parabola is at the origin.

34. The following figure depicts the cable of a suspension bridge. The cable is in the form of a parabola with a vertical axis. The horizontal highway on the bridge is 300 ft long. The longest of the vertical supporting wires is 100 ft; the shortest is 40 ft. Find the length of a supporting wire that is 50 ft from the middle.

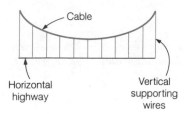

C

**35.** If $\overline{PQ}$ is a focal chord of the parabola $x^2 = 4py$, and the coordinates of $P$ are $(x_0, y_0)$, show that the coordinates of $Q$ are

$$\left( \frac{-4p^2}{x_0}, \frac{p^2}{y_0} \right)$$

**36.** Let $F$ and $V$ denote the focus and the vertex, respectively, of the parabola $x^2 = 4py$. If $\overline{PQ}$ is a focal chord of the parabola, show that

$$PF \cdot FQ = VF \cdot PQ$$

## THE ELLIPSE

### 12.2

*"The heavenly motions are nothing but a continuous song for several voices, to be perceived by the intellect, not by the ear."*

JOHANNES KEPLER (1571–1630)

In this section, we discuss the symmetric oval-shaped curve known as the *ellipse*. As Kepler discovered, and Newton later proved, this is the curve described by the planets in their motions around the sun.

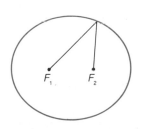

> ### The Ellipse
>
> An **ellipse** is the set of all points in the plane, the sum of whose distances from two fixed points is constant. Each fixed point is called a **focus** (plural: **foci**).

**FIGURE 1**

**FIGURE 2**   When the distance between the foci is small compared to the length of the string, the ellipse resembles a circle.

**FIGURE 3**   When the distance between the foci is nearly equal to the length of the string, the ellipse is relatively flat.

Subsequently, we will derive an equation describing the ellipse, just as we found an equation for the parabola in the previous section. But first let us consider some rather immediate consequences of the definition. In fact, we can learn a great deal about the ellipse even before we derive its equation.

There is a simple mechanical method for constructing an ellipse that arises directly from the definition of the curve. Mark the given foci, say $F_1$ and $F_2$, on a drawing board and insert thumbtacks at those points. Now take a piece of string whose length is greater than the distance from $F_1$ to $F_2$ and tie the ends of the string to the tacks. Next pull the string taut with a pencil point. Then if you move the pencil while keeping the string taut, the curve traced out will be an ellipse, as indicated in Figure 1. The reason the curve is an ellipse is that for each point on the curve, the sum of the distances from the foci is constant, the constant being the length of the string. By actually carrying out this construction for yourself several times, each time varying the distance between the foci or the length of the string, you can learn a great deal about the ellipse. For instance, when the distance between the foci is small compared to the length of the string, the ellipse begins to resemble a circle, as in Figure 2. On the other hand, when the distance between the foci is nearly equal to the length of the string, the ellipse becomes relatively flat, as in Figure 3.

We now derive one of the standard forms for the equation of an ellipse. As indicated in Figure 4, we assume that the foci are $F_1(-c, 0)$ and $F_2(c, 0)$ and that the sum of the distances from a point $P(x, y)$ on the ellipse to the foci is $2a$.

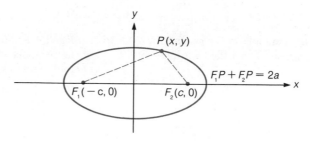

**FIGURE 4**

Since the point $P$ lies on the ellipse, we have

$$F_1P + F_2P = 2a$$

and therefore

$$\sqrt{(x + c)^2 + (y - 0)^2} + \sqrt{(x - c)^2 + (y - 0)^2} = 2a$$

or

$$\sqrt{(x - c)^2 + y^2} = 2a - \sqrt{(x + c)^2 + y^2} \qquad (1)$$

Now, by following a straightforward but lengthy process of squaring and simplifying (as outlined in detail in Exercise 37), we find that equation (1) becomes

$$(a^2 - c^2)x^2 + a^2y^2 = a^2(a^2 - c^2) \qquad (2)$$

To write equation (2) in a more symmetric form, we define the positive number $b$ by the equation

$$b^2 = a^2 - c^2 \qquad (3)$$

NOTE   For this definition to make sense, we need to know that the right-hand side of equation (3) is positive. (See Exercise 38 at the end of this section for details.)

Finally, using equation (3) to substitute for $a^2 - c^2$ in equation (2), we obtain

$$b^2x^2 + a^2y^2 = a^2b^2$$

which can be written

$$\frac{x^2}{a^2} + \frac{y^2}{b^2} = 1 \qquad (a > b) \qquad (4)$$

We have now shown that the coordinates of each point on the ellipse satisfy equation (4). Conversely, it can be shown that if $x$ and $y$ satisfy equation (4), then the point $(x, y)$ indeed lies on the ellipse. We refer to equation (4) as the **standard form** for the equation of an ellipse with foci $(-c, 0)$ and $(c, 0)$. For an ellipse in this form, it will always be the case that $a$ is greater than $b$; this follows from equation (3).

For purposes of graphing, we want to know the intercepts of the ellipse. To find the $x$-intercepts, we set $y = 0$ in equation (4) to obtain

$$\frac{x^2}{a^2} = 1$$

$$x^2 = a^2$$

$$x = \pm a$$

The $x$-intercepts are therefore $a$ and $-a$. In a similar fashion, you can check that the $y$-intercepts are $b$ and $-b$. Also (according to the symmetry tests in Section 3.2), note that the graph of equation (4) must be symmetric about both coordinate axes. Figure 5 shows the graph of the ellipse $(x^2/a^2) + (y^2/b^2) = 1$. (Calculator exercises at the end of this section will help convince you that the general shape of the curve in Figure 5 is correct.)

We can use Figure 5 to introduce some terminology for the ellipse. In Figure 5, line segment $\overline{AB}$ is called the **major axis** of the ellipse. The length of this major axis is $2a$. Line segment $\overline{CD}$ is the **minor axis**; its length is $2b$. The **eccentricity** of an ellipse is a number between 0 and 1; it is defined by the equation

$$\text{eccentricity} = \frac{c}{a} = \frac{\sqrt{a^2 - b^2}}{a}$$

Figure 6 displays examples of ellipses of various eccentricities. As you can see from the figure, the eccentricity provides a numerical measure of how much the ellipse deviates from being a circle. The closer the eccentricity is to zero, the more the ellipse resembles a circle. In the other direction, as the eccentricity approaches 1, the ellipse becomes increasingly flat. In the box that follows, we summarize these remarks. (In the box, note that we've used the letter $e$ to denote the eccentricity. This is the conventional choice, even though the same letter is used with a very different meaning in the context of exponential functions and logarithms.)

**FIGURE 5**

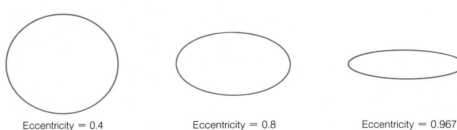

Eccentricity = 0.4          Eccentricity = 0.8          Eccentricity = 0.967

**FIGURE 6**   Eccentricity is a number between 0 and 1 that describes the shape of an ellipse. The narrowest ellipse in this figure has the same proportions as the orbit of Halley's comet. By way of contrast, the eccentricity of Earth's orbit is 0.0017; if an ellipse with this eccentricity were included in Figure 6, it would appear indistinguishable from a circle.

| PROPERTY SUMMARY | The Ellipse $\dfrac{x^2}{a^2} + \dfrac{y^2}{b^2} = 1 \quad (a > b)$ |
|---|---|

Foci: $(\pm c, 0)$, where $c^2 = a^2 - b^2$

Length of major axis: $2a$

Length of minor axis: $2b$

Eccentricity: $e = \dfrac{c}{a} = \dfrac{\sqrt{a^2 - b^2}}{a}$

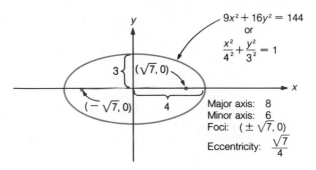

**EXAMPLE 1** Find the lengths of the major and minor axes of the ellipse $9x^2 + 16y^2 = 144$. Specify the coordinates of the foci and the eccentricity. Graph the ellipse.

**Solution** To convert the equation $9x^2 + 16y^2 = 144$ to standard form, we divide both sides by 144. This yields

$$\frac{9x^2}{144} + \frac{16y^2}{144} = \frac{144}{144}$$

$$\frac{x^2}{16} + \frac{y^2}{9} = 1$$

$$\frac{x^2}{4^2} + \frac{y^2}{3^2} = 1$$

This is the standard form. Comparing this equation with $\dfrac{x^2}{a^2} + \dfrac{y^2}{b^2} = 1$, we see

that $a = 4$ and $b = 3$. Thus, the major and minor axes are 8 and 6 units, respectively. Next, to determine the foci, we use the equation $c^2 = a^2 - b^2$.

FIGURE 7

We have

$$c^2 = 4^2 - 3^2 = 7$$
$$c = \sqrt{7}$$

(We choose the positive square root because $c > 0$.) It follows that the co-ordinates of the foci are $(-\sqrt{7}, 0)$ and $(\sqrt{7}, 0)$. We now calculate the eccentricity using the formula $e = c/a$. This yields $e = \sqrt{7}/4$. Figure 7 shows the graph along with the required information.

**EXAMPLE 2**  The foci of an ellipse are $(-1, 0)$ and $(1, 0)$. The eccentricity is $\frac{1}{3}$. Find the equation of the ellipse (in standard form), and specify the lengths of the major and minor axes.

Solution  Since the foci are $(\pm 1, 0)$, we have $c = 1$. Using the equation $e = c/a$ with $e = \frac{1}{3}$ and $c = 1$, we obtain

$$\frac{1}{3} = \frac{1}{a} \qquad \text{and therefore} \qquad a = 3$$

Recall now that $b^2$ was defined by the equation $b^2 = a^2 - c^2$. In view of this, we have

$$b^2 = 3^2 - 1^2 = 8$$
$$b = \sqrt{8} = 2\sqrt{2}$$

The equation of the ellipse in standard form is therefore

$$\frac{x^2}{3^2} + \frac{y^2}{(2\sqrt{2})^2} = 1$$

Furthermore, since $a = 3$ and $b = 2\sqrt{2}$, the lengths of the major and minor axes are 6 and $4\sqrt{2}$ units, respectively.

Suppose now that we translate the graph of $(x^2/a^2) + (y^2/b^2) = 1$ by $h$ units in the $x$-direction and $k$ units in the $y$-direction, as shown in Figure 8. Then (according to our work on translation in Section 4.3), the equation of the translated ellipse is

$$\frac{(x-h)^2}{a^2} + \frac{(y-k)^2}{b^2} = 1 \tag{5}$$

Equation (5) is another **standard form** for the equation of an ellipse. As indicated in Figure 8, the **center** of this ellipse is the point $(h, k)$. (In general, the

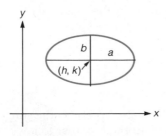

FIGURE 8

center of an ellipse can be defined as the midpoint of the line segment joining the foci.) In the next example, we use the technique of completing the square to convert an equation for an ellipse to standard form. Once the equation is in standard form, the graph is readily obtained.

**EXAMPLE 3**  Determine the center, foci, and eccentricity of the ellipse

$$4x^2 + 9y^2 - 8x - 54y + 49 = 0$$

Graph the ellipse.

**Solution**  We will convert the given equation to standard form by using the technique of completing the square. We have

$$4x^2 - 8x + 9y^2 - 54y = -49$$
$$4(x^2 - 2x) + 9(y^2 - 6y) = -49$$
$$4(x^2 - 2x + 1) + 9(y^2 - 6y + 9) = -49 + (4)(1) + 9(9)$$
$$4(x - 1)^2 + 9(y - 3)^2 = 36$$

Dividing this last equation by 36, we obain

$$\frac{(x - 1)^2}{9} + \frac{(y - 3)^2}{4} = 1$$

or

$$\frac{(x - 1)^2}{3^2} + \frac{(y - 3)^2}{2^2} = 1$$

This last equation represents an ellipse with center at $(1, 3)$ and with $a = 3$ and $b = 2$. We can calculate $c$ using the formula $c^2 = a^2 - b^2$:

$$c^2 = 3^2 - 2^2 = 5$$
$$c = \sqrt{5}$$

Since the center of this ellipse is $(1, 3)$, the foci are therefore $(1 + \sqrt{5}, 3)$ and $(1 - \sqrt{5}, 3)$. Finally, the eccentricity is $c/a$, which is $\sqrt{5}/3$. Figure 9 shows the graph of this ellipse.

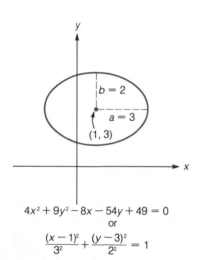

$$4x^2 + 9y^2 - 8x - 54y + 49 = 0$$
or
$$\frac{(x - 1)^2}{3^2} + \frac{(y - 3)^2}{2^2} = 1$$

**FIGURE 9**

In developing the equation $\dfrac{x^2}{a^2} + \dfrac{y^2}{b^2} = 1$, we assumed that the foci of the ellipse were located on the $x$-axis at the points $(-c, 0)$ and $(c, 0)$. If, instead, the foci are located on the $y$-axis at the points $(0, c)$ and $(0, -c)$, then the same method we used in the previous case will show the equation of the ellipse to be

$$\frac{x^2}{b^2} + \frac{y^2}{a^2} = 1 \qquad (a > b)$$

We still assume that $2a$ represents the sum of the distances from a point on the ellipse to the foci. In the box that follows, we summarize the situation for the ellipse with foci $(0, c)$ and $(0, -c)$.

| PROPERTY SUMMARY | The Ellipse $\dfrac{x^2}{b^2} + \dfrac{y^2}{a^2} = 1 \quad (a > b)$ |
|---|---|

Foci:   $(0, \pm c)$, where $c^2 = a^2 - b^2$

Length of major axis:   $2a$

Length of minor axis:   $2b$

Eccentricity:   $e = \dfrac{c}{a} = \dfrac{\sqrt{a^2 - b^2}}{a}$

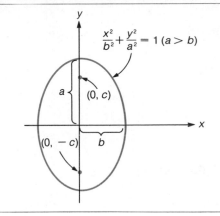

We now have two standard forms for the ellipse:

Foci on the $x$-axis at $(\pm c, 0)$:   $\dfrac{x^2}{a^2} + \dfrac{y^2}{b^2} = 1$     $(a > b)$

Foci on the $y$-axis at $(0, \pm c)$:   $\dfrac{x^2}{b^2} + \dfrac{y^2}{a^2} = 1$     $(a > b)$

Because $a$ is greater than $b$ in both standard forms, it is always easy to determine by inspection whether the foci lie on the $x$-axis or the $y$-axis. Consider, for instance, the equation $\dfrac{x^2}{5^2} + \dfrac{y^2}{6^2} = 1$. In this case, since $6 > 5$, we have $a = 6$. And since $6^2$ appears under $y^2$, we conclude that the foci lie on the $y$-axis.

**EXAMPLE 4**   The point $(5, 3)$ lies on an ellipse, and the endpoints of the major axis of the ellipse are $(0, \pm 2\sqrt{21})$. Find the equation of the ellipse; write the answer both in standard form and in the form $Ax^2 + By^2 = C$.

**Solution**   Since the endpoints of the major axis are $(0, \pm 2\sqrt{21})$, the standard form in this case is $(x^2/b^2) + (y^2/a^2) = 1$. Furthermore, in view of the coordinates of the endpoints, we have $a = 2\sqrt{21}$. Therefore,

$$\dfrac{x^2}{b^2} + \dfrac{y^2}{(2\sqrt{21})^2} = 1 \qquad \text{or} \qquad \dfrac{x^2}{b^2} + \dfrac{y^2}{84} = 1$$

Now, since the point $(5, 3)$ lies on the ellipse, its coordinates must satisfy this last equation. We then have

$$\dfrac{5^2}{b^2} + \dfrac{3^2}{84} = 1$$

$$\dfrac{25}{b^2} + \dfrac{3}{28} = 1$$

$$\dfrac{25}{b^2} = \dfrac{25}{28}$$

From this last equation, we see that $b^2 = 28$, and therefore,

$$b = \sqrt{28} = 2\sqrt{7}$$

Now that we've determined $a$ and $b$, we can write the equation of the ellipse in standard form. It is

$$\frac{x^2}{(2\sqrt{7})^2} + \frac{y^2}{(2\sqrt{21})^2} = 1$$

As you should verify for yourself, this equation can also be written in the equivalent form

$$3x^2 + y^2 = 84$$

## EXERCISE SET 12.2

### A

*In Exercises 1–24, graph the ellipses. In each case, specify the lengths of the major and minor axes, the foci, and the eccentricity. For Exercises 13–24, also specify the center of the ellipse.*

1. $4x^2 + 9y^2 = 36$
2. $4x^2 + 25y^2 = 100$
3. $x^2 + 16y^2 = 16$
4. $9x^2 + 25y^2 = 225$
5. $x^2 + 2y^2 = 2$
6. $2x^2 + 3y^2 = 3$
7. $16x^2 + 9y^2 = 144$
8. $25x^2 + y^2 = 25$
9. $15x^2 + 3y^2 = 5$
10. $9x^2 + y^2 = 4$
11. $2x^2 + y^2 = 4$
12. $36x^2 + 25y^2 = 400$

13. $\dfrac{(x-5)^2}{5^2} + \dfrac{(y+1)^2}{3^2} = 1$

14. $\dfrac{(x-1)^2}{2^2} + \dfrac{(y+4)^2}{3^2} = 1$

15. $\dfrac{(x-1)^2}{1^2} + \dfrac{(y-2)^2}{2^2} = 1$

16. $\dfrac{x^2}{4^2} + \dfrac{(y-3)^2}{2^2} = 1$

17. $\dfrac{(x+3)^2}{3^2} + \dfrac{y^2}{1^2} = 1$

18. $\dfrac{(x-2)^2}{2^2} + \dfrac{(y-2)^2}{2^2} = 1$

19. $3x^2 + 4y^2 - 6x + 16y + 7 = 0$
20. $16x^2 + 64x + 9y^2 - 54y + 1 = 0$
21. $5x^2 + 3y^2 - 40x - 36y + 188 = 0$
22. $x^2 + 16y^2 - 160y + 384 = 0$
23. $16x^2 + 25y^2 - 64x - 100y + 564 = 0$
24. $4x^2 + 4y^2 - 32x + 32y + 127 = 0$

### B

*In Exercises 25–32, find the equation of the ellipse satisfying the given conditions. Write the answer both in standard form and in the form $Ax^2 + By^2 = C$.*

25. Foci $(\pm 3, 0)$; endpoints of major axis $(\pm 5, 0)$
26. Foci $(0, \pm 1)$; endpoints of major axis $(0, \pm 4)$
27. Endpoints of major axis $(\pm 4, 0)$; eccentricity $\frac{1}{4}$
28. Foci $(0, \pm 2)$; endpoints of minor axis $(\pm 5, 0)$
29. Foci $(0, \pm 2)$; endpoints of major axis $(0, \pm 5)$
30. Endpoints of major axis $(\pm 10, 0)$; endpoints of minor axis $(0, \pm 4)$
31. Center at the origin; major axis on the $x$-axis; length of major axis twice the length of minor axis; $(1, \sqrt{2})$ lies on ellipse
32. Eccentricity $\frac{3}{5}$; one endpoint of minor axis $(-8, 0)$; center at the origin
33. The area $A$ of an ellipse $(x^2/a^2) + (y^2/b^2) = 1$ is given by the formula $A = \pi ab$. Use this formula to compute the area of the ellipse $5x^2 + 6y^2 = 60$.
34. Find a positive number $b$ such that the line $y = x - b$ intersects the ellipse $x^2 + 3y^2 = 12$ in exactly one point. Then, on the same set of axes, graph the ellipse and the line.

ⓒ 35. Solve the equation $(x^2/3^2) + (y^2/2^2) = 1$ to obtain $y = \dfrac{\pm\sqrt{36 - 4x^2}}{3}$. Then complete the following table and use the results to graph the given equation. (Use the fact that the graph must be symmetric about the $y$-axis.)

| $x$ | 0 | 0.5 | 1.0 | 1.5 | 2.0 | 2.5 | 3 |
|---|---|---|---|---|---|---|---|
| $y$ | $\pm 2$ | | | | | | 0 |

**☐ 36.** Solve the equation $(x^2/1^2) + (y^2/4^2) = 1$ for $y$ to obtain $y = \pm 4\sqrt{1 - x^2}$. Then complete the following table and use the results to graph the given equation. (After plotting the points obtained from your table, use the fact that the graph must be symmetric about the $y$-axis.)

| $x$ | 0 | 0.1 | 0.2 | 0.3 | 0.4 | 0.5 | 0.6 |
|-----|---|-----|-----|-----|-----|-----|-----|
| $y$ | $\pm 4$ | | | | | | |

| $x$ | 0.7 | 0.8 | 0.9 | 1.0 |
|-----|-----|-----|-----|-----|
| $y$ | | | | 0 |

**37.** This exercise outlines the steps needed to complete the derivation of the equation $(x^2/a^2) + (y^2/b^2) = 1$.
  **(a)** Square both sides of equation (1) on page 612. After simplifying, you should obtain

$$a\sqrt{(x + c)^2 + y^2} = a^2 + xc$$

  **(b)** Square both sides of the equation in part (a). Show that the result can be written

$$a^2 x^2 - x^2 c^2 + a^2 y^2 = a^4 - a^2 c^2$$

  **(c)** Verify that the equation in part (b) is equivalent to

$$(a^2 - c^2)x^2 + a^2 y^2 = a^2(a^2 - c^2)$$

  **(d)** Using the equation in part (c), replace the quantity $a^2 - c^2$ with $b^2$. Then show that the resulting equation can be rewritten $(x^2/a^2) + (y^2/b^2) = 1$, as required.

**38.** In the text we defined the positive number $b^2$ by the equation $b^2 = a^2 - c^2$. For this definition to make sense, we need to show that the quantity $a^2 - c^2$ is indeed positive. This can be done as follows. First, recall that in any triangle, the sum of the lengths of two sides is always greater than the length of the third side. Now apply this fact to triangle $F_1 P F_2$ in Figure 4 to show that $2a > 2c$. Conclude from this that $a^2 - c^2 > 0$, as required.

**☐ 39.** As you know, there is a simple expression for the circumference of a circle of radius $a$, namely, $2\pi a$. However, there is no similar type of elementary expression for the circumference of an ellipse. (The circumference of an ellipse can be computed to as many decimal places as required using the methods of calculus.) Nevertheless, there are some interesting elementary formulas that allow us to approximate the circumference of an ellipse quite closely. Three such formulas follow, along with the names of their discoverers and the dates of discovery. Each formula yields an approximate value for the circumference of the ellipse $(x^2/a^2) + (y^2/b^2) = 1$.

$$C_1 = \pi[a + b + \tfrac{1}{2}(\sqrt{a} - \sqrt{b})^2]$$
  Giuseppe Peano, 1887

$$C_2 = \pi[3(a + b) - \sqrt{(a + 3b)(3a + b)}]$$
  Srinivasa Ramanujan, 1914

$$C_3 = \frac{\pi}{2}[a + b + \sqrt{2(a^2 + b^2)}]$$
  R. A. Johnson, 1930

Use these formulas to complete the following table of approximations for the circumference of the ellipse $(x^2/5^2) + (y^2/3^2) = 1$. Round off the values of $C_1$, $C_2$, and $C_3$ to six decimal places. In order to complete the right-hand column of the table, you need two facts. First, the actual circumference of the ellipse, rounded off to six decimal places, is 25.526999. Second, percentage error in an approximation is given by

$$\frac{|(\text{true value}) - (\text{approximation})|}{\text{true value}} \times 100$$

Round off the percentage errors to two significant digits. Which of the three approximations is the best?

|       | APPROXIMATION OBTAINED | PERCENTAGE ERROR |
|-------|------------------------|------------------|
| $C_1$ |                        |                  |
| $C_2$ |                        |                  |
| $C_3$ |                        |                  |

**C**

**40.** In Section 12.1, we defined the parabola in terms of a focus–directrix property. This exercise shows that the ellipse also has a type of focus–directrix property.

Refer to the accompanying figure. Let $P$ be a point on the ellipse $(x^2/a^2) + (y^2/b^2) = 1$, and suppose that the coordinates of $P$ are $(x, y)$.
  **(a)** Show that $F_1 P = a + ex$. *Hint:* If you try to do this from scratch, it can involve a rather lengthy calculation. Begin instead with the equation $a\sqrt{(x + c)^2 + y^2} = a^2 + xc$ [which appears in Exercise 37(a)], and divide both sides by $a$.

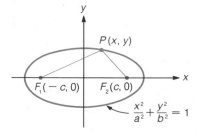

**(b)** Show that $F_2P = a - ex$. *Hint:* Make use of the result in part (a), along with the fact that the sum of $F_1P$ and $F_2P$ is, by definition, $2a$.

**(c)** In the following figure, $P(x, y)$ denotes an arbitrary point on the ellipse $(x^2/a^2) + (y^2/b^2) = 1$, and $R$

is the foot of the perpendicular drawn from $P$ to the line $x = a/e$. Show that

$$\frac{F_2P}{PR} = e$$

(This is the *focus–directrix property* of the ellipse. The line $x = a/e$ is the directrix in this case.) *Hint:* This result is easy to prove if you make use of the formula for $F_2P$ in part (b).

## THE HYPERBOLA

### 12.3

In the previous section, we defined an ellipse as the set of points $P$ such that the sum of the distances from $P$ to two fixed points is constant. By considering the difference instead of the sum, we are led to the definition of the hyperbola.

> **The Hyperbola**
>
> A **hyperbola** is the set of all points in the plane, the difference of whose distances from two fixed points is a positive constant. Each fixed point is called a **focus**.

As with the ellipse, we label the foci $F_1$ and $F_2$. Before obtaining an equation for the hyperbola, we can see the general features of the curve by using two sets of concentric circles, with centers $F_1$ and $F_2$, to locate points satisfying the

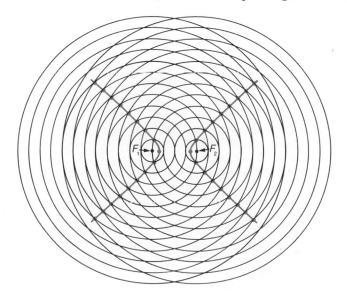

FIGURE 1

definition of a hyperbola. In Figure 1, we've plotted a number of points $P$ such that either $F_1P - F_2P = 3$ or $F_2P - F_1P = 3$. By joining these points, we obtain the graph of the hyperbola shown in Figure 1.

Unlike the parabola or the ellipse, the hyperbola is composed of two distinct parts, or **branches**. As you can check, the left branch in Figure 1 corresponds to the equation $F_2P - F_1P = 3$, while the right branch corresponds to the equation $F_1P - F_2P = 3$. Figure 1 also reveals that the hyperbola possesses two types of symmetry. First, it is symmetric about the line passing through the two foci $F_1$ and $F_2$; this line is referred to as the **focal axis** of the hyperbola. Second, the hyperbola is symmetric about the line that is the perpendicular bisector of the segment $\overline{F_1F_2}$.

To derive an equation for the hyperbola, let us initially assume that the foci are located at the points $F_1(-c, 0)$ and $F_2(c, 0)$, as indicated in Figure 2. We will use $2a$ to denote the positive constant referred to in the definition of the hyperbola. By definition, then, $P(x, y)$ lies on the hyperbola if and only if

$$|F_1P - F_2P| = 2a$$

or, equivalently,

$$F_1P - F_2P = \pm 2a$$

If we use the formula for the distance between two points, this last equation becomes

$$\sqrt{(x + c)^2 + (y - 0)^2} - \sqrt{(x - c)^2 + (y - 0)^2} = \pm 2a$$

or

$$\sqrt{(x + c)^2 + y^2} - \sqrt{(x - c)^2 + y^2} = \pm 2a$$

We can simplify this equation by carrying out the same procedure that we used for the ellipse in the previous section. As Exercise 42 asks you to verify, the resulting equation is

$$(c^2 - a^2)x^2 - a^2y^2 = a^2(c^2 - a^2) \tag{1}$$

Before further simplifying equation (1), we point out that the quantity $c^2 - a^2$ [which appears twice in equation (1)] is positive. To see why this is so, refer back to Figure 2. In triangle $F_1F_2P$ (as in any triangle), the length of any side is less than the sum of the lengths of the other two sides. Thus

$$F_1P < F_1F_2 + F_2P \qquad \text{and} \qquad F_2P < F_1F_2 + F_1P$$

and therefore

$$F_1P - F_2P < F_1F_2 \qquad \text{and} \qquad F_2P - F_1P < F_1F_2$$

These last two equations tell us that

$$|F_1P - F_2P| < F_1F_2$$

Therefore, in view of the definitions of $2a$ and $2c$, we have

$$0 < 2a < 2c$$

or

$$0 < a < c$$

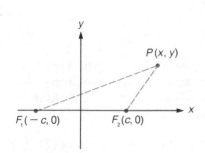

$P(x, y)$

$F_1(-c, 0)$    $F_2(c, 0)$

**FIGURE 2**

From this last inequality we conclude that $c^2 - a^2$ is positive, as we wished to show.

Now, since $c^2 - a^2$ is positive, we may define the positive number $b$ by the equation

$$b^2 = c^2 - a^2$$

With this notation, equation (1) becomes

$$b^2x^2 - a^2y^2 = a^2b^2$$

Dividing by $a^2b^2$, we obtain

$$\frac{x^2}{a^2} - \frac{y^2}{b^2} = 1 \tag{2}$$

We have now shown that the coordinates of every point on the hyperbola satisfy equation (2). Conversely, it can be shown that if the coordinates of a point satisfy equation (2), then the point satisfies the original definition of the hyperbola. Equation (2) is the **standard form** for the equation of a hyperbola with foci $F_1(-c,0)$ and $F_2(c,0)$.

The intercepts of the hyperbola are readily obtained from equation (2). To find the $x$-intercepts, we set $y$ equal to zero to obtain

$$\frac{x^2}{a^2} = 1$$

$$x^2 = a^2$$

$$x = \pm a$$

Thus, the hyperbola crosses the $x$-axis at the points $(-a,0)$ and $(a,0)$. On the other hand, the curve does not cross the $y$-axis, for if we set $x$ equal to zero in equation (2), we obtain $-y^2/b^2 = 1$ or

$$y^2 = -b^2 \qquad (b > 0)$$

Since the square of any real number $y$ is nonnegative, this last equation has no solution. Therefore, the graph does not cross the $y$-axis. Finally, let us note that (according to the symmetry tests in Section 3.2) the graph of equation (2) must be symmetric about both coordinate axes.

Before graphing the hyperbola $(x^2/a^2) - (y^2/b^2) = 1$, we point out the important fact that the two lines $y = (b/a)x$ and $y = -(b/a)x$ are asymptotes for the curve. We can see why as follows. First we solve equation (2) for $y$:

$$\frac{x^2}{a^2} - \frac{y^2}{b^2} = 1$$

$$b^2x^2 - a^2y^2 = a^2b^2 \qquad \text{multiplying by } a^2b^2$$

$$-a^2y^2 = a^2b^2 - b^2x^2$$

$$y^2 = \frac{b^2x^2 - a^2b^2}{a^2} = \frac{b^2(x^2 - a^2)}{a^2}$$

$$y = \pm\frac{b}{a}\sqrt{x^2 - a^2} \tag{3}$$

**TABLE 1**

| $x$ | $\sqrt{x^2 - 5^2}$ |
|---|---|
| 100 | 99.875 |
| 1000 | 999.987 |
| 10000 | 9999.999 |

Now, as $x$ grows arbitrarily large, the value of the quantity $\sqrt{x^2 - a^2}$ becomes closer and closer to $x$ itself. Table 1 provides some empirical evidence for this statement in the case when $a = 5$. (A formal proof of the statement properly belongs to calculus.) In summary then, we have the approximation $\sqrt{x^2 - a^2} \approx x$ as $x$ grows arbitrarily large. So, in view of equation (3), we have

$$y = \pm\frac{b}{a}\sqrt{x^2 - a^2} \approx \pm\frac{b}{a}x \qquad \text{as } x \text{ grows arbitrarily large}$$

In other words, the two lines $y = \pm(b/a)x$ are asymptotes for the hyperbola.

A simple way to sketch the two asymptotes and then graph the hyperbola is as follows. First draw the rectangle with vertices $(a, b)$, $(-a, b)$, $(-a, -b)$, and $(a, -b)$, as indicated in Figure 3(a). The slopes of the diagonals in this rectangle are $b/a$ and $-b/a$. Thus by extending these diagonals as in Figure 3(b), we obtain the two asymptotes $y = \pm(b/a)x$. Now, since the $x$-intercepts of the hyperbola are $a$ and $-a$, we can sketch the curve as shown in Figure 3(c).

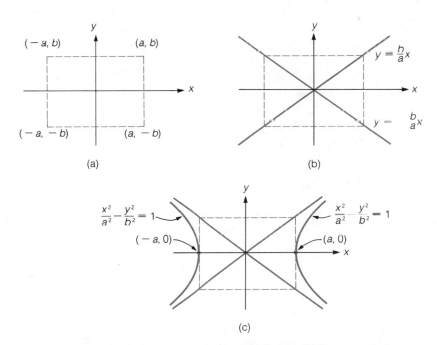

(a)

(b)

(c)

**FIGURE 3**  Steps in graphing the hyperbola $(x^2/a^2) - (y^2/b^2) = 1$ and its asymptotes

For the hyperbola in Figure 3(c), the two points $(\pm a, 0)$, where the curve meets the $x$-axis, are referred to as **vertices**. The midpoint of the line segment joining the two vertices is called the **center** of the hyperbola. (Equivalently, we can define the center as the point of intersection of the two asymptotes.) For the hyperbola in Figure 3(c), the center coincides with the origin. The line segment joining the vertices of a hyperbola is the **transverse axis** of the hyperbola. For reference now, in the box that follows we summarize our work up to this point on the hyperbola. (Note: several new terms describing the hyperbola are also given in the box.)

| PROPERTY SUMMARY | The Hyperbola $\dfrac{x^2}{a^2} - \dfrac{y^2}{b^2} = 1$ |
|---|---|

1. The **foci** are the points $F_1(-c, 0)$ and $F_2(c, 0)$. The hyperbola is the set of points $P$ such that $|F_1P - F_2P| = 2a$.

2. The **focal axis** is the line passing through the foci.

3. The **vertices** are the points at which the hyperbola intersects its focal axis. In Figure 4, these are the two points $V_1(-a, 0)$ and $V_2(a, 0)$.

FIGURE 4

4. The **center** is the point on the focal axis midway between the foci. The center of the hyperbola in Figure 4 is the origin.

5. The **transverse axis** is the line segment joining the two vertices. In Figure 4, the length of the transverse axis $\overline{V_1V_2}$ is $2a$.

6. The **conjugate axis** is the line segment perpendicular to the transverse axis, passing through the center and extending a distance $b$ on either side of the center. In Figure 4, this is the segment $\overline{AB}$.

7. The **eccentricity** $e$ is defined by $e = c/a$, where $c^2 = a^2 + b^2$.

In general, if $A$, $B$, and $C$ are positive numbers, then the graph of an equation of the form

$$Ax^2 - By^2 = C$$

will be a hyperbola of the type shown in Figure 4. Example 1 shows why this is so.

**EXAMPLE 1**   Graph the hyperbola $16x^2 - 9y^2 = 144$ after determining the following: vertices, foci, eccentricity, lengths of the transverse and conjugate axes, and asymptotes.

**Solution**  First convert the given equation to standard form by dividing through by 144. This yields

$$\frac{x^2}{9} - \frac{y^2}{16} = 1 \qquad \text{or} \qquad \frac{x^2}{3^2} - \frac{y^2}{4^2} = 1$$

By comparing this with the equation $(x^2/a^2) - (y^2/b^2) = 1$, we see that $a = 3$ and $b = 4$. The value of $c$ can be determined by using the equation $c^2 = a^2 + b^2$. We have

$$c^2 = 3^2 + 4^2 = 25$$
$$c = 5$$

Now that we know the values of $a$, $b$, and $c$, we can list the required information:

Vertices:  $(\pm 3, 0)$      Length of transverse axis ($= 2a$):   6

Foci:  $(\pm 5, 0)$      Length of conjugate axis ($= 2b$):   8

Eccentricity:  $\dfrac{5}{3}$      Asymptotes:  $y = \pm\dfrac{4}{3}x$

The graph of the hyperbola is shown in Figure 5.

$16x^2 - 9y^2 = 144$
or
$\dfrac{x^2}{3^2} - \dfrac{y^2}{4^2} = 1$

**FIGURE 5**

The same method that is used to derive the equation for the hyperbola with foci $(\pm c, 0)$ can be used when the foci are instead located on the $y$-axis at the points $(0, \pm c)$. The equation of the hyperbola in this case is

$$\frac{y^2}{a^2} - \frac{x^2}{b^2} = 1$$

We summarize the basic properties of this hyperbola in the following box.

| PROPERTY SUMMARY | The Hyperbola $\dfrac{y^2}{a^2} - \dfrac{x^2}{b^2} = 1$ |
|---|---|

Foci:   $(0, \pm c)$, where $c^2 = a^2 + b^2$

Vertices:   $(0, \pm a)$

Asymptotes:   $y = \pm (a/b)x$

Length of transverse axis:   $2a$

Length of conjugate axis:   $2b$

Eccentricity:   $e = c/a$

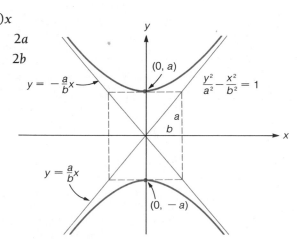

**EXAMPLE 2**   Use the technique of completing the square to show that the graph of the following equation is a hyperbola.

$$9y^2 - 54y - 25x^2 + 200x - 544 = 0$$

Graph the hyperbola and specify the center, the vertices, the foci, the length of the transverse axis, and the equations of the asymptotes.

**Solution**   $9(y^2 - 6y \quad) - 25(x^2 - 8x \quad) = 544$   factoring

$9(y^2 - 6y + 9) - 25(x^2 - 8x + 16) = 544 + 81 - 400$   completing the squares

$9(y - 3)^2 - 25(x - 4)^2 = 225$

$\dfrac{(y - 3)^2}{5^2} - \dfrac{(x - 4)^2}{3^2} = 1$   dividing by 225

Now, the graph of this last equation is obtained by translating the graph of the equation

$$\frac{y^2}{5^2} - \frac{x^2}{3^2} = 1 \tag{4}$$

4 units to the right and 3 units up. So first we analyze the graph of equation (4). The general form of the graph is shown in the box just prior to this example. In our case, we have $a = 5$, $b = 3$, and

$$c = \sqrt{a^2 + b^2}$$
$$= \sqrt{5^2 + 3^2} = \sqrt{34}\ (\approx 5.8)$$

Consequently, the vertices [for equation (4)] are $(0, \pm 5)$; the foci are $(0, \pm\sqrt{34})$; and the asymptotes are $y = \pm\frac{5}{3}x$. Figure 6 shows the graph of this hyperbola. Finally, by translating the graph in Figure 6 to the right 4 units and up 3 units, we obtain the graph of the original equation, as

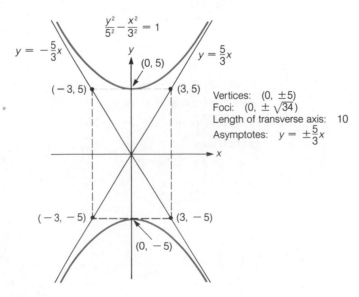

Vertices:  $(0, \pm 5)$
Foci:  $(0, \pm\sqrt{34})$
Length of transverse axis:  10
Asymptotes:  $y = \pm\frac{5}{3}x$

**FIGURE 6**

shown in Figure 7. You should verify for yourself that the information
accompanying Figure 7 is correct.

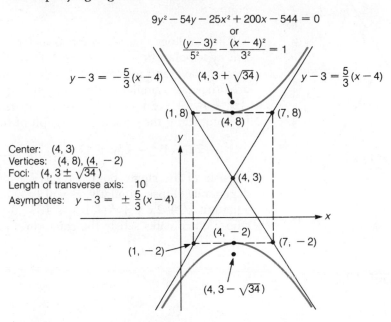

$$9y^2 - 54y - 25x^2 + 200x - 544 = 0$$
or
$$\frac{(y-3)^2}{5^2} - \frac{(x-4)^2}{3^2} = 1$$

$y - 3 = -\frac{5}{3}(x-4)$  $(4, 3 + \sqrt{34})$  $y - 3 = \frac{5}{3}(x-4)$

$(1, 8)$  $(4, 8)$  $(7, 8)$

Center: $(4, 3)$
Vertices: $(4, 8), (4, -2)$
Foci: $(4, 3 \pm \sqrt{34})$
Length of transverse axis: $10$
Asymptotes: $y - 3 = \pm \frac{5}{3}(x-4)$

$(4, 3)$

$(4, -2)$

$(1, -2)$  $(7, -2)$

$(4, 3 - \sqrt{34})$

FIGURE 7

If you reread the example just completed, you'll see that it was not necessary
to know in advance that the given equation represented a hyperbola. Rather,
this fact emerged naturally after we completed the square. Indeed, completing
the square is a useful technique for identifying the graph of any equation of
the form

$$Ax^2 + By^2 + Cx + Dy + E = 0$$

**EXAMPLE 3**  Identify the graph of the equation

$$4x^2 - 32x - y^2 + 2y + 63 = 0$$

Solution  As before, we complete the squares:

$$4(x^2 - 8x) - (y^2 - 2y) = -63$$
$$4(x^2 - 8x + 16) - (y^2 - 2y + 1) = -63 + 4(16) - 1$$
$$4(x - 4)^2 - (y - 1)^2 = 0$$

Since the right-hand side of this last equation is 0, dividing both sides by 4
will not bring the equation into one of the standard forms. Indeed, if we factor
the left-hand side of the equation as a difference of two squares, we obtain

$$[2(x - 4) - (y - 1)][2(x - 4) + (y - 1)] = 0$$
$$(2x - y - 7)(2x + y - 9) = 0$$

$2x - y - 7 = 0$  $\quad\quad$  $2x + y - 9 = 0$
$-y = -2x + 7$  $\quad\quad\quad\quad$  $y = -2x + 9$
$y = 2x - 7$

$4x^2 - 32x - y^2 + 2y + 63 = 0$
or
$(2x - y - 7)(2x + y - 9) = 0$

**FIGURE 8**

Thus the given equation is equivalent to the two linear equations $y = 2x - 7$ and $y = -2x + 9$. These two lines, taken together, constitute the graph. See Figure 8.   ▪▪

The two lines that we graphed in Figure 8 are actually the asymptotes for the hyperbola $4(x - 4)^2 - (y - 1)^2 = 1$. (Verify this for yourself.) For that reason, the graph in Figure 8 is referred to as a **degenerate hyperbola**. There are other cases similar to this that can arise in graphing equations of the form $Ax^2 + By^2 + Cx + Dy + E = 0$. For instance, as you can check for yourself by completing the squares, the graph of the equation

$$x^2 - 2x + 4y^2 - 16y + 17 = 0$$

consists of the single point $(1, 2)$. We refer to the graph in this case as a **degenerate ellipse**. Similarly, as you can check by completing the squares, the equation $x^2 - 2x + 4y^2 - 16y + 18 = 0$ has no graph; there are no points whose coordinates satisfy the equation.

## EXERCISE SET 12.3

### A

*For Exercises 1–24, graph the hyperbolas. In each case in which the hyperbola is nondegenerate, specify the following: vertices, foci, lengths of transverse and conjugate axes, eccentricity, and equations of the asymptotes. In Exercises 11–24, also specify the centers.*

1. $x^2 - 4y^2 = 4$
2. $y^2 - x^2 = 1$
3. $y^2 - 4x^2 = 4$
4. $25x^2 - 9y^2 = 225$
5. $16x^2 - 25y^2 = 400$
6. $9x^2 - y^2 = 36$
7. $2y^2 - 3x^2 = 1$
8. $x^2 - y^2 = 9$
9. $4y^2 - 25x^2 = 100$
10. $x^2 - 3y^2 = 3$
11. $\dfrac{(x - 5)^2}{5^2} - \dfrac{(y + 1)^2}{3^2} = 1$
12. $\dfrac{(x - 5)^2}{3^2} - \dfrac{(y + 1)^2}{5^2} = 1$
13. $\dfrac{(y - 2)^2}{2^2} - \dfrac{(x - 1)^2}{1^2} = 1$
14. $\dfrac{(y - 3)^2}{2^2} - \dfrac{x^2}{1^2} = 1$
15. $\dfrac{(x + 3)^2}{4^2} - \dfrac{(y - 4)^2}{4^2} = 1$
16. $\dfrac{(x + 1)^2}{5^2} - \dfrac{(y + 2)^2}{3^2} = 1$
17. $x^2 - y^2 + 2y - 5 = 0$

18. $16x^2 - 32x - 9y^2 + 90y - 353 = 0$
19. $x^2 - y^2 - 4x + 2y - 6 = 0$
20. $x^2 - 8x - y^2 + 8y - 25 = 0$
21. $y^2 - 25x^2 + 8y - 9 = 0$
22. $9y^2 - 18y - 4x^2 - 16x - 43 = 0$
23. $x^2 + 7x - y^2 - y + 12 = 0$
24. $9x^2 + 9x - 16y^2 + 4y + 2 = 0$

### B

25. Let $P(x, y)$ be a point in the first quadrant on the hyperbola $(x^2/2^2) - (y^2/1^2) = 1$. Let $Q$ be the point in the first quadrant with the same $x$-coordinate as $P$ and lying on an asymptote to the hyperbola. Show that $PQ = \frac{1}{2}(x - \sqrt{x^2 - 4})$.

26. The distance $PQ$ in Exercise 25 represents the vertical distance between the hyperbola and the asymptote. Complete the following table to see numerical evidence that this separation distance approaches zero as $x$ gets larger and larger. (Round off each entry to one significant digit.)

| $x$ | 10 | 50 | 100 | 500 | 1000 | 10000 |
|-----|----|----|-----|-----|------|-------|
| $PQ$ |  |  |  |  |  |  |

*In Exercises 27–36, determine the equation of the hyperbola satisfying the given conditions. Write each answer in the form* $Ax^2 - By^2 = C$ *or in the form* $Ay^2 - Bx^2 = C$.

27. Foci $(\pm 4, 0)$; vertices $(\pm 1, 0)$

28. Foci $(0, \pm 5)$; vertices $(0, \pm 3)$

29. Asymptotes $y = \pm\frac{1}{2}x$; vertices $(\pm 2, 0)$

30. Asymptotes $y = \pm x$; foci $(0, \pm 1)$

31. Asymptotes $y = \pm\sqrt{10}x/5$; foci $(\pm\sqrt{7}, 0)$

32. Length of transverse axis 6; eccentricity $\frac{4}{3}$; center $(0,0)$; focal axis horizontal

33. Vertices $(0, \pm 7)$; passing through the point $(1, 9)$

34. Eccentricity 2; foci $(\pm 1, 0)$

35. Length of transverse axis 6; length of conjugate axis 2; foci on the $y$-axis; center at the origin

36. Asymptotes $y = \pm 2x$; passing through $(1, \sqrt{3})$

37. Show that the two asymptotes of the hyperbola $x^2 - y^2 = 16$ are perpendicular to each other.

38. (a) Verify that the point $P(6, 4\sqrt{3})$ lies on the hyperbola $16x^2 - 9y^2 = 144$.
    (b) In Example 1, we found that the foci of this hyperbola were $F_1(-5, 0)$ and $F_2(5, 0)$. Compute the lengths $F_1P$ and $F_2P$ where $P$ is the point $(6, 4\sqrt{3})$.
    (c) Verify that $|F_1P - F_2P| = 2a$.

39. (a) Verify that the point $P(5, 6)$ lies on the hyperbola $5y^2 - 4x^2 = 80$.
    (b) Find the foci.
    (c) Compute the lengths of the line segments $\overline{F_1P}$ and $\overline{F_2P}$, where $P$ is the point $(5, 6)$.
    (d) Verify that $|F_1P - F_2P| = 2a$.

40. Let $e_1$ and $e_2$ denote the eccentricities of the hyperbolas $(x^2/A^2) - (y^2/B^2) = 1$ and $(y^2/B^2) - (x^2/A^2) = 1$, respectively. Verify that
$$e_1^2 e_2^2 = e_1^2 + e_2^2$$

41. (a) If the asymptotes of the hyperbola $x^2/a^2 - y^2/b^2 = 1$ are perpendicular, show that $a = b$. What is the eccentricity in this case?
    (b) Show that the asymptotes of the hyperbola $(x^2/a^2) - (y^2/a^2) = 1$ are perpendicular. What is the eccentricity of this hyperbola?

42. Derive equation (1) in this section from the equation that precedes it.

43. Let $P(x, y)$ be a point on the right-hand branch of the hyperbola $(x^2/a^2) - (y^2/b^2) = 1$. As usual, let $F_2$ denote

the focus located at $(c, 0)$. The following steps outline a proof of the fact that the length of the line segment $\overline{F_2P}$ in this case is given by
$$F_2P = xe - a$$

(a) Explain why
$$\sqrt{(x + c)^2 + y^2} - \sqrt{(x - c)^2 + y^2} = 2a$$

(b) In the preceding equation, add the quantity $\sqrt{(x - c)^2 + y^2}$ to both sides and then square both sides. Show that the result can be written
$$xc - a^2 = a\sqrt{(x - c)^2 + y^2}$$
or
$$xc - a^2 = a(F_2P)$$

(c) Divide both sides of the preceding equation by $a$ to show that $xe - a = F_2P$, as required.

44. Let $P(x, y)$ be a point on the right-hand branch of the hyperbola $(x^2/a^2) - (y^2/b^2) = 1$. As usual, let $F_1$ denote the focus located at $(-c, 0)$. Show that
$$F_1P = xe + a$$
   *Hint:* Use the result in Exercise 43 along with the fact that the right-hand branch is defined by the equation $F_1P - F_2P = 2a$.

45. Let $P$ be a point on the right-hand branch of the hyperbola $x^2 - y^2 = k^2$. If $d$ denotes the distance from $P$ to the center of the hyperbola, show that
$$d^2 = (F_1P)(F_2P)$$

46. Let $P(x, y)$ be a point on the right-hand branch of the hyperbola $(x^2/a^2) - (y^2/b^2) = 1$. From $P$, a line segment is drawn perpendicular to the line $x = a/e$, meeting this line at $D$. If $F_2$ denotes (as usual) the focus located at $(c, 0)$, show that
$$\frac{F_2P}{PD} = e$$

The line $x = a/e$ is called the **directrix** of the hyperbola corresponding to the focus $F_2$. *Hint:* Use the expression for $F_2P$ developed in Exercise 43.

47. Consider the hyperbola $(x^2/a^2) - (y^2/b^2) = 1$, and let $D$ be the point where the asymptote $y = (b/a)x$ meets the directrix $x = a/e$. If $F$ denotes the focus corresponding to this directrix, show that the line segment $\overline{FD}$ is perpendicular to the asymptote.

## CHAPTER 12 SUMMARY OF PRINCIPAL TERMS AND FORMULAS

| TERMS AND FORMULAS | PAGE REFERENCE | COMMENT |
|---|---|---|
| 1. Conic sections | 604 | These are the curves that are formed when a plane intersects the surface of a right circular cone. As indicated in Figure 1 in the chapter introduction, these curves are the circle, the ellipse, the hyperbola, and the parabola. |
| 2. Parabola | 604, 606 | A parabola is the set of all points in the plane equally distant from a fixed line and a fixed point not on the line. The fixed line is called the **directrix**, and the fixed point is called the **focus**. |
| 3. Axis (of a parabola) | 606 | This is the line drawn through the focus of the parabola, perpendicular to the directrix. |
| 4. Vertex (of a parabola) | 606 | This is the point at which the parabola intersects its axis. |
| 5. Focal chord (of a parabola) | 607 | A focal chord of a parabola is a line segment passing through the focus, with endpoints on the parabola. |
| 6. Focal width (of a parabola) | 607 | The focal width of a parabola is the length of the focal chord that is perpendicular to the axis of the parabola. For a given value of $p$, the two parabolas $x^2 = 4py$ and $y^2 = 4px$ have the same focal width; it is $4p$. |
| 7. Ellipse | 611 | An ellipse is the set of all points in the plane, the sum of whose distances from two fixed points is constant. Each fixed point is called a **focus** of the ellipse. |
| 8. Eccentricity (of an ellipse) | 613 | The eccentricity is a number that measures how much the ellipse deviates from being a circle. See Figure 6 in Section 12.2. The eccentricity $e$ is defined by the formula $e = c/a$, where $c$ and $a$ are defined by the following conventions. The distance between the foci is denoted by $2c$. The sum of the distances from a point on the ellipse to the two foci is denoted by $2a$. |
| 9. Center (of an ellipse) | 616 | This is the midpoint of the line segment joining the foci. |
| 10. Major and minor axes (of an ellipse) | 613 | In Figure 5, Section 12.2, the major and minor axes are the line segments $\overline{AB}$ and $\overline{CD}$, respectively. For the ellipse $(x^2/a^2) + (y^2/b^2) = 1$, the lengths of the major and minor axes are $2a$ and $2b$, respectively. |
| 11. $\dfrac{x^2}{a^2} + \dfrac{y^2}{b^2} = 1 \quad (a > b)$ | 614 | This is the standard form for the equation of an ellipse with foci $(\pm c, 0)$. The specifications for this ellipse are summarized in the box on page 614. |
| $\dfrac{x^2}{b^2} + \dfrac{y^2}{a^2} = 1 \quad (a > b)$ | 617 | This is the standard form for the equation of an ellipse with foci $(0, \pm c)$. The specifications for this ellipse are summarized in the box on page 617. |

| TERMS AND FORMULAS | PAGE REFERENCE | COMMENT |
|---|---|---|
| 12. Hyperbola | 620 | A hyperbola is the set of all points in the plane, the difference of whose distances from two fixed points is a positive constant. The two fixed points are the **foci**, and the line passing through the foci is the **focal axis**. |
| 13. $\dfrac{x^2}{a^2} - \dfrac{y^2}{b^2} = 1$ | 624 | This is the standard form for the equation of a hyperbola with foci $(\pm c, 0)$. |
| $\dfrac{y^2}{a^2} - \dfrac{x^2}{b^2} = 1$ | 625 | This is the standard form for the equation of a hyperbola with foci $(0, \pm c)$. |
| 14. Asymptotes for the hyperbola | 622 | As defined in earlier chapters, a line is an **asymptote** for a curve if the distance between the line and the curve approaches zero as we move farther and farther out along the line. The asymptotes for the hyperbola $\dfrac{x^2}{a^2} - \dfrac{y^2}{b^2} = 1$ are the two lines $y = \pm\dfrac{b}{a}x$. For the hyperbola $\dfrac{y^2}{a^2} - \dfrac{x^2}{b^2} = 1$, the asymptotes are $y = \pm\dfrac{a}{b}x$. |
| 15. Focal axis (of a hyperbola) | 624 | This is the line passing through the foci. |
| 16. Vertices (of a hyperbola) | 624 | The two points at which the hyperbola intersects its focal axis are called **vertices**. See Figure 4 on page 624 and the boxed figure on page 625. |
| 17. Center (of a hyperbola) | 624 | This is the point on the focal axis midway between the foci. See Figure 4 on page 624 and the boxed figure on page 625. |
| 18. Transverse axis | 624 | This is the line segment joining the two vertices of a hyperbola. See Figure 4 on page 624 and the boxed figure on page 625. |
| 19. Conjugate axis | 624 | This is the line segment perpendicular to the transverse axis of the hyperbola, passing through the center and extending a distance $b$ $(=\sqrt{c^2 - a^2})$ on either side of the center. See Figure 4 on page 624 and the boxed figure on page 625. |
| 20. Eccentricity (of a hyperbola) | 624 | For both of the standard forms for the hyperbola, the eccentricity $e$ is defined by $e = c/a$. |

## CHAPTER 12 REVIEW EXERCISES

NOTE  The ten exercises in this set with colored problem numbers constitute a chapter test.

*In Exercises 1–4, find the equation of the parabola satisfying the given conditions. In each case, assume that the vertex is* $(0,0)$.

1. (a) The focus is $(4,0)$.     (b) The focus is $(0,4)$.

2. The focus lies on the $x$-axis, and the curve passes through the point $(3,1)$.

3. The parabola is symmetric about the $y$-axis, the $y$-coordinate of the focus is positive, and the length of the focal chord perpendicular to the $y$-axis is 12.

4. The focus of the parabola is the center of the circle $x^2 - 8x + y^2 + 15 = 0$.

*In Exercises 5–7, find the equation of the ellipse satisfying the given conditions. Write your answer in the form $Ax^2 + By^2 = C$.*

5. Foci $(\pm 2, 0)$; endpoints of major axis $(\pm 8, 0)$

6. Foci $(0, \pm 1)$; endpoints of minor axis $(\pm 4, 0)$

7. Eccentricity $\frac{4}{5}$; one end of minor axis $(-6, 0)$; center at the origin

8. For any point $P$ on an ellipse, the sum of the distances from $(1, 2)$ and $(-1, -2)$ is 12. Find the equation of the ellipse.  *Hint:* Use the distance formula and the definition of an ellipse.

*In Exercises 9–12, find the equation of the hyperbola satisfying the given conditions. Write each answer in the form $Ax^2 - By^2 = C$ or in the form $Ay^2 - Bx^2 = C$.*

9. Foci $(\pm 6, 0)$; vertices $(\pm 2, 0)$

10. Asymptotes $y = \pm 2x$; foci $(0, \pm 3)$

11. Eccentricity 4; foci $(\pm 3, 0)$

12. Length of transverse axis 3; eccentricity $\frac{5}{4}$; center $(0, 0)$; focal axis horizontal

*In Exercises 13–18, graph the parabolas, and in each case specify the vertex, the focus, the directrix, and the focal width.*

13. $x^2 = 10y$      14. $x^2 = 5y$

15. $x^2 = -12(y - 3)$      16. $x^2 = -8(y + 1)$

17. $(y - 1)^2 = -4(x - 1)$      18. $(y + 3)^2 = 2(x - 1)$

*In Exercises 19–24, graph the ellipses, and in each case specify the center, the foci, the lengths of the major and minor axes, and the eccentricity.*

19. $x^2 + 2y^2 = 4$      20. $4x^2 + 9y^2 = 144$

21. $49x^2 + 9y^2 = 441$      22. $9x^2 + y^2 = 9$

23. $\dfrac{(x - 1)^2}{5^2} + \dfrac{(y + 2)^2}{3^2} = 1$      24. $\dfrac{(x + 3)^2}{3^2} + \dfrac{y^2}{3^2} = 1$

*In Exercises 25–30, graph the hyperbolas. In each case specify the center, the vertices, the foci, the equations of the asymptotes, and the eccentricity.*

25. $x^2 - 2y^2 = 4$      26. $4x^2 - 9y^2 = 144$

27. $49y^2 - 9x^2 = 441$      28. $9y^2 - x^2 = 9$

29. $\dfrac{(x - 1)^2}{5^2} - \dfrac{(y + 2)^2}{3^2} = 1$      30. $\dfrac{(y + 3)^2}{3^2} - \dfrac{x^2}{3^2} = 1$

*In Exercises 31–44, use the technique of completing the square to graph the given equation. If the graph is a parabola, specify the vertex, axis, focus, and directrix. If the graph is an ellipse, specify the center, foci, and lengths of the major and minor axes. If the graph is a hyperbola, specify the center, vertices, foci, and equations of the asymptotes. Finally, if the equation has no graph, say so.*

31. $3x^2 + 4y^2 - 6x + 16y + 7 = 0$

32. $y^2 - 16x - 8y + 80 = 0$

33. $y^2 + 4x + 2y - 15 = 0$

34. $16x^2 + 64x + 9y^2 - 54y + 1 = 0$

35. $16x^2 - 32x - 9y^2 + 90y - 353 = 0$

36. $x^2 + 6x - 12y + 33 = 0$

37. $5x^2 + 3y^2 - 40x - 36y + 188 = 0$

38. $x^2 - y^2 - 4x + 2y - 6 = 0$

39. $9x^2 - 90x - 16y^2 + 32y + 209 = 0$

40. $x^2 + 2y - 12 = 0$

41. $y^2 - 25x^2 + 8y - 9 = 0$

42. $x^2 + 16y^2 - 160y + 384 = 0$

43. $16x^2 + 25y^2 - 64x - 100y + 564 = 0$

44. $16x^2 - 25y^2 - 64x + 100y - 36 = 0$

45. Let $F_1$ and $F_2$ denote the foci of the hyperbola $5x^2 - 4y^2 = 80$.
   (a) Verify that the point $P$ with coordinates $(6, 5)$ lies on the hyperbola.
   (b) Compute the quantity $(F_1P - F_2P)^2$.

46. Show that the coordinates of the vertex of the parabola $Ax^2 + Dx + Ey + F = 0$ are given by

$$x = -\frac{D}{2A} \quad \text{and} \quad y = \frac{D^2 - 4AF}{4AE}$$

47. If the equation $Ax^2 + Cy^2 + Dx + Ey + F = 0$ represents an ellipse or a hyperbola show that the center is the point

$$\left( -\frac{D}{2A}, -\frac{E}{2C} \right)$$

48. The figure shows the parabola $x^2 = 4py$ and a circle with center at the origin and diameter $3p$. If $V$ and $F$ denote the vertex and focus of the parabola, respectively, show that the common chord of the circle and parabola bisects the line segment $\overline{VF}$.

# 13 ADDITIONAL TOPICS IN ALGEBRA

**INTRODUCTION**   A strong background in algebra is an important prerequisite for courses in calculus and in probability and statistics. In this final chapter we develop several additional topics that help provide a foundation in those areas of study. We begin in Section 13.1 with the principle of mathematical induction. This gives us a framework for proving statements about the natural numbers. In Section 13.2 we discuss the binomial theorem, which is used to analyze and expand expressions of the form $(a + b)^n$. As you'll see, the proof of the binomial theorem uses mathematical induction. Section 13.3 introduces the related (but distinct) concepts of sequences and series. Then in the next two sections (Sections 13.4 and 13.5), we study arithmetic and geometric sequences and series. The last topic introduced in Section 13.5 concerns finding sums of infinite geometric series. In a sense, this is an appropriate topic for our last chapter, for it is closely related to the idea of a *limit*, which is the starting point for calculus. Finally, in the last two sections of this chapter (Sections 13.6 and 13.7), we introduce the important concepts of permutations, combinations, and probability.

## MATHEMATICAL INDUCTION

### 13.1

*Mathematical induction is not a method of discovery but a technique of proving rigorously what has already been discovered.*

DAVID M. BURTON in his text *The History of Mathematics, An Introduction* (Boston: Allyn and Bacon, 1985)

*In Pascal's treatise* [Traité du triangle arithmétique, *written in 1653*] . . .
*appears one of the earliest acceptable statements of the method of mathematical induction.*

HOWARD EVES in his text *An Introduction to the History of Mathematics*, 5th ed. (Philadelphia: Saunders College Publishing, 1983)

Is mathematics an experimental science? The answer to this question is both yes and no, as the following example illustrates. Consider the problem of determining a formula for the sum of the first $n$ odd natural numbers:

$$1 + 3 + 5 + \cdots + (2n - 1)$$

We begin by doing some calculations in the hope that this may shed some light on the problem. Table 1 shows the results of calculating the sum of the first $n$ odd natural numbers for values of $n$ ranging from 1 to 5. Upon inspecting the table, we observe that each sum in the right-hand column is the square of the corresponding entry in the left-hand column. For instance, for $n = 5$,

**TABLE 1**

| $n$ | $1 + 3 + 5 + \cdots + (2n - 1)$ |
|---|---|
| 1 | $1 = 1$ |
| 2 | $1 + 3 = 4$ |
| 3 | $1 + 3 + 5 = 9$ |
| 4 | $1 + 3 + 5 + 7 = 16$ |
| 5 | $1 + 3 + 5 + 7 + 9 = 25$ |

we see that

$$\underbrace{1 + 3 + 5 + 7 + 9}_{\text{five terms}} = 5^2$$

Now let us try the next case where $n = 6$ and see if the pattern persists. That is, we want to know if it is true that

$$\underbrace{1 + 3 + 5 + 7 + 9 + 11}_{\text{six terms}} = 6^2$$

As you can easily check, this last equation is true. Thus, based on the experimental, or empirical, evidence, we are led to the following conjecture.

CONJECTURE    The sum of the first $n$ odd natural numbers is $n^2$. That is, $1 + 3 + 5 + \cdots + (2n - 1) = n^2$, for each natural number $n$.

At this point, the "law" we've discovered is indeed really only a conjecture. After all, we've checked it only for values of $n$ ranging from 1 to 6. It is conceivable at this point (although we may feel it is unlikely) that the conjecture is false for certain values of $n$. For the conjecture to be useful, we must be able to prove that it holds without exception for *all* natural numbers $n$. In fact, we will subsequently prove that this conjecture is valid. But before explaining the method of proof to be used, let us look at one more example.

Again, let $n$ denote a natural number. Then consider the following question: Which quantity is the larger, $2^n$ or $(n + 1)^2$? As before, we begin by doing some calculations. This is the experimental stage of our work. According to Table 2, the quantity $(n + 1)^2$ is larger than $2^n$ for each value of $n$ up through $n = 5$. Thus, we make the following conjecture.

CONJECTURE    $(n + 1)^2 > 2^n$ for all natural numbers $n$.

Again, we note that this is only a conjecture at this point. Indeed, if we try the case where $n = 6$, we find that the pattern does not persist. That is, when $n = 6$, we find that $2^n$ is 64, while $(n + 1)^2$ is only 49. So in this example, the conjecture is not true in general; we have found a value of $n$ for which it fails.

The preceding examples show that experimentation does have a place in mathematics, but we must be careful with the results. Where experimentation leads to a conjecture, proof is required before the conjecture can be viewed as a valid law. For the remainder of this section, we shall discuss one such method of proof, that of *mathematical induction*.

In order to state the principle of mathematical induction, we first introduce some notation. Suppose that for each natural number $n$ we have a statement $P_n$ to be proved. Consider, for instance, the first conjecture we arrived at:

$$1 + 3 + 5 + \cdots + (2n - 1) = n^2$$

If we denote this statement by $P_n$, then we have, for example, that

$P_1$ is the statement that $1 = 1^2$

$P_2$ is the statement that $1 + 3 = 2^2$

$P_3$ is the statement that $1 + 3 + 5 = 3^2$

**TABLE 2**

| $n$ | $2^n$ | $(n + 1)^2$ |
|---|---|---|
| 1 | 2 | 4 |
| 2 | 4 | 9 |
| 3 | 8 | 16 |
| 4 | 16 | 25 |
| 5 | 32 | 36 |

With this notation, we can now state the **principle of mathematical induction**.

> ### Principle of Mathematical Induction
>
> Suppose that for each natural number $n$, we have a statement $P_n$. Furthermore, suppose that the following two conditions hold:
>
> 1. $P_1$ is true.
> 2. For each natural number $k$, if $P_k$ is true, then $P_{k+1}$ is true.
>
> Then all the statements are true; that is, $P_n$ is true for all natural numbers $n$.

The idea behind mathematical induction is a simple one. Think of each statement $P_n$ as the rung of a ladder to be climbed. Then we can make the analogy shown in Table 3.

TABLE 3

| MATHEMATICAL INDUCTION | LADDER ANALOGY |
|---|---|
| Hypotheses $\begin{cases} \text{1. } P_1 \text{ is true.} \\[6pt] \text{2. If } P_k \text{ is true} \\ \quad \text{then } P_{k+1} \text{ is} \\ \quad \text{true, for any } k. \end{cases}$ | Hypotheses $\begin{cases} 1'.\ \text{You can reach the} \\ \quad \text{first rung.} \\[4pt] 2'.\ \text{If you are on the} \\ \quad k\text{th rung you can} \\ \quad \text{reach the } (k+1)\text{st} \\ \quad \text{rung, for any } k. \end{cases}$ |
| Conclusion $\begin{cases} \text{3. } P_n \text{ is true for} \\ \quad \text{all } n. \end{cases}$ | Conclusion $\begin{cases} 3'.\ \text{You can climb the} \\ \quad \text{entire ladder.} \end{cases}$ |

According to the principle of mathematical induction, we can prove that a statement or formula $P_n$ is true for all $n$ if we carry out the following two steps.

STEP 1   Show that $P_1$ is true.

STEP 2   Assume that $P_k$ is true, and on the basis of this assumption, show that $P_{k+1}$ is true.

In step 2, the assumption that $P_k$ is true is referred to as the **induction hypothesis**. (In computer science, step 1 is sometimes referred to as the **initialization step**.) Let us now turn to some examples of proof by mathematical induction.

**EXAMPLE 1**   Use mathematical induction to prove that

$$1 + 3 + 5 + \cdots + (2n - 1) = n^2$$

for all natural numbers $n$.

Solution   Let $P_n$ denote the statement that $1 + 3 + 5 + \cdots + (2n - 1) = n^2$. Then we want to show that $P_n$ is true for all natural numbers $n$.

STEP 1   We must check that $P_1$ is true. But $P_1$ is just the statement that $1 = 1^2$, which is true.

STEP 2   Assuming that $P_k$ is true, we must show that $P_{k+1}$ is true. Thus, we assume that

$$1 + 3 + 5 + \cdots + (2k - 1) = k^2 \tag{1}$$

That is the induction hypothesis. We must now show that

$$1 + 3 + 5 + \cdots + (2k - 1) + [2(k + 1) - 1] = (k + 1)^2 \tag{2}$$

To derive equation (2) from equation (1), we add the quantity $[2(k + 1) - 1]$ to both sides of equation (1). (The motivation for this stems from the observation that the left-hand sides of equations (1) and (2) differ only by the quantity $[2(k + 1) - 1]$.) We obtain

$$1 + 3 + 5 + \cdots + (2k - 1) + [2(k + 1) - 1] = k^2 + [2(k + 1) - 1]$$
$$= k^2 + 2k + 1$$
$$= (k + 1)^2$$

This last equation is what we wanted to show. And having now carried out steps 1 and 2, we conclude by the principle of mathematical induction that $P_n$ is true for all natural numbers $n$.

**EXAMPLE 2**   Use mathematical induction to prove that

$$2^3 + 4^3 + 6^3 + \cdots + (2n)^3 = 2n^2(n + 1)^2$$

for all natural numbers $n$.

**Solution**   Let $P_n$ denote the statement that

$$2^3 + 4^3 + 6^3 + \cdots + (2n)^3 = 2n^2(n + 1)^2$$

Then we want to show that $P_n$ is true for all natural numbers $n$.

STEP 1   We must check that $P_1$ is true. $P_1$ is the statement that

$$2^3 = 2(1^2)(1 + 1)^2 \qquad \text{or} \qquad 8 = 8$$

Thus, $P_1$ is true.

STEP 2   Assuming that $P_k$ is true, we must show that $P_{k+1}$ is true. Thus, we assume that

$$2^3 + 4^3 + 6^3 + \cdots + (2k)^3 = 2k^2(k + 1)^2 \tag{3}$$

We must now show that

$$2^3 + 4^3 + 6^3 + \cdots + (2k)^3 + [2(k + 1)]^3 = 2(k + 1)^2(k + 2)^2 \tag{4}$$

Adding $[2(k + 1)]^3$ to both sides of equation (3) yields

$$2^3 + 4^3 + 6^3 + \cdots + (2k)^3 + [2(k + 1)]^3 = 2k^2(k + 1)^2 + [2(k + 1)]^3$$
$$= 2k^2(k + 1)^2 + 8(k + 1)^3$$
$$= 2(k + 1)^2[k^2 + 4(k + 1)]$$
$$= 2(k + 1)^2(k^2 + 4k + 4)$$
$$= 2(k + 1)^2(k + 2)^2$$

We have now derived equation (4) from equation (3), as we wished to do. Having carried out steps 1 and 2, we conclude by the principle of mathematical induction that $P_n$ is true for all natural numbers $n$.

**EXAMPLE 3**  As indicated in Table 4, the number 3 is a factor of $2^{2n} - 1$ when $n = 1, 2, 3,$ and 4. Use mathematical induction to show that 3 is a factor of $2^{2n} - 1$ for all natural numbers $n$.

**Solution**  Let $P_n$ denote the statement that 3 is a factor of $2^{2n} - 1$. We want to show that $P_n$ is true for all natural numbers $n$.

STEP 1  We must check that $P_1$ is true. But $P_1$ in this case is just the statement that 3 is a factor of $2^{2(1)} - 1$; that is, 3 is a factor of 3, which is surely true.

STEP 2  Assuming that $P_k$ is true, we must show that $P_{k+1}$ is true. Thus, we assume that

$$3 \text{ is a factor of } 2^{2k} - 1 \tag{5}$$

and we must show that

$$3 \text{ is a factor of } 2^{2(k+1)} - 1$$

The strategy here will be to rewrite the expression $2^{2(k+1)} - 1$ in such a way that the induction hypothesis, statement (5), can be applied. We have

$$\begin{aligned} 2^{2(k+1)} - 1 &= 2^{2k+2} - 1 \\ &= 2^2 \cdot 2^{2k} - 1 \\ &= 4 \cdot 2^{2k} - 4 + 3 \\ &= 4(2^{2k} - 1) + 3 \end{aligned} \tag{6}$$

Now, look at the right-hand side of equation (6). By the induction hypothesis, 3 is a factor of $2^{2k} - 1$. Thus, 3 is a factor of $4(2^{2k} - 1)$, from which it certainly follows that 3 is a factor of $4(2^{2k} - 1) + 3$. In summary, then, 3 is a factor of the right-hand side of equation (6). Consequently, 3 must be a factor of the left-hand side of equation (6), which is what we wished to show. Having now completed steps 1 and 2, we conclude by the principle of mathematical induction that $P_n$ is true for all natural numbers $n$. In other words, 3 is a factor of $2^{2n} - 1$ for all natural numbers $n$.

There are instances in which a given statement $P_n$ is false for certain initial values of $n$, but true thereafter. An example of this is provided by the statement

$$2^n > (n+1)^2$$

As you can easily check, this statement is false for $n = 1, 2, 3, 4,$ and 5. But, as Example 4 shows, the statement is true for $n \geq 6$. In Example 4, we adapt the principle of mathematical induction by beginning in step 1 with a consideration of $P_6$ rather than $P_1$.

**EXAMPLE 4**  Use mathematical induction to prove that

$$2^n > (n+1)^2 \qquad \text{for all natural numbers } n \geq 6$$

**Solution**  STEP 1  We must first check that $P_6$ is true. But $P_6$ is simply the assertion that

$$2^6 > (6+1)^2 \qquad \text{or} \qquad 64 > 49$$

Thus, $P_6$ is true.

**TABLE 4**

| $n$ | $2^{2n} - 1$ |
|---|---|
| 1 | $3 \ (= 3 \cdot 1)$ |
| 2 | $15 \ (= 3 \cdot 5)$ |
| 3 | $63 \ (= 3 \cdot 21)$ |
| 4 | $255 \ (= 3 \cdot 85)$ |

STEP 2   Assuming that $P_k$ is true, where $k \geq 6$, we must show that $P_{k+1}$ is true. Thus, we assume that

$$2^k > (k+1)^2 \qquad \text{where } k \geq 6 \tag{7}$$

We must show that

$$2^{k+1} > (k+2)^2$$

Multiplying both sides of inequality (7) by 2 gives us

$$2(2^k) > 2(k+1)^2 = 2k^2 + 4k + 2$$

This can be rewritten

$$2^{k+1} > k^2 + 4k + (k^2 + 2)$$

However, since $k \geq 6$, it is certainly true that

$$k^2 + 2 > 4$$

We therefore have

$$2^{k+1} > k^2 + 4k + 4 \qquad \text{or} \qquad 2^{k+1} > (k+2)^2$$

as we wished to show. Having now completed steps 1 and 2, we conclude that $P_n$ is true for all natural numbers $n \geq 6$.

## EXERCISE SET 13.1

### A

*In Exercises 1–18, use the principle of mathematical induction to show that the statements are true for all natural numbers.*

1. $1 + 2 + 3 + \cdots + n = \frac{1}{2}n(n+1)$

2. $2 + 4 + 6 + \cdots + 2n = n(n+1)$

3. $1 + 4 + 7 + \cdots + (3n-2) = \frac{1}{2}n(3n-1)$

4. $5 + 9 + 13 + \cdots + (4n+1) = n(2n+3)$

5. $1^2 + 2^2 + 3^2 + \cdots + n^2 = \frac{1}{6}n(n+1)(2n+1)$

6. $2^2 + 4^2 + 6^2 + \cdots + (2n)^2 = \frac{1}{3}2n(n+1)(2n+1)$

7. $1^2 + 3^2 + 5^2 + \cdots + (2n-1)^2 = \frac{1}{3}n(2n-1)(2n+1)$

8. $2 + 2^2 + 2^3 + \cdots + 2^n = 2^{n+1} - 2$

9. $3 + 3^2 + 3^3 + \cdots + 3^n = \frac{1}{2}(3^{n+1} - 3)$

10. $e^x + e^{2x} + e^{3x} + \cdots + e^{nx} = \dfrac{e^{(n+1)x} - e^x}{e^x - 1} \quad (x \neq 0)$

11. $1^3 + 2^3 + 3^3 + \cdots + n^3 = [\frac{1}{2}n(n+1)]^2$

12. $2^3 + 4^3 + 6^3 + \cdots + (2n)^3 = 2n^2(n+1)^2$

13. $1^3 + 3^3 + 5^3 + \cdots + (2n-1)^3 = n^2(2n^2 - 1)$

14. $1 \cdot 2 + 3 \cdot 4 + 5 \cdot 6 + \cdots + (2n-1)(2n)$
$$= \frac{1}{3}n(n+1)(4n-1)$$

15. $1 \cdot 3 + 3 \cdot 5 + 5 \cdot 7 + \cdots + (2n-1)(2n+1)$
$$= \frac{1}{3}n(4n^2 + 6n - 1)$$

16. $\dfrac{1}{1 \times 3} + \dfrac{1}{2 \times 4} + \dfrac{1}{3 \times 5} + \cdots + \dfrac{1}{n(n+2)}$
$$= \dfrac{n(3n+5)}{4(n+1)(n+2)}$$

17. $1 + \dfrac{3}{2} + \dfrac{5}{2^2} + \dfrac{7}{2^3} + \cdots + \dfrac{2n-1}{2^{n-1}} = 6 - \dfrac{2n+3}{2^{n-1}}$

18. $1 + 2 \cdot 2 + 3 \cdot 2^2 + 4 \cdot 2^3 + \cdots + n \cdot 2^{n-1}$
$$= (n-1)2^n + 1$$

19. Show that $n \leq 2^{n-1}$ for all natural numbers $n$.

20. Show that 3 is a factor of $n^3 + 2n$ for all natural numbers $n$.

21. Show that $n^2 + 4 < (n+1)^2$ for all natural numbers $n \geq 2$.

22. Show that $n^3 > (n+1)^2$ for all natural numbers $n \geq 3$.

### B

23. Let $f(n) = \dfrac{1}{1 \cdot 2} + \dfrac{1}{2 \cdot 3} + \dfrac{1}{3 \cdot 4} + \cdots + \dfrac{1}{n(n+1)}$.

(a) Complete the following table.

| $n$ | 1 | 2 | 3 | 4 | 5 |
|------|---|---|---|---|---|
| $f(n)$ | | | | | |

(b) On the basis of the results in the table, what would you guess to be the value of $f(6)$? Compute $f(6)$ to see if this is correct.

(c) Make a conjecture about the value of $f(n)$, and prove it using mathematical induction.

24. Let $f(n) = \dfrac{1}{1 \times 3} + \dfrac{1}{3 \times 5} + \dfrac{1}{5 \times 7} + \cdots$

$$+ \dfrac{1}{(2n-1)(2n+1)}.$$

(a) Complete the following table.

| $n$ | 1 | 2 | 3 | 4 |
|-----|---|---|---|---|
| $f(n)$ | | | | |

(b) On the basis of the results in the table, what would you guess to be the value of $f(5)$? Compute $f(5)$ to see if your guess is correct.

(c) Make a conjecture about the value of $f(n)$, and prove it using mathematical induction.

25. Suppose that a function $f$ satisfies the following conditions: $f(1) = 1$ and

$$f(n) - f(n-1) + 2\sqrt{f(n-1)} + 1 \qquad \text{for } n \geq 2$$

(a) Complete the table.

| $n$ | 1 | 2 | 3 | 4 | 5 |
|-----|---|---|---|---|---|
| $f(n)$ | | | | | |

(b) On the basis of the results in the table, what would you guess to be the value of $f(6)$? Compute $f(6)$ to see if your guess is correct.

(c) Make a conjecture about the value of $f(n)$ when $n$ is a natural number, and prove the conjecture using mathematical induction.

26. This exercise demonstrates the necessity of carrying out both step 1 and step 2 before considering an induction proof valid.

(a) Let $P_n$ denote the statement that $n^2 + 1$ is even. Check that $P_1$ is true. Then give an example showing that $P_n$ is not true for all $n$.

(b) Let $Q_n$ denote the statement that $n^2 + n$ is odd. Show that step 2 of an induction proof can be completed in this case, but not step 1.

27. A **prime number** is a natural number that has no factors other than itself and 1. For technical reasons, 1 is not considered a prime. Thus, the list of the first seven primes looks like this:

$$2, 3, 5, 7, 11, 13, 17$$

Let $P_n$ be the statement that $n^2 + n + 11$ is prime. Check that $P_n$ is true for all values of $n$ less than 10. Check that $P_{10}$ is false.

28. Prove that if $x \neq 1$,

$$1 + 2x + 3x^2 + \cdots + nx^{n-1} = \dfrac{1 - x^n}{(1-x)^2} - \dfrac{nx^n}{1-x}$$

for all natural numbers $n$.

29. If $r \neq 1$, show that $1 + r + r^2 + \cdots + r^{n-1} = \dfrac{r^n - 1}{r - 1}$ for all natural numbers $n$.

30. Use mathematical induction to show that $x^n - 1 = (x - 1)(1 + x + x^2 + \cdots + x^{n-1})$ for all natural numbers $n$.

31. Prove that 5 is a factor of $n^5 - n$ for all natural numbers $n \geq 2$.

32. Prove that 4 is a factor of $5^n + 3$ for all natural numbers $n$.

33. Use mathematical induction to show that $x - y$ is a factor of $x^n - y^n$ for all natural numbers $n$.

*Suggestion for step 2:* Verify and then use the fact that $x^{k+1} - y^{k+1} = x^k(x - y) + (x^k - y^k)y$.

*In Exercises 34 and 35, use mathematical induction to prove that the formulas hold for all natural numbers $n$.*

34. $\log_{10}(a_1 a_2 \cdots a_n)$
$$= \log_{10} a_1 + \log_{10} a_2 + \cdots + \log_{10} a_n$$

35. $(1 + p)^n \geq 1 + np$, where $p > -1$

## THE BINOMIAL THEOREM

**13.2**

*A mathematician, like a painter or a poet, is a maker of patterns.*

G. H. HARDY (1877–1947)

If you look back at Section 1.8, you will see that two of the special products listed there are

$$(a + b)^2 = a^2 + 2ab + b^2$$

and

$$(a + b)^3 = a^3 + 3a^2b + 3ab^2 + b^3$$

Our present goal is to develop a general formula, known as the *binomial theorem*, for expanding any product of the form $(a + b)^n$, when $n$ is a natural number.

We begin by looking for patterns in the expansion of $(a + b)^n$. To do this, let's list the expansions of $(a + b)^n$ for $n = 1, 2, 3, 4,$ and 5. (Exercises 1 and 2 ask you to verify these results simply by repeated multiplication.)

$$(a + b)^1 = a + b$$
$$(a + b)^2 = a^2 + 2ab + b^2$$
$$(a + b)^3 = a^3 + 3a^2b + 3ab^2 + b^3$$
$$(a + b)^4 = a^4 + 4a^3b + 6a^2b^2 + 4ab^3 + b^4$$
$$(a + b)^5 = a^5 + 5a^4b + 10a^3b^2 + 10a^2b^3 + 5ab^4 + b^5$$

After surveying these results, we note the following patterns.

| PROPERTY SUMMARY | Patterns Observed in $(a + b)^n$ for $n = 1, 2, 3, 4, 5$ | |
|---|---|---|
| | GENERAL STATEMENT | EXAMPLE |
| | There are $n + 1$ terms. | There are $4\,(=3 + 1)$ terms in the expansion of $(a + b)^3$. |
| | The expansion begins with $a^n$ and ends with $b^n$. | $(a + b)^3$ begins with $a^3$ and ends with $b^3$. |
| | The sum of the exponents in each term is $n$. | The sum of the exponents in each term of $(a + b)^3$ is 3. |
| | The exponents of $a$ decrease by 1 from term to term. | $(a + b)^3 = a^{③} + 3a^{②}b + 3a^{①}b^2 + a^{⓪}b^3$ |
| | The exponents of $b$ increase by 1 from term to term. | $(a + b)^3 = a^3b^{⓪} + 3a^2b^{①} + 3ab^{②} + b^{③}$ |
| | When $n$ is even, the coefficients are symmetric about the middle term | The sequence of coefficients for $(a + b)^4$ is $1, 4, 6, 4, 1$. |
| | When $n$ is odd, the coefficients are symmetric about the two middle terms. | The sequence of coefficients for $(a + b)^5$ is $1, 5, 10, 10, 5, 1$. |

The patterns we have just observed for $(a + b)^n$ persist for all natural numbers $n$. (This will follow from the binomial theorem, which is proved at the end of this section.) Thus, for example, the form of $(a + b)^6$ must be as follows:

$$(a + b)^6 = a^6 + \underline{?}a^5b + \underline{?}a^4b^2 + \underline{?}a^3b^3 + \underline{?}a^2b^4 + \underline{?}ab^5 + b^6$$

The problem now is to find the proper coefficient for each term. To do this, we need to discover additional patterns in the expansion of $(a + b)^n$.

We have already written out the expansions of $(a + b)^n$ for values of $n$ ranging from 1 to 5. Let us now write only the coefficients appearing in those expansions. The resulting triangular array of numbers is known as **Pascal's**

**triangle.*** For reasons of symmetry, we begin with $(a + b)^0$ rather than $(a + b)^1$.

$$(a + b)^0 \dots\dots\dots\dots\dots\dots 1$$
$$(a + b)^1 \dots\dots\dots\dots\dots 1 \quad 1$$
$$(a + b)^2 \dots\dots\dots\dots 1 \quad 2 \quad 1$$
$$(a + b)^3 \dots\dots\dots 1 \quad 3 \quad 3 \quad 1$$
$$(a + b)^4 \dots\dots\dots 1 \quad 4 \quad 6 \quad 4 \quad 1$$
$$(a + b)^5 \dots\dots 1 \quad 5 \quad 10 \quad 10 \quad 5 \quad 1$$

The key observation regarding Pascal's triangle is this: *Each entry in the array* (other than the 1's along the sides) *is the sum of the two numbers diagonally above it*. For instance, the 6 that appears in the fifth row is the sum of the two 3's diagonally above it. Using this observation, we can form as many additional rows as we please. The coefficients for $(a + b)^n$ will then appear in the $(n + 1)$st row of the array.[†] For instance, to obtain the row corresponding to $(a + b)^6$, we have

sixth row, $(a + b)^5$:   1   5   10   10   5   1

seventh row, $(a + b)^6$:   1   6   15   20   15   6   1

Thus, the sequence of coefficients for $(a + b)^6$ is 1, 6, 15, 20, 15, 6, 1. This answers the question raised earlier about the expansions of $(a + b)^6$. We have

$$(a + b)^6 = a^6 + 6a^5b + 15a^4b^2 + 20a^3b^3 + 15a^2b^4 + 6ab^5 + b^6$$

For analytical work or for larger values of the exponent $n$, it is inefficient to rely on Pascal's triangle. For this reason, we point out another pattern in the expansions of $(a + b)^n$.

> In the expansion of $(a + b)^n$, the coefficient of any term after the first can be generated as follows. In the *previous* term, multiply the coefficient by the exponent of $a$ and then divide by the number of that previous term.

To see how this observation is used, let's compute the second, third, and fourth coefficients in the expansion of $(a + b)^6$. To compute the coefficient of the second term, we go back to the first term, which is $a^6$. We have

$$\text{coefficient of second term} = \frac{1 \cdot 6}{1} = 6$$

**FIGURE 1** "Pascal's" triangle—by Chu-Shi-Kie, A.D. 1303

---

* The array is named after Blaise Pascal, a seventeenth-century French mathematician and philosopher. However, as Figure 1 indicates, the Pascal triangle was known to Chinese mathematicians centuries earlier.

† That these numbers actually are the appropriate coefficients follows from the binomial theorem, which is proved at the end of this section.

Thus, the second term is $6a^5b$, and consequently, we have

coefficient of third term $= \dfrac{6 \cdot 5}{2} = 15$

where the annotations point to: coefficient of second term, exponent of $a$ in second term (from the $6 \cdot 5$), and number of second term (from the $2$).

Continuing now with this method, you should check for yourself that the coefficient of the fourth term in the expansion of $(a + b)^6$ is 20. *Note:* We now know that the first four coefficients are 1, 6, 15, and 20. By symmetry, it follows that the complete sequence of coefficients for this expansion is 1, 6, 15, 20, 15, 6, 1. No additional calculation for the coefficients is necessary.

**EXAMPLE 1**   Expand $(2x - y^2)^7$.

Solution   First we write the expansion of $(a + b)^7$ using the method explained just prior to this example, or using Pascal's triangle. As you should check for yourself, the expansion is

$$(a + b)^7 = a^7 + 7a^6b + 21a^5b^2 + 35a^4b^3 + 35a^3b^4 + 21a^2b^5 + 7ab^6 + b^7$$

Now we make the substitutions $a = 2x$ and $b = -y^2$. This yields

$$[2x + (-y^2)]^7 = (2x)^7 + 7(2x)^6(-y^2) + 21(2x)^5(-y^2)^2$$
$$+ 35(2x)^4(-y^2)^3 + 35(2x)^3(-y^2)^4$$
$$+ 21(2x)^2(-y^2)^5 + 7(2x)(-y^2)^6 + (-y^2)^7$$
$$= 128x^7 - 448x^6y^2 + 672x^5y^4 - 560x^4y^6 + 280x^3y^8$$
$$- 84x^2y^{10} + 14xy^{12} - y^{14}$$

This is the required expansion. Notice how the signs alternate in the final answer; this is characteristic of all expansions of the form $(a - b)^n$.

In preparation for the binomial theorem, we introduce two notations that are used not only in connection with the binomial theorem, but in many other areas of mathematics as well. The first of these notations is $n!$ (read "$n$ factorial").

| Definition of $n!$ | |
|---|---|
| | EXAMPLES |
| $n! = 1 \cdot 2 \cdot 3 \cdots n$ | $3! = 1 \cdot 2 \cdot 3 = 6$ |
| where $n$ is a natural number | $\dfrac{6!}{4!} = \dfrac{6 \cdot 5 \cdot 4 \cdot 3 \cdot 2 \cdot 1}{4 \cdot 3 \cdot 2 \cdot 1}$ |
| $0! = 1$ | $= 6 \times 5 = 30$ |

**EXAMPLE 2**   Simplify the expression $\dfrac{(n + 1)!}{(n - 1)!}$.

Solution   $\dfrac{(n + 1)!}{(n - 1)!} = \dfrac{(n + 1)(n)[(n - 1)!]}{(n - 1)!}$

$= (n + 1)n = n^2 + n$

The second notation that we introduce in preparation for the binomial theorem is $\binom{n}{k}$. (This notation is often read "$n$ choose $k$"; you'll see the reason for this in a later section when we discuss *permutations* and *combinations*.)

---

**Definition of the Binomial Coefficients $\binom{n}{k}$**

EXAMPLE

Let $n$ and $k$ be nonnegative integers with $k \le n$. Then the binomial coefficient $\binom{n}{k}$ is defined by

$$\binom{n}{k} = \frac{n!}{k!\,(n-k)!}$$

$$\binom{5}{2} = \frac{5!}{2!(5-2)!}$$

$$= \frac{5!}{2!\,3!}$$

$$= \frac{5 \cdot 4 \cdot 3 \cdot 2 \cdot 1}{(2 \cdot 1)(3 \cdot 2 \cdot 1)}$$

$$= \frac{5 \cdot 4}{2 \cdot 1} = 10$$

---

The binomial coefficients are so named because they are indeed the coefficients in the expansion of $(a + b)^n$. More precisely, the relationship is this:

The coefficients in the expansion of $(a + b)^n$ are the $n + 1$ numbers

$$\binom{n}{0}, \binom{n}{1}, \binom{n}{2}, \ldots, \binom{n}{n}.$$

Subsequently, we will see why this statement is true. For now, however, let us look at an example. Consider the binomial coefficients $\binom{3}{0}, \binom{3}{1}, \binom{3}{2},$ and $\binom{3}{3}$. According to our statement, these four quantities should be the coefficients in the expansion of $(a + b)^3$. Let us check:

$$\binom{3}{0} = \frac{3!}{0!(3-0)!} = \frac{3!}{1(3!)} = 1$$

$$\binom{3}{1} = \frac{3!}{1!(3-1)!} = \frac{3 \cdot 2 \cdot 1}{1(2 \cdot 1)} = 3$$

$$\binom{3}{2} = \frac{3!}{2!(3-2)!} = \frac{3 \cdot 2 \cdot 1}{(2 \cdot 1)1} = 3$$

$$\binom{3}{3} = \frac{3!}{3!(3-3)!} = \frac{3!}{3!\,0!} = 1$$

The values of $\binom{3}{0}, \binom{3}{1}, \binom{3}{2},$ and $\binom{3}{3}$ are thus 1, 3, 3, and 1, respectively. But these last four numbers are indeed the coefficients in the expansion of $(a + b)^3$, as we wished to check.

We are now in a position to state the binomial theorem, after which we will look at several applications. Finally, at the end of this section, we will use mathematical induction to prove the theorem. In the statement of the theorem that follows, we are assuming that the exponent $n$ is a natural number.

**Binomial Theorem**

$$(a + b)^n = \binom{n}{0}a^n + \binom{n}{1}a^{n-1}b + \binom{n}{2}a^{n-2}b^2 + \cdots + \binom{n}{n-1}ab^{n-1} + \binom{n}{n}b^n$$

One of the uses of the binomial theorem is in identifying specific terms in an expansion without computing the entire expansion. This is particularly helpful when the exponent $n$ is relatively large. Looking back at the statement of the binomial theorem, there are three observations we can make. First, the coefficient of the $r$th term is $\binom{n}{r-1}$. For instance, the coefficient of the third term is $\binom{n}{3-1} = \binom{n}{2}$. The second observation is that the exponent for $a$ in the $r$th term is $n - (r - 1)$. For instance, the exponent for $a$ in the third term is $n - (3 - 1) = n - 2$. Finally, we observe that the exponent for $b$ in the $r$th term is $r - 1$, the same quantity that appears in the lower position of the corresponding binomial coefficient. For instance, the exponent for $b$ in the third term is $r - 1 = 3 - 1 = 2$. We summarize these three observations with the following statement.

The $r$th term in the expansion of $(a + b)^n$ is

$$\binom{n}{r-1}a^{n-r+1}b^{r-1}$$

**EXAMPLE 3** Find the fifteenth term in the expansion of $\left(x^2 - \dfrac{1}{x}\right)^{18}$.

**Solution** Using the values $r = 15$, $n = 18$, $a = x^2$, and $b = -1/x$, we have

$$\binom{n}{r-1}a^{n-r+1}b^{r-1} = \binom{18}{15-1}(x^2)^{18-15+1}\left(\frac{-1}{x}\right)^{15-1}$$

$$= \binom{18}{14}x^8 \cdot \frac{1}{x^{14}}$$

$$= \frac{18 \times 17 \times 16 \times 15 \times (14!)}{14!(4 \times 3 \times 2 \times 1)}x^{-6}$$

$$= \frac{18 \times 17 \times 16 \times 15}{4 \times 3 \times 2 \times 1}x^{-6}$$

$$= 3060x^{-6} \qquad \text{as required}$$

**EXAMPLE 4**   Find the coefficient of the term containing $x^4$ in the expansion of $(x + y^2)^{30}$.

Solution   Again we use the fact that the $r$th term in the expansion of $(a + b)^n$ is $\binom{n}{r-1}a^{n-r+1}b^{r-1}$. In this case, $n$ is 30 and $x$ plays the role of $a$. The exponent for $x$ is then $n - r + 1$ or $30 - r + 1$. To see when this exponent is 4, we write

$$30 - r + 1 = 4$$
$$r = 27$$

The required coefficient is therefore $\binom{30}{27-1}$. We then have

$$\binom{30}{26} = \frac{30!}{26!(30-26)!}$$
$$= \frac{30 \times 29 \times 28 \times 27}{4 \times 3 \times 2 \times 1}$$

After carrying out the indicated arithmetic, we find that $\binom{30}{26} = 27{,}405$. This is the required coefficient.

**EXAMPLE 5**   Find the coefficient of the term containing $a^9$ in the expansion of $(a + 2\sqrt{a})^{10}$.

Solution   The $r$th term in this expansion will be

$$\binom{10}{r-1}a^{10-r+1}(2\sqrt{a})^{r-1}$$

We can rewrite this as

$$\binom{10}{r-1}a^{10-r+1}(2^{r-1})(a^{1/2})^{r-1}$$

or

$$2^{r-1}\binom{10}{r-1}a^{10-r+1+(r-1)/2}$$

This shows that the general form of the coefficient we wish to find is $2^{r-1}\binom{10}{r-1}$. We now need to determine $r$ when the exponent of $a$ is 9. Thus, we require that

$$10 - r + 1 + \frac{r-1}{2} = 9$$
$$-r + \frac{r-1}{2} = -2$$
$$-2r + r - 1 = -4 \qquad \text{multiplying by 2}$$
$$r = 3$$

The required coefficient is now obtained by substituting $r = 3$ in the expression $2^{r-1}\begin{pmatrix} 10 \\ r-1 \end{pmatrix}$. Thus, the required coefficient is

$$2^2\begin{pmatrix} 10 \\ 2 \end{pmatrix} = \frac{4 \cdot 10!}{2!(10-2)!} = 2 \cdot 10 \cdot 9 = 180$$

There are three simple identities involving the binomial coefficients that will simplify our proof of the binomial theorem.

IDENTITY 1 $\quad \begin{pmatrix} r \\ 0 \end{pmatrix} = 1 \quad$ for all nonnegative integers $r$

IDENTITY 2 $\quad \begin{pmatrix} r \\ r \end{pmatrix} = 1 \quad$ for all nonnegative integers $r$

IDENTITY 3 $\quad \begin{pmatrix} k \\ r \end{pmatrix} + \begin{pmatrix} k \\ r-1 \end{pmatrix} = \begin{pmatrix} k+1 \\ r \end{pmatrix} \quad$ for all natural numbers $k$ and $r$ with $r \le k$

All three of these identities can be proved directly from the definitions of the binomial coefficients, without the need for mathematical induction. The proofs of the first two are straightforward, and we omit them here. The proof of identity 3 runs as follows:

$$\begin{pmatrix} k \\ r \end{pmatrix} + \begin{pmatrix} k \\ r-1 \end{pmatrix} = \frac{k!}{r!(k-r)!} + \frac{k!}{(r-1)!(k-r+1)!}$$

$$= \frac{k!}{r(r-1)!(k-r)!} + \frac{k!}{(r-1)!(k-r+1)(k-r)!}$$

Now, the common denominator on the right-hand side of the last equation is $r(r-1)!(k-r+1)(k-r)!$ Thus, we have

$$\begin{pmatrix} k \\ r \end{pmatrix} + \begin{pmatrix} k \\ r-1 \end{pmatrix} = \frac{k!(k-r+1)}{r(r-1)!(k-r+1)(k-r)!} + \frac{k!\,r}{r(r-1)!(k-r+1)(k-r)!}$$

$$= \frac{k!(k-r+1) + k!\,r}{r(r-1)!(k-r+1)(k-r)!}$$

$$= \frac{k!(k-r+1+r)}{r(r-1)!(k-r+1)(k-r)!} = \frac{k!(k+1)}{r!(k-r+1)!}$$

$$= \frac{(k+1)!}{r![(k+1)-r]!}$$

$$= \begin{pmatrix} k+1 \\ r \end{pmatrix} \quad \text{as required}$$

Taken together, the three identities show why the $(n+1)$st row of Pascal's triangle consists of the numbers $\begin{pmatrix} n \\ 0 \end{pmatrix}, \begin{pmatrix} n \\ 1 \end{pmatrix}, \begin{pmatrix} n \\ 2 \end{pmatrix}, \ldots, \begin{pmatrix} n \\ n \end{pmatrix}$. Identities 1 and 2 tell us that this row of numbers begins and ends with 1. Identity 3 is then just a statement of the fact that each entry in the row, other than the initial and final 1, is generated by adding the two entries diagonally above it.

We conclude this section by using mathematical induction to prove the binomial theorem. The statement $P_n$ that we wish to prove for all natural numbers $n$ is this:

$$(a + b)^n = \binom{n}{0}a^n + \binom{n}{1}a^{n-1}b + \cdots + \binom{n}{n-1}ab^{n-1} + \binom{n}{n}b^n$$

First, we check that $P_1$ is true. The statement $P_1$ asserts that

$$(a + b)^1 = \binom{1}{0}a^1 + \binom{1}{1}a^0 b$$

However, in view of identities 1 and 2, this last equation becomes

$$(a + b)^1 = 1 \cdot a + 1 \cdot b$$

which is surely true. Now let us assume that $P_k$ is true and, on the basis of this assumption, show that $P_{k+1}$ is true. The statement $P_k$ is

$$(a + b)^k = \binom{k}{0}a^k + \binom{k}{1}a^{k-1}b + \cdots + \binom{k}{k-1}ab^{k-1} + \binom{k}{k}b^k$$

Multiplying both sides of this equation by the quantity $(a + b)$ yields

$$(a + b)^{k+1} = (a + b)\left[\binom{k}{0}a^k + \binom{k}{1}a^{k-1}b + \cdots + \binom{k}{k-1}ab^{k-1} + \binom{k}{k}b^k\right]$$

$$= a\left[\binom{k}{0}a^k + \binom{k}{1}a^{k-1}b + \cdots + \binom{k}{k-1}ab^{k-1} + \binom{k}{k}b^k\right]$$

$$+ b\left[\binom{k}{0}a^k + \binom{k}{1}a^{k-1}b + \cdots + \binom{k}{k-1}ab^{k-1} + \binom{k}{k}b^k\right]$$

$$= \binom{k}{0}a^{k+1} + \binom{k}{1}a^k b + \cdots + \binom{k}{k-1}a^2 b^{k-1} + \binom{k}{k}ab^k$$

$$+ \binom{k}{0}a^k b + \binom{k}{1}a^{k-1}b^2 + \cdots + \binom{k}{k-1}ab^k + \binom{k}{k}b^{k+1}$$

$$= \binom{k}{0}a^{k+1} + \left[\binom{k}{1} + \binom{k}{0}\right]a^k b + \cdots + \left[\binom{k}{k} + \binom{k}{k-1}\right]ab^k + \binom{k}{k}b^{k+1}$$

We can now make some substitutions on the right-hand side of this last equation. The initial binomial coefficient $\binom{k}{0}$ can be replaced by $\binom{k+1}{0}$, for both are equal to 1 according to identity 1. Similarly, the binomial coefficient $\binom{k}{k}$ appearing at the end of the equation can be replaced by $\binom{k+1}{k+1}$, since both are equal to 1 according to identity 2. Finally, we can use identity 3 to simplify each of the sums in the brackets. We obtain

$$(a + b)^{k+1} = \binom{k+1}{0}a^{k+1} + \binom{k+1}{1}a^k b + \cdots + \binom{k+1}{k}ab^k + \binom{k+1}{k+1}b^{k+1}$$

But this last equation is just the statement $P_{k+1}$; that is, we have derived $P_{k+1}$ from $P_k$, as we wished to do. The induction proof is now complete.

## EXERCISE SET 13.2

### A

In Exercises 1 and 2, verify each statement directly, without using the techniques developed in this section.

1. (a) $(a + b)^2 = a^2 + 2ab + b^2$
   (b) $(a + b)^3 = a^3 + 3a^2b + 3ab^2 + b^3$
       Hint:  $(a + b)^3 = (a + b)(a + b)^2$

2. (a) $(a + b)^4 = a^4 + 4a^3b + 6a^2b^2 + 4ab^3 + b^4$
       Hint:  Use the result in Exercise 1(b) and the fact that $(a + b)^4 = (a + b)(a + b)^3$.
   (b) $(a + b)^5 = a^5 + 5a^4b + 10a^3b^2 + 10a^2b^3 + 5ab^4 + b^5$

In Exercises 3–28, carry out the indicated expansions.

3. $(a + b)^9$
4. $(a - b)^9$
5. $(2A + B)^3$
6. $(1 + 2x)^6$
7. $(1 - 2x)^6$
8. $(3x^2 - y)^5$
9. $(\sqrt{x} + \sqrt{y})^4$
10. $(\sqrt{x} - \sqrt{y})^4$
11. $(x^2 + y^2)^5$
12. $(5A - B^2)^3$
13. $[1 - (1/x)]^6$
14. $(3x + y^2)^4$
15. $[(x/2) - (y/3)]^3$
16. $(1 - z^2)^7$
17. $(ab^2 + c)^7$
18. $[x - (1/x)]^8$
19. $(x + \sqrt{2})^8$
20. $(4A - \frac{1}{2})^5$
21. $(\sqrt{2} - 1)^3$
22. $(1 + \sqrt{5})^4$
23. $(\sqrt{2} + \sqrt{3})^5$
24. $(\frac{1}{2} - 2a)^6$
25. $(2\sqrt[3]{2} - \sqrt[3]{4})^3$

26. $(x + y + 1)^4$ Suggestion:  Rewrite the expression as $[(x + y) + 1]^4$.

27. $(x^2 - 2x - 1)^5$ Suggestion:  Rewrite the expression as $[x^2 - (2x + 1)]^5$.

28. $[x^2 - 2x - (1/x)]^6$

In Exercises 29–38, evaluate or simplify each expression.

29. $5!$

30. (a) $3! + 2!$
    (b) $(3 + 2)!$

31. $\binom{7}{3}\binom{3}{2}$

32. $\dfrac{20!}{18!}$

33. (a) $\binom{5}{3}$

    (b) $\binom{5}{4}$

34. (a) $\binom{7}{7}$

    (b) $\binom{7}{0}$

35. $\dfrac{(n + 2)!}{n!}$

36. $\dfrac{n[(n - 2)!]}{(n + 1)!}$

37. $\binom{6}{4} + \binom{6}{3} - \binom{7}{4}$

38. $(3!)! + (3!)^2$

39. Find the fifteenth term in the expansion of $(a + b)^{16}$.

40. Find the third term in the expansion of $(a - b)^{30}$

41. Find the one hundredth term in the expansion of $(1 + x)^{100}$

42. Find the twenty-third term in the expansion of $[x - (1/x^2)]^{25}$

43. Find the coefficient of the term containing $a^4$ in the expansion of $(\sqrt{a} - \sqrt{x})^{10}$.

44. Find the coefficient of the term containing $a^4$ in the expansion of $(3a - 5x)^{12}$.

45. Find the coefficient of the term containing $y^8$ in the expansion of $[(x/2) - 4y]^9$.

46. Find the coefficient of the term containing $x^6$ in the expansion of $[x^2 + (1/x)]^{12}$.

47. Find the coefficient of the term containing $x^3$ in the expansion of $(1 - \sqrt{x})^8$.

48. Find the coefficient of the term containing $a^8$ in the expansion of $[a - (2/\sqrt{a})]^{14}$.

49. Find the term that does not contain $A$ in the expansion of $[(1/A) + 3A^2]^{12}$.

50. Find the coefficient of $B^{-10}$ in the expansion of $[(B^2/2) - (3/B^3)]^{10}$.

### B

51. Show that the coefficient of $x^n$ in the expansion of $(1 + x)^{2n}$ is $(2n)!/(n!)^2$.

52. Find $n$ so that the coefficients of the eleventh and thirteenth terms in $(1 + x)^n$ are the same.

53. (a) Complete the following table.

| $k$ | 0 | 1 | 2 | 3 | 4 | 5 | 6 | 7 | 8 |
|---|---|---|---|---|---|---|---|---|---|
| $\binom{8}{k}$ | | | | | | | | | |

(b) Use the results in part (a) to verify that
$$\binom{8}{0} + \binom{8}{1} + \binom{8}{2} + \cdots + \binom{8}{8} = 2^8.$$

(c) By taking $a = b = 1$ in the expansion of $(a + b)^n$, show that
$$\binom{n}{0} + \binom{n}{1} + \binom{n}{2} + \cdots + \binom{n}{n} = 2^n$$

## C

54. Two real numbers $A$ and $B$ are defined as follows:

$$A = \sqrt[99]{99!} \qquad B = \sqrt[100]{100!}$$

Which number is larger, $A$ or $B$? *Hint:* Compare $A^{9900}$ and $B^{9900}$

55. This exercise outlines a proof of the identity
$$\binom{n}{0}^2 + \binom{n}{1}^2 + \binom{n}{2}^2 + \cdots + \binom{n}{n}^2 = \binom{2n}{n}$$

(a) Verify that
$$(1 + x)^n \left(1 + \frac{1}{x}\right)^n = \frac{(1 + x)^{2n}}{x^n} \qquad (1)$$

(This requires only basic algebra, not the binomial theorem.)

(b) Show that the coefficient of the term independent of $x$ on the right side of equation (1) is $\binom{2n}{n}$.

(c) Use the binomial theorem to expand $(1 + x)^n$. Then show that the coefficient of the term independent of $x$ on the left side of equation (1) is
$$\binom{n}{0}^2 + \binom{n}{1}^2 + \binom{n}{2}^2 + \cdots + \binom{n}{n}^2$$

## INTRODUCTION TO SEQUENCES AND SERIES

**13.3**

This section and the next two sections in this chapter deal with numerical sequences. We will begin with a somewhat informal definition of this concept. Then, after looking at some examples and terminology, we will present a more formal definition. A **numerical sequence** is an ordered list of numbers. Here are four examples:

| | |
|---|---|
| example A: | $1, \sqrt{2}, 10$ |
| example B: | $2, 4, 6, 8, \ldots$ |
| example C: | $1, \frac{1}{2}, \frac{1}{4}, \frac{1}{8}, \ldots$ |
| example D: | $1, 1, 1, 1, \ldots$ |

The individual entries in a numerical sequence are called the **terms** of the sequence. In this chapter, the terms in each sequence will always be real numbers, so for convenience, we will drop the adjective *numerical* and refer simply to *sequences*. (It is worth pointing out, however, that in more advanced courses, sequences are studied in which the individual terms are functions.) Any sequence possessing only a finite number of terms is called a **finite sequence**. Thus, the sequence in example A is a finite sequence. On the other hand, examples B, C, and D are examples of what we call **infinite sequences**; each contains infinitely many terms. As example D indicates, it is not necessary that all the terms in a sequence be distinct. In this chapter, all the sequences we discuss will be infinite sequences.

In examples B, C, and D, the three dots are read "and so on." In using this notation, we are assuming that it is clear what the subsequent terms of the sequence are. Toward this end, we often specify a formula for the $n$th term in a sequence. Example B in this case would appear this way:

$$2, 4, 6, 8, \ldots, 2n, \ldots$$

A letter with subscripts is often used to denote the various terms in a sequence. For instance, if we denote the sequence in example B by $a_1, a_2, a_3, \ldots$, then we have $a_1 = 2$, $a_2 = 4$, $a_3 = 6$, and, in general, $a_n = 2n$.

**EXAMPLE 1**   Consider the sequence $a_1, a_2, a_3, \ldots$ in which the $n$th term $a_n$ is given by

$$a_n = \frac{n}{n+1}$$

Compute the first three terms of the sequence, as well as the one thousandth term.

Solution   To obtain the first term, we replace $n$ by 1 in the given formula. This yields

$$a_1 = \frac{1}{1+1} = \frac{1}{2}$$

The other terms are similarly obtained. We have

$$a_2 = \frac{2}{2+1} = \frac{2}{3}$$

$$a_3 = \frac{3}{3+1} = \frac{3}{4}$$

$$a_{1000} = \frac{1000}{1000+1} = \frac{1000}{1001}$$

In the example just concluded, we were given an explicit formula for the $n$th term of the sequence. The next example shows a different way of specifying a sequence, in which each term except the first is defined in terms of the previous term. This is an example of a **recursive definition**. (Recursive definitions are particularly useful in computer programming.)

**EXAMPLE 2**   Compute the first three terms of the sequence $b_1, b_2, b_3, \ldots$, which is defined recursively by

$$b_1 = 4$$
$$b_n = 2(b_{n-1} - 1) \qquad \text{for } n \geq 2$$

Solution   We are given the first term: $b_1 = 4$. To find $b_2$, we replace $n$ by 2 in the formula $b_n = 2(b_{n-1} - 1)$ to obtain

$$b_2 = 2(b_1 - 1) = 2(4 - 1) = 6$$

Thus, $b_2 = 6$. Next we use this value of $b_2$ in the formula $b_n = 2(b_{n-1} - 1)$ to obtain $b_3$. Replacing $n$ by 3 in this formula yields

$$b_3 = 2(b_2 - 1) = 2(6 - 1) = 10$$

We have now found the first three terms of the sequence: $b_1 = 4$, $b_2 = 6$, and $b_3 = 10$.

If you think about the central idea behind the concept of a sequence, you can see that a function is involved: For each input $n$, we have an output $a_n$. For this reason, the formal definition of a sequence is phrased as follows.

> **Definition:   Sequence**
>
> A **sequence** is a function whose domain is the set of natural numbers.

If we denote the function by $f$ for the moment, then $f(1)$ would denote what we have been calling $a_1$, the first term of the sequence. Similarly, $f(2) = a_2$, $f(3) = a_3$, and so on.

**EXAMPLE 3**   Consider the sequence whose $n$th term is

$$a_n = \frac{1}{2^n}$$

Graph this sequence for $n = 1, 2, 3,$ and $4$.

**Solution**   First, we compute the required values of $a_n$, as shown in the following table.

| $n$ | 1 | 2 | 3 | 4 |
|-----|---|---|---|---|
| $a_n$ | $\frac{1}{2}$ | $\frac{1}{4}$ | $\frac{1}{8}$ | $\frac{1}{16}$ |

Now, locating the inputs $n$ on the horizontal axis and the outputs $a_n$ on the vertical axis, we obtain the graph shown in Figure 1.

We will often be interested in the sum of certain terms of a sequence. Consider, for example, the sequence

$$10, 20, 30, 40, \ldots$$

in which the $n$th term is $10n$. The sum of the first four terms in this sequence is

$$10 + 20 + 30 + 40 = 100$$

More generally, the sum of the first $n$ terms in this sequence is indicated by

$$10 + 20 + 30 + \cdots + 10n$$

This expression is an example of a **finite series**, which simply means a sum of a finite number of terms.

We can indicate the sum of the first $n$ terms of the sequence $a_1, a_2, a_3, \ldots$ by

$$a_1 + a_2 + a_3 + \cdots + a_n$$

Another way to indicate this sum uses the so-called **sigma notation**, which we now introduce. The capital Greek letter sigma is written $\Sigma$. We define the notation $\displaystyle\sum_{k=1}^{n} a_k$ by the equation

$$\sum_{k=1}^{n} a_k = a_1 + a_2 + a_3 + \cdots + a_n$$

For example, $\displaystyle\sum_{k=1}^{3} a_k$ stands for the sum $a_1 + a_2 + a_3$, the idea in this case being to replace the subscript $k$ successively by 1, 2, and 3 and then add the

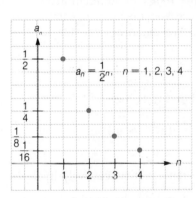

**FIGURE 1**   The graph of the sequence $a_n = \dfrac{1}{2^n}$, $n = 1, 2, 3, 4$

results. For a more concrete example, let us evaluate the expression $\sum_{k=1}^{4} k^2$. We have

$$\sum_{k=1}^{4} k^2 = 1^2 + 2^2 + 3^2 + 4^2$$
$$= 1 + 4 + 9 + 16 = 30$$

There is nothing special about the choice of the letter $k$ in the expression $\sum_{k=1}^{4} k^2$. For instance, we could equally well write

$$\sum_{j=1}^{4} j^2 = 1^2 + 2^2 + 3^2 + 4^2 = 30$$

The letter $k$ in the expression $\sum_{k=1}^{4} k^2$ is called the **index of summation**. Similarly, the letter $j$ appearing in $\sum_{j=1}^{4} j^2$ is the index of summation in that case. As we have seen, the choice of the letter used for the index of summation has no effect on the value of the indicated sum. For this reason, the index of summation is referred to as a *dummy variable*. The next two examples provide further practice with the sigma notation.

**EXAMPLE 4**   Express each of the following sums without sigma notation.

(a) $\sum_{k=1}^{3} (3k - 2)^2$     (b) $\sum_{i=1}^{4} ix^{i-1}$     (c) $\sum_{j=1}^{5} (a_{j+1} - a_j)$

**Solution**   (a) The notation $\sum_{k=1}^{3} (3k - 2)^2$ directs us to replace $k$ successively by 1, 2, and 3 in the expression $(3k - 2)^2$ and then add the results. We thus obtain

$$\sum_{k=1}^{3} (3k - 2)^2 = 1^2 + 4^2 + 7^2 = 66$$

(b) The notation $\sum_{i=1}^{4} ix^{i-1}$ directs us to replace $i$ successively by 1, 2, 3, and 4 in the expression $ix^{i-1}$ and then add the results. We have

$$\sum_{i=1}^{4} ix^{i-1} = 1x^0 + 2x^1 + 3x^2 + 4x^3$$
$$= 1 + 2x + 3x^2 + 4x^3$$

(c) To expand $\sum_{j=1}^{5} (a_{j+1} - a_j)$, we replace $j$ successively by 1, 2, 3, 4, and 5 in the expression $(a_{j+1} - a_j)$ and then add. We obtain

$$\sum_{j=1}^{5} (a_{j+1} - a_j) = (a_2 - a_1) + (a_3 - a_2) + (a_4 - a_3) + (a_5 - a_4) + (a_6 - a_5)$$
$$= a_2 - a_1 + a_3 - a_2 + a_4 - a_3 + a_5 - a_4 + a_6 - a_5$$

Combining like terms, we have

$$\sum_{j=1}^{5} (a_{j+1} - a_j) = a_6 - a_1$$

Sums such as $\sum_{j=1}^{5} (a_{j+1} - a_j)$ are known as **collapsing** or **telescoping sums**.

**EXAMPLE 5**   Use sigma notation to rewrite each sum.

(a) $\dfrac{x}{1!} + \dfrac{x^2}{2!} + \dfrac{x^3}{3!} + \cdots + \dfrac{x^{12}}{12!}$   (b) $\dfrac{x}{2!} + \dfrac{x^2}{3!} + \dfrac{x^3}{4!} + \cdots + \dfrac{x^n}{(n+1)!}$

**Solution**   (a) Since the exponents on $x$ run from 1 to 12, we choose a dummy variable, say $k$, running from 1 to 12. Also, we notice that if the numerator of a given term in the sum is $x^k$, then the corresponding denominator is $k!$. Consequently, the sum can be written

$$\frac{x}{1!} + \frac{x^2}{2!} + \frac{x^3}{3!} + \cdots + \frac{x^{12}}{12!} = \sum_{k=1}^{12} \frac{x^k}{k!}$$

(b) Since the exponents on $x$ run from 1 to $n$, we choose a dummy variable, say $k$, running from 1 to $n$. (Note that both of the letters $n$ and $x$ would be inappropriate here as dummy variables.) Also, we notice that if the numerator of a given term in the sum is $x^k$, then the corresponding denominator is $(k+1)!$. Thus, the given sum can be written

$$\frac{x}{2!} + \frac{x^2}{3!} + \frac{x^3}{4!} + \cdots + \frac{x^n}{(n+1)!} = \sum_{k=1}^{n} \frac{x^k}{(k+1)!}$$

# EXERCISE SET 13.3

## A

*In Exercises 1–20, compute the first five terms in each sequence. (In Exercises 1–12, where the nth term is given by an explicit formula, for instance, $a_n = n/(n+1)$, assume that the equation holds for all natural numbers n.)*

1. $a_n = \dfrac{n}{n+1}$   2. $a_n = n^3$

3. $a_n = (n-1)^2$   4. $b_n = n!$

5. $b_n = [1 + (1/n)]^n$   6. $b_n = \dfrac{n^n}{n!}$

7. $u_n = (-1)^n$   8. $a_n = (-1)^n \sqrt{n^n}$

9. $a_n = (-1)^{n+1}/n!$   10. $S_n = 1/[(n+1)!]$

11. $a_n = 3n$   12. $b_n = 3$

13. $a_1 = 1;\ a_n = (1 + a_{n-1})^2,\ \ n \geq 2$

14. $a_1 = 2;\ a_n = \sqrt{a_{n-1}^2 + 1},\ \ n \geq 2$

15. $a_1 = 2;\ a_2 = 2;\ a_n = a_{n-1}a_{n-2},\ \ n \geq 3$

16. $F_1 = 1;\ F_2 = 1;\ F_n = F_{n-1} + F_{n-2},\ \ n \geq 3$

17. $a_1 = 1;\ a_{n+1} = na_n,\ \ n \geq 1$

18. $a_1 = 1;\ a_2 = 2;\ a_n = a_{n-1}/a_{n-2},\ \ n \geq 3$

19. $a_1 = 0;\ a_n = 2^{a_{n-1}},\ \ n \geq 2$

20. $a_1 = 0;\ a_2 = 1;\ a_n = \tfrac{1}{2}(a_{n-1} + a_{n-2}),\ n \geq 3$

*In Exercises 21–26, graph the sequences for the indicated values of n.*

21. $a_n = (-1)^n,\ n = 1, 2, 3, 4$

22. $a_n = 2n - 1,\ n = 1, 2, 3$

23. $a_n = 1/n,\ n = 1, 2, 3, 4$

24. $a_n = (-1)^n/n,\ n = 1, 2, 3, 4$

25. $a_n = (n-1)/(n+1),\ n = 1, 2, 3$

26. $a_n = [1 + (1/n)]^n,\ n = 1, 2, 3$

In Exercises 27–32, *find the sum of the first five terms in the sequence whose nth term is as given.*

**27.** $a_n = 2^n$

**28.** $b_n = 2^{-n}$

**29.** $a_n = n^2 - n$

**30.** $b_n = (n-1)!$

**31.** $a_n = (-1)^n/n!$

**32.** $a_1 = 1; \quad a_n = \dfrac{1}{n-1} - \dfrac{1}{n+1} \quad (n \geq 2)$

**33.** Find the sum of the first five terms of the sequence defined recursively by $a_1 = 1; a_2 = 2; a_n = a_{n-1}^2 + a_{n-2}^2 \quad (n \geq 3)$.

**34.** Find the sum of the first four terms of the sequence defined recursively by $a_1 = 2; a_n = \dfrac{n}{a_{n-1}} \quad (n \geq 2)$.

**35.** Find the sum of the first four terms of the sequence defined recursively by $a_1 = 2; a_n = (a_{n-1})^2 \quad (n \geq 2)$.

**36.** Find the sum of the first six terms of the sequence defined recursively by $a_1 = 1; a_2 = 2; a_{n+1} = a_n a_{n-1}$ $(n \geq 2)$.

In Exercises 37–48, *express each of the sums without using sigma notation. Simplify your answers where possible.*

**37.** $\displaystyle\sum_{k=1}^{3} (k-1)$

**38.** $\displaystyle\sum_{k=1}^{5} k$

**39.** $\displaystyle\sum_{k=4}^{5} k^2$

**40.** $\displaystyle\sum_{k=2}^{6} (1-2k)$

**41.** $\displaystyle\sum_{n=1}^{3} x^n$

**42.** $\displaystyle\sum_{n=1}^{3} (n-1)x^{n-2}$

**43.** $\displaystyle\sum_{n=1}^{4} \frac{1}{n}$

**44.** $\displaystyle\sum_{n=0}^{4} 3^n$

**45.** $\displaystyle\sum_{j=1}^{9} \log_{10} \frac{j}{j+1}$

**46.** $\displaystyle\sum_{j=2}^{5} \log_{10} j$

**47.** $\displaystyle\sum_{j=1}^{6} \left( \frac{1}{j} - \frac{1}{j+1} \right)$

**48.** $\displaystyle\sum_{j=1}^{5} (x^{j+1} - x^j)$

In Exercises 49–58, *rewrite the sums using sigma notation.*

**49.** $5 + 5^2 + 5^3 + 5^4$

**50.** $5 + 5^2 + 5^3 + \cdots + 5^n$

**51.** $x + x^2 + x^3 + x^4 + x^5 + x^6$

**52.** $x + 2x^2 + 3x^3 + 4x^4 + 5x^5 + 6x^6$

**53.** $\frac{1}{1} + \frac{1}{2} + \frac{1}{3} + \cdots + \frac{1}{12}$

**54.** $\frac{1}{1} + \frac{1}{2} + \frac{1}{3} + \cdots + \frac{1}{n}$

**55.** $2 - 2^2 + 2^3 - 2^4 + 2^5$

**56.** $\dbinom{10}{3} + \dbinom{10}{4} + \dbinom{10}{5} + \cdots + \dbinom{10}{10}$

**57.** $1 - 2 + 3 - 4 + 5$

**58.** $\frac{1}{2} - \frac{1}{4} + \frac{1}{8} - \frac{1}{16} + \frac{1}{32} - \frac{1}{64} + \frac{1}{128}$

**B**

**59.** The *Fibonacci numbers* $F_1, F_2, F_3, \ldots$ are defined as follows:

$$F_1 = 1; \quad F_2 = 1; \quad F_{n+2} = F_n + F_{n+1} \quad (n \geq 1)$$

**(a)** Verify that the first ten Fibonacci numbers are as follows.

| $F_1$ | $F_2$ | $F_3$ | $F_4$ | $F_5$ | $F_6$ | $F_7$ | $F_8$ | $F_9$ | $F_{10}$ |
|---|---|---|---|---|---|---|---|---|---|
| 1 | 1 | 2 | 3 | 5 | 8 | 13 | 21 | 34 | 55 |

**(b)** Complete the following table.

| $n$ | $F_1 + F_2 + F_3 + \cdots + F_n$ | $F_{n+2} - 1$ |
|---|---|---|
| 1 | | |
| 2 | | |
| 3 | | |
| 4 | | |
| 5 | | |

**(c)** Use mathematical induction to prove that

$$F_1 + F_2 + F_3 + \cdots + F_n = F_{n+2} - 1$$

for all natural numbers $n$.

**(d)** Use mathematical induction to prove that

$$F_n \geq n \qquad \text{for all natural numbers } n \geq 5$$

**(e)** Use mathematical induction to show that

$$F_1^2 + F_2^2 + F_3^2 + \cdots + F_n^2 = F_n F_{n+1}$$

for all natural numbers $n$.

**(f)** Use mathematical induction to show that

$$F_{n+1}^2 = F_n F_{n+2} + (-1)^n$$

for all natural numbers $n$. *Hint for Step 2:* Add $F_{k+1} F_{k+2}$ to both sides of the equation in the induction hypothesis. Then factor $F_{k+1}$ from the left-hand side and factor $F_{k+2}$ from the first two terms on the right-hand side.

**(g)** This part of the exercise requires a knowledge of matrix multiplication. Show that

$$\begin{pmatrix} 1 & 1 \\ 1 & 0 \end{pmatrix}^n = \begin{pmatrix} F_{n+1} & F_n \\ F_n & F_{n-1} \end{pmatrix} \qquad \text{for } n \geq 2$$

## ARITHMETIC SEQUENCES AND SERIES

**13.4**

One of the most natural ways to generate a sequence is to begin with a fixed number $a$ and then repeatedly add a fixed constant $d$. This yields the sequence

$$a, \ a + d, \ a + 2d, \ a + 3d, \ldots$$

Such a sequence is called an **arithmetic sequence** or **arithmetic progression**. Notice that the difference between any two consecutive terms is the constant $d$. We call $d$ the **common difference**. Here are several examples of arithmetic sequences:

example A:     $1, 2, 3, \ldots$

example B:     $3, 7, 11, 15, \ldots$

example C:     $10, 5, 0, -5, \ldots$

In example A, the first term is $a = 1$ and the common difference is $d = 1$. For example B, we have $a = 3$. The value of $d$ in this example is found by subtracting any two consecutive terms; thus, $d = 4$. Finally, in example C, we have $a = 10$ and $d = -5$. Notice that when the common difference is negative, the terms of the sequence decrease.

There is a simple formula for the $n$th term in an arithmetic sequence

$$a, \ a + d, \ a + 2d, \ a + 3d, \ldots$$

In this sequence, notice that

$$a_1 = a + 0d$$
$$a_2 = a + 1d$$
$$a_3 = a + 2d$$
$$a_4 = a + 3d$$

Following this pattern, it appears that the formula for $a_n$ should be

$$a_n = a + (n - 1)d$$

Indeed, this is the correct formula, and Exercise 32 asks you to verify it using mathematical induction.

---

**$n$th Term of an Arithmetic Sequence**

The $n$th term of an arithmetic sequence $a, a + d, a + 2d, \ldots$ is given by

$$a_n = a + (n - 1)d$$

---

**EXAMPLE 1**   Determine the one hundredth term of the arithmetic sequence

$$7, \ 10, \ 13, \ 16, \ldots$$

**Solution**   The first term is $a = 7$, and the common difference is $d = 3$. Substituting these values in the formula $a_n = a + (n - 1)d$ yields

$$a_n = 7 + (n - 1)3 = 3n + 4$$

To find the one hundredth term, we replace $n$ by 100 in this last equation to obtain

$$a_{100} = 3(100) + 4 = 304 \qquad \text{as required}$$

**EXAMPLE 2**  Determine the arithmetic sequence in which the second term is $-2$ and the eighth term is 40.

**Solution**  We are given that the second term is $-2$. Using this information in the formula $a_n = a + (n-1)d$, we have

$$-2 = a + (2-1)d = a + d$$

This gives us one equation in two unknowns. We are also given that the eighth term is 40. Therefore,

$$40 = a + (8-1)d = a + 7d$$

We now have a system of two equations in two unknowns:

$$\begin{cases} -2 = a + \phantom{7}d \\ \phantom{-}40 = a + 7d \end{cases}$$

Subtracting the first equation from the second gives us

$$42 = 6d$$
$$7 = d$$

To find $a$, we replace $d$ by 7 in the first equation of the system. This yields

$$-2 = a + 7$$
$$-9 = a$$

We have now determined the sequence, since we know that the first term is $-9$ and the common difference is 7. The first four terms of the sequence are $-9, -2, 5,$ and 12.

Next we would like to derive a formula for the sum of the first $n$ terms of an arithmetic sequence. Such a sum is referred to as an **arithmetic series**. If we use $S_n$ to denote the required sum, we have

$$S_n = a + (a + d) + \cdots + [a + (n-2)d] + [a + (n-1)d] \qquad (1)$$

Of course, we must obtain the same sum if we add the terms from right to left rather than left to right. That is, we must have

$$S_n = [a + (n-1)d] + [a + (n-2)d] + \cdots + (a + d) + a \qquad (2)$$

Let us now add equations (1) and (2). Adding the left-hand sides is easy; we obtain $2S_n$. Now we add the corresponding terms on the right-hand sides. For the first terms, we have

$$\underset{\substack{\text{first term} \\ \text{in equation (1)}}}{a} + \underset{\substack{\text{first term} \\ \text{in equation (2)}}}{[a + (n-1)d]} = 2a + (n-1)d$$

Next we add the second terms:

$$\underset{\substack{\text{second term} \\ \text{in equation (1)}}}{(a + d)} + \underset{\substack{\text{second term} \\ \text{in equation (2)}}}{[a + (n-2)d]} = 2a + d + (n-2)d = 2a + (n-1)d$$

Notice that the sum of the second terms is again $2a + (n - 1)d$, the same quantity we arrived at with the first terms. As you can check, this pattern continues all the way through to the last terms. For instance,

$$\underbrace{[a + (n - 1)d]}_{\substack{\text{last term} \\ \text{in equation (1)}}} + \underset{\substack{\uparrow \\ \text{last term} \\ \text{in equation (2)}}}{a} = 2a + (n - 1)d$$

We conclude from these observations that by adding the right-hand sides of equations (1) and (2), the quantity $2a + (n - 1)d$ is added a total of $n$ times. Therefore,

$$2S_n = n[2a + (n - 1)d]$$

$$S_n = \frac{n}{2}[2a + (n - 1)d]$$

This gives us the desired formula for the sum of the first $n$ terms in an arithmetic sequence. There is an alternate form of this formula, which now follows rather quickly:

$$S_n = \frac{n}{2}[2a + (n - 1)d]$$

$$= \frac{n}{2}\{a + [a + (n - 1)d]\}$$

$$= \frac{n}{2}(a + a_n) = n\left(\frac{a + a_n}{2}\right)$$

This last form of the formula is easy to remember. It says that the sum of an arithmetic series is obtained by averaging the first and last terms and then multiplying this average by $n$, the number of terms. For reference, we summarize both formulas as follows.

---

**Formulas for the Sum of an Arithmetic Series**

$$S_n = \frac{n}{2}[2a + (n - 1)d]$$

$$S_n = n\left(\frac{a + a_n}{2}\right)$$

---

**EXAMPLE 3**    Find the sum of the first 30 terms of the arithmetic sequence $2, 6, 10, 14, \ldots$.

**Solution**    We have $a = 2$, $d = 4$, and $n = 30$. Substituting these values in the formula $S_n = \frac{n}{2}[2a + (n - 1)d]$ then yields

$$S_{30} = \frac{30}{2}[2(2) + (30 - 1)4]$$

$$= 15[4 + 29(4)]$$

$$= 1800 \qquad \text{(Check the arithmetic)}$$

**EXAMPLE 4** In a certain arithmetic sequence, the first term is 6 and the fortieth term is 71. Find the sum of the first 40 terms and also the common difference for the sequence.

Solution We have $a = 6$ and $a_{40} = 71$. Using these values in the formula

$$S_n = n\left(\frac{a + a_n}{2}\right)$$

yields

$$S_{40} = \frac{40}{2}(6 + 71) = 20(77)$$

$$= 1540$$

The sum of the first 40 terms is thus 1540. The value of $d$ can now be found by using the formula $S_n = \frac{n}{2}[2a + (n - 1)d]$. We obtain

$$1540 = \frac{40}{2}[2(6) + (40 - 1)d]$$

$$= 20(12 + 39d)$$

$$= 240 + 780d$$

$$1300 = 780d$$

$$d = \frac{1300}{780} = \frac{5}{3}$$

The required value of $d$ is therefore $\frac{5}{3}$.

**EXAMPLE 5** Show that $\sum_{k=1}^{50} (3k - 2)$ represents an arithmetic series, and compute the sum.

Solution There are two different ways to see that $\sum_{k=1}^{50} (3k - 2)$ is an arithmetic series. One way is simply to write out the first few terms and look at the pattern. We then have

$$\sum_{k=1}^{50} (3k - 2) = 1 + 4 + 7 + 10 + \cdots + 148$$

From this it is clear that we are indeed summing the terms in an arithmetic sequence in which $d = 3$ and $a = 1$.

Another, more formal way to show that $\sum_{k=1}^{50} (3k - 2)$ represents an arithmetic series is to prove that the difference between successive terms in the indicated sum is a constant. Now, the form of a typical term in this sum is $(3k - 2)$. Thus, the form of the next term must be $[3(k + 1) - 2]$. The difference between these terms is then

$$[3(k + 1) - 2] - (3k - 2) = 3k + 3 - 2 - 3k + 2 = 3$$

The difference therefore is constant, as we wished to show.

To evaluate $\sum_{k=1}^{50} (3k - 2)$, we can use either of the two formulas for the sum

of an arithmetic series. Using the formula $S_n = \dfrac{n}{2}[2a + (n - 1)d]$, we obtain

$$S_{50} = \tfrac{50}{2}[2(1) + 49(3)]$$
$$= 25(149) = 3725$$

Thus, the required sum is 3725. You should check for yourself that the same value is obtained using the formula $S_n = n(a + a_n)/2$.

## EXERCISE SET 13.4

### A

1. Find the common difference $d$ for each of the following arithmetic sequences.
   (a) $1, 3, 5, 7, \ldots$   (b) $10, 6, 2, \phantom{-}2, \ldots$
   (c) $\frac{2}{3}, 1, \frac{4}{3}, \frac{5}{3}, \ldots$
   (d) $1, 1 + \sqrt{2}, 1 + 2\sqrt{2}, 1 + 3\sqrt{2}, \ldots$

2. Which of the following are arithmetic sequences?
   (a) $2, 4, 8, 16, \ldots$   (b) $5, 9, 13, 17, \ldots$
   (c) $3, \frac{11}{5}, \frac{7}{5}, \frac{3}{5}, \ldots$   (d) $-1, -1, -1, -1, \ldots$
   (e) $-1, 1, -1, 1, \ldots$

*In Exercises 3–8, find the indicated term in each sequence.*

3. $10, 21, 32, 43, \ldots$; $a_{12}$

4. $7, 2, -3, -8, \ldots$; $a_{20}$

5. $6, 11, 16, 21, \ldots$; $a_{100}$

6. $\frac{2}{5}, \frac{4}{5}, \frac{6}{5}, \frac{8}{5}, \ldots$; $a_{30}$

7. $-1, 0, 1, 2, \ldots$; $a_{1000}$

8. $42, 1, -40, -81, \ldots$; $a_{15}$

9. The fourth term in an arithmetic sequence is $-6$, and the tenth term is 5. Find the common difference and the first term.

10. The fifth term in an arithmetic sequence is $\frac{1}{2}$, and the twentieth term is $\frac{7}{8}$. Find the first three terms of the sequence.

11. The sixtieth term in an arithmetic sequence is 105, and the common difference is 5. Find the first term.

12. Find the common difference in an arithmetic sequence in which $a_{10} - a_{20} = 70$.

13. Find the common difference in an arithmetic sequence in which $a_{15} - a_7 = -1$.

14. Find the sum of the first 16 terms in the sequence $2, 11, 20, 29, \ldots$.

15. Find the sum of the first 1000 terms in the sequence $1, 2, 3, 4, \ldots$.

16. Find the sum of the first 50 terms in an arithmetic series whose first term is $-8$ and whose fiftieth term is 139.

17. Find the sum: $\dfrac{\pi}{3} + \dfrac{2\pi}{3} + \pi + \dfrac{4\pi}{3} + \cdots + \dfrac{13\pi}{3}$.

18. Find the sum: $\dfrac{1}{e} + \dfrac{3}{e} + \dfrac{5}{e} + \cdots + \dfrac{21}{e}$.

19. Determine the first term of an arithmetic sequence in which the common difference is 5 and the sum of the first 38 terms is 3534.

20. The sum of the first 12 terms in an arithmetic sequence is 156. What is the sum of the first and twelfth terms?

21. In a certain arithmetic sequence, the first term is 4 and the sixteenth term is $-100$. Find the sum of the first 16 terms and also the common difference for the sequence.

22. The fifth and fiftieth terms of an arithmetic sequence are 3 and 30, respectively. Find the sum of the first ten terms.

23. The eighth term in an arithmetic sequence is 5, and the sum of the first ten terms is 20. Find the common difference and the first term of the sequence.

*In Exercises 24–26, find each sum.*

24. $\displaystyle\sum_{i=1}^{10} (2i - 1) = 1 + 3 + 5 + \cdots + 19$

25. $\displaystyle\sum_{k=1}^{20} (4k + 3)$   26. $\displaystyle\sum_{n=5}^{100} (2n - 1)$

27. The sum of three consecutive terms in an arithmetic sequence is 30, and their product is 360. Find the three terms. *Suggestion:* Let $x$ denote the *middle* term and $d$ the common difference.

28. The sum of three consecutive terms in an arithmetic sequence is 21, and the sum of their squares is 197. Find the three terms.

29. The sum of three consecutive terms in an arithmetic sequence is 6, and the sum of their cubes is 132. Find the three terms.

30. In a certain arithmetic sequence, $a = -4$ and $d = 6$. If $S_n = 570$, find $n$.

31. Let $a_1 = \dfrac{1}{1 + \sqrt{2}}$, $a_2 = -1$, and $a_3 = \dfrac{1}{1 - \sqrt{2}}$.

    (a) Show that $a_2 - a_1 = a_3 - a_2$.
    (b) Find the sum of the first six terms in the arithmetic sequence
    $$\frac{1}{1 + \sqrt{2}}, \; -1, \; \frac{1}{1 - \sqrt{2}}, \ldots$$

32. Using mathematical induction, prove that the $n$th term of the sequence
    $$a, \; a + d, \; a + 2d, \ldots$$
    is given by
    $$a_n = a + (n - 1)d$$

**B**

33. Let $b$ denote a positive constant. Find the sum of the first $n$ terms in the sequence
    $$\frac{1}{1 + \sqrt{b}}, \; \frac{1}{1 - b}, \; \frac{1}{1 - \sqrt{b}}, \ldots$$

34. The sum of the first $n$ terms in a certain arithmetic sequence is given by $S_n = 3n^2 - n$. Show that the $r$th term is given by $a_r = 6r - 4$.

35. Let $a_1, a_2, a_3, \ldots$ be an arithmetic sequence, and let $S_k$ denote the sum of the first $k$ terms. If $S_n/S_m = n^2/m^2$, show that
    $$\frac{a_n}{a_m} = \frac{2n - 1}{2m - 1}$$

36. If the common difference in an arithmetic sequence is twice the first term, show that
    $$\frac{S_n}{S_m} = \frac{n^2}{m^2}$$

37. The lengths of the sides of a right triangle form three consecutive terms in an arithmetic sequence. Show that the triangle is similar to the 3–4–5 right triangle.

38. Suppose that $1/a$, $1/b$, and $1/c$ are three consecutive terms in an arithmetic sequence. Show that:

    (a) $\dfrac{a}{c} = \dfrac{a - b}{b - c}$     (b) $b = \dfrac{2ac}{a + c}$

39. Suppose that $a$, $b$, and $c$ are three positive numbers with $a > c$. If $1/a$, $1/b$, and $1/c$ are consecutive terms in an arithmetic sequence, show that
    $$\ln(a + c) + \ln(a - 2b + c) = 2\ln(a - c)$$

## GEOMETRIC SEQUENCES AND SERIES

**13.5**

A **geometric sequence** or **geometric progression**, is a sequence of the form

$$a, \; ar, \; ar^2, \; ar^3, \ldots \qquad \text{where } a \text{ and } r \text{ are constants}$$

As you can see, each term after the first in a geometric sequence is obtained by multiplying the previous term by $r$. The number $r$ is called the **common ratio**, since the ratio of any term to the previous one is always $r$. For instance, the ratio of the fourth term to the third is $ar^3/ar^2 = r$. Here are two examples of geometric sequences:

$$1, \frac{1}{2}, \frac{1}{4}, \frac{1}{8}, \ldots$$

$$10, \; -100, \; 1000, \; -10000, \ldots$$

In the first example, we have $a = 1$ and $r = \frac{1}{2}$; in the second example, we have $a = 10$ and $r = -10$.

**EXAMPLE 1** In a certain geometric sequence, the first term is 2, the third term is 3, and the common ratio is negative. Find the second term.

**Solution** Let $x$ denote the second term, so that the sequence begins

$$2, \; x, \; 3, \ldots$$

By definition, the ratios $3/x$ and $x/2$ must be equal. Thus, we have

$$\frac{3}{x} = \frac{x}{2}$$
$$x^2 = 6$$
$$x = \pm\sqrt{6}$$

Now, the second term must be negative, since the first term is positive and the common ratio is negative. Thus, the second term is $x = -\sqrt{6}$.

The formula for the $n$th term of a geometric sequence is easily deduced by considering Table 1. The table indicates that the exponent on $r$ is one less than the value of $n$ in each case. On the basis of this observation, it appears that the $n$th term must be given by $a_n = ar^{n-1}$. Indeed, it can be shown by mathematical induction that this formula does hold for all natural numbers $n$. (Exercise 30 asks you to carry out the proof.) We summarize this result as follows.

TABLE 1

| $n$ | $a_n$ |
|-----|-------|
| 1 | $ar^0$ |
| 2 | $ar^1$ |
| 3 | $ar^2$ |
| 4 | $ar^3$ |
| $\vdots$ | $\vdots$ |

---

**$n$th Term of a Geometric Sequence**

The $n$th term of the geometric sequence $a, ar, ar^2, \ldots$ is given by

$$a_n = ar^{n-1}$$

---

**EXAMPLE 2**  Find the seventh term in the geometric sequence $2, 6, 18, \ldots$.

**Solution**  We can find the common ratio $r$ by dividing the second term by the first. Thus, $r = 3$. Now, using $a = 2$, $r = 3$, and $n = 7$ in the formula $a_n = ar^{n-1}$, we have

$$a_7 = 2(3)^6 = 2(729) = 1458$$

The seventh term of the sequence is therefore 1458.

Suppose that we begin with a geometric sequence $a, ar, ar^2, \ldots$ in which $r \neq 1$. If we add the first $n$ terms and denote the sum by $S_n$, we have

$$S_n = a + ar + ar^2 + \cdots + ar^{n-2} + ar^{n-1} \tag{1}$$

This sum is called a **finite geometric series**. We would like to find a formula for $S_n$. To do this, we multiply equation (1) by $r$ to obtain

$$rS_n = ar + ar^2 + ar^3 + \cdots + ar^{n-1} + ar^n \tag{2}$$

We now subtract equation (2) from equation (1). This yields (after combining like terms)

$$S_n - rS_n = a - ar^n$$
$$S_n(1 - r) = a(1 - r^n)$$
$$S_n = \frac{a(1 - r^n)}{1 - r} \qquad (r \neq 1)$$

This is the formula for the sum of a finite geometric series. We summarize this result in the box that follows.

---

**Formula for the Sum of a Geometric Series**

Let $S_n$ denote the sum $a + ar + ar^2 + \cdots + ar^{n-1}$, and assume that $r \neq 1$. Then

$$S_n = \frac{a(1 - r^n)}{1 - r}$$

---

**EXAMPLE 3**  Evaluate the sum $\dfrac{1}{2^1} + \dfrac{1}{2^2} + \dfrac{1}{2^3} + \cdots + \dfrac{1}{2^{10}}$.

**Solution**  This is a finite geometric series with $a = \frac{1}{2}$, $r = \frac{1}{2}$, and $n = 10$. Using these values in the formula for $S_n$ yields

$$S_{10} = \frac{\frac{1}{2}[1 - (\frac{1}{2})^{10}]}{1 - \frac{1}{2}}$$

$$= \frac{\frac{1}{2}[1 - \frac{1}{1024}]}{\frac{1}{2}} = 1 - \frac{1}{1024}$$

$$= \frac{1023}{1024} \qquad \text{as required}$$

We would now like to attach a meaning to certain expressions of the form

$$a + ar + ar^2 + \cdots$$

Such an expression is called an **infinite geometric series**. The three dots indicate (intuitively at least) that the additions are to be carried out indefinitely, without end. To see how to proceed here, let us look at some examples involving finite geometric series. In particular, we will consider the series

$$\frac{1}{2^1} + \frac{1}{2^2} + \frac{1}{2^3} + \cdots + \frac{1}{2^n}$$

for increasing values of $n$. The idea is to look for a pattern as $n$ grows ever larger. Let $S_1 = \dfrac{1}{2}$, $S_2 = \dfrac{1}{2^1} + \dfrac{1}{2^2}$, $S_3 = \dfrac{1}{2^1} + \dfrac{1}{2^2} + \dfrac{1}{2^3}$, and, in general,

$$S_n = \frac{1}{2^1} + \frac{1}{2^2} + \cdots + \frac{1}{2^n}$$

Then we can compute $S_n$ for any given value of $n$ by means of the formula for the sum of a finite geometric series. From Table 2, which displays the results of such calculations, it seems clear that as $n$ grows larger and larger, the value of $S_n$ grows ever closer to 1. More precisely (but leaving the details for calculus), it can be shown that the value of $S_n$ can be made as close to 1 as we please, provided only that $n$ is sufficiently large. For this reason, we say that the *sum of the infinite geometric series* $\dfrac{1}{2^1} + \dfrac{1}{2^2} + \dfrac{1}{2^3} + \cdots$ is 1. That is,

$$\frac{1}{2^1} + \frac{1}{2^2} + \frac{1}{2^3} + \cdots = 1$$

**TABLE 2**

| $n$ | $S_n = \dfrac{1}{2} + \dfrac{1}{2^2} + \cdots + \dfrac{1}{2^n}$ |
|---|---|
| 1 | 0.5 |
| 2 | 0.75 |
| 5 | 0.96875 |
| 10 | 0.999023437... |
| 15 | 0.999969482... |
| 20 | 0.999999046... |
| 25 | 0.999999970... |

We can arrive at this last result another way. First we compute the sum of the finite geometric series $\frac{1}{2^1} + \frac{1}{2^2} + \cdots + \frac{1}{2^n}$. As you can check, the result is

$$S_n = 1 - \left(\frac{1}{2}\right)^n$$

Now, as $n$ grows larger and larger, the value of $\left(\frac{1}{2}\right)^n$ gets closer and closer to zero. Thus, as $n$ grows ever larger, the value of $S_n$ will more and more resemble $1 - 0$, or 1.

Now let us repeat our reasoning to obtain a formula for the sum of the infinite geometric series

$$a + ar + ar^2 + \cdots \qquad \text{where } |r| < 1$$

First consider the finite geometric series

$$a + ar + ar^2 + \cdots + ar^{n-1}$$

The sum $S_n$ in this case is

$$S_n - \frac{a(1 - r^n)}{1 - r}$$

We want to know how $S_n$ behaves as $n$ grows ever larger. This is where the assumption $|r| < 1$ is crucial. Just as $\left(\frac{1}{2}\right)^n$ approaches zero as $n$ grows larger and larger, so will $r^n$ approach zero as $n$ grows larger and larger. Thus, as $n$ grows ever larger, the sum $S_n$ will more and more resemble

$$\frac{a(1 - 0)}{1 - r} = \frac{a}{1 - r}$$

For this reason, we say that the sum of the infinite geometric series is $\dfrac{a}{1 - r}$.

We will make free use of this result in the subsequent examples. However, a more rigorous development of infinite series properly belongs to calculus.

---

**Formula for the Sum of an Infinite Geometric Series**

Suppose that $|r| < 1$. Then the sum $S$ of the infinite geometric series $a + ar + ar^2 + \cdots$ is given by

$$S = \frac{a}{1 - r}$$

---

**EXAMPLE 4**   Find the sum of the infinite geometric series $1 + \frac{2}{3} + \frac{4}{9} + \cdots$.

Solution   In this case, we have $a = 1$ and $r = \frac{2}{3}$. Thus,

$$S = \frac{a}{1 - r}$$

$$= \frac{1}{1 - \frac{2}{3}} = \frac{1}{\frac{1}{3}} = 3$$

The sum of the series is 3.

**EXAMPLE 5** Find a fraction equivalent to the repeating decimal of $0.2\overline{35}$.

**Solution** Let $S = 0.2\overline{35}$. Then we have

$$S = 0.2353535\ldots$$

$$= \frac{2}{10} + \frac{35}{1000} + \frac{35}{100,000} + \frac{35}{10,000,000} + \cdots$$

Now, the expression following $\frac{2}{10}$ on the right-hand side of this last equation is an infinite geometric series in which $a = \frac{35}{1000}$ and $r = \frac{1}{100}$. Thus,

$$S = \frac{2}{10} + \frac{a}{1-r}$$

$$= \frac{2}{10} + \frac{\frac{35}{1000}}{1 - \frac{1}{100}} = \frac{233}{990} \qquad \text{(Check the arithmetic!)}$$

The given decimal is therefore equivalent to 233/990.

# EXERCISE SET 13.5

## A

1. Find the second term in a geometric sequence in which the first term is 9, the third term is 4, and the common ratio is positive.

2. Find the fifth term in a geometric sequence in which the fourth term is 4, the sixth term is 6, and the common ratio is negative.

3. The product of the first three terms in a geometric sequence is 8000. If the first term is 4, find the second and third terms.

*In Exercises 4–8, find the indicated term for the geometric sequence given.*

4. $9, 81, 729, \ldots; a_7$

5. $-1, 1, -1, 1, \ldots; a_{100}$

6. $\frac{1}{2}, \frac{1}{4}, \frac{1}{8}, \ldots; a_9$

7. $\frac{2}{3}, \frac{4}{9}, \frac{8}{27}, \ldots; a_8$

8. $1, -\sqrt{2}, 2, \ldots; a_6$

9. Find the common ratio in a geometric sequence in which the first term is 1 and the seventh term is 4096.

10. Find the first term in a geometric sequence in which the common ratio is $\frac{4}{3}$ and the tenth term is $\frac{16}{9}$.

11. Find the sum of the first ten terms of the sequence $7, 14, 28, \ldots$.

12. Find the sum of the first five terms of the sequence $-\frac{1}{2}, \frac{3}{10}, -\frac{9}{50}, \ldots$.

13. Find the sum: $1 + \sqrt{2} + 2 + \cdots + 32$.

14. Find the sum of the first 12 terms in the sequence $-4, -2, -1, \ldots$.

*In Exercises 15–17, evaluate each sum.*

15. $\sum_{k=1}^{6} \left(\frac{3}{2}\right)^k$

16. $\sum_{k=1}^{6} \left(\frac{2}{3}\right)^{k+1}$

17. $\sum_{k=2}^{6} \left(\frac{1}{10}\right)^k$

*In Exercises 18–22, determine the sum of each infinite geometric series.*

18. $\frac{1}{4} + \frac{1}{4^2} + \frac{1}{4^3} + \cdots$

19. $\frac{2}{3} - \frac{4}{9} + \frac{8}{27} - \cdots$

20. $\frac{9}{10} + \frac{9}{100} + \frac{9}{1000} + \cdots$

21. $1 + \frac{1}{1.01} + \frac{1}{(1.01)^2} + \cdots$

22. $-1 - \frac{1}{\sqrt{2}} - \frac{1}{2} - \cdots$

*In Exercises 23–27, express each repeating decimal as a fraction.*

23. $0.555\ldots$

24. $0.\overline{47}$

25. $0.1\overline{23}$

26. $0.050505\ldots$

27. $0.\overline{432}$

## B

28. The lengths of the sides in a right triangle form three consecutive terms of a geometric sequence. Find the common ratio of the sequence. (There are two distinct answers.)

29. The product of three consecutive terms in a geometric sequence is $-1000$, and their sum is 15. Find the com-

mon ratio. (There are two answers.)  *Suggestion:* Denote the terms by $a/r$, $a$, and $ar$.

30. Use mathematical induction to prove that the $n$th term of the geometric sequence $a, ar, ar^2, \ldots$ is $ar^{n-1}$.

31. Show that the sum of the following infinite geometric series is $\frac{3}{2}$:

$$\frac{\sqrt{3}}{\sqrt{3}+1} + \frac{\sqrt{3}}{\sqrt{3}+3} + \cdots$$

32. Let $A_1$ denote the area of an equilateral triangle, each side of which is 1 unit long. A second equilateral triangle is formed by joining the midpoints of the sides of the

first triangle. Let $A_2$ denote the area of this second triangle. This process is then repeated to form a third triangle with area $A_3$; and so on. Find the sum of the areas $A_1 + A_2 + A_3 + \cdots$.

33. Let $a_1, a_2, a_3, \ldots$ be a geometric sequence such that $r \neq 1$. Let $S = a_1 + a_2 + a_3 + \cdots + a_n$, and let $T = \frac{1}{a_1} + \frac{1}{a_2} + \cdots + \frac{1}{a_n}$. Show that $\frac{S}{T} = a_1 a_n$.

34. Suppose that $a$, $b$, and $c$ are three consecutive terms in a geometric sequence. Show that $\frac{1}{a+b}$, $\frac{1}{2b}$, and $\frac{1}{c+b}$ are three consecutive terms in an arithmetic sequence.

## PERMUTATIONS AND COMBINATIONS

### 13.6

*I have all the words. What I am seeking is the perfect order of the words in the sentence.*

JAMES JOYCE (1882–1941)

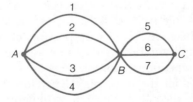

**FIGURE 1**

Suppose that you flip a coin twice, keeping track of the result each time. How many possible outcomes are there for this experiment? Using $H$ and $T$ to denote heads and tails, respectively, we can list the possible outcomes this way:

$$HH \qquad HT \qquad TH \qquad TT$$

For instance, $HH$ describes the experiment in which we obtain two heads in succession. Looking at this list then, we can say that there are four possible outcomes for the experiment.

Now let's ask the same question for an experiment in which we flip a coin three times in succession. How many possible outcomes are there? One way to organize the counting procedure uses the *tree diagram* in Figure 1. A tree diagram shows us every possibility at each stage in the experiment. Each path through the tree (from left to right) corresponds to a possible outcome for the experiment. For instance, the topmost path corresponds to the case in which we obtain three heads in succession ($HHH$). As you can see in Figure 1, there are eight possible outcomes for our experiment, corresponding to the eight branches on the right side of the tree.

**EXAMPLE 1** There are four roads from town $A$ to town $B$ and three roads from town $B$ to town $C$, as indicated schematically in Figure 2. Draw a tree diagram to find out how many routes there are from $A$ to $C$ via $B$.

**FIGURE 2**

Solution   Our tree diagram in Figure 3 has 12 branches on the right side. (Count them!) So there are 12 routes from *A* to *C* via *B*.   ▪▪

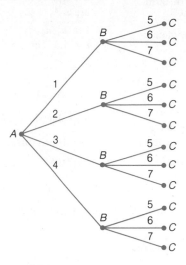

**FIGURE 3**

Rather than actually counting all 12 branches in Figure 3, we could have reasoned as follows. There are four possible ways to go from *A* to *B*, and for each of these ways, there are then three ways to go from *B* to *C*. Thus there are 4 *times* 3, or 12, ways to go from *A* to *C*. The *multiplication principle*, which we now state, formalizes this idea.

### The Multiplication Principle

Suppose that a task involves two consecutive operations. If the first operation can be performed in *M* ways, and the second operation can be performed in *N* ways, then the entire task can be performed in *MN* ways.

**EXAMPLE 2**   How many different two-digit numbers can be formed using the digits 1, 3, 5, 7, and 9 if:   **(a)** no digit is repeated (within a given number)?   **(b)** repetitions are allowed?

Solution   **(a)**   We can think of the task of forming a two-digit number as composed of two consecutive operations: first choose a tens digit, then choose a units digits. There are five choices for the tens digit. Then, after choosing a tens digit (assuming that no repetitions are allowed), there will be four choices left for the units digit. It follows now from the multiplication principle that there are $5 \times 4 = 20$ possible choices. That is, there are 20 different two-digit numbers that can be formed.

   **(b)**   With repetitions allowed, there are five choices for the tens digit and five choices for the units digit. Therefore (using the multiplication principle again) there are $5 \times 5 = 25$ different two-digit numbers that can be formed.   ▪▪

The multiplication principle can be extended to tasks involving three or more consecutive operations. This idea is used in the next example.

**EXAMPLE 3**   You buy three books at a book sale: one by Hemingway, one by Joyce, and one by Fitzgerald. How many ways are there to arrange these on a (single) bookshelf?

Solution   There are three positions we have to fill on the shelf: left, middle, and right. Starting with the left position, there are three choices. Then, once we make a choice for that left position, there are two choices left for the middle position. And finally, after choosing a book for that middle position, we'll have but one choice for the right position. Thus, according to the multiplication principle, there are $3 \times 2 \times 1 = 6$ ways to arrange the three books.

Let's list the six possible arrangements from Example 3. Using $H$ for Hemingway, $J$ for Joyce, and $F$ for Fitzgerald, the six arrangements are:

HJF    JHF    FJH

HFJ    JFH    FHJ

Each of the six arrangements here is called a *permutation* of the set $\{H, J, F\}$. In general, by a **permutation** of a set of $n$ distinct objects we just mean a listing or arrangement of those objects in a definite order. For instance, if we take the set $\{a, b, c, d\}$, then two (of the many possible) permutations of this set are *abcd* and *dcba*.

**EXAMPLE 4**   How many permutations are there of the set $\{a, b, c, d\}$?

Solution   To form a permutation, we have to assign a letter to each of four positions:

___  ___  ___  ___

We have four choices for the first position. Then, after filling that position, we'll have three letters left from which to choose for the second position. After carrying that out, there will be only two letters from which to choose for the third position. And finally, with the first three positions filled, there will be but one letter left for the fourth position. In summary we have

___  ___  ___  ___
4 choices   3 choices   2 choices   1 choice

Therefore (by the multiplication principle) the total number of permutations or arrangements is $4 \times 3 \times 2 \times 1 = 24$. Notice that this is 4!. (Recall from Section 13.2 that $n!$ stands for the product of the first $n$ natural numbers.)

In the example just concluded, we saw that there were 4! permutations of a set with four elements. By following the same kind of reasoning (and using mathematical induction), it can be shown that there are $n!$ permutations of a set with $n$ elements. For reference we state this fact in the box that follows.

The number of permutations of a set with $n$ elements is $n!$

**EXAMPLE 5** A regular deck of playing cards contain 52 cards. How many arrangements (permutations) of the deck are there?

**Solution** The number of permutations of a set with 52 elements is 52!. With the aid of a calculator we find that

$$52! \approx 8 \times 10^{67}$$

We've said that a permutation of a set with $n$ elements is a listing or arrangement of all $n$ of those elements in some definite order. Sometimes, though, we want to consider arrangements that use less than all $n$ of the elements. For example, let's take the set $\{H, J, F\}$ that we used earlier and this time list the possible arrangements using only two elements at a time:

$$
\begin{array}{lll}
HJ & JH & FH \\
HF & JF & FJ
\end{array}
$$

Each arrangement here is called a **permutation of three objects taken two at a time**. The number of such permutations is denoted by $P(3,2)$. So we have $P(3,2) = 6$, since there are six arrangements. More generally, we have the following definition.

---

**Definition**

By a **permutation of $n$ objects taken $r$ at a time**, we mean a listing or an arrangement of $r$ of the objects in a definite order, where $r \leq n$. The number of such arrangements is denoted by

$$P(n, r)$$

---

**EXAMPLE 6** Using the set $\{a, b, c, d\}$, list the permutations of these four letters taken two at a time. Evaluate $P(4, 2)$.

**Solution**
| | | | |
|---|---|---|---|
| Using $a$ and $b$: | $ab, ba$ | Using $b$ and $c$: | $bc, cb$ |
| Using $a$ and $c$: | $ac, ca$ | Using $b$ and $d$: | $bd, db$ |
| Using $a$ and $d$: | $ad, da$ | Using $c$ and $d$: | $cd, dc$ |

We've listed 12 permutations of the four given letters taken two at a time. (Convince yourself that we've listed all possibilities.) So $P(4, 2) = 12$.

We could have evaluated $P(4, 2)$ in Example 6 by reasoning as follows. To form a permutation of four letters taken two at a time, we have to fill two positions:

First letter        Second letter

There are four choices for the first letter. Then, after making a choice, we're left with three choices for the second letter. Therefore (according to the multiplication principle) there are $4 \times 3 = 12$ possible choices, in agreement with the result in Example 6. By using this type of reasoning, we can obtain the following general expression for the number of permutations of $n$ objects taken $r$ at a time. (Exercise 54 asks you to prove that the two expressions for $P(n, r)$ in the box are equal.)

$$P(n,r) = n(n-1) \cdots (n-r+1) = \frac{n!}{(n-r)!}$$

**EXAMPLE 7**   (a) Evaluate $P(8,5)$, the number of permutations of eight objects taken five at a time.
(b) Evaluate $P(8,8)$.

Solution   (a) In the first expression for $P(n,r)$ in the box, let $n = 8$ and $r = 5$. This yields

$$P(8,5) = 8 \cdot (7) \cdots (8-5+1)$$
$$= 8 \cdot (7)(6)(5)(4) = 6720$$

So there are 6720 permutations of eight objects taken five at a time.

(b) $P(8,8) = 8 \cdot (7) \cdots (8-8+1)$
$$= 8 \cdot (7) \cdots (1) = 8! = 40{,}320$$

So there are 8! or 40,320 permutations of eight objects taken eight at a time. (Note that a permutation of eight objects taken eight at a time is just a permutation of those eight objects.)   ▪

**EXAMPLE 8**   From a set of 10 pictures, five are to be selected and hung in a row. How many arrangements are possible?

Solution   We want the number of permutations of 10 objects taken five at a time. This number is

$$P(10,5) = (10)(9) \cdots (10-5+1)$$
$$= (10)(9)(8)(7)(6)$$
$$= 30{,}240$$

There are, therefore, 30,240 possible arrangements.   ▪

**EXAMPLE 9**   How many possible arrangements are there of the letters $\{a,b,c,d,e\}$ in which $a$ and $b$ are next to each other?

Solution   First think of $a$ and $b$ as a single object and count the number of arrangements of the *four* objects $(ab)$, $c$, $d$, and $e$. This number is

$$P(4,4) = 4! \qquad \text{(Why?)}$$
$$= 4 \cdot (3)(2)(1) = 24$$

Now, in each of these 24 arrangements, the letters $a$ and $b$ can appear in one of two ways, either $ab$ or $ba$. Consequently (by the multiplication principle), the number of possible arrangements of $\{a,b,c,d,e\}$ in which $a$ and $b$ are adjacent is $24 \times 2 = 48$.   ▪

**EXAMPLE 10**   How many possible arrangements are there of the letters $\{a,b,c,d,e\}$ in which $a$ and $b$ are not next to each other?

Solution   In this type of counting problem, the direct approach is not the simplest. Instead, we reason as follows. The number of arrangements in which $a$ and $b$

are not adjacent equals the total number of possible arrangements of the five letters, minus the number of arrangements of the five letters in which $a$ and $b$ *are* adjacent. So (using the result in Example 9), the required number of arrangements is

$$P(5,5) - 48 = 5 \cdot (4)(3)(2)(1) - 48$$
$$= 120 - 48 = 72$$

Therefore there are 72 ways to arrange the five given letters such that $a$ and $b$ are not adjacent.

In working with permutations, we've seen that the order in which the objects are listed is an essential factor. Now we want to introduce the idea of a *combination*, in which the order of the objects is not a consideration. Our work will be based on the following definition.

---

**Definition**

Suppose that $r$ objects are selected from a set of $n$ objects, where $r \leq n$. Then this selection or subset of $r$ objects is called a **combination of the $n$ objects taken $r$ at a time**. The number of such combinations is denoted

$C(n,r)$

---

To illustrate this definition, let's take the set $\{H, J, F\}$. As we saw earlier, there are six permutations of these three letters taken two at a time:

| | |
|---|---|
| Permutations involving $H$ and $J$: | $HJ, JH$ |
| Permutations involving $J$ and $F$: | $JF, FJ$ |
| Permutations involving $H$ and $F$: | $HF, FH$ |

Now, the two permutations $HJ$ and $JH$ use exactly the same letters. So, corresponding to these two permutations, we have the single combination or set of letters $\{H, J\}$. Also, whether we write this set as $\{H, J\}$ or $\{J, H\}$, it's still the same combination of letters. Similarly, corresponding to the two permutations $JF$ and $FJ$ we have the single combination $\{J, F\}$. And finally, corresponding to the two permutations $HF$ and $FH$, we have the combination $\{H, F\}$. Thus, although we have six permutations of the three letters taken two at a time, there are only three combinations of the letters taken two at a time. We can summarize these results by writing

$$P(3,2) = 6 \quad \text{and} \quad C(3,2) = 3$$

**EXAMPLE 11** Using the set $\{1, 2, 3, 4\}$, list both the combinations and the permutations of these four numbers taken three at a time. Evaluate $C(4,3)$ and $P(4,3)$.

**Solution**

| COMBINATION | PERMUTATIONS GENERATED FROM THE COMBINATION | | | | | |
|---|---|---|---|---|---|---|
| $\{1,2,3\}$ | 123 | 132 | 213 | 231 | 312 | 321 |
| $\{1,2,4\}$ | 124 | 142 | 241 | 214 | 412 | 421 |
| $\{1,3,4\}$ | 134 | 143 | 341 | 314 | 413 | 431 |
| $\{2,3,4\}$ | 234 | 243 | 342 | 324 | 423 | 432 |

We've found four combinations. (Convince yourself that there are no other distinct combinations using three of the four given digits.) Thus $C(4,3) = 4$. Next, let's evaluate $P(4,3)$. Assuming that the listing we've made is complete, a direct count shows that $P(4,3) = 24$. You can check this result for yourself using the formula for $P(n,r)$ that was given earlier. Actually, there is yet another way to calculate $P(4,3)$, and we'll show it here because it will help you to follow the reasoning in the next paragraph. Each of our combinations contains three numbers. So, corresponding to each combination there are 3! or six permutations. Thus, by the multiplication principle, we have $P(4,3) = 6 \times C(4,3) = 6 \times 4 = 24$.

We can derive a formula for $C(n,r)$ as follows. First of all, the number of permutations of $n$ objects taken $r$ at a time is

$$P(n,r) = n(n-1) \cdots (n-r+1)$$

We can think of the task of forming any one of these permutations as composed of two consecutive operations:

1. Select $r$ of the $n$ objects.
2. Then arrange these $r$ objects in some definite order.

Since operation 1 can be performed in $C(n,r)$ ways, and then operation 2 can be performed in $r!$ ways, the multiplication principle tells us that

$$P(n,r) = C(n,r) \cdot r!$$

Therefore

$$C(n,r) = \frac{P(n,r)}{r!} = \frac{n(n-1) \cdots (n-r+1)}{r!}$$

We can simplify this last fraction. The technique is to multiply both the numerator and the denominator by the quantity $(n-r)!$. This gives us

$$C(n,r) = \frac{n(n-1) \cdots (n-r+1)[(n-r)!]}{r![(n-r)!]}$$

The numerator of this last fraction is just $n!$, so we have

$$C(n,r) = \frac{n!}{r!(n-r)!}$$

Do you recognize the fraction on the right-hand side of this last equation? It's the binomial coefficient $\binom{n}{r}$ that we defined in connection with our earlier work on the binomial theorem. In summary, then, we have the following two expressions for $C(n,r)$. (*Note:* In the second expression, there are $r$ factors in the numerator and $r$ factors in the denominator.)

$$C(n,r) = \binom{n}{r} = \frac{n!}{r!(n-r)!}$$
$$= \frac{n(n-1) \cdots (n-r+1)}{r!}$$

**EXAMPLE 12** There are 100 senators in the United States Senate. How many four-member committees are possible?

Solution Since we are interested in the members of a committee rather than the order in which they are chosen or seated, we use combinations rather than permutations. We want the number of combinations of the 100 senators taken four at a time. This number is

$$C(100,4) = \binom{100}{4} = \frac{100!}{4!(100-4)!}$$

$$= \frac{100(99)(98)(97)(96!)}{4!(96!)}$$

$$= \frac{100(99)(98)(97)}{4!}$$

$$= 3{,}921{,}225 \qquad \text{(using a calculator)}$$

FIGURE 4

So there are 3,921,225 possible four-member committees.

**EXAMPLE 13** Figure 4 shows a decagon (a ten-sided figure) with all its diagonals. How many diagonals are there?

Solution Each pair of vertices determines either a diagonal or a side of the decagon. So the total number of diagonals and sides is the same as the number of pairs of vertices. This latter number is $C(10,2)$, the number of combinations of the ten vertices taken two at a time. Evaluating $C(10,2)$, we have

$$C(10,2) = \binom{10}{2} = \frac{10!}{2!(10-2)!} = \frac{10(9)(8!)}{2(8!)} = 45$$

Thus the total number of sides and diagonals is 45. Since 10 of these are sides, we're left with $45 - 10 = 35$ diagonals. (For a different approach to this problem, see Exercise 53.)

**EXAMPLE 14** On a certain committee there are eight Republicans and seven Democrats. How many possibilities are there for a five-member subcommittee that consists of three Republicans and two Democrats?

Solution Think of the task of forming a five-member subcommittee as composed of two consecutive operations: First choose three Republicans; then choose two Democrats. The number of ways to choose three Republicans is $C(8,3)$, and the number of ways to choose two Democrats is $C(7,2)$. Therefore, according to the multiplication principle, the total number of subcommittees here is

$$C(8,3) \cdot C(7,2) = \binom{8}{3} \cdot \binom{7}{2}$$

$$= \frac{8!}{3!(8-3)!} \cdot \frac{7!}{2!(7-2)!}$$

$$= \frac{8(7)(6)}{3(2)} \cdot \frac{7(6)}{2} = 1176 \qquad \text{(Check the arithmetic)}$$

So there are 1176 possible subcommittees consisting of three Republicans and two Democrats.

# EXERCISE SET 13.6

## A

1. There are three roads from town $A$ to town $B$ and two roads from town $B$ to town $C$. By drawing a tree diagram, find out how many routes there are from $A$ to $C$ via $B$. (Check your answer using the multiplication principle.)

2. There are five roads from town $A$ to town $B$ and three roads from town $B$ to town $C$. Draw a tree diagram to find out how many routes there are from $A$ to $C$ via $B$. (Check your answer using the multiplication principle.)

3. There are three roads from town $A$ to town $B$, two roads from town $B$ to town $C$, and two roads from town $C$ to town $D$. Draw a tree diagram to find out how many routes there are from town $A$ to town $D$ via $B$ and $C$.

4. At the beginning of this section we noted that if a coin is flipped twice, there are four possible outcomes. Draw a tree diagram illustrating this.

5. How many different two-digit numbers can be formed using the digits 2, 4, 6, and 8 if no digit is repeated (within a given number)?

6. Follow Exercise 5, but allow repetitions.

7. How many three-digit numbers can be formed using the digits 1, 2, 3, 4, and 5 if no digit can be repeated (within a given number)?

8. How many two-digit numbers that are even can be formed using the digits 2, 4, 5, 7, and 9 if no digit is repeated (within a given number)?

9. How many three-digit numbers that are less than 600 can be formed using the digits 3, 4, 5, 6, 7, and 8 if repetitions are allowed?

10. In a certain state, the license plates contain three digits followed by three letters. (Repetitions of digits or letters on a given plate are allowed.) How many different license plates are possible?

11. Follow Exercise 10, but assume that repetitions are not allowed.

12. In how many ways can five books be arranged on a bookshelf?

13. In how many ways can 10 books be arranged on a bookshelf?

14. How many permutations are there of the set $\{a, e, i, o, u\}$?

15. If the number of permutations of a certain set of letters is 24, how many letters must there be in the set?

16. Five framed photographs are going to be hung on the wall in one row. If one particular photograph must always be in the middle, how many arrangements are possible?

For Exercises 17–20, evaluate each expression.

17. $P(8, 1)$    18. $P(8, 2)$

19. $P(6, 3)$    20. $P(11, 3)$

21. Show that $P(n, n-1) = n!$.

22. Show that $P(n, 1) = n$.

23. Which is larger, $P(100, 50)$ or $P(101, 51)$?

24. Which is larger, $P(100, 50)$ or $P(101, 52)$?

25. How many possible arrangements are there of the six letters $\{a, b, c, d, e, f\}$ in which $b$ and $c$ are next to each other? (Use the method shown in Example 9.)

26. How many arrangements are there of the six letters $\{a, b, c, d, e, f\}$ in which $b$ and $c$ are not next to each other? (Use the method shown in Example 10.)

27. How many four-digit numbers can be formed using the digits $\{2, 7, 1, 8\}$ such that 7 and 1 are next to each other? (No repetitions.)

28. How many four-digit numbers can be formed using the digits $\{2, 7, 1, 8\}$ such that 7 and 1 are not next to each other? (No repetitions.)

29. How many three-digit numbers can be formed from the ten digits $\{0, 1, 2, \ldots, 9\}$ if: (a) repetitions are allowed? (Don't forget, the first digit cannot be 0.) (b) repetitions are not allowed?

30. How many three-digit numbers can be formed using the digits $\{1, 2, 3, 4, 5, 6\}$ if: (a) repetitions are not allowed and each number ends with 32? (b) repetitions are not allowed and the number formed in each case is odd? (c) repetitions are not allowed and each number formed is divisible by 5?

31. How many three-letter code words can be formed using the letters in the word *equations* if: (a) repetitions are allowed? (b) repetitions are not allowed? (c) repetitions are not allowed and each code word begins with $q$? (d) repetitions are not allowed and each code word begins with a vowel and ends with a consonant?

32. Using the set $\{a, b, c, d\}$, list both the combinations and the permutations of these four letters taken two at a time. Evaluate $C(4, 2)$ and $P(4, 2)$.

For Exercises 33–38, evaluate each expression.

33. (a) $C(12, 11)$    34. (a) $C(15, 14)$
    (b) $C(12, 1)$         (b) $C(15, 1)$

35. (a) $C(8, 3)$    36. (a) $C(10, 8)$
    (b) $C(8, 5)$         (b) $C(10, 2)$

37. (a) $C(12,5)$
    (b) $C(12,7)$

38. (a) $C(15,13)$
    (b) $C(15,2)$

39. Simplify $C(n, n-2)$.

40. Simplify $C(n, 1)$.

41. There are eight people on a certain committee. How many three-member subcommittees are possible?

42. How many five-member subcommittees are possible from a committee with 10 members?

43. There are 10 people in a certain club.
    (a) How many different six-member committees are possible?
    (b) Of the 10 people in the club, suppose that four are men and six are women. How many six-member committees are possible if each committee must be composed of three men and three women?

44. Suppose you are one of 12 people in a certain club.
    (a) How many five-member committees can be formed that include you?
    (b) How many five-member committees can be formed that exclude you?

45. Use the method shown in Example 13 to find out how many diagonals can be drawn in a hexagon. (A hexagon is a six-sided polygon.)

46. Use the method shown in Example 13 to count the number of diagonals that can be drawn in a regular 14-sided polygon.

47. Suppose that we are given eight points in a plane, and no three of the points lie on the same straight line. How many different triangles can be formed in which the vertices are chosen from these eight points?

48. Suppose that we are given 12 points lying on a certain circle. How many triangles can be formed if the vertices are chosen from the 12 points?

**B**

49. You have one penny, one nickel, one dime, and one quarter. How many different sums of money can you form using one or more of these coins? *Hint:* You need to compute $C(4,1) + C(4,2) + C(4,3) + C(4,4)$.

50. You have a \$1 bill, a \$5 bill, a \$10 bill, a \$20 bill, and a \$100 bill. How many different sums of money can you form using one or more of, these bills? (Study the hint in the previous exercise.)

51. A group consists of nine Republicans and eight Democrats. How many different eight-member committees can be formed that contain (a) six Republicans and two Democrats? (b) at least six Republicans?

52. (a) From a group of 12 people, a committee with eight members will be formed. Then, from those eight, a subcommittee with three people will be chosen. How many ways are there to carry out these two consecutive operations?
    (b) From a group of 12 people, a committee of three will be chosen. Later, five additional people are picked to increase the committee size to eight. How many ways are there to carry out these two consecutive operations?
    (c) Use parts (a) and (b) to explain why
$$\binom{12}{8}\binom{8}{3} = \binom{12}{3}\binom{9}{5}$$
    (d) Verify the equality in part (c) by direct calculation (involving ratios of factorials).
    (e) Show that $\binom{n}{k}\binom{k}{r} = \binom{n}{r}\binom{n-r}{k-r}$, where $n \geq k \geq r$.
    *Suggestion:* You can do this either by direct calculation or by repeating the reasoning used in parts (a), (b), and (c) with the quantities 12, 8, and 3 replaced by $n$, $k$, and $r$, respectively.

53. In Example 13 we computed the number of diagonals in a decagon by considering combinations. Here is a different way to approach the problem. First observe that there are seven diagonals emanating from each of the 10 vertices. Why does this imply that the total number of diagonals is $\dfrac{7 \cdot 10}{2} = 35$?

54. Show that
$$n(n-1) \cdots (n-r+1) = \frac{n!}{(n-r)!}$$

## INTRODUCTION TO PROBABILITY

**13.7**

*The true logic of this world is in the calculus of probabilities.*

JAMES CLERK MAXWELL

In order to define the mathematical notion of probability, we first introduce the idea of a *sample space* for an experiment. As a particularly simple first example, we can think of flipping a coin as an experiment in which there are two possible *outcomes*: heads (*H*) or tails (*T*). The set $\{H, T\}$ consisting of all possible outcomes for this experiment is called a **sample space**. Intuitively, at least, we believe that each outcome in this sample space is equally likely to occur.

As another example of this terminology, let's suppose that an experiment consists of tossing a single die and then noting the number of dots showing on the top face. Then a sample space for this experiment is the set

$$\{1, 2, 3, 4, 5, 6\}$$

Here, as in the previous example, our sample space consists of all possible outcomes for the experiment. And again, intuitively at least, we believe that each outcome is equally likely to occur. In this section, all the sample spaces that we consider are finite; that is, they contain finitely many outcomes.

Sometimes we wish to focus on a particular outcome, or a particular set of outcomes, in an experiment. For instance, in tossing the die we might be interested in whether the die shows an even number of dots. In this context, we refer to the set $\{2, 4, 6\}$ as an *event* in the sample space $\{1, 2, 3, 4, 5, 6\}$. More generally, any subset of a given sample space is referred to as an **event**. We shall use the notation $n(E)$ to denote the number of outcomes in an event $E$. For example, if $E$ is the event "obtaining an even number" that we've just discussed, then $n(E) = 3$, because there are three outcomes in the set $E = \{2, 4, 6\}$.

**EXAMPLE 1** Use set notation to specify each of the following events in the die-tossing experiment using the sample space $\{1, 2, 3, 4, 5, 6\}$.

Event *A*: obtaining an odd number
Event *B*: obtaining a number less than 3
Event *C*: obtaining a 6
Event *D*: obtaining a number less than 7
Event *E*: obtaining a 7

Solution

| | | | |
|---|---|---|---|
| Event *A*: | $\{1, 3, 5\}$ | Event *D*: | $\{1, 2, 3, 4, 5, 6\}$ |
| Event *B*: | $\{1, 2\}$ | Event *E*: | $\varnothing$ |
| Event *C*: | $\{6\}$ | | |

Two of the events we've listed here deserve special mention. First, notice that event *D* is the entire sample space. This is simply because every outcome in the sample space is a number less than 7. At the other extreme we have event *E*, "obtaining a 7." There are no outcomes in the sample space corresponding to this event. So (remembering that an event is a subset of the sample space), we say that event *E* is the so-called **empty set**, that is, the set with no elements. By convention, the empty set is denoted by the symbol $\varnothing$.

**EXAMPLE 2** Using the events $A$, $B$, $C$, $D$, and $E$ in Example 1, compute $n(A)$, $n(B)$, $n(C)$, $n(D)$, and $n(E)$.

**Solution** We need only count the number of outcomes in each set. The results are

$$n(A) = 3 \qquad n(D) = 6$$
$$n(B) = 2 \qquad n(E) = 0$$
$$n(C) = 1$$

For each event $E$ in a finite sample space, we are going to define a number $P(E)$ called the *probability* of the event. Informally speaking, the number $P(E)$ is a rating of how likely it is that the event $E$ will occur in any trial of the experiment. As you'll see, the number $P(E)$ always satisfies the condition $0 \leq P(E) \leq 1$.

---

**Definition:   Probability of an Event**

Let $S$ be a finite sample space in which each outcome is equally likely to occur. If $E$ is an event in $S$, then the **probability** of $E$, denoted by $P(E)$, is defined by

$$P(E) = \frac{n(E)}{n(S)} = \frac{\text{number of outcomes in } E}{\text{number of outcomes in } S}$$

---

**EXAMPLE 3** Find the probability for each of the events $A$, $B$, $C$, $D$, and $E$ discussed in Examples 1 and 2.

**Solution** $P(A) = P(\text{obtaining an odd number}) = \dfrac{n(A)}{n(S)} = \dfrac{3}{6} = \dfrac{1}{2}$

$P(B) = P(\text{obtaining a number less than 3}) = \dfrac{n(B)}{n(S)} = \dfrac{2}{6} = \dfrac{1}{3}$

$P(C) = P(\text{obtaining a 6}) = \dfrac{n(C)}{n(S)} = \dfrac{1}{6}$

$P(D) = P(\text{obtaining a number less than 7}) = \dfrac{n(D)}{n(S)} = \dfrac{6}{6} = 1$

$P(E) = P(\text{obtaining a 7}) = \dfrac{n(E)}{n(S)} = \dfrac{0}{6} = 0$

Notice that each of the probabilities that we computed in Example 3 is a number between 0 and 1, as mentioned previously. Also note that a probability of 0 is assigned to the "impossible" event, whereas a probability of 1 is assigned to the "certain" event.

**EXAMPLE 4** Two (distinct) letters are selected at random from the set $\{a, b, c, d, e\}$. What is the probability that one of the letters is $c$?

**Solution** Let the sample space $S$ consist of all combinations of the five given letters taken two at a time. For instance, $ab$, $ac$, $ad$, and $ae$ are four of the outcomes in this sample space. We needn't list every outcome in $S$ in order to find $n(S)$; for, from our work in the previous section, we know that

$$n(S) = C(5,2) = \frac{5!}{2!3!} = 10$$

Now we let $E$ be the event "$c$ is selected." We can specify $E$ as follows:

$$E = \{ac, bc, dc, ec\}$$

Thus $n(E) = 4$ and, consequently,

$$P(E) = \frac{n(E)}{n(S)} = \frac{4}{10} = \frac{2}{5}$$

So the probability that $c$ is one of the two letters selected is $\frac{2}{5}$. This is also expressed by saying there is a 40% chance that one of the two letters is $c$. ▪

**EXAMPLE 5** Suppose that five faulty transistors are accidentally packaged with 10 reliable transistors. What is the probability that three transistors chosen at random from this package all will be reliable?

Solution Let the sample space $S$ consist of the combinations of the 15 transistors taken three at a time. Then we have

$$n(S) = C(15,3) = \frac{15!}{3!12!} = 455$$

Next let $E$ denote the event "choose three reliable transistors." Since these three must be chosen from the 10 reliable ones, we have

$$n(E) = C(10,3) = \frac{10!}{3!7!} = 120$$

Thus

$$P(E) = \frac{n(E)}{n(S)} = \frac{120}{455} = \frac{24}{91} \qquad \text{as required}$$

To continue our development of probability a bit further, it will be convenient to recall two basic ideas from set theory. Suppose that $A$ and $B$ are two sets. Then the **union** of these two sets, written $A \cup B$, is defined to be the set of elements that are in $A$ *or* in $B$ (or in both). For example, if

$$A = \{2,3,4,5\} \qquad \text{and} \qquad B = \{4,5,6,7\}$$

then

$$A \cup B = \{2,3,4,5,6,7\}$$

The **intersection** of two sets $A$ and $B$, written $A \cap B$, is defined as the set of elements that belong to both $A$ *and* $B$. For example, using $A$ and $B$ as just defined, we have

$$A \cap B = \{4,5\}$$

because 4 and 5 are the only elements common to both sets. As another example, if $E = \{1,2\}$ and $F = \{3,4\}$, then $E \cap F = \varnothing$, because there are no elements common to the two sets. Two sets that have no elements in common are said to be **disjoint**.

If $A$ and $B$ are events (remember, that means subsets of a sample space), then $A \cup B$ is the event "$A$ or $B$," whereas $A \cap B$ is the event "$A$ and $B$." Here's an example. Let $S$ be the sample space $\{1,2,3,4,5,6\}$ that we've used for the die-tossing experiment. And let the events $A$ and $B$ be as follows:

$A$:  obtaining an even number

$B$:  obtaining a number greater than four

Then we have

$A \cup B =$ "obtaining an even number" *or*
"obtaining a number greater than four"
$= \{2,4,6\} \cup \{5,6\}$
$= \{2,4,5,6\}$

whereas

$A \cap B =$ "obtaining an even number" *and*
"obtaining a number greater than four"
$= \{2,4,6\} \cap \{5,6\}$
$= \{6\}$

**EXAMPLE 6**  A coin is flipped three times and the result is recorded each time. Let $A$, $B$, and $C$ denote the following events.

$A$:  obtaining exactly two heads

$B$:  obtaining no heads

$C$:  obtaining more than one tail

Compute the following probabilities:

(a)  $P(A)$, $P(B)$, and $P(C)$      (b)  $P(A$ or $B)$      (c)  $P(B$ or $C)$

Solution  Our sample space $S$ will be as follows. (See Figure 1 in the previous section if you're not certain how we've obtained the eight outcomes in $S$.)

$S = \{HHH, HHT, HTH, HTT, THH, THT, TTH, TTT\}$

(a)  Using the definitions given for $A$, $B$, and $C$, we have

$A = \{HHT, HTH, THH\}$
$B = \{TTT\}$
$C = \{HTT, THT, TTH, TTT\}$

Thus

$$P(A) = \frac{n(A)}{n(S)} = \frac{3}{8} \qquad P(B) = \frac{n(B)}{n(S)} = \frac{1}{8} \qquad P(C) = \frac{n(C)}{n(S)} = \frac{4}{8} = \frac{1}{2}$$

**(b)** The event "A or B" is $A \cup B$, so we have

$$A \cup B = \{HHT, HTH, THH, TTT\}$$

and consequently

$$P(A \cup B) = \frac{n(A \cup B)}{n(S)} = \frac{4}{8} = \frac{1}{2}$$

**(c)** Likewise, the event "B or C" is $B \cup C$, so we have

$$B \cup C = \{TTT, HTT, THT, TTH\}$$

and consequently

$$P(B \cup C) = \frac{n(B \cup C)}{n(S)} = \frac{4}{8} = \frac{1}{2} \qquad \text{as required}$$

We can use the results in Example 6 to introduce a rule that simplifies certain probability computations. If you review the results in Example 6, you'll see that $P(A \cup B) = \frac{1}{2}$, $P(A) = \frac{3}{8}$, and $P(B) = \frac{1}{8}$. Thus we have

$$P(A \cup B) = P(A) + P(B) \qquad \text{(because } \tfrac{1}{2} = \tfrac{3}{8} + \tfrac{1}{8}\text{)}$$

That is,

$$P(A \text{ or } B) = P(A) + P(B) \tag{1}$$

So (in this case at least), we find that the probability of the event "A or B" is the sum of the individual probabilities of A and B. Furthermore, it can be shown that equation (1) remains valid for any two events (in a given sample space) that have no outcomes in common. *Terminology:* Two events with no outcomes in common are said to be **mutually exclusive**. For instance, the events A and B in Example 6 are mutually exclusive. (Why?)

The remarks in the previous paragraph can be summarized this way: If E and F are mutually exclusive, then $P(E \text{ or } F) = P(E) + P(F)$. This formula is *not* valid, however, in cases in which the two events fail to be mutually exclusive. To see an example of this, let's take the events B and C in Example 6. In that example we found that $P(B \cup C) = \frac{1}{2}$, $P(B) = \frac{1}{8}$, and $P(C) = \frac{1}{2}$. Consequently we have

$$P(B \cup C) \neq P(B) + P(C) \qquad \text{(because } \tfrac{1}{2} \neq \tfrac{1}{8} + \tfrac{1}{2}\text{)}$$

The difference here, as we've said, has to do with the fact that B and C are not mutually exclusive. As you can check, B and C do have an outcome in common, namely TTT. Because of this, when we compute $P(B \cup C)$, the outcome TTT is counted but once; but when we compute $P(B) + P(C)$, the outcome TTT is counted twice (once with B and once again with C). This is essentially the reason that we obtain different values for the quantities $P(B \cup C)$ and $P(B) + P(C)$. By following this same type of reasoning (in which we keep track of how many times an outcome is counted) the following general formula can be derived. Notice that the formula agrees with equation (1) in the case when $A \cap B = \varnothing$ (that is, when A and B are mutually exclusive).

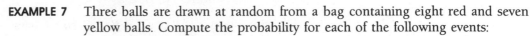

**The Addition Theorem for Probabilities**

If $A$ and $B$ are events in a sample space $S$, then

$$P(A \cup B) = P(A) + P(B) - P(A \cap B)$$

**EXAMPLE 7** Three balls are drawn at random from a bag containing eight red and seven yellow balls. Compute the probability for each of the following events:

(a) drawing three red balls;    (b) drawing three yellow balls;
(c) drawing either three red or three yellow balls.

**Solution** (a) The total number of possible outcomes in the experiment is $C(15,3)$, the number of ways of choosing three balls from a set of 15. We have

$$C(15,3) = \frac{15!}{3!\,12!} = \frac{15 \cdot 14 \cdot 13}{3 \cdot 2} = 455$$

Next, we need to know how many outcomes are in the event "drawing three red balls." This number is

$$C(8,3) = \frac{8!}{3!\,5!} = \frac{8 \cdot 7 \cdot 6}{3 \cdot 2} = 56$$

Consequently we have

$$P(\text{drawing 3 red balls}) = \frac{56}{455} = \frac{8}{65}$$

(b) Following the reasoning used in part (a), we find that

$$P(\text{drawing 3 yellow balls}) = \frac{C(7,3)}{C(15,3)} = \frac{35}{455} \qquad \text{(Check the arithmetic)}$$

$$= \frac{1}{13}$$

(c) Let $A$ and $B$ denote the events described in parts (a) and (b), respectively. Then $A \cup B$ is the event "drawing either three red or three yellow balls," and we have

$$P(A \cup B) = P(A) + P(B) - P(A \cap B)$$

$$= \frac{8}{65} + \frac{1}{13} - P(\varnothing)$$

$$= \frac{8}{65} + \frac{5}{65} - 0 = \frac{13}{65} = \frac{1}{5}$$

So the probability of drawing either three red or three yellow balls is $\frac{1}{5}$.

**EXAMPLE 8** A card is selected from a well-shuffled deck of 52 cards. What is the probability that it is a jack or a heart?

Solution    In this experiment there are 52 possible outcomes. Let $A$ denote the event "selecting a jack," and let $B$ denote the event "selecting a heart." Then $P(A \cup B)$ is the probability of the event "selecting a jack or a heart," and we have

$$P(A \cup B) = P(A) + P(B) - P(A \cap B)$$
$$= \frac{4}{52} + \frac{13}{52} - \frac{1}{52} \quad \text{(Why?)}$$
$$= \frac{16}{52} = \frac{4}{13} \quad \text{as required}$$

## EXERCISE SET 13.7

### A

*Exercises 1–20 refer to an experiment that consists of tossing a single die and then noting the number of dots showing on the top face. As in the text, assume that the sample space for this experiment is*

$$S = \{1, 2, 3, 4, 5, 6\}$$

*Let A, B, C, D, E, F, and G be the following events.*

A:  *obtaining an even number*
B:  *obtaining a number greater than 3*
C:  *obtaining a 1*
D:  *obtaining a number greater than 1*
E:  *obtaining an odd number*
F:  *obtaining a number less than 8*
G:  *obtaining an 8*

1. Use set notation to specify the following events.
   (a) $A$   (b) $B$   (c) $A \cup B$   (d) $A \cap B$

2. Use set notation to specify the following events.
   (a) $C$   (b) $D$   (c) $C \cup D$   (d) $C \cap D$

3. Use set notation to specify the following events.
   (a) $E$   (b) $F$   (c) $E \cup A$   (d) $E \cap A$

4. Use set notation to specify the following events.
   (a) $G$   (b) $G \cup B$   (c) $G \cap B$   (d) $G \cup F$

5. Are events $A$ and $B$ mutually exclusive?

6. Are events $A$ and $C$ mutually exclusive?

7. Are events $C$ and $D$ mutually exclusive?

8. Are events $B$ and $E$ mutually exclusive?

*For Exercises 9–20, compute each of the indicated quantities.*

9. (a) $n(A)$   (b) $n(B)$   (c) $n(A \cup B)$
10. (a) $n(C)$   (b) $n(D)$   (c) $n(C \cup D)$
11. (a) $n(E)$   (b) $n(F)$   (c) $n(G)$
12. $n(B \cap E)$   13. (a) $P(A)$   (b) $P(B)$
14. (a) $P(C)$   (b) $P(D)$

15. (a) $P(E)$   (b) $P(F)$   16. $P(G)$
17. (a) $P(A \cup B)$   (b) $P(A) + P(B)$
18. (a) $P(B \cup C)$   (b) $P(B) + P(C)$
19. (a) $P(B \cup E)$   (b) $P(B) + P(E)$
20. (a) $P(D) + P(E) - P(D \cap E)$   (b) $P(D \cup E)$

*Exercises 21–40 refer to the experiment described in Example 6, in which a coin is flipped three times and the result is recorded each time. As in Example 6, let the sample space $S$ be*

$$S = \{HHH, HHT, HTH, HTT, THH, THT, TTH, TTT\}$$

*Let A, B, C, D, E, and F be the following events.*

A:  *obtaining exactly two tails*
B:  *obtaining two heads on the first two tosses*
C:  *obtaining heads on the first toss*
D:  *obtaining no heads*
E:  *obtaining tails on both the second and third tosses*
F:  *obtaining tails on the first toss*

Compute each of the following probabilities.

21. $P(A)$   22. $P(B)$   23. $P(C)$
24. $P(D)$   25. $P(E)$   26. $P(F)$
27. $P(A \cup B)$   28. $P(A \cap B)$   29. $P(A \cup C)$
30. $P(A \cap C)$   31. $P(C \cup F)$   32. $P(C \cap F)$
33. $P(E \cup F)$   34. $P(E \cap F)$   35. $P(B \cup C)$
36. $P(B \cap C)$   37. $P(D \cap A)$   38. $P(D \cup A)$
39. $P(D \cup E)$   40. $P(D \cap E)$

41. Two letters are selected at random (and without replacement) from the set $\{a, b, c, d, e, f\}$. What is the probability that one of the letters is $d$?

42. The numbers 1, 2, 3, and 4 are arranged in random order to form a four-digit number. What is the probability that the number is greater than 4000?

43. Suppose that two faulty transistors are accidentally packaged with a dozen reliable ones.
    (a) If one transistor is chosen at random from the package, what is the probability that it will be reliable?
    (b) If five transistors are chosen at random from the package, what is the probability that they are all reliable?

44. Suppose that five cartons of spoiled milk are inadvertently placed on a grocery shelf along with 15 cartons of fresh milk.
    (a) If three cartons are selected at random, what is the probability that they all contain unspoiled milk?
    (b) What is the probability that three cartons selected at random all contain spoiled milk?

*In Exercises 45–48, use the addition theorem to compute the required probabilities.*

45. If $P(A) = \frac{1}{3}$, $P(B) = \frac{1}{4}$, and $A \cap B = \emptyset$, compute $P(A \cup B)$.

46. If $P(A) = \frac{2}{5}$, $P(B) = \frac{3}{10}$, and $P(A \cap B) = \frac{1}{5}$, compute $P(A \cup B)$.

47. If $P(A \cup B) = \frac{7}{9}$, $P(A) = \frac{5}{9}$, and $P(B) = \frac{4}{9}$, compute $P(A \cap B)$.

48. If $P(A) = \frac{1}{3}$, $P(A \cup B) = \frac{2}{3}$, and $P(A \cap B) = \frac{1}{6}$, compute $P(B)$.

49. Two balls are drawn at random from a bag containing six red and 10 black balls. Compute the probability for each of the following events.
    (a) Drawing two red balls
    (b) Drawing two black balls
    (c) Drawing either two red or two black balls

50. A card is selected at random from a deck of 52 cards. Compute the probability of each of the following events.
    (a) Selecting a 4 or 2
    (b) Selecting a red card or an ace
    (c) Selecting an ace or a picture card

51. Your mathematics instructor claims that the probabilities of your getting A's on the first exam, on the second exam, and on both exams are $\frac{1}{2}$, $\frac{3}{10}$, and $\frac{1}{10}$, respectively. According to the figures, what is the probability that you will get an A on at least one exam? *Hint:* Let $G$ and $H$ denote the events "A on first exam" and "A on second exam," respectively. Then you want to compute $P(G \text{ or } H)$.

52. Two cards are drawn at random from a regular 52-card deck. Find the probability that either both are black or both are twos.

53. Five cards are drawn at random from a regular deck of 52 cards. Compute the probability that all five cards are clubs.

**B**

54. An experiment consists of selecting at random one of the 100 points in the $x$-$y$ plane shown in the accompanying figure. Compute the probability for each of the following events.
    (a) The point lies on the line $y = x$.
    (b) The point lies on the line $y = 2x$.
    (c) The point lies on the circle $x^2 + y^2 = 25$.
    (d) The point lies within (but not on) the circle $x^2 + y^2 = 25$.
    (e) The point lies on or within the circle $x^2 + y^2 = 25$.
    (f) The point lies outside the circle $x^2 + y^2 = 25$.
    (g) The point lies inside the circle $x^2 + y^2 = 25$ but not on the graph of $y = |x - 3|$.
    (h) The point lies either on the circle $x^2 + y^2 = 25$ or on the line $y = x$.
    (i) The point lies on the $x$-axis or on the circle $x^2 + y^2 = 25$.
    (j) The sum of the $x$- and $y$-coordinates of the point is 18.
    (k) The sum of the $x$- and $y$-coordinates of the point is 21.
    (l) The $y$-coordinate of the point is greater than the square of the $x$-coordinate.

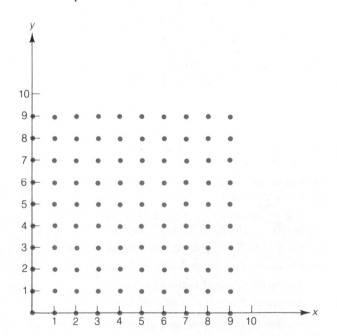

## CHAPTER 13    SUMMARY OF PRINCIPAL TERMS AND FORMULAS

| TERMS OR FORMULAS | PAGE REFERENCE | COMMENT |
|---|---|---|
| 1. Principle of mathematical induction | 635 | This can be stated as follows. Suppose that for each natural number $n$ we have a statement $P_n$. Suppose $P_1$ is true. Also suppose that $P_{k+1}$ is true whenever $P_k$ is true. Then according to the **principle of mathematical induction**, all the statements are true; that is, $P_n$ is true for all natural numbers $n$. The principle of mathematical induction has the status of an axiom; that is, we accept its validity without proof. |
| 2. Pascal's triangle | 641 | Pascal's triangle refers to the triangular array of numbers displayed on page 641. Additional rows can be added to the triangle according to the following rule: Each entry in the array (other than the 1's along the sides) is the sum of the two numbers diagonally above it. The numbers in the $n$th row of Pascal's triangle are the coefficients of the terms in the expansion of $(a + b)^{n-1}$. |
| 3. $n!$ (read: $n$ factorial) | 642 | This denotes the product of the first $n$ natural numbers. For example, $4! = (4)(3)(2)(1) = 24$. |
| 4. $\binom{n}{k}$ (read: $n$ choose $k$) | 643 | Let $n$ and $k$ be nonnegative integers with $k \leq n$. Then the **binomial coefficient** $\binom{n}{k}$ is defined by $$\binom{n}{k} = \frac{n!}{k!(n-k)!}$$ |
| 5. Binomial theorem | 644 | The **binomial theorem** is a formula that allows us to analyze and expand expressions of the form $(a + b)^n$. If we use the sigma notation, then the statement of the binomial theorem that appears on page 644 can be abbreviated to read as follows: $$(a + b)^n = \sum_{k=0}^{n} \binom{n}{k} a^{n-k} b^k$$ |
| 6. Sequence | 649 | In the context of the present chapter, a **sequence** is an ordered list of real numbers: $a_1, a_2, a_3, \ldots$. The numbers $a_k$ are called the **terms** of the sequence. A sequence can also be defined as a *function* whose domain is the set of natural numbers. |
| 7. $\sum_{k=1}^{n} a_k$ | 651 | This expression stands for the sum $a_1 + a_2 + a_3 + \cdots + a_n$. The letter $k$ in the expression is referred to as the **index of summation**. |
| 8. Arithmetic sequence | 655 | A sequence in which the successive terms differ by a constant is called an **arithmetic sequence**. The general form of an arithmetic sequence is $$a, \ a + d, \ a + 2d, \ a + 3d, \ldots$$ In this sequence, $d$ is referred to as the **common difference**. |

| TERMS OR FORMULAS | PAGE REFERENCE | COMMENT |
|---|---|---|
| 9. $a_n = a + (n-1)d$ | 655 | This is the formula for the $n$th term of an arithmetic sequence. |
| 10. $S_n = \dfrac{n}{2}[2a + (n-1)d]$ <br><br> $S_n = n\left(\dfrac{a + a_n}{2}\right)$ | 657 | These are the formulas for the sum of the first $n$ terms of an arithmetic sequence. |
| 11. Geometric sequence | 660 | A sequence in which the ratio of successive terms is constant is called a **geometric sequence**. The general form of a geometric sequence is <br><br> $a,\ ar,\ ar^2, \ldots$ <br><br> The number $r$ is referred to as the **common ratio**. |
| 12. $a_n = ar^{n-1}$ | 661 | This is the formula for the $n$th term of a geometric sequence. |
| 13. $S_n = \dfrac{a(1 - r^n)}{1 - r}$ | 662 | This is the formula for the sum of the first $n$ terms of a geometric sequence. |
| 14. $S = \dfrac{a}{1 - r}$ | 663 | This the formula for the sum of the infinite geometric series $a + ar + ar^2 + \cdots$, where $|r| < 1$. |
| 15. Multiplication principle | 666 | Suppose that a task involves two consecutive operations. If the first operation can be performed in $M$ ways, and the second operation can be performed in $N$ ways, then the entire task can be performed in $MN$ ways. |
| 16. Permutation | 667 | A **permutation** of a set of $n$ objects is a listing or arrangement of those objects in a definite order. |
| 17. $P(n,r) = n(n-1)\cdots(n-r+1)$ <br><br> $\quad = \dfrac{n!}{(n-r)!}$ | 668 | $P(n,r)$ denotes the number of permutations of $n$ objects taken $r$ at a time $(r \le n)$. |
| 18. $C(n,r) = \dfrac{n!}{r!(n-r)!}$ <br><br> $\quad = \dfrac{n(n-1)\cdots(n-r+1)}{r!}$ | 670 | $C(n,r)$ denotes the number of ways of selecting $r$ objects from a set of $n$ objects $(r \le n)$. Each selection or subset of $r$ objects is called a **combination** of the $n$ objects taken $r$ at a time. |
| 19. Sample space; event | 675 | The sample space for an experiment is the set of all possible outcomes of the experiment. A subset of the **sample space** is called an **event**. The notation $n(E)$ is used to denote the number of outcomes in an event $E$. |
| 20. $P(E)$ | 676 | Let $S$ be a finite sample space in which each outcome is equally likely to occur. If $E$ is an event in $S$, then the probability of $E$, denoted by $P(E)$, is defined by <br><br> $P(E) = \dfrac{n(E)}{n(S)} = \dfrac{\text{number of outcomes in } E}{\text{number of outcomes in } S}$ |
| 21. The addition theorem for probabilities | 680 | If $A$ and $B$ are events in a sample space $S$, then <br><br> $P(A \cup B) = P(A) + P(B) - P(A \cap B)$ |

## CHAPTER 13 REVIEW EXERCISES

NOTE  The fifteen exercises in this set with colored problem numbers constitute a chapter test.

*In Exercises 1–10, use the principle of mathematical induction to show that the statements are true for all natural numbers.*

1. $5 + 10 + 15 + \cdots + 5n = \frac{5}{2}n(n+1)$

2. $10 + 10^2 + 10^3 + \cdots + 10^n = \frac{10}{9}(10^n - 1)$

3. $1 \cdot 2 + 2 \cdot 3 + 3 \cdot 4 + \cdots + n(n+1)$
$$= \tfrac{1}{3}n(n+1)(n+2)$$

4. $\dfrac{1}{2} + \dfrac{2}{2^2} + \dfrac{3}{2^3} + \cdots + \dfrac{n}{2^n} = 2 - \dfrac{2+n}{2^n}$

5. $1 + 3 \cdot 2 + 5 \cdot 2^2 + 7 \cdot 2^3 + \cdots$
$$+ (2n - 1) \cdot 2^{n-1} = 3 + (2n - 3) \cdot 2^n$$

6. $\dfrac{1}{1 \cdot 4} + \dfrac{1}{4 \cdot 7} + \dfrac{1}{7 \cdot 10} + \cdots$
$$+ \dfrac{1}{(3n-2)(3n+1)} = \dfrac{n}{3n+1}$$

7. $1 + 2^2 \cdot 2 + 3^2 \cdot 2^2 + 4^2 \cdot 2^3 + \cdots + n^2 \cdot 2^{n-1}$
$$= (n^2 - 2n + 3)2^n - 3$$

8. $9$ is a factor of $n^3 + (n+1)^3 + (n+2)^3$.

9. $3$ is a factor of $7^n - 1$.

10. $8$ is a factor of $9^n - 1$.

*In Exercises 11–20, expand the given expressions.*

11. $(3a + b^2)^4$

12. $(5a - 2b)^3$

13. $(x + \sqrt{x})^4$

14. $(1 - \sqrt{3})^6$

15. $(x^2 - 2y^2)^5$

16. $\left(\dfrac{1}{a} + \dfrac{2}{b}\right)^3$

17. $\left(1 + \dfrac{1}{x}\right)^5$

18. $\left(x^3 + \dfrac{1}{x^2}\right)^6$

19. $(a\sqrt{b} - b\sqrt{a})^4$

20. $(x^{-2} + y^{5/2})^8$

21. Find the fifth term in the expansion of $(3x + y^2)^5$.

22. Find the eighth term in the expansion of $(2x - y)^9$.

23. Find the coefficient of the term containing $a^5$ in the expansion of $(a - 2b)^7$.

24. Find the coefficient of the term containing $b^8$ in the expansion of $\left(2a - \dfrac{b}{3}\right)^{10}$.

25. Find the coefficient of the term containing $x^3$ in the expansion of $(1 + \sqrt{x})^8$.

26. Expand $(1 + \sqrt{x} + x)^6$.  *Suggestion:*  Rewrite the expression as $[(1 + \sqrt{x}) + x]^6$.

*In Exercises 27–32, verify each assertion by computing the indicated binomial coefficients.*

27. $\dbinom{2}{0}^2 + \dbinom{2}{1}^2 + \dbinom{2}{2}^2 = \dbinom{4}{2}$

28. $\dbinom{3}{0}^2 + \dbinom{3}{1}^2 + \dbinom{3}{2}^2 + \dbinom{3}{3}^2 = \dbinom{6}{3}$

29. $\dbinom{4}{0}^2 + \dbinom{4}{1}^2 + \dbinom{4}{2}^2 + \dbinom{4}{3}^2 + \dbinom{4}{4}^2 = \dbinom{8}{4}$

30. $\dbinom{2}{0} + \dbinom{2}{1} + \dbinom{2}{2} = 2^2$

31. $\dbinom{3}{0} + \dbinom{3}{1} + \dbinom{3}{2} + \dbinom{3}{3} = 2^3$

32. $\dbinom{4}{0} + \dbinom{4}{1} + \dbinom{4}{2} + \dbinom{4}{3} + \dbinom{4}{4} = 2^4$

*In Exercises 33–38, compute the first four terms in each sequence. Also, in Exercises 33–35, graph the sequences for $n = 1$, 2, 3, and 4.*

33. $a_n = \dfrac{2n}{n+1}$

34. $a_n = \dfrac{3n-2}{3n+2}$

35. $u_n = (-1)^n\left(1 - \dfrac{1}{n+1}\right)$

36. $a_0 = 4;\ a_n = 2a_{n-1},\ n \ge 1$

37. $a_0 = -3;\ a_n = 4a_{n-1},\ n \ge 1$

38. $a_0 = 1;\ a_1 = 2;\ a_n = 3a_{n-1} + 2a_{n-2},\ n \ge 2$

*In Exercises 39 and 40, evaluate each sum.*

39. (a) $\displaystyle\sum_{k=1}^{3} (-1)^k(2k+1)$  (b) $\displaystyle\sum_{k=0}^{8} \left(\dfrac{1}{k+1} - \dfrac{1}{k+2}\right)$

40. (a) $\displaystyle\sum_{j=1}^{4} \dfrac{(-1)^j}{j}$  (b) $\displaystyle\sum_{n=1}^{5} \left(\dfrac{1}{n} - \dfrac{1}{n+1}\right)$

*In Exercises 41 and 42, rewrite each sum using sigma notation.*

41. $5/3 + 5/3^2 + 5/3^3 + 5/3^4 + 5/3^5$

42. $1/2 - 2/2^2 + 3/2^3 - 4/2^4 + 5/2^2 - 6/2^6$

*In Exercises 43–46, find the indicated term in each sequence.*

43. $5, 9, 13, 17, \ldots;\ a_{18}$

44. $5, \frac{9}{2}, 4, \frac{7}{2}, \ldots;\ a_{20}$

45. $10, 5, \frac{5}{2}, \frac{5}{4}, \ldots;\ a_{12}$

46. $\sqrt{2} + 1, 1, \sqrt{2} - 1, 3 - 2\sqrt{2}, \ldots;\ a_{10}$

47. Determine the sum of the first 12 terms of an arithmetic sequence in which the first term is 8 and the twelfth term is $\frac{43}{2}$.

**48.** Find the sum of the first 45 terms in the sequence $10, \frac{29}{3}, \frac{28}{3}, 9, \ldots$.

**49.** Find the sum of the first 10 terms in the sequence $7, 70, 700, \ldots$.

**50.** Find the sum of the first 12 terms in the sequence $\frac{1}{3}, -\frac{2}{9}, \frac{4}{27}, -\frac{8}{81}, \ldots$.

**51.** In a certain geometric sequence, the third term is 4 and the fifth term is 10. Find the sixth term, given that the common ratio is negative.

**52.** For a certain infinite geometric series, the first term is 2 and the sum is $\frac{18}{11}$. Find the common ratio $r$.

*In Exercises 53–56, find the sum of each infinite geometric series.*

**53.** $\frac{3}{5} + \frac{3}{25} + \frac{3}{125} + \cdots$     **54.** $\frac{7}{10} + \frac{7}{100} + \frac{7}{1000} + \cdots$

**55.** $\frac{1}{9} - \frac{1}{81} + \frac{1}{729} - \cdots$

**56.** $1 + \dfrac{1}{1 + \sqrt{2}} + \dfrac{1}{(1 + \sqrt{2})^2} + \cdots$

**57.** Find a fraction equivalent to $0.\overline{45}$.

**58.** Find a fraction equivalent to $0.2\overline{13}$.

*In Exercises 59–62, verify each equation using the formula for the sum of an arithmetic series:*

$$S_n = \frac{n}{2}[2a + (n - 1)d]$$

*(The formulas given in Exercises 59–62 appear in* Elements of Algebra *by Leonhard Euler, first published in 1770.)*

**59.** $1 + 2 + 3 + \cdots + n = n + n(n - 1)/2$

**60.** $1 + 3 + 5 + \cdots$ (to $n$ terms) $= n + 2n(n - 1)/2$

**61.** $1 + 4 + 7 + \cdots$ (to $n$ terms) $= n + 3n(n - 1)/2$

**62.** $1 + 5 + 9 + \cdots$ (to $n$ terms) $= n + 4n(n - 1)/2$

<u>c</u> **63.** In this exercise, you will use the following (remarkably simple) formula for approximating sums of powers of integers:

$$1^k + 2^k + 3^k + \cdots + n^k \approx \frac{(n + \frac{1}{2})^{k+1}}{k + 1} \qquad (1)$$

[This formula appears in the article, "Sums of Powers of Integers," by B. L. Burrows and R. F. Talbot, published in the *American Mathematical Monthly*, **91** (1984): 394.]

(a) Use formula (1) to estimate the sum $1^2 + 2^2 + 3^2 + \cdots + 50^2$. Round off your answer to the nearest integer.

(b) Compute the exact value of the sum in part (a) using the formula $\sum_{k=1}^{n} k^2 = \frac{1}{6}n(n + 1)(2n + 1)$. (This formula can be proved using mathematical induction.) Then compute the percent error for the

approximation obtained in part (a). The percent error is given by

$$\frac{|\text{actual value} - \text{approximate value}|}{\text{actual value}} \times 100$$

(c) Use formula (1) to estimate the sum $1^4 + 2^4 + 3^4 + \cdots + 200^4$. Round off your answer to six significant digits.

(d) The following formula for the sum $1^4 + 2^4 + \cdots + n^4$ can be proved using mathematical induction:

$$\sum_{k=1}^{n} k^4 = \frac{n(n + 1)(2n + 1)(3n^2 + 3n - 1)}{30}$$

Use this formula to compute the sum in part (c). Round off your answer to six significant digits. Then use this result to compute the percent error for the approximation in part (c).

<u>c</u> **64.** According to *Stirling's formula* [named after James Stirling (1692–1770)], the quantity $n!$ can be approximated as follows:

$$n! \approx \sqrt{2\pi n}\left(\frac{n}{e}\right)^n$$

In this formula, $e$ is the constant $2.718\ldots$ (discussed in Section 6.2). Use a calculator to complete the following table. Round off your answers to five significant digits. As you will see, the numbers in the right-hand column approach 1 as $n$ increases. This shows that, in a certain sense, the approximation improves as $n$ increases.

| $n$ | $n!$ | $\sqrt{2\pi n}\,(n/e)^n$ | $\dfrac{n!}{\sqrt{2\pi n}\,(n/e)^n}$ |
|---|---|---|---|
| 10 | | | |
| 20 | | | |
| 30 | | | |
| 40 | | | |
| 50 | | | |
| 60 | | | |
| 65 | | | |

**65.** How many different two-digit numbers can be formed using the digits 2, 4, 6, and 8? (Repetitions are allowed.)

**66.** How many permutations are there of the set $\{A, B, C, D, E\}$?

**67.** Using the set $\{a, b, c, d\}$, list the permutations of these four letters taken three at a time. Evaluate $P(4, 3)$.

**68.** Follow Exercise 67, using combinations rather than permutations, and evaluate $C(4,3)$.

**69.** From a collection of 12 pictures, eight are to be selected and hung in a row. How many arrangements are possible?

**70.** How many possible arrangements are there of the letters $a, b, c, d, e, f$ in which $a$, $b$, and $c$ appear consecutively (but not necessarily in that order)?

**71.** How many possible arrangements are there of the letters $a, b, c, d, e, f$ in which $e$ and $f$ are not next to each other?

**72.** How many four-member committees are possible in a club with 15 members?

**73.** On a certain committee there are nine males and 10 females. How many possibilities are there for a four-member subcommittee consisting of two females and two males?

**74.** Given 10 points in a plane, no three of which lie on the same line, find the number of different triangles that can be formed when the vertices are selected from the given points.

**75.** Suppose that three faulty batteries are accidentally shipped along with nine reliable batteries. What is the probability that two batteries selected at random from this shipment will both be reliable?

**76.** The accompanying figure shows a 12-sided figure with all its diagonals. How many diagonals are there?

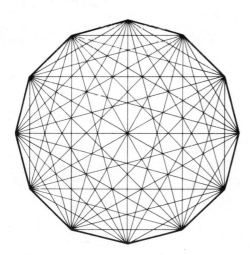

**77.** Two balls are selected at random from a bag containing five red and seven green balls. Compute the probability for each of the following events.
(a) Drawing two red balls
(b) Drawing two green balls
(c) Drawing either two red or two green balls

*For Exercises 78–88, consider an experiment in which two dice are tossed. Assume that one die is red and the other is white. Then the outcomes can be recorded as ordered pairs. Let us agree that the first element in any ordered pair corresponds to the number of dots showing on the red die and the second element corresponds to the number of dots showing on the white die. For example, $(5,4)$ denotes the outcome "5 on red and 4 on white." Compute the probabilities for the following events.*

**78.** Obtaining a sum of 2     **79.** Obtaining a sum of 3

**80.** Obtaining a sum of 5     **81.** Obtaining a sum of 7

**82.** Obtaining a sum of 8

**83.** Obtaining a sum of either 7 or 11

**84.** Obtaining "doubles" (i.e., the same number of dots on both dice)

**85.** Obtaining a sum of at least 10

**86.** Obtaining a sum that is either even or less than 5

**87.** Obtaining any sum other than 7 and 8

**88.** Obtaining either doubles or a sum greater than 9

**89.** An integer between 1 and 15, inclusive, is selected at random. Compute the probabilities for the following events.
(a) Selecting an even number
(b) Selecting an odd number
(c) Selecting either a number greater than 10 or an even number
(d) Selecting either a multiple of 2 or a multiple of 3

**90.** (a) How many three-digit numbers are there with no repeated digits? (The first digit cannot be zero.)
(b) A three-digit number is selected at random. What is the probability that it has no repeated digits?

**91.** (a) How many three-digits numbers are there with one or more repeated digits?
(b) What is the probability that a three-digit number selected at random will contain one or more repeated digits?

**92.** The sum of three consecutive terms in a geometric sequence is 13, and the sum of the reciprocals is $\frac{13}{9}$. What are the possible values for the common ratio? (There are four answers.)

**93.** The nonzero numbers $a$, $b$, and $c$ are consecutive terms in a geometric sequence, and $a + b + c = 70$. Furthermore, $4a$, $5b$, and $4c$ are consecutive terms in an arithmetic sequence. Find $a$, $b$, and $c$.

**94.** The nonzero numbers $a$, $b$, and $c$ are consecutive terms in a geometric sequence, and $a$, $2b$, and $c$ are consecutive terms in an arithmetic sequence. Show that the common ratio in the geometric sequence must be either $2 + \sqrt{3}$ or $2 - \sqrt{3}$.

95. If the numbers $\dfrac{1}{b+c}$, $\dfrac{1}{c+a}$, and $\dfrac{1}{a+b}$ are consecutive terms in an arithmetic sequence, show that $a^2$, $b^2$, and $c^2$ are also consecutive terms in an arithmetic sequence.

96. If $a$, $b$, and $c$ are consecutive terms in a geometric sequence, prove that $\dfrac{1}{a+b}$, $\dfrac{1}{2b}$, and $\dfrac{1}{c+b}$ are consecutive terms in an arithmetic sequence.

97. (a) Find a value for $x$ such that $3+x$, $4+x$, and $5+x$ are consecutive terms in a geometric sequence.
    (b) Given three numbers $a$, $b$, and $c$, find a value for $x$ (in terms of $a$, $b$, and $c$) such that $a+x$, $b+x$, and $c+x$ are consecutive terms in a geometric sequence.

98. If $\ln(A+C) + \ln(A+C-2B) = 2\ln(A-C)$, show that $1/A$, $1/B$, and $1/C$ are consecutive terms in an arithmetic sequence.

# APPENDIX

## SIGNIFICANT DIGITS AND CALCULATORS

**A.1**

Many of the numbers that we use in scientific work and in daily life are approximations. In some cases the approximations arise because the numbers were obtained through measurements or experiments. Consider, for example, the following statement from an astronomy textbook:

The diameter of the Moon is 3476 km

We interpret this statement as meaning that the actual diameter $D$ is closer to 3476 km than it is to either 3475 km or 3477 km. In other words,

$$3475.5 \text{ km} \leq D \leq 3476.5 \text{ km}$$

The interval $[3475.5, 3476.5]$ in this example provides information about the accuracy of the measurement. Another way to indicate accuracy in an approximation is by specifying the number of *significant digits* it contains. The measurement 3476 km has four significant digits. In general, the number of significant digits in a given number is found as follows.

| Significant Digits | | |
|---|---|---|
| | | EXAMPLES |
| The number of significant digits in a given number is determined by counting the digits from left to right, beginning with the left-most nonzero digit. | NUMBER | NUMBER OF SIGNIFICANT DIGITS |
| | 1.43 | 3 |
| | 0.52 | 2 |
| | 0.05 | 1 |
| | 4837 | 4 |
| | 4837.0 | 5 |

Numbers obtained through measurements are not the only source of approximations in scientific work. For example, to five significant digits, we have the following approximation for the irrational number $\pi$:

$$\pi \approx 3.1416$$

This statement tells us that $\pi$ is closer to 3.1416 than it is to either 3.1415 or 3.1417. In other words,

$$3.14155 \leq \pi \leq 3.14165$$

Table 1 provides some additional examples of the ideas we've introduced.

TABLE 1

| NUMBER | NUMBER OF SIGNIFICANT DIGITS | RANGE OF MEASUREMENT |
|---|---|---|
| 37 | 2 | $[36.5, 37.5]$ |
| 37.0 | 3 | $[36.95, 37.05]$ |
| 268.1 | 4 | $[268.05, 268.15]$ |
| 1.036 | 4 | $[1.0355, 1.0365]$ |
| 0.036 | 2 | $[0.0355, 0.0365]$ |

There is an ambiguity involving zero that can arise in counting significant digits. Suppose that someone measures the width $w$ of a rectangle and reports the result as 30 cm. How many significant digits are there? If the value 30 cm was obtained by measuring to the nearest 10 cm, then only the digit 3 is significant, and we can conclude only that the width $w$ lies in the range $25 \text{ cm} \leq w \leq 35 \text{ cm}$. On the other hand, if the 30 cm was obtained by measuring to the nearest 1 cm, then both the digits 3 and 0 are significant and we have $29.5 \text{ cm} \leq w \leq 30.5 \text{ cm}$.

By using *scientific notation* we can avoid the type of ambiguity discussed in the previous paragraph. As explained in Section 1.5, a number written in the form

$$b \times 10^n \qquad \text{where } 1 \leq b < 10 \text{ and } n \text{ is an integer}$$

is said to be expressed in scientific notation. For the example in the previous paragraph, then, we would write

$$w = 3 \times 10^1 \text{ cm} \qquad \text{if the measurement were to the nearest 10 cm}$$

and

$$w = 3.0 \times 10^1 \text{ cm} \qquad \text{if the measurement were to the nearest 1 cm}$$

As the figures in Table 2 indicate, for a number $b \times 10^n$ in scientific notation, the number of significant digits is just the number of digits in $b$. (This is one of the advantages in using scientific notation; the number of significant digits, and hence the accuracy of the measurement, is readily apparent.)

TABLE 2

| MEASUREMENT | NUMBER OF SIGNIFICANT DIGITS | RANGE OF MEASUREMENT |
|---|---|---|
| Mass of the earth: | | |
| $6 \times 10^{27} \text{ g}$ | 1 | $[5.5 \times 10^{27} \text{ g}, 6.5 \times 10^{27} \text{ g}]$ |
| $6.0 \times 10^{27} \text{ g}$ | 2 | $[5.95 \times 10^{27} \text{ g}, 6.05 \times 10^{27} \text{ g}]$ |
| $5.974 \times 10^{27} \text{ g}$ | 4 | $[5.9735 \times 10^{27} \text{ g}, 5.9745 \times 10^{27} \text{ g}]$ |
| Mass of a proton: | | |
| $1.67 \times 10^{-24} \text{ g}$ | 3 | $[1.665 \times 10^{-24} \text{ g}, 1.675 \times 10^{-24} \text{ g}]$ |

There are a number of exercises in this text in which a calculator either is required or is extremely useful. Although you could work many of these exercises using tables instead of a calculator, this author recommends that you purchase a scientific calculator. *Scientific calculator* is a generic term describing a calculator with (at least) the following features or functions beyond the usual arithmetic functions:

1. Memory
2. Scientific notation
3. Powers and roots
4. Logarithms (base ten and base $e$)
5. Trigonometric functions
6. Inverse functions

Because of the variety and differences in the calculators that are available, no specific instructions are provided in this appendix for operating a calculator. When you buy a calculator, read the owner's manual carefully and work through some of the examples in it. In general, learn to use the memory capabilities of the calculator so that, as far as possible, you don't need to write down the results of the intermediate steps in a given calculation.

Many of the numerical exercises in the text ask that you round off the answers to a specified number of decimal places. Our rules for rounding off are as follows.

### Rules for Rounding Off a Number (With More Than $n$ Decimal Places) to $n$ Decimal Places

1. If the digit in the $(n + 1)$st decimal place is greater than 5, increase the digit in the $n$th place by 1. If the digit in the $(n + 1)$st place is less than 5, leave the $n$th digit unchanged.

2. If the digit in the $(n + 1)$st decimal place is 5 and there is at least one nonzero digit to the right of this 5, increase the digit in the $n$th decimal place by 1.

3. If the digit in the $(n + 1)$st decimal place is 5 and there are no nonzero digits to the right of this 5, then increase the digit in the $n$th decimal place by 1 only if this results in an even digit.

The examples in Table 3 illustrate the use of these rules.

TABLE 3

| NUMBER | ROUNDED TO ONE DECIMAL PLACE | ROUNDED TO THREE DECIMAL PLACES |
|--------|------------------------------|---------------------------------|
| 4.3742 | 4.4 | 4.374 |
| 2.0515 | 2.1 | 2.052 |
| 2.9925 | 3.0 | 2.992 |

These same rules can be adapted for rounding off a result to a specified number of significant digits. As examples of this, we have

| | | | |
|---|---|---|---|
| 2347 | rounded off to two significant digits is 2300 | $= 2.3 \times 10^3$ |
| 2347 | rounded off to three significant digits is 2350 | $= 2.35 \times 10^3$ |
| 975 | rounded off to two significant digits is 980 | $= 9.8 \times 10^2$ |
| 0.985 | rounded off to two significant digits is 0.98 | $= 9.8 \times 10^{-1}$ |

**FIGURE 1**

In calculator exercises that ask you to round off your answers, it's important that you postpone rounding until the final calculation is carried out. For example, suppose that you are required to determine the hypotenuse $x$ of the right triangle in Figure 1 to two significant digits. Using the Pythagorean Theorem, we have

$$x = \sqrt{(1.36)^2 + (2.46)^2}$$
$$= 2.8 \qquad \text{(using a calculator and rounding off the final result to two significant digits)}$$

On the other hand, if we first round off each of the given lengths to two significant digits, we obtain

$$x = \sqrt{(1.4)^2 + (2.5)^2}$$
$$= 2.9 \qquad \text{(to two significant digits)}$$

This last result is inappropriate, and we can see why as follows. As Table 4 shows, the maximum possible values for the sides are 1.365 and 2.465, respectively.

Thus, the maximum possible value for the hypotenuse must be

$$\sqrt{(1.365)^2 + (2.465)^2} = 2.817\ldots \qquad \text{(calculator display)}$$
$$= 2.8 \qquad \text{(to two significant digits)}$$

This shows that the value 2.9 is indeed inappropriate, as we stated previously.

An error often made by people working with calculators and approximations is to report a final answer with a greater degree of accuracy than the data warrant. Consider, for example, the right triangle in Figure 2. Using the Pythagorean Theorem and a calculator with an eight-digit display, we obtain

$$h = 3.6055513 \text{ cm}$$

**TABLE 4**

| NUMBER | RANGE OF MEASUREMENT |
|--------|----------------------|
| 1.36   | [1.355, 1.365]       |
| 2.46   | [2.455, 2.465]       |

**FIGURE 2**

This value for $h$ is inappropriate, since common sense tells us that the answer should be no more accurate than the data used to obtain that answer. In particular, since the given sides of the triangle apparently were measured only to the nearest tenth of a centimeter, we certainly should not expect any improvement in accuracy for the resulting value of the hypotenuse. An appropriate form for the value of $h$ here would be $h = 3.6$ cm. In general, for calculator exercises in this text that do not specify a required number of decimal places or significant digits in the final results, you should use the following guidelines.

---

**Guidelines for Computing with Approximations**

1. *For adding and subtracting:* Round off the final result so that it contains only as many decimal places as there are in the data with the fewest decimal places.

2. *For multiplying and dividing:* Round off the final result so that it contains only as many significant digits as there are in the data with the fewest significant digits.

3. *For powers and roots:* In computing a power or a root of a real number $b$, round off the result so that it contains as many significant digits as there are in $b$.

## SOME PROOFS FOR SECTION 1.2

**A.2**

Some of the fundamental algebraic properties of the real number system are listed on page 6 in Section 1.2. In this appendix we use those properties to prove the following theorem.

| **Theorem** |
| --- |
| Let $a$ and $b$ be real numbers. Then |

(a) $a \cdot 0 = 0$      (d) $a(-b) = -ab$
(b) $-a = (-1)a$      (e) $(-a)(-b) = ab$
(c) $-(-a) = a$

PROOF OF PART ($a$)

$$a \cdot 0 = a \cdot (0 + 0) \qquad \text{additive identity property}$$
$$= a \cdot 0 + a \cdot 0 \qquad \text{distributive property}$$

Now since $a \cdot 0$ is a real number, it has an additive inverse $-(a \cdot 0)$. Adding this to both sides of the last equation, we obtain

$$a \cdot 0 + [-(a \cdot 0)] = (a \cdot 0 + a \cdot 0) + [-(a \cdot 0)]$$
$$a \cdot 0 + [-(a \cdot 0)] = a \cdot 0 + \{a \cdot 0 + [-(a \cdot 0)]\} \qquad \text{associative property of addition}$$
$$0 = a \cdot 0 + 0 \qquad \text{additive inverse property}$$
$$0 = a \cdot 0 \qquad \text{additive identity property}$$

Thus $a \cdot 0 = 0$, as we wished to show.

PROOF OF PART ($b$)

$$0 = 0 \cdot a \qquad \text{using part (a) and the commutative property of multiplication}$$
$$= [1 + (-1)]a \qquad \text{additive inverse property}$$
$$= 1 \cdot a + (-1)a \qquad \text{distributive property}$$
$$= a + (-1)a \qquad \text{multiplicative identity property}$$

Now, by adding $-a$ to both sides of this last equation, we obtain

$$-a + 0 = -a + [a + (-1)a]$$
$$-a = (-a + a) + (-1)a \qquad \text{additive identity property and associative property of addition}$$
$$= 0 + (-1)a \qquad \text{additive inverse property}$$
$$= (-1)a \qquad \text{additive identity property}$$

This last equation asserts that $-a = (-1)a$, as we wished to show.

PROOF OF PART ($c$)

$$-(-a) + (-a) = 0 \qquad \text{additive inverse property}$$

By adding $a$ to both sides of this last equation, we obtain

$$[-(-a) + (-a)] + a = 0 + a$$

$-(-a) + (-a + a) = a$      associative property of addition and additive identity property

$-(-a) + 0 = a$      additive inverse property

$-(-a) = a$      additive identity property

This last equation states that $-(-a) = a$, as we wished to show.

PROOF OF PART (*d*)

$a(-b) = a[(-1)b]$      using part (b)

$= [a(-1)]b$      associative property of multiplication

$= [(-1)a]b$      commutative property of multiplication

$= (-1)(ab)$      associative property of multiplication

$= -(ab)$      using part (b)

Thus $a(-b) = -ab$, as we wished to show.

PROOF OF PART (*e*)

$(-a)(-b) = -[(-a)b]$      using part (d)

$= -[b(-a)]$      commutative property of multiplication

$= -[-(ba)]$      using part (d)

$= ba$      using part (c)

$= ab$      commutative property of multiplication

We've now shown that $(-a)(-b) = ab$, as required.

# $\sqrt{2}$ IS IRRATIONAL

## A.3

*The proof is by* reductio ad absurdum, *and* reductio ad absurdum, *which Euclid loved so much, is one of a mathematician's finest weapons.*

G. H. HARDY (1877–1947)

We will use an *indirect proof* to show that the square root of two is an irrational number. The strategy is as follows.

1. We suppose that $\sqrt{2}$ is a rational number.
2. Using (1) and the usual rules of logic and algebra, we derive a contradiction.
3. On the basis of the contradiction in (2), we conclude that the supposition in (1) is untenable; that is, we conclude that $\sqrt{2}$ is irrational.

In carrying out the proof, we'll assume as known the following three statements.

If $x$ is an even natural number, then $x = 2k$, for some natural number $k$.

Any rational number can be written in the form $a/b$, where the integers $a$

and $b$ have no common integral factors other than $\pm 1$. (In other words, any fraction can be reduced to lowest terms.)

If $x$ is a natural number and $x^2$ is even, then $x$ is even.

Our indirect proof now proceeds as follows. Suppose that $\sqrt{2}$ were a rational number. Then we would be able to write

$$\sqrt{2} = \frac{a}{b} \qquad \text{where } a \text{ and } b \text{ are natural numbers with no common factor other than 1} \qquad (1)$$

Since both sides of equation (1) are positive, we can square both sides to obtain the equivalent equation

$$2 = \frac{a^2}{b^2}$$

or

$$2b^2 = a^2 \qquad (2)$$

Since the left-hand side of equation (2) is an even number, the right-hand side must be even. But if $a^2$ is even, then $a$ is even, and so

$$a = 2k, \text{ for some natural number } k$$

Using this last equation to substitute for $a$ in equation (2), we have

$$2b^2 = (2k)^2 = 4k^2$$

or

$$b^2 = 2k^2$$

Hence (reasoning as before) $b^2$ is even, and therefore $b$ is even. But then we have that both $b$ and $a$ are even, contrary to our hypothesis that $b$ and $a$ have no common factor other than 1. We conclude from this that equation (1) cannot hold; that is, there is no rational number $a/b$ such that $\sqrt{2} = a/b$. Thus $\sqrt{2}$ is irrational, as we wished to prove.

# TABLES

**TABLE 1** Squares and Square Roots

| NO. | SQ. | SQ. RT. | NO. | SQ. | SQ. RT. |
|---|---|---|---|---|---|
| 1 | 1 | 1.000 | 51 | 2,601 | 7.141 |
| 2 | 4 | 1.414 | 52 | 2,704 | 7.211 |
| 3 | 9 | 1.732 | 53 | 2,809 | 7.280 |
| 4 | 16 | 2.000 | 54 | 2,916 | 7.348 |
| 5 | 25 | 2.236 | 55 | 3,025 | 7.416 |
| 6 | 36 | 2.449 | 56 | 3,136 | 7.483 |
| 7 | 49 | 2.646 | 57 | 3,249 | 7.550 |
| 8 | 64 | 2.828 | 58 | 3,364 | 7.616 |
| 9 | 81 | 3.000 | 59 | 3,481 | 7.681 |
| 10 | 100 | 3.162 | 60 | 3,600 | 7.746 |
| 11 | 121 | 3.317 | 61 | 3,721 | 7.810 |
| 12 | 144 | 3.464 | 62 | 3,844 | 7.874 |
| 13 | 169 | 3.606 | 63 | 3,969 | 7.937 |
| 14 | 196 | 3.742 | 64 | 4,096 | 8.000 |
| 15 | 225 | 3.873 | 65 | 4,225 | 8.062 |
| 16 | 256 | 4.000 | 66 | 4,356 | 8.124 |
| 17 | 289 | 4.123 | 67 | 4,489 | 8.185 |
| 18 | 324 | 4.243 | 68 | 4,624 | 8.246 |
| 19 | 361 | 4.359 | 69 | 4,761 | 8.307 |
| 20 | 400 | 4.472 | 70 | 4,900 | 8.367 |
| 21 | 441 | 4.583 | 71 | 5,041 | 8.426 |
| 22 | 484 | 4.690 | 72 | 5,184 | 8.485 |
| 23 | 529 | 4.796 | 73 | 5,329 | 8.544 |
| 24 | 576 | 4.899 | 74 | 5,476 | 8.602 |
| 25 | 625 | 5.000 | 75 | 5,625 | 8.660 |
| 26 | 676 | 5.099 | 76 | 5,776 | 8.718 |
| 27 | 729 | 5.196 | 77 | 5,929 | 8.775 |
| 28 | 784 | 5.292 | 78 | 6,084 | 8.832 |
| 29 | 841 | 5.385 | 79 | 6,241 | 8.888 |
| 30 | 900 | 5.477 | 80 | 6,400 | 8.944 |
| 31 | 961 | 5.568 | 81 | 6,561 | 9.000 |
| 32 | 1,024 | 5.657 | 82 | 6,724 | 9.055 |
| 33 | 1,089 | 5.745 | 83 | 6,889 | 9.110 |
| 34 | 1,156 | 5.831 | 84 | 7,056 | 9.165 |
| 35 | 1,225 | 5.916 | 85 | 7,225 | 9.220 |
| 36 | 1,296 | 6.000 | 86 | 7,396 | 9.274 |
| 37 | 1,369 | 6.083 | 87 | 7,569 | 9.327 |
| 38 | 1,444 | 6.164 | 88 | 7,744 | 9.381 |
| 39 | 1,521 | 6.245 | 89 | 7,921 | 9.434 |
| 40 | 1,600 | 6.325 | 90 | 8,100 | 9.487 |
| 41 | 1,681 | 6.403 | 91 | 8,281 | 9.539 |
| 42 | 1,764 | 6.481 | 92 | 8,464 | 9.592 |
| 43 | 1,849 | 6.557 | 93 | 8,649 | 9.644 |
| 44 | 1,936 | 6.633 | 94 | 8,836 | 9.695 |
| 45 | 2,025 | 6.708 | 95 | 9,025 | 9.747 |
| 46 | 2,116 | 6.782 | 96 | 9,216 | 9.798 |
| 47 | 2,209 | 6.856 | 97 | 9,409 | 9.849 |
| 48 | 2,304 | 6.928 | 98 | 9,604 | 9.899 |
| 49 | 2,401 | 7.000 | 99 | 9,801 | 9.950 |
| 50 | 2,500 | 7.071 | 100 | 10,000 | 10.000 |

**TABLE 2** Exponential Functions

| $x$ | $e^x$ | $e^{-x}$ | $x$ | $e^x$ | $e^{-x}$ |
|---|---|---|---|---|---|
| 0.00 | 1.0000 | 1.0000 | 1.5 | 4.4817 | 0.2231 |
| 0.01 | 1.0101 | 0.9901 | 1.6 | 4.9530 | 0.2019 |
| 0.02 | 1.0202 | 0.9802 | 1.7 | 5.4739 | 0.1827 |
| 0.03 | 1.0305 | 0.9704 | 1.8 | 6.0496 | 0.1653 |
| 0.04 | 1.0408 | 0.9608 | 1.9 | 6.6859 | 0.1496 |
| 0.05 | 1.0513 | 0.9512 | 2.0 | 7.3891 | 0.1353 |
| 0.06 | 1.0618 | 0.9418 | 2.1 | 8.1662 | 0.1225 |
| 0.07 | 1.0725 | 0.9324 | 2.2 | 9.0250 | 0.1108 |
| 0.08 | 1.0833 | 0.9231 | 2.3 | 9.9742 | 0.1003 |
| 0.09 | 1.0942 | 0.9139 | 2.4 | 11.023 | 0.0907 |
| 0.10 | 1.1052 | 0.9048 | 2.5 | 12.182 | 0.0821 |
| 0.11 | 1.1163 | 0.8958 | 2.6 | 13.464 | 0.0743 |
| 0.12 | 1.1275 | 0.8869 | 2.7 | 14.880 | 0.0672 |
| 0.13 | 1.1388 | 0.8781 | 2.8 | 16.445 | 0.0608 |
| 0.14 | 1.1503 | 0.8694 | 2.9 | 18.174 | 0.0550 |
| 0.15 | 1.1618 | 0.8607 | 3.0 | 20.086 | 0.0498 |
| 0.16 | 1.1735 | 0.8521 | 3.1 | 22.198 | 0.0450 |
| 0.17 | 1.1853 | 0.8437 | 3.2 | 24.533 | 0.0408 |
| 0.18 | 1.1972 | 0.8353 | 3.3 | 27.113 | 0.0369 |
| 0.19 | 1.2092 | 0.8270 | 3.4 | 29.964 | 0.0334 |
| 0.20 | 1.2214 | 0.8187 | 3.5 | 33.115 | 0.0302 |
| 0.21 | 1.2337 | 0.8106 | 3.6 | 36.598 | 0.0273 |
| 0.22 | 1.2461 | 0.8025 | 3.7 | 40.447 | 0.0247 |
| 0.23 | 1.2586 | 0.7945 | 3.8 | 44.701 | 0.0224 |
| 0.24 | 1.2712 | 0.7866 | 3.9 | 49.402 | 0.0202 |
| 0.25 | 1.2840 | 0.7788 | 4.0 | 54.598 | 0.0183 |
| 0.30 | 1.3499 | 0.7408 | 4.1 | 60.340 | 0.0166 |
| 0.35 | 1.4191 | 0.7047 | 4.2 | 66.686 | 0.0150 |
| 0.40 | 1.4918 | 0.6703 | 4.3 | 73.700 | 0.0136 |
| 0.45 | 1.5683 | 0.6376 | 4.4 | 81.451 | 0.0123 |
| 0.50 | 1.6487 | 0.6065 | 4.5 | 90.017 | 0.0111 |
| 0.55 | 1.7333 | 0.5769 | 4.6 | 99.484 | 0.0101 |
| 0.60 | 1.8221 | 0.5488 | 4.7 | 109.95 | 0.0091 |
| 0.65 | 1.9155 | 0.5220 | 4.8 | 121.51 | 0.0082 |
| 0.70 | 2.0138 | 0.4966 | 4.9 | 134.29 | 0.0074 |
| 0.75 | 2.1170 | 0.4724 | 5.0 | 148.41 | 0.0067 |
| 0.80 | 2.2255 | 0.4493 | 5.5 | 244.69 | 0.0041 |
| 0.85 | 2.3396 | 0.4274 | 6.0 | 403.43 | 0.0025 |
| 0.90 | 2.4596 | 0.4066 | 6.5 | 665.14 | 0.0015 |
| 0.95 | 2.5857 | 0.3867 | 7.0 | 1096.6 | 0.0009 |
| 1.0 | 2.7183 | 0.3679 | 7.5 | 1808.0 | 0.0006 |
| 1.1 | 3.0042 | 0.3329 | 8.0 | 2981.0 | 0.0003 |
| 1.2 | 3.3201 | 0.3012 | 8.5 | 4914.8 | 0.0002 |
| 1.3 | 3.6693 | 0.2725 | 9.0 | 8103.1 | 0.0001 |
| 1.4 | 4.0552 | 0.2466 | 10.0 | 22026 | 0.00005 |

**TABLE 3**   Logarithms to the Base Ten

| x | 0 | 1 | 2 | 3 | 4 | 5 | 6 | 7 | 8 | 9 |
|---|---|---|---|---|---|---|---|---|---|---|
| 1.0 | .0000 | .0043 | .0086 | .0128 | .0170 | .0212 | .0253 | .0294 | .0334 | .0374 |
| 1.1 | .0414 | .0453 | .0492 | .0531 | .0569 | .0607 | .0645 | .0682 | .0719 | .0755 |
| 1.2 | .0792 | .0828 | .0864 | .0899 | .0934 | .0969 | .1004 | .1038 | .1072 | .1106 |
| 1.3 | .1139 | .1173 | .1206 | .1239 | .1271 | .1303 | .1335 | .1367 | .1399 | .1430 |
| 1.4 | .1461 | .1492 | .1523 | .1553 | .1584 | .1614 | .1644 | .1673 | .1703 | .1732 |
| 1.5 | .1761 | .1790 | .1818 | .1847 | .1875 | .1903 | .1931 | .1959 | .1987 | .2014 |
| 1.6 | .2041 | .2068 | .2095 | .2122 | .2148 | .2175 | .2201 | .2227 | .2253 | .2279 |
| 1.7 | .2304 | .2330 | .2355 | .2380 | .2405 | .2430 | .2455 | .2480 | .2504 | .2529 |
| 1.8 | .2553 | .2577 | .2601 | .2625 | .2648 | .2672 | .2695 | .2718 | .2742 | .2765 |
| 1.9 | .2788 | .2810 | .2833 | .2856 | .2878 | .2900 | .2923 | .2945 | .2967 | .2989 |
| 2.0 | .3010 | .3032 | .3054 | .3075 | .3096 | .3118 | .3139 | .3160 | .3181 | .3201 |
| 2.1 | .3222 | .3243 | .3263 | .3284 | .3304 | .3324 | .3345 | .3365 | .3385 | .3404 |
| 2.2 | .3424 | .3444 | .3464 | .3483 | .3502 | .3522 | .3541 | .3560 | .3579 | .3598 |
| 2.3 | .3617 | .3636 | .3655 | .3674 | .3692 | .3711 | .3729 | .3747 | .3766 | .3784 |
| 2.4 | .3802 | .3820 | .3838 | .3856 | .3874 | .3892 | .3909 | .3927 | .3945 | .3962 |
| 2.5 | .3979 | .3997 | .4014 | .4031 | .4048 | .4065 | .4082 | .4099 | .4116 | .4133 |
| 2.6 | .4150 | .4166 | .4183 | .4200 | .4216 | .4232 | .4249 | .4265 | .4281 | .4298 |
| 2.7 | .4314 | .4330 | .4346 | .4362 | .4378 | .4393 | .4409 | .4425 | .4440 | .4456 |
| 2.8 | .4472 | .4487 | .4502 | .4518 | .4533 | .4548 | .4564 | .4579 | .4594 | .4609 |
| 2.9 | .4624 | .4639 | .4654 | .4669 | .4683 | .4698 | .4713 | .4728 | .4742 | .4757 |
| 3.0 | .4771 | .4786 | .4800 | .4814 | .4829 | .4843 | .4857 | .4871 | .4886 | .4900 |
| 3.1 | .4914 | .4928 | .4942 | .4955 | .4969 | .4983 | .4997 | .5011 | .5024 | .5038 |
| 3.2 | .5051 | .5065 | .5079 | .5092 | .5105 | .5119 | .5132 | .5145 | .5159 | .5172 |
| 3.3 | .5185 | .5198 | .5211 | .5224 | .5237 | .5250 | .5263 | .5276 | .5289 | .5302 |
| 3.4 | .5315 | .5328 | .5340 | .5353 | .5366 | .5378 | .5391 | .5403 | .5416 | .5428 |
| 3.5 | .5441 | .5453 | .5465 | .5478 | .5490 | .5502 | .5514 | .5527 | .5539 | .5551 |
| 3.6 | .5563 | .5575 | .5587 | .5599 | .5611 | .5623 | .5635 | .5647 | .5658 | .5670 |
| 3.7 | .5682 | .5694 | .5705 | .5717 | .5729 | .5740 | .5752 | .5763 | .5775 | .5786 |
| 3.8 | .5798 | .5809 | .5821 | .5832 | .5843 | .5855 | .5866 | .5877 | .5888 | .5899 |
| 3.9 | .5911 | .5922 | .5933 | .5944 | .5955 | .5966 | .5977 | .5988 | .5999 | .6010 |
| 4.0 | .6021 | .6031 | .6042 | .6053 | .6064 | .6075 | .6085 | .6096 | .6107 | .6117 |
| 4.1 | .6128 | .6138 | .6149 | .6160 | .6170 | .6180 | .6191 | .6201 | .6212 | .6222 |
| 4.2 | .6232 | .6243 | .6253 | .6263 | .6274 | .6284 | .6294 | .6304 | .6314 | .6325 |
| 4.3 | .6335 | .6345 | .6355 | .6365 | .6375 | .6385 | .6395 | .6405 | .6415 | .6425 |
| 4.4 | .6435 | .6444 | .6454 | .6464 | .6474 | .6484 | .6493 | .6503 | .6513 | .6522 |
| 4.5 | .6532 | .6542 | .6551 | .6561 | .6571 | .6580 | .6590 | .6599 | .6609 | .6618 |
| 4.6 | .6628 | .6637 | .6646 | .6656 | .6665 | .6675 | .6684 | .6693 | .6702 | .6712 |
| 4.7 | .6721 | .6730 | .6739 | .6749 | .6758 | .6767 | .6776 | .6785 | .6794 | .6803 |
| 4.8 | .6812 | .6821 | .6830 | .6839 | .6848 | .6857 | .6866 | .6875 | .6884 | .6893 |
| 4.9 | .6902 | .6911 | .6920 | .6928 | .6937 | .6946 | .6955 | .6964 | .6972 | .6981 |
| 5.0 | .6990 | .6998 | .7007 | .7016 | .7024 | .7033 | .7042 | .7050 | .7059 | .7067 |
| 5.1 | .7076 | .7084 | .7093 | .7101 | .7110 | .7118 | .7126 | .7135 | .7143 | .7152 |
| 5.2 | .7160 | .7168 | .7177 | .7185 | .7193 | .7202 | .7210 | .7218 | .7226 | .7235 |
| 5.3 | .7243 | .7251 | .7259 | .7267 | .7275 | .7284 | .7292 | .7300 | .7308 | .7316 |
| 5.4 | .7324 | .7332 | .7340 | .7348 | .7356 | .7364 | .7372 | .7380 | .7388 | .7396 |
| x | 0 | 1 | 2 | 3 | 4 | 5 | 6 | 7 | 8 | 9 |

**TABLE 3**  Logarithms to the Base Ten (*continued*)

| x | 0 | 1 | 2 | 3 | 4 | 5 | 6 | 7 | 8 | 9 |
|---|---|---|---|---|---|---|---|---|---|---|
| 5.5 | .7404 | .7412 | .7419 | .7427 | .7435 | .7443 | .7451 | .7459 | .7466 | .7474 |
| 5.6 | .7482 | .7490 | .7497 | .7505 | .7513 | .7520 | .7528 | .7536 | .7543 | .7551 |
| 5.7 | .7559 | .7566 | .7574 | .7582 | .7589 | .7597 | .7604 | .7612 | .7619 | .7627 |
| 5.8 | .7634 | .7642 | .7649 | .7657 | .7664 | .7672 | .7679 | .7686 | .7694 | .7701 |
| 5.9 | .7709 | .7716 | .7723 | .7731 | .7738 | .7745 | .7752 | .7760 | .7767 | .7774 |
| 6.0 | .7782 | .7789 | .7796 | .7803 | .7810 | .7818 | .7825 | .7832 | .7839 | .7846 |
| 6.1 | .7853 | .7860 | .7868 | .7875 | .7882 | .7889 | .7896 | .7903 | .7910 | .7917 |
| 6.2 | .7924 | .7931 | .7938 | .7945 | .7952 | .7959 | .7966 | .7973 | .7980 | .7987 |
| 6.3 | .7993 | .8000 | .8007 | .8014 | .8021 | .8028 | .8035 | .8041 | .8048 | .8055 |
| 6.4 | .8062 | .8069 | .8075 | .8082 | .8089 | .8096 | .8102 | .8109 | .8116 | .8122 |
| 6.5 | .8129 | .8136 | .8142 | .8149 | .8156 | .8162 | .8169 | .8176 | .8182 | .8189 |
| 6.6 | .8195 | .8202 | .8209 | .8215 | .8222 | .8228 | .8235 | .8241 | .8248 | .8254 |
| 6.7 | .8261 | .8267 | .8274 | .8280 | .8287 | .8293 | .8299 | .8306 | .8312 | .8319 |
| 6.8 | .8325 | .8331 | .8338 | .8344 | .8351 | .8357 | .8363 | .8370 | .8376 | .8382 |
| 6.9 | .8388 | .8395 | .8401 | .8407 | .8414 | .8420 | .8426 | .8432 | .8439 | .8445 |
| 7.0 | .8451 | .8457 | .8463 | .8470 | .8476 | .8482 | .8488 | .8494 | .8500 | .8506 |
| 7.1 | .8513 | .8519 | .8525 | .8531 | .8537 | .8543 | .8549 | .8555 | .8561 | .8567 |
| 7.2 | .8573 | .8579 | .8585 | .8591 | .8597 | .8603 | .8609 | .8615 | .8621 | .8627 |
| 7.3 | .8633 | .8639 | .8645 | .8651 | .8657 | .8663 | .8669 | .8675 | .8681 | .8686 |
| 7.4 | .8692 | .8698 | .8704 | .8710 | .8716 | .8722 | .8727 | .8733 | .8739 | .8745 |
| 7.5 | .8751 | .8756 | .8762 | .8768 | .8774 | .8779 | .8785 | .8791 | .8797 | .8802 |
| 7.6 | .8808 | .8814 | .8820 | .8825 | .8831 | .8837 | .8842 | .8848 | .8854 | .8859 |
| 7.7 | .8865 | .8871 | .8876 | .8882 | .8887 | .8893 | .8899 | .8904 | .8910 | .8915 |
| 7.8 | .8921 | .8927 | .8932 | .8938 | .8943 | .8949 | .8954 | .8960 | .8965 | .8971 |
| 7.9 | .8976 | .8982 | .8987 | .8993 | .8998 | .9004 | .9009 | .9015 | .9020 | .9025 |
| 8.0 | .9031 | .9036 | .9042 | .9047 | .9053 | .9058 | .9063 | .9069 | .9074 | .9079 |
| 8.1 | .9085 | .9090 | .9096 | .9101 | .9106 | .9112 | .9117 | .9122 | .9128 | .9133 |
| 8.2 | .9138 | .9143 | .9149 | .9154 | .9159 | .9165 | .9170 | .9175 | .9180 | .9186 |
| 8.3 | .9191 | .9196 | .9201 | .9206 | .9212 | .9217 | .9222 | .9227 | .9232 | .9238 |
| 8.4 | .9243 | .9248 | .9253 | .9258 | .9263 | .9269 | .9274 | .9279 | .9284 | .9289 |
| 8.5 | .9294 | .9299 | .9304 | .9309 | .9315 | .9320 | .9325 | .9330 | .9335 | .9340 |
| 8.6 | .9345 | .9350 | .9355 | .9360 | .9365 | .9370 | .9375 | .9380 | .9385 | .9390 |
| 8.7 | .9395 | .9400 | .9405 | .9410 | .9415 | .9420 | .9425 | .9430 | .9435 | .9440 |
| 8.8 | .9445 | .9450 | .9455 | .9460 | .9465 | .9469 | .9474 | .9479 | .9484 | .9489 |
| 8.9 | .9494 | .9499 | .9504 | .9509 | .9513 | .9518 | .9523 | .9528 | .9533 | .9538 |
| 9.0 | .9542 | .9547 | .9552 | .9557 | .9562 | .9566 | .9571 | .9576 | .9581 | .9586 |
| 9.1 | .9590 | .9595 | .9600 | .9605 | .9609 | .9614 | .9619 | .9624 | .9628 | .9633 |
| 9.2 | .9638 | .9643 | .9647 | .9652 | .9657 | .9661 | .9666 | .9671 | .9675 | .9680 |
| 9.3 | .9685 | .9689 | .9694 | .9699 | .9703 | .9708 | .9713 | .9717 | .9722 | .9727 |
| 9.4 | .9731 | .9736 | .9741 | .9745 | .9750 | .9754 | .9759 | .9763 | .9768 | .9773 |
| 9.5 | .9777 | .9782 | .9786 | .9791 | .9795 | .9800 | .9805 | .9809 | .9814 | .9818 |
| 9.6 | .9823 | .9827 | .9832 | .9836 | .9841 | .9845 | .9850 | .9854 | .9859 | .9863 |
| 9.7 | .9868 | .9872 | .9877 | .9881 | .9886 | .9890 | .9894 | .9899 | .9903 | .9908 |
| 9.8 | .9912 | .9917 | .9921 | .9926 | .9930 | .9934 | .9939 | .9943 | .9948 | .9952 |
| 9.9 | .9956 | .9961 | .9965 | .9969 | .9974 | .9978 | .9983 | .9987 | .9991 | .9996 |
| x | 0 | 1 | 2 | 3 | 4 | 5 | 6 | 7 | 8 | 9 |

**TABLE 4**  Natural Logarithms

| $n$ | $\ln n$ | $n$ | $\ln n$ | $n$ | $\ln n$ |
|---|---|---|---|---|---|
|  |  | 4.5 | 1.5041 | 9.0 | 2.1972 |
| 0.1 | −2.3026 | 4.6 | 1.5261 | 9.1 | 2.2083 |
| 0.2 | −1.6094 | 4.7 | 1.5476 | 9.2 | 2.2192 |
| 0.3 | −1.2040 | 4.8 | 1.5686 | 9.3 | 2.2300 |
| 0.4 | −0.9163 | 4.9 | 1.5892 | 9.4 | 2.2407 |
| 0.5 | −0.6931 | 5.0 | 1.6094 | 9.5 | 2.2513 |
| 0.6 | −0.5108 | 5.1 | 1.6292 | 9.6 | 2.2618 |
| 0.7 | −0.3567 | 5.2 | 1.6487 | 9.7 | 2.2721 |
| 0.8 | −0.2231 | 5.3 | 1.6677 | 9.8 | 2.2824 |
| 0.9 | −0.1054 | 5.4 | 1.6864 | 9.9 | 2.2925 |
| 1.0 | 0.0000 | 5.5 | 1.7047 | 10 | 2.3026 |
| 1.1 | 0.0953 | 5.6 | 1.7228 | 11 | 2.3979 |
| 1.2 | 0.1823 | 5.7 | 1.7405 | 12 | 2.4849 |
| 1.3 | 0.2624 | 5.8 | 1.7579 | 13 | 2.5649 |
| 1.4 | 0.3365 | 5.9 | 1.7750 | 14 | 2.6391 |
| 1.5 | 0.4055 | 6.0 | 1.7918 | 15 | 2.7081 |
| 1.6 | 0.4700 | 6.1 | 1.8083 | 16 | 2.7726 |
| 1.7 | 0.5306 | 6.2 | 1.8245 | 17 | 2.8332 |
| 1.8 | 0.5878 | 6.3 | 1.8405 | 18 | 2.8904 |
| 1.9 | 0.6419 | 6.4 | 1.8563 | 19 | 2.9444 |
| 2.0 | 0.6931 | 6.5 | 1.8718 | 20 | 2.9957 |
| 2.1 | 0.7419 | 6.6 | 1.8871 | 25 | 3.2189 |
| 2.2 | 0.7885 | 6.7 | 1.9021 | 30 | 3.4012 |
| 2.3 | 0.8329 | 6.8 | 1.9169 | 35 | 3.5553 |
| 2.4 | 0.8755 | 6.9 | 1.9315 | 40 | 3.6889 |
| 2.5 | 0.9163 | 7.0 | 1.9459 | 45 | 3.8067 |
| 2.6 | 0.9555 | 7.1 | 1.9601 | 50 | 3.9120 |
| 2.7 | 0.9933 | 7.2 | 1.9741 | 55 | 4.0073 |
| 2.8 | 1.0296 | 7.3 | 1.9879 | 60 | 4.0943 |
| 2.9 | 1.0647 | 7.4 | 2.0015 | 65 | 4.1744 |
| 3.0 | 1.0986 | 7.5 | 2.0149 | 70 | 4.2485 |
| 3.1 | 1.1314 | 7.6 | 2.0281 | 75 | 4.3175 |
| 3.2 | 1.1632 | 7.7 | 2.0412 | 80 | 4.3820 |
| 3.3 | 1.1939 | 7.8 | 2.0541 | 85 | 4.4427 |
| 3.4 | 1.2238 | 7.9 | 2.0669 | 90 | 4.4998 |
| 3.5 | 1.2528 | 8.0 | 2.0794 | 95 | 4.5539 |
| 3.6 | 1.2809 | 8.1 | 2.0919 | 100 | 4.6052 |
| 3.7 | 1.3083 | 8.2 | 2.1041 |  |  |
| 3.8 | 1.3350 | 8.3 | 2.1163 |  |  |
| 3.9 | 1.3610 | 8.4 | 2.1282 |  |  |
| 4.0 | 1.3863 | 8.5 | 2.1401 |  |  |
| 4.1 | 1.4110 | 8.6 | 2.1518 |  |  |
| 4.2 | 1.4351 | 8.7 | 2.1633 |  |  |
| 4.3 | 1.4586 | 8.8 | 2.1748 |  |  |
| 4.4 | 1.4816 | 8.9 | 2.1861 |  |  |

**TABLE 5**  Values of Trigonometric Functions

| ANGLE $\theta$ | | | | | | | | | |
| --- | --- | --- | --- | --- | --- | --- | --- | --- | --- |
| Degrees | Radians | $\sin\theta$ | $\csc\theta$ | $\tan\theta$ | $\cot\theta$ | $\sec\theta$ | $\cos\theta$ | | |
| 0° 00′ | .0000 | .0000 | No value | .0000 | No value | 1.000 | 1.0000 | 1.5708 | 90° 00′ |
| 10 | 029 | 029 | 343.8 | 029 | 343.8 | 000 | 000 | 679 | 50 |
| 20 | 058 | 058 | 171.9 | 058 | 171.9 | 000 | 000 | 650 | 40 |
| 30 | 087 | 087 | 114.6 | 087 | 114.6 | 000 | 1.0000 | 621 | 30 |
| 40 | 116 | 116 | 85.95 | 116 | 85.94 | 000 | .9999 | 592 | 20 |
| 50 | 145 | 145 | 68.76 | 145 | 68.75 | 000 | 999 | 563 | 10 |
| 1° 00′ | .0175 | .0175 | 57.30 | .0175 | 57.29 | 1.000 | .9998 | 1.5533 | 89° 00′ |
| 10 | 204 | 204 | 49.11 | 204 | 49.10 | 000 | 998 | 504 | 50 |
| 20 | 233 | 233 | 42.98 | 233 | 42.96 | 000 | 997 | 475 | 40 |
| 30 | 262 | 262 | 38.20 | 262 | 38.19 | 000 | 997 | 446 | 30 |
| 40 | 291 | 291 | 34.38 | 291 | 34.37 | 000 | 996 | 417 | 20 |
| 50 | 320 | 320 | 31.26 | 320 | 31.24 | 001 | 995 | 388 | 10 |
| 2° 00′ | .0349 | .0349 | 28.65 | .0349 | 28.64 | 1.001 | .9994 | 1.5359 | 88° 00′ |
| 10 | 378 | 378 | 26.45 | 378 | 26.43 | 001 | 993 | 330 | 50 |
| 20 | 407 | 407 | 24.56 | 407 | 24.54 | 001 | 992 | 301 | 40 |
| 30 | 436 | 436 | 22.93 | 437 | 22.90 | 001 | 990 | 272 | 30 |
| 40 | 465 | 465 | 21.49 | 466 | 21.47 | 001 | 989 | 243 | 20 |
| 50 | 495 | 494 | 20.23 | 495 | 20.21 | 001 | 988 | 213 | 10 |
| 3° 00′ | .0524 | .0523 | 19.11 | .0524 | 19.08 | 1.001 | .9986 | 1.5184 | 87° 00′ |
| 10 | 553 | 552 | 18.10 | 553 | 18.07 | 002 | 985 | 155 | 50 |
| 20 | 582 | 581 | 17.20 | 582 | 17.17 | 002 | 983 | 126 | 40 |
| 30 | 611 | 610 | 16.38 | 612 | 16.35 | 002 | 981 | 097 | 30 |
| 40 | 640 | 640 | 15.64 | 641 | 15.60 | 002 | 980 | 068 | 20 |
| 50 | 669 | 669 | 14.96 | 670 | 14.92 | 002 | 978 | 039 | 10 |
| 4° 00′ | .0698 | .0698 | 14.34 | .0699 | 14.30 | 1.002 | .9976 | 1.5010 | 86° 00′ |
| 10 | 727 | 727 | 13.76 | 729 | 13.73 | 003 | 974 | 981 | 50 |
| 20 | 756 | 756 | 13.23 | 758 | 13.20 | 003 | 971 | 952 | 40 |
| 30 | 785 | 785 | 12.75 | 787 | 12.71 | 003 | 969 | 923 | 30 |
| 40 | 814 | 814 | 12.29 | 816 | 12.25 | 003 | 967 | 893 | 20 |
| 50 | 844 | 843 | 11.87 | 846 | 11.83 | 004 | 964 | 864 | 10 |
| 5° 00′ | .0873 | .0872 | 11.47 | .0875 | 11.43 | 1.004 | .9962 | 1.4835 | 85° 00′ |
| 10 | 902 | 901 | 11.10 | 904 | 11.06 | 004 | 959 | 806 | 50 |
| 20 | 931 | 929 | 10.76 | 934 | 10.71 | 004 | 957 | 777 | 40 |
| 30 | 960 | 958 | 10.43 | 963 | 10.39 | 005 | 954 | 748 | 30 |
| 40 | .0989 | .0987 | 10.13 | .0992 | 10.08 | 005 | 951 | 719 | 20 |
| 50 | .1018 | .1016 | 9.839 | .1022 | 9.788 | 005 | 948 | 690 | 10 |
| 6° 00′ | .1047 | .1045 | 9.567 | .1051 | 9.514 | 1.006 | .9945 | 1.4661 | 84° 00′ |
| 10 | 076 | 074 | 9.309 | 080 | 9.255 | 006 | 942 | 632 | 50 |
| 20 | 105 | 103 | 9.065 | 110 | 9.010 | 006 | 939 | 603 | 40 |
| 30 | 134 | 132 | 8.834 | 139 | 8.777 | 006 | 936 | 573 | 30 |
| 40 | 164 | 161 | 8.614 | 169 | 8.556 | 007 | 932 | 544 | 20 |
| 50 | 193 | 190 | 8.405 | 198 | 8.345 | 007 | 929 | 515 | 10 |
| 7° 00′ | .1222 | .1219 | 8.206 | .1228 | 8.144 | 1.008 | .9925 | 1.4486 | 83° 00′ |
| | | $\cos\theta$ | $\sec\theta$ | $\cot\theta$ | $\tan\theta$ | $\csc\theta$ | $\sin\theta$ | Radians | Degrees |
| | | | | | | | | ANGLE $\theta$ | |

**TABLE 5** Values of Trigonometric Functions (*continued*)

| ANGLE $\theta$ | | | | | | | | | |
|---|---|---|---|---|---|---|---|---|---|
| Degrees | Radians | $\sin\theta$ | $\csc\theta$ | $\tan\theta$ | $\cot\theta$ | $\sec\theta$ | $\cos\theta$ | | |
| 7° 00′ | .1222 | .1219 | 8.206 | .1228 | 8.144 | 1.008 | .9925 | 1.4486 | 83° 00′ |
| 10 | 251 | 248 | 8.016 | 257 | 7.953 | 008 | 922 | 457 | 50 |
| 20 | 280 | 276 | 7.834 | 287 | 7.770 | 008 | 918 | 428 | 40 |
| 30 | 309 | 305 | 7.661 | 317 | 7.596 | 009 | 914 | 399 | 30 |
| 40 | 338 | 334 | 7.496 | 346 | 7.429 | 009 | 911 | 370 | 20 |
| 50 | 367 | 363 | 7.337 | 376 | 7.269 | 009 | 907 | 341 | 10 |
| 8° 00′ | .1396 | .1392 | 7.185 | .1405 | 7.115 | 1.010 | .9903 | 1.4312 | 82° 00′ |
| 10 | 425 | 421 | 7.040 | 435 | 6.968 | 010 | 899 | 283 | 50 |
| 20 | 454 | 449 | 6.900 | 465 | 827 | 011 | 894 | 254 | 40 |
| 30 | 484 | 478 | 765 | 495 | 691 | 011 | 890 | 224 | 30 |
| 40 | 513 | 507 | 636 | 524 | 561 | 012 | 886 | 195 | 20 |
| 50 | 542 | 536 | 512 | 554 | 435 | 012 | 881 | 166 | 10 |
| 9° 00′ | .1571 | .1564 | 6.392 | .1584 | 6.314 | 1.012 | .9877 | 1.4137 | 81° 00′ |
| 10 | 600 | 593 | 277 | 614 | 197 | 013 | 872 | 108 | 50 |
| 20 | 629 | 622 | 166 | 644 | 6.084 | 013 | 868 | 079 | 40 |
| 30 | 658 | 650 | 6.059 | 673 | 5.976 | 014 | 863 | 050 | 30 |
| 40 | 687 | 679 | 5.955 | 703 | 871 | 014 | 858 | 1.4021 | 20 |
| 50 | 716 | 708 | 855 | 733 | 769 | 015 | 853 | 1.3992 | 10 |
| 10° 00′ | .1745 | .1736 | 5.759 | .1763 | 5.671 | 1.015 | .9848 | 1.3963 | 80° 00′ |
| 10 | 774 | 765 | 665 | 793 | 576 | 016 | 843 | 934 | 50 |
| 20 | 804 | 794 | 575 | 823 | 485 | 016 | 838 | 904 | 40 |
| 30 | 833 | 822 | 487 | 853 | 396 | 017 | 833 | 875 | 30 |
| 40 | 862 | 851 | 403 | 883 | 309 | 018 | 827 | 846 | 20 |
| 50 | 891 | 880 | 320 | 914 | 226 | 018 | 822 | 817 | 10 |
| 11° 00′ | .1920 | .1908 | 5.241 | .1944 | 5.145 | 1.019 | .9816 | 1.3788 | 79° 00′ |
| 10 | 949 | 937 | 164 | .1974 | 5.066 | 019 | 811 | 759 | 50 |
| 20 | .1978 | 965 | 089 | .2004 | 4.989 | 020 | 805 | 730 | 40 |
| 30 | .2007 | .1994 | 5.016 | 035 | 915 | 020 | 799 | 701 | 30 |
| 40 | 036 | .2022 | 4.945 | 065 | 843 | 021 | 793 | 672 | 20 |
| 50 | 065 | 051 | 876 | 095 | 773 | 022 | 787 | 643 | 10 |
| 12° 00′ | .2094 | .2079 | 4.810 | .2126 | 4.705 | 1.022 | .9781 | 1.3614 | 78° 00′ |
| 10 | 123 | 108 | 745 | 156 | 638 | 023 | 775 | 584 | 50 |
| 20 | 153 | 136 | 682 | 186 | 574 | 024 | 769 | 555 | 40 |
| 30 | 182 | 164 | 620 | 217 | 511 | 024 | 763 | 526 | 30 |
| 40 | 211 | 193 | 560 | 247 | 449 | 025 | 757 | 497 | 20 |
| 50 | 240 | 221 | 502 | 278 | 390 | 026 | 750 | 468 | 10 |
| 13° 00′ | .2269 | .2250 | 4.445 | .2309 | 4.331 | 1.026 | .9744 | 1.3439 | 77° 00′ |
| 10 | 298 | 278 | 390 | 339 | 275 | 027 | 737 | 410 | 50 |
| 20 | 327 | 306 | 336 | 370 | 219 | 028 | 730 | 381 | 40 |
| 30 | 356 | 334 | 284 | 401 | 165 | 028 | 724 | 352 | 30 |
| 40 | 385 | 363 | 232 | 432 | 113 | 029 | 717 | 323 | 20 |
| 50 | 414 | 391 | 182 | 462 | 061 | 030 | 710 | 294 | 10 |
| 14° 00′ | .2443 | .2419 | 4.134 | .2493 | 4.011 | 1.031 | .9703 | 1.3265 | 76° 00′ |
| | | $\cos\theta$ | $\sec\theta$ | $\cot\theta$ | $\tan\theta$ | $\csc\theta$ | $\sin\theta$ | Radians | Degrees |
| | | | | | | | | ANGLE $\theta$ | |

**TABLE 5** Values of Trigonometric Functions (*continued*)

| ANGLE $\theta$ | | | | | | | | |
|---|---|---|---|---|---|---|---|---|
| Degrees | Radians | $\sin\theta$ | $\csc\theta$ | $\tan\theta$ | $\cot\theta$ | $\sec\theta$ | $\cos\theta$ | |
| 14° 00′ | .2443 | .2419 | 4.134 | .2493 | 4.011 | 1.031 | .9703 | 1.3265 | 76° 00′ |
| 10 | 473 | 447 | 086 | 524 | 3.962 | 031 | 696 | 235 | 50 |
| 20 | 502 | 476 | 4.039 | 555 | 914 | 032 | 689 | 206 | 40 |
| 30 | 531 | 504 | 3.994 | 586 | 867 | 033 | 681 | 177 | 30 |
| 40 | 560 | 532 | 950 | 617 | 821 | 034 | 674 | 148 | 20 |
| 50 | 589 | 560 | 906 | 648 | 776 | 034 | 667 | 119 | 10 |
| 15° 00′ | .2618 | .2588 | 3.864 | .2679 | 3.732 | 1.035 | .9659 | 1.3090 | 75° 00′ |
| 10 | 647 | 616 | 822 | 711 | 689 | 036 | 652 | 061 | 50 |
| 20 | 676 | 644 | 782 | 742 | 647 | 037 | 644 | 032 | 40 |
| 30 | 705 | 672 | 742 | 773 | 606 | 038 | 636 | 1.3003 | 30 |
| 40 | 734 | 700 | 703 | 805 | 566 | 039 | 628 | 1.2974 | 20 |
| 50 | 763 | 728 | 665 | 836 | 526 | 039 | 621 | 945 | 10 |
| 16° 00′ | .2793 | .2756 | 3.628 | .2867 | 3.487 | 1.040 | .9613 | 1.2915 | 74° 00′ |
| 10 | 822 | 784 | 592 | 899 | 450 | 041 | 605 | 886 | 50 |
| 20 | 851 | 812 | 556 | 931 | 412 | 042 | 596 | 857 | 40 |
| 30 | 880 | 840 | 521 | 962 | 376 | 043 | 588 | 828 | 30 |
| 40 | 909 | 868 | 487 | .2944 | 340 | 044 | 580 | 799 | 20 |
| 50 | 938 | 896 | 453 | .3026 | 305 | 045 | 572 | 770 | 10 |
| 17° 00′ | .2967 | .2924 | 3.420 | .3057 | 3.271 | 1.046 | .9563 | 1.2741 | 73° 00′ |
| 10 | .2996 | 952 | 388 | 089 | 237 | 047 | 555 | 712 | 50 |
| 20 | .3025 | .2979 | 357 | 121 | 201 | 048 | 546 | 683 | 40 |
| 30 | 054 | .3007 | 326 | 153 | 172 | 048 | 537 | 654 | 30 |
| 40 | 083 | 035 | 295 | 185 | 140 | 049 | 528 | 625 | 20 |
| 50 | 113 | 062 | 265 | 217 | 108 | 050 | 520 | 595 | 10 |
| 18° 00′ | .3142 | .3090 | 3.236 | .3249 | 3.078 | 1.051 | .9511 | 1.2566 | 72° 00′ |
| 10 | 171 | 118 | 207 | 281 | 047 | 052 | 502 | 537 | 50 |
| 20 | 200 | 145 | 179 | 314 | 3.018 | 053 | 492 | 508 | 40 |
| 30 | 229 | 173 | 152 | 346 | 2.989 | 054 | 483 | 479 | 30 |
| 40 | 258 | 201 | 124 | 378 | 960 | 056 | 474 | 450 | 20 |
| 50 | 287 | 228 | 098 | 411 | 932 | 057 | 465 | 421 | 10 |
| 19° 00′ | .3316 | .3256 | 3.072 | .3443 | 2.904 | 1.058 | .9455 | 1.2392 | 71° 00′ |
| 10 | 345 | 283 | 046 | 476 | 877 | 059 | 446 | 363 | 50 |
| 20 | 374 | 311 | 3.021 | 508 | 850 | 060 | 436 | 334 | 40 |
| 30 | 403 | 338 | 2.996 | 541 | 824 | 061 | 426 | 305 | 30 |
| 40 | 432 | 365 | 971 | 574 | 798 | 062 | 417 | 275 | 20 |
| 50 | 462 | 393 | 947 | 607 | 773 | 063 | 407 | 246 | 10 |
| 20° 00′ | .3491 | .3420 | 2.924 | .3640 | 2.747 | 1.064 | .9397 | 1.2217 | 70° 00′ |
| 10 | 520 | 448 | 901 | 673 | 723 | 065 | 387 | 188 | 50 |
| 20 | 549 | 475 | 878 | 706 | 699 | 066 | 377 | 159 | 40 |
| 30 | 578 | 502 | 855 | 739 | 675 | 068 | 367 | 130 | 30 |
| 40 | 607 | 529 | 833 | 772 | 651 | 069 | 356 | 101 | 20 |
| 50 | 636 | 557 | 812 | 805 | 628 | 070 | 346 | 072 | 10 |
| 21° 00′ | .3665 | .3584 | 2.790 | .3839 | 2.605 | 1.071 | .9336 | 1.2043 | 69° 00′ |
| | | $\cos\theta$ | $\sec\theta$ | $\cot\theta$ | $\tan\theta$ | $\csc\theta$ | $\sin\theta$ | Radians | Degrees |
| | | | | | | | | ANGLE $\theta$ | |

**TABLE 5**  Values of Trigonometric Functions (*continued*)

| ANGLE $\theta$ | | | | | | | | | |
|---|---|---|---|---|---|---|---|---|---|
| Degrees | Radians | $\sin \theta$ | $\csc \theta$ | $\tan \theta$ | $\cot \theta$ | $\sec \theta$ | $\cos \theta$ | | |
| 21° 00′ | .3665 | .3584 | 2.790 | .3839 | 2.605 | 1.071 | .9336 | 1.2043 | 69° 00′ |
| 10 | 694 | 611 | 769 | 872 | 583 | 072 | 325 | 1.2014 | 50 |
| 20 | 723 | 638 | 749 | 906 | 560 | 074 | 315 | 1.1985 | 40 |
| 30 | 752 | 665 | 729 | 939 | 539 | 075 | 304 | 956 | 30 |
| 40 | 782 | 692 | 709 | .3973 | 517 | 076 | 293 | 926 | 20 |
| 50 | 811 | 719 | 689 | .4006 | 496 | 077 | 283 | 897 | 10 |
| 22° 00′ | .3840 | .3746 | 2.669 | .4040 | 2.475 | 1.079 | .9272 | 1.1868 | 68° 00′ |
| 10 | 869 | 773 | 650 | 074 | 455 | 080 | 261 | 839 | 50 |
| 20 | 898 | 800 | 632 | 108 | 434 | 081 | 250 | 810 | 40 |
| 30 | 927 | 827 | 613 | 142 | 414 | 082 | 239 | 781 | 30 |
| 40 | 956 | 854 | 595 | 176 | 394 | 084 | 228 | 752 | 20 |
| 50 | 985 | 881 | 577 | 210 | 375 | 085 | 216 | 723 | 10 |
| 23° 00′ | .4014 | .3907 | 2.559 | .4245 | 2.356 | 1.086 | .9205 | 1.1694 | 67° 00′ |
| 10 | 043 | 934 | 542 | 279 | 337 | 088 | 194 | 665 | 50 |
| 20 | 072 | 961 | 525 | 314 | 318 | 089 | 182 | 636 | 40 |
| 30 | 102 | .3987 | 508 | 348 | 300 | 090 | 171 | 606 | 30 |
| 40 | 131 | .4014 | 491 | 383 | 282 | 092 | 159 | 577 | 20 |
| 50 | 160 | 041 | 475 | 417 | 264 | 093 | 147 | 548 | 10 |
| 24° 00′ | .4189 | .4067 | 2.459 | .4452 | 2.246 | 1.095 | .9135 | 1.1519 | 66° 00′ |
| 10 | 218 | 094 | 443 | 487 | 229 | 096 | 124 | 490 | 50 |
| 20 | 247 | 120 | 427 | 522 | 211 | 097 | 112 | 461 | 40 |
| 30 | 276 | 147 | 411 | 557 | 194 | 099 | 100 | 432 | 30 |
| 40 | 305 | 173 | 396 | 592 | 177 | 100 | 088 | 403 | 20 |
| 50 | 334 | 200 | 381 | 628 | 161 | 102 | 075 | 374 | 10 |
| 25° 00′ | .4363 | .4226 | 2.366 | .4663 | 2.145 | 1.103 | .9063 | 1.1345 | 65° 00′ |
| 10 | 392 | 253 | 352 | 699 | 128 | 105 | 051 | 316 | 50 |
| 20 | 422 | 279 | 337 | 734 | 112 | 106 | 038 | 286 | 40 |
| 30 | 451 | 305 | 323 | 770 | 097 | 108 | 026 | 257 | 30 |
| 40 | 480 | 331 | 309 | 806 | 081 | 109 | 013 | 228 | 20 |
| 50 | 509 | 358 | 295 | 841 | 066 | 111 | .9001 | 199 | 10 |
| 26° 00′ | .4538 | .4384 | 2.281 | .4877 | 2.050 | 1.113 | .8988 | 1.1170 | 64° 00′ |
| 10 | 567 | 410 | 268 | 913 | 035 | 114 | 975 | 141 | 50 |
| 20 | 596 | 436 | 254 | 950 | 020 | 116 | 962 | 112 | 40 |
| 30 | 625 | 462 | 241 | .4986 | 2.006 | 117 | 949 | 083 | 30 |
| 40 | 654 | 488 | 228 | .5022 | 1.991 | 119 | 936 | 054 | 20 |
| 50 | 683 | 514 | 215 | 059 | 977 | 121 | 923 | 1.1025 | 10 |
| 27° 00′ | .4712 | .4540 | 2.203 | .5095 | 1.963 | 1.122 | .8910 | 1.0996 | 63° 00′ |
| | | $\cos \theta$ | $\sec \theta$ | $\cot \theta$ | $\tan \theta$ | $\csc \theta$ | $\sin \theta$ | Radians | Degrees |
| | | | | | | | | ANGLE $\theta$ | |

**TABLE 5**  Values of Trigonometric Functions (*continued*)

| | ANGLE $\theta$ | | | | | | | | |

| Degrees | Radians | $\sin\theta$ | $\csc\theta$ | $\tan\theta$ | $\cot\theta$ | $\sec\theta$ | $\cos\theta$ | | |
|---|---|---|---|---|---|---|---|---|---|
| 27° 00′ | .4712 | .4540 | 2.203 | .5095 | 1.963 | 1.122 | .8910 | 1.0996 | 63° 00′ |
| 10 | 741 | 566 | 190 | 132 | 949 | 124 | 897 | 966 | 50 |
| 20 | 771 | 592 | 178 | 169 | 935 | 126 | 884 | 937 | 40 |
| 30 | 800 | 617 | 166 | 206 | 921 | 127 | 870 | 908 | 30 |
| 40 | 829 | 643 | 154 | 243 | 907 | 129 | 857 | 879 | 20 |
| 50 | 858 | 669 | 142 | 280 | 894 | 131 | 843 | 850 | 10 |
| 28° 00′ | .4887 | .4695 | 2.130 | .5317 | 1.881 | 1.133 | .8829 | 1.0821 | 62° 00′ |
| 10 | 916 | 720 | 118 | 354 | 868 | 134 | 816 | 792 | 50 |
| 20 | 945 | 746 | 107 | 392 | 855 | 136 | 802 | 763 | 40 |
| 30 | .4974 | 772 | 096 | 430 | 842 | 138 | 788 | 734 | 30 |
| 40 | .5003 | 797 | 085 | 467 | 829 | 140 | 774 | 705 | 20 |
| 50 | 032 | 823 | 074 | 505 | 816 | 142 | 760 | 676 | 10 |
| 29° 00′ | .5061 | .4848 | 2.063 | .5543 | 1.804 | 1.143 | .8746 | 1.0647 | 61° 00′ |
| 10 | 091 | 874 | 052 | 581 | 792 | 145 | 732 | 617 | 50 |
| 20 | 120 | 899 | 041 | 619 | 780 | 147 | 718 | 588 | 40 |
| 30 | 149 | 924 | 031 | 658 | 767 | 149 | 704 | 559 | 30 |
| 40 | 178 | 950 | 020 | 696 | 756 | 151 | 689 | 530 | 20 |
| 50 | 207 | .4975 | 010 | 735 | 744 | 153 | 675 | 501 | 10 |
| 30° 00′ | .5236 | .5000 | 2.000 | .5774 | 1.732 | 1.155 | .8660 | 1.0472 | 60° 00′ |
| 10 | 265 | 025 | 1.990 | 812 | 720 | 157 | 646 | 443 | 50 |
| 20 | 294 | 050 | 980 | 851 | 709 | 159 | 631 | 414 | 40 |
| 30 | 323 | 075 | 970 | 890 | 698 | 161 | 616 | 385 | 30 |
| 40 | 352 | 100 | 961 | 930 | 686 | 163 | 601 | 356 | 20 |
| 50 | 381 | 125 | 951 | .5969 | 675 | 165 | 587 | 327 | 10 |
| 31° 00′ | .5411 | .5150 | 1.942 | .6009 | 1.664 | 1.167 | .8572 | 1.0297 | 59° 00′ |
| 10 | 440 | 175 | 932 | 048 | 653 | 169 | 557 | 268 | 50 |
| 20 | 469 | 200 | 923 | 088 | 643 | 171 | 542 | 239 | 40 |
| 30 | 498 | 225 | 914 | 128 | 632 | 173 | 526 | 210 | 30 |
| 40 | 527 | 250 | 905 | 168 | 621 | 175 | 511 | 181 | 20 |
| 50 | 556 | 275 | 896 | 208 | 611 | 177 | 496 | 152 | 10 |
| 32° 00′ | .5585 | .5299 | 1.887 | .6249 | 1.600 | 1.179 | .8480 | 1.0123 | 58° 00′ |
| 10 | 614 | 324 | 878 | 289 | 590 | 181 | 465 | 094 | 50 |
| 20 | 643 | 348 | 870 | 330 | 580 | 184 | 450 | 065 | 40 |
| 30 | 672 | 373 | 861 | 371 | 570 | 186 | 434 | 036 | 30 |
| 40 | 701 | 398 | 853 | 412 | 560 | 188 | 418 | 1.0007 | 20 |
| 50 | 730 | 422 | 844 | 453 | 550 | 190 | 403 | .9977 | 10 |
| 33° 00′ | .5760 | .5446 | 1.836 | .6494 | 1.540 | 1.192 | .8387 | .9948 | 57° 00′ |
| | | $\cos\theta$ | $\sec\theta$ | $\cot\theta$ | $\tan\theta$ | $\csc\theta$ | $\sin\theta$ | Radians | Degrees |

| | ANGLE $\theta$ |

**TABLE 5**  Values of Trigonometric Functions (*continued*)

| ANGLE θ | | | | | | | | |
|---|---|---|---|---|---|---|---|---|
| Degrees | Radians | sin θ | csc θ | tan θ | cot θ | sec θ | cos θ | |
| 33° 00' | .5760 | .5446 | 1.836 | .6494 | 1.540 | 1.192 | .8387 | .9948 | 57° 00' |
| 10 | 789 | 471 | 828 | 536 | 530 | 195 | 371 | 919 | 50 |
| 20 | 818 | 495 | 820 | 577 | 520 | 197 | 355 | 890 | 40 |
| 30 | 847 | 519 | 812 | 619 | 511 | 199 | 339 | 861 | 30 |
| 40 | 876 | 544 | 804 | 661 | 501 | 202 | 323 | 832 | 20 |
| 50 | 905 | 568 | 796 | 703 | 492 | 204 | 307 | 803 | 10 |
| 34° 00' | .5934 | .5592 | 1.788 | .6745 | 1.483 | 1.206 | .8290 | .9774 | 56° 00' |
| 10 | 963 | 616 | 781 | 787 | 473 | 209 | 274 | 745 | 50 |
| 20 | .5992 | 640 | 773 | 830 | 464 | 211 | 258 | 716 | 40 |
| 30 | .6021 | 644 | 766 | 873 | 455 | 213 | 241 | 687 | 30 |
| 40 | 050 | 688 | 758 | 916 | 446 | 216 | 225 | 657 | 20 |
| 50 | 080 | 712 | 751 | .6959 | 437 | 218 | 208 | 628 | 10 |
| 35° 00' | .6109 | .5736 | 1.743 | .7002 | 1.428 | 1.221 | .8192 | .9599 | 55° 00' |
| 10 | 138 | 760 | 736 | 046 | 419 | 223 | 175 | 570 | 50 |
| 20 | 167 | 783 | 729 | 089 | 411 | 226 | 158 | 541 | 40 |
| 30 | 196 | 807 | 722 | 133 | 402 | 228 | 141 | 512 | 30 |
| 40 | 225 | 831 | 715 | 177 | 393 | 231 | 124 | 483 | 20 |
| 50 | 254 | 854 | 708 | 221 | 385 | 233 | 107 | 454 | 10 |
| 36° 00' | .6283 | .5878 | 1.701 | .7265 | 1.376 | 1.236 | .8090 | .9425 | 54° 00' |
| 10 | 312 | 901 | 695 | 310 | 368 | 239 | 073 | 396 | 50 |
| 20 | 341 | 925 | 688 | 355 | 360 | 241 | 056 | 367 | 40 |
| 30 | 370 | 948 | 681 | 400 | 351 | 244 | 039 | 338 | 30 |
| 40 | 400 | 972 | 675 | 445 | 343 | 247 | 021 | 308 | 20 |
| 50 | 429 | .5995 | 668 | 490 | 335 | 249 | .8004 | 279 | 10 |
| 37° 00' | .6458 | .6018 | 1.662 | .7536 | 1.327 | 1.252 | .7986 | .9250 | 53° 00' |
| 10 | 487 | 041 | 655 | 581 | 319 | 255 | 969 | 221 | 50 |
| 20 | 516 | 065 | 649 | 627 | 311 | 258 | 951 | 192 | 40 |
| 30 | 545 | 088 | 643 | 673 | 303 | 260 | 934 | 163 | 30 |
| 40 | 574 | 111 | 636 | 720 | 295 | 263 | 916 | 134 | 20 |
| 50 | 603 | 134 | 630 | 766 | 288 | 266 | 898 | 105 | 10 |
| 38° 00' | .6632 | .6157 | 1.624 | .7813 | 1.280 | 1.269 | .7880 | .9076 | 52° 00' |
| 10 | 661 | 180 | 618 | 860 | 272 | 272 | 862 | 047 | 50 |
| 20 | 690 | 202 | 612 | 907 | 265 | 275 | 844 | .9018 | 40 |
| 30 | 720 | 225 | 606 | .7954 | 257 | 278 | 826 | .8988 | 30 |
| 40 | 749 | 248 | 601 | .8002 | 250 | 281 | 808 | 959 | 20 |
| 50 | 778 | 271 | 595 | 050 | 242 | 284 | 790 | 930 | 10 |
| 39° 00' | .6807 | .6293 | 1.589 | .8098 | 1.235 | 1.287 | .7771 | .8901 | 51° 00' |
| | | cos θ | sec θ | cot θ | tan θ | csc θ | sin θ | Radians | Degrees |

| ANGLE θ |

**TABLE 5**   Values of Trigonometric Functions (*continued*)

| ANGLE $\theta$ | | | | | | | | | |
|---|---|---|---|---|---|---|---|---|---|
| Degrees | Radians | $\sin\theta$ | $\csc\theta$ | $\tan\theta$ | $\cot\theta$ | $\sec\theta$ | $\cos\theta$ | | |
| 39° 00′ | .6807 | .6293 | 1.589 | .8098 | 1.235 | 1.287 | .7771 | .8901 | 51° 00′ |
| 10 | 836 | 316 | 583 | 146 | 228 | 290 | 753 | 872 | 50 |
| 20 | 865 | 338 | 578 | 195 | 220 | 293 | 735 | 843 | 40 |
| 30 | 894 | 361 | 572 | 243 | 213 | 296 | 716 | 814 | 30 |
| 40 | 923 | 383 | 567 | 292 | 206 | 299 | 698 | 785 | 20 |
| 50 | 952 | 406 | 561 | 342 | 199 | 302 | 679 | 756 | 10 |
| 40° 00′ | .6981 | .6428 | 1.556 | .8391 | 1.192 | 1.305 | .7660 | .8727 | 50° 00′ |
| 10 | .7010 | 450 | 550 | 441 | 185 | 309 | 642 | 698 | 50 |
| 20 | 039 | 472 | 545 | 491 | 178 | 312 | 623 | 668 | 40 |
| 30 | 069 | 494 | 540 | 541 | 171 | 315 | 604 | 639 | 30 |
| 40 | 098 | 517 | 535 | 591 | 164 | 318 | 585 | 610 | 20 |
| 50 | 127 | 539 | 529 | 642 | 157 | 322 | 566 | 581 | 10 |
| 41° 00′ | .7156 | .6561 | 1.524 | .8693 | 1.150 | 1.325 | .7547 | .8552 | 49° 00′ |
| 10 | 185 | 583 | 519 | 744 | 144 | 328 | 528 | 523 | 50 |
| 20 | 214 | 604 | 514 | 796 | 137 | 332 | 509 | 494 | 40 |
| 30 | 243 | 626 | 509 | 847 | 130 | 335 | 490 | 465 | 30 |
| 40 | 272 | 648 | 504 | 899 | 124 | 339 | 470 | 436 | 20 |
| 50 | 301 | 670 | 499 | .8952 | 117 | 342 | 451 | 407 | 10 |
| 42° 00′ | .7330 | .6691 | 1.494 | .9004 | 1.111 | 1.346 | .7431 | .8378 | 48° 00′ |
| 10 | 359 | 713 | 490 | 057 | 104 | 349 | 412 | 348 | 50 |
| 20 | 389 | 734 | 485 | 110 | 098 | 353 | 392 | 319 | 40 |
| 30 | 418 | 756 | 480 | 163 | 091 | 356 | 373 | 290 | 30 |
| 40 | 447 | 777 | 476 | 217 | 085 | 360 | 353 | 261 | 20 |
| 50 | 476 | 799 | 471 | 271 | 079 | 364 | 333 | 232 | 10 |
| 43° 00′ | .7505 | .6820 | 1.466 | .9325 | 1.072 | 1.367 | .7314 | .8203 | 47° 00′ |
| 10 | 534 | 841 | 462 | 380 | 066 | 371 | 294 | 174 | 50 |
| 20 | 563 | 862 | 457 | 435 | 060 | 375 | 274 | 145 | 40 |
| 30 | 592 | 884 | 453 | 490 | 054 | 379 | 254 | 116 | 30 |
| 40 | 621 | 905 | 448 | 545 | 048 | 382 | 234 | 087 | 20 |
| 50 | 650 | 926 | 444 | 601 | 042 | 386 | 214 | 058 | 10 |
| 44° 00′ | .7679 | .6947 | 1.440 | .9657 | 1.036 | 1.390 | .7193 | .8029 | 46° 00′ |
| 10 | 709 | 967 | 435 | 713 | 030 | 394 | 173 | .7999 | 50 |
| 20 | 738 | .6988 | 431 | 770 | 024 | 398 | 153 | 970 | 40 |
| 30 | 767 | .7009 | 427 | 827 | 018 | 402 | 133 | 941 | 30 |
| 40 | 796 | 030 | 423 | 884 | 012 | 406 | 112 | 912 | 20 |
| 50 | 825 | 050 | 418 | .9942 | 006 | 410 | 092 | 883 | 10 |
| 45° 00′ | .7854 | .7071 | 1.414 | 1.000 | 1.000 | 1.414 | .7071 | .7854 | 45° 00′ |
| | | $\cos\theta$ | $\sec\theta$ | $\cot\theta$ | $\tan\theta$ | $\csc\theta$ | $\sin\theta$ | Radians | Degrees |

| | | | | ANGLE $\theta$ | |
|---|---|---|---|---|---|

# ANSWERS TO SELECTED EXERCISES

## CHAPTER 1
### FUNDAMENTAL CONCEPTS

#### 1.1 Notation and Language

1. $x^5y^2$   3. $(x+1)^3$   5. (a) $4x$
(b) $3(x^2+1)$ or $3x^2+3$
7. (a) $3a+2b$   (b) $3a^2+2b^2$
9. (a) $(2a+1)^3(2b+1)^2$
(b) $(2a+1)^3(2b+1)^2$   11. 22   13. 25
15. 36   17. 21   19. 42   21. 2   23. 6
25. (a) 256   (b) 512   (c) 81   27. $2^{34}$
29. $x+2$   31. $4(a+b^2)$   33. $x^2+y^2$
35. $x+2x^2$   37. $\frac{1}{3}(x+y+z)$
39. $\frac{1}{3}(x^2+y^2+z^2)$   41. $2xy-1$
43. area: $9\pi y^2$; circumference: $6\pi y$
45. $V=36\pi x^3$   47. (a) $V=\pi r^3$
(b) $V=64\pi\,\text{cm}^3$   49. $10.8\,\text{cm}^3$
51. $86.8\,\text{cm}^3$   53. (a) $\frac{3}{2}$   (b) $\frac{3}{2}$
55. (a) $V=12w^3$   (b) $S=40w^2$
57. (a) 1.12   (b) 1.12   59. (a) 1000
(b) $\frac{5}{8}$

#### 1.2 Properties of the Real Numbers

1. 832   3. 4410   5. (a) $A^2+2AB+B^2$
(b) $A^2-B^2$   (c) $A^2-2AB+B^2$
7. (a) $x^2+4xy+4y^2$   (b) $x^2-4y^2$
(c) $x^2-4xy+4y^2$   9. (a) $Ax^2-Ax+A$
(b) $x^3+1$
11. commutative property of addition
13. additive inverse property
15. associative property of addition
17. associative property of multiplication
19. identity property for addition
21. closure property with respect to addition
23. distributive property   25. $-23$
27. 2   29. $-1$   31. 2   33. $\frac{31}{15}$
35. $-5/x$   37. $2y^2/(x+y)$
39. $-\frac{49}{36}$   41. $\frac{6}{35}$   43. $\frac{3}{5}$   45. (a) $\frac{22}{37}$
(b) $-1$   47. $-1$   49. $\frac{5}{7}$   51. 16
53. 0   55. 1   57. 145
63. $y=-1.6$ (to two significant digits)
65.

(a)
| $x$ | 1 | 1.250 | 1.500 | 1.750 | 2.000 | 2.250 |
|---|---|---|---|---|---|---|
| $y$ | 0 | .2880 | .3704 | .3848 | .3750 | .3567 |

(b) $x=1.750$   (c) $y=.3849$

#### 1.3 Sets of Real Numbers

1. (a) nat., int., rat.   (b) int., rat.
3. (a) rat.   (b) irr.   5. (a) nat., int., rat.
(b) rat.   7. (a) rat.   (b) rat.   9. irr.
11. nat., int., rat.   13. (a) $\frac{27}{5}$   (b) $\frac{49}{9}$
15. (a) $\frac{99}{100}$   (b) 1   17. (a) $\frac{17}{10}$   (b) $\frac{16}{9}$
19.
21.

23. false   25. true   27. false
29.
(a)
(b)

31.
(a)
(b)

33.

35.

37. $\frac{19}{99}$   39. $\frac{103}{330}$   41. (a) 3.863703
(b) 3.162277   (c) 1.847759
(d) 1.414213

#### 1.4 Absolute Value

1. 3   3. 6   5. 2   7. 0   9. 0   11. 17
13. 1   15. 0   17. 25   19. $-1$   21. 0
23. 3   25. $\sqrt{2}-2$   27. $x-3$
29. $t^2+1$   31. $\sqrt{3}+4$   33. $-2x+7$
35. 1   37. $3x+11$   39. $|x-4|=8$

41. $|x-1|=\frac{1}{2}$   43. $|x-1|\geq\frac{1}{2}$
45. $|y+4|<1$   47. $|y|<3$
49. $|x^2-a^2|<M$
51.
53.
55.
57.
59.
61.
63.
(a)
(b)
(c) The interval in (b) does not include 2.
65. (a) 1   (b) $-6$   (c) $-6$

#### 1.5 Integer Exponents. Scientific Notation

1. $-10$   3. $\frac{2}{3}$   5. 2   7. $(x+y)^2$
9. $(x+y)^2+3$   11. $[\frac{1}{2}(x^2-2y^3)]^2$
13. $|x-1|^3$   15. (a) $a^{15}$   (b) $(a+1)^{15}$
(c) $(a+1)^{15}$   17. (a) $y^{11}$
(b) $(y+1)^{11}$   (c) $(y+1)^{18}$
19. (a) $x^2+3$   (b) $1/(x^2+3)$   (c) 12
21. (a) $t^6$   (b) $1/t^6$   (c) $(t^2+3)^6$
23. (a) $x^4/y^5$   (b) $y^5/x^4$   (c) $y^{10}/x^8$
25. (a) $4x^6$   (b) $16x^6$   (c) 4   27. 1
29. 1   31. $\frac{11}{100}$   33. 5   35. 1
37. $1/a^6b^3$   39. $a^4b^2/c^6$   41. $y^{12}/x^6z^{12}$
43. $y^{20}/x^6z^{16}$   45. $1/a^8b^{12}c^{16}$   47. $y^6$
49. 576   51. 12   53. $9.29\times10^7$
55. $6.68\times10^4$   57. $2.5\times10^{19}$
59. (a) $8.6688\times10^1$
(b) $1.0604772\times10^4$   (c) $8.9424\times10^4$

**61.** (a) $1 \times 10^{-9}$ (b) $1 \times 10^{-18}$
(c) $1 \times 10^{-24}$ **63.** 1 **65.** $x^2$ **67.** $9^{10}$
**69.** 8 min

## 1.6 nth Roots

**1.** false **3.** true **5.** true **7.** true
**9.** true **11.** (a) $-4$ (b) undefined
**13.** (a) $\frac{2}{5}$ (b) $-\frac{2}{5}$ **15.** (a) undefined
(b) undefined **17.** (a) $\frac{4}{3}$ (b) $-\frac{3}{5}$
**19.** (a) $-2$ (b) 2 **21.** (a) $3\sqrt{2}$
(b) $3\sqrt[3]{2}$ **23.** (a) $7\sqrt{2}$ (b) $-2\sqrt[3]{2}$
**25.** (a) $\frac{5}{2}$ (b) $\frac{2}{5}$ **27.** (a) $3\sqrt{2}$
(b) $3\sqrt[3]{2}$ **29.** (a) $-4\sqrt{2}$ (b) $5\sqrt[4]{2}$
**31.** (a) 0.3 (b) 0.2 **33.** $-14\sqrt{6}$
**35.** $2\sqrt{2}$ **37.** (a) $6x$ (b) $-6y$
**39.** (a) $ab\sqrt{ab}$ (b) $a^2b^2$
**41.** $6ab^2c^2\sqrt{2ac}$ **43.** $2ab\sqrt[4]{b}$
**45.** $3ab\sqrt{2a}$ **47.** $2a^4\sqrt[3]{2b^2}/c^3$
**49.** $a^2\sqrt[6]{5}/b$ **51.** $4\sqrt{7}/7$ **53.** $\sqrt{2}/4$
**55.** $\frac{1}{4}(1-\sqrt{5})$ or $\frac{1}{4}(\sqrt{5}-1)$
**57.** $-2-\sqrt{3}$ **59.** $61\sqrt{5}/5$ **61.** $\sqrt[3]{5}/5$
**63.** $\sqrt[4]{27}$ **65.** $\sqrt[4]{8a^3b^3}/2ab^2$
**67.** $1/(\sqrt{x}+\sqrt{5})$ **69.** $1/(\sqrt{2+h}+\sqrt{2})$
**71.** (a) Both equal 1.645751
**73.** $(a+b)/(a-b)$ **75.** $\frac{1}{4}(\sqrt{2}+2-\sqrt{6})$

## 1.7 Rational Exponents

**1.** 4 **3.** $\frac{1}{6}$ **5.** undefined **7.** 5 **9.** 2
**11.** 4 **13.** $-2$ **15.** $-10$ **17.** $\frac{1}{7}$
**19.** undefined **21.** $\frac{1}{216}$ **23.** 25 **25.** $-1$
**27.** $\frac{255}{16}$ **29.** $-\frac{28976}{243}$ **31.** $6a^{7/12}$
**33.** $2(4^{1/4})a^{1/12}$ or $2^{3/2}a^{1/12}$ **35.** $x^2+1$
**37.** (a) $2^{1/3}3^{5/6}$ (b) $\sqrt[6]{972}$
**39.** (a) $2^{7/12}3^{1/3}$ (b) $\sqrt[12]{10368}$
**41.** (a) $x^{2/3}y^{4/5}$ (b) $\sqrt[15]{x^{10}y^{12}}$
**43.** (a) $x^{(a+b)/3}$ (b) $\sqrt[3]{x^{a+b}}$
**45.** $(x+1)^{2/3}$ **47.** $(x+y)^{2/5}$ **49.** $2x^{1/6}$
**51.** $x^{1/6}y^{1/8}$ **53.** $9^{10/9}$
**55.**

| $n$ | 2 | 5 | 10 | 100 |
|---|---|---|---|---|
| $n^{1/n}$ | 1.4142 | 1.3797 | 1.2589 | 1.0471 |

| $n$ | $10^3$ | $10^4$ | $10^5$ | $10^6$ |
|---|---|---|---|---|
| $n^{1/n}$ | 1.0069 | 1.0009 | 1.0001 | 1.0000 |

**57.** (a) $2^{3/2}$ (b) $5^{1/2}$ (c) $2^{1/2}$
(d) $(\frac{1}{2})^{1/3}$ (e) $10^{1/10}$
**59.** (a) $2^{1/2}$ (b) $(\sqrt{2})^2$ **61.** $(-0.5)^{1/3}$

## 1.8 Polynomials

**1.** (a) all real numbers
(b) nonnegative real numbers
**3.** (a) all real numbers
(b) all real numbers
**5.** (a) nonnegative real numbers
(b) positive real numbers except 1
**7.** (a) degree: 3;
coefficients: $4, -2, -6, -1$
(b) degree: 2; coefficients: $a, b, c$
**9.** (a) degree: 1; coefficients: 6
(b) degree: 3; coefficients: $-1, 1, 6$
**11.** $20x^2 + 2x + 1$ **13.** $-2x - 2$
**15.** $-4x^2 - 12x - 8$ **17.** $4x^2 - 19x - 14$
**19.** $-ax^2 + 4bx$ **21.** $2x^3 - 8x^2 - 10x$
**23.** $x^2 - 3x + 2$ **25.** $2x^2 + 6x + 4$
**27.** $x^4 + 4x^3 + 3x^2$ **29.** $4x^2y^2 - 8xy + 3$
**31.** $x^2 + 2 + (1/x^2)$
**33.** $a + b + 4\sqrt{a+b} + 3$
**35.** $x - 3x^{1/2} + 2$
**37.** $y^3 - y^2 - 11y - 10$
**39.** $x^2 - 4y^2 + 4yz - z^2$
**41.** (a) $x^2 - y^2$ (b) $x^4 - 25$
**43.** (a) $A^2 - 16$ (b) $a^2 + 2ab + b^2 - 16$
**45.** (a) $x^2 - 16x + 64$
(b) $4x^4 - 20x^2 + 25$
**47.** (a) $x + 2\sqrt{xy} + y$
(b) $2x + y - 2\sqrt{x^2 + xy}$
**49.** (a) $a^3 + 3a^2 + 3a + 1$
(b) $27x^6 - 54x^4a^2 + 36x^2a^4 - 8a^6$
**51.** (a) $x^3 + 1$ (b) $x^3 + 8$
**53.** (a) quotient: $x + 7$; remainder: 6
(b) quotient: $x^2 + x + 7$; remainder: 6
(c) quotient: 1; remainder: $6x$
**55.** quotient: $x^2 - 3$; remainder: 14
**57.** quotient: $x^5 + 2x^4 + 4x^3 + 8x^2$
$+ 16x + 32$; remainder: 0
**59.** quotient: $a$; remainder: $b - ac$

**61.** (a)

| $x$ | $\dfrac{x^2-16}{x-4}$ |
|---|---|
| 3.9 | 7.9 |
| 3.99 | 7.99 |
| 3.999 | 7.999 |
| 3.9999 | 7.9999 |
| 3.99999 | 7.99999 |
| 4.1 | 8.1 |
| 4.01 | 8.01 |
| 4.001 | 8.001 |
| 4.0001 | 8.0001 |
| 4.00001 | 8.00001 |

(b) 8 **67.** (a) $ar^2 + br + c$
(b) $ar^3 + br^2 + cr + d$
(c) $ar^4 + br^3 + cr^2 + dr + e$

**69.** (a)

| $n$ | 1 | 2 | 3 | 4 | 5 | 6 |
|---|---|---|---|---|---|---|
| $n^2$ | 1 | 4 | 9 | 16 | 25 | 36 |
| $n+1$ | 2 | 3 | 4 | 5 | 6 | 7 |

(b) no

## 1.9 Factoring

**1.** (a) $(x+8)(x-8)$ (b) $7x^2(x^2+2)$
(c) $z(11+z)(11-z)$
(d) $(ab+c)(ab-c)$
**3.** (a) $(x+3)(x-1)$
(b) $(x-3)(x+1)$ (c) irreducible
(d) $(-x+3)(x+1)$ or $-(x-3)(x+1)$
**5.** (a) $(x+1)(x^2-x+1)$
(b) $(x+6)(x^2-6x+36)$
(c) $8(5-x^2)(25+5x^2+x^4)$
(d) $(4ax-5)(16a^2x^2+20ax+25)$
**7.** $2x(1-x)(1+x)$
**9.** $x^3(10+x)(10-x)$
**11.** $x^2(2x-3)(x+3)$ **13.** $x(2x-5)^2$
**15.** $(xz+t)(xz+y)$
**17.** $(a^2+b^2)(t^2-c)$
**19.** $x(x-18)(x+5)$
**21.** $(4x+3y)(x-8y)$ **23.** irreducible
**25.** $(1+x+y)(1-x-y)$
**27.** $(x-1)(x+1)(x^2+1)(x^4+1)$
**29.** $(x+1)^3$ **31.** $(3x+4)^3$
**33.** $(2a-b)(4a^2+2ab+b^2)$
$\times (2a+b)(4a^2-2ab+b^2)$
**35.** $(x-3)(x+3)(x-4)(x+4)$
**37.** irreducible **39.** $x(x+17)(x-15)$
**41.** $(x+a)(x^2-xa+a^2+1)$
**43.** $(x-y-a)(x-y+a)$
**45.** $x(7x-3)(3x+13)$
**47.** $(5-2x+3y)(5+2x-3y)$
**49.** (a) 199 (b) 296 (c) 1999
**51.** $(ax+y)(x+by)$
**53.** $(a+2)^2(9a-8)(u-2)$
**55.** $-x(x+1)^{1/2}$
**57.** $x(x+1)^{-3/2}$ or $x/(x+1)^{3/2}$
**59.** $2x(x-1)(x-2)^{-4}$ or
$2x(x-1)/(x-2)^4$
**61.** $(x-t)(x+y+z+t)$
**63.** $3(b-a)(c-b)(c-a)$
**65.** (a) $(x^2-3)^2$
(b) $(x^2-3x-3)(x^2+3x-3)$

## 1.10 Fractional Expressions

**1.** $x - 3$ **3.** $1/(x-2)(x^2+4)$
**5.** $1/(x-2)$ **7.** $3b/2a$
**9.** $(a^2+1)/(a-1)$
**11.** $(x^2+xy+y^2)/(x-y)^2$ **13.** 2
**15.** $(x^2+x)/(x+4)(x-2)$
**17.** $(x^2-xy+y^2)/(x-y)(x+y)$

19. $(x - 3y)/(x + 3y)$  21. $(4x - 2)/x^2$
23. $(36 - a^2)/6a$
25. $(4x + 11)/(x + 3)(x + 2)$
27. $(3x^2 + 6x - 6)/(x - 2)(x + 2)$
29. $(6ax^2 - 4ax + a)/(x - 1)^3$
31. $(2x^2 - 3)/(x - 3)(x + 3)(x - 2)$
33. $8/(x - 5)$  35. $2a/(a + b)$
37. $-9/(x + 5)(x - 4)^2$
39. $(-p^2 - pq - q^2)/p(2p + q)(p - 5q)$
41. 0  43. 0  45. $-1/(x + a)$  47. 1
49. $(x + 4)/(3 - 2x)$  51. $a - 1$
53. $-1/2(2 + h)$  55. $(a + x)/x^2a^2$
57. 0  59. $(a^2 - b^2)/2$
61. $(px + q)/(x - a)(x - b)$  63. 0
65. $-1$  67. (a) $\frac{1}{42}$; same

## 1.11 The Complex Number System

1.

| $i^2$ | $i^3$ | $i^4$ | $i^5$ | $i^6$ | $i^7$ | $i^8$ |
|---|---|---|---|---|---|---|
| $-1$ | $-i$ | 1 | $i$ | $-1$ | $-i$ | 1 |

3. (a) real: 4; imaginary: 5
(b) real: 4; imaginary: $-5$
(c) real: $\frac{1}{2}$; imaginary: $-1$
(d) real: 0; imaginary: 16
5. $c = 4, d = -3$  7. (a) $14 - 4i$
(b) $-4 - 8i$  9. (a) $19 - 17i$
(b) $19 - 17i$  (c) $\frac{11}{26} - \frac{23}{26}i$  (d) $\frac{11}{25} + \frac{23}{25}i$
11. (a) $11 - i$  (b) $11 - 7i$  (c) 4
13. $4 - 2i$  15. $30 - 19i$  17. 13
19. $-191 - 163i$  21. $19 - 4i$
23. $-70 + 84i$  25. $539 + 1140i$
27. $-46 + 9i$  29. $\frac{6}{97} + \frac{35}{97}i$  31. $\frac{6}{97} - \frac{35}{97}i$
33. $-\frac{5}{13} + \frac{12}{13}i$  35. $-4$  37. $\frac{1}{26} + \frac{5}{26}i$
39. $-i$  41. $12i$  43. $-3\sqrt{5}i$  45. $-35$
47. $12\sqrt{2}i$  49. (a) $(a + c) + (b + d)i$
(b) $(a - c) + (b - d)i$
(c) $(ac - bd) + (bc + ad)i$
(d) $\dfrac{ac + bd}{c^2 + d^2} + \dfrac{bc - ad}{c^2 + d^2}i$
57. real: $(2a^2 - 2b^2)/(a^2 + b^2)$;
imaginary: 0  59. 0

## Chapter 1 Review Exercises

1. $(a - 4b)(a + 4b)$
3. $(1 - a)(7 + 4a + a^2)$  5. $x(ax + b)^2$
7. $(4x + 1)(2x + 1)$
9. $a^4x^4(1 - xa)(1 + xa)(1 + x^2a^2)$
11. $(2 + a)^3$  13. $z^3(4xy + 1)(xy - 1)$
15. $(1 - x)(1 + x + x^2)(1 + x)(1 - x + x^2)$
17. $(ax + b - 2abx)(ax + b + 2abx)$
19. $(a - b)(a + b + c + ab)$  21. 8
23. 1  25. $\frac{1}{16}$  27. $\frac{1}{9}$  29. $ab^3c^4$
31. $2a^2b^4$  33. $5\sqrt[3]{2}$  35. $5ab\sqrt{6b}$
37. $|t|$  39. $2|x|$  41. (a) $x^{13/12}$
(b) $x^{12}\sqrt{x}$  43. (a) $t^{17/30}$  (b) $\sqrt[30]{t^{17}}$

45. $2\sqrt{3}$  47. $\frac{1}{3}(\sqrt{6} + \sqrt{3})$
49. $(a^2 + \sqrt{a^4 - x^4})/x^2$
51. $\frac{1}{12}(2\sqrt{3} + 3\sqrt{2} - \sqrt{30})$
53. $1/(\sqrt{t} + a)$  55. $1/(\sqrt[3]{x^2} + 2\sqrt[3]{x} + 4)$
57. $9x^2 + 6x + 1$  59. $9x^4 + 6x^2y^2 + y^4$
61. $8a^3 + 36a^2 + 54a + 27$
63. $1 - 27a^3$  65. $x - y$
67. $x + 3x^{2/3}y^{1/3} + 3x^{1/3}y^{2/3} + y$
69. $x^4 - 6x^3 + 11x^2 - 6x + 1$  71. $5 + 6i$
73. 6  75. $\frac{1}{2}(1 - \sqrt{3}i)$  77. $3\sqrt{2} - 4\sqrt{2}i$
81. (a) both are $2\sqrt{10}$  (b) 3
83. distributive property (used twice),
associative property of addition
85. distributive property  87. $|x - 6| = 2$
89. $|a - b| = 3$  91. $|x| > 10$  93. $\sqrt{6} - 2$
95. $x^4 + x^2 + 1$  97. (a) $-2x + 5$
(b) 1  (c) $2x - 5$  99. (a) true
(b) true  (c) true  (d) true
101.
103.
105. $1.4 \times 10^{-3}$  107. $1.2001 \times 10^1$
109. $-37$  111. $-12$
113. $x^3 + 2x^2 - x - 2$  115. $3x + 2$
117. quotient: $x^2 - x$; remainder: 4
119. quotient: $x^2 - 2$; remainder: 3
121.
123.
125.
(a)
(b)
127. $\frac{1259}{99}$
129.

| $x$ | $1 + \dfrac{x}{2}$ | $\sqrt{1 + x}$ |
|---|---|---|
| 0.1 | 1.05 | 1.048808 |
| 0.01 | 1.005 | 1.004987 |
| 0.001 | 1.0005 | 1.000499 |

131. (a) 0.447214 (to six decimal places)
(b) 0.447214 (to six decimal places)
137. $x^2$

## CHAPTER 2
## EQUATIONS AND INEQUALITIES

## 2.1 Linear Equations

1. $-1$  3. 3  5. $-11$  7. 12  9. 3
11. $-1$  13. $-6$  15. $-7$  17. $-3$

19. $-15$  21. 9  23. $-\frac{13}{7}$  25. 8
27. all real numbers except 5 and $-5$
29. no solution  31. no solution  33. 3
35. $6, 4$  37. $5, -\frac{10}{3}$  39. 0.31
41. $x = (b + 1)/a$  43. $x = 1$
45. $x = 1/(a + b)$  47. $x = ab/(b + c)$
49. $x = (a^2 - b^2)/2a$  51. $x = -3$
53. $x = (a^2 + ab + b^2)/(a - b)$
55. $x = b$  57. $x = (a + b)/2$
59. $r = d/(1 - dt)$
61. $h = (S - 2\pi r^2)/2\pi r$
63. $r = (S - a)/(S - l)$
67. $x = 2pq/(p + q)$

## 2.2 Quadratic Equations

1. yes  3. no  5. no  7. $6, -1$
9. $\pm 10$  11. $\frac{6}{5}$  13. $-\frac{1}{5}, \frac{3}{5}$  15. $1, -3$
17. $\frac{1}{2}, 6$  19. $-13, 12$  21. $\pm 7$
23. $\pm\frac{1}{2}$  25. $\pm 8$  27. $-1, -9$
29. $\frac{1}{2}(1 \pm \sqrt{21})$  31. $\frac{1}{4}(-3 \pm \sqrt{41})$
33. $-\frac{1}{6}, -\frac{5}{2}$  35. $\frac{1}{4}(1 \pm \sqrt{41})$
37. $\frac{1}{6}(6 \pm \sqrt{42})$  39. $-1.47, -1.53$
41. two  43. two  45. one  47. two
49. 36  51. $\pm 2\sqrt{5}$  53. $\frac{5}{2}, -4$
55. $\frac{1}{3}, \frac{1}{2}$  57. 1  59. $-3 \pm 2\sqrt{2}$
61. $x = 1, -1$
63. $A = 0, B = 0; A = 1, B = -2$
69. $x = (a + b)/ab, -2/(a + b)$

## 2.3 Applications

1. $12, -5$  3. $13, 4$  5. 21  7. 8  9. 5
11. $12, 6$  13. 78  15. 44
17. $5\,cm, 12\,cm, 13\,cm$  19. $2.5\,cm$
21. $20\,cm$  23. $4.8\,in., 5.2\,in.$
25. $10\,cm, 30\,cm, 30\,cm$  27. 8 years
29. 10 years  31. 6.6%
33. $8000 account: 7.5%;
$9000 account: 8.5%
35. $80\,cm^3$ of 10%, $120\,cm^3$ of 35%
37. 15 lb  39. 8 tons  41. 17 cm
43. $\frac{1}{2}(3 - \sqrt{5})$  45. $\frac{1}{2}(1 + \sqrt{5})$
47. (a) 6 sec  (b) 1 sec (rising), 5 sec
(falling)  49. $12\frac{3}{8}$ min ($= 12$ min 22 sec)
51. 55 mph  53. 350 mph  55. 15 miles
57. $5117  59. $\frac{1}{24}$ mile/minute
61. $\dfrac{2v_1v_2}{v_1 + v_2}$ mph  63. $\dfrac{100(d + D)}{dr + DR}$ years
65. $\dfrac{a^2 - 2ab + 2b^2}{2(a - b)}$ in.
67. $\dfrac{b}{\sqrt{a^2 - 1}}$ units $\left(\text{or } \dfrac{b\sqrt{a^2 - 1}}{a^2 - 1}\text{ units}\right)$
69. 4.62 cm  71. $ab/(a + b)$

## 2.4 Other Types of Equations

1. $0, 16$  3. $5$  5. $0, \pm 2$  7. $1, 15, -15$
9. $0, \frac{5}{2}$  11. $0, -3, 1$  13. $0, \frac{1}{4}, -6$
15. $\pm\sqrt{3}$  17. $\pm 1$  19. $\pm\sqrt{6}/3$
21. $0, \pm\sqrt{6}, \pm 2$  23. $\pm\sqrt{-\frac{1}{2} + \frac{1}{2}\sqrt{5}}$
25. $\pm\sqrt{-\frac{3}{2} + \frac{1}{2}\sqrt{17}}$  27. $-2, 1$  29. $\frac{1}{3}, \frac{1}{4}$
31. $\frac{3}{5}, 4$  33. $\pm\frac{1}{3}$  35. $\pm 27$  37. $\pm 3$
39. $1 + \sqrt[3]{7}$  41. $-4$  43. $\pm 1, \pm\frac{1}{27}$
45. $-1$  47. $\pm 2, \pm 3$  49. $-4, -\frac{20}{9}$
51. no solution  53. $2$  55. no solution
57. $\pm 1.80$  59. $0.47$  61. $a + b$
63. $a^2 b^2/(2a + b)^2$  65. $(1 \pm \sqrt{21})/2$
67. $-a/15$  69. $b^2/4a, 9a^3/b^2$
71. $\frac{1}{2}(\sqrt{3} - 9)$  73. $0.21, 1.56, -1.21,$
$-2.56$  75. $0, a(2^m - 1)/(2^m + 1)$

## 2.5 Inequalities. More about Absolute Value

1. (a)

(b)

3. (a)

(b)

5.

7. $(-\infty, -1)$  9. $[\frac{1}{3}, \infty)$  11. $(9, \infty)$
13. $[-\frac{10}{3}, \infty)$  15. $(-\infty, 1)$  17. $[0, \infty)$
19. $(-\infty, -3)$  21. $(-\infty, -5]$
23. $(-\infty, \frac{5}{2})$  25. $[-\frac{17}{23}, \infty)$  27. $[\frac{42}{5}, \infty)$
29. $[4, 6]$  31. $(-22, -10)$  33. $[-\frac{1}{2}, 1]$
35. $(3.98, 3.998)$  37. $(0.49, 0.51)$
39. $\$2000$  41. $-13° \le F \le 887°$
43. $[\frac{5}{16}, \frac{5}{8}]$  45. $(-6, 6)$
47. $(-\infty, -2] \cup [2, \infty)$  49. $(-\infty, \infty)$
51. $(0, 8)$  53. $(-\frac{22}{3}, 4)$  55. $(\frac{39}{50}, \frac{41}{50})$
57. $(-10, 14)$  59. $[\frac{2}{5}, \frac{58}{25}]$
61. (a) $(-2h, h)$  (b) $(h, -2h)$

63.

| $a$ | $b$ | $\sqrt{ab}$ (G.M.) | $\dfrac{a+b}{2}$ (A.M.) | $\sqrt{\dfrac{a^2+b^2}{2}}$ (R.M.) | LARGEST | SMALLEST |
|---|---|---|---|---|---|---|
| 1 | 2 | 1.4142 | 1.5 | 1.5811 | R.M. | G.M. |
| 1 | 3 | 1.7320 | 2.0 | 2.2361 | R.M. | G.M. |
| 1 | 4 | 2.0000 | 2.5 | 2.9155 | R.M. | G.M. |
| 2 | 3 | 2.4495 | 2.5 | 2.5495 | R.M. | G.M. |
| 3 | 4 | 3.4641 | 3.5 | 3.5355 | R.M. | G.M. |
| 9 | 10 | 9.4868 | 9.5 | 9.5131 | R.M. | G.M. |
| 99 | 100 | 99.4987 | 99.5 | 99.5012 | R.M. | G.M. |
| 999 | 1000 | 999.4999 | 999.5 | 999.5001 | R.M. | G.M. |

65. (a)

| $x$ | $y$ | $\dfrac{x}{y} + \dfrac{y}{x}$ | $\dfrac{x}{y} + \dfrac{y}{x} \ge 2$ |
|---|---|---|---|
| 1 | 1 | 2.0000 | true |
| 2 | 3 | 2.1667 | true |
| 3 | 5 | 2.2667 | true |
| 4 | 7 | 2.3214 | true |
| 5 | 9 | 2.3556 | true |
| 9 | 10 | 2.0111 | true |
| 49 | 50 | 2.0004 | true |
| 99 | 100 | 2.0001 | true |

71. (c) $x = 5$; 5 ft by 5 ft

## 2.6 More on Inequalities

1. $(-3, 2)$  3. $(-\infty, 2) \cup (9, \infty)$
5. $(-\infty, 4] \cup [5, \infty)$
7. $(-\infty, -4] \cup [4, \infty)$  9. no solution
11. $(-7, -6) \cup (0, \infty)$
13. $[-15, 0] \cup [15, \infty)$  15. $(-\infty, \infty)$
17. $(-\infty, -\frac{3}{4}) \cup (-\frac{2}{3}, 0)$
19. $(-\infty, -\frac{1}{2} - \frac{1}{2}\sqrt{5}) \cup (-\frac{1}{2} + \frac{1}{2}\sqrt{5}, \infty)$
21. $[4 - \sqrt{14}, 4 + \sqrt{14}]$
23. $[-4, -3] \cup [1, \infty)$
25. $(-\infty, -6) \cup (-5, -4)$
27. $(-\infty, -\frac{1}{3}) \cup (\frac{1}{3}, 2) \cup (2, \infty)$
29. $(-\infty, -\frac{2}{3}]$
31. $(-\infty, -\sqrt{5}] \cup [-2, 2] \cup [\sqrt{5}, \infty)$
33. $(-\infty, \frac{3}{2} - \frac{1}{2}\sqrt{11}) \cup (\frac{3}{2} + \frac{1}{2}\sqrt{11}, 4)$
35. $(-2, -1) \cup (1, \infty)$  37. $(-1, 1]$
39. $(-\infty, \frac{3}{2}) \cup [2, \infty)$
41. $(-\infty, -1) \cup (0, 9)$
43. $(-\frac{7}{2}, -\frac{4}{3}) \cup (-1, 0) \cup (1, \infty)$
45. $(-5, \infty)$  47. $[-1, 1) \cup (2, 4]$
49. $(-1, 2 - \sqrt{5}) \cup (1, 2 + \sqrt{5})$
51. $(-\frac{3}{2}, 0)$  53. $(-\infty, -1) \cup (0, \infty)$
55. (a) $(-\infty, -1] \cup [5, \infty)$
(b) $(-\infty, -1) \cup (5, \infty)$

57. $(-\infty, -2] \cup [2, \infty)$
59. $(-\infty, 0) \cup (\frac{1}{2}, \infty)$  61. $(\frac{1}{5}, \frac{4}{5})$
63. $(0, 2)$
65. (a)

| $x$ | $x^2 + 1000$ | $2x^2 + x$ | $\dfrac{x^2 + 1000}{2x^2 + x}$ |
|---|---|---|---|
| 1 | 1001 | 3 | 333.67 |
| 2 | 1004 | 10 | 100.4 |
| 5 | 1025 | 55 | 18.64 |
| 10 | 1100 | 210 | 5.24 |
| 100 | 11,000 | 20,100 | 0.55 |
| 200 | 41,000 | 80,200 | 0.51 |
| 10,000 | 100,001,000 | 200,010,000 | 0.50 |

(b) $(-\frac{1}{2}, 0) \cup (2000, \infty)$  (c) 2001
67. $(-\infty, \frac{3}{8}a + \frac{5}{8}b)$

## Chapter 2 Review Exercises

1. true  3. false  5. false  7. true
9. false  11. false  13. $\frac{1}{3}$  15. $-\frac{37}{11}$
17. $-\frac{11}{3}$  19. $5$  21. $8$  23. $-6, -2$
25. $1, -\frac{1}{3}$  27. $\frac{1}{3}, -\frac{1}{2}$  29. $-6, 4$
31. $\frac{14}{11}, -1$  33. $10, -1$  35. $\frac{2}{3}, 3$
37. $-2 \pm \sqrt{5}$  39. $\frac{1}{2}(-1 \pm \sqrt{3})$
41. $\pm\sqrt{3}$  43. $\pm 2, \pm\sqrt{3}$
45. $0, \pm\sqrt{1 + \sqrt{3}}$  47. $-6, -8$
49. $\pm\frac{27}{8}, \pm 1$  51. $6561, 256$  53. $-7$
55. $\frac{1}{4}$  57. $2$  59. $16$  61. $10$  63. $2$
65. $x = (de - b)/(a - ce)$  67. $x = a + b$
69. $x = a$  71. $x = 1/2y$
73. $x = -2/a, (a + b)/ab$
75. $x = -a, -b$
77. $x = (a-b)^2/(a+b)^2, (a+b)^2/(a-b)^2$
79. $(-\infty, 3]$  81. $(-2, 1)$  83. $(\frac{27}{10}, \frac{31}{10})$
85. $[-\frac{1}{2}, \frac{1}{2}]$  87. $(-\frac{41}{10}, -\frac{39}{10})$
89. $(-\infty, -2] \cup [3, \infty)$  91. $(-8, 5)$
93. $(3 - \sqrt{10}, 3 + \sqrt{10})$  95. $[-8, \infty)$
97. $(-5, -3) \cup (3, 5)$  99. $(-2, \infty)$
101. $(-\infty, -1) \cup [1, 9]$
103. $(-\infty, -\frac{1}{2}) \cup [\frac{5}{6}, \infty)$
105. $(-2, -1 - \frac{1}{3}\sqrt{3}]$
$\cup (-1, -1 + \frac{1}{3}\sqrt{3}] \cup (0, \infty)$
107. $(0, 25]$  109. $[-\frac{1}{2}, \frac{1}{2}]$
111. $(-\infty, 3 - 2\sqrt{2}) \cup (3 + 2\sqrt{2}, \infty)$
113. Kona: 168 lb; Colombian: 112 lb
115. $47$  117. 12:20 P.M.
119. $-1 + \sqrt{2}$ cm  121. $x > 16$ cm
123. $21, 22, 23$  125. 7.5 mph; 1 hr, 17 min
127. $\frac{1}{4}[-(a + b) + \sqrt{a^2 + 6ab + b^2}]$
129. second ball (the one thrown from 100 ft)

**131.** $\sqrt{2}/2$ cm
**135.** (c) $x = \frac{1}{2}(-1 \pm \sqrt{4a-3})$,
$\frac{1}{2}(1 \pm \sqrt{4a+1})$
(d) $x = \frac{1}{2}(-1 + \sqrt{4a-3})$

(b) midpoint of $\overline{AC}$: $(0, \frac{7}{2})$;
midpoint of $\overline{BD}$: $(0, \frac{7}{2})$   (c) same
**23.** (a) 128   (b) 96   (c) 4/3
**25.** $a = \sqrt{2}$; $b = \sqrt{3}$; $c = 2$; $d = \sqrt{5}$;
$e = \sqrt{6}$; $f = \sqrt{7}$; $g = 2\sqrt{2}$
**35.** (a) $-2 \pm 4\sqrt{2}$   (b) no; the lines
$y = 3$ and $y = 1$ are too far apart

**7.** $y$-intercept: $-6$; $x$-intercepts: $-1, 6$
**9.** $y$-intercept: $-1$; $x$-intercepts:
$\frac{1}{2}(-1 \pm \sqrt{5})$
**11.** $y$-intercept: 3; $x$-intercept: 2
**13.** $y$-intercept: $-2$; $x$-intercept: $\frac{10}{3}$
**15.** $y$-intercept: $-8$; $x$-intercept: 2
**17.** (a)

# CHAPTER 3
# COORDINATES AND GRAPHS

## 3.1 Rectangular Coordinates

**1.**

## 3.2 Graphs and Symmetry

**1.** $x$-intercept: 4; $y$-intercept: 3

(b)

**3.** (a)

(b) 6   **5.** (a) 5   (b) 13
**7.** (a) 10   (b) 9   **9.** 4
**11.** (a) $(4, \frac{1}{2})$   (b) $(-6, 7)$   **13.** (a), (b)
**15.** area is 0; the three points are collinear
**17.** (a) $(6, 5)$   (b) $(\frac{1}{2}, -\frac{3}{2})$
(c) $(1, -4)$   **19.** $(11, -8)$
**21.** (a)

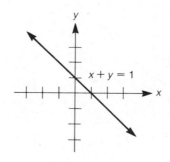

**3.** $x$-intercept: 2; $y$-intercept: $-4$

**5.** $x$-intercept: 1; $y$-intercept: 1

(c)

**19. (a)**

**(b)**

**(c)**

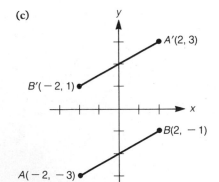

**21.** y-axis  **23.** y-axis
**25.** x-axis, y-axis, origin  **27.** none
**29.** none  **31.** x-intercept: 0;
y-intercept: 0; symmetry: y-axis

**33.** x- and y-intercepts: none; symmetry: origin

**35.** x-intercept: 0; y-intercept: 0; symmetry: y-axis

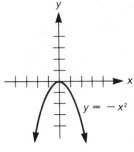

**37.** x- and y-intercepts: none; symmetry: y-axis

**39.** x-intercept: 0; y-intercept: 0; symmetry: y-axis

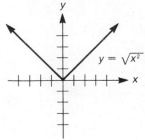

**41.** x-intercept: 1; y-intercept: 1; symmetry: none

**43.** x-intercept: 2; y-intercept: $-4$; symmetry: none

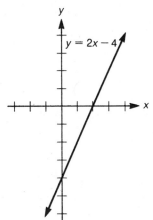

**45.** x-intercepts: $\frac{1}{4}(-1 \pm \sqrt{33})$; y-intercept: $-4$; symmetry: none

**47.** (a) $x$-intercept: 2; $y$-intercept: $-6$; symmetry: none

(b) $x$-intercept: 2; $y$-intercept: 6; symmetry: none

**49.** center: $(3,1)$; radius: 5; $x$-intercepts: $3 \pm 2\sqrt{6}$; $y$-intercepts: $5, -3$

**51.** center: $(0,0)$; radius: $\sqrt[4]{2}$; $x$-intercepts: $\pm\sqrt[4]{2}$; $y$-intercepts: $\pm\sqrt[4]{2}$

**53.** center: $(-4,3)$; radius: 1; $x$-intercepts: none; $y$-intercepts: none

**55.** center: $(-3, \frac{1}{3})$; radius: $\sqrt{2}$; $x$-intercepts: $-3 \pm \frac{1}{3}\sqrt{17}$; $y$-intercepts: none   **57.** $A = \pi c^2$; $C = 2\pi c$

**59.** $(x + \frac{1}{2})^2 + (y - 2)^2 = \frac{169}{4}$

**61.** $(x + 3)^2 + (y - 2)^2 = 64$

**63.** (a)

(b) $0 \le s \le 16$

(c) $16 \le s \le 64$

| $x$ | 0 | 0.5 | 1 | 1.5 | 2 | 2.5 | 3 | 3.5 | 4 | 4.5 | 5 | 5.5 | 6 |
|---|---|---|---|---|---|---|---|---|---|---|---|---|---|
| $y$ | 0 | $-13.4$ | $-26$ | $-37.1$ | $-46$ | $-51.9$ | $-54$ | $-51.6$ | $-44$ | $-30.4$ | $-10$ | $17.9$ | $54$ |

**65.**

In the interval $[0, 1]$, the graph of $y = x^{100}$ is practically a flat line on the $x$-axis.

**67.** (a) 500 bacteria   (b) 1.5 hours
(c) 3.5 hours   (d) $3 \le t \le 4$

**69.** $x$-intercepts: $0, \pm 3\sqrt{3}$; $y$-intercept: 0; symmetry: origin

**71.**

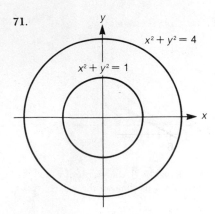

$x^2 + y^2 = 4$

$x^2 + y^2 = 1$

**73.**

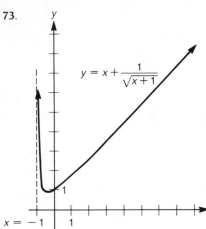

$y = x + \dfrac{1}{\sqrt{x+1}}$

$x = -1$    1

**75.** $(x + 2)^2 + (y - 3)^2 = 20$

**3.3 Slope**

**1. (a)** $-2$  **(b)** $3$  **(c)** $-\frac{7}{3}$  **(d)** $3$
**3. (a)** $1$

$(1, 1)$

$\Delta y$

$(-1, -1)$

$\Delta x$

**(b)** 0

$(-8, 5)$

$\Delta y$    $\Delta x$    $(0, 5)$

**(c)** $-1$

$(-1, 1)$

$\Delta y$

$\Delta x$    $(1, -1)$

**(d)** $-1$

$(a, b)$

$b$

$\Delta y$

$a$    $\Delta x$    $(b, a)$

$a$    $b$

**5.** $m_3 < m_2 < m_4 < m_1$   **7.** no   **9.** yes
**11. (a)** $\Delta x = 2; \Delta y = 2; m = 1$
**(b)** $\Delta x = -2; \Delta y = -2; m = 1$
**13. (a)** $m = 2$

$y = 2x - 3$

**(b)** $m = 2$

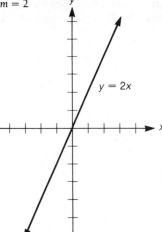

$y = 2x$

**(c)** $m = 2$

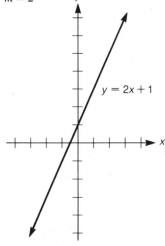

$y = 2x + 1$

**15.** $m = 5$; $m = $ speed

$d$

$5$

$d = 5t$

$1$    $t$

**17.** (a) $\frac{4}{5}$ ft/sec   (b) 0 cm/sec   (c) 8 mph
**19.** slope = 0.5; mar. cost = $0.50/album

**27.** *Monter* means "to climb."
**29.**

| $x$ | 1.1 | 1.01 | 1.001 | 1.0001 | 1.00001 |
|---|---|---|---|---|---|
| $y$ | 1.048809 | 1.004988 | 1.000500 | 1.000050 | 1.000005 |
| $\Delta x$ | 0.1 | 0.01 | 0.001 | 0.0001 | 0.00001 |
| $\Delta y$ | 0.048809 | 0.004988 | 0.0005 | 0.00005 | 0.000005 |
| slope | 0.48809 | 0.4988 | 0.5 | 0.5 | 0.5 |

The slope of $\overline{PT}$ seems to approach 0.5, so we would estimate the tangent line slope at $T$ to be 0.5.
**31.** $\left(-\frac{1}{2}, -\frac{1}{8}\right)$   **33.** $\left(8, \frac{1}{8}\right)$

## 3.4 Equations of Lines

**1.** (a) $y = -5x - 9$   (b) $y = 4x - 20$
(c) $y = \frac{1}{3}x + \frac{4}{3}$   (d) $y = -x + 1$
**3.** (a) $y = 2x$   (b) $y = -2x - 4$
(c) $y = \frac{1}{7}x - \frac{11}{7}$   **5.** $x = -3$   **7.** $y = 4$
**9.** $y$-axis   **11.** (a) $y = -4x + 7$
(b) $y = 2x + \frac{3}{2}$   (c) $y = -\frac{4}{3}x + 14$

**13.** slope: $-7$; $y$-intercept: 4

**15.** slope: 1; $y$-intercept: $-1$

**17.** slope: $-\frac{3}{2}$; $y$-intercept: 3

**19.** slope: $\frac{1}{2}$; $y$-intercept: 3

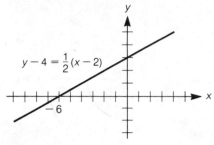

**21.** (a) $y = 4x + 11$

(b) $y = \frac{1}{2}x - \frac{5}{4}$

(c) $y = -\frac{5}{6}x + 5$

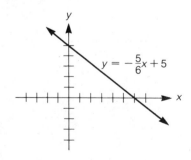

**(d)** $y = \frac{3}{4}x + \frac{3}{2}$

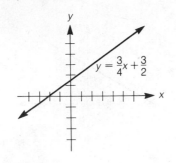

**(e)** $y = 4x - 2$

**23.** $y = x + 2$

**25.** $y - 4 = 0$

**27. (a)**

**(b)**

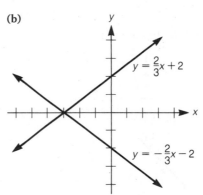

**29. (a)** neither  **(b)** parallel
**(c)** perpendicular  **(d)** perpendicular
**(e)** neither  **(f)** parallel
**31.** $y = \frac{2}{5}x + \frac{12}{5}$; $2x - 5y + 12 = 0$
**33.** $y = -\frac{4}{3}x + \frac{16}{3}$; $4x + 3y - 16 = 0$
**35.** $y = \frac{3}{5}x + 11$; $3x - 5y + 55 = 0$
**37. (a)**

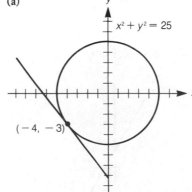

**(b)** $y = -\frac{4}{3}x - \frac{25}{3}$  **39. (a)** $530
**(b)** $538  **(c)** $8/fan; $m = 8$
**41. (a)** 5 ft/sec  **(b)** 75 feet
**(c)** 80 feet  **43.** $\frac{441}{10} = 44.1$ square units
**45.**  **(b)** $7x + 6y - 42 = 0$

**47. (a)**

**(b)** $y = -2x + 5$  **(c)** $(4, -3)$
**(d)** $3\sqrt{5}$  **51. (a)** $x = 410 - \frac{2}{3}P$
**(b)** 230 units  **(c)** $307.50 per unit
**55. (a)** $3x^2 + 3y^2 + 3x - 11y = 0$
**(b)** $3x^2 + 3y^2 + 3x - 11y = 0$
**(c)** $(0, \frac{4}{3})$  **(h)** $(-\frac{2}{3}, 2)$

## Chapter 3 Review Exercises

**1.** $y = -2x - 6$  **3.** $y = \frac{1}{4}x - \frac{5}{2}$
**5.** $y = 2x + 8$  **7.** $y = -2$  **9.** $y = x + 1$
**11.** $y = \frac{9}{4}x + \frac{7}{2}$  **13.** $y = -2x$
**15.** $y = \frac{3}{4}x$  **17.** $y = -x + 1$ or
$y = \frac{1}{2}x - 2$  **19.** 13  **21.** $(15, 6)$
**23.** $x$-axis: no; $y$-axis: yes; origin: no
**25.** $x$-axis: no; $y$-axis: no; origin: yes
**27.** $x$-axis: no; $y$-axis: no; origin: yes
**29.** $x$-axis: no; $y$-axis: no; origin: yes
**31.** $x$-axis: no; $y$-axis: no; origin: yes
**33.** $x$-axis: yes; $y$-axis: no; origin: no
**35.** $x$-axis: yes; $y$-axis: yes; origin: yes
**37.** $x$-axis: yes; $y$-axis: no; origin: no
**39.** $x$-axis: yes; $y$-axis: no; origin: no
**41.** $x$-axis: no; $y$-axis: yes; origin: no
**43.**

**45.**

**47.**

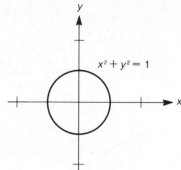

$x^2 + y^2 = 1$

**49.**

$y = |x| + 1$

**51.**

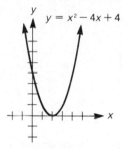

$y = x^2 - 4x + 4$

**53.**

$y = |x - 2| + 2$

**55.**

$y = -3x$

**57.**

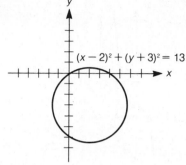

$(x - 2)^2 + (y + 3)^2 = 13$

**59.**

$y = \dfrac{x^2 - 16}{x - 4}$

**61.**

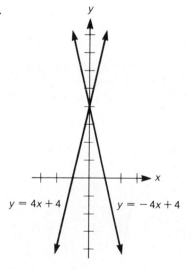

$y = 4x + 4$    $y = -4x + 4$

**63.** 19  **65.** $m = x^2 - 2x + 4$
**67.** all are $\sqrt{c^2 + b^2}$  **69.** (a) $\sqrt{130}$
(b) $-\frac{11}{3}$  (c) $(\frac{7}{2}, -\frac{1}{2})$  **71.** (a) 2
(b) $\sqrt{3}/3$  (c) $(0,0)$  **73.** $\frac{49}{6}$
**75.** (a) \$12/tire  (b) \$20,500
(c) \$20,512  **79.** 0  **85.** $13\sqrt{5}/5$
**87.** $17\sqrt{13}/13$

# CHAPTER 4
# FUNCTIONS

## 4.1 The Definition of a Function

**1.** (a) $(-\infty, \infty)$  (b) $(-\infty, \frac{1}{5}]$
(c) $(-\infty, \frac{1}{5}) \cup (\frac{1}{5}, \infty)$
**3.** (a) $(-\infty, 4) \cup (4, \infty)$
(b) $(-\infty, -2) \cup (-2, 2) \cup (2, \infty)$
(c) $(-\infty, -2) \cup (2, \infty)$
**5.** (a) $(-\infty, -1] \cup [5, \infty)$
(b) $(-\infty, -1) \cup (5, \infty)$  (c) $(-\infty, \infty)$
**7.** (a) $(-\infty, 1) \cup (1, \infty)$
(b) $(-\infty, 1) \cup (1, \infty)$  **9.** (a) $[4, \infty)$
(b) $(-\infty, \infty)$  **11.** $f, g, F, H$
**13.** (a) $\text{range}_f = \{1, 2, 3\}$; $\text{range}_g = \{2, 3\}$;
$\text{range}_F = \{1\}$; $\text{range}_H = \{1, 2\}$
(b) $\text{range}_g = \{i, j\}$; $\text{range}_F = \{i, j\}$;
$\text{range}_G = \{k\}$  **15.** (a) $y = (x - 3)^2$
(b) $y = x^2 - 3$  (c) $y = (3x)^2$
(d) $y = 3x^2$  **17.** (a) $-1$  (b) 1  (c) 5
(d) $-\frac{5}{4}$  (e) $z^2 - 3z + 1$
(f) $r^2 - x - 1$  (g) $a^2 - a - 1$
(h) $x^2 + 3x + 1$  (i) 1  (j) $4 - 3\sqrt{3}$
(k) $1 - \sqrt{2}$  (l) 2  **19.** (a) $12x^2$
(b) $6x^2$  (c) $3x^4$  (d) $9x^4$  (e) $3x^2/4$
(f) $3x^2/2$  **21.** (a) 1  (b) $-7$  (c) $-3$
(d) $-\frac{7}{18}$  (e) $-2x^2 - 4x - 1$
(f) $1 - 2x^2 - 4xh - 2h^2$
(g) $-4xh - 2h^2$  (h) $-4x - 2h$
**23.** (a) domain: $(-\infty, 2) \cup (2, \infty)$;
range: $(-\infty, 2) \cup (2, \infty)$  (b) $\frac{1}{2}$  (c) 0
(d) 1  (e) $\dfrac{2x^2 - 1}{x^2 - 2}$  (f) $\dfrac{2 - x}{1 - 2x}$

(g) $\dfrac{2a - 1}{a - 2}$  (h) $\dfrac{2x - 3}{x - 3}$
**25.** (a) $d(1) = 80$; $d(\frac{3}{2}) = 108$;
$d(2) = 128$; $d(t_0) = -16t_0^2 + 96t_0$
(b) $t = 0, 6$  (c) $\frac{1}{4}(12 \pm \sqrt{143})$
**27.** $g(3) = 1$; $g(x + 4) = |x|$
**29.** (a) $2x + h$  (b) $4x + 2h - 3$
(c) $3x^2 + 3xh + h^2$
**31.** (a) $-1/(x - 1)(a - 1)$
(b) $-1/2(x - 1)$
(c) $-1/(x - 1)(x + h - 1)$
(d) $-1/2(2 + h)$  **33.** (a) $\frac{11}{5}$
(b) $\pm 2$  (c) $0, 1$  (d) $\frac{3}{2}, -1$
**35.** (a) \$125.51  (b) \$363.76
**37.**

(a)

| $n$ | 2 | 3 | 4 |
|---|---|---|---|
| $g(n)$ | 1.4142 | 1.4422 | 1.4142 |

| $n$ | 5 | 6 | 7 | 8 |
|---|---|---|---|---|
| $g(n)$ | 1.3797 | 1.3480 | 1.3205 | 1.2968 |

**(b)** 15   **39. (a)** $f(a) = 0$; $f(2a) = \frac{1}{3}$;
$f(3a) = \frac{1}{2}$; no   **43.** $a = \frac{1}{2}, b = -\frac{3}{2}$   **49.** $-\frac{1}{3}$
**51. (a)** 0   **(b)** 1   **(c)** 2   **(d)** 6   **(e)** $-1$
**(f)** $-2$   **(g)** $-6$   **(h)** $\frac{1}{2}$   **53.** 0
**55. (a)** $x$   **(b)** $\frac{22}{7}$
**57.** $G$ assigns more than one value; let $G$
assign the nearest prime less than or equal
to $x$.   **59.** $P(1) = 41$; $P(2) = 43$; $P(3) = 47$;
$P(4) = 53$; $x = 41$

**(c)**

$h(x) = x^3 + 1$

## 4.2   The Graph of a Function

**1. (a)** positive   **(b)** $f(-2) = 4$; $f(1) = 1$;
$f(2) = 2$; $f(3) = 0$   **(c)** $f(2)$   **(d)** $-3$
**(e)** 3   **(f)** domain $= [-2, 4]$;
range $= [-2, 4]$   **3. (a)** $g(-2)$   **(b)** 5
**(c)** $f(2) - g(2)$   **(d)** $-2, 3$
**(e)** range of $f$
**5. (a)** $f(3) = 3$; $f(\pi) = 3$; $f(\frac{39}{10}) = 3$;
$f(-\frac{3}{2}) = -2$   **(b)** the integers

**7. (a)**

$f(x) = x^3$

**(b)**

$g(x) = |x^3|$

**(d)**

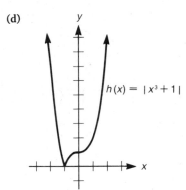

$h(x) = |x^3 + 1|$

**9. (a)** $[-2, 6]$
**(b)** $G(7) < G(0) < G(9) < G(2) < G(3)$
**(c)** $(3, 6)$   **(d)** $(7, -2)$
**11. (a)** yes   **(b)** no   **(c)** no   **(d)** yes
**13.** $m = 7$

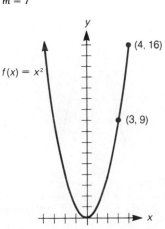

$f(x) = x^2$   (4, 16)

(3, 9)

**15.** $(1, T(1))$ and $(4, T(4))$

**17. (a)**

$k(x) = \sqrt{1 - x^2}$

**(b)**

$m(x) = \sqrt{1 - x^2}$

**(c)**

$n(x) = |x|$

**(d)**

$z(x) = x^2$

**19.**

$f(x)$

**21.**

**23.**

**25. (a)**

**(b)**

**27.**

**29.**

**31.**

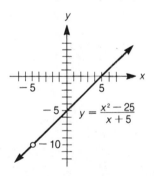

**33. (a)** $(-\infty, -5) \cup (-5, \infty)$

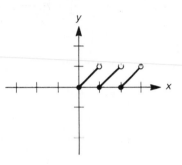

**(b)**

**35.** $t = 5$

**37.**

**39.**

**41.**

**43. (a)**

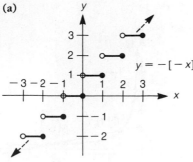

$y = -[-x]$

**(b)** The function is the same as in part (a)

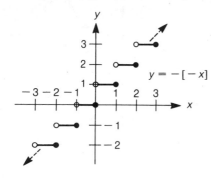

$y = -[-x]$

**45.** $B(x + h, f(x + h))$
**47. (a)**

**(b)** between 10 and 11   **(c)** $g(x) = x^3$
**(d)** **(i)** 2.1; **(ii)** 2.01; **(iii)** 2.001;
approaches 2

**49. (a)** 48 ft

$f(t) = 16t^2$

**(b)** 80 ft

$f(t) = 16t^2$

**(c)** 64 ft/sec = slope

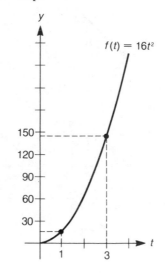

$f(t) = 16t^2$

**(d)** **(i)** 33.6; **(ii)** 32.016; **(iii)** 32.00016;
approaches 32 ft/sec

## 4.3 Techniques in Graphing

**1.**

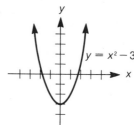

$y = x^2 - 3$

**3.**

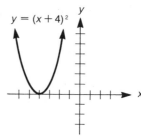

$y = (x + 4)^2$

**5.**

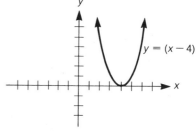

$y = (x - 4)^2$

**7.**

$y = -x^2$

**9.**

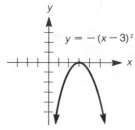

$y = -(x - 3)^2$

**11.**

$y = \sqrt{x-3}$

**13.**

$y = -\sqrt{x+1}$

**15.**

$y = \dfrac{1}{x+2} + 2$

**17.**

$y = (x-2)^3$

**19.**

$y = -x^3 + 4$

**21.**

$y = [-x]$

**23.**

$y = -[-x]$

**25.**

$y = |x-5|$

**27.**

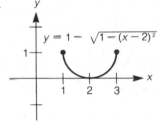

$y = 1 - |x-5|$

**29.**

$y = \dfrac{1}{x+3}$

**31.**

$y = -\dfrac{1}{x+3}$

**33.**

$y = \sqrt{1-(x-2)^2}$

**35.**

$y = 1 - \sqrt{1-(x-2)^2}$

**37. (a)**

$y = -f(x)$

**(b)**

$y = f(-x)$

**(c)**

$y = -f(-x)$

39. (a)

$y = g(-x)$

(b)

$y = -g(x)$

(c)

$y = -g(-x)$

41. (a)

$y = 10^{-x}$

(b)

$y = -10^x$

(c)

$y = -10^{-x}$

(d)

$y = 10^{x-1}$

(e)

$y = 10^x + 1$

$y = 1$

(f)

$y = -1$

$y = -10^{x-1} - 1$

43. (a)

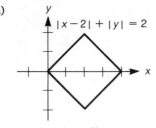

$|x-2| + |y| = 2$

(b)

$|x+2| + |y| = 2$

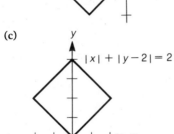

(c)

$|x| + |y-2| = 2$

(d)

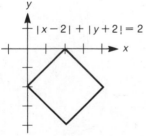

$|x| + |y+2| = 2$

(e)

$|x-2| + |y+2| = 2$

(f)

$|x+2| + |y+2| = 2$

**45. (a)**

**(b)**

**(c)**

**(d)**

**47. (a)** $(a+3,b)$  **(b)** $(a,b-3)$
**(c)** $(a+3,b-3)$  **(d)** $(a,-b)$
**(e)** $(-a,b)$  **(f)** $(-a,-b)$
**49.**

**51.**

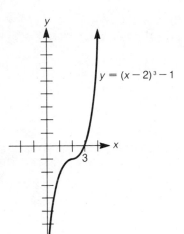

**53.** $[1,2)$

### 4.4 Methods of Combining Functions

**1. (a)** $x^2-x-7$  **(b)** $-x^2+5x+5$
**(c)** 5  **3. (a)** $x^2-2x-8$
**(b)** $-x^2+2x+8$
**5. (a)** $4x-2$  **(b)** $4x-2$  **(c)** $-4$
**7. (a)** $\dfrac{-x^4+22x^2-4x-80}{2x^3-x^2-18x+9}$  **(b)** $-\frac{80}{9}$
**9. (a)** $-x^5+9x^3+2x^2-18$
**(b)** $-x^5+9x^3-2x^2-18$  **(c)** $-24$
**11. (a)** $-6x-14$  **(b)** $-74$
**(c)** $-6x-7$  **(d)** $-67$
**13. (a)** $(f\circ g)(x)=-x$;
$(f\circ g)(-2)=2$; $(g\circ f)(x)=2-x$;
$(g\circ f)(-2)=4$
**(b)** $(f\circ g)(x)=9x^2-3x-6$;
$(f\circ g)(-2)=36$;
$(g\circ f)(x)=-3x^2+9x+14$;
$(g\circ f)(-2)=-16$
**(c)** $(f\circ g)(x)=\frac{1}{3}(1-x^4)$;
$(f\circ g)(-2)=-5$; $(g\circ f)(x)=1-\frac{1}{81}x^4$;
$(g\circ f)(-2)=\frac{65}{81}$  **(d)** $(f\circ g)(x)=2^{x^2+1}$;
$(f\circ g)(-2)=32$; $(g\circ f)(x)=2^{2x}+1$;
$(g\circ f)(-2)=\frac{17}{16}$
**(e)** $(f\circ g)(x)=3x^5-4x^2$;
$(f\circ g)(-2)=-112$;
$(g\circ f)(x)=3x^5-4x^2$;
$(g\circ f)(-2)=-112$
**(f)** $(f\circ g)(x)=x$; $(f\circ g)(-2)=-2$;
$(g\circ f)(x)=x$; $(g\circ f)(-2)=-2$
**15. (a)** $(-x+7)/6x$  **(b)** $(-t+7)/6t$
**(c)** $\frac{5}{12}$  **(d)** $(1-6x)/7$  **(e)** $(1-6y)/7$
**(f)** $-\frac{11}{7}$
**17. (a)** $M(7)=\frac{13}{5}$; $M[M(7)]=7$

**(b)** $x$  **(c)** 7  **19. (a)** 1  **(b)** $-3$
**(c)** $-1$  **(d)** 2  **(e)** 2  **(f)** $-3$
**21. (a)** $(T\circ I)(x)=4x^3-3x^2+6x-1$;
$(I\circ T)(x)=4x^3-3x^2+6x-1$
**(b)** $(G\circ I)(x)=ax^2+bx+c$;
$(I\circ G)(x)=ax^2+bx+c$
**(c)** $(f\circ I)(x)=f(x)=(I\circ f)(x)$

**23.**

| $x$ | 0 | 1 | 2 | 3 | 4 |
|---|---|---|---|---|---|
| $(f\circ g)(x)$ | 1 | 3 | 2 | und. | 2 |

| $x$ | $-1$ | 0 | 1 | 2 | 3 | 4 |
|---|---|---|---|---|---|---|
| $(g\circ f)(x)$ | 0 | 0 | 3 | 4 | 2 | und. |

**25. (a)**

**(b)**

**27. (a)** domain: $[0,\infty)$; range: $[-3,\infty)$

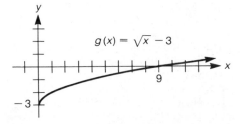

**(b)** domain: all real numbers; range: all real numbers

$f(x) = x - 1$

**(c)** domain: $[0, \infty)$; range: $[-4, \infty)$

$(f \circ g)(x) = \sqrt{x} - 4$

**(d)** domain: $[1, \infty)$; range: $[-3, \infty)$
$g[f(x)] = \sqrt{x - 1} - 3$
**(e)**

$(g \circ f)(x) = \sqrt{x - 1} - 3$

**29.** Let $f(x) = x^4$ and $g(x) = 3x - 1$; then $C(x) = (f \circ g)(x)$.
**31.** **(a)** Let $f(x) = \sqrt[3]{x}$ and $g(x) = 3x + 4$; then $F(x) = (f \circ g)(x)$.
**(b)** Let $f(x) = |x|$ and $g(x) = 2x - 3$; then $G(x) = (f \circ g)(x)$.
**(c)** Let $f(x) = x^5$ and $g(x) = ax + b$; then $H(x) = (f \circ g)(x)$.
**(d)** Let $f(x) = 1/x$ and $g(x) = \sqrt{x}$; then $T(x) = (f \circ g)(x)$.
**33.** **(a)** $(b \circ c)(x)$ **(b)** $(a \circ d)(x)$
**(c)** $(c \circ d)(x)$ **(d)** $(c \circ b)(x)$
**(e)** $(c \circ a)(x)$ **(f)** $(a \circ c)(x)$
**(g)** $(b \circ d)(x)$ or $(d \circ b)(x)$
**35.** $(C \circ f)(t) = 2\pi/(t^2 + 1)$;
$(C \circ f)(3) = \pi/5$ ft
**37.** **(a)** $100 + 450t - 25t^2$ **(b)** \$1225
**(c)** no **39.** $f(x) = (x + 6)/4$
**41.** $a = -\frac{1}{2}, b = \frac{1}{2}$ **43.** **(a)** $2x + 2a - 2$
**(b)** $4x + 4a - 4$ **45.** **(a)** $\frac{1}{2}x^2 + 1$
**(b)** $\frac{1}{2}(x + 1)^2$ **(c)** $\frac{1}{4}x^2 + 1$
**(d)** $\frac{1}{4}(x + 1)^2$ **(e)** $\frac{1}{2}(x^2 + 1)$
**47.** **(a)** $(g \circ f \circ h)(x)$ **(b)** $(h \circ g \circ f)(x)$
**(c)** $(f \circ g \circ h)(x)$ **(d)** $(h \circ f \circ g)(x)$

**49.** **(a)** $(x + 1)/\sqrt{x^2 + 2x + 2}$
**(b)** $10\sqrt{101}/101$

**51.** **(a)**

| $\circ$ | $i$ | $a$ | $b$ | $c$ |
|---|---|---|---|---|
| $i$ | $i$ | $a$ | $b$ | $c$ |
| $a$ | $a$ | $i$ | $c$ | $b$ |
| $b$ | $b$ | $c$ | $i$ | $a$ |
| $c$ | $c$ | $b$ | $a$ | $i$ |

**(b)** yes **(c)** $a^2 = i$; $b^2 = i$; $c^2 = i$; $c^3 = c$
**(d)** yes **(e)** $\{i\}$ **(g)** yes

## 4.5 Inverse Functions

**3.** **(a)** 4 **(b)** $-1$ **(c)** $\sqrt{2}$ **(d)** $t + 1$
**(e)** $f(0) = 1$; $f^{-1}(1) = 0$
**(f)** $f(-1) = -2$; $f^{-1}(-2) = -1$
**5.** **(a)** $\frac{1}{3}(x + 1)$
**(c)**

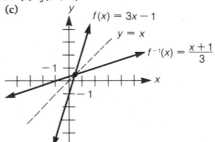

$f(x) = 3x - 1$
$y = x$
$f^{-1}(x) = \dfrac{x + 1}{3}$

**7.** **(a)** $x^2 + 1$
**(c)**

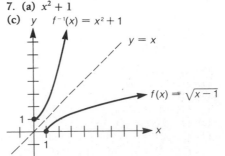

$f^{-1}(x) = x^2 + 1$
$y = x$
$f(x) = \sqrt{x - 1}$

**9.** **(a)** domain of $f$: $(-\infty, 3) \cup (3, \infty)$;
range of $f$: $(-\infty, 1) \cup (1, \infty)$
**(b)** $(3x + 2)/(x - 1)$
**(c)** domain of $f^{-1}$: $(-\infty, 1) \cup (1, \infty)$;
range of $f^{-1}$: $(-\infty, 3) \cup (3, \infty)$;
domain of $f$ = range of $f^{-1}$;
range of $f$ = domain of $f^{-1}$
**11.** $\sqrt[3]{\frac{1}{2}(x - 1)}$

**13.** $f[f^{-1}(x)] = x$; $f^{-1}[f(x)] = x$

$f(x) = 3x - 4$
$y = x$
$f^{-1}(x) = \dfrac{x + 4}{3}$
$(2, 2)$

**15.** **(a)** $\sqrt[3]{x + 1} + 3$
**(b)**

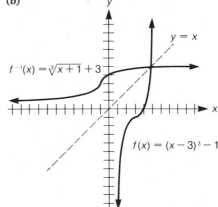

$f^{-1}(x) = \sqrt[3]{x + 1} + 3$
$y = x$
$f(x) = (x - 3)^3 - 1$

**17.** **(a)**

$y = g^{-1}(x)$

**(b)**

$y = g^{-1}(x) - 1$

**(c)**

$$y = g^{-1}(x - 1)$$

**(d)**

$$y = g^{-1}(-x)$$

**(e)**

$$y = -g^{-1}(x)$$

**(f)**

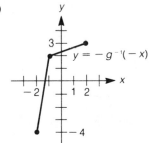

$$y = -g^{-1}(-x)$$

19. no   21. yes   23. yes   25. no

27. no   29. yes   33. (a) $\dfrac{b - dx}{cx - a}$

(b) $\dfrac{(a^2 + bc)x + (ab + bd)}{(ac + cd)x + (bc + d^2)}$   (c) $\dfrac{ab + bd}{bc + d^2}$

(d) $x$   (e) 0   35. 0   39. yes

41. $m = \frac{2}{3}$, $b = 1$

45. (b)

$$\left( \frac{2mb - (m^2 - 1)a}{m^2 + 1}, \frac{2ma + (m^2 - 1)b}{m^2 + 1} \right)$$

## 4.6 Variation

1. (a) $y = kx$   (b) $A = k/B$
3. (a) $x = kuv^2$   (b) $z = kA^2B^3$
5. (a) $F = k/r^2$   (b) $V^2 = k(U^2 + T^2)$
7. (a) $-2$   (b) $A = -2/B$   (c) $-\frac{8}{5}$
9. $-24$
11. six times the original value of $A$
13. eight times the original value of $x$
15. (a) $k = 4\pi$; $S = 4\pi r^2$
(b) $S = 12\pi \text{ cm}^2$
17. 48 times the original force
19. 122.5 m   21. (a) $k = 2.05$; $V = 2.05/P$
(b) 2.05 liters   23. (a) 6 times the original
value   (b) $\sqrt{2}$ times the original velocity
25. 110.6 lb   27. (a) F   (b) C   (c) D
(d) A   (e) B   29. $\frac{3}{4}L$
31. $3\sqrt{2}$ ft $\approx 4.24$ ft
35.   Mercury: 87.9693 days
   Venus: 224.7007 days
   Earth: 365.2564 days
   Mars: 686.9786 days
   Jupiter: 4336.6159 days
   Saturn: 10826.9994 days
   Uranus: 30873.7244 days
   Neptune: 60300.6863 days
   Pluto: 91814.3739 days
37. (a) yes   (b) no   (c) yes

## Chapter 4 Review Exercises

1. $x$-intercepts: $\pm 2$; $y$-intercept: 4

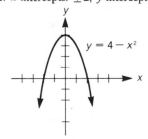

$$y = 4 - x^2$$

3. $x$-intercept: $-1$; $y$-intercept: none

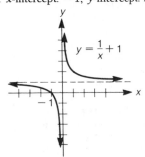

$$y = \frac{1}{x} + 1$$

5. $x$-intercept: $-2$; $y$-intercept: 2

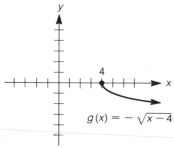

$$y = \frac{1}{x + 1} + 1$$

7. $x$-intercept: 4; $y$-intercept: none

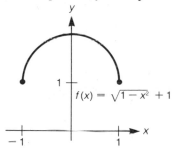

$$g(x) = -\sqrt{x - 4}$$

9. $x$-intercept: none; $y$-intercept: 2

$$f(x) = \sqrt{1 - x^2} + 1$$

11. $x$-intercept: $-16$; $y$-intercept: 4

$$y = 4 - \sqrt{-x}$$

13. $x$-intercept: $-3$; $y$-intercept: 9

$$(f \circ g)(x) = (x + 3)^2$$

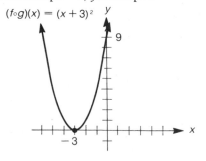

**15.** *x*-intercept: $-1$; *y*-intercept: 1

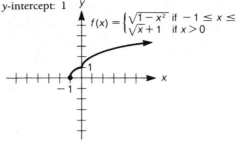

$$f(x) = \begin{cases} \sqrt{1-x^2} & \text{if } -1 \le x \le 0 \\ \sqrt{x} + 1 & \text{if } x > 0 \end{cases}$$

**17.** *x*-intercept: 1, 3; *y*-intercept: 1

$$f(x) = \begin{cases} |x-1| & \text{if } 0 \le x \le 2 \\ |x-3| & \text{if } 2 < x \le 4 \end{cases}$$

**19.** *x*-intercept: $\pm 1$; *y*-intercept: $\pm 1$

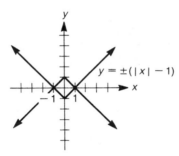

$y = \pm(|x| - 1)$

**21.** *x*-intercept: $\sqrt[3]{2}$; *y*-intercept: $-2$

$g^{-1}(x) = x^3 - 2$

**23.** *x*-intercept: none; *y*-intercept: none

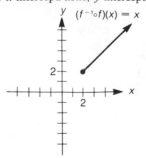

$(f^{-1} \circ f)(x) = x$

**25.** $-\frac{1}{2}$   **27.** $(0, -1)$   **29.** $(-\infty, \infty)$
**31.** $(-\infty, -\frac{3}{2}) \cup (-\frac{3}{2}, \frac{1}{3}) \cup (\frac{1}{3}, \infty)$
**33.** $(-\infty, \infty)$   **35.** $[-\sqrt{5}, \sqrt{5}]$
**37.** all noninteger real numbers
**39.** $(-\infty, 2) \cup (2, \infty)$
**41.** $(-\infty, \frac{2}{5}) \cup (\frac{2}{5}, \infty)$
**43.** $(-\infty, -1) \cup (-1, \infty)$
**45.** $(g \circ f)(x)$   **47.** $(g \circ G)(x)$
**49.** $(F \circ g \circ g)(x)$
**51.** $(f \circ G \circ g \circ g \circ g)(x)$   **53.** $2 + \sqrt{2}$
**55.** $t^2 - t$   **57.** $1 - 4x$   **59.** $1 - 2x - 2h$
**61.** $x^2 + x - 1$   **63.** $2xh + h^2 - h$
**65.** $x - 1$   **67.** $x^4 - 2x^3 + x$   **69.** $-11$
**71.** $-6x^2 + 4x + 1$   **73.** $-2$
**75.** $7/(x+4)(a+4)$   **77.** $x$   **79.** $\frac{13}{3}$

**81.** $(1-x)/2$   **83.** $-\frac{1}{2}$   **85.** $x + 1$
**87.** domain of $f$: $[-6, 8]$; range of $f$: $[-2, 4]$
**89.** $f(-\frac{5}{2})$   **91.** 1   **93.** $[-6, -5]$
**95.** no   **97.** $-2$   **99.** $[0, 4]$   **101.** 5
**103. (a)** 2   **(b)** 3   **(c)** 2   **(d)** 3
**105.** $-4$   **107.** $[3, 5]$   **109. (a)** $(-6, 5)$
**(b)** $(0, 0)$   **(c)** $(7, 5)$   **(d)** $(0, 5)$
**(e)** $(10, 0)$   **(f)** $(-10, 0)$   **111.** $\frac{9}{16}$
**113.** increased by 5.41%   **115.** 3
**117. (a)** $F[G(x)] = 1/x^2$; $G[F(x)] = 1/x^2$
**(b)** no   **119. (a)** E   **(b)** C   **(c)** L
**(d)** A   **(e)** J   **(f)** G   **(g)** B   **(h)** M
**(i)** K   **(j)** D   **(k)** I   **(l)** H   **(m)** F

# CHAPTER 5
# POLYNOMIAL FUNCTIONS AND RATIONAL FUNCTIONS

## 5.1 Linear and Quadratic Functions

**1.** $f(x) = \frac{2}{3}x + \frac{2}{3}$   **3.** $g(x) = \sqrt{2}x$
**5.** $f(x) = x - \frac{7}{2}$   **7.** $f(x) = \sqrt{3}$   **9.** yes
**11.** $V(t) = -2375t + 20{,}000$
**13. (a)** $V(t) = -12{,}000t + 60{,}000$

**(b)**

| END OF YEAR | YEARLY DEPRECIATION | ACCUMULATED DEPRECIATION | VALUE (V) |
|---|---|---|---|
| 0 | 0 | 0 | 60,000 |
| 1 | 12,000 | 12,000 | 48,000 |
| 2 | 12,000 | 24,000 | 36,000 |
| 3 | 12,000 | 36,000 | 24,000 |
| 4 | 12,000 | 48,000 | 12,000 |
| 5 | 12,000 | 60,000 | 0 |

**15.** vertex: $(-2, 0)$; axis: $x = -2$;
minimum: 0 (at $x = -2$); *x*-intercept: $-2$;
*y*-intercept: 4

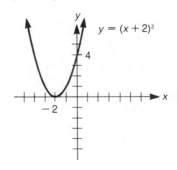

$y = (x+2)^2$

**17.** vertex: $(-2, 0)$; axis: $x = -2$;
minimum: 0 (at $x = -2$); $x$-intercept: $-2$;
$y$-intercept: 8

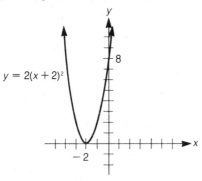

$y = 2(x+2)^2$

**19.** vertex: $(-2, 4)$; axis: $x = -2$;
maximum: 4 (at $x = -2$);
$x$-intercepts: $-2 \pm \sqrt{2}$; $y$-intercept: $-4$

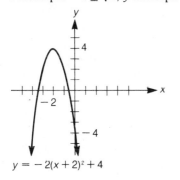

$y = -2(x+2)^2 + 4$

**21.** vertex: $(2, -4)$; axis: $x = 2$;
minimum: $-4$ (at $x = 2$); $x$-intercepts: 0, 4;
$y$-intercept: 0

$f(x) = x^2 - 4x$

**23.** vertex: $(0, 1)$; axis: $x = 0$;
maximum: 1 (at $x = 0$); $x$-intercepts: $\pm 1$;
$y$-intercept: 1

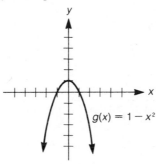

$g(x) = 1 - x^2$

**25.** vertex: $(1, -4)$; axis: $x = 1$;
minimum: $-4$ (at $x = 1$);
$x$-intercepts: 3, $-1$; $y$-intercept: $-3$

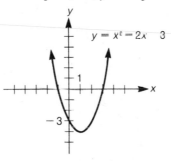

$y = x^2 - 2x - 3$

**27.** vertex: $(3, 11)$; axis: $x = 3$;
maximum: 11 (at $x = 3$);
$x$-intercepts: $3 \pm \sqrt{11}$; $y$-intercept: 2

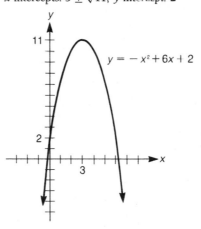

$y = -x^2 + 6x + 2$

**29.** (a) $x = 1$; minimum
(b) $x = -\frac{1}{16}$; minimum
(c) $x = \frac{1}{16}$; maximum   (d) $x = \frac{49}{8}$;
maximum   (e) $x = 0$; minimum
(f) $x = \frac{3}{4}$; maximum   **31.** 5 units
**33.** (a) minimum = 8 at $x = 3$
(b) minimum = 4 at $x = 3$
(c) minimum = 64 at $x = \pm\sqrt{3}$
**35.** (a) maximum = 4 at $x = 2$
(b) maximum = $2\sqrt[3]{2}$ at $x = 2$
(c) maximum = 16 at $x = \pm\sqrt{2}$
**37.** $y = 7.91x - 10.88$
**39.** $V(t) = -4{,}050t + 26{,}450$
**41.** $y = (x + 4)^2 + 3 = x^2 + 8x + 19$
**43.** $y = (x - 1)^2 = x^2 - 2x + 1$
**45.** $(x + 3)^2 + (y + 4)^2 = 25$
**47.** $x = \frac{1}{2}(a + b)$
**49.** (a) $p(x) = -2x + 500$
(b) $R(x) = -2x^2 + 500x$
**51.** (a) 233.0 sec; 0.82% error
(b) 227.6 sec; 0.52% error
(c) 225.19 sec; 0.36% error
(d) 208.42 sec; yes: 0.75% error

## 5.2  Applied Functions: Setting up Equations

**1.** (a) $P(x) = 2x + 2\sqrt{144 - x^2}$
(b) $A(x) = x\sqrt{144 - x^2}$
**3.** (a) $D(x) = \sqrt{x^4 + 3x^2 + 1}$
(b) $M(x) = (x^2 + 1)/x$
**5.** (a) $A(y) = \pi y^2/4$   (b) $A(y) = \pi^2 y^2/16$
**7.** (a) $P(x) = 16x - x^2$
(b) $S(x) = 2x^2 - 32x + 256$
(c) $D(x) = x^3 - (16 - x)^3$ or
$D(x) = (16 - x)^3 - x^3$
(d) $A(x) = 8$; the answer does not depend
on $x$   **9.** $R(x) = -\frac{1}{4}x^2 + 8x$
**11.**

(a)

| $x$ | 1 | 2 | 3 | 4 | 5 | 6 | 7 |
|---|---|---|---|---|---|---|---|
| $P(x)$ | 17.88 | 19.49 | 20.83 | 21.86 | 22.49 | 22.58 | 21.75 |

(b) 22.58, corresponding to $x = 6$
(c) $P \approx 22.63$

**13.** (a) $h(s) = \sqrt{3}s$  (b) $A(s) = \sqrt{3}s^2$
(c) $4\sqrt{3}$ cm  (d) $\frac{1}{4} \cdot 25\sqrt{3}$ in.$^2$
**15.** $V(r) = 2\pi r^3$  **17.** (a) $h(r) = 12/r^2$
(b) $S(r) = 2\pi r^2 + (24\pi/r)$
**19.** $V(S) = S\sqrt{S\pi}/6\pi$
**21.** $A(x) = \frac{1}{2}x\sqrt{400 - x^2}$
**23.** $d(x) = (x + 4)\sqrt{x^2 + 25}/x$
**25.** $x = 25$ cm; $L = 25$ cm

| $x$ | 5 | 10 | 20 | 24 | 24.8 | 24.9 | 25 | 25.1 | 25.2 | 45 |
|---|---|---|---|---|---|---|---|---|---|---|
| $A(x)$ | 225 | 400 | 600 | 624 | 624.96 | 624.99 | 625 | 624.99 | 624.96 | 225 |

**27.** (a) Table 1:  $x = 2$ yields the largest area

| $x$ | 1 | 2 | 3 | 4 |
|---|---|---|---|---|
| $A$ | 7.5 | 12 | 10.5 | 0 |

Table 2:  $x = 2.25$ yields the largest area

| $x$ | 1.75 | 2.00 | 2.25 | 2.50 | 2.75 |
|---|---|---|---|---|---|
| $A$ | 11.3203 | 12.0000 | 12.3047 | 12.1875 | 11.6016 |

Table 3:  $x = 2.30$ yields the largest area

| $x$ | 2.15 | 2.20 | 2.25 | 2.30 | 2.35 |
|---|---|---|---|---|---|
| $A$ | 12.2308 | 12.2760 | 12.3047 | 12.3165 | 12.3111 |

(b) $x = 2.30$; area $= 12.3168$
**29.** (a) $V(r) = \sqrt{3}\pi r^3/3$  (b) $S(r) = 2\pi r^2$
**31.** (a) $r(h) = 3h/\sqrt{h^2 - 9}$
(b) $h(r) = 3r/\sqrt{r^2 - 9}$
**33.** $A(x) = [4x^2 + \pi(14 - x)^2]/16\pi$
**35.** $A(r) = r(1 - 4\pi r)/4$
**37.** $A(x) = \pi x^2/3$
**39.** $A(h) = \pi R^2 - h\sqrt{2Rh - h^2}$
**41.** $V(x) = 4x^3 - 28x^2 + 48x$
**43.** (a) $A(r) = 32r - 2r^2 - \frac{1}{2}\pi r^2$
(b) It will open downward and pass
through the origin.
**45.** (a) $y(s) = 3s/\sqrt{1 - s^2}$
(b) $s(y) = y/\sqrt{y^2 + 9}$
(c) $z(s) = 3/\sqrt{1 - s^2}$
(d) $s(z) = \sqrt{z^2 - 9}/z$
**47.** (a) $m = (a^2 + 1)/a$
**49.** $A(x) = 2x - \frac{1}{4}x^2$
**51.** $A(m) = (2m^2 - 8m + 8)/(m^2 - 4m)$
**53.** $A(m) = -(ma - b)^2/2m$

## 5.3 Maximum and Minimum Problems

**1.** $\frac{25}{4}$  **3.** $\frac{1}{2}$  **5.** $\frac{25}{4}$ m by $\frac{25}{4}$ m
**7.** 1250 in.$^2$  **9.** (a) 18  (b) $\frac{23}{4}$  (c) $\frac{47}{8}$
(d) $\frac{95}{16}$  **11.** (a) 16 ft; 12 ft
(b) maximum height is 16 ft at $t = 1$ sec
(c) $\frac{7}{4}$ sec, $\frac{1}{4}$ sec  **13.** $(\frac{7}{2}, 1 + \frac{1}{2}\sqrt{6})$;
minimum distance is $\sqrt{7}/2$  **15.** (a) $\frac{1}{2}$

(b) $\frac{1}{4}$  **17.** 125 ft by 250 ft  **19.** $x = 40$
**21.** $x = 60$; revenue $= \$900$; price $= \$15$
**23.** (a) $\frac{36}{13}$  (b) $6\sqrt{13}/13$; this is the
square root of the answer in part (a).
**25.** (a) $\frac{225}{2}$  **27.** $x = \sqrt{2}/2$; area is $\frac{1}{2}$
**29.** $\frac{49}{12}$  **33.** 100 yd by 150 yd
**37.** (a) $p(x) = -2x + 500$
(b) maximum revenue is $\$31,250$; $p = \$250$
**39.** $\pm\sqrt{2}/2$  **41.** $\pm\sqrt{3}$  **43.** $b = \sqrt{2}$
**47.** (a) $(3.5, 3.43)$

| $x$ | 0.5 | 1 | 1.5 | 2 | 2.5 | 3 | 3.5 |
|---|---|---|---|---|---|---|---|
| $y$ | 24 | 12 | 8 | 6 | 4.8 | 4 | 3.43 |
| $xy$ | 12 | 12 | 12 | 12 | 12 | 12 | 12 |
| $x + y$ | 24.5 | 13 | 9.5 | 8 | 7.3 | 7 | 6.93 |

(b) $2\sqrt{3}, 2\sqrt{3}$
**49.** (a) $x = 0.7$

| $x$ | 0.4 | 0.5 | 0.6 | 0.7 | 0.8 | 0.9 |
|---|---|---|---|---|---|---|
| $f(x)$ | $-0.3500$ | $-0.5000$ | $-0.5667$ | $-0.5857$ | $-0.5750$ | $-0.5444$ |

$x = 0.7$

| $x$ | 0.68 | 0.69 | 0.70 | 0.71 | 0.72 | 0.73 |
|---|---|---|---|---|---|---|
| $f(x)$ | $-0.5847$ | $-0.5854$ | $-0.5857$ | $-0.5858$ | $-0.5856$ | $-0.5851$ |

(b) $x = \sqrt{2}/2$
(c) $f(\sqrt{2}/2) = -2 + \sqrt{2} \approx -0.585787$

## 5.4 Polynomial Functions

**1.** yes  **3.** yes  **5.** no
**7.** $x$-intercept: none; $y$-intercept: 5

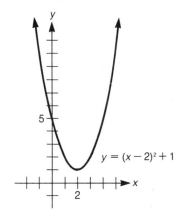

$y = (x - 2)^2 + 1$

**9.** $x$-intercept: 1; $y$-intercept: $-1$

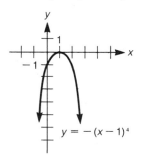

$y = -(x - 1)^4$

**11.** $x$-intercept: $4 + \sqrt[3]{2}$;
$y$-intercept: $-66$

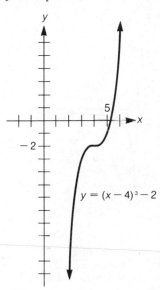

$y = (x - 4)^3 - 2$

**13.** $x$-intercept: $-5$; $y$-intercept: $-1250$

$y = -2(x + 5)^4$

**15.** $x$-intercept: $-1$; $y$-intercept: $\frac{1}{2}$

$y = \frac{1}{2}(x + 1)^5$

**17.** $x$-intercept: 0; $y$-intercept: 0

$y = -(x - 1)^3 - 1$

**19.**

$y = (x - 2)(x - 1)(x + 1)$

**21.**

$y = 2x(x + 1)(x + 3)$

**23.**

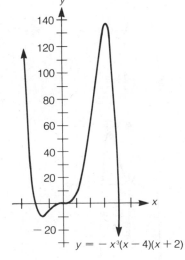

$y = x^3(x + 2)$

**25.**

$y = 2(x - 1)(x - 4)^3$

**27.**

$y = (x + 1)^2(x - 1)(x - 3)$

**29.**

$y = -x^3(x - 4)(x + 2)$

**31.**

$y = -4x(x-2)^2(x+2)^3$

**33.** From left to right, they are: $f(x) = x$, $g(x) = x^2$, $h(x) = x^3$, $F(x) = x^4$, $G(x) = x^5$, $H(x) = x^6$.  **35.** $[0, 0.68]$  **37.** no
**39.** yes, at the point $(100, 100)$
**41.**

$y = x^3(2-x)$

**43.**

$y = x(x+2)(x-1)$

**45.**

$y = x^2(2+x)(2-x)$

## 5.5 Graphs of Rational Functions

**1.** domain: $(-\infty, 3) \cup (3, \infty)$; $x$-intercept: $-5$; $y$-intercept: $-\frac{5}{4}$  **3.** domain: $(-\infty, -2) \cup (-2, 3) \cup (3, \infty)$; $x$-intercepts: $9, -1$; $y$-intercept: $\frac{3}{2}$
**5.** domain: $(-\infty, 0) \cup (0, \infty)$; $x$-intercepts: $-2, 2, 1$; $y$-intercept: none
**7.** $x$-intercept: none; $y$-intercept: $\frac{1}{4}$; horizontal asymptote: $y = 0$; vertical asymptote: $x = -4$

$y = \dfrac{1}{x+4}$

$x = -4$

**9.** $x$-intercept: none; $y$-intercept: $\frac{3}{2}$; horizontal asymptote: $y = 0$; vertical asymptote: $x = -2$

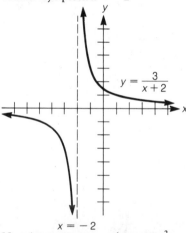

$y = \dfrac{3}{x+2}$

$x = -2$

**11.** $x$-intercept: none; $y$-intercept: $\frac{2}{3}$; horizontal asymptote: $y = 0$; vertical asymptote: $x = 3$

$y = \dfrac{-2}{x-3}$

$x = 3$

**13.** $x$-intercept: 3; $y$-intercept: 3; horizontal asymptote: $y = 1$; vertical asymptote: $x = 1$

$y = \dfrac{x-3}{x-1}$

$x = 1$

**15.** $x$-intercept: $\frac{1}{2}$; $y$-intercept: $-2$; horizontal asymptote: $y = 2$; vertical asymptote: $x = -\frac{1}{2}$

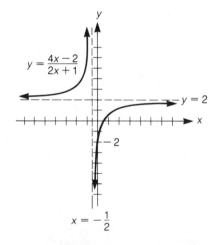

$y = \dfrac{4x-2}{2x+1}$

$y = 2$

$x = -\dfrac{1}{2}$

**17.** $x$-intercept: none; $y$-intercept: $\frac{1}{4}$; horizontal asymptote: $y = 0$; vertical asymptote: $x = 2$

$y = \dfrac{1}{(x-2)^2}$

$x = 2$

**19.** $x$-intercept: none; $y$-intercept: 3; horizontal asymptote: $y = 0$; vertical asymptote: $x = -1$

$$y = \frac{3}{(x+1)^2}$$

$x = -1$

**21.** $x$-intercept: none; $y$-intercept: $\frac{1}{8}$; horizontal asymptote: $y = 0$; vertical asymptote: $x = -2$

$$y = \frac{1}{(x+2)^3}$$

$x = -2$

**23.** $x$-intercept: none; $y$-intercept: $-\frac{4}{125}$; horizontal asymptote: $y = 0$; vertical asymptote: $x = -5$

$$y = \frac{-4}{(x+5)^3}$$

$x = -5$

**25.** $x$-intercept: 0; $y$-intercept: 0; horizontal asymptote: $y = 0$; vertical asymptotes: $x = -2, x = 2$

$$y = \frac{-x}{(x+2)(x-2)}$$

$x = -2 \qquad x = 2$

**27.** $x$-intercept: 0; $y$-intercept: 0; horizontal asymptote: $y = 0$; vertical asymptotes: $x = -3, x = 1$

$$y = \frac{x}{(x-1)(x+3)}$$

$x = -3 \qquad x = 1$

**29.** $x$-intercepts: $-2, 4$; $y$-intercept: $-\frac{8}{3}$; horizontal asymptote: $y = 1$; vertical asymptotes: $x = 1, x = 3$; crosses horizontal asymptote at $x = \frac{11}{2}$

$$y = \frac{(x-4)(x+2)}{(x-1)(x-3)}$$

$y = 1$

$-2$

$x = 1 \quad x = 3$

**31.** $x$-intercept: $-1$; $y$-intercept: $\frac{1}{3}$; horizontal asymptote: $y = 1$; vertical asymptotes: $x = 1, x = 3$; crosses horizontal asymptote at $x = \frac{1}{3}$

$$y = \frac{(x+1)^2}{(x-1)(x-3)}$$

$y = 1$

$-9$

$x = 3$

$x = 1$

**35. (a)**

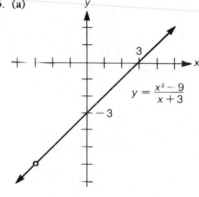

$$y = \frac{x^2 - 9}{x + 3}$$

$3$

$-3$

**(b)**

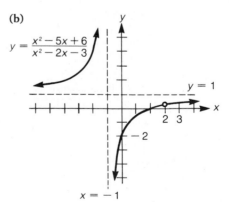

$$y = \frac{x^2 - 5x + 6}{x^2 - 2x - 3}$$

$y = 1$

$2 \quad 3$

$-2$

$x = -1$

**(c)**

$$y = \frac{(x-1)(x-2)(x-3)}{(x-1)(x-2)(x-3)(x-4)}$$

$x = 4$

**37.** $x$-intercept: 0; $y$-intercept: 0; horizontal asymptote: $y = 0$; vertical asymptote: $x = 3$; low point: $\left(-3, -\frac{1}{12}\right)$

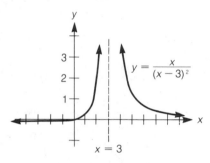

$$y = \frac{x}{(x-3)^2}$$

$x = 3$

**39. (b)**

| $x$ | $x + 4$ | $\dfrac{x^2 + x - 6}{x - 3}$ |
|---|---|---|
| 10 | 14 | 14.8571 |
| 100 | 104 | 104.0619 |
| 1000 | 1004 | 1004.0060 |

| $x$ | $x + 4$ | $\dfrac{x^2 + x - 6}{x - 3}$ |
|---|---|---|
| $-10$ | $-6$ | $-6.4615$ |
| $-100$ | $-96$ | $-96.0583$ |
| $-1000$ | $-996$ | $-996.0600$ |

**(c)** vertical asymptote: $x = 3$; $x$-intercepts: $-3, 2$; $y$-intercept: 2

**(d)**

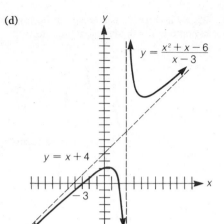

$$y = \frac{x^2 + x - 6}{x - 3}$$

$y = x + 4$

$-3$

$x = 3$

**(e)** $(3 + \sqrt{6}, 7 + 2\sqrt{6})$, $(3 - \sqrt{6}, 7 - 2\sqrt{6})$

**41.**

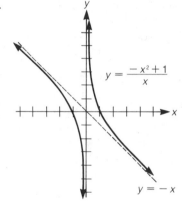

$$y = \frac{-x^2 + 1}{x}$$

$y = -x$

## Chapter 5 Review Exercises

**1.** $f(x) = x + 2$   **3.** $f(x) = \frac{3}{8}x - \frac{5}{2}$
**5.** $f(x) = -\frac{3}{4}x + \frac{11}{4}$
**7.** vertex: $(-1, -4)$; $x$-intercepts: $1, -3$; $y$-intercept: $-3$

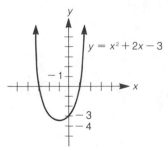

$$y = x^2 + 2x - 3$$

$-1$

$-3$
$-4$

**9.** vertex: $(\sqrt{3}, 6)$; $x$-intercepts: $\sqrt{3} + \sqrt{6}$, $\sqrt{3} - \sqrt{6}$; $y$-intercept: 3

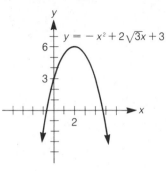

$$y = -x^2 + 2\sqrt{3}x + 3$$

6

3

2

**11.** vertex: $(2, 12)$; $x$-intercepts: $0, 4$; $y$-intercept: 0

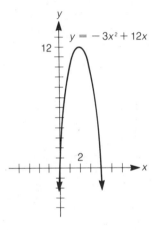

$$y = -3x^2 + 12x$$

12

2

**13.** 5   **15.** $\frac{3}{4}$   **17. (a)** At $t = v_0/32$ sec, maximum height is $v_0^2/64$ ft.   **(b)** $v_0/16$ sec
**19. (a)** $d = \sqrt{x^4 - 3x^2 + 4}$
**(b)** $\left(-\sqrt{6}/2, \frac{3}{2}\right)$   **21.** $b = \frac{17}{3}, -11$
**23.** 1   **25.** $\frac{225}{4}$ cm$^2$   **27.** $a = 1$
**29.** $x = 400$ units; $p = \$80$
**31.** $x$-intercepts: $-4, 2$; $y$-intercept: $-8$; vertex: $(-1, -9)$

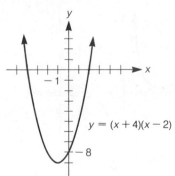

$-1$

$y = (x + 4)(x - 2)$

$-8$

33. $x$-intercept: $-5$; $y$-intercept: $-125$

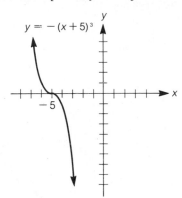

35. $x$-intercepts: $-1, 0$; $y$-intercept: $0$

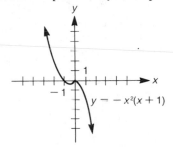

37. $x$-intercepts: $0, 2, -2$; $y$-intercept: $0$

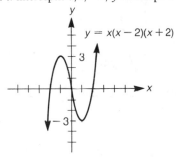

39. $x$-intercept: $-\frac{1}{3}$; $y$-intercept: none;
horizontal asymptote: $y = 3$;
vertical asymptote: $x = 0$

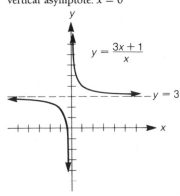

41. $x$-intercept: none; $y$-intercept: $-1$;
horizontal asymptote: $y = 0$;
vertical asymptote: $x = 1$

43. $x$-intercept: $2$; $y$-intercept: $\frac{2}{3}$;
horizontal asymptote: $y = 1$;
vertical asymptote: $x = 3$

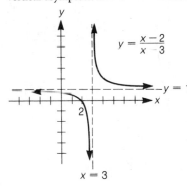

45. $x$-intercept: $1$; $y$-intercept: $\frac{1}{4}$;
horizontal asymptote: $y = 1$;
vertical asymptote: $x = 2$

47. (a) $1$   (b) $\sqrt{13}$   (c) $b = \pm\sqrt{2}/2$
49. $k = 6$
51. $(-\infty, 4 - 2\sqrt{3}] \cup [4 + 2\sqrt{3}, \infty)$
53. $A(m) = m/2$
55. $A(x) = (1 - x)\sqrt{1 - x^2}$

# CHAPTER 6
# EXPONENTIAL AND
# LOGARITHMIC FUNCTIONS

## 6.1 Exponential Functions

1. $10^9$   3. $10^{27}$   5. $(-\infty, \infty)$
7. $(-\infty, \infty)$
9.

11.

13.

**15.**

**17.** domain: $(-\infty, \infty)$; range: $(-\infty, 1)$;
$x$-intercept: 0; $y$-intercept: 0;
asymptote: $y = 1$

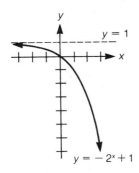

**19.** domain: $(-\infty, \infty)$; range: $(1, \infty)$;
$x$-intercept: none; $y$-intercept: 2;
asymptote: $y = 1$

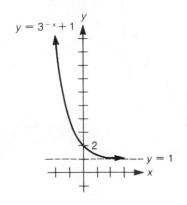

**21.** domain: $(-\infty, \infty)$; range: $(0, \infty)$;
$x$-intercept: none; $y$-intercept: $\frac{1}{2}$;
asymptote: $y = 0$

**23.** domain: $(-\infty, \infty)$; range: $(1, \infty)$;
$x$-intercept: none; $y$-intercept: 4;
asymptote: $y = 1$

**25.** $x = -\frac{1}{3}$   **27.** $x = \frac{3}{2}, 1$
**31. (a)**

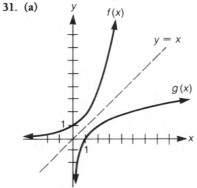

**(b)** domain: $(0, \infty)$; range: $(-\infty, \infty)$;
$x$-intercept: 1; $y$-intercept: none;
asymptote: $x = 0$
**33. (a)** 1.4   **(b)** 1.15   **(c)** 1.5

**35.**

| $x$ | $\log_{10} x$ |
|---|---|
| 1 | 0 |
| 2 | 0.3 |
| 3 | 0.48 |
| 4 | 0.6 |
| 5 | 0.7 |
| 6 | 0.78 |
| 7 | 0.85 |
| 8 | 0.9 |
| 9 | 0.95 |
| 10 | 1 |

## 6.2 The Exponential Function $y = e^x$

**1.** domain: $(-\infty, \infty)$;  range: $(0, \infty)$;
$x$-intercept: none;  $y$-intercept: 1;
asymptote: $y = 0$

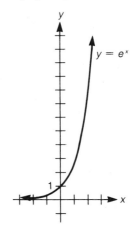

**3.** domain: $(-\infty, \infty)$;
range: $(-\infty, 0)$;  $x$-intercept: none;
$y$-intercept: $-1$  asymptote: $y = 0$

**5.** domain: $(-\infty, \infty)$; range: $(1, \infty)$;
$x$-intercept: none; $y$-intercept: 2;
asymptote: $y = 1$

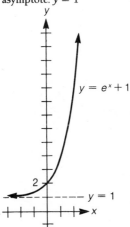

**7.** domain: $(-\infty, \infty)$; range: $(1, \infty)$;
$x$-intercept: none; $y$-intercept: $e + 1$;
asymptote: $y = 1$

**9.** domain: $(-\infty, \infty)$; range: $(-\infty, 0)$;
$x$-intercept: none; $y$-intercept: $-1/e^2$;
asymptote: $y = 0$

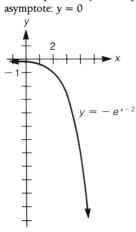

**11.** domain: $(-\infty, \infty)$; range: $(-\infty, e)$;
$x$-intercept: 1; $y$-intercept: $e - 1$;
asymptote: $y = e$

**13.**

**15.** 2  **17.** $\sqrt[3]{4}$  **19. (a)** $\sqrt{3}$
**(b)** 486,000  **21.** 6400 bacteria
**23.** 0.55 g  **25.** 97.15%
**29.**

**31.**

| $n$ | $1 + \dfrac{1}{n}$ | $\left(1 + \dfrac{1}{n}\right)^n$ |
|---|---|---|
| 1 | 2 | 2.0000 |
| 10 | 1.1 | 2.5937 |
| 100 | 1.01 | 2.7048 |
| 1000 | 1.001 | 2.7169 |
| 10,000 | 1.0001 | 2.7181 |
| 100,000 | 1.00001 | 2.7183 |

**33. (a)** 3.22 billion  **(b)** too low by
about 1.6 billion people
**35. (a)**

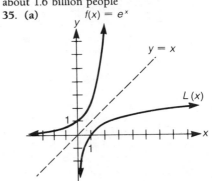

**(b)** domain: $(0, \infty)$; range: $(-\infty, \infty)$;
$x$-intercept: 1; asymptote: $x = 0$
**(c) (i)** $x$-intercept: 1; asymptote: $x = 0$

**(ii)** $x$-intercept: $-1$; asymptote: $x = 0$

**(iii)** $x$-intercept: 2; asymptote: $x = 1$

## 6.3 Logarithmic Functions

1. (a) no (b) yes (c) yes
3. (a) $\dfrac{4x+1}{2-3x}$ (b) $\dfrac{3x+4}{2x-1}$ (c) $\dfrac{1}{2}$
(d) $-4$ 5. $(-1,3)$ and $(6,-1)$
7. (a) $\log_3 9 = 2$ (b) $\log_{10} 1000 = 3$
(c) $\log_7 343 = 3$ (d) $\log_2 \sqrt{2} = \frac{1}{2}$
9. (a) $2^5 = 32$ (b) $10^0 = 1$
(c) $e^{1/2} = \sqrt{e}$ (d) $3^{-4} = \frac{1}{81}$ (e) $t^v = u$
11. $\log_5 30$ 13. (a) $\frac{3}{2}$ (b) $-\frac{5}{2}$ (c) $\frac{3}{2}$
15. (a) $2$ (b) $\frac{1}{5}$ 17. (a) $(0,\infty)$
(b) $(-\infty,\frac{3}{4})$ (c) $(-\infty,0) \cup (0,\infty)$
(d) $(0,\infty)$ (e) $(-\infty,-5) \cup (5,\infty)$
19. domain: $(0,e) \cup (e,\infty)$;
range: $(-\infty,0) \cup (0,\infty)$
21. (a) domain: $(-1,\infty)$;
range: $(-\infty,\infty)$; $x$-intercept: 0;
$y$-intercept: 0; asymptote: $x = -1$

(b) domain: $(-1,\infty)$; range: $(-\infty,\infty)$;
$x$-intercept: 0; $y$-intercept: 0;
asymptote: $x = -1$

23. (a) domain: $(0,\infty)$; range: $(-\infty,\infty)$;
$x$-intercept: 1; $y$-intercept: none;
asymptote: $x = 0$

(b) domain: $(-\infty,0)$; range: $(-\infty,\infty)$;
$x$-intercept: $-1$; $y$-intercept: none;
asymptote: $x = 0$

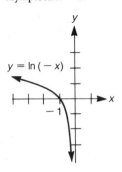

(c) domain: $(-\infty,0)$; range: $(-\infty,\infty)$;
$x$-intercept: $-e$; $y$-intercept: none;
asymptote: $x = 0$

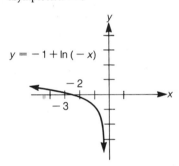

25. (a) 4 (b) $-1$ (c) $\frac{1}{2}$
27. $\log_{10} 25 \approx 1.40$
29. $\pm\sqrt{1 + \log_{10} 4} \approx \pm 1.27$
31. $\frac{1}{2}(\ln 10 - 3) \approx -0.35$
33. $\frac{1}{4}(1 - \ln 12.405) \approx -0.38$
35. (a) $-1.54 \times 10^{-10}$
(b) 99.9999846%
37. (a) $\ln\frac{1}{2} \approx -0.6931$
(b) 3.32 years 39. (a) 0.51 hr
(b) 2.89 hr
41. $x$-intercept: $e$; asymptote: $x = 0$

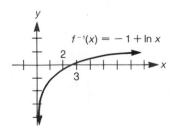

43. The shaded area is less than that of
the rectangle (shown dotted on the graph),
which has an area of 2 square units.

45. $\ln 6 \approx 1.79$ 47. $10^{30}$
49. $\dfrac{\ln (A_0/B_0)}{k_2 - k_1}$; this quantity represents
the amount of time it takes for two
radioactive substances to decay to the
same mass.
51. (a) $P(10) = 4$; $P(18) = 7$; $P(19) = 8$
(b)

| $x$ | $P(x)$ | $\dfrac{x}{\ln x}$ | $\dfrac{P(x)}{x/\ln x}$ |
|---|---|---|---|
| $10^2$ | 25 | 21.715 | 1.151 |
| $10^4$ | 1229 | 1,085.736 | 1.132 |
| $10^6$ | 78,498 | 72,382.414 | 1.084 |
| $10^8$ | 5,761,455 | 5,428,681.024 | 1.061 |
| $10^9$ | 50,847,534 | 48,254,942.43 | 1.054 |
| $10^{10}$ | 455,052,512 | 434,294,481.9 | 1.048 |

## 6.4 Properties of Logarithms

1. 1 3. $\frac{1}{2}$ 5. 4 7. 0 9. 4
11. $\log_{10} 60$ 13. $\log_5 20$ 15. (a) $\ln 6$
(b) $\ln \left(\frac{3}{16384}\right)$ 17. $\log_b \left[\dfrac{4(1+x)^3}{(1-x)^{3/2}}\right]$
19. $\log_{10} \left[\dfrac{27\sqrt{x+1}}{(x^2+1)^6}\right]$
21. (a) $2\log_{10} x - \log_{10}(1+x^2)$
(b) $2\ln x - \frac{1}{2}\ln(1+x^2)$
23. (a) $\frac{1}{2}\log_{10}(3+x) + \frac{1}{2}\log_{10}(3-x)$
(b) $\frac{1}{2}\ln(2+x) + \frac{1}{2}\ln(2-x)$
$-\ln(x-1) - \frac{3}{2}\ln(x+1)$
25. (a) $\frac{1}{2}\log_b x - \frac{1}{2}$
(b) $\ln(1+x^2) + \ln(1+x^4) + \ln(1+x^6)$
27. (a) $a + 2b + 3c$
(b) $\frac{1}{2}(1 + a + b + c)$
(c) $1 + a - \frac{1}{2}b - \frac{1}{2}c$
(d) $2 + 2a - 4b - \frac{1}{3}c$ (e) $5a + 5b - c$
29. $\frac{1}{2}(\ln 5 - \ln 2 + 1)$
31. $\ln 2 - \ln 3 - 1$ 33. $(\ln 9)/(\ln 2)$

35. $\dfrac{\ln 10}{\ln 5 - \ln 2} = \dfrac{\ln 5 + \ln 2}{\ln 5 - \ln 2}$  37. $\frac{1}{2}$

39. 3  41. $\frac{203}{99}$  43. 3  45. 7

47. (a) $x = 10^y/[3(10^y) - 1]$

(b) $x = (1 - y)/2$

49. $(\log_{10} 5)/(\log_{10} 2)$

51. $(\log_{10} 3)/(\log_{10} e)$

53. $(\log_{10} 2)/(\log_{10} b)$

55. $(\ln 6)/(\ln 10)$  57. $1/(\ln 10)$

59. $[\ln(\ln x) - \ln(\ln 10)]/(\ln 10)$

61. (a) true  (b) true  (c) true

(d) false  (e) true  (f) false  (g) true

(h) false  (i) true  (j) false  (k) false

(l) true  (m) true .75. 0.123456  81. $x^3$

83. $\frac{1}{2}$  85. $\beta e^{\alpha/3}$  93. $\dfrac{e^{2x} - 1}{2e^x}$

95.

| $x$ | $10^2$ | $10^3$ | $10^6$ | $10^{20}$ | $10^{50}$ | $10^{99}$ |
|---|---|---|---|---|---|---|
| $y$ | 0.4234 | 0.6589 | 0.9654 | 1.3428 | 1.5573 | 1.6918 |

## 6.5 Applications

1. $1009.98  3. 8.45%  5. $767.27

7. (a) $3869.68  (b) $4006.39

9. 13 quarters  11. $3487.50

13. $2610.23  15. 5.83%  17. first

19. (a) 14 years  (b) 13.86 years

(c) 1.01%  21. $26.5 trillion

23. (a) 14 years

(b)

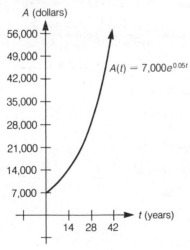

$A(t) = 7,000e^{0.05t}$

25.

|  | 1975 POPULATION (BILLIONS) | % POPULATION IN 1975 | GROWTH RATE (% PER YEAR) | YEAR 2000 POPULATION (BILLIONS) | % OF WORLD POPULATION IN 2000 |
|---|---|---|---|---|---|
| World | 4.090 | 100 | 1.8 | 6.414 | 100 |
| More Dev. | 1.131 | 27.7 | 0.6 | 1.314 | 20.5 |
| Less Dev. | 2.959 | 72.3 | 2.1 | 5.002 | 78.0 |

27.

|  | 1975 POPULATION (BILLIONS) | GROWTH RATE (% PER YEAR) | YEAR 2000 POPULATION (BILLIONS) | % INCREASE IN POPULATION |
|---|---|---|---|---|
| Low | 4.043 | 1.5 | 5.883 | 45.5 |
| Medium | 4.090 | 1.8 | 6.414 | 56.8 |
| High | 4.134 | 2.0 | 6.816 | 64.9 |

29. (a) 0.0200  (b) 170,853,155

(c) slower  31. (a) 0.68%  (b) 0.96%

(c) 19,751,512  (d) higher

33. (a)

Time (years)

(b)

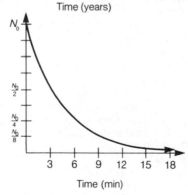

Time (min)

35. (a) $-0.0533$  (b) 58.7%; 0.5% of $N_0$

37. (a) $-0.0248$  (b) 279 years

(c) 280 years  39. (a) 2020  (b) 2035

(c) 2019  41. (a) 200 years

(b) 132 years  43. (a) $E = (10^{11.4})(10^{1.5M})$

(b) 31.6  45. $-1.4748 \times 10^{-11}$

41. 4.181 billion years old

49. 15,505 years old

## Chapter 6 Review Exercises

1. horizontal asymptote: $y = 0$; vertical asymptote: none; $x$-intercept: none; $y$-intercept: 1

$y = e^x$

3. horizontal asymptote: none; vertical asymptote: $x = 0$; $x$-intercept: 1; $y$-intercept: none

$y = \ln x$

**5.** horizontal asymptote: $y = 1$; vertical asymptote: none; $x$-intercept: none; $y$-intercept: 3

**7.** horizontal asymptote: $y = 0$; vertical asymptote: none; $x$-intercept: none; $y$-intercept: 1

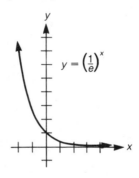

**9.** horizontal asymptote: $y = 1$; vertical asymptote: none; $x$-intercept: none; $y$-intercept: $e + 1$

**11.** horizontal asymptote: none; vertical asymptote: none; $x$-intercept: 0; $y$-intercept: 0

**13.** 4  **15.** 3  **17.** $\frac{2}{3}$  **19.** $\sqrt[3]{3}$
**21.** $(1 - 2\ln 3)/10$  **23.** $\frac{200}{99}$  **25.** 2
**27.** $(0, \infty)$  **29.** 1  **31.** $\frac{1}{2}$  **33.** $\frac{1}{5}$
**35.** $-1$  **37.** 16  **39.** 4  **41.** 2  **43.** 2
**45.** $\frac{9}{14}$  **47.** $2a + 3b + (c/2)$
**49.** $8a + 4b$  **51.** 2, 3  **53.** 2, 3
**55.** $-2, -3$
**57.** (a) third quadrant

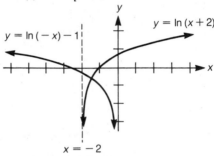

(b) $-1.46$
**59.** $k = (\ln\frac{1}{2})/T$  **61.** 6.25%
**63.** $[d\ln(c/b)]/(\ln\frac{1}{2})$ days  **65.** $\log_{10} 2$
**67.** $\ln 10$  **69.** $\ln x^a y^b$
**71.** $\frac{1}{2}\ln(x - 3) + \frac{1}{2}(x + 4)$
**73.** $3\log_{10} x - \frac{1}{2}\log_{10}(1 + x)$
**75.** $\frac{1}{3}\log_{10} x - \frac{2}{3}$
**77.** $3\ln(1 + 2e) - 3\ln(1 - 2e)$
**79.** $(\ln 2)/[\ln(1 + \frac{1}{100}R)]$ years  **81.** 9.92%
**83.** $(100\ln 2)/R$ years  **85.** (a) $7\frac{3}{4}$ years
(b) $t = (\ln n)/\ln(1 + \frac{1}{400}R)$
**87.**(a) $(0, \infty)$  (b) $[1, \infty)$
**89.** (a) $(-\infty, -3) \cup (-3, 5) \cup (5, \infty)$
(b) $(-\infty, -3) \cup (5, \infty)$
**91.** $(-\infty, -1) \cup (1, \infty)$  **93.** $-0.7$
**95.** $-2.2$  **97.** 4.3  **99.** 2.7  **101.** 1.6

**103.** (a)

| REGION | 1990 POPULATION (MILLIONS) | GROWTH RATE (%) | 2025 POPULATION (MILLIONS) |
|---|---|---|---|
| North America | 275.2 | 0.7 | 351.6 |
| Soviet Union | 291.3 | 0.7 | 372.2 |
| Europe | 499.5 | 0.2 | 535.7 |
| Nigeria | 113.3 | 3.1 | 335.3 |

(b) 222.0 million  (c) 193.5 million
(d) Our results support his projection.
**105.** (a) 1.0986  (b) 0.00112%
(c) $\ln\sqrt{3} \approx 0.5493$; $\ln 9 \approx 2.1972$;
$\ln\frac{1}{3} \approx -1.0986$

# CHAPTER 7
# TRIGONOMETRIC FUNCTIONS OF ANGLES

## 7.1 Trigonometric Functions of Acute Angles

**1.** (b) $\sin\theta = \frac{15}{17}$; $\cos\beta = \frac{15}{17}$
(c) $\cos\theta = \frac{8}{17}$; $\sin\beta = \frac{8}{17}$
(d) $\tan\theta = \frac{15}{8}$; $\csc\theta = \frac{17}{15}$; $\sec\theta = \frac{17}{8}$;
$\cot\theta = \frac{8}{15}$  (e) $\tan\beta = \frac{8}{15}$; $\csc\beta = \frac{17}{8}$;
$\sec\beta = \frac{17}{15}$; $\cot\beta = \frac{15}{8}$
**3.** (a) $\sin A = 2/\sqrt{13}$; $\cos A = 3/\sqrt{13}$;
$\tan A = \frac{2}{3}$  (c) $\cos B = 2/\sqrt{13}$
**5.** $\sin A = \frac{5}{13}$; $\cos A = \frac{12}{13}$; $\tan A = \frac{5}{12}$;
$\sec A = \frac{13}{12}$; $\csc A = \frac{13}{5}$; $\cot A = \frac{12}{5}$
**7.** (a) $\sin B = \frac{4}{5}$; $\cos A = \frac{4}{5}$;
(b) $\cos B = \frac{3}{5}$; $\sin A = \frac{3}{5}$
(c) $(\tan A)(\tan B) = 1$
**9.** (a) $\sin A = \sqrt{5}/4$; $\cos A = \sqrt{11}/4$
(b) $\tan B = \sqrt{11}/\sqrt{5}$; $\sin B = \sqrt{11}/4$;
$\cos B = \sqrt{5}/4$  **23.** (a) $\sin\theta = \dfrac{2x\sqrt{4x^2 + 9}}{4x^2 + 9}$;
$\cos\theta = \dfrac{3\sqrt{4x^2 + 9}}{4x^2 + 9}$; $\tan\theta = \dfrac{2x}{3}$
(b) $\sin^2\theta = \dfrac{4x^2}{4x^2 + 9}$; $\cos^2\theta = \dfrac{9}{4x^2 + 9}$;
$\tan^2\theta = 4x^2/9$
(c) $\sin(90° - \theta) = \dfrac{3\sqrt{4x^2 + 9}}{4x^2 + 9}$;
$\cos(90° - \theta) = \dfrac{2x\sqrt{4x^2 + 9}}{4x^2 + 9}$;
$\tan(90° - \theta) = 3/2x$
**25.** (a) $\sin\beta = \dfrac{\sqrt{16x^2 - 1}}{4x}$; $\cos\beta = \dfrac{1}{4x}$;
$\tan\beta = \sqrt{16x^2 - 1}$
(b) $\csc\beta = \dfrac{4x\sqrt{16x^2 - 1}}{16x^2 - 1}$;

$\sec \beta = 4x$;  $\cot \beta = \dfrac{\sqrt{16x^2 - 1}}{16x^2 - 1}$

**(c)** $\sin(90° - \beta) = 1/4x$;

$\cos(90° - \beta) = \dfrac{\sqrt{16x^2 - 1}}{16x^2 - 1}$;

$\tan(90° - \beta) = \sqrt{16x^2 - 1}/4x$

**27.** $\sin B = \sqrt{33}/7$;  $\tan B = \sqrt{33}/4$;

$\sec B = \frac{7}{4}$;  $\csc B = 7\sqrt{33}/33$;

$\cot B = 4\sqrt{33}/33$  **29.** $\cos \theta = \sqrt{13}/5$;

$\tan \theta = 2\sqrt{39}/13$;  $\sec \theta = 5\sqrt{13}/13$;

$\csc \theta = 5\sqrt{3}/6$;  $\cot \theta = \sqrt{39}/6$

**31.** $\sin A = \dfrac{2\sqrt{3} - \sqrt{6}}{6}$;

$\cos A = \dfrac{2\sqrt{3} + \sqrt{6}}{6}$;  $\sec A = 2\sqrt{3} - \sqrt{6}$;

$\csc A = 2\sqrt{3} + \sqrt{6}$;  $\cot A = 3 + 2\sqrt{2}$

**33.** $\cos \theta = \dfrac{\sqrt{4 - x^2}}{2}$;  $\tan \theta = \dfrac{x\sqrt{4 - x^2}}{4 - x^2}$;

$\sec \theta = \dfrac{2\sqrt{4 - x^2}}{4 - x^2}$;  $\csc \theta = \dfrac{2}{x}$;

$\cot \theta = \sqrt{4 - x^2}/x$  **35.** $\sin \theta = \sqrt{1 - x^4}$;

$\tan \theta = \sqrt{1 - x^4}/x^2$;  $\sec \theta = 1/x^2$;

$\csc \theta = \dfrac{\sqrt{1 - x^4}}{1 - x^4}$;  $\cot \theta = \dfrac{x^2\sqrt{1 - x^4}}{1 - x^4}$

**37. (a)** $\cos 30° \approx .86603$;

$\cos 45° \approx .70711$;  $\sin 60° \approx .86603$;

**(b)** $\cos 30° \approx .86603$;  $\cos 45° \approx .70711$;

$\sin 60° \approx .86603$  **39.** $\csc 25° \approx 2.36620$;

$\sec 25° \approx 1.10338$;  $\cot 25° \approx 2.14451$

## 7.2 Right-Triangle Applications

**1.** $BC = 30$ cm;  $AC = 30\sqrt{3}$ cm

**3.** $AB = 32\sqrt{3}/3$ cm;  $BC = 16\sqrt{3}/3$ cm

**5.** $AC \approx 11.5$ cm;  $BC \approx 9.6$ cm

**7. (a)** $9\sqrt{3} \approx 15.59$ ft  **(b)** 9 ft

**9.** 34 million miles  **11. (a)** 1.5 in.²

**(b)** 11.28 cm²  **13.** 1175.6 cm²

**15.** 10,660 ft  **17.** 136 m

**19.** $18(\sqrt{3} - 1)$ cm

**21. (a)** $\angle OBA = 90°$;  $\angle OAB = \theta$;

$\angle BAP = 90° - \theta$;  $\angle BPA = \theta$

**(b)** $AO = \sin \theta$;  $AP = \cos \theta$;  $OB = \sin^2 \theta$;

$BP = \cos^2 \theta$  **23. (a)** 53.1°  **(b)** 67.4°

**(c)** 48.6°  **25.** $AC = 4 \sec \theta + 5 \csc \theta$

**27. (a)** $DE = \sin \theta$  **(b)** $OE = \cos \theta$

**(c)** $CF = \tan \theta$  **(d)** $OC = \sec \theta$

**(e)** $AB = \cot \theta$  **(f)** $OB = \csc \theta$

**29. (b)** 1080 miles

**31. (b)**

| $n$ | 5 | 10 | 50 | 100 | 1000 | 5000 | 10,000 |
|---|---|---|---|---|---|---|---|
| $A_n$ | 2.38 | 2.94 | 3.1333 | 3.1395 | 3.141572 | 3.1415918 | 3.1415924 |

**(c)** As $n$ gets larger, $A_n$ becomes closer to the area of the circle, which is $\pi$.

## 7.3 Trigonometric Functions of General Angles

**1. (a)** reference angle: 70°

**(b)** reference angle: 60°

**(c)** reference angle: 60°

**(d)** reference angle: 60°

**3. (a)** 1  **(b)** 1  **(c)** $-1$  **(d)** 1

**5. (a)** $-1$  **(b)** undefined  **(c)** $-1$

**(d)** 1  **7. (a)** $\cos \theta = \frac{3}{4}$

**(b)** $\sin \theta = \sqrt{7}/4$;  $\tan \theta = \sqrt{7}/3$

**(c)** $\cos(\beta - 90°)$  **9. (a)** $-\sqrt{2}/2$  **(b)** $\frac{1}{2}$

**(c)** $-\sqrt{3}/3$  **(d)** $\sqrt{2}/2$  **11. (a)** $-1$

**(b)** $-\sqrt{3}/3$  **(c)** $-\sqrt{3}/3$  **(d)** 1

**13.** $\sin(-30°) = -\frac{1}{2}$;  $\cos(-30°) = \sqrt{3}/2$;

$\tan(-30°) = -\sqrt{3}/3$;

$\sec(-30°) = 2\sqrt{3}/3$;  $\csc(-30°) = -2$;

$\cot(-30°) = -\sqrt{3}$

**15. (a)**

| $\theta$ | 0° | 90° | 180° | 270° | 360° | 450° | 540° | 630° | 720° |
|---|---|---|---|---|---|---|---|---|---|
| $\sin \theta$ | 0 | 1 | 0 | $-1$ | 0 | 1 | 0 | $-1$ | 0 |
| $\cos \theta$ | 1 | 0 | $-1$ | 0 | 1 | 0 | $-1$ | 0 | 1 |

**(b)**

| $\theta$ | 30° | 60° | 90° | 120° | 150° | 180° |
|---|---|---|---|---|---|---|
| $\sin \theta$ | $\dfrac{1}{2}$ | $\dfrac{\sqrt{3}}{2}$ | 1 | $\dfrac{\sqrt{3}}{2}$ | $\dfrac{1}{2}$ | 0 |

| $\theta$ | 210° | 240° | 270° | 300° | 330° | 360° |
|---|---|---|---|---|---|---|
| $\sin \theta$ | $-\dfrac{1}{2}$ | $-\dfrac{\sqrt{3}}{2}$ | $-1$ | $-\dfrac{\sqrt{3}}{2}$ | $-\dfrac{1}{2}$ | 0 |

**17.** $\cos \theta = -2\sqrt{6}/5$;  $\tan \theta = -\sqrt{6}/12$;

$\sec \theta = -5\sqrt{6}/12$;  $\csc \theta = 5$;

$\cot \theta = -2\sqrt{6}$  **19.** $\sin \theta = -\frac{4}{5}$;

$\tan \theta = \frac{4}{3}$;  $\sec \theta = -\frac{5}{3}$;  $\csc \theta = -\frac{5}{4}$;

$\cot \theta = \frac{3}{4}$  **21.** $35\sqrt{3}/4$ cm²  **23.** 302.7 cm²

**25.** $90°, 450°, -270°$ (other answers possible)  **27.** (a) $\sin 10° \approx 0.17$; $\cos 10° \approx 0.99$  (b) $\sin 20° \approx 0.34$; $\cos 20° \approx 0.94$  (c) $\sin 30° \approx 0.50$; $\cos 30° \approx 0.87$  (d) $\sin 40° \approx 0.64$; $\cos 40° \approx 0.77$  (e) $\sin 50° \approx 0.77$; $\cos 50° \approx 0.64$  (f) $\sin 70° \approx 0.94$; $\cos 70° \approx 0.34$  (g) $\sin 80° \approx 0.99$; $\cos 80° \approx 0.17$  (c) $\sin 100° \approx 0.99$; $\cos 100° \approx -0.17$  (i) $\sin 130° \approx 0.77$; $\cos 130° \approx -0.64$  **29.** (a) $\frac{1}{4}(\sqrt{2}-\sqrt{6})$

(b) $-\frac{1}{4}\sqrt{10+2\sqrt{5}}$  (c) $\frac{1}{2}\left(\dfrac{\sqrt{2}-\sqrt{6}}{\sqrt{2}+\sqrt{3}}\right)$

(d) $\frac{1}{4}(\sqrt{2}-\sqrt{6})$  (e) $\frac{1}{4}\sqrt{10+2\sqrt{5}}$

(f) $-\frac{1}{4}\sqrt{10+2\sqrt{5}}$

**31.** (a) $P(\cos\theta,\sin\theta)$; $Q(\cos\phi,\sin\phi)$
**35.** (a) Since $OP$ and $OQ$ are both radii of the same circle, they are equal. Thus $\triangle OPQ$ is isosceles and the base angles $PQO$ and $QPO$ are equal, and so $\angle PQO = 30°$.
(b) They are opposite interior angles and thus equal.  (c) Since $\angle QCO$ and $\angle PCO$ are $90°$, then $\angle QOC$ and $\angle POC$ are each $60°$. Thus we have side–angle–side equality and so the two triangles are congruent.
(d) The lengths $CQ = CP$, since they are corresponding parts of congruent triangles. Hence the $x$-coordinates are equal but of opposite sign.

## 7.4 Algebra and the Trigonometric Functions

**1.** (a) $1 - 2\cos\theta + \cos^2\theta$
(b) $1 + 2\sin\theta\cos\theta$  (c) $\dfrac{\cos\theta\sin\theta + 1}{\sin\theta}$
**3.** (a) $(\tan\theta - 6)(\tan\theta + 1)$
(b) $(\sin B + \cos B)(\sin B - \cos B)$
(c) $(\cos A + 1)^2$  **5.** $-2\cos\theta - 1$
**7.** $-1$  **9.** $\sin A + \cos A$  **11.** $\csc\theta$
**13.** $\cos^2 B$  **15.** $\cos A + 4$  **17.** $\cot A$
**19.** $0$  **21.** $0$  **23.** $1$

## Chapter 7 Review Exercises

**1.** $\sqrt{2}/2$  **3.** $-\sqrt{3}$  **5.** $-2$  **7.** $-1$
**9.** $\sqrt{2}/2$  **11.** $1$  **13.** $-2\sqrt{3}/3$  **15.** $2$
**17.** (a) $\sin^2 33° + \sin^2 57° \neq \sin^2(33° + 57°)$
(b) $1$  **19.** $a = \frac{1}{2}$; $c = \sqrt{3}/2$  **21.** $b = \frac{35}{2}$
**23.** $\sin A = \sqrt{55}/8$; $\cot A = 3\sqrt{55}/55$
**25.** area $= 5$  **27.** $\sin^2 A + \cos^2 B = \frac{1}{5}$
**29.** $a = 16\sqrt{3}/3$  **31.** $c = 9$;
area $= 9\sqrt{19}/4$  **35.** $\cos\theta = \dfrac{p^4 - q^4}{p^4 + q^4}$;

$\tan\theta = \dfrac{2p^2 q^2}{p^4 - q^4}$  **37.** $\sin A \cos A$
**39.** $\sin^2 A$  **41.** $\cos A + \sin A$
**43.** $\sin A \cos A$  **45.** $2\sin A \cos A$
**47.** $\sqrt{1 - a^2}$  **49.** $\sqrt{1 - a^2}$  **51.** $-a$
**53.** $-a$  **55.** $-\sqrt{1 - a^2}$
**57.** $\sin\theta - \cos\theta$  **59.** $\alpha = 45°$; $\beta = 30°$
**61.** $\sin\theta = \frac{4}{5}$; $\tan\theta = \frac{4}{3}$
**63.** $\tan\theta = -\frac{24}{7}$  **65.** $\cot\theta = \frac{5}{12}$
**67.** $\tan(90° - \theta) = -3\sqrt{7}/7$
**69.** $\cos(180° - \theta) = -\frac{7}{9}$  **73.** $2521\ \text{cm}^2$
**109.** (a) $\sin\theta$  (b) $\cos\theta$  (c) $\tan\theta$
(d) $\sec\theta$  (e) $1 - \cos\theta$  (f) $\sin\theta\tan\theta$
**111.** (a) $\sqrt{3}/2$  (b) $\sqrt{3}$  (c) $1$
**113.** (a) $\sqrt{3}/2$  (b) $-\sqrt{2}$
**115.** $(3\tan\theta + 1)(2\tan\theta + 3)$
**117.** $-2\tan\theta - 1$  **119.** $24\sqrt{2}\ \text{cm}^2$
**121.** $50(\tan 55° - \tan 25°)$
**125.** $BP = \cos^2\theta$; $AB = \cos\theta\sin\theta$

# CHAPTER 8
## TRIGONOMETRIC FUNCTIONS OF REAL NUMBERS

## 8.1 Radian Measure

**1.** (a) $\pi/3$ radians  (b) $5\pi/4$ radians
(c) $\pi/5$ radians  (d) $5\pi/2$ radians
(e) $0$ radians  **3.** (a) $15°$  (b) $270°$
(c) $1080°$  (d) $18°$  (e) $90°$
(f) $540°/\pi$  **5.** smaller

**7.**

| $\theta$ | $0$ | $\pi/2$ | $\pi$ | $3\pi/2$ | $2\pi$ |
|---|---|---|---|---|---|
| $\cos\theta$ | $1$ | $0$ | $-1$ | $0$ | $1$ |

**9.** (a) positive  (b) positive
**11.** (a) negative  (b) positive  **13.** $4\pi$ ft
**15.** $\pi/2$ cm  **17.** $\frac{1}{6}$
**19.** (a) $12\pi$ radians/sec  (b) $144\pi$ cm/sec

(c) $72\pi$ cm/sec  **21.** (a) $6\pi$ rad/sec
(b) $150\pi$ cm/sec  (c) $75\pi$ cm/sec
**23.** (a) $50\pi/3$ rad/sec  (b) $750\pi$ cm/sec
(c) $375\pi$ cm/sec  **25.** (a) $175\pi/12$ cm$^2$
(b) $\pi/6$ in.$^2$  **27.** $50\pi/3 - 25\sqrt{3}$ cm$^2$
**29.** $57.3°$
**31.** $x = 22619/3\pi$ rev $\approx 600$ rev
**33.** (a) $0.000073$ rad/sec  (b) $1040$ mph
**35.** (a) yes  (b) $\theta = \pi/2$

## 8.2 Trigonometric Functions of Real Numbers

**1.**

| $\theta$ | $\sin\theta$ | $\cos\theta$ | $\tan\theta$ | $\csc\theta$ | $\sec\theta$ | $\cot\theta$ |
|---|---|---|---|---|---|---|
| $0$ | $0$ | $1$ | $0$ | undef. | $1$ | undef. |
| $\pi/6$ | $\frac{1}{2}$ | $\sqrt{3}/2$ | $\sqrt{3}/3$ | $2$ | $2\sqrt{3}/3$ | $\sqrt{3}$ |
| $\pi/4$ | $\sqrt{2}/2$ | $\sqrt{2}/2$ | $1$ | $\sqrt{2}$ | $\sqrt{2}$ | $1$ |
| $\pi/3$ | $\sqrt{3}/2$ | $\frac{1}{2}$ | $\sqrt{3}$ | $2\sqrt{3}/3$ | $2$ | $\sqrt{3}/3$ |
| $\pi/2$ | $1$ | $0$ | undef. | $1$ | undef. | $0$ |
| $2\pi/3$ | $\sqrt{3}/2$ | $-\frac{1}{2}$ | $-\sqrt{3}$ | $2\sqrt{3}/3$ | $-2$ | $-\sqrt{3}/3$ |
| $3\pi/4$ | $\sqrt{2}/2$ | $-\sqrt{2}/2$ | $-1$ | $\sqrt{2}$ | $-\sqrt{2}$ | $-1$ |
| $5\pi/6$ | $\frac{1}{2}$ | $-\sqrt{3}/2$ | $-\sqrt{3}/3$ | $2$ | $-2\sqrt{3}/3$ | $-\sqrt{3}$ |
| $\pi$ | $0$ | $-1$ | $0$ | undef. | $1$ | undef. |

**3.** $\cos\theta = -\frac{4}{5}$; $\tan\theta = \frac{3}{4}$
**5.** $\tan t = -\sqrt{39}/13$
**7.** $\sec\alpha = \frac{13}{5}$; $\cos\alpha = \frac{5}{13}$; $\sin\alpha = \frac{12}{13}$
**13.** $(\sqrt{7}\cos\theta)/7$  **15.** (a) $-\frac{2}{3}$  (b) $\frac{1}{4}$
(c) $\frac{1}{5}$  (d) $-\frac{1}{5}$  **17.** (a) $-\frac{1}{3}(1 + 2\sqrt{2})$
(b) $1$  **19.** (a) $\sqrt{2}/2$  (b) $\sqrt{3}/2$
(c) $1$  **21.** $\cos^2\theta$  **23.** $\sin^2\theta$

**37.**

| $\theta$ | $1 - \dfrac{\theta^2}{2}$ | $\cos\theta$ |
|---|---|---|
| $0.1$ | $0.995$ | $0.995004\ldots$ |
| $0.2$ | $0.980$ | $0.980066\ldots$ |
| $0.3$ | $0.955$ | $0.955336\ldots$ |

**41.** (a) $0, \pi, 2\pi, 3\pi$  (b) all real numbers except $k\pi$, where $k$ is any integer
(c) all real numbers except the integers

## 8.3 The Addition Formulas

**1.** (a) $\sin 3\theta$  (b) $1$  **3.** (a) $\cos 5u$
(b) $\cos u$  **5.** $\sin B$  **7.** $\cos\theta$  **9.** $-\cos\theta$
**11.** $\sin\theta$  **13.** $\frac{1}{4}(\sqrt{6} - \sqrt{2})$
**15.** $\frac{1}{4}(\sqrt{6} + \sqrt{2})$  **17.** $\sqrt{2}\sin s$

**19.** $\sqrt{3}\sin\theta$  **21. (a)** $\frac{56}{65}$  **(b)** $\frac{16}{65}$
**(c)** $\frac{33}{65}$  **(d)** $-\frac{63}{65}$  **23. (a)** $-\frac{5}{13}$  **(b)** $\frac{119}{169}$
**25.** $\sin(s-t) = -\sqrt{2}/10$;
$\cos(s+t) = -\sqrt{2}/10$

**35. (a)**

| $t$ | 1 | 2 | 3 | 4 |
|---|---|---|---|---|
| $f(t)$ | 1.5 | 1.5 | 1.5 | 1.5 |

**(b)** $f(t) = 1.5$

## 8.4 Further Identities

**1. (a)** $\tan(\theta+t) = \frac{4}{3}$
**(b)** $\tan(\theta-t) = \frac{44}{117}$
**3. (a)** $\tan\left(\theta + \frac{\pi}{4}\right) = 7$
**(b)** $\tan\left(\theta - \frac{\pi}{4}\right) = -\frac{1}{7}$
**5. (a)** $\sin 2\theta = \frac{24}{25}$  **(b)** $\cos 2\theta - \frac{7}{25}$
**(c)** $\tan 2\theta = \frac{24}{7}$  **7. (a)** $\sin 2\beta = \frac{24}{25}$
**(b)** $\cos 2\beta = -\frac{7}{25}$  **(c)** $\tan 2\beta = -\frac{24}{7}$
**9. (a)** $\sin(\theta/2) = \sqrt{10}/10$
**(b)** $\cos(\theta/2) = 3\sqrt{10}/10$
**(c)** $\tan(\theta/2) = 1/3$
**11. (a)** $\sin(\beta/2) = \sqrt{5}/5$
**(b)** $\cos(\beta/2) = 2\sqrt{5}/5$
**(c)** $\tan(\beta/2) = 1/2$  **13. (a)** $-3\sqrt{7}/8$
**(b)** $-\frac{1}{8}$  **(c)** $\frac{1}{4}\sqrt{8+2\sqrt{7}}$
**(d)** $\frac{1}{4}\sqrt{8-2\sqrt{7}}$  **15. (a)** $4\sqrt{2}/9$
**(b)** $-\frac{7}{9}$  **(c)** $\sqrt{6}/3$  **(d)** $-\sqrt{3}/3$
**17. (a)** $\frac{1}{2}\sqrt{2-\sqrt{3}}$  **(b)** $\frac{1}{2}\sqrt{2+\sqrt{3}}$
**(c)** $2-\sqrt{3}$  **19. (a)** $\frac{1}{2}\sqrt{2+\sqrt{3}}$
**(b)** $-\frac{1}{2}\sqrt{2-\sqrt{3}}$  **(c)** $-2-\sqrt{3}$
**21.** $\sin 2\theta = \frac{2}{25}x\sqrt{25-x^2}$;
$\cos 2\theta = 1 - \frac{2}{25}x^2$
**23.** $\sin 2\theta = \frac{1}{2}(x-1)\sqrt{3+2x-x^2}$;
$\cos 2\theta = (1+2x-x^2)/2$
**25.** $\sin^4\theta = (3 - 4\cos 2\theta + \cos 4\theta)/8$
**27.** $\sin^4(\theta/2) = (3 - 4\cos\theta + \cos 2\theta)/8$
**55. (b)** $\frac{1}{2}(\sqrt{3}+\sqrt{2})$
**67. (a)** $-1.1918$; $-1.1918$; they are equal
**(b)** $-1.3764$; $-1.3764$; they are equal

## 8.5 Graphs of the Sine and Cosine Functions

**1.** period: 2; amplitude: 1
**3.** period: 4; amplitude: 6
**5.** period: 4; amplitude: 2
**7.** period: 6; amplitude: $\frac{3}{2}$

**9. (a)** amplitude: 2; period: $2\pi$;
$x$-intercepts: $0, \pi, 2\pi$; high point: $(\pi/2, 2)$;
low point: $(3\pi/2, -2)$

**(b)** amplitude: 1; period: $\pi$; $x$-intercepts:
$0, \pi/2, \pi$; high point: $(3\pi/4, 1)$; low point:
$(\pi/4, -1)$

**11. (a)** amplitude: 1; period: $\pi$;
$x$-intercepts: $\pi/4, 3\pi/4$; high points:
$(0, 1), (\pi, 1)$; low points: $(\pi/2, -1)$

**(b)** amplitude: 2; period: $\pi$; $x$-intercepts:
$\pi/4, 3\pi/4$; high points: $(0, 2), (\pi, 2)$;
low points: $(\pi/2, -2)$

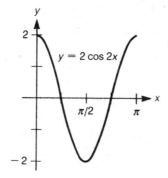

**13. (a)** amplitude: 3; period: 4;
$x$-intercepts: $0, 2, 4$; high point: $(1, 3)$;
low point: $(3, -3)$

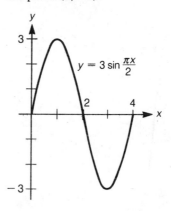

**(b)** amplitude: 3; period: 4; $x$-intercepts:
$0, 2, 4$; high point: $(3, 3)$, low point: $(1, -3)$

**15. (a)** amplitude: 1; period: 1;
$x$-intercepts: $\frac{1}{4}, \frac{3}{4}$; high points: $(0, 1), (1, 1)$;
low points: $(\frac{1}{2}, -1)$

**(b)** amplitude: 4; period: 1; x-intercepts: $\frac{1}{4}$, $\frac{3}{4}$; high point: $(\frac{1}{2}, 4)$; low points: $(0, -4)$, $(1, -4)$

**17.** amplitude: 1; period: $\pi$; x-intercept: $3\pi/4$; high point: $(\pi/4, 2)$; low point: $(3\pi/4, 0)$

**19.** amplitude: 1; period: 6; x-intercepts: 0, 6; high point: $(3, 2)$; low points: $(0, 0)$, $(6, 0)$

**21.** amplitude: 1; period: $\pi$; phase shift: $\pi/4$; x-intercepts: $\pi/4$, $3\pi/4$, $5\pi/4$; high point: $(\pi/2, 1)$; low point: $(\pi, -1)$

**23.** amplitude: 1; period: $\pi$; phase shift: $\pi/2$; x-intercepts: $3\pi/4$, $5\pi/4$; high points: $(\pi/2, 1)$, $(3\pi/2, 1)$; low point: $(\pi, -1)$

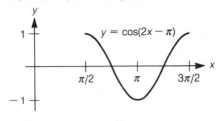

**25.** amplitude: 3; period: $4\pi$; phase shift: $-\pi/3$; x-intercepts: $-\pi/3$, $5\pi/3$, $11\pi/3$; high point $(2\pi/3, 3)$; low point: $(8\pi/3, -3)$

**27.** amplitude: 4; period: $2\pi/3$; phase shift: $\pi/12$; x-intercepts: $\pi/4$, $7\pi/12$; high points: $(\pi/12, 4)$, $(3\pi/4, 4)$; low point: $(5\pi/12, -4)$

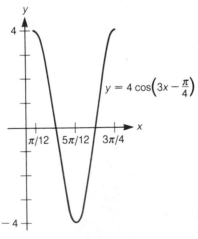

**29.** amplitude: $\frac{1}{2}$; period: 4; phase shift: $2\pi$; x-intercepts: $2\pi$, $2\pi + 2$, $2\pi + 4$; high point: $(2\pi + 1, \frac{1}{2})$; low point: $(2\pi + 3, -\frac{1}{2})$

**31.** amplitude: 1; period: $\pi$; phase shift: $\pi/6$; x-intercepts: $\pi/6$, $7\pi/6$; high point: $(2\pi/3, 2)$; low points: $(\pi/6, 0)$, $(7\pi/6, 0)$

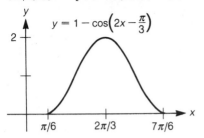

**33. (b)** amplitude: $\frac{1}{2}$; period: $\pi$

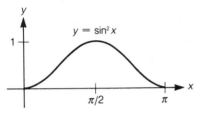

**35.** amplitude: $\frac{1}{2}$; period: $\pi$

**37. (a)** amplitude: 3; period: 6

**(b)** The weight is farthest from the origin when $|s| = 3$. This occurs when $t = 0, 3, 6, 9, 12$ sec.
**(c)** $t = \frac{3}{2}, \frac{9}{2}, \frac{15}{2}, \frac{21}{2}$ sec
**(d)** amplitude: $\pi$; period: 6

**(e)**  $t = 0, 3, 6, 9, 12 \sec;\ s = 3$ or $-3$
**(f)**  $3 < t < 6,\ 9 < t < 12$
**(g)**  $v$ is greatest when $t = \frac{3}{2}, \frac{15}{2}$; then
$v \approx 3.14\,\text{ft/sec}$ and $s = 0$; $v$ is least when
$t = \frac{9}{2}, \frac{21}{2}$; then $v \approx -3.14\,\text{ft/sec}$ and $s = 0$
**(h)**

**39.  (b)**

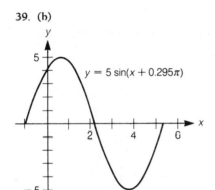

## 8.6  Graphs of the Tangent and the Reciprocal Functions

**1.  (a)**  $x$-intercept: $-\pi/4$; $y$-intercept: 1;
asymptotes: $x = -3\pi/4, x = \pi/4$

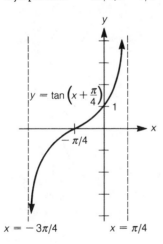

**(b)**  $x$-intercept: $-\pi/4$; $y$-intercept: $-1$;
asymptotes: $x = -3\pi/4, x = \pi/4$

**3.  (a)**  $x$-intercept: 0; $y$-intercept: 0;
asymptotes: $x = -3\pi/2, x = 3\pi/2$

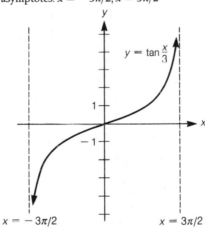

**(b)**  $x$-intercept: 0; $y$-intercept: 0;
asymptotes: $x = -3\pi/2, x = 3\pi/2$

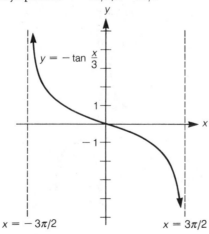

**5.**  $x$-intercept: 0; $y$-intercept: 0;
asymptotes: $x = -1, x = 1$

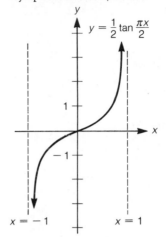

**7.**  $x$-intercept: 1; $y$-intercept: none;
asymptotes: $x = 0, x = 2$

**9.**  $x$-intercept: $3\pi/4$; $y$-intercept: 1;
asymptotes: $x = \pi/4, x = 5\pi/4$

**11.** *x*-intercept: $\pi/4$; *y*-intercept: none;
asymptotes: $x = 0, x = \pi/2$

**13.** *x*-intercept: none; *y*-intercept: $-\sqrt{2}$;
asymptotes: $x = -3\pi/4, x = \pi/4, x = 5\pi/4$

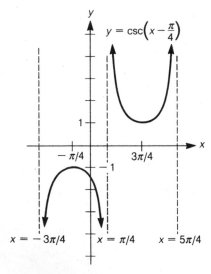

**15.** *x*-intercept: none; *y*-intercept: none;
asymptotes: $x = -2\pi, x = 0, x = 2\pi$

**17.** *x*-intercept: none; *y*-intercept: none;
asymptotes: $x = -1, x = 0, x = 1$

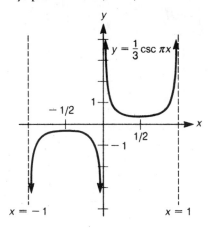

**19.** *x*-intercept: none; *y*-intercept: $-1$;
asymptotes: $x = -\pi/2, x = \pi/2, x = 3\pi/2$

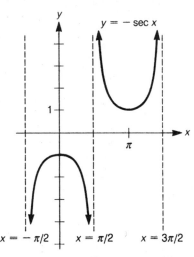

**21.** *x*-intercept: none; *y*-intercept: $-1$;
asymptotes: $x = \pi/2, x = 3\pi/2, x = 5\pi/2$

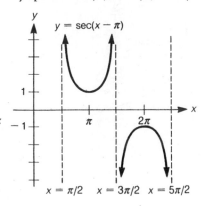

**23.** *x*-intercept: none; *y*-intercept: 3;
asymptotes: $x = -1, x = 1, x = 3$

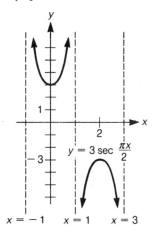

**25.** period: $\pi$; phase shift: $-\pi/6$;
*x*-intercept: none; *y*-intercept: $4\sqrt{3}/3$;
asymptotes: $x = -2\pi/3, x = -\pi/6, x = \pi/3$

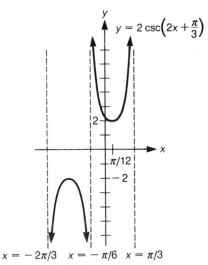

**27.** *x*-intercept: 0; *y*-intercept: 0;
asymptotes: $x = -\pi/2, x = \pi/2$

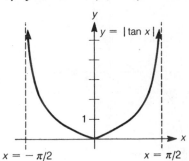

**29.** (a) $x$-intercept: none; $y$-intercept: 1; asymptotes: $x = -\pi/2, x = \pi/2$

(b) $x$-intercept: none; $y$-intercept: 1; asymptotes: $x = -\pi/2, x = \pi/2$

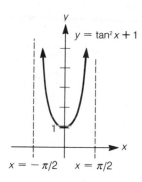

**31.** (b) $2\pi$

## 8.7 Trigonometric Equations

**1.** yes **3.** no **5.** $\theta = (\pi/3) + 2\pi k$ or $\theta = (2\pi/3) + 2\pi k$, where $k$ is any integer
**7.** $\theta = (7\pi/6) + 2\pi k$ or $\theta = (11\pi/6) + 2\pi k$, where $k$ is any integer
**9.** $\theta = \pi + 2\pi k$, where $k$ is any integer
**11.** $\theta = (\pi/3) + \pi k$, where $k$ is any integer
**13.** $x = \pi k$, where $k$ is any integer
**15.** $\theta = (\pi/2) + \pi k$, $\theta = (2\pi/3) + 2\pi k$, or $\theta = (4\pi/3) + 2\pi k$, where $k$ is any integer
**17.** $t = \pi k$, where $k$ is any integer
**19.** $x = (\pi/6) + 2\pi k$, $x = (5\pi/6) + 2\pi k$, or $x = (3\pi/2) + 2\pi k$, where $k$ is any integer
**21.** $t = (\pi/4) + 2\pi k$ or $t = (3\pi/4) + 2\pi k$, where $k$ is any integer **23.** $0°, 120°, 240°$
**25.** $75°, 105°, 195°, 225°, 315°, 345°$
**27.** $60°, 180°, 300°$
**29.** $30°, 120°, 210°, 300°$
**31.** $14.5°, 165.5°$ **33.** $116.6°, 296.6°$
**35.** $128.2°, 231.8°$ **37.** $0, \pi$ **39.** $2\pi/3, \pi$

**41.** $(3\pi/16) + (k\pi/4)$, where $k$ is any integer **43.** $\pi/8, 3\pi/8, 5\pi/8, 7\pi/8, 9\pi/8, 11\pi/8, 13\pi/8, 15\pi/8, 0, \pi, 2\pi$
**45.** $\pi/10, \pi/2, 9\pi/10, 13\pi/10, 17\pi/10$
**47.** $\pi/3, 4\pi/3$ **49.** $60.45°$

## 8.8 The Inverse Trigonometric Functions

**1.** $\pi/3$ **3.** $\pi/3$ **5.** $-\pi/6$ **7.** $\pi/4$
**9.** undefined **11.** $\frac{1}{4}$ **13.** $\frac{3}{4}$ **15.** $-\pi/7$
**17.** $\pi/2$ **19.** 0 **21.** $\frac{4}{3}$ **23.** $\sqrt{2}/2$
**25.** $\frac{12}{5}$ **27.** $\frac{1}{2}$ **29.** $2\sqrt{2}/3$ **31.** (a) 0.85 or $48.59°$ (b) 0.84 or $48.19°$ (c) 1.26 or $72.34°$ (d) 1.56 or $89.21°$ **33.** $\sqrt{2}$
**37.** $\frac{1}{4}\sin^{-1}(3x/2) - \frac{3}{2}x\sqrt{4 - 9x^2}$
**39.** $\tan^{-1}(\frac{1}{2}x - \frac{1}{2}) - 2/\sqrt{5 - 2x + x^2}$
**41.** $\frac{8}{17}$ **43.** $31\sqrt{218}/1090$ **45.** $-\sqrt{2}/2$

## Chapter 8 Review Exercises

**1.** $135°$ **3.** $900°$ **5.** $150°$ **7.** $360°/\pi$
**9.** $2\pi$ radians **11.** $\pi/180$ radians
**13.** $\pi/24$ radians **15.** $\frac{1}{180}$ radians
**17.** $-1$ **19.** $2\sqrt{3}/3$ **21.** $-\sqrt{3}$ **23.** $\frac{1}{2}$
**25.** 1 **27.** $-2$ **29.** $s = 2\pi$ cm; $A = 16\pi$ cm² **31.** $A = \frac{1}{2}$ cm²
**33.** $r = 20/\pi$ cm; $A = 40/\pi$ cm²
**35.** $r = 3$ cm; $\theta = 4$ radians
**37.** $7\pi/18$ radians **39.** 0.841
**41.** $-1.000$ **43.** 0.012 **49.** $5\cos\theta$
**51.** $10\tan\theta$ **53.** $\sqrt{5}/5\cos\theta$ **55.** $\sqrt{10}/10$
**57.** $\frac{527}{625}$ **59.** $2\sqrt{29}/29$ **61.** $\sqrt{2}/10$
**63.** $\frac{1}{2}(\pi - 2)$ cm² **65.** (a) $1.618 = 1.618$
(b) $1.614 = 1.614$ (c) $1.609 = 1.609$
**75.** $-\frac{16}{65}$ **77.** $\sin x$ **79.** 1 **81.** 0
**83.** $\cos x$ **85.** $-3$ **87.** $\frac{1104}{1105}$
**89.** $y = 4\sin x$ **91.** $y = -2\cos 4x$
**93.** $\sqrt{2}$
**95.** $x$-intercepts: $\pi/8, 3\pi/8$, high point: $(\pi/4, 3)$; low points: $(0, -3), (\pi/2, -3)$

**97.** $x$-intercepts: $\frac{1}{2}, \frac{5}{2}, \frac{9}{2}$; high point: $(\frac{3}{2}, 2)$; low point: $(\frac{7}{2}, -2)$

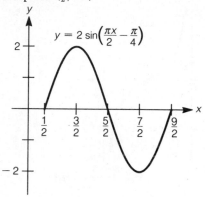

**99.** $x$-intercepts: $\frac{5}{2}, \frac{9}{2}$; high points: $(1, 3)$, $(7, 3)$; low point: $(4, -3)$

**101.** $x$-intercepts: $0, \pi/4, \pi/2$; high point: $(3\pi/8, \frac{1}{2})$; low point: $(\pi/8, -\frac{1}{2})$

**103.** (a)

**(b)**

$y = \cot \dfrac{\pi x}{4}$

$x = 4$

**105. (a)**

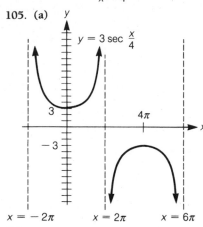

$y = 3 \sec \dfrac{x}{4}$

$x = -2\pi$   $x = 2\pi$   $x = 6\pi$

**(b)**

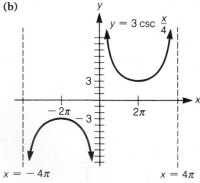

$y = 3 \csc \dfrac{x}{4}$

$x = -4\pi$   $x = 4\pi$

**107.**

$y = \tan(x - 2)$

$x = 2 - \dfrac{\pi}{2}$   $x = 2 + \dfrac{\pi}{2}$

**145.** $\pi/3, 2\pi/3, 4\pi/3, 5\pi/3$   **147.** $0, 3\pi/2$
**149.** $0, \pi, 7\pi/6, 11\pi/6$
**151.** $\pi/3, 2\pi/3, 4\pi/3, 5\pi/3$
**153.** $\pi/4, \pi/2, 3\pi/4, 5\pi/4, 3\pi/2, 7\pi/4$
**155.** $\pi/6, \pi/3, 2\pi/3, 5\pi/6, 7\pi/6, 4\pi/3, 5\pi/3,$
$11\pi/6$   **157.** $\pi/2, 3\pi/4, \pi, 7\pi/4$   **161.** $3\pi/4$
**163.** $0$   **165.** $\pi/3$   **167.** $-\pi/4$   **169.** $1$
**171.** $\sqrt{3}/2$   **173.** $\sqrt{3}/3$   **175.** $-\frac{3}{5}$
**177.** $\frac{24}{25}$   **179.** $\pi/7$   **181.** $\sqrt{2}/2$
**191. (a)** $0.0625$

# CHAPTER 9
# ADDITIONAL TOPICS IN TRIGONOMETRY

## 9.1 The Law of Sines and the Law of Cosines

**1.** $AC = 4\sqrt{6}$ cm   **3.** $BC = 20 \sin 50°$ cm
**5.** $a \approx 9.7$ cm; $c \approx 16.4$ cm
**7.** $A \approx 63.3°$; $C \approx 50.7°$; $c \approx 25.5$ cm
**9. (a)** $\angle B = 45°$ or $135°$
**(b)** $\angle A = 150°$
**(c)** $\angle D \approx 14.5°$ or $165.5°$
**11. (c)** $C = 105°$; $c \approx 1.93$
**(d)** $C = 15°$; $c \approx 0.52$
**13.** $a = \dfrac{2 \sin 70°}{\sin 20°}$ cm; $b = \dfrac{2 \sin 50°}{\sin 20°}$ cm;
$c = \dfrac{2 \sin 50° \sin 70°}{\sin 20° \sin 85°}$ cm;
$d = \dfrac{2 \sin 50° \sin 15°}{\sin 20° \sin 85°}$ cm
**15. (a)** $x = 21$ cm   **(b)** $x \approx 5.4$ in.
**17.** $\cos A = \frac{113}{140}$; $\cos B = \frac{29}{40}$; $\cos C = -\frac{5}{28}$
**19.** $5.9$ units   **21. (a)** $a \approx 4.2$ cm
**(b)** $C \approx 29.3°$   **(c)** $B \approx 110.7°$
**23.** $PD = \sqrt{2 - \sqrt{2}}$
**25. (b)** $a = 5, b = 3, c = 7$ (with $m = 2,$
$n = 1$)   **27.** $x \approx 881.5$ ft

## 9.2 Vectors in the Plane, Part 1: A Geometric Approach

**1.** $|\overrightarrow{PQ}| = \sqrt{34}$

**3.** $|\overrightarrow{SQ}| = \sqrt{10}$

**5.** $|\overrightarrow{OP}| = \sqrt{10}$

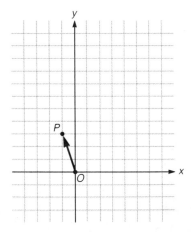

**7.** $|\overrightarrow{PQ} + \overrightarrow{QS}| = |\overrightarrow{PS}| = 6\sqrt{2}$

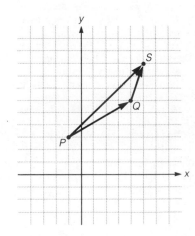

**9.** $|\overrightarrow{OP} + \overrightarrow{PQ}| = |\overrightarrow{OQ}| = 2\sqrt{13}$

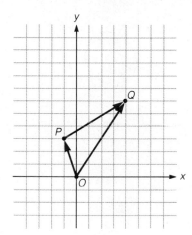

**15.** $|\overrightarrow{SR} + \overrightarrow{PO}| = 9$

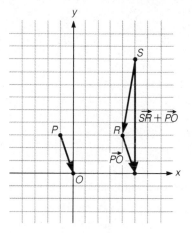

**21.** $|\overrightarrow{OP} + \overrightarrow{OR}| = 3\sqrt{5}$

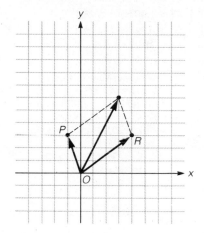

**11.** $|(\overrightarrow{OS} + \overrightarrow{SQ}) + \overrightarrow{QP}| = |\overrightarrow{OP}| = \sqrt{10}$

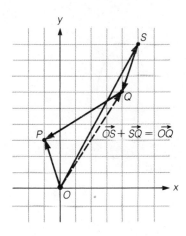

**17.** $|\overrightarrow{OP} + \overrightarrow{RQ}| = \sqrt{37}$

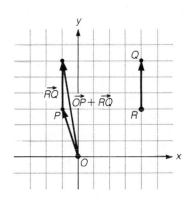

**23.** $|\overrightarrow{RP} + \overrightarrow{RS}| = 2\sqrt{13}$

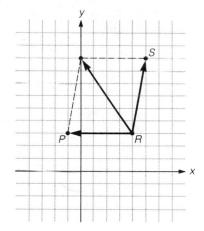

**13.** $|\overrightarrow{OP} + \overrightarrow{QS}| = 6$

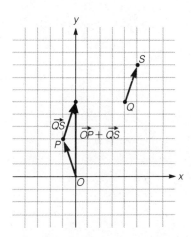

**19.** $|\overrightarrow{SQ} + \overrightarrow{RO}| = \sqrt{61}$

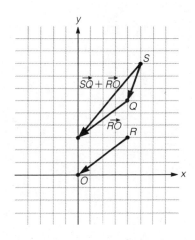

**25.** $|\overrightarrow{SO} + \overrightarrow{SQ}| = 6\sqrt{5}$

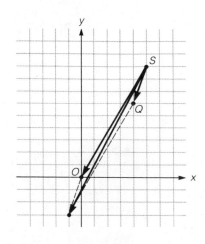

**27.** $|\mathbf{F} + \mathbf{G}| = \sqrt{41}\,\text{N}$; $\theta \approx 38.7°$

**29.** $|\mathbf{F} + \mathbf{G}| = 9\sqrt{2}\,\text{N}$; $\theta = 45°$

**31.** $|\mathbf{F} + \mathbf{G}| \approx 7.90\,\text{N}$; $\theta \approx 24.1°$

**33.** $|\mathbf{F} + \mathbf{G}| \approx 6.92\,\text{N}$; $\theta \approx 34.67°$

**35.** $|\mathbf{F} + \mathbf{G}| \approx 39.20\,\text{N}$; $\theta \approx 21.46°$

**37.** $|\mathbf{F} + \mathbf{G}| \approx 38.96\,\text{N}$; $\theta \approx 29.44°$

**39.** $V_x \approx 13.86\,\text{cm/sec}$; $V_y = 8\,\text{cm/sec}$

**41.** $F_x \approx 3.62\,\text{N}$; $F_y \approx 13.52\,\text{N}$

**43.** $V_x \approx -0.71\,\text{cm/sec}$; $V_y \approx 0.71\,\text{cm/sec}$

**45.** $F_x \approx -1.02\,\text{N}$; $F_y \approx 0.72\,\text{N}$

**47.** ground speed $\approx 301.04$ mph; drift angle $\approx 4.76°$; bearing $\approx 25.24°$

**49.** ground speed $\approx 293.47$ mph; drift angle $\approx 8.82°$; bearing $\approx 91.18°$

**51.** perpendicular $\approx 9.83$ lb; parallel $\approx 6.88$ lb

**53.** perpendicular $\approx 11.82$ lb; parallel $\approx 2.08$ lb

**55.** **(a)** initial point: $(-1,2)$; terminal point: $(2,-3)$ **(b)** initial point: $(-1,2)$; terminal point: $(2,-3)$

### 9.3 Vectors in the Plane, Part 2: An Algebraic Approach

**1.** $|\langle 4,3\rangle| = 5$

**3.** $|\langle -4,2\rangle| = 2\sqrt{5}$

**5.** $|\langle \frac{3}{4}, -\frac{1}{2}\rangle| = \sqrt{13}/4$

**7.** $\langle 1,4\rangle$ **9.** $\langle -1,1\rangle$ **11.** $\langle 8,-5\rangle$

**13.** $\langle 7,7\rangle$ **15.** $\langle 24,22\rangle$ **17.** $\sqrt{130}$

**19.** $2\sqrt{17} - \sqrt{13} - \sqrt{37}$ **21.** $\langle 13,6\rangle$

**23.** $\langle 14,21\rangle$ **25.** $\langle -3,-1\rangle$

**27.** $\langle 23,12\rangle$ **29.** $\langle -9,0\rangle$ **31.** $-48$

**33.** $3\mathbf{i} + 8\mathbf{j}$ **35.** $-8\mathbf{i} - 6\mathbf{j}$ **37.** $19\mathbf{i} + 23\mathbf{j}$

**39.** $\langle 1,1\rangle$ **41.** $\langle 5,-4\rangle$

**43.** $\langle \frac{1}{5}\sqrt{5}, \frac{2}{5}\sqrt{5}\rangle$ **45.** $\langle \frac{2}{5}\sqrt{5}, -\frac{1}{5}\sqrt{5}\rangle$

**47.** $\frac{8}{145}\sqrt{145}\mathbf{i} - \frac{9}{145}\sqrt{145}\mathbf{j}$

**49.** $u_1 = \sqrt{3}/2$, $u_2 = \frac{1}{2}$

**51.** $u_1 = -\frac{1}{2}$, $u_2 = \sqrt{3}/2$

**53.** $u_1 = -\sqrt{3}/2$, $u_2 = \frac{1}{2}$

**59.** **(a)** $|\overrightarrow{PQ}| = \sqrt{|\overrightarrow{OP}|^2 + |\overrightarrow{OQ}|^2 - 2(x_1 x_2 + y_1 y_2)}$

### 9.4 Introduction to Polar Coordinates

**1.** **(a)** $(-\frac{3}{2}, \frac{3}{2}\sqrt{3})$ **(b)** $(2\sqrt{3}, -2)$

**(c)** $(2\sqrt{3}, -2)$ **3.** **(a)** $(0,1)$

**(b)** $(0,1)$ **(c)** $(\frac{1}{2}\sqrt{2+\sqrt{2}}, \frac{1}{2}\sqrt{2-\sqrt{2}})$

**5.** $(\sqrt{2}, \frac{5}{4}\pi)$ **7.** $(x-1)^2 + y^2 = 1$

**9.** $x^4 + x^2 y^2 - y^2 = 0$

**11.** $x^6 - 9x^4 + 3x^4 y^2 + 18x^2 y^2 + 3x^2 y^4 - 9y^4 + y^6 = 0$

**13.** $\frac{1}{4}x^2 + \frac{1}{8}y^2 = 1$ **15.** $x + 2y = 1$

**17.** $r = 2/(3\cos\theta - 4\sin\theta)$

**19.** $r = \tan^2\theta\sec\theta$ **21.** $r^2 = \csc 2\theta$

**23.** $r^2 = 9/(9\cos^2\theta + \sin^2\theta)$

**25.**

$r = 3\cos\theta$

**27.**

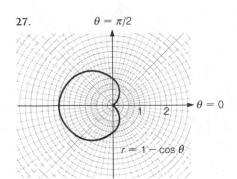

$r = 1 - \cos\theta$

**29.**

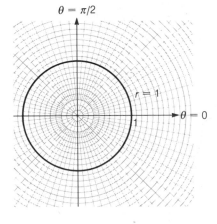

$r^2 = 2\sin 2\theta$

**31.**

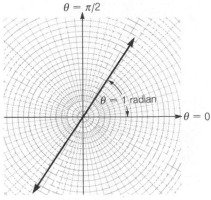

$r = 1$

**33.**

$\theta = \pi/2$, $\theta = 1$ radian, $\theta = 0$

**35.**

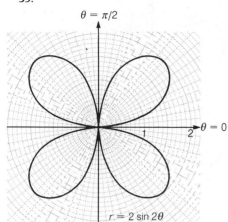

$r = 2 \sin 2\theta$

**41.**

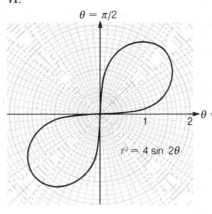

$r^2 = 4 \sin 2\theta$

**51. (c)**

$r^2 = 2k^2 \cos 2\theta$

**37**

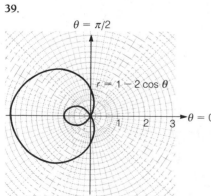

$r = 2 + \sin \theta$

**43.**

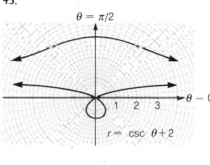

$r = \csc \theta + 2$

**45.**

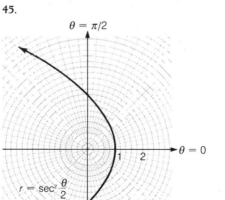

$r = \sec^2 \frac{\theta}{2}$

**39.**

$r = 1 - 2 \cos \theta$

**47.** $\left(x - \dfrac{a}{2}\right)^2 + \left(y - \dfrac{b}{2}\right)^2 = \dfrac{a^2 + b^2}{4}$;

center $= \left(\dfrac{a}{2}, \dfrac{b}{2}\right)$; radius $= \dfrac{\sqrt{a^2 + b^2}}{2}$

## 9.5 Trigonometric Form for Complex Numbers

**1.**

$4 + 2i$

**3.**

$-5 + i$

**5.**

$1 - 4i$

**7.**

$-i$

**9.** $\sqrt{2} + \sqrt{2}i$  **11.** $-2\sqrt{3} + 2i$
**13.** $-1 - i$  **15.** $\sqrt{3}i$
**17.** $(\sqrt{6} - \sqrt{2}) + (\sqrt{6} + \sqrt{2})i$
**19.** $\cos(\pi/6) + i\sin(\pi/6)$
**21.** $2[\cos(2\pi/3) + i\sin(2\pi/3)]$
**23.** $4[\cos(7\pi/6) + i\sin(7\pi/6)]$
**25.** $6[\cos(3\pi/2) + i\sin(3\pi/2)]$
**27.** $\frac{1}{2}[\cos(11\pi/6) + i\sin(11\pi/6)]$
**29.** $3 + 3\sqrt{3}i$  **31.** $1 - \sqrt{3}i$
**33.** $3\sqrt{2}[\cos(2\pi/7) + i\sin(2\pi/7)]$
**35.** $\frac{3}{2}\sqrt{2} + \frac{3}{2}\sqrt{2}i$  **37.** $\sqrt{3} + i$  **39.** $1$
**41.** $\frac{243}{2} - \frac{243}{2}\sqrt{3}i$  **43.** $\frac{1}{128}\sqrt{2} + \frac{1}{128}\sqrt{2}i$
**45.** $-4i$  **47.** $-1 - i$  **49.** $\frac{1}{2} + \frac{1}{2}\sqrt{3}i$
**51.** $128 + 128i$  **53.** $-128 + 128\sqrt{3}i$
**55.** $3i, -\frac{3}{2}\sqrt{3} - \frac{3}{2}i, \frac{3}{2}\sqrt{3} - \frac{3}{2}i$
**57.** $1, \frac{1}{2}\sqrt{2} + \frac{1}{2}\sqrt{2}i, i, -\frac{1}{2}\sqrt{2} + \frac{1}{2}\sqrt{2}i, -1,$
$-\frac{1}{2}\sqrt{2} - \frac{1}{2}\sqrt{2}i, -i, \frac{1}{2}\sqrt{2} - \frac{1}{2}\sqrt{2}i$
**59.** $4, -2 + 2\sqrt{3}i, -2 - 2\sqrt{3}i$
**61.** $3, \frac{3}{2} + \frac{3}{2}\sqrt{3}i, -\frac{3}{2} + \frac{3}{2}\sqrt{3}i, -3,$
$-\frac{3}{2} - \frac{3}{2}\sqrt{3}i, \frac{3}{2} - \frac{3}{2}\sqrt{3}i$  **63.** 92,236,816
**65.** $0.95 + 0.31i, i, -0.95 + 0.31i,$
$-0.59 - 0.81i, 0.59 - 0.81i$
**67.** Using addition formulas:
$\frac{1}{2}(\sqrt{6} - \sqrt{2}) + \frac{1}{2}(\sqrt{6} + \sqrt{2})i,$
$-\frac{1}{2}(\sqrt{6} + \sqrt{2}) + \frac{1}{2}(\sqrt{6} - \sqrt{2})i,$
$\frac{1}{2}(\sqrt{2} - \sqrt{6}) - \frac{1}{2}(\sqrt{2} + \sqrt{6})i,$
$\frac{1}{2}(\sqrt{2} + \sqrt{6}) + \frac{1}{2}(\sqrt{2} - \sqrt{6})i;$ using

half-angle formulas: $\sqrt{2 - \sqrt{3}} + \sqrt{2 + \sqrt{3}}i,$
$-\sqrt{2 + \sqrt{3}} + \sqrt{2 - \sqrt{3}}i,$
$-\sqrt{2 - \sqrt{3}} - \sqrt{2 + \sqrt{3}}i,$
$\sqrt{2 + \sqrt{3}} - \sqrt{2 - \sqrt{3}}i$
**69.** (a) $1, -\frac{1}{2} + \frac{1}{2}\sqrt{3}i, -\frac{1}{2} - \frac{1}{2}\sqrt{3}i$
**71.** $-1$  **73.** $1$

## Chapter 9 Review Exercises

**1.** $C = 55°$; $a \approx 12.6\,\text{cm}$; $b \approx 19.5\,\text{cm}$
**3.** (a) $B \approx 62.4°$; $C \approx 65.6°$; $c \approx 9.2\,\text{cm}$
(b) $B \approx 117.6°$; $C \approx 10.4°$; $c \approx 1.8\,\text{cm}$
**5.** $c \approx 7.7\,\text{cm}$; $A \approx 108.5°$; $B \approx 47.5°$
**7.** $C \approx 106.6°$; $B \approx 48.2°$; $A \approx 25.2°$
**9.** $9.21\,\text{cm}$  **11.** $32.48\,\text{cm}^2$
**13.** $55.23\,\text{cm}^2$  **15.** $7.89\,\text{cm}$  **17.** $15.43\,\text{cm}$
**19.** $15\,\text{cm}$  **21.** $11\,\text{cm}$
**23.** magnitude: $25\,\text{N}$; direction: $53.1°$
**25.** $v_x \approx 41.0\,\text{cm/sec}$; $v_y \approx 28.7\,\text{cm/sec}$
**27.** weight: $36.4\,\text{lb}$; perpendicular: $33.2\,\text{lb}$
**29.** $b = \pm\sqrt{15}$  **31.** $\langle 10, 9 \rangle$  **33.** $\langle 12, 7 \rangle$
**35.** $48$  **37.** $\langle 12, 8 \rangle$  **39.** $\langle -6, 2 \rangle$
**41.** $\langle -7, -6 \rangle$  **43.** $7i - 6j$
**45.** $u_1 = 3\sqrt{13}/13, u_2 = 2\sqrt{13}/13$
**47.** $u_1 = (\sqrt{6} + \sqrt{2})/4, u_2 = (\sqrt{6} - \sqrt{2})/4$

**49.** (a)

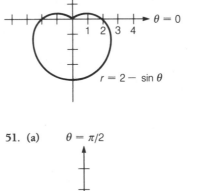

$r = 2 - 2\cos\theta$

(b)

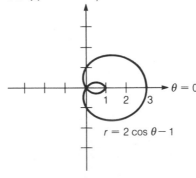

$r = 2 - \sin\theta$

**51.** (a)

$r = 2\cos\theta - 1$

(b)

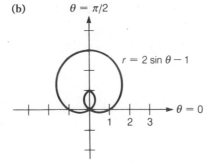

$r = 2\sin\theta - 1$

**53.** (a)

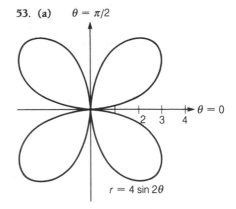

$r = 4\sin 2\theta$

(b)

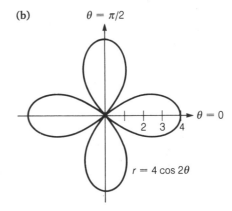

$r = 4\cos 2\theta$

**55.** (a)

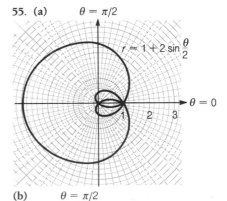

$r = 1 + 2\sin\dfrac{\theta}{2}$

(b)

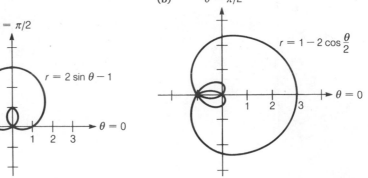

$r = 1 - 2\cos\dfrac{\theta}{2}$

57. $\frac{3}{2} + \frac{3}{2}\sqrt{3}i$  59. $2^{-1/4} - 2^{-1/4}i$
61. $1[\cos(\pi/3) + i\sin(\pi/3)]$
63. $6[\cos(5\pi/4) + i\sin(5\pi/4)]$
65. $5\sqrt{2} + 5\sqrt{2}i$  67. $\sqrt{2} - \sqrt{2}i$
69. $3[\cos(5\pi/9) + i\sin(5\pi/9)]$
71. $\frac{27}{2} + \frac{27}{2}\sqrt{3}i$  73. $512 - 512\sqrt{3}i$
75. $1, \frac{1}{2} + \frac{1}{2}\sqrt{3}i, -\frac{1}{2} + \frac{1}{2}\sqrt{3}i, -1,$
$-\frac{1}{2} - \frac{1}{2}\sqrt{3}i, \frac{1}{2} - \frac{1}{2}\sqrt{3}i$
77. $-\frac{1}{2}\sqrt{4 + 2\sqrt{2}} + \frac{1}{2}\sqrt{4 - 2\sqrt{2}}i,$
$-\frac{1}{2}\sqrt{4 + 2\sqrt{2}} - \frac{1}{2}\sqrt{4 - 2\sqrt{2}}i$
79. $1.06 + 0.17i, 0.17 + 1.06i,$
$-0.95 + 0.49i, -0.76 - 0.76i,$
$0.49 - 0.95i$
81. $\cos 3\theta = \cos^3 \theta - 3\cos\theta\sin^2\theta,$
$\sin 3\theta = 3\cos^2\theta\sin\theta - \sin^3\theta$

## CHAPTER 10
## SYSTEMS OF EQUATIONS

### 10.1 Systems of Two Linear Equations in Two Unknowns

1. $a, c, d$  3. yes  5. no  7. yes
9. $(-6, \frac{1}{2})$  11. $(-\frac{5}{2}, \frac{13}{2})$  13. $(\frac{1}{5}, \frac{7}{10})$
15. $(-\frac{60}{13}, \frac{60}{13})$  17. $(\sqrt{6}, 1)$  19. $(-\frac{2}{9}, \frac{23}{27})$
21. $(-\frac{283}{242}, -\frac{3}{121})$  23. $(-\frac{226}{25}, -\frac{939}{50})$
25. $(\frac{5}{8}, 0)$
27. $\left(\dfrac{4\sqrt{3} - \sqrt{10} - 5\sqrt{2}}{4}, \dfrac{-3 - \sqrt{5}}{2}\right)$
29. $b = 3, c = 4$  31. $A = -28, B = -22$
33. $\frac{49}{20}$ square units  35. 80 cm³ of the 10% solution, 120 cm² of the 35% solution
37. 8 lb of the $5.20 coffee, 8 lb of the $5.80 coffee
39. $\left(\dfrac{ab}{a + b}, \dfrac{ab}{a + b}\right), a \neq \pm b$
41. $\left(\dfrac{a + b}{ab}, -\dfrac{1}{ab}\right), ab \neq 0$
43. $(-\frac{1}{11}, \frac{1}{9})$  45. $(5, 2)$  47. $(1.32, -1.62)$
49. $(4.04, 0.44)$  51. 59  53. 12 in., 5 in.
55. $(60, 40)$  57. $(0, a)$  59. $(1/b, 1/a)$
61. $-\frac{29}{7}, 2$

### 10.2 Gaussian Elimination

1. $(-3, -2, -1)$  3. $(-\frac{1}{60}, -\frac{2}{15}, \frac{3}{5})$
5. $(\frac{25}{36}, \frac{5}{9}, -\frac{1}{3})$  7. $(x, x/8, 0)$, where $x$ is any real number  9. $(-1, 0, 1, -5)$
11. $(\frac{11}{3}, \frac{8}{3}, \frac{17}{3})$  13. $(1, 0, 1)$  15. $(2, 3, 1)$
17. no solution
19. $\left(-\dfrac{z + 1}{7}, \dfrac{5z + 5}{7}, z\right)$, where $z$ is any real number
21. $\left(\dfrac{11 - 5z}{7}, \dfrac{-3z - 6}{7}, z\right)$, where $z$ is any real number  23. $(4, 1, -3, 2)$

25. $\left(\dfrac{17 - 17z}{5}, \dfrac{8z - 3}{5}, z\right)$, where $z$ is any real number
27. $\left(\dfrac{12 + 10w}{11}, \dfrac{146 + 19w}{55}, \dfrac{61w + 159}{55}, w\right)$, where $w$ is any real number
29. $\left(-\dfrac{5z}{12}, \dfrac{2z + 3}{3}, z\right)$, where $z$ is any real number  31. (a) $A = \frac{1}{4}, B = -\frac{1}{4}$
(b) $A = \frac{1}{2}, B = \frac{1}{2}$  33. $A = \frac{4}{3}, B = \frac{5}{3}$
35. (a) $A = \frac{3}{8}, B = -\frac{3}{8}$
(b) $A = \frac{3}{64}, B = -\frac{3}{64}, C = -\frac{3}{8}$
37. $A = \frac{1}{3}, B = -\frac{1}{3}$, and $C = \frac{2}{3}$
39. $A = 4, B = 4$
41. $A = \frac{1}{2}, B = \frac{1}{2}, C = -\frac{1}{2}, D = \frac{1}{2}$
43. $17x^2 + 17y^2 - 49x + 65y - 166 = 0$
45. $x = \ln a, y = \ln 2a, z = \ln(a/2)$
49. (a) $A = 1/(a - b), B = 1/(b - a)$
(b) $A = (ap + q)/(a - b),$
$B = (bp + q)/(b - a)$
51. $A = (a^2 + pa + q)/(a - b)(a - c),$
$B = (b^2 + pb + q)/(b - a)(b - c),$
$C = (c^2 + pc + q)/(c - a)(c - b)$
53. 60 miles

### 10.3 Matrices

1. (a) $2 \times 3$  (b) $3 \times 2$  3. $5 \times 4$
5. coefficient matrix: $\begin{pmatrix} 2 & 3 & 4 \\ 5 & 6 & 7 \\ 8 & 9 & 10 \end{pmatrix}$
augmented matrix: $\begin{pmatrix} 2 & 3 & 4 & 10 \\ 5 & 6 & 7 & 9 \\ 8 & 9 & 10 & 8 \end{pmatrix}$
7. coefficient matrix: $\begin{pmatrix} 1 & 0 & 1 & 1 \\ 1 & 1 & 0 & 2 \\ 0 & 1 & 1 & 1 \\ 2 & -1 & -1 & 0 \end{pmatrix}$
augmented matrix: $\begin{pmatrix} 1 & 0 & 1 & 1 & -1 \\ 1 & 1 & 0 & 2 & 0 \\ 0 & 1 & 1 & 1 & 1 \\ 2 & -1 & -1 & 0 & 2 \end{pmatrix}$
9. $(-1, -2, 3)$  11. $(-5, 1, 3)$
13. $(3, 0, -7)$  15. $(8, 9, -1)$
17. $\left(\dfrac{9z + 5}{19}, \dfrac{31z - 6}{19}, z\right)$, for any real number $z$  19. $(2, -1, 0, 3)$
21. no solution
23. $\begin{pmatrix} 3 & 2 \\ 2 & 4 \end{pmatrix}$  25. $\begin{pmatrix} 6 & 4 \\ 4 & 8 \end{pmatrix}$
27. $\begin{pmatrix} 11 & -2 \\ 11 & 1 \end{pmatrix}$  29. $\begin{pmatrix} 2 & 3 \\ -1 & 4 \end{pmatrix}$
31. not defined  33. $\begin{pmatrix} 10 & -2 \\ -8 & 0 \\ 4 & 6 \end{pmatrix}$

35. $\begin{pmatrix} 2 & 4 & 11 \\ -12 & 16 & 19 \\ 14 & 12 & 43 \end{pmatrix}$
37. $\begin{pmatrix} -9 & 10 & 10 \\ 4 & -8 & -12 \\ 10 & 4 & 21 \end{pmatrix}$
39. not defined
41. $\begin{pmatrix} 0 & 0 & 0 \\ 0 & 0 & 0 \\ 0 & 0 & 0 \end{pmatrix}$  43. $\begin{pmatrix} 4 & 2 \\ 2 & 5 \end{pmatrix}$
45. not defined
47. $\begin{pmatrix} 1 & 18 \\ -6 & 13 \end{pmatrix}$  49. $\begin{pmatrix} -16 & 75 \\ -25 & 34 \end{pmatrix}$
51. (a) $\begin{pmatrix} -13 & 1 & 40 \\ 43 & 17 & 0 \\ 89 & 61 & 60 \end{pmatrix}$
(b) $\begin{pmatrix} -13 & 1 & 40 \\ 43 & 17 & 0 \\ 89 & 61 & 60 \end{pmatrix}$
(c) $\begin{pmatrix} -52 & -82 & 61 \\ 87 & 141 & 0 \\ 216 & 318 & 165 \end{pmatrix}$
(d) $\begin{pmatrix} -52 & -82 & 61 \\ 87 & 141 & 0 \\ 216 & 318 & 165 \end{pmatrix}$
53. (a) $\begin{pmatrix} 16 & 20 \\ 24 & 28 \end{pmatrix}$
(b) $\begin{pmatrix} 18 & 26 \\ 18 & 26 \end{pmatrix}$  (c) $\begin{pmatrix} 14 & 14 \\ 30 & 30 \end{pmatrix}$
(d) $\begin{pmatrix} 18 & 26 \\ 18 & 26 \end{pmatrix}$  55. (a) $\begin{pmatrix} -x \\ y \end{pmatrix}$
(b) $\begin{pmatrix} x \\ -y \end{pmatrix}$  (c) $\begin{pmatrix} -1 & 0 \\ 0 & -1 \end{pmatrix}$,
reflection about origin  57. (a) $f(A) = -2$; $f(B) = 29; f(AB) = -58$; yes

### 10.4 The Inverse of a Square Matrix

1. $\begin{pmatrix} 4 & -1 \\ -5 & 2 \end{pmatrix}$  3. $\begin{pmatrix} 3 & 0 & -2 \\ 0 & 5 & 6 \\ 1 & 4 & -7 \end{pmatrix}$
5. $\begin{pmatrix} -5 & 9 \\ 4 & -7 \end{pmatrix}$  7. $\begin{pmatrix} -\frac{6}{23} & \frac{1}{23} \\ \frac{5}{23} & \frac{3}{23} \end{pmatrix}$
9. no $A^{-1}$ exists.  11. $\begin{pmatrix} 2 & -3 \\ 1 & 3 \end{pmatrix}$
13. $\begin{pmatrix} 2 & -1 \\ -3 & 2 \end{pmatrix}$  15. $\begin{pmatrix} \frac{6}{11} & 1 \\ -\frac{1}{11} & 0 \end{pmatrix}$
17. does not exist  19. $\begin{pmatrix} -1 & 2 & -3 \\ 2 & 1 & 0 \\ 4 & -2 & 5 \end{pmatrix}$

**21.** $\begin{pmatrix} 5 & -\frac{10}{3} & 1 \\ 0 & \frac{1}{3} & 0 \\ 4 & -\frac{8}{3} & 1 \end{pmatrix}$ **23.** $\begin{pmatrix} 2 & 1 & 4 \\ 3 & 2 & 5 \\ 0 & -1 & 1 \end{pmatrix}$

**25.** does not exist **27.** (a) $x = -1, y = 1$
(b) $x = -132, y = 48$ **29.** (a) $x = 2$,
$y = -1, z = 4$ **(b)** $x = 1, y = 1, z = -2$

**31.** (a) $\begin{pmatrix} 1 & 0 & 0 \\ 0 & 1 & 0 \\ 0 & 0 & 1 \end{pmatrix}$, equal to $I_3$

(b) $x = \frac{1}{2}, y = -1, z = 1$

**35.** (a) $A^{-1} = \begin{pmatrix} -\frac{5}{2} & \frac{3}{2} \\ 2 & -1 \end{pmatrix}$;

$B^{-1} = \begin{pmatrix} 7 & -8 \\ -6 & 7 \end{pmatrix}$;

$B^{-1}A^{-1} = \begin{pmatrix} -\frac{67}{2} & \frac{37}{2} \\ 29 & -16 \end{pmatrix}$

(b) $\begin{pmatrix} -\frac{67}{2} & \frac{37}{2} \\ 29 & -16 \end{pmatrix}$, same as (a)

## 10.5 Determinants and Cramer's Rule

**1.** (a) 29 (b) $-29$ **3.** (a) 0 (b) 0
**5.** $-1$ **7.** $-60$ **9.** 9 **11.** (a) 314
(b) 674 (c) part (b) **13.** (a) 0 (b) 0
(c) 0 (d) 0 **15.** 0 **17.** $-3$ **19.** 6848
**21.** 17120 **23.** $(y - x)(z - x)(z - y)$
**25.** $xy$ **27.** 20 **29.** 120 **33.** $(1,1,2)$
**35.** $(2, -3, 6)$ **37.** $(0,0,0)$
**39.** $(13 - \frac{11}{3}y, y, 13 - 4y)$, for any real
number $y$ **41.** $(1, 0, -10, 2)$
**43.** $x = 4, -4, -1$ **55.** (a) 1 (b) 1
**59.** $\left( \dfrac{k(k - b)(k - c)}{a(a - b)(a - c)}, \dfrac{k(k - a)(k - c)}{b(b - a)(b - c)}, \right.$
$\left. \dfrac{k(k - a)(k - b)}{c(c - a)(c - b)} \right)$

**63.** (b) $\begin{pmatrix} -\frac{9}{61} & \frac{7}{61} \\ \frac{1}{61} & \frac{6}{61} \end{pmatrix}$

## 10.6 Nonlinear Systems of Equations

**1.** $(0,0), (3,9)$ **3.** $(2\sqrt{6}, 1), (-2\sqrt{6}, 1)$
**5.** $(-1, -1)$ **7.** $(2,3), (2, -3)$,
$(-2, 3), (-2, -3)$ **9.** $(1, 0), (-1, 0)$
**11.** $(-\frac{1}{8} + \frac{1}{8}\sqrt{65}, \frac{1}{2} + \frac{1}{2}\sqrt{65})$,
$(-\frac{1}{8} - \frac{1}{8}\sqrt{65}, \frac{1}{2} - \frac{1}{2}\sqrt{65})$
**13.** $(\frac{1}{17}\sqrt{17}, 1), (\frac{1}{17}\sqrt{17}, -1)$,
$(-\frac{1}{17}\sqrt{17}, 1), (-\frac{1}{17}\sqrt{17}, -1)$
**15.** $(1, 0), (4, -\sqrt{3})$ **17.** $(2, 4)$
**19.** $(100, 1000), (100, \frac{1}{1000}), (\frac{1}{100}, 1000)$,
$(\frac{1}{100}, \frac{1}{1000})$
**21.** $\left( \dfrac{2\ln 2 - 5\ln 3}{\ln 2 - \ln 3}, \dfrac{3\ln 2}{\ln 2 - \ln 3} \right)$
**25.** $(1/a, 1/b)$ **27.** $(9, 14), (14, 9)$
**29.** $\dfrac{p - \sqrt{2d^2 - p^2}}{2}, \dfrac{p + \sqrt{2d^2 - p^2}}{2}$

**31.** $(5, 4), (5, -4)$ **33.** $\left( \dfrac{p^2}{A}, \dfrac{q^2}{A}, \dfrac{r^2}{A} \right)$,
$\left( \dfrac{p^2}{A}, \dfrac{-q^2}{A}, \dfrac{-r^2}{A} \right)$, where $A = \sqrt{p^2 + q^2 + r^2}$

**35.** 9 cm, 40 cm **37.** 3 cm, 20 cm
**39.** $(1, 2), (-1, -2), (2, 1), (-2, -1)$
**41.** $(\frac{1}{2} + \frac{1}{2}\sqrt{13}, -\frac{1}{2} + \frac{1}{2}\sqrt{13})$,
$(-\frac{1}{2} + \frac{1}{2}\sqrt{13}, \frac{1}{2} + \frac{1}{2}\sqrt{13})$,
$(\frac{1}{2} - \frac{1}{2}\sqrt{13}, -\frac{1}{2} - \frac{1}{2}\sqrt{13})$,
$(-\frac{1}{2} - \frac{1}{2}\sqrt{13}, \frac{1}{2} - \frac{1}{2}\sqrt{13})$
**43.** $(2, 3), (-2, -3), (1, 2), (-1, -2)$
**45.** $(e^{9/2}, e^3)$

## 10.7 Systems of inequalities Linear Programming

**1.** (a) no (b) yes

**3.**

**5.**

**7.**

**9.**

**11.**

**13.**

**15.**

**17.**

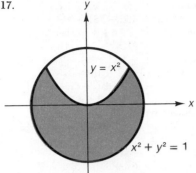

$y = x^2$

$x^2 + y^2 = 1$

**19.**

$y = |x|$

$y = 1$

**21.**

$y = 1 - x^2$

**23.** convex: yes; bounded: yes;
vertices: $(0,0), (7,0), (3,8), (0,5)$

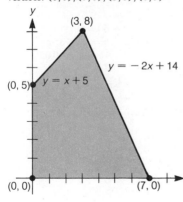

$(3, 8)$

$y = -2x + 14$

$(0, 5)$  $y = x + 5$

$(0, 0)$  $(7, 0)$

**25.** convex: yes; bounded: yes; ·
vertices: $(0,0), (0,4), (3,5), (8,0)$

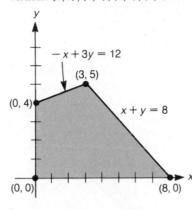

$-x + 3y = 12$

$(3, 5)$

$(0, 4)$  $x + y = 8$

$(0, 0)$  $(8, 0)$

**27.** convex: yes; bounded: no;
vertices: $(2,7), (8,5)$

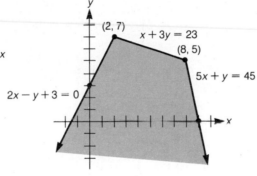

$(2, 7)$  $x + 3y = 23$

$(8, 5)$

$5x + y = 45$

$2x - y + 3 = 0$

**29.** convex: yes; bounded: no;
vertex: $(6,0)$

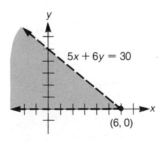

$5x + 6y = 30$

$(6, 0)$

**31.** convex: yes; bounded: yes;
vertices: $(0,0), (0,5), (6,0)$

$(0, 5)$

$5x + 6y = 30$

$(0, 0)$  $(6, 0)$

**33.** convex: yes; bounded: yes;
vertices: $(5,30), (10,30), (20,15), (20,20)$

$(5, 30)$

$(10, 30)$

$(20, 20)$

$(20, 15)$

**35.** max $= 9$ at $(3,4)$;  min $= -5$ at $(5,0)$
**37.** max $= 4$ at $(1,2)$;  min $= 0$ at $(0,0)$
**39.** max $= 113$ at $(6,6)$;  min $= 27$ at $(2,1)$

**41.** max $= 188$ at $(6,7)$;  min $= 100$ at
$(0,0)$   **43. (a)** max $= 819$ at $(1,8)$;
min $= 0$ at $(0,0)$    **(b)**  max $= 826$ at $(6,7)$;
min $= 0$ at $(0,0)$    **45.** $W_1$ to $D_1$: 0 cars;
$W_2$ to $D_1$: 40 cars; $W_1$ to $D_2$: 25 cars;
$W_2$ to $D_2$: 0 cars; min. cost \$10,150
**47.** \$766.20: 833 units of $A$, 333 units of $B$
**49.** \$23,700: 150 acres of cherry,
450 acres of regular
**51.** \$4000 in stock $A$, \$8000 in stock $B$
**53.** \$4000 in $A$, \$1000 in $B$, \$5000 in $C$
**55.** 10 lb of $A$, 4 lb of $B$

**Chapter 10 Review Exercises**

**1.** $(3, -5)$   **3.** $(-1, 4)$   **5.** $(-3, 15)$
**7.** $\left(-\frac{18}{5}, \frac{8}{5}\right)$   **9.** $\left(\frac{2}{3}, -\frac{1}{5}\right)$
**11.** $(-12, -8)$   **13.** $\left(\frac{1}{3}, -\frac{1}{4}\right)$
**15.** $\left(\dfrac{-1}{a^2 - 3a + 1}, \dfrac{1-a}{a^2 - 3a + 1}\right)$, where
$a^2 - 3a + 1 \neq 0$, or $a \neq (3 \pm \sqrt{5})/2$
**17.** $(a^2, 1 - a^2)$   **19.** $(a^2 - b^2, a^2 + b^2)$
**21.** $\left(\dfrac{pq(p+q)}{p^2 + q^2}, \dfrac{p^3 - q^3}{p^2 + q^2}\right)$, where
$(p,q) \neq (0,0)$

**23.** $\left(\dfrac{a}{a-b}, \dfrac{b}{a+b}\right)$, where $a \neq \pm b$,

$9a - 8b \neq 0$ **25.** $(2,3,4)$
**27.** $(-1,-2,0)$ **29.** no solution
**31.** $(x, 6 - 2x, -1)$, for any real number $x$
**33.** no solution **35.** $(2b - z, a - b, z)$, for any real number $z$ **37.** $(4, 3, -1, 2)$
**39.** $A = \frac{1}{20}, B = -\frac{1}{20}.$ **41.** $A = 2, B = -2$
**43.** $A = -\frac{5}{4}, B = \frac{5}{4}$
**45.** $A = \frac{1}{16}, B = -\frac{1}{16}, C = -\frac{1}{4}$
**47.** $A = 3, B = 1, C = 0$
**49.** $A = \frac{1}{48}, B = -\frac{1}{48}, C = \frac{1}{6}$
**51.** $A = 0, B = 1, C = a$
**53.** $A = b, B = -a$ **55.** 34 **57.** $-56$
**59.** 0 **61.** 24 **67.** $a = 2, b = 1$
**69.** (a) $(\frac{2}{3}, 2)$ (b) $(\frac{2}{3}, 2)$
(c) $(\frac{2}{3}, 2)$ (d) all are 2
**73.** $(x + \frac{17}{6})^2 + (y + \frac{8}{3})^2 = \frac{245}{36}$

**75.** $\begin{pmatrix} 10 & -2 \\ 4 & 26 \end{pmatrix}$ **77.** $\begin{pmatrix} 8 & 4 \\ 4 & 32 \end{pmatrix}$

**79.** $\begin{pmatrix} 4 & -13 \\ 7 & 41 \end{pmatrix}$ **81.** $\begin{pmatrix} -3 & -14 \\ -4 & 3 \end{pmatrix}$

**83.** $\begin{pmatrix} 1 & 1 \\ 1 & 7 \end{pmatrix}$ **85.** $\begin{pmatrix} 1 & -11 \\ 6 & 36 \end{pmatrix}$

**87.** $\begin{pmatrix} 4 & 3 \\ 10 & 33 \end{pmatrix}$ **89.** $\begin{pmatrix} -42 & 58 \\ 5 & 20 \end{pmatrix}$

**91.** undefined **93.** undefined

**95.** $\begin{pmatrix} 4 & -1 \\ 2 & 12 \end{pmatrix}$ **97.** $\begin{pmatrix} -4 & 13 \\ -7 & -41 \end{pmatrix}$

**101.** (a) $\begin{pmatrix} -9 & 5 \\ 2 & -1 \end{pmatrix}$

(b) $x = -47, y = 10$

**103.** (a) $\begin{pmatrix} 10 & -2 & 5 \\ 6 & -1 & 4 \\ 1 & 0 & 1 \end{pmatrix}$

(b) $x = 16, y = 15, z = 4$

**105.** $\begin{pmatrix} -65 & 20 & 9 & 133 \\ 3 & -1 & 0 & -6 \\ 26 & -8 & -4 & -53 \\ -23 & 7 & 3 & 47 \end{pmatrix}$

**107.** $D = -20, D_x = 12, D_y = 13,$
$D_z = -31$; solution: $(-\frac{3}{5}, -\frac{13}{20}, \frac{31}{20})$
**109.** $D = 0$; no solution **111.** $D = 0$;
solution: $(-5 - 4y, y, -8 - 5y)$,
for any real number $y$
**113.** $D = 149, D_x = 596, D_y = 149,$
$D_z = -149, D_w = 447$;
solution: $(4, 1, -1, 3)$ **115.** $(0,0), (6,36)$
**117.** $(3,0), (-3,0)$
**119.** $(\frac{5}{2}\sqrt{2}, \frac{1}{2}\sqrt{14}), (-\frac{5}{2}\sqrt{2}, \frac{1}{2}\sqrt{14}),$
$(\frac{5}{2}\sqrt{2}, -\frac{1}{2}\sqrt{14}), (-\frac{5}{2}\sqrt{2}, -\frac{1}{2}\sqrt{14})$
**121.** $(-\frac{1}{2} + \frac{1}{2}\sqrt{5}, \frac{1}{2}\sqrt{-2 + 2\sqrt{5}})$
**123.** $(\frac{1}{4}\sqrt{-2 + 2\sqrt{17}}, -\frac{1}{4} + \frac{1}{4}\sqrt{17}),$
$(-\frac{1}{4}\sqrt{-2 + 2\sqrt{17}}, -\frac{1}{4} + \frac{1}{4}\sqrt{17})$

**125.** $(\sqrt{2}, 3\sqrt{2}), (-\sqrt{2}, -3\sqrt{2}),$
$(\frac{7}{3}\sqrt{22}, \frac{31}{33}\sqrt{22}), (-\frac{7}{33}\sqrt{22}, -\frac{31}{33}\sqrt{22})$
**127.** $(5,2), (5,-4), (1,2), (1,-4)$
**129.** $as/(a + b)$ and $bs/(a + b)$
**131.** $(24, 60, 120)$
**133.** $\frac{1}{2}(\sqrt{n + 2m} + \sqrt{n - 2m})$ and
$\frac{1}{2}(\sqrt{n + 2m} - \sqrt{n - 2m});$
$\frac{1}{2}(\sqrt{n - 2m} - \sqrt{n + 2m})$ and
$\frac{1}{2}(-\sqrt{n - 2m} - \sqrt{n + 2m})$
**135.** convex: no; bounded: no

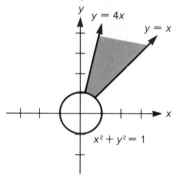

**137.** convex: yes; bounded: yes

**139.** convex: no; bounded: no

**141.** $1100 with 50 small, 25 large

---

## CHAPTER 11
## ROOTS OF POLYNOMIAL EQUATIONS

### 11.1 More on Division of Polynomials

**1.** quotient: $x - 1$; remainder: $-7$;
$x^2 - 6x - 2 = (x - 5)(x - 1) - 7$
**3.** quotient: $4x - 5$; remainder: 0;
$4x^2 - x - 5 = (x + 1)(4x - 5) + 0$
**5.** quotient: $6x^2 + 19x + 78$;

remainder: 313; $6x^3 - 5x^2 + 2x + 1 =$
$(x - 4)(6x^2 + 19x + 78) + 313$
**7.** quotient: $x^2 + 2x + 4$; remainder: 7;
$x^3 - 1 = (x - 2)(x^2 + 2x + 4) + 7$
**9.** quotient: $x^4 - 2x^3 + 4x^2 - 8x + 16$;
remainder: $-33$; $x^5 - 1 =$
$(x + 2)(x^4 - 2x^3 + 4x^2 - 8x + 16) - 33$
**11.** quotient: $x^3 - 10x^2 + 40x - 160$;
remainder: 642; $x^4 - 6x^3 + 2 =$
$(x + 4)(x^3 - 10x^2 + 40x - 160) + 642$
**13.** quotient: $x^2 + 6x + 57$; remainder:
576; $x^3 - 4x^2 - 3x + 6 =$
$(x - 10)(x^2 + 6x + 57) + 576$
**15.** quotient: $x^2 - 6x + 30$;
remainder: $-150$;
$x^3 - x^2 = (x + 5)(x^2 - 6x + 30) - 150$
**17.** quotient: $54x^2 + 9x - 21$; remainder:
$0; 54x^3 - 27x^2 - 27x + 14 =$
$(x - \frac{2}{3})(54x^2 + 9x - 21) + 0$
**19.** quotient: $x^3 + 3x^2 + 12x + 36$;
remainder: 120; $x^4 + 3x^2 + 12 =$
$(x - 3)(x^3 + 3x^2 + 12x + 36) + 120$
**21.** $q(x) = ax + (ar + b)$;
$R(x) = ar^2 + br + c$
**23.** $q(x) = 2x$; $R(x) = \frac{1}{3}$
**25.** $q(x) = 3x^2 - \frac{3}{2}x + \frac{3}{4}$;
$R(x) = \frac{1}{8}$ **27.** 4
**31.** $q(x) = x + (-4 + i)$; $R(x) = -4i$
**33.** $q(x) = x + (-1 + i)$; $R(x) = 0$

### 11.2 Roots of Polynomial Equations: The Remainder Theorem and the Factor Theorem

**1.** yes **3.** yes **5.** yes **7.** yes **9.** yes
**11.** no **13.** (a) yes (b) no
**15.** (a) 1, 2 (multiplicity 3), 3
(b) 1 (multiplicity 3) (c) 5
(multiplicity 6), $-1$ (multiplicity 4)
(d) 0 (multiplicity 5), 1 **17.** $-170$
**19.** $-9$ **21.** $-3\sqrt{2} - 2$ **23.** $-22$
**25.** $-3, 4, 3$ **27.** $1, -1 + \sqrt{6}, -1 - \sqrt{6}$
**29.** $-2, \frac{2}{3}, 3$ **31.** $-\frac{3}{2}, \frac{1}{2}(1 + \sqrt{5}),$
$\frac{1}{2}(1 - \sqrt{5})$ **33.** $0, 5$ **35.** $-4, 3$
**37.** $-9, 1 + \sqrt{2}, 1 - \sqrt{2}$
**39.** $x^3 - 4x^2 - 17x + 60 = 0$
**41.** $x^3 + 8x^2 + 13x + 6 = 0$ **43.** $c = 5,$
$d = -2$ **45.** 20.44 **47.** (a) $-0.05$
(b) 0.07 **49.** yes **51.** yes **53.** $a = -1,$
$b = -1$ **55.** $\pm\frac{3}{2}\sqrt{2}$

### 11.3 The Fundamental Theorem of Algebra

**1.** (a) yes (b) yes (c) yes (d) no
**3.** $(x - (-1))(x - 3)$
**5.** $4(x - \frac{1}{4})(x - (-6))$

7. $(x - (-\sqrt{5}))(x - \sqrt{5})$
9. $(x - (5 + i))(x - (5 - i))$
11. $x^3 + x^2 - 5x + 3$
13. $x^4 - 16$   15. $x^6 + 10x^4 - 87x^2 + 144$
17. $-\frac{5}{42}x^2 + \frac{25}{42}x + \frac{30}{7}$
19. $\frac{1}{30}x^3 - \frac{19}{30}x + 1$
21. $x^2 + (i + \sqrt{3})x + i\sqrt{3} = 0$
23. $x^2 - 3x - 54 = 0$
25. $x^2 - 2x - 4 = 0$
27. $x^2 - 2ax + a^2 - b = 0$
29. $(x + 2 - 2i)(x + 2 + 2i)(x - 2 - 2i)$
$\times (x - 2 + 2i)$
33. $4, 3, -3$

## 11.4 Rational and Irrational Roots

1. $\pm 1, \pm \frac{1}{2}, \pm \frac{1}{4}, \pm 3, \pm \frac{3}{2}, \pm \frac{3}{4}$
3. $\pm 1, \pm \frac{1}{2}, \pm \frac{1}{4}, \pm \frac{1}{8}, \pm 3, \pm \frac{3}{2}, \pm \frac{3}{4}, \pm \frac{3}{8}, \pm 9,$
$\pm \frac{9}{2}, \pm \frac{9}{4}, \pm \frac{9}{8}$
5. $\pm 1, \pm \frac{1}{2}, \pm 2, \pm 3, \pm \frac{3}{2}, \pm 6$
13. $1, -1, -3$   15. $-\frac{1}{4}, -\sqrt{5}, \sqrt{5}$
17. $-1, -\frac{2}{3}, -\frac{1}{3}$
19. $1, -1, \frac{1}{2}(-1 + \sqrt{97}), \frac{1}{2}(-1 - \sqrt{97})$
21. $1$ (multiplicity 4)   23. $\frac{1}{2}, 6, -4$
25. $\frac{1}{4}, \frac{1}{5}, -\frac{1}{2}\sqrt{6}, \frac{1}{2}\sqrt{6}$
27. (a) upper: 2; lower: $-1$
(b) upper: 2; lower: $-1$   (c) upper: 6;
lower: $-2$   29. 0.68 and 0.69
31. 2.88 and 2.89   33. 4.31 and 4.32
35. $-2.15$ and $-2.14$
37. $-5.27$ and $-5.26$
39. (b) $A$ has a factor in common with $C$,
so the result does not apply.   (c) We
assume $p/q$ is a root.
41. $x = 0.32$

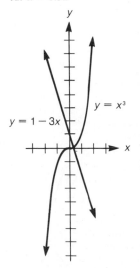

43. $x = 1.49$   45. $p = 3, x = 1$
47. (a) 1.27   (b) 1.272

## 11.5 Conjugate Roots and Descartes's Rule of Signs

1. $7 + 2i$   3. $5 - 2i, 3$
5. $-2 - i, -3$ (multiplicity 2)
7. $6 + 5i, -\frac{1}{4}$   9. $4 - \sqrt{2}i, -\frac{3}{2}i, \frac{3}{2}i$
11. $10 - 2i, 1 + \sqrt{5}, 1 - \sqrt{5}$
13. $\frac{1}{3}(1 - \sqrt{2}i), \frac{2}{5}$
15. $3 + 2i, -1 - i, -1 + \sqrt{2}, -1 - \sqrt{2}$
17. $x^2 - 2x - 5 = 0$
19. $x^2 - \frac{4}{3}x - \frac{2}{3} = 0$   21. 2 complex, 1
negative real   23. 4 complex, 1 negative
real   25. 2 complex, 1 positive real, 1
negative real   27. either 1 positive real
and 2 negative real or 1 positive real and 2
complex   29. 1 positive real, 1 negative
real, 6 complex   31. 1 positive real, 8
complex   33. 1 positive real, 1 negative
real, 6 complex   35. 1 positive real, 1
negative real, 4 complex
39. $f(x) = x^4 + 2x^2 + 49$

## Chapter 11 Review Exercises

1. quotient: $x^3 + 3x^2 + 7x + 21$;
remainder: 71; $x^4 - 2x^2 + 8 =$
$(x - 3)(x^3 + 3x^2 + 7x + 21) + 71$
3. quotient: $2x^2 - 13x + 46$;
remainder: $-187$; $2x^3 - 5x^2 - 6x - 3 =$
$(x + 4)(2x^2 - 13x + 46) - 187$
5. quotient: $5x - 20$; remainder: 0;
$5x^2 - 19x - 4 = (x + 0.2)(5x - 20) + 0$
7. quotient: $x - (3 - 2i)$; remainder: 0;
$x^2 - 6x + 13 =$
$(x - (3 + 2i))(x - (3 - 2i)) + 0$
9. (a) Let $p(x)$ and $d(x)$ be polynomials,
where $d(x) \neq 0$. Then there are unique
polynomials $q(x)$ and $R(x)$ such that
$p(x) = d(x) \cdot q(x) + R(x)$, where either
$R(x) = 0$ or the degree of $R(x)$ is less than
the degree of $d(x)$.   (b) When a
polynomial $f(x)$ is divided by $x - r$, the
remainder is $f(r)$.   11. 2   13. $-5a^4$
15. 0.331   17. (a) Let $f(x)$ be a
polynomial. If $f(r) = 0$, then $x - r$ is a
factor of $f(x)$. Conversely, if $x - r$ is a factor
of $f(x)$, then $f(r) = 0$.   (b) Every
polynomial equation of the form
$a_nx^n + a_{n-1}x^{n-1} + \cdots + a_1x + a_0 = 0$
$(n \geq 1, a_n \neq 0)$ has at least one root among
the complex numbers (This root may be a
real number.)   19. $5, -5$
21. $f(x) = 3x^2 + 21x - 24$
23. $f(x) = (x - 2)(x - 3i)^3(x - (1 + \sqrt{2}))^2$
27. (a) $\pm 1, \pm 2, \pm 3, \pm 4, \pm 6, \pm 8, \pm 12,$
$\pm 24$,   (c) $x = 1$
(d) none   29. $\pm 1, \pm 2, \pm 3, \pm 6, \pm 9, \pm 18$
31. $\pm p, \pm 1$   33. $2, -\frac{3}{2}, -1$

35. $\frac{5}{2}, -1 + \sqrt{3}, -1 - \sqrt{3}$
37. $\frac{2}{3}, \frac{1}{2}(-1 + \sqrt{3}i), \frac{1}{2}(-1 - \sqrt{3}i)$
39. $-1, -\sqrt{7}$ (multiplicity 2), $\sqrt{7}$
(multiplicity 2)   41. 2 (multiplicity 2), 5
43. $-3, 1 + \sqrt{6}, 1 - \sqrt{6}$
45. $6(x - \frac{4}{3})(x - (-\frac{5}{2}))$
47. $(x - 4)[x - (-\sqrt[3]{5})]$
$\cdot [x - \frac{1}{2}(\sqrt[3]{5} + i\sqrt{3}\sqrt[3]{25})]$
$\cdot [x - \frac{1}{2}(\sqrt[3]{5} - i\sqrt{3}\sqrt[3]{25})]$
49. $2 - 3i, 2 + 3i, 3$
51. $1 + \sqrt{2}i, 1 - \sqrt{2}i, \sqrt{7}, -\sqrt{7}$
53. (a) $\frac{3}{2}$   (b) $\frac{3}{2}, \frac{1}{2}(-1 + \sqrt{3}i),$
$\frac{1}{2}(-1 - \sqrt{3}i)$
55. (c) $-1.88$ and $-1.87$   57. 1 positive
real, 1 negative real, 2 complex   59. 1
positive real, 2 complex   61. 1 negative
real, 2 complex   63. 1 positive real, 1
negative real, 2 complex
65. $-3i, 3i, 1 + \sqrt{3}, 1 - \sqrt{3}, -2$   67. 0.82
and 0.83   69. F   71. T   73. F   75. F
77. $x^2 - 8x + 11 = 0$
79. $x^4 - 12x^3 + 35x^2 + 60x - 200 = 0$
81. $x^3 + 6x + 20 = 0$
83. $0, -1$

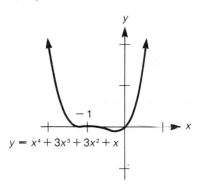

$y = x^4 + 3x^3 + 3x^2 + x$

85. $1, -4, -3$

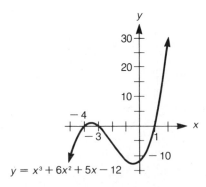

$y = x^3 + 6x^2 + 5x - 12$

87. $x^4 - 4x^3 - 4x^2 + 16x - 8 = 0$

## CHAPTER 12
## THE CONIC SECTIONS

### 12.1 The Parabola

**1.** focus: $(0, 1)$; directrix: $y = -1$; focal width: 4

**3.** focus: $(-2, 0)$; directrix: $x = 2$; focal width: 8

**5.** focus: $(0, -5)$; directrix: $y = 5$; focal width: 20

**7.** focus: $(-7, 0)$; directrix: $x = 7$; focal width: 28

**9.** focus: $(0, \frac{3}{2})$; directrix: $y = -\frac{3}{2}$; focal width: 6

**11.** focus: $(0, \frac{7}{16})$; directrix: $y = -\frac{7}{16}$; focal width: $\frac{7}{4}$

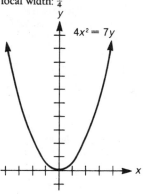

**13.** vertex: $(2, 3)$; focus: $(3, 3)$; directrix: $x = 1$; focal width: 4

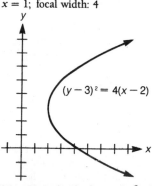

**15.** vertex: $(4, 2)$; focus $(4, \frac{9}{4})$; directrix: $y = \frac{7}{4}$; focal width: 1

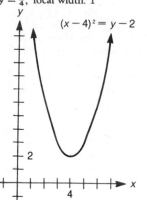

**17.** vertex: $(0, -1)$; focus: $(\frac{1}{4}, -1)$; directrix: $x = -\frac{1}{4}$; focal width: 1

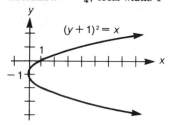

**19.** vertex: $(3, 0)$; focus: $(3, \frac{1}{8})$; directrix: $y = -\frac{1}{8}$; focal width: $\frac{1}{2}$

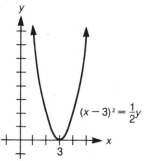

**21.** vertex: $(4, 1)$; focus: $(4, \frac{9}{8})$; directrix: $y = \frac{7}{8}$; focal width: $\frac{1}{2}$

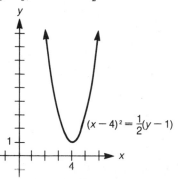

**23.** line of symmetry: $y = 1$

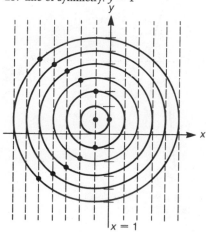

**25.** $x^2 = 12y$   **27.** $y^2 = 128x$
**29.** $y^2 = -9x$   **31.** (a) $y = \frac{3}{4}x + 2$
(b) $(-2, \frac{1}{2})$   (c) $\frac{25}{2}$
(d) $(x - 3)^2 + (y - \frac{17}{4})^2 = \frac{625}{16}$
(e)

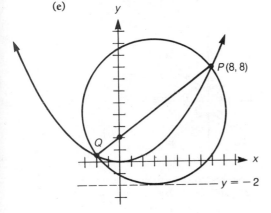

**33.** 11.25 ft

## 12.2 The Ellipse

**1.** major axis: 6; minor axis: 4; foci:
$(\pm\sqrt{5}, 0)$; eccentricity: $\sqrt{5}/3$

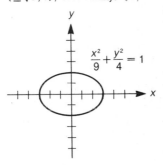

**3.** major axis: 8; minor axis: 2;   foci:
$(\pm\sqrt{15}, 0)$; eccentricity: $\sqrt{15}/4$

**5.** major axis: $2\sqrt{2}$; minor axis: 2; foci:
$(\pm1, 0)$; eccentricity: $\sqrt{2}/2$

**7.** major axis: 8; minor axis: 6; foci:
$(0, \pm\sqrt{7})$; eccentricity: $\sqrt{7}/4$

**9.** major axis: $2\sqrt{15}/3$; minor axis:
$2\sqrt{3}/3$; foci: $(0, \pm2\sqrt{3}/3)$; eccentricity:
$2\sqrt{5}/5$

**11.** major axis: 4; minor axis: $2\sqrt{2}$;
foci: $(0, \pm\sqrt{2})$; eccentricity: $\sqrt{2}/2$

**13.** center: $(5, -1)$; major axis: 10;
minor axis: 6; foci: $(9, -1)$, $(1, -1)$;
eccentricity: $\frac{4}{5}$

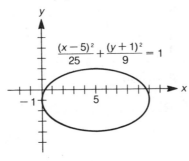

**15.** center: $(1, 2)$; major axis: 4;
minor axis: 2; foci: $(1, 2 \pm \sqrt{3})$;
eccentricity: $\sqrt{3}/2$

**17.** center: $(-3, 0)$; major axis: 6;
minor axis: 2; foci: $(-3 \pm 2\sqrt{2}, 0)$;
eccentricity: $2\sqrt{2}/3$

**19.** center: $(1, -2)$; major axis: 4;
minor axis: $2\sqrt{3}$; foci: $(2, -2)$, $(0, -2)$;
eccentricity: $\frac{1}{2}$

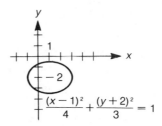

**21.** degenerate ellipse: the point $(4, 6)$

**23.** no graph **25.** $16x^2 + 25y^2 = 400$,
$$\frac{x^2}{25} + \frac{y^2}{16} = 1$$

**27.** $15x^2 + 16y^2 = 240, \dfrac{x^2}{16} + \dfrac{y^2}{15} = 1$

**29.** $25x^2 + 21y^2 = 525, \dfrac{x^2}{21} + \dfrac{y^2}{25} = 1$

**31.** $x^2 + 4y^2 = 9, \dfrac{x^2}{9} + \dfrac{y^2}{9/4} = 1$

**33.** $2\sqrt{30}\pi$

**35.**

| $x$ | 0 | 0.5 | 1.0 | 1.5 | 2.0 | 2.5 | 3.0 |
|---|---|---|---|---|---|---|---|
| $y$ | $\pm 2$ | $\pm 1.97$ | $\pm 1.89$ | $\pm 1.73$ | $\pm 1.49$ | $\pm 1.11$ | 0 |

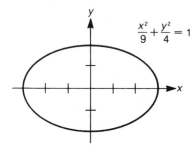

$$\frac{x^2}{9} + \frac{y^2}{4} = 1$$

**39.**

| | APPROXIMATION OBTAINED | PERCENTAGE ERROR |
|---|---|---|
| $C_1$ | 25.531776 | 0.019 |
| $C_2$ | 25.526986 | 0.000049 |
| $C_3$ | 25.519489 | 0.029 |

$C_2$ is the best approximation.

## 12.3 The Hyperbola

**1.** vertices: $(\pm 2, 0)$; transverse axis: 4;
conjugate axis: 2; asymptotes: $y = \pm\frac{1}{2}x$;
foci: $(\pm\sqrt{5}, 0)$; eccentricity: $\sqrt{5}/2$

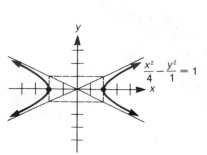

$$\frac{x^2}{4} - \frac{y^2}{1} = 1$$

**3.** vertices: $(0, \pm 2)$; transverse axis: 4;
conjugate axis: 2; asymptotes: $y = \pm 2x$;
foci: $(0, \pm\sqrt{5})$; eccentricity: $\sqrt{5}/2$

$$\frac{y^2}{4} - \frac{x^2}{1} = 1$$

**5.** vertices: $(\pm 5, 0)$; transverse axis: 10;
conjugate axis: 8; asymptotes: $y = \pm\frac{4}{5}x$;
foci: $(\pm\sqrt{41}, 0)$; eccentricity: $\sqrt{41}/5$

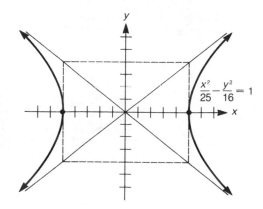

$$\frac{x^2}{25} - \frac{y^2}{16} = 1$$

**7.** vertices: $(0, \pm\sqrt{2}/2)$; transverse axis:
$\sqrt{2}$; conjugate axis: $2\sqrt{3}/3$; asymptotes:
$y = \pm\frac{1}{2}\sqrt{3}x$; foci: $(0, \pm\sqrt{30}/6)$;
eccentricity: $\sqrt{15}/3$

$$\frac{y^2}{\frac{1}{2}} - \frac{x^2}{\frac{1}{3}} = 1$$

**9.** vertices: $(0, \pm 5)$; transverse axis: 10;
conjugate axis: 4; asymptotes: $y = \pm\frac{5}{2}x$;
foci: $(0, \pm\sqrt{29})$; eccentricity: $\sqrt{29}/5$

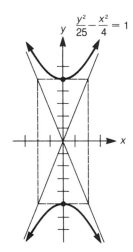

$$\frac{y^2}{25} - \frac{x^2}{4} = 1$$

**11.** center: $(5, -1)$; vertices: $(10, -1)$, $(0, -1)$; transverse axis: 10; conjugate axis: 6; asymptotes: $y = \frac{3}{5}x - 4$, $y = -\frac{3}{5}x + 2$; foci: $(5 \pm \sqrt{34}, -1)$; eccentricity: $\sqrt{34}/5$

**21.** center: $(0, -4)$; vertices: $(0, -9)$, $(0, 1)$; transverse axis: 10; conjugate axis: 2; asymptotes: $y = 5x - 4$, $y = -5x - 4$; foci: $(0, -4 \pm \sqrt{26})$; eccentricity: $\sqrt{26}/5$

$$\frac{(y+4)^2}{25} - \frac{x^2}{1} = 1$$

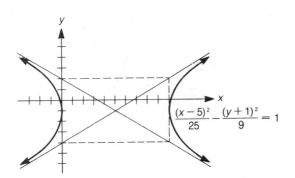

$$\frac{(x-5)^2}{25} - \frac{(y+1)^2}{9} = 1$$

**13.** center: $(1,2)$; vertices: $(1,4)$, $(1,0)$; transverse axis: 4; conjugate axis: 2; asymptotes: $y = 2x$, $y = -2x + 4$; foci: $(1, 2 \pm \sqrt{5})$; eccentricity: $\sqrt{5}/2$

**17.** center: $(0,1)$; vertices: $(\pm 2, 1)$; transverse axis: 4; conjugate axis: 4; asymptotes: $y = x + 1$, $y = -x + 1$; foci: $(\pm 2\sqrt{2}, 1)$; eccentricity: $\sqrt{2}$

**23.** degenerate hyperbola: graph consists of the two lines $y = x + 3$ and $y = -x - 4$

$$\frac{(y-2)^2}{4} - \frac{(x-1)^2}{1} = 1$$

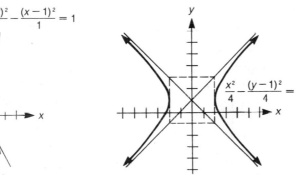

$$\frac{x^2}{4} - \frac{(y-1)^2}{4} = 1$$

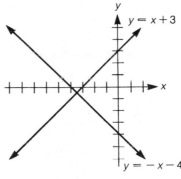

$y = x + 3$

$y = -x - 4$

**27.** $15x^2 - y^2 = 15$   **29.** $x^2 - 4y^2 = 4$
**31.** $2x^2 - 5y^2 = 10$   **33.** $y^2 - 32x^2 = 49$
**35.** $y^2 - 9x^2 = 9$
**39.** **(b)** $F_1 = (0,6)$, $F_2 = (0,-6)$
**(c)** $F_1P = 5$, $F_2P = 13$
**41.** **(a)** $e = \sqrt{2}$   **(b)** $e = \sqrt{2}$

**15.** center: $(-3,4)$; vertices: $(1,4)$, $(-7,4)$; transverse axis: 8; conjugate axis: 8; asymptotes: $y = x + 7$, $y = -x + 1$; foci: $(-3 \pm 4\sqrt{2}, 4)$; eccentricity: $\sqrt{2}$

**19.** center: $(2,1)$; vertices: $(-1,1)$, $(5,1)$; transverse axis: 6; conjugate axis: 6; asymptotes: $y = x - 1$, $y = -x + 3$; foci: $(2 \pm 3\sqrt{2}, 1)$; eccentricity: $\sqrt{2}$

## Chapter 12 Review Exercises

**1.** **(a)** $y^2 = 16x$   **(b)** $x^2 = 16y$
**3.** $x^2 = 12y$   **5.** $15x^2 + 16y^2 = 960$
**7.** $25x^2 + 9y^2 = 900$   **9.** $8x^2 - y^2 = 32$
**11.** $240x^2 - 16y^2 = 135$
**13.** vertex: $(0,0)$; focus: $(0, \frac{5}{2})$; directrix: $y = -\frac{5}{2}$; focal width: 10

$$\frac{(x+3)^2}{16} - \frac{(y-4)^2}{16} = 1$$

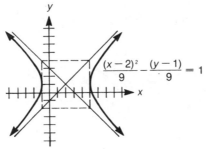

$$\frac{(x-2)^2}{9} - \frac{(y-1)}{9} = 1$$

$x^2 = 10y$

**15.** vertex: $(0,3)$; focus: $(0,0)$; directrix: $y = 6$; focal width: 12

$$x^2 = -12(y - 3)$$

**17.** vertex: $(1,1)$; focus: $(0,1)$; directrix: $x = 2$; focal width: 4

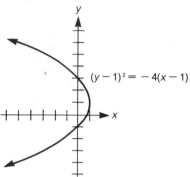

$$(y - 1)^2 = -4(x - 1)$$

**19.** center: $(0,0)$; foci: $(\pm\sqrt{2},0)$; major axis: 4; minor axis: $2\sqrt{2}$; eccentricity: $\sqrt{2}/2$

$$\frac{x^2}{4} + \frac{y^2}{2} = 1$$

**21.** center: $(0,0)$; foci: $(0, \pm 2\sqrt{10})$; major axis: 14; minor axis: 6; eccentricity: $2\sqrt{10}/7$

$$\frac{x^2}{9} + \frac{y^2}{49} = 1$$

**23.** center: $(1, -2)$; foci: $(5, -2)$, $(-3, -2)$; major axis: 10; minor axis: 6; eccentricity: $\frac{4}{5}$

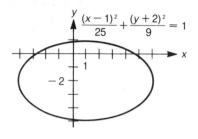

$$\frac{(x - 1)^2}{25} + \frac{(y + 2)^2}{9} = 1$$

**25.** center: $(0,0)$; vertices: $(\pm 2, 0)$; foci: $(\pm\sqrt{6}, 0)$; asymptotes: $y = \pm\frac{1}{2}\sqrt{2}x$ eccentricity: $\sqrt{6}/2$

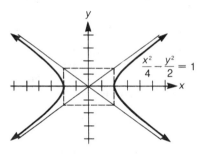

$$\frac{x^2}{4} - \frac{y^2}{2} = 1$$

**27.** center: $(0,0)$; vertices: $(0, \pm 3)$; foci: $(0, \pm\sqrt{58})$; asymptotes: $y = \pm\frac{3}{7}x$; eccentricity: $\sqrt{58}/3$

$$\frac{y^2}{9} - \frac{x^2}{49} = 1$$

**29.** center: $(1, -2)$; vertices: $(6, -2)$, $(-4, -2)$; foci: $(1 \pm \sqrt{34}, -2)$; asymptotes: $y = \frac{3}{5}x - \frac{13}{5}, y = -\frac{3}{5}x - \frac{7}{5}$ eccentricity: $\sqrt{34}/5$

$$\frac{(x - 1)^2}{25} - \frac{(y + 2)^2}{9} = 1$$

**31.** Ellipse with center: $(1, -2)$; foci: $(0, -2)$, $(2, -2)$; major axis: 4; minor axis: 4; minor axis: $2\sqrt{3}$

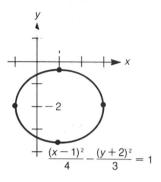

$$\frac{(x - 1)^2}{4} - \frac{(y + 2)^2}{3} = 1$$

**33.** Parabola with vertex: $(4, -1)$; axis: $y = -1$; focus: $(3, -1)$; directrix: $x = 5$

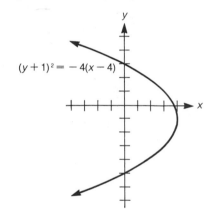

$$(y + 1)^2 = -4(x - 4)$$

**35.** Hyperbola with center: $(1, 5)$; vertices: $(-2, 5)$, $(4, 5)$; foci: $(-4, 5)$, $(6, 5)$; asymptotes: $y = \frac{4}{3}x + \frac{11}{3}, y = -\frac{4}{3}x + \frac{19}{3}$

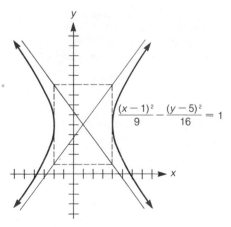

$$\frac{(x - 1)^2}{9} - \frac{(y - 5)^2}{16} = 1$$

37. Degenerate ellipse: the point $(4,6)$

39. Degenerate hyperbola: the lines $3x - 4y = 11$ and $3x + 4y = 19$

41. Hyperbola with center: $(0, -4)$; vertices: $(0,1)$, $(0,-9)$; foci: $(0, -4 \pm \sqrt{26})$; asymptotes: $y = \pm 5x - 4$

$$\frac{(y+4)^2}{25} - \frac{x^2}{1} = 1$$

43. No graph   45. (b) 64

CHAPTER 13
ADDITIONAL TOPICS IN ALGEBRA

13.1 Mathematical Induction

23. (a)

| $n$ | 1 | 2 | 3 | 4 | 5 |
|-----|---|---|---|---|---|
| $f(n)$ | $\frac{1}{2}$ | $\frac{2}{3}$ | $\frac{3}{4}$ | $\frac{4}{5}$ | $\frac{5}{6}$ |

(b) $\frac{6}{7}$   (c) $f(n) = n/(n+1)$

25. (a)

| $n$ | 1 | 2 | 3 | 4 | 5 |
|-----|---|---|---|---|---|
| $f(n)$ | 1 | 4 | 9 | 16 | 25 |

(b) 36   (c) $f(n) = n^2$

13.2 The Binomial Theorem

3. $a^9 + 9a^8b + 36a^7b^2 + 84a^6b^3 + 126a^5b^4 + 126a^4b^5 + 84a^3b^6 + 36a^2b^7 + 9ab^8 + b^9$   5. $8A^3 + 12A^2B + 6AB^2 + B^3$
7. $1 - 12x + 60x^2 - 160x^3 + 240x^4 - 192x^5 + 64x^6$
9. $x^2 + 4x\sqrt{xy} + 6xy + 4y\sqrt{xy} + y^2$
11. $x^{10} + 5x^8y^2 + 10x^6y^4 + 10x^4y^6 + 5x^2y^8 + y^{10}$
13. $1 - \frac{6}{x} + \frac{15}{x^2} - \frac{20}{x^3} + \frac{15}{x^4} - \frac{6}{x^5} + \frac{1}{x^6}$
15. $\frac{x^3}{8} - \frac{x^2y}{4} + \frac{xy^2}{6} - \frac{y^3}{27}$
17. $a^7b^{14} + 7a^6b^{12}c + 21a^5b^{10}c^2 + 35a^4b^8c^3 + 35a^3b^6c^4 + 21a^2b^4c^5 + 7ab^2c^6 + c^7$
19. $x^8 + 8\sqrt{2}x^7 + 56x^6 + 112\sqrt{2}x^5 + 280x^4 + 224\sqrt{2}x^3 + 224x^2 + 64\sqrt{2}x + 16$
21. $5\sqrt{2} - 7$   23. $89\sqrt{3} + 109\sqrt{2}$
25. $12 - 24\sqrt[3]{2} + 12\sqrt[3]{4}$
27. $x^{10} - 10x^9 + 35x^8 - 40x^7 - 30x^6 + 68x^5 + 30x^4 - 40x^3 - 35x^2 - 10x - 1$
29. 120   31. 105   33. (a) 10   (b) 5
35. $n^2 + 3n + 2$   37. 0   39. $120a^2b^{14}$
41. $100x^{99}$   43. 45   45. 294,912
47. 28   49. the fifth term

53. (a)

| $k$ | 0 | 1 | 2 | 3 | 4 | 5 | 6 | 7 | 8 |
|-----|---|---|---|---|---|---|---|---|---|
| $\binom{8}{k}$ | 1 | 8 | 28 | 56 | 70 | 56 | 28 | 8 | 1 |

13.3 Introduction to Sequences and Series

1. $\frac{1}{2}, \frac{2}{3}, \frac{3}{4}, \frac{4}{5}, \frac{5}{6}$   3. $0, 1, 4, 9, 16$
5. $2, \frac{9}{4}, \frac{64}{27}, \frac{625}{256}, \frac{7776}{3125}$
7. $-1, 1, -1, 1, -1$
9. $1, -\frac{1}{2}, \frac{1}{6}, -\frac{1}{24}, \frac{1}{120}$
11. $3, 6, 9, 12, 15$
13. $1, 4, 25, 676, 458329$
15. $2, 2, 4, 8, 32$
17. $1, 1, 2, 6, 24$
19. $0, 1, 2, 4, 16$
21. $a_n$

23. $a_n$

25. $a_n$

27. 62   29. 40   31. $-\frac{19}{30}$   33. 903
35. 278   37. 3   39. 41   41. $x + x^2 + x^3$
43. $\frac{25}{12}$   45. $-1$   47. $\frac{6}{7}$
49. $\sum_{i=1}^{4} 5^i$   51. $\sum_{i=1}^{6} x^i$   53. $\sum_{k=1}^{12} (1/k)$
55. $\sum_{i=1}^{5} (-1)^{i+1}2^i$   57. $\sum_{i=1}^{5} (-1)^{i+1}i$

59. (b)

| $n$ | $F_1 + F_2 + \cdots + F_n$ | $F_{n+2} - 1$ |
|-----|----------------------------|---------------|
| 1 | 1 | 1 |
| 2 | 2 | 2 |
| 3 | 4 | 4 |
| 4 | 7 | 7 |
| 5 | 12 | 12 |

13.4 Arithmetic Sequences and Series

1. (a) 2   (b) $-4$   (c) $\frac{1}{3}$   (d) $\sqrt{2}$
3. 131   5. 501   7. 998
9. $d = \frac{11}{6}, a_1 = -\frac{23}{2}$   11. $-190$
13. $-\frac{1}{8}$   15. 500,500   17. $91\pi/3$
19. $\frac{1}{2}$   21. $S_{16} = -768, d = -\frac{104}{15}$
23. $d = \frac{6}{5}, a = -\frac{17}{5}$   25. 900
27. $2, 10, 18$ or $18, 10, 2$
29. $-1, 2, 5$ or $5, 2, -1$
31. (b) $-6 - 9\sqrt{2}$
33. $n[2 + (n-3)\sqrt{b}]/2(1 - b)$

13.5 Geometric Sequences and Series

1. 6   3. 20;100   5. 1   7. $\frac{256}{6561}$
9. $4, -4$   11. 7161   13. $63 + 31\sqrt{2}$

15. $\frac{1995}{64}$   17. 0.011111   19. $\frac{2}{5}$
21. 101   23. $\frac{5}{9}$   25. $\frac{61}{495}$   27. $\frac{16}{37}$
29. $-\frac{1}{2}, -2$

## 13.6 Permutations and Combinations

1. 6 routes

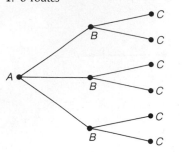

3. 12 routes

5. 12   7. 60   9. 108   11. 11,232,000
13. 3,628,800   15. 4   17. 8
19. 120   23. $P(101, 51)$   25. 240
27. 12   29. (a) 900   (b) 648
31. (a) 729   (b) 504
(c) 56   (d) 140   33. (a) 12   (b) 12
35. (a) 56   (b) 56   37. (a) 792
(b) 792   39. $n(n-1)/2$   41. 56
43. (a) 210   (b) 80   45. 9   47. 56
49. 15 sums   51. (a) 2352   (b) 2649

53. To avoid counting each diagonal twice,
we must divide by 2; hence there are
$\frac{1}{2}(7 \cdot 10) = 35$ diagonals.

## 13.7 Introduction to Probability

1. (a) $\{2,4,6\}$   (b) $\{4,5,6\}$
(c) $\{2,4,5,6\}$   (d) $\{4,6\}$
3. (a) $\{1,3,5\}$   (b) $\{1,2,3,4,5,6\}$
(c) $\{1,2,3,4,5,6\}$   (d) $\varnothing$   5. no
7. yes   9. (a) 3   (b) 3   (c) 4
11. (a) 3   (b) 6   (c) 0
13. (a) $\frac{1}{2}$   (b) $\frac{1}{2}$   15. (a) $\frac{1}{2}$   (b) 1
17. (a) $\frac{2}{3}$   (b) 1   19. (a) $\frac{5}{6}$   (b) 1
21. $\frac{3}{8}$   23. $\frac{1}{2}$   25. $\frac{1}{4}$   27. $\frac{5}{8}$   29. $\frac{3}{4}$
31. 1   33. $\frac{5}{8}$   35. $\frac{1}{2}$   37. 0   39. $\frac{1}{4}$
41. $\frac{1}{3}$   43. (a) $\frac{6}{7}$   (b) $\frac{36}{91}$   45. $\frac{7}{12}$
47. $\frac{2}{9}$   49. (a) $\frac{1}{8}$   (b) $\frac{3}{8}$   (c) $\frac{1}{2}$
51. $\frac{7}{10}$   53. $\frac{33}{66640}$

## Chapter 13 Review Exercises

11. $81a^4 + 108a^3b^2 + 54a^2b^4 + 12ab^6 + b^8$
13. $x^4 + 4x\sqrt[3]{x} + 6x^3 + 4x^2\sqrt{x} + x^2$
15. $x^{10} - 10x^8y^2 + 40x^6y^4 - 80x^4y^6$
$+ 80x^2y^8 - 32y^{10}$

17. $1 + \dfrac{5}{x} + \dfrac{10}{x^2} + \dfrac{10}{x^3} + \dfrac{5}{x^4} + \dfrac{1}{x^5}$

19. $a^4b^2 - 4a^3b^2\sqrt{ab} + 6a^3b^3$
$- 4a^2b^3\sqrt{ab} + a^2b^4$

21. $15xy^8$   23. 84   25. 28
33. $1, \frac{4}{3}, \frac{3}{2}, \frac{8}{5}$

35. $-\frac{1}{2}, \frac{2}{3}, -\frac{3}{4}, \frac{4}{5}$

37. $-3, -12, -48, -192$

39. (a) $-5$   (b) $\frac{9}{10}$   41. $\sum\limits_{k=1}^{5} (5/3^k)$

43. 73   45. $\frac{5}{1024}$   47. 177

49. 7,777,777,777   51. $-5\sqrt{10}$   53. $\frac{3}{4}$
55. $\frac{1}{10}$   57. $\frac{5}{11}$   63. (a) 42929
(b) 42925; 0.00932% error
(c) $6.48040 \times 10^{10}$
(d) $6.48027 \times 10^{10}$; $2 \times 10^{-3}$% error
65. 16   67. $abc$; $acb$; $bac$; $bca$; $cab$; $cba$;
$abd$; $adb$; $bad$; $bda$; $dab$; $dba$; $acd$; $adc$;
$cad$; $cda$; $dac$; $dca$; $bcd$; $bdc$; $cbd$; $cdb$; $dbc$;
$dcb$;   $P(4, 3) = 24$   69. 19,958,400
71. 480   73. 1620   75. $\frac{6}{11}$
77. (a) $\frac{5}{33}$   (b) $\frac{7}{22}$   (c) $\frac{31}{66}$   79. $\frac{1}{18}$
81. $\frac{1}{6}$   83. $\frac{2}{9}$   85. $\frac{1}{6}$   87. $\frac{25}{36}$
89. (a) $\frac{7}{15}$   (b) $\frac{8}{15}$   (c) $\frac{2}{3}$   (d) $\frac{2}{3}$
91. (a) 252   (b) $\frac{7}{25}$
93. 40, 20, 10 or 10, 20, 40
97. (a) No such value for $x$ exists.

(b) $x = \dfrac{b^2 - ac}{a + c - 2b}$ if $a + c \neq 2b$